Recent Advances in Multidisciplinary Applied Physics

Proceedings of the
First International Meeting on Applied Physics (APHYS-2003)

October 13-18th, 2003, Badajoz, Spain

Elsevier Internet Homepage - http://www.elsevier.com
Consult the Elsevier homepage for full catalogue information on all books, major reference works, journals, electronic products and services.

Elsevier Titles of Related Interest

ADVANCES IN IMAGING AND ELECTRON PHYSICS - series
Series Editor: Peter Hawkes
ISSN: 1076-5670
Volume 138 (2005) ISBN: 0120147807
www.elsevier.com/locate/isbn/0120147807

STUDIES IN APPLIED MECHANICS - series
Series Editors: I. Elishakoff
Volume 49 (2004) ISBN: 0-08-044348-6
www.elsevier.com/locate/isbn/0080443486

Related Journals:
Elsevier publishes a wide-ranging portfolio of high quality research journals, encompassing the applied physics and mechanics field. A sample journal issue is available online by visiting the Elsevier web site (details at the top of this page). Leading titles include:

Current Applied Physics
Journal of the Mechanics and Physics of Solids
Applied Surface Science
Communications in Nonlinear Science and Numerical Simulation
Physica D: Nonlinear Phenomena
Optics Communications
Physica A: Statistical Mechanics and its Applications
Thin Solid Films
International Journal of Engineering Science

All journals are available online via ScienceDirect: www.sciencedirect.com

To contact the Publisher
Elsevier welcomes enquiries concerning publishing proposals: books, journal special issues, conference proceedings, etc. All formats and media can be considered. Should you have a publishing proposal you wish to discuss, please contact, without obligation, the publisher responsible for Elsevier's mechanics and mechanical engineering programme:

Lynne Honigmann
Publisher
Elsevier Ltd
The Boulevard, Langford Lane — Phone: +44 1865 843462
Kidlington, Oxford — Fax: +44 1865 843987
OX5 1GB, UK — E.mail: l.honingman@elsevier.com

General enquiries, including placing orders, should be directed to Elsevier's Regional Sales Offices – please access the Elsevier homepage for full contact details (homepage details at the top of this page).

Recent Advances in Multidisciplinary Applied Physics

Proceedings of the
First International Meeting on Applied Physics (APHYS-2003)

October 13-18th, 2003, Badajoz, Spain

Editor

A. Méndez-Vilas

FORMATEX Research Center, Badajoz, Spain

2005

ELSEVIER

Amsterdam • Boston • Heidelberg • London • New York • Oxford
Paris • San Diego • San Francisco • Singapore • Sydney • Tokyo

ELSEVIER B.V.
Radarweg 29
P.O. Box 211, 1000 AE Amsterdam
The Netherlands

ELSEVIER Inc.
525 B Street, Suite 1900
San Diego, CA 92101-4495
USA

ELSEVIER Ltd.
The Boulevard, Langford Lane
Kidlington, Oxford OX5 1GB
UK

ELSEVIER Ltd.
84 Theobalds Road
London WC1X 8RR
UK

First edition 2005

ISBN: 0080446485

∞ The paper used in this publication meets the requirements of ANSI/NISO Z39.48-1992 (Permanence of Paper).
Printed in Great Britain by MPG Books Ltd, Bodmin, Cornwall.

Foreword

This Book contains a collection of papers presented at the 1st International Meeting on Applied Physics (APHYS-2003), held in Badajoz (Spain), from 15th to 18th October 2003
(URL: http://www.formatex.org/aphys2003/aphys2003.htm).

APHYS-2003 was born as an attempt to create a new international forum on Applied Physics in Europe. Since Applied Physics is not really a branch of Physics, but the application of all the branches of Physics to the broad realms of practical problems in Science, Engineering and Industry, this conference was a truly inter-disciplinary event. The organizers called for papers relating Physics with other sciences such as Biology, Chemistry, Information Science, Medicine, etc, or relating different Physics areas, and aimed at solving practical problems, from an transnational perspective. In other words, the Conference was specifically interested in reports applying the methods, the training, and the culture of Physics to research areas usually associated with other scientific and engineering disciplines.

It was extremely rewarding that over 800 researchers, from over 65 countries, attended the conference, where more than 1000 research papers were presented. We feel really proud of this excellent response obtained (in number and quality), for this first edition of the conference.

We are very grateful to all the members of the Organizing Committee, for the hard work done for the preparation of the Conference (which began one year before the conference start), and to the members of the International Advisory Committee, for the valuable contribution to the evaluation of submitted works. Also we would like to thank Prof. F. Guiberteau from the University of Extremadura (Spain), Prof. R. Tannenbaum from the Georgia Institute of Technology (USA), Dr. M.L.González-Martín, from University of Extremadura (Spain), Dr. R. Chacón, from University of Extremadura (Spain), Prof. A. Martín-Sánchez, from University of Extremaudra (Spain), for their help as sessions Chairpersons, and to the referees for the excellent work done in the revision of submitted papers (more than 600 research papers).

We were honoured to count on the following Plenary Speakers. We are extremely grateful to all of them, which delivered excellent enthusiastic Lectures on some cutting-edge areas of Applied Physics.

Prof. Adam Curtis, Director of the Centre of Cell Engineering at the University of Glasgow, UK
"Nanobiotechnology - Interactions of Cells with Nanofeatured Surfaces and with Nanoparticles"

Prof. Allan S. Hoffman, Department of Bioengineering, University of Washington, USA

Prof. Lars Persson, Retired Scientist of the Swedish Radiation Protection Authority, Sweden
"Radiation Protection of Nuclear Workers - Ethical Issues"

Prof. K Alan Shore, University of Wales ,Bangor, UK. Head of the School of Informatics, Director of the ICON Centre of Excellence and Chair of the Institute of Physics (Wales)
"Chaotic Data Encryption for Optical Communications"

Finally, we would like to thank the Department of Physics of the University of Extremadura, and the Foundation for the Development of Science and Technology in Extremadura (FUNDECYT), for their support, and the Regional Government (Junta de Extremadura / Consejería de Educación, Ciencia y Tecnología), as well as INNOVA Instrumentación, for sponsoring the Conference.

Thanks are also due to the following companies, which chose APHYS-2003 for presenting some of their most interesting products for the conference audience:

- Innova Instrumentacion
- Attocube Systems - Positioning for the nanoworld
- Owis GmBH
- Shaefer Techniques
- Nanotec Electronica

- Caburn, the Vacuum Components Company
- Alfa Aesar - Johnson Matthey
- ATOS
- Kurt J. Lesker Company
- Advanced Design Consulting USA, INC
- EDP Sciences
- World Scientific Publishers
- American Scientific Publishers - Encyclopedia of Nanoscience and Nanotechnology

Without the efforts of everyone, the Conference could not have been such a success and this Book could not have been published. We hope to see you in APHYS-2006!

A. Méndez-Vilas
Editor
FORMATEX Research Centre, Badajoz, Spain
Phone/Fax: +34 924 258 615
e-mail: amvilas@formatex.org

Local Organising Committee

International Scientific Advisory Committee

Surfaces. Interfaces. Nanosciences. Nanotechnology. Imaging Techniques

Prof. Klaus Kern, Max-Planck-Institute for Solid State Research, Nanoscale Science Department, Germany

Prof. Raul A. Baragiola, Director of the Laboratory for Atomic and Surface Physics, Engineering Physics, University of Virginia, USA

Prof. Federico Rosei. INRS-EMT Université du Québec, Varennes (QC), Canada

Prof. Sven Tougaard, Physics Department, University of Southern Denmark, Denmark

Dr. M.L. Gonzalez-Martin, Biosurfaces Group, Physics Department, University of Extremadura, Spain

Dr. M.J. Nuevo, Physics Department, University of Extremadura, Spain

Prof. Bronislaw Janczuk, Dept. of Interfacial Phenomena, Faculty of Chemistry, Maria Curie-Sklodowska Universit, Poland

Dr. Randy Headrick, Department of Physics, University of Vermount, USA

Dr. A. Patrick Gunning, Institute of Food Research, Norwich Research Park, UK

Dr. Alberto Diaspro, LAMBS (Laboratory for Advanced Microscopy, Bioimaging and Spectroscopy), Department of Physics, University of Genova, Italy

Prof. Buddy D. Ratner, Director, University of Washington Engineered Biomaterials, Washington Research Foundation, Endowed Professor of Bioengineering and Professor of Chemical Engineering, University of Washington, USA.

Prof. Igor Yaminsky, Head of the Scanning Probe Microscopy Group, Moscow State University & Advanced Technologies Center, Moscow, Russia

Dr. José Angel Martín Gago, Department of Surface Physics and Engineering, Madrid Institute of Materials Science, CSIC, Spain

Dr. Terry McMaster, H.H. Wills Physics Laboratory, University of Bristol, UK

Materials Science. Applied Solid State Physics/Chemistry

Prof. Isaac Hernández-Calderón, Physics Department, CINVESTAV, Mexico

Dr. Vasco Teixeira, Head of the Research Group GRF-Functional Coatings Group, University of Minho, Portugal

Prof. María Teresa Mora, Group of Materials Physics I (GFMI), Autonomous University of Barcelona, Spain

Dr. José Manuel Saniger Blesa, Center of Applied Sciences and Technological Development (Applied Research and Sensors), National Autonomus University of Mexico, Mexico

Prof. Fernando Guiberteau, Materials Science and Metallurgical Engineering Area, University of Extremadura, Spain

Dr A. Pajares Vicente, Condensed Matter Physics Area, University of Extremadura, Spain

Prof. Miguel Avalos Borja, Department of Nanostructures of the Centre of Condensed Matter Sciences, National Autonomous University of Mexico, Mexico

Prof. Giovanni Giacometti, Università di Padova, Dipartmento di Chimica Fisica, Italy

Prof. Marcelo Knobel, Laboratório de Materiais e Baixas Temperaturas, Departamento de Física da Matéria Condensada, Instituto de Física "Gleb Wataghin", Universidade Estadual de Campinas, Brazil

Prof. Norman K.Y. Cheng, Electrical and Computer Engineering Department, University of Illinois at Urbana-Champaign, USA

Computational Physics. Non-linear Physics

Prof. Francisco Jiménez-Morales, Condensed Matter Physics Department, Faculty of Physics, University of Sevilla, Spain.

Dr. Ricardo Chacón, Assoc. Prof. of Applied Physics, School of Industrial Engineering, University of Extremadura, Spain

Prof. M.A. Jaramillo, School of Industrial Engineering, University of Extremadura, Spain.

M.J. Nuevo, Assoc. Prof. of Theoretical Physics, Physics Department, University of Extremadura, Spain

Prof. Victor M. Pérez García, School of Industrial Engineering, University of Castilla-La Mancha, Spain

Dr. Alberto Pérez Muñuzuri, Faculty of Physics, University of Santiago de Compostela, Spain

Prof. Jason A.C. Gallas, Institute of Physics, University of Rio Grande do Sul, Brazil

Dr. Francisco Balibrea Gallego, Faculty of Mathematics, University of Murcia, Spain

Dr. Pedro J. Martínez, Technical School of Industrial Engineering, University of Zaragoza (Group of Theory and Simulation of Complex Systems)

Dr. Juan José Mazo, Faculty of Sciences, University of Zaragoza and CSIC, Spain (Group of Theory and Simulation of Complex Systems).

Applied Optics. Optoelectronics. Photonics

Prof. Michael Gal, School of Physics, The University of NSW, Sydney, Australia

Dr. Manuel Martinez-Corral, 3D Diffraction & Imaging Group, Department of Optics. University of Valencia, Spain

Prof. Augusto Beléndez Vázquez, Departamento de Física, Ingeniería de Sistemas y Teoría de la Señal, Universidad de Alicante, Spain

Dr David Binks, Laser Photonics Group, Dept. of Physics and Astronomy, University of Manchester, U.K

Dr. M. Isabel Suero Lopez, ORION Optics and Physics Education Research Group, Physics Department, University of Extremadura, Spain

Dr. A. Luis Perez Rodriguez, ORION Optics and Physics Education Research Group, Physics Department, University of Extremadura, Spain

Prof. Sergei Turitsyn, Photonics Research Group, Aston University, Birmingham, UK

Prof. Ivan Chambouleyron, Institute of Physics "Gleb Wataghin", UNICAMP, Brazil

Biophysics. Biomedical Engineering

Prof. Oswaldo Baffa, Departamento de Física e Matemática, FFCLRP-Universidade de São Paulo, Brasil

Dr Steven James Swithenby, Department of Physics and Astronomy, Group of Bio-magnetism, The Open University, UK

Dr. Manuel Monleon Pradas, Director of the Center of Biomaterials, Dept. de Termodinamica Aplicada, ETSII, Universidad Politecnica de Valencia, Spain

Dr. Jose Luis González Carrasco, Departamento de Metalurgia Física Centro, Nacional de Investigaciones Metalúrgicas, Madrid, Spain

Dr. J.C. Knowlesm, Reader in Biomaterials and Head of Department, Department of Biomaterials, Eastman Dental Institute, University College London, UK

Prof. Yu-Li Wang, Biomedical Engineering and Medical Physics, University of Massachusetts Medical School, USA

Radiation Physics/Chemistry and Processing. Radioactivity. Radiation Protection. Medical Physics

CONTENTS

Applied Physics (APHYS 2003)
A. Méndez-Vilas (Ed.)

Absorption of RF Radiation in Confocal and Uniformly Shelled Ellipsoidal Biological Cell Models

J.L. Sebastián, S. Muñoz San Martín, M. Sancho and J.M. Miranda

Departamento de Física Aplicada III, Facultad de Física.
Universidad Complutense de Madrid, Ciudad Universitaria 28040 Madrid. Spain.

Abstract. This paper presents a detailed calculation of the electric field distribution induced in biological cell models exposed to a RF radiation. The study shows the importance of using realistic cell shapes with the proper geometry and electrical properties to study the mechanisms of direct cellular effects from RF exposure. For this purpose, the electric field distribution within confocal and shelled ellipsoidal cell models is calculated by using a finite element technique with adaptive meshing. The cell models are exposed to linearly polarized electromagnetic plane waves of frequencies 900 and 2450 MHz. The results show that the amplification of the electric field within the membrane of the confocal shape cell is more significant than that observed in shelled cell geometries. The results show the dependence of the induced electric field distribution on frequency, electrical properties of membrane and cytoplasm and the orientation of the cell with respect to the applied field.

INTRODUCTION

Exposure of a biological cell to RF fields can produce a variety of profound biochemical and biophysical responses. However, any possible cell response is directly related to the internal field distribution, and in particular to the modification of the field strength across the cell membrane induced by the external RF radiation. Weak electric field effects have generally been attributed, at least as a primary event, to field interaction with either membrane or glycocalix constituents. The magnitude of transmembrane voltage and the deposited energy are basic issues [1,2] for understanding the relation between the exposition to fields and the subsequent physiological reactions at the cell level. Considering that the membrane is a site of high field amplification, it is uncertain how the detailed geometry and electrical properties of the cell can affect the exactness of the predictions in the electric behaviour. Therefore, in order to determine the mechanism of the basic interaction of RF fields with a biological structure, the knowledge of the electric field distribution within the cell membrane is of primary importance.

The analytical approach used by many researchers to find the cell internal field strength has severe limitations since an explicit solution of Laplace equation requires a geometry consisting in one or several uniform media separated by interfaces which coincide with a surface of a constant coordinate, within a certain set of coordinate types. Therefore, it turns out that only numerical methods can give a sufficiently precise estimation of field values in realistic cell anatomies. However, conventional computational methods have difficulties in dealing with a very thin membrane in a shelled structure. It is for this reason that up to now, geometric configurations representing more realistic cell shapes, such as ellipsoids [3], erythrocytes or rods with a uniform membrane thickness have not been studied.

Detailed numerical calculations of the electric field within a mammalian cell with basic spherical and cylindrical geometries have been already carried out by the authors [4,5] in previous works. The results indicated the important role played by the geometry of the cell model in the electric field determination. These studies made also clear that in order to have a good insight into the possible mechanisms of the action of electromagnetic fields, including athermal effects, more realistic models than cylindrical and confocal ellipsoidal geometries should be used.

This paper analyzes the influence on the internal electric field distribution of the geometry and electrical properties of confocal and shelled ellipsoidal shape cells exposed to a RF plane wave. The frequencies of the RF radiation used in this work are 900 MHz and 2450 MHz and both orientations of the E field (electric and magnetic polarizations) with respect to the cell model have been considered. The numerical technique used to calculate the field distribution is based on the well known finite elements (FE) theory. However, the efficiency and precision of this technique have been improved by using perfectly matched layers (PML) in the boundary conditions of the radiation region and an adaptive mesh in the mesh sizes for the different cell layers. In order to analyze the influence of the cell electrical properties, different values of permittivity and conductivity for both membrane and cytoplasm have been considered.

CELL MODELS

Figure 1 shows the geometrical dimensions for the confocal and shelled ellipsoidal cell geometries considered in this work to model a mammalian cell. The values of the major and minor semi-axis for both geometries were kept constant to 3.5μ and 1μm respectively. For the confocal ellipsoid, the membrane is described within the same coordinate system, which is determined by the foci of the ellipsoidal surfaces. For this geometry, the maximum values of the non uniform thickness of the membrane along the major and minor semi-axis are 10 nm and 5.8 nm respectively. For the shelled ellipsoidal model, the membrane has a uniform thickness of 10 nm. The cell structure is formed by two layers, cytoplasm and membrane, and the cell is considered immersed in an external continuous medium formed by electrolytes in free water with the dielectric properties of physiological saline.

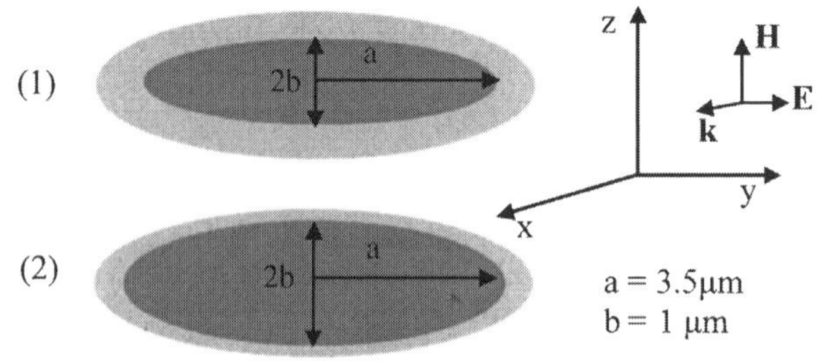

FIGURE 1. Ellipsoidal cell models, 1) shelled, with uniform thickness membrane and 2) confocal.

Table 1 shows the electrical parameters for the non conductive membrane and the highly polarizable cytoplasm considered for the cell as well as those for the external medium. At the higher frequency of 2450 MHz, the values used for the permittivity and conductivity of the cell layers have been taken to be the same as those used originally by Liu and Cleary [6], whereas for the lower frequency of 900 MHz the values of the complex permittivity of the cytoplasm and external medium have been found using the variations that correspond to the dielectric dispersion of bound and free water [7]. The cytoplasm is a physiological saline solution with a protein volume fraction of 0.26, whereas the membrane is made of a phospholipids bilayer that has no conductivity and a frequency-independent relative permittivity of 11.3.

TABLE 1. Electrical Parameters for the Different Layers of the Cell Model at the two frequencies used in this study.

Layer	Parameter	f = 900 MHz	f = 2450 MHz
Cytoplasm	ε	50.2	48.699
$a = 3.5\mu m$ $b = 1\mu m$ $c = 0.5\mu m$	σ (S/m)	0.992	1.417
	tgδ	0.395	0.214
Membrane	ε	11.3	11.3
Uniform Shell $d = 10$nm	σ (S/m)	0	0
Major semiaxis 10 nm Minor semiaxis 5.8 nm	tgδ	0	0
External Medium	ε	71.78	70.87
	σ (S/m)	1.947	2.781
	tgδ	0.542	0.288

In order to know the influence of the cell electric properties on its internal field strength, the permittivity and conductivity of the membrane and cytoplasm have been varied within reasonable ranges found in the literature [8]. For the membrane, the relative permittivity has been varied from 2 to 22 for both geometries, whereas the conductance has been kept constant to a negligible value at both frequencies. For the cytoplasm, the relative permittivity and conductivity have been varied from 30 to 70 and from 0.8 to 1.2 S/m respectively.

The radiation region in which the cell is immersed is filled with a continuous medium formed by electrolytes in free water with the dielectric properties of physiological saline.

ELECTRIC FIELD CALCULATION

In order to have a sufficiently precise estimation of the electric field distribution in realistic cell geometries it is necessary to apply numerical methods. But up to date very few studies of this type have been reported [4], the main reason being the difficulty they have to face in handling regions of very different size scales for the cell diameter and for the membrane thickness. As the numerical solution of Laplace equation in the form of finite differences involves a kind of polynomial approximation in nodes of a convenient grid, the existence of very small domains makes it necessary to use a very dense grid or alternatively sophisticated non-uniform meshing methods. As the cell dimensions (7 μm) are much smaller than the wavelength at the working frequencies (~4 cm), it is reasonable to assume that the cell is exposed to a uniform field propagating along the *x*-axis with E and H field components linearly polarized along the *y* and *z* axes respectively as shown in Figure 1.

The external field strength is E = 1 V/m and the electric field intensity within the different layers of the cell is found by using a finite element (FE) technique. The full Maxwell equations are then solved considering a discretization of the geometry into tetrahedral elements. An adaptive mesh is used so that the size of the basic tetrahedron is varied for the different regions. Therefore, the number of tetrahedra in the membrane had to be considerably higher (~10000 for the membrane) than the corresponding number for the region occupied by the cytoplasm (~8000). In order to keep the computational resource requirements reasonable, the computational domain is truncated to a radiation region (external medium) in which the cell is immersed and surrounded by finite thickness absorbing layers, called perfectly matched layers (PML). This provides a reflectionless interface between the region of interest and the PML layers at all incident angles [(9)]. The dimensions of the radiation region (of the order of 4 wavelengths in the external medium) are adjusted so that a good compromise between accuracy and reasonable computing times is obtained. The accuracy of this technique is conditioned by the smaller size of the mesh single element. The field values at the mesh nodes are calculated by an iterative method and a solution is found when a convergence criterion limit is accomplished. In all analysis, the convergence error limit was smaller than 10^{-6} for a result to be considered correct and the computing times for 2 GHz speed microprocessor were of the order of 200 minutes.

In order to validate the FE numerical technique used to obtain the electric field values, a comparison between the FE results and an analytical solution obtained by using a quasistatic approximation was made for the confocal ellipsoids cell model. For the ellipsoidal model and extension of the solution for a homogeneous dielectric ellipsoid [(10)] to a multi-layer case was used. The use of this approximation is fully justified as explained earlier. A discrepancy of less than 2% was found between the solutions obtained with the FE technique for the specified convergence criteria (smaller than 10^{-6}) and the quasistatic solution. This figure also gives an estimation of the uncertainties on the calculated values.

In the analysis, the orientation of the cell models exposed to RF was varied so that both electric polarization (where the electric vector is aligned with the minor ellipsoid semi-axis) and magnetic polarization (when the magnetic vector is aligned with the minor ellipsoid semi-axis) were considered.

ANALYSIS OF THE RESULTS

Figures 2a and 2b show that at 900 MHz, the values of the electric field intensity within the membrane and cytoplasm are higher than the values found at 2450 MHz. For a uniformly shelled cell the field strength value is lower than the value found for a confocal geometry. Also, the values of the E field for electric polarization are higher than for magnetic polarization, thus showing the important role of the polarization of the incident field.

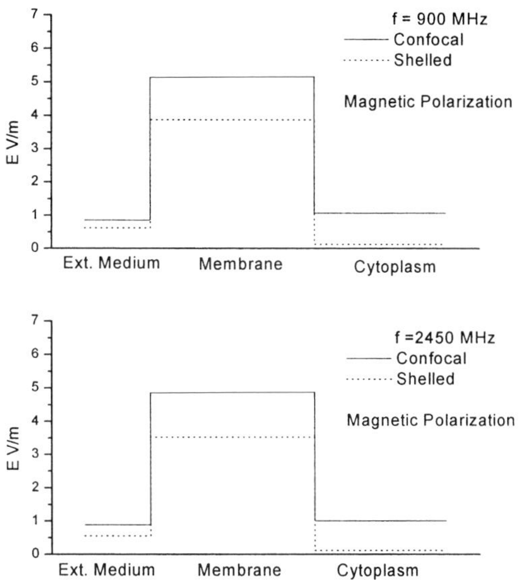

(a)

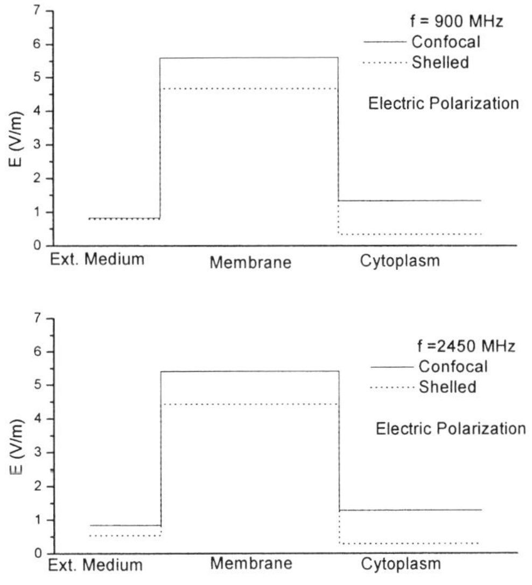

(b)

FIGURE 2. Electric field amplitude (V. m^{-1}) distribution within the membrane and cytoplasm at 900 and 2450 MHz. The electric field is represented along the minor semiaxis of the cell, which is parallel to a) the magnetic field (magnetic polarization) and b) the electric field direction (electric polarization).

As for the influence of the electrical parameters of the different cell parts, Figures 3a and 3b show that the electric field intensity is reduced as the membrane permittivity increases. The results show that the confocal geometry presents again higher values for the electric intensity than the corresponding values for the uniformly shelled model. A comparison of both figures shows that the values of the E field for electric polarization are higher than for magnetic polarization. For the results of Figures 3a and 3b, the cytoplasm and external medium electrical parameters have been kept constant to the values specified in Table I.

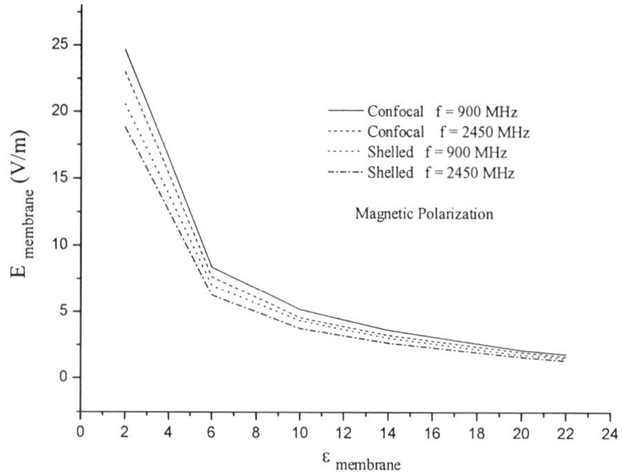

(a)

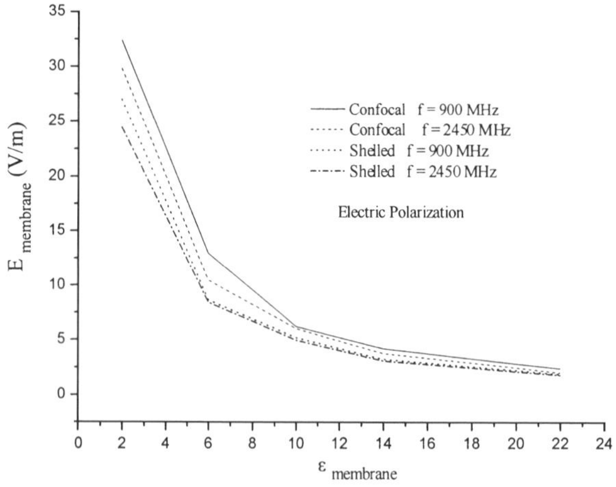

(b)

FIGURE 3. Electric field intensity within the membrane of the shelled and confocal ellipsoids as a function of the membrane permittivity for a) magnetic polarization and b) electric polarization.

Figures 4a and 4b show the influence of the cytoplasm permittivity for magnetic and electrical polarizations respectively. It has been found that the influence of the conductivity of the cytoplasm is not as strong as its permittivity, being the maximum differences between the highest and lowest electric field intensity values within the membrane of 2%.

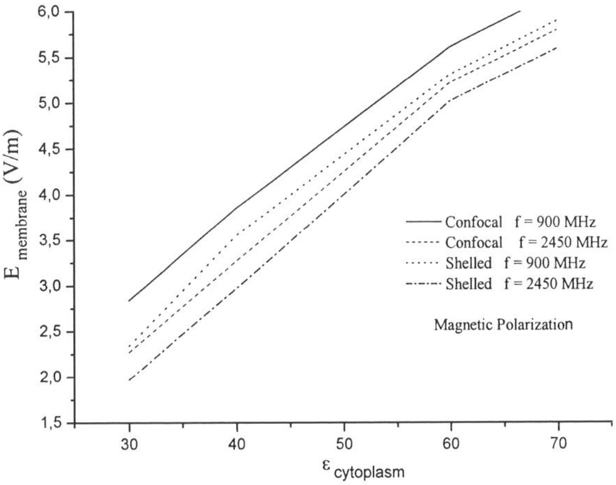

(a)

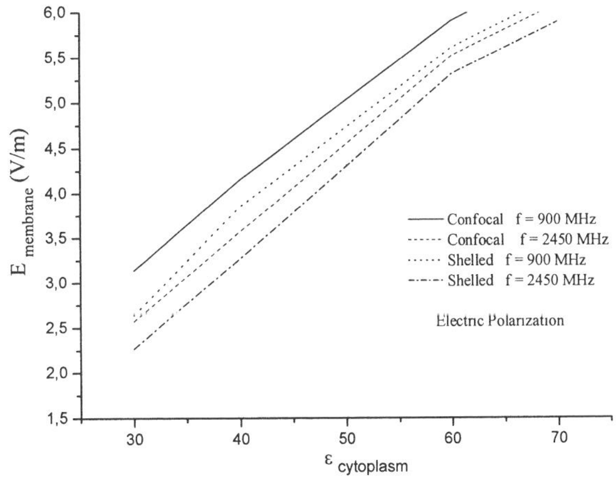

(b)

FIGURE 4. Electric field intensity within the membrane of the shelled and confocal ellipsoids as a function of the cytoplasm permittivity for a) magnetic polarization and b) electric polarization.

CONCLUSIONS

The results presented above clearly show the influence of the geometry, electrical properties and orientation of the cell on the induced field at the cell membrane, as a possible target of harmful effects of field exposure the internal electric field distribution. To calculate the field distribution, an accurate FE computational approach with adaptive meshing and improved boundary conditions has been used. The observed variations of the field within the membrane with the permittivity and conductivity call for the need of accurate theoretical and experimental determination of electrical parameters of cells.

The comparison of our results with other values previously published for spheres and ellipsoids shows the importance of using a realistic shape for modeling a cell, at the expense of mathematical complexity and computing time. As it has been shown, many studies which have used confocal ellipsoids as a model for erythrocytes are rough approximations if a precise simulation of the bioeffects in cells is desired.

REFERENCES

1. Gimsa J. and Wachner D., *On the analytical description of the transmembrane voltage induced on spheroidal cells with zero membrane conductance*, Biophys. J., **30,** 463-466 (2001).
2. Lin J.C., Guy, A.W. and Johnson C.C., *Power deposition in a spherical model of man exposed to 1-20 MHz electromagnetic fields*, IEEE Trans Microwave Theory Tech., **23**, 246-253 (1975).
3. Gimsa J. and Wachner D., *Analytical description of the transmembrane voltage induced on arbitrarily oriented ellipsoidal an cylindrical cells*, Biophys. J., **81**, 1888-1896 (2001).
4. Sebastian J.L., Muñoz San Martin, Sancho M. and Miranda J.M., *Analysis of the influence of the cell geometry orientation and cell proximity effects on the electric field distribution from direct RF exposure*, Phys. Med. Biol. **46**, 213-225 (2001).
5. Muñoz San Martin S., Sebastian J.L., Sancho M. and Miranda J.M., A study of the electric field distribution in erythrocyte and rod shape cells from direct RF exposure, Phys. Med. Biol. 48, 1649-1659 (2003).
6. Liu L.M and Cleary SF, Absorbed energy distribution from radiofrequency electromagnetic radiation in mammalian cell model: effect of membrane-bound water, Bioelectromagnetics, 16, 160-171 (1995).
7. S. Gabriel R., Lau W. and Gabriel C., The dielectric properties of biological tissues III: Parametric models for the dielectric spectrum of tissues. Phys, Med. Biol., 41, 2271-2293 (1996).
8. Gimsa J. and Wachner D., A unified resistor-capacitor model for impedance, dielectrophoresis, electrorotation and induced transmenbrane potential, Biophys. J 75, 1107-1116 (1998).
9. Becache E. and Joly P., On the analysis of Berenger's perfectly matched layers for Maxwell equations, INRIA Raport No. 4164, 2001.
10. Stratton J., Electromagnetic Theory, 513-573, McGraw-Hill, New York, (1941).

Applied Physics (APHYS 2003)
A. Méndez-Vilas (Ed.)

CHERNOBYL CLEAN-UP WORKERS: 17 YEARS OF FOLLOW-UP IN LATVIA

N. Mironova-Ulmane, A. Pavlenko
Solid State Physics Institute, University of Latvia, 31 Miera St., Salaspils, LV-2961, LATVIA

M. EGLITE, E. CURBAKOVA, T. ZVAGULE, N. KURJANE
Medical Academy of Latvia, 16 Dzirciema St., Riga, LV-1007, LATVIA

R.BRUVERE, N. GABRUSEVA
Biomedical Research and Study Centre, 1 Ratsupites St., LV-1067 Riga, Latvia

About 5000 Latvian inhabitants took part at clean up works after accident on Chernobyl Nuclear Power Plant during 1986–1991. Clinical follow-up programme showed that clean-up workers having higher morbidity rate compare to the general population with prevalence of poly–symptomatic sicknesses caused by depression of immune system. Dose reconstruction made by Electron Paramagnetic Resonance (EPR) indicated underestimation of officially documented doses by factor 2-10. Blood measurement performed by Inductively Coupled Plasma Mass Spectrometry (ICP MS) indicated elevated concentration of lead, cadmium and thorium. It was assumed that toxic compounds, physical stress, incorporated long-lived radionuclides are the principal factors affecting health of clean–up workers.

INTRODUCTION

It is already more then 17 years after the accident at Chernobyl Nuclear Power Plant. There are huge number of publications devoted to monitoring of health impact to European population [1,2] analysis of radionuclide distribution and its transfer in the environment [2,3] estimation of absorbed doses to personal taking part in mitigating of accident and etc [4].

At the moment it is very well investigated topic demonstrating decreasing of public and scientific interest all over the world. However, for Republic of Belarus, Russian Federation and Ukraine and parts of some European countries is still subject of great consequence because of the environment and human impact caused by this accident. For instance, about 5000 Latvian inhabitants or 0.45% of all population took part in clean-up activities from 1986 till 1991. Nowadays they present a group of chronically sick people that requires further observation, examination and scientific investigation as well as proper treatment and rehabilitation [5].

MATERIALS AND METHODS

The Cohort

Latvian inhabitants taking part in mitigating activities after the Chernobyl accident is an unique cohort for scientific studies because they had radiation exposure during an exact period of time and then had been moved to not-contaminated region.

The 1320 workers worked close to the reactor and turbine hall, 1130 worked at the installations next to the reactor site, 2213 worked at general environment. These workers were mainly at the age 18-45 at the time of work in Chernobyl, Table 1. People employed between 1986 and 1987 worked 1 to 3 months in average, while those employed later worked for 4 to 6 months in average. The workers employed in 1988 to 1989 were mainly occupied with building of town near the Chernobyl reactor and were not involved in the works on reactor itself, therefore their doses are lower.

Table 1. Distribution of Chernobyl NPP accident clean-up workers by age and time of work in Chernobyl

Age / Year	<18	18-20	21-25	26-30	31-35	36-40	41-45	>45
1986	5	122	712	635	540	387	191	46
1987	2	30	257	347	382	238	74	40
1988	0	2	6	57	227	163	60	9
1989	0	2	4	8	45	47	11	4
1990	0	0	0	0	3	2	2	0
after 1990	0	0	0	0	0	0	3	1
Total	7	156	979	1047	1196	837	341	100

Part of the cohort has officially documented exposure records ranging form 0.01 to 0.5 Gy, Table 2.

Table 2. Doses of radiation received by Latvian Chernobyl NPP clean-up workers

Doses of radiation, Gy	Amount of clean up workers
0.5 and >	3
0.4-0.49	2
0.3-0.39	4
0.2-0.29	515
0.1-0.19	909
0.01-0.1	1218
Not measured	2012
	Total: 4663

Physical methods

Electron Paramagnetic Resonance

EPR retrospective dosimetry is based on measurements of amount of radiation induced radicals in hydroxyapatite (HAP) $Ca_{10}(PO_4)_2(OH)$ [6], which present in mineralized tissues like tooth enamel, dentine and bone. Tooth enamel is more suitable for EPR measurements taking into account high HAP content, slow metabolism and high sensitivity to ionising radiation [7].

For the given studies teeth extracted during medical treatment mostly molars and pre-molars have been selected for measurements. The crown was mechanically separated from the root. The dentine was removed with a hard alloy dental drill keeping enamel as the most suitable for EPR measurements, as in [8]. The enamel was crushed into coarse chips using an agate mortar and pestle.

The samples were kept in alcohol-ethyl mixture for several days for degreasing and dried in air. Following such processing, the mechanically induced radicals decayed and the amplitude of the background signal was reduced. The samples were stored for a few more weeks in silica gel for decreasing of water content.

EPR experiments on teeth enamel have been made at room temperature by an X-band cavity PE-1306 spectrometer. The measurements were compared to the standard samples MgO crystals doped with Mn^{2+}or Cr^{3+}.

Two types of radioactive sources were used for sample irradiation: a ^{137}Cs gamma ray source with energy 0.6662 MeV, and a ^{90}Sr-^{90}Y beta particles source with maximum energies 0.546 MeV and 2.28 MeV respectively. Five measurements, one initial and four after cumulative irradiation were used to achieve linear fit in steps of 500 mGy. The dose was reconstructed by additive dose method [5].

Inductively Coupled Plasma Mass Spectrometry

ICP MS technique has been used for trace element measurements in blood of clean-up workers. Venous blood was collected during medical examination in plastic syringes using stainless steel needles and stored at freezer before the test.

Human blood has very high salt and protein content as well as high viscosity [9], which leads to blocking of the nebulizer. For analysing purposes blood samples have been digested in 10 ml nitric acid and diluted in de-ionised water.

ICP MS measurements have been made using ELAN 6000 (Perkin–Elmer SCIEX, Concord, Ontario, Canada). The instrument was operated in the standard configuration in clean room condition.

Clinical observation

The clinical studies were conducted in the Centre of Occupational and Radiological Medicine, P. Sradins Clinical Hospital. The Latvia State Register for clean-up workers and their children has been created and includes information about 4663 Chernobyl clean-up workers, 97 persons evacuated from Chernobyl and 1250 children from the families of clean-up workers (born after the accident). Register contains data about time and period of stay in Chernobyl, as well as data about place and kind of work, radiation dose received, protective measures used, food consumed, disease and health status. The data about death cases of clean-up workers are registered.

There is a screening system of clean-up workers in every district of Latvia. The local hospitals make a follow-up of every clean-up worker at least once a year including following examinations: physical examination, full blood analysis, biochemical analyzes, EKG, chest x-ray. The registration and more detailed examination of their health condition are carried out in the P. Stradins Clinical Hospital at the Centre of Occupational and Radiological Medicine where the following examinations are performed: estimation of the function and structure of the thyroid gland, analysis in dynamic and correction of the immune status, levels of the heavy metals in the biosamples, endoscopic examinations when necessary (fibrogastroscopy, fibrocolonoscopy, bronchoscopy), functional tests (dopplerography, veloergometry, respiratory function), sonography of thyroid gland and abdominal organs.

The results of medical investigations have been compared with a control group of 237 employees of Ministry of Internal Affairs of corresponding age having no occupational radiation exposure. For more detailed medical tests the control group was subdivided according to profession, duration of service, individual habits, Table 3.

Table 3. Subdivision of control groups according to age, duration of service, individual habits and absorbed dose.

Profession	Clean-up workers	Control
Age of investigated cohort		
Drivers	46.23 ±1.4	45.79±2.98
Other professions	46.10 ±1.18	43.79±1.05
Average age at (2000-2001)	46.2±0.91	44.07±0.99
Work experience		
Drivers	16.25 ±1.54	19.12±2.72
Other professions	15.12 ±0.98	14.66±0.94
Duration of service	15.57±0.85	15.28±0.92
Average absorbed dose, mGy		
Drivers	183.9±1.76	
Other professions	178.0±1.2	
Average dose	180.5±1.06	
Amount of investigated people		
Drivers	85	33
Smokers	146	76
Alcohol users	145	141
Handicaps	192	11

RESULTS AND DISCUSSION

Latvian clean-up workers provide an opportunity to investigate mainly the short-term impact of ionizing radiation and toxic materials on the human health. After working in Chernobyl for a limited period of time (mainly 3-6 months) they have been residing in territories free from radioactive contamination.

In the previous studies [10] the content of microelements and macro element calcium in the Chernobyl clean-up worker teeth in comparison with practically healthy man teeth, obtained from dentist has been measured. It was found presence of strontium (40-250 μg/g), thorium (0-200 μg/g) in clean up workers teeth. In teeth of healthy individuals

these elements have not been detected. Calcium content in clean-up workers teeth varied from 25% to 33% compare to 36% in healthy individual.
The doses measured by EPR on Chernobyl clean-up workers enamel are always higher as documented individual doses [5,10]. The teeth dose included all external exposure from natural background radiation, medical examinations, clean-up works and internal exposure from natural and artificially incorporated radionuclides in the teeth. Some authors claiming evolution of EPR signal related to dosimetry by Ultraviolet light [11]. This probably is not the case for given studies taking into account the position of extracted teeth in mouth and solar activity in Latvia.
The main radionuclide discovered in Latvian clean-up workers teeth was ^{90}Sr [12], beta-emitting radionuclide with long physical half-life. It is considered like bone-seeking radionuclide uniformly distributed throughout the volume of mineral bone [13] and therefore, through the other calcified tissues, teeth among them. The biological half-life for ^{90}Sr is about 33 years for cortical bone and 5.6 years for trabecular bone [14]. The maximum range of beta particles in calcified tissues from ^{90}Sr is 1 mm and form its daughter product ^{90}Y is around 3 mm therefore the organs of preliminary concern are cells near bone surface and marrow. The long biological half-life, dense ionization, target organs of irradiation and high concentration in calcified tissues make ^{90}Sr the radionuclide heavily affected health of clean-up workers.
Blood analysis performed by ICP MS method showed elevated concentration of toxic elements in blood of clean up workers, Table 4. The lead concentration for workers Nr. 2, 4 and 7 is much higher then average value for European population. For example, for European countries using unleaded fuel the average lead concentration is around 10-230 μg/l of blood [15].

For given analysis the element concentrations have been compared with the blood of individual from the same region (P), where the lead containing petroleum still in use. For this individual the lead concentration in blood was 245 μg/l. As mentioned in [16] the usage of leaded petroleum is the main source of lead exposure to population, however increase of lead concentration in 5 times for worker 2 and 4 times for worker 7 could be attributed mainly to occupational exposure. In general, lead content in blood of Chernobyl clean-up workers at least two times higher compared to general Latvian population [17].
In clinical picture Chernobyl clean-up workers most commonly have complaints of headache (70%), dizziness and weakness of memory, prostration, lowering of working abilities (76%). Characteristic attack with loses of consciousness, impotence is observed. There are complaints of changeable arterial blood pressure, excessive nervousness, local cramps and pains in the bones (70-72%) [18].
The disease prevalence rate of clean-up workers is much higher that that in the age matched population group gradually increases even 10-15 years after the work in Chernobyl [19]. Diseases of nervous, digestive, circulatory system, mental disorders and diseases of muscles and connective tissue are the most frequent in the clean-up workers group, Table 5.

Every clean-up worker has more than one disease [20]. Furthermore, the disease prevalence rate with the most frequent diseases was significantly higher in the clean-up workers who suffered with seizures of unconsciousness in comparison with the control group.

Table 4. Concentration of chemical elements in blood of clean-up workers and healthy individual (P), (μg/l).

Worker / Analyte	1	2	3	4	5	6	7	P	Reference values
Al	62006	10450	9899	6109	5306	5325	2821	44724	2000-10200*
Co	1	2	1	1	13	1	1	2	20-40*
Ni	148	142	79	47	14	460	121	487	300-1100*
Cu	4193	2779	121 4	3736	3287	2740	2361	4314	570-1850*
Zn	8047	7289	624 4	6460	9535	7207	6470	7217	700-1140*
As	336	517	340	351	303	359	281	653	460-4000*
Rb	1412	1300	160 5	1547	1431	1669	1468	2618	155-315*
Sr	156	236	59	84	115	71	145	221	No data
Mo	3	5	5	2	1	1	3	8	300-1300*
Cd	0	16	39	3	0	0	6	1	0.7-6**
Sn	12	80	107	91	68	69	53	5	1000*
Cs	3	2	3	23	5	3	19	8	450-1180**
Ba	1247	870	451	485	5224	505	761	1198	No data
La	2	4	5	4	2	4	4	9	4-6*
Hg	10	7	9	0	7	7	0	10	2-9**
Pb	100	1201	122	839	196	175	575	245	50-200*
Th	7	50	156	4	2	4	2	17	No data
U	3	11	0	5	3	4	5	4	NO data

* taken from [9]
**taken from [15]

From 1996 to 2000, 43 Chernobyl accident clean-up workers with suspected thyroid cancer were operated. The thyroid cancer was actually found in 21% of the cases. There were 6 papillary adenocarcinomas and 3 follicular adenocarcinomas found. The first thyroid cancer was discovered in 1996 – after ten years of latent period. Six patients having thyroid cancer worked at Chernobyl in 1986 during so called "iodine period" [21]. The morphological modifications of thyroid tissue in Chernobyl clean-up workers were follicle atrophy with epithelium sclerosis, which is more frequent than for other men in control group. Thyroid cancer occurs 10,6 times more frequently in the Chernobyl accident clean-up workers and at in the earlier age than in the rest of population (selection from 1990 till 2000).

Table 5. Incidence (%) of general diseases between all diagnosis in hospitalized patient per 1000 (age group 45-59 years)

Diseases	Groups	1992	1994	1996	1997	1998	1999	2000
Diseases of digestive system	Clean-up workers	17.2	18.6	21.8	19.2	16.1	14.2	14.2
	Inhabitants	No data	No data	10.3	10.3	9.8	9.15	10.2
Diseases of nervous and sensory organs	Clean-up workers	15.3	16	18.4	22.1	31.3	27.5	24.5
	Inhabitants	No data	No data	5.2	5.1	5.1	5.2	4.8
Mental disorders	Clean-up workers	13.2	13.8	11.7	10.5	7.2	7	10.4
	Inhabitants	No data	No data	9.7	8.5	9	9.2	9.5
Diseases of circulatory system	Clean-up workers	8.2	9.7	9	9	6.7	6.5	8.4
	Inhabitants	No data	No data	16.9	17.1	16.9	17	16.8
Diseases of the musculoskeletal system and connective tissues	Clean-up workers	16.7	18	19.1	17.2	13.6	12.7	12.4
	Inhabitants	No data	No data	9.8	10	10.2	10.6	10.6

At 2001 the status of whole immune system for 59 Chernobyl clean-up workers with most common thyroid diseases – euthyroid nodular and diffuse goiter has been tested [22]. There were 47 healthy blood donors taken as a control. The level of immunoglobulins (IgA, IgG and IgM), the number of peripheral blood leukocytes (PBL), lymphocytes (Ly), monocytes (Mo), T-lymphocytes and their subpopulations (CD3+, CD4+, CD8+), B-lymphocytes (CD19+), NK cells (CD16+), classical and alternative pathway activity of complement (CH50, APH50), the C3 split product C3d, and neutrophil phagocytosis were determined in the peripheral blood serum, Table 6.

Significantly decreased number of CD16+ cells (natural killer), of CD4+ and CD8+ T-lymphocytes, a reduced neutrophil phagocytic activity as well as significant complement activation in Chernobyl clean-up workers was found compare with control group. In addition, the number of CD3+ and CD4+ cells was significantly higher in patients with nodular goiter when compared with patients with diffuse goiter. Levels of IgG and numbers of monocytes were significantly decreased in those persons who worked in Chernobyl during the first 2 months after the accident (maximum radiation exposure).

These results indicate that even after a period of 15-17 years an impact of radiation and toxic compounds on immune system can be observed that could be possible reason for thyroid disorders as well as for other somatic [22, 23] and oncologycal diseases.

Table 6. Immunological profile of Chernobyl clean-up workers with and without thyroid diseases.

Indices	Chernobyl clean-up workers		Control group
	with thyroid pathology	without thyroid pathology	
PBL	6415.2±165.7 n=59	6514±93.3 n=326	6153.2±298.1 n=47
Ly, %	26.4±0.9 n=59	28.19±0.44 n=326	28.7±2.07 n=47
Mo, %	5.5±0.28 n=59	4.9±0.15 n=326	5.1±0.2 n=47
CD3+, %	57.3±2.3* n=59	58.6±0.97* n=171	75.7±1.37* n=47
CD4+, %	41.5±1.1* n=59	39.2±0.64* n=296	46.6±1.89* n=47
CD8+, %	25.8±0.9* n=59	26.49±0.46* n=270	30.46±1.17* n=47
CD16+, %	14.2±0.8**/* n=44	17.08±0.44**/* n=239	23.21±2.21* n=47
CD19+, %	14.4±1.4 n=40	16.5±1.4 n=129	18.36±0.6 n=47
IgA, mg%	274±11.18 n=59	280.9±6.23 n=318	279±15 n=47
IgG, mg%	1140.6±35.6**/* n=59	1051.7±15.2**/* n=323	1209±32* n=47
IgM, mg%	135.4±6.5* n=59	138.15±4.06* n=318	119±7* n=47
Neutrophil phagocyte activity, %	41.57±11.7* n=57	30.9±2.06* n=91	78.3±1.7* n=47
CH50, %	93.5±4.15** n=24	71.52±5.1** n=37	89.03±1.5 n=21
APH50, %	100.2±9.76 n=15	96.47±7.41 n=15	98.1±1.63 n=20
C3d, mU/l	83.04±11.72* n=24	78.74±10.9* n=31	37.2±1.6* n=17

CH50, normal level in blood serum 80-120 %
APH50, normal level in blood serum 80-120 %
C3d, normal level in blood serum <55mU/l
PBL, normal range 4500-9500
Ly, normal range 20-40%
Mo, normal range 4-8%

CONCLUSION

Latvian inhabitants participated in clean-up activities after Chernobyl accident represent group of chronically sick people with prevalence of neurological, gastrointestinal and musculoskeletal systems diseases. Observed clean-up workers have disturbances of the immune functions leading to higher rate of somatic and oncologycal diseases compare to general Latvian population.
Retrospective EPR dosimetry on teeth enamel of Chernobyl clean-up workers demonstrated underestimation of officially documented doses at factor 2-10 claiming the incorporated ^{90}Sr as possible source for internal irradiation.
Elevated concentration of toxic elements in blood discovered by ICP MS was an additional factor affected health status of clean-up workers.

REFERENCES

1. World Health Organization. International Programme on the Health Effects of the Chernobyl Accident. *Health consequences of the Chernobyl accident: results of the IPHECA pilot projects and related national programmes*: Scientific report. ISBN: 5-88429-0 (1996)
2. International Atomic Energy Agency. *Testing of Environmental Transfer Models Using Chernobyl Fallout From the Iput River, Catchment Area, Bryansk Region, Russian Federation.* ISBN 92-0-104003-2. (2003)
3. International Atomic Energy Agency. *Present and Future Environmental Impact of the Chernobyl Accident.* IAEA TECDOC Series No. 1240. (2001)
4. United Nations Scientific Committee on the Effects of Atomic Radiation UNSCEAR 2000 *Report to the General Assembly. Sources and Effects of Ionizing Radiation. Annex J: Exposures and effects of the Chernobyl accident.* p.453-551
5. Mironova-Ulmane, N., Pavlenko, A., Zvagule, T., Kärner, T., Bruvere, R. and Volarte, A. *Retrospective dosimetry for Latvian workers at Chernobyl.* Radiat. Prot. Dosim. **96**(1) 37-240 (2001)
6. International Commission on Radiation Units and Measurements. *Retrospective Assessment of Exposures to Ionising Radiation.* (ICRU Report 68). Journal of UCRU 2(2), 2002
7. International Atomic Energy Agency. *Use of Electron Paramagnetic Resonance Dosimetry With Tooth Enamel for Retrospective Dose Assessment.* IAEA-TECDOC-1331. (2002)
8. Liidija, G. and Wieser, A. *Electron Paramagnetic Resonance o Human Tooth Enamel at High Gamma Ray Doses.* Radiat. Prot. Dosim.**101**(1-4) 503-506 (2002)
9. Versieck, J., Cornelis, R. *Trace element in human plasma or serum.* (CRC Press). ISBN 0-8493-6810-3. (2000)
10. Mironova, N., Eglite, M., Churbakova, E., Zvagule, T., Riekstina, D. *Electron spin resonance and instrumental neutron activation analyses of Chernobyl nuclear power plant accident clean-up worker teeth.* Proc. Latvian Acad. Sci. **52** 194-196 (1998)
11. Liidja, G., Past, J., Puskar, J. and Lippmaa, E. *Paramagnetic resonance in tooth enamel created by ultraviolet light.* Appl Radiat Isot. **47**(8) 785-758 (1996)
12. Bruvere, R., Mironova-Ulmane, N., Feldmane, G., Volrate, A., Zvagule, T. *β-radioactivity of teeth and the ability of leukocytes to produce interferons in Chernobyl incident clean-up workers.* Proc.VI Int. Conf. Med. Phys., Physica Medica. **15** 139 (1999)
13. Romanyukha, A.A., Seltzer, S.M., Desrosiers, M., Ignatiev, E.A., Ivanov, D.V., Bayankin, S., Degteva, M.O., Eichmiller, F.C., Wieser, A. and Jacob, P. *Correction factors in the EPR dose reconstruction for residents of the Middle and Lower Techa riverside.* Health Phys. **81**(5) 554-566 (2001)

14. The International Commission on Radiological Protection. *Limits for Intakes of Radionuclides by Workers: Part 1.* ICRP Publication No 30.Annals ICRP 2 (3/4) ISBN 0 08 022638 8 (1979)
15. Barany, E., Bergdhall, I.A., Schutz, A., Skerfving, S. and Oskarsson, A. *Inductively Coupled Plasma Mass Spectrometry for Direct Multi-Element Analysis of Diluted Human Blood and Serum.* J. Analyt. Atom. Spec.**12** 1005-1009 (1997)
16. Truckenbrodt, R., Winter, L.and Schaller, K.H. *Effect of occupational lead exposure on various elements in the human blood. Effects on calcium, cadmium, iron, copper, magnesium, manganese and zinc levels in the human blood, erythrocytes and plasma in vivo.* Zentralbl. Bakteriol. Mikrobiol. Hyg. **179**(3) 187-197 (1984)
17. Kurjane, N., Bruvere, R., Shitova, O., Romanova, T., Jaunalksne, I., Kirschfink, M. and A.Sochnevs. *Analysis of the immune status in Latvian Chernboyl clean-up workers with non-oncological thyroid diseases.* Scand. J. Immunol. **54** 528-533 (2001)
18. Curbakova, E., Dzerve, B. and Eglite, M. *Health status and follow-up of the Chernobyl NPP accident liquidators in Latvia.* Proc. 1st Int. Conf. Rad. Conseq. Chernobyl. Acc., Minsk. Report Eur 16 544, pp. 929-934 (Brussels: European Commission) (1996)
19. Zvagule, T., Mironova-Ulmane, N., Pavlenko, A., Kartner, T., Bruvere, R., Garbuseva, N., Volrate, A. *Retrospective Dosimetry and Clinical Follow-up program for Chernobyl Accident Clean-up workers in Latvia.* Proc. IRPA Reg. Cong. Rad. Prot. CD-ROM 6o-03 ISBN 953-96133-3-7 (2001)
20. Curbakova, E., Fabtuha, T., Zvagule, T., Eglite, M., Jekabsone, I. and Eglite, A. *The health status of Chernobyl nuclear power plant accident liquidators in Latvia.* Proc. Latvian Acad. Sci., **52** 187-191 (1998)
21. Kurjane, N., Groma, V., Orļikovs, G., Ritenberga, R., Skudra, M., Lemane, R., Lemanis, A., Čurbakova, E. and Socnevs, A. *Thyroid disorders in Chernobyl clean-up workers from Latvia.* Proc. Latv. Acad. Sci. **53**(6) 315-321 (1999)
22. Bruvere, R., Gabruseva, N., Volrate, A., Heisele, O., Feldmane, G., Zvagule, T., Balodis, V. *Functional deficiency of the immune system of the Chernobyl accident clean-up workers residng in Latvia.* Proc. Latv.an Acad. Sci., **57** 17-21 (2003)
23. Bruvere, R., Heisele, O., Zvagule, T. and Curbakova, E. *The immune state of Latvia's inhabitants involved in the clean-up of radioactivity in Chernobyl.* Acta Medica Baltica. **1** 30-37 (1994)

THREE-DIMENSIONAL DOSE DISTRIBUTION AROUND AIR AND BONE IN TISSUE MEASURED BY MRI-BASED POLYMER GEL DOSIMETER

Yoichi Watanabe and Rob Mooij
Department of Radiation Oncology, Columbia University
622 W168th St., New York, NY 10032, USA

Gerard M. Perera
Department of Medical Physics, Memorial Sloan-Kettering Cancer Center
1275 York Ave., New York, NY 10021, USA

Abstract - Heterogeneity correction in dose calculation is necessary for radiation therapy treatment plans. Dosimetric measurements of the effects are hampered if the detectors are large and their radiological characteristics are not equivalent to water. Gel dosimetry can solve these problems. Furthermore, it provides detailed three-dimensional (3D) dose distributions. We used a cylindrical phantom filled with BANG-3 polymer gel to measure 3D dose distributions in inhomogeneous media. The phantom has a cavity, in which water-equivalent or bone-equivalent solid blocks can be inserted. The irradiated phantom was scanned with an MRI scanner. Dose distributions were obtained by calibrating the gel for a relationship between the absorbed dose and the spin-spin relaxation rate. We observed 3D dosimetric structures around the heterogeneity. The dose distributions were unique to the photon energy (6MV or 18MV), the field size and the type of heterogeneity material (air or bone).

INTRODUCTION

Heterogeneity in a medium could cause a significant change in absorbed dose in comparison to the dose in a homogeneous medium. If the effects are not considered correctly, it may result in undesirable dose delivery for radiation therapy. Therefore, the dose distributions in tissue with heterogeneity such as air-gap and bone are of great interest to medical physicists and radiation oncologists. There are many studies presenting measurements in inhomogeneous media for clinically relevant beam energies. However, the geometrical set-up of the measurements is mostly limited to one-dimensional (1D) slab geometry, for which heterogeneity is created as a slab sandwiched between tissue-equivalent materials. There are two-dimensional (2D) dose measurements using radiographic films, thermoluminescent dosimeters (TLD), and ionization chambers. Full three-dimensional (3D) dose measurements in inhomogeneous media have rarely been attempted. Therefore, valid 3D measurements are urgently needed to evaluate computational methods, which may accurately estimate the dose in 3D inhomogeneous media. Polymer gel dosimeter may provide true 3D absorbed dose distributions with high spatial resolution. There are some investigations, which apply polymer gel or Fricke gel to dose measurements in inhomogeneous media[1, 2, 3, 4, 5].

There exists electronic disequilibrium (EDE) near heterogeneity due to the difference in the kinetic energy transferred from incident photons to secondary electrons and absorbed dose[6]. Some calculation methods may be able to accurately predict the dose in the EDE regions and experimental verification of those methods, in particular, in 2D or 3D, is required. Hence, we are particularly interested in measuring the dose distributions in the EDE region. In this article we will present results of dose measurements with MRI-based polymer gel dosimetry. The dependence of 2D dose distributions near the heterogeneity on the type of heterogeneity (air or bone), the photon beam energy, and the field size are studied.

MATERIAL AND METHODS

Gel dosimetry

We used the BANG-3 polymer gel manufactured by MGS Research, Inc. (Guilford, CT, USA). When the polymer gel is irradiated, polymerization progresses, leading to absorbed dose-dependent structural variation[7]. This physical change can be detected by using a magnetic resonance imaging (MRI) scanner, an optical scanner, an X-ray computed-tomography (CT) device, or an ultrasound device. For this study we used MRI to measure spin-spin relaxation time (T2). The spin-spin relaxation rate (R2), which is the inverse of T2, is proportional to the absorbed dose. Therefore, MRI data can be converted to absorbed dose distributions with adequate calibration.

Phantoms

A heterogeneity phantom containing BANG-3 polymer gel was designed and manufactured. The phantom is a 16 cm diameter and 15 cm long cylinder with two hexagonal end plates for stable horizontal positioning. It has a 6 cm diameter and 6 cm deep cylindrical cavity on one of the flat ends. This cavity can be filled with cylindrical inserts made of solid bone-equivalent and water-equivalent materials to simulate tissue heterogeneity. Figure 1 is a photograph of the phantom. Note that the container has an inlet of about 2 cm diameter, through which the gel is poured into the container. The phantom container is made of oxygen-impermeable Barex plastic (BP, Naperville, ILL, USA). The thickness of the Barex wall is about 0.7 mm.

Figure1. Photograph of a heterogeneity phantom. The cavity for heterogeneity inserts at one end of the cylindrical phantom is visible. A part of a white water-equivalent block can be seen inside the cavity.

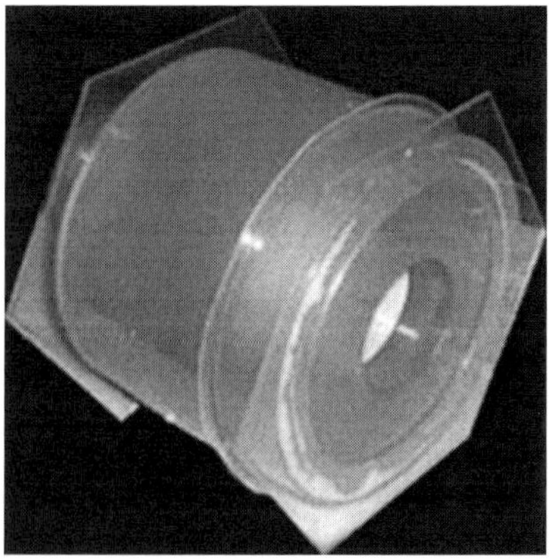

Polymer gel contained in 2.5-cm diameter and 9-cm long cylindrical glass vials was used to obtain a calibration relation between absorbed dose and R2 value. The polymer gel for the calibration was taken from the same batch as that used to fill the heterogeneity phantom.

The BANG-3 polymer gel simulates the radiological characteristics of water. The water-equivalent material includes some heavier elements such as Mg and Al; but the over all characteristics such as mass density and electron density are similar to water. The bone-equivalent material simulates human bones, in particular, the cortical bone. Refer to a recent publication(5) for more details on the materials.

Experimental procedure

A heterogeneity phantom was mounted in the Leksell stereotactic frame. It was irradiated with photon beams of a Varian linear accelerator. It was set up at 95 cm to the top surface of the phantom. Eight calibration vials containing polymer gel were irradiated with a 6 MV photon beam to 0.3, 0.6, 0.8, 1.0, 1.3, 1.6, 2.0 and 2.5 Gy. One vial was left unirradiated as a control.

A heterogeneity phantom was placed in a head coil and scanned with a 1.5 T SIGNA MRI scanner (GE Medical Systems, Waukesha, WI, USA). We used the Hahn spin-echo pulse sequence. The repetition time was 2 s. The echo times were 20 and 100 ms. The echo times were determined to receive sufficiently large signal for expected spin-spin relaxation time (T2) of longer than 40 ms (or R2 $< 25\ s^{-1}$). The field of view was 24 cm by 24 cm. Acquisition matrix was 256 by 256. Hence, the pixel size was 0.937 mm by 0.937 mm on the image plane. The number of acquisitions per scan was set to 2. The receiver bandwidth was 32 kHz. The frequency encoding was made in the direction of the static magnetic field. The slice thickness was 5 mm with no overlap between adjacent slices. We chose the transverse axis perpendicular to the cylinder axis for the slice select axis. All calibration vials were mounted in a homemade mounting block made of Styrofoam. The block was MRI-scanned independently from the heterogeneity phantom, but using the same imaging parameters.

Analysis methods

MRI data were transferred to a workstation for analyses. With in-house software the spin-spin relaxation rate was calculated. The MR signal decay was modeled by an exponential equation for the calculation(8). A relation between R2 and absorbed dose was derived from the calibration vial data. We converted the data from R2 to absorbed dose and displayed dose distributions with MATLAB (MathWorks, Inc., Natick, MA, USA).

Irradiation experiments

We studied the effects of air-gap and bone on dose distributions. Air heterogeneity was simulated by placing a 3-cm thick water-equivalent insert in the top part of the phantom cavity and leaving a 3-cm thick air-gap. For the bone heterogeneity experiment, the 3-cm thick cavity was filled with a bone-equivalent material. The inserts were removed for MRI scanning and replaced with water-equivalent inserts to minimize MR image artifacts. For irradiation we used photon beams of 6 MV and 18 MV energies. The field sizes were 4x4 cm^2 and 10x10 cm^2 square fields.

RESULTS

Calibration

Figure 2 shows the data points (filled circles) of a calibration experiment. The figure also includes regression curves: a linear equation and a quadratic equation. The quadratic equation fits the experimental data slightly better than the linear equation as evident by the goodness-of-fit value, R^2, which is closer to unity. For this study we used a linear equation by considering that there is about 5% uncertainty with the measured R2 values. It is noted that the calibration was done with a 6MV photon beam; but the data hold for a 18 MV photon beam because the sensitivity of the BANG-3 polymer gel is not energy dependent in the MeV range.[9]

Figure 2. Absorbed dose [Gy] vs. spin-spin relaxation rate R2 [1/s]. Linear (thick solid) and quadratic (thin solid) curves obtained by regression analyses are displayed together with the original data points (filled circles). The goodness-of-fit values for linear and quadratic equations were 0.9859 and 0.9977, respectively.

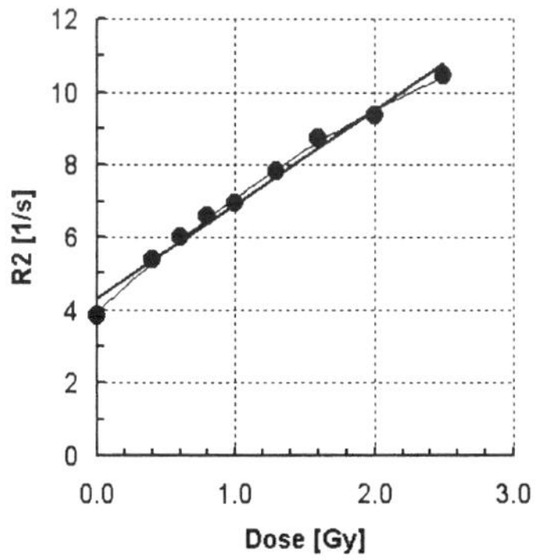

Air heterogeneity

Isodose distributions on a sagittal plane are presented for 6 MV and 18 MV photon beams in Figures 3 and 4. The field size was 10x10 cm^2 for those measurements. In this article photon beams enter the phantom from the left.

Figure3. Isodose distributions on a sagittal plane: 6MV photon beam, 10x10 cm^2 field size, and air-gap. (a) Entire view. (b) Magnification plot of the region near the top-right corner of the air-gap.

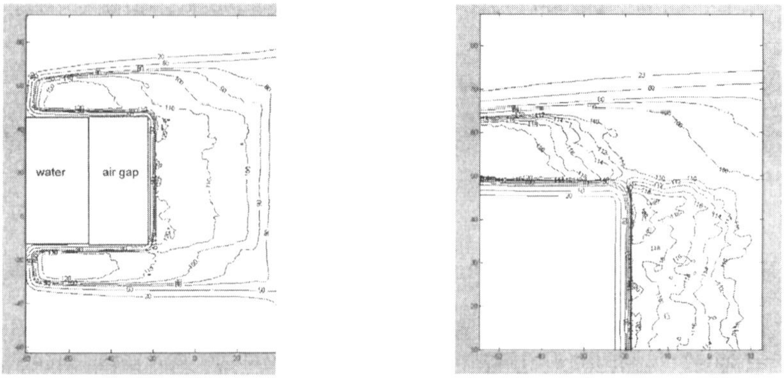

a

b

In Figure 3 (a) the dose along the line transverse to the beam axis at 1 cm downstream from the air-gap indicates that the dose on the downstream side of the air-gap is 10% higher than the dose at locations not beyond the air-gap. Figure 3 (b) is a magnification of the right top corner of the air-gap. The figure indicates a bifurcation of the isodose lines at the corner. The dose is the smallest in the 45-degree outward direction with respect to the beam direction. It is also noted that there is no dose build-up region around the air-gap, whereas there is the common 1.5-cm thick build-up near the beam entrance.

Figure4. Isodose distributions on a sagittal plane: 18MV photon beam, 10x10 cm^2 field size, and air-gap. (a) Entire view. (b) Magnification plot of the region near the top-right corner of the air-gap.

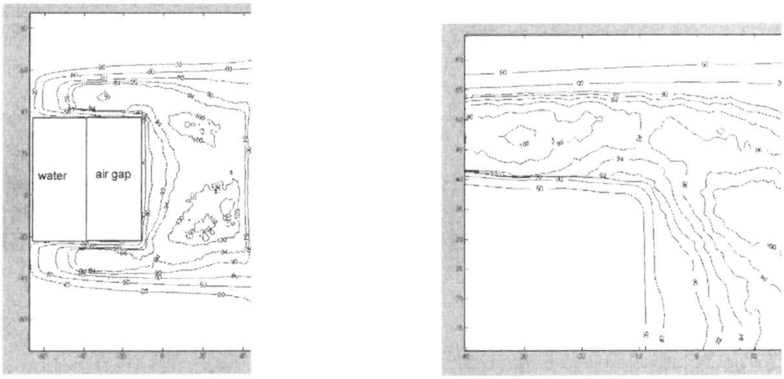

a b

Figure 4 (a) illustrates the dose distribution for the 18 MV photon beam. The isodose distribution is very different from that for the 6 MV case. The dose increase beyond the air-gap is not observable as clearly as the 6MV case. We can identify a dose build-up region on the downstream side of the air-gap and on the sides near the bottom face of the air-gap. The build-up depth on the downstream side of the air-gap is the largest at the center of the air-gap (or near the beam axis) and it is about 3 cm. The 94% isodose line in Figure 4 (b) shows that the width of the lateral build-up region on the side of the air-gap is about 0.5 cm.

The field size relative to the width of air-gap affects the dose distribution. Figure 5 shows an isodose distribution on a sagittal plane for a 4x4 cm^2 field and an 18 MV photon beam. Because there is no contribution of electrons generated in the gel at the sides of the air-gap for such a small field, the air-gap leads to a full dose build-up at the downstream side of the air-gap. The build-up depth is approximately 3.5 cm. The result represents a well-known depth dose profile beyond an air-gap for a slab geometry case[10, 11].

Bone heterogeneity

Figures 6 and 7 show measured isodose distributions in a phantom with bone heterogeneity for 6 MV and 18 MV photon beams, respectively.

High-density material such as bone reduces the amount of photons transmitting through the medium and leads to a decrease in absorbed dose on the downstream side of the heterogeneity. Figure 6 (a) shows that the dose beyond the bone is 10 % smaller than the dose in the region that is not behind the bone at the 1 cm from the bone-gel boundary. Figure 6 (b) shows more clearly the isodose distribution near the top-right corner of the bone insert. The 90% isodose line extends toward the downstream side in the beam direction. Lower isodose lines, i.e., 85% and 82%, extend into the region downstream from the bone.

Figure5. Isodose distributions on a sagittal plane: 18MV photon beam, 4x4 cm^2 field size, air-gap.

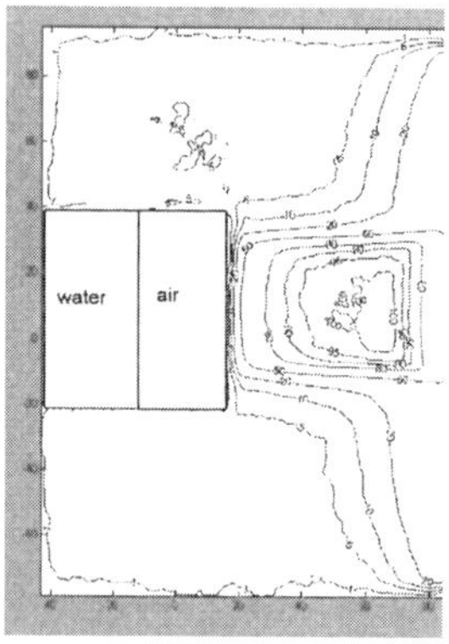

Figure6. Isodose distributions on a sagittal plane: 6MV photon beam, 10x10 cm^2 field size, and bone heterogeneity. (a) Entire view. (b) Magnification plot of the region near the top-right corner of the bone.

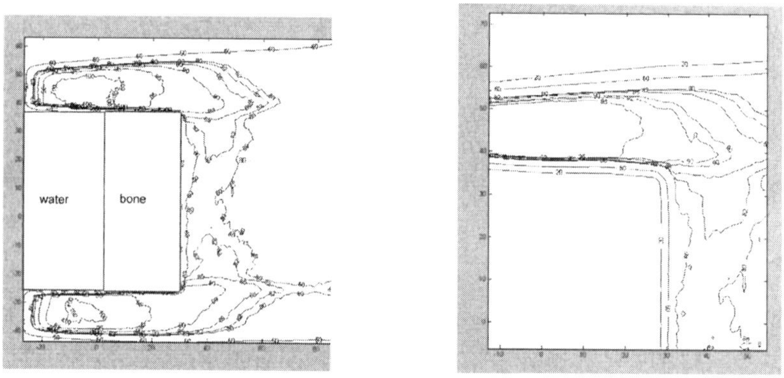

a b

The isodose distribution for the 18 MV photon beam is different from the 6 MV case. Although there is a dose decrease behind the bone, there is now a noticeably dose build-up region. The build-up depth is about 3 cm. There are dose build-up regions in the lateral corners of the bone as seen in Figure 7 (b). A comparison of Figure 7 with Figure 4 indicates a similarity of the dose distributions of the bone heterogeneity to that of the air-gap. A difference between air-gap and bone heterogeneities for the 18MV photon beam can be recognized on the transverse plane at a 3 cm depth from the face of heterogeneity, where the bone case indicates a dose decrease, but the air-gap leads to a dose increase near the beam axis.

Figure7. Isodose distributions on a sagittal plane: 18MV photon beam, 10x10 cm^2 field size, and bone heterogeneity. (a) Entire view. (b) Magnification plot of the region near the top-right corner of the bone.

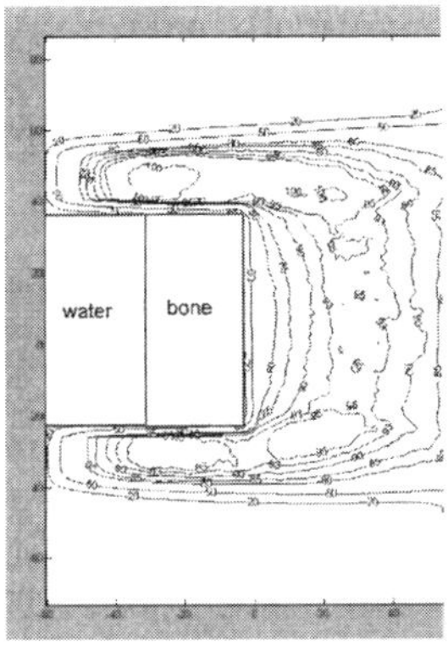

a

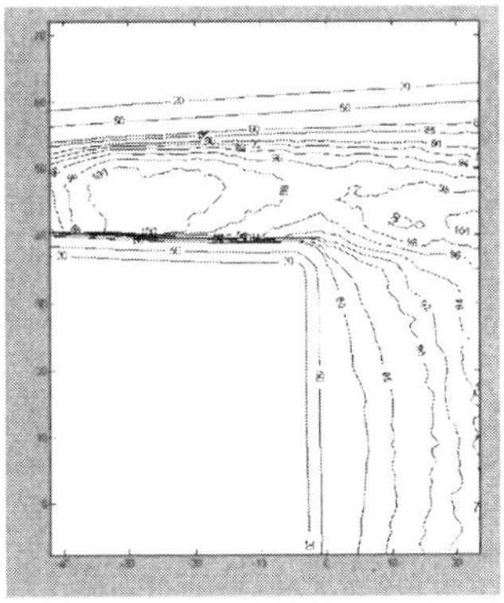

b

DISCUSSION

The dosimetric patterns observable in the measured isodose distributions of Figures 3 to 7 are generated by intricate interplay between the energy carried by photons and electrons. When a water-like medium is irradiated with a high energy photon beam, i.e., the energy ranging from 1 to 20 MeV, the photon energy is transferred to electrons through one of the following interaction mechanisms: photo-electric interactions, Compton interactions, and pair productions. The major interaction process with 6 MV and 18 MV photon beams is the Compton interaction, with which a secondary electron is liberated from an atom, whereas the incident photon is scattered with less kinetic energy than before the collision. The energy carried by the secondary electron is called KERMA. The energy absorption in a medium is due to those secondary electrons. At a shallow depth in a medium (or build-up region) the photon energy is transferred to electrons. Since those electrons do not loose the energy instantly, there is imbalance between KERMA and the absorbed energy (or dose) in the shallow depth. Such a phenomenon is called "electronic disequilibrium" (EDE)(6).

Dosimetric patterns in the downstream side of heterogeneity are complex for electron beams(6). Localized hot and cold spots are developed near the corner of an air-gap or high-Z heterogeneity. In the EDE region the electron transport processes dominate the formation of specific dosimetric patterns. Hence, patterns similar to those of the pure electron beams could be observed.

Figure 3 (a) shows that the dose in the region downstream from the air-gap is significantly larger than in the region below the gel for the 6 MV photon beam because of less attenuation of photons in the air-gap. Near the bottom corner of the air-gap there are more electrons scattered inward (or toward the beam axis) from the gel on the side of the air-gap than electrons scattered outward from the air-gap and the water-equivalent insert above the air-gap; hence, there is a sharp dose gradient along the line transverse to the beam axis. The electrons produced in the sidewall of the air-gap penetrate into the air and reach the bottom corner of the air-gap, creating the localized 120% hot spots near the air-gel interface.

The build-up depth of an 18 MV photon beam is 3.5 cm; hence, full electronic equilibrium condition is not established in the 3-cm thick water-insert above the air-gap. The bottom face of the air-gap acts as a beam entrance surface. Consequently, there is a dose build-up region around the air-gap as seen in Figure 4 (b). This also explains why the build-up depth on the air-gap interface is about 3 cm instead of 3.5 cm at the beam entrance. More Compton electrons are emitted in the forward direction as the photon energy increases. Hence, few secondary electrons produced in the gel on the side of the air-gap travel inward. As a result, there is no sharp dose gradient along the transverse direction in the region downstream from the air-gap.

The bone-equivalent material has about two times more electrons per unit volume. Hence, photons are attenuated twice more in the bone than in the gel. At the same time, twice more electrons are generated in the bone than in the gel. These electrons travel outward and deposit their energy in the region downstream from the gel on the side of the bone. This leads to a formation of localized hot spots. Figure 6 shows not a localized hot spot but a 90% isodose line locally extended toward the downstream side for the 6 MV photon beam. Such a hot spot is evident for the 18 MV case in Figure 7, which displays localized 100% hot spots. The 82% and 95% isodose lines extend into the center of the beam as seen in Figure 6 (a) and Figure 7 (a), respectively. These might be created by the photons scattered inward from the outer portion of the field.

The formation mechanisms of a build-up region near high-Z material in slab geometry are discussed by Werner et al[12, 13]. They predict a build-up region on the downstream side of aluminum, which is radiologically similar to bone, only for low energy photons, i.e., a 6 MV photon beam. However, they expect a dose build-down for higher energy photons such as an 18 MV photon beam. Therefore, our results for the 18 MV case disagree with their prediction. Our results may be explained by a difference of particle transport processes in 3D geometry and 1D (or slab) geometry. Because of the cylindrical shape of the phantom, more photons and electrons are scattered inward and travel to the center of the field than in the slab geometry. The maximum effects of the scattered photons take place at some distance from the bone-gel interface. This might lead to the dose build-up in 3D.

The physical explanations given above are less quantitative; but our experimental observations reveal an interesting interplay between photons and electrons in the EDE regions and warrant further investigation.

CONCLUSIONS

In this article we have presented initial results of three-dimensional dose measurements in heterogeneous media using BANG-3 polymer gel dosimeter. We have successfully demonstrated that the polymer gel dosimeter can be used to measure dosimetric structures of isodose lines in the electronic disequilibrium region near heterogeneity.

We have made qualitative comparison of dose distributions by varying the photon energy, the type of heterogeneity (air or bone), and the field size. However, the validity of the measurements, in particular, the accuracy of measured dose distributions has not been assessed in this study. Hence, the evaluation of the measurement accuracy and precision and more systematic analyses of heterogeneity effects will be undertaken in the future.

ACKNOWLEDGEMENTS

The work was partially supported by the RSNA Research and Education Foundation Seed Grant Program. We acknowledge Dr.M.Maryanski for his collaboration in the early phase of the project.

REFERENCES

1. Hepworth, S. J., McJury, M., Oldham, M., Morton, M. J. and Doran, S. J., *Dose mapping of inhomogeneities positioned in radiosentive polymer gels*, Nucl. Instrum. Methods Phys. Res. A 422(1-3) 756-760 (1999).
2. Watanabe, Y., Perera, G. M., Mooij, R. B. and Maryanski, M., *Three-Dimensional Dose Distribution in a Heterogeneous Phantom Measured by Polymer Gel Dosimeter (abstract)*, Med. Phys. 29(6) 1371 (2002).
3. Gum, F., Scherer, J., Bogner, L., Solleder, M., Rhein, B. and Bock, M., *Preliminary study on the use of an inhomogeneous anthropomorphic Fricke gel phantom and 3D magnetic resonance dosimetry for verification of IMRT treatment plans*, Phys. Med. Biol. 47(7) N67-77 (2002).
4. Vergote, K., De Deene, Y., Claus, F., De Gersem, W., Van Duyse, B., Paelinck, L., Achten, E., De Neve, W. and De Wagter, C., *Application of monomer/polymer gel dosimetry to study the effects of tissue inhomogeneities on intensity-modulated radiation therapy (IMRT) dose distributions*, Radiotherapy and Oncology 67(1) 119-128 (2003).
5. Watanabe, Y., Mooij, R., Perera, G. M. and Maryanski, M. J., *Heterogeneity phantoms for visualization of 3D dose distributions by MRI-based polymer gel dosimetry*, Med Phys 31(5) 975-84 (2004).
6. Kahn, F. M., *The Physics of Radiation Therapy*. (Baltimore, MD: Williams&Wilkins). (1994).
7. Maryanski, M. J., Schulz, R. J., Ibbott, G. S., Gatenby, J. C., Xie, J., Horton, D. and Gore, J. C., *Magnetic resonance imaging of radiation dose distributions using a polymer-gel dosimeter*, Phys. Med. Biol. 39(9) 1437-55 (1994).
8. Haake, E. M., Brown, R. W., Thompson, M. R. and Venkatesan, R., *Magnetic Resonance Imaging: Physical Principles and Sequence Design*. (New York, NY: Wiley-LISS). (1999).
9. Maryanski, M. J., Ibbott, G. S., Eastman, P., Schulz, R. J. and Gore, J. C., *Radiation therapy dosimetry using magnetic resonance imaging of polymer gels*, Med Phys 23(5) 699-705 (1996).
10. Epp, E. R., Lougheed, M. N. and McKay, J. W., *Ionization build-up in upper respiratory air passages during teletherapy with Co-60 radiation*, Br. J. Radiol. 31(4) 361-367 (1958).
11. Shahine, B. H., Al-Ghazi, M. S. A. L. and El-Khatib, E., *Experimental evaluation of interface doses in the presence of air cavities compared with treatment planning algorithms*, Med. Phys. 26(3) 350-355 (1999).
12. Werner, B. L., Das, I. J. and Salk, W. N., *Dose perturbations at interfaces in photon beams: secondary electron transport*, Med. Phys. 17(2) 212-26 (1990).
13. Werner, B. L., Das, I. J., Khan, F. M. and Meigooni, A. S., *Dose perturbations at interfaces in photon beams*, Med. Phys. 14(4) 585-95 (1987).

Applied Physics (APHYS 2003)
A. Méndez-Vilas (Ed.)

DETERMINATION OF THE *V* FUNCTION FOR CR-39 BY ATOMIC FORCE MICROSCOPE

K.N. Yu*, J.P.Y. Ho , D. Nikezic, C.W.Y.Yip

Department of Physics and Materials Science, City University of Hong Kong,
Tat Chee Avenue, Kowloon Tong, Kowloon, Hong Kong. P.R. CHINA

ABSTRACT

In the present work, the V function (= V_t/V_b, ratio between the track etch rate and the bulk etch rate) of alpha particles in the CR-39 detector was determined. CR-39 detectors were irradiated with alpha particles with different energies in the range 0.5 to 5.1 MeV. After irradiation, the detectors were etched in a 6.25 N aqueous solution of NaOH maintained at 70 °C by a water bath for 15 min, which was much shorter than the normally employed etching time. Atomic Force Microscopy (AFM) was used to measure the track lengths in order to calculate V. The corresponding alpha energies were converted into ranges in the CR-39 detector using the SRIM software. Our V values were based on very short etching time and direct measurements, so the determined V function should be more accurate.

INTRODUCTION

The use of solid-state nuclear track detectors (SSNTDs) has already become a well-known technique which has been applied in many branches of science. It has always been desirable to build a model for track development in these SSNTDs. On one hand, the processes involved in track development can be better understood if a satisfactory model exists. On the other hand, relevant parameters, such as the lengths of the major and minor axes and the depths, of tracks formed under different conditions can be predicted using the model. The most widely accepted model involves two etch rates, namely, the track etch rate V_t (i.e., along a track in the SSNTD) and the bulk etch rate V_b (i.e., the in undamaged areas of the SSNTD). The track etch rate of α particles in the CR-39 SSNTD will be the focus of the present paper.

Most current studies on V_t for alpha particles in CR-39 employ long etching time [1-7] or diameter measurements by optical microscope [6,7]. However, measurements of V_t by means of the optical microscope are either difficult or indirect. For example, V_t is indirectly derived from the growth of the track diameter D, with the understanding that D is somehow related to V_t, but it has been noticed that the dependence of D on V_t is weak and not explicit so multi-parametric fitting is needed to determine V_t [8]. In most studies, tracks were obtained by etching for at least 4 h. The accuracies of V_t values obtained from such long etching are arguable since it is well established that V_t is not a constant throughout the alpha-particle trajectory in the detector. As track lengths and diameters are the results of integration of V_t, it is understood that the derived V_t will become more inaccurate if measurements are made from tracks with a longer etching time.

Naturally, therefore, one will try to minimize the etching time to obtain the most accurate V_t. However, a short etching time will also mean very shallow tracks, which would be very difficult to measure. In the present paper, Atomic Force Microscopy (AFM) is employed to measure the depth of these tracks under very short etching time (15 min) [9-14]. As AFM has very good spatial resolution and is capable of measuring very minute structures, it can be exploited to measure the parameters of small and shallow tracks for very short etching time.

METHODOLOGY

The CR-39 detectors used in the present study were purchased from Page Mouldings (Pershore) Limited (Worcestershire, England). The original dimensions of a sheet of the detector are 30cm × 47cm × 0.1cm (thickness). The detectors for our studies were cut to a size of 1.5×1.5 cm^2. The CR-39 detectors were irradiated with alpha particles with different energies in the range from 0.5 to 5.1 MeV under normal incidence through a collimator.

The alpha source employed in the present study was a planar ^{241}Am source (main alpha energy = 5.49 MeV under vacuum). Normal air was used as the energy absorber to control the final alpha energies incident on the detector. A relationship between the alpha energy and the air distance traveled by an alpha particle (with initial energy of 5.49 MeV from ^{241}Am) was therefore needed. This relationship was obtained by measuring the energies for alpha particles passing different distances through normal air using α spectroscopy systems (ORTEC Model 5030) with Passivated Implanted Planar Silicon (PIPS) detectors of areas of 300 mm^2.

After irradiation, the detectors were etched in a 6.25 N aqueous solution of NaOH maintained at 70 °C by a water bath for 15 min, which was much shorter than the normally employed etching time.

AFM was employed to capture the surface topography and the depths of the tracks. The AFM used in the present project was the Autoprobe CP model from Park Scientific Instruments (1171 Borregas Avenu, Sunnyavale, CA 94089). The probe of the AFM employed was an Ultralever, with an opening angle of 10° and a length of 4 μm. Contact mode operation was used where high-resolution image was expected. A constant force of 13.2 nN was applied on the tip and the scan rate was 1 Hz. The surfaces of the detectors were imaged directly in air and room temperature.

During measurements, the Ultralever scanned the studied surface many times, with a 256×256 and 512×512 pixel resolution for a scanning area of 5×5 μm^2 and 10×10 μm^2, respectively, to ensure that the Ultralever will pass across any one track at least 15 times, and we could be reasonably confident that the Ultralever would scan across or at least close to the deepest point of the tracks so that the track depths were effectively correct. From the cross-sectional profiles containing the alpha tracks recorded by the AFM, the track lengths can be obtained to determine V_t for the corresponding ranges in the CR-39 detector.

The V function is usually expressed as a function of the residual range in the detector. By establishing a Range-Energy relationship for alpha particles in the CR-39 detector, the al ranges of alpha particles with known energies can be conveniently cbtained. This onship can be found using the SRIM software [15].

ULTS AND DISCUSSION

A two-dimensional image showing the tracks (created by 3.50 MeV alpha particles) in CR-39 detector recorded by the AFM with a scanning area of 10×10 μm^2 and a esolution of 512×512 pixels is shown in Fig. 1(a). The depths of the structures shown in ig. 1(a) are represented by the gray levels. More detailed information on the structures can be obtained by drawing lines on the images to generate the corresponding track profiles. The corresponding cross-sectional track profile across the upper horizontal line shown in Fig. 1(a) is shown in Fig. 1(b). The track depth is calculated from the coordinates of the intersection points between the two vertical cursors and the line profile. It is noted that the track depth is less than half a micron, but is still readily observable by the AFM. The track lengths, together with the bulk etch rate V_b [16] will give the V_t values and the V ratios.

Fig. 1. (a) Two-dimensional image recorded by AFM for tracks created by 3.50 MeV lpha-particles in CR-39 and etched for 15 min (with scanning area 10×10 μm^2 and a resolution of 512×512 pixels). (b) Cross-sectional profile of the upper line in (a). The vertical cursors are used for measurements of the height from the coordinates of the intersection points between the cursors and the line profile. The results are output in another dialogue box which is not shown here.

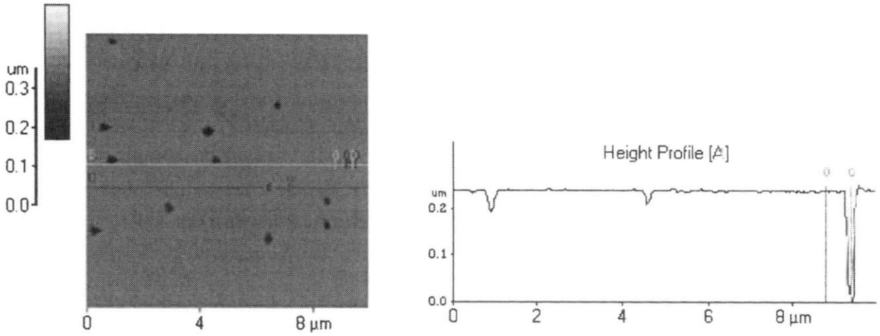

(a) (b)

By using the Microcal™ Origin™ software, the function best fitting our experimental data were found as

$$V = e^{(-0.0680R+1.1784)} - e^{(-0.6513R+1.1784)} + 1 \qquad (1)$$

where R is the residual range in μm. The best fit together with the experimental data are shown in Fig. 2(a). This function was obtained by fitting a parametric equation with iterations, where the parametric equation has been modified from the V function of the Besançon team [7] (hereafter referred to as the Brun's function):

$$V = e^{(-a_1 R + a_4)} - e^{(-a_2 R + a_3)} + e^{a_3} - a^{a_4} + 1 \qquad (2)$$

and the original parameters were: $a_1 = 0.1$, $a_2 = 1$, $a_3 = 1.27$ and $a_4 = 1$. A comparison between the Brun's function and our function is shown in Fig. 2(b).

Fig. 2. (a) The relationship between the V ratio and the residual range. Open circles: Experimental data; Solid line: best fit line. (b) The V functions from the present project (solid line) and from Brun et al. (1999) (dotted line).

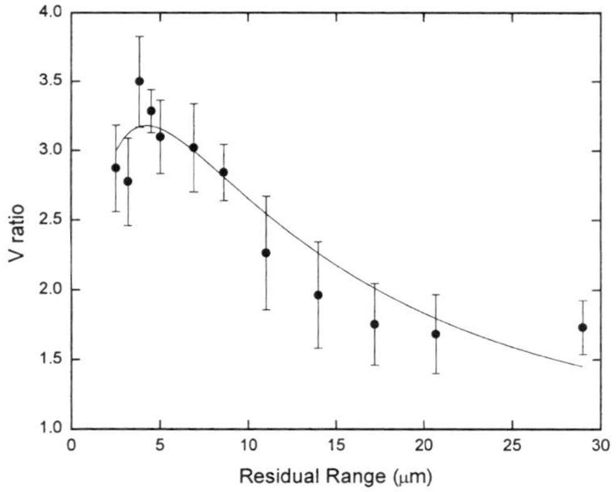

(a)

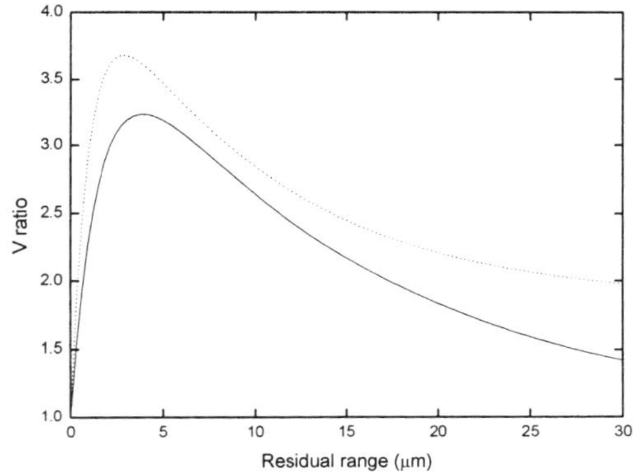

(b)

In the derivation of the Brun's function, CR-39 detectors (Track Analysis systems Ltd, UK) were irradiated perpendicularly by alpha particles with energies ranging from 0.2 to 4.5 MeV. Etching was conducted in 7N NaOH at 70°C. The etching duration ranged between 4 and 7 h. The track diameters were measured using an OLYMPUS BH2 microscope linked with an image analyzing system. The spatial resolution of this device was ~0.4 μm at a magnification of 1000.

From Fig. 2(b), we can see that the trends for the two V functions are generally commensurate with one another, both of them showing a 'Bragg peak'. However, the width of the Bragg peak, the maximum V value and the corresponding R for the maximum V value are slightly different. The V values obtained in the present project were found to be systematically lower than those predicted by the Brun's function, which was based on measurements of track-opening diameters and relatively long etching durations (in the order of hours).

In conclusion, our present work have been based on direct measurements of the track etch rate (use of Atomic Force Microscopy) as well as the very short etching time involved (in the order of 15 min). These lead to more accurate determination of the V function.

ACKNOWLEDGMENT

The present research is supported by the CERG grant CityU 1206/02P from the Research

... K. and Rößler P. *Studies on the variation of the ... alpha particle trajectories in CR-39.* Radiat. Meas. 25, 157-158

2. Dörschel, B., Hartmann, H. and Kadner K. *Variation of the track etch rates along the alpha particle trajectories in two types of CR-39.* Radiat. Meas. 26, 51-57 (1996).
3. Dörschel, B., Fülle, D., Hartmann, H., Hermsdorf, D., Kadner, K. and Radlach, Ch. *Measurement of track parameters and etch rates in proton-irradiated CR-39 detectors and simulation of neutron dosemeter responses.* Radiat. Prot. Dosim. **69**(4), 267-274 (1997).
4. Dörschel, B., Fülle, D., Hartmann, H., Hermsdorf, D., Kadner, K. and Radlach, Ch. *Dependence of the etch rate ratio on the energy loss in proton irradiated CR-39 detectors and recalculation of etch pit parameters.* Radiat. Prot. Dosim. 71(2), 99-106 (1997).
5. Dörschel, B., Bretschneideer, R., Hermsdorf, D., Kadner, K. and Kühne, H. *Measurement of the track etch rates along proton and alpha particle trajectories in CR-39 and calculation of the detection efficiency.* Radiat. Meas. 31, 103-108 (1999).
6. Green, P.F., Ramli, A.G., Al-Najjar S.A.R., Abu-Jarad, F. and Durrani, S.A. *A study of bulk-etch rates and track0etch rates in CR-39.* Nucl. Instr. Meth. 203, 551-559 (1982).
7. Brun, C., Fromm, M., Jorffroy, M., Meyer, P., Groetz J.E., Dörschel, B., Hermsdorf, D., Bretschneideer, R., Kadner K. and Kühne H. *Inter-comparison study of the detection characteristics of the CR-39 SSNTD for Light ions: Present status of the Besancon-Dresden approaches.* Radiat. Meas. **31** 89-98 (1999).
8. Fromm, M., Membrey, F., Rahamay, El. and Chambaudet, A. *Principle of light ions micromapping and dosimetry using CR-39 polymeric detector: modelized and experimental uncertainties.* Nucl. Tracks Radiat. Meas. 21, 357-365 (1993).

9. He, Y.D., Hancox, C.I. and Solarz, M. *Measurement of general etch rate for CR-39 plastic at short distance scale*. Nucl. Instr. Meth. B. 132, 109-113 (1997).

10. Palmino, F., Klein, D. and Labrune, J.C. *Observation of nuclear track in organic material by atomic force microscopy in real time during etching*. Radiat. Meas. 31, 209-212 (1999).

11. Rozlosnik, N., Glavák, C.S., Pálfalvi, L., Sajó-bohus, L. and Birattari, C. *Investigation of nuclear reaction products by atomic force microscopy*. Radiat. Meas. 28, 277-280 (1997).

12. Vázquez-López, C., Fragoso, R., Golzarri, J.I., Castillo-Mejía, F., Fujii, M. and Espinosa, G. *The atomic force microscope as a fine tool for nuclear track studies*. Radiat. Meas. 34, 189-191 (2001).

13. Yamamoto, M., Yasuda, N., Kaizuka, Y., Yamagishi, M., Kanai, T., Ishigure, N., Furukawa, A., Kurano, M., Miyahara, N., Nakazawa, M., Doke, T. and Ogura, K. *CR-39 sensitivity analysis on heavy ion bean with atomic force microscope*. Radiat. Meas. 28, 227-230 (1997).

14. Yamamoto, M., Yasuda, [illegible]

15. Ziegler, J.F. SRIM [illegible]

16. Ho, J.P.Y., Yip, C.W.Y., Nikezic, D. and [illegible] *rate of CR-39 detector*. Radiat. Meas. 36, 141-143 (2003).

Applied Physics (APHYS 2003)
A. Méndez-Vilas (Ed.)

THE ROLE OF ^{238}Pu/$^{239+240}$Pu ACTIVITY RATIOS AS ISOTOPIC SIGNATURE OF PLUTONIUM ORIGIN IN ENVIRONMENTAL SAMPLES: QUALITY ASSURANCE IN Pu DETERMINATION BY ALPHA-PARTICLE SPECTROMETRY

R. GARCÍA-TENORIO*, I. VIOQUE*, G. MANJÓN* and F. EL-DAOUSHY†

**E.T.S. Arquitectura, Departamento de Física Aplicada II, Universidad de Sevilla. Av. Reina Mercedes 2, E-41012-Sevilla, Spain.*
†Department of Physics, Uppsala University, Box 530, S-75121 Uppsala, Sweden

Abstract. In general, the validation of any method for Pu-isotopes determination in environmental samples by alpha-spectrometry is performed through the analysis of reference materials or participating in intercomparison exercises. However, in most cases only $^{239+240}$Pu activities are considered, and no information is given concerning to ^{238}Pu. A complete validation of the method requires additional tests in order to be also confident for ^{238}Pu. In this sense, if the origin of the plutonium in an environmental sample is known, ^{238}Pu/$^{239+240}$Pu activity ratios can be used as the demanded additional quality assurance test. The value of this ratio is usually known in any ecosystem where the origin of Pu is well determined. We have participated in several $^{239+240}$Pu intercomparison exercises and, additionally, we have analysed some samples, which are contaminated by a well-defined source of plutonium, for the complete validation of the method used in our laboratory for Pu–isotopes determination by alpha-spectrometry.

INTRODUCTION

Results of Pu-activity concentrations in environmental samples, determined by alpha-particle spectrometry, should be checked in order to detect the presence of any possible alpha contamination in the final source and in their corresponding spectrum. Alpha-emissions of some natural radionuclides (i.e. ^{228} Th, ^{234}U, ^{210}Po) could increase in the alpha-spectra the count rate in the ROIs corresponding to the different Pu-isotopes alpha emitters (included the radiochemical tracer) if the radiochemical procedure cannot efficiently remove these interferents from the analysed sample.

Usually, the quality of the technique is checked by inter-comparison exercises or using reference materials. However such possibility cannot assure in general the quality of all the information that can be derived from a Pu alpha-spectrum, because usually only are referenced or intercompared the $^{239+240}$Pu levels in these samples. In this paper an additional quality procedure, based on the determination of the $^{238}Pu/^{239+240}Pu$ activity ratio in specific environmental samples is proposed, in order to validate also the quality of the technique for ^{238}Pu determinations. In these specific samples, it is only necessary to know the source-term of the Pu contaminating the analysed sample. These additional validation exercises are needed attending to the general low ^{238}Pu concentrations found in environmental samples, and the distortion that can be produced in its determination if, for example, some traces of Th are present in the measured source (alpha-emissions of ^{228}Th cannot be resolved from the ^{238}Pu emissions in the alpha-spectrum).

The value of the $^{238}Pu/^{239+240}Pu$ activity ratio is in general unambiguously related to the source term causing the presence of the Pu in the environmental sample analysed, and for that reason it is used as an isotopic signature of the origin of these isotopes. Indeed a typical activity ratio of 0.03 is obtained in bulk samples collected over all the Northern Hemisphere [1,2,3,4] when they are only affected by global fallout (weapon tests), whereas the ratio is about 0.2 over all the Southern Hemisphere [5,6], when the analysed samples are only affected by the global Pu sources affecting this part of the world: weapon tests fallout and the input of ^{238}Pu caused by the burnt of the SNAP-9A satellite.

But, additionally, some local or regional areas were affected by accidents or controlled discharges that released Pu-isotopes to the close environments. In each of these areas, the $^{238}Pu/^{239+240}Pu$ activity ratio has a characteristic particular value. For instance, the $^{238}Pu/^{239+240}Pu$ activity ratio in the ecosystem affected by the Palomares accident (Spain), where two thermonuclear bombs burnt up, is 0.02 [7] indicating the contamination with weapon-grade plutonium; whereas this ratio is about 0.2 in marine samples affected by Sellafield (U.K.) reprocessing plant [8] and 0.47 in samples heavily contaminated by the Pu releases produced by the Chernobyl accident.

The main objective of this work was to check a radiochemical procedure to isolate plutonium isotopes in order to determine their activity concentrations in environmental samples by alpha-spectrometry. Two type of samples were used: $^{239+240}Pu$ intercomparison samples (for $^{239+240}Pu$ validation), and samples where the source-term of Pu is known. The expected $^{238}Pu/^{239+240}Pu$ activity ratios, in these later samples, have been used as quality parameter of this radiochemical procedure for ^{238}Pu determinations.

EXPERIMENTAL

A radiochemical method, previously applied to marine samples [9], was improved for the proper determination of Pu isotopes in environmental samples (soils, sediments and peats) by alpha-particle spectrometry [10]. Samples were ashed at 550ºC for 24 hours. Plutonium was dissolved by leaching using 8M HNO_3 and co-precipitated with Fe^{3+}. Interfering alpha-emitters (U- and Th-isotopes, ^{210}Po and ^{241}Am) were separated by ion exchange purification. Finally, plutonium was electrodeposited in order to obtain a thin source to be measured by alpha-spectrometry.

Measurements were done with an alpha-spectrometry chain (Alpha Analyst, Canberra) with eight chambers. Alpha-particles were detected by passivated implanted planar silicon (PIPS) detectors. Alpha-spectra were analysed with Alpha Analyst software. An ultra-low background was carefully kept during the measurement period in order to obtain low detection limits.

RESULTS AND DISCUSSION

$^{239+240}Pu$ intercomparison material was used in order to check partially the goodness of radiochemical method and the alpha-spectrometry technique ($^{239+240}Pu$ validation). The activity concentrations of $^{239+240}Pu$, determined by us in these samples, were compared to the intercomparison values. The results of these comparisons are presented in Table 1. The sample LDP was a peat sample collected in Cumbria (U.K.) on the frame of a European project. The rest of the samples were provided by the Consejo de Seguridad Nuclear (Spanish Nuclear Safety Organism). Thus, a peat sample, two soil samples and a fish sample were analysed in the exercise. A good agreement can be deduced in all the cases.

TABLE 1. $^{239+240}Pu$ (Bq/kg) activity concentration obtained in intercomparison materials. Two or three aliquots were analysed in every sample.

Key	Kind of sample	Intercomparison value $^{239+240}Pu$ (Bq/kg)	Calculated value $^{239+240}Pu$ (Bq/kg)
LDP (33)#3 - 1	Peats	55 ± 4	55 ± 2
LDP (33)#3 - 2			51 ± 4
CSN98 - 1	Soil	1.040 ± 0.100	0.95 ± 0.06
CSN98 - 2			1.00 ± 0.05
CSN98 - 3			1.01 ± 0.06
CSN2000 - 1	Soil	0.102 ± 0.005	0.104 ± 0.008
CSN2000 - 2			0.096 ± 0.010
CSN2002-1	Fish	0.121 ± 0.011	0.12 ± 0.01
CSN2002-2			0.11 ± 0.01

Additionally, and for ^{238}Pu validation, $^{238}Pu/^{239+240}Pu$ activity ratios were calculated in some of the previously analysed samples. Other samples, mostly collected in the Southern Hemisphere were also treated. In all cases, the origin of the Pu contaminating the sample was known. The results are listed in Table 2, where they are ordered according to the terrestrial latitude of the sampling station. In each case the result obtained by us (column named "this work") can be compared to the expected one attending to the Pu source term affecting the sample and taken from the data published in the current literature (column named "reported value"). The majority of the samples were mainly affected by global fallout and their $^{238}Pu/^{239+240}Pu$ activity ratios present certain latitude dependence. In these cases, the activity ratio was 0.03-0.04 in the Northern Hemisphere and about 0.2 in the Southern Hemisphere, as it was expected. Additionally a sample mainly affected by Chernobyl accident debris was analysed. In this case the activity ratio we have determined was 0.46, whereas the expected value was 0.47.

A good agreement is observed in all the cases, taking into account the uncertainties of the results, providing confidence (together with the previous $^{239+240}Pu$ tests) in the quality of our experimental technique for alpha-emitters Pu determination in environmental samples by alpha-particle spectrometry.

TABLE 2. $^{238}Pu/^{239+240}Pu$ activity ratio in environmental samples affected by a defined source-term of Pu. Most of them were affected by global fallout sources of Pu and collected at different Earth latitudes. Only one sample was mainly affected by Chernobyl debris. Results are compared to data reported in the literature in samples collected at the same latitude, and affected only by the commented sources.

Location	Pu source-term	Kind of sample	This work $^{238}Pu/^{239+240}Pu$	Reported value $^{238}Pu/^{239+240}Pu$	Ref.
Northern Hemisphere					
Sweden (55°N)	Global Fallout	Soil	0.034 ± 0.008	0.034 ± 0.007	[1]
Sweden (55°N)	Global Fallout	Sediment	0.034 ± 0.003	0.034 ± 0.007	[1]
Sweden (55°N)	Global Fallout	Peats	0.034 ± 0.004	0.034 ± 0.007	[1]
Cumbria (50°N)	Global Fallout	Peats	0.05 ± 0.01	0.04 ± 0.02	[11]
Spain (35°-40°N)	Global Fallout	Soil	0.035 ± 0.010	0.036 ± 0.004	[3]
Ukraine	Chernobyl	Soil	0.46 ± 0.08	0.47 ± 0.07	[12]
Southern Hemisphere					
Pacific (10°-30°S)	Global Fallout	Fish	0.19 ± 0.06	0.22 ± 0.15	[1]
Perú (13°S)	Global Fallout	Soils	0.15 ± 0.02	0.22 ± 0.15	[1]
Argentina (37°S)	Global Fallout	Sediment	0.46 ± 0.13	0.18 ± 0.07	[1]
Antarctica (60°-70°S).	Global Fallout	Lichen	0.23 ± 0.02	0.22 ± 0.02	[5]

CONCLUSIONS

The possible use of the $^{238}Pu/^{239+240}Pu$ activity ratio, as a partial quality test of the radiochemical procedure applied for Pu determination in environmental samples by alpha-spectrometry analysis, was explored. The $^{238}Pu/^{239+240}Pu$ activity ratios obtained in samples contaminated by a well defined source-term of Pu were compared to the expected ones attending to the Pu origin. A good agreement was observed with independence of the kind of environmental sample analysed and the source-term of the Pu contamination. As the activity concentration of $^{239+240}Pu$ determined in intercomparison materials has validated previously the procedure for the determination of these isotopes, the good results obtained in the $^{238}Pu/^{239+240}Pu$ activity ratio tests validate the ^{238}Pu determinations by our experimental technique.

REFERENCES

1. Hardy, E. P., Krey, P. W. and Volchok, H. L. *Nature* 241, 444-446 (1973).
2. Bunzl, K. and Kracke, W. *J. Radioanal. and Nucl. Chem., Articles*, 115 (1) 13-21 (1987).
3. Holm, E., Fukai, R. and Whithead, N. E. "Radiocaesium and transuranium elements in the Mediterranean Sea: sources, inventories and environmental levels" *Proceedings of International Conference on Environmental Radioactivity in the Mediterranean Sea*, Barcelona, Spain, May 1988, 601-615 (1988).
4. Yamamoto, M., Tamauchi, Y., Chatani, K., Igarashi, S., Komura, K., Ueno, K. and Sakanoue, M. *J. Radioanal. Nucl. Chem.* 147 (1) 165-176 (1991).
5. Mietelski, J. W., Gaca, P. and Olech, M. A. *J. Radioanal. and Nucl. Chem.*, 245 (3) 527-537 (2000).
6. Mitchel, P. I., Sánchez-Cabeza, J. A., Ryan, T. P., McGarry, A. T. and Vidal-Quadras, Ross, P., Holm, E., Persson, R. B. R., Aarkrog, A. and Nielsen, S. P. *J. Environ. Radioac.*, 24, 235-251 (1994).

7. Manjón, G., García-León, M., Ballestra, S. and López, J. J. *J. Environ. Radioact.* 28 (2) 171-189 (1995).
8. Mitchell, P. I., Sánchez-Cabeza, J. A., Vidal-Quadras, A., García-León, M. and Manjón, G. "The impact on Irish coastal waters of long-lived radioactive waste discharges to the Irish Sea", in *The Irish Sea: A Source at Risk.* Chapter 12, edited by. John C. Sweeney. Special Publication n° 3. Geographical Society of Ireland (1988).
9. Ballestra, S., Gastaud, J. and López, J. J. "Radiochemical procedures used at IAEA-ILMR Monaco for measuring artificial radionuclides resulting from Chernoby accident" in the book: *Low-level Measurements of Man-made Radionuclides in the Environment,* edited by. M. García-León and G. Madurga. World Scientific Publishing Co. 395-415 (1991).
10. Vioque, I., Manjón, G., García-Tenorio, R. and El-Daoushy, F. *Analyst* 127, 530-535 (2002).
11. Cawse, P. A. "Studies of environmental radioactivity in Cumbria, part 4. Caesium-137 and plutonium in soils of Cumbria and the Isle of Man". *AERE-R9851 Report*, Harwell (1980).
12. Aarkrog, A. *J. of Environ. Radioact.* 6, 151-162 (1988).

Applied Physics (APHYS 2003)
A. Méndez-Vilas (Ed.)

Making Predictions On The Evolution Of Radioactive Spots In The Ocean. Validation In The Baltic Sea

M. Toscano-Jimenez[1], R. Garcia-Tenorio[2] and J. M. Abril[3]

[1] *Departamento de Física Aplicada III. ES Ingenieros, Universidad de Sevilla. Avda. Descubrimientos s/n. 41092-Sevilla. Spain. e-mail: mtoscano@esi.us.es*

[2] *Departamento de Física Aplicada II, E.T.S. Arquitectura, Universidad de Sevilla Avda. Reina Mercedes 2, E-41012-Sevilla, Spain.*

[3] *Departamento de Física Aplicada I, E.U.I.T. Agrícola, Universidad de Sevilla. Carretera de Utrera km 1, 41013.- Sevilla, Spain.*

Abstract. A 3D dispersion model has been constructed in order to simulate the dispersion in marine scenarios of conservative nuclear contaminants originated in accidents or industrial releases. The model has been validated taken the Baltic Sea as the physical scenario, and selecting the ^{137}Cs with origin in the Chernobyl accident as the substance that experimented the dispersion.

In this paper special emphasis is given to the description of the original approach developed for the independent treatment in the model of the advective and diffusive transport affecting any contaminant in marine ecosystems. This original approach is based on the use of Monte Carlo techniques.

INTRODUCTION

It is well known that the exponential increase in industrialisation during the last century has produced the consequent incorporation of hazardous materials in several compartments of our globe. One of the most affected ecosystem by these anthropogenic intrusions has been the marine environment, where the contaminants, once incorporated, undergo a series of dispersion and transport processes.

A wide range of studies have been performed in order to obtain greater understanding of the behaviour of different contaminants in the marine environment [1]. This knowledge, when associated to a specific contaminant, facilitates an accurate prediction of its evolution in a specific marine scenario through the use of mathematical models simulating its behaviour. These predictions can be of vital importance in order to know, in advance, the impact of any contaminant input to the analysed marine compartment.

In this field, we have focussed in the construction of a 3D mathematical model simulating, in marine scenarios, the dispersion processes experienced by anthropogenic radionuclides originating from accidents or industrial releases. This model have been formulated by combining several original approaches and by using different tools in oceanography, and schematically is formed as the composition of three sub-models which respectively simulates the advection, diffusion and sedimentation processes which affect the behaviour of any nuclide in the marine compartment analysed.

In this paper we will specifically detail and justify the original approach developed for the independent treatment of the advective and diffusive transport of any substance in the sea. Through this approach, which is based on the use of Monte Carlo techniques, the mathematical treatment of the dispersion processes is enormously simplified.

The value of the methodology used for simulate the diffusion processes has been checked by implementing the advective and diffusion sub-models for the analysis in the Baltic sea of the dispersion of ^{137}Cs for one year, and by setting initial conditions as those which existed just after the accident of Chernobyl nuclear reactor in spring 1986.

The radionuclide and the time interval of study was selected due to the following reasons: a) the very-well documented existence of a clear and unevenly distributed ^{137}Cs contamination in the Baltic Sea just after the accident (presence of hot spots), b) the existence of very detailed information about the ^{137}Cs concentrations in the Baltic water columns in spring 1987 [1], which indicates that since the accident the Cs levels have experienced clearly identifiable spatial and temporal evolutions, and c) the possibility of considering as a good approximation that ^{137}Cs behaviour is conservative in the marine environment, avoiding the necessity to implement the sedimentation sub-model.

THE MODEL

The structure of the model implemented for this work can be described as follows: it is formed by two sub-models (advection and diffusion sub-models) together with the numerical and mixing algorithms applied to the individual cells into which the scenario was divided. By implementing these two sub-models, we take into consideration that the water circulation pattern in the Baltic Sea can be described superimposing eddy-like motions with variable intensities and scales on the mean circulation field, in such a way that the velocity $\vec{v}$ of a general point of our system can be described as the sum of its annual mean velocity $\vec{v}^m$ plus its fluctuation $\vec{v}'$.

In its particularisation to the Baltic Sea, the horizontal resolution of the model was selected as 20 km, while the vertical scale has been divided into six layers in order to properly reproduce the dispersion effects caused by the vertical sea-current profiles.

Our advection/circulation sub-model reproduces the mean annual current-velocity fields ($\vec{v}^m$) for the six layers into which the Baltic Sea has been divided, following the well-recognised study carried out in this scenario [3]. We have therefore produced a 3D field of mean annual currents, by applying interpolations to the Funkquist results in order to fit them to our horizontal resolution. The thicknesses of the six layers considered in the Funkquist study, as well as in our model, are the following: [0,5]m, [5,10]m, [10,20]m, [20,40]m, [40,60]m and [60m,bottom].

On the other hand, the modelling of the random movements $\vec{v}'$ superimposed over the mean circulation pattern $\vec{v}^m$, is carried out by the diffusion sub-model. Due to the originality of the approach used in its formulation, a detailed description is given below.

Horizontal diffusion

Typical experimental measurements carried out in the Baltic Sea show that temporal current spectra have strong fluctuations with respect to a mean annual value. These fluctuations have contributions with periods shorter than two days and contributions with periods longer than two days as can be deduced from the application of appropriate numerical filters to the current spectra taken at different stations in the Baltic Sea.

The last statements are demonstrated by observing the Figure 1, where a representative hourly mean current spectrum is shown [4]. The filtering of this original data gets the daily mean current spectrum , removing almost all the fluctuations shorter than 2 days , while the oscillations with longer periods remain. These last oscillations can be considered as fluctuations of the mean annual flow displayed by the advection sub-model.

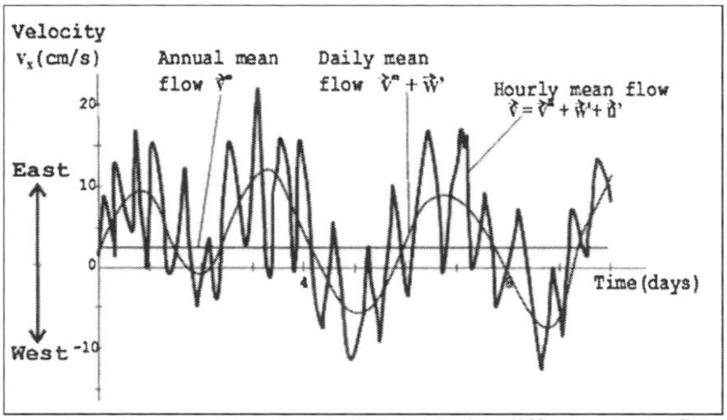

FIGURE 1. A representative current spectrum [4], with the different mean flows indicated in the manuscript.

In consequence, the fluctuations with regard to the annual mean velocity, at any point of the analysed marine ecosystem, can be formulated as the sum of two terms: $\vec{v}' = \vec{u}' + \vec{w}'$, where $\vec{u}'$ represents the short-term fluctuations (shorter than two days) and $\vec{w}'$ represents the long-term fluctuations (longer than two days).

The simulation of the horizontal short-term fluctuation velocities $\vec{u}'_x$ and $\vec{u}'_y$ at any point of the Baltic Sea is carried out taking into account that their variances are expressed as:

$$\sigma(u'_x) \equiv \sqrt{\overline{(u'_x)^2}} = \sqrt{\frac{K_x}{T_x}} \qquad \sigma(u'_y) \equiv \sqrt{\overline{(u'_y)^2}} = \sqrt{\frac{K_y}{T_y}} \qquad (1)$$

where K_x and K_y denote the horizontal diffusion coefficients while T_x and T_y represent the Eulerian integral timescales in horizontal directions.

Straightforward determinations of these variances can be carried out at any point of the analysed ecosystem by using average experimental values of K_x, K_y, T_x and T_y fixed after the analysis of the measurements taken at stations which are distributed in the Baltic Sea [4]. Indeed, we have adopted those of average horizontal-diffusion coefficients which are compiled in Table 1 for the six layers into which the Baltic has been sliced.

TABLE 1. Horizontal diffusion coefficients, K_x and K_y (m^2/s), used in the diffusion sub-model.

$\{K_x, K_y\}\, m^2/s$	30	10	8	6	4
$depth(m)$	[0,10]	[10,20]	[20,40]	[40,60]	[60,bottom]

The selection of these values reflects their close and reasonably uniform correlation with the depth (independent of any geographic zone inside the Baltic sea), as well as the existence of a clear and generalised isotropy on the horizontal scale (i.e. $K_x = K_y$).

However, the analysis of the experimental values of T_x and T_y indicates their moderate uniformity over the Baltic Sea, with their values slightly oscillating around an average of 2 hours for all stations and at all depths. This smooth uniformity has allowed the assignation of defined values for these parameters at any point of the Baltic Sea by simple interpolation of their experimental values determined at several stations.

Once the K_x, K_y, T_x and T_y 3D maps have been constructed in the Baltic Sea, and consequently the variances of the short-term fluctuation components of the velocity are known, the diffusion sub-model determines the values of u'_x and u'_y at any point of the system by the simple assignment of values through the Monte Carlo Method.

In relation with the analysis of the horizontal long-term fluctuation velocities w'_x and w'_y, we can indicate that their simulation at any point of the Baltic Sea is also based on observations made at several stations distributed over the Baltic Sea which show the clear correlation between the w'_x and w'_y experimental values and the annual mean velocities v_x^m and v_y^m. For this reason, as good approximation, the following expressions for the variances of w'_x and w'_y are assumed in the diffusion sub-model:

$$\left\{\begin{array}{l} \overline{w'_x} = \overline{w'_y} = 0 \\ \sigma(w'_x) \equiv \sqrt{\overline{(w'_x)^2}} = F(v_x^m) = \lambda \,|\, v_x^m \,| \\ \sigma(w'_y) \equiv \sqrt{\overline{(w'_y)^2}} = F(v_y^m) = \lambda \,|\, v_y^m \,| \end{array}\right\} \quad (2)$$

where λ can take a value in the interval 1 to 2.5, as can be also deduced by analysing different current spectra taken in most of the stations distributed over the Baltic Sea.

Once these variances are determined, the calculations of the w'_x and w'_y 3D maps are performed in a straightforward way through the application of the Monte Carlo method in the same way that was carried out for the determinations of u'_x and u'_y.

Vertical diffusion

According to the experimental results obtained, a proper simulation of the dispersion processes over the Baltic Sea also imply the necessity to consider diffusion processes in the vertical scale. In this case, and following the same methodology applied in the analysis of the horizontal diffusion, it is only necessary the modelling of the vertical component of the short-term fluctuation velocities, u'_z, because the long-term component, w'_z, can be considered negligible in the Baltic Sea according to the fact that the vertical dimensions in this scenario are significantly lower than the horizontal ones.

The construction through modelling of the 3D u'_z map is performed by using the same approach applied in the horizontal scale, and consequently its determination at any point of the system simply requires the construction of the average vertical-diffusion coefficient (K_z) map and the vertical Eulerian integral time scale (T_z) map.

For the construction of the K_z map associated to our selected scenario (the Baltic Sea) it has been taken into consideration two important facts: a) the vertical diffusion coefficients are very much dependent on the depth, and b) the non-existence of a clear dependence of the vertical diffusion coefficient values on the geographical situation in the Baltic Sea. These average values oscillate around an average annual value for a fixed depth and can be associated to every location of the system analysed.

These facts, simplify the construction of the K_z map in the Baltic Sea, being possible to use the average annual values of K_z which are compiled in the Table 2, and which are based on experimental values determined in several stations placed in the Baltic Sea [5].

TABLE 2. Vertical diffusion coefficients, K_z (cm^2/s), used in the diffusion sub-model.

$K_z\ (cm^2/s)$	1.00	0.50	0.20	0.20	0.05
$depth(m)$	5	10	20	40	60

The construction of the T_z map is still easier. In fact, it is a copy of the previously obtained map for T_x and T_y since experimental results over the Baltic sea show that, at any selected point, the value of T_z is the same as those corresponding T_x and T_y.

VALIDATION

The model implemented ad-hoc for this work, and formed by the described advection and diffusion sub-models together with numerical and mixing algorithms, has allowed the simulation of the evolution of the ^{137}Cs concentration in the Baltic Sea for the time interval June 1986-June 1987. The reasons to select the scenario, as well as the radionuclide and the time interval to be simulated, were detailed previously.

For the performance of this validation, two experimental spatial distribution maps of ^{137}Cs concentrations have been reconstructed in the Baltic Sea from experimental values found in several radiological publications [2].

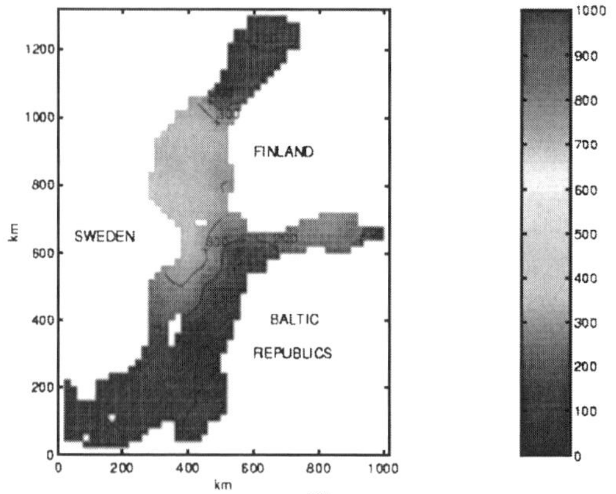

FIGURE 2(a) Experimental ^{137}Cs specific activity distribution (Bq/m^3) at the surface of the Baltic Sea in summer 1987.2(a)

The first spatial ^{137}Cs distribution, corresponds with the situation existing in the Baltic Sea in June 1986 (starting date of the simulation) and has been implemented to the model as initial conditions for the simulation, while the second one, Figure 2(a) , corresponds to the situation existing in the Baltic Sea in June 1987 (final date of simulation) and has been used for validation by comparing it with the simulated spatial distribution obtained after running of the model.

The model produces ^{137}Cs spatial distributions for the six layers in which the Baltic Sea was divided. In this paper, and for the sake of simplicity, we only show the modelled ^{137}Cs distribution maps obtained for the superficial layer by using different values of the parameter λ.

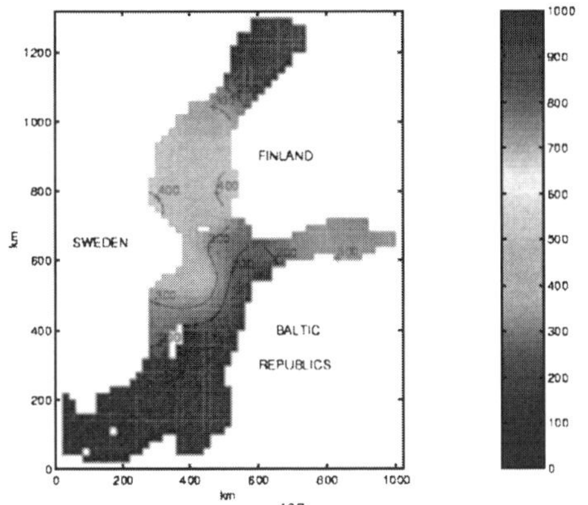

FIGURE 2(b) Modelled ^{137}Cs specific activity distribution (Bq/m^3) at the surface of the Baltic Sea in summer 1987 (λ = 1.5)

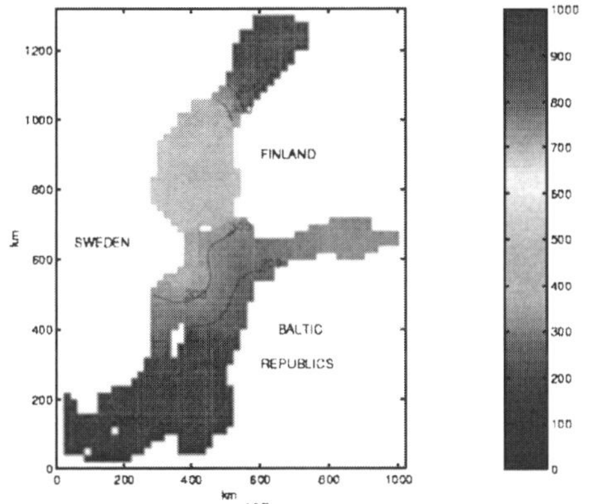

FIGURE 2(c) Modelled ^{137}Cs specific activity distribution (Bq/m^3) at the surface of the Baltic Sea in summer 1987 (λ = 2.5)

These modelled ^{137}Cs distribution maps, for λ = 1.5 and λ = 2.5 are shown in Figure 2(b) and Figure 2(c) which should be compared with the experimental distribution shown in Figure 2a. This comparison, performed only here in a qualitative way, shows the existence of a very satisfactory agreement between the modelled and experimental distribution maps, thereby guaranteeing the validity and strength of the approach formulated in this paper for the simulation of the dispersion of conservative substances in marine scenarios.

CONCLUSIONS

A 3D mathematical model simulating in a specific marine scenario the dispersion processes experienced by anthropogenic conservative radionuclides originating from accidents or industrial releases have been constructed and validated, with the aim to verify the goodness of an original approach formulated for the independent treatment of the advective and diffusive transport of any substance in the sea.

REFERENCES

1. J. Ribbe, S.H. Müller-Navarra and H. Nies. *Journal of Environmental Radioactivity*, **14**, 55-72 (1991).

2. Z.G. Gritchenko, L.M. Ivanova, T.E. Orlova, N.A. Tishkova, V.P. Toporkov and I. Tochilov "Radiation situation in the Baltic Sea in 1986 in sea water and sediments" in *Three years observations of the levels of some radionuclides in the Baltic Sea after the Chernobyl accident*, edited by Helsinki Commission, Baltic Sea environment proceedings 31, Helsinki, Finland, 1989, pp.10-30.

3. L. Funkquist, and L. Gidhagen, *SMHI-Report RHO* **39** (1984).

4. L. Gidhagen, L. Funkquist and R. Murthy, *SMHI-Report RO* **1** (1986).

5. A. Voipio, *The Baltic Sea,* Elsevier, Amsterdam, 2002, pp. 162-167

Applied Physics (APHYS 2003)
A. Méndez-Vilas (Ed.)

HYPERION NET - A Distributed Measurement System for Monitoring Background Ionizing Radiation

Aleksandar Žigić, Djordje Šaponjić, Nenad Jevtić, Branislava Radenković, Vojislav Arandjelović

Institute of Nuclear Sciences "VINČA" – Electronics Department, University of Belgrade,
Mike Alasa 12-14, 11000 Belgrade, Serbia and Montenegro

Abstract. The distributed monitoring system - HYPERION NET, based on the concept of FieldBus technology and FieldBus Development Tools and using a powered EIA RS-485 full-duplex bus, has been developed for the purpose of monitoring the background ionizing radiation. The net may comprise up to 255 transmitters which may be segmented in order to allow the use of suitable repeaters if a large area is covered by the monitoring system. A segment may contain up to 32 transmitters. The intelligent GM transmitter for background ionizing radiation is based on GM tube SI-8B and is used for measuring the absorbed dose of soft β and γ radiation in the air in the range 0.087 to 720 μGy/h. The transmitter makes use of an advanced count rate measurement algorithm capable of suppressing the statistical fluctuations of the measured quantity.

INTRODUCTION

The modern, information technology, approach to the measurement of physical quantities, control of technological processes, environmetal/object monitoring aimed at protection of humans and/or environment requires the execution of multiple measurements at many remote sites. A measuring system organized in this way is called distributed measuring system. A distributed measuring system allows the acquisition of sufficent information giving a complete insight in the state of a complex system and gives the opportunity of monitoring and prediciting the developments within the monitored system. A distributed measuring system is in essence a network of measuring stations which allow the measurement, processing, and storing of the acquired data, generation of the measurement reports, and interfacing to the higher hierarchical levels of the total system. The distributed measuring system for the measurement of ionizing radiation HYPERION is a specific measuring net because it has to function under exceptional operating conditions. HYPERION net is devoted to the measurement of ionizing radiation over a wide area covered by the distributed measuring system, over a wide range of dose rates of the ionizing radiation, and under all climatic conditions.

THE ARCHITECTURE AND ORGANIZATION OF HYPERION NET

The distributed meausrement system HYPERION, comprising several hierarchical levels, makes use of a powered bus transmitter and is organized as a device network. Each individual level performs its task in the process of the measurement, acquisition, and processing of the measured data. The lowest level of the system is the net of sensors, in the FieldBus terminology denoted as "Device Network" [1, 2]. A constituent part of the device net is an intelligent GM transmitter (IGMT); it measures the level of ionizing radiation, performs local processing of the measured data, and, on request, forwards the data to local node C1 using standard digital communication interface. Within the device net the master-slave relation is clearly defined in that local node C1 is master and all intelligent GM transmitters in the net are slaves. The data collected by the local node are processed and stored in the local data base, they are available on display to the operator and in the form of packages they are exchanged with the higher hierarchical level – central node C. The exchange of data between local node C1 and central node C is carried out by using standard digital Ethernet protocol via Internet. The data collected from local nodes C1, C2, ..., Cn are: processed by central node C, stored in the central data base of the HYPERION net, displayed to the operator in the central node, and organized in a WEB site, in HTML document format, for the purpose of further vertical connection to higher hierarchical levels or public information services.

DESCRIPTION OF THE INTELLIGENT GM TRANSMITTER

The basic measuring element in the distributed measuring system HYPERION is an intelligent GM transmitter. It performs the following tasks: measurement of the levels of □ and soft □ ionizing radiation, processing of the measured data, selftesting for the purpose of identifying the state of the transmitter, and communication to the higher hierarchical level, local node C1. Block diagram of the intelligent GM transmitter is shown in Fig. 1.

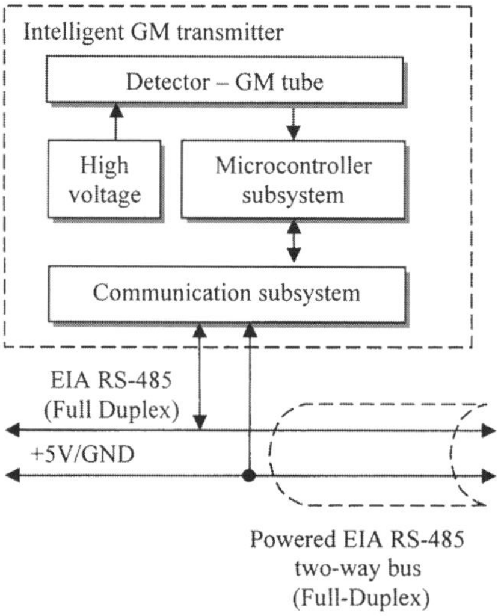

FIGURE 1. Block diagram of the intelligent GM transmitter.

The intelligent GM transmitter is made of two SI-8B GM tubes supplied by a 400V d.c. power supply [3]. The pulses from the two detectors are collected and processed by the microcontroller subsystem. The speed and accuracy of the applied preset count measurement method have been improved by implementatrion of a new algorithm for suppressing statistical fluctuations [4]. The microcontroller subsystem [5] also performs the autotest of the IGMT functions, e.g. control of the high voltage supply, operating temperature of the transmitter, etc. This subsystem prepares the processed measurement data and through the communication subsystem forwards them to local node C1.

THE INTERCONNECTION OF IGMT AND LOCAL NODE C1 AND THE DEVICE NET

The IGMTs are connected to local node C1 by a common bus and constitute a segment of the device net clusterred around local node C1. The communication between IGMT and C1 is realized as a standard two-way digital EIA RS-485 interface with additional lines required for the IGMTs supply [6,7]. Fig. 2 shows the block diagram of the interconnection between a local node C1 and n IGMTs. For the purpose of designing device nets comprising large number of IGMTs ($n < 255$) and meeting the requirements imposed by EIA RS-485 standard the segment of the net between two repeaters may contain up to 32 IGMTs. The configuration and parametrization of IGMT is carried out using point-to-point communication mode implementing the well established concept of FDT Technology [8].

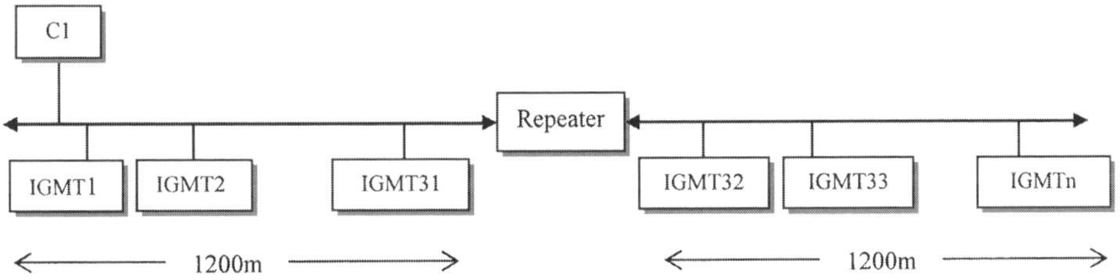

FIGURE 2. Block diagram of the device net and interconnection of a local node C1 to a large number of IGMTs

The communication protocol for the C1 – IGMT system is defined as follows:

A. The process of reading measured data from IGMT

- The protocol is cyclic, i.e. the local node polls IGMTs in the device net, point-to-multipoint.
- The communication is initiated by the master, local node C1, who sends two bytes formatted SYNC+ADR (SYNC = 7Eh, ADR – IGMT address) to identify the IGMT requested to send measured data.
- The identified IGMT returns measured data to the master, local node C1, seven bytes formatted ALARM+DATA+CHECK_SUM (ALARM – autotest status, one byte , DATA – mean time between successive pulses, four bytes, CHECK_SUM – general parity check).

B. The process of parametrization of IGMT table

- The protocol implies point-to-point communication of the configuring PC and IGMT.

• The communication is initiated by the master, the configuring PC, who sends one byte CALW (CALW=2Ah).

• Upon receipt of the CALW byte, IGMT is ready to receive the configuring data. The data contain identification, measurement, and calibration parameters.

C. The process of reading the identification, measurement, and calibration parameters from IGMT table

• The protocol is not cyclic, i.e. the local node reads the table of parameters of the IGMT in the device net upon request by the operator, the configuration is point-to-multipoint.

• The communication is initiated by the master, local node C1, who sends two bytes formatted CALR+ADR (CALR = 2Dh, ADR – IGMT address) for identification of the IGMT requested to send its parameter table.

• The identified IGMT returns 28 bytes from its EEPROM table and, for the purpose of ensuring the synchronization master-slave, upon receipt of each byte local node C1 returns one ACK byte for confirmation (ACK=2Eh).

DESCRIPTION OF THE LOCAL NODE AND THE CONCEPT OF VIRTUAL MEASURING POINT

The local node, as the central point of the lowest hierarchical level of the distributed HYPERION system, i.e. of the device net, is designed on the basis of the concept of virtual measuring point [9]. The functions of the local node are: acquisition of data from IGMTs, processing of these data, storing the processed data in a local data base, displaying locally the measured data, generation of reports on the measurements carried out by the portion of the device net it controls, and forwarding the acquired data to the higher hierarchical level, central node C. A general purpose PC can be used as local node C1. The necessary adjustment required for connecting the PC to the device net is the use of adapter EIA RS-232 to EIA RS-485 full duplex with a +5V supply for the bus. The adapter allows two-way communication with the device net at the speed of 9600 Baud at distances up to 1200m without repeaters. The local node is equipped with an autonomous power supply (UPS). The data acquired by the device net are forwarded to the higher hierarchical level, central node C, by the standard digital protocol TCP/IP[10].

INTERCONNECTION OF THE LOCAL NODES WITH THE CENTRAL NODE AND THE CONTROLLER NET

The local nodes, Ci (i=1,2,...,m), are connected to the central node, C, by using net topologies LAN, MAN, or WAN, or by Internet, and constitute a net of controllers. The communication between a local node Ci and central node C is realized by using standard digital protocol TCP/IP. Fig. 3 shows a block diagram of the interconnections of local nodes Ci with central node C.

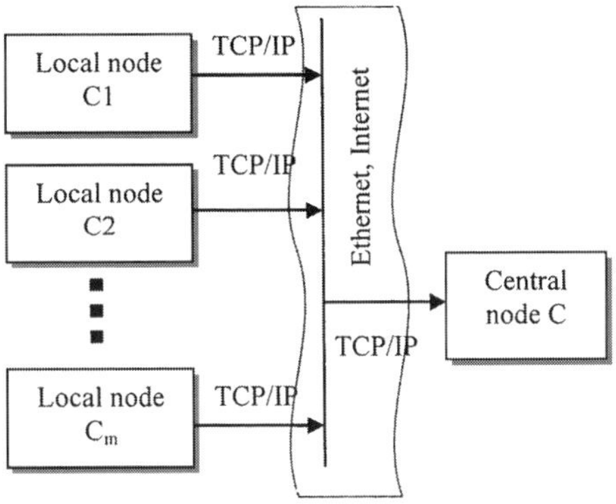

Figure 3 The interconnections of local nodes Ci with central node C

The use of the standard digital protocol TCP/IP ensures the integrity of data sent from local nodes Ci to central node C. The distributed measuring system HYPERION gives the possibilty that the local nodes could have dynamic IP addresses, whereas central node C must have a fixed IP address.

DESCRIPTION OF CENTRAL NODE C AND THE ORGANIZATION OF THE SOFTWARE

Central node C is the central point of the middle hierarchical level of the HYPERION system i.e. the net of controllers. The functions of the central node are: collection of the measured data from the local nodes, processing and storing the processed data in the central data base, displaying of the status of the measuring system and displaying of the measured data to the operator, generation of reports on the measurements carried out by the complete HYPERION net, presentation of the collected data by WEB technology using HTML data format [11] in the manner that they are accessible to the higher hierachical levels: Management Networks or public information services. A general purpose PC can be used for the realization of central node C of the minimum configuration of HYPERION measurement system. If the performance of the system is to be upgraded and the reliability and integrity of data increased, it is recommendable to separate the functions of acquisition and processing (server of central node C) from the function of storing data in the data base (data base server) and WEB site (WEB server). The central node is equipped with one UPS in order to ensure the autonomy of the power supply. The software of the central node comprises: the application for acquisition of data from the local nodes based on TCP/IP, the application data base, and the application WEB server. For the minimum configuration of the system it is required that the data acquisition and data base applications are executed by using two physically separate hard discs. As theWEB server application the Apache server is used. Configuring the Apache sever is carried out locally by the operator of the central node.

CONCLUDING REMARKS

The measuring system HYPERION is a modern measuring system based on Internet technologies which allow full integration of this distributed measuring system in the information systems of higher hicrarchical levels. The system can provide complete information regarding the levels of ionizing radiation over a teritory and constitutes a basis of an early warning system.

The radiological protection of people and human environment is very important in the modern society full of unintentional and intenional radiation hazards. Any defense against radiation risks requires radiation measurements. The proposed distributed, Internet-based, monitoring measurement system is shown to represent a suitable sollution for informing of the background ionizing radiation level the appropriate Government institutions, interested scientific community and the public, accurately and in due time.

The accuracy and timeliness are guaranteed by the implementation of robust, reliable, low-noise sensors, intelligent transmitters with self-diagnostics, tree-like structure for data collection from all measuring sites, from local nodes to the central node, and then, further to human communities concerned about one of the important parameters of their environment – the level of ionizing radiation.

REFERENCES

1. Profibus - Technical Description, Process Field Bus Foundation, 1999.
2. HART – Field Communication Protocol, Application Guide, HART Communication Foundation, 1999.
3. SI-8B GM Counter, Data Sheet, Techsnabexport, 1989.
4. Arandjelović, V., Koturović, A. and Vukanović, R., "A Software Method for Suppressing Statistical Fluctuations in Preset Count Digital-Rate Meter Algorithms", IEEE Trans. on Nucl. Sci. 49, 2561-2566, (2002).
5. AT89S8252 – 8 Bit Microcontroller with 8K Bytes Flash , Data Sheet, Atmel, 1997.
6. ADM485 – +5V Low Power EIA RS-485 Transciever, Analog Devices, Rev. 0, 2001.
7. Haseloff, E., Beckemeyer, H. and Zipperer, J., Data Transmission Design Seminar, Reference Manual, Texas Instruments, 1998.
8. T. Hadlich, T., and Szczepanski, T.," FDT – The New Concept for FieldBus Communication" Control Engineering Europe, 9, 33-37, (2002).

9 LabWindows/CVI – User Manual, National Instruments, 1998.

10 Murhammer, .M.W., Atakan, Bretz, O.S. Pugh, L.R. Suzuki, K. Wood, D.H. "TCP/IP Tutorial and Technical Overview", IBM Internation Technical Support Organization, 1998.

11 Mullen, R.,"HTML 4 Programmer's Reference", The Coriolis Group, 1997.

Applied Physics (APHYS 2003)
A. Méndez-Vilas (Ed.)

Automatic Diagnostic Of Plasmas With Finite Positive Ion Temperature

J. Ballesteros, J. I. Fernández Palop, M. A. Hernández, R. Morales Crespo and S. Borrego del Pino

Department of Physics, University of Córdoba, Campus Univ. de Rabanales, Edif. C-2, 14071 Córdoba, SPAIN. e-mail: fa1bapaj@uco.es

Abstract. A LabView Virtual Instrument (VI) to diagnostic cold plasmas, in which the positive ion temperature is not negligible compared to the electron one, is developed. The VI automatically measures the *I-V* characteristic of a cylindrical Langmuir probe immersed in the plasma. By using the measured characteristic, the VI obtains the electron energy distribution function (EEDF) and other parameters characterizing the plasma by using several methods one of them including the influence of the positive ion thermal motion. The comparison among these results ensures the goodness of the results and stands out the influence of the positive ion temperature. The measurement process and treatment is very quick, about 0.5 s, so temporal evolution in the plasma conditions can be pursued. Finally, the program is developed in the easy to use and portable LabView environment, so it can be easily adapted to other platforms.

INTRODUCTION

Cold plasmas are widely used in several industrial processes, such as Plasma etching and Plasma Assisted Chemical Vapour Deposition (PACVD). In this kind of plasmas the positive ion temperature, T_i, is similar to the electron one, T_e, so it can not be neglected because it influences notably the process. Therefore, an easy and quick method to control the plasma, developing *in situ* measurements of the plasma parameters is needed. This work develops a Virtual Instrument (VI) in the easy of use and portable LabView environment to diagnostic and control plasmas considering the influence of the positive ion thermal motion. The program performs the automatic measurement of the *I-V* characteristic of a cylindrical Langmuir probe. By using this characteristic, several magnitudes characterizing the plasma are automatically obtained. All this magnitudes are obtained by using different diagnostic methods each one applied on different zones of the characteristic [1,2]. This ensures the goodness of the results and the control of the technological process. The influence of the positive ion temperature is considered in the diagnostic process by using a theoretical radial model developed by the authors [3-5]. The time spent in the whole measurement and data treatment is always less than 0.5 s. Moreover, the program can repeat the diagnosis process in other places of the discharge, in order to measure the spatial homogeneity of the plasma. So, spatial and temporal evolutions in the plasma conditions can be determined, ensuring the control of the technical process. The

LabView environment has been chosen because it is a programming language of Virtual Instrumentation that works on the most usual hardware/software platforms. This allows the portability of the program without significant modifications.

THEORETICAL MODEL CONSIDERING THE POSITIVE ION TEMPERATURE

The influence of the positive ion temperature in the electron density is considered by using a theoretical radial model, developed by the authors, of the positive ion sheath surrounding a cylindrical or spherical Langmuir probe immersed in a plasma [3-5]. This model proposes appropriate functions to fit the theoretical I_+-V characteristic of the probe (I_+ being the positive ion current collected by the probe, per unit length in the case of a cylindrical probe) as a function of the probe radius, r_p, the biasing probe potential , V_p (referred to the plasma potential, V_{plasma}) and the ratio between T_i and T_e, $\beta = T_i/T_e$. This fitting is very useful, because these are common experimental parameters. For cylindrical probes, these fitting functions are:

$$I(x_p, y_p, \beta) = \sum_{i=0}^{1} h_i(x_p, \beta) y_p^{i/2}, \text{ with } h_i(x_p, \beta) = \sum_{j=0}^{2} h_{ij}(\beta) x_p^j, \text{ being } h_{ij} = c_{ij} + d_{ij}\beta.$$

In these equations the following dimensionless parameters have been used:

$$x_p = \frac{r_p}{\lambda_D}, \quad y_p = -\frac{eV_p}{K_B T_e} \quad \text{and} \quad I(x_p, y_p, \beta) = \frac{I_+(x_p, y_p, \beta) r_p e}{2\pi\varepsilon_0}\left(\frac{m_+}{2K_B^3 T_e^3}\right),$$

m_+ being the positive ion mass, and e the elementary charge.

The coefficients c_{ij} and d_{ij} for cylindrical probes are given in Table 1, with x_p in the range [1,10], y_p in the range [1,150] and β <0.3. The coefficients for higher ranges and for the case of spherical probes can be found in a work developed by the authors [5].

For given y_p and β values, the Sonin plot can be drawn. This is a useful representation of the positive ion current collected by the probe because it leads to obtain the electron density, n_e. In the Sonin plot, the following dimensionless positive ion current

TABLE 1. Fitting coefficients for the I-V characteristic of a cylindrical probe. $x_p \in [1,10]$, $y_p \in [1,150]$ and $\beta < 0.3$.

$c_{00} = -3.30218 \cdot 10^{-1}$	$c_{01} = 1.76216 \cdot 10^{-1}$	$c_{02} = 4.60087 \cdot 10^{-1}$
$c_{10} = -2.37501 \cdot 10^{-1}$	$c_{11} = 9.92961 \cdot 10^{-1}$	$c_{12} = 2.22827 \cdot 10^{-2}$
$d_{00} = 9.08056 \cdot 10^{-1}$	$d_{01} = -9.29967 \cdot 10^{-1}$	$d_{02} = 5.07961 \cdot 10^{-1}$
$d_{10} = -2.80119 \cdot 10^{-1}$	$d_{11} = 6.23191 \cdot 10^{-1}$	$d_{12} = -4.23166 \cdot 10^{-3}$

$$I'(x_p, y_p = -15, \beta) = \frac{I_+(x_p, -15, \beta)}{er_p n_e} \sqrt{\frac{m_+}{2\pi K_B T_e}},$$

is represented versus

$$I'(x_p, y_p = -15, \beta) \cdot x_p^2 = \frac{I_+(x_p, -15, \beta) e r_p}{\varepsilon_0} \sqrt{\frac{m_+}{2\pi K_B^3 T_e^3}}.$$

As can be shown, the last magnitude does not depend on n_e. By using the experimental values of V_{plasma} and T_e, the value for the abscissa of the Sonin plot is known and by cross plotting, the corresponding value for the ordinate can be obtained and the experimental n_e value can be determined.

Figure 1 shows the Sonin plot used to determine the electron density for $y_p = -15$ (this high negative value is chosen to ensure that the current collected by the probe is only due to the positive ions) considering two β values: β=0.2 and β=0. An example of the cross plotting of a real experimental measurement has been drawn. As can be seen, $I'(x_p,-15,0)$ is lower than $I'(x_p,-15,0.2)$. This is reasonable since the cold positive ions have less energy to reach the probe [3-6]. Therefore, the positive ion temperature must be considered in the plasmas where $\beta \neq 0$ because this condition notably influences the process.

Moreover, it is important to remark that this diagnosis technique is very adequate because it uses the ion saturation zone of the *I-V* characteristic. In this zone, the current drained by the probe is very small diminishing the perturbation that the measurement causes in the plasma [7].

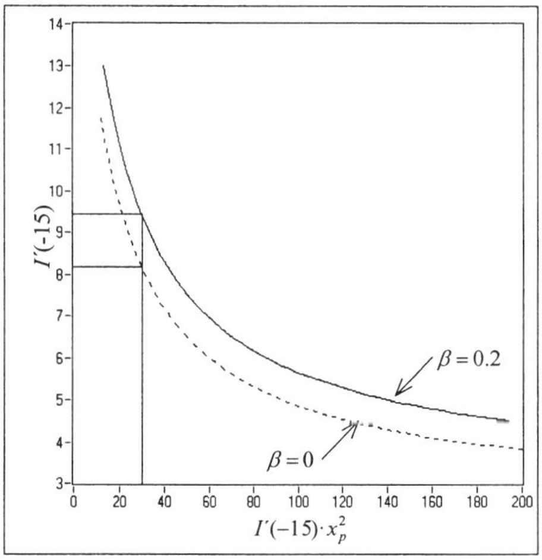

FIGURE 1. Sonin plot.

EXPERIMENTAL DEVICE

The experimental device, in which the program has been tested is shown in Fig. 2. This is very adequate for this study because the plasma generated by it is very stable and the electron temperature is low [1,8], $T_e(K) \in [1000,3000]$, T_i is usually considered similar to the room temperature, 300 K (although it is greater in PACVD devices reaching even 1000 K). The discharge device consists on a large pyrex cylinder (31 cm inner diameter and 40 cm height) with two electrodes connected to a very stable high voltage power supply which generates the discharge in a low pressure Argon gas.

The *I-V* characteristic of a cylindrical Langmuir probe, designed following the adequate conditions [1,9], is measured by using an inexpensive analog to digital (A/D) converter card, with 8 differential A/D channels, 12 bits resolution, $2 \cdot 10^5$ samples per second and several digital input/output. Two voltage dividers are used to adapt the probe voltage to the A/D card input ranges. A LabView program controls the whole experiment. First the VI asks for the kind of gas, its pressure and the probe dimensions while the probe is highly biased in order to be cleaned by electron impact. When the "begin" switch is pressed, the VI measures the discharge current and voltage by using a multimeter. Then, it orders the card to generate a TTL pulse which

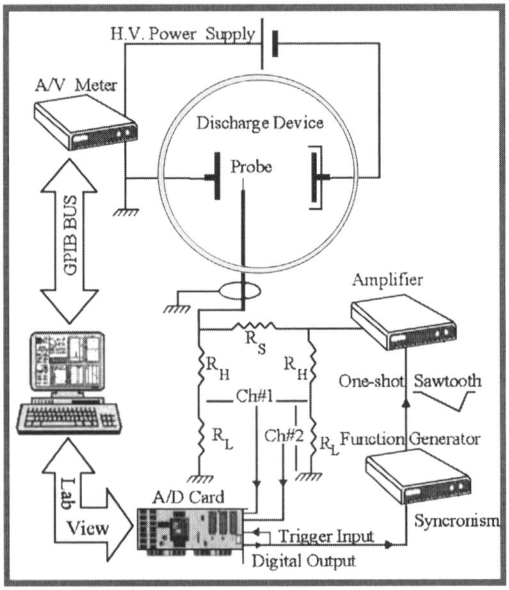

FIGURE 2. Experimental setup.

simultaneously triggers the A/D converter and the function generator to generate a one-shot saw tooth which is amplified to bias the probe. This process is developed twice: first with channel#1 to measure the potential drop in R and finally with channel#2 to measure the potential drop in R_L, see Fig. 2. Both data series are easily related with the *I-V* characteristic. In our case, each channel measures 2000 data for every *I-V* characteristic in 10 ms. Simultaneously measurement of both channels is not needed because the data acquired with channel#2 do not depend on the plasma. In this way, the other six A/D channels of the card could be used to develop similar measurements of the *I-V* characteristic of other six probes placed in several points of the discharge. In this way, information about the spatial homogeneity of the plasma in the discharge device could be obtained without extra cost.

DATA TREATMENT

Figure 3 shows the Virtual Instrument Scope that the user sees in the screen of the computer. This VI is developed not only to acquire the data, but to treat them too. In this way, the VI has two input modes. In the "Meas" (measure) mode the data are acquired from the experimental device as described in the previous section. In the "Oper" (operate) mode the VI reads previously measured data, saved to a file by the VI itself. Once the data are read/acquired, the rest of the operations developed by the VI are similar in both modes. These operations are the following:

Smoothing the data. As will be seen, in order to obtain the EEDF we must calculate the second derivative of the experimental *I-V* characteristic. This derivative increases the noise inherent to the measurement process, and so the measured data must be previously smoothed:

Channel#1 data are smoothed by using an iterative convolution method with the instrument function which was demonstrated to be very adequate for this task [1,2,10-13]. If the instrument function is assumed to be a Gaussian distribution function, the whole process can be carried out in a single step by performing a convolution with the following function:

$$g_n(x)=\sum_{k=1}^{n}\binom{n}{k}(-1)^{k+1}\frac{\alpha}{\sqrt{\pi k}}e^{-\alpha^2x^2/k}, \tag{1}$$

$\sigma=1/\alpha\sqrt{2}$ being the standard deviation of the Gaussian distribution function and n the number of iterations.

Channel#2 data correspond to a linear zone ramp of the saw-tooth polarizing the probe. So, they are smoothed by a linear fitting by using the least square method.

Once the data are smoothed, we obtain the *I-V* characteristic is obtained and also its first and second derivative, the plasma potential, V_{plasma} (from the condition $d^2I/dV^2=0$) the floating potential, V_{float} (from the condition $I=0$) and the electron energy distribution function (EEDF) [1,2]. This EEDF is obtained from the Druyvenstein formula [14]:

$$[f_E(E)]_{E=-eV_p} = -\frac{4}{S_p e^2}\sqrt{\frac{-m_e V_p}{2\ e}}\ \frac{d^2 I}{dV_p^2}, \qquad V_p \le 0\ , \tag{2}$$

where $f_E(E)$ is the EEDF, S_p the probe area and I the current collected by the probe.

Figure 3 shows the plots of the measured and smoothed I-V characteristic (their comparison indicates if it has occurred some problem in the smoothing process) the first and second derivatives, the EEDF and $\ln(d^2I/dV^2)$ vs V. This last plot must be observed because it allows us to see if the EEDF approaches to a Maxwellian distribution function, in this case it should has a linear behavior. This is important because several of the diagnostic methods used in this work are based on plasmas whose EEDF approaches to a Maxwellian. Obviously, other diagnostic method should be programmed for plasmas with different EEDFs.

The mean electron energy, E_{mean}, is calculated by direct integration from the EEDF:

$$E_{mean} = \int_0^\infty E f_E(E)dE\ .$$

This expression does not depend on the specific EEDF.

The electron temperature is obtained from the slope, m, of the plot of $\ln I$ vs V in the retarding zone ($V<V_{plasma}$). Assuming a Maxwellian EEDF, $T_e=e/mK_B$ [1]. This temperature must be similar to $2E_{mean}/3K_B$ if the EEDF approaches to a Maxwellian one.

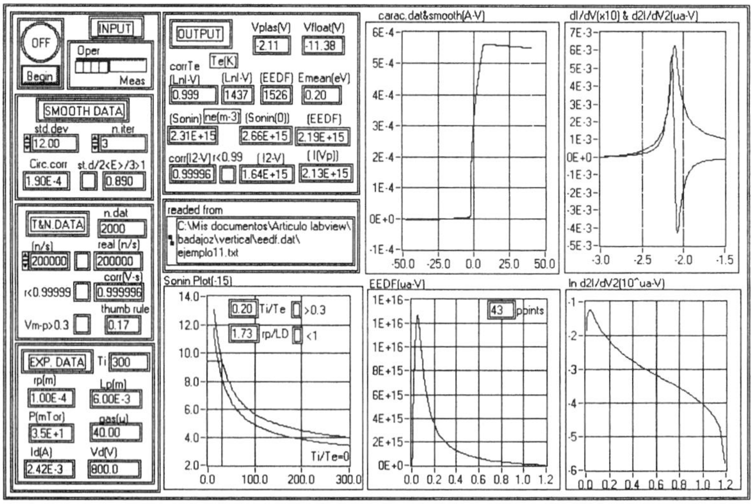

FIGURE 3. Virtual Instrument Scope

The electron density, n_e, is calculated by using several methods:

By direct integration of the EEDF:

$$n_e = \int_0^{\infty} f_E(e)dE \,.$$

This expression does not depend on the specific EEDF. For the following methods the EEDF must approach to a Maxwellian one:

From the slope of the plot of I^2-V in the electron saturation zone $(V > V_{plasma} + 2K_BT_e/e)$ [1].

From the current collected by the probe when biased at the plasma potential [1].

From the Sonin plot in the case β=0.

These four values are obtained for the case of cold positive ions and must be similar in this case. This must be emphasized since these methods use different zones of the I-V characteristic.

The influence of the positive ion temperature in the diagnostic of n_e is considered by using the Sonin plot for the corresponding β value.

All these values appears in the Virtual Instrument Scope illustrated in Fig. 3. The VI includes luminous warning and error indicators to test the goodness of the measured data and of the smoothing parameters. In this way, the following parameters are controlled:

The linear behavior of the saw tooth (the linear correlation coefficient must satisfy the condition r>0.99999).

The linear behavior of the plot I^2-V (r>0.99).

The probe contamination is controlled by using the thumb rule [1,2] (the difference between the plasma potential and the maximum of the second derivative, V_{m-p}, must be lower than $0.3E_{mean}$). This contamination is due to a slow acquisition time [9].

The adequacy of the smoothing parameters is controlled with the condition $\sigma < 2E_{mean}/3$ [12]. Moreover, the distortion introduced by the smoothing is controlled by using a circular correlation coefficient [1,2].

The conditions $1 < r_p/\lambda_D < 10$ and $\beta < 0.3$ are controlled to accomplish with the fitting conditions.

All these processes, read/acquire and data treatment are carried out in less than 0.5 s. Therefore, temporal evolutions and drifts in the plasma conditions can be measured ensuring the control of the technical process.

As the last step, the VI ask if the data must be saved, this lets us to save only the data to be treated later.

This VI program can be downloaded for free from the web site http://www.uco.es/~fa1fepai/research.html. The program is developed for LabView 5.1 or later. Some experimental data files have been included to be treated in the "Oper" mode.

ACKNOWLEDGMENTS

This work has been supported by the Spanish Ministerio de Educación y Cultura (Secretaría de Estado de Univesidades, Investigación y Cultura, Dirección General de Enseñanza Superior, Programa Sectorial de Promoción General del Conocimiento). Project No. PB96-0985.

REFERENCES

1. Fernández Palop, J. I., Ballesteros, J., Colomer, V., and Hernández, M.A., Rev. Sci. Instrum. **66**, 4625 (1995).
2. Ballesteros, J., Fernández Palop, J.I., Hernández, M. A., Morales Crespo, R., and Borrego del Pino, S., Rev. Sci. Instrum.**75**, 90 (2004).
3. Fernández Palop, J. I., Ballesteros, J., Colomer, V., and Hernández, M.A., J. Phys. D: Appl. Phys. D **29**, 2832 (1996).
4. Morales Crespo, R., Fernández Palop, J.I., Hernández, M. A., and Ballesteros, J., J. Appl. Phys. **94**, 4788 (2003).
5. Morales Crespo, R., Fernández Palop, J.I., Hernández, M. A., and Ballesteros, J., J. Appl. Phys. **95**, 2982 (2004).
6. Ballesteros, J., Fernández Palop, J. I., Colomer, V. and Hernández, M. A., "XXII Conference on Phenomena in Ionized Gases" (Edited by Becker, K. H., Carr, W. E. and Kunhardt, E. E.) (Stevens Institute of Technology, Hoboken, 1995) pp.195-196.
7. Ballesteros, J., Hernández, M. A., Dengra, A., and Colomer, V., Contrib. Plasma Phys. **31**, 595 (1991).
8. Fernández Palop, J. I., Ballesteros, J., Colomer, V., and Hernández, M.A., J. Appl. Phys. **80**, 4282 (1996).
9. Godyak, V. A:, "Plasma Surface Interactions and Processing of Materials" (Edited by Auciello, O. *et al*) (Dordrecht, Kluwer, 1990). pp. 95-134.
10. Monterde, M. P., Haines, M. G., Dangor, A. E., Malik, A. K., and Fearn, D. G., J. Phys. D: Appl. Phys. **30**, 842 (1997).
11. Demidov, V. I., Ratynskaia, S. V., and Rypdal, K., Rev. Sci. Instrum. **73**, 3409 (2002).
12. Kortshagen, U., Maresca, A., Orlov, K., and Heil, B., Appl. Surf. Sci. **192**, 244, (2002).
13. Estamate, E. and Ohe, K., J. Appl. Phys. **84**, 2450 (1998).
14. Druyvesteyn, M.J., Z. Phys. **64**, 781 (1930).

Applied Physics (APHYS 2003)
A. Méndez-Vilas (Ed.)

Biophysical Device For The Treatment Of Neurodegenerative Diseases

I. Montiel [(1)], J.L. Bardasano [(2)], J.L. Ramos [(2)]

(1) Instituto Nacional de Técnica Aeroespacial (INTA).Systems and Equipment Department. Ctra de Ajalvir Km. 4, 28850 Torrejón de Ardoz. Madrid. Spain. E-mail: montielsi@inta.es
(2) Universidad de Alcalá (UA), Medical Specialities Department, Alcalá de Henares, 28871, Madrid, SPAIN, e-mail:JoseLuis.Bardasano@uah.es, Joseluis.Ramos@uah.es.

Abstract. The medical treatment of Neurodegenerative Diseases is becoming an issue for contemporary society. The causes of these diseases are not known yet and the Chemical-Pharmacological remedies are not working properly as it is demanded. Some publications exist which show a possible relationship between neurodegenerative diseases and electromagnetic fields. Although epidemiological studies show difficulties to isolate the links between the different parameters involved, some results could guide us to relate the effects of Electromagnetic Fields with these kind of diseases. Based in the possibility of a break in the organism synchronicity it is possible to adjust the endogenous signals with an artificial one in order to force the body synchronism. This synchronicity is reached by different mechanisms among which the natural electromagnetic fields could play the main role. Here we present a prototype of a new non-invasive device which objective is to synchronise the organism to the lost natural electromagnetic signal.
Keywords: Biomedical Engineering, Schumann Resonance, Bioelectromagnetism, Neurodegenerative diseases, Electromagnetic Fields.

INTRODUCTION

The human organism is a complex electrochemical system being extremely sensitive to the surrounding environment. This environment has interacted with living beings for a long time by the means of fields like the Earth magnetic field (30 μT to 70 μT), the Earth electric field (120 a 150 V/m), the electric field present in the storms (< 10 kV/m) and electromagnetic fields like sun light and cosmic rays [10].

These fields are in some part responsible for the physiology of living matter because life has evolved under the constraint of the environmental characteristics.

When new fields are added, living beings try to adapt to them, evolving in a way to improve their survivability. But in some cases, that adaptation is not possible and different diseases can be produced.

Nowadays, electromagnetic pollution is a serious fact to analyze. The electronic devices we are using at home or in the office can produce electromagnetic fields of quite a high level depending on their proximity. The mobile telephony is introducing levels of electromagnetic fields in frequencies that had been almost completely clean before.

All that man-made noise in the electromagnetic spectrum can degrade the natural signals we have grown-up with and cause some of our homeostasis mechanisms to be altered in a negative way.

Some treatments for neurodegenerative diseases have been proposed [8,12-14] using magnetic fields of Extremely Low Frequency (ELF) and intensity in the picoTesla (pT) range. The beneficial effect caused by these fields has not been clearly explained before.

Our intention is to explain these effects as a de-synchronization of the mechanisms used by the organism to regulate its function, i.e. through endogenous oscillators. Based on this explanation, the disease could be caused by the failure of the body natural electromagnetic sensors.

A new prototype of a device is then presented which mission is to provide a signal for the organism to use it to synchronize itself through its endogenous oscillators.

BIOPHYSICAL TREATMENTS

The work of Cohen [6] is the origin of most of the neurodegenerative disease treatments by the use of electromagnetic fields of recent development. Cohen measured the brain magnetic activity through the Superconducting Quantum Interference Device (SQUID) magnetometer. This device is capable to provide a very high sensitivity and has given this new insight into the extremely weak magnetic fields originated by the organism.

These measurements showed the different magnetic profile from a healthy brain to the pathologic one. The idea was to apply a slightly higher level field to the brain in order to readjust the profiles to the healthy ones, but always maintaining the physiological levels of the body. Promising results have been obtained through that proceeding [1,2,8,12,13,14].

Jacobson [8] has proposed a physical explanation to these treatments with magnetic fields. He has found a resonance named after his name by equating the force exerted on any particle with mass to the value of the Lorentz force produced by a static magnetic field on a moving particle. By means of this equation he founds different resonant frequencies for every ion. His aim is to produce vibrations on the particles in order to re-establish their healthy status of vibration.

Nevertheless, there is not any clear explanation of the reason why these electromagnetic fields should improve the health of the patient.

SYNCHRONIZATION

The human organism has biological oscillators to set the rhythms needed for its physiological activity. These oscillators require external references to synchronize with, which are known as Zeitgebers [4].

The main oscillator seems to be located in the Suprachiasmatic Nucleus of the hypothalamus. There are more endogenous oscillators in the hypothalamus considered as secondary ones. The Pineal Gland is also considered as an oscillator with an important role by synchronizing different hormonal functions within the light-darkness cycle [3].

The geo-magnetic fields could play a part in this synchronization sensed by the pineal gland as the main candidate to be an electromagnetic sensor although a distributed array of sensors based in the biomagnetite should not be discarded neither. It has been shown that magnetic fields provoke changes in the behavior and depressive alterations in human beings [4].

On the other hand, there have been shown some links between neurodegenerative diseases and electromagnetic fields [7, 16]. In these cases, a possible explanation may be that the noise induced by these man-made electromagnetic fields could interfere with the natural ones. Then, it would not be possible to synchronize with the signal obtained by the electromagnetic body sensors, being that the cause of the disease.

This could be an important factor to cause the neurodegenerative diseases.

The resonating cavity formed by the nearly perfectly conducting terrestrial surface and the Ionosphere can be excited by broadband electromagnetic impulses, like those from lightning flashes.

The electromagnetic signals appearing due to that excitation are called the Schumann Resonance and their frequencies range from 4 to 50 Hz [5,15,16,17]. The average fundamental mode of resonance is around 7.8 Hz, and the rest of modes are 14, 20, 26, 33, 39, and 45 Hz with slight diurnal variation.

The human brain waves behave similarly in frequency to Schumann Resonance. We can see the frequency band correspondence between the Electroencephalogram (EEG) and the Schumann Resonance in the following table.

TABLE 1. Schumann Resonance and EEG band.

EEG Band	Frequency (Hz)	Schumann Resonance	EEG Band
Delta	0.5 - 4	4	Delta
Theta	4 - 8	4 , 7.8	Theta
Alpha	8 – 14	7.8, 14	Alpha
Beta	14 -30	14, 20, 26	Beta

INTERACTION MECHANISM

Based on the hypothesis presented here, it is possible to design a device to force the brain to synchronize to the Schumann Resonance frequency. On these basis we can explain the treatments used before, based on ELF magnetic fields and picoTesla intensity levels.

The body would use these fields as synchronizing signals. But in the case that the sensitivity of the body sensors was highly degraded, the use of a signal from a different kind could be the solution.

We propose here the use of a microwave frequency signal modulated by the ELF frequency to take advantage of the ferromagnetic resonance effect to generate a 7.8 Hz magnetoacoustic wave. That vibration could be enough to synchronize the endogenous oscillators to the natural Schumann Resonance.

This mechanism would be based in the biomagnetite present in the human brain discovered by Kirschvink in 1992 [9]. In order to achieve this, we have to design a system capable of introducing microwave energy inside the brain of the patients. This is done through the design of an antenna matched to the head.

BIOMEDICAL DEVICE

A complete system has been built comprising a signal generator and the antenna applicator. The signal generator is an oscillator that provides the microwave carrier frequency around 10 GHz. It modulates this carrier with a 7.8 Hz signal. We can see the aspect of the generator in figure 1.

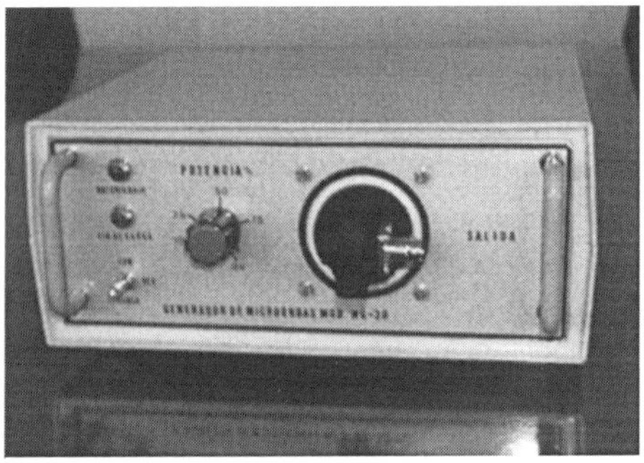

FIGURE 1. Signal Generator

A microstrip antenna has been designed taken into account the different permitivities of the tissues involved to achieve the match to the head of the patient. The geometry of the antenna can be seen in figure 2. The measure of the antenna match is shown in figure 3. It can be seen that the match is very good, allowing the signal to avoid the barrier presented by skin, bone and brain.

FIGURE 2. Antenna

The power used is very low, never reaching the $1 mW/cm^2$ limit imposed by the European Commission Recommendation (1999), in order to protect patients from thermal effects. Special care has to be taken to avoid illuminating the eyes.

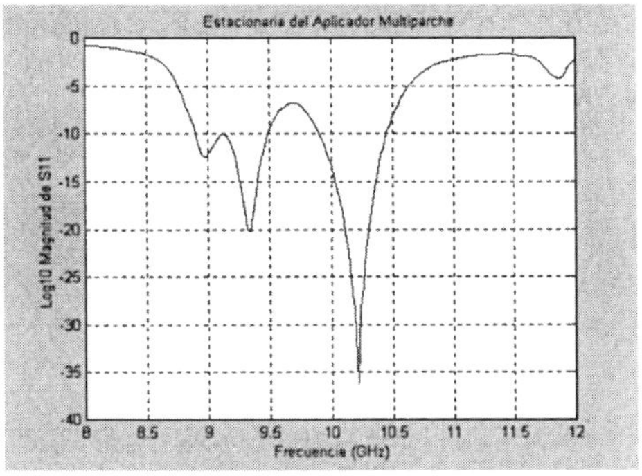

FIGURE 3. Antenna S11 parameter

EXPERIMENTAL RESULTS

A preliminary use of this technique has been performed by Dra. Vanessa de San Gregorio Cidón who has treated a limited number of Alzheimer Disease (AD) (6) and Multiple Sclerosis (MD) (2) patients for a period of six months. The conclusion is that this treatment shows very promising improvements in many of the cognitive functions such us the comprehension capability, short term memory and in the general status of the AD patients. A decrease in the symptoms of the MD has been seen too. More complete clinical tests have to be done to characterize the performance of this biophysical device.

CONCLUSIONS

We have presented a new biophysical device for the treatment of neurodegenerative diseases. It is based in the use of a microwave signal carrier in X band modulated in the ELF band. The aim of the device is to synchronize the organism to a signal that replaces the natural signal responsible of the healthy function of the human being. This 10 GHz carrier frequency is new for this kind of treatments.

The preliminary results obtained using this system are very promising although more research and clinical tests have to be done. Moreover, with this understanding it is possible to justify the reason why treatments using magnetic fields in the ELF band and picoTesla intensities can produce beneficial effects to the patients.

This work is protected under patent nº P200300891

ACKNOWLEDGMENTS

We have to thank the support from the INTA Systems and Equipment Department, the UA Medical Specialities Department, the Fundación Europea de Bioelectromagnetismo (FEB) and the work done by Dra. Vanessa de San Gregorio which has provided a preliminary experimental insight into this matter.

REFERENCES

1. Anninos, P.A., Tsagas, N., Sandyk, R. and Derpapas, K., Magnetic Stimulation in the Treatment of Partial Seizures. *Interational Journal on Neuroscience*. 60: 141-147, 1991.

2. Bardasano J.L., Ramos J.L., Picazo M.L..- Dispositivo para el tatamiento no invasivo con campos electromagnéticos ELF (intensidad pT). *En Bioelectromagnetismo y Salud Pública. Efectos, prevención, diagnóstico y tratamiento*. Instituto de Bioelectromagnetismo (IBASC) Universidad de Alcalá de Henares (UAH): 395-400, 1997.

3. Bardasano, J.L., Elorrieta J.I.. Bioelectromagnetismo. Ciencia y Salud. Serie McGraw-Hill de Divulgación Científica. 2000.

4. Betés de Toro M., Bardasano J.L., Picazo M.L. "Osciladores circadianos, psicofármacos y campos magnéticos". En Bioelectricidad, Cronobiología y Glándula Pineal. Instituto de Bioelectromagnetismo (IBASC) Universidad de Alcalá de Henares (UAH). 1993.

5. Bliokh, P. V., A. P. Nikolaenko, and Y. F. Filippov, "Schumann Resonances in the Earth-Ionosphere Cavity", Peter Perigrinus, London, 1980.

6. Cohen D. "Detection of the brain´s electrical activity with a superconducting magnetometer". *Science*. 175:664, 1972.

7. Feytching , M., Pedersen, N., Svedberg,P., Floderus, B. & Gatz, M. "Dementia and occupational exposure to magnetic fields." *Scandinavian Journal of Work, Environment and Health*, In Press. 1998.

8. Jacobson J.I, Yamanashi S. "An initial physical mechanism in the treatment of neurologic disorders with externally applied picoTesla magnetic fields." *Neurological Research*, Vol.17. April, 1995.

9. Kirschvink J.L., Kobayasi-Kirschvink A, & Woodford B.J. "Magnetite biomineralization in the human brain." *Proceedings of the National Acaddemy of Sciences*. Vol 89: 7683-7687. 1992.

10. Llanos,C. "Medida de CEM Próximos a Líneas Eléctricas de Alta Tensión." En *Encuentros Sanitarios en la Sociedad Actual (2002)* C. 2: 97 – 108.

11. Montiel I. "Diseño y construcción de una antena para el tratamiento de enfermedades neurodegenerativas mediante CEM." Tesis Doctoral. UA. 2003.

12. Sandyk, R. "Successful treatment of multiple sclerosis with magnetic fields." *International Journal on Neuroscience*, 66:237-250.1992-a

13. Sandyk, R. "Weak magnetic fields as a novel therapeutic modality in Parkinson's desease." *International Journal on Neuroscience*, 66: 1-15.1992.

14. Sandyk R. "Alzheimer's disease: Improvement of visual memory and visuoconstructive performance by treatment with Picotesla range Magnetic Fields." *International Journal on Neuroscience*, Vol.76. pp 185-225. 1994.

15. Sentman, D. D., "Magnetic polarization of Schumann resonances", *Radio Science, 22*, 595-606, 1987.

16. Sentman, D. D. and B. J. Fraser, "Simultaneous observations of Schumann resonances in California and Australia: Evidence for intensity modulation by the local height of the D region", *J. Geophys. Res., 96*, 15973-15984, 1991.

17. Schumann, W. O., "Uber die strahlungslosen Eigenschwingungen einer leitenden Kugel, die von einer Luftshicht und einer Ionosphärenhulle umgeben ist", *Z. Naturforsch., 7a*, 149, 1952.

18. Sobel, E., Dunn, M., Davanipour, Z., Qian, Z. & Chui, H.C. "Elevated risk of Alzheimer's disease among workers with likely electromagnetic field exposure". Neurology, 47, 1477-1481. 1996.

Applied Physics (APHYS 2003)
A. Méndez-Vilas (Ed.)

New Facility at the Portuguese Research Reactor for Irradiation with Fast Neutrons

J.G. Marques*†, A.C. Fernandes*†, I.C. Gonçalves*, and A.J.G. Ramalho*

**Reactor Português de Investigação, Instituto Tecnológico e Nuclear, Estrada Nacional 10, P-2686-953 Sacavém, Portugal*

†CFNUL, Av. Prof. Gama Pinto nr. 2, P-1649-003 Lisboa, Portugal

Abstract. A dedicated irradiation facility was built in the Portuguese Research Reactor and used in the irradiation of electronic components with fast neutrons. Fast neutron fluxes of up to $1.5x10^9$ n/cm^2/s are achievable in the facility. The irradiation facility is described and the characterization of the neutron and photon fields is presented.

INTRODUCTION

The performance of electronic components under irradiation is a concern for the nuclear industry, the space community and the high-energy physics community. In many situations the use of radiation hard components is not an option due to the high costs involved and standard commercial components are used instead. However, the use of these components complicates the radiation hardness assurance process since, contrary to radiation hard parts, there is little information on the actual implementation of the circuit. Only testing can give an indication on the radiation tolerance of the component and indicate which malfunctions can occur in a radiation environment [1].

In this work we describe a fast neutron irradiation facility built in the Portuguese Research Reactor (RPI). This facility has been used to irradiate electronic components for cryogenic thermometry at the Large Hadron Collider (LHC) facility at CERN [2]. Fast neutron fluxes of up to $1.2x10^9$ n/cm^2/s are achievable. The irradiation goal is to achieve during one week of operation of the reactor (about 60 hr) a fast neutron fluence that corresponds to the expected values for 10 years of operation in the LHC. Several parameters of the components are monitored during the irradiation using a PC--controlled data acquisition system.

DESCRIPTION OF THE FACILITY

The RPI is a 1 MW pool-type reactor with a maximum thermal neutron flux of $3x10^{13}$ n/cm^2/s. Its irradiation facilities include 7 beam tubes, a thermal column and a rabbit system. It was built by AMF Atomics Inc., in the period of 1959/61 and its design is similar, e.g., to the FRM I reactor ("Atomic Egg") in Munich, Germany.

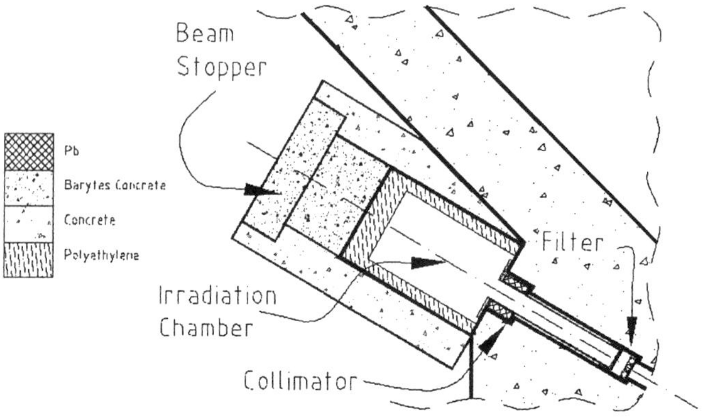

FIGURE 1. Horizontal cut of the fast neutron facility. For simplification the outer radiation shielding is not shown complete.

Most irradiations in pool-type reactors are done inside watertight containers placed in the pool. However, in this case reasonably large containers would be required to accommodate the boards with the components to be irradiated. Moreover, the necessary length of the cables to connect to the measuring instruments compromises the necessary signal-to-noise ratio. Thus, this approach was abandoned in favor of a facility built around a beam tube.

Figure 1 shows a horizontal cut of the installed facility. A dry irradiation chamber with 100 x 60 x 60 cm (l x w x h) was created at the end of beam tube E4. The chamber was prolonged inside the beam tube, through the introduction of a 100 cm long cylinder with 15 cm diameter. The inner face of this prolongation is approximately 160 cm away from the core. When the facility is not in use the end portion of the beam tube is flooded, the water acting as a neutron and photon shield. The attenuation for fission neutrons provided by a water shield 160 cm thick is $7x10^{-8}$ [3]. The measured attenuation for the photon dose rate is $1x10^{-2}$. The neutron beam size is 15 cm, as defined by the diameter of the beam tube close to the core.

The shielding of the facility is a combination of polyethylene lined with Cd and concrete. Provision for passage of cables to the measuring instruments was made. Insertion and removal of the circuits is done with the reactor stopped and the beam tube flooded.

The reactor core is a set of fuel assemblies mounted on a grid plate with a 9x6 pattern (Fig. 2). It is reflected by a thermal column on one side and by Beryllium and water on the remaining sides. The fuel is of the "Materials Test Reactor" type, with flat plates, using uranium enriched to 93% in the ^{235}U isotope. The core was rearranged in order to increase the fast neutron flux in the tube E4. In previous core configurations the entrance of this tube was partially covered with Be reflectors and the remaining surface with water. A threefold increase in the fast neutron flux was obtained placing in the core an aluminum block covering three grid positions facing the entrance of the tube, as shown in Fig. 2. In order to obtain a neutron / photon ratio

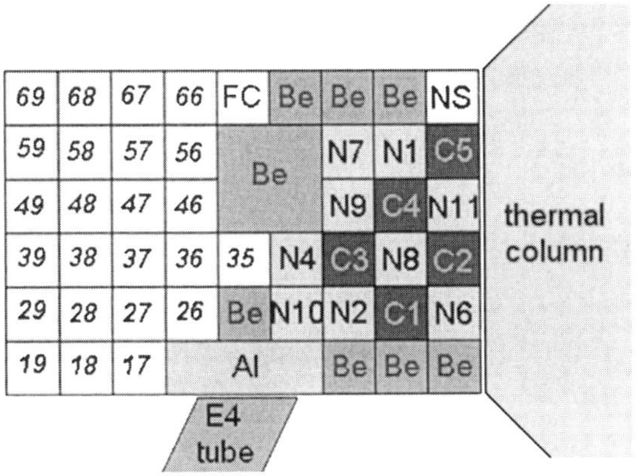

FIGURE 2. Schematic view of RPI's core configuration N2-P1/6 (NS=SbBe neutron source; FC=fission chamber; Ni=standard fuel assemblies; Ci=fuel assemblies with control rods; Be=Beryllium reflectors; Al=aluminum block in front of beam tube E4). Free grid positions are labeled in italic.

close to the one expected for the location of the cryogenic temperature processing units at the LHC, a Pb filter was placed inside the irradiation tube. The 4 cm thick Pb filter installed attenuates fast neutrons approximately by a factor of two and photons by a factor of ten. The thermal neutron component of the beam was strongly reduced by a 0.7 cm thick Boral (Al-B_4C) disk placed next to the Pb. The power deposited in the Pb and Boral filter is less than 3 W.

The components under test are mounted on several boards (8 x 8 cm), inside assembling boxes. Up to 18 boxes can be accommodated in the current box holder, which has provisions for cooling the boards using a small flow of compressed air. Connection to the measuring instruments is done through four cables with 25 conductors. Normally each irradiation campaign runs from Monday to Friday with approximately 13 hours of irradiation followed by 11 hours of stand-by per day, due to the two-shift per day operation of the reactor, and the central box receives a neutron fluence of 5×10^{13} n/cm^2.

CHARACTERIZATION OF THE NEUTRON AND PHOTON FIELDS

The neutron spectrum in the irradiation facility is essentially a leakage spectrum in a water moderated fission reactor, with a reduced thermal component due to the Boral filter. In the case of ^{235}U the fraction of fission neutrons emitted per unit energy about E, $\chi(E)$, can be described by a modified Maxwellian distribution in the range of 0.2 to 12 MeV, commonly called a Watt-Cranberg distribution [4]:

$$\chi(E) = 0.453 \exp(-1.036E) \sinh(\sqrt{2.29E}). \qquad (1)$$

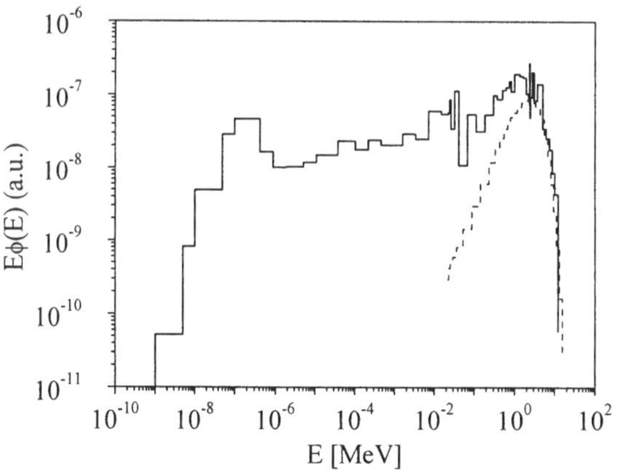

FIGURE 3. MCNP simulation of the neutron spectrum in the irradiation chamber (full line). A Watt-Cranberg distribution, normalized at 2 MeV, is also shown for comparison (dashed line).

The energy distribution of the fission neutrons is highly asymmetric. Most neutrons have energies between 1 and 2 MeV, although there are still neutrons with energy in excess of 10 MeV.

Figure 3 shows a Monte-Carlo simulation, using the MCNP-4C code [5] of the neutron spectrum inside the irradiation chamber, at the point closest to the core. This simulation uses a detailed three-dimensional model of the core validated with extensive measurements in the reactor core [6]. A Watt-Cranberg distribution, as described by Equation (1), normalized to the value at 2 MeV, is also shown for comparison. The fast neutron component of the spectrum is well described by this distribution, as expected. The thermal component is well described by a Maxwellian distribution but shifted to 323 K, due to the effect of the Boral filter. The epithermal portion of the spectrum follows closely a 1/E dependence.

Figure 4 shows measured values of the fast neutron flux, for three situations: free beam, i.e., no materials being irradiated, a standard irradiation and a high flux irradiation. The neutron fluxes were measured with Ni detectors. They are based on the averaged neutron cross section for the $^{58}Ni(n,p)^{58}Co$ reaction in a ^{235}U fission spectrum [7]. In the case of the actual irradiations of circuits, two Ni detectors were placed at the ends of each box; the values in Fig. 4 are the average of the two measured values and represent the flux at the center of the box. Fig. 4 shows that the free-beam values can be used to predict the initial fast flux; the materials being irradiated then dominate the attenuation. The attenuation of the fast neutron flux by the circuits shows an effective removal coefficient of 0.05 cm^{-1}, which is about half of the one of water for fission neutrons [3].

The photon dose rate in the facility was obtained with thermoluminescent dosemeters (TLD): TLD-700 (^7LiF:Mg,Ti) and Aluminum oxide (Al_2O_3:Mg,Y). The measurements were performed at a reduced power of 50 kW from the inner face of the irradiation chamber until 135 cm downstream. An ionization chamber was also placed

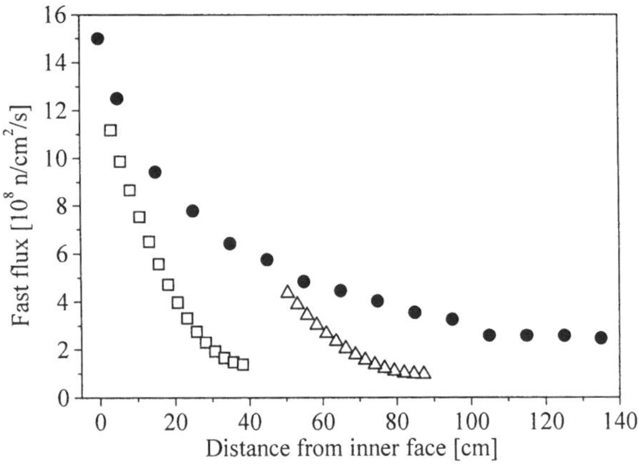

FIGURE 4. Measured values of the fast neutron flux in three situations: free beam (circle); a standard irradiation (triangle); a high flux irradiation (square).

at 70 cm from the inner face of the irradiation chamber and the measured value was in good agreement with the one obtained via TLD. Figure 5 shows the profile of photon dose rate at 1 MW in the free beam condition and with the assembling boxes mounted (but without circuits). The values with the assembling boxes are only 10-15% higher than the free beam values. The ionization chamber was subsequently used to control the gamma dose during the actual irradiation of the circuits at 1 MW power.

The characterization of the circuits is done by the Complutense University of Madrid in collaboration with LHC. A recent report of results can be found elsewhere [8].

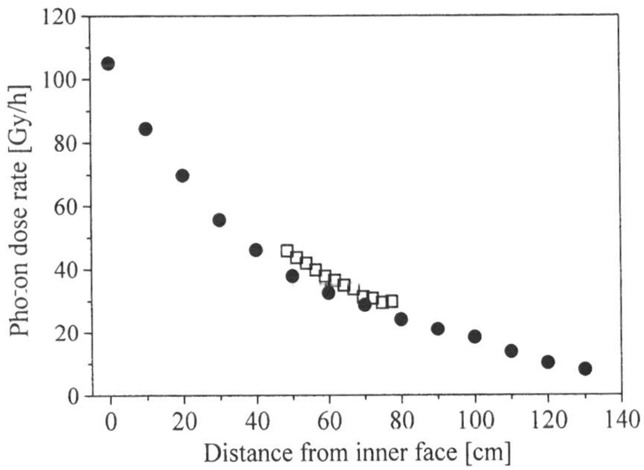

FIGURE 5. Measured values of the photon dose rate: free beam (circle); with boxes (square).

CONCLUSIONS

A dedicated irradiation facility was built in the Portuguese Research Reactor and used in the irradiation of electronic components with fast neutrons. Fast neutron fluxes of up to 1.5×10^9 n/cm^2/s are achievable in the facility. The photon dose rate is in the 20-40 Gy/h range in the standard irradiation conditions, maintaining a neutron/photon ratio close to the one expected at the location of the cryogenic temperature processing signals at the LHC.

ACKNOWLEDGMENTS

The authors are grateful to the operating staff of the Portuguese Research Reactor for their assistance. This work was partially supported by FCT, Portugal, through projects CERN/P/FIS/43773/2001, POCTI/49559/FNU/2002 and a PhD grant (ACF).

REFERENCES

1. F. Faccio, Proceedings 6th Workshop on Electronics for LHC Experiments, CERN-LHCC-2000-041, CERN, Geneva, 2000, pp. 50-65.
2. J. Casas-Cubillos, P. Gomes, K.N. Henrichsen, U. Jordnung and M.A. Rodriguez-Ruiz, CERN-LHC-Project-Report-333, CERN, Geneva, 1999.
3. J. Shapiro, *Radiation Protection*, Harvard University Press, Cambridge, 1990.
4. S. Glasstone and A. Sesonske, *Nuclear Reactor Engineering*, Chapman and Hall, New York, 1994.
5. J.F. Briesmeister, LA-12626-M, LANL, Los Alamos, 1990.
6. A.C. Fernandes, I.C. Gonçalves, N.P. Barradas and A.J. Ramalho, *Nuclear Technology* **143**, 358-363 (2003).
7. J.H. Baard, W.L. Zijp and H.J. Nolthenius, *Nuclear Data Guide for Reactor Neutron Metrology*, Kluwer Academic Publishers, Dordrecht, 1989.
8. J.A. Agapito, N.P. Barradas, F.M. Cardeira, J. Casas, A.P. Fernandes, F.J. Franco, P. Gomes, I.C. Goncalves, A.H. Cachero, J. Lozano, J.G. Marques, A. Paz, M.J. Prata, A.J.G. Ramalho, M.A. Rodriguez Ruiz, J.P. Santos and A. Vieira, Proceedings 7th Workshop on Electronics for LHC Experiments, CERN-LHCC-2001-034, CERN, Geneva, 2001, pp. 117-121.

Applied Physics (APHYS 2003)
A. Méndez-Vilas (Ed.)

Patient Doses from Conventional Diagnostic Radiology Procedures in Serbia and Montenegro

Olivera Ciraj, Srpko Markovic, Dusko Kosutic

VINCA Institute of Nuclear Sciences
P.O.Box 522, 11001 Belgrade, Serbia and Montenegro

Abstract. The objective of this work is to assess the patient doses for most frequent X-ray examinations in Serbia and Montenegro. A total 491 procedures, for 11 different examination categories were analyzed. Using X-ray tube output data, entrance surface dose for each plane radiography was calculated, as well as effective dose for each patient. Except for chest PA examination, all estimated doses are less than stated reference levels for plane film examinations. For fluoroscopy examinations total kerma-area product was measured and contributions from fluoroscopy and radiography were assessed. The study of kerma-area product reference doses confirms that dose level for complex fluoroscopy investigations are closely related to technique and individual patient variation, in terms of fluoroscopy time and number of radiography exposures. Survey data are aimed to help in development of national quality control and radiation protection programme for medical exposures.

INTRODUCTION

X-ray examinations are an established tool of medical diagnosis. Their widespread use means that, on the average, health system in Serbia and Montenegro provides annually 880 examinations per 1000 inhabitants [1]. Patients can undoubtedly derive enormous benefit from these examinations, although the ionizing nature of the X-rays means that their use is not entirely without risk. For this reason, all exposures to diagnostic X-rays need to be justified and optimized in terms of benefit and risk [2]. One of the basic requirements for such requirement is knowledge of patient doses. Regular patient dosimetry is recommended to evaluate the potential for optimization of radiation protection of patients. Although mandatory by the low, systematic recording of patient exposure is not yet part of radiology practice in Serbia and Montenegro. Also, there are no established national diagnostic reference levels. Following Directive 97/43 Euroatom, establishment of patient dose measurement, diagnostic reference levels and measures to reduce patient dose have become mandatory [3]. There are reference levels for patient doses only in simple examinations and few published papers on reference levels in complex examinations [4, 5]. In recent period, in Radiation and Environmental Protection Laboratory of the VINČA Institite of Nuclear Sciences an effort have been made to collect data on patient doses during standard radiological examinations, as a part of Quality Assurance Programme [6]. The patient dose survey was performed in order to examine the situation and evaluate how the International Commission on Radiological

Protection principle of optimization could be implemented in practice. The purpose of this work is to estimate patient doses for simple radiographic examinations and barium meal procedure. Further analysis of patient doses (including image quality aspect) are in progress and will be reported subsequently.

PATIENT DOSE SURVEY

The extent of dose survey must be limited and measurements have to be confined to most frequent X-ray examinations which give a large collective dose to the population. In that sense, measurements were concentrated on high-frequency examinations. Initially, measurements have been performed in five non-specialized local hospitals, performing annually more than 150 000 examinations. In summary, 56 barium meal and 435 conventional procedures were studied, so at least 10 patients were observed for each examination type. The examinations were carried in three X-ray rooms, equipped with three-phase, 6-pulse X-ray unit, in a room equipped with three-phase, 12-pulse unit and in two rooms with high frequency units. Only later, high frequency units are using Automatic Exposure Control (AEC). Using established Quality Control Protocol [7,8], all X-ray tubes and generators were tested before start of patient dose survey using calibrated Barracuda Multimeter (Barracuda, R100, RTI Electronics AB, Goteborg, Sweden) and RMI set of quality control tools (RMI, Middleton, USA), calibrated in traceable Secondary Standard Dosimetry Laboratory at the VINČA Institute of Nuclear Sciences, Belgrade. For each examination studied, personal data and technical parameters were collected according to a questionnaire designed by the patient dosimetry protocol. In questionnaire the data collected were:

- Radiological room and equipment;
- Patient sex, age , weight and height;
- Type of procedure;
- Analysis for each patient for simple examination (kV, mAs, kerma-area product, film size and focal spot – film distance);
- Analysis for complex examination for each patient (mean kV for fluoroscopy, mean mA for fluoroscopy, kV for radiography, mAs for radiography, kerma-area product for fluoroscopy and radiography, fluoroscopy time, size of film).

At the end of procedure, the quality of each film was verified by radiologist. In this paper, the survey is summarized in terms of mean doses, medians and associated range, to illustrate the often-wide distributions of observed doses for each type of examination. This will provide a useful baseline for future measurements on patient doses.

Entrance Surface Dose

Various dosimetry quantities are applied in patient dosimetry with respect to actual examination type and equipment performance [9]. It is important that patient dose measurements are time-effective and not disturb the patient and staff during examination. Only a brief outline of the method employed is given here. Full details of patient dosimetry techniques are given elsewhere [10, 11]. After having evaluated several options available, it was decided to use indirect method for dose assessment, i.e. air kerma measurements for plane film examinations and kerma-area product measurement for complex examinations. Dosimetry in diagnostic radiology is in air kerma domain [12]. Although absorbed dose and air kerma are almost equal in the diagnostic energy range, but vary if the medium is different, air kerma is easier to measure accurately, due to practical problem associated with achieving electronic equilibrium in the field [13]. The dosimetry method involve a measurement of X-ray tube output (Y_D), e.g. air kerma at defined geometry for a range of tube voltages, followed by the use of backscatter factor (*BSF*) data and geometry corrections to determine the entrance surface dose. This methodology enables a relatively large number of patient dose estimates from a small number of measured parameters, the measurements are part of a quality assurance programme and they are useful for the estimation of low surface doses. The knowledge of the tube output, tube voltage, tube current, exposure time and focal spot-skin distance, enables to deduce the air kerma at the point corresponding to the position of the patient's skin. Entrance surface dose is air kerma measured in the primary beam in the entrance plane of the patient including backscatter radiation. Backscatter factor values for diagnostic X-ray beam qualities are reported in literature [12]. Then entrance surface dose is given by [11]:

$$ESD = \frac{Y_D \cdot mAs \cdot D^2}{\left(L - (d + b)\right)^2} \cdot BSF \tag{1}$$

where Y_D is X-ray tube output at distance D normalized by mAs (μGy/mAs), *mAs* is the product of the tube current and exposure time, L is focus-film distance and b and d film-table top distance and patient thickness, respectively. To calculate entrance surface dose, X-ray tube output Y_D was measured at distance of 1 m for X-ray tube voltage in range (50-120) kVp, in 10 kVp steps. Patient thickness was deduced from the recorded patient weight and height. For each patient entrance surface dose was calculated using real examination data, according to Eq. (1).

Kerma –Area Product

Kerma-area product is integral of the air kerma (K_{air}) over the area A, a perpendicular to the beam axis. Thus:

$$KAP = \int_A K_{air}(A)dA \tag{2}$$

In this quantity, radiation backscattered from the patient is excluded. By including a measure of the area of the beam as well as the dose, kerma-area product provides an indication of the level of beam collimation achieved which is a very important method for patient dose reduction. The kerma-area product meter essentially integrates the air kerma over the whole beam area and for any number of exposures. Kerma-area product was determined using KERMAX-Plus transmission ionizing chamber fitted to an X-ray tube light-beam diaphragm. The chamber was calibrated against reference dosimeter (Barracuda, R100, RTI Electronics AB, Goteborg, Sweden) on both X-ray units enrolled into the survey. The energy response of chamber was better than ±8% related to 100 kV [14]. To allow practical estimation of effective dose, United Kingdom's National Radiological Protection Board conversion factors have been used [15]. Entrance surface doses and kerma-area product as input parameters, have allowed organ equivalent dose and effective dose assessment for each patient and actual radiation quality.

PATIENT DOSE ASSESSMENT

To obtain an estimation of typical dose to an average patient, the measurements have been performed on a representative sample of adult patients with mean weight of 70 kg. Patients of extreme body weight have been excluded from the survey.

TABLE 1. Characteristics of the patients and radiological procedures in terms of sample size, patients weight (with standard deviations) and applied tube voltage (with range in parenthesis) for plane film and complex examinations

Plane film examinations										
	Cervical spine AP	**Cervical spine LAT**	**Pelvis AP**	**Th spine AP**	**Lumbal spine AP**	**Lumbal spine LAT**	**Chest PA**	**Chest LAT**	**Skull PA**	**Skull LAT**
Hospital 1										
Sample size	12	23	13	5	15	25	46	4	10	7
m (kg)	77±9	82±11	86±8	65±9	69±12	71±11	72±10	67±7	73±7	73±6
U (kV)	77 (63-90)	82 (66-102)	82 (66-102)	84 (77-90)	83 (70-96)	117 (85-141)	91 (70-117)	117 (102-133)	84 (77-102)	82 (71-96)
Hospital 2										
Sample size	12	12	10	6	10	10	41	-	11	11
m (kg)	70±14	70±14	73±11	70±15	77±17	77±17	73±13	-	68±10	67±11
U (kV)	61 (55-65)	60 (55-60)	64 (40-75)	70 (65-75)	70 (65-75)	80 (65-85)	64 (55-70)	-	65 (60-70)	59 (55-65)
Hospital 3										
Sample size	17	17	8	8	16	16	10	-	-	-
m (kg)	74±8	74±8	69±9	71±8	72±10	72±10	72±11	-	-	-
U (kV)	79 (75-85)	79 (75-90)	73 (67-85)	81 (80-85)	67 (65-75)	81 (75-90)	82 (75-87)	-	-	-

Complex examination -Barium meal		
	Hospital A	**Hospital B**
Sample size	29	27
m (kg)	71±9	72±9
Radiography U(kV)	100 (77-102)	76 (55-90)
Fluoroscopy time (s)	199±81 (73-294)	283±93 (103-440)
Number of images	3±2 (2-8)	3±1 (1-6)

Table 1. shows the characteristics of the patients and technical parameters selected for various examinations types. It summarizes sample size, patient weight and applied X-ray tube voltage for plane film and complex examinations. The broad classification of used projections is: anterio-posterior (AP), posterior-anterior (PA) and lateral (LAT) projections. Table 2. summarizes entrance surface dose mean values with

standard deviation and medians for plane film examinations in three hospitals. Estimated effective dose mean vales are also given for each examination type.

TABLE 2. Means (with standard deviations) and medians of entrance surface doses (mGy) and mean effective doses (mSv) for different radiographic examinations in three hospitals

	Hospital 1			Hospital 2			Hospital 2		
Procedure	**Mean**	**Median**	**Effective dose (mSv)**	**Mean***	**Median**	**Effective dose (mSv)**	**Mean**	**Median**	**Effective dose (mSv)**
Cervical spine AP	0.5±0.2	0.5	0.02	1.0±0.4	0.9	0.04	2.4±0.6	2.5	0.12
Cervical spine LAT	0.4±0.3	0.3	<0.01	1.0±0.5	0.8	<0.01	1.7±1.0	0.8	0.02
Pelvis AP	1.7±0.9	1.6	0.23	2.13±1.3	1.7	0.30	2.4±0.3	2.3	0.35
Thoracic spine AP	0.9±0.5	1.16	0.09	2.0±0.2	1.9	0.16	1.7±0.4	2.0	0.18
Lumbal spine AP	1.6±1.0	1.4	0.21	2.7±0.8	2.5	0.27	4.0±0.3	4.1	0.36
Lumbal spine LAT	2.2±1.0	1.9	0.06	5.9±1.8	5.3	0.10	5.2±0.8	5.3	0.85
Chest PA	0.2±0.14	0.19	0.03	0.6±0.2	0.6	0.05	0.4±0.2	0.3	0.04
Chest LAT	0.3±0.2	0.3	0.03	-	-	-	-	-	-
Skull PA	1.0±0.7	0.9	0.01	1.3±0.4	1.2	0.01	-	-	-
Skull LAT	0.9±0.6	0.9	0.01	1.0±0.3	0.9	0.01	-	-	-

Table 3. Compiles kerma-area product values for barium meal procedure. Analysis of results indicates that fluoroscopy is the main contributor to total dose.

TABLE 3. Total kerma-area product (with standard deviation), associated range and median (Gy cm^2) and mean effective doses (mSv) for barium studies of upper gastrointestinal tract for 56 patients in two hospitals. Percentage of total kerma-area product due to fluoroscopy and radiography is also shown.

		Kerma-area product (Gy cm^2)							Effective
Hospital	No of patients	Total	Max	Min	Median	3rd quartile	F (%)	R (%)	dose (mSv)
A	29	8.4±5.4	24.5	2.2	7.2	10.7	81±7	19±7	1.7±1.1
B	27	24.3±11.6	45.9	5.2	22.1	31.1	75±11	25±11	4.8±2.3

DISCUSSION AND CONCLUSIONS

Great variations in patient doses were found in this survey. Some reasons for the variations became apparent, as speed class of film-screen combination, which was 200-400, and manual exposure control settings. The typical technical factors used vary by a wide range. For instance, loading factors extend (55-117) kVp and (1-62) mAs for chest radiography (Table 1). In spite of observed fluctuations in applied workload (tube current and exposure time product), there is a tendency of smaller product of tube current and exposure time for high tube voltage. This combination provides lower entrance surface dose, which is case in Hospital 1. Besides tube dialed exposure parameters, other equipment related, technologically limited, factors also affect patient dose. These are three phase generators, insufficient beam filtration and manual exposure control setting. Distributions observed for various dose quantities are typical skewed, with mean values generally greater than corresponding medians, so small numbers of patients receive high doses. Since the survey was not extensive and the median value is not influenced by the values that lie outside the main part of distribution as the mean value, it can be argued that the median is very helpful in typical practice assessment. The entrance surface dose to patients in diagnostic radiology is a dose descriptor to quantify diagnostic reference doses for simple radiographic examination. Diagnostic reference levels (DRL) are part of the quality criteria as laid down in the European Guidelines on Quality Criteria for Diagnostic Radiographic Images [16]. They are also recommended by the International Commission on Radiological Protection [2] and by the International Atomic Energy Agency, as guidance doses [17]. Diagnostic reference dose values provide quantitative guidance to identify relatively poor and inadequate use of technique and need for appropriate corrective action. They are usually based on the third quartile values of large patient dose surveys [10]. The adopted reference levels in Serbian legislation are those proposed by International Atomic Energy Agency, but only for simple examinations. As it is presented in Table 2, obtained doses for plane radiography examinations in hospital 1 are well below the stated reference levels. The explanation for relatively low doses is in good radiographic technique applied, which supports high X-ray tube voltages and sufficient beam filtration. The mean and median values in Hospital 2 and Hospital 3 for chest PA examination are greater than reference value of 0.3 mGy. The explanation for relatively high doses lies down soft radiation qualities applied during the examination. Applied tube voltage in Hospital 2 and 3 was significantly lower than 90 kVp which in combination with insufficient tube filtration has resulted in increased patient doses. For chest film, high physiological contrast among lung and bone tissue is well transformed into long gray scale at high tube voltage values. It keeps down the relative number of photoelectric events in bone and leads to lower overall patient dose. Although the assessed doses for other examinations were well below reference level, general practice if far from good radiographic technique. In addition to chest X-ray examinations, the optimization of practice for other X-ray examinations is also necessary.

Kerma-area product is reference dosimetric parameter in complex examinations. The KAP measurements results for barium meal procedure obtained here are comparable with other survey results [18, 19]. The findings from the present study showed that optimization of technical and clinical factors may leas to a substantial patient dose reduction. In fact, fluoroscopy and radiography have been performed at higher X-ray tube voltages and proper beam filtration in hospital A which resulted with three times lower doses than in hospital B. Also, the importance of Automatic Exposure Control settings is enormous. The average kerma-area product obtained here is lower than results of other surveys, mostly due to relatively small number of images made during the barium meal procedure.

Patient doses are determined by multitude factors which interact in very complicated manner. It is very important to perform real patient dose measurements in hospitals. Besides quantitative data obtained, these results allow better understanding how different working habits and examination technology influence the patient doses and make medical staff aware of patient doses and their responsibility for optimization of daily practice. In that sense, the survey results are link between patient dosimetry, as a first step in optimization of radiation protection, and Quality Assurance Programme in diagnostic radiology. Since patient doses using kerma-area product meter have not been extensively investigate, the idea is to extend the methodology presented here to other complex examinations and to more radiology departments, in order to establish diagnostic reference levels on national scale. Reference dose levels for diagnostic radiology examinations provide the benchmark for comparison X-ray exposures from different facilities, in order to reduce patient doses and maintain good image.

The results of this survey will be important input for a new radiation protection measures within this field. The mean values of dose estimates presented for each type of examination may be representative for average adult patient undergoing these examinations. These data may be used to assess collective dose to the population arising from diagnostic X-rays and to evaluate the radiation risk from the various radiological procedures.

ACKNOWLEDGMENTS

This work has been supported by Ministry of Sciences, Technologies and Development of the Republic of Serbia through Projects No. 2016 (Physics of Radiation Protection).

REFERENCES

1. Ciraj, O. Košutić, D. Marković, S. *Patient Dosimetry in Diagnostic Radiology and Dose Measurement in Practice*, Proceedings of Scientific Meeting Applied Physics in Serbia, Serbian Academy of Sciences and Art, Belgrade, May 2002, ISBN 86-7025-391-4, 317-320 (2002)
2. International Commission on Radiological Protection, *Recommendations of the ICRP*. (Publication 60), Annals ICRP 21, (1991)
3. European Commission. *On health protection of individuals against the dangers of ionizing radiation in relation to medical exposure.* Council Directive 97/43/Euroatom of June 30 1997, (1997)
4. Ruiz-Cruces, R. and Ruiz, F. *Patient doses from barium procedures.* Br J Radiol 73:752-61 (2002)
5. Warren-Forward, H.M. and Haddaway, M.J. *Dose-area product readings for fluoroscopic and plain film examinations, including an analysis of the source of variation for barium enema examinations.* Br J Radol 71:961-967 (1998)
6. Ciraj, O. Košutić, D. Marković, S. *Quality Assurance in Diagnostic Radiology with X-rays.* Proceedings of Scientific Meeting Applied Physics in Serbia, Serbian Academy of Sciences and Art, Belgrade, May 2002, 321-324 (2002)
7. Institution of Physics and Engineering in Medicine and Biology. *Measurement of the performance characteristics of diagnostic x-ray systems used in medicine. Part I x-ray tubes and generators.* Report No 32, IPEM (1995)
8. The Institute of Physics and Engineering in Medicine. *Recommended Standards for the Routine Performance Testing of Diagnostic X-ray Imaging Systems.*(Report No. 77). IPEM, (1997)
9. National Radiological Protection Boared. *Doses to Patients from Medical X-ray Examinations in the UK*, NRPB, (2000)
10. National Radiological Protection Boared. *Guidelines on Patient Dose to Promote the Optimization of Protection for diagnostic Medical Exposures.* (NRPB Vol.10 No.1). (1999)
11. Faul kner, K. Broadhead, D.A. Harrison, R.M. *Patient dosimetry measurement methods*, Appl Rad and Isotopes, 50:113-123 (1999)
12. Petoussi -Henss, N. Zankl, M. Drexler, G. Panzer,W. Regulla, D. *Calculation of backscatter factors for diagnostic radiology using Monte Carlo methods.* Phys. Med. Biol. 43: 2237-2250(1998)

13. Kase, K. bjarngard, B. Attix, F. *The dosimetry of ionizing radiation.* Volume 1. (Academic Press) ISBN 0-12-392651-3 (1985)
14. Internatio nal Electrotechnical Commission. *Medical Electrical equipment –Dose Area product meters.* IEC 2000-01. (2000)
15. Hart,D.,Jones,D.G.,WallB.F. *Normalized Organ Doses for Medical X-ray Examinations Calculated Using Monte Carlo Techniques.* NRPB SR262 (1994)
16. European Commission. *European Commission. European Guidelines on Quality Criteria for Diagnostic Radiographic Images.* EUR 16260 EN.(1996)
17. International Commission on Radiological Protection. *Recommendations of the ICRP,* Publication 73 (1997)
18. Maccia, C. and Benedittini, M. *Doses to patients from diagnostic radiology inFrance.* Health Phys 54:397-408 (1988)
19. United Nations Scientific Committee on the Effects of Atomic Radiation. *Source and Effects of Ionizing Radiation.* Report to the General Assembly. United Nations, New York (2000)

Applied Physics (APHYS 2003)
A. Méndez-Vilas (Ed.)

Analysis of the Background Levels of Tritium in Precipitation in Valladolid (Spain)

Miriam Pequeño*♦, Marta G.-Talavera*, Raul López*♦, Luis Deban♦, Eulogio L. García⅄, Rafael Pardo♦ and Victor Peña*

*LIBRA, Edificio I+D, Campus Miguel Delibes, University of Valladolid, Paseo Belen, 47011, Valladolid, SPAIN.
♦Department of Analytical Chemistry, Faculty of Sciences, University of Valladolid, C/ Dr. Mergelina s/n, 47005 Valladolid, SPAIN
⅄ Department of Physics, Faculty of Agrarian and Environmental Sciences, University of Salamanca, Avda. Filiberto Villalobos, 119-129, 37007, Salamanca, SPAIN

Abstract. An accurate knowledge of background tritium levels in precipitation is important for a number of applications using tritium as a radiotracer. In this paper, we present a study of the levels of tritium in precipitation in Valladolid (Spain) based on measurements of samples collected during a period of three years. Seasonal variations in the ^{3}H activity have been detected. The anomalous values with respect to the general trend are discussed in terms of the synoptic conditions associated to the precipitation event.

INTRODUCTION

Tritium is a cosmogenic radionuclide ubiquitous in the environment, coming both from natural and anthropogenic sources. Natural tritium is formed in the upper atmosphere from the bombardment of nitrogen by the flux of neutrons in cosmic radiation. Subsequently, it is directly incorporated into the water molecule ($^{1}H^{3}HO$) being introduced into the hydrological cycle. On the other hand, the anthropogenic inventory is largely due to the vast amounts of this isotope released to the stratosphere as a result of the nuclear weapons testing from 1952 to 1963. Besides, nuclear power plants and other industries discharge considerable volumes of tritium to the environment under controlled conditions.

The interest of tritium regarding radiation protection issues is stimulated by its physical half-life (12.43 y), the continued production from existing nuclear facilities and the potential of even greater releases from thermonuclear reactors [1]. Furthermore, ^{3}H is used as a tracer to provide a better understanding of many natural processes, mainly in hydrology, oceanography and atmospheric dynamics. Tritium is the most applied radionuclide for dating of ground waters, especially for distinguishing pre-bomb and post-bomb recharge. Other applications include the authentication of wines and derivatives [2] or the reconstruction of the historical tritium exposures using tree rings [3].

But to develop any of those studies or to detect accidental releases from industrial or energetic activities, the background levels of this isotope in precipitation must be accurately characterized in the region under study. ^3H concentration in precipitation is a function of geomagnetic latitude, with greater production at higher latitudes [4]. Besides, yearly cycles have been reported at several locations, with maxima in the spring-summer and minima in the winter, related to air exchange between the stratosphere and the troposphere [5].

In this paper, we present a study of the levels of tritium in precipitation in Valladolid (Spain) based on measurements performed by liquid scintillation spectrometry. Seasonal variations in the ^3H activity, in agreement with those reported in other stations [6,7], have been detected. The anomalous values with respect to the general trend are explained in terms of the synoptic conditions associated to the precipitation event.

SAMPLING PROCEDURE

Rainwater samples were collected periodically, from spring 2001 till autumn 2003, at a station in the city of Valladolid. Generally, samples were collected once per month, except for major rain events, when they were collected immediately afterwards. Sampling was performed with a polyethylene collector.

Valladolid is a medium-size city of 350.000 inhabitants. Several industries are situated in the outskirts of the city but none of them generates tritium – the closest nuclear power plant is situated 215 km northeast. The weather is typically continental but with marked Mediterranean characteristics, with a high temperature range between the hottest and the coldest months in the year and scarce precipitation (400 mm/y).

MEASURING METHODS

Electrolytic Enrichment

Tritium levels in natural waters are usually too low to be measured directly by means of liquid scintillation. Therefore, an electrolytic enrichment process was undertaken prior to measurement. Water samples were first distilled, in order to remove interfering substances and to minimize chemical variability among samples. Afterwards, an electrolytic enrichment process in basic medium was carried out, adding 1.0 g of Na_2O_2 in 500 ml of water - the use of a higher concentration of Na_2O_2 would accelerate the process but would increase corrosion in the electrodes. The electrolytic enrichment cells consist of two electrodes of Ni and mild steel, respectively, and follow the design proposed by Östlund [8]. Because isotopic effects are higher at low temperature, the cells are installed within a refrigerator system that keeps temperature stable ($\pm$1°C) at 1°C. Once the electrolysis process is finished, the sample is neutralized by adding $PbCl_2$ and distilled again to reduce quenching agents.

Measurement Of Tritium By Liquid Scintillation

Tritium analyses were performed in an ultra low-level background QUANTULUS 1220 detector by Wallac. The electrolytic enriched samples were disposed in polyethylene vials and mixed with the scintillation cocktail Optiphase Hisafe 3 (Wallac). The optimal ratio sample to cocktail was set to 8:12, taking into account minimum detectable activities and a proper mixture sample-cocktail. To avoid chemiluminiscence, samples were stored for at least 12 h in the dark before measuring.

Since the measuring efficiency for a given sample depends on its quenching level, this variable has to be quantified in order to determine accurately the tritium activity. In particular, we measured the quenching for each sample by means of its SQP(E) parameter, which is obtained by external irradiation of the vial using an ^{152}Eu source. The efficiency must be determined in a range covering the SQP(E) values common in real samples. To do so, we prepared eight calibration vials by spiking free-tritium water with a calibration standard provided by Wallac and adding different quantities of a quenching agent - CCl_4 (Merk) - to each vial. The measured SQP(E) values for each calibration sample were fitted to the known measuring efficiencies by means of a least-squares fit to a polynomial function (R^2=0.9976). Such function was then used to obtain the efficiency of the samples given the corresponding SQP(E) parameter.

The tritium activity for each sample was calculated according to eq. 1:

$$A(UT) = \frac{CPM_s - CPM_b}{60 \cdot Ef \cdot m \cdot Z \cdot 0.118} \qquad (1)$$

CPM_s and CPM_b denoting counts per minute from sample and background, respectively; 60 being the conversion factor to Bq; *Ef*, the counting efficiency; *m*, sample mass (kg); *N*, the volume reduction factor; *Z*, the electrolytic enrichment factor and; 0.118, the conversion factor to TU.

Associated uncertainties were calculated taking into account the individual uncertainty sources [9]. The final uncertainty was then estimated by the error-propagation formula:

$$\sigma(A) = \sqrt{\sum_{i=1}^{n} \left(\frac{\partial A}{\partial x_i}\right)^2 \cdot \sigma(x_i)^2} \qquad (2)$$

RESULTS

In Fig. 1, we have represented the tritium activities during the above mentioned period. The mean annual tritium level in precipitation in Valladolid is 4.11 TU. Deviations from this average value reach up to 50 %. As can be seen from the figure, the data follow a seasonal trend with maxima in spring and minima in the winter. This behaviour has been reported at different stations in the North Hemisphere. It is known to be due to a weakening of the tropopause between 30 y 60° N every spring, causing a leakage of stratospheric vapour with a higher content of tritium [10].

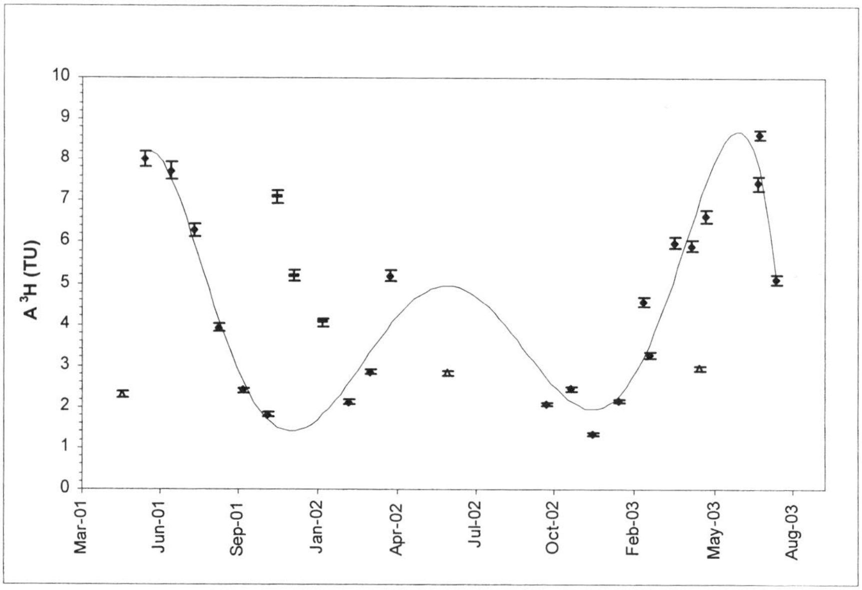

FIGURE 1. Trend in the ^3H concentration in precipitation from spring 2001 till autumn 2003. The solid line was obtained by a polynomial least-squares fit to the experimental data points (♦). The dash (–) and the square (□) points was related to atypical synoptic situation (snow, hail and thunderstorm).

Nevertheless, there are several data points markedly differing from this general trend. Most of them correspond to the winter 2001-2002, which presented extremely atypical synoptic flow features compared to the usual winter conditions at Valladolid [11]. The typical winter synoptic situation is shown in Fig.2. It corresponds to advections of maritime masses entering inland from the NW.

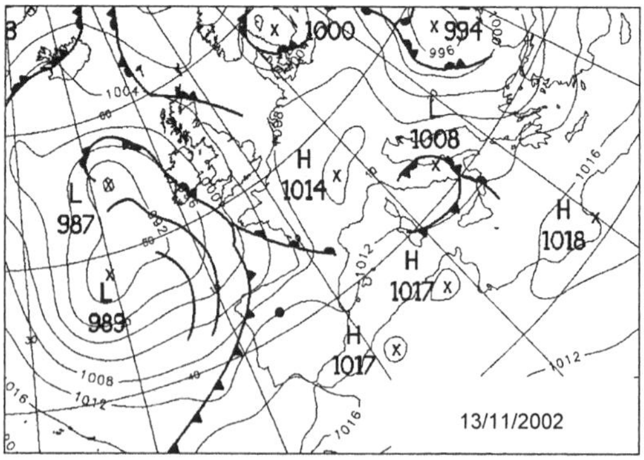

FIGURE 2. Synoptic situation relative to a Zonal circulation (Polar Air)

On the other hand, the precipitation events from November 2001 correspond to the synoptic map shown in Fig. 3. At low levels, there was a cold northerly flow over the Iberian Peninsula, bringing a continental polar air mass of elevated tritium content. Besides, precipitation fell as snow, which presents higher tritium levels than rain because the frozen phases, i.e. hail and snow, are not subject to a rapid exchange of isotopes between the rain droplets and ambient vapour [12].

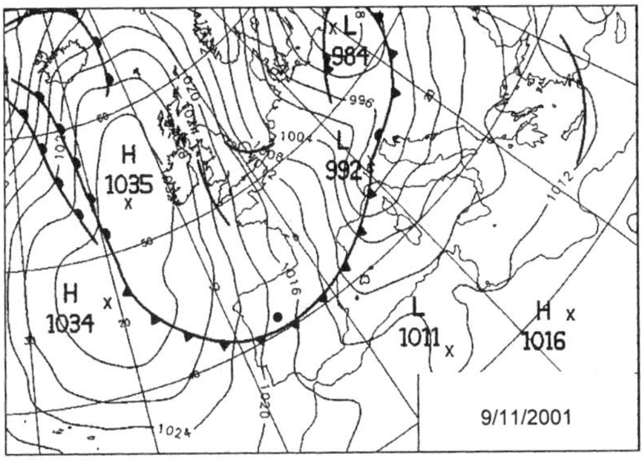

FIGURE 3. Synoptic situation relative to a meridian circulation North-South (Arctic Air)

The situations occurred during December 2001 and January 2002, are associated to the advection of a mass of tropical maritime air entering from the SE (see Fig. 4). The high moisture content of the mass and the longest trajectory inland, compared to air masses from the NW, favours the uptake of tritium, whose concentration in air is higher over the continent than over the oceans.

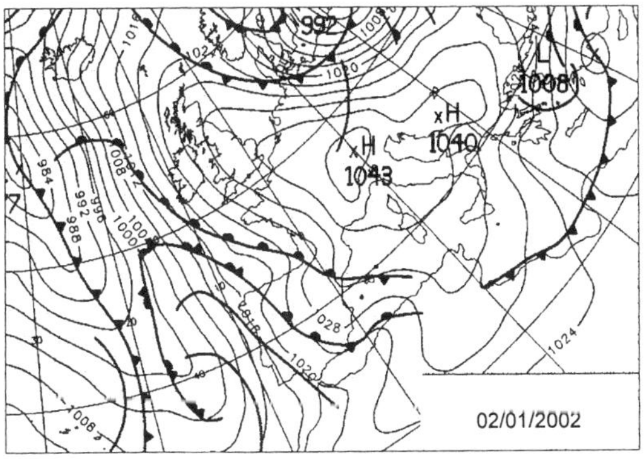

FIGURE 4. Synoptic situation relative to southwest wind (Maritime Tropical Air)

Moreover, two rain samples presented anomalously low tritium values. They come from thunderstorms produced in the early spring. The three conditions to produce a thunderstorm are potential instability, high levels of moisture in the atmospheric boundary layer and forced lifting [13]. But these factors can be related to many different synoptic features. The reason for the low tritium content is still under investigation.

CONCLUSIONS

We have measured the tritium activities in precipitation for a period of three years in the city of Valladolid. Large variations with respect to the seasonal value have been found. In general, there is a seasonal fluctuation with maxima in late spring-summer and minima during the winter. Unusually high values found in the winter have been interpreted based on the synoptic conditions. The reason for anomalously values in thunderstorm is still under investigation. These results are of great importance for studies related to the dating of aquifers and authentications of wines, which are currently carried out in the region.

ACKNOWLEDGMENTS

The authors are grateful to the *Junta de Castilla y León* and European Social Fund for financial support and to the *Instituto Nacional de Meteorología* and the web *Infomet* for providing information.

REFERENCES

1. Edlund, O. Normal releases from fusion processes and environmental radiation doses. Radiation Protection Dosimetry. **16**(1 -2), 27-30 (1986).
2. Medina, B. Wine Authenticity. Food Authentication, Blackie Academic and Professional, London, 60-107 (1996)
3. Kozák, K., Rank, D., Biró, T., Rajner, V., Golder, F. and Staudner, F. Retrospective Evaluation of Tritium Fallout by Tree-ring Analysis. J. Environ. Radioactivity, **19** 67-77 (1993)
4. Ferronsky, V. and Polyakoc I. Environmental Isotopes in the Hidrosphere. New York. Wiley and Sons, 1982.
5. Mayo A. Isotopes, 1991.
6. Brown, R.M. A Review of Tritium Dispersal in the Environment. Proceedings of a Workshop on Tritium and Advanced Fuel in Fusion Reactors. Bologna 557-576 (1989)
7. Rank, D. Environmental Tritium in Hydrology: Present State. (1992)
8. Östlund, H.G. and Werner E. The electrolytic enrichment of tritium and deuterium for natural tritium measurements. Tritium in the Physical and Biological Sciences, IAEA, Vienna, **1**, 95-104 (1962)
9. Rozanski K. and Gröning M. Quantifying uncertainties of tritium assay in water samples using electrolytic enrichment and liquid scintillation spectrometry.
(http://www.iaea.or.at/programmes/rial/pci/isotopehydrology/docs/intercomparison/ViennaH3-v12.htm)
10. Gat, J.R. The isotopes of hydrogen and oxygen in precipitation. Handbook of Environmental Isotope Geochemistry (P. Fritz and J.Ch. Fontes, eds.). **1**, 22-48 (1980)
11. Sánchez, J. Situaciones atmosféricas en España. Ministerio de Obras Públicas, Transportes y Medio Ambiente (1993)
12. Ehhalt, D.H. Vertical profiles and transport of HTO in the troposphere. J. Geophys. **76**, 7351-7367. (1971)
13. Van Delten. The synoptic setting of thunderstorms in Western Europe. Atmospheric Research. **56,** 89-110 (2001).

Applied Physics (APHYS 2003)
A. Méndez-Vilas (Ed.)

Continuous Monitoring Of Environmental Radioactivity In Belgrade

Dejan Jokovic*, Radomir Banjanac*, Aleksandar Dragic*, Vladimir Udovicic*, Bratimir Panic*, Ivan Anicin¶, Jovan Puzovic¶

**Center for Applied and Technical Physics, Institute of Physics, Pregrevica 118, 11080 Belgrade, Serbia and Montenegro*
¶Faculty of Physics, University of Belgrade, St. trg 16, 11000 Belgrade, Serbia nad Montenegro

Abstract. Two identical plastic scintillators, shaped prismatic (50x23x5)cm and similar to NE102, have been developed for continuous cosmic-ray measurements. The first detector is situated on the ground level and the second one at the depth of 25 m.w.e. in the low-level underground laboratory [1]. All recorded spectra are mainly the cosmic-ray muon ΔE spectra. Here, the low-energy (below 6 MeV) region of the spectra is analyzed in order to continuously monitor variations in environmental radioactivity, as well as to estimate long term stability of used electronics.

INTRODUCTION

Preliminary measurements made during the year 2002. were basically analyzed in the spectrum region of single muon detection originated from cosmic rays [2]. Intensity variations of detected cosmic rays and determination of muon flux may be correlated with processes on the Sun only if fluctuations of all measurement parameters are eliminated. These fluctuations, caused by electrical power supply, ambiental temperature etc., make a shift in acquisited spectrum yielding additional error in muon flux measurements. Analysis of the eventual shifting in the spectra has been made during the whole 2002. by observing the low energy part of the spectra. Energy calibration sets lower limit of muon counting integration on 6 MeV; this means mainly ambiental gamma rays ($E \leq 2.7$ MeV) are detected in the spectrum region below 6 MeV. Gamma rays dominantly interact with plastic scintillators through Compton scattering yielding continuous spectrum. Detection of particles from the cosmic rays (muons, gammas, electrons) in the energy region below 6 MeV can be neglected; this energy is limit line of detection of the ambiental gamma rays and the cosmic ray muons, respectively.

EXPERIMENTAL SETUP

Environmental gamma rays are detected by two identical plastic scintillator detectors of prismatic (50cm x 23cm x 5cm) shape. Detectors are produced by the High Energy Physics Laboratory of JINR, Dubna, and are similar to NE102. The detectors as well as some electronic modules (such as MPI) are assembled in our laboratory. The experimental scheme with description of used modules is presented in figure 1.

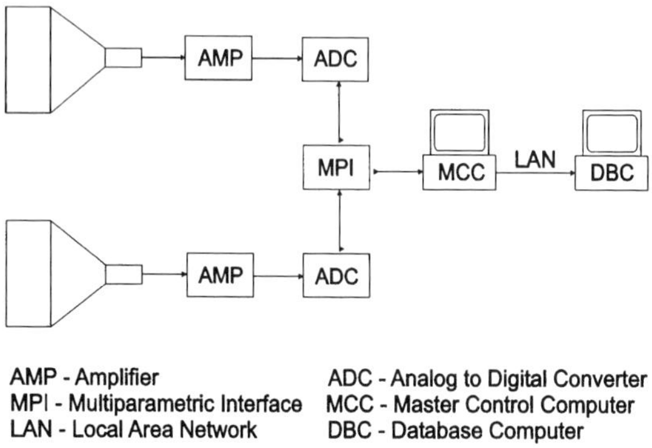

FIGURE 1. Experimental scheme

Each detector lies horizontally on its largest side and single 5 cm photomultiplier watches its long side (50cm x 5cm) via a correspondingly shaped light guide. The 4K channel spectra are registered every 5 minutes, with 270 seconds dedicated to measurements and 30 seconds being allowed for the recording of the spectrum and some quick interventions on the system. A typical ΔE spectrum of the underground level detector is presented in figure 2.

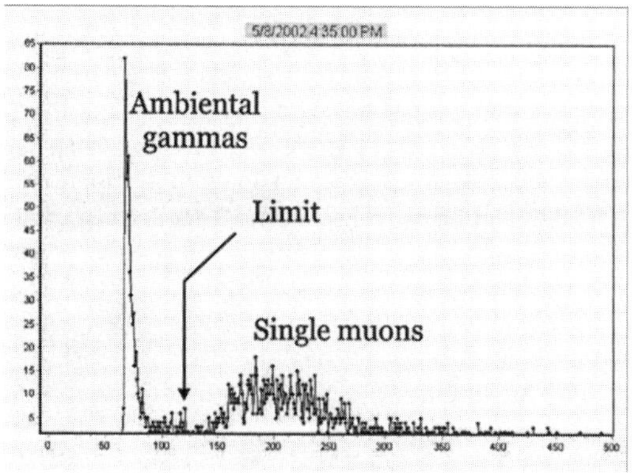

FIGURE 2. A typical 5 minutes spectrum of the underground detector.

Monte Carlo simulation of this spectrum, developed for muon interactions only, agrees very well with the experimental spectrum in the muon region (figure 3). Difference in the low energy part of the experimental spectrum from the simulated one is attributed to the environmental gamma background.

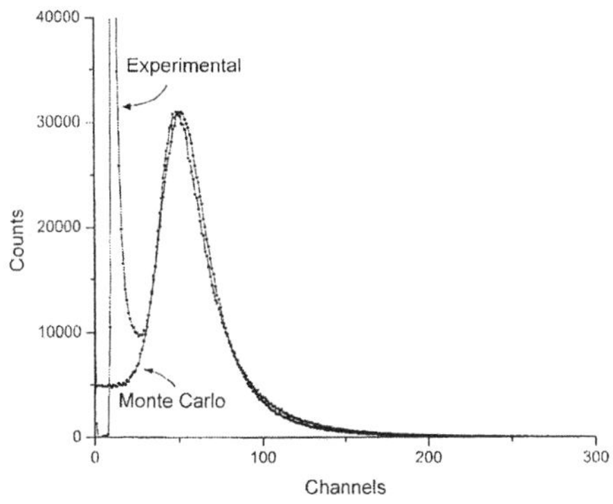

FIGURE 3. The experimental spectrum compared to the Monte Carlo simulated spectrum.

RESULTS AND DISCUSSION

The data from the two detectors for the year 2002 are studied for fluctuations in environmental radioactivity intensity. Figure 4. shows data from the underground detector, averaged over period of 1 hour for entire year 2002, plotted in form of deviation from the mean value, while figure 5. shows data from the ground level detector. For comparison, data from muon flux measurements are also plotted. The change of intensity is bigger in the data from the ground detector.

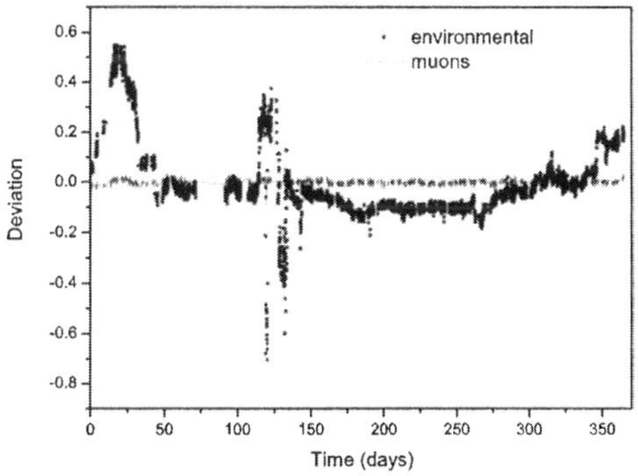

FIGURE 4. Annual intensity variations of underground environmental radioactivity for the year 2002.

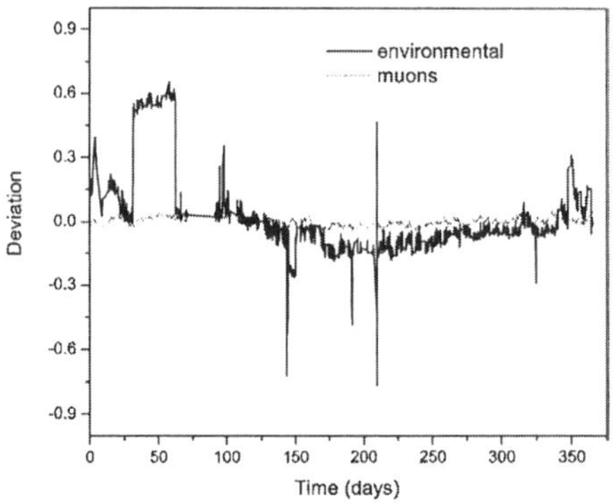

FIGURE 5. Annual intensity variations of above-ground environmental radioactivity for the year 2002

Although our measurements are continuous, some data are missing due to electric power or computer system failures, etc. This problem introduced difficulties in the analysis, especially in comparison of different observation period. Problem of long term stability of high voltage (and temperature in the ground detector environment) makes it difficult to take full advantage of spectral energy loss measurements. For that purpose, the optical pulsing system for regular checking of the calibration is under construction.

ACKNOWLEDGMENTS

This work has been supported by the Ministry of Science, Technologies and Development of the Republic of Serbia, project no. 1461.

REFERENCES

1. Antanasijevic, R., Anicin, I., Bikit, I., Banjanac, R., Dragic, A., Joksimovic, D., Krmpotic, Dj., Udovicic, V., and Vukovic, J., *Radiat. Meas.* **31**, 371-374 (1999).
2. Puzovic, J., Dragic, A., Udovicic, V., Jokovic, D., Banjanac, R., and Anicin, I., *Proceedings of the 28th International Cosmic Ray Conference,* Tsukuba, Japan, 2003, pp. 1199-1202.

Applied Physics (APHYS 2003)
A. Méndez-Vilas (Ed.)

Variations of ^{7}Be and ^{210}Pb (1992-1999) In Air At a Sampling Station On The Mediterranean Coast

C. DUEÑAS, M.C. FERNANDEZ, J. CARRETERO*, E. LIGER* AND S. CAÑETE

Department of Applied Physics I, University of Málaga, Campus de Teatinos s/n, 29071
* *Department of Applied Physics II, University of Málaga, Campus de Teatinos s/n, 29071.* MÁLAGA, SPAIN.

Abstract. The activity concentrations in air of ^{7}Be and ^{210}Pb have been investigated at Málaga (South-East Spain) from 1992 to 1999. In this period, the measurements yielded an average ^{7}Be activity concentrations at ground level of $4.16 \cdot 10^{-3}$ Bq m^{-3} and an average of $0.54 \cdot 10^{-3}$ Bq m^{-3} for ^{210}Pb. The variation of the data with time was studied by time series analysis and seasonal patterns were identified. Activity concentrations of ^{7}Be and ^{210}Pb were greatly affected by the meteorological conditions showing pronounced differences between seasons. Maximum activity concentrations in air were systematically observed in spring-summer of each year for ^{7}Be and in summer for ^{210}Pb. Minimum activity concentrations were observed in autumn for ^{7}Be and in autumn-winter for ^{210}Pb. In the long term, annual variations of approximately a factor of 1.7 were observed in ^{7}Be activity concentrations and a factor of 1.8 for ^{210}Pb. The study reveals that much of the variability in the concentration activities is explained by temperature, rainfall, relative humidity and wind speed.

1. INTRODUCTION

Long-term measurements of cosmogenic and atmospheric radionuclides such as ^{7}Be and ^{210}Pb provide important data in studying global atmospheric processes and comparing environmental impact of radioactivity from man-made sources to natural ones (1) y (2). ^{7}Be and ^{210}Pb are natural radionuclide tracers of aerosols originating over a range of altitudes in the atmosphere. ^{7}Be ($T_{½}$ = 53 days) is produced by cosmic rays impact on nitrogen and oxygen atoms in the stratosphere and the troposphere. ^{210}Pb ($T_{½}$ = 22.3 years) decays from ^{222}Rn ($T_{½}$ = 3.8 days) which is emitted from continents. Taken together, ^{7}Be and ^{210}Pb yield information about vertical motions in the atmosphere and the scavenging of aerosols. Several features make ^{7}Be and ^{210}Pb highly suitable tracers for improving general circulation model (GCM) aerosol simulation (3). Their sources are known, global in extent and relatively steady in time. The altitudinally distinct sources of ^{7}Be and ^{210}Pb suggest that they may allow assessment of the relative importance of stratospheric and tropospheric transport pathways. We report eight years measurements (1992-1999) of ^{7}Be and ^{210}Pb concentrations in surface air. Using these data, the present research was undertaken with the following principal goals: a) To study the variations of the data and b). To analyse seasonal patterns and to identify the main meteorological parameters that are responsible for the variations of these concentrations.

2 .MATERIAL AND METHODS

Aerosol samples were collected weekly in cellulose nitrate filters, 47 mm diameter (collection efficiency 99.99% for 0.8 μm pore size) with an air sampler

(RADECO, mod.AVS-28A) at a flow rate of 30 L min^{-1}. The air sampler was lodge in an all weather sampling station and situated 10m above the ground, on the roof of the Faculty of Science building, University of Málaga (4° 28′8″W; 36°43′40″N). Measurements of ^{7}Be and ^{210}Pb in each sample were carried out by non-destructive -ray spectrometry by means of its 477.6 keV and 46 keV -ray respectively, using a REGe-detector made by CANBERRA (relative efficiency about 30% to the efficiency of a 3″×3″ NaI(Tl) at 25 cm distance; resolution 2 keV for 1.33 MeV -ray of ^{60}Co).

3. RESULTS AND DISCUSSION

3.1Characterization of data

The results from individual measurements of ^{7}Be and ^{210}Pb concentrations were analyzed to derive the statistical estimates characterizing the distributions. Table 1 provides arithmetic mean (AM) and related statistical information such as standard deviation (SD), geometric mean (GM), dispersion factor of geometric mean (DF), maximum and minimum values and coefficient of variation (CV). These values are given in mBq m^{-3}.

TABLE 1. Statistical parameters.

	AM	GM	DF	SD	Max	Min	CV
Be-7	4.16	4.07	1.4	0.15	6.0	2.5	20.4
Pb-210	0.54	0.52	0.6	0.026	0.92	0.28	27.3

Normal distribution for ^{7}Be is significant at the 0.1 level. Otherwise, ^{210}Pb concentration appears approximately log-normal (significant level less than 0.1). Assuming these types of distribution, the GM for the ^{210}Pb data and the AM for ^{7}Be data should be used to characterize average values. A range of values of 3.5 mBq m^{-3} and a mean value of 4.16 mBq m^{-3} were found for ^{7}Be activity. A range of 0.64 mBq m^{-3} and a geometric mean of 0.52. mBq m^{-3} were found for ^{210}Pb activity.

3.2Temporal variations and some meteorological factors affecting variation in concentrations.

In Fig. 1 the box and whiskers diagrams of ^{7}Be and ^{210}Pb concentrations along the eight years studied are represented . From a visual inspection of the data in the box and whiskers diagrams, annual changes seem to be produced and are commonly attributable to different factors such as temperature, atmospheric stability or frequency and amount of precipitation. Several studies have been performed on the relationship between the meteorological parameters and the concentration of diverse radionuclides in air.

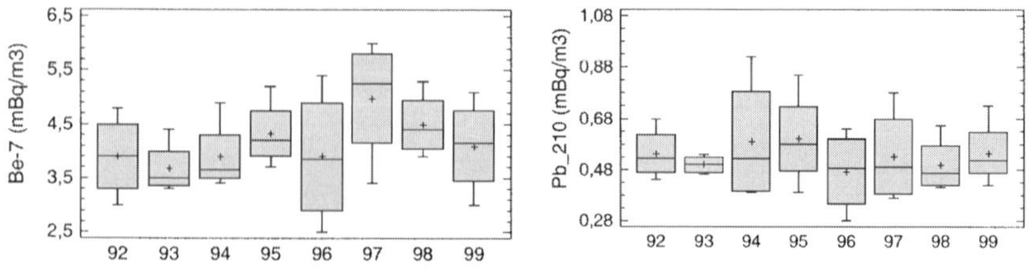

FIGURE 1. Annual variations of the activities for the 1992-99 period.

In Fig. 2 the box and whisker diagrams of the maximum air temperature, relative humidity and wind speed are plotted during eight years and also the amount precipitation per year. Examining the diagrams we conclude that there are differences among the studied years, mainly concerning the precipitation and wind speed. Firstly, we observe the variations in the amounts of precipitation during the eight period. The amounts of precipitation during all years, exception 1996 and 1997 are lower than the 550 mm average of the annual rainfall in Málaga over a 56 years period while in 1996 and 1997 the total annual rainfall was higher than this average.

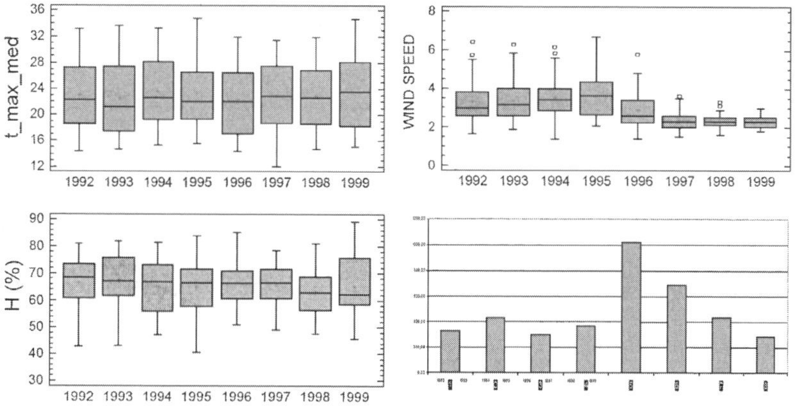

FIGURE 2. Box and whiskers diagrams for some meteorological parameters over the studied period

As meteorology plays an important role in the dispersion and transport of pollutants, we have performed a study to identify which meteorological parameters are strongly associated with the fluctuations of weekly concentrations. During the period of this study, meteorological data (wind speed, temperature, pressure, rainfall, and relative humidity) were supplied by the nearby weather station. First we performed a simple regression of ^{7}Be and ^{210}Pb concentrations and some meteorological factors and then we carried out a multiple regression in order to determine the extent to which the variations in concentrations might be attributed to the combination of these meteorological parameters. In our analysis we used average of the maximum temperature (*T*), rainfall (*r*), average relative humidity (H), average pressure (*p*), average wind speed (*v*). In Tables 2 the correlation coefficient between ^{7}Be and ^{210}Pb concentrations respectively and those parameters are summarized.

TABLE 2. Correlation coefficients between concentrations and meteorological factors

	T_max	Rainfall	Rel_Humidity	Wind Speed	Pressure
Be-7	0.502	-0.459	-0.570	0.087	-0.343
Pb-210	0.589	-0.472	-0.211	-0.290	-0.006

The study of correlation reveals a pronounced positive correlation with the average of the maximum temperature and a negative one with the other meteorological factors.

Temperature is the variable most strongly correlated to the activities. High temperatures are often associated to upward convection currents in the atmosphere .On the contrary , the correlation with the humidity is negative because air temperature and air humidity show an opposite behaviour. A multiple regression analysis was carried out in order to find the factors that influence the ^{7}Be concentrations. The variables were obtained choose the maximum temperature and relative humidity, obtaining the equation:

$$\text{Be-7 (mBq/m}^3) = (0.066 \pm 0.030)\ T\_max - (0.066 \pm 0.024)\ H + (7.251 \pm 1.835) \quad (1)$$

The validity of analysis of the regression equation were taken into account the relative error of the coefficient of each independent variable, the standard error of the estimate, the R-squared value and the p-value of regression. The R-squared for Eq.(1) is 42% and the p-value is less than 0.01. So, there is a statistically significant relationship between the variables at 99%confidence level The study of the correlations for ^{210}Pb reveals a correlation with the maximum temperature and pressure. A multiple regression analysis was carried out then in order to find the factor ^{7}Be, we have obtained the equation 2:

$$\text{Pb-210 (mBq/m}^3) = (0.024 \pm 0.005)\ T\ max + (9.7\ 10^{-5} \pm 1.2\ 10^{-7})\ P - (9.714 \pm 4.603) \quad (2)$$

The R-squared is and the p-value is less than, so there is a statistically significant relationship between the variables at the 95% confidence level. The correlation of ^{7}Be and ^{210}Pb air concentrations with meteorological parameters is very difficult, mainly due to the relative short times of atmospheric reload, less than 2 days (4) y (5) compared with the sampling time. This could be the best explication for the relatively poor correlation obtained .

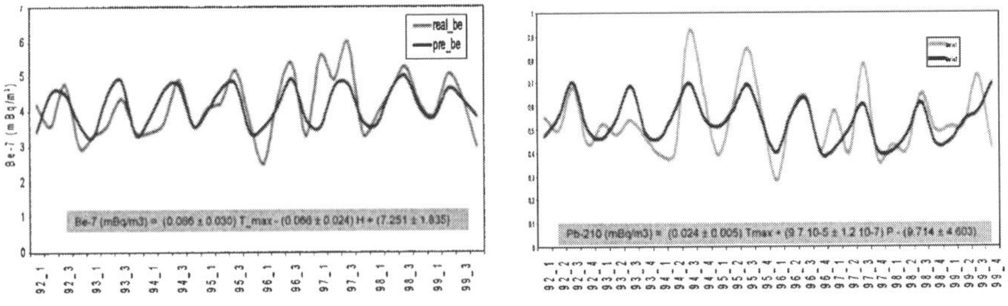

FIGURE 3.- Observed and predicted data for Be-7 and Pb-210

In Fig. 2 we plotted the observed and predicted data obtained applying Eqs (1) and (2).There are a better concordance between the calculated values and those experimentally observed for ^{7}Be than ^{210}Pb.

CONCLUSIONS

The results presented in this paper identify several meteorological parameters which are strongly associated with the fluctuations of weekly concentrations of ^{7}Be and ^{210}Pb. These results are useful to provide information on the atmosphere from a coastal Mediterranean area where data are insufficient now days.

REFERENCES

1. Dueñas, C., M.C. Fernández, J. Carretero, E. Liger and S. Cañete . *J.Geophys. Res.*, **106**, D24,34,059-34,065 (2001).
2. Dueñas, C., M.C. Fernández, J. Carretero, E. Liger and S. Cañete . *J. Geophys. Res.*, **108**, D11,4336-4345 (2003).
3. Koch D.M., D.J. Jacob and W.C. Graustein *J. Geophys. Res.*,**101**, 18,151 (1996).
4. Bergametti,G.,Dutot A.L. Buat-Ménart, P. Losno,R. and Remoudaki,E *Tellus*,**41B**, pp.353-361 (1989).
5. Caillet, S. Arpagaus, P.,Monna,F. and Dominif. *J. Environ. Radioact.*,**53**, pp.241-256 (2001).

The Radioactivity Levels and Physical-Chemical Properties In Public Water Supplies Of Malaga

C. DUEÑAS*, M.C. FERNÁNDEZ*, E. LIGER**, S. CAÑETE* AND M. CASTRILLO*

*Department of Applied Physics I, University of Málaga** Department of Applied Physics II, University of Málaga, Campus de Teatinos s/n, 29071Málaga, Spain.

A. FERNÁNDEZ AND R. PÉREZ

Empresa Municipal de Aguas de Málaga S.A. (EMASA)

Abstract. The measurement of radioactivity in drinking water permits us to determine the exposure of population to radiation from the habitual consumption of water. An intensive study of the water supply in the city of Malaga has been carried out in order to determine the gross alpha, gross beta activities and natural and artificial radionuclides by gamma spectrometry. A data base on natural and artificial radioactivity in water was produced. Results indicated that 90% of the water sampled contains a gross alpha radioactivity of less than 0.10Bq/L and 100% gross beta of less than 1 Bq/L, limit of activity recommended by the Spanish Regulatory Organization. For most of the samples, there is not regular correspondence between gross alpha and gross beta activities concentrations. The water samples have been analyzed during three periods. In order to quantify the influence of the origin of water on its radioactivity content, these samples were also classified in 2 categories; surface waters(from rivers and reservoirs) and subterranean waters (from wells and springs).Because of the different origin of the water samples, the following physic-chemical parameters have been obtained: (pH, conductivity, dry residue, Ca^{+2},K^{+}, SO_4^{-2}) A more detailed analysis of the results from individual measurements shows that there is certain degree of correlation between the gross alpha and gross beta activities and some physical-chemical parameters.

1. INTRODUCTION

Radioactive elements can enter the body via food, water, or the air. Given the special condition of water as a universal solvent, it is potentially an important carrier of dissolved radionuclides, which are absorbed by the population when consumed. This fact is of importance considering that the average consumption of water per day of an adult has been established as 2 L.

An exhaustive analysis of gross alpha and beta activities as well as the concentration of gamma emitting radionuclides has been carried out, as in compliance with the Real Decreto 140/2003, 7 of February, whereby standard levels are recommended for background natural radioactivity as well as for that present in water used by humans. From a radiological point of view, these levels are such that water can be considered to be drinkable without the need of any other kind of radiological examination.

Equally, a series of chemical factors (Standard Method 20th edition) have been analysed to determine the distribution of radionuclides in the aqueous environment such as pH, conductivity, hardness, carbonates .....

2. MATERIALS AND METHODS

The existence of water with a significant activity in natural radionuclides is relatively probable according to the origin of the water analysed. In this study, 14 points were sampled which were classified as surface waters (from rivers and reservoirs) and subterranean waters (from wells and springs). These are the sources which constitute the public water supply of Málaga. Málaga has a stable population of 600.000 inhabitants, rising to as many as 1.000.000 in the summer season. The study of the water from wells and springs is fundamental in this case due to the existence of long dry periods which in turn oblige drilling for subterranean water to supplement the available drinking water. The water samples were collected in the same places at three different times (4th quarter of 2001, 1st and 2nd quarters of 2002) with the aim of presenting a spatial and temporal average of the concentrations.

The analytical procedure used to determine the gross alpha activity level is that of coprecipitation of a volume of $500cm^3$ of sample water. This method consists of the selective precipitation of radium isotopes followed by a coprecipitation of the actinoids [1]. Both having been precipitated, they are separated by filtration and the level of total alpha activity is measured by a solid SZn(Ag) scintillation counter. The beta activity is measured with a gas-flow proportional counter with a precipitate obtained by evaporation until nearly dry of a maximum volume of sample water of $150cm^3$ as determined by the conductivity values obtained from each of the samples analysed, and is carried out on a steel planchet of dimensions appropriate to our detector. An analysis has also been carried out of radioisotopes by gamma spectrometry using an intrinsic Ge detector type ReGe. Afterwards, with a multi-channel analyser, a gamma emission spectrum is obtained, which being previously calibrated for energy and efficiency, allows us to obtain the activity of each of the emission sources that are within the limits of detection.

3. RESULTS

In order to quantify the influence of the origin of the water samples on the radioactive content, the box and whisker diagrams have been used to show the gross alpha and beta activity level corresponding to the classification of water as either superficial or subterranean Fig1. In these box and whisker plot it can be observed that the gross level of alpha activity is greater in subterranean waters (sub) as compared to the surface waters (sup). With respect to the total level of beta activity, no marked difference exists. Fig.1 shows us that the range of the level of the total alpha and beta activity is greater in subterranean waters than in superficial waters since the filtration of water through the ground raises the degree of mineralisation of the analysed waters depending on the geological characteristics of the respective aquifer. The geological structure of Malaga is characterised by predominantly sedimentary rock (carbonated and detritical) which has a low concentration of radioactive elements [2] [3] .

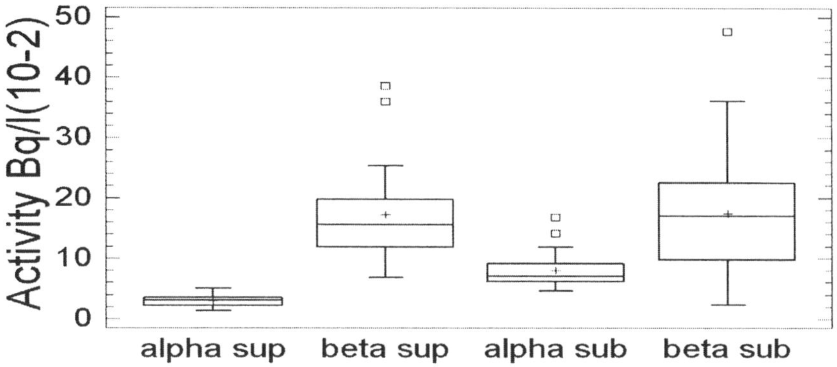

Fig. 1.- Box and Whisker diagrams of the gross alpha and gross beta activities (Bq/L) from samples waters according to the origin.

With the object of studying the temporal variation of these two groups of water, Fig.2 shows the corresponding histograms of the gross alpha and gross beta activity level for the quarterly periods analysed. No appreciable change can be observed in most of the waters measured. All maintain a notable consistency in their concentrations with a repetition of their values for different samples. This is possibly explained by the small variation in rainfall during the period analysed: 135.2 mm 3th quarter and 179.6 mm in the 4th quarter of 2001, and 137.0 mm and 80.7 mm in the 1st and 2nd quarters 2002 respectively.

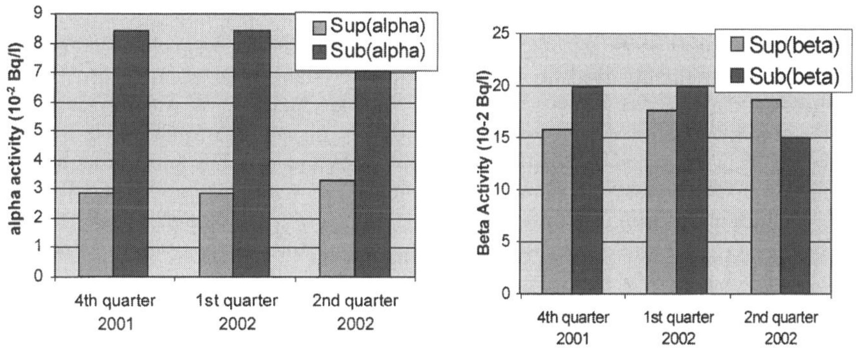

FIGURE 2.- Histograms of the gross alpha and gross beta activity level (10^{-2} Bq/L)

In Table 1 some of the physical-chemical parameters established in each of the water types analysed are shown. To study the influence of these parameters have on the alpha and beta activity level, we have made a lineal regression adjustment between these activity levels and the chemical parameters considered (which are not autocorrelated between themselves).

waters	pH	Hardness	SO_4^{2-} (mg/)	Cl^- (mg/)	Ca^{2+} (mg/)	Na^+ (mg/)	K^+ (mg/)
surface 4th 2001	8.0	282	159	84	61	61	5
subterranean 4th 2001	7.6	164	173	164	98	100	4
surface 1st 2002	8.1	372	161	92	79	66	4
subterranean 1st 2002	7.6	447	114	160	97	110	3
surface 2nd 2002	8.0	332	171	169	80	119	5
subterranean 2nd 2002	7.6	474	188	162	102	113	4

Table 1. Average the physical-chemical parameters established in each of the water types analysed.

Table 2 are laid out the most significant statistics of the lineal regression analysis. However, two aspects stand out from an observation of this table [4].

• The first chemical parameter selected is the hardness of the water, and we find that the correlation coefficient, r, is always positive when we establish the relationship between the gross alpha and gross beta activity level with the hardness of water since as the degree of mineralisation of the water is increased, the dissolving of radioactive isotopes in the water is enhanced, especially of those of the Uranium-238 family due to their solubility in water as compared to Thorium-232. One of the most abundant radionuclides in the drinking water is Ra-226 [5]. This natural radionuclide is of relative importance in the Earth's crust. It is generally found forming composites which are highly soluble in water.

• The second of the chemical parameters that exerts an important influence is pH, since there is an acceptable correlation with the alpha activity level, with negative values for the correlation coefficient which indicates an inverse relationship between the two parameters. This confirms the hypothesis that a slightly alkaline pH favours a possible elimination of the Ra-226 from the solution. However the correlation coefficient between pH and the gross beta activity level is not significant. There is however an acceptable correlation coefficient between the beta activity level and the stable K^+ concentration in the analysed waters, which allows us to affirm that this K^+ has a marked influence on the total beta activity level.

During the sampling period activity measurements were made with gamma spectroscopy for the natural isotopes Ac-238 (Th-232 series), Pb-214 (U-238 series) and of K-40 and the artificial Cs-137 and Co-58. The average activity during the sampling periods (4th quarter of 2001, 1st and 2nd quarters of 2002) corresponding to the natural radioisotopes are below the normal level of environmental radioactivity, for both the superficial and subterranean waters, with no artificial radionuclides having been detected in any of the analysed waters.

Independ variable	Depend variable alpha activity: Intercept	slope	C. Coef	C. level	Depend variable beta activity: Intercept	slope	C. Coef	C. level
Hardness	1.6 ± 0.7	$(8.5 \pm 1.8)10^{-3}$	0.60	99%	5.5 ± 3.0	0.03 ± 0.007	0.57	99%
pH	38 ± 8	-4.2 ± 1.0	0.55	99%	----	----	---	N S
K	---	-----	----	N. S	8.4 ± 2.8	2.4 ± 0.6	0.52	99%

Table 2. Summary of the statistical parameters obtained from the lineal regression analysis between gross alpha and gross beta activities which some of the physical-chemical parameters.

CONCLUSIONS

-Gross alpha activity is greater in subterranean waters than the surface waters. With respect to the gross beta activity, no marked difference exists.
-Surface and subterranean waters can be considered drinkable from a radiological a point of view .
-The lineal regression analysis between gross alpha and gross beta activities with some of the physical-chemical parameters enable, in most cases, a quantitative prediction of the radioactivity background natural waters in the interval of the chemical parameters values

REFERENCES

1. Procedimientos Radioquímicos: Índice de actividad alfa total en aguas por coprecipitación. J.A. Suárez, Ll. Pujol y Mª A. de Pablo. *Cuadernos de Investigación* 39. CEDEX, 2000.
2. ^{226}Ra and ^{224}Ra in waters in Spain. C. Dueñas, M.C. Fernández ,J.A González, J. Carretero and M. Pérez. *Toxicological and Environmental Chemistry*, 39, 71-79. (1993).
3.-Natural Radioactivity levels in bottled water in Spain. C. Dueñas, M.C. Fernández , J. Carretero, E. Liger. *Water Research.* **31**, nº 8, 1997, pp 1919-1924.
4. Factors determining the radioactivity levels of waters in the province of Cáceres (Spain). A. Baeza, L.M. del Rio, A. Jiménez, C. Miró and J.M. Paniagua. *Appl. Radiat. Isot.* , **46**, nº 10, pp1053-1059. (1995).
5. ^{226}Ra and ^{222}Rn concentrations and doses in bottled waters in Spain. C. Dueñas, M.C. Fernández , J. Carretero, E. Liger and S. Cañete. *Journal of Environmental Radioactivity*, **45**, 283-290. (1999).

Applied Physics (APHYS 2003)
A. Méndez-Vilas (Ed.)

Spectral Analysis and Measurements of Exposure to Magnetic Fields (up to 32 kHz) in Private and Public Transport

Jesús M. Paniagua, Antonio Jiménez, and M. Montaña Rufo

Department of Physics, University of Extremadura, Avda. de la Universidad s/n, 10071 Cáceres, SPAIN

Abstract. During recent years there has been a recognition of the growing public as well as scientific interest in the biological effects of electric and magnetic fields. Transport, for instance, is not free from electric and magnetic fields because electricity is used both as the power source and as a control mechanism generating oscillating magnetic fields with a wide range of frequencies. We here present the results of a study of people's levels of exposure to the magnetic fields that exist in different modes of private and public transport. The study included the determination of the intensity and frequency of the fields, and a comparison with the *ICNIRP* reference levels.

INTRODUCTION

The numerous studies that have been carried out aimed at determining whether there is any relationship between exposure to low frequency magnetic fields and adverse effects on health have aroused considerable public interest, and led both national and international organisms to draw up a series of health measures to protect the population against non-ionizing radiation. In Spain, these measures consist of certain basic restrictions and reference levels that coincide with the recommendations of the *International Commission on Non-Ionizing Radiation Protection (ICNIRP)* [1]. The said reference levels depend on the frequency of the magnetic field, so that it is necessary to measure this as well as the field intensity to perform an appropriate dosimetric evaluation, especially in environments in which there exist elements generating alternating magnetic fields of many frequencies.

One such environment is transport, whether propelled electrically or by fossil fuels. There are alternating magnetic fields produced by the vehicle's motor, control systems, air conditioning, sound, video, communications, etc., as well as those produced by magnetized metal in the wheels [2].

We here present the results of a study of people's levels of exposure to the magnetic fields that exist in different modes of private and public transport. The study included the determination of the intensity and frequency of the fields, and a comparison with the *ICNIRP* reference levels.

METHODS

We measured the magnetic fields with a WG EFA-200 meter. This device makes isotropic measurements, permits continuous data logging, and performs a fast Fourier transform spectral analysis. It can be configured to make measurements with band-pass filtering and also broad band.

The measurements were made in the following means of transport: car, bus, train, tram, plane, and ship. The position was chosen to be one that a traveler would normally occupy. We logged the intensity every 5 seconds, and performed the spectral analysis using the frequency bands 5 Hz – 2 kHz and 40 Hz – 32 kHz. In order to obtain information on the "background" recorded by the meter, we also carried out both types of analysis at sites distant from power lines and other sources of magnetic fields.

We shall present the results of the measurements as a series of statistical parameters, including the arithmetic mean, the geometric mean, and the median. The arithmetic mean is equivalent to the time-weighted average. The geometric mean and the median are central statistics that are less sensitive to atypical data such as the brief exposure to high magnetic fields that often appear in continuous records. The statistical parameters will be presented as calculated after subtracting the mean value of the background.

RESULTS AND DISCUSSION

Prior to the measurements in the modes of transport, we took readings of the magnetic flux density and gathered background spectra in the aforementioned frequency bands. In the 5 Hz – 2 kHz spectrum there existed high signals at frequencies below 20 Hz, with the remainder of the spectrum being practically flat, as was the 40 Hz – 32 kHz spectrum. This leads to broad band measurements having a relatively high background – 0.132 μT and 0.144 μT, respectively – with respect to the meter's background of 0.014 μT when, for example, the band-pass filter corresponding to the 50 Hz frequency is used. Nonetheless, it was necessary to make broad band measurements since we did not know *a priori* the emission frequencies that would appear in the modes of transport to be studied.

Table 1 shows some of the statistical data for the different measurements together with the corresponding journey times. Those for the journey by ship (Helsinki – Tallin) are not included since neither the continuous log of the magnetic flux densities nor the spectral analyses showed any difference with the background recorded by the meter.

With respect to the other means of transport, the mean magnetic flux densities after subtracting the background were: 0.11 μT for the car, 0.29 μT for the bus and the train, 0.32 μT for the tram, and 1.04 μT for the airplane. It was in the car, however, that we detected the greatest maximum value, 10.94 μT, although considering the interval between the 5 and 95 percentiles, for which the extreme high and low values are excluded, most of the levels lay within a smaller range, 0.01 to 0.30 μT. The coefficient of variation (the relative standard deviation) was very high in almost all cases, from 410 % for the car to 30 % for the airplane. This indicates that the frequency distributions have a strong positive skew except in the case of the airplane, as is also reflected in the greater value of the arithmetic mean relative to the geometric mean and the median.

TABLE 1. Logging time and statistical parameters of the magnetic flux density measured in different means of transport.

Transport	Car	Bus	Train	Tram	Plane
Logging time (min)	155	45	141	15	51
Average (μT)	0.11	0.29	0.29	0.32	1.04
Median (μT)	0.06	0.21	0.09	0.19	0.99
Geom. Mean (μT)	0.05	0.20	0.13	0.12	0.98
Std. dev. (μT)	0.45	0.21	0.36	0.41	0.31
Minimum (μT)	< 0.01	< 0.01	0.02	< 0.01	0.01
Maximum (μT)	10.94	0.98	1.82	2.46	2.14
5 % - 95 % interval (μT)	0.01 – 0.30	0.04 – 0.75	0.04 – 0.95	< 0.01 – 1.01	0.67 – 1.85

It is not always straightforward to compare this type of data between similar studies since the frequency ranges measured are not always the same, and the magnetic flux density may vary with position within the vehicle and with time. It is not surprising therefore that the literature values for the flux densities in means of transport cover a very wide range [3-6]. The values obtained in the present work lie within those ranges. The values of the magnetic flux densities in the plain, the train, and the tram are higher than the literature values for residential zones [7-9], and are closer to those of workplaces [10,11].

An appropriate dosimetric estimate of the exposure to magnetic fields requires a knowledge of the frequency involved. This is no easy task since this frequency may vary during the course of the journey due to many causes.

Of the cases that we studied, the most stable values of the frequency and the magnetic flux density corresponded to the airplane. Figure 1 shows conjointly the magnetic flux density and the frequency of the main emission versus the time of duration of the Stockholm-Helsinki flight, with data logged every 5 seconds. The frequency scale in the figure is chosen so that the curves do not overlap and can be distinguished better. One observes that there were maxima in the magnetic flux density at take-off and landing. The greatest value after subtracting the background was 2.14 μT at take-off. The frequency remained at around 400 Hz throughout most of the flight (the power frequency widely adopted in aircraft on-board systems), although it rose to 600 Hz at certain moments. Figure 2 shows a spectrum in the 5 Hz – 2 kHz range in which the 400 Hz peak is clearly observed together with the 3rd and 5th harmonics. Odd harmonics of the principal frequency continue to be observed in the 40 Hz – 32 kHz spectrum, but no new emissions.

The train measurements were made on a journey from Valencia to Madrid on an electrified line. Figure 3 shows the temporal variation of the magnetic flux density during the journey. One observes that the values fluctuated widely – the maxima reached 1.82 μT, but the mean value was relatively low. The spectra showed no clear peaks at specific frequencies. For approximately 30 minutes, the mean value rose to some 1.1 μT (one sees this in the figure from minute 106 of the journey). This rise coincided with the appearance of an emission with a principal frequency of about

24 kHz (Figure 4). As well as this dominant peak, one observes in the figure other smaller peaks at 100 Hz, 250 Hz, and some harmonics of the latter.

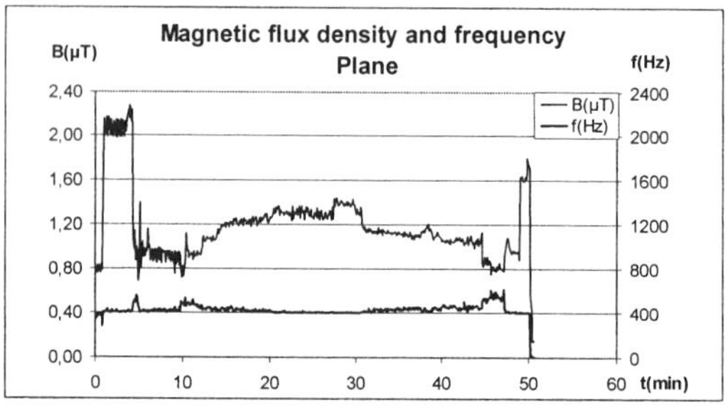

FIGURE 1. Temporal variation of the magnetic flux density and its main frequency during a airplane journey. The magnetic flux densities during take-off and landing are marked.

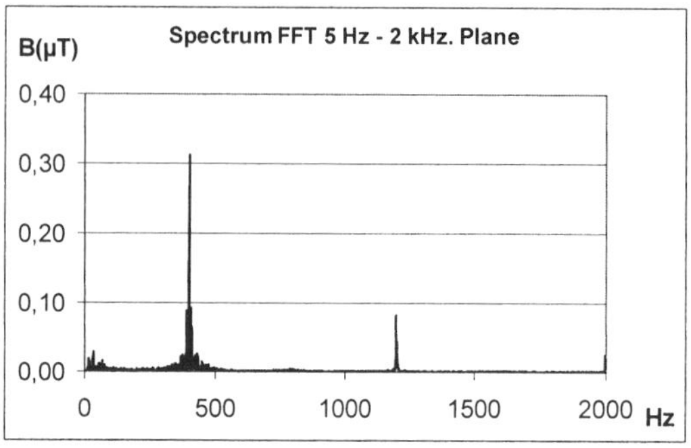

FIGURE 2. Magnetic field spectrum recorded during the airplane journey (5 Hz – 2 kHz).

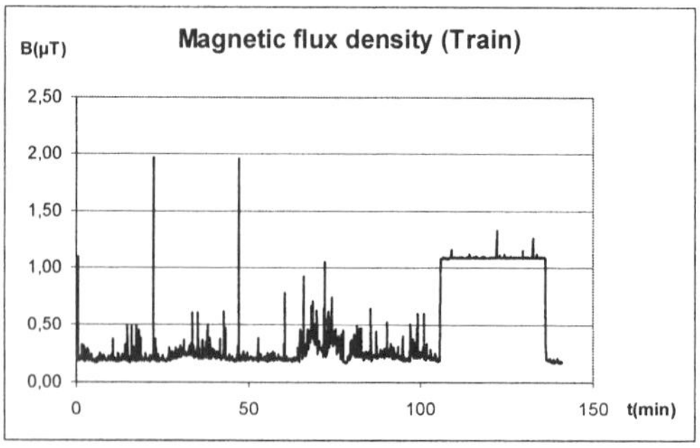

FIGURE 3. Temporal variation of the magnetic flux density and its main frequency during a train journey on an electrified line.

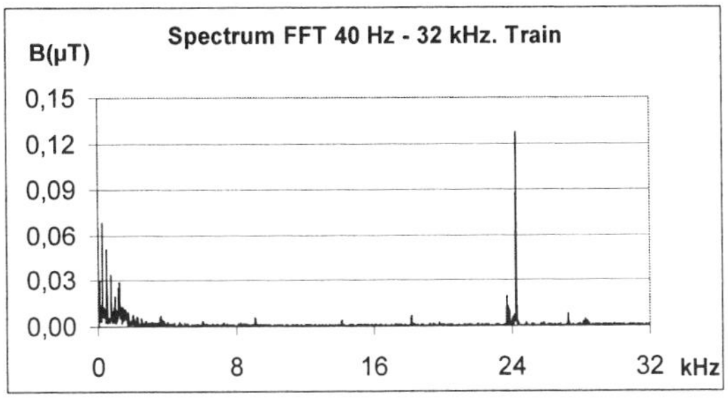

FIGURE 4. Magnetic field spectrum recorded during the train journey on an electrified line (40 Hz – 32 kHz).

The tram measurements (both broad band and spectral analyses) of the magnetic flux density were made in Helsinki during a short journey of 15 minutes. The broad band measurements showed fluctuations that reached a maximum value of 2.46 μT. The spectra had a complex structure with many peaks, the most intense being those corresponding to 150 and 300 Hz.

In the case of the bus – a city bus in Cáceres (Spain) – the spectrum also presented several peaks, the most intense at 23 Hz, followed by another at 183 Hz. Other less intense peaks appeared up to 600 Hz. No emissions were detected beyond this frequency.

In the car, on a journey Cáceres – Madrid (Spain), we detected low frequency emissions related to the rotation of the magnetized wheels and to the ventilation systems. In particular, the principal frequency detected was 27 Hz, and there was another smaller peak at around 500 Hz when the air conditioning was on.

Table 2 presents the values of the magnetic flux densities and of its most intense frequencies, together with the *ICNIRP* reference level and the percentage of that level represented by the measured values. One observes that the greatest value corresponded to the airplane: with its high magnetic flux density values and the 400 Hz frequency, the mean value of 1.04 μT represents 8.32 % of the reference norm. The next in order of importance was the train in which relatively high values were recorded – greater than 0,9 μT for approximately 30 minutes coinciding with the appearance of the 24 kHz peak. The rest of the time, there was no clearly dominant frequency, and the values were relatively low – below 0.1 μT.

TABLE 2. Comparison of the mean values of the magnetic flux density with the *ICNIRP* reference levels.

Transport	Car	Bus	Train	Tram	Plane
Frequency (Hz)	27	23	24000	150	400
Average (μT)	0.11	0.29	0.29	0.32	1.04
Reference Level (μT)	185	217	6.25	33.7	12.5
%	0.06	0.20	4.64	0.96	8.32

Assuming the least favourable case that this is the principal frequency throughout the journey, the reference level is 6.25 μT, and the measured mean value represents 4.64 % of that level. In the other three – tram, bus, and car – with their low magnetic flux density levels and low frequencies, the *ICNIRP* reference levels are higher, and hence the corresponding percentages are lower: 0.96 %, 0.20 %, and 0.06 %, respectively.

CONCLUSIONS

The mean values of the magnetic flux densities in the transportation environment found in the present work ranged from 0.11 μT in the car to 1.04 μT in the airplane. These values are within the range of published values for means of transport, and are closer to those of workplaces than of residential environments.

We analysed the principal frequencies of the magnetic fields measured in the 5 Hz – 32 kHz range, finding that they varied from very low frequencies – 23 and 27 Hz in the bus and the car, respectively – to values of 24 kHz in one section of the journey in the train.

On the basis of these values of the magnetic flux densities and the frequencies, we made a dosimetric assessment of the exposure of the traveler to magnetic fields. Given the complexity of the spectra and, in some cases, the great variability of the flux densities and the frequencies, however, further studies need to be carried out for an appropriate dosimetric evaluation.

REFERENCES

1. ICNIRP. International Commission on Non-Ionizing Radiation Protection, *Health Phys.* **74**, 494-522 (1998).
2. Milham S., Hatfield J.B. and Tell R., *Bioelectromagnetics* **20**, 440-445 (1999).
3. Dietrich F.M. and Jacobs W.L., *Survey and assessment of electric and magnetic field (EMF) public exposure in the transportation environment.* US Department of Transportation, Federal Railroad Administration. Report No. PB99-130908 (1999).
4. Wenzl T., Am. *Ind. Hyg. Assoc. J.* 667-671 (1997).
5. Hamalainen, A.M., Hietanen, M., Juunti, P., and Juutilainen, J., "Exposure to magnetic fields at work and public areas at the Finnish Railways", in *Proceedings*, Second World Congress for Electricity and Magnetism in Medicine and Biology, Bologna, 1997.
6. Kim Y.S., and Cho Y.S., "Environmental exposure to magnetic fields in public transit systems", in *Proceedings*, Bioelectromagnetics Society Annual Meeting. Long Beach, California, 1999.
7. Merchant C..J. Renew D.C., and Swanson J., *J. Radiol. Prot.* **14**, 77-87 (1994).
8. Swanson J., *J. Radiol. Prot.* **17**, 111-113 (1997).
9. Clinard F., Milan C., Harb M., Carli P.M., Bonithon-Kopp C., Moutet J.P., Faivre J., and Hillon P., *Bioelectromagnetics* **20**, 319-326 (1999).
10. Kim Y.S., and Cho Y.S., *J. Occup Health* **43**, 141-149 (2001).
11. Sakurazawa H., Iwasaki A., Higashi R., Nakayama T., and Kusaka Y. *J. Occup. Health* **45**, 104-110 (2003).

Applied Physics (APHYS 2003)
A. Méndez-Vilas (Ed.)

Photophysical Properties and Phototoxicity Effect of Supramolecular Sensitizers

H. Kolarova [1], M. Huf[1], R. Bajgar[1], J. Mosinger[2,3], M. Modrianský[1], M. Strnad[1]

[1]*Department of Biophysics, Faculty of Medicine, Laboratory of Growth Regulators, Palacky University, Hněvotínská 3, 775 15 Olomouc, Czech Republic, e–mail:* kol@tunw.upol.cz, [2]*Department of Inorganic Chemistry, Faculty of Sciences, Charles University in Prague, Hlavova 2030, 128 43 Prague, e–mail:* mosinger@natur.cuni.cz, [3]*Institute of Inorganic Chemistry, Academy of Sciences of the Czech Republic, 250 68 Rez, Czech Republic*

In this study we tested the cellular uptake and the phototoxicity of *meso*-tetrakis(4-sulfonatophenyl)porphyrin (TPPS4) and its metallocomplex ZnTPPS4 in the presence or absence of 2-hydroxypropyl-cyclodextrins (hpCDs) on G361human melanoma cells. The dependence of concentration of porphyrin sensitizers in combination with laser irradiation on photodamage of G361 cells was investigated by in vitro methods. Viability of cells was determined by means of molecular probes for fluorescence microscopy. The quantitative changes of cell viability in relation to sensitizers concentrations and irradiation doses were proved by fluorometric measurement. G361 cells are sensitive to photodynamic damage for both of the tested sensitizers. TPPS4 and ZnTPPS4 especially in the supramolecular complex with nontoxic cyclodextrin carriers represent efficient sensitizers with high phototoxicity to G361 human melanoma cells.

INTRODUCTION

Photodynamic therapy of cancer (PDT) uses the interaction of sensitizers and light to destroy cancerous cells and tumors [1, 2, 3,]. The photochemical interactions of sensitizer, light, and molecular oxygen produce cytotoxic singlet oxygen ($^1O_2$) and other forms of active oxygen, such as hydroxyl radical etc. The tumor is destroyed either by $^1O_2$ (generated *via* energy transfer from excited sensitizer to triplet oxygen, type II mechanism) or radical products (generated *via* electron transfer from excited sensitizer, type I mechanism). The selectivity of tumor damage depends on specific retention of a sensitizer in the tumor tissue after systemic administration, combined with directed illumination [3, 4]. The cellular effects include plasma membrane, lysosomes and mitochondria damage leading to tumor ablation [5, 6]. In this work we studied, cellular uptake and the phototoxicity of *meso*-tetrakis(4-sulfonatophenyl)porphyrin (TPPS4) and its metallocomplex ZnTPPS4, [7, 8] in the presence or absence of 2-hydroxypropyl-cyclodextrins (hpCDs) as nontoxic carriers on G361human melanoma cells.

MATERIAL AND METHODS

Absorption Spectra

The absorption spectra of 10 μM sensitizers (TPPS4, ZnTPPS4) in the absence or presence of in hundred-fold concentration excess of hpCDs were measured in DMEM medium on spectrophotometer Unicam UV 550.

Cell Uptake

The G361 cells (ATTC, USA) were divided in the amount of 10^4 to each well (Dynatech plates 8 x 12, flat bottom) and filled in DMEM with 10% FCS. The sensitizer was added into the holes in concentrations of 0; 0.1; 0.3; 1; 3; 10; 30 and 100 μM in the absence or presence of hpCDs in a hundred-fold concentration excess compared to the sensitizer. The cells were incubated in a thermobox (37 °C, 5% CO_2). After 1; 3; 6; 10; 16; 24 and 48 hours of incubation, the medium above the adhering cells was removed. Each emptied hole was 2x washed with 120 μl of DMEM. After washing 100 μl of DMEM was added into each hole and self-fluorescence (TPPS4 excitation at 415 nm, emission at 645nm, $ZnTPPS_4$ excitation at 413 nm, emission at 606 nm) in G361 cells were measured with respect to individual sensitizer by Perkin-Elmer LS50B luminometer equipped with well plate reader accessory (Perkin-Elmer Corp., Norwalk, CT). The whole plate was read once with a read time of 0.2 s for each well. We found these settings optimal, increasing the read time per well and/or adjusting slit widths did not improve the signal to background ratio. Subsequently, from each of the holes 10 μl of medium was withdrawn and the volume was replaced by the same one of 20 % SDS. The holes were mildly shaken and incubated for 5 minutes; then their fluorescence was measured again.

Phototoxicity of Sensitizers

Twice washed trypsinated G361 cells were divided in the amount of 10^4 to each well and filled in DMEM with 10% FCS in a total volume of 80 μl. After 24 hours of cultivation at 37°C in 5% CO_2 the 20 μL of sensitizer was added. Cells were cultivated with sensitizers at concentrations ranging from 0.1 to 125 μg/ml. The total volumes of 100 μl (cells with additives) were cultivated for 24 hours. The controls contained cells in the cultivation medium only. After 24 hours of cultivation the cells were subsequently irradiated by a halogen lamp (24V/250W) at a dose of 0.5 to 150 J/cm^2. The halogen lamp has continuous irradiance spectrum (from 360 to 2700 nm) with maximum in VIS and NIR region. Irradiance was measured by Radiometer RK 2500 (Meopta Prerov, Czech Republic). Morphological changes in cells have been evaluated using inversion fluorescent microscope and image analysis. The quantitative changes of cell viability in relation to sensitizers concentrations and irradiation doses were proved by fluorimetric measurement with fluoroscan Ascent (Labsystems). Viability of cells was determined by means of molecular probes for fluorescence microscopy (LIVE/DEAD kit) [9].

RESULTS

Absorption Spectra

Fig. 1 shows the absorption spectra of the selected porphyrins in DMEM. The wavelength maximum of Soret band for free TPPS4 is at 415 nm. ZnTPPS4 shows a drift of maximum wavelength aproximately 2 nm to the value 413 nm. As a rule, the presence of hpCDs cause the red shift of Soret band.

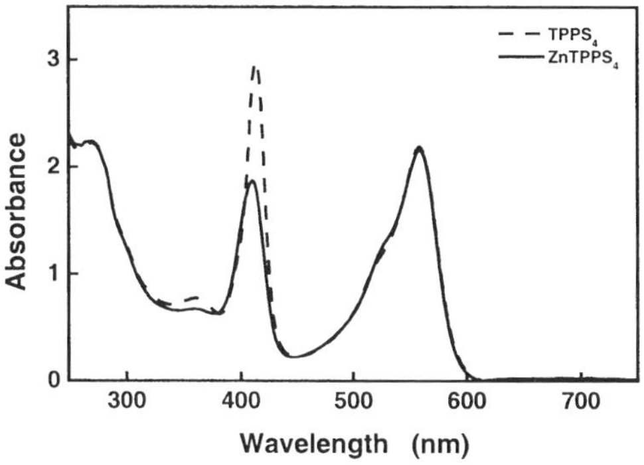

FIGURE 1. Absorption spectra of 10 μM sensitizers in growth medium.

Cell Uptake

Fig. 2 shows that the uptake of the sensitizer ZnTPPS4 at the given time interval is markedly higher than the uptake of TPPS4. The presence of the hpCD carrier did not affect the accumulation of TPPS4, but significantly affect uptake of ZnTPPS4. The highest uptake was found for sensitizer ZnTPPS4 in combination with hpβCD. The presence of the hpCD significantly increases the level of an accumulation of ZnTPPS4 in cells after a longtime period of incubation and gives no saturation character even after 50 hours of incubation. This is in contrary of free ZnTPPS4 that reaches saturation after 24 hours (Fig. 3).

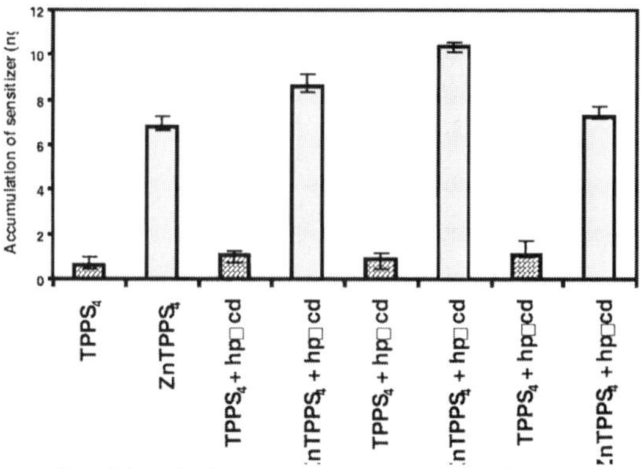

FIGURE 2. The uptake of sensitizers in the absence or presence of hpCDs into G361 cells.

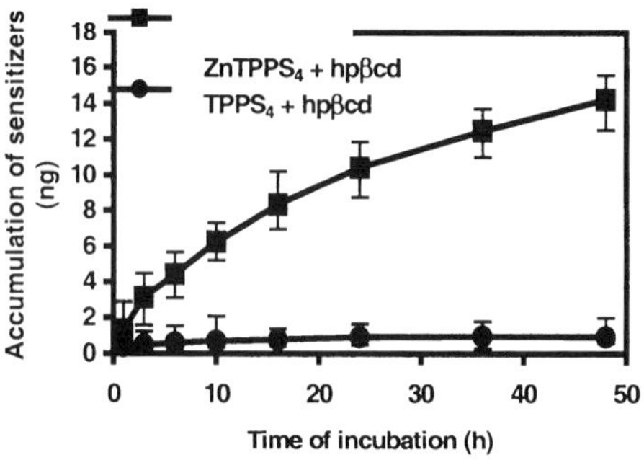

FIGURE 3. Time dependent uptake of TPPS4 and ZnTPPS4 sensitizers bound to hpβCD.

Phototoxicity of Sensitizers

Viability studies have shown, that the optimum phototoxic effect observed in G361 melanoma cells was obtained in the presence of light dose of 40 J/cm^2 and 25 μg/ml ZnTPPS4 in a hundred-fold concentration excess of hpβCD. The radiation dose of visible light used in our study is, however, non-toxic in the absence of sensitizer, and the concentration of photosensitizer without light exposure is also non-toxic. Microscopical study (Fig. 4) shows morphological changes in cell cultures after PDT treatment with ZnTPPS4 at concentration of 25 μg/ml, hundred-fold concentration excess of hpβCD and dose of light irradiation 40 J/cm^2. After this procedure all cells were dead (Fig.4B, 4D). Fig. 4A presents live control human meanoma cells G361 which were not irradiated. Fig. 4C demonstrates detection of live control cells by green fluorescence of Calcein AM and Fig. 4D shows detection of dead cells by red fluorescence of Ethidium Homodimer in photodamaged cells after PDT. Living and dead cells were counted by software Olympus Micro Image for each well.

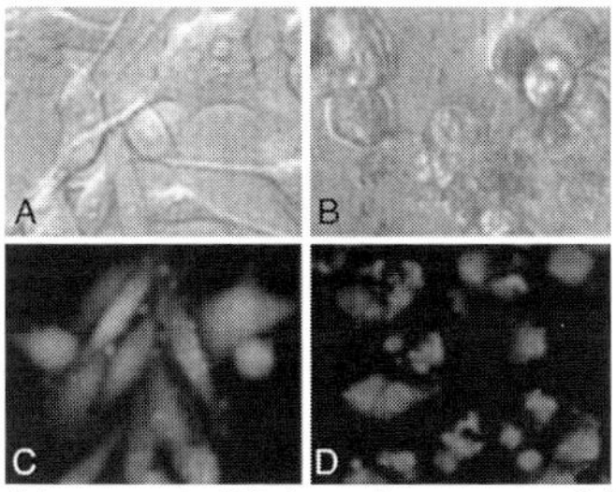

FIGURE 4. G361 cells observed in transmited light - microscope and fluorescence microscope

CONCLUSION

Efficiency of PDT is affected by various factors including photophysical properties of a sensitizer, wavelength of the activation light, depth of the light penetration in the biological tissue, tissue response on singlet oxygen, etc. [1, 2]. The measurement of the uptake of sensitizers into the G361cells shows the difference between the free sensitizers and bound to hpCD carriers. The kinetic of ZnTPPS4 uptake is higher than for TPPS4. While the presence of hpCDs does not notably affect the uptake of TPPS4, in the case of ZnTPPS4 cyclodextrin carriers hpβCD cause a significant magnification in accumulation of the sensitizer. The most effect on the level of distribution of the sensitizers in G361cells was found for ZnTPPS4 in combination with hpβCD. G361 cells are sensitive to photodynamic damage by all of the tested sensitizers in the presence or absence of hpCDs. ZnTPPS4 is more phototoxic than TPPS4. In conclusion, ZnTPPS4 and TPPS4 especially in the supramolecular complexes with hpCDs carriers represent efficient sensitizers with high phototoxicity to G361 human melanoma cells.

ACKNOWLEDGMENTS

This work was supported by the grant project of Grant Agency No. 203/02/1483 and Ministry of Education MSM No. 153100008.

REFERENCES

1. Sibata, Ch., Conlussi, V.C., Oleinick, N.L., Kinsella, T.J. *Exp.Op. Pharm.*, **2 (6)**, 917 (2001).
2. Brown, S.B., Brown, J.E.,Vcron, D.I. *Expert Opinion on Investigative Drugs*. **8(12)**, 1967 (1999).
3. Lui, H., Anderson, R. R. Dermatologics Clinics **11**, 1(1993) .
4. Moor, A.C.E. *Journal of Photochemistry and Photobiology B: Biology* **57**, 1(2000)
5. Moan, J. , Berg, K. *Photochemistry and Photobiology* **55**, 931(1992).
6. Dahle, J., Steen, H.B., Moan, J. *J.l Photoch. Photobiol. B: Biology* **70(3)**, 363 (1999).
7. Mosinger, J., et. al. *Journal of Photochemistry and Photobiology A. : Chemistry* **130**, 13 (2000).
8. Mosinger, J., Kliment V., Sejbal, J., Kubát, P.,Lang, K. *Journal of Porphyrins and Phthalocyanines* **6**, 513 (2002).
9. Kolarova, H., Kubínek, R., Strnad, M. *Internet Journal of Photochemistry and Photobiology* web site http://www.photochem.photobiol.com (Accessed 15.09.98.)(1998).

Applied Physics (APHYS 2003)
A. Méndez-Vilas (Ed.)

Determination of *Punica granatum* L. Carpellary Membrane Elastic Properties Using Atomic Force Microscopy

M.C. Millan*, M. Gasque*, F.J. Garcia-Diego*, J. Curiel* and G. Ruiz†

* *Department of Applied Physics, Polytechnic University of Valencia, Camino de Vera s/n. 46022 Valencia, SPAIN. E-mail: mcmillan@fis.upv.es*
† *Department of Physics and Computer Architecture, Miguel Hernández University, Crta. Beniel, km 3,2. 03300 Orihuela, SPAIN. E-mail: gabriel.ruiz@umh.es*

Abstract. Atomic Force Microscopy (AFM) is used to observe the *Punica granatum* L. carpellary membrane topography and also to determine its elastic properties. Mechanical properties are studied to improve the knowledge of the carpellary membrane, and how its relative value of elasticity varies from one zone to another. AFM is the only technique that allows to determine local elastic properties. AFM makes possible the measurement of the force between the probe and the sample, at any point of the sample, as a function of the displacement, where the displacement is varied using a piezoelectric crystal. Contact mode is used for these determinations. AFM three-dimensional images are obtained and compared with SEM images.

INTRODUCTION

Atomic Force Microscopy (AFM) has developed quickly since 1986 [1]. AFM is a non-destructive imaging technique with nanometre resolution that has been adopted for the analysis of biological material. No requirement for Au, Pd or C sputter coating or dehydration is needed for AFM analysis; thus sample alteration is minimum. The analysis of mechanical and elastic properties of cells is very useful for understanding their function [2]. AFM has also been used to determine elastic properties since 1989 [3]. The study of elastic properties is becoming increasingly important, especially with a view to the research with soft materials like biological specimens [4].

Usually, when images and elastic properties are obtained with AFM, sample size, tip characteristics and tip artefacts must be taken into account. Tip calibration, using non-destructive techniques [5] is not necessary for this study due to the roughness of the area analysed and the kind of tip used. Moreover, in the surface analysed, tip artefacts do not add valuable information. When corrugated sample surfaces are analysed, the study of tip artefacts is fundamental [6]. It can be appreciated certain image distortion function of the sharpness of the tip and the sample. The level of distortion accepted for an image depends on the subsequent use of data or the purpose of the study [7], [8].

Hertz model, used to study elastic properties, describes elastic deformation of the surface in contact under load [9]. Some relations between the applied force and the displacement have been developed using this model and AFM techniques. Some of the values needed for the parameters used in Hertz's model are not obtained in AFM experiments, although the values given by other authors for these parameters are valid (i.e. some references indicate a value of 0,5 for the Poisson ratio in biological material but it is not evaluated for different types of materials). AFM can be used to study elastic properties by collecting force-distance curves from the sample surface provided that viscous contributions are small. By collecting arrays of force-distance curves, called force volume, maps of mechanical properties can be obtained showing differences in local elastic properties. Some authors prefer to determine relative microelastic properties [10]. For this reason, a relative value of elasticity can be evaluated if only the values obtained in the AFM experiment are taken into account. This value allows the identification of the variations on points of the surface (carpellary membrane) or when zones are homogeneous, the evaluation of the measurements and experimental errors. The aim of this work is the study of the measurement repeatability with the instrumentation and methodology used starting from the variability of the obtained data.

Image study is completed using a Cryogenic Scanning Electron Microscopy (CryoSEM). CryoSEM analysis of carpellary membrane enables to observe the carpellary membrane without alteration of its chemistry or texture. These flat images can be compared with three-dimensional AFM images. Samples are analysed by CryoSEM after AFM studies, but not reversely.

This work is centred on pomegranate *(Punica granatum* L.*)* carpellary membrane topography and its elastic properties. Studies on pomegranate have already been developed in previous research carried out in our laboratory [11].

THEORY

The AFM works by detecting interactions between a tip on a cantilever and the surface of the sample. In the calculation, it was considered that a sphere approximates the shape of the tip end. Then, using Hertz mechanics [9] the force F on the cantilever is given by:

$$F = \frac{4\sqrt{RE}}{3\left(1-\upsilon^2\right)} \cdot \delta^{3/2} \qquad (1)$$

where δ is the indentation, E is the elastic modulus, ν is the Poisson ratio and R is the radius of the tip end. The sample is assumed to be much softer than the tip (Etip=150 GPa). This equation can be expressed using a factor k (function of E and ν):

$$F = \frac{4\sqrt{R}}{3\pi k} \cdot \delta^{3/2} \qquad (2)$$

In order to compare the elastic properties at different positions, some authors have obtained a model based on a simple relationship that relates the work (w) done by the AFM cantilever during an indentation to the k. This model is called FIEL (Force Integration to Equal Limits) [10]. Using this model, the relationship between the energy at two points is:

$$\frac{w_1}{w_2} = \left(\frac{k_1}{k_2}\right)^{2/3} \quad (3)$$

MATERIAL AND METHODS

Sample Preparation

In this study, different portions of fresh carpellary membrane of pomegranate are used. For AFM mounting, carpellary membrane portions were fixed with a double-side adhesive tape on metallic stubs.

For the images obtained in cryogenic scanning electron microscope (Jeol scanning microscope 5400), sections of carpellary membrane were quickly frozen by contact with a liquid nitrogen-cooled. The frozen sample was introduced in the microscope and was sputtered with gold for its observation Fig.1.

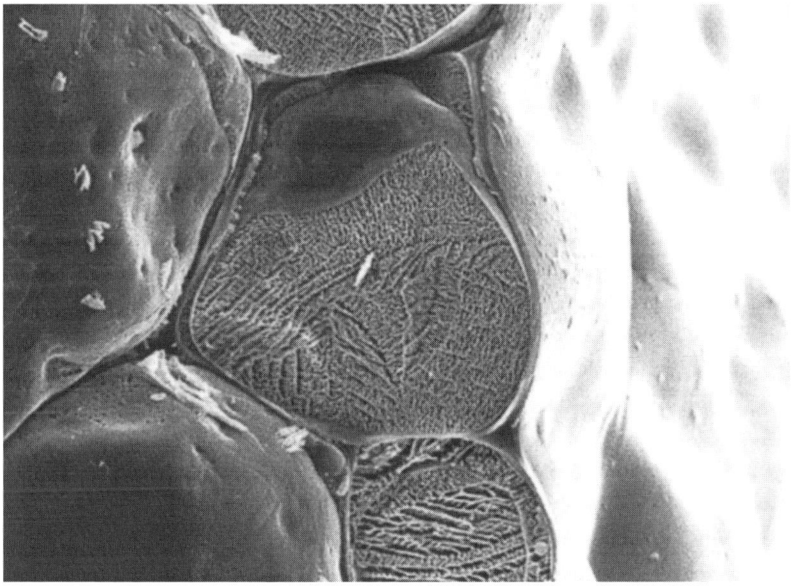

FIGURE 1: Cryofracture of carpellary membrane area observed using cryosem.

Instrumentation

The AFM measurements were performed, under air at room temperature, using a Nanoscope III atomic force microscope from Digital Instruments Inc. with a V-shaped cantilever with spring constant of 0,15 N/m (manufacturer's value). The tip used in these experiments was a pyramidal standard silicon nitride (nominal radius minor than 20 nm). Veeco Metrology Group manufactured tip and cantilever. The same probe was used in all measurements.

The force-distance curves, retracting and approaching curves were recorded in contact mode. Slower scan rate minimises the viscous contribution. During force mapping, force curves were taken continuously while the tip was scanned laterally over the sample. Lateral scan sizes were usually of 10 μm and the spacing between each force curve was of 0,67 μm. The total number of points analysed was 15 x 15 (225).

Software Analysis

The FIEL mapping of data were done with software developed in our laboratory using the LabVIEW environment. This program follows these steps:

1. It opens the 225 data files each one corresponding with a point of our sample and calculates the media of the approaching and retracting curves because they do not exhibit hysteresis for all points measured as can be seen in Fig. 2 for one point.

2. It takes the minimum value of all the maxim force each file. This will be the first value of force for the calculation of FIEL values.

3. It makes the lineal regression of each file in the interval where the deflection of the cantilever is different from cero.

4. With this regression it interpolates the value of the separation distance for the force obtained in step two and takes this point as the first one on the force-distance diagram.

5. The FIEL values are determined by a simple numerical integration.

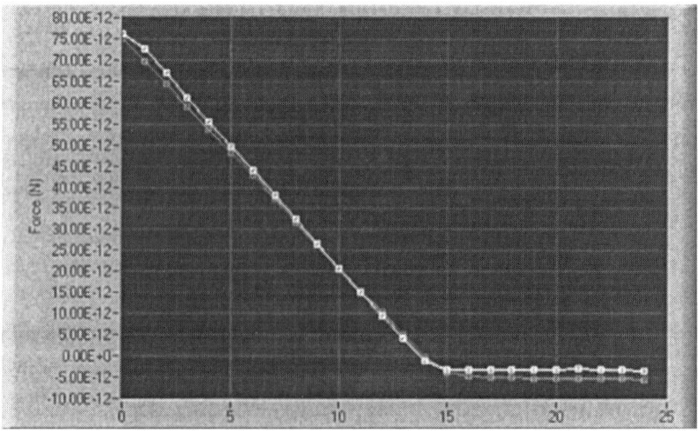

FIGURE 2: Approaching and retracting curves for one measured point.

6. It creates a text file with the FIEL values and the regression analysis data for each point.

Data were analysed in order to verify if they showed a normal distribution and also to obtain their statistic parameters. Statistical analysis were performed using the SPSS V.11.5 package for Windows (SPSS Inc., Chicago, IL).

RESULTS AND DISCUSSION

The topography of a 50x50 μm carpellary membrane area was observed and characterised using AFM (Fig. 3) and CryoSEM (Fig. 1). As the area under study was rather homogeneous, a less extensive area of 10x10 μm was selected to obtain the force-volume curves (Fig. 4). These curves allow to obtain the work at each point using the software previously described (see analysis software). In order to obtain a mapping of the k ratio, every work data was divided by the average value (835.384 work units). After this, it was applied the exponent aforementioned (see theory) to these k ratios.

Data showed a normal distribution as indicated the Kolmogorov-Smirnov test for one sample (P>0,05). The frequency distribution of the relation $k/k_{average}$ is shown in TABLE 1. The coefficient of variation of this distribution is lower than 6%.

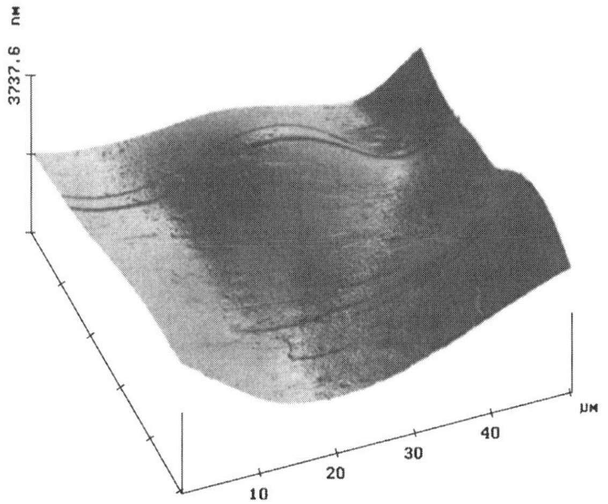

FIGURE 3: Topography of a 50x50 μm carpellary membrane area observed using AFM.

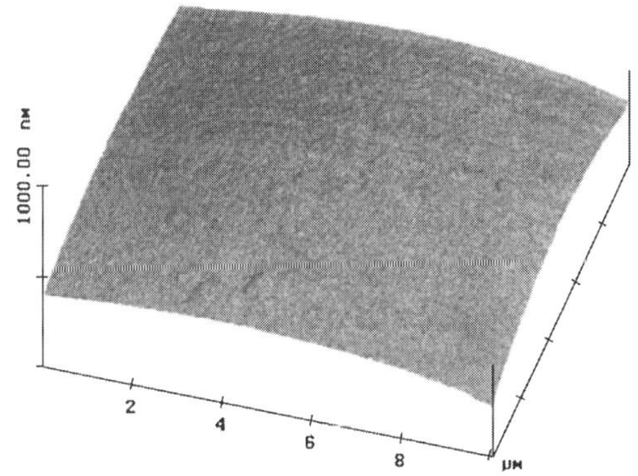

FIGURE 4: Topography of a 10x10 μm carpellary membrane area selected to obtain the force-volume curves observed using AFM.

TABLE 1. Frequency Distribution of the Relation $K/K_{average}$ and the Kolmogorov-Smirnov Test For One Sample.

Kolmogorov-Smirnov test for one sample		$k/k_{average}$
Number of samples		225
Normal parameters [a, b]	mean	1.000552
	Typical deviation	5,78E-2
Most extreme differences	Absolute	0.039
	Positive	0.039
	Negative	-0.033
Z of Kolmogorov-Smirnov		0.583
Asint. Sig. (bilateral)		0.885

a Contrast distribution is the Normal.
b Calculated from data.

CONCLUSIONS

Results show that AFM is a good technique to determine elastic properties in homogeneous biological samples. It presents a high repeatability. The measurement system repeatability indicates the variation obtained when independent measurements on the same sample are done using the same method, the same material, the same equipment, the same user, and when measurements are taken in time intervals as short as possible.

This repeatability is due on the one hand to the contribution of the homogeneity of the sample in its different points (important in some biological samples), and on the other to the low error of the measurement procedure. In future works we will try to discern between these two factors.

ACKNOWLEDGEMENTS

The authors appreciate the assistance of Mr. Jose Luis Moya Lopez at the Microscopy Service, Polytechnic University of Valencia, Spain.

REFERENCES

1. G. Binnig and C.F. Quate, *Physical Review Letters* **56 (9)**, 930-933 (1986).
2. E.K. Dimitriadis et al., *Biophysical Journal* **82**, 2798-2810 (2002).
3. N.A. Burnham and R.J. Colton, *Journal Vacuum Science & Technology A* **7**, 2906-2913 (1989).
4. N.J. Tao, S.M. Lindsay and S. Lees, *Biophysical Journal* **63**, 1166-1169 (1992).
5. S. Xu and M.F. Arnsdorf, *Journal of Microscopy* **173 (3)**, 199-210 (1994).
6. U.D. Schwarz et al., *Journal of Microscopy* **173 (3)**, 183-197 (1994).
7. K.L. Westra and D.J. Thomson, *Journal Vacuum Science & Technology B* **12 (6)**, 3176-3181 (1994).
8. K.L. Westra and D.J. Thomson, *Journal Vacuum Science & Technology B* **13 (2)**, 344-349 (1995).
9. H. Hertz, *J. Reine Angew. Math.* **92**, 156-171 (1881).
10. E.A-Hassan et al., Biophysical Journal **74**, 1564-1578 (1998).

11. G. Ruiz, "Contribución al estudio morfológico y analítico del granado (*Punica granatum* L.) mediante microscopía electrónica de barrido y microanálisis de rayos-X." PhD., Valencia, 2001.

Applied Physics (APHYS 2003)
A. Méndez-Vilas (Ed.)

Micro-fabrication with nanoparticles: Assembling DNA molecules by a focused laser

Masatoshi Ichikawa*, and Kenichi Yoshikawa
Department of Physics, Graduate School of Science, Kyoto University & CREST, Kyoto 606-8502, Japan

Yukiko Matsuzawa
Department of Ecological Engineering, Toyohashi University of Technology, Toyohashi, Aichi 441-8580, Japan

Yoshiyuki Koyama
Department of Home Economics, Otsuma Women's University, 12 Sanban-cho, Chiyoda-ku, Tokyo 102-8357, Japan

Abstract

We report a novel method to assemble colloidal particles to a linear rod structure. Giant DNA molecules individually folded by a sticky condensing agent can be colligating into a linear micrometer-sized chain under trapping field with a focused IR laser. Using this method, the macromolecules are assembled in a sequential order, and then a complex structure is fabricated on a glass plate through mechanical and optical manipulations. This approach may be useful for the further development of micro-manufacture using macromolecules as nano-elements.

A focused laser traps a dielectric particle on the laser focus [1]. The phenomenon, which is called laser trapping or optical tweezers, has been actively applied in biophysical and colloidal experiments as a powerful tool to manipulate a nanoparticle [2]. Recently, spontaneous assembling of giant molecules or colloidal particles under optical trapping potential has been investigated [3,4]. In the present paper, we report the process of linear assembling of sticky colloidal particle under the focused laser and demonstrate a fabrication on the substrate.

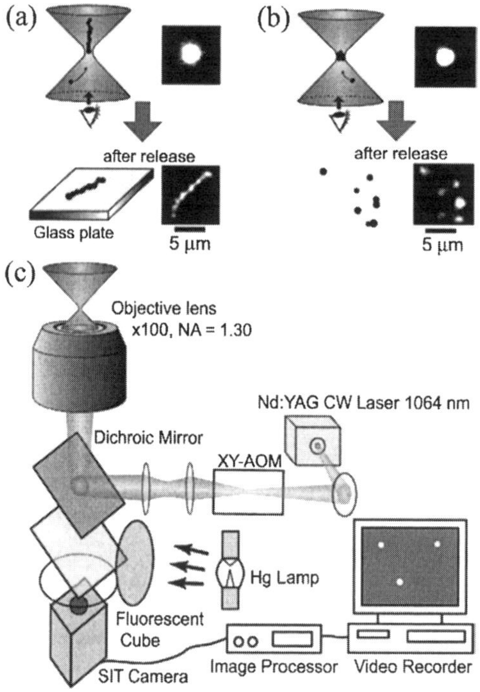

Fig. 1. Experimental setup and the results. (Reference [4]) (a) The folded DNA particles with Chol-PEG-A assemble to micrometer-sized linear bar under laser trapping field. Laser power for optical trapping is about 500 mW. (b) The DNA particles complex with PEG-A are once collected by the optical tweezers. Since its non-sticky or weak repulsive nature of folded DNA, each particle re-disperses after laser off. (c) Schematics of experimental system. XY-AOM is an acousto-optic modulator. XY-AOM was used in arraying assemblies on the plate.

A naked DNA thermally fluctuates in water solution with random coil shape, and is collapsed to a globular particle by adding various reagents [5]. Under dilute condition of DNA, individual DNA molecules are compacted through first order phase transition [6], and behave as colloidal particles. A single DNA molecule with about 166 kbp (56 μm) folded by multivalent-cations, polyethylene glycol derivatives with both cholesteryl- and amino-pendant groups (Chol-PEG-A) or either amino-pendant groups (PEG-A), are prepared as a sticky particle and as a non-sticky particle, respectively. Behavior of the folded DNA particles corresponds to dispersion of colloidal particles with about 50 to 100 nm in radii [7].

The experimental setup is shown in Fig. 1. We observed assembling phenomena of DNA particles complex with Chol-PEG-A. The particle dispersed in aqueous solution is attracted to the beam focus along the light line, since the optical trapping potential is proportional to the light intensity. When another floating particle has been caught in the focus, the falling particle conflicts to the other one, and thus the particles cling together. Repetition of collisions and coalitions around the focus generates a linear structure as shown in Fig. 1(a). Meanwhile, DNA particles collapsed by PEG-A also gather on the focus, but do not form a linear structure.

To make clear essential components for linear assembling, we have confirmed them through Brownian dynamics simulations. Optical tweezers are roughly composed of two forces, attractive trapping force and one directional pushing force. First one is called optical gradient force or dipole trapping force, which originates from induced dipole interaction and attracts a dielectric object with the direction toward higher light intensity. The latter, which derives from scattering force, etc., pushes particles along the light line with the direction following the laser beam.

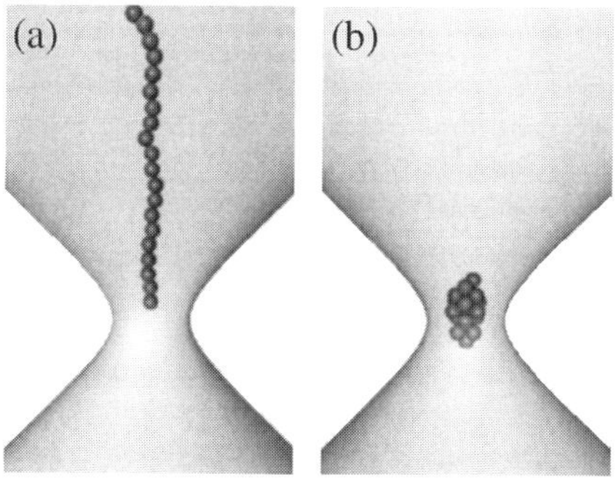

Fig. 2. Typical results of the simulations. Optical cones in the snap shots depict converged Gausian beam with 1 micrometer in beam waist diameter. The laser beams with 1 W power are incident from bottom side of the each panel. The particle is 100 nm in radius and is 1.60 in refractive index. (a) An assembly of sticky particles formed under optical gradient force with one directional pushing force is shown. The pushing force is three times larger than that of the Rayleigh scattering. (b) An aggregate of non-sticky particles is shown. The optical trapping force and the pushing force is same as (a).

Figures 2(a) and 2(b) shows simulation results corresponding to Fig. 1 experiments, respectively. The assembly shown as in Fig. 2(a) is generated under optical trapping potential with uni-directional pushing forces. The linear product tails form the focus following the light. Figure 2(b) shows an assembly of no-sticky particles with only excluded volume interaction under the same ambient condition as Fig. 2(a). The product on the focus is similar to the experiments in Fig. 1(b). The cluster forms dense packing in the small area by means of the strong collective force.

The results of the experiments and the corresponding simulations indicate strong pushing forces, for example, Rayleigh scattering and absorbing, convection induced by the laser heating, play an important roll in making the straight bars. Briefly, we explain the effects of directional force and the process of linear assembling from simulative and experimental observations: Particles are falling to the focus along the laser light line because a trapping potential is proportional to the light intensity. Thus the particles around the focus are arranged their collisions to be made linear bar with taking them along the beam line. A linear object in optical tweezers tends to stand before the light from the focus. In addition the one directional pushing force enhance a probability of one side approach of particles. Therefore the collision point between the particle and the bar is fixed near the focus, i.e. the edge of the bar on the focus. The linear assembly grows with the direction before the beam.

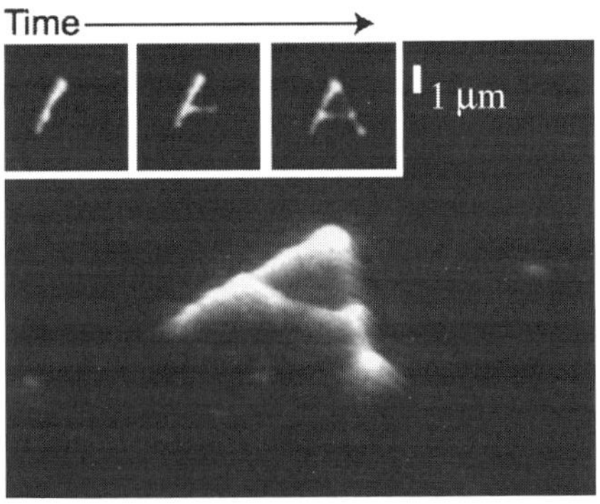

Fig.3. "A" prefabrication built on the substrate. The character is fabricated from three bars of the DNA assemblies. The magnified panel is quasi-3D image of fluorescence.

Figure 3 shows the demonstration of prefabrication on the glass substrate made from assembled DNA bars. The DNA bars are first generated by the above method, and are transported over the substrate. Then the bars are put in desired position through optical and mechanical manipulation.

In conclusion, optical trapping potential with light pressure generates a linear assembly from sticky nanoparticles. We also have demonstrated selective or sequential assembling by use of the method, and have made a complex structure from the assembled bars on the glass plate through the optical and mechanical manipulation [4,8]. Such a novel method is expected to be applicable to assemble nanoparticles to a desired structure.

References

[1] Ashkin, A., Dziedzic, J. M., Bjorkholm, J. E. & Chu, S., Observation of a single-beam gradient force optical trap for dielectric particles. *Opt. Lett.* **11**, 288-290 (1986).

[2] Grier, D. G., A revolution in optical manipulation. *Nature* **424**, 810-816 (2003).

[3] Hofkens, J., Hotta, J., Sasaki, K., Masuhara, H. & Iwai, K., Molecular Assembling by the Radiation Pressure of a Focused Laser Beam: Poly(N-isopropylacrylamide) in Aqueous Solution, *Langmuir* **13**, 414-419 (1997).

[4] Ichikawa, M., Matsuzawa, Y., Koyama, Y. & Yoshikawa, K., Molecular Fabrication: Aligning DNA Molecules as Building Blocks. *Langmuir* **19**, 5444-5447 (2003).

[5] Bloomfield, V. A., Condensation of DNA by multivalent cations: condensation on mechanism. *Biopolymers* **31**,1471-1481 (1991).

[6] Minagawa, K., Matsuzawa, Y., Yoshikawa, K., Khokhlov, A. R. & Doi, M., Direct Observation of the Coil-Globule Transition in DNA Molecules. *Biopolymers*, **34**, 555-558 (1994).

[7] Yoshikawa, K., Yoshikawa, Y., Koyama, Y. & Kanbe, T., Highly Effective Compaction of Long Duplex DNA Induced by Polyethylene Glycol with Pendant Amino Groups. *J. Am. Chem. Soc.*, **119**, 6473-6477 (1997).

[8] Matsuzawa, Y., Hirano, K., Mizuno, A., Ichikawa, M. & Yoshikawa, K., Geometric manipulation of DNA molecules with a laser. *Appl. Phys. Lett.*, **81**, 3494-3496 (2002).

Applied Physics (APHYS 2003)
A. Méndez-Vilas (Ed.)

METROLOGIC ASSESSMENT OF HIGH POWER LASER GENERATED SURFACE ROUGHNESS BY CONFOCAL LASER SCANNING MICROSCOPY

C. MOLPECERES[1,2], S. LAUZURICA[2], J.A. PORRO[1,2], J.L. OCAÑA[1,2]

(1) ETSIIMLAS. Dept. of Applied Physics. ETSI Industriales. Universidad Politécnica de Madrid. C/ José Gutiérrez Abascal, 2. 28006 Madrid. SPAIN
(2) Centro Láser UPM. Universidad Politécnica de Madrid Ctra. de Valencia, km. 7,300. 28031 Madrid. SPAIN

TOPICS+KEYWORDS: Optical Microscopy, Laser Material Processing, Laser Scanning Confocal Microscopy, Surface Roughness, Quality Assessment, Metrology.

ABSTRACT
This work presents the results of different techniques, based on Confocal Laser Scanning Microscopy (CLSM), applied to the study, quality control and assessment of most typical laser industrial processes. Topometry and roughness measurements in laser cutting and welding, pattern morphology analysis in laser marking, ablation control in laser cleaning, and surface characterization in laser shock processing are relevant examples of CLSM techniques applied to laser fabrication quality assessment, offering also a valuable tool for the understanding of basic physical phenomena involved in laser-matter interaction in the field of major laser industrial applications.

1. INTRODUCTION

CLSM has been used for years as an important tool for optical characterization in the fields of biology (Shuman et al., 1989), medicine (Voort et al., 1989), and, more recently, material science (Becker et al., 2001; Lindseth & Bardal, 1999; Draeger & Case, 2002). In this last topic, high power laser applications cover a huge range of material processing and machining techniques, ranging from nowadays standard cutting and welding techniques to the very new developments in surface modification procedures as laser shock processing (LSP). In all these areas, laser as industrial machine is a very precise tool, with a very high degree of flexibility in its usage, able to competitively solve a variety of industry demands, though also needing sophisticated techniques for quality control.

An important set of instrumental techniques for laser processes assessment is commonly used depending on the particular application and study depth requirements. Metallurgical analysis, profilometry, residual stresses measurements, microphotography, hardness measurement, contact roughness measurement, etc. are widely used together with different microscopic techniques (optical and electronic) lacking, however, of a more extensive use of CLSM techniques that appear as especially well suited to some of these analysis. Major advantages of CLSM applied to laser machining characterization rely on non-contact surface topography assessment (including standardized roughness measurements), and 3D dimensional measurements with high aspect ratios. Additionally correlation between topometry data and surface generation mechanisms (Draeger & Case, 2002), offers a better comprehension of the very complex physical phenomena arising in laser matter interaction processes.

2. EQUIPMENT DESCRIPTION

A Leica ICM 1000 CLSM, using only reflected light, has been used for this work. Equipment operation wavelength (635 nm corresponding to diode laser emission) together with objective numerical aperture and pinhole diameter give final resolution in axial direction.

Different objectives have been used (Table 1) aiming to cover almost completely standard dimensional assessment of laser material processing techniques (nowadays limited to spatial resolutions of about 100 nm in micromachining applications). Long working distances objectives (suffixed L in Table 1) have been set up for complicated 3D workpieces analysis.

Objective Magnification	10X	20X	20X L	50X L	50X	100X
Numerical Aperture	0.25	0.40	0.40	0.5	0.75	0.90
Resolution XY (nm)	780.8	488.0	488.0	390.4	260.3	216.9
Resolution XZ (nm)	6915.6	2630.4	2630.4	1639.1	648.6	389.3
Scan Area (μm x μm)	1000 x 800	500 x 400	500 x 400	200 x 160	200 x 160	100 x 80

Table 1. Magnification, numerical aperture, resolution and scanning area of CLSM objectives used in this work. All data correspond to nominal parameters.

Special emphasis must be done in topometry and roughness evaluation. In confocal microscopic techniques, 3D images are obtained moving the focus plane and acquiring single images (optical slices) that can be put together building up a three dimensional stack of images that can be digitally processed (Wilson, 1992). In this case all of these images have been treated with noise filtering techniques to eliminate highest and lowest intensity values of the topographical images (Wendt et al., 2002), according to normalized standards for roughness measurements (ISO 11562:1996)

Commercial CLSM are normally calibrated to fulfil nominal specifications, but for industrial assessment an in-situ calibration (including if required a trazability plan) should be recommended. For this work different roughness standards (Fig. 1) with trazability certificate have been used for each objective, and all measurements have been performed with environmental temperature and humidity control.

Following sections summarize the results obtained with this equipment in some characteristic laser applications carried out at *Centro Laser UPM* (Universidad Politécnica de Madrid). An introduction to each involved laser process is included for a better comprehension of this particular research frame.

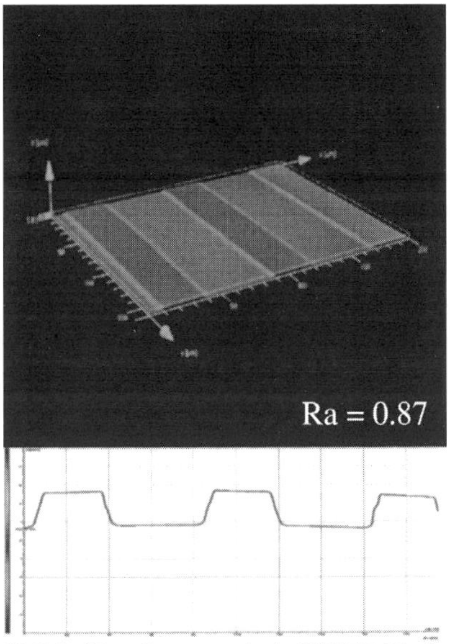

Fig.1 Topographical image CLSM of Roughness standard. 112/557 (Taylor-Hobson). (R_a = 0.87 ±0.07 □m certified value)

3. LASER CUTTING

Laser cutting represents, from the economical point view, the main laser industrial application. Quality assessment in laser cutting is compulsory in different industrial processes and CLSM gives appropriate solutions to deal with some complex problems characteristics of this application.

Being mainly a thermal process, laser cutting is based on absorption of laser light in the material surface leading to partial molten and evaporation of the material in a narrow kerf. In order to remove molten material from the interaction zone an assist gas is used, and relative movement between laser beam and workpiece leads to cut kerf formation. Physical mechanisms involved in laser cutting are very complex and still there is no a complete description of the whole process dynamics (Olsen, 1998).

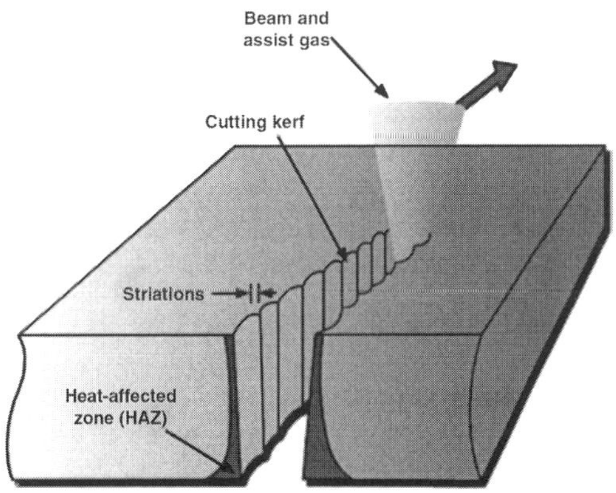

Fig. 2 Conceptual drawing of characteristic laser cut striations origin

As consequence of laser cutting intrinsic dynamics, a typical striation, overall for metallic materials, is formed in the cut front (Fig. 2). Such a striation is related with molten layer dynamics and is strongly dependent on the main process variables considered in laser cutting (velocity, laser power and assist gas variables). Some work (Biermann, 1988; Schüocker, 1987) have be done in order to explain striation formation and morphology (striation wavelength and bending, influence of oxidation on striation pattern , etc) , but there is no doubt about the fact that striation analysis is important both from the point of view of final laser cut quality (defines morphology of the cut front and roughness of this area) and theoretical modelling validation (Ocaña, 1994), being this last fact of great importance for a better comprehension of the physics involved and in order to offer predictive tools for industry in laser cutting processes.

CLSM appears as an adequate alternative to standard microscopy and contact roughness measurement techniques for cut front characterization. In fact, CLSM analysis of cut front topography, allows a detailed study of striations, both qualitative and quantitative, with an easy reading of roughness, bending of lines, striations wavelength and detailed morphology of the striation pattern. This capacity gives and invaluable help in quality assessment and models development in such an important segment of laser industrial applications.

Figure 3 shows confocal striation pattern images analysis from 3mm and 8 mm cast iron probes cut with a 1500 W CO_2 laser (cw operating). Striation wavelength values are shown (directly related with cutting velocity, closer striation lines relates to higher cutting velocities) and also roughness determination in perpendicular direction to striation pattern. Particular morphology of these striations and straight cylindrical surfaces aspect of the presented results are indicative of appropriate processes parameters selection and very good cut quality.

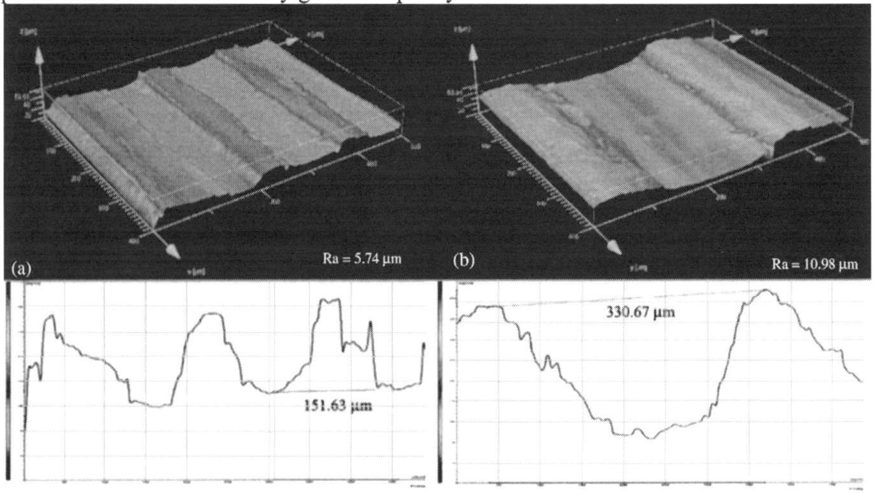

Fig. 3 Confocal 3D reconstruction and perpendicular sections of cut edge striation in cw laser cutting of cast iron probes a) 3 mm and b) 8 mm thickness. Roughness and striation wavelength values are shown in the images.

4. LASER WELDING

Laser welding refers to a set of procedures for material joining (metallic and non metallic) considerably different from the physical point of view. Main laser welding process in metallic materials avoid the use of filler material and, depending on material phase transitions produced during the irradiation, laser conduction welding or *keyhole* (deep penetration) laser welding is obtained. The last one is the most interesting from the technological point of view leading to very good quality welds beads. In laser keyhole welding, formation of a plasma filled capillary increase drastically the energy absorption and leads to a very high aspect ratio welds (Fig. 4) but implying a strong sensibility on quality from the process variables (Dawes 1992.).

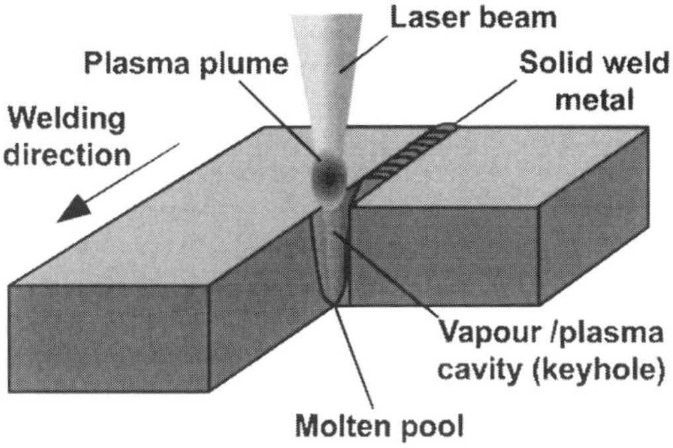

Fig. 4 Keyhole laser welding

Confocal microscopy helps mainly, and with remarkable advantage, in welding bead morphology characterization (Fig. 5), which is particularly important from the point of view of laser welding modelling assessment (Ocaña, 1994). Laser keyhole welding phenomenology implies non-trivial hydrodynamical problems like plasma plume and molten pool dynamics (being both phenomena strongly related to final bead morphology). Usually for narrow welds beads, a chevron-like aspect of weld seam (Fig. 5) is observed that is strongly correlated with keyhole parameters (welding speed, and keyhole stability mainly). In particular, an accurate estimation of material raised over surface level and accurate morphology determination of bead geometry are very important for laser welding quality assessment. An extraordinary set of different details can be observed in CLSM images, as fluid lines geometry, geometric defects of the bead, cracks on surface, etc. giving the opportunity of a very complete weld quality analysis.

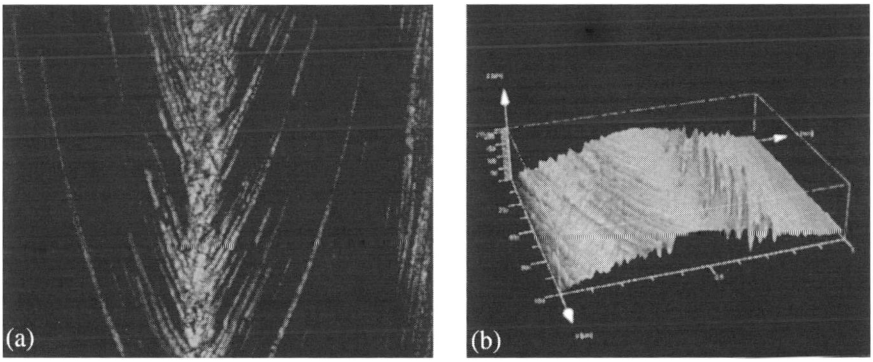

Fig 5. a) Extended Focus image of welding seam with a 10X magnification. b) Topographical CLSM image of a laser weld in titanium (bead-on-plate essay) with a 10X magnification.

5. LASER ABLATION

Laser ablation process covers a great diversity of particular applications in which mass removal of laser irradiated materials is the essence of the final required process. Even more than previously mentioned cutting and welding processes, physical mechanisms involved in laser ablation are extremely complex (Von Allmen 1987) depending on the particular ablation technique considered (molten material ablation, vaporization phase ablation, sublimation techniques, non thermal ablation, etc.). Quality control in laser ablation implies surface final state characterization, including, if possible, estimation of ablated mass, walls morphology in ablation fronts and layer behaviour in multilayer laser ablation. That is also the case in laser cleaning (Meja 2001), and other process as laser lithography (Lamda Physics 2001) and surface modification (Kaplan 1998) very well known in the frame of micromachining techniques. Even for fully commercial laser ablation applications as laser marking and engraving CLSM offers complete assessment including marking depth, walls slope, pattern homogeneity, etc. (Fig. 6).

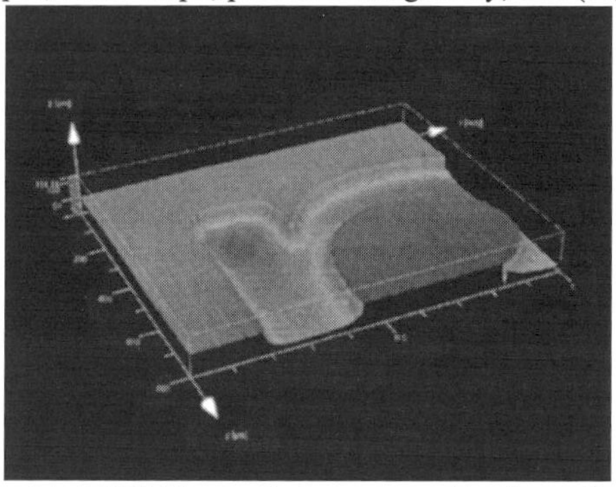

Fig. 6 Topographical CLSM image of a laser text marking in anodized aluminium

Another good example of CLSM capabilities in laser ablation assessment is laser paint stripping process in ship building industry. Laser advantages with respect to other paint removal techniques as sand blasting and water jets are an enhanced cleanliness and environmental impact and an improved final surface quality allowing a better surface maintenance. Figure 7 shows confocal reconstruction and realistic images of the step-like transition zone between clean material and residual painted areas in laser stripping of shipbuilding steel coated with two layers of different epoxy paints (transitions zone appears in white colour in Fig. 7 digital images). A sharp step is related with non-thermal ablation effects and non well defined steps prove strong thermal interaction. Images show clearly that operation in cw for high power diode lasers leads to strong thermal interaction mechanism (corresponding to soft height gradients) while pulse laser dynamic (with short peak power pulses) leads to important gradients that implies a less thermal affection of base material and better layer removal control.

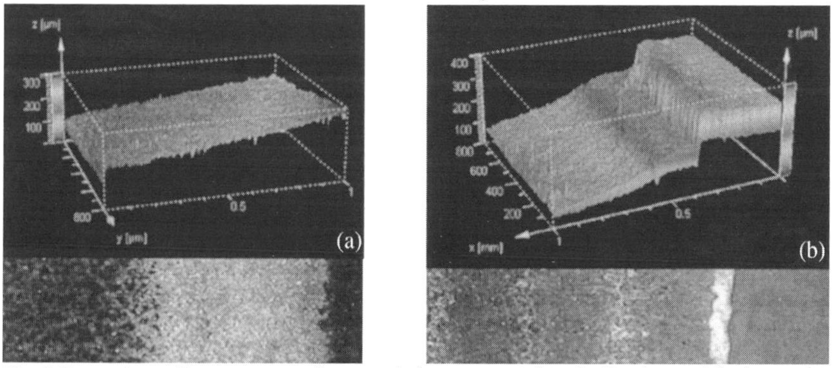

Fig. 7 Comparison of laser source performance in laser generated tracks (border area) for laser paint stripping in ship building industry. a) cw high power (2700 W) diode laser b) Pulsed Nd:YAG laser source (400 W mean power).

6. LASER SHOCK PROCESSING

LSP is a very new technique for material properties improvement (mainly fatigue and wear resistance) based on the generation of residual compression stresses in metals by means of laser generated shock waves (Fairand & Clauer, 1979; Ocaña et al., 2001). Q-switch high power lasers are used to irradiate material surface, leading to a quickly expanding plasma formation. Plasma expansion induces, by momentum transfer, high pressures that lead to shock wave formation and propagation, giving the resulting stress field inside the material. Overlapping laser pulses onto surface material defines the final treated area and is the responsible mechanism together with irradiation parameters of final surface morphology. In practice this processes competes with well established surface modification techniques as shock peening, but LSP offers better surface finishing and deeper (until 1 mm) treated zones inside the material. Assessment of surface topometry and roughness is essential in LSP processes, mainly considering that current research lines (Ocaña, 2001) add new elements that can affect final surface morphology (implying important roughness changes), as the employment of dielectric confinement media for plasma expansion in order to increase the effective pressure applied to the material and deposition of absorbing materials layers in target surface for better energy coupling.

Fig. 8 a) 3D FEM code overlapping simulation and topographical CLSM images of LSP processes in aluminum with confinement media b) air and c) water.

In figure 8 an aluminum LSP treated surface as appear in a 3D finite element code simulation (Ocaña, 1994) and confocal images from same material with confinement media air a) and water b) is shown. Measured values of roughness (indicated in the figures) for treated targets proves a clear difference between these two treatments.

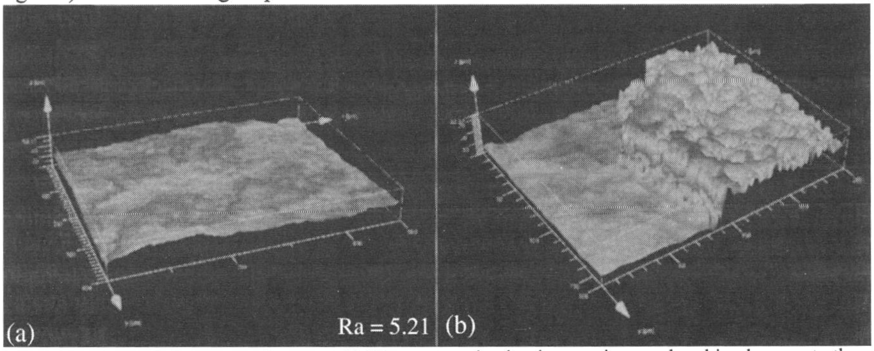

Fig. 9 Topographical CLSM images of LSP treatment in aluminum using an absorbing layer onto the material a) 20X magnification b) in the interface 50X magnification.

Roughness values for air are close to those encountered in shot peening applications; however values for water confinement LSP indicate a clear improvement in final surface cosmetic quality. Fig 9 shows a LSP treatment in aluminum but using an absorbing layer onto the material. With this technique deeper compression stress field can be achieved but with significant lack of surface cosmetic quality.

CONCLUSIONS

High power laser applications usually demand a great variety of techniques for quality assessment. Authors, with a significant experience in development, control and assessment of laser industrial applications, consider CLSM, as demonstrated by the examples presented in this work, an adequate alternative to complement or even substitute conventional dimensional measurement techniques and roughness evaluation in different fields of the wide range of processes that industrial high power laser techniques covers nowadays.

ACKNOWLEDGEMENTS

Work partially supported by MCYT PU 2002-36

REFERENCES

Becker, J.M., Grousson, S., Jourlin, M. (2001) *Surface state analysis by means of confocal microscopy Cement & Concrete Comp.* 23. 255-259

Biermann, S. (1988) *Proc. Laser Treatment of Materials.* ECLAT'88. 20-2

Dawes, C. (1992) *Laser welding.* Abington Publishing.

Draeger, D.J., Case, E.D. (2002) *Characterization of channels in zirconia ceramics using laser scanning confocal microscopy.* J. of Material Science Lett. 21. 787-781

Excimer Laser Technology Handbook: Laser Sources, Optics, Systems and Applications. Ed. Dirk Basting. LAMBDA PHYSIK. (2001)

Fairand, B. P. & Clauer, A.H. (1979) *Laser generation of high-amplitude stress waves in materials.* J. Appl. Phys. 50. No. 3, 1497-1502

Kaplan, A. (1998) *Precision Ablation Processing.* Handbook of the Eurolaser Academy. Ed. D. Schüocker. Chapman & Hall.

Lindseth, I., Bardal, A. (1999) *Quantitative topography measurements of rolled aluminium surfaces by atomic force microscopy and optical methods.* Surf. Coatings and Tech. 111 276-286

Meja, P. & Autric, M. (2001) *Laser cleaning of technological surfaces.* Proc. LANE 2001. Ed. M. Geiger. Meisenbach-Verlag.

Ocaña, J.L., Molpeceres, C., Morales, M., Moreno, M. (2001) *Laser Shock Processing as a Method for the Improvement of the Fatigue Resistance of Metals and Metal Alloys: Numerical Simulation and Experimental Assessment.* In Laser Assisted Net Shape Engineering 3. M. Geiger, A. Otto, Eds. Meisenbach-Verlag, 199-210

Ocaña, J.L. (1994) *Review on the physics and calculational methods for the modelling of the laser-matter interaction in high intensity laser processing applications*, Lasers in engineering Vol 3.

Olsen, A. (1998) *Cutting Handbook of the Eurolaser Academy.* Ed. D. Schüocker. Chapman & Hall.

Schüocker, D. (1987). *The physical mechanisms and theory of laser cutting in* Industrial Annual Laser Handbook. 65-79.

Shuman, H., Murray, J.M., DiLullo, C. (1989). *Confocal microscopy, an overview.* Biotechniques 7. 154-163

Von Allmen, M. (1987). *Laser Beam Interaction with Materials.* Springer Verlag, Berlin.

Voort, HTM, Brakenhoff, G.J., Baarslag MW (1989) *Three dimensional visualization methods for confocal microscopy.* J. Microsc. 153:123-132

Wendt, U., Stiebe-Lange, K. & Smid, M, (2002) *On the influence of imaging conditions and algorithms on the quantification of surface topography.* J. Microsc. 207, 169-179

Wilson, T (1992) *Confocal Microscopy.* Academic Press, London.

Applied Physics (APHYS 2003)
A. Méndez-Vilas (Ed.)

Characterization Of Membrane Distillation Membranes By Tapping Mode Atomic Force Microscopy

Mohamed Khayet

Department of Applied Physics I, Faculty of Physics, University Complutense of Madrid, Avda. Complutense s/n, 28040, Madrid, Spain. Tel. & Fax: 34 (91) 3945191.
E-mail address: khayetm@fis.ucm.es

Abstract. The surface structure of polytetrafluoroethylene and polyvinylidene fluoride flat sheet membranes used in membrane distillation were studied by tapping mode atomic force microscopy (TM-AFM). The membranes are porous and hydrophobic. The mean pore size, nodule size, pore size distribution, pore density, surface porosity and roughness parameters were determined by TM-AFM analysis. The obtained pore sizes of the membranes were fitted to the log-normal probability distribution function. The characteristics of the same membranes obtained by other techniques were compared to those obtained in this study by means of TM-AFM.

INTRODUCTION

The separation process membrane distillation (MD) offers a superior potential for production of high-purity water than the pressure-driven membrane separation processes (i.e. microfiltration, ultrafiltration, nanofiltration and reverse osmosis) [1,2]. MD has also been used for the concentration of aqueous solutions containing non-volatile solutes such as ions and colloids. Moreover, the removal of trace organic compounds (VOCs) from waste water by MD is well recognized [3].

MD employs polymeric microporous and hydrophobic membranes that act only as support for vapor-liquid interface and do not contribute in the separation performance. The driving force in this process is the vapor pressure difference between both sides of membrane pores. Four configurations were applied to establish the transmembrane vapor pressure, namely, direct contact membrane distillation (DCMD), sweeping gas membrane distillation (SGMD), vacuum membrane distillation (VMD) and air gap membrane distillation (AGMD). These embodiments were explained in Refs. [1,4].

Direct contact membrane distillation (DCMD) configuration refers to a thermally driven transport of vapor through the membrane pores. Both the heated feed and the cold permeate aqueous solutions are maintained always in contact with each side of the membrane. Due to the hydrophobic nature of the membrane, aqueous solutions cannot penetrate inside dry membrane pores unless a transmembrane hydrostatic pressure exceeds the "liquid entry pressure of water (*LEPw*)", which is characteristic of each membrane. Under this condition, liquid-vapor interfaces are formed at the entrances of each pore. Therefore, since a vapor pressure difference is maintained by a temperature difference between both sides of the membrane pores, molecules evaporate from the feed liquid/vapor interface, cross the pores in vapor phase and condense on the liquid/vapor interface kept at lower temperature.

In MD literature, various physical methods were used to characterize the MD membranes. The techniques used can be divided in two main groups, the one which is related to membrane permeation (i.e. gas flow test, liquid displacement, bubble point), and the other which permits to obtain directly the morphological properties of the membranes (i.e. scanning electron microscopy, SEM; transmission electron microscopy, TEM; field emission scanning electron microscopy, FESEM) [3-8]. Atomic force microscopy (AFM) study of MD membranes has not been reported yet although this technique permits to determine useful characteristics in MD and no special sample preparation is needed to obtain directly three-dimensional topographical images of the membrane surface. On the contrary, heavy metal coating required in SEM and TEM give some artifacts and tend to damage the polymeric membranes. In other words, a more true surface morphology of a membrane could be observed by AFM. In fact, AFM was first applied to polymer membrane surface by Albrecht et al. [9] in 1988 soon after it was invented by Binning et al. [10]. The technique has been applied, since then, extensively for studying various types of membranes giving information about surface morphology of both flat sheet and hollow fiber membranes [11,12]. Recent review article on characterization of synthetic membranes by AFM was given by Khulbe and Matsuura [13].

In the present study, the surface topography and membrane structure of two flat sheet MD membranes are analyzed by TM-AFM. The membrane characteristics needed in MD process such as the mean pore size, pore size distribution, pore density and surface porosity are determined by TM-AFM analysis and compared with those determined by other membrane characterization techniques.

EXPERIMENTAL

Two microporous hydrophobic flat-sheet membranes, polytetrafluoroethylene (TF200, Gelman) supported by a polypropylene net and polyvinylidene fluoride (GVHP, Millipore), were used. Their principal characteristics as specified by the manufacturers are the following: TF200 (nominal pore size = 0.2 μm; thickness = 178 μm; void volume = 80 %; *LEPw* = 2.82 bar); GVHP (nominal pore size = 0.22 μm; thickness = 125 μm; void volume = 75 %).

The above membranes have been structurally characterised by a tapping mode atomic force microscope on a Nanoscope III equipped with 1553D scanner from Digital Instruments, Santa Barbara, CA, USA. The mode of operation to take the AFM pictures was explained in previous works [11,12]. In this study, small pieces of approximately 0.5 x 0.5 cm^2 in area were cut from each membrane sample and fixed over a magnetic holder. All the TM-AFM images were made in air at room temperature over different locations of each membrane sample. Scans were made on areas of 6 μm x 6 μm and areas of 3 μm x 3 μm were selected randomly for analysis.

To obtain the pore sizes and nodule sizes, cross-sectional line profiles were selected to traverse the obtained TM-AFM images and the diameters of nodules (i.e. high peaks) or pores (i.e. low valleys) were measured by means of a pair of cursors. The horizontal distance between each pair of cursors was taken as the diameter of the nodule or pore.

The AFM analysis software program allowed computation of various statistics related to the surface roughness on predetermined scanned membrane area. The definition of the roughness parameters is listed later on. It should be emphasised here that the roughness parameters depend on the curvature and size of the TM-AFM tip as well as on the treatment of the captured surface data (plane fitting, flattening, filtering, etc.). Therefore, the roughness parameters obtained from TM-AFM images should not be considered as absolute roughness values. In the present study, the same tip was used to take the TM-AFM images of both membranes and the captured pictures were treated in the same way.

THEORETICAL

Determination Of Mean Pore Size And Pore Size Distribution By TM-AFM Analysis

The surfaces of the membranes were compared in terms of the pore size, nodule size and the roughness parameters obtained from TM-AFM study. The measured pore sizes were arranged in ascending order and the corresponding median ranks (50 %) were calculated using the following equation [11,12].

$$\chi = \left(\frac{i-0.3}{n+0.4}\right)100 \tag{1}$$

where i is the order number of the measured pore size arranged in ascending order and n is the total number of the measured pores.

The median ranks were plotted on the ordinate against the pore sizes. If the obtained pore sizes fit to a log-normal distribution, a straight line on log-normal probability paper will be obtained. From the log-normal plot, the mean pore size (μ_p) and the corresponding geometric standard deviation (σ_p) can be calculated. μ_p will correspond to 50 % of the cumulative number of pores and σ_p can be obtained from the ratio of 84.13 % of the cumulative number of pores to that of 50 %. Therefore, the pore size distribution can be expressed by the following probability density function.

$$\frac{df(d_p)}{d(d_p)} = \frac{1}{d_p \ln\sigma_p (2\pi)^{1/2}} \exp\left(-\frac{(\ln d_p - \ln\mu_p)^2}{2(\ln\sigma_p)^2}\right) \tag{2}$$

where d_p is the membrane pore size.

In addition, the surface pore density, which is the number of pores per unit membrane area (ρ_s) can be obtained directly from the AFM analysis software program. The corresponding surface porosity, ε_s, may be determined from Eq. (3) using the pore size distribution obtained from the TM-AFM images.

$$\varepsilon_s = \frac{\rho_s\, \pi}{4} \sum_{j=1}^{n} f_j\, d_j^{\,2} \tag{3}$$

where f_j is the fraction of the number of pores with size d_j.

It must be pointed out that ε_s is different to the void volume, ε [5].

$$\varepsilon = \varepsilon_s\, \tau \tag{4}$$

where τ is the pore tortuosity. This factor may be estimated if ε is known.

Surface Roughness Parameters

The mean roughness of each membrane, R_a, the root mean-square of Z data, R_q, and the average difference in height between the five highest peaks and the five lowest valleys, R_z were determined. R_z was calculated relative to the mean plane, which is a plane about which the image data has a minimum variance.

The mean roughness, R_a, represents the mean value of the surface relative to the center plane for which the volumes enclosed by the images above and below this plane are equal.

$$R_a = \frac{1}{L_x L_y} \int_0^{L_x}\int_0^{L_y} |f(x,y)|\, dxdy \tag{5}$$

where $f(x,y)$ is the surface profile relative to the center plane, L_x and L_y are the dimensions of the surface in the x and y directions, respectively.

The root mean-square roughness, R_q, is the standard deviation of the Z values within the specific area and can be calculated using the following expression.

$$R_q = \sqrt{\frac{\sum (Z_i - Z_m)^2}{N_p}} \tag{6}$$

where Z_i is the current Z value, Z_m is the average of the Z values and N_p is the number of points within a given area.

RESULTS AND DISCUSSION

The three-dimensional TM-AFM pictures of GVHP and TF200 membranes are presented in Fig. 1.

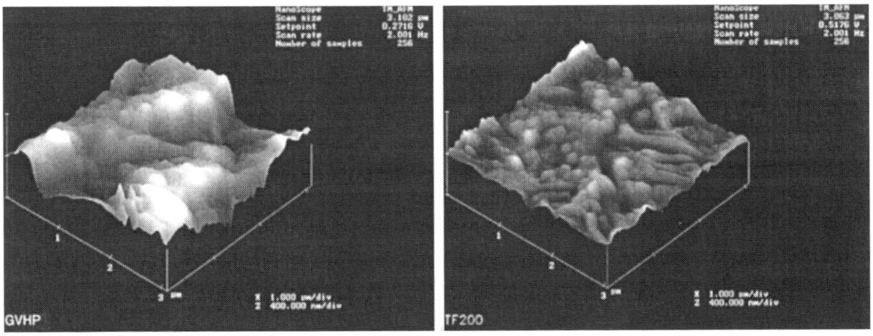

FIGURE 1. Three-dimensional TM-AFM images of GVHP and TF200 membranes.

As expected, a difference in the morphology of the membrane surfaces can be observed. The nodules are seen as bright high peaks whereas the pores are seen as dark depressions. The surfaces of both membranes are not smooth and the nodule-like structure and nodule aggregates are present at the surface of the membranes [14]. The pore sizes and the nodule sizes were measured as stated earlier and the TM-AFM image analysis was done by the built-in computer program. It was observed that the pores were not necessarily circular and therefore the average of their length and width was recorded as pore size.

The median ranks were plotted versus the measured pore sizes on a log-normal probability paper as can be observed in Fig. 2. Straight lines have been fitted to the data with reasonably high correlation coefficient ($r^2 > 0.965$ for GVHP and $r^2 > 0.957$ for TF200).

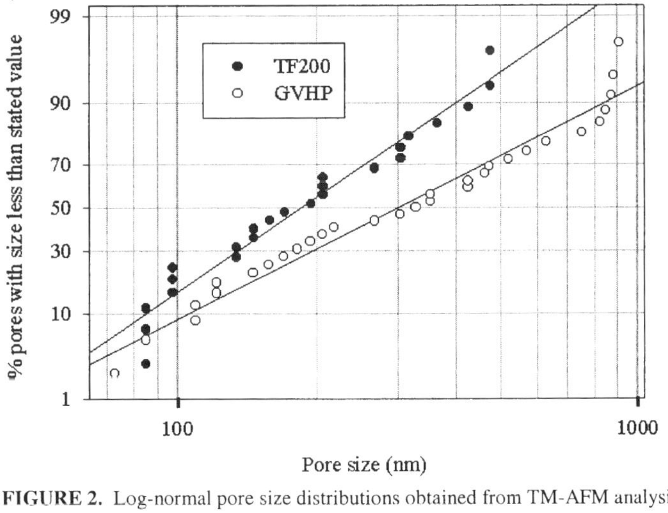

FIGURE 2. Log-normal pore size distributions obtained from TM-AFM analysis.

μ_p and σ_p were calculated for each membrane according to the method described earlier. The results are presented in Table 1. The cumulative pore size distribution and

μ_p and σ_p were calculated for each membrane according to the method described earlier. The results are presented in Table 1. The cumulative pore size distribution and the probability density function curves were generated from Eq. (2). Figure 3 shows the results. When compared to the membrane TF200, it can be observed that the pore size distribution curve of the GVHP membrane shifts to the right and its pore size distribution is lower and broader around the mean pore size. In other words, GVHP membrane exhibits larger pore sizes than TF200 membrane. For the same membranes, from the liquid displacement technique, Martínez et al. [5,6] found lower mean pore size (i.e. 0.32 μm) for GVHP and larger mean pore size (i.e. 0.31 μm) for TF200 membrane than those obtained from TM-AFM analysis.

Recently, the surface and the bulk pore sizes of various polyetherimide and polyvinylidene fluoride membranes were determined by TM-AFM study, solute transport and gas permeation test [15]. It was observed that the pore sizes at the membrane surface are larger than the bulk membrane pore sizes and the mean pore sizes determined by TM-AFM analysis were 2.1 and 1.7 times larger than those determined from the gas permeation test for flat sheet and hollow fiber membranes, respectively. Therefore, it may be stated that the pores of GVHP membrane measured by TM-AFM have maximum openings at the surface entrance and a few small pores could easily be misinterpreted as one large pore when they are amalgamated, resulting in an overestimation of the pore sizes. On the contrary, this is not the case for TF200 membrane, which exhibits different structure and smaller membrane thickness.

Table 1. MD membrane characteristics: membrane thickness, δ; void volume, ε; mean pore size, μ_p; geometric standard deviation, σ_p; nodule sizes (minimum, ν_L; average, ν_M; maximum, ν_H).

Membrane	δ (μm)	ε (%)	TM-AFM analysis				
			μ_p (nm)	σ_p	ν_L (nm)	ν_M (nm)	ν_H (nm)
GVHP [a]	117.7	70.1	449.79	1.32	44.9	415.2	1273.0
TF200 [b]	54.8	68.7	250.72	1.23	18.4	92.6	153.9

[a] Membrane supplied by Millipore with the trade name Durapore.
[b] Membrane supplied by Gelman with the trade name TF200. Total thickness, 165.2 μm.

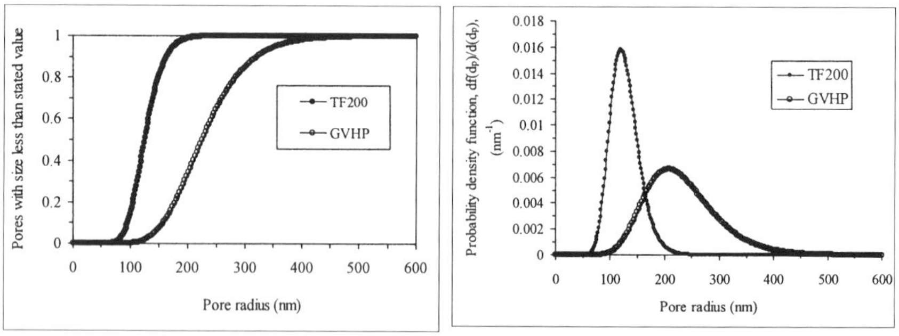

FIGURE 3. Cumulative distributions of pore sizes (a) and probability density function curves (b) generated from the pore sizes measured from the TM-AFM images of GVHP and TF200 membranes.

From the field emission scanning electron microscopy (FESEM), Phattaranawik et al. [7] obtained mean pore sizes of 251 nm for GVHP membrane and almost the same pore size (i.e.253 nm) as the one reported in Table 1 for TF200 membrane. The low value obtained for GVHP membrane may due to pore contraction during metal-coating of the membrane sample, which is required to take the FESEM images. In addition, as can be seen in Table 1, σ_p value of the membrane GVHP is higher than that of TF200 membrane. However, based on FESEM study lower σ_p values, 1.04 and 0.94, were reported for GVHP and TF200 membranes, respectively [7].

Table 1 also shows the minimum, maximum and the average nodule sizes of both membranes. The GVHP membrane surface contains nodule aggregates and super nodular aggregates [14], while the nodule sizes are lower for TF200 membrane. This result may be attributed to the different techniques and polymers used for membrane preparation.

The pore density was obtained by counting the number of pores on the TM-AFM images and the surface porosity was calculated from Eq. (3) using the pore size distribution (Fig. 3). Both the pore density and surface porosity data are higher for TF200 membrane (Table 2). It must be pointed out that the effective porosity obtained by other authors from the liquid displacement technique is higher for TF200 membrane than for GVHP membrane [5,6]. However, the measured void volume of both membranes were almost the same as presented in Table 1 [3]. It is worth quoting that the porosity of the membrane GVHP is specified as 75 % by the manufacturer, while that of TF200 is 80 %. Moreover, the GVHP membrane is thicker than TF200 membrane. This may induce lower effective membrane porosity for GVHP as found from the liquid displacement method [5,6].

For the two MD membranes, the tortuosity factor was calculated from Eq. (4). The obtained τ value of GVHP is higher (Table 2). This may be attributed partly to the higher membrane thickness of the membrane GVHP. In MD studies, a pore tortuosity factor of 2 is frequently assumed [7]. A value as high as 3.9 was also reported for GVHP membrane [16].

The roughness parameters were evaluated from TM-AFM images on different parts of the same membrane sample and the mean values are reported in Table 2. As expected, the GVHP membrane is rougher than TF200. Generally, the membrane roughness parameters become higher with larger pore size and nodule size. This may be explained since the roughness parameters depend on the Z values. In this case GVHP membrane exhibits larger nodules and pore sizes. The same relationship between roughness parameter and pore size was also observed in previous studies for both flat sheet and hollow fiber membranes [11,12].

Table 2. Membrane pore density, ρ_s; surface porosity, ε_s; pore tortuosity, τ; and roughness parameters.

Membrane	ρ_s (μm^{-2})	ε_s (%)	τ	R_a (nm)	R_q (nm)	R_z (nm)
GVHP	1.35	25.28	2.77	81.7	113.8	1052.0
TF200	6.25	33.97	2.02	44.2	59.6	441.8

CONCLUSIONS

Morphological characterization of microporous polyvinylidene fluoride (GVHP) and polytetrafluoroethylene (TF200) membranes used in MD were carried out by tapping mode atomic force microscopy. It was observed that the membrane surfaces are not smooth and GVHP membrane is rougher than TF200 membrane due to the larger nodules and pore sizes of GVHP. The log-normal distribution is appropriate to describe the pore size distribution of MD membranes. The pore tortuosity of the membrane GHVP is higher due to its higher membrane thickness. The membranes characteristics obtained from TM-AFM study may be used for the prediction of the permeability of MD membranes.

ACKNOWLEDGMENTS

Financial support from the University Complutense of Madrid is acknowledged.

REFERENCES

1. Khayet, M., Godino, M.P., and Mengual, J.I., *Int. J. of Nuclear Desalination* **1**, 30-46 (2003).
2. Mengual, J.I., and Peña, L., *Colloid & Interface Sci.* **1**, 17-29 (1997).
3. Khayet, M., and Matsuura, T., *Ind. Eng. Chem. Res.* **40**, 5710-5718 (2001).
4. Lawson, K.W., and Lloyd, D.R., *J. Membr. Sci.* **124**, 1-25 (1997).
5. Martínez, L., Florido-Díaz, F.J., Hernández, A., and Prádanos, P., *J. Membr. Sci.* **203**, 15-27 (2002).
6. Martínez, L., Florido-Díaz, F.J., Hernández, A., and Prádanos, P., *Sep. & Pur. Tech.* **33**, 45-55 (2003).
7. Phattaranawik, J., Jiraratananon, R., and Fane, A.G., *J. Membr. Sci.* **215**, 75-85 (2003).
8. Youn, I.J., Jeong, J., Kim, T., and Lee, J.C., *J. Membr. Sci.* **145**, 265-269 (1998).
9. Albrecht, T.R., and Quate, C.F., *J. Vac. Sci. Technol. A* **6**, 271-275 (1988).
10. Binning, G., Quate, C.F., and Gerber, C., *Phys. Rev. Lett.* **56**, 930-933 (1986).
11. Khayet, M., Feng, C.Y., and Matsuura, T., *J. Membr. Sci.* **213**, 159-180 (2003).
12. Khayet, M., *Chem. Eng. Sci.* **58**, 3091-3104 (2003).
13. Khulbe, K.C., and Matsuura, T., *Polymer* **41**, 1917-1935 (2000).
14. Kesting, R.E., *J. Appl., Polym. Sci.* **41**, 2739-2752 (1990).
15. Khayet, M., and Matsuura, T., *Desalination* **158**, 57-64 (2003).
16. Fernández-Pineda, C., Izquierdo-Gil, M.A., and García-Payo, M.C., *J. Membr. Sci.* **198**, 33-49 (2002).

Applied Physics (APHYS 2003)
A. Méndez-Vilas (Ed.)

Software For Practical Training In Medical Biophysics

Jiri Zahora, Josef Hanus

Charles University in Prague, Medical Faculty in Hradec Kralove, Department of Medical Biophysics, Simkova 870, P.O. Box 38, 500 38 Hradec Kralove, Czech Republic

Abstract. The course of medical biophysics at our medical faculty consists of three parts. The first one and the most extended one is medical biophysics, the second one is introduction to medical data processing and the last one is biostatistics. Mainly the conventional forms of teaching are used – lectures, seminars and practical training. At present we have six practical trainings: Audiometry, Principle of computed tomography, Noninvasive measurement of electrical heart activity and blood pressure, Magnitude measurement of microscopic objects, Mechanical properties of Nitinol and Diagnostic use of ultrasound. To increase efficiency of the teaching process we have decided to combine and to apply all three parts of our course in every practical training. The base of all practical trainings is some medical or biophysical problem. Students must learn how to work with measuring device, they must collect data and they must process them and prepare final report. So we have decided to make the best account of personal computers, of software for data acquisition and statistical data processing and of local area network (LAN). For some tasks we use commercial software, when it was necessary we have developed our own one. For all practical trainings we use personal computers connected to LAN. Computers are provided for measuring with either AC/DC converters and sensors (Mechanical properties of Nitinol, Noninvasive measurement of electrical heart activity) or they communicate with measuring unit (audiometer, digital camera, ultrasonograph, spectrometer). We use some commercial software as DP Soft by Olympus for image processing and remote controlling of digital camera, MS Excel for statistical data processing or MS Word for creating final reports. Software for remote controlling of audiometer and database of patients was customized to be consistent with our special demands. For measuring of mechanical properties of Nitinol we have developed software for data acquisition by means of LabVIEW (National Instruments). We have developed also a number of teaching programs for biophysics (Construction of electrical heart axes, Propagation and reflection of ultrasound waves through interfaces between tissues, Blood circulation, Model of multithermocouple probe in the temperature gradient, Principle of osmosis, ...) and for biostatistics (Theoretical models of random quantities, STATPROMED – interactive hypertext textbook of biostatistics). For developing of this software we use Authorware (Macromedia). Resulting practical tasks are very complex. Besides of investigating and cognition of basic biophysical phenomena and principles the students have possibility to learn and to train new modern ways of data acquisition and data processing. They also obtain experiences with various types of software and their effective exploitation.

INTRODUCTION

Various types of software are at present available and used in probably all parts of medicine. Physicians may need to work with large scale of software. For example they use hospital information systems or picture archives or with various databases, information search systems and bibliographical systems. Both may be local or on Internet. They must work with software for controlling medical devices, with e-mail clients, etc. The communication with insurance companies is usually in electronic way too. That is the reason why students of medical faculties must collect experiences in this area from the first days of their studies. Main educational aims of our practical course are of course focused on biophysics and biostatistics, but the other important goal is to obtain practical experiences in effective use of various types of software.

SOFTWARE FOR PRACTICAL TRAINING IN MEDICAL BIOPHYSICS

About software, we are using commercial software and where it is necessary we are developing our own one. All software can be divided into three groups from the point of view of the way, how it is used. The first group is software used during the assignments by students, the second one used for preparation the assignments by teachers. The last one can be entitled auxiliary. As to the first group, it is software for controlling measurement, for data acquisition, for data processing, for creating final reports and various teaching programs. The second group contains software for creating various study texts for students (PowerPoint), authoring systems for development of courseware (Macromedia Authorware) and finally a development environment to design software for controlling measurement units (LabVIEW). Under the last group comes operating system, e-mail clients, databases ...

INTEGRATED SET OF PRACTICAL ASSIGNMENTS IN MEDICAL BIOPHYSICS

Few years ago we have decided to develop an integrated system of practical assignments in medical biophysics [1., 2.]. The basic idea of this integrated system is to combine in every assignment biophysical problems and statistical problems with computer-aided data acquisition and data processing. For every assignment we have one computer which provides controlling of measurement, measurement, data collecting and storage, data processing and creating final report. All computers are connected to local area network, which enables for example sharing of student texts or files with data for farther elaboration. There are also at student's disposal descriptions of assignments and sample final reports on the intranet. At present we have six practical assignments.

Audiometry

Description

Audiometry is qualitative and quantitative hearing examination method using special equipment called audiometer. The basic audiometric examination is pure tone thresholds audiography for air conduction and bone conduction. In clinical practice it is called **"hearing thresholds audiogram"**. Students must record the relative audiogram (Fig 1) in frequency range 125 Hz – 8000 Hz, in steps by octave (125 Hz, 250 Hz, 500 Hz, 1 kHz, 2 kHz, 4 kHz, 8 kHz).

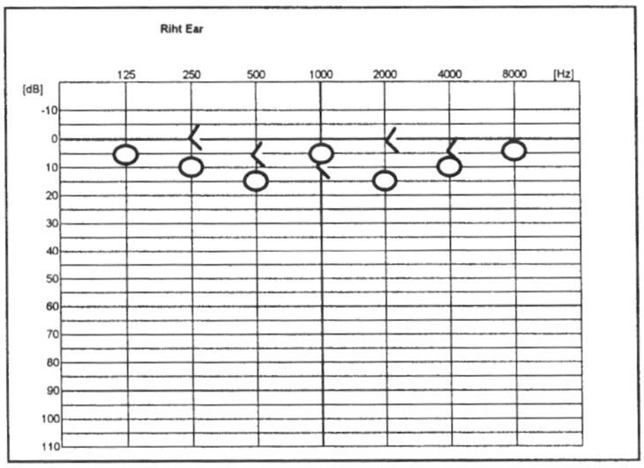

FIGURE 1. Relative audiogram for the right ear, bone conduction and air conduction.

We are using audiometer controlled by computer, so the students obtain experiences in this clinical test, they will learn how to work with such equipment and with patient database, the learn how to evaluate results.

Software

The software, which is used in this practical assignment, is partly commercial but it is adjusted for usage in teaching process. It has two parts – one is controlling the audiometer, the second one is the patient database and picture (audiogram) database. The MS Excel and Word is used to create final report.

Principle Of Computed Tomography

Description

In this practical assignment students measure absorption coefficients of a model of tissue layer. They use a source of gamma rays and spectrometer. Another task is theoretical. They receive a virtual model of layer, which consists of nine volume elements, it means, that they work with matrix 3x3 and must solve system of nine linear equations. The result is 2D reconstruction of structure, which is simulated.

Software

The spectrometer is equipped with simple software for transfer of data into computer by means of port RS 232. MS Excel is used for further data processing. We have developed a special computational program, which is used for preparation of input data for the theoretical task.

Noninvasive Measurement Of Electrical Heart Activity And Blood Pressure

Description

Here the ECG is recorded and evaluated and blood pressure measured both in the rest and after load. The digital electrocardiograph controlled by computer is used. Record of I., II. and

III. bipolar lead is used. Students should determine from the record the heart rate and the mean electrical axis in the rest and after the load and elaborate results by using statistics. These values are compared with the physiological values.

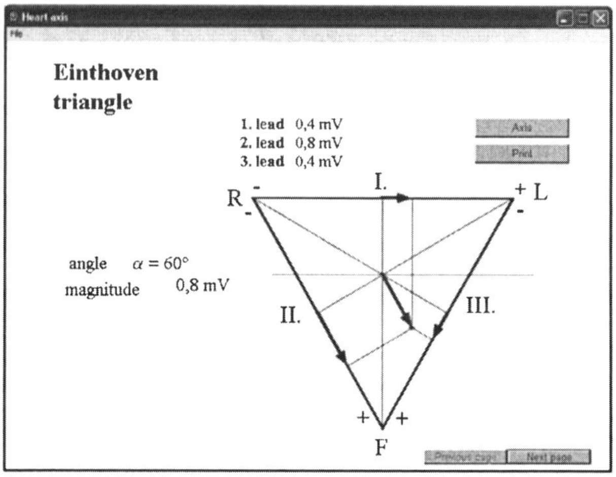

FIGURE 2. Construction of an electrical heart axis.

Software

Students must use software for controlling process of recording and for evaluating records and with patient database. As to the construction of an electrical heart axis they can use teaching program, which can learn them, how to do it and it can be used also for processing real data. This program was developed by means of Macromedia Authorware.

Magnitude Measurement Of Microscopic Objects

Description

The goal of this assignment is to measure diameters of erythrocytes in the sample and to decide, if it corresponds with the physiological value. The t-test is used. The measuring device is the combination of microscope and digital camera.

Software

The commercial software DP Soft is used for controlling the camera, taking the picture and its further processing. The students will obtain experiences with software for controlling digital camera and image analyses. They learn how to calibrate the software, how to measure linear dimensions, how to define objects and how to find their number in certain area. Collected data (diameters of erythrocytes) are statistically evaluated by means of MS Excel.

Mechanical Properties Of Nitinol

Description

In this assignment the students have the possibility to study mechanical properties of Nitinol and their change after its phase change from the austenitic phase to the martensitic phase. This phenomenon is demonstrated using a spring made of Nitinol wire. The spring constant is measured so that the length of the spring and the corresponding force is recorded. The spring constant is measured twice, once at low temperature (room temperature), second time at high temperature (about 50°C) and its change is investigated. The measuring device consists of system of sensors, AD/DA converter and controlling software.

Software

Controlling software was developed by means of LabVIEW. It was designed as the virtual device. Users can use only the control panel, which is similar to the real one. This program has basic functions as possibility to set period of reading data, number of data readings, export of data to the text file or as the Excel sheet. The force, the length of the spring and its temperature is measured concurrently. The Excel is used for further data processing.

Diagnostic Use Of Ultrasound

Description

The students must investigate simple model of blood vessel system. They must measure by means of sonograph with linear probe geometrical parameters (diameters of vessels, their position) and velocity of blood flow. Blood is simulated by water. Then they must create 3D reconstruction of the model and to verify the equation of continuity.

Software

We are using software, which is a part of sonograph. It enables to control the sonograph and to analyze the image. It is possible to do various calculations, to create reports. The patient data management and image archiving is available too.

Another Software And Electronic Resources

We have also developed a set of teaching programs, which are available for students. For biophysics we have programs Construction of electrical heart axes, Propagation and reflection of ultrasound waves through interfaces between tissues, Blood circulation, Model of multithermocouple probe in the temperature gradient, Principle of osmosis, Imaging by means of converging lens and for biostatistics Theoretical models of random quantities, interactive hypertext textbook of biostatistics STATPROMED. All lectures are available as PowerPoint presentations too.

CONCLUSION

Each student has during the practical assignment possibility to work with the measuring device, he/she must use corresponding software. All computers in our laboratory are connected to the local area network and all students have their personal folders, which are available from computers in the computer laboratory. At present there is about fifty computers. The study texts, instructions and sample reports are accessible from laboratory and from computer laboratory at any time. The same software for creating final reports is installed in both laboratories. Finally, the reports are to be sent to the teachers by e-mail.

DISCUSSION

We have good experiences with this integrated set of computer-aided practical assignments. Students can work with modern software, which they will probably meet in their future clinical practice (or similar type of software). Now is this set of practical assignments based on local area network, but the technologies based on Internet services are every year more important. So in the near future we want to include them into our practical course too. We suppose that the electronic materials will be available on web site. We also are developing the administrative web system for checking attendance, testing, collecting final reports.

ACKNOWLEDGMENT

The work was supported by state-aided research project of CR No. 111500004 and by grant agency FRVS No. 2964/H/03.

REFERENCES

1. Tompkins, W. J., "Using electronic technologies to teach biomedical engineering", *Biomed. Eng.* **7**, No. 5, 509-514 (1995).
2. Zahora, J. and Hanus, J., "Courseware for medical biophysics" in 2nd European Medical and Biological Engineering Conference EMBEC'02, edited by H. Hutten et al., IMBE Proceedings, December 04-08, 2002 Vienna, Austria, pp. 704-705, ISBN 3-901351-62-0, ISSN 1680-0737.

Applied Physics (APHYS 2003)
A. Méndez-Vilas (Ed.)

REACTIVE ELEMENT EFFECT STUDIED BY LASER ABLATION

R. GUERRERO-PENALVA M. G. MORENO-ARMENTA, M.H. FARIAS AND L.COTA ARAIZA

Instituto Tecnológico de Tijuana, Centro de Graduados e Investigación, Apartado Postal 1166, Tijuana, BC 22000, México. penalva@telnor.net
Centro de Ciencias de la Materia Condensada de la UNAM, Apartado Postal 2681, Ensenada, BC 22800, México. moreno@ccmc.unam.mx, Fax +++52-6-1744603

Abstract. A significant improvement in corrosion resistance of the protecting oxide of alloys has been observed when adding small amounts of reactive elements, such as yttrium, this effect has been called reactive element effect (REE). The general mechanism of the REE has not been determined yet. In this work, we study the growth of an yttrium oxide film and its interaction with the phases η and α that constitutes the alloy Zn-22Al-2Cu named Zinalco™. The alloy's surface was coated by a pulsed laser deposition technique. The deposit is controlled and characterized by x-ray photoelectron spectroscopy. The mechanism by which the reactive element produce its effects in this alloy is explained by the preferential interaction among the active sites related to the zinc rich phase and enhancing aluminum movement toward the surface where it is oxidized and the protection film formed.

INTRODUCTION

According to some recent review articles [1], adding small amounts of reactive elements such as yttrium oxide caused an improvement in oxidation and exfoliation resistance of some alloys. This effect was named reactive element effect (REE). The studies have been performed on alloys under a broad commercial use with the common characteristic of aluminum or chromium as alloying metal. These last two metals are the best in forming protective oxide films. Little or nothing has been published about the REE produced improvements on alloys based on magnesium or zinc. Several particular mechanisms, which can be applied to certain specific cases, have been proposed. However, the general mechanism of the REE has not been determined yet. The REE can be obtained by incorporating either metallic yttrium or a dispersion of oxide particles. The beneficial effect could be obtained by means of surface coatings with the reactive element. Several surface deposition techniques have been used to obtain the REE, such as: ion implantation, fusion of the reactive element with substrate by laser cladding, immersion of reactive element in nitrate aqueous solutions, sol-gel method by immersion or electrophoresis, reactive sputtering and electrochemical deposition from organic nitrate solutions. To our best knowledge, REE studies by means of pulsed laser deposition (PLD) have been investigated only in our laboratory [2]. This technique, although relatively expensive for commercial purposes, presents advantages for studying the REE. The amount of deposited material can be easily controlled, and the films can be characterized in situ.

EXPERIMENTAL PROCEDURE

Samples were prepared slicing a zinalco bar, polished until mirror-like surface, acetone degreased in ultrasonic bath and introduced in in a Riber LDM-32 ultra high vacuum (UHV)system equipped for growing thin films by PLD and characterization in situ by x-ray photoelectron spectroscopy (XPS) with Cameca MAC-3 analyzer and magnesium Kα radiation as x-ray source. Under UHV, sample surface preparation was completed by bombarding it with 5 to 10 eV argon plasma, generated by microwave electron cyclotron resonance (ECR) using an ASTeX model AX 2000 system. We call the samples prepared in this way "standard samples".

PLD was utilized as depositing technique with a generator of KrF (λ=248 nm) excimer, model Lextra 2000 from Lambda Physik. Delivered energy per pulse is about 5 joules/cm^2 with a 1 Hz frequency. Target to substrate distance was set to 6 cm. An Y_2O_3 sintered wafer was utilized as target in order to generate the depositing material. Substrate was a zinalco standard sample at a temperature of 100° C, rotating at 120 rpm during deposition. On standard prepared surfaces, a Y_2O_3 film was grown in situ immediately after bombarding the sample with argon ions and pumping out the vacuum chamber down to a base pressure of $5x10^{-9}$ Torr. The surface coating consisted of the deposition of Y_2O_3 generated by groups of 5 laser pulses, followed by in situ XPS characterization, up to a total of 80 pulses. After 80 pulses, the XPS substrate signal completely disappeared and only the yttrium oxide signals was detected.

Film thickness, *d*, was calculated using a widely accepted method by Hill et al. [3], where they consider the electron emission angle with respect to the surface normal, θ (20° as defined by our equipment geometry), the signal intensity ratio between substrate with film and substrate of infinite thickness, I_s/I_s^{∞}, and the intensity ratio between film and film of infinite thickness, I_p/I_p^{∞}. Instrumental factors, common to substrate and adsorbate get canceled. Since the kinetic energy of the measured signals are very close: E_k (Y 3d) = 1129 eV, E_k (Al 2s) = 1168 eV and E_k (Zn 3p) = 1196 eV, the mean-free-paths are similar and the method can be used in order to evaluate the film thickness.

Control samples and coated-samples by were thermally treated at 120° C under atmospheric conditions during three 8 h periods. After each heating period the samples were cooled down to room temperature and the surface composition was monitored by means of XPS.

RESULTS AND DISCUSSION

The Zn-22Al-2Cu alloy has been studied and it is known to be composed mainly by two different solid phases: η and α. The η phase is a solid solution composed mainly by zinc with dissolved aluminum, which keeps the hexagonal compact structure of zinc; while the α phase is rich in aluminum with dissolved zinc and keeps the face centered cubic structure of aluminum. The two phases are arranged in a lamellar microstructure with alternate regions of phases η and α, covered by an oxides layer, mainly Al_2O_3 and some spots over the η phase where the concentration of zinc oxide predominates. In those regions where there is enough aluminum to form a continuous oxide film, the surface is well protected and the corrosion is drastically reduced. However, over some parts of the η regions, rich in zinc, the surface contains zinc and aluminum oxides and the barrier against corrosion is not so efficient. The corrosion progresses in these areas, producing growths that spread out above the aluminum oxide layer until all the surface is eventually covered by these zinc oxide rich corrosion products [4]. The bombardment of the surface with low energy ECR argon plasma is strong enough to remove the less well-adhered zinc oxides and hydroxides that spread over the aluminum oxide layer, leaving an aluminum oxide enriched surface with some spots still showing zinc oxide products. This resulting surface is used as a clean standard surface, ready for carrying out in situ PLD deposits.

The XPS spectra obtained during the deposition of the Y_2O_3 film over a standard zinalco substrate let us keep track of the intensity of the signals from the substrate and the over-layer as a function of the number of laser pulses sent to the target. As expected, the yttrium signal increases with the number of laser pulses, while the aluminum and zinc signals gradually diminish, since they become covered by yttrium oxide, although not to the same relative speed. We made a correlation of the increment in Y_2O_3 thickness as a function of the laser pulses number and the type of substrate, and found out that the coating growths 1.72 times faster over zinc than over aluminum. This suggests that the adsorption and film growth over each phase is not the same.

Figure 1 shows normalized relative atomic intensities (I) and exponential fittings of the deposited film and substrate from XPS results, as a function of film thickness. In order to observe the growth rate of yttrium oxide over zinc or aluminum we compare the I_{Zn} and I_Y divided by $(I_{Zn} + I_Y)$, and I_{Al} and I_Y divided by $(I_{Al} + I_Y)$ respectively. It can be observed the exponential crossover between the Zn and Y related signals in Fig. 1(a) to occur at a film thickness of about 9.5 Å, while the corresponding crossover between the Al and Y signals in Fig. 1(b) happens at around 18 Å. In the case of the zinc rich phase the crossing point of the normalized intensities at 9.5 Å tends to the point at which the interactions between the substrate atoms and atoms of the film are stronger than the interaction between the neighboring film atoms [5]. In the case of the aluminum the crossing point at around the 18 Å, indicating that the interaction with the substrate is weaker than the interaction between deposited atoms.

To realize this fact, we need to take into account the substrate and the deposit material. We already know that in the substrate there are two types of well defined phases available, and although the surface was cleaned by bombardment with an ECR argon plasma, the surface that interact with the laser plasma still was covered by zinc and aluminum oxides and hydroxides. The incoming plasma composition, generated by the pulsed laser radiation hitting the Y_2O_3 target and expand in the vacuum, was already

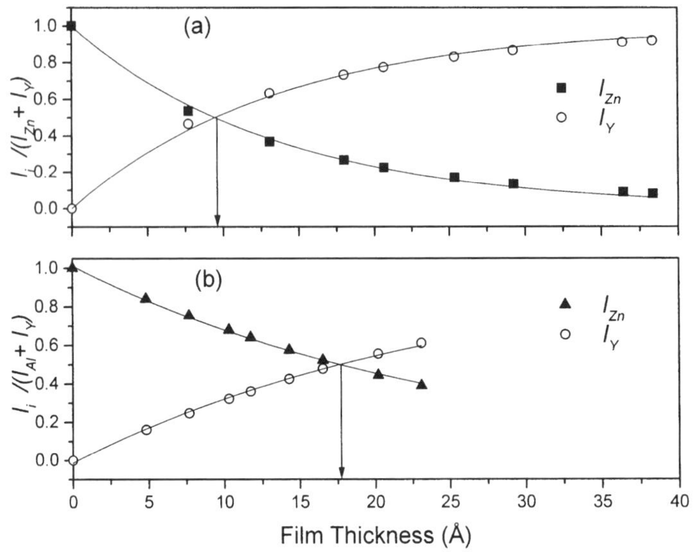

FIGURE 1. Exponential fit from the normalized relative atomic intensities from the substrate and over-layer as a function of the Y_2O_3 film thickness: a) over zinc, b) over aluminum.

analyzed [6] through optic emission spectra, finding out that the main radiative species were metallic ions of Y^+ and diatomic ions of O_2^-. The kinetic energy distribution in the plasma [7] falls in the 10 to 100 eV ranges. Knowing that the substrate rotates during deposition, we can assume a homogeneous distribution of incoming material. The Y^+ ions arrives to the surface with enough kinetic energy to break all kinds of bonds and diffuse along the surface in search of active sites, in order to minimize their energy. The whole oxide layer can be broken although not with the same probability. The easiest bonds to brake are that between Zn-O because they have a dissociation energy D^0 = -1.64 eV [8] which is weaker than those of bonds between Al-O D^0 = -5.27 eV.

The laser pulse that hit the target produces plasma that expands in the direction normal to the target plane. It creates a material flow traveling toward the substrate. If the atoms, ions or molecules that arrive onto the substrate (zinalco surface) detach some material without interacting with it, this material would tend to redeposit because the dragging effect of the incoming plasma. As a result of this process the surface would tend to increase the concentration of the easiest detached material. The experimental evidence shows contrarily that the zinc intensity diminishes faster than the aluminum; it means that although the zinc is detached easily the interaction of yttrium ions or yttrium oxide with the zinc oxides is more probable. Furthermore the formation of bonds Y-Al D^0=+0.21 eV is highly unfavorable (Al-Al D^0=-1.37eV), whereas the bonds Y-Zn, D^0=-0.60 eV (Zn-Zn D^0=-0.30eV), are thermodynamically favored. It can cause that some yttrium would be implanted in the zinc metallic matrix. The kinetic energy of yttrium ions is higher than the dissociation energy of the probable species involved in the experiment and the energy dissipation is very fast, in the experimental conditions so we have a system out of equilibrium that freeze rapidly giving as a result species not necessarily in its minimal energy configuration and the bonds between yttrium and zinc can be possible. According to the experimental results observed, there is a greater reduction in the intensity of the zinc signals, meaning a better adhesion of the yttrium oxide to the zinc substrate, this could be explained based on the major probability for cracking the Zn-O and Zn-Zn bonds which abundance is higher in the η phase.

In the thermal treatment of the coated samples the XPS spectra show that after the first eight hours of annealing, any change can be detected and the surface remains composed of yttrium oxide. After 16 hours, the film surface was formed only by aluminum and yttrium oxides, with no zinc present yet and after a total of 24 hours the zinc appears but in very low concentration, whereas the control sample do not show appreciable increment in the aluminum concentration. This yttrium oxide preference for active sites related to the zinc, could function as an anchor for the zinc atoms limiting their mobility and leaving the aluminum atoms readily available to migrate towards the surface and react with the oxygen to form a protective aluminum oxide film.

CONCLUSIONS

The PLD and XPS techniques could be used advantageously in the reactive element effect studies, because with these method is possible to control small amounts of deposited material and the thickness can be calculated accurately, in order to detect the preferentially adsorbing sites in the different phases of a binary alloy.

The mechanism by which the reactive element produces its results is by interaction preferentially to the active sites related to the zinc rich phase impeding its movements and allowing the free aluminum movement toward the surface where it is oxidized and the protection film formed.

We are working now in first principles calculations about the interaction of yttrium with the existent phases in the alloy to confirm our results and get more insight in the REE.

ACKNOWLEDGMENTS

We would like to thank Dr. G. Torres-Villaseñor for providing the zinalco bar used in this work and to Ing. Israel Gradilla and M.C. Eloisa Aparicio for valuable technical assistance. This work has been partially supported by a COSNET-México research project No. 737.99-P.

REFERENCES

1. B. Pint, *Oxid. Metal.*, 45 (1-2) (1996) 1.

2. R. Guerrero, M.H. Farias and L. Cota, *App. Surf. Sci.*, 185 (3-4) (2002) 248.
3. Hill JM et al., *Chem. Phys. Lett.* 44 (1976) 225.
4. R. Guerrero, M.H. Farias, L. Cota,. *App. Surf. Sci.*, 195 (1-4) (2002) 137.
5. H. Luth, *Surfaces and Interfaces of solid Materials*, Third edition, Springer-Verlag. (1995).
6. V. Tsaneva, et al., *Vacuum*, 48 (10) (1997) 803.
7. K. R. Singh and J. Narayan, *Phys. Rev. B*, 41 (13) (1990) 8843.
8. *CRC Handbook of Chemistry and Physics*. 83RD Edition, CRC Press, LLC, 2002.

Applied Physics (APHYS 2003)
A. Méndez-Vilas (Ed.)

Analysis of Sea-Land Breeze Around the City of Huelva (Spain)

J. A. Adame Carnero*¶, J.P. Bolívar Raya* and B.A. De la Morena¶

**Atmospheric Sounding Station. El Arenosillo - INTA, Crta. Huelva – Matalascañas km 33. 21130. Mazagón. Huelva. SPAIN. adamecj@inta.es*

¶Department of Applied Physics, University of Huelva. Campus de El Carmen, s/n. 21071.Huelva. SPAIN.

Abstract. In the surroundings of the city of Huelva there are three important industrial complexes, which emit several pollutants. The sea-land breeze influence in the atmospheric processes of the dispersion and it plays an important role in phenomena such as the recirculation of pollutants. In order to know the phenomenon of the sea-land breeze in this region we have previously studied the wind system with the data registered during 1999 to 2002 in Punta del Sebo (situated in an industrial centre, 4 km from the city of Huelva). We have calculated different monthly parameters to characterise the breeze, i.e. the frequency, medium speed, medium direction and hour when the marine breeze starts.

INTRODUCTION

The land-sea breeze circulation is one the most interesting mesoscale atmospheric phenomena observed in coastal regions. The land-sea breeze is in close connection with air pollution at coastal areas. The sea breeze grows during the day in the warm season and the features that the sea-land breeze present dominates the behaviour of photochemical pollutants [3,4]. The Huelva city, situated at SW of Spain (see figure 1) is one coastal region, characterised by high solar radiation level, which cause that the mesoscale atmospheric flows are frequents. In the environs of the Huelva city, three important industrial complexes are located, which emit primary pollutants to the atmosphere (oxides of nitrogen and sulphur, volatile organic compounds and particles).

These substances are put under processes of dispersion and chemistry transformation in the atmosphere, which originated new secondary pollutants like ozone. In order to interpret the behaviour of these pollutants it is necessary to know the atmospheric flows in the region, and in a coastal zone like this, the breeze regime.

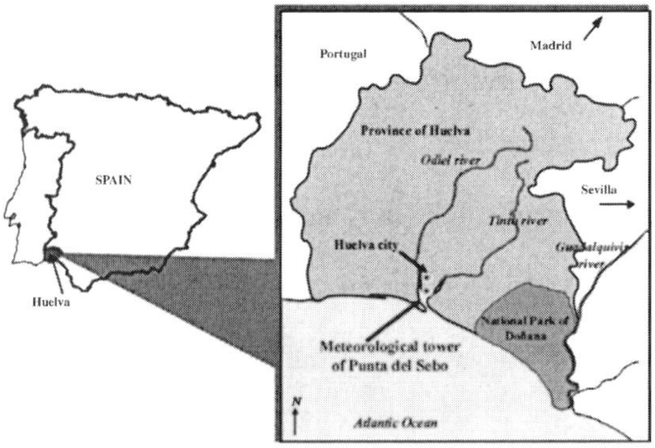

FIGURE 1. Location of Huelva city and meteorological tower of Punta del Sebo.

STUDIED AREA, PERIOD, DATA AND METHODOLOGY TO ANALYSE THE SEA-LAND BREEZE

The wind data was collected in the meteorological tower of Punta del Sebo, observatory situated at coastline of Huelva (see figure 1), very near of Huelva city. We have studied the wind data during the period 1999-2002. These data have been taken with sensors to height 56 m over the ground and with a temporary resolution of 15 minutes. It's no easy to extract the days of breeze of wind data historical, perhaps the most complicated it is the establishment of a criteria to make this selection. Some authors use criteria like the change in wind direction, weak gradient of pressure or positive gradient of temperature in the beginning of the sea-breeze [5]. At present, we have no data in our region to apply these criteria, then we have used the wind normal component to the coast v_b, to select the days which the wind present sea-land breeze typical variations [8]. This parameter is formulated according to equation 1 and your geometric meaning is show in figure 2.

$$v_b = v\cos(f - f') \qquad (1)$$

where v is wind speed, f is the perpendicular angle to the coast and f' is direction wind.

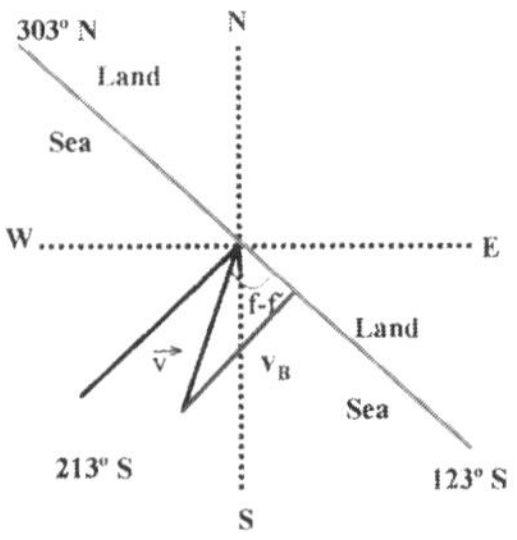

FIGURE 2. Parameter v_b and Coastline of Huelva.

To calculate the v_b is necessary to know the perpendicular direction to the coast; in the case of the Huelva the coastline is 303°N-123°S and the perpendicular direction is 213° S (see figure 2). We have used next criteria to select the days with sea-land breeze.

a) $v_b > 0$ between 15 and 17 hour (UTC), which indicates that we have sea breeze during this diurnal period.
b) $v_b < 0$ between 2 and 5 hour (UTC), which indicates that we have land breeze during this nocturnal period.
c) $v > 3$ m/s between 15 and 17 hour (UTC), this criteria show that the sea breeze is well stablished.

RESULTS

Analysis of Wind Data

First step to know the days of breeze in a region consists to analyse the wind data historical. We have calculated the wind directions roses for each season of the year. The directions predominant in the months of winter and autumn have mainly north component (NE and ENE), and with smaller frequency wind directions of the third quadrant, see figure 3.a.

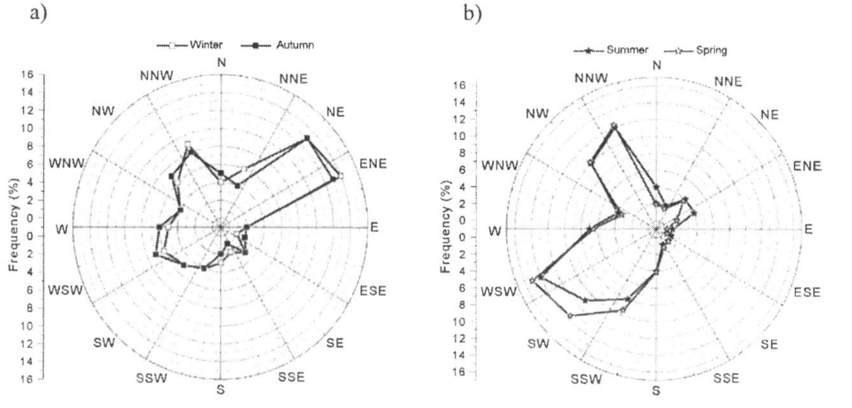

FIGURE 3. a) Rose of wind directions to the winter (January, February and March) and autumn (October, November and December). Calm: winter 12% and autumn 14%. b) Rose of wind directions to the spring (April, may and June) and summer (July, august and September). Calm: spring 10% and summer 11%.

This is a synoptic-scale wind originated by the air masses and fronts that enter from the north of Iberian Peninsula in this time of the year. The spring and summer are characterised by wind directions of WSW and NNW, see figure 3.b. The wind in these months has a double origin, the fronts coming from Atlantic ocean and the mainly is in diurnal (wind direction NW-NNW) and nocturnal (wind direction SW-WSW) flows of the breeze [1].

Analysis of Sea-Land Breeze

After to apply the mentioned criteria to wind data, we has been obtained the sea-land breeze days. The aim to characterise the breeze we have calculated next six parameters:

f_B: monthly frequency of sea-land breeze days.

v_{iBD}: wind speed medium when the sea breeze starts.

d_{iBD}: wind direction medium when the sea breeze starts.

H_{iBD}: hour when the sea breeze front moves inland.

H_{fBD}: hour when the sea breeze front finish to moves inland.

L_{BD}: coefficient for estimation the distance of sea breeze front moves inland [1], but it is not a real measurement of this distance, only a approximation. To calculate this coefficient we has used the equation 2.

$$L_{BD} = \left(\sum \sqrt{u^2 + v^2}\right)\Delta t \tag{2}$$

where u and v are the components north-south and east-west of the wind vector, and Δt is the time averaging interval of the data, in our case are 15 minutes. The value obtained for these parameters appear in next table.

TABLE. Parameters to Characterize the Sea-Land Breeze Days.

Month	f_B	v_{iBD} (m/s)	d_{iBD} (°)	H_{iBD} (UTC)	H_{fBD} (UTC)	L_{BD} (km)
January	8.1	2.2	241	11.40	17.20	70
February	26.8	3.1	221	12.00	18.15	86
March	35.5	4	201	12.20	18.45	96
April	22.1	4.9	205	11.30	19.30	129
May	38.2	4.9	208	9.40	20.15	137
June	37.9	5	213	8.40	20.20	200
July	50.5	5	202	9.40	20.15	167
August	50.0	4.5	204	9.25	20.10	155
September	36.7	4.7	210	11.00	19.30	120
October	23.4	3.9	216	12.00	18.50	103
November	12.9	3.4	211	11.20	18.20	103
December	5.6	4	208	13.00	17.50	84

The sea-land breeze occurs the more frequently in summer months with a value of 50% in august, very similar to the obtained in other points of the Spanish Mediterranean coast [6,7].

The sea breeze starts with medium wind speed between 2 and 5 m/s, wind speed greater in summer when the insolation is stronger. The sea breeze starts with wind direction almost perpendicular to the coastline, between 202° and 240°.

The table 1 shows the hour at which sea breeze begins and finish, these parameters present a clear seasonally. During the summer the sea breeze starts two or three hours after sunrise in Huelva and it ceases at 20h. In these months sea breezes reaches its maximum duration, between 10-11 hour. In addition, they also present greater distances of penetration inland. June and July shows the maximum values with distances 50-60% greater than winter-autumn months.

CONCLUSIONS

The sea-land breeze in Punta del Sebo (Huelva) is a very frequent wind in summer months and it is more difficult to observe in winter-autumn months because the synoptic-scale wind is superimposed. However, during the 50% of the summer days there is a well developed sea-land breeze, these circulations begin two or three hours after sunrise in this months with wind speed of 4-5 m/s and wind directions almost perpendicular to he coast. The distance of penetration inland presents differences of 50-60% between summer and winter months.

ACKNOWLEDGMENTS

We thank to the Consejería de Medio Ambiente of Junta de Andalucía for wind data.

REFERENCES

1. Adame, J.A., Bolívar J.P., De la Morena, B., and Cachorro, V. "La dinámica de la baja atmósfera en el litoral de la provincia de Huelva: resultados preliminares", in *VIII Congreso de Ingeniería Ambiental*, Bilbao, 2003.
2. Allwine, K. and Whiteman, C. "Single-Station integral measures of atmospheric stagnation, recirculation and ventilation". *Atmospheric Environment* **28,** 713-721 (1994).
3. Gangoiti, G., Millán, M., Salvador, R. and Mantilla, E. "Longe-range transport and recirculation of pollutants in the western Mediterranean during the Project Regional Cycles of air pollution in the west-central Mediterranean area". *Atmospheric Environment* **35**, 6267-6276 (2001).
4. Heping, L. and Chan, C.L. "An investigation of air-pollutant patterns under sea-land breezes during a severe air-pollution episode in Hong-Kong". *Atmospheric Environment* **36**, 591-601 (2002).
5. Hiroshi, Y. "Statistical analyses of the sea breeze pattern in relation to general weather conditions". *Journal of the Meteorological Society of Japan* **59**, n° 1, 98–107 (1981).
6. Martín, F. and Palomino, I. "Análisis de las brisas en la costa Atlantico-Andaluza y su penetración en el valle del Gualquivir" in *XXV Bienal de la Real Sociedad Española de Física.* 1995. pp 71-72.
7. Redaño, A., Cruz, J., and Lorente J. "Main features of sea breeze in Barcelona". *Meteorology Atmospheric Physics* **46,** 175–179 (1991).
8. Salvador, R. "Análisis y modelización de los procesos atmosféricos durante condiciones de brisa en la costa Mediterránea Occidental: Zona de Castellón". *PhD. Thesis.* Universidad Politecnica de Cataluña (1999).

Applied Physics (APHYS 2003)
A. Méndez-Vilas (Ed.)

Analysis of the Electromagnetic Behaviour of a Variable-Waveform-Supplied Iron Core Inductor, Modelled with Finite Elements

Consuelo Gragera Peña(1), M. Isabel Milanés Montero(2) and Enrique Romero Cadaval (3)

Department of Electronic and Electromechanical Engineering
University of Extremadura, Avda. de Elvas s/n, 06071 Badajoz, SPAIN.
(1)cgragera@unex.es, (2)milanes@unex.es, (3)eromero@unex.es

Abstract. This paper presents a model which allows simulating approximately the behaviour of a ferromagnetic core inductor. The permeability (μ) of ferromagnetic materials depends on the magnetic field intensity applied to them (H) and their history. This phenomenon causes a non-linear behaviour on the core, which has been analysed using the Finite Element Method (FEM). ANSYS© has been chosen as FEM software. The study has been particularized to the real case of an E-I core inductor made of ferromagnetic sheets. The model has been implemented in parametric form using the APDL (Ansys Parametric Design Language). The main application of the model is the analysis of the hysteresis curve variables evolution (B and H) with diverse amplitudes and waveforms in the current that flows through the inductor. Special attention will be paid to the study of the inductor behaviour when the current has a high harmonic content. On the other hand, the work has allowed the validation of the inductor design used in the experimentation.

1. INTRODUCTION

The analysis of a ferromagnetic core inductor under sinusoidal conditions is widely known. However, its behaviour when the current has a waveform different from the conventional in an electrical power system is not so evident.

It is very common that inductors belong to circuits with electronic converters, such as controlled or non-controlled rectifiers. In those situations the current that flows through the inductors will have a high harmonic content. The aim of this paper is to analyse the inductor behaviour in these cases.

As the resolution of non-linear equations is necessary for the analysis of the non-linear behaviour of the core, it has been used a Finite Element Method for solving that equations.

2. EXPERIMENTAL PROTOTYPE

The study has been particularized to the real case of an E-I core inductor made of ferromagnetic sheets (as Figure 1 shows), whose design magnitudes are perfectly known (geometric characteristics and nominal electromagnetic parameters).

The inductor has been built for comparing simulation with experimental results and it has been designed using an application in DELPHI 5.0 developed in the School of Industrial Engineering (shown on Figure 2). It has been winded an additional coil for measuring the flux or magnetic induction to permit the hysteresis curve visualization. The geometric and electromagnetic parameters of the prototype are on the right side of figures 1 and 2.

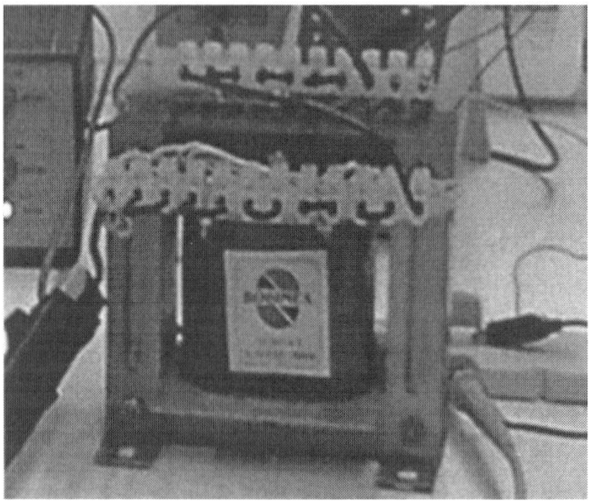

Nominal Parameters:

Imax = 4.5 A

I = 2.5 A

N (number of turns): 240

Lnominal : 300 mH

Internal resistance: 0.855 Ω

Copper section: 1.13mm2

FIGURE 1. E-I core inductor built for empirical experiments.

3. MODEL IN ANSYS

As it can be seen on Figure 2, the core is axisymmetric, so it is convenient to model only the middle of the complete geometry. The 2-D axisymmetric model, which is displayed in Figure 3, allows the plot of the magnetic flux density lines, the analysis of the electromagnetic magnitudes evolution (magnetomotrice force, magnetic induction, etc.) and the study of the magnetic permeability variation in the material, which produces a modification in the inductance value.

In the study it has been neglected the dispersion flux outside the inductor, but this variable has been taken into account because of the airgap between the E and I parts of the core. It has been chosen the Plane53 element for the analysis (as can be seen on Figure 3), because it allows the harmonic and transient analysis in Ansys, which is the objective of our study. It can also be observed in that figure the 4 materials into which the model has been divided.

The model has been implemented in parametric form using the APDL (Ansys Parametric Design Language) in order to permit its optimization later.

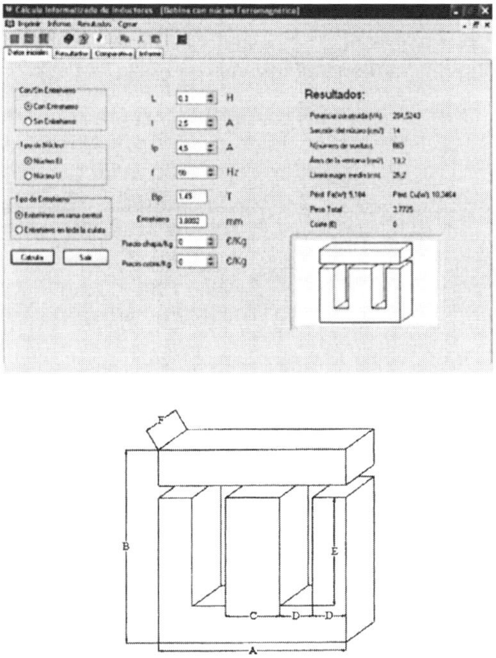

Geometric parameters:

Iron cross section: 14 cm^2

$C = 4.2$ cm

$D = 2.1$ cm

$E = 6.3$ cm

$B = E + 2\,D = 10.5$ cm

$F = 3.6$ cm

FIGURE 2. Application in Delphi used for the inductor design.

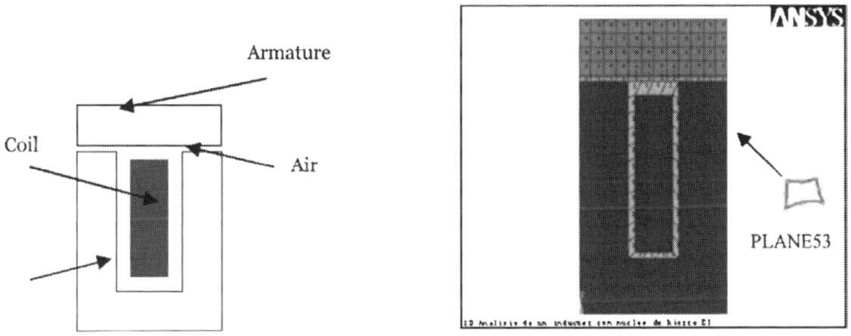

FIGURE 3. 2-D axisymmetric E-I core inductor Finite Element Model.

4. SIMULATION AND EXPERIMENTAL RESULTS UNDER SINUSOIDAL CONDITIONS.

For validating the model, simulation results have been compared with empirical results obtained in the laboratory with the inductor built for experiments.

In the study under sinusoidal conditions, the coil has been with excited with a sinusoidal voltage. Experimental results are shown on Figure 4 (a). Simulation results obtained with the Finite Element Model are displayed on Figure 4 (b), which contain the magnetic flux lines in two different instants (t = 0 and t = 15 ms = 3T/4) and Figure 4 (c), that shows the magnetic flux in the same instants.

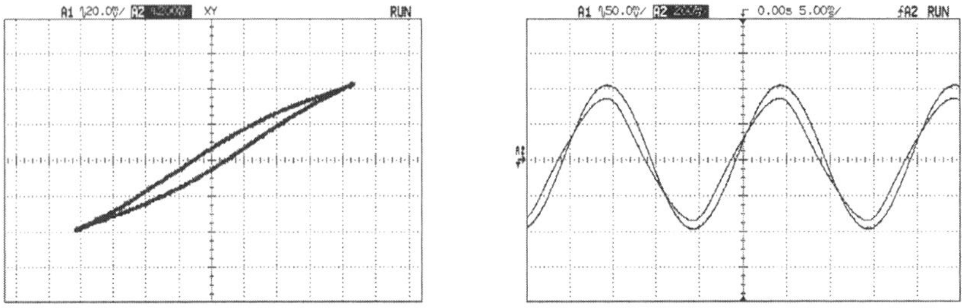

(a)

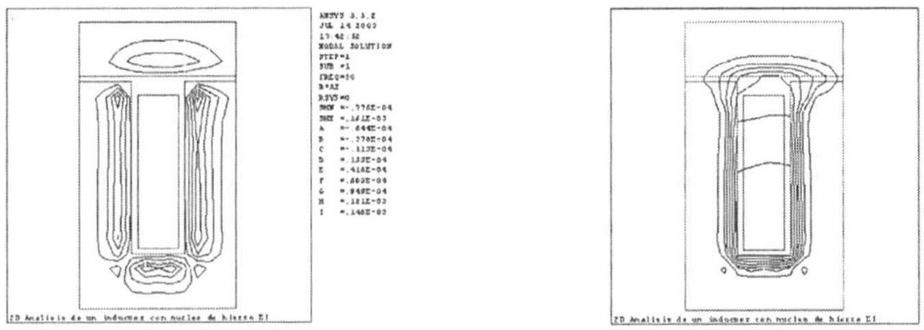

(b)

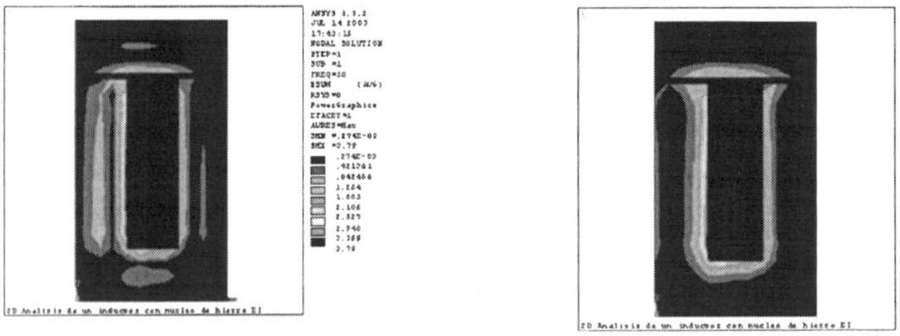

(c)

FIGURE 4. (a) Experimental results (left: hysteresis curve, right: magnetic induction and current); (b) Simulation results: magnetic flux lines (left: t = 0; right: t = 15 ms); (c) Simulation results: magnetic induction (left: t = 0; right: t = 15 ms).

5. SIMULATION AND EXPERIMENTAL RESULTS UNDER NON-SINUSOIDAL CONDITIONS

It has been also proved the model validity when the excitation is different from sinusoidal conditions. As an example, figure 5 shows the experimental and simulation results when the coil is excited with a square waveform voltage.

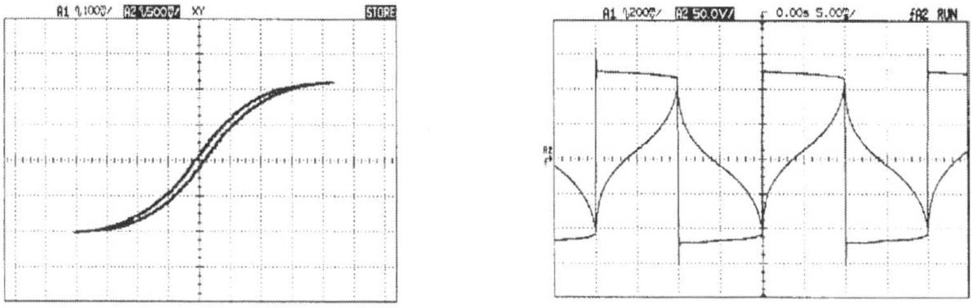

(a)

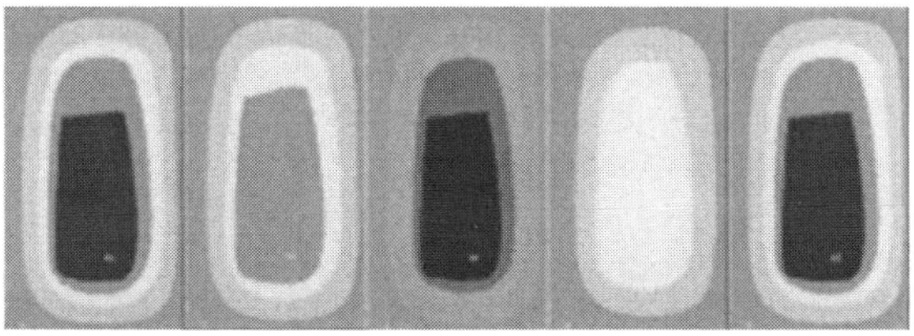

(b)

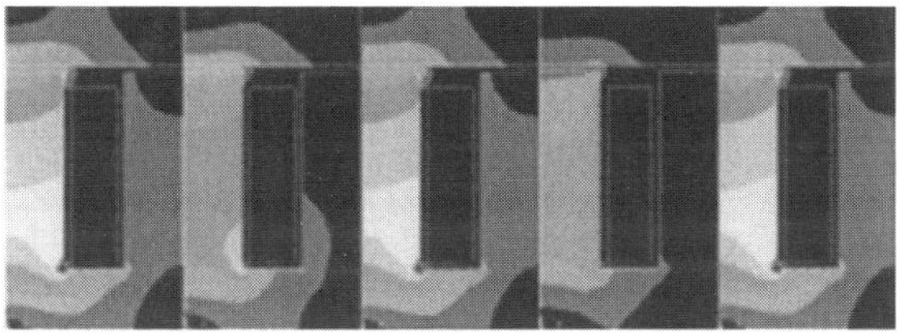

(c)

FIGURE 5. (a) Experimental results (left: hysteresis curve, right: voltage and current); (b) Simulation results: magnetic flux lines evolution in one period; (c) Simulation results: magnetic induction evolution in one period.

6. CONCLUSIONS

It has been developed a model for studying the behaviour of a ferromagnetic inductor. Simulation and experimental results for an E-I core inductor have been shown. The model has also allowed the validation of the inductor design testing that the electromagnetic magnitudes which were used as initial data for the coil calculus are closed to the real values obtained in the laboratory, and to the simulation results got with the Finite Element Model. The authors are working towards the improvement of the model for eliminating some simplifications made in this initial study.

REFERENCES

1. Barrero, F.; Milanés, M.I.; Romero, E.; Montero J.M., "Estudio del efecto de los armónicos y de alimentaciones conmutadas en inductores con núcleo ferromagnético", in *XIIIRGIIE Prooceedings*. Vigo. 2003.
2. R. Prieto, Cobos, J.A., "Modelado de componentes magnéticos mediante técnicas de elementos finitos" in *Métodos Numéricos en Ingeniería*, SEMNI, Sevilla. 1999.
3. J. Sarrate, R. Clarisó, "El Método de los Elementos Finitos en problemas electromagnéticos: planteamiento y aplicaciones" in *Revista Internacional de Métodos Numéricos para Cálculo y Diseño en Ingeniería*. Vol 17, 1, 219-248. 2001.
4. J.K. Sykulski, C. W. Trowbridge, "Computational Electromagnetics: Some Conjetures for Industrial and Teaching Requeriments" in *COMPEL Int. Journal*, vol. 17, no. 1/2/3, pp. 36-46. MCB University Press. 1998.

Applied Physics (APHYS 2003)
A. Méndez-Vilas (Ed.)

The PHOTONS - AERONET Network Stations in Spain

V.E. Cachorro[1], C. Toledano[1], R. Vergaz[1], A.M. de Frutos[1], M. Sorribas[2], J.M. Vilaplana[2], B. A. de la Morena[2]

[1] *GOA-UVA / Depart. of Optics and Applied Physics. University of Valladolid 47011, Valladolid, Spain, chiqui@baraja.opt.cie.uva.es*
[2] *INTA / Atmospheric Sounding Station "El Arenosillo". 21130, Mazagón,Huelva, Spain, sorribasmm@inta.es*

Abstract. The Atmospheric Optics Group of the University of Valladolid, Spain (GOA-UVA) in collaboration with the Instituto Nacional de Técnica Aeroespacial (INTA) has been measuring direct solar irradiance and sky radiance with a Cimel sun-photometer belonging to PHOTONS-AERONET since February 2000 up to date. The site for measurements "El Arenosillo" is located in the southwest coast of Spain (37.1° N, 6.7° W). Since January 2003 another Cimel sunphotometer is operating in Palencia (41.9°N, 4.5°W), located in a continental area of north-central Spain. The characteristics and protocols for measurements, data transmission and processing of the sunphotometers according to AERONET standards are described. Problems due to calibration were detected and solved by development of a new correction method (KCICLO). The method provides *in situ* correction for Aerosol Optical Depth (AOD) measurements as well as it explains the AOD fictitious diurnal cycle detected in the data series.

SITES OF MEASUREMENTS

The University of Valladolid and Instituto Nacional de Técnica Aeroespacial maintain a Framework agreement for collaboration in different subjects. Aerosol studies by remote sensing techniques at El Arenosillo started in 1996, using a LiCor-1800 spectrometer, and continued with a Cimel-318 radiometer since February 2000.

El Arenosillo is located in the coastal area of the Gulf of Cádiz, in Huelva (37.1° N, 6.7° E, sea level). This area is very interesting for aerosol studies. The maritime character is influenced by air masses of different origin, and also frequent African desert dust events occur, giving rise to a complex mixed aerosol type. The Cimel was installed in collaboration with the Laboratoire d'Optique Atmospherique (LOA, Univ.of Lille, http://loacli.univ-lille1.fr/photons/sites.html). LOA supports PHOTONS network, with more than 25 aerosol monitoring sites around Europe and Africa.

AERONET (AErosol RObotic NETwork, http://aeronet.gsfc.nasa.gov) program, created in 1998 and supported by Goddard Space Flight Center of NASA, is a worldwide federation of ground-based remote sensing aerosol networks established by AERONET and PHOTONS, and expanded by other agencies, institutes and universities. More than 200 sites make part of AERONET. At the moment there are two AERONET sites in Spain: El Arenosillo and Palencia.

According to AERONET protocols, the instruments (labeled with numbers) are deployed for field operation around six months. Then the instrument is sent for calibration and a new instrument is deployed. Four different instruments have been operating at El Arenosillo up to date: Cimel number #48, #114, #45 and Cimel#50 operating right now (see Table 1).

TABLE 1. History of Cimel instruments at El Arenosillo and Palencia.

El Arenosillo	Start	Finish	Database
Cimel #48	16/02/2000	9/07/2000	Level 2.0
Cimel #114	10/07/2000	21/07/2001	Level 1.5
Cimel #45	22/07/2001	27/01/2003	Level 2.0
Cimel #50	27/01/2003	Up to date	Level 1.5
Palencia			
Cimel #243	23/01/2003	Up to date	Level 1.5

Another Cimel sunphotometer is operating at Escuela Técnica Superior de Ingenierías Agrarias, in Palencia (41.9ºN, 4.5ºW, elevation 740m.) since February 2003 (table 1). This town is located in north-central Spain, in an uncontaminated continental area.

AERONET CIMEL DATA PROCESSING

The goal of AERONET is to assess aerosol optical properties and validate satellite retrievals of aerosol optical properties (Holben et al. 1998). AERONET standard instrument is the Cimel-318. Direct solar irradiance and sky radiance are measured at four wavelengths: 440nm, 670nm 870nm and 1020nm. Aerosol Optical Depth (AOD) is obtained from the direct solar measurements. The Cimel-318 has 3 polarized filters at 870nm and another filter at 936nm for water vapor determination. Polarization ratio, particles size distribution and some other aerosol parameters are retrieved from sun and sky measurements, using inversion algorithms (Dubovik and King, 2000).

Three levels of data are available from AERONET database: Level 1.0 (unscreened), Level 1.5 (cloud-screened), and Level 2.0 (cloud-screened and quality-assured). This Level 2.0 is reached once the instrument is sent for a post calibration.

Data are transmitted half hourly from the sunphotometer to METEOSAT and then retransmitted to the ground receiving station in Darmstadt. The data are retrieved for routine checks and processing in LOA and GSFC by Internet linkage, resulting in near real-time acquisition from the site. The AOD (Level 1) is available in the AERONET web site approximately in 2 hours.

Instruments are calibrated both before and after deployment in the field, for determining the coefficients needed to convert the instrument output digital number to aerosol optical depth (AOD), precipitable water, and radiance ($W/m^2/sr/nm$). Field instruments are returned to GSFC or LOA for intercomparison with reference instruments approximately every 6 to 12 months in order to maintain accurate calibration (1-2%). The AERONET reference Cimel are calibrated by the Langley technique at Mauna Loa Observatory. For the sky radiance measurements, calibration is performed using a calibrated integrating sphere to an accuracy of 5%.

Aerosol Optical Depth and Ångström Parameters Retrieval

Total optical thickness for the atmosphere column is obtained from the well-known Beer-Bouguer-Lambert law:

$$I(\lambda) = I_0(\lambda) \cdot e^{-\tau(\lambda)\cdot m} \tag{1}$$

Finding the total optical thickness:

$$\tau = -\frac{1}{m} Ln(I/I_0) \tag{2}$$

where m is the optical air mass (function of the solar zenith angle), I_0 is the calibration coefficient and I is the measured direct irradiance. There is no absorption by other atmospheric components (water vapor, O_2, or CO_2) at the 4 Cimel wavelengths except for the ozone at 670 nm. Aerosol Optical Depth (τ_a) is obtained from the total optical thickness once discounted the Rayleigh optical thickness τ_R contribution and the ozone optical thickness τ_{o3} for the 670nm filter, according to:

$$\tau_a = \tau - \tau_R - \tau_{O3} \tag{3}$$

From the Ångström expression, it is possible to determine α parameter by a logarithmic fitting of the AOD versus the wavelength:

$$\tau_a(\lambda) = \beta \cdot \lambda^{-\alpha} \tag{4}$$

The result is a representative α value for a given spectral range, often 440-870nm. This parameter describes the behavior of the aerosol optical thickness with wavelength, which is related to the aerosol size distribution function.

QUALITY OF DATA SERIES: KCICLO METHOD CORRECTION

In the analysis of the data series from Cimel #114 some problems were detected, i.e. negative AOD values and a pronounced diurnal cycle for AOD measurements in the 870nm, and 1020nm filters. These data did not reach Level 2.0 in AERONET database (table 1). We have developed a new processing for these data. The procedure is based in the realization of in situ correction method, KCICLO. The new processing achieved excellent results, showing that there were problems with the pre-calibration.

According to Romero and Cuevas (2002), we can define K as the ratio constant between the current calibration constant I_0', and the true calibration constant I_0:

$$K = \frac{I_0'}{I_0} \tag{5}$$

The corresponding erroneous retrieved Aerosol Optical Depth τ'_a can be derived:

$$\tau'_a = \frac{LnI'_0 - LnI}{m} - \tau_R = \frac{LnI_0 + LnK - LnI}{m} - \tau_R = \tau_a + \frac{1}{m} LnK \tag{6}$$

Assuming as a first approximation that 1/m=cos(SZA), where SZA is the solar zenith angle, expression (6) gives the cosine o near parabolic shape of the observed false diurnal cycle on the measured AOD values (τ'_a) if the true AOD value τ_a remains constant during the day. Obviously this requirement is very hard in the atmosphere and hence we find more o less this perfect shape:

$$\tau'_a = \tau_a + \frac{1}{m} LnK \tag{7}$$

The error added to the real AOD is modulated by 1/m, thus is higher at noon (m near 1) and lower for high SZA. It will be positive if the ratio constant K is greater than 1 (LnK>0), and the current calibration constant is overestimated. Then the AOD increases in the morning and decreases in the afternoon giving a convex curve shape. If K is less than 1 (LnK negative) a concave curve shape can be observed for the day, and the calibration constant is underestimated.

Reprocessing by KCICLO Method

Note that equation (7) is a straight line between the measured AOD values (τ'_a) and 1/m, with the slope given by LnK and the intercept being τ_a. Therefore we can determine the slope by a linear regression, then K, and finally the true calibration constant I_0 if the earlier I'_0 was known.

A sufficient number of clear-sky and stable days are selected, with a clear convex or concave behaviour. Obviously the selected days must fulfill a set of requirements: an air mass range, a minimum of data points for the fit, standard deviation of the fit with a given threshold, etc. A linear regression following expression (7) is applied to the previously selected days. A set of K values is obtained for each filter (figure 2).

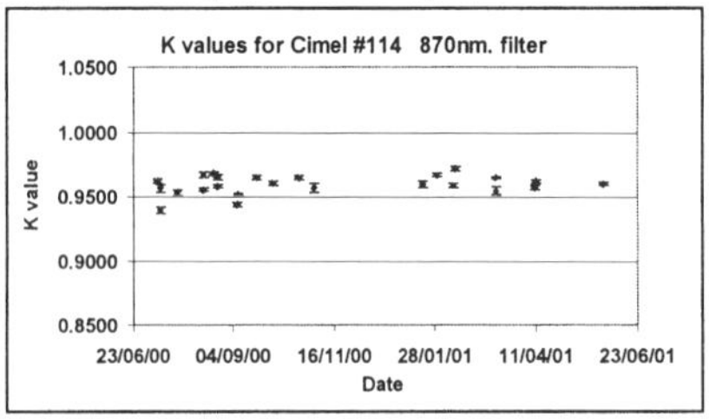

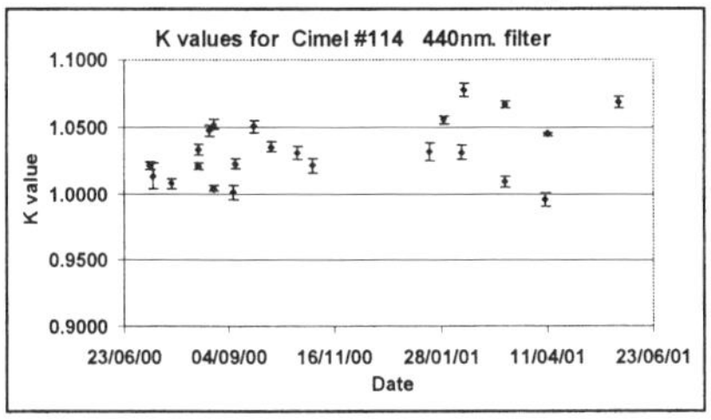

FIGURE 1. K values of the KCICLO fits for Cimel #114 at El Arenosillo at 440nm. and 870nm. filters.

The average K value for each filter (see table 2) is used for correcting Cimel #114 data series. The difference between the current (pre-calibration) and the true calibration constant for each filter is given in the last row of table 2.

TABLE 2. K ratio, new calibration coefficients and associated errors. Cimel #114 El Arenosillo

Filter	1020nm	870nm	670nm	440nm
CN'_0 Pre.	16967.223	18390.122	19254.920	14655.182
K ±STD	0.9494± 0.0018	0.9593± 0.0015	0.9956± 0.0019	1.032± 0.005
New CN_0(error%)	17872 (0.19)	19171 (0.16)	19341 (0.19)	14200 (0.46)
Diff.-Pre(%)	-5.0	-4.0	-0.4	3.2

The precision of the KCICLO correction is given by the STD of K, which ranges from 0.2-0.5%, an order of magnitude less than the correction we must do on the calibration constant. In figure 3 the AERONET retrieval for AOD and the corrected AOD with KCICLO method are shown. The false diurnal cycle in 870nm and 1020nm filters disappeared, as well as the negative values produced for the concave shape, that where rejected in the AERONET processing (Toledano, 2003).

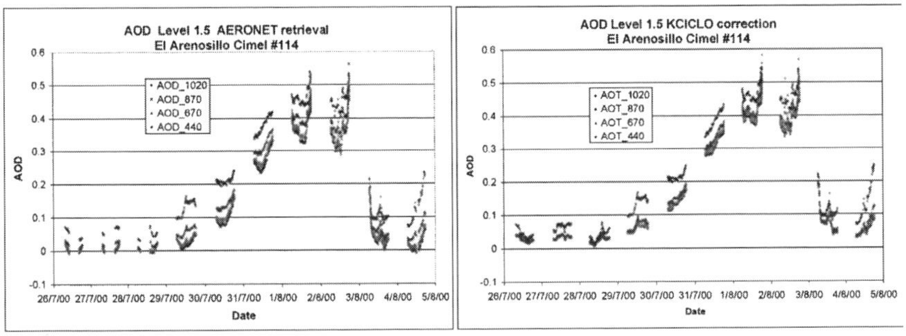

FIGURE 2. AOD results for AERONET(left) and for KCICLO correction (right). El Arenosillo data.

The proposed KCICLO method for AOD data correction is a contribution for in situ calibration methods. The method is based on the detection of a false AOD diurnal cycle. The requirement of sufficient clear days can be a drawback for this methodology.

ACKNOWLEDGMENTS

Thanks are due to the AERONET and PHOTONS teams, and to CICYT (Spanish Commission for Research and Technology Development) for the financial support of the project REN2000-0903-CO8-04 CLI.

REFERENCES

1. Dubovik, O., M.D. King (2000) A flexible inversion algorithm for retrieval of aerosol optical properties from sun and sky radiance measurements, J. Geophys. Res., 105, 20673-20696.
2. Holben, B. N., T.F. Eck, I. Slutsker, D. Tanré, D., J.P. Buis, A. Setzer., E.F. Vermote, J.A. Reagan, Y.J. Kaufman, T. Nakajima, F. Lavenu, I. Jankowiak, and A. Smirnov, "AERONET – A federated instrument network and data archive for aerosol characterization". Remote Sensing of Environment, 66(1), 1-16, 1998.
3. Romero, P.M. and E. Cuevas, Variación diurna del espesor óptico de aerosoles: ¿Ficción o realidad?. Proceeding of 3ª Asamblea Hispano Portuguesa de Geofísica y Geodesia. Tomo II, (S13)1252-1256, Valencia, España, 2002.
4. Toledano, C., Validación y análisis de la serie de datos del espesor óptico de aerosoles del fotómetro Cimel en El Arenosillo. Predoctoral Work of the Third Cycle. University of Valladolid, Spain, 2003.

Applied Physics (APHYS 2003)
A. Méndez-Vilas (Ed.)

Application of ICP-MS for Measuring Soil Metal Cations from Sequential Extraction

[1¶]H. Barros, [1]J.M. Abril, [3]A. Ludicina and [2]A. Delgado

[1]Dpto. Física Aplicada I. [2]Dpto Cs. Agroforestales. EUITA. Universidad de Sevilla, Crta Utrera km 1. 41013.Sevilla. SPAIN. [3]Dpto. Agronomia, Coltivazioni Erbacee e Pedologia. Università di Palermo. Viale delle Scienze. 90128 Palermo ITALY.

Abstract. The present work is part of a collaboration between Palermo and Seville Universities to investigate the cations incorporated into calcite and its effect in the trace metal mobility in soils. The study area is located in the central part of Sicily. Calcite is usually greatly dispersed and has a major influence on the pH of soils and therefore on trace element behaviour. A great affinity for carbonates has been observed for a wide range of cations including common diagenetic tracers and important environmental contaminants. Although many studies have been investigate trace elements in soils, only few has considered the metal cations incorporated into calcite. In order to investigate the metal cations in soil carbonate, 1M MgCl was used to remove the exchangeable Ca. Sub-samples were treated with acid acetic 1M to quantify the total amount of Ca and metal cations in soil carbonate. And finally others sub-samples were used to perform a kinetic extraction with Na-citrate 0,2M. The Inductively Coupled Plasma Mass Spectrometry (ICP-MS) was selected as a suitable technique to reliable and simultaneously measure trace elements reducing matrix effects. The equipment used was a X7 SERIES ICP-MS from Thermo Elemental (U.K). The present work shows in detail the methodology used to determine, in acid acetic and Na-citrate solutions, the following isotopes: Cr, Mn, Fe, Co, Ni, Sr, Ba, Pb and U. We have adapted the U.S EPA Method 6020 for our special needs. It is important to remark that a careful procedure has been carry out to reproduce the sample matrix in all the calibration and QC standards, and to properly select the internal standards since the complexity of the matrix.

INTRODUCTION

ICP-MS has become a widely spread analysis method because it's capacity of ultra trace measurements, it's relatively low cost per sample and fast analysis, when compared with other analytical methods. This technique is useful in geological and soil science studies, where often a low level determination is required for a large number of samples. An interesting problem, as seem from an agrological perspective, is the fate of metallic cations in soils, which is influenced by the physic-chemical and mineralogical properties of the substratum.

In arid and semi-arid regions, often, Ca-carbonate is found in large amounts, either spread in the soil matrix or cumulated in particular horizons (calcic and/or petrocalcic). When it is spread within the soil then exerts a strong influence in the pH and in the behaviour of metal cations [1]. These last, besides to influence on the carbonate's precipitation process, can coprecipitate with it becoming part of the new

¶ Corresponding author: haydn@us.es

structure, or can be occluded in the oxides (mainly of Fe and Mn), or can even be adsorbed onto the carbonate's surface [2].

A wide range of metallic cations, including common diagenetic tracers (Mn, Sr) and important environmental pollutants (Ni, Zn, Co, Cd, Pb), can substitute for Ca in the calcite structure [3]. Although many studies has investigated the presence of metallic cations, only few of them has considered the calcite's content in the soil. Some authors [4] have observed that secondary $CaCO_3$ shows a high content of Sr or Co, others [5] couldn't find a correlation between metallic cations and $CaCO_3$ in soils, particularly in Sicily where many soils has along its profile a variable content of carbonates, which represent a possible sink/source of metallic cations.

The scope of this work is to bring up all the necessary technical resources to investigate, in a sequence of forestall soils from central Sicily, which metallic cations are associated to the soil carbonate and to quantify it. With this purpose it was developed a method to measure metallic cations by ICP-MS, these method makes a compromise between the particular chemical matrix and the technical requirements to allow a low level determination for such cations.

MATERIAL AND METHODS

The Area Under Study

The chosen area on study was the Forest Complex Mustigarufi [6-7] (central Sicily). It were selected five soil pedons along a ground elevation which is formed from different lithotypes comprising from the Tortonian to the Olocenean. The sequence of such lithotype along the elevation (described from the top) is the following: co-alluvial deposits, clays with gypsum and/or gypsum, marly clays, marly and/or sandy clays and alluvial deposits (terraced).

Analytical Methods

The soil samples were desalted with Millipore water and washed twice with $MgCl_2$ 1M for removing the exchangeable cations. A set of samples was treated with 1M acetic acid in an ultrasonic bath to promote the selective dissolution of calcium carbonate [8], and another sub-samples were used to perform a kinetic extraction with 0,2M Na-citrate (5, 15, 30, 120 minutes, 1 day and 1 and 2 week). The obtained solutions were analysed by AAS (Atomic Absorption Spectrometry) for Ca determination and ICP-MS for: Cr, Mn, Fe, Co, Ni, Sr, Ba, Pb and U. It was used an X7 SERIES ICP-MS (Thermo Elemental, UK) and we developed an adaptation of the 6020 US EPA method, restricted to the isotopes of our interest and modified for matrix matching the calibration solutions, as detailed in what follows.

The selection of a given isotope depends on the interferences that can be found in the spectra. The most standard and simplified methods suggest the use of the simplest matrix (or diluted as much as possible), however a compromise have to be achieved in order to avoid losses in the count rate and still allowing special chemical procedures.

For that reason, the samples were measured separately and all the calibration solutions and QC samples were prepared in the corresponding matrix to eliminate any difference between patterns and samples to be analysed.

It was perform a previous mass scan for each unknown solution to determine the most appropriate isotopes, matrix interferences and the acquisition method to be chosen. In general, peak jumping was used for quantitative analysis. In Table 1 it is shown the summary of the set up parameters and in Table 2 the analytes with the additional interferences introduced by the used matrixes (MgCl, acetic acid and Na-citrate in the presence of a high content of Ca).

TABLE 1. ICP-MS System details.

ICP-MS	X7 SERIES ICP-MS, Thermo Elemental (U.K)
Lens	Infinity
Interface	Environmental – Ni cones
Detector	Simultaneous AutoRange Plus
	Peak jumping - 60 sweeps - 10 ms dwell/peak (58 s Total)
Spraychamber	Peltier cooled impact based, 3 °C
ICP	Solid State 27 MHz - 1200 (500*) W
Nebulizer	Concentric
Torch	1.5 mm (quartz) - 12 (14*) L min^{-1} Coolant flow - 0.7 (1*) L min^{-1} Aux flow

* Set up values for Cool Plasma Screen

TABLE 2. Analytes.

Element	Monitor	Used for	Ionisation potential	Possible matrix associated interferences
Sc	45	Internal standard	6.54	-
Cr	52,53	Analyte	6.77	$^{12}C+^{40}Ar$, $^{12}C+^{40}Ca$, $^{16}O+^{37}Cl$
Mn	55	Analyte	7.43	$^{15}N+^{40}Ca$, $^{12}C+^{43}Ca$
Fe¶	54,56	Analyte	7.87	$^{16}O+^{40}Ar$, $^{16}O+^{40}Ca$, $^{12}C+^{44}Ca$
Co	59	Analyte	7.86	$^{36}Ar+^{23}Na$, $^{16}O+^{43}Ca$
Ni	60,62	Analyte	7.63	$^{16}O+^{44}Ca$, $^{36}Ar+^{24}Mg$, $^{12}C+^{48}Ca$, $^{14}N+^{48}Ca$
Sr	86,88	Analyte	5.7	-
In	115	Internal standard	5.79	-
Ba	137,138	Analyte	5.21	-
Pb	206,207,208	Analyte	6.21	-
Bi	209	Internal standard	7.29	-
U	238	Analyte	6.08	-

¶ The main interferences for Fe were reduced using Cool Plasma Screen (see text)

Interferences in ICP-MS spectra becomes important particularly in samples with complicated matrix, these interferences are caused for matrix or plasma gas ions.

Although the selected chemical treatment was incorporated into the calibration patterns, it was performed a verification following the standard addition technique. This is generally used when matrix effects cannot be adequately corrected for by either sample dilution or matrix matching the calibration solutions (when matrix elimination is not possible, as is the case). It consist in the addition of a small amount of analyte to the unknown sample, which is then re-analyzed. This procedure was repeated three times. The obtained curve intercept point corresponds to the signal of the analyte in the unspiked sample. The slope of the additions curve is the analyte sensitivity in the matrix, and can be applied to subsequent unknown samples. Interferences are not eliminated but can be corrected for using blank subtraction.

For Fe the Cool Plasma mode was used. It consists of a Ni screen, covered with a quartz bonnet, which is located around the outside of the torch and within the load coil. This screen is grounded once the plasma has been ignited. The main use of the Plasma Screen option is to allow the plasma to be operated at low power (500 W) which drastically reduces species such as $^{40}Ar^{16}O$ (which overlaps with ^{56}Fe).

RESULTS AND DISCUSSION.

The 6020 US EPA method [11] was adapted as previous description. The standard addition calibration has shown that, in the standard mode of operation, the matrix matching calibration solution was a suitable method for the matrix considered. An exception was Fe, which was separately measured in the Cool Plasma mode.

This method can be used in the determination of trace metallic elements in selective or sequential extractions. The detection limits results in higher values due to the instability associated to the matrix suppression/enhanced effects. As an example the 6020 MDL for ^{52}Cr is 0.007 ppb and in the present work 0.02. However in the rest of the cases the measured concentration resulted above the found MDL.

The carbonate content varies with the Ca amount and with its origin (primary or secondary). Comparing the calcium extracted with acetic acid and the content in calcium carbonate, it can be stated that the applied extraction [8] allowed the selective dissolution of soil's carbonates. In fact, the correlation between Ca (extracted by acetic acid) and the total $CaCO_3$ is r=0,96, excluding horizons rich in gypsum. Only Cr was not detected, while others shows an order, from the higher contents: Fe $\approx$ Sr > Mn > Co > Ni > U; and Ba and Pb shows a variable content ($0 < Ba < 132,7$; $0 < Pb < 16,4$).

From the analysis of the results it can be seen that the Ca extracted by acetic acid is correlated with Fe (r 0.701), Co (r 0.922), Ni (r 0.906), Sr (r 0.900), but not with Mn (r 0.480) and U (r 0.436). However, if the correlation coefficients are calculated taking into account the Ca coming from primary and secondary $CaCO_3$ separately, then the results are widely diverse.

Even more evident is the influence of the carbonization in metal cations and specially the related to Mn, Fe and Sr. In particular, it have to be noted the relation between Ca and Sr (usual substitute of Ca). In fact, separating the horizon richer in primary $CaCO_3$ from those richer in secondary $CaCO_3$, the equation of linear correlation between Sr and Ca from primary carbonate is $y_1 = 6{,}03x_1 + 204$ while related to secondary carbonate is $y_2 = 4{,}41x_2 + 216$ (Fig. 1). The two different slopes (6,03 and e 4,41 respectively) means that in the case of secondary carbonate the amount of calcium is higher than in the case of primary carbonate and a such difference is justified admitting that in the formation of this secondary carbonate there is a source of external calcium, as can be for instance the gypsum [10].

Contrarily, in the case of Co and Ni, the effect or carbonization seems to be lower, despite the direct relation between these elements and Ca, and the consequent high correlation between themselves (r 0,98). This last means that the carbonization influence in the same way these two metal cations, which can be expected since both elements have a similar chemical character.

On the basis of the previous results, it can be stated that the carbonate, in the studied soils, represents a source of metal cations, and the carbonization process exerts certain degree of influence on the soil mobility.

FIGURE 1. Calcium and strontium in primary (y_1) and secondary (y_2) carbonate.

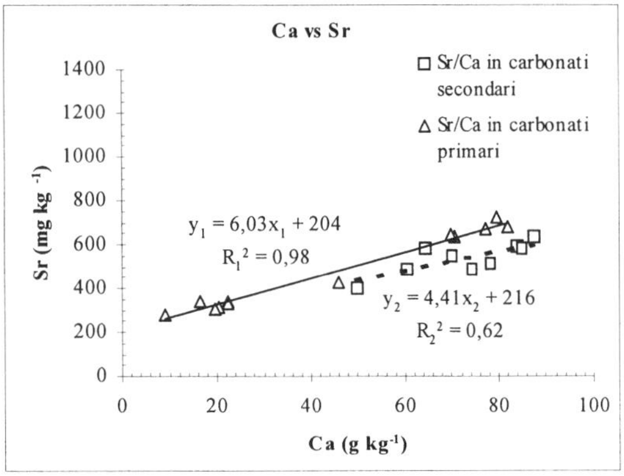

REFERENCES

1. Birkeland P. W. Soils and Geomorphology. Oxford University Press, New York. 1999.
2. Kabata-Pendias A. & Pendias H. Trace elements in soils and plants. CRC press, 2nd ed. 1992.
3. Temmam M., Paquette J., Vali H. Mn and Zn incorporation into calcite as a function of chloride aqueous concentration. *Geochim. et Cosmochim. Acta.* **64** 2417. 2000.
4. Vochten R. C. & Geyes J. G. Pyrite and calcite in septarian concretions from the Rupelian clay at Rumst (Belgium) and their geochemical composition. *Chem. Geol.* **14** 123. 1974.
5. Palumbo B., Angelone M., Bellanca A., Dazzi C., Hauser S., Neri R., Wilson J. Influence of inheritance and pedogenesis on heavy metal distribution in soils of Sicily, Italy. *Geoderma.* **95** 247. 2000.
6. Dazzi C. & Monteleone S. Soils and soil-landform relationships along an elevational transect in a gypsiferous hilly area in central Sicily, Italy. *Proceedings 7th International Meeting of Soils with Mediterranean Type of Climate*, Valenzano, Bari. 2001.
7. Dazzi C. & Scalenghe R. Soils with gypsic horizon in mediterranean climate: a case study. *Proceedings 17th World Congress of Soil Science*, Bangkok 2002.
8. De Paolo D.J., Kyte F.T., Marshall B.D., O'Neil J.R. and Smit J. Rb-Sr, Sm-Nd, K-Ca, O and H isotopic study of Cretaceous-Tertiary boundary sediments, Caravaca, Spain: Evidence for an oceanic impact site. *Earth Planetary Science Letters.* **64** 356. 2003.
9. Laudicina V.A., Pisciotta A., Parello F. and Dazzi C. Differenziazione e quantificazione dei carbonati litogenici e pedogenici di Gypsisuoli forestali attraverso l'analisi isotopica. *Convegno annuale della Società Italiana della Scienza del Suolo.* Siena, 9-12 June 2003.
10. FAO. Management of gypsiferous soils. Soil Bulletin n.62, Rome. 1990.
11. Thermo Elemental. ICP-MS Application Notes. S311AN Issue 1. June 2001.

Applied Physics (APHYS 2003)
A. Méndez-Vilas (Ed.)

Fractal Plotter: Visual Tools for Non-Linear Dynamical Systems Study

G. Álvarez, F. Montoya, M. Romera, G. Pastor

Instituto de Física Aplicada, C.S.I.C., Serrano 144, 28006, Madrid, SPAIN.
E-mail: gonzalo@iec.csic.es

Abstract. A new computer program for the visual study of non-linear dynamical systems is presented. The program, called Fractal Plotter, offers an integrated graphic environment allowing for the graphical representation of the most common maps. It also offers a complete set of tools for the computation of mathematical operations closely related to the study of chaotic physics. The program is written in C++ and runs on any Windows platform.

INTRODUCTION

Non-linear phenomena are a rapid developing research field in applied physics. As a consequence, it is of vital importance the development of software tools to help researchers in their work. We have created a program, called Fractal Plotter, consisting of a suite of tools aimed at the generation, representation, manipulation and further study of a wide range of different chaotic maps including, amongst the most widely studied in the literature, the Mandelbrot set, Julia sets, Myrberg maps, exponential maps, etc.

Fractal Plotter offers an integrated graphic environment to aid the researcher in the analysis of topics such as hyperbolic components, Misiurewicz points and symbolic sequences. It also includes some tools for the computation of specific mathematical operations involving symbolic sequences extensively used in 1D quadratic maps: hyperbolic components centre determination, Gray sequences conversion, computation and arithmetic operations with MSS- and Fourier-harmonics, heredity computation, composition rules, external rays and rational form conversion, etc.

The program is written in C++ and runs on any Windows platform.

THE PROGRAM

The two main menus of the program are **Graphics** and **Tools**. The former offers all the functions to generate the different maps. The latter offers utilities to compute mathematical functions.

Graphics

The **Graphics** menu constitutes the kernel of the program. This menu contains the commands to open the dialog boxes allowing for the parameter definition and generation of the different maps supported by the program.

Parable

This command plots over the parable $f(x) = x^2 + c$ the orbit followed by the initial point x_0 for the given parameter value *c*, using the graphic iteration method described by Feigenbaum [1]. It is also possible to represent the curves corresponding to higher order iterations of the function: $f(x) = x^2 + c$, $f^2(x)$, $f^3(x)$, ..., $f^n(x)$.

Mandelbrot Map

This command plots the complex Mandelbrot set map, one of the most intensively studied in the literature.

Once the Mandelbrot set image has been plotted, a Misiurewicz point symbolic sequence [2], along with its preperiod and period can be obtained pressing the **Misiurewicz** button.

Fractal Plotter provides four different methods to plot a Mandelbrot set:

1. **Shadows**: the image is plotted in color, using different colors according to the number of iterations before diverging to infinite.

2. **Black&White**: only points belonging to the Mandelbrot Set are plotted, in black.

3. **Milnor**: the distance estimator method introduced by Milnor [3] is used. This method draws points that do not belong to the Mandelbrot set but are very close to it. Filaments appear and a lot of midgets that could not be seen with the B&W method can be now localized thanks to the help of filaments.

4. **Escape lines**: the escape lines method consists of plotting those points that diverge at exactly *N* iterations. It is an improvement in regard to the B&W and the Milnor methods because many more midgets can be observed [4] and their period can be easily calculated with the naked eye.

Fig. 1 shows the same Mandelbrot set region plotted using the four different methods.

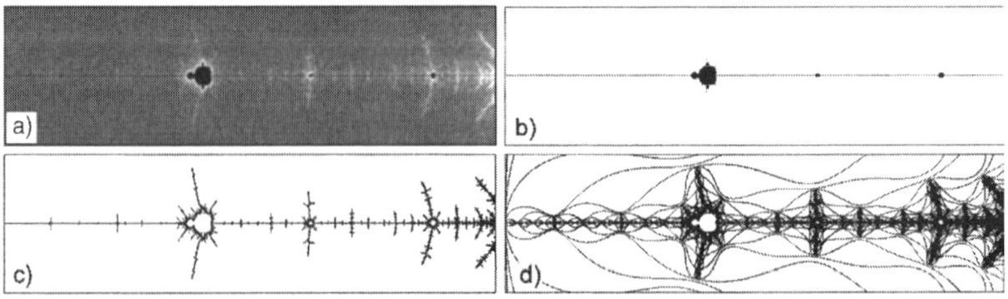

FIGURE 1. Mandelbrot set images plotted by Fractal Plotter using different plotting methods: a) shadows; b) Black&White; c) Milnor; and d) escape lines. Note that the last method allows for finding the greatest number of midgets with the naked eye.

Furthermore, Fractal Plotter easily locates hyperbolic components with a given rotation number. When introducing the rotation number and pressing the **Find** button, a small dot is plotted on the tangency point between the main cardioid and the hyperbolic component looked for.

Once the image has been plotted, or even while it is being plotted, it is possible to know the period of a disk or cardioid by simply clicking on it while holding the *Control* key pressed. The information about its symbolic sequence and center coordinates is also provided. If instead of holding the *Control* key, the *Shift* key is held, then in addition to that information, the disk/cardioid is colored in red if Myrberg method [5] converges and in white if not.

Julia Map

This command plots the Julia Set. If the **Iterate** button is pressed, the orbit followed by the initial point $z_0 = 0$ for the given parameter value is plotted. If the checkbox **Lines** is checked, then the points are linked by segments.

For Julia maps only two plotting methods are considered: shadows and B&W.

Baker-Rippon-Devaney (BRD) maps

This command plots Julia- and Mandelbrot-like exponential maps, such as $z_{n+1} = ce^{z_n}$, $z_{n+1} = e^{cz_n}$, and $z_{n+1} = e^{z_n^2+c}$. We have generically denoted these maps as Baker-Rippon-Devaney (BRD) [6]. In Fig. 2 a Mandelbrot-like set is plotted upon increasing number of iterations

Tools

Under this menu we have grouped many different tools to compute calculations on the study of non-linear dynamical systems.

Lip

Given a symbolic sequence corresponding to an orbit in the real Mandelbrot map, it informs whether it is admissible (legal inverse path) or not.

Ancestral path

Given a symbolic sequence corresponding to an orbit in the real Mandelbrot map, it returns its ancestral path [7].

Centers

Given a symbolic sequence corresponding to a superstable periodic orbit in the real Mandelbrot map, it calculates the coordinates of the centre (the parameter value) whenever the Myrberg method converges [5].

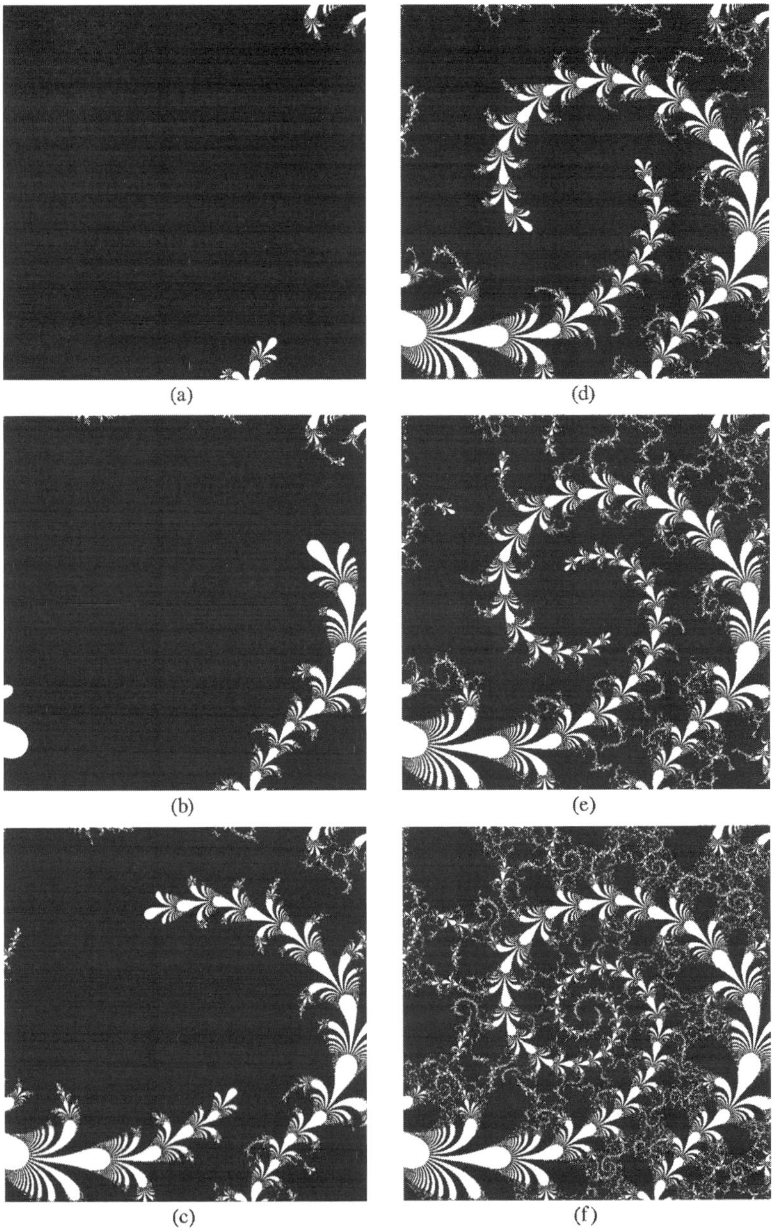

FIGURE 2. BRD Mandelbrot-like set centered at $c = -2.437924149711 + 1.939831341204i$, corresponding to Misiurewicz point $M_{8,1}$. The six frames correspond to 15, 20, 25, 30, 35, and 50 iterations respectively.

Gray

Given a symbolic sequence corresponding to a superstable periodic orbit in the real Mandelbrot map, it calculates its generalized Gray number and, conversely, given a Gray number, it computes the corresponding symbolic sequence [8].

Harmonics

Given a symbolic sequence corresponding to a superstable periodic orbit in the real Mandelbrot map, it calculates its harmonics and antiharmonics, both Fourier and MSS [9].

Composition

Given two symbolic sequences corresponding to superstable periodic orbits in the real Mandelbrot map, it calculates the result of their composition, both leftward and rightward [9].

External arguments

It computes the conversions between the different forms of representation of hyperbolic components: symbolic sequence and external arguments (represented both as a binary expansion and in rational form), both for hyperbolic components and Misiurewicz points [10].

ACKNOWLEDGMENTS

This research was supported by Ministerio de Ciencia y Tecnología, Proyecto TIC2001-0586.

REFERENCES

1. M. J. Feigenbaum, *J. Stat. Phys.* **19**, 25-52 (1978).
2. Romera, M., Pastor, G., and Montoya, F., *Physica A* **232**, 517-535 (1996).
3. J. Milnor, "Self-similarity and hairiness in the Mandelbrot Set", in *Computers in Geometry and Topology*, edited by Tangora, Lect. Notes Pure Appl. Math 114, 211-257 (1989).
4. Pastor, G., Romera, M., and Montoya, F., *Chaos, Solitons & Fractals* **7**, 565-584 (1996).
5. Alvarez, G., Romera, M., Pastor, G., and Montoya, F., *Chaos, Solitons and Fractals* **9**, 1997 2005 (1998).
6. Romera, M., Pastor, G., Alvarez, G., and Montoya, F., *Computers & Graphics* **24**, 115-131 (2000).
7. Romera, M., Pastor, G., Alvarez, G., and Montoya, F., *Physical Review E* **58**, 7214-7218 (1998).
8. Alvarez, G., Romera, M., Pastor, G., and Montoya, F., *Electronics Letters* **34**, 1304-1306 (1998).
9. Pastor, G., Romera, M., and Montoya, F., *Physic Review E* **56**, 1476-1483 (1997).
10. Pastor, G., Romera, M., Alvarez, G., and Montoya, F., *Physica D* **171**, 52–71 (2002).

Applied Physics (APHYS 2003)
A. Méndez-Vilas (Ed.)

Parameterizing non-linear magnetic cores for PSpice Simulation

R. García-Gil, J. M. Espí, J. Jordán, S. Casans and J. Castelló

Department of Electronics Engineering, University of Valencia, C/ Dr. Moliner 50, 46100 Burjassot, SPAIN

Abstract. In this paper the main Jiles & Atherton parameters of a magnetic material, according to their particular working condition (mainly frequency, temperature and waveform) are calculated by means of a systematic procedure that use PSpice and MathCad. The modelled magnetic material is used for power electronics circuit simulation, which uses magnetic materials, working at frequencies in the region of kH, to construct transformers and inductors. Magnetic materials are the most critical components in electronics simulators due to their non-linear behaviour and the accuracy in the simulation can only be obtained through a proper magnetic material modelling. This paper examines the modelling of the magnetic core hysteresis for high-frequency applications. As example of application, a buck converter with its inductor working near saturation is implemented and its experimental waveforms are compared with that obtained by simulation.

INTRODUCTION

When engineers try to simulate power converters, a main drawback exists due to the non-linear magnetic cores. Usually, simulation with linear inductors or transformers does not give good results, because saturation and hysteresis are not included in the model. In these cases, the accuracy in the power circuit simulation can only be obtained through a proper magnetic modelling, which means to model the hysteresis or B-H loop of the magnetic material. But, the B-H loop of a magnetic component is highly dependent on frequency, waveform (not very important for ferrite materials that are mostly used in power electronics applications), temperature, excitation level, etc. Although the electronics simulation package like PSpice includes in its magnetic library a list of non-linear components, rarely these parameters are coincident with our working conditions or simply they are not specified. In all of these cases it is necessary the modelling of non-linear magnetic cores for the particular application.

Most of the literature [1 - 3] describes the B-H loop obtained at low frequency, but the inductors and transformers used in power electronics normally work at frequencies in the region of kHz, to reduce volume and weight of these components that normally are the most voluminous part in the power equipment. In this work we present a systematic procedure to obtain the main PSpice parameters used to model non-linear magnetic cores at high-frequencies.

In the final section, the modeled magnetic material will be used to simulate the behavior of a real inductor in a power supply.

JILES & ATHERTON MODEL

Based on the magnetic domain theory, the Jiles & Atheton model (J-A) [4, 5] starts with the magnetic material response without hysteresis losses, named anhysteretic curve (**M_{an}**) and defined by the Langevin equation as follows,

$$M_{an} = M_s\left(\coth\left(\frac{H_e}{A}\right) - \frac{A}{H_e}\right) = M_s\left(\coth\left(\frac{H+\alpha M}{A}\right) - \left(\frac{A}{H+\alpha M}\right)\right). \tag{1}$$

where $H_e = H + \alpha M$ is the effective field experienced by the domains, **M** is the total magnetization, **H** is the applied magnetic field, M_s is the saturation magnetization, A is the domain density and α the domain coupling [1].

Hysteresis is then introduced as the consequence of a frictional force which opposes to the block domain-wall motion. According to [4, 5] the differential equation for the total magnetization **M** results from the reversible and irreversible domain wall motion and can be expressed as follows,

$$\frac{dM}{dH} = (1-C)\frac{M_{an} - M_{irr}}{k\delta - \alpha(M_{an} - M_{irr})} + C\frac{dM_{an}}{dH} \tag{2}$$

Equation (2) describes the hysteresis loop of any magnetic material. The parameter k represents the loss coefficient, M_{irr} the irreversible magnetization, the parameter δ takes the value +1 when $dH/dt>0$ and -1 when $dH/dt<0$ and $C \in [0, 1]$ is the reversibility coefficient [1].

The parameters M_s, A, α, C and k are the J-A parameters that use PSpice to model non-linear magnetic cores, together with the geometrical parameters AREA (mean magnetic cross-section), PATH (mean magnetic path length), PACK (staking factor) and GAP (gap length).

From equations (1) and (2), the five J-A parameters can be calculated from points and slopes of the experimental B-H loop [1, 2] as follows,

1) **The saturation magnetization (M_s),**

$$M_s = B_{max}/\mu_o \tag{3}$$

where B_{max} will be obtained from the data-sheet of the magnetic material.

2) **Initial susceptibility (χ_{in}):** Represent the slope at the origin of the initial magnetization curve. Substituting H=0 and M=0 in equation (2) and considering that near the origin only a reversible wall motion exists ($M_{irr}=0$),

$$\chi_{in} = \left.\frac{dM}{dH}\right|_{\substack{H=0\\M=0}} = \frac{M_s C}{3A} \tag{4}$$

3) **Maximum differential susceptibility (χ_{max}):** It is defined as the slope of the magnetization loop at the coercive point (H_c) and is calculated from equation (2) with M=0, $H=H_c$ and $\delta=1$,

$$\chi_{max} = \left.\frac{dM}{dH}\right|_{Hc(M=0)} = (1-C)\frac{M_{an}(H_c)}{k - \alpha M_{an}(H_c)} + C\left.\frac{dM_{an}}{dH}\right|_{H=Hc} \tag{5}$$

4) **Remanent differential susceptibility (χ_r):** It is defined as the slope of the magnetization loop at the remanent point (M_r) and is calculated from equation (2) with $M=M_r$, $H=0$ and $\delta=-1$,

$$\chi_r = \left.\frac{dM}{dH}\right|_{Mr(H=0)} = (C-1)\frac{M_{an}(M_r)-M_r}{k+\alpha[M_{an}(M_r)-M_r]} + C\left.\frac{dM_{an}}{dH}\right|_{Mr} \tag{6}$$

5) **Differential susceptibility at the loop tip**: It is defined as the slope at the (H_m, M_m) loop tip of the measured hysteresis loop and is calculated from equation (2) with the supposition that near saturation $dM/dH \sim dM_{an}/dH$, $M_{irr} \sim M_m$ and $\delta=1$

$$\chi_m = \left.\frac{dM}{dH}\right|_{\substack{Mm\\Hm}} = \frac{M_{an}(M_m)-M_m}{k-\alpha[M_{an}(M_m)-M_m]} \tag{7}$$

Using equations (3) to (7) the five J-A parameters can be calculated from experimental measurement of: the coercive magnetic field (H_c) and their slope (χ_c), the remanent magnetization ($M_r=B_r/\mu_o$) and their slope (χ_r), the saturation magnetization ($M_s=B_s/\mu_o$), the coordinates of the loop tip (H_m, M_m) and their slope (χ_m).

The effect of the variation of the k, A and α parameters in the B-H loop is shown in Fig. 1.

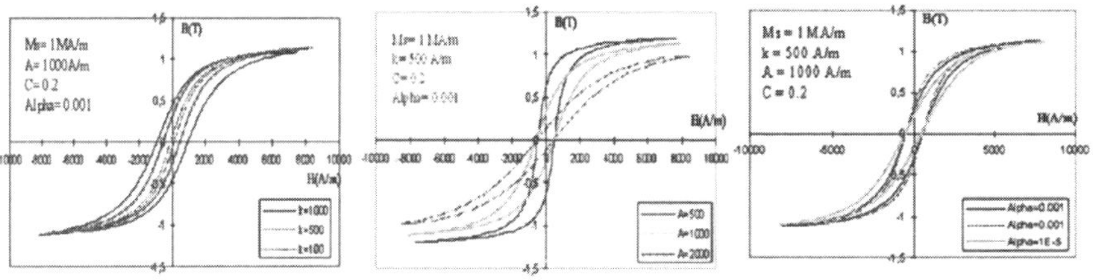

(a) (b) (c)

FIGURE 1. Effect on the theoretical B-H loop of the (a) k parameter, (b) the A parameter, (c) and the α parameter.

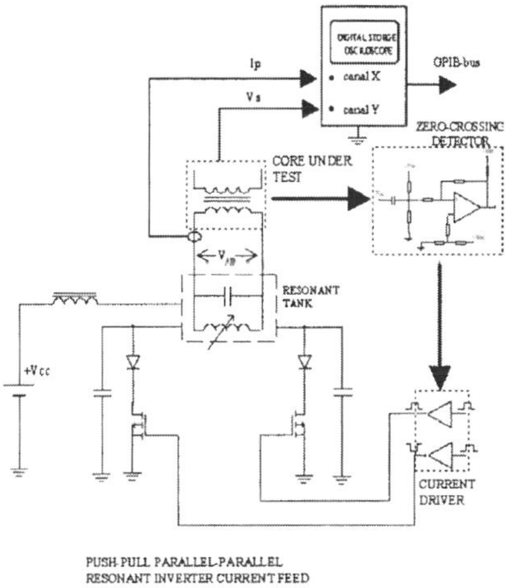

FIGURE 2. Block diagram of the experimental setup used to obtain the experimental B-H loop.

EXPERIMENTAL VALIDATION OF THE MODEL

The experimental bench for the magnetic material characterization is shown in Fig. 2. A power supply generating a sinusoidal waveform of variable frequency is applied to the selected magnetic material. A reduced geometry of this magnetic material with primary and secondary winding is used to bring the component into saturation with low power requirements. Measuring the primary current and the secondary voltage and by using the Ampere and the Faraday laws, the experimental B-H loop can be obtained. Later on, this experimental loop is used to extract the points and slopes described in previous paragraph. With these experimental points and slopes, solve the equation system (3), (4), (5), (6) and (7) with the mathematical package MathCad. We must give initial points from which Matcad looks for the zeros of the function to solve these equations. As initial values are taken those used for PSpice as default values (M_S=1 MA/m, A = 1000 A/m, α=0.001, C= 0.2, k= 500 A/m) and the equation system is calculated with minimum error. The obtained results are used as entry for a new calculation. The iterative procedure finish when the error between measured $M_{measured}$ and modeled $M_{modeled}$ magnetization is beside the considered error. The error function used is,

$$\varepsilon = \left[\sum_{samples} \frac{\left(B_{experimental} - B_{modeled}\right)^2}{N} \right]^{1/2} \tag{8}$$

The selected procedure is used to model different magnetic materials used for power electronics application at different working conditions (mainly temperature in the range of 25 to 50 °C and frequency in the range of 50 to 200 kHz). As an example, in Fig. 3 and 4 there are presented the measured hysteresis curve and the best fitting curve generated by the J-A model for the N_47 material at 110 kHz and 30°C and the H7C4 material at 140 kHz and 40°C, respectively. As can be shown, the obtained parameters fit well the hysteresis loop.

These five parameters together with the selected geometry (AREA, PATH and PACK) are then introduced in the Parts Editor of PSpice. The resultant model is saved in the magnetic library for further simulations.

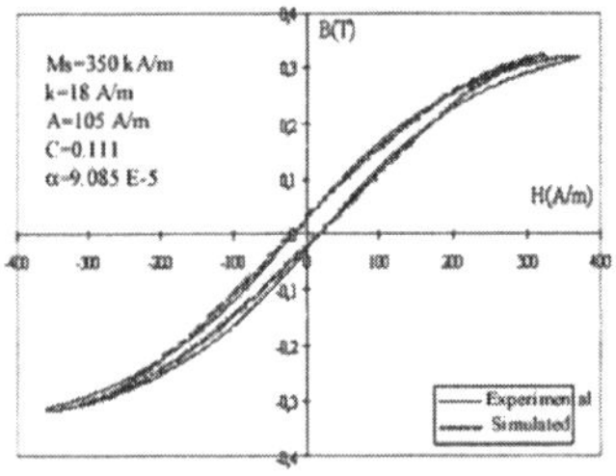

FIGURE 3. Experimental and simulated B-H loop for the N-47 material at 110 kHz and 30°C.

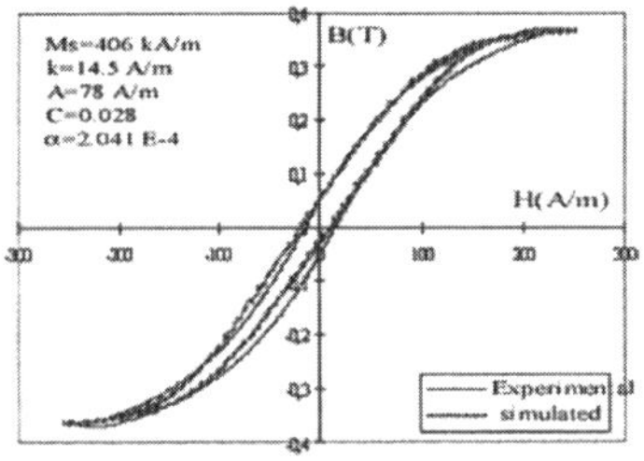

FIGURE 4. Experimental and simulated B-H loop for the H7C4 material at 140kHz and 40°C.

SIMULATION OF THE MODELED MAGNETIC IN A POWER SUPPLY

Another series of tests have been performed in order to check the capability of the model to predict the waveform of the power supply shown in Fig. 5 (a). In this example the magnet (inductor) is fed by a triangular waveform of 110 kHz. The selected material is the ferrite N-47, which for this working frequency has been modeled in Fig. 3. The selected model is as follows,

.model KE30_47 core (Alpha=1.283E-4 Ms=342k A=64 C=0.271 k=17 AREA=0.6 PATH=4.4 GAP=0)

As a conclusion, Fig. 5(b) and (c) show the experimental and simulated waveform for the inductor current, when the converter is working near saturation, showing a good agreement between them. In addition, Fig. 5 (d) shows the simulated waveform obtained by most of the designers using an ideal inductor, which is very different of that obtained experimentally.

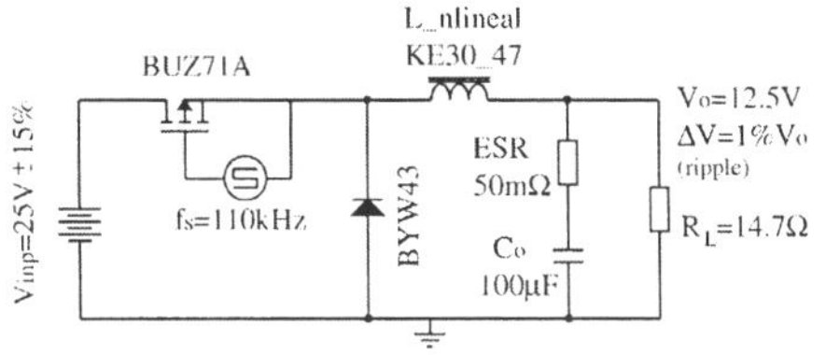

(a) Buck converter

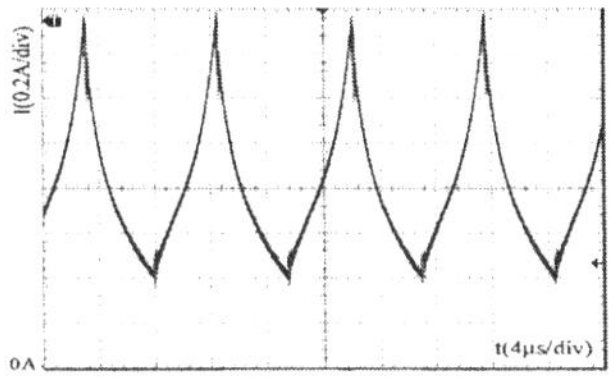

(b) Experimental waveform.

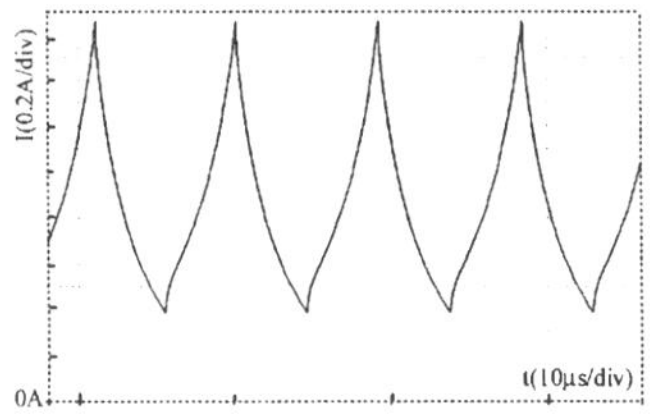

(c) Simulation with the modeled inductor.

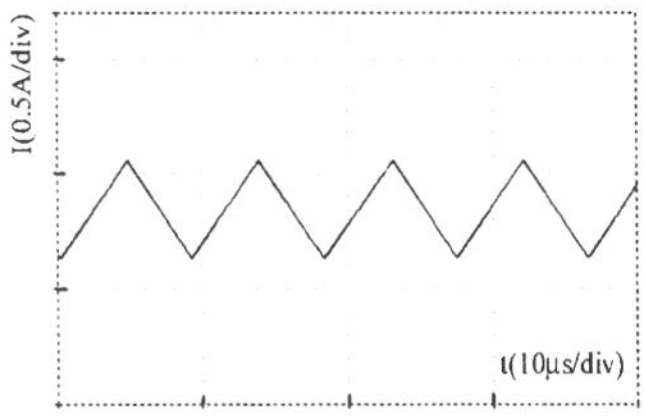

(d) Simulation with the ideal inductor

FIGURE 5. Comparison between experimental and simulated (modeled and ideal) waveform.

CONCLUSION

In this paper the main J&A parameters of a magnetic material at high-frequencies has been obtained. These parameters have been adjusted to fit the experimental curve. Finally, the modelled magnetic core has been used to successfully simulate the real behaviour of a power supply when working near saturation.

In addition, the magnetic modelling allows for core optimisation, which permits to reduce the volume and weight of the power supply.

REFERENCES

1. Benabou, A., Clénet, S., and Piriou, F., *Journal of Magnetism and Magnetic Materials* **261**, 139-160 (2003).
2. Del Moral Hernández, E., Muranaka, C. S., and Cardoso, J. R., *Physica B* **275**, 212-215 (2000).
3. Angeli, M., Cardelli, E., and Della Torre, E., *Physica B* **275**, 154-158 (2000).
4. Jiles, D. C., and Atherton, D. L., *Journal of Magnetism and Magnetic Materials* **61**, 48-60 (1986).
5. Jiles, D. C., Thoelke, J. B., and Devine, M. K., *IEE Transactions on Magnetics* **28**, 27-35 (1992).
6. Garcia, R., Carrasco, J. A, Espí, J. M., Dede, E. J., and Castelló, J., *EPE Journal* **11(2)**, 13-22 (2001).

Applied Physics (APHYS 2003)
A. Méndez-Vilas (Ed.)

Obtaining the Electrical Model of a Power Transformer by Means of Finite Element Software

R. García-Gil, J. M. Espí, S. Casans, J. Jordán and J. Castelló

Dpto. Ingeniería Electrónica, Universidad de Valencia, Apartado 46100 Burjassot, Valencia, Spain

Abstract. The behaviour of transformers and inductors working at high-frequencies and high power density can dramatically differ from the ideal or low-frequency operation. Their performance is very dependent on the magnetic material, operating frequency and physical construction. In this paper, finite element analysis (FEA) is used to obtain an accurate characterization of the electromagnetic behaviour of these magnetic components.

INTRODUCTION

Transformers and inductors are essential and important elements of power systems. Normally, they are analysed as ideal or linear components, but their behaviour when working at high-frequencies and high power density can be dramatically different from your ideal or low-frequency operation. The non-ideal proximity and skin effects increase the AC resistance of the windings at high frequencies, and consequently increase the dissipated power by Joule effect, heating the component. Other parasitic parameters are the leakage inductance and the equivalent capacitance of the windings, which become more and more important in the performance of the magnetic component in transient studies; i.e. the increasing of the leakage inductance leads to over-voltage in the switches of a power supply that reduce efficiency and can lead to its destruction.

Determination of these parasitic components is very useful to obtain an accurate model of the power transformer for electrical simulation. However computation of the equivalent parameters is difficult and unreliable due to the complexity of the geometry of the core and windings. Moreover, all of these parasitic elements are very dependent on the magnetic material, operating frequency and physical construction.

When increasing working frequency and power density, the reduction of these parasitic parameters is mandatory and different core geometries and winding strategies have to be analysed. Although the qualitative effects of the winding technique is very well known by most designers, quantifying them is difficult without building the component. Due to component complexity, overcoat at high-power application (more than 500 kW) at which windings are not only copper wires but water cooled tubes [1], they are difficult to machine. For this reason, we propose to analyse these parasitic elements by means of FEA software. In present work, the FLUX2D software [2] was used and a systematic procedure based on the open and short-circuit analysis is presented to obtain these parameters.

As an example, in this paper FLUX2D was used to compare two winding strategies at two different frequencies to analyze how this can affect to the parasitic elements.

ELECTRICAL PARAMETERS FOR THE TRANSFORMER

The electrical model of the power transformer used in the present work is shown in Fig. 1. The circuit is consisted of the series resistance R_p and R_s (both AC and DC) of the primary and the secondary winding, the leakage inductance of the primary L_{dp} and secondary L_{ds} windings, the magnetizing inductance L_μ, the resistance R_μ that represents core losses due to hysteresis phenomenon of the magnetic material and the equivalent stray capacitance of the windings C_{str}. In fact, C_{str} represents the combination of the stray capacitance due to the turn-to-turn electrical coupling in the primary (C_p) and the secondary windings (C_s) and the electrical coupling between the primary and secondary windings (C_{ps}) [3].

These parameters can be obtained by the open and short-circuit analysis, as described bellow.

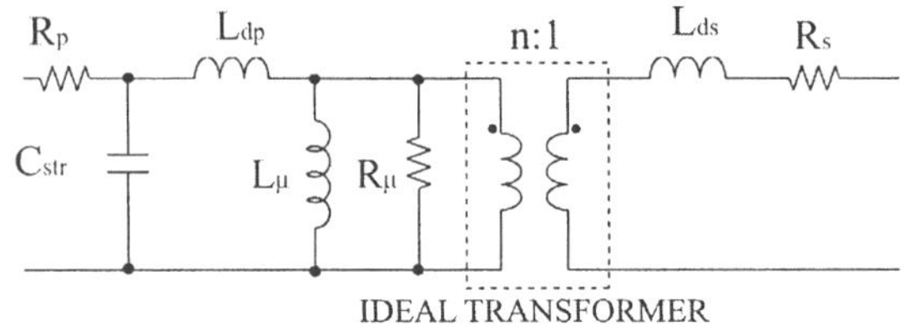

FIGURE 1. Equivalent circuit of the power transformer.

RESULTS AND DISCUSSION

We have made computations with FLUX2D in order to obtain the electrical parameters of the power transformer. As an example, in present work, we analyze the effect of the winding strategy in all of these parameters.

The analysis has been performed for a magnetic core with U+I geometry, as shown in Fig. 2. The selected magnetic material has the following characteristics: μ_i= 6545, B_{sat}= 510mT @ H= 3000A/m, thermal conductivity 4W/m°C and specific heat $0.36x10^7$ J/m3°C. The transformer has been excited by means of a sinusoidal waveform of 500 V with a frequency in the range of 10 to 500 kHz. It is composed by a 20 turns winding on the primary and a single turn on the secondary. Windings are made of squared copper and they have an inner hole for water cooling.

Two winding strategies were considered. In the first one, the primary and secondary are wounded on different columns of the transformer, as can be shown in Fig. 2. In the second technique the primary and secondary are wounded on the same column (sandwich structure).

The topology to be analyzed is actually 3D, but it can be reduced to three 2D simulations in a similar way as described in [4]. The simulation procedure is based on the division of the magnetic component in three parts, each one producing field distribution in different planes of the space. The resultant 2D geometries are depicted in Fig. 2.

In order to get accurate computation, dense meshing was used in the regions between windings and magnetic material, where high current density and high flux density are expected.

With this geometry we made a magnetodynamic computation that allows obtaining non-uniform current distribution on inductors, equiflux isovalues and the distribution of the flux density. Beside, we got also currents, voltages, active and reactive power values in each region. As an example in Fig. 3 we represent the flux density **B** in the transformer.

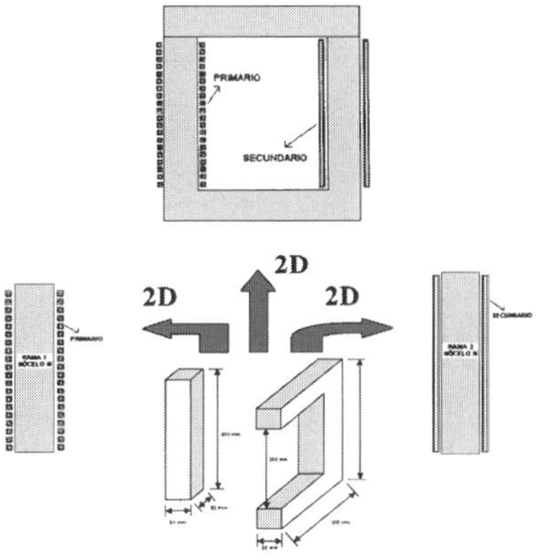

FIGURE 2. Reduction of the 3D geometry to three 2D geometries.

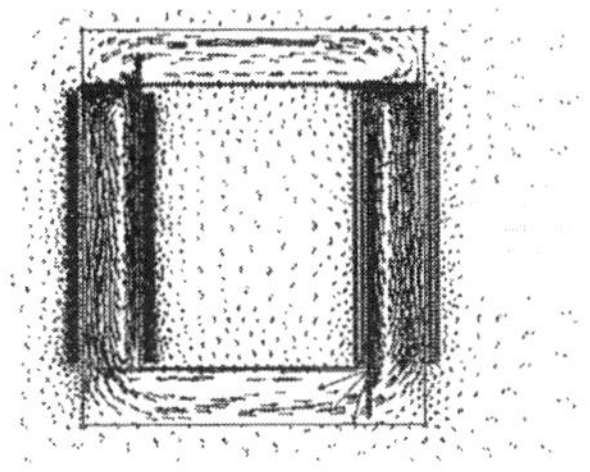

FIGURE 3. Flux density in the transformer (B_{max}= 53 mT)

No load analysis

By considering the secondary in open circuit, the resultant equivalent circuit is shown in Fig. 4(a), where R_μ is supposed to be high enough to be removed as compared with the impedance of L_μ. This fact means that core losses associated with hysteresis phenomenon of the magnetic material are expected to be low. Actually, this parameter can be computed directly by simulation with FLUX2D using the command P_iron.

Under no load condition we compute the current and voltage at primary side that will give information on the input impedance $Z_{p\_open} = V_1 / I_1$. This impedance can be calculated from circuit of Fig. 4(a) as follows,

$$Z_{p\_open}(f) = \frac{\left(L_\mu + L_{dp}\right)\cdot 2\pi \cdot f \cdot j}{1-\left(L_\mu + L_{dp}\right)\cdot C_{str} \cdot 4\pi^2 \cdot f^2} + R_p \tag{1}$$

where f is the working frequency.

Moreover, the value (L_μ + L_{dp}) can be directly computed by simulation with compute / inductance command. The resistance of the primary winding (R_p) will be calculated from the computed active power at the primary side. The last one is directly the AC+DC resistance, because computations with FEA software take into account the proximity and skin effect.

Similar analysis can be performed with the primary in open circuit. The resultant equivalent circuit is shown in Fig. 4(c) and the impedance to analyse is named Z_{s_open} which can be calculated by elementary circuit analysis. The resultant equation will be a function of the operating frequency, the transformer turns ratio n and the electrical parameters L_μ, L_{dp}, L_{ds}, C_{str}. This impedance can be obtained computing with FLUX2D the current and voltage at the secondary side $Z_{s\_open} = V_2 / I_2$ for this load condition. Mathematical description has been omitted for simplicity.

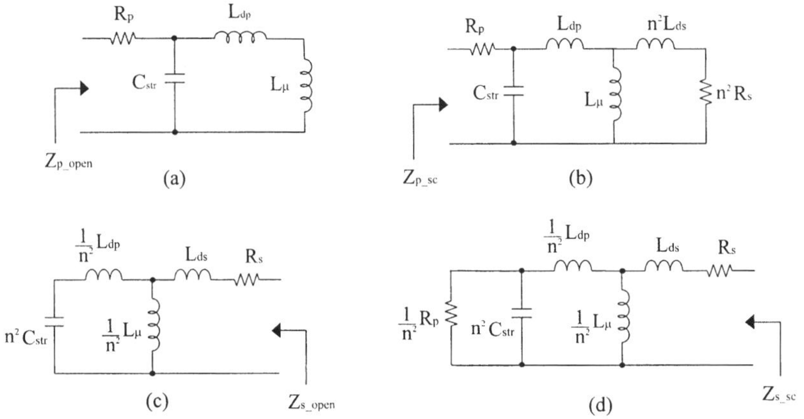

FIGURE 4. Equivalent circuit of the transformer (a) with the secondary in open circuit; (b) with the secondary in short-circuit; (c) with the primary in open circuit; (d) with the primary in short-circuit.

Short-circuit analysis

By considering the secondary in short-circuit, the resultant equivalent circuit is shown in Fig. 4(b). Under this load condition we compute the current and voltage at the primary side to determine the impedance Z_{p_sc}. Also, the Z_{s_sc} can be calculated from computation of the voltage and current at the secondary side when the primary is in short-circuit condition, see Fig. 4(d). By elementary circuit analysis it can be obtained Z_{p_sc}, Z_{s_sc} as a function of the operating frequency, the transformer turns ratio n and the electrical parameters L_μ, L_{dp}, L_{ds}, C_{str}.

In all these analysis it is essential to ensure that, for the test conditions, the magnet is far from saturation.

Finally, the four unknown L_μ, L_{dp}, L_{ds}, C_{str} will be obtained as a solution from the system of equations composed by Z_{p_open}, Z_{s_open}, Z_{p_sc} and Z_{s_sc}.

TABLE 1. Electrical parameters of the power transformer for winding technique 1.

	L_μ (mH)	R_{prim} (mΩ)	C_{prim}(pF)	L_{dp} (μH)	L_{ds} (nH)
100 kH	1.31	81.4	172	39.4	109.5
500 kH	0.61	253.6	42	34.9	118.1

TABLE 2. Electrical parameters of the power transformer for winding technique 2.

	L_μ (mH)	R_{prim} (mΩ)	C_{prim}(pF)	L_{dp} (μH)	L_{ds} (nH)
100 kH	1.31	31	215	0.47	3.2
500 kH	0.493	59.3	71	49.1	4.3

From Table I and II can be concluded that while configuration 1 reduce the leakage inductance, this one increases the primary capacitance. Depending on the particular application it will be more interesting to reduce the leakage inductance (i.e. power rectification) or the parasitic capacitance (i.e. power supply for induction heating). In both cases the resistance increases with the frequency, due to skin and proximity effect, but it is larger for the configuration 1. The current tends to flow through windings close to primary, so current will be better distributed in configuration 2, reducing the total AC resistance and, consequently, the dissipated power. Therefore, the way to reduce AC resistance is applying “interleaving” or “sandwich” strategy between windings.

CONCLUSION

The electrical parameters of a power transformer have been obtained by finite element analysis software. Determination of these parameters is really useful to obtain an accurate model for computer simulation.

Two different winding strategies have been analysed. From these analysis several general conclusion have been obtained. Reduction of the AC resistance can be obtained placing secondary windings near the primary (sandwich strategy), because current will flow thought the winding close to primary in order to minimize the energy. I nterleaving of winding reduce leakage inductance, but increase the stray capacitances.

REFERENCES

1. García-Gil, R., “Optimization of the Electrical Parameters in a Power Transformer”, Flux Magazine **33**, pp. 11-12, 2000.
2. FLUX2D v 7.30 User’s Guide (1998)
3. Lu, H. Y., Zhu, J. G., Ramsden, V. S., and Hui, S. Y. R., “Measurement and Modeling of Stray Capacitance in High Frequency Transformers” in 30th Annual IEEE Power Electronics Specialists Conference, PESC’99 (2), Charleston, South Carolina, 1999, pp. 763-768.
4. Prieto, R., Cobos, J. A., García, O., Alou, P. and Uceda, J., “Model of Integrated Magnetics by means of a “Double 2D” Finite Element Analysis Technique” in 30th Annual IEEE Power Electronics Specialists Conference, PESC’99 (1), Charleston, South Carolina, 1999, pp. 598-603.
5. Prieto, R., Cobos, J. A., García, O., Alou, P. and Uceda, J., “Using Paralle Windings in Planar Magnetic Componentes” in 32nd Annual IEEE Power Electronics Specialists Conference, PESC’99 (4), Vancouver, Canada, 2001, pp. 2055-2060.

Applied Physics (APHYS 2003)
A. Méndez-Vilas (Ed.)

Forced Low-Frequency Cell Oscillations In Human Blood Suspensions

A. RAMÍREZ and A. ZEHE

Benemérita Universidad Autónoma de Puebla, Facultad de Cs. de la Electrónica, Apdo. Post. # 1505, 72000 Puebla, Pue., México. e-mail: aramirs@siu.buap.mx

Abstract. The physical mechanics of cell membranes is an essential aspect of cell deformation and shape oscillations, when exposed to an external electric a.c. field. Dielectrodeformation is related to the induction of a dipole moment due to electric charges on the opposite boundaries of the cell. Dielectric responses of induced dipoles are resonance phenomena, which are expected to occur at intrinsic cell frequencies due to the forced cell oscillations. It has been shown, that higher harmonics of oscillation frequencies go along with the possible appearance of double- or even triple-peak formation in low-frequency impedance spectra. Literature data of experimental low-frequency impedance spectroscopy confirm this finding.

1. INTRODUCTION

Dielectric relaxation spectroscopy in frequency or time domain has gained considerable attraction for studying dielectric properties of biological cell- and particle suspensions [1]. Biological systems are complex. Blood, for instance, is an aqueous solution of many substructures, different in composition, geometrical arrangement and size. The interaction with an external electric field involves multiple relaxation processes, including interfacial polarization around the cells.

The interface polarization in the case of biological cells is related to the complex dielectric permittivity of the cell structural parts, which has led to the physical presentation of a cell by shelled models [1,2]. Multilayer ellipsoidal cell models exposed to electric and magnetic fields up to 100 MHz have been studied [3]. Given the spherical shape of lymphocytes with a thin cell membrane and a spherical nucleus, which by itself is covered by a thin nuclear envelope, a double-shell model has been assumed and characterized by both time domain dielectric spectroscopy and computer modeling [2]. It is known, that the conductivity and the permittivity of nucleoplasm is about twice that of cytoplasm. This fact together with the volume differences of both can play an important role in the interpretation of bioimpedance spectra. While lymphocytes possess a spherical double-shell structure, human erythrocytes are single-shelled and of discoid shape. Nevertheless are dielectric responses in an external electrical field frequently treated within the framework of a spherical model. The highly oblate spheroid with semi axes $a=b>>c$ is transformed under the action of an electrical field in a three-axial ellipsoid with semi axes $a>b>>c$ (a is parallel to the external field). A shelled cell model with one or several membranes embracing electrolytes with mobile dipoles or charge carriers implies also

the possibility for a mechanically vibrating oscillator. Given the elevated oscillator mass, this could happen only at far lower than microwave frequencies.
In the present study frequency-domain dielectric spectroscopy in the low frequency range of 100 Hz...1 MHz was considered in order to test the viability of an oscillatory excitation of cell structures by a low intensity external electric field.

2. DIELECTRIC MODEL OF FORCED CELL RESONANCE

Forced vibrations of an oscillator result, when an external oscillatory force of frequency ω is applied to a particle subject to an electric field. The double-shell lymphocyte bears similarities in the physical concept of motion, where the external force acts on both, the inner (nucleoplasm) sphere and the outer (cytoplasm) sphere. Given the different geometrical and dielectric characteristics, the response of each might be different (Fig.1a). In the case of erythrocytes only one shell exists and the oscillator model could be simpler, although the more complex shape makes it more difficult in the experimental approach (Fig.1b). Let us suppose, that a shelled spherical cell containing a certain dipole charge density is exposed to an harmonically variable electric field. It is easy to imagine, that the elastic properties of the nuclear envelope and the cell membrane provide for a restoring force after a deformation due to the external field action on the mobile ions inside the sphere plasmas has occurred.

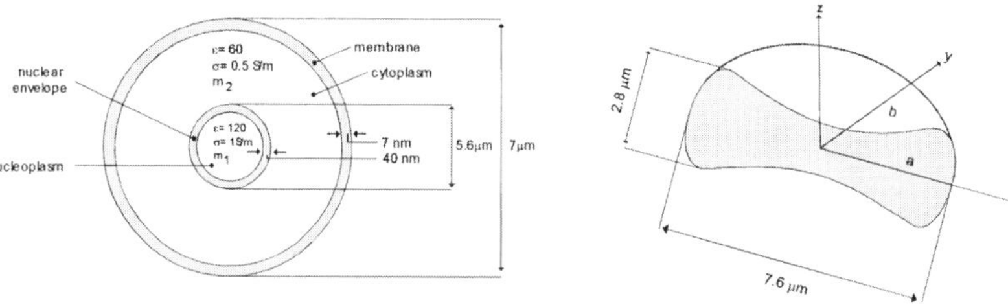

Figure 1(a): Double-shelled lymphocyte cell as proposed by Ermolina et al. [2]; ε is the dielectric constant, σ - specific electric conductivity, m_1 - mass of the inner nucleus (m_1=9·10^{11}g), m_2- mass of the outer shell region (m_2=6·10^{11}g), being m_c= m_1+m_2 the total cell mass. **(b):** Discoid shape of an erythrocyte of diameter 2R=7.6μm and thickness at the rim 2L=2.8 μm. *a, b* represent semi axes *(a=b=R)* in a spheroidal approach with small semi axis c in direction of z.

The periodic deformation of cells in an electric field $E_0(\omega)$ with frequency ω is determined by the distribution of the electrical forces applied and the mechanical forces generated in the deformed membranes and in the adjacent layers of the cell. The mechanical forces include elastic and viscous shear stresses in the membranes. The axis relation $q=R/L$ with L the longer half axis of the field-deformed ellipsoid of radius R (i.e. $q=1$ corresponds to a spheroid) characterizes both, the elongation (contraction) deformation and the forces, which determine the deformation [4].
The electric ponderomotive force is caused by the applied time-dependent field $E_0(t)$, and contains the dielectric properties of the cell:

$$F_{pond}(\omega,q) = V \cdot F(\omega)\left|E_{loc}(\omega,q)\right|^2 \cdot \frac{\partial f_L(q)}{\partial q}. \quad (1)$$

Here $f_L(q)$ is the depolarizing factor in direction of the main axis L parallel to the external field vector E_0. The local field $E_{loc}(\omega, q)$ is the value of the applied field inside the ellipsoid:

$$E_{loc}(\omega,q) = \Phi(\omega,q)\cdot E_o;\ \Phi(\omega,q) = \frac{\varepsilon_m^*(\omega)}{\varepsilon_m^*(\omega) + [\varepsilon_p^*(\omega) - \varepsilon_m^*(\omega)]\cdot f(q)} = \alpha(q)\cdot\varepsilon_m^*(\omega) \quad (2)$$

The factor $\alpha(q)$ has been calculated previously for the general case of a dielectric object with arbitrary shape [5]. The frequency dependence of the ponderomotive electric field force is given by $F(\omega)$

$$F(\omega) = \frac{\varepsilon_0\varepsilon_m}{4}\left[(\frac{\varepsilon_p}{\varepsilon_m} - 1)^2(\omega\tau_e)^2 + (\frac{\sigma_p}{\sigma_m})^2 - \frac{2\varepsilon_p}{\varepsilon_m} + 1\right]\cdot\left[1 + (\omega\tau_e)^2\right]^{-1} \quad ;\ \tau_e = \varepsilon_0\varepsilon_m/\sigma_m. \quad (3)$$

Here ε_p and ε_m are the static relative permittivities of the cell and the suspending medium, σ_p and σ_m are the static conductivities, and ε_0 is the vacuum permittivity.
The elastic force of the cell membrane is given by

$$F_{ela}(q) = -\tfrac{1}{2}\cdot S\mu(1 - q^2) \quad (4)$$

with S the surface area of the object and μ the shear elasticity modulus.
The friction force F_{fr} $(q, dq/dt)$ between the cell and the surrounding medium is

$$F_{fr}(q, dq/dt) = -S\cdot\eta_s q^2\frac{dq}{dt} \quad (5)$$

with η_s the surface viscosity coefficient of the membrane.
Due to the small mass of the considered object and low velocities of deformation dq/dt, it is safe to exclude the inertia term in the equation of cell motion. All forces acting on the cell will then fulfill the condition of instantaneous equilibrium; i.e., its sum must be equal to zero. Introducing $\tau_0 = \eta_s/\mu$ as time of viscoelastic relaxation of the membrane, and summing up the forces given by ecs. (3,4,5), it follows

$$\left[1 - F(\omega)\cdot\left|\Phi(\omega,q)\right|^2\cdot\left|\frac{\partial f(q)}{\partial q}\right|\cdot\frac{VE_0^2}{\mu S}\right]\cdot q^{-2} = 1 - 2\tau_0\frac{dq}{dt}. \quad (6)$$

This equation describes the time-dependent elongation (contraction) of the ellipsoid, depending on geometric (V, S), electrical $(\varepsilon_p, \sigma_p)$ and mechanical (μ, η_s) cell parameters, as well as on experimental properties of the medium $(\varepsilon_m, \sigma_m)$. As can be appreciated in ec. (1), the cell deformation is caused not only by the electric field amplitude E_0, but by the dimensionless force parameters $(VE_o^2/\mu S)^{1/2}$. If we apply a time-dependent electric field $E_0(t) = E_0[1+\delta\cos\Omega t]$ of frequency Ω, modulation depth δ, and consider the periodic changes of the axis relation $q(t)$ as forced oscillations of the ellipsoid, a solution of ec. (1) is found, which contains such oscillations not only at the frequency Ω but also at its higher harmonics. Details of the calculations are given in [4]. For small-signal modulation $\delta \ll 1$ of the electric field amplitude, one gets

$$q(t) \sim \frac{\cos(\Omega t - \varphi_1)}{\left\{1 + [\Omega \tau(E_o)]^2\right\}^{1/2}} + \frac{\delta}{2} \frac{\cos(2\Omega t - \varphi_2)}{\left\{1 + [2\Omega \tau(E_o)]^2\right\}^{1/2}} + \cdot\,; \tau(E_o) = \tau_o/q_o\left(1 + Bx_0^2/q_0^2\right);\ x_o^2 = VE_o^2/\mu S \quad (7)$$

φ_1 and φ_2 are phase angles of the harmonics, and related to tan $\varphi_n = n \cdot \Omega\tau(E_o)$.

3. COMPARISON TO EXPERIMENTAL SPECTRA

If by any reason forced oscillations of cells give rise to a dissipative process at a peak frequency Ω of the external field, one should expect peak repetitions at 2Ω, and possibly at 3Ω. Indeed would such a peak sequence in a spectrum be indicative, that a forced oscillation of cells has occurred.

In contrast to the relaxation type response of dielectric objects with permanent dipoles, the dielectric response of induced dipoles are resonant phenomena, where energy resonance occurs at frequencies ω_0, which are easily calculated from the expressions for $\chi'(\omega)$ and $\chi''(\omega)$, and result in $\Omega_0 \cong [2Ne^2\Delta\Omega/\varepsilon_0 m\chi']^{1/3}$.

Here N is the number of induced dipoles per unit volume, m is the particle (oscillator) mass and e the elementary charge. $(\Omega_0^2-\Omega^2)=(\Omega_0+\Omega)(\Omega_0-\Omega)$ was substituted by $2\Omega_0\Delta\Omega$ as a first-order approximation for $\Omega \approx \Omega_0$.

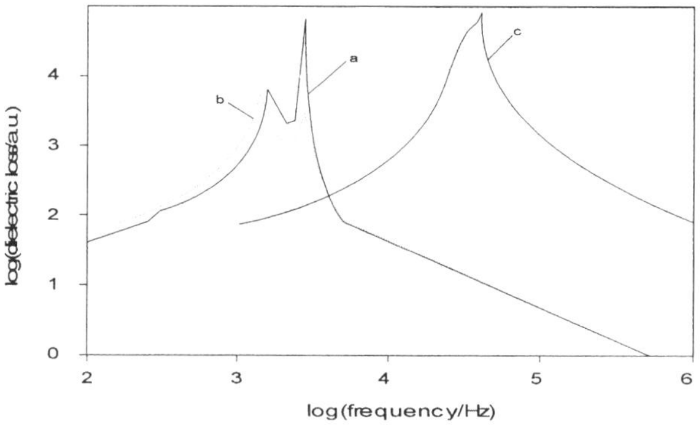

Figure 2: Spectra of dielectric loss D vs. frequency ν [1/s] of human blood, as published by Vázquez et al. [6]. The sample is described to be composed of (a) 0.1 ml blood + 20 ml H_2O; (b) 0.1 ml blood + 20 ml H_2O + 0.5 ml ethanol, introduced into a parallel-plate capacitor cell. (c) is a different donor.

In order to arrive at numerical values, we apply for $N \cong 10^{24}\ m^{-3}$, $m = 1.5 \cdot 10^{-10}$ g, $\chi' = 120$, and $\Delta\Omega = 200\ s^{-1}$, which correspond to realistic experimental parameters of a certain type of human blood cells. Here $\Delta\Omega$ is the width of the single resonance line, and $\Omega_0 \cong 1.7\ kHz$ is the experimentally determined [6] and by ec. (7) reproduced resonance frequency.

A second peak could then be found at 3.4 kHz, which is actually seen in Fig. 2 of the mentioned spectra.

4. DISCUSSION

The low-frequency dispersion occurring below a few kHz is often assigned to counter-ion displacement about molecular structural parts or cell membranes. Moreover are biological tissues inhomogeneous and show considerable variability in structure and composition, and hence in the dielectric response. Although such variations are natural and may be due to physiological processes or other functional requirements, they make the meaningful interpretation of the measuring results an involved task. Additionally are relaxation processes of the sample material often obscured by electrode polarization. The implications of these effects deserve particular consideration.

The appearance of pronounced peaks in low-frequency impedance spectra of human blood suspensions seems to support the forced oscillatory behavior of cells with the induction of dipole moments corresponding to the oscillation frequency of the external field. The dielectric responses of induced dipoles are resonant phenomena. The difference to other macromolecular structures arises from the fact, that the cell electrolytes are constrained in elastic membranes, which make them interact as a whole with the harmonically oscillating electric field. The elastic deformation of each single cell under the action of the slowly varying external field from spherical to oblate with harmonically forth and back changing the direction of the internal polarization vector makes the charges oscillate between two preferred localization sites, or the dipoles to assume one of two discrete orientations in space. Indeed would this picture represent an array of polarizable macrostructures with mutual interactions only by way of the counter-ions. The precise effect of the latter is not yet clear.

Certainly, electrode polarization and thus the generation of electrode artifacts deserves attention. The measurements of conductive materials in the frequency range 1 Hz to 10 MHz are not so straight-forward as in higher frequency regions. Electrode polarization is a manifestation of molecular charge organization, which in the presence of water molecules and hydrated ions occur at the sample-electrolyte interface. As discussed previously [7], the effect increases with increasing sample conductivity, and its consequences are more pronounced on the capacitance of the ionic solutions and biological samples. It has been found that in the case of biological material the poorly conducting cells shield part of the electrodes from the ionic current and reduce polarization effects.

ACKNOWLEDGEMENTS

Financial support of SEP-SESIC, México, under contract #2003-21-001-023 is gratefully acknowledged.

REFERENCES

1. Irimajiri, I., Hanai, T., Inouye, A., *J. Theor. Biol.* **78,** 251-269 (1979).
2. Remolina, I., Polevaya, Yu., Feldman, Yu., *Eur. Biophys. J.* **29**, 141-145 (2000).
3. Sebastián, J.L., Muñoz, S., Sancho, M., Miranda, J.M., *Phys. Med. Biol.* **46**, 213-225 (2001).

4. Kononenko, V.L., Kasymova, M.R., *Biol. Membrany*, **8**, 297-313 (1991).
5. Zehe, A., Ramírez, A., *Rev. Mex. Fís.* **48**, 427-431 (2002).
6. Vázquez, J., Starostenko, O., Martínez, J., Gutiérrez, F., Proceedings 1° Congreso Latino-Americano de Ingeniería Biomédica, México, 1998, pp. 686-688.
7. Schwan, H.P., *Ann. Biomed. Engineering* **20**, 269-288 (1992).

VERY LOW GAMMA-RAY ACTIVITY PORTABLE INSTRUMENT FOR THE DETERMINATION OF %Pb ON Pb-Zn ORE SURFACE

MIHAI BORSARU, MARK BERRY, CRAIG SMITH AND ANDREW ROJC
[1]CSIRO-Exploration and Mining, P.O. Box 883, Kenmore, Queensland 4069, Australia,

ABSTRACT

A Surface Analyzer instrument for the determination of %Pb in Zn-Pb ore was developed and tested in the laboratory. The instrument is based on the backscattered γ-γ technique in a 2π geometry and employs two γ-ray microsources: 1.1 MBq ^{133}Ba and 37 kBq ^{137}Cs. The instrument weighs less than 2 kg and the time of measurement is 15 sec per sample.

Keywords: Backscatter gamma-gamma; In situ analysis; Surface analyzer; %Pb

INTRODUCTION

Geophysical techniques are well established in the oil, gas, uranium, coal and mineral industries. Today they are largely used in mining applications in all stages: exploration, mine development and mine production. Nucleonic instrumentation based on nuclear techniques, which represent a subset of this group, is used for on-stream analysis and borehole logging applications in the coal mining industry and is slowly making in-roads in the metalifferous mining industry.

One drawback for the nucleonic instruments is the reluctance of some industries or mines to employ nuclear techniques due to the fact that they use radioactive sources; either gamma-ray or neutron sources. To overcome this problem, research was carried out in the last 9 years to develop portable nucleonic instruments that use very low activity gamma-ray sources, below 2 MBq. Such instruments are light in weight because they do not need special shielding to protect the user from radiation and the user is not exposed to unacceptable levels of radiation. One instrument was developed for the determination of ash on the coal face (Borsaru et al., 1997; Borsaru et al., 2001). It employs two γ-ray sources: a 1.1 MBq ^{133}Ba source as the primary source of radiation and a 37 kBq ^{137}Cs for gain stabilization. The technique for the determination of ash on the coal face is the backscattered γ-γ technique in a 2π geometry. Another instrument was reported for the determination of ash in coal stockpiles (Borsaru et al., 2002). It employs two γ-ray sources: a 1.1 MBq ^{133}Ba source and a 370 kBq ^{137}Cs source. The instrument is based on the backscattered γ-γ technique in a 4π geometry. Both instruments are handheld and the measuring devise weighs less than 1.5 kg.

The present work deals with the determination of %Pb on the Pb-Zn ore surface by the backscattered γ-γ technique in a 2π geometry using two γ-ray sources: ^{133}Ba and ^{137}Cs of very low activity. The instrument does not require any special radiation shielding, is handheld (1.5 kg weight) and takes less than 15 seconds per measurement. The surface analyzer has been commercialized.

METHODOLOGY

Principle of %Pb determination. The lead grade is determined from the backscattered γ-ray spectrum. Figure 1 shows the backscattered spectra recorded on the 10 samples of crushed Zn-Pb ore available for the present work. A Cd-Cu filter was placed on the plastic cover of the instrument in front of the ^{133}Ba source to cut the 80 keV and 32 keV γ-rays released by ^{133}Ba. However, some 80 and 32 keV γ-rays are backscattered by the plastic cover and appear in the recorded spectrum. The peaks at ~28 keV in the backscattered spectra do not show variation with the lead content in the samples and are produced by the plastic cover of the instrument. The peaks at ~ 80 keV show good variation with the lead content of the samples and are the 73 keV and 87 keV lead X-rays generated in the samples be the multiscattered γ-rays produced by ^{133}Ba. The count-rates measured in the energy region from ~110 to ~170 keV show good variation with the lead content in samples and they go in opposite direction with %Pb. The explanation for this behavior is that the photoelectric absorption cross-section per atom as function of photon energy E is given by $\sigma(E) \approx \frac{Z^{4.5}}{E^n}$, where $2.5 \le n \le 3.5$ and Z is relates to the number of protons in the atom. Therefore the count-rates in high grade Pb samples (high Z average) are lower than count-rates in low grade lead samples (lower Z average).

Fig. 1. Backscattered spectra collected by the surface analyzer on Zn-Pb samples with Cd-Cu filter.

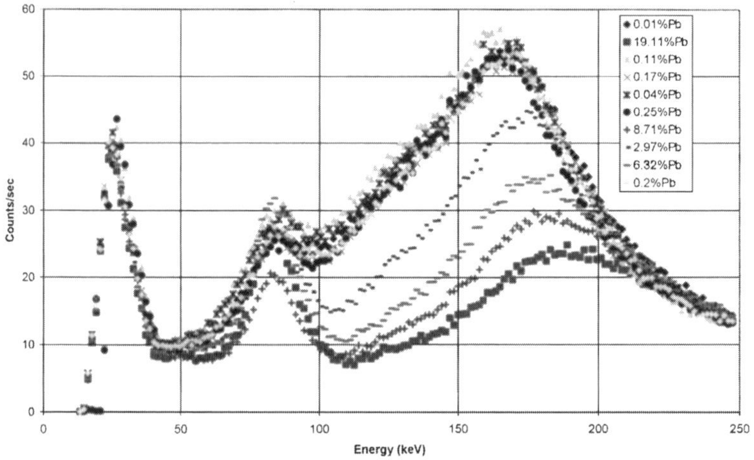

The lead content in the present work is determined by the count-rates recorded in these energy regions.

The instrument. The instrument used for the determination of %Pb in the present work was initially developed for the determination of ash on the coal face (M. Borsaru et al., 1997; M. Borsaru et al. 2001). The work described in this paper is an extension of its applications.

The primary source of γ-radiation is ^{133}Ba of activity 1.1 MBq. The prominent γ-rays released by the source have energies of 300 keV, 350 keV, 80 keV and 32 keV. Gain stabilization is essential for obtaining high accuracy of measurement and is provided by a ^{137}Cs γ-ray source of activity 37 kBq. Figure 2 shows the source-detector-shielding configuration designed for a 2π surface measurement. Lead provides the shielding between the 37 dia x 25 mm NaI(Tl) scintillation detector and the two γ-ray sources. A Cu-Cd filter lines the lead shielding to cut down the 73 and 87 keV X-rays generated in lead by the backscattered γ-rays. Such X-rays would act as a variable background in the backscattered spectrum and would worsen the accuracy. The distance between the ^{133}Ba source and the end of the lead shielding, which houses the detector, is 25 mm. The instrument consists of the handheld measuring device that weighs ~1.5 kg and the display unit that weighs ~2.5 kg. They are connected by a coiled cable ~2 m long. The scintillation detector, the preamplifier and HV generator are located in the handheld device and a 40 x 2 line Alpha-numeric LCD is used for output of %Pb content in the display unit. This unit also houses the acquisition electronics, CPU, LCD and battery pack.

Fig. 2. Sources-detector-shielding configuration of the surface analyzer

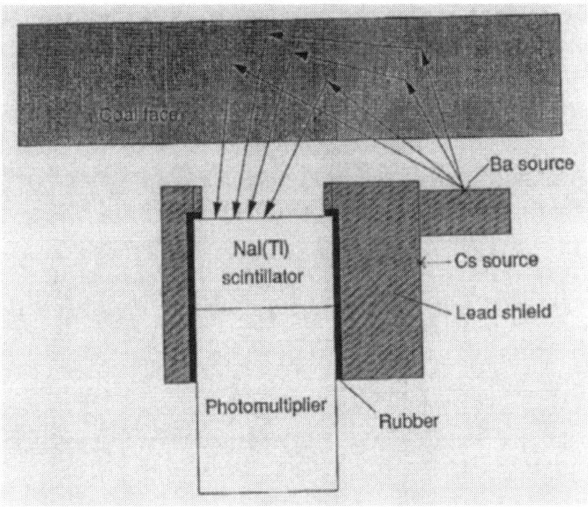

Data analysis. Linear regression analysis is used for the determination of %Pb. Windows are set in different energy regions of the backscattered spectrum, as mentioned above, and a calibration for the determination of %Pb is obtained by fitting a linear regression model of the form:

$$\%Pb = a_0 + a_1X_1 + a_2X_2$$

where a_0, a_1, and a_2 are constants and X_1 and X_2 are variables corresponding to count-rates recorded in the energy windows or ratios of the count rates.

LABORATORY INVESTIGATION

10 samples were available for the present work. The samples were prepared from crushed Zn-Pb ore to ~40 mm particle size. Each sample was contained in a 4 litre plastic bucket and the measurements were taken on the surface of the ore. The elemental composition of the samples is given in Table 1. A block of Zn-Pb ore was

also selected from the mine to correspond to each of the samples selected. Its average Pb content should be, theoretically, close to the Pb content of the samples. However, the Pb distribution in the ore blocks is not homogeneous. For this reason, the crushed samples were used to test the accuracy given by the instrument for the determination of %Pb. The instrument was calibrated on the crushed samples and later used to predict the %Pb on the surface of the ore blocks. One face of each ore block was polished and the material collected was analyzed in the laboratory to estimate the lead content of the ore blocks. Since the lead content of the crushed samples can be estimated with high accuracy by sampling and laboratory analysis, the determination of the lead content of the ore blocks was not used to assess the accuracy of this technique.

Two measurements were taken on each sample on different areas of the sample, thus artificially doubling the number of data points in the regression analysis. Two sets of measurements were taken on the 10 samples of crushed Zn-Pb ore. In the first set of measurements a Cd-Cu filter was placed on the surface of the surface analyzer to stop the 80 and 35 keV γ-rays released by the ^{133}Ba source reaching the Zn-Pb ore. Another set of measurements was taken without the Cd-Cu filter so that all the γ-rays reach the samples. The backscattered spectra measured in the two sets of measurements are shown in Figure 1 and Figure 3 respectively. The count-rates for the 80 keV peak in Figure 3 is much higher than the count-rates shown in Figure 1. This is due to the fact that the 80 keV peaks for the measurements taken without

Table 1 Elemental composition of the Zn-Pb samples

Sample No	Density	Pb	Zn	S	Fe_2O_3	SiO_2	Al_2O_3	BaO	CaO	MgO	K_2O	MnO
1	2.81	0.11	0.24	1.05	4.76	63.0	17.5	0.27	2.74	1.73	6.24	0.65
2	2.94	0.17	0.17	5.72	9.13	46.8	21.6	1.05	1.94	2.46	7.31	1.26
3	4.5	19.11	38.9	28.3	8.61	4.72	0.54	1.09	0.32	0.21	0.19	0.19
4	2.80	0.04	0.53	1.49	5.99	71.8	13.6	0.11	0.18	1.42	4.42	0.9
5	2.97	0.25	5.46	10.2	12.7	61.5	6.8	0.11	0.08	0.61	2.27	1.06
6	3.61	8.71	16.3	17.1	6.81	22.6	6.0	14.4	0.01	0.46	1.66	0.12
7	3.19	2.97	11.7	16.0	15.5	31.5	8.64	0.24	3.44	1.07	3.03	3.51
8	3.31	6.32	18.2	14.2	8.13	37.0	10.3	0.55	0.29	0.87	3.31	0.49
9	2.95	0.2	0.52	3.50	10.47	58.5	17.2	0.56	0.54	2.71	4.84	0.77
10	2.96	0.01	0.08	1.74	6.66	50.9	16.0	0.72	0.48	1.24	5.09	7.98

the Cd-Cu filter are a superposition of the 80 keV X-rays excited in the lead atoms of the ore samples and the near 80 keV γ-ray peaks produced by the backscattered 80 keV γ-rays released by the primary source of radiation ^{133}Ba. The data analysis was carried out by setting up windows in the gamma ray spectra at energies encompassing the 80 keV peak and above this peak where the spectra show good sensitivity between the count-rates and the %Pb in the samples.

The data collected with the Cd-Cu filter were also processed by regression analysis considering the count-rates in the net peak areas of the 80 keV peaks only, instead of the total count-rates in the energy windows set around the peaks. This entailed the subtraction of exponential backgrounds from the total count-rates recorded in the energy windows set around the 80 keV peaks. Table 2 shows the RMS deviations between the laboratory assays and the surface analyzer predictions for the laboratory measurements. The range of %Pb in the samples, as shown in Table 1, was between 0.01 and 19.1 %Pb with a standard deviation of 6.04 %Pb. The regression equations and the energy windows selected for the calibrations are shown in Table 3.

Fig. 3. Backscattered spectra collected by the surface analyzer on Zn-Pb samples without Cd-Cu filter.

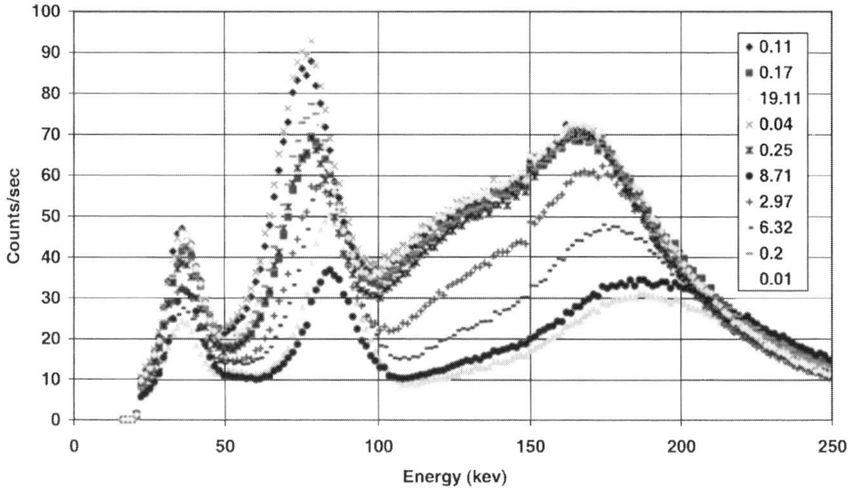

Table 2 Regression analysis results

Type of measurement and data analysis	RMS deviation (wt%)	Correlation coefficient
Measurements taken without Cd-Cu filter and no background subtraction for the 80 keV peak in regression analysis	0.8	0.99
Measurements taken with Cd-Cu filter and no background subtraction for the 80 keV peak in regression analysis	1	0.99
Measurements taken with Cd-Cu filter and background subtraction for the 80 keV peak in regression analysis	0.6	0.99

Table 3 Regression equations for the determination of %Pb

Type of measurement and data analysis	Regression equation	Energy windows and ratios (keV)
Measurements taken without Cd-Cu filter and no background subtraction for the 80 keV peak in regression analysis	%Pb = - 4.3 + 18 x Rat A- - 0.0076 x Roi A	Rat A = (48 – 96) / / (120 – 165) Roi A = (52.5 – 91.5)
Measurements taken with Cd-Cu filter and no background subtraction for the 80 keV peak in regression analysis	%Pb = -6.8+35.5 x Rat B - - 0.012 x Roi B	Rat B = (52.5 - 112.5) / / (120 - 180) Roi B = (52.5 - 112.5)
Measurements taken with Cd-Cu filter and background subtraction for the 80 keV peak in regression analysis	%Pb = 2.6+77.38 x Rat C - - 1191.78 x Roi C	Rat C = (52.5 – 112.5) / / (120 – 180) Roi C = (52.5 – 112.5)

Table 2 shows that there is not much difference in the RMS deviations between the measurements taken with the Cd-Cu filter and without the filter when no background subtraction is used in the regression analysis. However, there is a significant improvement when the background subtraction is applied. Figure 4 shows a cross plot between the predicted %Pb by regression analysis with background subtraction as given in table 3 and the laboratory analysis assays. The measurements were taken with a Cd-Cu filter.

Fig. 4. %Pb (surface analyzer) vs. chemical analysis given by regression analysis

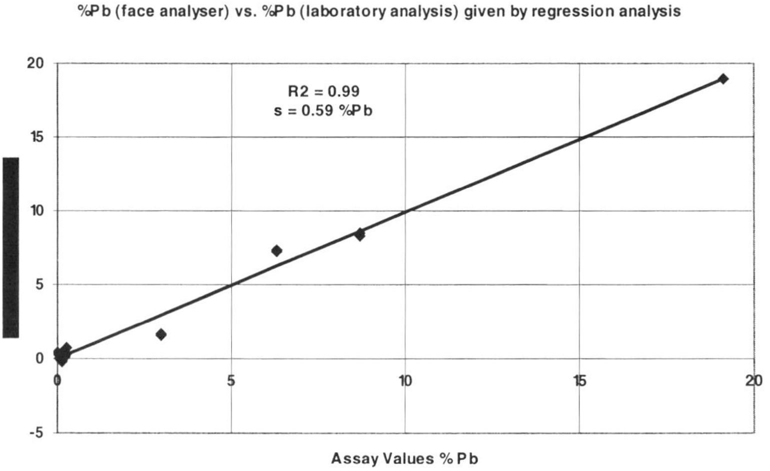

The regression equations were used to predict the %Pb on the surface of the blocks of Zn-Pb ore. Measurements were taken on different sides of the blocks (only one side was polished) that were large enough in size for the measurement. Because the blocks are not homogeneous in regard to the Pb distribution, an average for the %Pb prediction on different faces was taken. Table 4 gives the results of the surface analyzer's predictions and the chemical assays for the rocks. One should mention that the chemical assays for the Zn-Pb blocks are not accurate due to the fact that the %Pb of the blocks are estimated from the ore collected as a result of polishing one face.

Table 4 %Pb (laboratory assays) vs. surface analyzer predictions on the Zn-Pb rocks

Sample No	Chemical Assay (%Pb)	No Cd-Cu filter and no background subtraction (%Pb)	Cd-Cu filter and no background subtraction	Cd-Cu filter and background subtraction
1	0.1		0.4	0.2
2	0.06	0	0.1	1.5
3	9.96	8.7	7.7	7.6
4	0.01		0.3	0.4
5	0.03		0	0.4
6	7.06	9	8.6	8.6
7	0.33	0	0	1.3
8	4.21	4	3	1.4
9	0.05	1	0	0.7
10	0.05		0.1	0.9

The RMS deviations between laboratory assays and nuclear predictions shown in table 4 in columns 3 to 5, were 0.7, 0.7 and 1 respectively. They are comparable to the RMS deviations given by the regression equations.

SUMMARY AND CONCLUSIONS

This work demonstrates that the surface analyzer can determine the %Pb on the surface of Zn-Pb ores with acceptable accuracy. The two gamma-ray microsources used in the present instrument had a combined activity of 1.1 MBq. The instrument does not require extra shielding and it does not expose the user to unacceptable levels of radiation. The time of measurement is 15 seconds.

The primary source of radiation used in the instrument was ^{133}Ba. However, ^{137}Cs can be an alternative source to ^{133}Ba due to the fact that the accuracy for Pb determination did not change significantly when the 80 keV γ-ray from ^{133}Ba was cut by the Cd-Cu filter.

The depth of penetration for Pb determination given by this technique is larger than the penetration given by an XRF technique. This is due to the fact that the lead X-rays are excited by the γ-rays multiple scattered in the Zn-Pb ore.

REFERENCES

Borsaru M., Ceravolo C., Carson G. and Tchen T. (1997) Low radioactivity portable coal face ash analyzer. Appl. Radiat. Isot. 48, 715-720.

Borsaru M., Dixon R., Rojc A., Stehle R. and Jecny Z. (2001) Coal face and stockpile ash analyzer for the coal mining industry. Appl. Radiat. Isot. 55, 407-412.

Borsaru M., Charbucinski J., Rojc A., Thanh N.D. and Tuy N.T. (2002) Probe for the determination of ash in coal stockpiles. IRRMA-V, 5th International Topical Meeting on Industrial Radiation and Radioisotope Measurement Applications, Bologna, Italy, 9-14 June 2002. (In press. Nucl Instr. Meth. B)

Applied Physics (APHYS 2003)
A. Méndez-Vilas (Ed.)

Microscopic Optical Interferometry Study of the Cottrell Atmospheres in Si-Doped GaAs

M. A. González[1], L.F.Sanz[1], M. Avella[1], J. Jiménez[1], J. Adiego[2], P.F. Redondo[2], R. Frigeri[3]

[1]*Dep. Física de la Materia Condensada, ETSII, University of Valladolid.*
[2]*Dep. de Informática. Facultad de Informática, University of Valladolid, 47011 Valladolid, Spain*
[3]*CNR-IMEM Institute, Area delle Scienze 37/A, I-43010 Fontanini, Parma, Italy*
e-mail: mrebollo@eis.uva.es

Abstract. Diluted Sirtl (DSL) etching of GaAs constitutes a very powerful method to study the distribution of impurities and native defects in GaAs. The etching rate is governed by the concentration of the free holes at the solid liquid interface. As a result of the inhomogeneous distribution of both dopants and native defects the etched surface presents a complex topography. In particular different dislocations with their corresponding Cottrell atmospheres are revealed. These Cottrell atmospheres appear as hillocks or depressions depending of the defects present in the atmospheres. The study of the etching rate can be achieved by high resolution surface topography of relatively large areas of the etched surface. Microscopic optical interferometry is a powerful tool to study the surface topography with nanometric vertical resolution. The topography of DSL etched Si-doped GaAs grown with either Ga-rich or As-rich stoichiometries is studied by means of a home made microscope interferometer, which the main properties will be described. The topography data are compared with micro-photoluminescence (μ-PL) and cathodoluminescence (CL). CL and μ–PL provide complementary information to DSL etching. The combination of these characterization techniques allows to understand the nature of the Cottrell atmospheres in Si-doped n-type GaAs.

INTRODUCTION

The performance of GaAs devices are influenced by the presence of crystal defects (dislocations, precipitates, point defects etc) either native or produced during the technological processing. In particular dislocations play an important role in the reliability of devices. The interaction between dislocations and the crystal matrix introduce defects that form the Cottrell atmospheres, increasing the crystal inhomogeneity at micrometric scale. Doping of Si GaAs was found to be very effective to reduce the density of the dislocations without microprecipitation [1-2]. In this work we present a study of dislocations and related impurity atmospheres performed on as-grown Liquid Encapsulated Czhochralski (LEC) n-type GaAs Si doped samples. The analysis of the dislocations after DSL etching (Diluted Sirtl applied with light) was achieved by Phase Stepping Microscopy (PSM), Microphotoluminescence (μPL) and Cathodoluminescence (CL).

EXPERIMENTAL AND SAMPLES

The n-type Si doped GaAs crystals were grown by the LEC method. We studied two types of samples with an average carrier concentration above 10^{17} cm^{-3}. Sample No1 was grown from As-rich melt, while sample No 2 was grown on Ga-Rich melt. The samples were mechano-chemically polished with a PA-7 solution [3] and dipped into HCL-H_2O. Both samples were submitted to a Diluted Sirtl applied with light etching [4-5].

DSL is a selective etching allowing to reveal dislocations and their impurity atmospheres. Etching is based on a diluted Sirtl solution (CrO_3:HF:H_2O) The oxidative process for GaAs requires 6 holes (h^+) per molecule supplied by the solution

$$GaAs + 6h^+ \rightarrow Ga^{3+} + As^{3+} \quad (1)$$

To enhance the etching rate the sample is illuminated by a halogen lamp; the photogenerated holes accelerate the etching rate [5]. Therefore, the reduction of hole concentration at the surface decreases the etching rate and inversely the higher the free hole concentration, the faster the etching rate. Since the free hole concentration at the surface is governed by the concentration of shallow acceptors, the concentration of deep trap levels and the thickness of the space charge region thickness (concentration of shallow donors) is related doping and defect distributions.

The PSM technique is based on optical interferometry [6] and digital image processing providing a 3D surface reconstruction with nm in-depth resolution. The set-up has been developed at the University of Valladolid, where a new algorithm was designed [7]. This algorithm minimizes the errors in the calculation process of 3D map surface reconstruction.

High spatial resolution PL maps were performed at 82 K. An Ar^+ laser (λ = 488 nm) was used as the excitation source. The detection was done with an S1 photomultiplier tube. The laser beam is focused with a 50 x high numerical aperture long working distance microscope objective, which is also used to collect the emitted light. The sample was moved by means of a high precision motorized X-Y stage. The step size for the PL mapping was ~ 1μm. The set-up, and the control software were also home made. Cathodoluminiscence measurements were carried out with a GatanXiCLOne system mounted in the SEM (Scanning Electron Microscope). CL spectral images were obtained at liquid nitrogen temperature.

RESULTS

DSL photoetching combined with PSM allow to show the different surface topographies of Cottrell atmospheres revealed by DSL for the two types of samples. One can observe different types of grown-in (G) dislocations and stress induced glide grown-in dislocations (G-S). In this paper we study mainly the last kind of defects. In the As-rich crystal PSM topographies reveal hillocks at the start (s) and end (e) regions of the G-S dislocations, Fig 1.

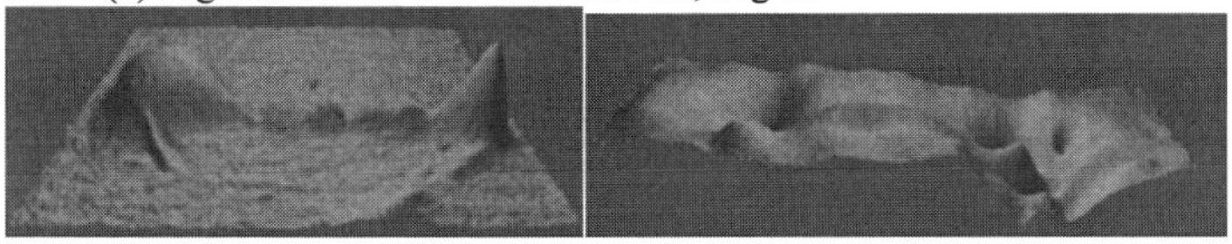

Figure 1 PSM 3d map of a typical GS dislocation a) As-rich (left) b) Ga-rich (right) samples

In Ga-rich samples, the topography of the Cottrell atmospheres revealed by DSL at the atmospheres s and e is basically a depression, with a peak inside the e atmosphere that corresponds to the outcrop of the dislocation Fig 1. The trace of the dislocations presents a similar topography for both samples, where it is revealed as a ridge.

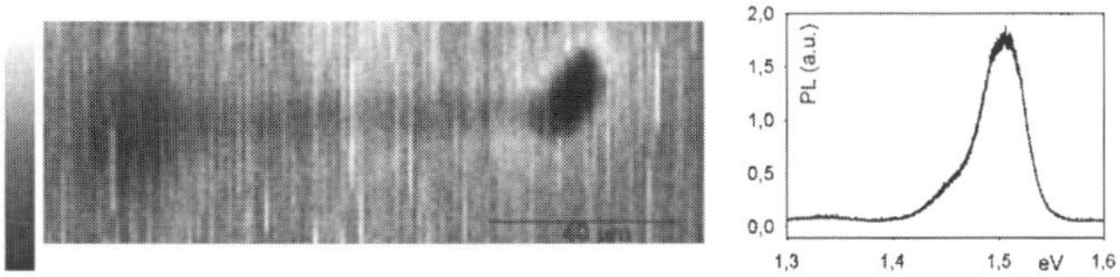

Figure 2. –Left: Monochromatic μ–PL map for at 825 nm of G-S dislocation at sample 1. –Right: PL spectra at the matrix at sample 2.

High spatial resolution cathodoluminescence and photoluminescence allow a qualitative description of the defects forming the Cottrell atmospheres around the dislocations. Monochromatic PL map at 825 nm for sample 1 is shown in Fig 2. The G-S Cottrell atmosphere exhibits lower luminescence intensity than the surrounding crystal matrix. Conversely, the PL intensity of the e atmosphere in sample 2 is higher than the matrix one. The PL spectra of both samples present similar features, exhibiting two main bands centred about 822 and 958 nm; and a shoulder at ~852 nm Fig 2.

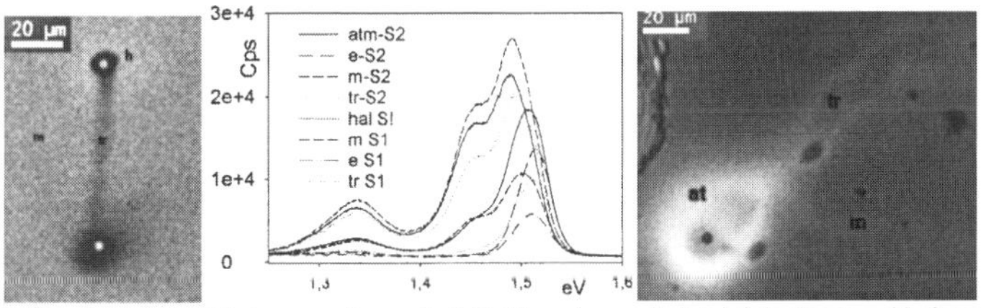

Figure 3. Panchromatic CL image of sample 1 (left) and sample 2 (right). Center: CL Spectra at e, halo, atmosphere, trace and matrix points of two samples

Figure 3 shows panchromatic CL image and local CL spectra at different points of two samples. The relative intensity of three bands changes at different regions of the Cottrell atmosphere as compared to the crystal matrix. The most remarkable effect concerns the shoulder at 852 nm, which is weakened at e atmosphere, vanishing close to the dislocation outcrop. It should be noted that only intrinsic and shallow level related transitions were observed, silicon complexes at lower energy were not detected [9].

DISCUSSION

Photoluminescence and cathodoluminescence contrast is determined by the distribution of shallow levels and deep non radiative recombination centres in GaAs. On the other hand, dislocations will getter point defects and/or impurity atoms, which constitutes the Cottrell atmosphere. The lower luminescence intensity emission from the Cottrell atmospheres suggests the formation of non-radiative recombination centers as a consequence of the interaction between the dislocation and the crystal matrix. The main non-radiative recombination deep center in GaAs is related to Arsenic antisite As_{Ga}. Since arsenic interstitials As_i are present in all type of crystals, both As and Ga rich Si doped GaAs. The formation of arsenic antisites around the dislocation can obey the following reaction:

$$V_{Ga} + As_i \rightarrow As_{Ga} \quad (2)$$

In Ga-Rich samples it exists also other non-radiative recombination center Ga_{As} originated by the following reaction:

$$V_{As} \rightarrow V_{Ga} + Ga_{As} \quad (3)$$

Therefore, different point defects are present in these samples V_{As}, V_{Ga}, Si_i, Si_{As}, Si_{Ga}. The distribution of which is modified as a consequence of the interaction between the dislocation and the matrix which is at the origin of the spectral changes seen in the luminescence spectra and the etching rate.

The first peak (B1) centered at about 1.50 eV Fig. 3 was attributed to (e, A^0) transitions with Si_{As} as an acceptor level [10]. Silicon is amphoteric in the GaAs host (donor at III site and acceptor at V site). Si_{Ga} donor level is very close to the conduction band and therefore D^o-VB transitions are not detected at 82 K. The maximum intensity of this peak yields at the matrix (sample 1) and at the halo around e (sample 2). This last result was in agreement with PSM and DSL measurements at sample 2 and explains the higher etch rate at the halo in Ga-rich samples.

The shoulder (B2) appears at about 1.46 eV and was ascribed to the V_{Ga}-As_i complex [11]. In both samples, this band vanishes at e-point because the concentration of those complex diminishes, when one of the components, V_{Ga} disappears due to reaction (2) and gettering Si impurities that preferentially enter in Ga sites.

The third band (B3) at ~1.33 eV was attributed to (e, A^0) transitions with V_{Ga} acceptors [11]. The features of this peak are similar to shoulder.

Relative intensities (B1/B2 and B1/B3) are represented in Table 1. It can be observed that the relative values are greater for sample 2. These results are coherent with the defects concentrations attempted to both samples: [Si_{As}] lower and [V_{Ga}] higher in sample 1 than in sample 2 according to the melt stoichiometries of both samples.

Table 1. Relative intensities of bands at samples 1 and 2

	Sample 1 (As-rich)			Sample 2 (Ga-rich)		
	halo	trace	matrix	atmosphere	trace	matrix
B1/B2	1.6	1.6	1.4	3.1	5.7	2
B1/B3	5.2	4	1.4	7.9	11	4.7

CONCLUSIONS

This work demonstrates the ability of the PSM microscope measurements combined with DSL treatments and other optical techniques to characterize crystal defects in LEC Si doped GaAs samples. The main characteristics of the Cottrell atmospheres of G-S dislocations in Ga and As-rich melts have been described.

ACKNOWLEDGEMENTS

The authors are indebted to the Italian-Spanish cooperation program. The Spanish group was funded by CICYT (project MAT98-0710).

REFERENCES

1. S.Miyazawa, *Mater Sci.Forum,* **10-12**, 1 (1986)
2. F.Fujii, N.Nakajima, T.Fukuda, K.Nitta, T.Komatsubara, *J.Electron. Mater.*, **16** 219 (1987)
3. J.C.Dyment, G.A.Rozgony, *J.Electrochem Soc.*, **118**, 1346 (1971)
4. C.Frigeri, J.L.Weyher, *J.Appl. Phys.*, **65** (12) (1989)
5. J.l.Weyher, J.Van de Ven, J.Cryst. Growth **63** (1983).
6. C.Frigeri, J.L.Weyher, J.Jiménez, P.Martin, , *J.Phys III France*, **7** (1997)
7. K.Creath. "Phase Measurements Interferometry Techniques". *Progress in Optics XXVI* edited by E.Wolf. Elsevier Science Publishers B.V. (Amsterdam) (1988)
8. R. Fernandez, J.Adiego. *Sistema de reconstrucción tridimensional de superficies por Interferometría óptica.* Memoria Proyecto Fin de Carrera. Universidad de Valladolid (1999)
9. O. Martínez, A. M. Ardila, M. Avella, J. Jimenez, F. Rossi, N. Armani, B. Gérard, E. Gil-Lafon. *J. Phys. C* (to be published).
10. J.L.Weyher, J.Van de Ven, C. Frigeri. *Semicond. Sci. Technol.*, 7 A294 (1992)
11. Pro perties of Gallium Arsenide, *EMIS Data Reviews Series* No 2, (INSPEC, The Institute of Electrical Engineers, London) (1990).

Applied Physics (APHYS 2003)
A. Méndez-Vilas (Ed.)

AN EDGE-BASED DATA STRUCTURE FOR NAVIER-STOKES EQUATIONS RESOLUTION

R. GOMEZ-MIGUEL

Department of Propulsion, INTA, 28850 Torrejon de Ardoz, SPAIN

Abstract. This work presents an edge-based data structure, that allows to compute solutions of Euler and Navier-Stokes equations, independently of the selected mesh type used to discretize the computational domain: The code developed, 'BERTA', is grid-transparent. Based in this structure a complete solution technique has been formulated, which allows to handle structured grids, block structured grids, and unstructured grids of tethrahedral or mixed elements without any modification. The contribution of the work lie in the possibility of unbinding the information needed to solve the equetions from the geometric data and original dimension of the mesh. Allocating original geometric data to the mesh edge, subsequent references to the original grid are not needed. An efficient data structure has been constructed, minimizing memory overhead and amount of gather/scatter, in comparation to others structures (element-based or face-based). Furthemore, the gains in computational efficiency afforded by the use of hybrid meshes over fully tetrahedral meshes are demostrated. A finite-volumen scheme has been used for the spatial discretization of the equations, and multi-stage time-stepping scheme for the time discretization. Viability of this data structure to solve termofluiddynamics problems wiht large generality has been demonstrated.

INTRODUCTION

From end of the century XIX it is had a general formulation of the fluid mechanics laws that describes the behavior of the continuous fluid flow. These laws are based on three consevation principles: mass conservation, momentumconservation and energy conservation.These equations are denominated Navier-Stokes, and are nolinear,nonunique and difficult to solve, they cannot be solved analytically except in special cases. The Computational Fluid Dinamics (CFD) solve the equations numerically. The field of CFD during old years has developed greatly, Both computer power and numerical algorithm are improving with the time. However the necessity to hadle flows and geometries more complex requires a continuous improvent of efficiency of algoritms.

SOLVER: DISCRETIZACIÓN OF THE EQUATIONS OF NAVIER-STOKES

The Navier-Stokes equations are discretized in integral form with conserved variables. The finite-volume node-based method is employed for the spatial discretization. A centered scheme of second order with addition of artificial viscosity is used.

The flow equations are avanced in time to obtain the steady-state solution using a multistage Runge-Kutta explicite scheme. Local time stepping and residual averaging are employed to accelerate the convergence.

Edge-based Data Structure.

While with the element-based data structure information is gathered from all the nodes of each element, operated on the element, and then scattered back to the nodes of the element. The edge-based algorithm gathers information from the two nodes of each edge, operates it on the edge, and then scatters it back to the nodes of the edge. A significant reduction in gather/scatter costs and memory requirements can be achieved by going from an element-based to an edge-based data structure. For a typical tetrahedral mesh with N nodes the number of cells, N_c is about $5.5N$ and the number of edges, N_e $7N$. An cell-based data structure requires $2x4xN_c$, i.e., $44N$ gather/scatter operations, while its edge-based counterpart needs $2x2xN$, i.e., $28N$. Note that a significant gather/scatter overhead reduction is achieved using an edge data structure, thus leading to a remarkable CPU saving.

Grid-Transparency, Dual Mesh, and Control-Volume.

The use of mixed grids raises the question of how the existence of different cell type affect the solution method. While it may be unavoidable to handle different cell types (arbitrary polyhedral elements) during the pre- and post-processing stages, we regard it as undesirable to handle arbitrary polyhedral cells differently during the flow solution stage. The resulting conditional statements adversaly affect program speed and result in untidy code. It is therefore sensible handle the different cell type in the same way wherever possible. The ideal case is a code which does not require any information on the local cell topology, this is independent of the cell type. Such a code is termed 'grid-transparent'. In the present projet, every effort has been made to ensure that the resulting code is grid-transparent. Grid-transparent algorithms are not allowed to refer to cell information. All knowledge of cell-topology is resticted to the pre- and post-processing stages.

Dual mesh is formed by constructing non-overlapping volumes, referred to as dual cells, around each node. The dual cells represent the control volume associated with the respective node. The dual mesh, for a $2D$ unstructured grid, is shown with dashed lines in Figure $1a$. The mesh is constructed by conecting the mid-points of the edges and the centroids of elements that constitute the grid and hencefort dividing each triangle/quadrilateral into three/four quadrilaterals of equal areas. The finite-volume around any node, say 0, is constituted by union of all the quadrilaterals which share that node. Analogously, the dual mesh for a $3D$ hybrid mesh. The BERTA pre-processing can handle mesh involving tetrahedral, hexahedral, and 'in-between' elements such as prims and pyramids. The dual mesh is constructed by dividing each cell into n (n is the number of vertex) hexaedron of equal volumes,by connecting the mid-edge

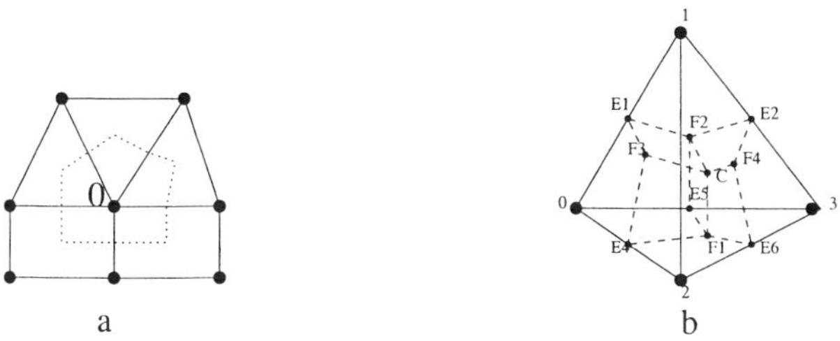

FIGURE 1. Dual Mesh a: $2D$ for a mixed mesh, b: for a tetrahedron.

points, face-centroids, and the centroid of the element. Figure $2b$ shows a thetraedron $0-1-2-3$ with the two hexaedral cells $0-E1-F2-C-F3-E4-F4-E5$ and $3-E2-F2-C-F4-E6-F1-E5$ that constitute a portion of the duals cells around the nodes 0 and 3, respectively. The control volume around a node 0 is thus constitued by a polyhedral hull which is the union of all such hexaedra that share that node.

Convective Terms Discretization.

The fluxes over the boundary of the control volume are integrated using a simple midpoint integration rule, where the value of the flux at each control volumen face, is computed by averaging the flow variables in the two control volumes on either side of the face. Such a scheme corresponds to a central difference discretization on structured meshes and requires the addition of artificial diffusive terms to ensure stability. By analogy with the blended second and fourth differences employed on strucrtured meshes [1] artificial dissipation is construted as a combination of an undivided Laplacian and biharmonic operator. First-order shocks-capturing diffusive terms are provided by the Lapalacian operator, which is turned off in regions of smooth flow where biharmonic dissipation ensure stability second-orden accuracy. This approach has been used extensively to compute inviscid and viscous two- and three-dimensional flows ([2], [3]). The discretization is simple and direct to implement in the edge-based data structure.

Viscous Discretization.

In discretizations of finite volume node based. The viscous terms are traditionally thought of as a sequence of two loops: one to construct (gradients at triangle or tetrahedron) centers and another to form the final residual contributions. However, the final discretes viscous terms obtained in this manner form a nearest neighbor stencil. The viscous terms for a vertix 0 depend only on values at 0 and at vertices k, such than k is joined to 0 by a mesh edges. Thus, an edge-based data structure may also be employed to assemble the viscous terms. This fact has previously been pointed out in several references [4], [5]. In Ref 4 , a complete derivation of the edge-based coefficients for Hessian matrix is given. In three dimensions, this would require the storage of nine coefficients

TABLE 1. *Data-structure efficiency*.

operator	cell-based struture (s/call)	edge-based struture (s/call)
convective	2.04	1.26
viscous	3.04	1.36

per edge, since the discrete Hessian is written as

$$\begin{bmatrix} u_{xx} & u_{xy} & u_{xz} \\ u_{yx} & u_{yy} & u_{yz} \\ u_{zx} & u_{zy} & u_{zz} \end{bmatrix} = \frac{1}{V_0} \sum_{k=1}^{n} \begin{bmatrix} \alpha_{xx} & \alpha_{xy} & \alpha_{xz} \\ \alpha_{yx} & \alpha_{yy} & \alpha_{yz} \\ \alpha_{zx} & \alpha_{zy} & \alpha_{zz} \end{bmatrix} (u_0 - u_k), \tag{1}$$

where Vol_0 represents the control volumen to vertex 0. However, the local edge-based coefficient matrix is symmetric about the diagonal. Thus, we need only store six coefficients per edge for the discretization of the viscous terms, and the implementation in the edge-based data structure is obvius. The edges coefficients for Hessian matrix are computed in the pre- processing, they only depend of mesh local geometry [6].

RESULTS

Data-structure efficiency

In order to illustrate the efficiency of structure three dimensional flow over a Gas Turbine T106 cascade has been computed. The mesh contains 100985 nodes, 550894 tetrahedron, and 671302 edges. The table 1 shows run time of convective and viscous operators for an element-based data structure and an edge-based data structure. The code has been run in a pentium III, 500Mhz.The gains in efficiency of edge-based data structure are 38.8% for convective operator and 55.3% for viscous operator.

Test case

ONERA M6 Wing ($Mach_\infty = 0.84$, $\alpha_\infty = 3.06$): This study is an example demostrating the application of the BERTA code for solving the three-dimesional inviscis case. Figure 2 shows pressure coefficients at four setions along the span of the wing, for BERTA's solution and the referenced experiment [7]. One may conclude that the comparisions are good overall.

Flat Plate Boundary Layer: An assessment of the accury of the scheme may be performed by examing the ability of the method to reproduce the wel-know compresible boundary-layer solution over a adiabatic flat plate. The computations were performed for a Mach number of 0.8 and a Reynolds number based on the plate length of 5000. An exact analytical selfsimilar solution for this flow may by obtained i by an application of the Illingworth transformation to incompressible Blasius selfsimilar solution. Figure 3 shows the calculated and exact boundary-layer profiles at the station $x = 0.6$ for a

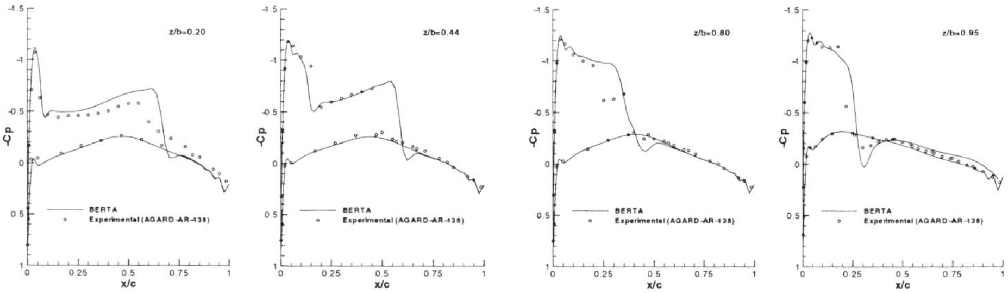

FIGURE 2. Pressure Coefficients.ONERA-M6

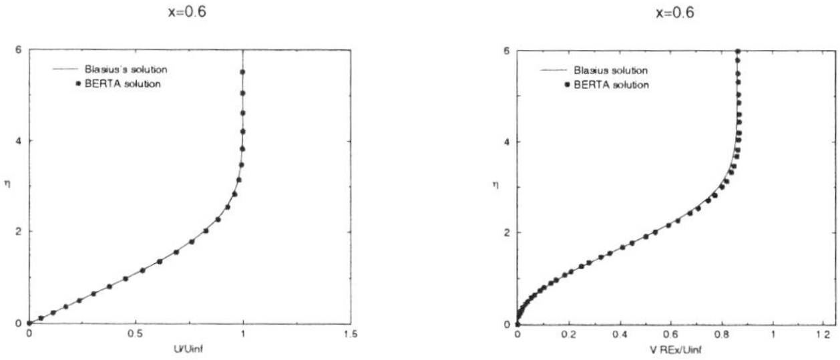

FIGURE 3. Flat Plate velocity profiles

plate of length unity. Excellent agreement between the computed and exact profiles of streamwise and normal velocity is observed.

CONCLUSIONS

BERTA code gets numerical solution of Navier-Stokes Equations, independently on the selected mesh type to discretize the computational domain: the solver is grid-transparent, and allows to solve a great variety of fluid-mechanics problems with the same tool. It minimizes both memory overhead and the amount of gather-scatter: it has a very good run time efficieny.

REFERENCES

1. Jameson, A., Schmidt, W., and Turkel, E., *AIAA Paper 81-1259* (1981).
2. Mavriplis, D. J., *AIAA J.*, **30**, 1753–61 (1992).
3. Mavriplis, D. J., and Venkatakrishan, V., *AIAA Paper 95-0345* (1995).
4. Barth, T. J., *AIAA Paper 91-0721* (1991).
5. Lou, H., Baum, J. D., Lohner, R., and Cabello, J., *AIAA Paper 93-0336* (1993).
6. Gomez-Miguel, R., *Ph. D. Disertation, UPM* (2001).
7. Schmitt, V., and Charpin, F., *AGARD-AR-138* (1979).

Radiofrequency Ablation on Heart-equivalent Phantom. Functionality Testing of Percutaneous Single-use Catheter.

Francesco Tessarolo, Paolo Ferrari, Renzo Antolini, Giandomenico Nollo

Department of Physics, University of Trento, & ITC-IRST, 38050 Povo, Trento, ITALY.
E-mail: tessaro@science.unitn.it

Abstract. Aim of the presented experimental and theoretical model was to characterize the power deposition process in cardiac catheter ablation. A dedicated set-up, based on a transparent temperature-sensitive gel-phantom was developed. A liquid crystal film was used to map the temperature distribution during RF deposition. An infrared acquiring system and a post-processing image software were used to define the heating zone and to quantify the device specific heating fingerprint. Experimental results were compared to a thermal conduction model with spherical symmetry, showing good agreement for ablation time $100<t<600$s. Differences between model and experimental results were explained by deviation of the real heating pattern from spherical geometry.

INTRODUCTION

Cardiac trans-catheter ablation is a valid alternative to drug therapy and cardiac surgery in the treatment of myocardial arrhythmias [1-3]. The deep comprehension of the radio-frequency (RF) heating mechanism and the characterization of the functional properties of the devices are mandatory to guarantee the expected clinical outcome. Furthermore, testing the functionality effectiveness of ablation catheters allows certifying performances and defining regeneration procedures. From this point of view functional testing methods and experimental models have to be non destructive and reproducible.

Aim of this work was to monitor the power deposition process occurring in simulated cardiac catheter ablation, identifying a quantitative procedure to assess device functionality. We developed an apparatus for simulating power deposition by RF cardiac catheters on a tissue-equivalent phantom. Phantom ablation presents the uncountable advantages of a physical and chemical properties fixed system with a simple and optimised geometry. Bi-phases phantoms, miming the myocardium-blood interface, were previously realized [4-7]. However, the discrete distribution of thermometric probes limited the spatial resolution in heating pattern reconstruction. Here is presented a high spatial definition method based on the thermal to colour conversion obtained by liquid crystal film [8]. This experimental set-up should allow an easy data interpretation which is fundamental for selecting intervals and boundary conditions of use and functionality. A theoretical model was developed to help the comprehension of experimental data and to assess the functional properties of new and reprocessed devices.

MATERIALS AND METHODS

The realized set-up consisted in a RF production unit, a signal control system and a temperature-sensitive phantom. A sinusoidal signal was amplified by a custom realized amplifier to 500 KHz, 130V max. The current intensity, flowing thought the phantom, was measured by monitoring the voltage across a 5Ω calibrated resistance. The released power P and the phantom impedance Z were monitored during all the simulated ablation.

The phantom consisted in a 140x72x50mm PMMA box, filled with a transparent gel obtained by mixing deionised water and 0.3% wt. Carbopol® Ultrez 21 (Noveon, Inc.) neutralized with 1%wt. of 18%wt. NaOH solution. To maximise clarity, the preparation was carried out at 60°C with boiled water and low share mixing. Physical gel characterization indicated a 3.3±0.1mS/cm conductivity at 35°C and a 4.2±0.1J/g specific heat, in accordance with myocardium values reported in literature [9]. The obtained viscosity allowed to consider negligible any convective flux during the simulated ablation. A 140x72mm grounded electrode was positioned on the bottom of the box.

A liquid crystal film (LCF) (Edmund Scientific 72-374) was placed parallel to the front box side and immersed in the gel. The distal portion of the catheter was immersed in the phantom, in contact with the LCF. The ablation induced LCF colour pattern evolution was recorded by capturing 640x480 pixels images in infrared (875 nm) and solar light. Off-line digital image analysis was performed by IMAQ (Labview™, National Instruments™) obtaining quantitative parameters as volume, area, and width of the heated zone.

Constant power mode ablations were performed by using Medtronic RF Conductr™MC catheters. Phantom temperature was set at T_0=22.5°C and the infrared sensitivity of LCF was T_s=24.6°C. Images were collected every 15s for 15 minutes of ablation.

THERMAL MODEL

Experimental data were obtained by the software recognition of the isotherm T_s. The isotherm represented the intersection between the LCF plane and the phantom surface at $T=T_s$. Although the problem presented a cylindrical symmetry, spherical approximation was possible because of the catheter tip reduced dimension. Considering that the power due to heating Joule effect decreases as $1/r^4$ (r distance from tip centre) [9], a thermal conduction model should be suitable for r greater than 2mm. Thus according to thermodynamics:

$$Q = mc\Delta T \tag{1}$$

where T is temperature, c the specific heat, and m the mass.
Considering a punctiform heating source immersed in a homogeneous and isotropic medium with density ρ and specific heat c, the infinitesimal heat required to raise the temperature from T_0 to T_s in a circular ring of radius r and width dr is:

$$dQ = \rho c 4 \pi r^2 dr \, (T_s - T_0) \tag{2}$$

In a constant power mode, given $dQ/dt = \mathrm{P}$:

$$\mathrm{P} = \rho c 4 \pi r^2 \frac{d}{dt} r(t)(T_s - T_0) \tag{3}$$

so:

$$r(t) = 3\sqrt[3]{\alpha}\, t^{1/3} \qquad \alpha = \frac{\mathrm{P}}{\rho c 4 \pi (T_s - T_0)} \tag{4}$$

Considering the specific parameters we were interested in, we obtain:

$$\text{width } (t) = 6\sqrt[3]{\alpha}\, t^{1/3} \tag{5}$$

$$\text{area } (t) = 9\pi \sqrt[3]{\alpha^2}\, t^{2/3} \tag{6}$$

$$\text{volume } (t) = 36 \pi \alpha t \tag{7}$$

RESULTS

Experimental data for volume, area and width vs. time, obtained from infrared images processing, are reported in Fig.1 and Fig. 2.

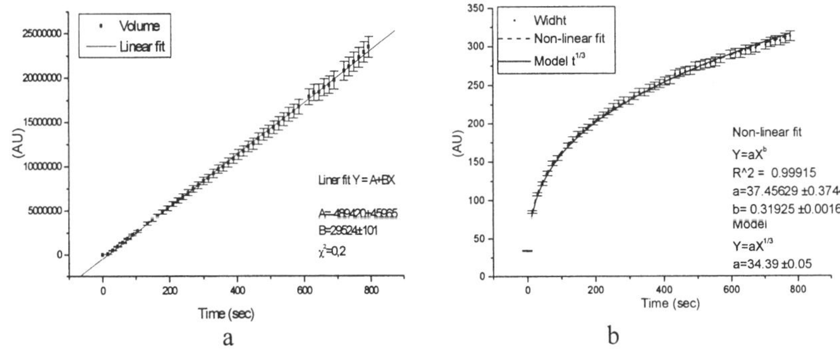

Figure 1. Time dependence of heating parameters. a: volume real data and linear fit. b: Data of heating pattern width. Non linear fit, and model curve are superimposed to real data. Fitting and model parameters are shown.

The linear regression on volume data, confirmed the validity of the first-order model and identified a 17±2s lag for LCF colour changes. This period of time, required by the phantom for reaching the threshold temperature T_S, was omitted in the analysis of area and width values. Resulting data were fitted with a non-linear regression using a two-independent parameters function $Y=aX^b$. Appling this parametric fit on width and area points we obtained respectively b=0.319±0.002 and b=0.730±0.007, while model predicted value were 0.333 and 0.667. The agreement between model and fit worsened for ablation time lower than 100 s and greater than 600s. Particularly for t<50s and t>600s the theoretical expected value under-estimated the heating area points. On the contrary, solar light images processing of the first 50s of ablation, was useful for detecting the influence of the electrode shape on the heating pattern. In Fig.3 electrodes features are compared to the respective LCF images detected in solar light for LG CRV Blazer II™ XP and Medtronic RF Conductr™ MC.

DISCUSSION

The model curve and the non-linear fits on data of heating area and width were in good agreement, confirming the reliability of the thermal conduction model over the 100-600s time ablation interval. In this range the model well forecasted the ablative development, peculiar of the given catheter type. This approach to catheter characterization could represent a quantitative functionality test. For t>600s a non spherical isotherm shape was noted, suggesting the presence of other heating sources. An inductive coupling was revealed indeed, between ablation and closest recording electrodes. This electrical coupling caused minor power sources, breaking the spherical symmetry of the heated zone. For t<50, Joule-effect was not negligible. This additional mechanism affecting the detected area accelerated the heating process in the starting phase of ablation. Furthermore, the non-spherical shape and the finite dimensions of the ablative electrode influenced the first 100s of power deposition determining a non-spherical pattern. The deviation from spherical symmetry constituted a useful tool for identifying the device peculiar heating fingerprint. This characterization of catheter performance is mandatory for relating ablation electrode design and power-delivery distinctive features as needed in electrode design optimisation and lesion size and depth foreseeing [10, 11].

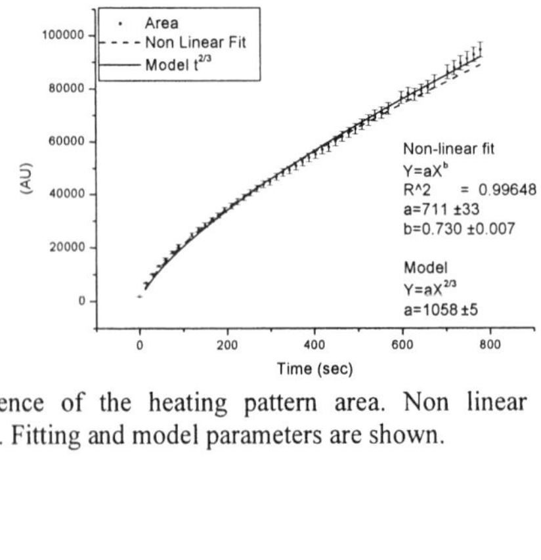

Figure 2. Time dependence of the heating pattern area. Non linear fit, and model curve are superimposed to real data. Fitting and model parameters are shown.

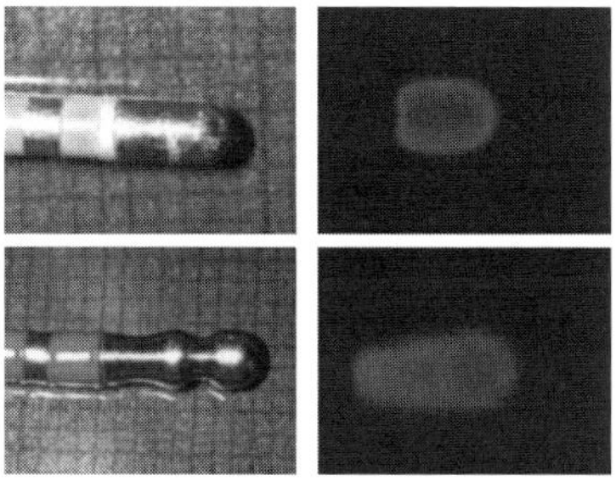

Figure 3. Electrode shape (left) compared to heating fingerprint (right). Images are obtained by Medtronic RF Conductr MC (top) and LG CRV Blazer II XP.

CONCLUSIONS

In this wok a tissue-equivalent phantom ablation, able to prove the functionality state and to quantify the catheter power deposition was presented. Experimental set-up and theoretical model allowed to identify device specific heating features and to comprehend the dissipative mechanism active in in-vivo cardiac catheter RF ablation. Real data shifts from model-predicted values have to be considered as warnings for malfunction or device functionality alteration. The procedure described here should represent an integral part of a testing protocol in pre-market approval or in reprocessing procedure of percutaneous cardiac ablation catheter.

REFERENCES

1. W.G. Stevenson, K.E. Eleison, D.C. Lefroy, P.L. Friedman, "Ablation Therapy for cardiac arrhythmias" *Am J Cardiol*, 1997, 80(8A), pp. 56G-66G.
2. G.N. Kay, V.J. Plumb, "The present role of radiofrequency catheter ablation in the management of cardiac arrhythmias" *Am J Med*, March 1996, vol.100, pp. 344-356.
3. W.S. Teo, R. Kam, Y.L. Lim, T.H. Koh, " Curative therapy of cardiac tachyarrhythmias with catheter ablation – A Review of the experience with the first 1000 patients" *Singapore Med J*, 1999, vol. 40(04).
4. H. Cao, V.R Vorperian, J. Tsai, S. Tugjitkusolmun, E.J. Woo, J.G. Webster, "Temperature measurement within myocardium during in vitro RF catheter ablation", *IEEE Trans. Biomed. Eng.*, vol. 47, no. 11, 2000.
5. I. Chang, B. Beard, " Precision test apparatus for evaluating the heating pattern of radiofrequency ablation devices", *Med. Eng. & Phys.* 24, 2002, pp. 633-640.
6. S.S. Hsu, L. Hoh, R.M. Rosembaum, A. Rosen, P. Walinsky, A.J. Greenspon, "A method for in vitro testing of cardiac ablation catheters", *IEEE Trans Microwave Theory and Techniques*, vol. 44, no.10, October 1996.
7. T.L. Wonnell, P. R. Stauffer, J.J. Langberg, "Evaluation of microwave and radiofrequency catheter ablation in a myocardium-equivalent phantom model", *IEEE Trans. Biomed. Eng.*, vol. 39 no. 10, October 1992.
8. L. Criostoforetti, R. Pontati, L. Cescatti, R. Antolini, "Quantitative colorimetric analysis of liquid crystal films for phantom dosimetry in microwave hyperthermia", *IEEE Trans. Biomed. Eng.*, 1993, vol. 40, pp. 1159- 1165.
9. S. Gabriel, R.W. Lau, C. Gabriel," The dielectric properties of biological tissues: II. Measurement in the frequency range 10 Hz to 20 GHz." *Phys. Med. Biol.*, 41, 1996, pp. 2251-2269.
10. D. Panescu, S. D. Fleishman, J.G. Swanson, J.G. Webster, " The influence of different electrode geometries on the current density distribution during radio-frequency ablation. A three-dimensional finite element study" *IEEE*, 1994.
11. S. D. Fleishman, J.G. Whayne, D. Panescu, D.K. Swanson, " In vitro study of lesion size dependence on electrode geometry during temperature-controlled radiofrequency" *IEEE*, 1994.

Applied Physics (APHYS 2003)
A. Méndez-Vilas (Ed.)

Determination of Synchronization of Electrical Activity in the Heart by Shannon Entropy Measure

Michela Masè, Luca Faes, Giandomenico Nollo, Renzo Antolini, and Flavia Ravelli

Department of Physics and INFM, University of Trento, Povo, Trento, ITALY

Abstract. In this paper we propose a new index of synchronization for the study of heart's electrical activity during atrial fibrillation (AF). The index relies on the measure of the time delays between correspondent activations in two atrial electrograms and on the characterization of their dispersion by a measure of Shannon Entropy. The algorithm was validated on simulated signals mimicking different degree of synchronization. Results showed the index was able to discriminate among different levels of organization, provided that it works on series of at least 50 activations (time resolution of almost 10 sec during AF). Moreover, we applied the algorithm to real bipolar electrograms, obtained from a multipolar basket catheter in right atrium in two patients during atrial fibrillation: this showed the index able to distinguish different levels of complexity in AF.

INTRODUCTION

Although atrial fibrillation (AF) has classically been described as a random phenomenon, due to the presence of multiple activation wavelets circulating in the atrial tissue [1], it is now recognized the existence of a structure in the electrical activity recorded during the arrhythmia: in fact, several factors (such as anatomy, refractoriness, etc.) should play a role in limiting the randomness of propagation mechanisms and "organizing" atrial fibrillation. Therefore the researchers' interest is now focused on obtaining a quantitative measure of organization during AF, this allowing an objective distinction of different kinds of AF and the development of aimed and specific treatments. Up to now many definitions of organization have been proposed with different conceptual and methodological approaches [2,3]. Among those, two-sites measures of synchronization are intrinsically more informative than single site ones. The available synchronization measures based on considering the time of the atrial activations in two sites [4] are sensitive to the distance between the two sites and do not provide information about the direction of wavefront propagation.

In this study, we propose an algorithm to evaluate the causal coupling between the electrical activity of two atrial sites as an indicator of rhythm organization. More specifically, we have defined an index of synchronization S, quantifying the distribution of the time delays between correspondent activations in the two signals by

Shannon Entropy. The method was validated by simulations of different levels of organization and applied to real signals acquired during AF.

METHODS

2.1 Description of the Index of Synchronization

The proposed algorithm quantifies the synchronization level of two atrial signals, recorded in different sites of the atrium. It is based on the assumption that, if two atrial sites are interested by the passage of the same activation-wave, the consecutive time delays between coupled activations in the two sites should have a small dispersion.

Once the activation time series have been estimated for both signals by means of a morphology-based algorithm [5], these times are coupled associating each activation of the first series with the subsequent closest activation of the second. A positive time delay for each couple is then defined as the difference between the two activation times. The time delay values are then organized in a histogram, made up by 5ms-large bins, starting from the shortest delay and going on adding bins until the longest delay is included in one of them. The spreading of the delays in different bins is characterized by a measure of Shannon Entropy (SE), defined as follows:

$$SE = -\sum_{i=1}^{N} p(i) \cdot \ln p(i) \tag{1}$$

where p(i) is the probability for the i-th bin, estimated as the ratio of the number of delays in the bin to their total number in the series, and N is the total number of bins.

Finally, the synchronization index S, is defined as follows:

$$S = \left(1 - \frac{SE}{\ln(ndelays)}\right) \tag{2}$$

where *ndelays* is the total number of delays in the histogram.

In this way index S presents values ranging from 0 to 1, increasing with increasing degree of synchronization: it reaches its maximum value 1, when all time delays lie in the same bin (SE=0), while it takes its minimum 0, when the spreading of time delays in the histogram is maximal (all time delays lie in different bins, SE=ln(*ndelays*)). The last condition is rarely satisfied also by independent series, because of a non-null probability of having at least two time delays in the same bin. For this reason, a statistical test based on surrogate data analysis is used to differentiate significant S values from background coupling values. For each original pair of activation time series, a set of surrogate pairs is obtained by randomly permuting the order of the FF intervals (i.e., the intervals between two consecutive activation times in the same series). The level above which the synchronization between the two series can be considered significant, with significance 0.05, is then obtained as the 95th percentile of the distribution of S on the surrogate pairs.

To provide information about the causal verse of synchronization (i.e. which site activates the other), the presented procedure is repeated switching the order of the two series, thus obtaining two synchronization values.

2.2 Verification of the Index by Simulation

To verify that our algorithm can discriminate among different degrees of synchronization between two signals, its performances were tested on simulated signals. The couples of simulated signals are formed by: (i) an activation time series, obtained by summing up simulated FF intervals extracted from an Erlang distribution of the third type (with minimum of 60 ms and mean of 180 ms); (ii) a second series, obtained from the first by adding a Gaussian random delay, with mean value of 50ms and different standard deviation (STD) chosen in the range (1, 3, 5, 7, 10, 12, 15, 20, 25, 30, 35, 40, 45 ms). The use of Erlang distribution, instead of gaussian, allows us to mimic the low bound of FF interval distribution due to refractory period, while the increasing standard deviation of the delays simulates different degrees of synchronization. The consistency of the obtained S values was assessed by surrogate data analysis, generating 10 pairs of surrogate series for each measure.

The dependence of the algorithm on the number of activation times included in the analyzed epoch was evaluated by computing the index S on strips of the simulated signals, with an increasing number of activations (from 5 to 200).

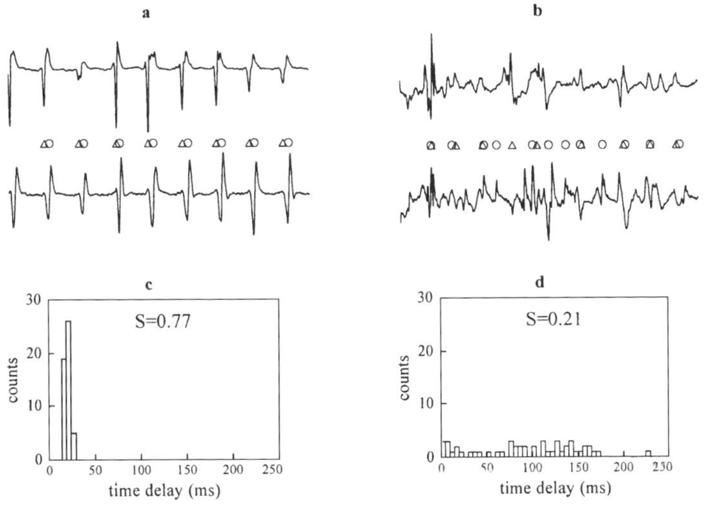

FIGURE 1. Working steps of the proposed algorithm running on strips of 50 activation real bipolar signals recorded during AF episodes with low (left) and high complexity (right). (a,b): Activation times estimation (triangles for first, circles for second signals). (c,d): Histogram distributions of time delays and corresponding synchronization values. See text for details.

RESULTS

As an example of the index S working on real signals, two couples of bipolar signals were considered. They were both recorded by a multipolar basket catheter in right atrium during AF, but presented different degree of complexity according to an expert cardiologist's visual scoring [6]. 50 activation long strips of the signals were considered. Figure 1 shows the algorithm running on the two couples of signals: after estimating the activation times for each signal and the associated time delays (a,b), we characterize their histograms (c, d) by S index. While in the less complex episode, the activation series are highly synchronized as all delays are gathered in three bins, in the more complex the coupling is lacking as the delays are highly spread.

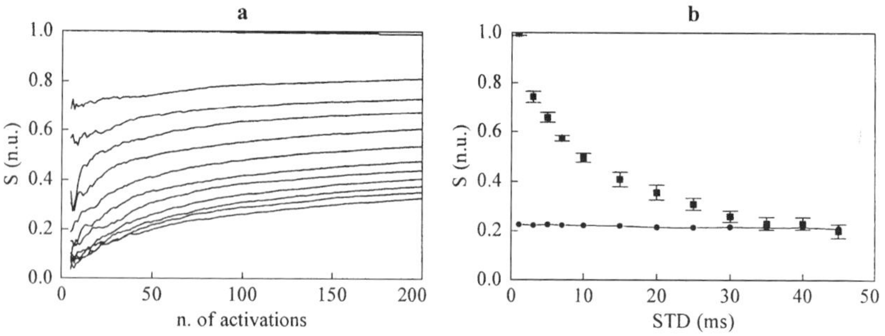

FIGURE 2. a. Dependence of the index S on the number of activations in simulated signals (average on 10 couples of signals). The simulated signals correspond, from top to bottom, to STD=1,3,5,7,10,15,20,25,30,35,40,45 ms. b. Values of index S for simulated signals (■) and surrogate counterparts (●) 50 activation long (average on 10 couples of signals). See text for details.

Figure 2a shows the dependence of S on the number of activations for simulated signals with different values of STD in the delay distributions. The index shows a high dispersion (not displayed in figure) and an oscillating behavior for small numbers of activations, while it shows smaller standard deviation and a slow monotonic variation for higher number of activations. However, it discriminates different classes of synchronization mimicked by different STD values, as the p values of the t-test become statistically significant for all pairs of signals with STD<35 ms and number of activations $\geq$ 50. This suggests to calculate S on series with at least 50 activations, implying a temporal resolution of almost 10 sec during AF.

Figure 2b shows the behavior of S calculated on 50 activation long simulated signals and on their surrogate counterparts. For simulated signals S decreases as STD increases, following the loss of synchronization in the couple of signals. Instead, S is quite constant for surrogate signals. The values for simulated signals are significantly ($p<0.001$) different from those obtained from their surrogate counterparts as long as STD<35 ms, while the significance of the index is lost for higher STD. This means that for STD<35 ms, we can distinguish different degrees of synchronization in the couples of signals, while for higher values of STD the signals behave as independent and can't be distinguished, in accordance with the t-test results for simulated signals.

DISCUSSION AND CONCLUSIONS

In this study we have proposed a new index, S, which measures the level of synchronization between couples of electrograms recorded in two sites of the human atrium during different heart rhythms. It is based on the assumption that if two sites are persistently activated by the passage of the same wavefront (i.e. are synchronized), the time delays between correspondent activations have a small dispersion.

The index, tested on simulated signals, shows its ability in distinguishing different imposed levels of synchronization, provided that the number of activations in the signal is at least 50 (time resolution of almost 10 sec during AF). Good performances have been obtained also for real signals: in fact S can distinguish AF episodes with different degrees of complexity [6]. The test with simulated signals shows a slight dependence of S from the number of activations, suggesting the use of signals' strips with the same number of activations, instead of the same temporal length, in comparing different rate arrhythmias (for example atrial flutter and fibrillation).

So far several methods have been proposed to measure synchronization during AF [2], with approaches going from chaos theory to frequency domain techniques. Their main problems concern long time for calculus and need of long data set. In comparison with these methods our algorithm presents a small computational overhead and a good time resolution. An approach similar to ours was used by Barbaro et al. [4], who considered time delays as a starting point too. They obtained a better time resolution (2 sec), but their algorithm could only be applied to sites belonging to the same splines of a basket catheter, implying a limitation in the number of sites that could be compared and a lacking use of the information such catheters provide. On the contrary, our algorithm can be applied to any couple of sites chosen and also supplies information about the causal verse of the excitation wavelets. This allows the study of synchronization for increasing site distances and the construction of spatially distributed synchronization maps, in addition to supplying clues about the direction of wavefront propagation.

The knowledge of the spatial behavior of synchronization could help to understand the mechanisms and sites, which play a main role in promoting and sustaining AF, and, consequently, to individuate specific treatments to interrupt it. In fact, our index should reflect the number of wavelets circulating in the atria, one of the main factors which determinate the stability or termination of AF.

REFERENCES

1. Moe, G. K., *Arch. Int. P. Ther* **140**, 83-188 (1962).
2. Sih, H. J., *Ann. Ist. Super. Sanità* **37**, 361-369 (2001).
3. Faes, L., Nollo, G., Antolini, R., Gaita, F., Ravelli, F., *IEEE Trans.Biomed.Eng.*, **49**, 1504-1513 (2002).
4. Barbaro, V., Bartolini, P., Calcagnini, G., Censi, F., and Michelucci, A., *Med.Biol.Eng Comput.* **40**, 56-62 (2002).
5. Sandrini, L., Faes, L., Ravelli, F., Antolini, R., and Nollo, G., *Comput. Cardiol.* **29**, 593-596 (2002).
6. Wells, J. L., Jr., Karp, R. B., Kouchoukos, N. T., MacLean, W. A., James, T. N., and Waldo, A. L., *Pacing Clin.Electrophysiol.* **1**, 426-438 (1978).

Applied Physics (APHYS 2003)
A. Méndez-Vilas (Ed.)

Neural Network Control In A Wastewater Treatment Plant

Miguel A. Jaramillo[1]; Juan C. Peguero[2], Enrique Martínez de Salazar[1], Montserrat García del Valle[1]

([1])Escuela de Ingenierías Industriales. ([2])Centro Universitario de Mérida. Universidad de Extremadura. Avda. de Elvas s/n. 06071 Badajoz SPAIN. e-mail: miguel@unex.es, jcpeg@unex.es, dsalazar@unex.es, montse@nernet.unex.es

Abstract. Control in a Wastewater Treatment Plant is a task mainly performed by technicians because the complexity of the dynamics of the biochemical processes performing the wastewater treatment and the delays in the measurement of several variables describing the system dynamics prevent from using classical automatic regulators. A Neural Network that learns the control actions carried out by the technicians controlling the plant is developed and then used as an automatic controller in a simulation of the plant dynamics. This one is represented by a set of nonlinear differential equations describing the time evolution of waste and bacterial population. The neural network try to keep the waste concentration as low as possible and the bacterial population between certain limits. Results obtained with simulation are compared with those of a real plant

INTRODUCTION

Wastewater Treatment Plants are facilities where municipal and industrial wastewaters are processed to eliminate as much pollution as possible. When these waters enter the plant large particles are rejected and greases and oils are skimmed off in a first stage. In a second one, organic pollution is removed with a biological treatment where bacteria "eat" the pollutants in the water reducing its concentration to an appropriate level. Then this water may be released to a natural stream. This second stage may be divided into two different processes. The first one is performed in an aeration tank where the incoming polluted water is mixed with a sludge made up of bacteria. After they have "eaten" most of the organic matter, water and sludge are driven to a settler tank where the second process is performed: the separation of these two components. The sludge flows downwards while the water stays at the top of the tank and flows over an overflow weir to be released to natural streams. The sludge is withdrawn from the tank and then divided into two streams: one is driven to the aeration tank to keep the sludge concentration at an appropriate level while the other is eliminated.

From the preceding description it may be inferred that the sludge feedback must play an important roll in the biological process control. In fact it may be considered as the most important control parameter of the plant dynamics. It allows the sludge concentration in the aeration tank to be kept at an appropriate level to ensure an adequate plant operation.

As the plant dynamics must describe biological processes, the formulation of a mathematical model is very difficult. Therefore a simplification must be applied in order to obtain a computational treatable structure along with a reasonable understanding of the plant dynamics. Fortunately a good approximation to the behavior of the plant may be obtained with the description of only the aeration tank, the main element of the whole system. The equations describing this reduced model are nonlinear and the use of classical linear control techniques is not possible. Moreover some of the system variables need several days to be obtained and, therefore, automatic controllers may not be included, as they need online variables values (or at least with small time delays) to operate. So the plant must be controlled by technicians that adjust the plant parameters. As they have only delayed values of the plant variables they must rely on his own experience to adjust the values of those parameters. This experience may be complemented with the direct observation of some properties of the fluids in the tanks, as their apparent dirtiness, that allows the operator to suppose an approximate concentration of bacteria and pollutants. The operator experience is then based on the study of the effects his actions produce once the actual values of the plant variables (several days delayed) have been obtained.

As wastewater treatment plant technicians provide a very effective control, the learning of their experience may define a very effective tool to provide an automatic control of the plant. Nevertheless, as these plants are continuously working, they may not be used to test the performance of such controllers, so it will be necessary to simulate both the plant and the control. This is the aim of the present work: the definition of the model of a wastewater treatment plant controlled by a system that reproduce the operator's knowledge. As this knowledge is based on personal experience it must be learned with artificial intelligence techniques. We have selected Neural Networks as that tool because the operator's experience is represented by a time series and they have proved their capabilities to learn a dynamical system behavior from its time evolution [1]-[2]-[3]. As all the information concerning the plant dynamics is available at the training time, the proposed model may learn the operator's actions relating then to the corresponding system variables. In this way the operator's knowledge is associated with the plant variables. Once the neural network has been trained it will be included in the plant model to test its capabilities in simulation.

PLANT DESCRIPTION

As it has been previously stated the system model will be restricted to the aeration tank, where the main processes of the plant take place. In this model the input variables are the "Influent Flowrate" *Qf* and the "Influent Pollution Concentration" *Sf*. The output variables are the "Sludge Concentration" in the aeration tank *Xva* and the "Output Pollution Concentration" *Se*. The "Sludge Concentration" in the settler, *Xvu*, is assumed to be a process parameter that will be provided at every time step. This parameter defines de concentration of the sludge in the settler, a portion of which will be recirculated to the aeration tank to keep the sludge concentration at an appropriate level. This recirculation is defined by a flowrate, *Qr*, that is adjusted by the operator supervising the plan dynamics. It is represented by a parameter; *r*, that provides the ratio between this "Recirculation Flowrate" and the "Influent Flowrate".

The system dynamics may be defined by a set of differential equation of the variables *Se* and *Xva*. No dependence on the temperature or the aeration will be considered, their values will be assumed to be optimal, a supposition that is quite close to reality because the great volume of the tanks in the plant ensure a quite constant value for temperature, while the aeration may be set to an adequate level by air pumping. So the proposed model is described by the following equations [4]-[5]:

$$\frac{dSe}{dt} = -\frac{Qf}{V} \cdot Se - k \cdot Xva \cdot Se + Sf \cdot \frac{Qf}{V} \tag{1}$$

$$\frac{dXva}{dt} = -\left[\frac{Qf}{V} \cdot (1+r) + kd\right] \cdot Xva - \frac{Qf}{V} \cdot Y \cdot Se + \frac{Qf}{V} \cdot r \cdot Xvu + \frac{Qf}{V} \cdot Y \cdot S \tag{2}$$

$$r = \frac{Qr}{Qf} \cong \frac{Xva}{Xvu - Xva} \tag{3}$$

CONTROL OF THE RECIRCULATION PARAMETER

As it has been previously stated the plant control is performed adjusting the value of the recirculation parameter *r*. To provide an easier manipulation it has been expressed as the sum of two elements: $r=ro+\delta r$. The first one is fixed and define a mean value of the recirculation, while δr represents the oscillations around that value. They have been selected so that δr has only values between +1 and –1 to ensure an optimal neural behavior, as it will be seen later. *ro* has been set to *ro*=1.74.

As this parameter is adjusted by the plant operator, whose experience is to be learned by a neural network, the time series describing its evolution may be used to perform the network training along with all the other time series of the plant variables and parameters. So, the neural network may learn the value the operator gaves to the recirculation, relating it to the values of the plant variables and parameters obtained at the same time. Al these data have been taken from the activity reports of an actual Wastewater Treatment Plant [6]. This neural network may be used as a controller in simulation, where direct values of the variables may be used to control the system dynamics.

The variables and parameters used as network inputs are *Sf(t-1)Qf(t-1)*, *Xva(t-1)*, *Se(t-1)*, *Xvu(t-1)* and $\delta r(t-1)$, while the output is $\delta r(t)$. We have considered the product of the input flowrate and its concentration instead of their separated values because that product represents the value of the total pollutant mass, which have a more direct influence on the sludge evolution than the separated values. Several simulations carried out both with the product and with the separated variables have proved that better results were obtained with the fist configuration.

NEURAL NETWORK STRUCTURE

The selected Neural Network is a Multilayer Preceptron with tow hidden layers with 15 and 10 neurons with sigmoids as output functions. The input one has 5 inputs (those defined in the previous section). The network output has one only neuron that provides the recirculation correction δr. It has an arctangent function as output to provide the desired variation of δr (between +1 and –1). We have also tested several one hidden layer structures but they all provided quite worse results than the two layers network proposed here.

The network training was performed with a "Backpropagation" [7] algorithm with the Levenberg-Marquardt optimization method to adapt the neuron sinaptic weights.

Two years of activities in the actual plant [6] were considered, 1998 for training and 1999 for simulation. These data were shared out into two groups, one for the first and last three months of the year (which we named wet months) and the other for the remaining six (which we named dry). This distribution try to ensure a more or less uniform temperature distribution through the six months that form each group to fulfill the aforementioned supposition that the temperature is kept to a constant value. So two different networks were generated. The plant model used equations (1) to (3) with the parameters of the same plant that provided the time series.

RESULTS AND COMMENTS

To check the good behavior of the proposed controller the simulation was repeated with δr obtained from the time series of the real plant. In Figures 1.- and 2.- the results obtained for two wet months, April and December, are presented. Very similar results were obtained with dry months. First of all we can see that the neural controller provides a very effective regulation of *Xva* and *Se*. They are kept at very good levels, very close to those the plant was designed for: about 2000 mg/l for *Xva* and about 7.5 mg/l or less for *Se*. It can be seen that the output pollutant concentration is very similar to that obtained when the control is performed by the time series describing the operator's actions, what proves the effectiveness of the neural network at learning the operator's skills. Differences are bigger for the sludge concentration but both dynamics are very similar. Moreover the simulation with the neural controller provides values that are closer to those assumed in the plant design than those provided by the time series of the recirculation. This is not a surprising fact because while the operator must adjust that parameter taking into account delayed measures of the plant variables and subjective observations, the neurocontroller uses precise values. If it were possible to measure that variables with a little delay in an actual plant the defined controller might be used to provide an automatic control, because as the time evolution of the plant variables is quite slow (it takes more than an hour to notice the effect of a modification in the input values) time delays close to half an hour will allow an almost real time control. Unfortunately devices providing those fast measures are not used in actual plants because they are too expensive.

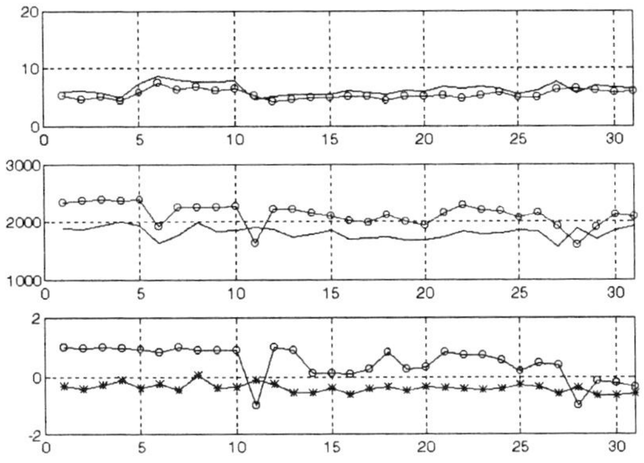

FIGURE 1. April 1999. Top to down *Se* (mg/l), *Xva* (mg/l) and δr (no dimension). "o": data obtained with the neurocrontroller. Solid line: data obtained from operator.

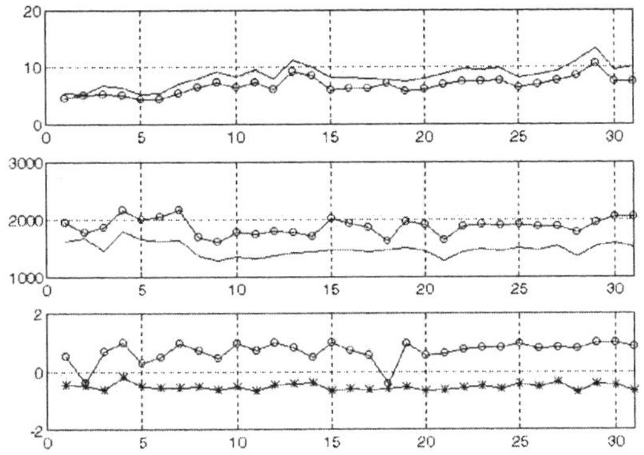

FIGURE 2. December 1999. Top to down *Se* (mg/l), *Xva* (mg/l) and δr (no dimension). "o": data obtained with the neurocrontroller. Solid line: data obtained from operator.

REFERENCES

1. Bishop, C. M., *Neural Networks for Patter Recognition*, Oxford, 2000.
2. Haykin S., *Neural Networks. A Comprehensive Foundation*, IEEE Press, 1994.
3. Hrycej T., *Neurocontrol, Towards an Industrial Control Methodology*, Wiley Interscience, 1997.
4. Ramalho R. S. *Tratamiento de Aguas Residuales*, Reverte, 1996.
5. Olsson, G.; Newell, B. *Wastewater Treatment System*, IWA, 1999.
6. E.D.A.R. Aguas de Mérida S. L., *Operation and Maintenance Reports. 1998 and 1999.*
7. Norgaard, M. *Neural Networks for Modelling and Control of Dynamic Systems,* Spinger-Verlag, 2000.

Applied Physics (APHYS 2003)
A. Méndez-Vilas (Ed.)

Time Series Prediction With Neural Networks. Application To Electric Energy Demand

M. A. Jaramillo, D. Carmona, E. González and J. A. Álvarez

School of Industrial Engineering University of Extremadura Avda. de Elvas s/n, 06071 Badajoz (Spain)
phone:+34 924 289600, fax:+34 924 289601, e-mail: miguel@unex.es, dcarmona@unex.es, evagzlez@unex.es, jalvarez@unex.es

Abstract. Electric energy demand forecasting represents a fundamental information to plan the activities of the companies that generate and distribute it. So a good prediction of its demand will provide an invaluable tool to plan their production and growth policies. This demand may be seen as a temporal series when its data are conveniently arranged. In this way the prediction of a future value may be performed studying the past ones. Neural networks have proved to be a very powerful tool to do this. They are mathematical structures that mimic that of the nervous system of living beings and are used extensively for system identification and prediction of their future evolution. In this work a neural network is presented to forecast the evolution of the monthly demand of electric consumption. Two strategies are proposed: the first uses a network that is trained once an then used to predict future values of the time series, while in the second the network is trained with all the past data every time a prediction is to be performed. The Spanish monthly consumption from 1975 to 2002 has been used to validate the models proposed. Errors smaller than 5% have been obtained in most of the predictions.

INTRODUCTION

Electric energy demand forecasting is a fundamental tool for production and distribution companies because it provides them a prediction of the market needs in order to fit their production to the society demand. Two kinds of forecasting may be performed: a short term one that deals with prediction of hourly or daily consumption and a long term one that works with monthly data.

As electric consumption data evolve along time they may be assumed to form a time series. Nevertheless as short time electric energy consumption is highly influenced by factors different from past consumption, forecasting also needs to take into account other information as temperature, humidity, hour of the day or day of the week, that have a remarkable influence on demand.

On the other hand, as long term forecasting deals with monthly data, the influence of the aforementioned factors is diluted in an overall value. So only the time series of electric consumption is needed to perform the prediction.

Among the several tools usually used in time series prediction, neural networks have shown to be a very effective one because of their flexibility and easy configuration, a fact that is hardly surprising if it is taken into account that recurrent neural networks may be considered as a special case of nonlinear autoregressive moving average models (NARMA), a very powerful tool for predicting the time series evolution [1]. So neural networks have become popular tools for the electric demand forecasting [2], [3], [4].

Nevertheless in spite of its potential interest for electric energy companies, long term forecasting has received little attention from researchers in contrast with the higher interest that short term one has had [5]. In this work a monthly demand prediction is carried out by a neural network, the Multilayer Feedforward Perceptron, one of the most popular neural structures.

The outline of this work is as follows. In Section 2 the normalization process of the input data is presented. Section 3 describes the structure of the neural network used to perform the forecasting. In Section 4 two forecasting strategies and their results are presented.

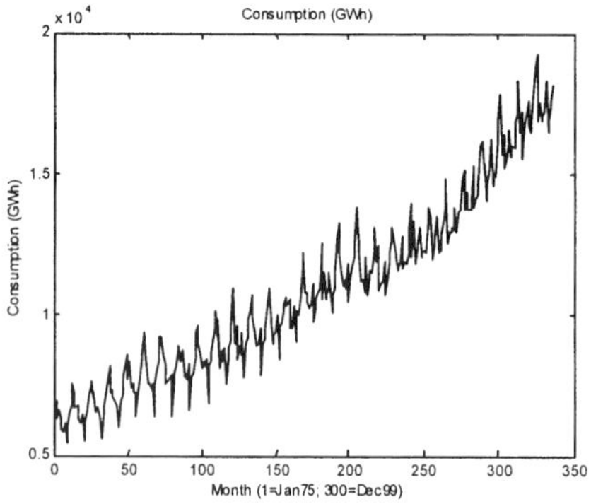

FIGURE 1. Monthly Spanish electric consumption from January 1975 to December 2002.

NORMALIZATION OF THE TIME SERIES

Time series of monthly electric consumption have a rising tendency (Figure.- 1) due to the influence of the economic and technological evolution that generates a growing demand of electric energy.

That rising trend represents an important problem for neural networks because the use of nonlinear saturating functions as neural outputs imposes boundaries to the network prediction. So the model will not be able to predict values of the time series falling out of that boundaries. Therefore it is necessary to define a procedure that eliminates that rising tendency from the time series but retains the monthly fluctuations. Nevertheless there is no beforehand way to define an optimal procedure to perform that normalization process, so several strategies must be tested in order to find out the one performing the best.

As the prediction is performed using only past values of the time series, a set of them, along with the data to be normalized, will be considered to do the normalization. Two options have been tested: the division of every datum by the sum of those elements or by their mean value. Several sets of past values have been used and the best results ware obtained with the mean value of eleven past elements and that considered to be normalized (twelve values) [6].

In the same way a denormalization process will be applied to the network output in order to obtain the actual value of the forecasting. As there is not a normalization constant associated to the predicted datum, the network output will be multiplied by the normalization constant of the last one used to perform the prediction.

NEURAL NETWORK STRUCTURE

As it has been previously stated the neural network structure used in this work is the well-known Multilayer Feedforward Perceptron. It has been widely used in time series prediction because of its ability to identify the time evolution of a dynamic system[7].

The network is formed by an input layer whose elements are the data to be processed, an output layer that provides the network output (in this work an only neuron) and one or several hidden layers that process the input data. Every layer (apart from the first one) is formed by a variable number of units named neurons that compute the weighted sum of all its inputs and a bias constant. The result is processed by an activation function that provides the neuron output:

$$x_j^i = \sum_k w_{jk}^i y_k^{i-1} + \theta_j^i \tag{1}$$

$$y_j^i = \sigma\left(x_j^i\right) \tag{2}$$

In these expressions x_j^i represents the activity of neuron j in layer i, w_{jk}^i the strength of the connections between this neuron and all those that are in the previous layer, y_j^i the neuron output while θ_j^i is a bias constant. The output function $\sigma\left(x_j^i\right)$ represents an important element of the neural network paradigm, because it supplies a nonlinear element to the model that allows these structures to identify the nonlinear behavior inherent to complex dynamics. The most frequently used as output function is the so called sigmoid for its "S-shape". Gaussians or linear functions may also be used. A combination of both linear and nonlinear functions is usually found in the literature, where a very common configuration is the use of nonlinear functions for the hidden layers and linear ones for the output. A combination of sigmoids in the hidden layer and a linear function in the output will be used in this work.

The learning capability of neural networks is provided by the adaptation of the input weights of every neuron. This process is performed presenting an input pattern and the desired output to the network modifying every weight until an error function reaches a minimum or falls below a fixed value. This procedure is repeated for each pattern to be learned, defining a strategy known as "Backpropagation" [8]. The way this adaptation process is carried out defines the learning strategy of the neural network. The well known Levenberg-Marquardt algorithm (a combination of the gradient descent and Newton methods for solving optimization problems) [8] has been selected to do it.

The number of neurons and hidden layers is an essential issue in the network design. Several structures have been tested to find out that performing the best [5]. It was a neural network with three hidden layers with 8, 4 and 8 neurons.

Finally, a corrective process has been applied to the resulting values because, although the neural network can learn the behavior of a time series, it cannot predict an unexpected change in the series tendency until enough values of the monthly demand have been presented to the network. To compensate for this effect a new variable is defined. It represents the difference between the denormalized output and an expected value that is obtained with the following expression:

$$C(i) = P(i) - R(i-12) \times \frac{R(i-1) - R(i-13)}{R(i-13)} \tag{3}$$

where C(i) is the corrective variable to be applied to the prediction P(i) and R(j) the real value of datum j.

This correction is weighted by a factor 0.6, a value that has been obtained by a trial and error procedure which provides the best results among all those tested between 0 and 1. So, a factor 0.6C(i) is added to the denormalized datum to obtain the network prediction.

SIMULATION RESULTS

The Spanish monthly consumption from January 1975 to December 2002 (a total of 336 values) has been used to validate the proposed model. All these data has been divided into two blocks: one for training (from January 1975 to December 1996, 264 months) and the other for validation (the remaining information, 72 months).

Once the network structure and the normalization process have been defined, the forecasting may be performed. Two options are to be considered, the first assumes to train the network with the first 264 elements and then predict the electric demand of the remaining months. In he second, the network is trained with the first 264 elements once and then predicts the demand of the following month. Now the actual value of consumption is added to the training group and the network is trained again with 265 elements. This process is repeated until all the predictions have been performed. These two options define two different strategies in the way the forecasting will be performed in the real world by an electric company. The first assumes that the network is trained once and then used to predict any month. The second assumes that the network will be trained with all the past values of consumption that the user has every time a prediction is to be performed.

The two options have been tested and their results are presented in Figure 2. As it can be seen both options present very similar results, but while the first one needs a few seconds to perform every prediction, the second requires many time to perform both the training and the prediction (up to seven hours have been needed to carry out the forecasting of the 72 month used for validation). Although monthly forecasting is not time dependent (the prediction may be performed once the consumption datum of the past month is available during the first days of the month to be forecasted) it looks little affective to spend so much time in a task that may be carried out in a few seconds and, indeed, with the same precision.

So we can conclude that the best option is to train the network once and predict every monthly demand with that network with no further training.

Apart from a few values, most of the predictions present an error less than 5%, a value that may be assumed as acceptable by electric companies.

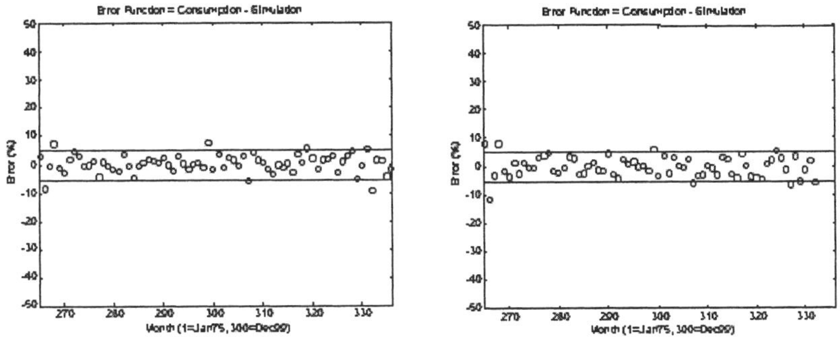

a) b)

FIGURE 2. a) Errors in the prediction when the network is trained only once. b) Errors when the network is trained before any prediction is performed.

REFERENCES

1. Connor, J. T., Martin, R. D., and Atlas, L. E.. "Recurrent Neural Networks and Time Series Prediction". IEEE Trans. on Neural Networks, Vol. 5, No. 2, pp. 240-254, March 1994.
2 Charytoniuk, W., Don Box, E., Lee, W. J., Chen, M.-S., Kostas, P. and Van Olinda, P.. "Neural-Network-Based Demand Forecasting in a Deregulated Environment". IEEE Trans. on Industry Applications, Vol. 36, No. 3, pp. 893-898, May/June 2000.
3. Doveh, F., Fegin, P., Greig, D. and Hyams, L.. "Experience with FNN Models for Medium Term Power Demand Predictions". IEEE Trans. on Power Systems, Vol. 14, No. 2, pp. 538-546, May 1999.
4. Derzga, I. and Rahman, S.. "Input Variable Selection for ANN-Based Short-Term Load Forecasting". IEEE Trans. on Power Systems, Vol. 13, No. 4, pp. 1238-1244, November 1998.
5. Carmona, D., Jaramillo, M. A., González, E., Álvarez, J. A.. "Electric energy demand forecasting with neural networks". IECON-2002. 28th Conference of the IEEE Industrial Electronics Society. Sevilla (Spain). 5-8 November 2002.
6. Jaramillo, M. A., Carmona, D., González, E., Álvarez, J. A.. "Reliability of the forecasting of the monthly demand of electric energy with neural networks". ICREPQ'03. International Conference on Renewable Energy and Power Quality. Vigo (Spain). 9-12 April, 2003.
7. Hornik, K., Stinchcombe M. and White, H. "Multilayer Feedforward Networks are Universal Approximators". Neural Networks, Vol. 2, 1989, pp. 359-366.
8. Bishop, C. M., "Neural Networks for Pattern Recognition". *Oxford University Press*, 1995.

Applied Physics (APHYS 2003)
A. Méndez-Vilas (Ed.)

Five Years Tumor Therapy with Heavy Ions at GSI Darmstadt

D. Schardt
for the Heavy-Ion Therapy Collaboration*

Gesellschaft für Schwerionenforschung, Biophysics Division, Planckstr. 1, D-64291 Darmstadt, Germany

**GSI Darmstadt, FZ Rossendorf, Radiological Clinic and DKFZ Heidelberg*

Abstract. Heavy-ion beams offer favourable conditions for the treatment of deep-seated local tumors. The well defined range and the small lateral beam spread make it possible to deliver the dose with millimeter precision. In addition, heavy ions have an enhanced biological efficiency in the Bragg peak region which is caused by the dense ionization and the resulting reduced cellular repair rate. Furthermore, heavy ions offer the unique possibility of in-vivo range monitoring by applying Positron-Emission-Tomography (PET) techniques. Taking advantage of these clinically relevant properties, more than 180 patients have been treated with carbon ions at GSI Darmstadt since December 1997 with promising results so far. A dedicated heavy-ion treatment center at the Radiological Clinic Heidelberg with a designed capacity of 1000 patients per year is under construction.

INTRODUCTION

The application of high-energy particle beams to cancer treatment represents an exceptional example of interdisciplinary scientific work involving the basic sciences of medicine, biology, and physics. The recent achievements resulting in a highly tumor conform radiation treatment with particle beams would not have been possible without the strong and fruitful interdisciplinary collaboration of scientists in the fields of oncology and radiation medicine, radiation biology, accelerator technology and engineering, as well as atomic and nuclear physics.

Heavy-ion beams offer favourable conditions for the treatment of deep-seated local tumors. In contrast to photon beams they have an inverted depth-dose profile (Bragg curve) and the narrow Bragg peak can be precisely adjusted to the desired depth by the beam energy. Ions heavier than protons like carbon in addition offer an enhanced biological efficiency in the Bragg peak region (Fig.1) which can be explained by the very dense ionization towards the end of the particle track.

Patient treatments with heavy charged particles started in 1954 at LBL Berkeley, first with protons and later with helium beams. Radiotherapy with heavier ions was initiated by Tobias et al. [1,2] at the BEVALAC facility at LBL. There most of the patient treatments (1975-1992) were performed with beams of ^{20}Ne. The beams were delivered to the patient by passive beam shaping systems including a scatterer or wobbler magnets and other passive modulating elements [3]. In 1994 the heavy-ion medical accelerator HIMAC [4] dedicated to radiotherapy started with carbon ions at NIRS Chiba (Japan), using similar technical concepts as those pioneered at Berkeley.

At GSI Darmstadt (Germany) a new concept [5] was developed, differing significantly from the previous designs at the BEVALAC and HIMAC: moving a narrow pencil beam over the target volume (raster scan) a tumor conform treatment can be achieved to a high degree, restricting the biologically most effective ions to the target volume and minimizing the dose to the surrounding normal tissue. In spite of the demanding technical concept the fully active rasterscan system has proven to operate reliably since the first patient treatment in December 1997. Within the Heavy-Ion Therapy Collaboration three other institutes take care of various parts of the project: The Radiological Clinic Heidelberg (all clinical aspects such as patient selection, diagnostic, dose calculation), the German Cancer Research Center DKFZ Heidelberg (patient immobilisation, treatment planning, dosimetry), and the Research Center FZ Rossendorf near Dresden (PET monitoring). Up to now more than 180 patients, most of them with tumors in the skull base region, were treated with carbon beams at GSI.

PHYSICAL AND RADIOBIOLOGICAL ASPECTS

The slowing-down process of heavy charged particles is governed by interactions with the atomic shell of the absorber nuclei (tissue) and well described by the Bethe-formula. The resulting depth-dose profile (Bragg curve) exhibits a flat plateau region and a distinct peak near to the end of range of the particles. This represents the major physical advantage of heavy charged particle beams in radiotherapy. For heavy ions, nuclear reactions along the penetration path lead to an attenuation of the primary beam flux and a build-up of lower-Z fragments with increasing penetration depth [6]. As the range of the particles scales with A/Z^2 the

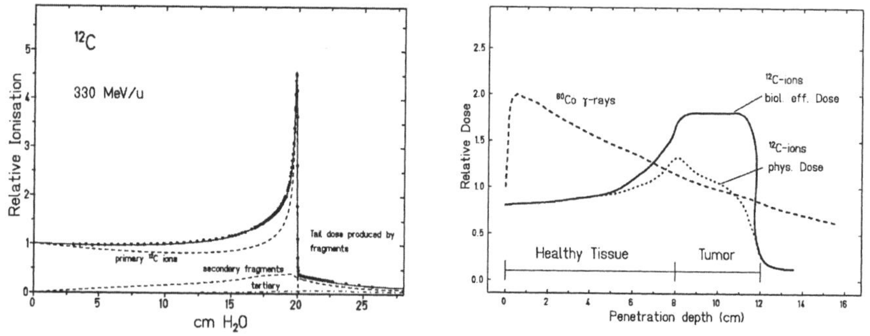

FIGURE 1. Left: Bragg curve for a carbon beam in water measured at GSI. The data points are compared to a model calculation [7] (solid line) and the dashed lines indicate the calculated contributions from the primary particles and from nuclear fragments.
Right: Depth-dose profiles of photon and carbon beams. The extended Bragg peak of carbon ions is generated by superposition of several Bragg curves with different particle energies. The increased biological effectiveness in the Bragg peak region is an advantage for the treatment of deep-seated tumors, especially in the vicinity of radiosensitive structures.

depth-dose profile of heavy-ion beams shows a characteristic fragment tail beyond the Bragg peak (Fig.1). For carbon ions the tail-dose contribution is relatively small and the production of the radioactive isotopes ^{10}C and ^{11}C enables an interesting application of PET-techniques for in-vivo range verification (see below).

The radiobiological efficieny of charged particles is mainly characterized by their local ionization density which can directly be correlated to the local density of DNA damage. As a result of extensive irradiation experiments with cell cultures at LBL, NIRS and GSI it was found that carbon beams meet the therapy requirements best possible [8]. At high energies in the entrance region carbon ions have a sufficiently low ionization density and act like photons, producing mostly repairable DNA damage. Towards the Bragg peak the ionization density increases significantly, resulting in irrepairable damages and high cell killing power. These findings were confirmed by measuring molecular DNA damage and repair along therapeutic beams [9].

The treatment planning program [10] developed at GSI and DKFZ Heidelberg is based on a biological model [11] which takes into account the variance of RBE in a mixed particle field (incl. fragmentation). The physical dose is adjusted to the RBE by an iterative procedure in order to achieve a homogeneous biological effect.

THE GSI TREATMENT UNIT

In order to have a maximum benefit of the physical and biological advantages of heavy-ion beams for the treatment of deep-seated tumors the concept of the GSI treatment unit includes a fully active beam delivery system [12]. A pencil-like beam with well-defined energy and corresponding depth of the Bragg peak is moved by fast horizontal and vertical scanning magnets slice-by-slice over the target volume. By choosing appropriate steps in the beam energy the whole target volume is irradiated according to the dose prescription (Fig.2).

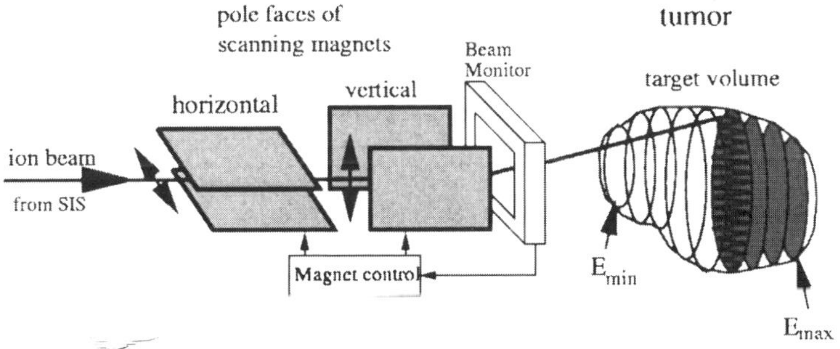

FIGURE 2. Principle of the magnetic scanning system at GSI. The target volume is irradiated by moving the ion beam (80-430 MeV/u ^{12}C) with fast magnets over each slice. The required beam energies - corresponding to the depth of the Bragg peak for each slice - are supplied on a pulse-to-pulse operation by the synchrotron (SIS) control system.

In comparison with passive beam delivery, the active system has a clear advantage as there are no restrictions in shaping the target volume and any prescribed 3-dimensional dose distribution can in principle be generated. Furthermore, beam losses and contamination by nuclear fragmentation in passive beam shaping elements in front of the patient are minimized in active systems.

An interesting positive aspect of the nuclear fragmentation effect discussed above is the formation of shortlived positron-emitting isotopes which can be used for an in-situ monitoring of their stopping points by Positron-Emission-Tomograpy (PET)-techniques [13]. This yields a verification of the correct treatment planning and beam delivery in each patient. Especially the penetration depths is monitored which is invaluable for treating tumors near critical structures. This feature is unique to heavy-ion beams as here the induced β^+-activity in tissue mainly stems from positron-emitting projectile fragments which have nearly the same end-of-range as the primary particles.

FIRST CLINICAL RESULTS

The treatment of radioresistant skull-base tumours with carbon ion beams seemed to be most promising because conventional radiotherapy with photons is often not applicable due the limited tolerance of neighbouring radiosensitive structures. As can be seen from a typical treatment plan shown in Fig. 3, the target volume for such tumors is close to the brain stem or the optical nerves that can be spared very well by the precision irradiation with ion beams. In addition, the enhanced biological effectiveness (RBE) of heavy ions ensures that a sufficiently high dose can be delivered to the target volume.

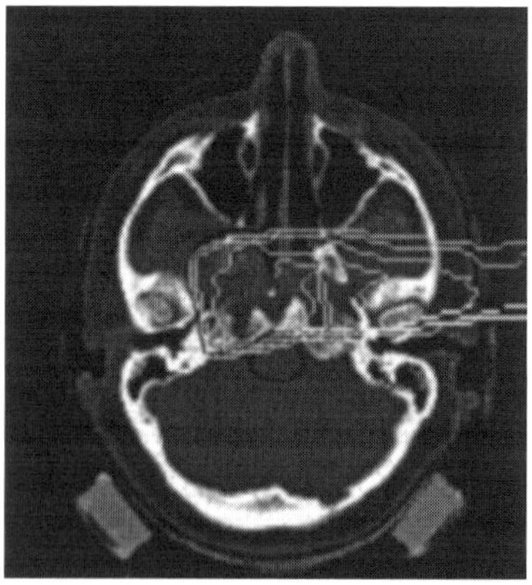

FIGURE 3. Carbon-ion treatment plan for a large tumor in the skull base. The target volume is close to the brain stem and optical nerve which can be spared very well by the carbon ion irradiation.

Since December 1997 more than 180 patients were treated at GSI within clinical pase I/II trials. First results were reported for 45 patients with chordomas, chondrosarcomas and other skull base tumors [14,15]. Most of these patients received a fractionated carbon ion irradiation in 20 consecutive days with a median total dose of 60 GyE which was well tolerated without severe side effects. No local recurrence within the treatment volume was observed. As a preliminary result local control rates at 3 years of 87% for chordomas and 100% for chondrosarcomas were found [16]. Ongoing clinical studies of patients with adenoid cystic carcinomas and sacral/spinal chordomas and chondrosarcomas show a good effectiveness and no severe toxicity.

Encouraged by the positive clinical results obtained in this pilot project the plans for a dedicated hospital-based ion treatment facility [17] are further pursued. The facility to be built near the Radiological Clinics in Heidelberg is planned for the treatment of 1000 patients per year and includes three treatment rooms, one of which is equipped with a rotating gantry. This facility is expected to become fully operational in 2007 and will be the first clinical heavy-ion treatment center in Europe.

REFERENCES

1. Tobias C. A. and Todd P.W., Heavy charged particles in cancer therapy, In *Radiobiology and Radiotherapy*. Institute Monograph No.24 (1967).
2. Petti P. L. and Lennox A. J., Hadronic radiotherapy, *Ann. Rev. Nucl. Part. Sci.* **44** (1994) pp.155-197.
3. Chu W.T., Ludewigt B. A. and Renner, T. R., Instrumentation for treatment of cancer using proton and light-ion beams, *Rev. Sci. Instrum.* **64** (1993) pp. 2055-2122.
4. Hirao Y., Ogawa H., Yamada S. et al., Heavy-ion synchrotron for medical use - HIMAC project at NIRS Japan, *Nucl. Phys.* **A538** (1992) pp. 541c-550c.
5. Kraft G., Gademann G. (Eds.), Einrichtung einer experimentellen Strahlen-therapie bei der Gesellschaft für Schwerionenforschung Darmstadt, *GSI-Report* **93-23** (1993).
6. Schall I., Schardt D., Geissel H. et al., Charge-changing nuclear reactions of relativistic light-ion beams ($5 \leq Z \leq 10$) passing through thick absorbers, *Nucl. Instr. and Meth.* **B11** (1996) pp.221-234.
7. Sihver L., Schardt D. and Kanai T., Depth-dose distributions of high-energy carbon, oxygen and neon beams in water, *Jpn. J. Med. Phys.* **18** (1998) pp. 1 - 21.
8. Weyrather W. K., Ritter S., Scholz M. and Kraft G., RBE for carbon track-segment irradiation in cell lines of differing repair capacity, *Int. J. Radiat. Biol.* **75** (1999) pp. 1357 - 1364.
9. Heilmann J., Taucher-Scholz G., Haberer T., Scholz M., Kraft G., Measurement of intracellular DNA double strand breaks, induction and rejoining along the tracks of carbon and neon particles in water, *Int. J. Radiat. Oncol. Biol. Phys.* **34** (1996) pp. 599-608.
10. Krämer M., Jäkel O., Haberer T., Kraft G., Schardt D. and Weber U., Treatment planning for heavy-ion radiotherapy, *Phys. Med. Biol.* **45**(2000)pp.3299-3317
11. Scholz M. and Kraft G., Calculation of heavy-ion inactivation probabilities based on track structure, X-ray sensitivity and target size, *Rad. Prot. Dosim.* **52** (1994) pp. 29-33.
12. Haberer T., Becher W., Schardt D. and Kraft G., Magnetic scanning system for heavy-ion therapy, *Nucl. Instr. and Meth. in Phys. Res.* **A330** (1993) pp. 296-305.
13. Enghardt W., Debus J., Haberer T. et al., The application of PET to quality assurance of heavy-ion tumor therapy, *Strahlenther. Onkol.* **175** (1999) Suppl. II, pp. 33-36.
14. Debus J., Haberer T., Schultz-Ertner D. et al., Fractionated carbon ion irradiation of skull base tumors at GSI. First clinical results and future perspectives, *Strahlenther. Onkol.* **176** (2000) pp. 211-216.
15. Schultz-Ertner D. et al., Radiotherapy for Chordomas and Low-Grade Chondrosarcomas of the Skull base with Carbon Ions, *Int. J. Radiation Oncology Biol. Phys.*53(2002)pp. 36-42.
16. Debus J. et al., GSI Scientific Report 2002 (2003) p.174
17. Groß K. D. and Pavlovic M. (Eds.), Proposal for a dedicated ion beam facility for cancer therapy, GSI Darmstadt 1998.

Applied Physics (APHYS 2003)
A. Méndez-Vilas (Ed.)

Ionizing Radiation as a Tool for Detoxification of Whole Effluents

S. I. Borrely; C. L. Duarte and M. H. O. Sampa
Instituto de Pesquisas Energéticas e Nucleares – IPEN
Centro de Tecnologia das Radiações – CTR
Av. Lineu Prestes, 2422, CEP 05508-000 – São Paulo, Brasil
sborrely@ipen.br

Abstract. Ionizing Radiation Technology (IRT) will be presented as an alternative for complex effluent detoxification. The acute toxicity reduction will be discussed as percentage evaluated for luminescent bacteria and microcrustacean (*Vibrio fischeri* and *Daphnia similis*, respectively). At Instituto de Pesquisas Energéticas e Nucleares, IPEN, the use of IRT has been considered mainly for industrial effluent detoxification. The results to be presented will show the efficiency of radiation processing to several kind of residues that have been submitted to electron beam irradiation allowing the selection of ideal radiation doses also considering the potential for toxicity reduction.

INTRODUCTION

The development of electron beam technology for environmental purposes has gained attention not only because of several important and hard contaminants reaching waters, soil and groundwater but also due to the capability of degrading organic molecules. Ionizing radiation for environmental area may be obtained from cobalt-60 (gamma radiation) or from Electron Beam Accelerator (EBA), which seams to be more suitable for environment applications.

Typically, electron-beam devices have been powered by conventional high-voltage sources. Conventional accelerator or high-voltage generators are available in voltages ranging from 100keV up to 10MeV. New concepts of accelerator for a machine which allows the use of solid insulation at high voltages, known as Nested High Voltage Generator, may indicate suitability for environmental application (Curry, et al., 1998). Other important development is centered into the devices which take the water or sewage to the electron beam region (Rela et al, 2000). Podzorova, et al (1998), developed electron irradiation of municipal wastewater in the aerosol flow in order to increase throughput by 20-50 times in comparison with liquid wastewater, rendering 500 m^3/day due the spray irradiation system, 0.3 MeV electron energy, beam power 15kW.

The degradation kinetics of hundred compounds have been reported by Zele et al (1998), nevertheless relatively few data with real samples is available (Borrely et al, 2000 Duarte, 2000; Nickelsen et al, 1998).

Due to difficulties for a complete chemical characterization of real effluents, another important parameter to be taken into account for evaluating the efficacy of radiation process is toxicity.

Focusing remainder toxicity and disinfection processes, it was evidenced that toxicity in the γ-irradiated effluent either decreased or was unchanged as compared to an undisinfected sample and that chlorination (with or without dechlorination), ozonation (O_3) and ultraviolet (UV) irradiation increased toxicity with statistic significance (Thompson and Blatchley, 1999). Biological assays have been routinely used to assess the toxicity of real effluents before and after irradiation (activities: industrial effluents, produced water – petrol, textile and sewage), Borrely et al (2000).

At the Instituto de Pesquisas Energéticas e Nucleares, IPEN, a ten year period was dedicatad for environmental irradiation applications. This project elected electron beam irradiation (EBI) for real effluents although few experiments were performed with gamma radiation. The experiments were carried out with sewage and sludge disinfection, organics sewage degradation, chemical and pharmaceutical organic degradation and detoxification. EBI will be presented here as an alternative for complex effluent detoxification.

METHOD

Electron beam irradiation was applied to sewage and industrial samples collected at Suzano Wastewater Treatment Plant, Suzano – São Paulo, using laboratory scale. Three sites were sellected for sampling: industrial effluents (a); hard mixture of sewage and industrial effluents (b); final biological effluent (c).

Sampling: Seven composite samples (six hour per fraction until getting 24 hours with four subsamples) were collected from each of the three sites studied at the same wastewater treatment plant, respecting the entrance flow rate. Three sites were included for this study totalizing 21 samples.

Irradiation: The samples were processed at an Electron beam accelerator (Dynamitron, RDI), 37.50kW power. The machine energy was fixed at 1.4 MeV, varaible current according to the desired radiation dose. The wastewater was irradiated contained in pyrex vessels (248mL), the wastewater thickness was 4.0mm in order to obtain total electron penetration into the water. The samples were conducted at electron beam region by a trail at 6.72 meters per minute.

Acute Toxicity Evaluation:

Acute toxicity evaluation (ATE): Two biological systems were applied for the study. The first was a luminescent bacteria and the second was a daphnid water flea. The marine bacteria ***Vibrio fischeri***, Microtox®, basic test protocol for 15 minutes exposure. The test was performed using the following sample concentrations: 45.45%, 22.72%, 11.36%, 5.68% and two controls. Microtox® test was the first acute evaluation carried out as an screening test, applied to each sample, even during the preliminary studies in order to select the radiation doses to be applied for each sampling site.

Daphnid Acute Toxicity Test was the second ATE performed with the water flea *Daphnia similis*, 24 hours exposure, measuring non motile neonates (ABNT, 1997). In general the sample testing concentrations varied from 0.1% to 10% for site 1, from 0.1% to 50% for site 2 and from 10% to 100% for the final effluent, site 3.

Both tests results were expressed as effective concentration that reduced the measured parameter by 50%, EC50 (%, v/v). These numbers (EC50) were transformed into toxic units (TU) as follow:

$$TU = 100/EC50 \qquad (01)$$

$$RDE (\%) = (TU_{untr} - TU_{tr}) \times 100 / TU_{untr} \qquad (02)$$

RDE (%) = Radiation dose efficiency (radiation dose applied to obtain a given percentage of toxicity reduction, reported as Toxic Unit, for radiation treated samples (TU_{tr}) and untreated samples (TU_{untr}));

RESULTS

Radiation effect on toxicity was evidenced by toxic unit percentage ranging from 82 to 100. For industrial effluents that required higher radiation dose than the others, 50kGy was selected as enough dose for the most of the collected samples. At Fig. 1.the radiation effect on the toxicity reduction was presented and it can be noted a very important reduction (82%, 90% and 54% for 50kGy, including highly toxic samples collected during six months.

Taking the two last bars of Fig. 1, when radiation process was less effective than to the others for toxicity removal. Assuming an average flow of 2.000 liters per day of a similar industrial effluent, if 50kGy would actually be applied, it should represent at least a reduction of total toxic charge from 166.000 liters to 78.000 liters per day.

At the second studied site, entrance of wastewater, it was tested doses from 10 kGy up to 50kGy, meaning an improvement (average toxicity reduction higher than 90%) for 10kGy in two different sampling. Nevertheless, some other samples required higher radiation dose to achieve such a good result, Figure 2. This fact can be related to the huge variety of organic and inorganic contained in the effluent which are also included in the solids residues.

The importance of evaluating the biological effects of effluents when working on treatment process is increasing due to the complexity of mixture of wastewater with that remained from industrial activities. This also justify the different sensitivity when a given effluent result in different inter-specific biological effects (ie daphnids and luminescent bacteria, for instance).

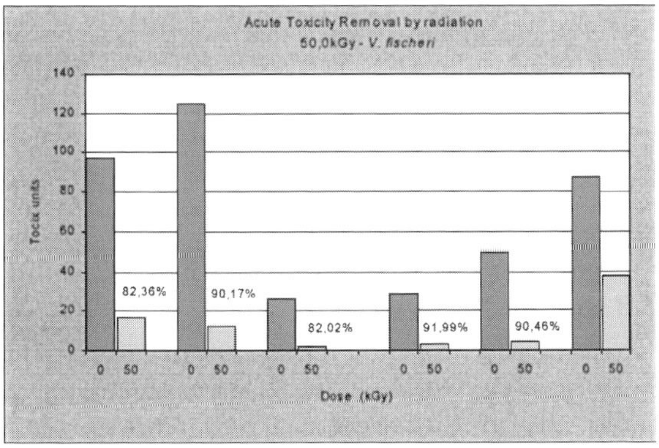

Figure 1 – Radiation effect on acute toxicity for 50kGy selected for industrial effluents

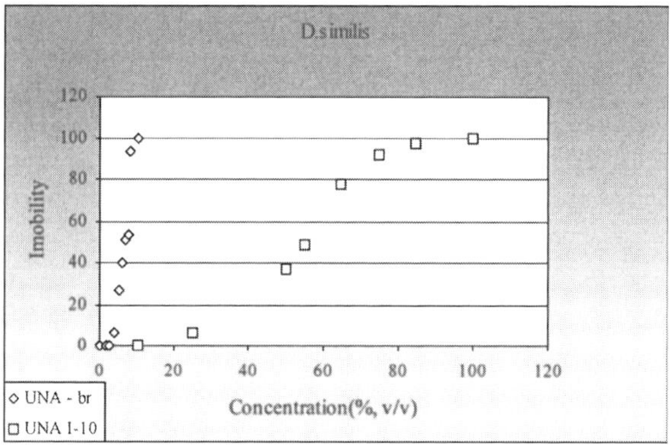

Figure 2 – The potential for causing adverse effect to aquatic organisms (daphnid) of treated and untreated samples. Radiation dose was 10kGy.

Fig. 2 may to evidence the benefic effect of radiation treatment to a given sample. It shows the samples concentration to reduce 50% of daphnid mobility before and after being treated with 10kGy. For untreated wastewater only 10% of sample can eliminate 100% biological activity while for treated effluent a 80% concentration was need to obtain the same adverse effect, at the same standardized condition.

Table 1 – Electron beam efficience for organic compounds decomposition

Name	Site 1 (a) (%)	Site 2 (b) (%)	Site 3 (c) and (d) (%)
Chloroform	87.45 ± 9.80	99.99	68.34 ± 21.21 94.10 ± 6.52
Dichloroethane	99.98 ±0.24	93.45 ± 7.06	85.30 ± 9.56 99.17 ± 1.41
MIC (*)	99.99	97.80 ± 1	72.10 ± 20.50 87.82 ± 17.18
Benzene	81.78 ± 12.0	-	-
Toluene	88.36 ±23.27	71.70 ± 40	76.91 ± 10 97.46 ± 3.60
Xylene	96.57 ± 5.55	99.99 ± 64.90	86.42 ± 15 99.99
Phenol	52.60 ± 21	-25.96 ± 64.90	43.0

At Table 1 it was summarized the efficiency of radiation for several organic compounds degradation on the same samples that were used for this toxicity studies (Duarte, 2000). The mainly radiation byproducts detected were organic acids. The improvement of waters is a function of radiation dose and the efficiency of degradation for a chemical is dependent upon the initial concentration.

was applied to real effluents.

CONCLUSIONS

Ionizing radiation technology can play an important role for complex mixtures of effluent. The radiation process can be applied combined to biological treatment in order to reduce toxic substances assuring the efficiency of biological process or can be used for an specific industrial effluent.

(1) The efficiency of the process for effluent from chemical industries reached about 86% for 50kGy.

(2) Less contaminated effluents required lower radiation doses which were from 5,0kGy to 20kGy.

(3) When radiation was combined to biological process an overall toxicity reduction was higher than 95%.

REFERENCES

ABNT. Água – Ensaio de toxicidade aguda com *Daphnia similis* Claus, 1876[(1)] (Cladocera, Crustacea). NBR 12713, 1997.

BORRELY, S.I., A.C. CRUZ, N.L.Del MASTRO, M.H.O. SAMPA, E.S.SOMESSARI. Radiation Processing of Sewage and Sludge. A Review. ***Progress in Nuclear Energy***. v 33, No. 1 / 2. p. 3-21. 1998.

BORRELY, S.I.; SAMPA, M.H.O.; BADARÓ-PEDROSO, C.; OIKAWA, H.; SILVEIRA, C.G.; CHERBAKIAN, E.H.; SANTOS, M.C.F. Radiation processing of wastewater evaluated by toxicity assays. ***Rad Phys. and Chem.***, v.57, 507-511. 2000.

CETESB L5.227 (1987). Bioensaio de Toxicidade Aguda com *Photobacterium phosphoreum*, Sistema Microtox. (Método de ensaio).

CURRY, R.D.; D. Johns; K. Rathbun; k. Unklesbay. The Nested High-Voltage Generator: An overview of the Technology and Applications. ***In: Environ. Applic. of Ioniz. Rad.*** p.33-46. 1998.

DUARTE, C.L. *Aplicação do Processo Avançado de Oxidação por feixe de elétrons na degradação de compostos orgânicos presentes em efluentes Industriais*. 1999. Tese (Doutorado) Instituto de Pesquisas Energéticas e Nucleares, São Paulo.

DUARTE, C.L.; SAMPA, M.H.O.; RELA, P.R.; OIKAWA, H.; CHERBAKIAN, E.H.; SENA, H.C.; ABE, H.; SCIANI, V. Application of electron beam irradiation combined to conventional treatmento to treat industrial effluents. ***Radiat. Phys. and Chem. (57***), p. 513-518. 2000.

MICROBICS CORPORATION. (1994). Microtox ® M500 Manual. A Toxocity Testing Hankbook, Version 3.

NICKELSEN,M.G.; D.C. KAJDI; W.J. COOPER; C.N.KURUCZ; T.D. WAITE; F.GENSEL; H. LORENZI H; U. SPARKA. Field Application of a Mobile 20-kW Electron-

Beam Treatment System on Contaminated Groundwater and Industrial Wastes. ***In: Environ. Applic. of Ioniz. Rad.*** 451-58. 1998.

RELA, P.R., SAMPA, M.H.O., DUARTE, C.L., COSTA, F.E., SCIANI, V. Development of an up-flow irradiation device for electron beam wastewater treatment. ***Radiat. Phys. And Chem***. 57, 657-660, 2000.

PODZOROPVA, E.A., PIKAEV, A.K., BELICHEV, V.A. and LYSENKO, S.L. New data on electron-beam treatmento fo municipal wastewater in the aerosol flow. ***Radiat. Phys. Chem***. Vol 52 Nos. 1-6, pp. 361-364.1998.

THOMPSON, J.E. and BLATCHLEY, E.R. Toxicity effects of γ- irradiated wastewater effluents. ***Wat. Res***. Vol. 33, No. 9, pp. 2053-2058. 1999.

ZELE, S.R. ; NICKELSEN, M.G.; COOPER, W.J.; KURUCZ, C.N.; WAITE, T.D. Modeling kinetis of benzene, phenol and toluene removal in aqueous solution using the high-energy electron-beam process. In: ***: Environ. Applic. of Ioniz. Rad.*** p.395-403, 1998.

Applied Physics (APHYS 2003)
A. Méndez-Vilas (Ed.)

Sequence of Phase Transitions of Li-Na Niobate Solid Solutions in the High Temperature Region.

B. Jiménez, R. Jiménez, A. Castro and L. Pardo
Instituto de Ciencia de Materiales de Madrid. CSIC. Cantoblanco. Madrid

Abstract. Dielectric, thermal expansion and mechanoelastic measurements as a function of the temperature in Lithium Sodium Niobate ($Li_XNa_{1-X})NbO_3$ ceramics in the low Lithium content side, X ≤ 12%, have been performed. The obtained results from the different techniques in the 300K-1000K-temperature interval show several reversible anomalies besides that corresponding to the main ferro-paraelectric phase transition at Tc. The anomalies above Tc show a ferroelastic character and those below Tc are of the ferroelectric type. The existence of ferroeleastic phases is non-dependent on the Li content. A secondary ferroelectric phase below Tc (T' ≈ 560K) stabilizes only for X ≥ 10%Li content and the anomaly for X < 5% Li content could better considered as a diffusive ferroelectric phase transition. The main ferro-paraelectric phase transition was attributed to Na displacements but the other observed phase transitions are better accounted for NbO_6 octahedra tilts.

INTRODUCTION

The low lithium content Lithium Sodium Niobate solid solutions have become very interesting materials for microwave, pyroelectric applications and nowadays for piezoelectric applications as a substitute for lead titanate-circonate compositions to reduce environmental pollution.

The Sodium Niobate (NN) is a base compound with perovskite type structure (ABX_3) that presents several structural phase transitions [1]. $Li_XNa_{(1-X)}NbO_3$ compositions, hereafter referred to as LNN(X), have also shown several phase transitions [2-6] in the 100K - 800K temperature interval. The compositions with Li content lower than 2% are antiferroelectric (AFE) [3] and those with a Li content higher than 2% are ferroelectric (FE) [4,5] at room temperature.

Dielectric measurements in the composition with X = 9% [7] present an unique peak in the dielectric constant, that corresponds to the ferro-paraelectric phase transition temperature Tc = 658K. Compositions with 10% ≤ X ≤ 15% show several peaks in dielectric constant around the main phase transition temperature on heating but an unique wide peak on cooling [4,5] is observed.

It has been proved in pure sodium niobate LNN(0) that composition fluctuations produce regions with different Na content (local non- stoichiometry) giving place to the coexistence of phases, named (FE) Q, (AFE) P phases, at room temperature. The transformation from (AFE) P phase to the (FE) Q phase can be stabilised in the whole volume material by applying a strong electric field or by substitution of 2.5% K atoms for Na.[8] The temperature also seems to modify the relation P/Q phases according the wide, small anomaly observed [9] in the dielectric constant of LNN (0) at about 460K By using optical techniques to follow the thermal evolution of the birefringence in single crystals with X = 2% Li content [10], ferroelastic domains have been observed in the paraelectric phase of this composition. According to these results, this composition should be ferroelastic from room temperature to 970K. Above this temperature the compound acquire the prototype cubic phase (m3m). Many studies on dielectric properties have also been carried out in LNN(x) compositions [5], but the high values of the a.c. electric conductivity hides the true behaviour of dielectric properties at high temperatures. To overcome this problem thermal expansion and mechanoelastic measurements that provide valuable information at high temperatures must be used [11,12].

The aim of this work is to deliver additional information on the character of the phase transitions that could be assigned to the anomalies observed by the different measuring techniques (dielectric, thermal expansion and mechanoelastic) in $0 \leq X \leq 10\%$ compositions in the 300K-1000K-temperature interval

EXPERIMENTAL PROCEDURE

Ceramics of lithium-sodium substituted niobates LNN(X) of nominal composition $Li_xNa_{(1-x)}NbO_3$, ($0 \leq X \leq 10\%$ Li) have been prepared by sintering of the precursors obtained by solid-state chemical reaction at temperatures of 1073 K of stoichiometric mixtures of the component oxides and carbonates. Ceramics with 90-93% densification were obtained by sintering pellets in air at 1513 K during 4 hours. X-ray diffraction patterns showed orthorhombic single phased materials without second phases.

Electrodes of 873 K sintered silver paste were used to electrode rectangular samples of 4 x 7 x 0.5 mm^3 dimensions. Dielectric measurements at different frequencies as a function of the temperature were performed with a HP4284A LCR Precision meter. The measurements were carried out during a complete heating-cooling thermal cycle.

Mechanoelastic measurements were achieved by the Three Points Bending (TPB) technique, as it is described elsewhere [13] on samples of 12.00x2.00x0.35mm^3. Using the TPB system in the appropriated condition disposition performed thermal expansion measurements.

The temperature rate in all the used experimental techniques was of 2 K/min between room temperature and 1000K from room temperature to 1000K-temperature interval.

EXPERIMENTAL RESULTS

The Figure 1 shows the dielectric constant as a function of the temperature for samples with X = 2%, 5% and 10% Li content in the cooling runs. The plot for the sample with X = 2% Li shows a wide reversible non-hysteretic maximum at 460K, below the main peak at Tc. The curve of the dielectric constant for sample with X = 10%Li on cooling presents an elbow at 535K (565K on heating) and the main peak at Tc = 595K (625K on heating). The behaviour is reversible with a thermal hysteresis of 30K.

In the Figure 2 we have plotted the Young's modulus, Y, and mechanical loss tangent, tanδ, as a function of the temperature for samples with X = 1%(a) and 8% Li (b). several peaks are observed above
the phase transition temperature. The temperature of these peaks is higher for high Li content sample. All the peaks are reversible and those of the paraelectric phase show small thermal hysteresis. The inset shows the behaviour for X = 10% Li content composition in the ferroelectric phase.

The Figure 3 depicts the thermal expansion in the phase transition temperature interval as function of the temperature for samples of 5%, 8% and 10% Li compositions. One anomaly in the samples with X = 5% and 8% Li, and two clear ones in the sample with X = 10% Li content are observed. Non- significant anomalies are observed above Tc.

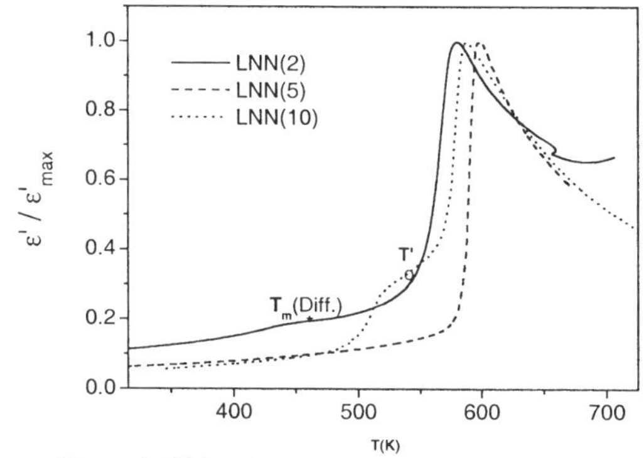

Figure 1. Dielectric constant as function of the temperature for samples LNN(2), LNN(5) and LNN(10).

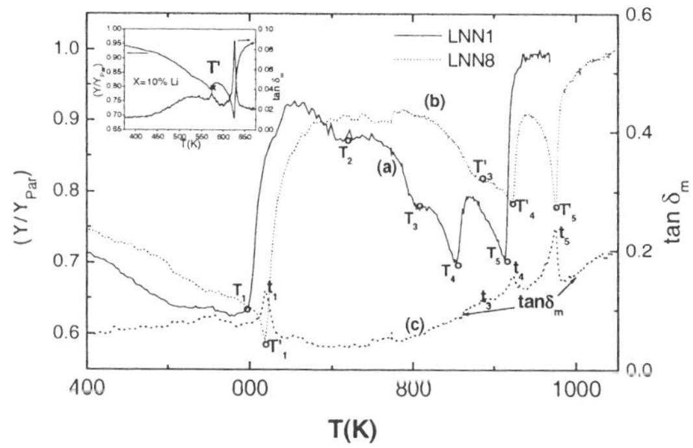

Figure 2. Relative Young's modulus, $Y/Y_{par.}$, for samples of LNN(1)(a) and LNN(8)(b) and mechanical losses tangent, $\tan\delta_m$, for sample LNN(8)(c) as a function of the temperature. Y_{par} is the Young's modulus in the paraelastic phase. T_n and t_n correspond to the phase transition temperatures. The inset shows the results for X = 10% Li content sample in the ferroelectric phase with the anomaly at T'.

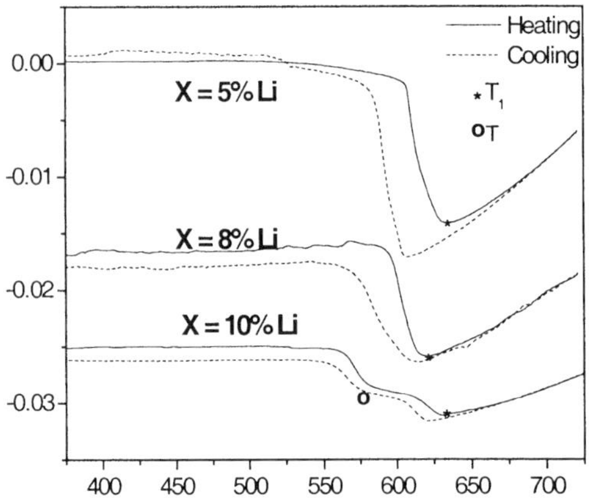

Figure 3. Thermal expansion behaviour in the structural phase transition temperature region for samples with different Li content. The linear thermal expansion has been subtracted from the complete experimental experimental data. T_1 corresponds to Tc and T' to the new FE-FE phase transition.

In the Table 1 we gather the temperatures of the anomalies in the Young's modulus for some of the studied samples obtained from the heating run curve of mechanoelastic measurements.

Table 1. Temperatures (K) of the anomalies above Tc obtained from elastic measurements.

X	**T1 = Tc**	**T2**	**T3**	**T4**	**T5**
0[1]	**643(D)**	**763**	**790**	**840**	**906**
1	604	707	807	855	913
5	624		851	892	957
8	620		871	921	978
10	627		908	946	999

[1]Data from the literature obtained from different measuring techniques.
(D) = From dielectric measurements.

DISCUSSION

The results obtained from the different used measuring techniques show that the substitution Li/Na provokes the gradual appearance of a ferroelectric phase for increasing Li content. The lower Li content compositions X < 5%, some residual AFE regions remain due to lack of compositional homogeneity. By increasing the temperature these regions experiment a phase transition probably to a ferroelectric Q–like phase at about 460K. The X = 5% Li composition contains a single ferroelectric phase with a ferro-paraelectric phase transition at Tc =624K. The X = 10% composition clearly shows two anomalies at 565K and 627K with important structural and dimensional changes, Fig. 3. The first anomaly should correspond to a FE-FE phase transition and the second one to a FE-PE phase transition. The X = 8% Li composition with a Tc = 616K could be considered as that for which the new FE phase starts to appear.

The mechanoelastic results of the Fig. 2 show a series of anomalies in the paraelectric phase besides the main one at Tc. Above Tc all the studied compositions have very small thermal hysteresis and non-significant dimensional changes. These characteristics with the optical results [3] lead to identify these anomalies as ferroelastic phase transitions [11] being the anomaly at T > 900K the corresponding to the ferroelastic-paraelastic phase transition.

The ferro-paraelectric phase transition with important dimensional changes is attributed to Na^+ cations displacement, the other observed phase transitions are attributed to Nb_2O_6 octahedra tilts [14].

In summary, the LNN(X) (X≤ 10% Li) compositions present a sequence of phase transitions in the room temperature to 1000K-temperature interval. At temperatures T < Tc the phase transitions have diffusive or ferroelectric/ferroelectric character, but at temperatures above Tc they are predominantly ferroelastic.

ACKNOWLEDGEMENT

This work was supported by the LEAF G5RD-CT-2001-00431 EU project. R. Jiménez acknowledges the Ramon y Cajal program of the Spanish S&T Ministry.

REFERENCES

1.- H.D. Megaw, Ferroelectrics, **7**, 87 (1974).

2.- R. Von der Mühll, A. Sadel, J. Ravez, and P. Hagenmüller, Solid State Comm. **31**, 151 (1979).

3.- A. Sadel, R. Von der Mühll, J. Ravez, J.P. Chaminade, and P. Hagenmüller, Solid State Comm. **44**, 345 (1982).

4.- M.A.L. Nobre and S. Lanfredi. J. Phys. Chem. of Solids, **62**, 1999(2001)

5.- M.A.L. Nobre and S. Lanfredi . J. Phys. : Conden. Matter, **12**, 7833(2000)

6.- J. Lacomte and E. Quemeneur, Bull. Soc. Chem. France, **12**, 2779(1974).

7.- Y. Song, H.F. Chen, F. Chen, D. Sun, P. Zhang and W. Zhong, J. Synthetic Crystals, **18**(2), 117(1989)

8.- A.C. Sakowski-Cowley, K. Lukaszewicz, and H.D. Megaw, Acta Cryst. **B25**, 851 (1969).

9.- I.P. Raevski and S.A. Prosandeev, J. Phys. Chem. Solids **63,** 1939 (2002).

10.- A. Sadel, R. Von der Mühll, and J. Ravez, J.Mat. Res.Bull. **18**, 45 (1983).

11.- B. Jiménez, A. Castro and L. Pardo. Appl. Phys. Letters, **82**, 3940 (2003)

12.- B. Jiménez and R. Jiménez, Pys. Rev. B, **66**, 4104 (2002).

13.- B. Jiménez and J.M. Vicente, J. Phys.:D, **31**, 446 (1998).

14.- Z.X. Shen, X.B. Wang, M.H. Kuok, and S.H. Tang, J. Raman Spectroscopy **29**, 279 (1998).

Applied Physics (APHYS 2003)
A. Méndez-Vilas (Ed.)

Study of kinetic friction of solid using driven lattice of quantized vortex in high-temperature superconductors – a new route to study microscopic tribology –

A. Maeda*†, Y. Inoue* and H. Kitano*†

*Department of Basic Science, Univeristy of Tokyo
3-8-1, Komaba, Meguro-ku, Tokyo, Japan, 153-8902
†CREST, Japan Science and Technology Corporation (JST), 4-1-8, Honcho, Kawaguchi, Saitama, Japan 332-0012

Abstract.
We show that quantized magnetic vortex lattice in high-temperature cuprate superconductors driven by dc or ac current is a very good model system for investigating physics of friction of solids and tribology in microscopic systems. Based on the *I*-*V* characteristic measurement and viscosity measurement using microwave techniques in $Bi_2Sr_2CaCu_2O_y$, kinetic friction of the vortex system was obtained as a function of temperature, magnetic field and driving current density. With increasing magnetic field and temperature, velocity dependence of kinetic friction bahaves as those at cleaner interfaces of solid. These result means that we can control the kinetic friction in the vortex system, and that systematic experiments are available in a reproducible manner with this system. Behavior of the kinetic friction at high velocities is also discussed.

INTRODUCTION

Control of friction is one of most important key technologies in microscopic movable systems. However, physics of kinetic friction has not been well understood more than 500 years. Empirically, phenomena of friction is summarized as the Coulomb-Amontons' law[1]. According to recent studies, the essential mechanism of static friction has been found to be the adhesion of atoms and molecules at the interface[1]. However, almost nothing has been clarified for kinetic friction, and several important questions have remained to be seen, such as (1) why kinetic friction is smaller than the static friction, (2) why kinetic friction does not depend on velocity, (3) why kinetic friction does show some dependences on velocity at low velocities, *etc*. It is noteworthy that an intersting scaling law between the velocity dependence of kinetic friction and the aging time dependence of the static friction was reported in the friction between thick papers[2].

To clarify the underlying mechanism of kinetic friction, good model systems, with wchich systematic experiments are available in cotroled and reproducible manners.

However, at present, it seems that such systems are lacking.

In contemporary condensed matter physics, it has become known that many electrons form a quantum condensate at low temperatures, where phase of the wave function of many electrons are coherent. In such systems, a new type of collective motion is available, and that can participate in the electric conduction. The most famous example is the charge density wave (CDW) in some quasi-one dimensional materials and the quasi two dimensional quatized vortex lattices in superconducting state under magnetic field. Indeed, the equation of motion for those systems[3, 4] are very similar to that for the friction at the solid interface[5]. This suggests that these systems can be good model systems for the study of kinetic friction at the solid interface.

In this paper, we investigate the kinetic friction deduced for vortex lattice of a high-temperature superconductor as a fuction of magnetic field, temperature, and driving current, and show that quantized magnetic vortex lattice of superconductor driven by dc or ac current is a very good model system for investigating tribology in microscopic systems.

EXPERIMENTAL

Fisrt, we should express the kinetic friction, F_k, by measurable physical quantities shown up in the physics of vortex dynamics of superconductor. According to Matsukawa and Fukuyama[5], in case of the friction at the solid interface, F_k was found to be equal to the summation of the interraction between an atom in the upper material and that in the lower material. That means F_k for the vortex system in superconductor can be expressed as follows[6].

$$F_k = j\Phi_0[1 - \frac{\rho}{\rho_\infty}], \tag{1}$$

where Φ_0 is the flux quantum, j is the current density, ρ and ρ_∞ is resistivity and the flux flow resistivity, respectively. We can know both of j and ρ by the usual dc I-V characteristics measurement, whereas ρ_∞ should be obtained in the microwave complex surface impedance measurement[7]. In dc I-V measurement, joule heating becomes serious at high current densities. Thus, I-V measurement using short rectangular pulses was also performed, details of which were described elsewhere[9].

For the flux-flow resistivity, we need both of real part and imaginary part of the microwave surface impedance, because of the large pinning frequency. With an aid of some general phenomenological models[8], we can extract flux flow resistivity. Detailes on the procedure were described in the literature[7].

Single crystals of $Bi_2Sr_2CaCu_2O_y$ (BSCCO) used in this study were made by ourselves using the so-called floating zone method.

RESULTS AND DISCUSSION

Figure 1 shows the kinetic friction, F_k, of vortices in BSCCO as a function of the velocity of vortices for various magnetic field at four different temperatures. All of the data shows that F_k increases with increasing vortex velocity. This is largely different from what we know for the kinetic friction at the solid interface in our daily life, as was described in the Coulmb-Amontons' law. However, it is very similar to what was predicted in a numerical simulation of the kinetic friction for a very clean surface[5]. There, friction starts with zero static friction, and F_k increase with increasing velocity. Furthermore, dependence of F_k on the velocity becomes less pronounced with increasing the interaction at the interface. In other words, if one makes the surface dirtier, the velocity dependence of F_k becomes closer to what we know in our daily life.

Coming back to the experimental data of BSCCO, at a fixed temperature, the increase of F_k becomes more rapid at lower magnetic field, and at lower temperatures, the increase of F_k becomes more rapid. Thus, it can be said that increasing magnetic field and temperature bring the vortex system to a situation that corresponds to a cleaner interface of the solid friction problem, where F_k starts from zero and continues to increase with increasing velocity. Since the pinning force of each vortex decreases with increasing magnetic field and temperature, the vortex system behaves as a more rigid system at higher magnetic fields and temperatures. This is what is observed in the above experiment.

Next, we focus on the asymptotic behavior of F_k at high current densities. Phenomenologically, F_k in the high velocity limit can be expressed as

$$F_k \propto j^{\alpha}, \tag{2}$$

where α is a j-independent exponent. The Coulmmb-Amontons' law means α=0. Theoretically, several predictions have been made for quasi-onedimensional CDW, as for the asymptotic behavior of F_k. One dimensional motion of a rigid body predicts $\alpha = -1/2$[10], whereas the inclusion of the thermal fluctuation leads to $\alpha = 1$[11]. On the other hand, a perturbation theory for elastic medium predicts $\alpha = (d-2)/2$ (d is the dimension of internal degrees of freedom)[12], and a numerical simulation shows that α ranges between 1 and 1/2, depending on the degree of disorder[13]. An experiment in the CDW also shows that α ranges between 1 and 1/2, depending on samples[9].

Figure 2 shows F_k as a function of the driving current density, j. All the data shows that F_k increases with increasing j, which means that the exponent α is between 1 and 1/2. In terms of the models for the CDW's, these results corresponds to the deformable motion with the three dimensional internal degrees of freedom under the presence of finite disorder[12]. However, we should extend the current density region up to higher values to confirm that we are in the high current region. Indeed, another one-dimensional numeical simulation predicts that F_k, after showing a peak, decreases and approaching a constant value with increasing vortex velocity[14]. Similar studies using thin films are under way.

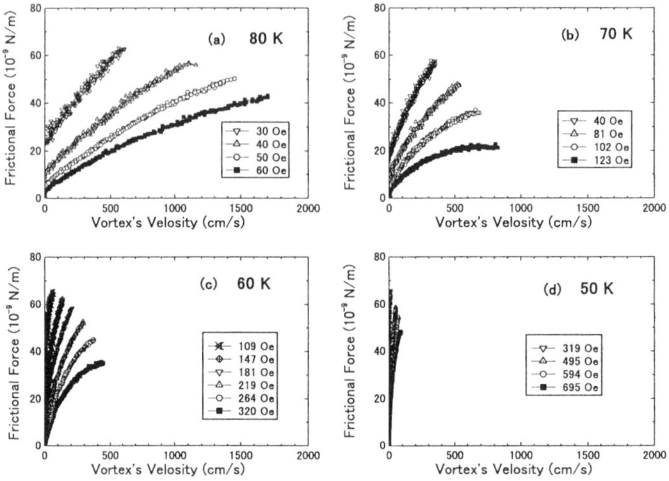

FIGURE 1. kinetic friction F_k as a function of velocity of vortices for various magnetic field at four tempertrues. (a) 80 K, (b) 70 K, (c) 60 K, and (d) 50 K.

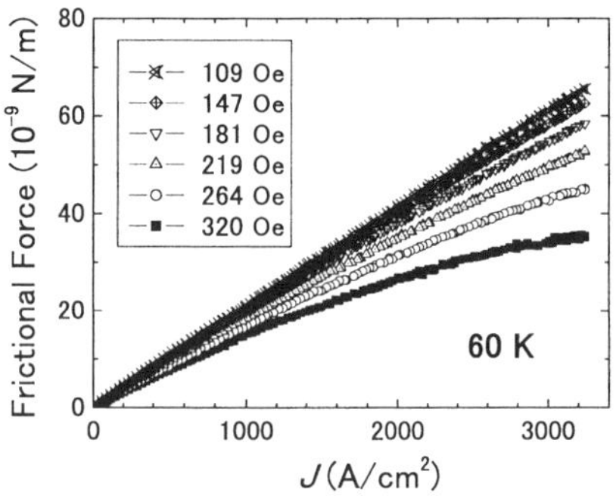

FIGURE 2. kinetic friction, F_k, as a function of driving current density, j, for various magnetic fields at 60 K.

Anyway, all of the data presented above show clearly that quantized magnetic vortex lattice in high-temperature superconductors driven by current is a very good model system for investigating tribology in microscopic systems. It is a very interesting issue how $F_k(v)$ changes if one introduces disorder into the crystals intentionally. This is also in progress.

CONCLUSION

Based on the I-V characteristic measurement and viscosity measurement using microwave techniques in $Bi_2Sr_2CaCu_2O_y$, kinetic friction of the vortex system was obtained as a function of voltex velocity for various magnetic fields and temperatures. With these data, we showed that the quantized magnetic vortex lattice in high-temperature superconductors driven by current is a very good model system for investigating tribology in microscopic systems, where controlled experiments are possible in a reproducible and systematic manners. Since all of the data presented here suggest that the vortex systems of high-T_c cuprate superconductors correspond to clean surfaces for the problem of the friction at the solid interface, it deserves an urgent study how $F_k(v)$ changes with introduction of vaiours types of disorder, that should contribute to an important progress for understanding the kinetic friction of solids.

ACKNOWLEDGMENTS

The authors thank profs. Hiroshi Matsukawa, Hidetoshi Fukuyama, and Franco Nori, and Dr. Sergei Savel'ev for fruitful discussions.

REFERENCES

1. Persson, B. N. J., *Sliding Friction – Physical Principles and Applications*, Springer, 1988.
2. Heslot, F., *et al.*, *Phys. Rev.*, **E49**, 4973 - 4988 (1994).
3. Grüner, G., *Rev. Mod. Phys.*, **60**, 1129 - 1181 (1988).
4. *For example*, Tinkham, M., *Introduction to Superconductivity* (2nd ed.), McGraw-Hill, 2000.
5. Matsukawa, H., and Fukuyama, H., *Phys. Rev.*, B49, 17286 - 17292 (1994).
6. Maeda, A., *et al.*: *unpublished*, Matsukawa, H., *private communication.*
7. Tsuchiya, Y., , Iwaya, K., Kinoshita, K., Hanaguri, T., Kitano, H., Maeda, A., Shibata. K., Nishizaki, T., and Kobayashi, K., *Phys. Rev.*, **B63**, 184517 - 184525 (2001)Cand references cited theirin, Maeda, A., *et al.*, *Physica*, **C362**, 127 - 133 (2001).
8. Martinoli, P. *et al.*, *Physica* **B165&166**, 1163 - 1164 (1990), Coffy, M. W., and Clem, J. R., *Phys. Rev. Lett.*, **67**, 386 - 389 (1991).
9. *For example,* Maeda, A., and Uchinokura, K., *J. Phys. Soc. Jpn.*, **59**, 234 - 252 (1990).
10. Grüener, G., Zawadowski, A., and Chaikin, P. M., *Phys. Rev. Lett.*, **46**, 511 - 514 (1981).
11. Ambegaokar, V., and Halperin, B. I., *Phys. Rev. Lett.* **22**, 1364 1366 (1969), Inui, M., Littlewood, P. B., and Coppersmith, S. N., *Phys. Rev. Lett.* **63**, 2421 - 2424 (1989).
12. Sneddon, L., Cross, M. C., and Fisher, D., *Phys. Rev. Lett.*, **49**, 292 - 295 (1982), and Sneddon, L., *Phys. Rev.*, **B29**, 719 - 724 (1984), 725 - 727 (1984).
13. Matsukawa, H., *J. Phys. Soc. Jpn.*, **56**, 1507 - 1521 (1987).
14. Savel'ev, S., and Nori, F., *unpublished.*

Applied Physics (APHYS 2003)
A. Méndez-Vilas (Ed.)

Novel Algorithms for Estimating Motion Characteristics within a Limited Sequence of Images

Oleg Starostenko*, Araceli Ramírez†, Alfred Zehe†, and Gennadiy Burlak¶

*Computer Science Department, University de las Américas, Cholula, Puebla, 72820, Mexico
†University Autónoma de Puebla, BUAP, Puebla, 72000, México
¶Autonomous State University of Morelos, Cuernavaca, 62210, México.

Abstract. Two novel algorithms for estimating the motion characteristics of the object in dynamic scene by processing a limited sequence of images are presented in this paper. The first algorithm is based on computation of space-temporal gradients of consecutive frames of video stream, another one has been designed for fast detection of motion by processing of principal corners of objects in real time applications. For quantitative estimation of motion characteristics, the novel segment and neighbours matching technique has been proposed. The method uses the concept of fuzzy sets and membership functions, which permits high-speed recognition and efficient interpretation of the patterns with significant level of noise and distortions. The introduced algorithms have been tested in order to evaluate their velocity, utility and efficiency.

INTRODUCTION

Each object within the image has its motion vector of three dimensions, which projection on the plane represents the image motion field [1]. The optical flow field is defined by the changing the brightness of the segments within the image. In general, the optical and motion field are different, but with a certain approximation they produce the similar responses [2]. The computing of motion field depends on conditions like noise of the image, sharp changing the brightness of segments, intersection of the different objects and its relation with the background, aperture problem for detection of the motion in straight monotonic borders, etc. [3]. In order to solve these problems and increase the speed of motion field construction without the loss of accuracy several techniques are proposed.

BASIC APPROACHES FOR MOTION FIELD CONSTRUCTION

Frequently, the efficient motion analysis methods are used for border detection and their translation into a sequence of images. The recent powerful well-known methods for border estimation can be subdivided as it shown below.

a) The gradient-based methods estimate the motion by analysis of the strong differences in brightness between analysed regions. These variations are modelled by differential equations represented by space and temporal gradients. The disadvantage of this approach is the dependence on noise, occlusions, and discontinuity. The motion field construction is exact but too complex [2], [4].

c) The correspondence methods are based on estimation of details of objects in motion [5], [6]. Details (regions) are classified during the analysis on high level (patterns, lines, forms) and low level (corners, simple patterns, colour variation). The high level approach is more complete but it requires image processing (filtering, pattern recognition) with corresponding computational cost. This approach strongly depends on rotation and scale variation of the objects within the images.

After analysis and test of some techniques the gradient based and block correspondence methods have been used as basic concepts of the proposed algorithms. The principal objective of proposed algorithms is the reduction of processed information by improving the well-known methods for motion analysis.

GRADIENT BASED ALGORITHM FOR BORDER DETECTION

The proposed algorithm is introduced by processing of frames presented in Fig.1.

a) For border detection, the Sobel filter is applied to normalised frame [7]:

$$\nabla f = \begin{pmatrix} G_x \\ G_y \end{pmatrix} = \begin{pmatrix} \partial f / \partial x \\ \partial f / \partial y \end{pmatrix} \tag{1}$$

where the value $f(x,y)$ defines the direction of the maximum variation of the gradient in the point (x,y). The computing of image gradient at each pixel position is done in digital form by discrete bi-dimensional convolution with 3x3 mask.

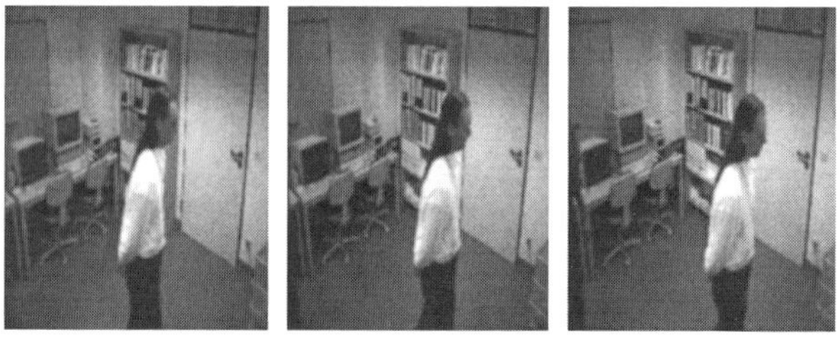

FIGURE 1. The sequence of images with objects in motion.

b) After the iterative process of mask displacement over all the pixels, the borders of the objects within each frame are estimated as it shown in Fig. 2.

c) For motion detection the following procedures are applied:

1. The difference between two frames is found taking into account the absolute values of early computed space gradients. Thus, the temporal gradient is obtained.

2. The result is multiplied by space gradient of the first frame.

3. The borders may be too weak. That is why, for their good presentation the binary threshold interval operator is applied to borders in motion.

d) Finally, the composite image is obtained by superimposing the binary image with borders in motion over the third frame as it shown in Fig. 2.

FIGURE 2. The extraction of the borders and composite image with borders in motion.

In the composite image the borders are presented as interrupted lines because the colour of the object in this region has the colour similar to background. That is considered as disadvantage of the proposed algorithm.

ALGORITHM FOR DETECTION OF PRINCIPAL CORNERS

One interesting method among the block correspondence techniques is the SUSAN method (Smallest Univalue Segment Assimilating Nucleus) which provides the analysis of strong differences in brightness. The method operates with a circular mask with the centre called nucleus [8]. If the brightness of pixels under the mask has the same value as the nucleus, they form the USAN area which is at minimum in the set of pixels with strong difference in brightness. This property of the USAN area may be used for detection of the borders and corners of 2D figures. Frequently, manipulation with principal corners of the objects sometimes is acceptable for computing of the motion characteristics as it shown in proposed algorithm.

a) Quite exact result is obtained with the experimentally established circular masks of radius 3.4 pixels. Using the equation:

$$C(r, r_0) = \exp - \left(\frac{I(r) - I(r_0)}{t} \right)^6 \tag{2}$$

the similarity of the pixels within the mask is computed with respect to the brightness of the nucleus producing the luminosity LUT (Look-up table). In equation *(2)* r_0 is the position of nucleus, r is the position of the pixels within the mask, $I(r)$ is the brightness of the analysed pixel, t defines the minimum acceptable difference in contrast and number of the detected principal corners. The power 6 is for good balance between stability of the threshold manipulation and computing of the USAN areas.

b) The average value of brightness of pixels within the mask is computed and assigned to the central pixel. Now, the USAN areas are computed. The result defines the brightness and position of the pixel considered as a corner if the USAN area is less than the half of mask area. Then the mask is displaced to next pixel.

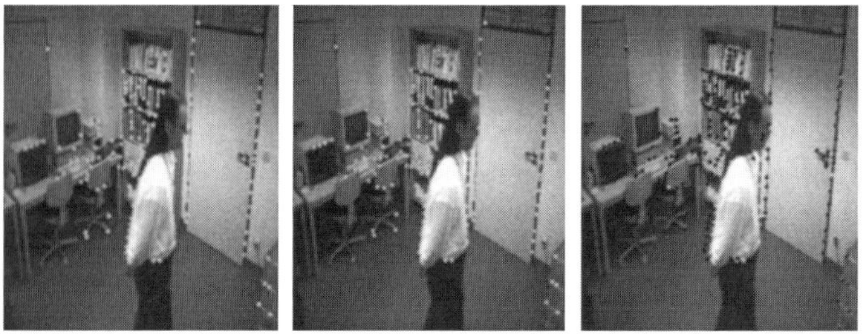

FIGURE 3. The result of superimposing the corners of two images over the second frame.

c) Finally, the wrong points (false corners) are detected using the gravity centre of the mask (medium value of the coordinates of x and y for USAN area).

d) The procedures of steps a) - c) are applied to the next frame and then two frames are superimposed as it shown in Fig. 3. The corners of static objects are at the same place and can be removed. The advantage of the proposed procedure is the quite simple way for detection of the objects in motion. It is important to note that there is no regularity in the appearance of the detected corners. The position of the corners depends on the brightness difference between the object and background. Another limitation of the algorithm is the additional errors due to non-linear motion which may be reduced by processing the frequently incoming frames, where the distance of translation is so short that it can be considered as linear motion.

MOTION CHARACTERISTICS ESTIMATION AND DISCUSSION

Usually, the motion field construction consists in searching the centre of analysed pattern at the time t and its correspondent centre at the time $(t-\Delta t)$. In this way, the progressive increment of stored processed patterns is avoided, but it introduces the accumulative error during the analysis of long sequence of the images. In this paper the Segment and neighbours matching method proposed by the authors is applied [9]. The method uses the concept of fuzzy sets and membership functions, which permits to accept the segment as candidate for inclusion to the selected set. The method permits to obtain the fuzzy skeleton of the frame applying the segment reduction technique substituting the non-arbitrary segments by fuzzy ones with predefined slope β (Fig.4b,c) as it shown for frame of Fig. 2 (Fig. 4a).

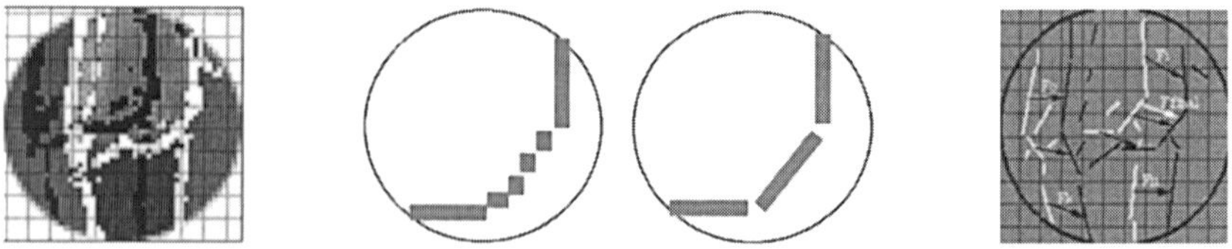

a) b) c) d)

FIGURE 4. a) the borders in motion, b) non-arbitrary segments, c) fuzzy segments, d) fuzzy segments displacement.

Finally, the displacement of the objects formed by fuzzy image is computed by estimation of the average displacement vector F_{final} (Fig. 4d). The displacement vectors of segments are obtained using the set of the fuzzy segments belonging with a certain error δ to the set with the similar slope $\gamma < \beta \pm \delta$. The centre of the segment can be used as significant point. Using the corner detection approach, their coordinates are taken for computing of displacement vectors.

The principal increment of the speed is obtained in the motion detection steps, where the algorithms operate with borders (only 3% of pixels of a frame are borders) or corners (less than 0.1% of pixels) in motion. The proposed approaches introduce the additional errors in case of the multiple objects in motion within same region. Therefore, the introduced algorithms are good for applications where the objects in motion have no occlusions or intersections. The block correspondence technique is noise sensitive but it may be reduces by operation with the reference patterns of the biggest size and on the step of substituting the non-arbitrary by fuzzy segments.

CONCLUSIONS

Two proposed algorithms permit the qualitative and quantitative analysis of the motion characteristics of the objects in dynamic scenes. In the proposed algorithms the quality of motion field construction is limited by borders or principal corners analysis, but it is sometimes enough for high-speed motion characteristics estimation with a small number of the objects in dynamic scene. The first algorithm operates with border in motion reducing the quantity of processed data. The second one is the fastest, but the principal problem is the correct selection of the thresholds defining the number of detected corners. The proposed Segment and neighbours matching method provides simple and fast motion field construction by analysis of the fuzzy segments and the membership functions for the object in motion.

REFERENCES

1. G. Kollios, V. Tsotras, Indexing animated objects using spatiotemporal access methods, IEEE Trans. on Knowledge and Data Engineering 13 (5) (2001) 758-777.
2. B. Jahne, H. Hauβecker, Computer Vision and Applications, Academic Press, USA, 2000.
3. L. Hongche, T. Hong, Accuracy vs efficiency trade-offs in optical flow algorithms, Comp. Vision and Image Understanding 72 (3) (1998) 271-286.
4. T. Hong, L. Hongche, Motion-model-based boundary extraction and a real-time implementation, Comp. Vision and Image Understanding 70 (1) (1998) 87-100.
5. C. Cheung, Fast motion estimation techniques for video compression, (Hong-Kong University: www.cityu.edu.hk, 1998).
6. I. Lagendijk, Motion estimation. information and communication theory group, (Delft University of Technology: www-ict.its.tudelft.nl/, 1999).
7. J. Sobel, Machine vision for three-dimensional scenes, Academic Press, New York, 1990.
8. S. Smith, J. Brady, SUSAN - a new approach to low level image processing, Int. Journal of Comp. Vision 23 (1) (1997) 45-78.
9. O. Starostenko, J.A. Neme, Automatic Complex Glyphs Recognition and Interpretation, Int. Journal IEICE Transaction, Japan E82-A (10) (1999) 2154-2160.

Applied Physics (APHYS 2003)
A. Méndez-Vilas (Ed.)

A Formalism for Quantum Computing and a Satisfiability Problem

C. Bautista*, M. Castro-Cardona* and A.F.K. Zehe†

*Facultad de Ciencias de la Computación. Benemérita Universidad Autónoma de Puebla. 14 Sur y Av San Claudio. Edif. 135. Ciudad Universitaria. Puebla, Pue. 72570 México
†Facultad de Ciencias de la Electrónica. Benemérita Universidad Autónoma de Puebla. 18 Sur y Av San Claudio. Edif. 129. Ciudad Universitaria. Puebla, Pue. 72570 México

Abstract. An efficient algorithm in a quantum computer for dealing with a related satisfiability (SAT) problem is shown. A bounded probability polynomial size algorithm that finds boolean formulas with n variables such that return more $1's$ than the number $2^n(1 - 1/\sqrt{2})$ is exposed (a quantum circuit of size $O(n)$). The algebraic framework used is the category theory. Our algebraic formalism includ es evolution and decoherence.

INTRODUCTION

There is little doubt, that nanotechnology will create the building blocks for practical quantum computers, consisting most probably in a set of coupled solid-state quantum systems, carefully put into place. On the other hand, efficient algorithms are required by the quantum information technology in order to meet the high expectations on speed and security.

Since the discovery of a polynomial *quantum* algorithm for factoring integers by Shor [1], quantum computation has drawn overwhelming attention from the science community, in particular from physicists and mathematicians interested in computer science.

Chuang et al. [2] proved the technological feasibility of building a quantum computer that can factorize the number 15. However a functional real quantum computer remains to be built, to the best of our knowledge. It is a general believe that it is just a technological problem that will be overcome [3, 4] some day.

A quantum computer is some kind of probabilistic machine with the additional feature of allowing superpositions of states which lead to a massive parallel processing and, as a consequence, to an amazing speedup of calculation. Excellent introductions and enjoyable readings, to quantum computation, are found in [11, 12].

Despite that at this moment, the most important problem in quantum computation is its physical realization, the design of quantum algorithms still is a very important issue. According with Shor [5] it may seem that computer scientists have to be accustomed to the very particular specifications of quantum computing, because most of these are far away from our daily experience (quantum mechanics).

The main objective of this paper is to show how quantum computation could deal with classical subjects of computing theory, like the SAT problems.

FORMALISM OF QUANTUM COMPUTATION

The mathematical framework of quantum computation is the Hilbert spaces theory. Although the setting for quantum mechanics is the infinite dimensional Hilbert space, the finite dimensional space is sufficient for dealing with quantum computation.

Actually, the building block for quantum computation theory is the Hilbert space $H = \mathbb{C}^2$, and the constructor is the tensorial product of such spaces, which is denoted $\otimes$.

We belive that category theory [6] provides the most economic way in concluding what the properties of $\otimes$ are. The use of category theory in computer science it is not a novelty here. Introductory readings about the role of category theory in computer science are founded in [7, 8]. The relationship between categories and quantum computation has been studied by Yu. Manin [9] and Abramsky and Coecke [10].

We denote by **FDVec** the category made of finite dimensional Hilbert spaces. Such category is an example of a *strict symmetrical monoidal category*. In particular,

$$\otimes : \mathbf{FDVec} \times \mathbf{FDVec} \to \mathbf{FDVec}$$

is a functor. Now, we denote with **qWord** the category with objects $\mathbb{C}^{2^n} = (\mathbb{C}^2)^{\otimes n}$, in **FDVec**, $n = 0, 1, \ldots$, and morphism $U(n) = \{U_i\}_{i=1}^m$ a finite collection of morphisms $U_i : \mathbb{C}^{2^n} \to \mathbb{C}^{2^n}$ in **FDVec** satisfying the so called completeness condition:

$$\sum_{i=1}^{m} U_i^* U_i = Id^{\mathbb{C}_{2^n}}$$

where $*$ stands for the complex conjugate transposition and $Id^{\mathbb{C}_{2^n}}$ is the identity morphisms on $\mathbb{C}^{2^n}$. If $U(n) = \{U_i\}_{i=1}^m$, $V(k) = \{V_j\}_{j=1}^l$ its composition is defined as

$$U(n) \circ V(k) = \{U_i V_j \,|\, i = 1 \ldots, m, j = 1, \ldots, l\}, \quad \text{iff } n = k \;,$$

where each $U_i V_j$ is a composition in **FDVec**. From

$$\begin{aligned}
\sum_{i,j} (U_i V_j)^* (U_i V_j) &= \sum_{i,j} V_j^* U_i^* U_i V_j \\
&= \sum_j V_j \sum_i U_i^* U_i V_j \\
&= \sum_j V_j^* Id V_j \\
&= Id
\end{aligned}$$

it follows that $U(n) \circ V(n)$ is a morphism in **qWord**.

The identities are $Id(n) = \{Id^{\mathbb{C}_{2^n}}\}$ which is a collection with a single element. Note that, in general, $U = \{U_1\} \in$ **qWord** if and only if U_1 is an unitary transformation.

Even more, if $U(n), V(k) \in$ **qWord** we can extend the tensorial product of **FDVec** to **qWord** by defining,

$$U(n) \otimes V(k) = \{U_i \otimes V_j\}_{i,j}$$

This linear morphism satisfies the completeness condition because

$$\begin{aligned}\sum_{i,j}(U_i\otimes V_j)^*(U_i\otimes V_j) &= \sum_{i,j}(U_i^*\otimes V_j^*)(U_i\otimes V_j)\\ &= \sum_{i,j}(U_i^*U_i)\otimes(V_j^*V_j)\\ &= (\sum_i U_i^*U_i)\otimes(\sum_j V_j^*V_j)\\ &= Id_{\mathbb{C}_{2^n}}\otimes Id_{\mathbb{C}_{2^k}}\\ &= Id_{\mathbb{C}_{2^{n+k}}}\ .\end{aligned}$$

Besides, the following equations imply that

$$\otimes : \mathbf{qWord}\times\mathbf{qWord}\rightarrow\mathbf{qWord}$$

is a functor:

$$\begin{aligned}(U(n) \ \otimes\ & V(k))\circ(W(n)\otimes X(k))\\ &= \{U_i\otimes V_j\}_{i,j}\circ\{W_a\otimes X_b\}_{a,b}\\ &= \{(U_i\otimes V_j)(W_a\otimes X_b)\}_{i,j,a,b}\\ &= \{(U_iW_a)\otimes(V_jX_b)\}_{i,j,a,b}\\ &= U_iW_{ai,a}\otimes\{V_jX_b\}_{j,b}\\ &= (U(n)\circ W(n))\otimes(V(k)\circ X(k))\end{aligned}$$

and

$$\begin{aligned}Id(n)\otimes Id(k) &= \{Id_{\mathbb{C}_{2^n}}\otimes Id_{\mathbb{C}_{2^k}}\}\\ &= \{Id_{\mathbb{C}_{2^{n+k}}}\}\\ &= Id(n+k)\end{aligned}$$

Therefore, we can say that *a closed quantum process* is a finite sequence

$$U_1(n),\ldots,U_l(n),U_{l+1}(n)$$

of morphisms in **qWord** such that $U_i(n)=\{U_{i,1}\}$, $i=1,\ldots l$, while *a quantum process with decoherence* is a finite sequence of morphisms in **qWord** such that it is not a closed quantum process.

One could say that decoherence (or *quantum noise*) is the process of the environment of constantly trying to look (measure) the state of our quantum system. This fact makes quantum computers quite sensitive to the environment. However, there exists ways to protect quantum information from decoherence [13].

Therefore, if $U:\mathbb{C}^{2^n}\rightarrow\mathbb{C}^{2^n}$, $V:\mathbb{C}^{2^n}\rightarrow\mathbb{C}^{2^n}$, $W:\mathbb{C}^{2^m}\rightarrow\mathbb{C}^{2^m}$ and $R:\mathbb{C}^{2^m}\rightarrow\mathbb{C}^{2^m}$ are in **qWord**, the following constructions are available:

$$UV,\ U\otimes W,\ U\otimes Id_{\mathbb{C}_{2^m}},\ Id_{\mathbb{C}_{2^m}}\otimes U$$

where Id is the identity map. Besides

$$(U \otimes W)(V \otimes R) = (UV) \otimes (WR) \ .$$

Finally, on vectors, $\otimes$ acts in a bilinear way:

$$(\alpha_1\phi + \alpha_2\varphi) \otimes (\beta_1\psi + \beta_2\chi) = \alpha_1\beta_1\phi \otimes \psi + \alpha_1\beta_2\phi \otimes \chi + \alpha_2\beta_1\varphi \otimes \psi + \alpha_2\beta_2\varphi \otimes \chi$$

The process in a quantum computer is the following

preparation$\rightarrow$ evolution $\rightarrow$ measurement

Preparation means setting the system in a known state: usually $|00\ldots0\rangle$. Where

$$|0\rangle = \begin{pmatrix} 1 \\ 0 \end{pmatrix}, \quad |1\rangle = \begin{pmatrix} 0 \\ 1 \end{pmatrix}$$

and $|\phi_1\phi_2\ldots\phi_n\rangle = |\phi_1\rangle \otimes |\phi_2\rangle \otimes \cdots \otimes |\phi_n\rangle$.

Evolution means applying the elements of a closed quantum process in **qWord** to the prepared state in order to obtain a final state (vector). While measurement means applying, to the final state $|\psi\rangle$, one element of an ad-hoc morphism in **qWord**. So, if we are using M_i in order to measure and the system has state (vector) $|\psi\rangle$, then the probability outcome p_i of the state $M_i|\psi\rangle$ after the measurement is

$$p_i = \langle M_i\psi | M_i\psi\rangle$$

and the state $|\psi\rangle$ collapses to

$$\frac{M_i\psi}{\sqrt{\langle M_i\psi|M_i\psi\rangle}}$$

[see 11].

The tensorial products of $|0\rangle$, $|1\rangle$ are called the basis of the calculation.

A SAT PROBLEM ON A QUANTUM COMPUTER

As usual, in order to enable the parallel abilities of the quantum computer, we are using the Hadamard gate H,

$$H = \begin{pmatrix} \frac{1}{\sqrt{2}} & \frac{1}{\sqrt{2}} \\ \frac{1}{\sqrt{2}} & -\frac{1}{\sqrt{2}} \end{pmatrix}.$$

Using the Dirac notation, $H|i\rangle = 1/\sqrt{2}(|0\rangle + (-1)^i|1\rangle)$, $i = 0, 1$. Then, if $|x\rangle$ is an element of the basis of calculation, it follows

$$H^{\otimes n}|x\rangle = \frac{1}{2^{n/2}} \sum_{y=0}^{2^n-1} (-1)^{x \cdot y}|y\rangle \qquad (1)$$

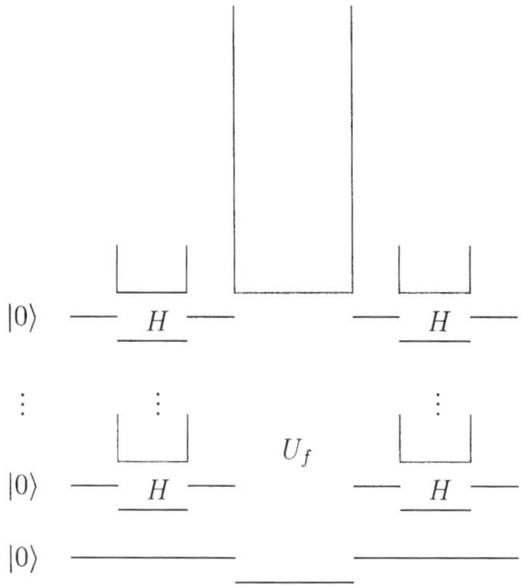

FIGURE 1. Our algorithm as a quantum circuit. Note that there is only one call to f: U_f.

where $x \cdot y$ is the scalar product bit by bit.

Suppose that f is a boolean formula with n literals, $f = f(x_1, \ldots, x_n)$. Since the SAT problem is, without question, a very important problem, we would like to know if there exists an assignation of the literals making f equal to 1. Equivalently, we would like to know if f is a contradiction or not.

We implement f in a reversible way:

$$U_f : \mathbb{C}^{2^n} \to \mathbb{C}^{2^n}, \quad |i\,j\rangle \mapsto |i\,j \oplus f(i)\rangle.$$

Now, let us consider the following quantum algorithm:

preparation : in the state $= |0\ldots 00\rangle = |0\rangle^{\otimes(n+1)}$;
evolution : $(H^{\otimes n} \otimes Id^{\mathbb{C}_2})U_f(H^{\otimes n} \otimes Id^{\mathbb{C}_2})$;
measurement : using the projective transformations relative to the basis of calculation.

The related quantum circuit is given at Fig. 1.

It is easy to see that, if f is a contradiction, then the final state $|\psi\rangle$ coincides to the initial state: $|\psi\rangle = |0\ldots 00\rangle$, and that, in such a case, the state $|\psi\rangle = |0\ldots 00\rangle$ results with probability equal to 1. Reciprocally, however, if the measure outcome is $|0\ldots 00\rangle$ then we cannot imply that f is a contradiction. For instance, for $f_0(x_1, x_2, x_3)$ which holds $f_0(x_1, x_2, x_3) = 1 \iff x_1 = 1, x_2 = 1$ and $x_3 = 1$, we have that the final state $|\psi\rangle = |0000\rangle$ appears with probability $\approx .77$, despite the fact that f_0 is satisfiable. Of course, this probability is high because f_0 is "near" to be a contradiction. Then a natural task is to search for a kind of boolean formulas that make the probability apparency of the state $|0\ldots 00\rangle$ low.

Straightforward calculations show that, in general, for a given boolean formula f, the final state $|\psi\rangle$, after running our algorithm, is

$$\begin{aligned} |\psi\rangle &= |0\ldots 00\rangle + \frac{1}{2^{n/2}} \sum_{i\,|\,f(i)=1} H^{\otimes n}|i\rangle(|1\rangle - |0\rangle) \\ &= |0\ldots 00\rangle + \frac{1}{2^n} \sum_{j=0}^{2^n-1} \sum_{i\,|\,f(i)=1} (-1)^{i \cdot j}|j\rangle(|1\rangle - |0\rangle) \end{aligned}$$

$$= (1 - \frac{u}{2^n})|0 \ldots 00\rangle + \frac{u}{2^n}|0 \ldots 01\rangle + \frac{1}{2^n} \sum_{j=1}^{2^n-1} \sum_{i \mid f(i)=1} (-1)^{i \cdot j} |j\rangle (|1\rangle - |0\rangle) \quad (2)$$

where u is the cardinality of the set $f^{-1}(1)$.

We would say that the probability outcome of $|0 \ldots 00\rangle$ is low if it is $< 1/2$, which is equivalent, because of (2), to $(1 - u/2^n)^2 < 1/2$, i.e., $u > 2^n(1 - 1/\sqrt{2})$. This means, if p stands for the probability outcome of the state $|0 \ldots 00\rangle$, then

$$p < 1/2 \iff \#f^{-1}(1) > 2^n(1 - 1/\sqrt{2})$$

where # stands for the cardinality set.

Therefore, we conclude that, if after running our quantum algorithm we get a state different from $|0 \ldots 00\rangle$ we can say, with high probability, that

$$\#f^{-1}(1) > 2^n(1 - 1/\sqrt{2}).$$

Note that such goal is achieved with only one query to the boolean formula f: the call to U_f in the step 2 of our algorithm; while a classic computer needs $\lceil 2^n(1 - 1/\sqrt{2}) \rceil$ queries to f, at least.

Finally we must say that it is impossible to solve the SAT problem in polynomial time [12, p. 41] by using only quantum parallelism.

Other works about satisfiability and quantum computing can be found in [14].

REFERENCES

1. P. W. Shor, Polynomial-time algorithms for prime factorization and discrete logarithms on a quantum computer. SIAM J. Comput. **26**, 1484–1509 (1997).
2. M. K. Lieven, K. Vandersypen, M. Steffan, G. Breyta, C. S. Yannoni, M.H. Sherwood, I. L. Chuang, Experimental realization of Shor's quantum factoring algorithm using nuclear magnetic resonance. Nature **414**, 883–887 (2001).
3. J. McCarthy, Problems and proyections in CS for the next 49 years. Journal of the ACM, **50**, 73–79 (2003).
4. Yu. A. Kitaev, A. H. Shen, M. N. Vyalyi, *Classical and Quantum Computation.* Graduate Studies in Mathematics. Vol. 47. American Mathematical Society, Rhode Island, 2002.
5. P. W. Shor, Why haven't more quantum algorithms have been found?. Journal of the ACM, **50**, 87–90 (2003).
6. S. MacLane, *Category Theory for the Working Mathematician,* Graduate Texts in Mathematics. Vol. 5. second edn., Springer, New York, 1998.
7. M. Barr, C. Wells, *Category Theory in Computing Science,* Prentice Hall International (UK), Hertfordshire, 1990.
8. R. F. C. Walters, *Categories and Computer Science,* Cambridge Computer Science Texts, Vol. 28. Cambridge University Press, Cambridge, 1991.
9. Yu. I. Manin, Classical computing, quantum computing, and Shor's factoring algorithm, http://arXiv.org/abs/quant-ph/9903008
10. S. Abramsky, B. Coecke, Physical traces: quantum vs classical information processing, Elsevier's Electronic Notes on Computer Science, **69**, (2002)

11. M. A. Nielsen, I. L. Chuang, *Quantum Computation and Quantum Information,* Cambridge University Press, Cambridge, 2000.
12. C.P. Williams, S.H. Clearwater, *Explorations in Quantum Computing,* Springer, New York, 1998.
13. D. Gottesman, "An introduction to quantum error correction" in *Quantum computation: A grand challenge for the twenty-first century and the millennium,* Proc. Sym. in Applied Math, Vol. 58. edited by S. J. Lomonaco Jr., American Mathematical Society, Rhode Island, 2002, pp. 221–235.
14. T. Hogg, Solving random satisfiability problems with quantum computers, http://arxiv.org/abs/quant-ph/0104048

Applied Physics (APHYS 2003)
A. Méndez-Vilas (Ed.)

A Novel High-Voltage Power Generator for Diesel Exhaust Gas Treatment

Mamoru Okumoto*, Shuiliang Yao*, Kazuhiko Madokoro*, Eiji Suzuki** and Tatsuaki Yashima*

**Chemical Research Group, Research Institute of Innovative Technology for the Earth, 9-2 Kizugawadai, Kizu-cho, Soraku-gun, Kyoto 619-0292, JAPAN*

***Faculty of Textile Science & Technology, Shinsyu University, 3-5-1, Tokida, Ueda City, Nagano 386-8577, JAPAN.*

Abstract. A high voltage pulse power generator was developed for particulate matter (PM) removal from a diesel engine using a dielectric barrier discharge (DBD) type of reactor. The influence of input power of the pulse power generator and geometry of the DBD reactor on power transform efficiency (ratio of input power from the pulse power generator to discharge power to the plasma reactor) was investigated. By applying pulse voltage from the pulse generator to the DBD reactor, decrease in PM concentration in the exhaust gas was confirmed. Maximum PM removal rate of 58 % was achieved at 480 W of input power of the pulse power generator. When input power of the pulse power generator was increased, the transform efficiency of the pulse power generator changed and converged to around 60 %.

INTRODUCTION

Recently, pulsed plasma technologies have been studied for natural gas conversion [1-3], removal of gaseous hazardous [4-6], and so on. Particularly, the high frequency (over 1 kHz) pulsed plasma attracts attention. Because of it can provide the excitation areas in short discharge duration ($1x10^{-8}$ to 10^{-6} s order) by high-energy injection rate and high repetition [1].

On the other hand, in order to apply the pulse power generator to the industrial chemical process, some improvements including the energy transfer efficiency are required. Therefore, performance evaluation and improvement of the pulse power generator are also studied [7-9]. Especially, development of a small size, low cost, and low loss energy type of pulse power generator is expected.

In this study, the authors developed a new high frequency pulse power generator for PM removal from diesel engines. The influence of input power of the pulse power generator and geometry of electrodes in the DBD reactor on power transform efficiency (ratio of input power from the pulse power generator to discharge power to the plasma reactor) was investigated. The Performance of the PM removal using the DBD reactor was also studied.

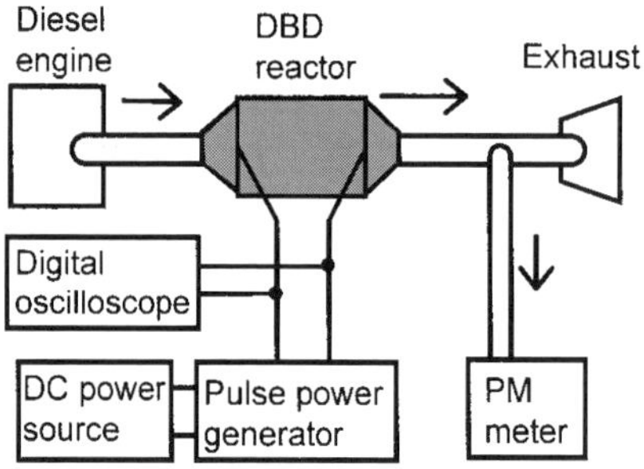

FIGURE 1. Experimental apparatus.

EXPERIMENTAL

Figure 1 shows a schematic of an experimental apparatus. Exhaust gases form a diesel engine (4-cylinder, 1974cc) were treated with a dielectric barrier discharge (DBD) type reactor. PM emission rate was measured using a PM meter (TEOM 1105 R&P). A pulse power generator driven by DC power source (24V DC, Peec, Japan) was used. Figure 2 shows the circuit diagram of the pulsed power generator. The second side of step-up transformer worked as electrical energy storage. Discharge voltage (V_D) and currents (I_D) were measured with a voltage divider (EP-50K, PEEC, Japan) and current transformer (model 2-1.0, Stangenes Industries Inc.), respectively. Signals from voltage and current transformers were recorded with a digital oscilloscope (Tektronix TDS 754D, Sony Tektronix) of a 500 MHz bandwidth.

The transform efficiency η_d of the pulse power supply is defined as Eq. (1).

$$\eta_d = \frac{P_D}{P_{in}} \times 100 \quad [\%] \tag{1}$$

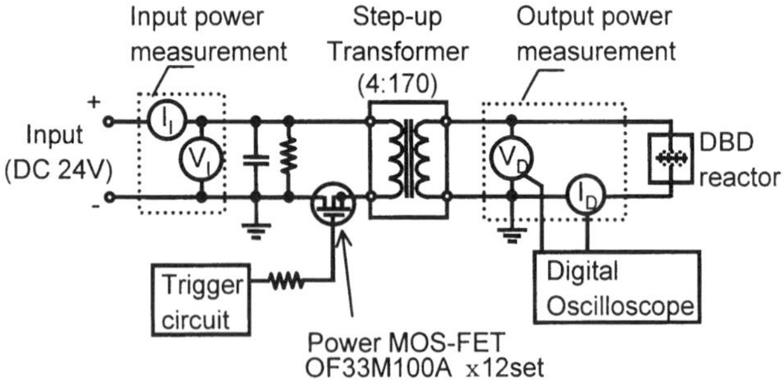

FIGURE 2. Circuit diagram of the pulse power generator.

P_{in} (input power to the pulse power generator), and P_D (discharge power to the DBD reactor) are calculated using Eqs. (2)-(3).

$$P_{in} = V_I \cdot I_I \text{ [W]} \tag{2}$$

Where V_I and I_I are input voltage [V] and input current [A], respectively.

$$P_D = f \times \sum_i \frac{V_{Di} + V_{Di+1}}{2} \frac{I_{Di} + I_{Di+1}}{2} (t_{i+1} - t_i) \text{ [W]} \tag{3}$$

Where, f is pulse frequency [Hz], V_{Di} and I_{Di} are discharge voltage [V] and cathode current [A] at discharge time t_i [s], respectively.

The PM removal rate was calculated as Eq. (4).

$$\text{PM removal rate} = \frac{\text{PM concentration after discharge tratment}}{\text{PM concentration before discharge treatment}} \times 100 \text{ [\%]} \tag{4}$$

Figure 3 shows the construction of the DBD reactor. A basic discharge layer consisted of two metal mesh electrodes (110 x 110 x 2 mm) placed on both sides of an alumina (Al_2O_3) plate (150 x 150 x 2 mm). 6-16 layers were connected in parallel and set into the reactor. The diesel engine was controlled at a fixed rotation rate (1000 rpm) and out put power (4 kW). The temperature of the exhaust gas at the inlet of the DBD reactor was 180 °C.

RESULTS

Figure 4 shows the transform efficiency of the pulse power generator as a function of input power to the pulse generator. Although the transform efficiency was variously distributed from 40% to 90% at lower input power, it was converged to 60% at higher input power. The number of layers had not influence on the

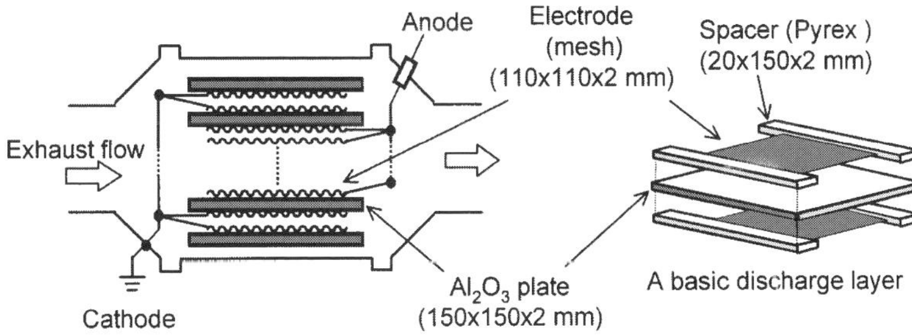

FIGURE 3. Construction of the BDB reactor.

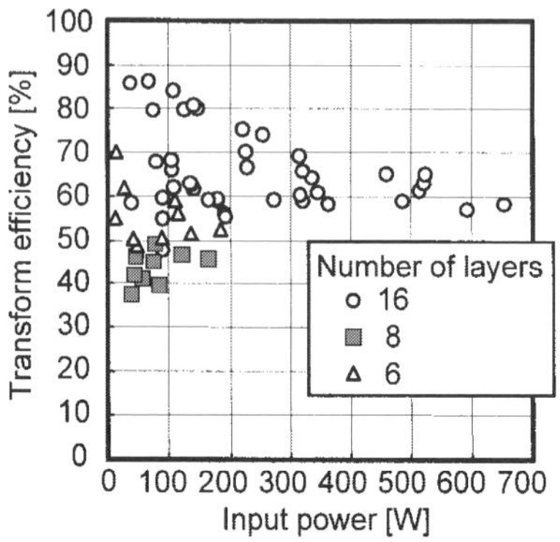

FIGURE 4. Transform efficiency as a function of input power (P_{in}).

transform efficiency. This result implied that the condition of discharge and flow of exhaust gas might affect to transform efficiency at lower input power. On the other hand, it is thought that convergence of the transfer efficiency in higher input power was related to increase of the influences of joule heat energy loss of transform line and/or mismatching of impedances, etc.

Figure 5 shows PM removal rate as a function of input power. The PM removal rate was increased by increasing input power of the pulse power generator. Maximum PM removal rate of 58% was achieved at 480 W of input power of the pulse power generator.

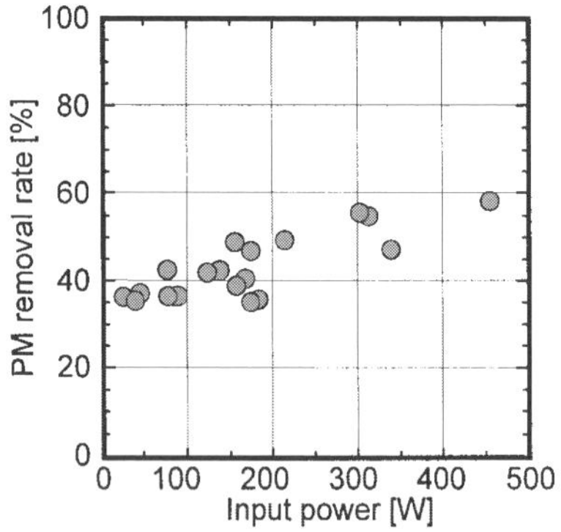

FIGURE 5. PM removal as a function of input power (P_{in}).

CONCLUSIONS

(1) The transform efficiency was variously distributed from 40% to 90% at lower input power, and converged to 60% at higher input power (over 300 W).
(2) The DBD reactor driven by the pulse power generator could be used for diesel PM removal. Maximum PM removal rate of 58% was achieved at 480 W of input power of pulse power generator.

ACKNOWREDGEMENTS

The Ministry of Education, Culture, Sports, Science and Technology supported this study. Yao is grateful to The New Energy and Industrial Technology Development Organization (NEDO) for a fellowship.

REFERENCES

1. S. L. Yao, T. Takemoto, F. Ouyang, A. Nakayama, E. Suzuki, A. Mizuno, M. Okumoto, *Energy Fuels*, **14**, 459-463 (2000).
2. C. Liu, A. Marafee, B. Hill, G. Xu, R. Mallinson, L. Lobban, *Ind. Eng. Chem. Res.*, **35,** 3295-3301 (1996).
3. L. M. Zhou, B. Xue, U. Kogelschatz, B. Eliasson, *Plasma Chem. Plasma Process.*, **18**, 375-393 (1998).
4. A. Mizuno, K. Shimizu, A. Chakurabarti, L. Dascalescu, S. Furuta, *IEEE Trans. Ind. Appl.*, **31,** 957-963 (1995).
5. S. Masuda, H. Nakao, *IEEE Trans. Ind. Appl.*, **26**, 374-383 (1990).
6. K. L. L. Vercammen, A. A. Berezin, F. Lox, J. S. Chang, *J. Adv. Oxid. Technology*, **2**, 312-329 (1997).
7. K. Yan, *Corona Plasma Generation*, Technische Universtieit Eindhoven, 2001; Chapter 3, pp. 47-128.
8. Y. S. Mok, *Plasma Chem. Plasma process.*, **20,** 353-364 (2000).
9. M. Okumoto, S. L. Yao, A. Nakayama, E. Suzuki, *Utilization of greenhouse gases*, ACS sym. series 852, 2003, pp.314-324.

Applied Physics (APHYS 2003)
A. Méndez-Vilas (Ed.)

Numerical Calculation of a Liquid Bridge Equilibrium Contour between Noncircular Supports

F. J. Acero †, J. M. Montanero ‡,

†Departamento de Física, Universidad de Extremadura, E-06071 Badajoz, Spain

‡Departamento de Electrónica e Ingeniería Electromecánica, Universidad de Extremadura, E-06071 Badajoz, Spain

March 11, 2004

Abstract

The equilibrium shape of a liquid bridge confined between two parallel plates of arbitrary shape is numerically calculated. Two alternative procedures are used: (i) an algorithm that minimizes the sum of the potential and surface energies, and (ii) a finite-difference scheme to integrate the Young-Laplace equation. Both methods preserve the imposed value for the liquid bridge volume. The comparison between the results obtained from the two methods shows an excellent agreement, which indicates the degree of accuracy of the calculations.

INTRODUCTION

A liquid bridge is a mass of liquid sustained by the action of the surface tension force between two parallel supporting disks. This fluid configuration has been extensively studied during the past decades from both theoretical and experimental points of view. The liquid bridge has been considered as a prototype to analyse mechanical and thermo-mechanical phenomena dominated by the surface tension. The equilibrium shapes [1, 2], their stability [3], the small amplitude dynamics [4], and the nonlinear breakup [5] of liquid bridges are classical consecutive steps in the theoretical study of their mechanical behaviour. On the experimental side, a liquid bridge can hold a significant liquid volume and provide a convenient way for its manipulation even under normal gravity conditions. In the Plateau-Rayleigh technique, a liquid bridge is surrounded by a liquid bath of similar density to partially compensate for the effect of the hydrostatic pressure along the interface [2]. In this way, one obtains a stable fluid configuration with a larger exposed surface area.

Apart from the fundamental point of view, a liquid bridge can be considered as the simplest idealisation of the configuration appearing in the floating zone technique used for crystal growth and purification of high melting point materials [6]. This confers to the study of liquid bridges a great interest not only in fluid mechanics but also in the field of material engineering. Most of the specialized literature deals with axisymmetric liquid bridges, although many papers considering non-axisymmetric configurations have been recently published. In these papers, the deviation of the liquid bridge contour from the axisymmetric shape is due to the action of lateral forces and/or eccentricity of the supporting disks [1, 2]. Works considering liquid bridges with noncircular supports are still very scarce. They have mainly focussed on the analytical [7, 8] and numerical [9] calculation of the stability limits. In the present contribution, we numerically calculate the equilibrium shape of a liquid bridge anchored to the edges of two parallel solid supports of arbitrary shape.

The equilibrium shape of a liquid drop can be calculated by minimizing the sum of the potential energy associated with gravity and inertial forces acting on it, and the energy associated with the interface. Formally, this calculation would lead to the well-known Young-Laplace equation, which represents the dynamical equilibrium between pressure, surface tension, gravity and inertial forces. Therefore, the equilibrium contour of the liquid drop can be alternatively obtained by integrating the Young-Laplace equation. In both procedures, the calculation has to take into account certain boundary conditions and preserve the value imposed for the drop volume. Here we use both approaches to obtain numerically the interface shape of a liquid bridge between noncircular solid supports. The minimization of the energy is performed by evolving the surface down the energy gradient [10], while the Young-Laplace equation is integrated by means of a finite difference scheme. Comparison between the results obtained from both procedures allows one to establish the accuracy of the methods.

FORMULATION OF THE PROBLEM

The fluid configuration considered is sketched in Fig. 1. It consists of a liquid bridge of volume $\mathcal{V}$ surrounded by either a gas or another liquid immiscible with the former, both contained in a vessel of volume $\mathcal{V}_c$. The liquid bridge is held between two parallel solid supports of arbitrary shape placed a distance L apart. Due to the sharpness of the support edges, one assumes that the liquid anchors perfectly to those edges, preventing motion of the contact line. The liquid bridge and the surrounding medium densities are ρ_1 and ρ_2, respectively, while the surface tension associated with the interface is σ. The liquid bridge is subjected to the action of both axial and lateral constant forces, g_a and g_l being their corresponding magnitudes per unit mass. The contour of the liquid bridge is characterized by the function $R(z,\theta)$, which measures the distance between a surface element and the z axis.

The energy $\mathcal{U}$ of the system is

$$\mathcal{U} = -\int_{\mathcal{V}_c} \rho(\mathbf{r})\ (\mathbf{g}\cdot\mathbf{r})\ d\mathcal{V} + \sigma\mathcal{S} + \mathcal{U}_0\,. \tag{1}$$

Here, $\rho(\mathbf{r})$ is the density field, $\mathcal{S}$ is the area of the interface, and $\mathcal{U}_0$ is an arbitrary constant.

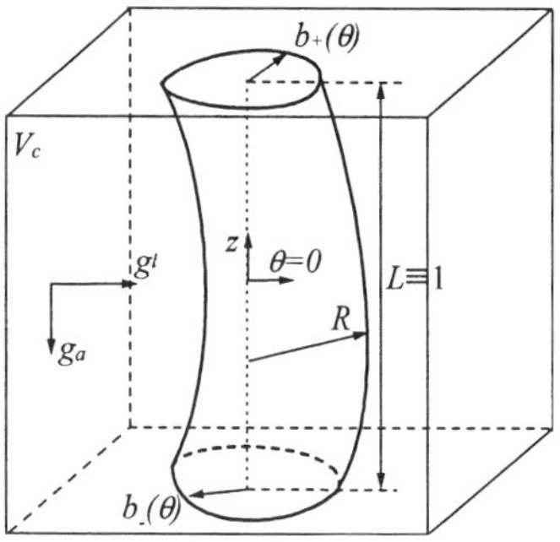

Figure 1: Sketch of the fluid configuration considered.

It is easy to verify that

$$\begin{aligned}\mathcal{U} &= -\int_{\mathcal{V}_c-\mathcal{V}} \rho(\mathbf{r})\,(\mathbf{g}\cdot\mathbf{r})\,d\mathcal{V} - \int_{\mathcal{V}} \rho_1\,(\mathbf{g}\cdot\mathbf{r})\,d\mathcal{V} + \int_{\mathcal{V}} \rho_2\,(\mathbf{g}\cdot\mathbf{r})\,d\mathcal{V} - \int_{\mathcal{V}} \rho_2\,(\mathbf{g}\cdot\mathbf{r})\,d\mathcal{V} \\ &\quad +\sigma\mathcal{S} + \mathcal{U}_0 \\ &= -\int_{\mathcal{V}_c} \rho_2\,(\mathbf{g}\cdot\mathbf{r})\,d\mathcal{V} - \int_{\mathcal{V}} \Delta\rho\,(\mathbf{g}\cdot\mathbf{r})\,d\mathcal{V} + \sigma\mathcal{S} + \mathcal{U}_0 \\ &= -\int_{\mathcal{V}} \Delta\rho\,(\mathbf{g}\cdot\mathbf{r})\,d\mathcal{V} + \sigma\mathcal{S} + \mathcal{U}_0', \end{aligned} \tag{2}$$

where $\Delta\rho \equiv \rho_1 - \rho_2$ and $\mathcal{U}_0'$ is a constant which does not depend on the liquid bridge shape.

In what follows, the space cylindrical coordinates (r, θ) and distances are made dimensionless using L as the characteristic length. The shapes (in terms of L) of the bottom and upper supporting solid surfaces are $b_-(\theta)$ and $b_+(\theta)$, respectively. These functions together with the set of dimensionless parameters $\{V,\ B_a,\ B_l\}$ uniquely characterize the above fluid configuration. Here, $V \equiv \mathcal{V}/L^3$ is the reduced volume, and $B_a \equiv \Delta\rho\ g_a L^2/\sigma$ and $B_l \equiv \Delta\rho\ g_l L^2/\sigma$ are the axial and lateral Bond numbers, respectively. The dimensionless energy $U \equiv \mathcal{U}/\sigma L^2$ is

$$\begin{aligned} U &= \int_V (B_a z - B_l r\ \cos\theta)\ dV + S \\ &= \tfrac{1}{2}B_a \int_{-1/2}^{1/2}\int_0^{2\pi} z\ R^2\ dz\ d\theta\ - \tfrac{1}{3}B_l \int_{-1/2}^{1/2}\int_0^{2\pi} \cos\theta\ R^3\ dz\ d\theta \\ &\quad + \int_{-1/2}^{1/2}\int_0^{2\pi} [R_\theta^2 + R^2 + (RR_z)^2]^{1/2}\ dz\ d\theta\ , \end{aligned} \tag{3}$$

where $S \equiv \mathcal{S}/L^2$ and the choice $\mathcal{U}_0' = 0$ has been made. In Eq. (3) and in what follows, a subscript denotes the partial derivative with respect to the corresponding coordinate.

The equilibrium contour $R(z,\theta)$ is the function that minimizes (3) with the boundary conditions

$$R(-1/2,\theta) = b_-(\theta)\ , \qquad R(1/2,\theta) = b_+(\theta)\ , \tag{4}$$

$$R(z,\theta) = R(z,\theta + 2\pi)\ , \tag{5}$$

and the volume constraint

$$V = \tfrac{1}{2}\int_{-1/2}^{1/2}\int_0^{2\pi} R^2\, dz\, d\theta\ . \tag{6}$$

Alternatively, the problem can be formulated in terms of the forces acting on the system. The Young-Laplace (capillary) equation

$$\nabla \cdot \mathbf{n} - B_a z + B_l R\cos\theta + P = 0 \tag{7}$$

expresses the dynamical equilibrium between that forces. In Eq. (7), $\mathbf{n}$ is the unitary vector normal to the interface, $\nabla \cdot \mathbf{n}$ is twice the (dimensionless) local mean curvature, and $P \equiv \Delta p_0 L/\sigma$ is the capillary pressure which is related to the difference Δp_0 between the outer pressure and the inner pressure at the origin of the coordinate system. In cylindrical coordinates

$$\begin{aligned}\nabla \cdot \mathbf{n} \;=\; & [R(1+R_z^2)(R_{\theta\theta} - R) + RR_{zz}(R^2 + R_\theta^2) - 2R_\theta(R_\theta + RR_zR_{z\theta})] \\ & \times [R^2(1+R_z^2) + R_\theta^2]^{-\frac{3}{2}}\ .\end{aligned} \tag{8}$$

Now, the equilibrium contour $R(z,\theta)$ is to be determined by integrating numerically the Young-Laplace equation (7) with the boundary conditions (4) and (5), and the volume constraint (6).

NUMERICAL APPROACHES

The SURFACE EVOLVER program [10] minimizes the energy of a fluid configuration subjected to constraints. The energy can include surface tension, gravity and other forms, while constraints can be geometrical conditions on boundary positions or constraints on integrated quantities such as body volumes. The program uses a finite element method to represent a surface as a union of simplexes. The minimization is carried out by evolving this representation down the energy gradient in a iteration process. At each iteration, the position of the simplexes is corrected to verify the constraints imposed. Also, it can be necessary to refine the mesh frequently.

The SURFACE EVOLVER program was used in Ref. [11] to obtain the stability limits of nonaxisymmetric liquid bridges between circular supports. Here, we have used it to calculate the equilibrium shapes between noncircular supports and subjected to the action of both axial and lateral gravity forces. This calculation is equivalent to find the function $R(z,\theta)$ that minimizes the energy U obtained from (3) with the constraints (4), (5) and (6). At each iteration, the algorithm provides the energy value

$$\Gamma \equiv \int_{-1/2}^{1/2}\int_0^{2\pi} [R_\theta^2 + R^2 + (RR_z)^2]^{1/2}\, dz\, d\theta \tag{9}$$

associated with the surface tension. In the iterative process, one assumes that the equilibrium shape has been reached if in an iteration the energy Γ changes of one part in 10^8 [11]. It must be pointed out that the use of this condition demonstrates to provide accurate results if the liquid bridge is actually stable, but it can lead to spurious equilibrium shapes beyond the stability limit. For this reason, the maximum value of the slenderness for which the liquid bridge is stable was slightly overestimated in some cases of Ref. [11].

As an alternative route, we have developed a finite difference scheme to integrate the Young-Laplace equation (7) with the conditions (4), (5) and (6). This procedure demonstrated to provide accurate results as compared with other theoretical approaches and experimental shapes for circular supports [2]. Recently, this method has been used to determine the stability limits of liquid bridges between an elliptical and a circular supporting disk subjected to axial gravity [9]. In the finite difference scheme, the continuum integration domain is replaced by a rectangular mesh with $I \times J$ nodes, I and J being the number of nodes along the θ and z axes, respectively. We distinguish between inner and boundary nodes, the later being those nodes on $z = \pm 1/2$. The partial derivatives in the local mean curvature (8) are replaced by central finite differences on the $I \times (J-2)$ inner points of the mesh. If in a finite difference a non-existent node is required then the periodic boundary condition (5) is used to replaced it with an inner node. Boundary condition (4) reduces to $2I$ conditions for the boundary nodes. The counterpart of integral (6) is obtained using the extended trapezoidal rule. The result is a nonlinear relation in which all the nodes are involved. Once the problem represented by (4)–(7) has been discretized, the goal is to solve a set of $N = I \times J + 1$ nonlinear equations to obtain R at the nodes and the capillary pressure P. In Refs. [1, 9] this set of equations was solved for different particular cases once the equations are linearized around a previously obtained solution. We here deal with the nonlinear problem by means of the Newton-Raphson method, thereby avoiding approximations associated with the linearization. The algorithm is found to be stable. Starting from the cylindrical solution slightly perturbed by a random function, the Newton-Raphson method converges to the sought values of R and P after several (less than 15) iterations. In all cases, the liquid bridge equilibrium contour is smooth so that the mesh size may be presumed not to play a relevant role. We will perform the calculations using a rectangular mesh with 266 nodes. In the final step, spline interpolation functions are constructed to provide $R(z, \theta)$ over the continuum integration domain. This numerical procedure has been automated and can be easily implemented inside other codes. Excellent agreement is found between our results and the stable configurations showed in Fig. 3 of Ref. [9], what indicates that the linearization used in that reference works, at least for the configurations considered in the figure.

RESULTS

Figure 2 shows a three-dimensional plot of a liquid bridge between solid supports of elliptic shape obtained from both the SURFACE EVOLVER and the finite difference scheme. The nodes of the meshes used in the calculations are also plotted. As can be observed, the action of the axial and lateral components of gravity produces a displacement of liquid that deforms the contour. Unfortunately, the SURFACE EVOLVER program does

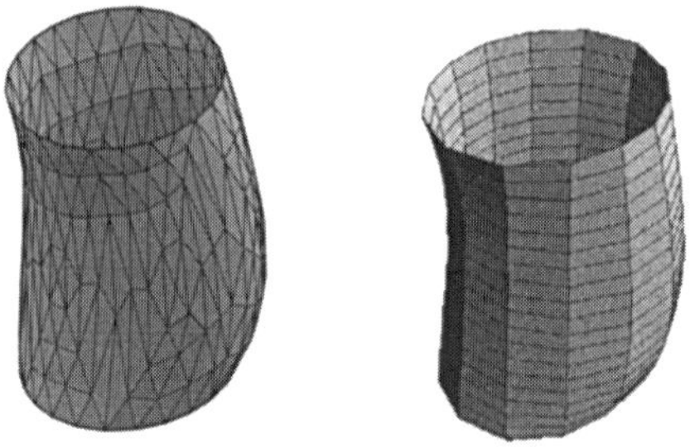

Figure 2: Three-dimensional plot of a liquid bridge obtained from SURFACE EVOLVER (left-hand plot) and the finite difference scheme (right-hand plot) for $b_- = b_+ = 0.3125(1.8\cos^2\theta + 1.2\sin^2\theta)^{1/2}$, $B_a = B_l = 1.5$ and $V = 0.47124$.

not allow one to print out values of the equilibrium contour $R(z,\theta)$, so that a detailed comparison between the results calculated from both methods is not feasible. In the present contribution, an only parameter is chosen to characterize the interface shape. From both the SURFACE EVOLVER program and the finite difference scheme, one can easily get the value of the (dimensionless) energy Γ associated to the interface. Tables 1–3 show the values of Γ obtained from both procedures for different choices of the set $\{b_-(\theta),\ b_+(\theta),\ V,\ B_a,\ B_l\}$. The agreement is excellent for all the cases considered. In fact, the discrepancies concern the fourth significant figure excepting close to the stability limit, where they increase due to discretization errors. The computing time expended to get the values showed in Tables 1–3 was similar using both methods, although the finite-difference scheme is more efficient close to the stability limit.

This work was supported by the Ministerio de Ciencia y Tecnología through Grant No. ESP2003-02859.

References

[1] A. Laverón-Simavilla and J. M. Perales: "Equilibrium shapes of nonaxisymmetric liquid bridges of arbitrary volume in gravitational fields and their potential energy," Phys. Fluids **7**, 1204 (1995).

[2] J. M. Montanero, G. Cabezas, J. Acero, J. M. Perales: "Theoretical and experimental analysis of the equilibrium contours of liquid bridges of arbitrary shape," Phys. Fluids **14**, 682 (2002).

[3] J. Meseguer, L. A. Slobozhanin and J. M. Perales: "A review on the stability of liquid bridges," Adv. Space Res. **16**, 5 (1995).

[4] J. A. Nicolás, J. M. Vega: "Linear oscillations of axisymmetric viscous liquid bridges," Z. Angew. Math. Phys. **51**, 701 (2000).

Table 1: Surface energy Γ for $b_- = b_+ = R_0(1.8\cos^2\theta + 1.2\sin^2\theta)^{1/2}$, $B_a = B_l = 1.5$ and different values of R_0 and $V^* \equiv V/(\pi R_0^2 L)$. The upper and lower values are obtained using the SURFACE EVOLVER and the finite-difference scheme, respectively.

R_0 \ V^*	1.32272	1.39621	1.49969	1.54317	1.61667	1.69015	1.76363	1.83712
0.3125	2.28772	2.34785	2.40884	2.47078	2.53376	2.59782	2.66301	2.72936
	2.28716	2.34741	2.40853	2.47061	2.53374	2.59796	2.66332	2.72987
0.29412	2.15098	2.20782	2.26516	2.32309	2.38167	2.44100	2.50112	2.56202
	2.15049	2.20746	2.26492	2.32298	2.38172	2.44119	2.50144	2.56252
0.27778	2.02977	2.08366	2.13774	2.19219	2.24609	2.30228	2.35806	2.41437
	2.02935	2.08335	2.13758	2.19212	2.24706	2.30247	2.35836	2.41486
0.26316	1.92160	1.97279	2.02401	2.07533	2.12686	2.17862	2.23069	2.28309
	1.92123	1.97255	2.02388	2.07532	2.12692	2.17884	2.23103	2.28356
0.25	1.82443	1.87322	1.92184	1.97041	2.01901	2.06770	2.11653	2.16555
	1.82414	1.87302	1.92174	1.97042	2.01913	2.06792	2.11687	2.16601
0.23809	1.73674	1.78328	1.82956	1.87565	1.92166	1.96763	2.01362	2.05968
	1.73648	1.78312	1.82949	1.87569	1.92181	1.96787	2.01397	2.06013
0.22727	1.65715	1.70165	1.74579	1.78966	1.83333	1.87689	1.92036	1.96381
	1.65694	1.70153	1.74576	1.78972	1.83349	1.87714	1.92073	1.96426
0.21739	1.58461	1.62724	1.66942	1.71127	1.75285	1.79422	1.83546	1.87659
	1.58445	1.62716	1.66943	1.71136	1.75303	1.79449	1.83581	1.87703

[5] J. Eggers: "Nonlinear dynamics and breakup of free-surface flows," Rev. Mod. Phys. **69**, 865 (1997).

[6] J. Meseguer, J. M. Perales, I. Martínez, N. A. Bezdenejnykh and A. Sanz: "Hydrostatic instabilities in floating zone crystal growth process," Crystal Growth Research **5**, 27 (1999).

[7] J. Meseguer, J. M. Perales and J. I. D. Alexander: "A perturbation analysis of the stability of long liquid bridges between almost circular supporting disks," Phys. Fluids **13**, 2724 (2001).

[8] M. Gómez, I. E. Parra and J. M. Perales: "Mechanical imperfections effect on the minimum volume stability limit of liquid bridges," Phys. Fluids **14**, 2029 (2002).

[9] A. Laverón-Simavilla, J. Meseguer, J. L. Espino: "Stability of liquid bridges between an elliptical and a circular supporting disk," Phys. Fluids **15**, 2830 (2003).

[10] K. Brakke: "The surface evolver," Experimental Math. **1**, 141 (1992).

[11] F. Zayas, J. I. D. Alexander, J. Meseguer and J. F. Ramus: "On the stability limits of long nonaxisymmetric cylindrical liquid bridges," Phys. Fluids **12**, 979 (2000).

Table 2: Surface energy Γ for $b_- = b_+ = (0.25\cos^2\theta + 0.09\sin^2\theta)^{1/2}$, $V = 0.47124$ and different values of B_a and B_l. The upper and lower values are obtained using the SURFACE EVOLVER and the finite-difference scheme, respectively.

B_l \ B_a	0.0	0.3	0.6	0.9	1.2	1.5
0.0	2.48537 2.48354	2.48557 2.48374	2.48616 2.48431	2.48715 2.48528	2.48858 2.48664	2.49033 2.48839
0.3	2.48595 2.48410	2.48622 2.48429	2.48674 2.48487	2.48773 2.48584	2.48913 2.48721	2.49093 2.48897
0.6	2.48769 2.48579	2.48789 2.48599	2.48849 2.48658	2.48950 2.48756	2.49091 2.48895	2.49273 2.49074
0.9	2.49061 2.48866	2.49089 2.48886	2.49143 2.48946	2.49246 2.49047	2.49392 2.49189	2.49577 2.49373
1.2	2.49476 2.49274	2.49497 2.49295	2.49560 2.49357	2.49667 2.49462	2.49819 2.49609	2.50010 2.49799
1.5	2.50021 2.49812	2.50042 2.49833	2.50107 2.49899	2.50219 2.50008	2.50379 2.50162	2.50577 2.50362
1.8	2.50707 2.50489	2.50727 2.50512	2.50795 2.50581	2.50912 2.50697	2.51082 2.50861	2.51298 2.51073
2.1	2.51545 2.51320	2.51566 2.51345	2.51637 2.51420	2.51760 2.51544	2.51941 2.5172	2.52185 2.51948
2.4	2.52555 2.52325	2.52576 2.52352	2.52651 2.52433	2.52784 2.52569	2.52979 2.52761	2.53248 2.53011
2.7	2.53762 2.53528	2.53783 2.53557	2.53864 2.53648	2.54009 2.53799	2.54223 2.54013	2.54519 2.54292
3.0	2.55199 2.54965	2.55222 2.54999	2.55312 2.55101	2.55474 2.55272	2.55713 2.55516	2.56049 2.55835

Table 3: The same as in Table 2.

B_l \ B_a	1.8	2.1	2.4	2.7	3.0
0.0	2.49255 2.49055	2.49519 2.49313	2.49827 2.49612	2.50179 2.49956	2.50577 2.50345
0.3	2.49317 2.49114	2.49582 2.49373	2.49890 2.49674	2.50244 2.5002	2.50649 2.50411
0.6	2.49502 2.49294	2.49770 2.49557	2.50084 2.49863	2.50444 2.50215	2.50851 2.50612
0.9	2.49813 2.49599	2.50089 2.49868	2.50411 2.50183	2.50782 2.50545	2.51202 2.50954
1.2	2.50256 2.50034	2.50545 2.50314	2.50879 2.50647	2.51266 2.51017	2.51705 2.51445
1.5	2.50837 2.50608	2.51145 2.50903	2.51497 2.51247	2.51905 2.51644	2.52370 2.52096
1.8	2.51567 2.51335	2.51903 2.51647	2.52278 2.52017	2.52715 2.52442	2.53215 2.52928
2.1	2.52466 2.52231	2.52838 2.52571	2.53243 2.52971	2.53716 2.53434	2.54263 2.53967
2.4	2.53558 2.53322	2.53969 2.53696	2.54416 2.54138	2.54939 2.54655	2.55546 2.55252
2.7	2.54875 2.5464	2.55323 2.55062	2.55840 2.55564	2.56427 2.56155	2.57117 2.56848
3.0	2.56463 2.56236	2.56971 2.56725	2.57580 2.57314	2.58262 2.58018	2.59077 2.58863

Applied Physics (APHYS 2003)
A. Méndez-Vilas (Ed.)

Hydrodynamic Lattice-Boltzmann Simulation of a Thermoplastic Fluid Film for Holographic Recording.

T. Belenguer Dávila*,†, G. Ramos Zapata*, E. Bernabeu Martínez†

*Laboratorio de Instrumentacón Espacial. INTA. Carretera de Ajalvir Km 4.5, 28850 Torrejón de Ardoz, Madrid, SPAIN.

† Facultad de Ciencias Físicas, Universidad Complutense de Madrid, Ciudad Universitaria, 28040 Madrid, SPAIN

Abstract. A multi-component lattice-Boltzmann model including external electric forces interactions is developed to study the behavior of a classical thermoplastic film which is used as a holographic recording material. The model incorporate also the spinodal decomposition originated in this type of polymeric fluids when a fast thermal pulse is used to develop the hologram. The surface deformation of the electrically charged thermoplastic film is simulated using the lattice-Boltzmann model and the main holographic parameters i.e. spatial frequency response (MTF) and diffraction efficiency are given and related to the hydrodynamic behavior of the fluid. The influence of the phase transition on the growth rate of the deformation and hence on the MTF response of the medium is studied and correlated with experimental results. The results obtained with our model are in agreement with those obtained using more complex theoretical analysis.

INTRODUCTION

Photo-thermoplastic holographic materials have been used in a number of different applications ranging from optical memories [1], adaptive data recording [2] to holographic interferometry [3]. Among them, the holographic interferometry is the field in which the thermoplastic materials have found most of its relevant applications. The main features that make such a material to be greatly attractive for holographic applications are (1) read-write-erase capabilities; (2) in-situ recording development, and fast read-out; and (3) virtually real time recording cycle.

Several theories have been developed in order to explain the behavior of a thermoplastic during holographic recording [4-7].

In recent years, the lattice Boltzmann method (LBM) has demonstrated to be an alternative for simulating fluid flows and modeling physics in fluid. This technique is particularly successful in fluid flow applications involving interfacial dynamics, complex boundaries and multiphase and multicomponents flows. These possibilities indicate that LBM could bc an excellent candidate for studying the fluid dynamic in thermoplastic recording media.

In this paper we summarize a comparison between the most important result of the classical theory of thermoplastic deformation and those obtained from the LBM, paying special attention in the spinodal decomposition

CLASSICAL APPROACH TO DYNAMICAL THEORY OF THERMOPLASTIC DEFORMATION

The photo-thermoplastic recording device consists basically on the following multilayer device: a conductive substrate, a photoconductor film and, finally, a thermoplastic layer (Figure 1). In the ref [7] is detailed the procedure to record a hologram in this kind of materials

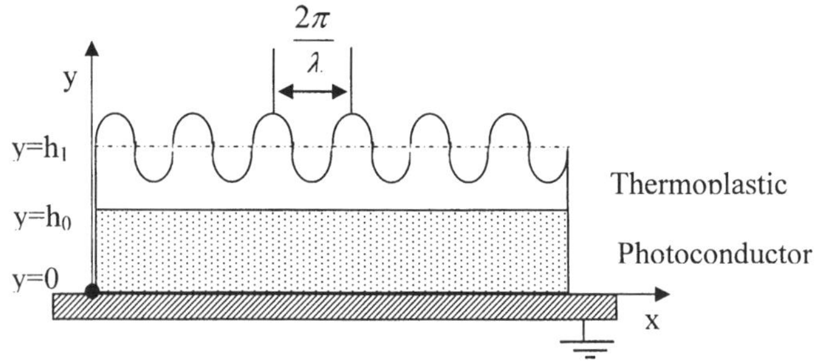

FIGURE 1. Schematic layout of a classical photo-thermoplastic recording device.

We have developed [8] a model to simulate the thermoplastic deformation of the thermoplastic layer following the classical approach.

Classical results

The most important curve obtained from the classical model is the deformation amplitude as a function of the spatial frequency for different thickness of the thermoplastic layer (fig.2). This curve is essentially the Modulation Transfer Function (MTF) of the thermoplastic response. The diffraction efficiency (η_{DE}) of the recorded grating is related to the deformation amplitude curve through the thin phase grating expression:

$$\eta_{DE} = J_1^2\left(\frac{2\pi}{\lambda_l}(n_{th}-1)A_m\right) \tag{1}$$

where J_1 is the Bessel function of first order, λ_l the wavelength of the recording-readout of the hologram , A_m the deformation amplitude obtained and n_{th} the thermoplastic layer refractive index. Hence, it can be concluded that to know the A-k curve is to have complete knowledge about the main holographic properties of the thermoplastic media.

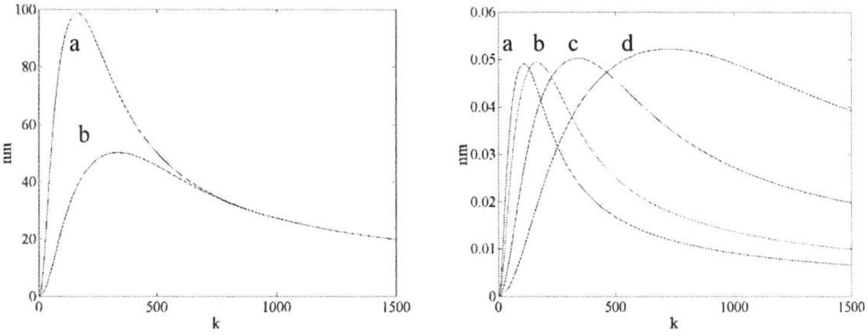

(a) (b)

FIGURE 2. Deformation amplitude as a function of spatial frequency, k: (a) A-k curve for two thickness of the thermoplastic layer ;a) 1 μm, b) 0.5 μm; (b) A-k curves for different thickness of the thermoplastic layer. The curves were divided by the thickness of the thermoplastic layer a) 4 μm, b) 2 μm, c) 1 μm, d) 0.5 μm.

The Data used in the present simulation are: T=0.04 N/m. ν=10 m^2/s, $P_0=10^5$ N/m^2, Thermoplastic dielectric constants: $\varepsilon_{th}=2.3\varepsilon_0$, Photoconductor dielectric constant $\varepsilon th=3\varepsilon_0$, ε_0 is the vacuum dielectric constant.

HYDRODYNAMIC LATTICE BOLTZMANN APPROACH

We have selected the D2Q7 scheme for our model that is based in a hexagonal grid with each node connected to its six nearest neighbors along a vector $\boldsymbol{e_i}$

Three immiscible component $Y_i(\mathbf{r},t)$ (yellow) ,$R_i(r,t)$ (red) and $B_i(r,t)$ (blue) with different relaxation parameters, τ_i ,are considered for each specie. The simplified discretized Boltzmann equation for a single component is given by

$$f^{\sigma}{}_i(\mathbf{r}+e_i,t+1)=f^{\sigma}{}_i(\mathbf{r},t)+\Theta^{\sigma}{}_i(f(\mathbf{r},t)) \tag{2}$$

where f^{σ}_i is the particle velocity distribution function along the e_i direction at site $\boldsymbol{r}$ and time t. Θ^{σ}_i: is the collision operator which represent the rate of change of f_i resulting from collision for each component. The left-hand side of Equation (14) describes the streaming o of the distribution function on the grid.

The density ρ and momentum density ρ$\boldsymbol{u}$ are defined as particle velocity moments of the distribution function, $f^{\sigma}{}_i$

$$\rho^{\sigma}(\mathbf{r},t)=\sum_{i\_0}^{5}f^{\sigma}{}_i(\mathbf{r},t) \tag{3}$$

The momentum of a single component σ is given by

$$\mathbf{u}(\mathbf{r},t)=\left(\sum_{i\_0}^{5}e_if^{\sigma}{}_i(\mathbf{r},t)\right)\Big/\left(\sum_{i\_0}^{5}f^{\sigma}{}_i(\mathbf{r},t)\right) \tag{4}$$

The collision operator the well-known BGK approach in which the distribution function is calculated using:

$$f^{\sigma}{}_i(x+e_{\sigma i},t+1)-f^{\sigma}{}_i(x,t)=-\frac{(f_{\sigma i}-f_{\sigma i}^{eq})}{\tau} \tag{5}$$

The Shan-Chen [11] approach for the velocity of each multicomponent node is considered assuming that the velocities of each component quickly equalize to a final velocity

We have used the modified Gunstensen-Rothmann [11] rules to take in to account the multi-componets behavior because this algorithm is able to maintain high different kinematic viscosity between components during phase separation.

The *color-blind* momentun density is used to calculate the overall density and velocity, in our case we have:

$$f_i(\mathbf{r},t) = Y_i(\mathbf{r},t) + R_i(\mathbf{r},t) + B_i(\mathbf{r},t) \qquad (6)$$

The equlibrium distribution of the global population is evaluated using Equation (18). The Surface tension is introduced in each component by considering a small mass and momentum perturbation. In our model the different relaxation parameters ,τ^{σ}, can be applied to each component simulating in this way different kinematic viscosities. When the thermoplastic simulation is considered as a multicomponent fluid, the population Red and Blue are initialized and special care have to be taken to define the viscosity of this mixed fluid.

Boundary conditions.

The inner wall of the thermoplastic coating is considered non-slip bounce-back boundary condition. The upper surface of the thermoplastic which is in contact with the air in real conditions, is simulated is in contact with a fluid with lower viscosity and lower density (air simulation). The upper wall of the domain is considered as an open boundary with re-injection of equilibrium population to prevent interactions with the wall. The boundary condition for right and left boundaries are considered periodic.

Electric forces.

The interaction between the electric forces and the fluid has been recently studied for liquid crystal application [15]. In our case the forces acting on each node of the lattice have been calculated using the distributions of the charge and the electric potential, equations (10)-(12), as

$$\vec{F} = q\vec{E} = -q\nabla V \qquad (7)$$

This interactions is included in the discretized Boltzmann equation as usual [13,14]

The Electric forces are applied in the interface between each component as real boundary conditions. The first step is to localize for each clock time of the lattice evolution the interface between populations simulating thermoplastic (Red and /or Blue) and air (Yellow). In the second step, the force is only applied in this interface following an exponential rise time until a specific value is reached. The force is maintained for a short period of time and after this, the force is decreased exponentially during time. This process is intended to simulate the charge decay occurred during the heat development pulse. The transversal and normal components of this forces with respect the interface layer are simulated functionally in full resemblance with the classical model of the electric field .

The model includes the possibility to consider the influence of the well-known *frost* effect occurred in this kind of materials when a fast heat pulse is used to develop the deformation. The material should be considered as a multicomponent polymer in which the rapid temperature quenching upon cooling produce phase separation in domains with different concentration of each component [3]. In the thermoplastic film the phase separation process is transformed in a thickness variation by effect of the electric field (see figure 3). It is important to mention that much of the work done to understand the deformations of thermoplastic film does not take in to account the effect of *frost* on the growing of the grating.

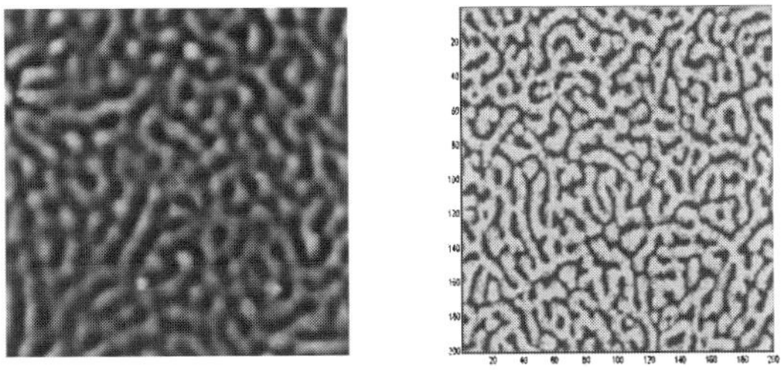

(a) (b)

FIGURE 3. Spinodal decomposition (*frost*) (a) measured with a AFM, (b) simulated by our LBM model

LBM results

We have simulated the thermoplastic layer using a bidimensional domain of 200x100 lattice nodes. The domain is initialized with near an equilibrium population for all the components. After a fixed number of time steps the forces, tangential and normal to the interface, are applied. The spatial period of the sinusoidal forces are changed simulating the recording process of a holographic image. When the forces are extinguished the population are let to evolve to the equilibrium distribution.

The contrast function, CF, defined as the difference between the maximum and the minimum of the deformation amplitude is computed during the evolution of the ensemble:

$$CF = \left| A^{l}_{\max} - A^{l}_{\min} \right| \tag{8}$$

This function is directly related to A-k curves of Figure 2.

The contrast function obtained from our LBM model is showed in Figure 4a. The comparison of both curves shows the good agreement obtained by the both models.

It is very interesting to analyze what is the influence of the phase separation during the recording of a holographic grating in the thermoplastic layer. In order to proceed with this, we have developed a simulation in which the thermoplastic layer is initialized with a density of Red and Blue particles by node equivalent to the monocomponent case. The applied forces are adapted to have the same momentum change by node that in the case before and the viscosity of the mixed components are considered. The contrast function obtained in presence of the spinodal decomposition

mechanism is found 10-20% better than the monocomponent case for low spatial frequencies, at medium and, mainly, at high frequencies the contrast function is considerably degraded showing a very variability versus time evolution of the simulation (see Figure 5).
We have verified that the spatial response of the contrast function is very dependent with the viscosity of the components and with the relative surface tension between species

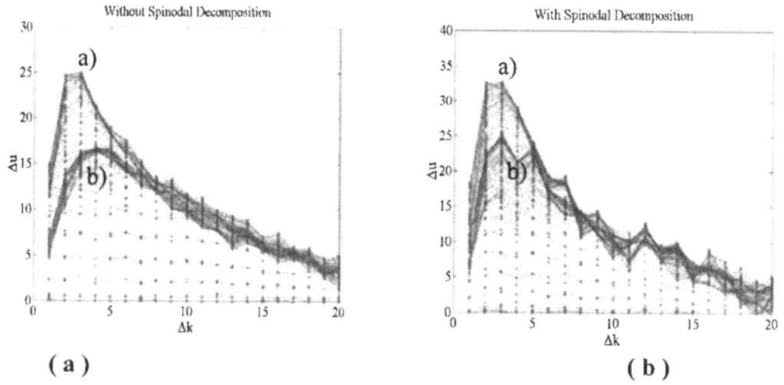

FIGURE 4. Deformation amplitude as a function of spatial frequency (in lattice units) obtained with our LBM model: a) 23 lattice units, and b) 56 lattice unit without spinodal decomposition;(b) Deformation amplitude as a function of spatial frequency (in lattice units) obtained with our LBM model: a) 23 lattice units, and b) 56 lattice unit witt spinodal decomposition
The Data used in the present simulation are: $A_{red}=A_{blue}=3*10^{-5}$, $A_{Yellow}=3*10^{-4}$, $\nu_{yellow}=0.5$, $\nu_{blue}=2$

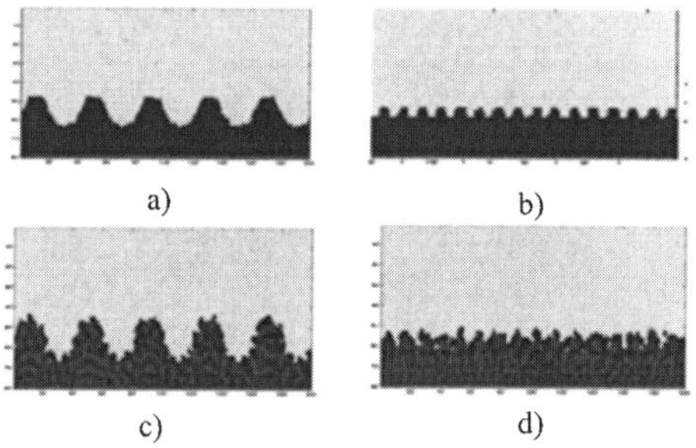

FIGURE 5. Deformation amplitude for two spatial frquencies: a) 40 lattice units, and b) 12 lattice units without spinodal decomposition; c) 40 lattice units, and d) 12 lattice units with spinodal decomposition.

SUMMARY

The LBM approach is suitable to understand the behavior of the complex fluids process occurred in the recording of holograms in thermoplastic film.

The effect of *frost*, not well understood in the literature, is naturally included in our model and hence the influence of this process in the main holographic parameter (A-k curve) seems to be double; to increase the amplitude of the deformation and to reduce the response of the media to high spatial frequencies.

The electric forces have been considered as a real boundary condition in the LBM scheme and this approach is considered enough to understand the main key parameters involved in the hydrodynamic behavior of the thermoplastic deformation.

REFERENCES

1. P. Gravey and J.Y. Moysan, Proc. SPIE **862**, 115 (1987)
2. Yu. A. Cherkasov, E.L. Aleksandrova,A.I. Rumyantsev, and M.V. Smirnov,J.Opt. Technol. 63 (4). April 1996
   Kono K., Ishizuka, T., Tsuda, H., and Kurosawa, A., Computer Phys. Comm. 129, 110- xxx (2000)
3. A.A. Friesem, Y. Katzir, Z. Rav-noy, B.Sharon Optiacal Engineering Vol **19** ,No. 5
4. H.F. Budd, J. Appl. Phys. **36**,1613 (1965)
5. U. Killat, J. Appl Phys. **46**, 5169 (1975)
6. U. Killat and D. Terrel, Opt Acta **24**, 441 (1977)
7. Z. Hirshfeld, A.A. Friesem, and Z. Rav-Noy, J. Appl. Phys. **52**, 605 (1981)
8. T. Belenguer et Al in preparation.
9. F. Carreño and E. Bernabeu J. Appl. Phys. **75** (9) 1994
10. Y. H. Qian, D'Humières and P. Lallemand, Europhy. Lett, **17** (1992)
11. A. K. Gunstensen, D. H. Rothman, S. Zaleski, and G. Zanetti, Phys. Rev. A,**43** (1991)
12. X. W. Shan and H. D. Chen , Phys. Rev. E, **49** (4) (1994)
13. Rothman, D. H., and Zaleski, S., *Lattice-Gas Cellular Autonomata. Simple Models of Complex Hydrodynamics*, Cambridge University Press, 1997.
14. Buick, J. M., and Greated, C. A., *Phys. Rev. E* **61**, (2000)
15. Denniston, C., Tóth, G., and Yeomans, J. M., *J. Statist. Phys.* **107**, 187(2002)

Applied Physics (APHYS 2003)
A. Méndez-Vilas (Ed.)

Diagnostics of a Pulsed Plasma Discharge

S. Yao*, M. Okumoto*, T. Yashima* and E. Suzuki†

*Chemical Research Group, Research Institute of Innovative Technology for the Earth, Kyoto 619-0292, Japan

†Department of Fine Material Engineering, Shinshu University, Ueda 386-8567, Japan

Abstract. A pulsed plasma discharge system was established for its diagnostics analysis. The pulsed plasma discharge system consisted of a pulse power supply and a point-to-point (PTP) type of reactor with a gas/liquid supply system. The discharge diagnostics was measured with a voltage probe, two current transformers, and an imaging spectrograph equipped with a high dynamic range streak camera and a dual cooled CCD camera. A mixture of CH_4 4.85%, O_2 9.71%, CO_2 9.71%, N_2 72.82% and H_2O 2.91% was supplied to the PTP reactor. The de-excitation rate constants of N (821.6 nm), H_α (656.2 nm), and O I (394~404 nm) were at an order of 10^7 s^{-1}. The formation rate of each species was also a function of the concentration of metastable atoms, and has an order of 10^8 s^{-1}. The authors proposed that the formation and de-excitation rates could be used for the estimation of dissociation and recombination rate of the source atoms or molecules with the help of the knowledge of kinetic simulation.

INTRODUCTION

Plasma technologies are now widely used in a variety of fields, such as microelectronics, material and chemical production, surface processing, decomposition of environmental pollutions. Particularly, non-thermal plasmas are of great interests and have been studied for gas cleanings of acid and greenhouse gases in combustion flue gases, such as NOx, SOx, VOC and PM removals, and for other applications such as methane conversion [1-6]. The authors recently developed a pulsed plasma discharge process for methane conversion and particulate matter removal from diesel exhaust gases [7, 8]. The wide application of plasma techniques requires both theoretical and practical studies to know the complicated processes in the plasma discharge. The information of atomic and molecular active components produced by the plasma discharges is important in further analysis of the plasma diagnostics and kinetic simulations [9, 10]. In a pulsed plasma discharge, electroluminescence is generally observed due to the induced and/or spontaneous emission of the excited species. After spectroscopic analysis of the electroluminescence, the presence of the excited species and their related ground state atoms and molecules and the kinetic rate constants of the quenching and de-excitation of the excited species can be estimated.

In this study, the rate constants of formation and quenching/de-excitation of N, H_α, and O I were measured in a pulse discharge duration over 821.6±5, 656.2±5, and 394~404 nm wavelength ranges, respectively.

EXPERIMENTAL

The experimental system mainly consisted of a pulse power supply (DP-17K35, Peec) and a point-to-point (PTP) type of reactor with a gas/liquid supply system. The discharge diagnostics was measured with a voltage probe, two current transformers, and an imaging spectrograph (C5094, Hamamatsu) equipped with a high dynamic range streak camera (C7700, Hamamatsu) and a dual mode cooled CCD camera (C4880, Hamamatsu). The pulse power supply had a pre-trigger signal (1 μs, 5V TTL) that was used as a fire signal to drive the streak camera. The analogue signals from the voltage probe, current transformers, and streak camera were recorded with a digital phosphor oscilloscope (TDS 7140, Tektronix). A mixture of CH_4 4.85%, O_2 9.71%, CO_2 9.71%, N_2 72.82% and H_2O 2.91% was supplied to the PTP reactor. A single pulse voltage of 10 kV-peak and about 20 ns rise-time was applied to the PTP gap at atmospheric pressure and room temperature.

RESULTS AND DISCUSSION

Properties of discharge voltage and currents

The discharge voltage and anode and cathode current were measured at a single pulse mode. The discharge voltage increased to a peak level of 10 kV in about 20 ns when the pulse voltage rose from 10% to 90% of peak level. The anode and cathode currents peaked to 76 A with a time lag of about 50 ns after the pulse voltage was applied. The most energy was injected over a time duration of 40 ns (from discharge time 100 ns to 140 ns). The peak energy injection was 0.45 MW.

Spectrum image

The spectrum of CH_4-O_2-CO_2-N_2-H_2O mixture in 380~870 nm region at various discharge times was shown in Fig. 1. Electroluminescence was observed from the time 100 ns, peaked in 130 ns to 140 ns and decreased. Several peaks due to N, O, and H_α de-excitation were found from the emission spectrum over a 500 ns time window. The positions of metastable species were confirmed from the spectra in a pulse discharge duration using pure N_2, O_2, CH_4, H_2, or CO_2 pulsed discharges.

Formation and de-excitation rates of metastable species

In consideration of excitation of an atom or molecule A by electron impact, the metastable A* is formed (Eq. (1)). The metastable species A* is then deactivated or quenched to A with or without luminescence (Eqs. (2) and (3)).

$$A + e \rightarrow A^* + e, \quad (1)$$

$$A^* \rightarrow A + h\nu, \quad (2)$$

$$A^* + X_i \rightarrow X_i^* + A. \quad (3)$$

The de-excitation rate of metastable spices A* can be represented as follows:

$$\frac{d[A^*]}{dt} = -(k + \sum_i k_i[X_i])[A^*], \quad \ln\left(\frac{[A^*]}{[A^*]_0}\right) = -(k + \sum_i k_i[X_i])t. \qquad (4)$$

Where, k and k_i are rate constants for Eqs. (2) and (3), respectively. $[X_i]$ is the concentration of specie X_i at time t. X_i^* is the product of X_i after interaction with A*.

The intensity I of the luminescence is proportional to the concentration of A*, $I = K[A^*]$, here K is a constant. Equation (4) is then given by

$$\ln\left(\frac{I}{I_0}\right) = -k't, \quad k' = k + \sum_i k_i[X_i]. \qquad (5)$$

The first order de-excitation rate coefficients (k') can be obtained from the slops of $\ln\left(I/I_0\right) \sim$ t plots.

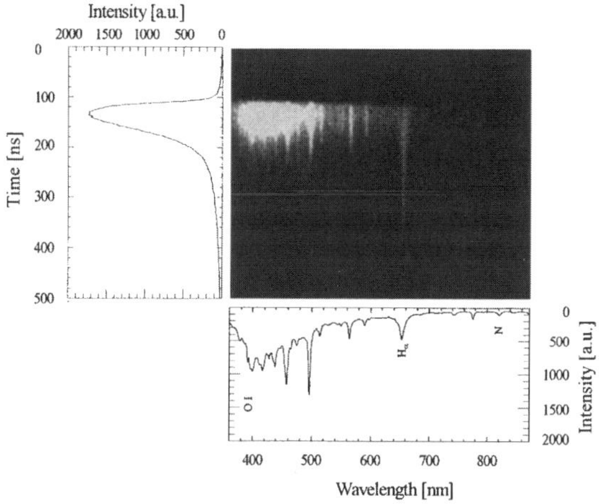

FIGURE 1. Spectrum of CH_4-O_2-CO_2-N_2-H_2O mixture in 380~870 nm at various discharge times.

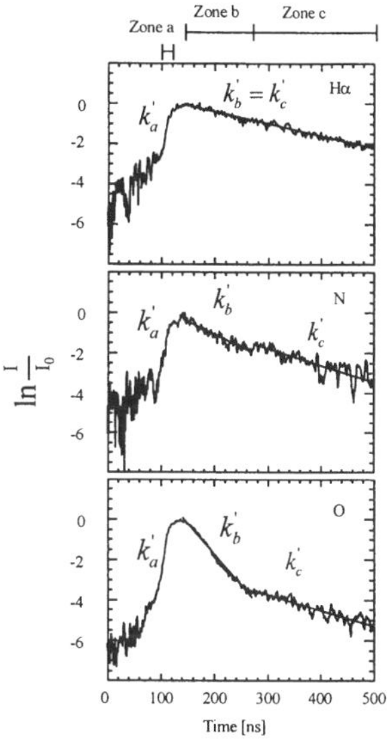

FIGURE 2. Relative luminescence changes of H, N, and O species.

The de-excitation rates of the metastable species of N, O, and H were then calculated from the luminescence intensity of each species over a 10 nm wavelength range in zones b and c (Table 1). As the rate constants of N and O species in the discharge duration from discharge time 140 ns to 270 ns, and after discharge time 270 ns were different, their values were given respectively.

The formation rate r of [A*] due to the excitation of A is then calculated using Eq. (6). From the results shown in Fig. 2, one can get Zone a in which the value of ln (I) is liner to discharge time t. This fact suggested that the formation rate r of A* within the Zone a is also a function of [A*] under the experimental conditions. For convince, one sets $r = k_a^{'}[A^*]$. $k_a^{'}$ values were then listed in Table 1.

$$\frac{d[A^*]}{dt} = r - k'[A^*], \quad \frac{d\ln[I]}{dt} \propto \frac{d\ln[A^*]}{dt} = \frac{r}{[A^*]} - k'. \tag{6}$$

Table 1. Rate Constants for Each Species.

Species	$k_a^{'}$, [s^{-1}] (Zone a)	$k_b^{'}$, [s^{-1}] (Zone b)	$k_c^{'}$, [s^{-1}] (Zone c)
H	1.09×10^8	$6.03\text{x}10^6$	$6.03\ \text{x}10^6$
N	9.82×10^8	$1.27\ \text{x}10^7$	$8.21\ \text{x}10^6$
O	1.13×10^8	$3.05\ \text{x}10^7$	$7.83\ \text{x}10^6$

Applications

In environmental protection studies using plasmas, the radicals such as N, H, O, OH are important, for example, formation of NO_2 and oxygen radicals that contributes particulate matter (C_xH_y) oxidation to CO_2 (Eq. (7)), formation of N that contributes NO reduction (Eq. (8)).

$NO + O \rightarrow NO_2, \quad NO_2 + C_xH_y + NO \rightarrow C_{x-1}H_y + NO + CO.$ (7)

$N + NO \rightarrow N_2 + O.$ (8)

Considering the activation reactions in atoms, molecules, and electrons during the pulsed plasma discharge, the scheme of the excitation and de-excitation, and reactions could be represented (Fig. 3). The rate constants of formation and de-excitation of A* can be measured with the method presented here. The rate of recombination of A with B can be estimated with the help of the knowledge of kinetic studies such as using Chemkin the kinetic simulation software. The dissociation rate of A_x ($x \geqq 1$) can then be calculated as the sum of excitation and recombination of A. Here, the recombination conditions are important for the kinetic simulation. The authors have built a model for estimation of the reaction temperature in the discharge channel [8]. A further study is under taken to modify the plasma reaction model in which the rate constant of excitation is measured using the present method.

CONCLUSION

The characteristics of a pulsed plasma discharge in a gas mixture of CH_4-O_2-CO_2-N_2-H_2O have been studied. The formation and de-excitation rate constants of N (821.6 nm), H_α (656.2 nm), and O I (394~404 nm) were calculated to be about $10^8\ s^{-1}$ and $10^7\ s^{-1}$, respectively. The formation rate of each species was also a function of the concentration of their metastable atoms. The authors proposed that the formation and de-excitation rates could be used for the estimation of dissociation and recombination rate of the source atoms or molecules with the help of the knowledge of kinetic simulation.

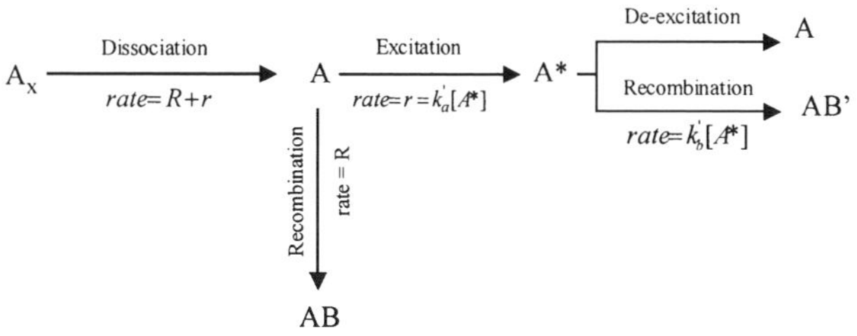

FIGURE 3. Rate estimation by measuring the de-excitation and recombination rate constant of A*. Here dissociation rate of A_x can be calculated from excitation rate r and recombination rate R.

ACKNOWLEDGEMENT

The Ministry of Education, Culture, Sports, Science and Technology supported this study. Yao is grateful to the New Energy and Industrial Technology Development Organization for a fellowship.

REFERENCE

1. K. L. L. Vercammen, A. A. Berezin, F. Lox, and J. S. Chang, *J. Adv. Oxid. Technol.* **2 (2)**, 312-329 (1997).
2. R. Dorai and M. J. Kushner, *J. Phys. D: Appl. Phys.* **35**, 2954-2968 (2002).
3. B. Eliasson and U. Kogelschatz, *IEEE Trans. Plasma Sci.* **19**, 1063–1077(1991).
4. B. M. Penetrante, *NATO ASI Series* **34(A)**, 65-89 (1993).
5. B. Sun, M. Sato, and J. S. Clements, *J. Phys. D: Appl. Phys.* **32**, 1908–1915 (1999).
6. C. G. Liu, A. Marafee, B. J. Hill, G. H. Xu, R. Mallinson, and L. Lobban, *Ind. Eng. Chem. Res.* **35 (10)**, 3295-3301 (1996).
7. S. Yao, M. Okumoto, J. Shimogami, K. Madokoro, T. Yashima, and E. Suzuki, *AIChE J.* in press.
8. S. Yao, E. Suzuki, and A. Nakayama, *Plasma Chem. Plasma Process.* **21 (4)**, 651-663 (2001).
9. M. Capitelli, C. M. Ferreira, B. F. Gordiets, and A. I. Osipov, *Plasma Kinetics in Atmospheric Gases*, Springer, 2000.
10. I. I. Fabrikant, O. B. Shpenik, A. V. Snegursky, and A. N. Zavilopulo, *Phys. Rep.* **159** (1&2), 1-97 (1988).

Applied Physics (APHYS 2003)
A. Méndez-Vilas (Ed.)

Critical Evaluation of Scattering Models Within the Full Band Monte Carlo Simulation Framework

M. Hjelm[1,2], H.-E Nilsson[1], A. Martinez[2]

[1]*Department of Information Technology and Media, Mid-Sweden University, SE-851 70 Sundsvall, Sweden*
[2]*Department of Microelectronics and Information Technology, KTH (Royal Institute of Technology), Electrum 229, SE-164 40 Kista, Sweden*

Abstract. The full band Monte Carlo (MC) simulation framework is regarded as the most accurate method available to study high-field carrier transport in semiconductors. Its potential has been demonstrated in a large number of studies over the years. In this work we focus on how the quantum mechanical uncertainty at high scattering rates affects the validity of Fermi's Golden Rule, which traditionally is the basis for the scattering handling in the MC method. Considering the uncertainty is important in for instance silicon carbide, which at moderate energies exhibits a scattering rate exceeding 10^{14} s^{-1}. The expression for time-dependent scattering rate is presented together with calculated rates for some initial states with acoustic as well as polar-optical phonon interaction. A first-order time-dependent algorithm for handling of scattering events in MC simulators is proposed.

INTRODUCTION

A MC simulation of a semiconductor is a computer model of the physical system constituted by charge carriers, crystal lattice (through the band structure), scattering mechanisms, electric field, and, if it is a device simulation, the electronic device with its contacts, borders, insulators, etc [1]. Normally, the handling of scattering mechanisms is based on Fermi´s Golden Rule, which is derived using first-order perturbation theory. In the derivation it is also assumed that the scattering events are taking place with a relatively long time interval, which motivates the usage of a δ function for the conservation of energy [2].

4H-SiC is a compound semiconductor with strong polar-optical scattering. For both holes and electrons the polar-optical scattering rate is about 10^{14} s^{-1} already at energies of about 0.15 eV. This means that about 10 % of the carriers, i.e. a significant proportion, undergo a scattering event within a time $\leq 10^{-15}$ s. According to the uncertainty relation this means an uncertainty in energy ≥ 0.6 eV.

We regard it as important to investigate the effect of this uncertainty on the energy distribution of semiconductors with high scattering rates. Especially is this true in impact ionization simulations, where a correct energy distribution is essential, and the uncertainty in energy may lead to increased probability for the carriers to be accelerated by the electric field to high energies by the electric field.

THEORY

We consider a carrier (electron or hole) in the initial state Ψ_a at the time $t = 0$. The particle is affected by a time-dependent perturbation with the matrix element $H_{b',a}$. As a result of the perturbation there is a probability $|c_{b'a}(t)|^2$ that the state is changed from the initial state a to the final state b'. This probability is according to the time-dependent perturbation theory

$$|c_{b'a}|^2 = \frac{2}{\hbar}|H_{b'a}|^2 \frac{t^2}{2} \cdot \frac{\sin^2(\omega_{b'a}t/2)}{(\omega_{b'a}t/2)^2}, \tag{1}$$

where $\omega_{b'a}$ is defined by

$$\hbar\omega_{b'a} = E_{b'} - E_a \pm \hbar\omega. \tag{2}$$

$E_{b'}$ and E_a are the final and initial state energy, respectively, while ω is the perturbation frequency. The + sign is for emission and the – for absorption. Assuming a constant matrix element, $|c_{b'a}(t)|^2$ is dominated by a peak for $|\omega_{b'a}| < 2\pi/t$. Clearly, this peak is broad for short times and its limit as *t* approaches infinity is a δ function, which is used in the derivation of the Golden Rule. In an MC simulator, we need to know the probability of change of state in an infinitely short time interval *dt*, i.e.

$$s_{b'a}(t) = \frac{d|c_{b'a}(t)|^2}{dt} = \frac{2}{\hbar^2}|H_{b'a}|^2 \cdot \frac{\sin(\omega_{b'a}t)}{\omega_{b'a}}. \tag{3}$$

Eq. 3 can be used in a precalculation step for a MC simulator to calculate the total scattering rate as a function of time. This calculation can be considered as a refinement of a numerical integration based on the Golden Rule, which is done for a large number of initial states (we are typically using around 10000 states for hexagonal SiC) in the irreducible wedge of the Brillouin zone (BZ), and where an even larger number of final states are used (we are typically using 100000 states in hexagonal SiC) [3]. The following observations can be done studying Eq. 3:

- The band structure contains a quasi-continuous range of final states with strongly varying density of states, which considerably influences the scattering rate as a function of energy.
- The energy range of possible final states is limited by the band gap, i.e. measuring the carrier energy relative the band gap, negative energies are not allowed. Assuming constant density of states and matrix element, this means that the scattering rate when time approaches zero is reduced to half the rate of long times.
- The rate has both positive and negative values, provided $\omega_{b'a} \neq 0$. In the interval $|\omega_{b'a}| < 2\pi/t$ it is positive for $|\omega_{b'a}| < \pi/t$, zero for $|\omega_{b'a}| = \pi/t$, and negative otherwise. We interpret the negative scattering rate as a probability of scattering from state b' to state a, i.e. a process opposite to the process when the rate is positive. This makes sense, since the negative rates only occur when $|c_{b'a}(t)|^2 > 0$.

TABLE 1. Scattering mechanisms, initial states, energies, and rates according to Golden Rule.

Interaction	Band	Initial k [2π/a]	Energy [meV]	Rate according to Golden Rule [s^{-1}]
Acoustic emission	3	[0, 0, 0.0229]	122	2.81×10^{12}
Acoustic emission	1	[0.1588,0.0289,0.0611]	250	6.68×10^{12}
Acoustic absorption	1	[0, 0, 0.0076]	0.5	2.35×10^{12}
Acoustic absorption	3	[0, 0, 0.0229]	122	2.83×10^{12}
Acoustic absorption	1	[0.1588,0.0289,0.0611]	250	2.07×10^{12}
Polar-optical emission	3	[0, 0, 0.0229]	122	1.22×10^{8}
Polar-optical emission	1	[0.1588,0.0289,0.0611]	250	9.74×10^{13}
Polar-optical absorption	1	[0, 0, 0.0076]	0.5	1.63×10^{12}
Polar-optical absorption	3	[0, 0, 0.0229]	122	1.58×10^{12}
Polar-optical absorption	1	[0.1588,0.0289,0.0611]	250	1.28×10^{12}

Calculation Results

The results presented below are based on the same band structure and matrix elements (including the overlap integral) as in Ref. 3. The results are obtained with a set of 1031625 states after scattering, all in the irreducible wedge of the BZ. In the calculation these states are transformed into every wedge that may be reached with a phonon-vector within the BZ. Table 1 gives a summary of the mechanisms, initial states, energies, and scattering rates calculated with the Golden Rule. For acoustic phonons the same dispersion relation as in Ref. 3 was used; for polar-optical phonons the energy was 120 meV.

Figure 1 shows two examples of the time evolution of the state calculated with Eq. 1 for a polar-optical emission scattering. The graph in a) shows a typical shape, which is found for most mechanisms and initial states, while the shape in b) represents a case representing a limited number of initial k-vectors. It demonstrates the relatively high probability to temporarily occupy states in regions with high density of states, although the total energy is not conserved. This effect can considerably influence the scattering probability for small times.

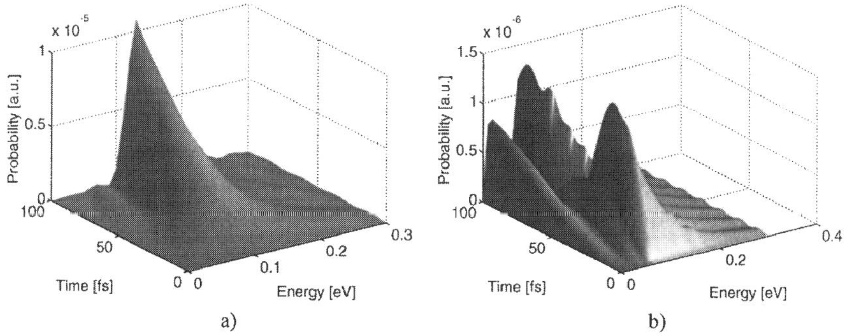

FIGURE 1. Time evolution in 4H-SiC of the probability of finding the electron at different energies as a result of the perturbation of a hole by a polar-optical phonon, energy = 0.120 eV. a) The initial state is k = [0.1588,0.0289,0.0611], band 1. b) The initial stat is k = [0,0,0.0229], band 3.

In Fig. 2 the scattering rate versus time is shown for acoustic and polar-optical phonons, both emission and absorption. The graphs for polar-optical phonons show a clear oscillatory shape, which appears in much smaller extent for acoustic phonons. We consider this as an effect of the matrix element dependence of $1/q^2$, where q is the phonon vector length, i.e. the scattering rate is much higher to a small portion of final states with a small energy range. For acoustic phonons the matrix element depends on q^2, which is counteracted by the overlap integral that in general is diminishing with the q vector length. Hence, the total rate is summed of a number of rates with different periods to different states, and no state is clearly dominating.

For acoustic phonons absorption with the initial energy 0.5 meV the time-dependent rate in the graph is considerably lower than according to the Golden Rule. We explain this with the small carrier energy, the loss of final states due to the broadening of the energy range, and that phonons with low energy are dominating. There is a high peak near 10 fs for polar-optical phonons emission from the third band near its minimum (in the Γ point), and the total rate afterwards oscillates around zero. In this case the initial carrier energy is only 2 meV higher than the phonon energy, and the rate calculated with the Golden Rule is very low due to the small number of possible final states. In the time-dependent calculation, considering the large energy broadening at short times leads to the high scattering rate.

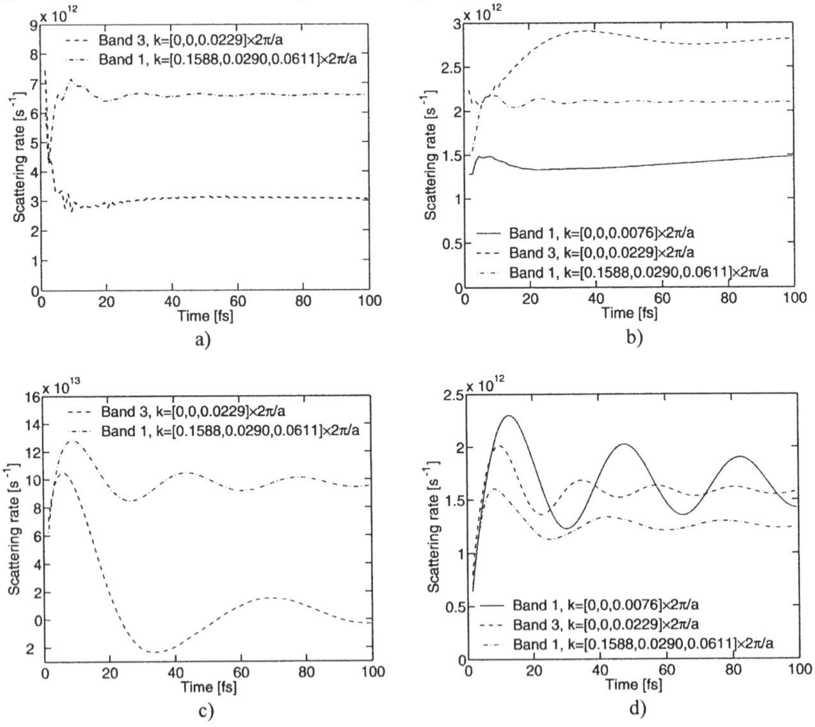

FIGURE 2. Scattering rate as a function of time for a) acoustic phonon emission, b) acoustic phonon absorption, c) polar-optical phonons emission, and d) polar-optical phonons absorption.

SKETCH TO A FIRST-ORDER ALGORITHM

An algorithm for selection of final state in an MC simulator can rely on an estimate of the time interval between the scattering events. When a scattering is handled, the time elapsed from the previous scattering event is known. In the simulator a random number is used to determine the time until the next event. Using this number an estimate of the time to the next scattering event can be made assuming constant scattering rate, i.e. disregarding the acceleration due to electric field. Because the time between the events is known, Eq. 1 can be used to produce the distribution of states. This means, that the anticipated energy distribution, when the next scattering handling is initialized, is used for the final state selection of the actual scattering. The detailed algorithm has to be based on the rejection technique, and can be considered as an extension of the algorithms presented in Ref. [4]. Since it in a simulator is difficult to handle a distribution with a very narrow peak, it is practical to have a minimum energy width where the final states are searched for long scattering time intervals.

CONCLUSIONS

Numerically integrated scattering rates versus time are presented for different initial *k*-vectors for acoustic and polar-optical phonon interaction. The rates are calculated with the band structure (and the corresponding overlap function) for 4H-SiC. There is a general agreement with calculations based on Fermi's Golden Rule. However, it is evident that, at certain points in the band structure, there are large differences due to the broadening of the energy spectrum at short times. At some of the *k*-points the particular shape of the density of states provides a splitting of the energy distributions into several peaks. There is also a clear oscillatory shape of the scattering rate for polar-optical scattering. An outline is presented for a first-order algorithm to be used in MC simulators for selection of state after scattering.

ACKNOWLEDGMENTS

We thank Dr. Ulf Lindefelt, Mid-Sweden University, for important discussions of the quantum mechanical basis of this paper. Furthermore, we gratefully acknowledge valuable financial support from the KK foundation, as well as the contribution from Sun Microsystems to the computational infrastructure at Mid-Sweden University.

REFERENCES

1. Jacoboni, C. and Lugli, P. *The Monte Carlo Method for Semiconductor Device Simulation* Vienna: Springer-Verlag, Vienna, 1989.
2. Lundstrom, M., *Fundamentals of carrier transport*, second edition, Cambridge: Cambridge Univer sity Press, 2000, pp. 41-45.
3. Hjelm, M., Nilsson, H.-E., Martinez, A., Brennan, K. F., Bellotti, E., *J. Appl. Phys.* **93**, 1099-1107 (2003).
4. Hjelm, M, and Nilsson, H.-E., *Simul. Pract. Theory* **9**, 321-332 (2002).

Applied Physics (APHYS 2003)
A. Méndez-Vilas (Ed.)

Magneto-acoustic and Barkhausen Emission in Wide Ribbons of One Magnetic Glass

Rafael J. López

Department of Physics , Universidad de Cantabria,39005,Santander,Spain

Abstract. The distribution function of the areas of Barkhausen jumps in wide ribbons of a magnetic glass (Metglas 2605SC) have been studied. The curves have been fitted to a power law with an exponential cutoff similar to that encountered in systems showing a self-organized criticality behavior. The high permeability and low coercivity for this very soft magnetic material requires the use of a low frequency (10^{-3}-10^{-5} Hz) of the applied AC magnetic field in the hysteresis loop in order to distinguish individual Barkhausen bursts. The size of the samples, 2 in. wide, allows piezoelectric sensors to be used in order to detect the Magneto-acoustic emission and the resonant waves produced by the magnetic field due to the large magnetoelastic coupling factor in this amorphous material.

INTRODUCTION

Barkhausen Emission (BE) pulse and Acoustic Emission detection techniques have been used as non-destructive techniques in several studies of the behavior of ferromagnetic materials[1]-[4] In the first case, the electromagnetic pulses caused by the sharp displacement of the walls of the magnetic domains are detected. In the second case, using piezoelectric sensors whose maximum sensitivity is in the hundreds of KHz., the elastic waves originated by the sudden changes of volume due to the magnetostriction of the material, also termed Magneto-acoustic Emission (MAE), are analyzed, normally in steels or nickel. Both techniques provide complementary information when studying the relation between BE and MAE with applied or residual stress, the effect of varying microstructures, structural defects,etc.

The industrial production of magnetic glass ribbons with high magnetomechanical coupling factors, K , led to new studies in the early 80s on the questions outlined above. One of the methods for determining the values of K is based on the measurement of the resonance and anti-resonance frequency for approximately one-dimensional specimens in which simple longitudinal oscillations arise. A coil surrounds the wire and is excited by a small AC current of varying frequency. The magnetic response of the material is detected by another small coil allowing the apparent permeability and its frequency dependence to be obtained [5]-[7]. The resonance frequencies measured are typically within the range of tens of KHz for amorphous ferromagnetic material wires or very narrow ribbons in the order of several centimeters.

At the same time, studies of the Barkhausen effect were intensified after the work of Per et al.[8] on the dynamics of disordered complex systems which show a behavior which has come to be known as self-organized criticality. The system of magnetic domains of ferromagnetic materials, easily controllable through the application of external magnetic fields, turned out to be an optimal candidate for verifying the validity of the theoretical hypotheses. Thus, there was a real 'avalanche' of work on this area in the 90s [9]-[19]. Ferromagnetic glass ribbons, with their high permeability and low coercivity were quickly incorporated into these studies analyzing the statistical properties of Barkhausen emission pulses [15]

The availability of amorphous ferromagnetic ribbons with widths in the order of some centimeters allows piezoelectric sensors normally used in MAE to be incorporated in the measurement techniques of the properties of these ribbons. The aim of this work is to study the information provided by these sensors in the frequency range of tens of KHz and to compare this data with that provided by inductive sensors when studying magnetoelastic resonances and Barkhausen pulse emission.

EXPERIMENTAL TECHNIQUES

The experimental set-up is made up of two independent installations in order to measure the electromagnetically induced signal and the magnetoelastic waves produced when there are variations in the magnetic field applied (Happl). This set-up is similar to a previous work [20].

The band of amorphous ferromagnetic material, Metglas 2605 SC from Allied Signal is 2 in. wide and 0.6 thousandths of an inch thick and was cut first in a square form and then in a fifteen cm. rectangular form. Given these dimensions, the generation of the magnetic field is performed either with 30 cm radius Helmholtz coils with the specimen placed in the centre and with the possibility of rotating on the horizontal plane, or with a magnetic yoke system with a coil wound around a bracket and the amorphous ribbon closing the magnetic circuit.

The feed signal is generated by a HP-3325B synthesizer and a bipolar Kepco source is used in cases where a higher power is required to magnetically saturate the specimen.

A 200 turn detection coil is wound around the sample and another identical one is wound in the opposite direction below the ribbon. In those cases where the sensor does not need to change its orientation as the ribbon rotates 90° with respect to the Happl. , a coil of 100 turns, with a diameter of 20 mm. and a thickness of 6 mm. is placed as a disk parallel to the specimen surface.

The signal can be sent from both types of coils either to a DSP Lock-in amplifier Model 7220 from EG&G for the analysis of the inductive resonances, or, for Barkhausen studies, to a low-noise preamplifier SR560 fed with batteries and connected to a digital oscilloscope, Tektronix TDS520,which stores the pulses .

The magnetoelastic resonances and the general acoustic emission pulse response are detected by means of piezoelectric sensors in the range of tens of KHz, an R6 from Physical Acoustics Corporation or a VS30 from Vallen-Systeme GmbH, together with a preamplifier P.A.C. 1220 with a gain of 60 dB and filters in the range of 20-100 KHz. Their signals can also be connected to another DSP Lock-in SR 850 or to the oscilloscope.

All of the tests were carried out at room temperature and with the specimen in the cast form, with no heat treatment. The manufacturing process leads to anisotropies along the longitudinal and transversal axes, underlined in the properties of the BE and MAE signals when the orientation of the specimen changes with respect to the Happl. field.

RESULTS

Figure 1 shows a comparison of the Barkhausen signals obtained for a square Metglas specimen located at the centre of the Helmholtz coils when the Happl. field is situated along the axis of the ribbon (X axis) or in a direction perpendicular to it (Y axis), tangent to the surface. The driving signal was triangular in shape and was in periods of 100 seconds. The detector coil was wound around the specimen and the maximum retention mode was used in the oscilloscope digitalising at125 Samples/s. The anistropy of the ribbon is highlighted by observing how the maximums are of a higher value for Happl. oriented along the X axis.

At this point, it is worth pointing out some of the main differences between the Barkhausen and AE signals associated with magnetostriction. The domain wall movements which imply variations in volume with a change in orientation of 180° should not originate elastic wave signals, while the inductive signal will be at the maximum. When the changes in orientation are of 90°, mechanical waves will appear in response to the magnetostriction of the material, but the induced signals will be of a lower amplitude.

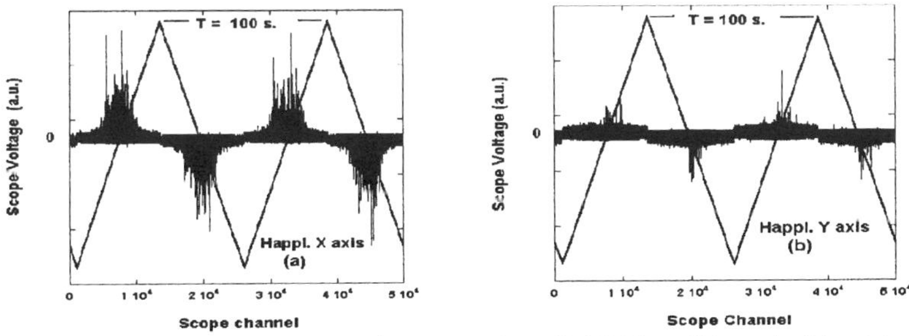

FIGURE 1. Barhausen signal obtained with a magnetic applied field parallel to axis ribbon **(a)**, or in a transversal direction **(b)**.

In order to discriminate these Barkhausen pulses individually, the Happl. was slowly varied, with frequencies of 10^{-3} to 10^{-5} Hz. To be able to perform the experiment in a reasonable time, one part of the cycle was performed at 10^{-2} Hz and when the field approaches values at which the BE begins to be detectable, the generator was set at a lower frequency. The yoke type device was used with the Metglas ribbon closing the magnetic circuit and feeding directly with the synthesizer which provides a signal of greater spectral purity, without the noise introduced by the bipolar source. In these soft magnetic materials, simple pulses are easy to interpret, but in general the avalanches are complex, even when dealing with very low frequencies.

_Figure 2 shows an example of the distribution of the areas of Barkhausen pulses. 2000 digitalised waves were taken at 2.5 MSamples/s. and at an interval of the Happl. in which the relative permeability remains approximately constant. The distribution function, the probability with which the area of a pulse is greater than a given value, was fitted to a potential-exponential function $P(x) = A\ X^{-B}\ \exp(-C\ X)$, where A, B and C are constants. The adjusted values of B are in the order of 0.1.

The study of the resonances of the magnetoelastic waves was carried out by subjecting the specimen to an alternate weak Happl field, lower than that required to reach magnetic saturation. One rectangular specimen was excited by a coil situated at one of its ends, with frequencies in the range of tens of KHz. Given the strong dependence of the magnetic properties of these amorphous ribbons on the applied stresses, a simultaneous measurement set-up was selected with piezoelectric and inductive sensors connected to two lock-ins. In this way, the possible effects derived from placing the AE sensor by gravity on the specimen should be evident in the two independent installations. The possible existence of intrinsic resonances in the detectors was verified by repeating the experiments performed on Metglas with a non-magnetic material.

Figure 3 shows a comparison of the results obtained using the two types of sensors at between 25 and 50 KHz with an interval between measurements of 125 Hz and connected to the two lock-ins. The similarity between the maximums can be

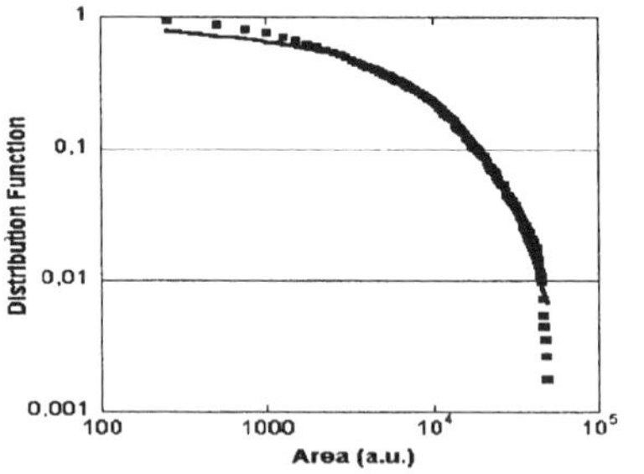

FIGURE 2. Cumulative distribution function of Barkhausen jump areas.

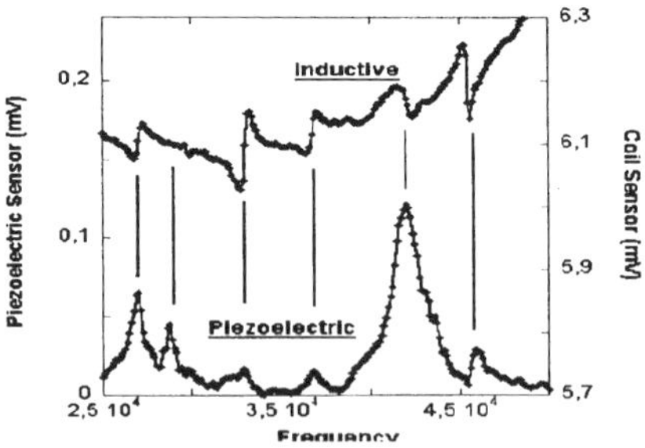

FIGURE 3. Simultaneous measurement of magnetoelastic resonances by using a coil and a piezoelectric sensor. Vertical lines for use as visual reference.

observed in the form of the peak of the piezoelectric sensor with the resonance-antiresonance characteristics of the inductive sensor. The AE sensor shows quite clear peaks with a good relation between the maximum value and the base line.

The lower frequency limit of the piezoelectric detection system used (20 KHz in our case) will determine the resonances which can be studied.

A more in-depth study will allow a modal analysis of this lamina considered as two-dimensional

CONCLUSIONS

- The anisotropies of wide Metglas 2605SC ribbons without heat treatment after manufacture can be identified with Barkhausen pulse detection techniques. These signals depend on the longitudinal or transversal direction of application of the magnetic field.
- Barkhausen pulses have complex profiles up to sweep frequencies of 10^{-5} Hz. The distribution functions of the pulse areas for this material can be satisfactorily adjusted by means of potential-exponential type curves.
- The width of the ribbons has made it possible to apply simultaneously piezoelectric and inductive sensors to the study of the magnetoelastic resonances from around 20 KHz. The signals from the piezoelectric sensors show clearly defined peaks in these resonances.
- An understanding of the effects of the magentoelastic resonances on the piezoelectric sensors can help in the interpretation of the classic Magneto-acoustic Emission results for other ferromagnetic materials.

REFERENCES

1. M.Shibata and K.Ono, Proc. of the Inst. of Acoustics Conference,Chelsea College,London (1979)
2. D.J.Buttle,C.B.Scruby,G.A.Briggs and J.P.Jakubovics , Proc.R. Soc.Lond. **A 414**, 469-497 (1987)
3. J.Kameda and R.Ranjan , Acta metall. **35**, n°7 , 1515 (1987)
4. S.Tyagi,JSteinberg,A.E.Lord, Jr. and P.M.Anderson III , Phys. Stat. Sol. (a) **64** ,443 (1981)
5. H.T.Savage and M.L.Spano , J.Appl.Phys. **53** (11) ,8092 (1982)
6. L.T.Kabacoff , J.Appl.Phys. **53** (11) ,8098 (1982)
7. P.M.Anderson III, J.Appl.Phys. **53** (11) ,8101 (1982)
8. P.Bak,C.Tang and K.Wiesenfeld, Phys.Rev.A **48** ,n°1, 364 (1988)
9. H.Yamazaki,Y.Iwamoto and H.Maruyama, Journal de Physique, **C8** ,Tome 49 , 1929 (1988)
10. L.P.Kadanoff,S.R.Nagel,L.Wu and S.Zhou, Phys.Rev A **39**,n°12 6524 (1989)
11. L.J.Swatzendruber,L.H.Bennet,H.Ettedgui and I.Aviram , J.Appl. Phys. **67**(9),5469 (1990)
12. B.Alessandro,C.Beatrice,G.Bertotti and A.Montorsi, J.Appl.Phys. **68** (6) , 2908 (1990)
13. X.Che and H.Suhl , Phys. Rev. Lett. **64** ,n° 14 , 1670 (1990)
14. O.Geoffroy and J.L.Porteseil, J.Magn.Magn.Mat. **97** , 205 (1991)
15. L.V.Meisel and P.J.Cote, Phys. Rev. B **46** n° 17 ,10822 (1992)
16. R.D.McMichael,L.J.Swartzendruber and L.H.Bennet, J.Appl.Phys. **73**(10),5848(1993)
17. J.S.Urbach,R.C.Madison and J.T.Markert, Phys. Rev. Lett. **75** n°2, 276(1995)
18. D.Spasojevic,S.Bukvic,S.Milosevic and H.Stanley, Phys. Rev. E **54** ,n°3 2531 (1996)
19. *KA..Dahmen,J.P.Sethna,MC..Kuntz and O.Perkovic, J.Magn.Magn.Mat 226-230,1287 (2001)*
20. R.Lopez Sanchez,M.Lopez,M.Armeite,R.Piotrkowski,J.Ruzzante,30thQNDE,Green Bay,(2003)

Template Mediated Nanofibrous Structure: Novel Chitosan/Polyethylene Glycol Scaffold for Tissue Engineering

J. WEN WANG* AND M. HSIUNG HON

Department of Materials Science and Engineering, National Cheng-Kung University, Tainan 70101, TAIWAN, ROC

Abstract. Biodegradable polymers have been widely used as scaffolding materials to regenerate new tissues. To mimic natural extra cellular matrix architecture, a novel three-dimensional fiber matrix with a fiber diameter ranging from 20 to 200 nm, has been created from chitosan/polyethylene glycol blending solution in this study. These nano-fiber matrixes were prepared from the polymer solutions by a method of template assisted process involving thermally induced gelation. The growths of these fibers were to be guided by infiltrating the polymer solution into the nanochannel of an anodic aluminum oxide (aao) template in contact with a silicon substrate with vacuum suction. The effects of polymer concentration, polymer molecular weight and channel diameter on the nano-scale structures were studied. In general, at a low chitosan/polyethylene glycol concentration (1wt%) and low chitosan molecular weight (150000), a directional nano-fiber matrix was formed. Under the conditions for nano-fiber matrix formation, the average fiber diameter (20–250 nm) was changed statistically with nanochannel diameter and the fiber length was changed with the polymer concentration and nanochannel diameter. This synthetic analogue of natural extra cellular matrix combined the advantages of biodegradable polymers and the nano-scale architecture of extra cellular matrix, and may provide a better environment for cell attachment.

INTRODUCTION

Chitosan is a biodegradable polymer that has been widely studied for biomedical, ecological, and industrial applications [1]. This polymer is known to be nontoxic, odorless, biocompatible in animal tissues, and enzymatically biodegradable [2]. However, its rigid crystalline structure, poor solubility in organic solvents and poor process ability has limited it to be utilized widely. In order to resolve these problems, chitosan-PEG polyblend was prepared by using PEG 6000 molecular weight in our group [3]. The result showed that by choosing PEG molecular length in appropriate range to blend with chitosan not only could improve the water affinity but also accelerate the degradation rate of polyblend. However, chitosan is easily dissolved in weak acids, therefore a modification on chitosan-PEG to obtain partially soluble devices is preferable. We have demonstrated that reaction of amino group of chitosan with aldehyde or keto group of low molecular weight reduced sugar (i.e. glucose and fructose) resulting in a partially soluble membrane in low pH environment. Thus, a scaffold made of glucose mediated chitosan/PEG gel should be mechanically stable and cable of functioning biologically in the implant site.

The typical scaffold fiber diameters approach 10 μ m, which is comparable to the diameter of a cell. However, in connective tissue, extracellular matrices (ECMs) are composed of ground substances and fibrous proteins. The fibrous proteins, collagens, embedded in ground substances maintain structure and mechanically stability. The collagen fibrous structure is organized in a fiber network composed of collagen fibers that are formed hierachically by nanometer-scale fibrils [4]. Therefore, ideally the dimensions of the building fibers of the scaffold should be on the same scale with those of natural ECM. Recently, synthetic biomimetic ECMs with nanofibrous structure similar to that of native tissue have been fabricated and cell adhension on the ECMs have been observed. It was found that the neurite cell anchors on the poly-(lactic acid) nanofiberous ECM in a similar way as that of native tissue[5].

Constituents of the natural (ECMs) exhibit fiber diameters that are in the range of 50-150 nm, a diameter far smaller than can be achieved with conventional processing strategies [6]. A nanostructure of biopolymer can be carried out both by chemically and electrochemically with a hard and soft template. Examples of hard templates include polycarbonate and anodized alumina or soft templates, such as surfactants, micelles, liquid crystals and polyacids are reported to be capable of directing the growth of biopolymer nanostructures with diameters smaller than 100 nm [7]. Physical methods, including electrospinning and mechanical stretching have also been used to prepare the nanofiber with diameters ranging from 100 to 900 nm[8].

In this study, the nano-fiber matrixes were prepared from the glucose mediated chitosan/PEG polymer gel by a method of template assisted process involving thermally induced gelation. The growths of these fibers were to be guided by infiltrating the polymer solution into nanochannel of an anodic aluminum oxide (AAO) template in contact with a silicon substrate with vacuum suction. At a low chitosan/polyethylene glycol concentration (1wt%) and low chitosan molecular weight (150000), a directional nano-fiber matrix was formed. The average fiber diameter (20–250 nm) was changed statistically with nanochannel diameter.

EXPERIMENTAL

In a typical synthesis, chitosan solutions with concentrations of 1 wt% were prepared by dissolving in 1% acetic acid. The mixture was stirred for 24 hr to obtain a perfectly transparent solution. Chitosan-PEG blend was prepared by mechanical stirring the filtered chitosan and PEG flakes in a weight percentage of 70:30 at room temperature. The 0.8wt% amount of glucose was then added with continued stirring until the solution was clear again at 120℃. After curing for 1hr at 68℃, the homogeneous aqueous solution of reaction mixture was prepared for fabricating nanofiber. Glucose mediated chitosan/PEG nanofibrous matrixes were fabricated by template-assisted process using the apparatus schematically shown in Fig 1. The AAO template with diameter 1.3 cm and 60μ m in thickness was adhered on silicon substrate by double adhesive then placed in a 20mL glass bottle. Polymer solution was infiltrating into the nanochannel of the AAO template in contact with a silicon substrate with vacuum suction for 0.5, 1 and 4hrs. The AAO template was removed by

0.5 M NaOH solution then rinsed with deionized distill water (DDW) twice. Brown glucose mediated chitosan/PEG nanofibrous matrixes are obtained by thermally drying at 68℃ for 1hr. Nanofibers are collected characterization with SEM, FT-IR and UV/vis spectroscopy.

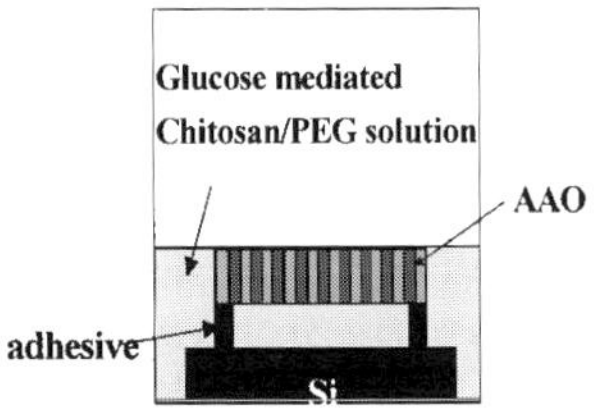

FIGURE 1. The apparatus designed for preparing glucose mediated chitosan/PEG nanofiber

RESULTS AND DISSCUSSION

Figure 2 shows the FT-IR spectra of chitosan, glucose mediated chitosan/PEG nanofiber. The spectrum of the chitosan exhibits an absorption around 1655cm^{-1} and 1590 cm^{-1}, which represent the amide I and amide II bands respectively. There are bands at 1420, 1380 and 1320 cm^{-1}, in addition to the usual C-H aliphatic band at 2880 cm^{-1}. The OH and NH_2 overlapping bands are found around 3310-3450 cm^{-1}. Glucose mediated chitosan/PEG nanofiber shows a deformation peak around 1460 cm^{-1}. It is proposed that the mechanism of crosslinking between the adhyde and the free amine on chitosan follows the Schiff's base reaction that results in C=N formation.

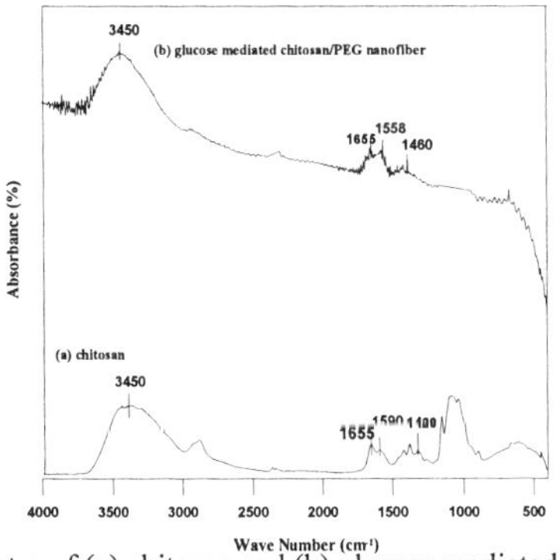

FIGURE 2. FT-IR spectra of (a) chitosan and (b) glucose mediated chitosan/PEG nanofiber

Figure 3 (a)(b) and (c) show a typical scanning electron microscopy (SEM, JEOL 5120) image of the glucose mediated chitosan/PEG nanofiber prepared in 200, 100 and 20 nm AAO template by vacuum suction. One directional orientation nanofibers with diameter in the range of 20 to 250 nm were obtained. The nanofibers tend to agglomerate into bundles and the tip of nanofibers clump together because of physical adsorption on the surface of the nanofibers. The diameters of the nanofibers are corresponding to the diameter of the nanochannel. The lengths of these nanofibers show that they don't span the complete thickness of the AAO template. According to our priminary study, the critical time for infiltrating the low molecular weight glucose mediated chitosan/PEG solution into AAO nanochannel is about 4hr and the vacuum suction pressure is fixed at 10^{-4} atm. After 4hr suction duration, polymer solution could not infiltrate the nanochannel further. This is attributed to the fact that AAO pore walls are negatively charged while the glucose mediated chitosan/PEG polymer sols are positively charged. Therefore, the sol particles preferentially adsorb and grow on the pore walls. But the grow process (gelation reaction) proceed too fast resulting in pore blockage at the AAO template surface before fibers growth within the nanochannel. It may be the reason for fiber length don't span the complete thickness of the template (60μ m).

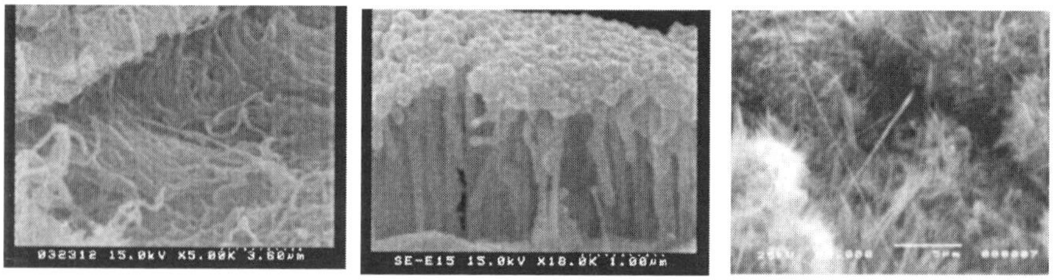

(a) (b) (c)

FIGURE 3. SEM image of glucose mediated chitosan/PEG nanofiber in (a)200, (b)100 and (c)20 nm AAO template by vacuum suction.

Figure 4 shows the UV/vis (spectrophotometer HITACHI U-2001) spectra of the starting glucose mediated chitosan/PEG powder and the final nanofiber in DDW. The adsorption band of nanofiber exhibits a blue-shifted in relation to the powder form of glucose mediated chitosan/PEG. It is clear that the nanostructure significantly change the characteristic visible spectrum of the glucose mediated chitosan/PEG powder. It also demonstrated that the glucose mediated chitosan/PEG nanofibers can be obtain by template assisted process with vacuum suction.

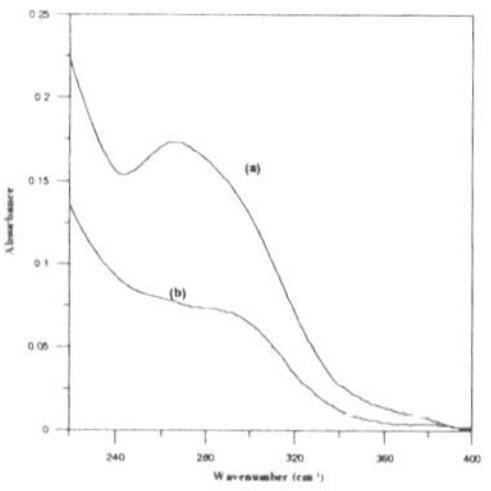

FIGURE 4. UV/visible spectra of the (a) starting glucose mediated chitosan/PEG powder and (b) the final nanofiber in DDW.

CONCLUSIONS

A novel glucose mediated chitosan/PEG nanofibers was prepared successfully by using a method of template-assisted process with vacuum suction. Nanofibers were synthesized in the pores of a nanoporous AAO membrane. The electron microscopy results showed that the template technique using an AAO membrane can control the length and diameter of the glucose mediated chitosan/PEG nanofibers, thus producing monodisperse nanofibers. The composition of the obtained nanofibers can also be estimated by FT-IR and UV/vis spectroscopy. The result shows that the mechanism of crosslinking between the adhyde and the free amine on chitosan follows the Schiff's base reaction that results in C=N formation and the nanofiber exhibits a blue-shifted in relation to its powder form. The designed length and diameter of the nanofiber can be a candidate for fabricating the ECMs.

ACKNOWLEDGMENTS

The authors gratefully acknowledge the financial support from the National Science Council of TAIWAN, ROC. (Grant No. 89-2216-E-006-072)

REFERENCES

1. Shigemasa Y., Morimoto M., Saimoto H., Okamoto Y., and Minami S., "Application of chitin and chitosan for biomaterials," in *Advance in Chitin Science*, edited by H. C. Chen. Lancaster:Technomic publishing, 1997, pp. 3-21.
2. Borzacchiello A., Ambrosio L., Netti P. A., Nicolais L., Peniche C., Gallardo A. and San Roman J., *J. Mater. Sci.-Mater. Med.***12**, 861-865 (2001).
3. Wang J. W. and Hon M. H., *J. Biomat. Poylmer. Sci. Polym. Ed.* **14**, 119-137 (2003).
4. Andrzej F., Wendy B. H. and Frank K. K., *J. Biomed. Mater. Res.* **57**, 48–58 (2001).
5. Kim B. S. and Mooney D. *J. Biomed. Mater. Res.* **41**, 322-332 (1998).
6. Gary E. W., Marcus E. C., David G. S., and Gary L. B., *Nano. Lett.* **3**, 213-216 (2003).
7. John C. H. and Charles R. M., *J. Mater. Chem.* **7**, 1075-1087 (1997).
8. Andrzej F., Wendy B. H. and Frank K. K., *J. Biomed. Mater. Res.* **57**, 48–58 (2001).

Applied Physics (APHYS 2003)
A. Méndez-Vilas (Ed.)

$CoSi_2$ formation with a thin Ti interlayer-Ti capping layer and Ti capping layer

A. Abdul Aziz[1], C.O. Lim[2], Z. Hassan[3], Z. Jamal[4]
[1] School of Physics, University Science Malaysia, 11800 Penang, Malaysia; lan@usm.my
[2] School of Physics, University Science Malaysia, 11800 Penang, Malaysia; co_lim@silterra.com
[3] School of Physics, University Science Malaysia, 11800 Penang, Malaysia; zai@usm.my
[4] School of Microelectronics Engineering, Northern Malaysia University College of Engineering, 02600 Perlis, Malaysia;zulazhar@kukum.edu.my

Abstract. In this work, sheet resistance measurements were used to characterize the progress of silicidation for $CoSi_2$. Two types of samples were prepared (a bi-layer of Ti/Co on Si; and a tri-layer of Ti/Co/thin Ti interlayer on Si). The influence of an additional thin Ti interlayer in between silicon and cobalt were investigated and compared to a Ti/Co/Si system. From XRD, it is shown that the presence of the thin Ti interlayer promotes preferential $CoSi_2$ crystal growth in the (111) direction. $CoSi_2$ grain size annealed at 750°C are the same from SEM top view images although annealing at 600°C yield bigger grains of CoSi with the presence of a thin Ti interlayer.

Keywords: Cobalt disilicide, sheet resistance, X-Ray Diffraction (XRD), Rutherford Backscattering (RBS), SEM top view.

Introduction

The preference to use $CoSi_2$ in the self-aligned silicide (SALICIDE) scheme is because of its low resistivity, good thermal stability and absence of nucleation problems in narrow lines on very large scale integrated circuit device processing **[1, 4, 5, 6, 8, 9]**. Techniques developed to obtain epitaxial $CoSi_2$ include molecular beam epitaxy (MBE), ion beam synthesis (IBS) and solid phase epitaxy (SPE). Although $CoSi_2$ may be formed epitaxially on Si due to its small lattice mismatch (-1.2%), the standard Co/Si SPE reaction often resulted in polycrystalline $CoSi_2$ **[2, 3, 7, 11, 13, 14]**.

The presence of a Ti interlayer is known to have promoted epitaxial growth of $CoSi_2$. As device geometry is continuously scaled down, demand for shallower junction increases as well. Issues like interface roughness, resistivity and thermal stability of the silicide layer are even more important **[1, 6, 10, 13, 14]**. Thus, formation of epitaxial $CoSi_2$ is favourable.

In this paper, the difference between a Ti cap SPE cobalt silicide and a thin Ti interlayer-Ti cap SPE cobalt silicide were investigated with sheet resistance measurements, Rutherford back scattering spectroscopy (RBS), X-ray diffraction (XRD) and SEM top view.

Experiment

In this work only p-type Si (100) substrate were used. Samples were dipped in buffered HF prior to metal deposition in dc magnetron sputtering Ti or Co chambers with base pressure at 10^{-9} Torr. Both bi-layer (150Å cobalt capped with 80Å titanium on silicon) and tri-layer (10Å of Titanium, 150Å cobalt capped with 80Å titanium on silicon) samples were deposited without breaking vacuum in a PVD cluster tool with sequential metal deposition processing. Isochronal RTP annealing (60s, N_2 ambient) was performed for both bi-layer and tri-layer samples to study the silicidation reaction. After annealing, the samples were carefully etched with H_2SO_4:H_2O_2 –based solution to remove the unreacted Ti layer. From the sheet resistance measurements, XRD, RBS and SEM top view were used to identify phases of cobalt silicide, preferential crystallographic orientation and grain size at 600°C and 750°C.

Results and discussions

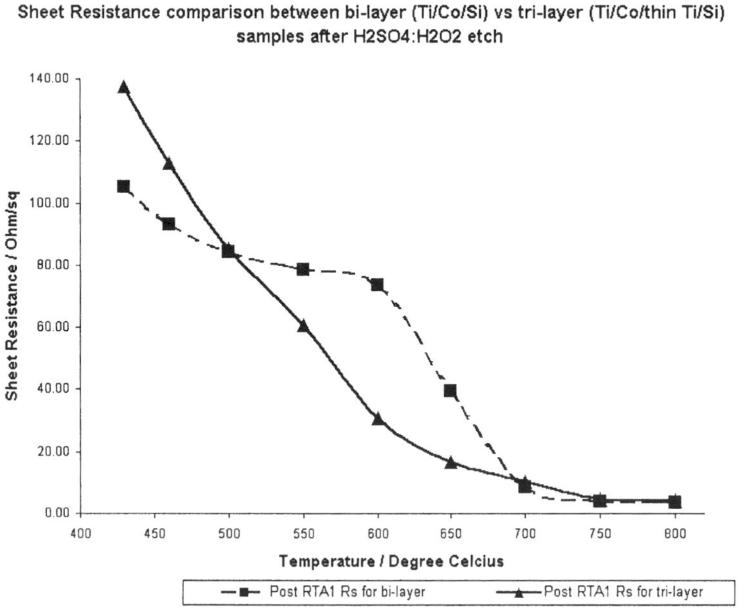

Fig. 1 : Sheet resistance versus annealing temperature range from 430°C to 800°C for a duration of 60 seconds each for both bi-layer and tri-layer samples.

From Fig. 1, it is obvious that the presence of a thin Ti between Co and Si in the tri-layer samples promote silicidation between 500°C and 700°C. Together with the XRD & RBS results, it is shown that high resistive CoSi forms at relatively low temperature (between 430°C and 600°C) as the precursor to $CoSi_2$ for both types of samples. At higher temperatures (more than 700°C), $CoSi_2$ formed as the low resistive phase. In between these two regions, there exists a region of rapid lowering of sheet resistance for the bi-layer samples. This region occur between 600°C and 700°C, indicating that between these two temperatures either a mixture of metastable CoSi and $CoSi_2$ co-exist whilst the thermal condition continues to favor $CoSi_2$ nucleation and formation **[1, 4, 7, 12–14]**.

For the tri-layer samples, no rapid lowering of sheet resistance effects was observed. The rapid lowering of sheet resistance is replaced by a continuous reduction in sheet resistance. At temperature region of 500°C to 700°C, sheet resistance for tri-layer samples are instead lower than the bi-layer samples. This observation confirms that the thin Ti interlayer lowered the $CoSi_2$ nucleation temperature. Samples from both bi-layer and tri-layer that were annealed at 600°C and 750°C respectively were identified for SEM top view, RBS and XRD.

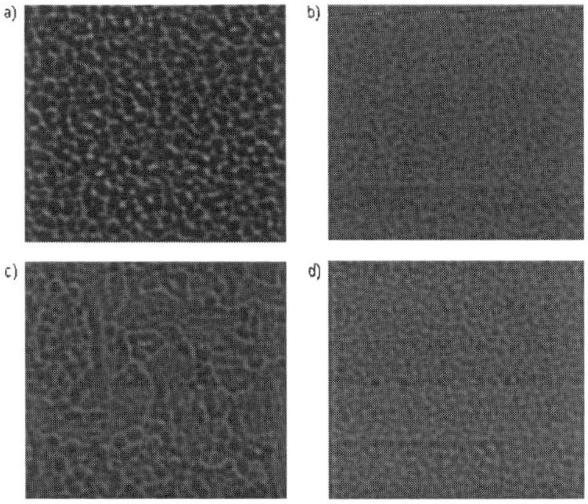

e)

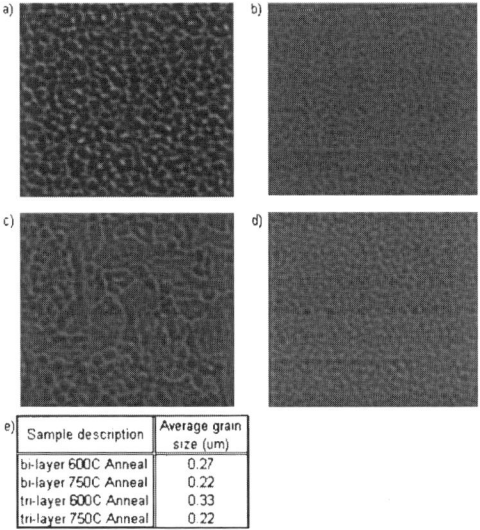

Sample description	Average grain size (um)
bi-layer 600C Anneal	0.27
bi-layer 750C Anneal	0.22
tri-layer 600C Anneal	0.33
tri-layer 750C Anneal	0.22

Fig. 2 : SEM top view images (50kX magnification with Ga^+) of (a) bi-layer 600°C Anneal (b) bi-layer 750°C Anneal (c) tri-layer 600°C Anneal (d) tri-layer 750°C Anneal (e) average grain size.

Fig. 2 (b) and 2 (d) shows that there is no effect from the thin Ti interlayer at 750°C anneal. The grain size on bi-layer and tri-layer sample type annealed at 750°C is similar. However, annealing at 600°C yields a bigger grain size at 0.33μm for a tri-layer sample. Comparing data on Fig. 2 (a) and 2 (c), it is observed that the presence of a thin Ti interlayer resulted in less nodular grains thus bigger grain size. This probably is due to the stronger preferential growth of CoSi in the (200) direction for the bi-layer annealed at 600°C (from XRD results). These SEM images (50kX magnification, 30kV, 51pA) were obtained by first sputtering the surface of the silicided samples with Ga+ ions.

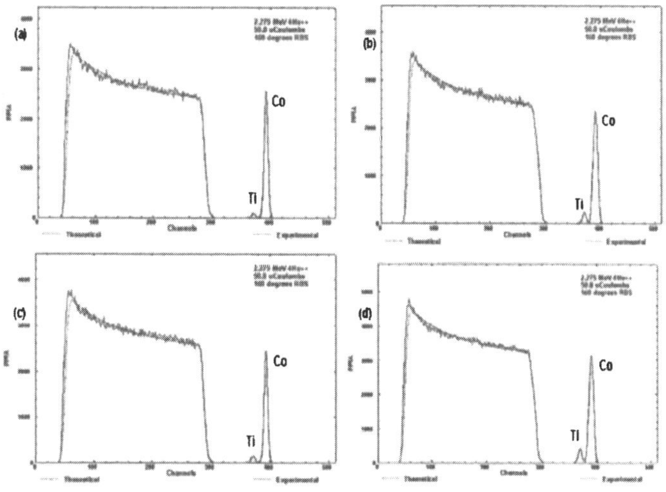

Fig. 3 : RBS Spectra (random) for (a) bi-layer 600°C, (b) bi-layer 750°C; (c) tri-layer 600°C (d) tri-layer 750°C. RBS Spectra performed at 2.275MeV $^4He^{2+}$, 50μC, Normal Detector Angle 160° RBS.

From Fig. 3, the RBS spectra showed that annealing at 600°C for bi-layer and tri-layer wafers strongly yield CoSi (Si/Co ratio at about 1), whereas when annealed at 750°C yield $CoSi_2$ (Si/Co ratio at about 2) for both bi-layer and tri-layer wafers.

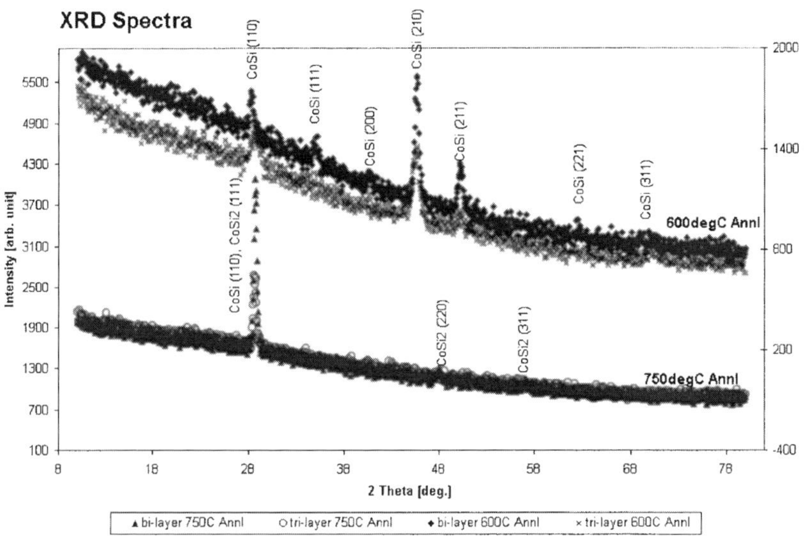

Fig. 4: XRD spectra

From Fig. 4, annealing either bi-layer or tri-layer wafers at 600°C yields CoSi and a 750°C anneal yields $CoSi_2$. No CoTi peak and no $CoSi_2$ peak were observed at 600°C for both samples. At annealing temperature of 600°C, the bi-layer sample showed a strong preferential growth of CoSi in the (210) direction besides the (110), (111) and (211) direction. Tri-layer sample also has a preferential growth of CoSi in the (210) direction albeit a weaker one relatively besides the (110), (211) and (311).

When samples are further annealed at 750°C, the bi-layer sample showed a strong preferential growth of $CoSi_2$ in the (111) direction and observation from fig. 4 also revealed the $CoSi_2$ existence in the (220) and (311) direction. This is not an issue for the tri-layer sample annealed at 750°C, showing preferential growth of $CoSi_2$ only in the (111) direction. CoSi peaks were also observed for both bi-layer and tri-layer sample when annealed at 750°C since CoSi phase is the most stable compound.

Thus from the XRD results, it can be deduced that the thin Ti interlayer influences strongly in the crystal orientation growth. Also both bi-layer and tri-layer silicided samples showed peaks for CoSi indicating that precipitation of CoSi occurred at 750°C.

4. Conclusion

Cobalt silicidation has been studied using 2 type of samples namely a bi-layer, Ti/Co/Si and a tri-layer film, Ti/Co/Ti/Si. Using a thin Ti interlayer, the nucleation temperature of $CoSi_2$ was reduced as shown with sheet resistance measurement. From the XRD results, the resultant CoSi phases differ between the silicided bi-layer and silicided tri-layer although both are by ratio relatively CoSi. $CoSi_2$ formed at 750°C for bi-layer sample was polycrystalline in nature but $CoSi_2$ formed at

750°C for tri-layer sample has a preferred crystal orientation **[2, 3, 11, 13, 14]**. SEM top view shows the nodular nature of the $CoSi_2$ grains for 750°C anneal regardless of bi-layer or tri-layer silicided samples.

Acknowledgements

The authors would like to acknowledge the support of *School of Physics, USM* under an IRPA RMK-8 Strategic Research Grant and also thank the management of *SilTerra (M) Sdn Bhd.* for the facilities to support this study.

References

[1] R.T.Tung and F. Schrey, Mat. Res. Soc. Symp. Proc., Vol. 402 (1996), p.173
[2] C. Detavernier, R. L. Van Meirhaeghe, F. Cardon, K. Maex, H. Bender and Shiyang Zhu, J. Appl. Phys. , 88 (2000), p. 133
[3] C. Detavernier, R. L. Van Meirhaeghe, F. Cardon and K. Maex, Thin Solid Films 386 (2001), p.19-26
[4] Sofia Hatzikonstantinidou, Peter Wilkman, Shi-Li Zhang and C. Sture Petersson, J. Appl. Phys. 80 (1996), p. 2
[5] Chang-Yong Kang, Dae-Gwan Kang and Joo-Wan Lee, J. Appl. Phys. 86 (1999), p. 9
[6] Dong Kyun Sohn, Ji-Soo Park, Byung Hak Lee, Jong-Uk Bae, Jeong Soo Byun and Jae Jeong Kim, Appl. Phys. Lett., 73 (1998), p. 16
[7] Karen Maex, Anne Lauwers, Paul Besser, Eiichi Kondoh, Muriel de Potter and An Steegen, IEEE Transactions on Electron Devices, Vol. 46, No.7, July 1999.
[8] J. Cardenas, S.-L. Zhang, B.G. Svensson and C.S. Petersson, J. Appl. Phys. 80 (1996), p. 2
[9] S.S. Lau, J.W. Mayer and K.N. Tu, J. Appl. Phys. 49 (1978), p. 7
[10] R.T. Tung and F. Schrey, Appl. Phys. Lett., 67 (1995), p.15
[11] C. Detavernier, R. L. Van Meirhaeghe, F. Cardon, K. Maex, B. Brijs and W. Vandervorst, Mat. Res. Soc. Symp. Vol. 611, 2000.
[12] W. L. Tan, K. L. Pey, Simon Y.M. Chooi and J.H. Ye, Mat. Res. Soc. Symp. Vol. 670, 2001.
[13] C. Detavernier, R. L. Van Meirhaeghe, F. Cardon and K. Maex, Mat. Res. Soc. Symp. Vol. 670, 2001.
[14] C. Detavernier, R. L. Van Meirhaeghe, K. Maex and F. Cardon, Mat. Res. Soc. Symp. Vol. 611, 2000.

Applied Physics (APHYS 2003)
A. Méndez-Vilas (Ed.)

Roughness Measurement by Speckle Correlation Interferometer with a Phase Shifting by Geometrical Phase Control

Daniel Gallego, Oscar López, M.C. Nistal and Vicente Moreno

Area of Optics. Department of Applied Physics. Universidade de Santiago de Compostela. 15782 Santiago de Compostela. SPAIN
E-mail: favmcr@usc.es

Abstract. Optical whole-field systems for surface roughness measurements (σ_R >1 μm) using speckle correlation technique, need the precise control of displacement to introduce the proper phase shift in the second image to be correlated with the reference one. In this way a fringe pattern is obtained after the subtraction image and a time-consuming computing post-processing, by a Fourier Transform, to measure the visibility is required. Here, we report a different technique to produce a phase shifted speckle second image, by means of the geometrical phase introduced by the rotation of the polarization state in the Poincaré sphere, using two $\lambda/4$ waveplates, one fix and one rotating, in one on the interferometers arms . With this method we observe a illuminated field which visibility can be easily evaluated by a simple and less time-consuming computer process that can finally related with surface roughness by means the use of the standard exponentially decreasing curves in function of the phase shifted. Every curve showed a different derivative depending of the rms surface roughness.

INTRODUCTION

In a very wide range of industrial processes (e.g.: paper mill), or research for testing techniques and materials (e.g: cement polishing in dentistry) and others, the quick and easy verification of the roughness of the surface material into allowed range of values is an item of very important consideration. The standard techniques like mechanical profilometry, or optical microscopy are considered useful for testing, with high resolution but just for few samples or small surfaces because are time consuming techniques even though using computerised systems.

However there are alternatives optical techniques, sharing the advantages over the mechanical ones, being a non-contact and non-destructive but besides to a wholefield technique which is very interesting in this case for saving time in the measurement process. Between these methods one the most used in industrial environments is based in speckle. The speckle (granulated aspect of the light reflected by a rough surface under laser illumination) is the result of the multiple interference originated by the back reflected light from the points of the surface.

This speckled distribution of the intensity is not related with the high of the *hills* (or deep of the *valleys*) of the rough surface, but with the visibility of the interference fringes produced by means of the Speckle Correlation as pointed out by Léger et al.

[1]. In practice is necessary of use an interferometric arrangement to observe these fringes and measure its visibility in function of the granularity of the surface. This system has been used by Leger et al. [1] and by Lehman et al. [2] with slight variations in the experimental set-up, is basically a two arms interferometer, an image capture system and a computing processor to calculate the Fourier Transform form the fringes observed in the image plane.

In this work we use a quite different method to introduce the controlled phase shift needed in any interferometric system (figure 1), using an on-axis illumination interferometer (Michelson) and avoiding the displacement of the optical elements to produce the fringes. As well as we use the *objective speckle* (without optical image system) and simplify the computer task (without FT calculation) just with a simple operation as a histogram of the image.

The main difference between this proposal respect to the previous ones is the way to introduce the phase shift using two waveplates (one fix and the other rotating) in one of the arms of the interferometer. This produce, between the entering beam (in this arm) and the exit beam, a phase shift related with solid angle closed by the path of the polarization state displacement in the Poincarè sphere. The phase introduced in this way is wide known as Geometric Phase (*Pancharatnam Phase*) [3].

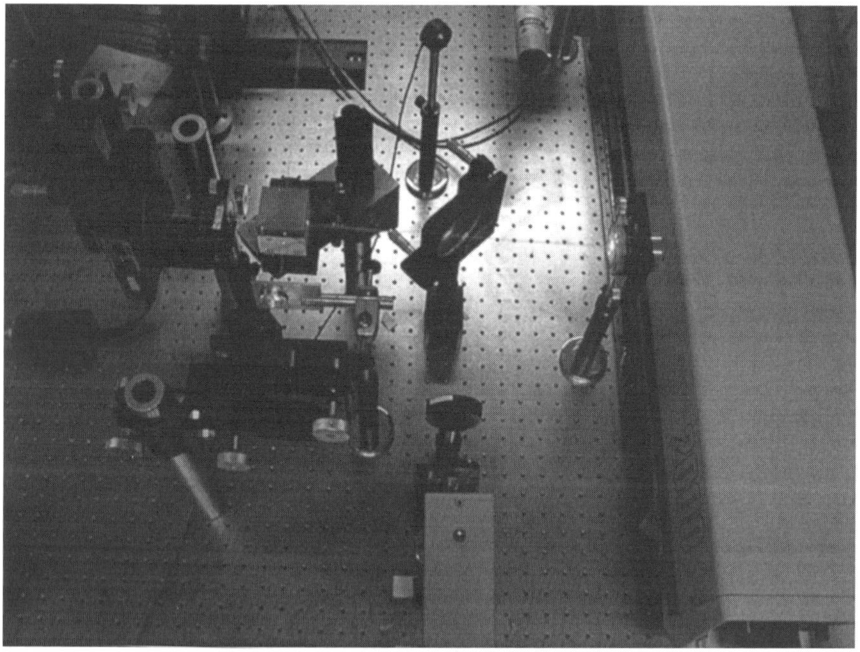

FIGURE 1. Photograph of the experimental set-up.

EXPERIMENTAL SET-UP AND PHASE SHIFTING

The technique we have used to introduce phase shifts speckle second image, is based on the geometrical phases generation formalism [3], [4], [5], [6] and [7] whose origin is found on the evolution of states over an Hilbert´s projective space. In our case, these states are the different polarization states experienced, particularly, by a light beam (Pancharatnam´s Phase), following an analogue technique to one introduced by [8], which consist in the variable geometrical phases producing by using a Michelson interferometer arrangement. The phase shift experienced by a light beam whose state of polarization is made to follow a closed (or even open [6]) circuit on the Poincare´s sphere is equal half of the solid angle subtended by the closed circuit at the center of the sphere.

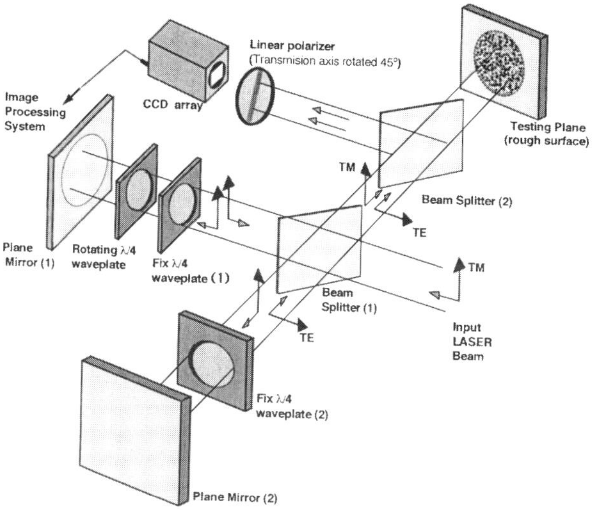

FIGURE 2. Schematic diagram of the experimental set-up.

The experimental arrangement is shown in the figure 2. The Michelson type interferometer is illuminated by an incident TM (*transversal magnetic*) or *vertical linear polarized* beam coming from a single mode argon laser beam, this input beam is divided into two equal beams by a 45° 50:50 beam-splitter (1), BS(1), having on antireflection coated face.

Beam 1 traverses a quarter wave-plate $\lambda/4$ (1), whose optical axis is fixed at 45° relative to the incident polarization, the coordinates system can be seen in the figure 2. This transform the linearly polarized light into a left circularly polarized beam that traverses a second quarter waveplate, whose axis makes an angle β with the original polarization, $\lambda/4(\beta)$, which is free to rotate about the beam direction. The light emerging from $\lambda/4(\beta)$ reaches a perpendicular mirror M(1), from which it is reflected and made to retrace its path recovering the initial polarization, with a phase increased a number of radians directly related with the rotated angle of the second waveplate.

Beam 2 traverses a quarter-waveplate $\lambda/4$ (2), whose optical axis is fixed at 45° relative to the incident polarization. This, of course, converts the linearly polarized light into a left circularly polarized beam, that travels to a perpendicular mirror M(2) and is reflected back traversing again $\lambda/4$ (2) returning to the system with a TE polarization (*horizontal linear polarized*) with a contribution of a constant factor phase increased.

The figure 3 shows the particular circuits over the unit Poincare´s sphere, whose three axes correspond to the three components of the Stokes vector that characterizes the state of polarization of the light. The poles B and D represent left and right circular polarizations, respectively, all points on the equatorial circle represent linearly polarized light.

In our experience, light leaving BS(1) in arm 1 is linearly polarized and may be represented by point A in figure 3. After cross over $\lambda/4$ (1), the light becomes left circularly polarized and is represented by point B on the North Pole of the sphere. After passing $\lambda/4$ (β), it is again linearly polarized and is represented by point C on the equatorial circle. Then, the light is reflected from M(1) on the plane mirror (1) and retraces its path trough the quarter-wave plates, first $\lambda/4$ (β) moves the beam to point D , and then $\lambda/4$ (1) to point A in the fig.3.

At this stage, the polarization state has made a closed circuit over the sphere and, of course, over a parameter space. In this closed circuit, it may be shown that the light has generated a phase shift whose value is equal to half the solid angle subtended by ABCDA at the center of the sphere. In our case, this geometrical or topological phase is given by $\phi_G(1) = 2\beta$.

Now, we consider the geometric phase contribution of the arm 2, indicated by a different circuit over the Poincare´s sphere: light leaving BS(1) in arm (2) is linearly polarized and also, of course, it is represented by point A in fig. 3. This beam, after cross over $\lambda/4$ (2) becomes left circularly polarized and is represented by point B. Then, the beam is reflected on mirror M(2) and retraces its path, when the beam cross over again the $\lambda/4$ (2), its state of polarization is orthogonal to the started one, and it is represented by point E in fig. 3. This second circuit over Poincare´s sphere suffered by beam of arm 2 is open, but, it has been shown by Samuel-Bhandari [4] that this open circuit can be closed by a geodesic which links with the starting point and, therefore, it has a geometrical phase whose quantitative contribution is given by the half of the solid angle subtended at the center of the sphere by the circuit ABEA. In our case, this topological contribution is a constant: $\phi_G(2) = 1/2\ \Omega_{(ABEA)} = 1/2(\pm\pi) = \pm\pi/2$.

The total geometric phase $\phi_{G(T)}$ is the difference of the two contributions, but since the arm 1 contribution is a variable contribution 2β, and the arm 2 contribution is constant ($\pm\pi/2$), we can always adjust the experimental arrangement in order to considerate only the relative geometrical phase contribution, that in our case is given by the variable geometrical phase $\phi_{G(T)}=2\beta$.

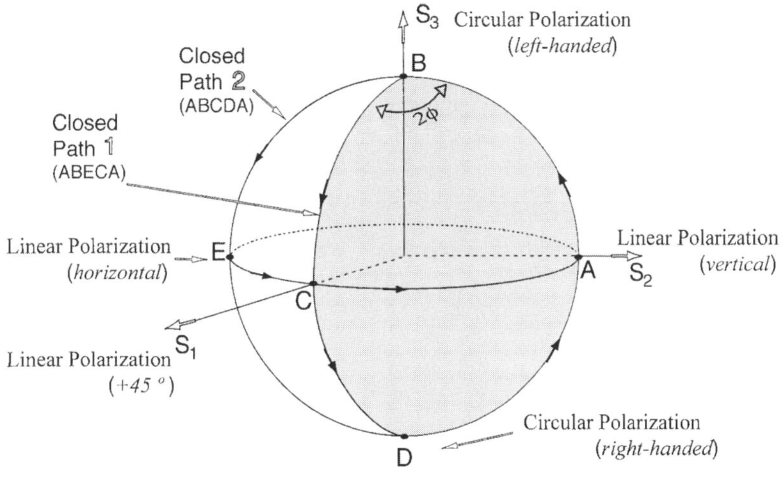

FIGURE 3. Outlines of the closed paths followed by the polarization states on the Poincarè sphere.

The BS(1) combines portions of the returning light from arm 1 and 2, both beams, with axial and parallel propagation, but with different polarization impinging the rough surface (after crossing a second beam splitter BS(2)) under not interfering conditions, and for this reason maintaining constant illumination level of the test surface. The reflecting beams from the test plane, traverses the second beam-splitter BS(2),then, a linear polarizer, whose transmission axis is rotated to 45° respect to the reference coordinate system, meets together the two portions of the beams and gives rise to an interference pattern that is directed to the detection system (CCD array) without crossing any imaging lens.

In this way we detect the *objective speckle* resulting from the interference (making the projection of the TM and TE reflected beams in the transmission axis of the linear polarizer). After, it can be made the computerised subtraction of the two experimental situations: before and after the introduction of the controlled phase shift of the both speckled patterns obtained by interference in the plane of the CCD array.

Finally we calculate the histogram of the subtracted image, filtering processing (for avoiding the electronic noise of the image), and obtaining a numerical value of the histogram width as proportional to the visibility of the resulting image.

The relation between the visibility of the fringes and the *standard deviation* σ (with $\sigma>\lambda$) of a normal distribution of the surface roughness has been established in previous theoretical and experimental works [1], [2] through a negative exponential like:

$$V \propto \exp\{-[C\phi\sigma]\}$$

Where C is a constant depending of the specific geometry of the experimental setup, and $\phi = 2\beta$ (the controlled phase shift introduced).

EXPERIMENTAL RESULTS

The experimental results in the figure 4, shown exponential decaying curves with different slopes in function of the roughness of the surfaces.

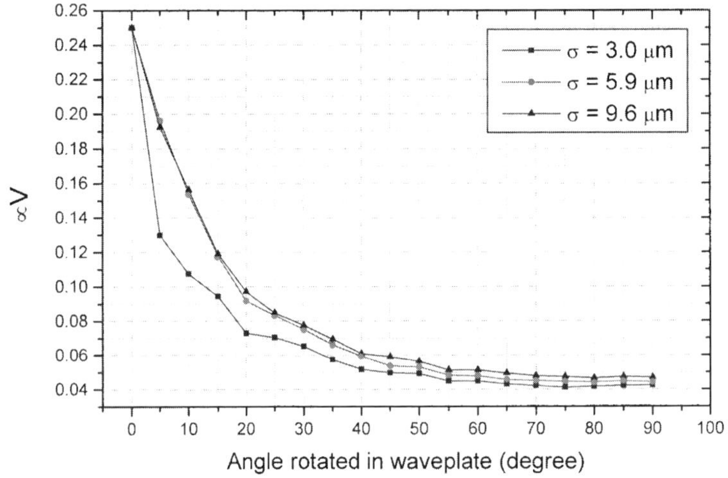

FIGURE 4. Experimental results. Decrease of visibility V by increasing the angle β for different values of σ.

CONCLUSIONS

The results obtained shown a great agreement with the published by other authors using quite different experimental arrangements, with several interesting advantages.

The use of controlled rotation stage is easier to maintain under alignment during all the measurement process than linear one, and introduce more precise phase shifting. As well as is to work with on-axis illumination could be possible to use the system to measure roughness in the regime of slightly rough surface ($\sigma<\lambda$) that cannot be done with two angled beams as demonstrated theoretically by Leger [1].

Because the no production of fringes to measure their visibility through out the Fourier Transform calculations, and once calibration has been done (as in the other proposals) the time needed for the computer to obtain the final data is lower in this case because we use just a simpler operation for an elementary image processor: an histogram.

The using a objective speckle pattern reflected by the rough surface, and simple optical elements (flat elements without the necessity of lenses) makes the system able to be compacted and used as portable device.

ACKNOWLEDGMENTS

We wish to thank F. Gómez for its useful comments and support of the image capture device system and A. Alonso for help in the experimental work. This work was supported by Xunta de Galicia. Galicia. Spain. Contract Number PIDT00TIC20601PR.

REFERENCES

1. Léger, D; Mathieu, E. and Perrin, J.C., "Optical surface roughness determination using speckle correlation technique". *Appl. Opt.* 14(4) 872-877 (1976).
2. M. Lehman, J.A. Pomarico, R.D. Torroba, *Optical Engineering*. Vol.34 No.4.1148-1152, 95 (1995).
3. S.Pancharatnam, "Generalized theory of interference, and its applications". *Indian Academy of Sciences*, Vol. 44 nº 5. sec A (1956), 247.
4. M.V.Berry,"Quantal phase factors accompanying adiabatic changes", *Proc.R.Soc.London* A 392 (1984) 45.
5. Y.Aharonov and J.Anandan, "Phase change During a cyclic Quantum Evolution". *Phys.Rev.Lett.*, vol 58, nº 16, (1987), 1593.
6. J.Samuel and R.Bhandari, "General setting for Berry´s phase". *Phys.Rev.Lett,* vol 60, nº 23, (1988), 2339.
7. J.Liñares and M.C.Nistal, "Geometric phases in multidirectional electromagnetic coupling theory". *Phys. Lett.* A 162 (1992), 7.
8. T.H.Chyba, L.J.Wang and L.Mandel, "Measurement of the Pancharatnam phase for a light beam", *Optics Lett.* Vol 13, nº 7 ,(1988) 562.

Structural Features in Granular and Amorphous Microwires

Julian Gonzalez*, Juan J. del Val*[$] and Arcady Zhukov[#]

*Departamento de Fisica Materiales, Universidad del País Vasco, Apdo. 1072, 20080 San Sebastian, SPAIN.

[$]also at Centro Mixto CSIC-UPV/EHU.

[#]Instituto de Ciencia de Materiales, CSIC, 28049 Cantoblanco, Madrid, SPAIN.

Abstract. The microstructural characteristics of granular and amorphous glass coated microwires are studied considering their geometry, which is connected to the fabrication conditions. In the case of granular microwires (Cu_xFe_{100-x} and Cu_xCo_{100-x}), the existing quantity of Cu, Fe or Co phases (with crystallite sizes around 40 nm) clearly depends of the ratio between the diameters of the metallic core and the external one of the layer. The scattering patterns of amorphous microwires consist on a complex superposition of halos (wide peaks) corresponding to structural correlations in the pyrex layer and in the metallic core. Their normalization to absolute units shows that the higher the size of the core the higher the intensity of the halos corresponding to the structural correlations in the core (and hence the number of structural correlations there). Small, although systematic, changes can be detected in the short range order in the glassy core.

INTRODUCTION

Glass coated metallic microwires are recently being a subject of interest owing to their potential applications as magnetic sensor devices [1]. Technological aspects of their fabrication related with their composition, structural, mechanical, electric and magnetic properties have been subject of great number of publications [1, 2]. These microwires present a metallic core with 1-20 μm diameter covered by an insulating Pyrex coating 2-20 μm thick. The metallic ferromagnetic core can be formed either by an amorphous atomic arrangement either by an alloy of inmiscible elements of magnetic and granular nature. It has been stated [3] that their structural and magnetic properties depend of their geometry through the parameter ρ, which is straightforwardly connected to the fabrication conditions of the microwires. ρ is defined as:

$$\rho = \frac{d}{D} \qquad (1)$$

where d and D are respectively the diameters of the internal metallic core and the external coating pyrex layer. The aim of this paper is to present a complete study of the influence of ρ paremeter on the structure of crystalline granular Cu-based microwires with different core compositions (Cu_xFe_{100-x} and Cu_xCo_{100-x}) as well as on the short range order of amorphous core microwires.

EXPERIMENTAL

Glass coated microwires with different core compositions were studied: a) granular Cu-based microwires: $Cu_{70}Fe_{30}$, $Cu_{63}Fe_{37}$, $Cu_{90}Co_{10}$ and $Cu_{80}Co_{20}$; b) amorphous microwires: $Fe_{75.5}B_{13}Si_{11}Mo_{0.5}$. In each composition four geometries ranging between ρ=0.05 and ρ=0.75 were choosen. The method of preparation of the samples with the Taylor-Ulitovsky technique was elsewhere described [4, 5].

The structural characteristics of the different samples were investigated by X-ray Diffraction (XRD) experiments which were carried out using CuKα radiation (λ = 0.154 nm) in a powder diffractometer provided with an automatic divergence slit and a secondary graphite monochromator. The accumulation time was much bigger in the case of amorphous microwires.

RESULTS AND DISCUSSION

Granular glass coated microwires

Fig. 1 shows four of the XRD patterns being evident the existence of inmiscible phases in all the cases.

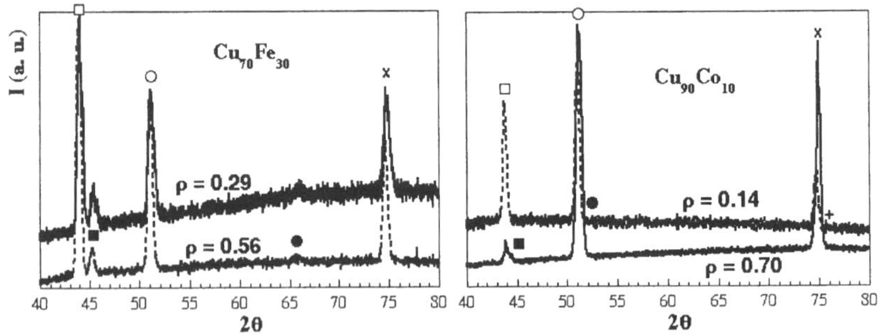

FIGURE 1. Examples of XRD patterns XRD patterns for $Cu_{70}Fe_{30}$, and $Cu_{90}Co_{10}$ samples with different ρ values. Open symbols and x correspond to Cu peaks; full symbols correspond to αFe peaks in Cu_xFe_{100-x} samples; full symbols and + correspond to Co peaks in Cu_xCo_{100-x} samples.

In all the studied samples the three peaks characteristic of the Cu phase (*fcc*, lattice parameter: 0.361 nm) corresponding to atomic spacings of 0.209 nm, 0.181 nm and 0.128 nm [6] are observed. Cu_xFe_{100-x} samples show well separated the peaks characteristic of the αFe phase (*bcc*, lattice parameter: 0.287 nm) corresponding to atomic spacings of 0.203 nm and 0.143 nm [6]. This last αFe peak is very small and is only well detected in the case of higher ρ values. Cu_xCo_{100-x} samples show the peaks characteristic of Co phase (*hcp*, lattice parameters: 0.251 nm and 0.407 nm) with atomic spacings of 0.205 nm, 0.177 nm and 0.125 nm [6]. We observe that, for all the samples, the higher the ρ value the neater and higher the intensity of the crystalline peaks and the smaller the background, which must be attributed to the proper one of the crystalline metallic core and to the amorphous halo resulting from the diffraction of the glass coating.

It is to be noted that the peaks corresponding to the Co phase are strongly overlapped to the ones of Cu phase. Even if only one maximum is observed, all the peaks showed a shoulder indicating that they are actually composed by two "elementary" contributions. These "double" peaks were deconvoluted by a fit to a sum of gaussian functions in order to fix their position, width and area. In all the cases the quality of the fits is very good [7].

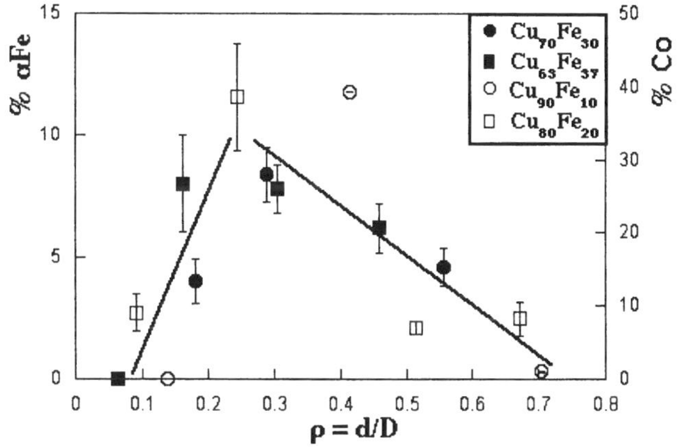

FIGURE 2. Relative content of αFe and Co phases plotted versus ρ for all the granular microwires.

The crystal size of each phase is evaluated from Scherrer's formula (correction of the instrumental broadening with the aid of a standard Si) and no systematic variation with the geometry has been found: crystals of sizes between 20 nm and 50 nm have been found for the three phases irrespectively on the alloy composition and the geometry. The content of each phase in each case is evaluated from the area of the crystalline peaks corresponding to a given existing phase compared to the total area of all the crystalline peaks in the XRD pattern. Focusing our attention on the obtained content for αFe and Co phases, we observe that these contents vary appreciably with the ρ parameter (see Fig. 2): irrespective of the composition, αFe and Co contents show a maximum around ρ = 0.25. This behaviour can be understood as consequence of the metastability of the obtained crystalline phases [1,4,5].

Amorphous glass coated microwires

XRD patterns of the three glas coated microwires with amorphous core ($Fe_{75.5}B_{13}Si_{11}Mo_{0.5}$) with three different geometries show large peaks located around 16 nm^{-1} and 30 nm^{-1} corresponding to the long structural correlation distances halos of the pyrex and the glassy metallic core respectively [8]. On the other hand, a smaller, wider and more complex peak appears at 40-60 nm^{-1} q-range that must be a mixture of the small intensity halos (long structural correlation distances) of the metallic core and the pyrex layer. The intensity of all these peaks changes significantly depending of the sample. The raw XRD patterns for the three samples were normalized to absolute (electronic) scale after being corrected from several effects that do not contain structural information [8], taking in account the composition of the samples, which is given by their geometry: background, polarization, absorption and incoherent scattering. The reduced intensity i(q) [8] was built from the coherent absolute intensity (I_{coh}) as follows

$$i(q) = \frac{I_{coh} - \sum_i x_i f_i^2}{\left[\sum_i x_i f_i\right]^2}, \qquad (2)$$

where f_i is the independent scattering factor of element i, present in atomic fraction x_i.The reduced intensity multiplied by q will enhance the peaks and make possible their separation (see Fig. 3):

Under these considerations, the three q.i(q) patterns were fit to a sum of two gaussian (for the large distances peaks) and two lorentzian (for the short distances peaks) functions. The results of the fits are very good in the large q region showing small deviations in the two sides of the large correlation distance corresponding to the core.

At this point, we will only consider the changes in the peaks associated to the structural correlations in the core. The two mean correlation distances, calculated from Bragg's law, are appreciably decreased in 1% for the sample with the thinest layer. The total area of the scattering peaks assigned to the core increases continuously. This fact indicates that the larger the ρ value the higher the electronic density or the number of diffracting structural units in the core, that is the number of structural correlations.

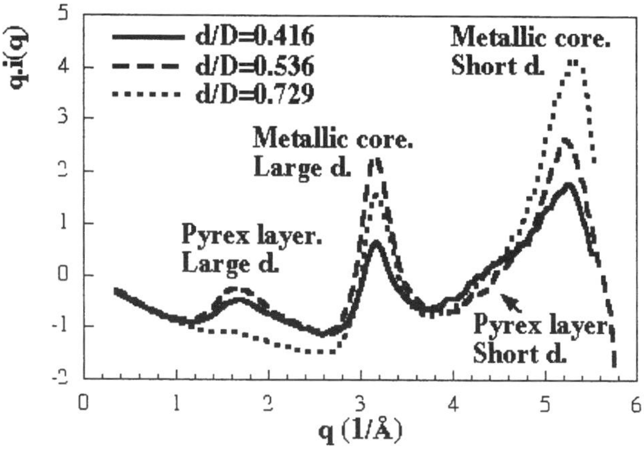

FIGURE 3. q.i(q) data obtained for the three glassy microwires.

CONCLUSIONS

We can conclude that the geometry of the microwires affects importantly the resulting microstructure of the core. This influence is neat in the case of the granular microwires where the ρ parameter drives the existing phases, being possible to define an optimal ρ value around 0.25. In the case of amorphous core microwires the influence of the geometry on the short range order characteristics is not so evident; even if the variations in the average correlation distances and in the area of the two peaks of the core are systematic, more studies with other amorphous microwires involving a larger ρ-range are needed in order to confirm these results.

ACKNOWLEDGEMENTS

The authors thank the spanish Ministerio de Ciencia y Tecnología for partial finantial support (project MAT2001-0082-C04-04).

REFERENCES

1. M. Vazquez and A. Zhukov, *J. Magn. Magn. Mat.* **160**, 223 (1996).
2. H. Chiriac and A. Ovari, *Progr. Mat. Sci.* **40**, 333 (1997).

3. J. J. del Val, A. Zhukov and J. Gonzalez, *J. Magn. Magn. Mat.* **249**, 126 (2002).

4. S. A. Baranov, V. N. Berzhanski, S. K. Kotov, V. S. Larin and A. V. Torcunov, *Phys. Met. Metall.*, **67**, 73 (1989).

5. V. S. Larin, A. V. Torcunov, A. Zhukov, J. Gonzalez, M. Vazquez and L. Panina, *J. Magn. Magn. Mat.* **249** 39 (2002).

6. *Powder Diffraction File, Inorganic Phases*, International Centre for Diffraction Data, JCPDS, USA 1985.

7. J. J. del Val, J. González and A. Zhukov, *Physica B: Cond. Matter* **299** 242 (2001).

8. H. P. Klug and L. E. Alexander, *X-ray Diffraction Procedures for Polycrystalline and Amorphous Materials*, Wiley, New York, 1974, p. 833.

Applied Physics (APHYS 2003)
A. Méndez-Vilas (Ed.)

Mechanical Alloying of $Fe_{100-x}B_x$ Compounds: A Structural Study

Carmen Miguel*, Julian Gonzalez*, Juan J. del Val*$ and Jesus M. Gonzalez#

*Departamento de Fisica Materiales, Universidad del País Vasco, Apdo. 1072, 20080 San Sebastian, SPAIN.

$also at Centro Mixto CSIC-UPV/EHU.

#Instituto de Ciencia de Materiales & Instituto de Magnetismo Aplicado, CSIC, 28049 Cantoblanco, Madrid, SPAIN.

Abstract. Decrystallization (amorphization) of $Fe_{100-x}B_x$ samples is made by means of mechanical alloying technique. X-ray diffraction has been used to study the decrystallization kinetics, whose time dependence is fitted by a single exponential decay. In all samples, milling times of around 500 h lead to very low crystallinity amounts. This residual crystalline phase corresponds to a mixture of FeB and Fe_2B phases, being the αFe phase almost disappeared at the same time. During the decrystallization, the grain size of the αFe phase changes appreciably from 50 nm down to 10 nm although FeB and Fe_2B phases maintain their grain size constant between 10 nm and 15 nm. Characterization of the samples in their final state with Vibrating Sample Magnetometer showed soft magnetic character in all the cases.

INTRODUCTION

Different techniques (ultrarapid quenching, vapor deposition, sputtering and solid state amorphization reactions (SSAR)) have been used to produce "transition metal-metalloid" amorphous materials. The first methods show problems of segregation when the components present very different melting points. Between the SSAR methods, Mechanical Alloying (MA) in a planetary mill avoids this inconvenience because the collisions of the balls with the other balls and with the walls of the containing vessel provide the necessary energy for the blending. The process is complex and involves deformation, fragmentation, cold welding and microdiffusion processes. As a consequence, the optimization of the parameters (type of mill, materials of the vessel and the balls, ball to powder ratio, quantity of the ductile element in the materials, etc...) and the monitoring of the technique are very difficult. Therefore it is difficult to get the same type of materials for a given initial mixture under different conditions and we can find in the literature that different phases are obtained [1-3] for similar compositions of $Fe_{100-x}B_x$ system introduced but different milling conditions. Even though Fe and B atoms show similar diffusion coefficients and the amorphization is not totally completed [4], it has been shown that the alloying is achieved by interdiffusion of both elements [5]. It is interesting to determine the final phases existing after the milling process because it exists some controversy in this question: some papers [2] point out that, after 600 hours of milling time, α-Fe, Fe_2B, FeB and amorphous phase are obtained although others [3] does not signal FeB phase to be present in the resulting material.

In the present work, a systematic study of the decrystallization (amorphization) process by MA of $Fe_{100-x}B_x$ samples has been developed in order to determine the final phases resulting from the process and to study the decrystallization kinetics. Moreover, structural and magnetic characterizations of the resulting samples in their final state have been performed in order to describe their properties.

EXPERIMENTAL

$Fe_{100-x}B_x$ (x = 25, 40, 50, 75) mixtures in atomic ratios obtained from powders of iron (99% purity, 60 nm grain size) and amorphous boron (95% purity, 0.9 nm grain size) were introduced into a WC vessel with two WC balls (ball to powder ratio: 14:1) and then put in a mill (PLANETARY MICRO MILL "PULVERISETTE-7") with rotation speed of 3467 rpm for 600-800 h. In order to avoid an excessive heating of the machine (following the method described in refs. [2, 3]), after periods of 225 min. of continuous milling, the process was interrupted during 15 min. The temperature of the system did not drop significantly during this period, avoiding possible structural reorganizations of the sample during these periods where the mill did not work.

At milling periods of approximately 50 or 100 hours, some powder was extracted and X-ray Diffraction (XRD) experiments were carried out in order to investigate the structural state of the sample at these stages. XRD measurements were made using CuKα radiation (λ = 0.154 nm) in a powder diffractometer provided with an automatic divergence slit and a secondary graphite monochromator.

Magnetic characterization of the final samples obtained by MA was carried out by means of Vibrating Sample Magnetometer (VSM) tecnique.

RESULTS AND DISCUSSION

Fig. 1 shows four XRD patterns for two of the samples. In the initial stages of the milling process (around 4 hours of milling) the patterns for all the samples only revealed three peaks characteristic of the α-Fe phase [6]. At short milling times (around 50 hours) new peaks appear with very low intensity: we will fix our attention on the two peaks that correspond to Fe_2B and FeB phases [6]. They prove that the alloy of the two elements was formed. At the same time, the α-Fe peaks begin to decrease in intensity and to broaden. At longer milling times (around 600 hours), the intensity of the peaks corresponding to the α-Fe phase is strongly diminished and is even lower than the intensity of the Fe_2B and FeB peaks which are not decreased. One could think, with a first view of Figure 1, that the intensity of these two last phases is increased with the milling process. We must consider that, at the same time that the intensity of the α-Fe phase peak is lowered and the two Fe_2B and FeB peaks are formed, the intensity of the background of the diffracted signal is continuously increased with the milling time. This fact indicates the formation of an amorphous phase in the material coexisting with the three above mentionned crystalline phases. In this sense we say that the sample is decrystallized during the milling process. In the case of the three other compositions the obtained behaviour is qualitatively the same.

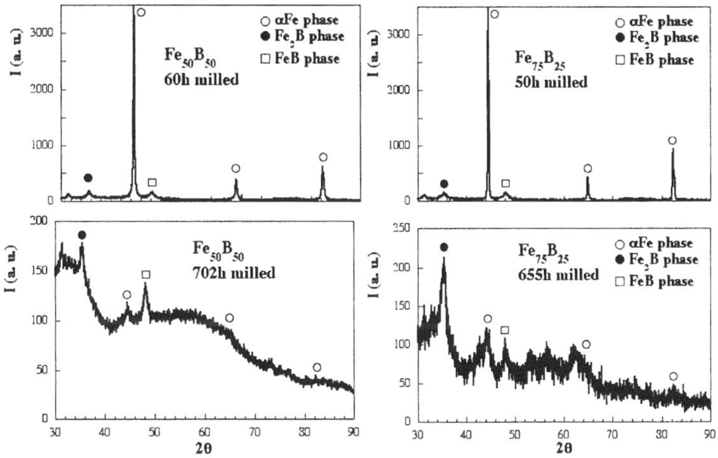

FIGURE 1. Raw XRD patterns for $Fe_{50}B_{50}$ and $Fe_{75}B_{25}$ samples obtained at different milling times.

At this point, one can study the crystallinity percentage (relative percentage of crystalline phase in the material) in each sample at the different milling times, as it is often done in semicrystalline materials, specially in polymers [7]. The samples in their final milled stage showed crystalline percentages lower than 6%, being the lowest (around 1%) for the sample $Fe_{50}B_{50}$. Nevertheless, we will pay our attention to the evolution of the main peak in each of the three crystalline phases before indicated assuming that the the higher the area of this main peak the higher the content of the corresponding phase. After substracting the intensity corresponding to the amorphous halo (which is extrapolated from the overall scattering curve) it has been found that the evolution with the milling time of the area of the αFe peak can be described by a single exponential decay in all the samples (see Fig. 2):

$$A \propto \exp[-t/\tau] \quad (1)$$

being τ the characteristic time for the decreasing of the concentration of the αFe crystalline phase, which is representative of the decrystallization (amorphization) process induced by the milling. On the contrary, the area of the Fe_2B and FeB phases remains approximately constant with the milling time.

On the other hand, the crystallite size D of the obtained phases was approximated by means of Scherrer formula [7. 8]. The results corresponding to the αFe peak, corrected from the instrumental broadening with the aid of a Si standard, show that, for all the samples, D decreases during the milling process from some 50-60 nm in the earlier stages down to some 10-20 nm at longer times. This last value is attained after 200-300 hours of milling. On the contrary, the peaks corresponding to Fe_2B and FeB phases show a D-value around 15 and 10 nm respectively, which is approximately constant during all the process and this fact occurs for all the samples.

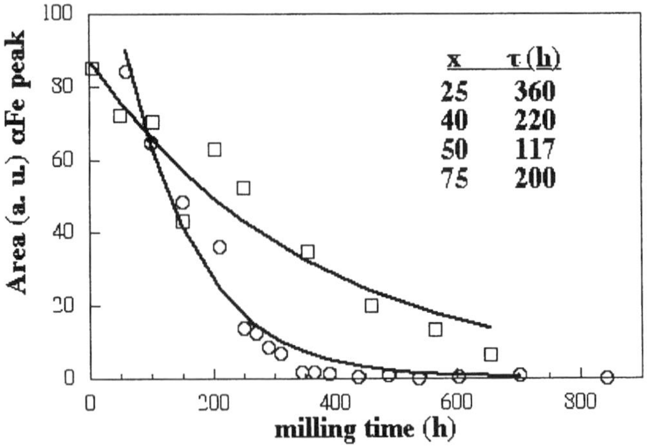

FIGURE 2. Plot of the area of the αFe, FeB peak versus the milling time for $Fe_{50}B_{50}$ (circles) and $Fe_{75}B_{25}$ (squares) samples. Solid lines show the fittings to a single exponential decay. The inset shows the τ-values for all the samples.

In order to characterize the properties of the samples in their final milled state, VSM measurements were carried out at room temperature and a soft magnetic character resulted for all the samples. Table 1 shows the obtained values for the coercitive field (Hc) and the saturation magnetization (Ms).

TABLE 1. Coercitive field (Hc) and Saturation magnetization (Ms) values obtained for all the milled samples.

Sample	$Fe_{25}B_{75}$	$Fe_{50}B_{50}$	$Fe_{60}B_{40}$	$Fe_{75}B_{25}$
Ms (emu/g)	3	3	40	35
Hc (Oe)	55	57	118	95

Higher Hc-values (around 100 Oe) were obtained for the B-rich samples in comparison to the 55 Oe measured in the Fe-rich ones (in any case 100 times higher than the obtained values for FeB samples reported in other works [2, 3]). This behaviour can be attributed to the extremely high stresses accumulated during the milling process. In fact, annealing of the samples at temperatures close to 750 K, allowing the internal stresses to relax, leads to a decreasing, in approximately a factor of 5, of the Hc-values. The results concerning the saturation magnetization show also a different behaviour between Fe-rich compositions (40 emu/g) and B-rich ones (3 emu/g), possibly because of the formation of a paramagnetic phase in the second ones.

CONCLUSIONS

The decrystallization (amorphization) of $Fe_{100-x}B_x$ alloys by means of MA technique is a complex process, as it is shown from XRD data. Simultaneously to a decreasing of the αFe phase content, Fe_2B and FeB are formed and their content seems to remain constant during all the process. The crystal size of these two formed phases is constant ($\approx$10 nm), whereas the size of the αFe crystals is decreased from 50 nm down to some 10 nm. The final milled samples show soft magnetic character (with strong internal stresses induced by the milling process) with different characteristics between Fe-rich samples and B-rich ones.

ACKNOWLEDGEMENTS

The authors thank the spanish Ministerio de Ciencia y Tecnología for partial finantial support (project MAT2001-0082-C04-04).

REFERENCES

1. C. Suryanarayana, *Progr. Mat. Sci* **46**, 1 (2001).
2. J. M. Gonzalez, L. Giri and A. K. Giri, *J. Magn. Magn. Mat.* **140-144**, 249 (1995).
3. A. K. Giri and J. M. Gonzalez, *IEEE Trans. Magn.* **31**, 4023 (1995).
4. A. L. Greer, *J. Non-Crystalline Solids*, **61**, 734 (1984).
5. J. Balogh, T. Kemeny, I. Vincze, L. Bujdoso, L. Toth and G. Vincze, *J. Appl. Phys.* **77** 4997 (1995).
6. *Powder Diffraction File, Inorganic Phases*, International Centre for Diffraction Data, JCPDS, U. S. A. 1985.
7. L. E. Alexander, *X-ray Diffraction Methods in Polymer Science*, Krieger, Malabar, 1985, p. 137.
8. R. Jenkins and R. L. Snyder, *Introduction to X-ray Powder Diffractometry*, Wiley, New York, 1996, p. 47.

Applied Physics (APHYS 2003)
A. Méndez-Vilas (Ed.)

Pericardial Biomechanical Adaptation to Low Frequency Noise Stress

Mariana Alves-Pereira*, João Joanaz de Melo*, and Nuno A. A. Castelo Branco ¶

*Dept. Environmental Sciences & Engineering, DCEA-FCT, New University of Lisbon, 2829-516 Caparica, Portugal mariana.pereira@oninet.pt
¶Center for Human Performance, Scientific Board, Apartado 173
2615 Alverca Codex, Portugal

Abstract. Low frequency noise (LFN) (≤500 Hz, including infrasound) exposure causes thickening of cardiac structures, namely the pericardium – a <0.5 mm thick, 3-layer, translucid sac that surrounds the heart. In LFN-exposed individuals, microscopy studies of their pericardia disclosed the existence of a 5-layered organ, up to 2.33 mm thick. The biomechanical properties which allow this abnormally large organ to coexist with a normal heart function (no diastolic dysfunction is exhibited by these patients) are discussed. The relative positioning of collagen and elastin fibers allow for an unusual reorganization of the extra layers of tissue so that the constant heart expansions and contractions are in no way impeded. The challenges of developing a biotensegrity-based mathematical model describing this new organ is still an open issue.

BACKGROUND

The pericardium is a fibrous sac that encases the heart, with the purpose of maintaining it in its normal position. External forces, due to respiration or changes in body posture, are absorbed by the pericardium so as to keep the heart and its cardiac rhythm intact. Consisting of three tissue layers – mesothelium, fibrosa and epipericardium – the pericardium is a highly organized mass of connective tissue, with a predominance of collagen fibers arranged in accordion-like bundles. Elastic fibers, much less numerous than collagen fibers, intersect the collagen bundles at right angles. This anatomical arrangement taken together with the viscoelastic properties of both collagen and elastin, provide the pericardium with the mechanical capability of protecting the integrity of the cardiac cycle. The thickness of the normal parietal leaflet of the pericardium is <0.5mm (1).

The mesothelium is in direct contact with the pericardial sac, and is formed by a one-layer thick sheet of mesothelial (cuboidal) cells (MC). Anchoring junctions (AJ) interconnect MC among themselves, through their cytoskeletal fibers. AJ that interconnect MC cytoskeletons through intermediate filaments are called desmosomes. Intermediate filament architecture consists of fibrous subunits associated side-by-side in overlapping arrays, forming a meshwork that extends throughout the cytoplasm, providing mechanical strength to the cell. Intermediate filaments do not rupture when stretched, and can withstand larger stresses and strains than actin filaments. Intermediate filaments are crucial to maintain cellular integrity.

Workers exposed to long-term (in years) low frequency noise (LFN) (≤500 Hz, including infrasound) can develop vibroacoustic disease (VAD) (2-4). Pericardial thickening in the hallmark of VAD (5). Pericarditis is an inflammatory condition whereby the pericardium thickens and restrains the cardiac diastolic motion. In VAD, despite pericardial thickening no constriction of cardiac diastole present, and no inflammatory process is observed. The question of how this is achieved prompted light and electron microscopy studies of pericardial fragments obtained from four VAD patients, with their informed consent, at the time they received cardiac surgery for other reasons (6). Since then, an additional eight pericardial fragments of LFN-exposed workers have been studied with light and electron microscopy (Table 1) (7,8).

The goal of this report is to a) qualitatively analyze the biomechanical and morphological changes observed in the pericardia of LFN-exposed workers; b) tentatively advance possible bio-mechanisms responsible for these changes; and c) bring this topic to the forum of biophysicists, bioengineers, and bio-mathematicians in an attempt to gain a deeper understanding of these biological processes.

METHODOLOGY

Data analyzed herein has been already been presented in medical and biological forums (Table 1), hence animal treatment and light and transmission electron microscopy (TEM) techniques are described in detail elsewhere (6-8). The qualitative analysis of the response of biological tissue to LFN is approached herein from a biomechanical standpoint, drawing upon previously published results

RESULTS

TABLE 1. Summary of VAD Patients' Age, Occupation and Pericardial Thickness

Case	Age	Occupation	Pericardial Thickness (mm)*
1	32	Aircraft Technician	1.35
2	37	Aircraft Pilot	2.04
3	39	Aircraft Pilot	1.22
4	41	Aircraft Pilot	2.19
5	44	Aircraft Pilot	2.23
6	48	Helicopter Pilot	1.40
7	51	Aircraft Technician	1.60
8	52	Truck Driver	1.67
9	53	Helicopter Pilot	1.03
10	57	Helicopter Pilot	2.02
11	60	Helicopter Pilot	1.06
12	62	Aircraft Technician	1.09

* Normal value: < 0.5 mm.

Instead of 3 layers of tissue, as expected in normal percardia, five distinct layers were identifiable (Fig.1): the fibrosa layer had thickened considerably, split in two, and now sandwiched a newly formed layer of loose tissue, i.e., containing blood vessels, fatty and nerve tissues (Fig. 2). Both the internal (closest to the mesothelium)

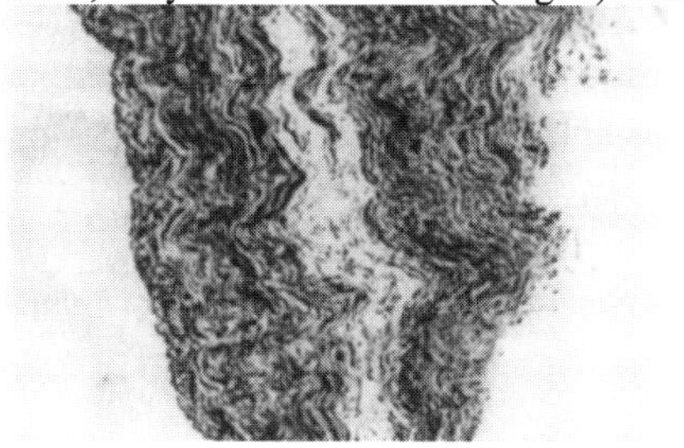

FIGURE 1. Light microscopy of a VAD patient pericardium, with pericardial sac on left. Five layers are identifiable: (from left) mesothelial, internal fibrosa, loose tissue, external fibrosa, and epipericardium. In both fibrous layers, wavy collagen bundles are visible. The loose tissue is rich in vessels. No inflammatory cellularity was identified in any of the five layers. (x200) (1)

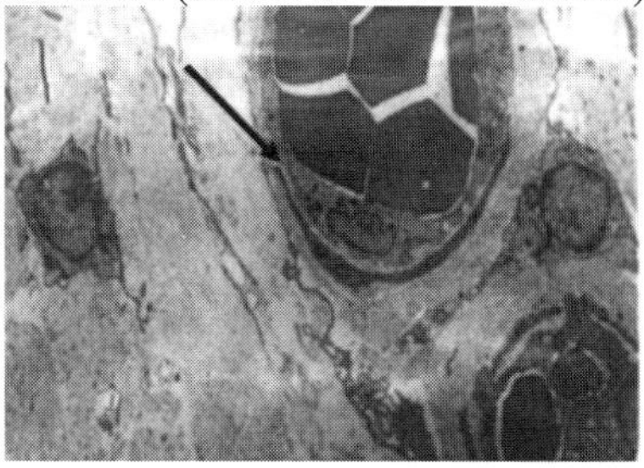

FIGURE 2. Light microscopy of the loose tissue layer of a 51-year-old VAD patient. Blood filled vessels with thickened walls (arrow). Two myofibroblasts, and numerous cytoplasmic extensions surround the bundles of collagen. (x2800).

and the external (closest to the epipericardium) exhibited the classical arrangement of accordion-like wave-forms of collagen bundles, intertwined with elastic fibers at seemingly regular intervals. No cilia were identified in any samples.

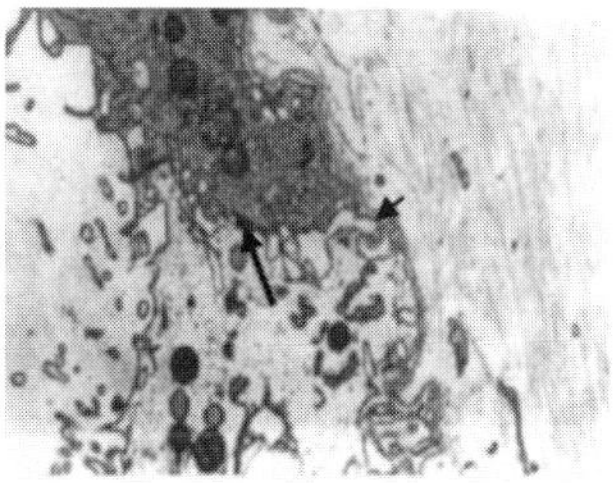

FIGURE 3. TEM of the mesothelial layer, with pericardial sac on left. MC have abundant microvilli and inter-cellular junctions possess numerous desmosomes (arrow). Nuclei shape is irregular reflecting cellular stress. Unusual gaps are seen in the lower portion of the MC layer (arrowhead). Cellular debris is visible in the submesothelial layer. (x2800)

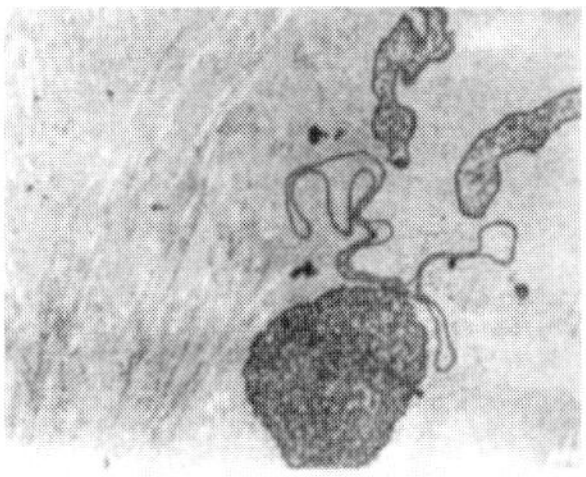

FIGURE 4. TEM of the loose tissue layer. Empty cell membrane in the vicinity of three fragments of nuclear material, embedded within multiple collagen bundles with different orientations. (x5300)

Cytoplasmic extensions, emanating from fibroblast cells also accompany the wavy collagen bundles (Figs. 5,6). Cellular death was observed in all layers, although more so in the internal fibrosa. This was not the classical, apoptotic death. Instead, cellular debris, consisting of seemingly live organelles, were scattered throughout all layers (Figs. 3,4), usually in the vicinity of elastic fibers, and frequently in the neighborhood of burst cytoplasmic extensions. In the MC layer, herniations of MC plasma membranes into the pericardial sac were frequently observed, some of which contained cellular organelles. No nuclear material was observed within these cytoplasmic extensions. Individual MC were seen protruding into the pericardial sac, in a process of surface extrusion of that cell.

MC with ruptured surfaces in direct contact with the pericardial sac were also observed. AJ between MC were very strong, sometimes formed by more than two,

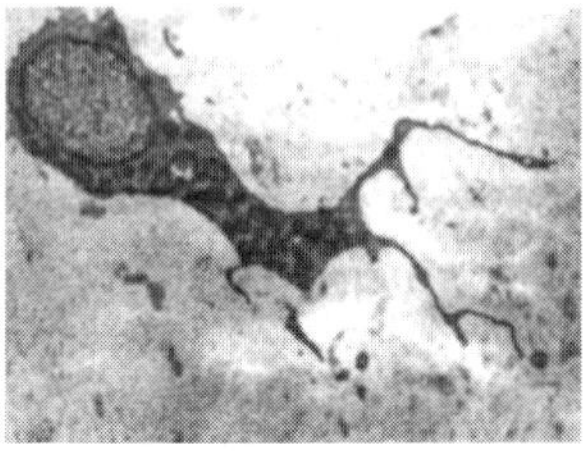

FIGURE 5. TEM of the internal fibrosa layer of a VAD patient. yofibroblast with numerous mitochondria, emanating multiple cytoplasmic extensions, and deeply intertwined with bundles of collagen. (x5300).

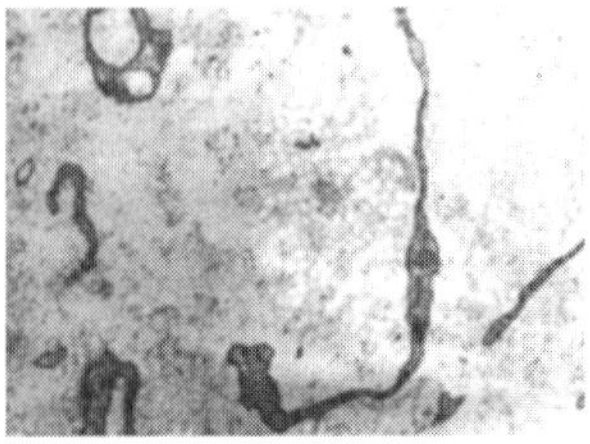

FIGURE 6. TEM of the external fibrosa layer of a VAD patient. Multiple view of fibroblast cytoplasmic extensions, intertwined with multiple collagen bundles. (x10000)

closely packed, desmosomes, and mostly located in the upper junctional area, closer to the pericardial sac (Fig. 5). Unusual gaps were seen in the lower section of the MC layer (Fig. 5), where desmosomes were less frequent. These gaps possess great plasticity, and are reversible occurrences.

DISCUSSION

The mechanical implications of this pericardial remodelling are far reaching. Although adaptation to LFN environments may be tentatively assumed, it is not trivial that not one single strand of cilia was identified in any of the multiple ultramicrographs of the 12 pericardial fragments. Ciliated structures under LFN stress are the object of an independent study (9). Hence, strictly speaking, biological adaptation may not be an adequate term for this process.

Taking together: a) the increased amount of collagen bundles, in wavy, accordion-like arrangements with b) different orientations in relation to each other, and c) with more than one elastic fiber accompanying the bundles at seemingly perpendicular angles, seems to suggest a pneumatic-like structure, designed to absorb abnormally large external forces. Most VAD patients exhibit labile hypertension, meaning that they suffer sudden and unusually violent tachycardia. This is a source of considerable cellular strain, exerted on the pericardium. The appearance of a soft tissue layer in between the thickened fibrosa layers, containing blood vessels, fatty and nerve tissue, as well as lymphatics, looks as if a layway station has been set up, in order to sustain this abnormally large organ. Thickened arteries have also been observed in LFN-exposed animal models (10), and in the autopsy of a deceased VAD patient (11).

Lastly, if the MC layer, a shearing surface, ceases to adequately function, the resultant adherences would interfere with the normal cardiac cycle. Therefore, maintaining the structural integrity of the one-cell deep mesothelial layer is of vital importance. By extruding older MC from the MC surface, the organism assures continued structural integrity of the MC layer, and a consequent intact cardiac cycle. This behavior could be explained by the principles of biotensigrity systems (12). Through desmosomes, all MC cytoskeletons of the pericardial mesothelium are interconnected through intermediate filaments. Given the violent and sudden tachycardia in these patients, MC surface integrity will strongly depend on the tensile strength of cytoskeletal intermediate filaments. Most probably, MC extrusion is energetically more efficient than reinforcement of cytoskeletal filaments. Spewing of cellular debris and extrusion of MC into the pericardial sac may explain why cellular features of apoptotic processes are not observed. Mechanical pericardial seizure can progressively limit the cardiac cycle. In conclusion, the authors are attempting to marshal bio-mathematicians,-physicists and –engineers to tackle the challenges offered by the response of biological tissues to LFN exposure.

ACKNOWLEDGMENTS

The authors thank the Portuguese Ministry of Defense (CIMO) for animal facilities, FLAD (Luso-American Foundation for Development) for continuous support, Prof. Nuno Grande (ICBAS) and Prof. Carlos Sá (CEMUP) for electron microscopy, Daniela Sousa Silva (CEMUP), António Costa e Silva and Emanuel Monteiro (ICBAS), and Carlos Lopes (CIMO) for technical support, and Pedro Castelo Branco for image treatment. Alves-Pereira & Joanaz de Melo also thank IMAR (Instituto do Mar) for hosting project POCTI/MGS/41089/2001 and FCT (Fundação para a Ciência e Tecnologia) for its funding.

REFERENCES

1. Shabetai, R.,"Diseases of the Pericardium," in *Hurst's The Heart, Arteries and Veins*, edited Schlant, R. C., and Wayne Alexander, R., New York; McGraw-Hill, 1994, pp. 1547-1649
2. Castelo Branco, N. A. A., and Rodriguez Lopez, E., , *Aviat. Space Environ. Med.***70**, A1-A6 (1999).
3. Castelo Branco, N. A. A., *Aviat. Space Environ. Med.***70**, A32-A39 (1999).
4. Castelo Branco, N. A. A., Rodriguez Lopez, E., Alves-Pereira, M., and Jones, D. R., *Aviat. Space Environ. Med.***70**, A145-A151 (1999).
5. Holt, B. D., "The Pericardium," in *Hurst's The Heart*, edited by Furster, V., Wayne Alexander, R., and Alexander, F., New York: McGraw-Hill, 2000, pp. 2061-82.
6. Castelo Branco, N. A. A., Águas, A., Sousa Pereira, A.,Fragata, J. I., Tavares, F., Grande, N., *Aviat. Space Environ. Med.***70**, A54-A62 (1999).
7. Castelo Branco, N. A. A., Fragata, J. I., Martins, A. P., Monteiro, E. and Alves-Pereira, M., *Proc. Inter. Commission. Biol. Effects Noise*, Rotterdam, Holland, 2003, pp. 376-377.
8. Castelo Branco, N. A. A., Castelo Branco, N. A. A., Fragata, J. I., Monteiro, E. and Alves-Pereira, M., *Proc. Intern. Comm. Biol. Effects of Noise*, Rotterdam, Holland, 2003, pp. 380-381.
9. Alves-Pereia, M., Castelo Branco, N. A. A., *Proc. Intern. Comm. Biol. Effects of Noise*, Rotterdam, Holland, 2003, pp. 366-367.
10. Castelo Branco, N. A. A., Alves-Pereira, M., Martins dos Santos, J., and Monteiro, E., "SEM and TEM Study of Rat Respiratory Epithelia Exposed to Low Frequency Noise," in *Science and Technology Education in Microscopy: An Overview*, edited by A. Mendez-Vilas, Badajoz, Spain: Formatex, 2003, pp. 505-533.
11. Castelo Branco, N. A. A., *Aviat. Space Environ. Med.***70**, A27-31 (1999).
12. Ingberg, D. E., *J Cell Sci* **104,** 613-627 (1993).

Applied Physics (APHYS 2003)
A. Méndez-Vilas (Ed.)

The Onset to Criticality in a Sheared Granular Medium

R. Lynch, D. Corcoran and F. Dalton

Department of Physics, University of Limerick, Limerick, Ireland.

Abstract. We have studied the temporal behaviour of an externally sheared granular medium in the transition from "mixed sliding and stick-slip" to "stick-slip" motion. These dynamic phases are morphologically distinct and are defined by the angular scale velocities of the system. The regime of stick-slip motion appears to be associated with a 2nd order rigidity phase transition, and the probability density distributions of slip event size are partial power laws with exponents that are robust to the changing scale velocities.

INTRODUCTION

Computational modeling has predicted that a second order phase transition in the shear modulus should exist for a compressed granular medium [1]. Recently, we presented experimental results supporting this prediction by showing that a granular bed under external shear exhibited stick-slip motion at low driving, and as expected of a critical phenomenon, the resultant fluctuations in slip size, duration and energy followed power law statistics [2]. Stick-slip motion has also been explored for solid-solid interaction [3]. In particular, at low driving, power law statistics for slip size distributions were observed only when the dynamic velocity (the maximum possible velocity of a slip) was low, else an exponential distribution of slip events resulted. The authors of this work proposed a dynamic phase diagram dependent on pulling, v_O, and dynamic, v_D, velocity phase parameters. Here we study the statistical behaviour of slipping events for an externally sheared granular medium in terms of this phase diagram, and their transition from a regime of mixed sliding and stick-slip, to stick-slip motion.

EXPERIMENTAL SYSTEM

The apparatus consists of an annular plate being rotated over a granular bed contained within a circular channel of width 75 mm. The plate has granular material attached to its base, to provide frictional contact, and is driven by a motor via a torsion spring. The apparatus is a modification of earlier work [2]. The boundary walls are more rigidly constructed and improvements to the bearings which couple the plate axle to the experiment rest frame, allow greater freedom of axle movement. During operation, if the plate sticks, the torque exerted by the torsion spring increases linearly in time until it is sufficient to overcome the frictional contact and the plate slips.

Driving and plate motion are recorded by rotary encoders, which have an angular resolution of 1.3 arc min. The plate is free to move vertically at the beginning of an experiment, but as the torque exerted by the spring increases, friction between the plate and the axle on which it is mounted, prevents further vertical motion. The apparatus thus self-determines a plate height with respect to the granular bed that allows for initial compaction or dilation. As a consequence, however, there is variation in the steady state torque of the torsion spring corresponding to a variation of the frictional state for the plate. We have studied the variation of the spring torque in time for a centrally positioned plate with width of 57 mm.

BEHAVIOURAL CLASSIFICATION

The results show two distinct types of characteristic plate motion which we refer to by type I and type II classifications (see Fig. 1). Both type I and type II are distinguishable in the temporal domain. Type I has predominantly symmetrical fluctuations of stick-increasing and slip-decreasing events while type II exhibits asymmetric fluctuations, which are "saw-tooth" in nature. Type II behaviour is only found at steady-state torques in the range 390-1200 N mm. At torques in the range 100-700 N mm, the plate can exhibit a mixture of both motions.

In fact the transition from the mixed type I+II behaviour to type II behaviour can also be identified by calculating the percentage of sliding to overall movement of the plate. When sliding decreases below 1% only type II behaviour is present, suggesting that type II behaviour is identifiable with stick-slip motion, while mixtures of type I+II are identifiable with mixed sliding and stick-slip motion.

The largest slips in type II fluctuations are at least an order of magnitude larger than

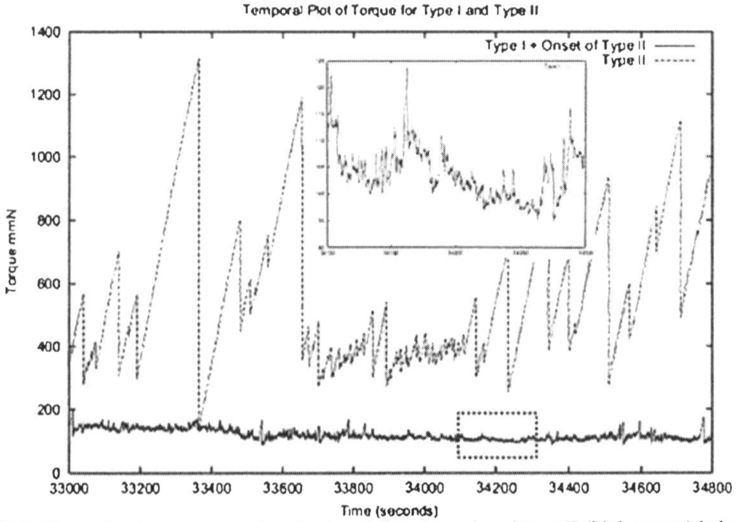

FIGURE 1. The spring torque versus time for type I (low torque) and type II (high torque) behaviours. Note the occasional type II event in the low torque state. Type I fluctuations are magnified in the inset.

those found in type I, and they have a preferred plate orientation. The preferred orientation suggests an experimental asymmetry such as a misalignment. In examples where the plate repeatedly switches between type I and type II modes, however, it does so without apparent regularity, and the preferred orientation of these large events changes. It would seem therefore that the preferred orientation is self-selected by the granular system. Such an effect could be caused by a slight tilt of the shearing plate with respect to the bed leaving a compressed medium in its wake, allowing the system to engage in a large slipping (resetting) event as the lowest plate position completes a cycle.

DYNAMIC PHASES OF SYSTEM

Experimental results have shown the dependence of the dynamic phase of the system on two scale parameters, the angular pulling velocity, ω_O, and the angular dynamic velocity, ω_D. The angular pulling velocity can be determined from the average turning speed of the drive axle. The angular dynamic velocity cannot be calculated directly from the experiment but it is taken as being proportional to the average torque being held by the torsion spring before a slip. To see this, consider that ω_D is the maximum possible angular velocity of a slip, which for an un-damped simple harmonic oscillator is $\omega_D = -\sqrt{k/I}\cdot\theta$, where k = spring constant, I = moment of inertia, θ = angle from equilibrium at the start of the slip [4]. Since the torque just before slipping is $\tau_S = -k\theta$ it follows that $\omega_D = \tau_S/\sqrt{k\cdot I}$. Therefore ω_D is proportional to τ_S, for constant k and I, and small system damping.

As described above, the set of shearing experiments can be classified into type I+II and type II behaviour. In Fig. 2, subject to these classifications, the experiments are

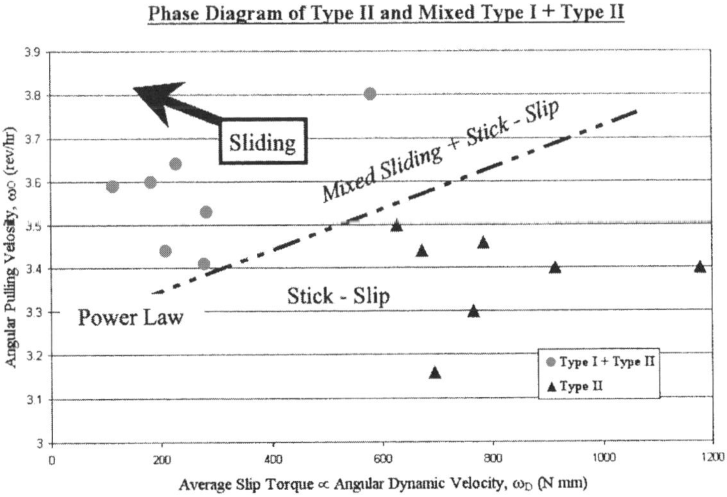

FIGURE 2. Dynamic phase diagram of sheared granular system.

plotted on a dynamic phase diagram and can be seen to form two distinct phase groupings. Increasing ω_D and decreasing ω_O causes a change from mixed sliding and stick-slip (type I+II) behaviour to stick–slip (type II) behaviour. The trend is consistent with predictions based on the dynamic phase diagram of a spring block system [3].

DISTRIBUTION OF SLIP SIZE OF EVENTS

The probability density distributions of slip event size have been found to follow partial power law decays (see Fig. 3). The power laws, and indeed the temporal dynamic behaviour are different to those obtained previously [5] and we identify these changes with a change in the dynamic phase space parameters.

The exponents of the power laws display a strong relationship with the steady state torque and consequently the steady state friction (Fig. 3 inset). For lower values of torque, corresponding to mixed type I + II behaviour, the exponents appear to increase linearly. The increase can be associated with the increasing appearance of type II fluctuations, which are distributed throughout type I as individual events and as isolated clusters of higher torque type II behaviour. From other work (not presented) it appears however that the linear change in exponent with increasing torque is a characteristic of the type I behaviour.

At the transition to type II behaviour, as can be seen in Fig. 3, the exponent saturates and becomes independent of the torque. We identify this behaviour with the 2nd order rigidity phase transition of a sheared granular medium [1]. Given that type II behaviour exhibits a power law which is robust over a wide range of torque (390-1200 N mm), it is possible that it represents a critical state with a finite basin of attraction,

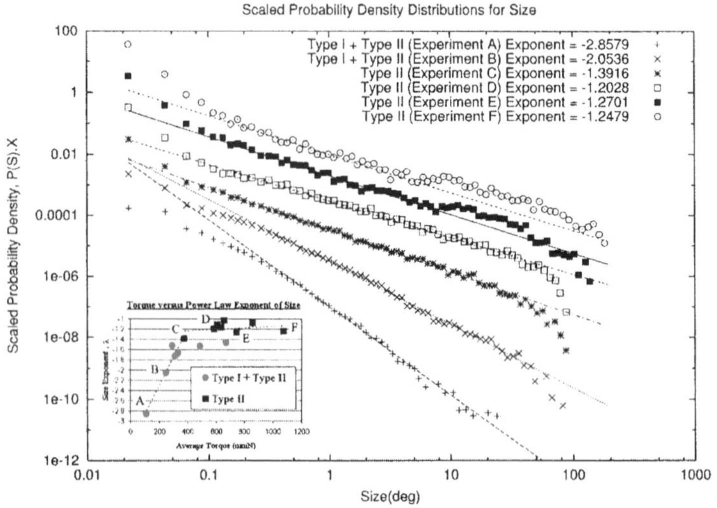

FIGURE 3. Probability Density Distributions of slip size displaced vertically for clarity. Inset shows the dependency, on Average Torque, of the Power Law Coefficient of Size.

what we have previously termed bounded self-organised criticality [5].

Closer examination of the type II PDDs of event size (Fig. 3) reveal that the distributions are altered as the average torque increases. In particular an excess of small events grows, the maximum slip size increases, and a broad large scale excess develops.

To understand these PDDs, we identify the critical state with the emergence of a stress chain network in the granular bed [6]. The eventual response of these stress chains to increased shear stress would be their reorganization [1] and this reorganization we believe is responsible for the observed scale invariant stick-slip motion. At increasing torque, corresponding to increased compression of the granular bed, it is expected that the granular stress chains would become denser [6]. The denser the supporting network, the smaller the possible rearrangements that could occur, allowing the power law adhering behaviour to extend to lower event sizes. This might then lead to events below the measurement resolution, and given the power law nature of the slip sizes, they would appear as an excess to the smallest events resolved. Therefore if the average torque/granular bed compression was to increase, an increase in the excess would be observed. At the same time, the denser stress chains, which would be more difficult to rearrange would allow an increase in the sustainable torque and resultant slip size of a resetting event. The broad excess of large events would then be a result of a scale-invariant spread of the maximum slip torque and the periodic nature of the resetting events.

CONCLUSION

We have identified two dynamic states for a sheared granular medium, type I, mixed sliding and stick-slip, and type II, stick-slip. The transition between these states is dependent on the angular dynamic, ω_D, and angular pulling, ω_O, scale velocities. The type II behaviour appears to be associated with a 2nd order rigidity phase transition, resulting in a robust partial power law exponent that is independent of the dynamic phase space parameters.

ACKNOWLEDGMENTS

We would like to thank Gerry Daly and acknowledge funding from the University of Limerick Foundation and Enterprise Ireland.

REFERENCES

1. E. Aharonov and D. Sparks, *Phys. Rev. E.* **60**, 6980-6896 (1999).
2. F. Dalton and D. Corcoran, *Phys. Rev. E.* **63**, 061312-1 -4 (2001).
3. A.Johansen, P.Dimon, C.Ellegaard, J.S.Larsen and H.H. Rugh, *Phys. Rev. E.* **48**, 4779-4790 (1993).
4. M. de Sousa Vieira, G.L. Vasconcelos, and S.R. Nagel, *Phys. Rev. E* **47**, R2221-R2224 (1993).
5. F. Dalton and D. Corcoran, *Phys. Rev. E.* **65**, 031310-1 –5 (2002).
6. H.A. Makse, D.L. Johnson and L.M. Schwartz, *Phys. Rev. Lett.* **84**, 4160-4163 (2000).

Applied Physics (APHYS 2003)
A. Méndez-Vilas (Ed.)

Electro-Optic Effect Induced In Glass Waveguides Containing A Charge-Trapping Layer

Yitao Ren, Carl Johan Marckmann, Rune Shim Jacobsen, and Martin Kristensen

COM, Technical University of Denmark, Building 345 west, DK-2800 Kgs.Lyngby, Denmark.

Abstract. Germanium-doped glass waveguides containing a thin silicon oxynitride layer as a charge trapper are thermally poled in air environment. The maximum electro-optic (EO) coefficient increases more than 20% compared to that induced in the waveguides without the trapping layer. The increased EO effect is attributed to the strength of the internal field by introducing the thin oxynitride layer in the waveguide structure. Our results demonstrate advantages of shaping the internal field to increase the optical nonlinearity in glass poling.

INTRODUCTION

A larger second-order nonlinear optical effect induced in poled silica or silica-based materials is of great interest because of its enormous potential to develop low-cost nonlinear silica components which can be integrated with optical waveguide components available, in particular electro-optic modulators for the telecom market. In our previous work [1], thermal poling was used to break the symmetrical structure of silica glass and to induce a stable electro-optic effect in germanium-doped silica channel waveguides. The explanations of the mechanism associated with the nonlinearity in poled silica are not fully satisfactory and intensive investigations are still in progress. Poling, according to the charge separation model [2, 3], induces an effective optical nonlinearity ($\chi^{(2)}_{eff}$) as a result of the interaction of the induced internal field (E_{int}) with the third-order nonlinearity ($\chi^{(3)}$) of silica material, $\chi^{(2)}_{eff} = 3\chi^{(3)}E_{\text{int}}$. Then $\chi^{(2)}_{eff}$ induced by poling in silica materials can be improved either strengthening E_{int} or increasing $\chi^{(3)}$. Here we report that more than 20% increase of the electro-optic (EO) coefficient is achieved by an optimized waveguide structure, where a charge-trapping layer is introduced to shape the distribution of the frozen-in field inside the structure. The results demonstrate directly that the charge-trapping effect leads to an improvement of the nonlinearity induced from the thermal poling. Results obtained earlier by Blazkiewicz *et al* [4] using a boron-doped layer in the poling of a twin-hole fiber were inconclusive, since the boron-doped layer can work both as charge trap and donor.

EXPERIMENTS

Waveguides containing a special charge-trapping layer were grown on n-type silicon wafers by thermal oxidation and plasma enhanced chemical vapor deposition (PECVD). The pure SiO_2 buffer layer from thermal oxidation helps to load a higher voltage during poling because of fewer defects introduced in the fabrication process referred to the PECVD process. The PECVD made core layer (~4 μm thick) consisted of germanium-doped silicon oxy-nitride, Ge:SiON (~5 at. % Ge and ~5 at. % N). Besides, a silicon oxy-nitride (SiON, ~9 at. % N and n=1.50 at 632.8 nm) layer of about 300 nm was deposited on the top of the core layer directly by a gas mixture of siliane (SiH_4), nitrous oxide (N_2O) and ammonia (NH_3) at a ratio of 1:94:21 to act as charge-trapping center increasing the charge density. Waveguide channels were formed by ultraviolet radiation from a KrF excimer laser. We used 3000 J/cm^2 pulse fluence of UV radiation for the channel formation. With the given core doping level, all formal channels from 4 to 10μm in width remained single mode. Waveguides broader than 10μm became multimode. These waveguides were poled thermally by a large static field applied at an elevated temperature from 345°C ~ 375°C in air environment. A negative bias referred to the silicon wafers is used deliberately to prevent more positive ions from entering into the samples. Due to increased conductivity from these ions in glass, a degradation of the electrode occurred leading to a failure of applying a higher poling voltage. The induced nonlinear effect of the poled samples was characterized using a fiber-based (single mode) Mach-Zehnder interferometer, and the linear electro-optic (LEO) coefficients were calculated by measuring the phase shift (compared with a reference phase shift from a $LiNbO_3$ phase modulator) from the waveguide samples.

RESULTS AND DISCUSSION

The measured dependence of the optical nonlinearity on the width of the channels is presented in figure 1, which shows that the channels of 10 μm exhibit the largest EO coefficients among all channels investigated. The result is attributed mainly to the overlap improvement between the excited mode in the channel and the space charge field, and is confirmed by a core confinement simulation. The calculated confinement of the UV-written core in the waveguide structure increases from 61% to 86% with enlarging of the channel width (figure 2).

The nonlinear properties of the waveguide samples, poled at 345°C ~ 375°C under -2.8 kV, were then characterized for the 10 μm channels and the results are shown in figure 4. An optimized EO coefficient (r_{TM}=0.093±0.001 pm/V, corresponding to $\chi^{(2)}$=0.230±0.002 pm/V) is achieved when the waveguides are poled at 357°C for 20 minutes under -2.8 kV (-208 V/μm). The maximum EO coefficient increased more than 20% compared to the waveguides without the silicon oxy-nitride trapping layer poled at the same conditions (-208 V/μm, 357°C). We find a nearly unchanged optimum poling temperature (T_{opt}=357°C) for our waveguides with and without the trapping layer. This is probably due to their similar structures and layer thickness.

However, the EO coefficients of the waveguide with the trapping layer are more sensitive to small changes of the poling temperature from the optimum.

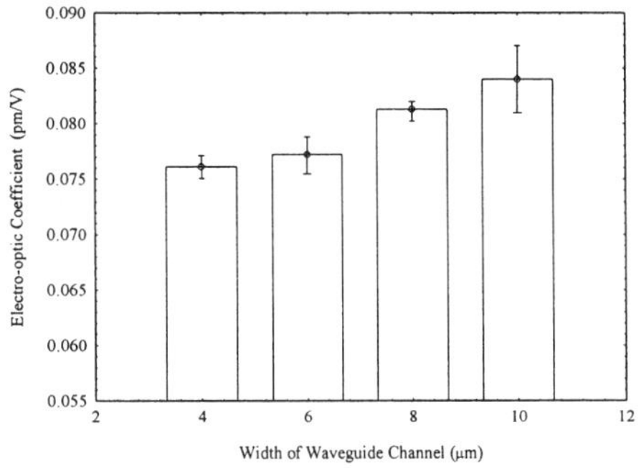

FIGURE 1. Electro-optic coefficient dependence on the waveguide widths (poled at –208 V/μm and 357°C).

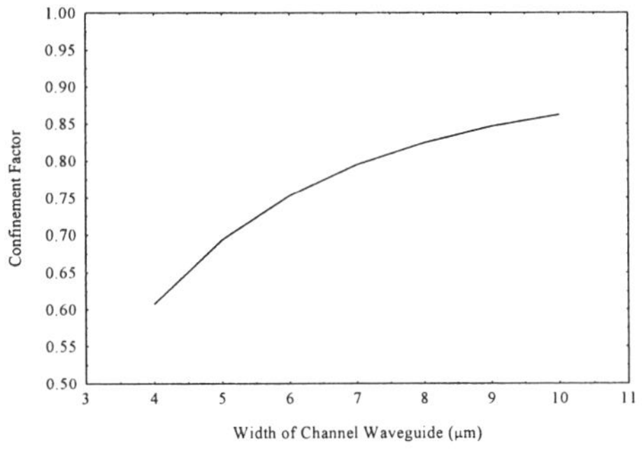

FIGURE 2. This is the Style for Figure Captions. Center this text if it doesn't run for more than one line.

The mobile charges (e.g. positive ions, electrons, etc…) in the glass are driven towards the electrodes when an external poling field is applied across the waveguides during the poling. Because the interfaces (between core and cladding layers) of the waveguides, acting as barriers formed by the conductivity difference between the layers [5,6], hinder the charges' motion in the waveguides, these charges accumulate at the interfaces and form two charge layers with different polarity on each side of the core. Consequently, the internal field (E_{int}) is built across the core layer when the whole poling process is finished. Trapping more charges at the interfaces will strengthen E_{int} and favor the induced nonlinear effect. Figure 3 shows EO coefficients measured from the poled waveguides with and without the trapping layer, which suggests that the distribution of the internal frozen-in field is changed and its intensity is reinforced after poling as a result of adding of the thin silicon oxy-nitride layer with charges trapped inside. At the same time the $\chi^{(3)}$ values remain nearly unchanged before and after poling for all the samples measured by a Bragg grating technique [7]. Increased optical nonlinearity and a gain of EO coefficient are therefore achieved via the stronger interaction between the internal field and the third-order nonlinearity of the waveguide material. However, the EO coefficients decrease significantly with any change of the temperature around the optimum, this may be due to the reduction of E_{int} from a loss of the trapped charges in the trapping layer. For the waveguides without the trapping layer, the internal frozen-in field is built mainly from the limited charge accumulation at the interfaces.

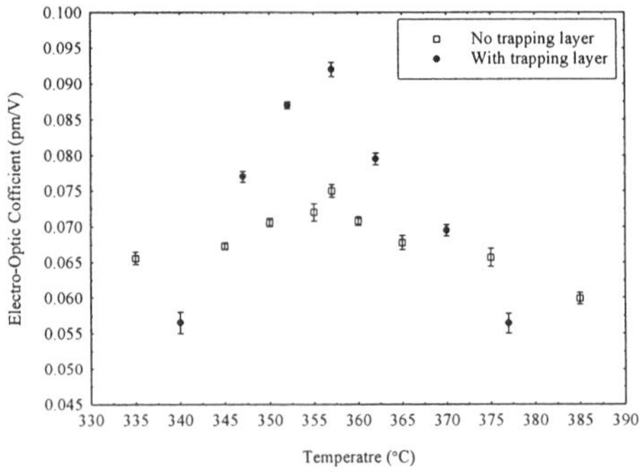

FIGURE 3. The dependence of EO coefficients on the poling temperature for the channel waveguides (TM mode) with and without charge-trapping layer. The samples were poled at –208 V/μm for 20 minutes in air environment.

Our results demonstrate an optimized charge distribution around the core by adding a charge-trapping layer in the waveguide structure and increased optical nonlinearity through the increased frozen-in field. The thin silicon oxy-nitride trapping layer does not show obvious influence on the optimized poling temperature. The increase of the EO effect from the charge-trapping layer improves the prospects to fabricate EO components on a chip. Further studies are in progress on the optimization of the charge distribution and the stability of the EO effect.

REFERENCES

1. Y. Ren, C. J. Marckmann, J. Arentoft, and M. Kristensen, *IEEE Photon. Technol. Lett.,* **14** (5), 639-641, (2002).
2. P. G. Kazansky, P. S. Russel, *Optics Communications*, **110**, 611-614 (1994).
3. R. A. Myers, N. Mukherjee and S. R. J. Brueck, *Opt. Lett.*, **16**, 1732-1734 (1991).
4. P. Blazkiewicz, W. Xu, D. Wong, S. Fleming, T. Ryan, *Journal of lightwave technology*, **19**, 1149-1154 (2001).
5. D. Faccio, A. Busacca, D. W. J Harwood, G. Bonfrate, V. Pruneri, P. G. Kazansky, *Optics Communications*, **196**, 187-190 (2001).
6. J. Arentoft, K. Pedersen, S. I. Bozhevolnyi, M. Kristensen, P. Yu, C. B. Nielsen, *Appl. Phys. Lett.*, **76**, 25-27 (2000).
7. C. J. Marckmann, Y. Ren, G. Genty, and M. Kristensen, *IEEE Photon. Technol. Lett.*, **14**, 1294-1296 (2002).

Applied Physics (APHYS 2003)
A. Méndez-Vilas (Ed.)

A Novel Method Of Measuring Light Absorption On A Self-Assembled Single Quantum Dot

B. Alén[1,4], F. Bickel[1], A. Hoegele[1], K. Karrai[1], R. J. Warburton[2], P. M. Petroff[3] and J. Martínez-Pastor[4]

[1]*Center for NanoScience, Sektion Physik, Ludwig-Maximilians Universität, D-80539 Munich, Germany.*
[2]*Department of Physics, Heriot-Watt University, Edimburgh EH14 4AS, UK.*
[3]*Materials Department, University of California, Santa Barbara, CA93106, United States of America.*
[4]*Materials Science Institute. University of Valencia, E46071 Valencia, Spain.*

Abstract. We present a novel method by wich excitonic interband optical transitions within single InAs self-assembled quantum dots can be directly observed in a transmission experiment. Due to the extremely high resolution of the tecnique, individual peaks associated to single exciton absorption resonances in single quantum dots can be spectrally resolved. Using this technique we investigate the oscillator strength, homogeneous linewidth and fine structure splitting in a collection of such individual resonances.

INTRODUCTION

The extremely small absorption cross section of InAs based single quantum dots has prevented up to date the possibility of performing absorption spectroscopy in this system, even if it has been previously achieved in single molecules [1], ions [2] or single GaAs/AlGaAs quantum well islands [3]. The main difficulty lies in the fact that the detection system has to extract the missing photons out of the orders of magnitude larger number of photons that impinges into the detector without interacting with the weak scatterer. Therefore, a prerequisite to improve the absorption signal is to reduce the spot size to allow the light to interact more efficiently with the single quantum dot. While the highest spatial resolution can be achieved by using a near field optical microscope, its operation is highly demanding and has been applied with limited success to the optical absorption of InAs quantum dots [4]. A further improve in the signal to noise ratio can be achieved by reducing the laser intensity fluctuations that dominate the noise level. In a temperature stabilised tunable diode laser, an important source for these fluctuations are the cavity mode inestabilities originated during the laser wavelength tuning. To avoid them, we can take advantage of any method capable of keeping constant the probing light energy while the probed resonance is tuned across it. This reversed sweep can be done, potentially, varying any external parameter like temperature, uniaxial or hydrostatic stress, electric field, magnetic field, etc, since all of them have a well known energy dispersion relation. However, in practice the suitable parameter has to satisfy some requirements like not disturbing the mechanical stability of the optical system or being available with enough precision and in a broad range to provide the desired spectral resolution and feasibility. In this work we present a practical realisation of such experimental scheme by which we have investigated the interband optical transitions between excited states of single self-assembled InAs quantum dots embedded in a field effect structure.

SAMPLE AND EXPERIMENT

The sample investigated in this work consists in a field effect device formed by a highly n-doped 20 nm-thick GaAs back-contact and a Ni-Cr metalized surface gate electrode. Both electrodes are separated by an intrinsic region where a single layer containing self-assembled InAs quantum dots was grown by molecular beam epitaxy in the Stranski-Krastanow mode. The dot layer is separated from the back-contact by a 25 nm-thick intrinsic GaAs layer which acts as tunneling barrier for the electrons. To further confine the carriers in the intrinsic region an AlAs/GaAs superlattice was also grown separating the dots from the gate electrode. The details of the growth has been published elsewhere [5]. In previous studies the central role played by the electric field in the electronic and optical properties of the dots was demonstrated. On the one hand, at negative voltages, the band bending produced by the field allows the electrons to tunnel into the dots determining their charging state. This produces several peaks in the capacitance-voltage curves when the fermi level of the structure crosses one empty electronic level in the conduction band. Indeed, varying the gate voltage V_g, the majority charging state of the dot ensemble can be controlled from N=0 (large negative V_g) to N=6, when the s-shell and p-shell of the conduction band are completely filled ($V_g \sim 0$). [6,7>footnote]. On the other hand, ensemble optical transmission experiments have shown that depending on the charging state the absorption signal can be totally quenched by Pauli-Blocking the electron state (case of the s-shell for $N \geq 2$, for instance), while, for a given charging state, the energy levels can be shifted by the quantum confined Stark effect if V_g is changed [8].

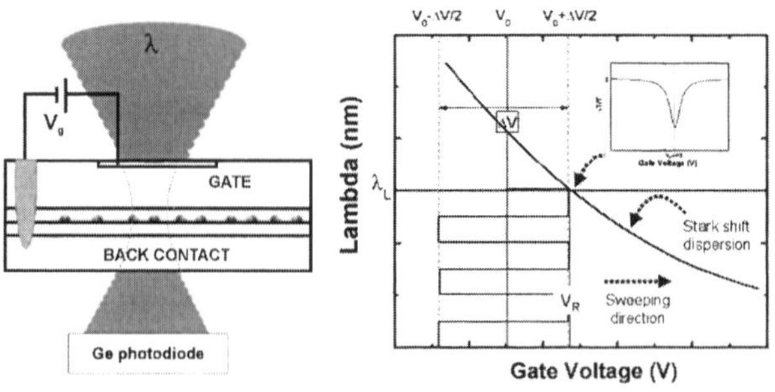

FIGURE 1. Experiment schematics (left) and stark shift modulation spectroscopy principle (right).

In Fig. 1 the schematics of the experiment are shown. The sample is piezoelectrically positioned in the focused spot (1.3 μm FWHM) of a confocal fiber-based microscope working at 4 K. Beneath the sample, an unbiased Ge photodiode detects the photons delivered by a stabilised narrow-band tunable laser, and transmitted through the sample. To avoid any cavity mode inestability, the laser wavelength is kept constant, being resonant, in this case, with the first excited state absorption optical transition centered at 1.17 eV as identified previously in ensemble transmission experiments. Now, in order to spectrally resolve the optical absorption, the voltage applied to the gate electrode is sweeped. It is well-known that the energy position of any given electronic level will be Stark shifted by the electric field. Hence, we can assume that there must be a voltage V_R where the interband optical transition is resonant with the incoming photons energy. At this voltage the absorption is non zero, although very low, and can be detected by a change in the transmission signal. In the right panel of Fig. 1, this principle is depicted. An additional advantage of the method, despite the granted laser stability, is the fact that the voltage itself can be modulated very easily allowing the use of phase sensitive amplifiers. This voltage modulation skips any signal dependence on the dynamic range of the photodector, a common drawback that is present when the beam is chopped externally. In the present experiment, the photon flux in the photodiode is almost constant excepting for the small number of photons taken by the electronic resonance itself, further improving the signal to noise ratio and allowing the detection of transmission changes up of 10 ppm. A detailed analysis of the Stark shift modulation principle and a complete description of the experiment set-up can be found elsewhere [9,10].

RESULTS AND DISCUSSION

Figure 2 shows two differential transmission spectra obtained at 4 K in different points of the sample. The voltage modulation peak amplitude, δV, was equal to 1 mV, which corresponds to an energy modulation $\delta\omega$=2.78 μeV and determines our spectral resolution [9]. The detected optical response do not resemble a lorentzian shape because we are using a modulation amplitude smaller than the homogeneous linewidth of the optical transition, $\Gamma=2\gamma$. In this regime, each single feature of the spectrum can be well described by a lorentzian derivative function given by:

$$\frac{\delta T}{T}=\frac{4\alpha_0\gamma^2\omega\delta\omega}{\left(\omega^2+\gamma^2\right)^2}. \qquad (1)$$

where α_0 stands for the maximum absorption strength or peak amplitude of the corresponding lorentzian curve. As observed in Fig. 2, the number and position of the peaks changes when the focused spot is displaced over the sample. This is an unequivocal evidence of the single quantum dot origin of the features observed, and is a direct consequence of the spatial and spectral inhomogeneity of the quantum dot ensemble. Furthermore, it should be noticed that given the spectral resolution of the technique, each peak in the spectra may be originated from non-degenerated single excitonic resonances. As an example, we can consider the typical values for energy splittings due to electron-hole exchange interaction in InAs quantum dots as recently observed in single quantum dot PL experiments [11]. Although their actual value depends on the particular dot, typical measured values are in the range of ten to hundred microelentronvolts, which count among the smallest fine structure splitting energies present in this system but are still larger than our resolution. This unique feature enables this method for the study of the oscillator strength and homogeneous linewidth of single interband optical transitions. In the case of excited state transitions, this investigation yields direct information about the carrier relaxation mechanisms which is model dependent using other indirect tecniques like excitation of the photoluminescence [12].

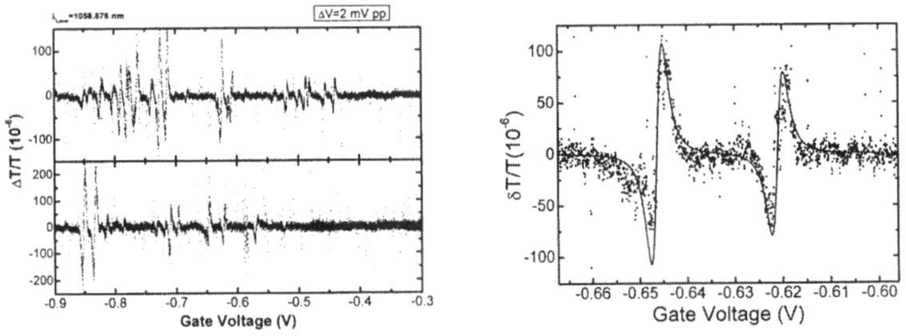

FIGURE 2. Differential transmission spectra obtained in two different points of the sample. Each spectrum shows several single excitonic absorption resonances from the different dots present in the spot (left). Two individual resonances together with the curve resulting of fitting the experimental data to equation 2 are shown in the right panel.

We have performed statistics on more than hundred resonances by measuring the differential absorption in different points of the sample and fitting the experimental data to equation 2. A typical fit is shown in the right panel of Fig. 2. Figure 3(a) shows the number of peaks as a function of the applied voltage which reveals a broad distribution centered at –0.65 V and covering a total spectral range of around 1.5 meV. In this voltage region most of the dots are singly or doubly charged, as demonstrates the capacitance-voltage characterization, although the individual charge state can not be determined just attending to its voltage position. This information, however, can be infered from the homogeneous linewidth distribution depicted in Fig. 3(b). This distribution is very steep having its maximum at around Γ=14 μeV, which corresponds to a typical excitonic dephasing rate of ~ 50 ps. This result is remarkable since is well known that the electron energy relaxation is a very efficient proccess that occurs in less than 10 ps by LO phonon emission. The fact that we systematically find dephasing rates larger that 40 ps is related to the multielectronic charge configuration in the conduction band. Indeed, with two electrons in the fundamental level of the conduction band, the relaxation of the electron from the excited to the ground state can not take place due to the Pauli exclusion principle. Hence, the exciton dephasing is determined by the relaxation of the hole that at 4 K occurs by emission of LA phonons which is much more slower [12]. This direcly proves that the majority of the observed resonances originate from dots doubly charged by electrons injected from the back contact. Nevertheless, as observed in Fig 3(b) there is a small number of absorption peaks whith larger linewidths up to ~ 60 μeV, which would correspond to optical transitions in dots singly charged or with no electrons at all.

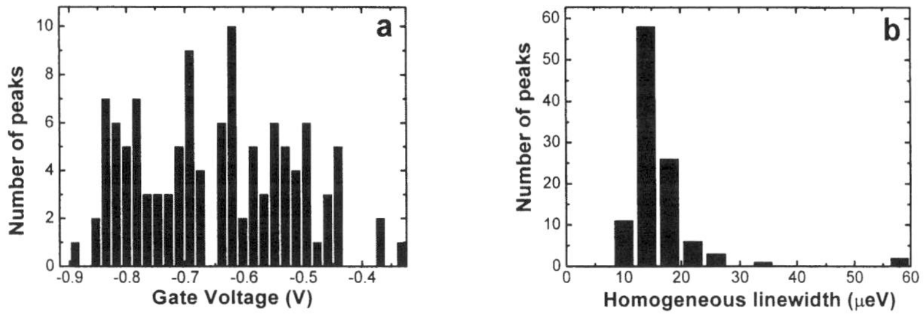

FIGURE 3. Statistics on more than hundred resonances have been performed. The distribution of peaks as a function of the applied voltage is shown on the left panel. On the right the homogeneous linewidth distribution for the different peaks is depicted.

To conclude, we have demonstrated that single excitonic resonances can be detected in an optical transmission experiment with extremely high signal to noise ratio and spectral resolution. The new tecnique developed opens the way to the direct investigation of fine structure splittings, oscillator strength and homogenous linewidth of charged and neutral excitons mapping the complete electronic structure of InAs self-assembled single quantum dots.

ACKNOWLEDGMENTS

We acknowledge useful discussions and technical help from C. Schulhauser. Financial support for this work was provided in Germany by the SFB348, in the UK by the EPSRC and in Spain by the CICYT. B. Alén wants also to acknowledge to the Marie Curie EU association for its financial support.

REFERENCES

1. D. J. Wineland *et al*, Opt. Lett., **6**, 389 (1987).
2. W. E. Moerner *et al*, Phys. Rev. Lett. **62**, 2535 (1989).
3. J. R. Guest *et al*, Phys. Rev. B **65**, 24310-1 (2002).
4. T. Matsumoto *et al*, Appl. Phys. Lett. **75**, 3246 (1999).
5. D. Leonard *et al*, Phys. Rev. B **50**, 11687 (1994).
6. A. Lorke *et al*, Microelectronic Engineering **47**, 95 (1999).
7. The s-shell and p-shell nomenclature refers to the electronic states of a parabolic in-plane potential. For a complete description of the degeneracies and coulomb interactions in this system see R. J. Warburton *et al*, Phys. Rev. B **58**, 16221 (1998).
8. R. J. Warburton *et al*, Phys. Rev. Lett. **79**, 5282 (1997).
9. B. Alén *et al*, Appl. Phys. Lett. **83**, 2235 (2003).
10. A. Hoegele *et al*, in *Proceedings of the MSS11-Eleventh International Conference in Modulated SemiconductorStructures*. Nara (Japan), 2003. To be published in Physica E as special issue.
11. B. Urbaszek *et al*, Phys. Rev. Lett. **90**, 247403-1 (2003).
12. B. Alén *et al*, in *Proceedings of the MSS11-Eleventh International Conference in Modulated SemiconductorStructures*. Nara (Japan), 2003. To be published in Physica E as special issue.

Applied Physics (APHYS 2003)
A. Méndez-Vilas (Ed.)

RADIATIVE EXCITON LIFETIMES IN INDIUM ARSENIDE SELF-ASSEMBLED QUANTUM WIRES

DAVID FUSTER*, JUAN MARTÍNEZ-PASTOR, JORDI GOMIS

Instituto de Ciencia de los Materiales, Universidad de Valencia, P.O. Box 22085, 46071 Valencia, Spain.

LUISA GONZÁLEZ* AND YOLANDA GONZÁLEZ*

**Instituto de Microelectrónica de Madrid (CNM-CSIC), Isaac Newton 8, 28760 Tres Cantos, Madrid, SPAIN.*

In this work we investigate the exciton recombination dynamics in InAs/InP semiconductor self-assembled quantum wires, by means of continuous wave and time resolved photoluminescence. The continuous photoluminescence results seem to indicate that the temperature quenching mechanism of the emission band is due to unipolar thermal escape of electrons towards the InP barrier. On the other hand, the time resolved photoluminescence study reveals that the temperature dependence of the radiative and non-radiative recombination times is characteristic of a dynamic between localized and free excitons in the quantum wires.

Introduction

Due to the increasing application potential of semiconductor quantum nanostructures, like quantum dots and quantum wires (QWr's), for the development of room temperature operating heterostructure lasers and high-speed optoelectronic devices, the carrier trapping and thermal escape processes are of considerable interest. In particular, self-assembled QWr's of InAs grown on InP (001) are good candidates for their use as the active region in optoelectronic devices working at the wavelengths of 1.30 and 1.55 μm [1], very useful for fiber telecommunications. In low-dimensional semiconductor structures, fluctuations of the nanostructure's size and/or the alloy composition result in a spatial variation of the material's band gap. The excitons trapped in such localization centers can remain stable with negligible lateral diffusion, mainly at low temperatures. The existence of localized and free accessible states for excitons will affect the thermal escape dynamics of carriers between the low-dimensional structures and the barrier [2].

In this work we present a study of the exciton dynamics in a set of two samples containing InAs/InP QWr's, carried out by means of continuous wave photoluminescence (PL) and time-resolved PL (TRPL) as a function of the temperature. The temperature evolution of the PL integrated intensity and decay time points out to an exciton recombination dynamics limited by localization effects.

Samples and experiment

The samples under study consist of a 100 nm thick InP buffer layer grown by solid source molecular beam epitaxy (MBE) on InP(001) substrate followed by a 1.6

monolayer (ML) thick InAs layer deposited at a substrate temperature T_S=515 °C, at a growth rate of 0.1 ML/s in a pulsed dynamic way (pulsed Indium cell sequence: 1s ON / 2s OFF). The deposited InAs thickness corresponds to the critical thickness in order to form the QWr, which is detected when the [1-10] reflection high-energy electron diffraction (RHEED) pattern shows the 2D-3D transition [1, 3]. Finally, a 20 nm thick cap layer is grown at a substrate temperature T_S=515 °C, by MBE (HT sample), and T_S=380 °C, by atomic layer molecular beam epitaxy (ALMBE) [4] (LT sample). For the ALMBE grown InP cap layer, the stoichiometry of the growth front is controlled by means of the reflectivity difference (RD) technique in order to obtain InP planar surfaces [5].

The continuous wave PL experiments were performed by using the 514.5 nm Ar^+ laser line as excitation source. The PL signal was dispersed by a 0.22 m focal length monochromator and synchronously detected with a cooled Ge photodiode. In time resolved experiments, sample excitation at 730 nm was done by a Ti:sapphire picosecond pulsed laser (Mira 900D, Coherent) pumped with a 5 W doubled Nd:YAG laser (Verdi, Coherent). The PL signal was dispersed by a single 0.5 m focal length imaging spectrometer and detected by a synchroscan streak camera (Hamamatsu C5680) with a type S1 cooled photocathode. The overall time response of the system in the widest temporal window (2 ns) was around 40-50 ps (full width at half maximum). In both kinds of experiments the samples were held in the cold finger of a closed-cycle cryostat to vary the temperature in the range 10 – 300 K, approximately.

Results

The continuous wave PL spectra at low temperature (12 K) for both samples are reported in Fig. 1.a. It is worth noting the fact that the PL band at room temperature is centered at the wavelengths of 1.30 and 1.55 μm for HT and LT samples, respectively, which makes this kind of samples potential candidates to fabricate optoelectronic devices for telecommunications. Several Gaussian lines can be deconvoluted from the PL band, which are related to the existence of wire height fluctuations. In this way, the energy difference between two adjacent Gaussian lines agrees approximately with a fluctuation of 1 ML in the QWr height [6, 7]. Considering the position of the most intense PL peaks shown in Fig. 1.a, we can estimate that the average wire height increases from 6 ML to 12 ML, from HT to LT sample. This important change in the wire height can be correlated with the different substrate temperature used for the growth of the InP cap layer (515 °C and 380 °C in samples HT and LT, respectively). In such a situation we can expect a different As/P exchange rate at the interface between the InAs QWr layer and the InP cap layer [8].

Typical PL decay curves from TRPL measurements at low temperature for the P2 component of HT sample are shown in Fig. 1.b for different excitation powers. An excitation power density of 40 kW/cm^2 has been used for the temperature evolution of the PL decay time, because we obtain a good detection signal over the whole temperature range. Above that excitation density the PL transient is not mono-exponential.

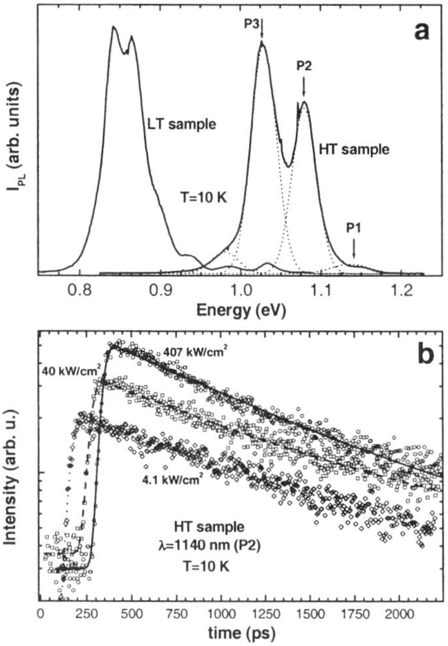

Fig. 1. (a) PL spectra measured at low temperature for HT and LT samples (continuous lines) and Gaussian lines (P1,P2,P3,P4) corresponding to the best multi-Gaussian fit to the PL band of the HT sample; (b) TRPL spectra of the P2 component in the HT sample at several excitation powers.

The evolution of the overall PL integrated intensity with the temperature is shown in Fig. 2. At room temperature the integrated intensity is two orders of magnitude lower than that measured at 12 K for both samples.

The TRPL study was limited to the HT sample because the quantum efficiency of the S1 photocathode in the streak camera drops exponentially above 1300 nm. However the phenomenology under continuous wave conditions is similar in both samples, which makes conclusions in this sample to be general. In Fig. 3.a we report the temperature dependence of the PL decay time for the P2 component in the HT sample. In contrast to the fast quenching of the PL band in the high temperature range, the PL decay time decreases slowly in the same temperature range. Such a different behavior has not been observed in quantum well heterostructures [9], and could be attributed to exciton localization effect in the InAs QWr, as explained below.

Discussion

The experimental temperature dependence of the PL integrated intensity shown in Fig. 2 can be accounted for by a Boltzmann model for excitonic recombination with two

Table I. Best fitting parameters to the experimental temperature dependence of the PL integrated intensity by using a Boltzmann model for HT and LT samples.

Sample	Γ_1 (s^{-1})	Γ_2 (s^{-1})	E_1(meV)	E_2(meV)
High T	$2.8 \cdot 10^9$	$8 \cdot 10^{12}$	20	140
Low T	$3.5 \cdot 10^9$	$20 \cdot 10^{12}$	21	145

quenching mechanisms [10] (the equation is given in Fig. 2 as an inset). I_0 is the integrated PL intensity at 0 K, E_1 and E_2 are the activation energies of the two quenching mechanisms, Γ_1 and Γ_2 the two related scattering rates and τ_0 the radiative recombination time of the excitons in QWr's, assumed to be constant, in a first approximation, and fixed to 1 ns [11]. The best fitting values for these parameters are shown on table I. The higher activation energy (E_2) is associated with the high temperature quenching of the PL band, and hence this parameter can be associated to the intrinsic non-radiative recombination mechanism in self-assembled QWr's. The best fitting parameters listed in table I correspond to the temperature evolution of the whole PL band. A similar study for the different components in sample HT, P1, P2 and P3 (see Fig. 1.a), provides activation energies E_1= 11, 10, 32 meV and E_2= 82, 89, 177 meV, respectively. Thermal escape of carriers, towards the barrier material (InP in our case), either ambipolar [10] or unipolar [9], is known as the most important temperature dependent non-radiative mechanism in quantum wells, and could be also appropriate in the case of QWr's. If the observed PL quenching mechanism was an ambipolar thermal escape (simultaneously electron + hole), we can estimate an average escape energy, difference between InP gap energy and average PL peak energy, to be about 360 and 550 meV for HT and LT samples, respectively. These values are well above the values obtained for E_2 in the fit to the Boltzmann model (see table I), which are close similar in both samples. On the other hand, if unipolar escape

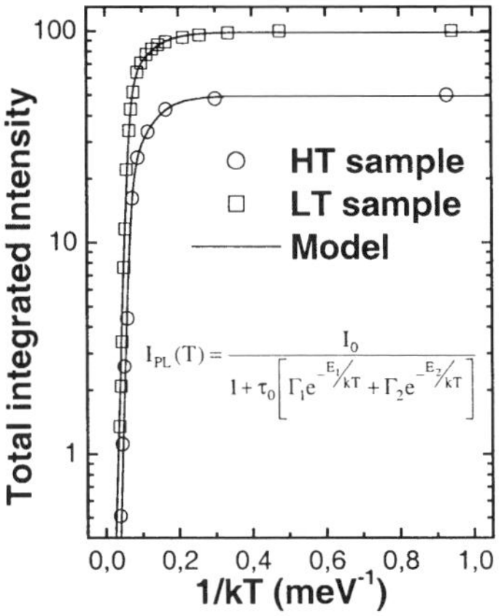

Fig. 2. Arrhenius plot of the PL integrated intensity for HT an LT samples and best fitting curves according to a Boltzmann model for excitonic recombination (equation is shown in the same plot).

is considered, the escape energy would be 380 to 290 meV for holes and 300 to 75 meV for electrons (the value is lower for smaller wires of the distribution) [7]. Therefore, unipolar thermal escape of electrons could be probably the non-radiavtive quenching mechanism limiting the emission of our samples at room temperature.

A separation of the radiative and non-radiative recombination times is possible under the assumption that the measured decay rate is given by the sum of the radiative and the non-radiative rates [12],

$$\frac{1}{\tau(T)} = \frac{1}{\tau_{rad}(T)} + \frac{1}{\tau_{nrad}(T)}. \qquad (1)$$

Hence, for the temperature-dependence PL intensity we can write

$$I(T) = I_0 \frac{\tau(T)}{\tau_{rad}(T)}. \qquad (2)$$

We assume a purely radiative recombination at low temperature (T=12 K) which yields $I_0 = I\,(12\ K)$. Then, we get for the radiative and the non-radiative recombination times,

$$\tau_{rad}(T) = \frac{I(12K)}{I(T)}\tau(T),$$
$$\tau_{norad}(T) = \frac{I(12K)}{I(12K) - I(T)}\tau(T). \qquad (3)$$

I(T) and τ(T) were determined experimentally (see Fig. 2 and Fig. 3.a). Figure 3.b shows the temperature evolution of the radiative exciton lifetime and non-radiative recombination time for the emission component P2. Radiative recombination dominates at low temperatures and the exciton lifetime increases at a rate about 8 ps/K, similar to the increase observed in quantum wells [13, 14]. Above 100 K the non-radiative channels are dominant, but the non-radiative time decreases at a slower rate than that expected for electron escape out the QWr's towards the InP barrier.

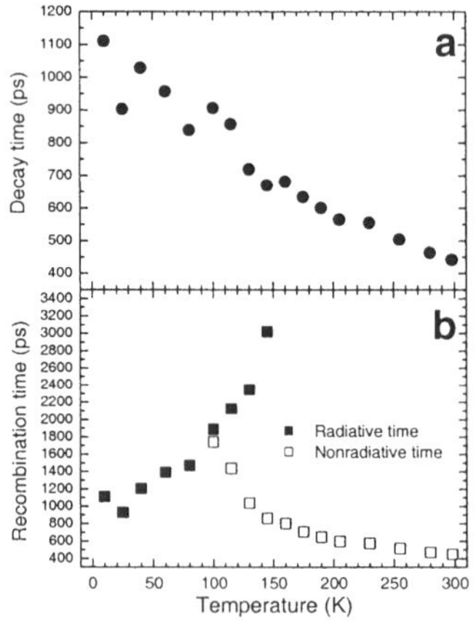

Fig. 3. Decay time (black circles) of the P2 component measured on the HT sample (a), and its separation in radiative (black squares) and non-radiative (white squares) contributions (b).

For a better interpretation of this unexpected behavior we can solve a rate equation model based on a three energy levels system: the InP barrier, free and localized exciton states (similar to the model proposed in Ref. 2). Free excitons (F) are formed by a generation rate g, corresponding to the carrier relaxation after arrival of the pulsed laser excitation. The free excitons can either recombine radiatively (radiative time τ_F), be captured into localized states at a rate $1/\tau_{FL}$, or escape towards the barrier at a rate Γ_{BF}, with a thermal activation energy E_A. The localized excitons (L) can either recombine radiatively with a lifetime τ_L or thermally transform into free excitons states. The equation rate system is then:

$$\frac{dF}{dt} = -\frac{F}{\tau_F} - (1 - L/N_L)\frac{F}{\tau_{FL}} - F\Gamma_{BF}\exp(-E_A/kT) + L\frac{\exp(-\Delta E_L/kT)}{\tau_{FL}} + g \qquad (4)$$

$$\frac{dL}{dt} = -\frac{L}{\tau_L} - L\frac{\exp(-\Delta E_L/kT)}{\tau_{FL}} + (1 - L/N_L)\frac{F}{\tau_{FL}} \qquad (5)$$

where ΔE_L is the activation energy for the thermally induced transition from localized to free exciton states, N_L the total density of localized states, and $(1\text{-}L/N_L)$ the factor that takes into account possible saturation effects. The PL transient would be determined by:

$$I(t) = \frac{F}{\tau_F} + \frac{L}{\tau_L} \qquad (6)$$

The PL integrated intensity gives account of the QWr's exciton depopulation when the temperature increases. The main channel for QWr's depopulation is the thermal escape towards the barrier which has the activation energy E_A. Such energy would correspond to the value E_2 obtained above with the Boltzmann model. An Arrhenius plot of the non-radiative time for the different PL components in HT sample is shown in Fig. 4. We obtain activation energies smaller than those obtained with the PL integrated intensity for the different Gaussian components (E_2= 82, 89, 177 meV). This disagreement can be explained within the model described above if the radiative recombination time for free excitons (τ_F) is considered longer than that corresponding to the localized excitons (τ_L). In this way, the activation energies deduced from results in Fig. 4 would correspond to the thermal transfer energy from localized to free exciton states. Given that the population of localized exciton states depends on the population of the free exciton states, according to equation (5), we expect the continuous wave PL to be quenched with the activation energy associated to the unipolar thermal escape of electrons towards the InP barrier. Moreover, this transition energy depends on the PL band component, the higher the exciton optical emission energy, the smaller free-localized exciton states transition energy. This is just we obtain from the results shown in Fig. 4: the activation energy for localized-to-free exciton transfer increases from 2 to 30 meV from the higher (P1) to the lower energy (P3) Gaussian components. On the other hand, from the fits performed on these results, we find τ_{FL} = 218, 225, 210 ps for the P1,P2 and P3 components, respectively. These values are comparable to the value of Γ_1 (357 ps, see table I) obtained above with the Boltzmann model. Indeed, the values found there for the E1 activation energies, 11, 10 and 32 meV for the three Gaussian components of the HT sample, could be now clearly associated to ΔE_L. Evidently, the values found from TRPL results (Fig. 4) are more direct and reliable, because the continuous wave PL intensity in the Boltzmann model contained the term $1/(1+\tau_0)$, where τ_0 in this model would be the radiative lifetime of localized excitons, assumed to be temperature independent [10].

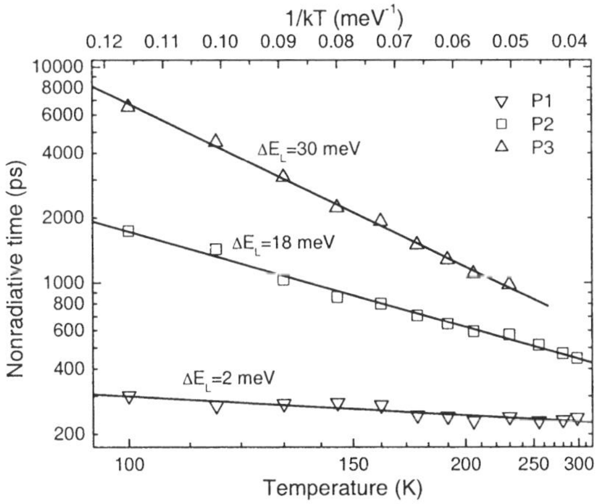

Fig. 4. Arrhenius plot of the non-radiative time for three of the Gaussian components in the HT sample (symbols) and the best linear fits (lines).

Summary

We have developed a process for controlling the optical emission wavelength of InAs QWr's growth on InP (001) based on the control of the As/P exchange by changing the growth temperature of the InP cap layer. The PL emission study reveals that the quenching mechanism limiting the emission of these samples at room temperature seems to be due to unipolar thermal escape of electrons towards the InP barrier. The temperature dependence of the radiative and non-radiative recombination time obtained in TRPL experiments indicates a possible influence of exciton localization effects in the exciton recombination dynamics. Within a three level rate equation model, we have put in evidence such exciton dynamics and estimated the value of the transition energy between free and localized exciton states, which depends on the excitonic optical transition energy.

Acknowledgments

This work was partially supported by Nanoself Project No. TIC2002-04096-C03 and by Nanomat of the EC Growth Program, Contract no. G5RD-CT-2001-00545.

References

[1] L. González, J. M. García, R. García, F. Briones, J. Martínez-Pastor, and C. Ballesteros, Appl. Phys. Lett. 76, 1104 (2000).
[2] K. Herz, G. Bacher, A. Forchel, H. Straub, G. Brunthaler, W. Fashinger, G. Bauer, and C. Vieu, Phys. Rev. B 59, 2888 (1999).
[3] H. R. Gutiérrez, M. A. Cotta, and M. M. G. de Carvalho, Appl. Phys. Lett. 79, 3854 (2001).
[4] F. Briones, L. González and A. Ruiz, Appl. Phys. A 49, 729-737 (1989).
[5] P. A. Postigo. PhD thesis.
[6] A. Rudrá, R. Houdré, J. F. Carlin, and M. Ilegems, J. Cryst. Growth 136, 278 (1994).
[7] B. Alén, J. Martínez-Pastor, A. García-Cristobal, L. González and J. M. García, Appl. Phys. Lett. 78, 4025 (2001).
[8] M. U. González, J. M. García, L. González, J. P. Silveira, Y. González, J. D. Gómez, F. Briones, Appl. Surf. Sci. 188 (2002) 188-192.
[9] M. Gurioli, J. Martinez-Pastor, M. Colocci, C. Deparis, B. Chastaingt, and J. Massies, Phys. Rev. B 46, 6922 (1992).
[10] E. M. Daly, T. J. Glynn, J. D. Lambkin, L. Considine, and S. Walsh, Phys. Rev. B 52, 4696 (1995).
[11] B. Ohnesorge, M. Albrecht, J. Oshinowo, A. Forchel, and Y. Arakawa, Phys. Rev. B 54, 11532 (1996).
[12] M. Gurioli, A. Vinattieri, M. Colocci, C. Deparis, J. Massies, G. Neu, A. Bosacchi, and S. Franchi, Phys. Rev. B 44, 3115 (1991).
[13] B. K. Ridley, Phys. Rev. B 41, 12190 (1990).
[14] J. Feldmann, G. Peter, E. O. Göbel, P. Dawson, K. Moore, C. Foxon, and R. J. Elliott, Phys. Rev. Lett. 59, 2337 (1987).

Applied Physics (APHYS 2003)
A. Méndez-Vilas (Ed.)

RADIATIVE EXCITON LIFETIMES ON DIFFERENT SHAPE SELF-ASSEMBLED SEMICONDUCTOR NANOSTRUCTURES

JORDI GOMIS, JUAN MARTINEZ PASTOR, BENITO ALEN, ENRIQUE NAVARRO,

Instituto de Ciencia de los Materiales, Universidad de Valencia, P.O. Box 22085, 46071 Valencia, Spain.

DANIEL GRANADOS and JORGE M. GARCIA

Instituto de Microelectrónica de Madrid (CNM-CSIC), Isaac Newton 8, 28760 Tres Cantos, Madrid, SPAIN.

PHILIPPE ROUSSIGNOL

Laboratoire de Physique de la Matière Condensée de l'Ecole Normale Supérieure, 24 rue Lhomond, 75231 Paris Cedex 05, France.

Abstract

In this work we present previous results of continuous wave and time resolved photoluminescence experiments performed on different shape InAs nanostructures grown on (001)-GaAs substrates: elongated dash-like, camel-like two-hump and ring-like nanostructures, evolving from InAs quantum dot layers by changing growth conditions of the GaAs overgrowth. The different shape and lateral sizes of the nanostructures produce changes in their electronic structure, as revealed from their photoluminescence and excitation photoluminescence spectra. Non resonant excitation gives slower transients (larger decay times), whose characteristic decay times are representative of the ground exciton lifetimes for QDh (950 ps), QC (750 ps) and QR (800 ps) ensembles. Resonant excitation in QC and QR nanostructures (camel and ring) is dominated by a fast transfer of carriers towards the electronic states of an InGaAs defect alloy layer that is formed during the GaAs overgrowth. For near resonant excitation on the sample containing QDh nanostructures a strong filtering effect is observed on the number of nanostructures to be populated within the illumination area. This leads to a maximum decay time at the PL energy whose first exciton excited state is being resonantly excited, because other emitting QDh nanostructures are being bpopulated through tunnelling processes, probably.

Introduction

Nowadays, increasing interest exists in the investigation of new methods to control the size and shape of self-assembled quantum dots (QD) to tune their optical properties. In particular, morphological changes of QD grown by molecular beam epitaxy (MBE) due to a thin GaAs cap have been reported [1]. This approach is now a powerful technique by which to obtain self-assembled nanostructures like quantum rings (QR) [2-4], even if only used by a few number of groups. Recently, Granados & García [5] have shown that the final morphology of the nanostructures depends strongly on the starting shape of the QD and on detailed growth conditions such as the substrate temperature for the GaAs capping process (necessary for optoelectronic devices) and the molecular species of the As flux. In this way, they have obtained two other different shapes for self-assembled nanostructures between QD and QR shapes, namely dash- (QDh) and camel-like (QC) nanostructures [5]. In this work we present a study of the exciton recombination dynamics observed in those kinds of InAs/GaAs nanostructures at low temperature. Continuous wave photoluminescence (PL) and time-resolved photoluminescence (TRPL) are used to characterize them.

Samples and experiment

The primary QDs are grown by depositing 1.7 monolayers (ML) of InAs onto a (001)-GaAs substrate (after a GaAs buffer layer) at 540 °C, under an As_2 beam equivalent pressure of 3-4x10^{-6} mbar. The InAs deposition takes place in a growth sequence of 0.1 ML InAs (at 0.06 ML/s) plus a 2 s pause under As flux. At the end of this sequence, the QDs are annealed 1 min to enhance the size distribution and to obtain low density ensembles (10^8 – 10^{10} cm^{-2}), which is useful either for improved AFM analysis or optical characterisation. Subsequently, a thin GaAs cap layer is grown (at a rate of 1 ML/s) at a different substrate temperature, T_{CAP}, depending on the sample, and annealed for 1 min. The samples for Atomic Force Microscopy (AFM) characterisation are cooled down immediately and removed from the chamber. The samples for PL measurements are obtained by capping these transformed nanostructures with a thicker GaAs layer, where the first 20 nm of GaAs is grown at T_{CAP} and after the substrate temperature is increased up to 595 °C. More details of the sample growth and AFM characterisation can be found on Ref. 5.

The final sample morphology was characterized by contact mode AFM (some micrographs are shown as insets in Fig. 1). The nanostructures called here as QDh are elongated islands whose typical lateral size is around 160x40x2 nm, QC has the aspect of camel-like two-hump islands (100x50x2 nm each "hump") and QR is similar to previous reported ring islands (100x90x1.5 nm) [1-5].

Continuous wave Photoluminescence (PL) and PL excitation (PLE) of the samples was performed with a standard optical set-up, using an Ar^+ pumped continuous wave Ti:sapphire for excitation. The light was dispersed by a double 0.6 m focal length monochromator and synchronously detected by a Si-APD. The sample was held in a He-liquid immersion cryostat. In time resolved PL (TRPL) experiments, sample excitation at different wavelengths was done by a Ti:sapphire picosecond pulsed laser (Mira 900D, Coherent) pumped with a 5 W doubled Nd:YAG laser (Verdi, Coherent). The PL signal was dispersed by a single 0.5 m focal length imaging spectrometer and detected by a synchroscan streak camera (Hamamatsu C5680) with a type S1 cooled photocathode. The overall time resolution in the widest temporal window (2 ns) was around 40-50 ps. The sample was held in the cold finger of a closed-cycle cryostat. The experimental results of PL and TRPL experiments shown here correspond to low excitation density conditions, for which any excited state induced PL line or saturation effects (because of the finite number of states per nanostructure) are observed. These effects have been well characterized, but out of the scope of this paper.

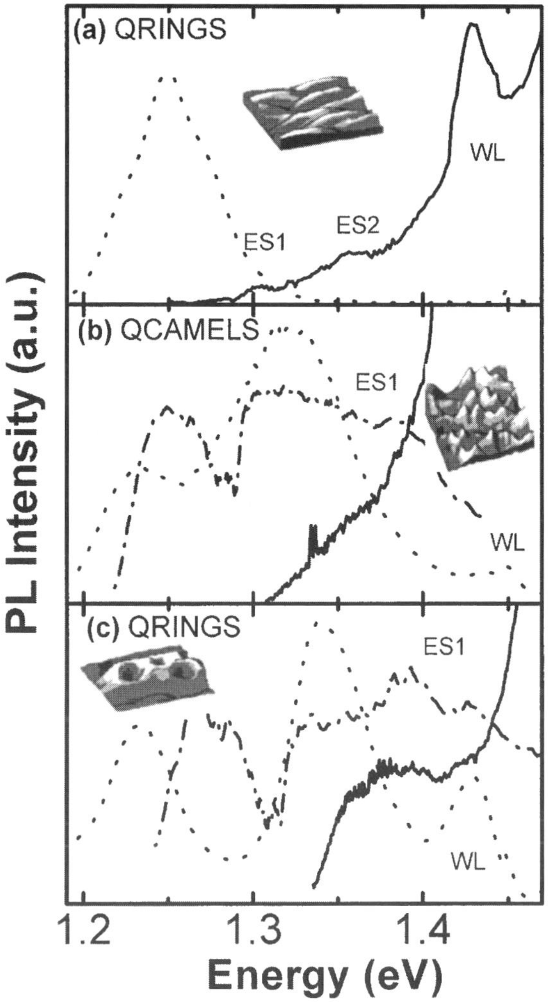

Figure 1. PL spectra (dotted lines), PLE detected at the PL maximum (continuous lines) and low energy detected PLE (dotted-dashed lines) for QDh (a), QC (b) and QR (c) at 10 K. The contribution of excited states (ES) and wetting layer (WL) absorption is indicated in the plots.

Results and Discussion

Figure 1 shows PL and PLE spectra measured at low temperature (and excitation above the wetting layer absorption edge) in the three samples. The main PL band is peaked at around 1.25, 1.33 and 1.34 eV, approximately, for QDh, QC and QR nanostructures, respectively, in accordance to their different average lateral and vertical dimensions. The wavelength tuning capability of the growth strategy is thus accomplished, as pointed out in Ref. 5.

As regards the optical quality, a better uniformity is observed for QR nanostructures, which exhibit a single PL band around 50 meV wide. It is worth noting that QDh nanostructures exhibit an overall PL bandwidth slightly larger than that of QR, but three Gaussian components can be perfectly deconvoluted within the band, that is, three families of QDh with appreciably different lateral sizes (the height being the main dimension to determine their corresponding emission energies). Furthermore, a low energy PL band peaked at around 1.23 eV is observed in the samples containing QR and QC ensembles, which is not the case for the sample containing QDh. The emission from the wetting layer (WL) is observed at 1.43-145 eV in the three samples.

On the other hand, also the electronic structure and exciton wavefunction expected for each nanostructure should be different. This is revealed by PL spectra under stronger excitation densities than that used in the PL spectra of Fig. 1 and PLE spectra. In the case of QDh nanostructures, as shown in Fig. 1(a), two resonances are observed at around 55 and 112 meV above detection (in the main peak of the PL spectrum, i.e., 1.25 eV), and a third more intense one corresponding to WL absorption edge. These two resonances are also observed in PL spectra measured under high excitation density conditions, which suggests the carrier filling of excited states to be the origin of those two resonances. When PLE is detected on the low or high energy parts of the PL spectrum (1.22 or 1.28 eV, for example), the energies found for those resonances changes slightly, accordingly to the change in the confinement energy of carriers with the average lateral size for each QDh family of the total ensemble. In the other two kinds of nanostructures less evident traces of excited state resonances are found at around 20-30 and 40-50 meV above the ground state transition for QC and QR, respectively, when combining PLE [Figs. 1(b) and 1(c)] and PL under high excitation density. Any well resolved resonance related to WL absorption is observed in PLE for these two samples [Figs. 1(b) and 1(c)], as compared to the sample containing QDh [Fig. 1(a)], but only an abrupt intensity increase of the PLE signal up to the GaAs absorption edge. A possible explanation for this behaviour can be the close distance between the excited states of the nanostructure's ensemble and the ground state of the WL. Another common feature of these two samples is the appearance of a low energy PL band, as outlined above, which is observed to be connected to the nanostructure's ensemble. In fact, the PLE spectra recorded by detecting at 1.22 - 1.23 eV [dashed-dotted lines in Figs. 1(b) and 1(c)] exhibit a clear absorption step when excitation is performed inside the PL band associated to the nanostructure's ensemble. The observation of that low energy PL band and its connection with QC and QR ensembles would indicate the presence of a sink of carriers, as evidenced in resonant TRPL experiments discussed below.

The PL transients under non resonant excitation (765 nm, above the GaAs absorption edge) are typically mono-exponential under low excitation density (1 – 10 W/cm^2), as shown in Fig. 2. If higher excitation density is used, exhaustion of available states per nanostructure will occur and the PL transient exhibit an intensity saturation at short times (the higher the excitation density the larger the temporal region of saturation), as also observed in QD [6,7]. A systematic study of these saturation effects is out of the scope of the paper and will be published elsewhere.

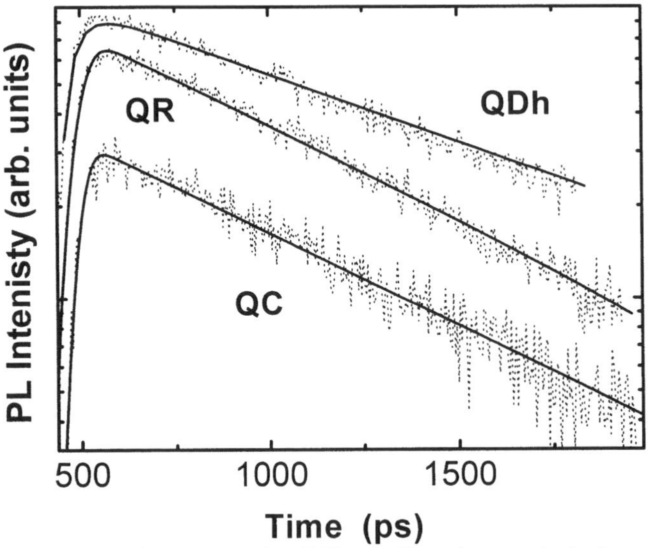

Figure 2. PL transients measured at 10 K at the PL peak energy in the three samples under low excitation density conditions. The wavelength of the excitation light was 765 nm.

Figure 3 shows the decay times measured (from PL transients) in the three samples under low excitation conditions in the whole emission band. The decay time measured at the PL peak energy for QDh nanostructures (950 ps) is appreciably larger than that for QC and QR, which are very similar between them (750 and 800 ps, respectively). This is expected from the more different electronic structure in the later cases with respect to that in QDh. The larger lateral size of QDh nanostructures will determine a weaker three dimensional confinement of carriers and hence a smaller oscillator strength (larger exciton lifetime) of their corresponding excitonic optical transitions. At the same time, the value found for QDh are not very far from decay times found in literature for larger size pyramidal InAs QD nanostructures (ensemble and single dots) [6-9], whose emission energy is below 1.1 eV. The effect of size confinement is confirmed within each ensemble of nanostructures, given the correlation between their lateral sizes and energies of the ground excitonic optical transitions. The decay time decreases from low to high PL detection energy, that is, from bigger to smaller nanostructures within the ensemble. For energies on the lowest energy side of the PL band the decay time suffers a fast decrease, contrary to the behaviour explained before. This effect can be due to defect nanostructures contributing to the low energy PL tail, for which non radiative channels (through interfaces, probably) can be more effective than radiative recombination.

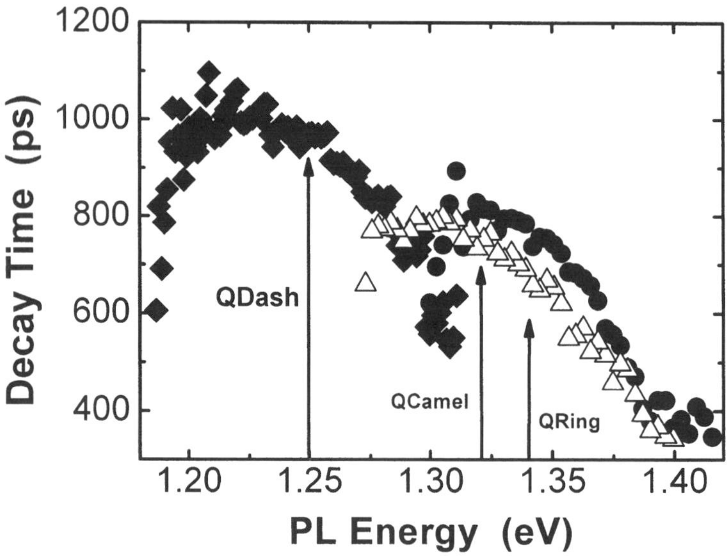

Figure 3. Decay times measured at 10 K inside the PL band at low excitation density for QDh (solid rhombus), QC (hollow triangle) and QR (solid circle) nanostructure's ensembles. The arrows indicate the position of the PL peak energy for each nanostructure's ensemble.

Finally, radiative recombination at low temperatures have been also examined under near resonant excitation. Three characteristic wavelengths have been used for this: 878 nm (WL absorption edge, approximately), 928 nm (ES2 optical transition) and 951 nm (ES1 optical transition). Surprinsingly, near resonant excitation in the samples containing QC and QR nanostructures exhibit a fast PL transient, from which any decay time can be deduced because it is practically equal to the temporal response function of our measure system (about 40-50 ps full width at half maximum). If one performs continuous wave PL spectra excited at similar wavelengths, the PL peak signal decreases abruptly and the spectra are dominated by the low energy PL band at around 1.23 eV, just because of the connection between excitonic states discussed above [dashed-dotted PLE curves on Figs. 1(b) and (c)]. Furthermore, the PL transient measured at 1.23 eV is as fast as that recorded at 1.32 eV (QC), or 1.34 eV (QR). This phenomenology is difficult to understand, because any intrinsic reason can be argued for justifying it. An heuristic explanation can be the fast transfer of carriers between the states of both nanostructures and "defects" (label for identifying the zones responsible of the PL band at 1.23 eV). At these defects the carriers should be able to fastly recombine non radiatively (similar to the explanation given above for the abrupt reduction of the PL decay times on the lowest energy PL tail) . The microscopic origin of such sink of carriers must be looked for in the surface dynamics during the capping growth. It is well known the In-Ga intermixing during QD formation, giving rise to smooth (InGaAs)-interfaces between InAs dots and GaAs capping layer. Indeed, these interfaces are not homogenous in all directions (can be more localized in the QD base) and strongly depends on the growth conditions [5]. Recent structural studies by Scanning Tunneling Microscopy on QR samples have demonstrated the existence of such InGaAs interface to be located at the top of the QR. The InGaAs layers can confine carriers because embedded between GaAs buffer (+InAs layer) and cap layers and be responsible of the low energy PL band and strong coupling with the states in the nanostructures.

In the sample containing QDh nanostructures the growth conditions are more similar to that used for dots [5] and no such a sink of carriers is formed. Decay times measured in this sample under near resonant excitation are plotted in Fig. 4. We observe how the decay time at the PL peak detection energy decreases from 950 ps to 770 ps, between 765 and 951 nm of excitation wavelength, respectively. It is also worth noting the strong reduction of the emission bandwidth when coming into more strictly resonant conditions, namely 951 nm, as observed in continuous wave PL. At this excitation wavelength, resonant with the ES1 optical transition, only a few number of nanostructures within the illumination area are being populated, because ES1 correspond to the first excited state of the nanostructures responsible of the main PL peak at 1.25 eV, approximately. This is the reason why a maximum decay time of 770 ps takes place at 1.25 eV and a fast decrease from this value is observed for lower and higher PL energies (hollow circles in Fig. 4). The nanostructures whose exciton ground state is either below or above 1.25eV can be populated through direct or phonon assisted tunnelling, for example, thus exhibiting an effective decay time shorter than their corresponding ground state exciton lifetime (better represented by curves in Fig. 3). For excitation at 928 nm (solid-hollow circles in Fig. 4) the filter effect on the number of QDh to be populated is no as strict as in the later case, because the internal dynamics (direct and phonon assisted tunnelling within the ensemble) is more efficient. It can be also added the fact that the ES2 transition associated to the 1.25 eV emitting QDh can be the ES1 one for higher energy emitting QDh, namely those emitting at 1.28 eV, approximately. This is the reason why we observe a clear decay time maximum at that energy. For excitation at the WL absorption (solid circles in Fig. 4) or above (solid rhombus in Fig. 3) any filter effect is longer observed and the entire QDh ensemble within the illumination area is being populated. In these cases any maxima are observed in the decay time curve as a function of the PL energy, now reflecting the variation of the ground state exciton lifetime (inverse of the oscillator strength) with the optical transition energy.

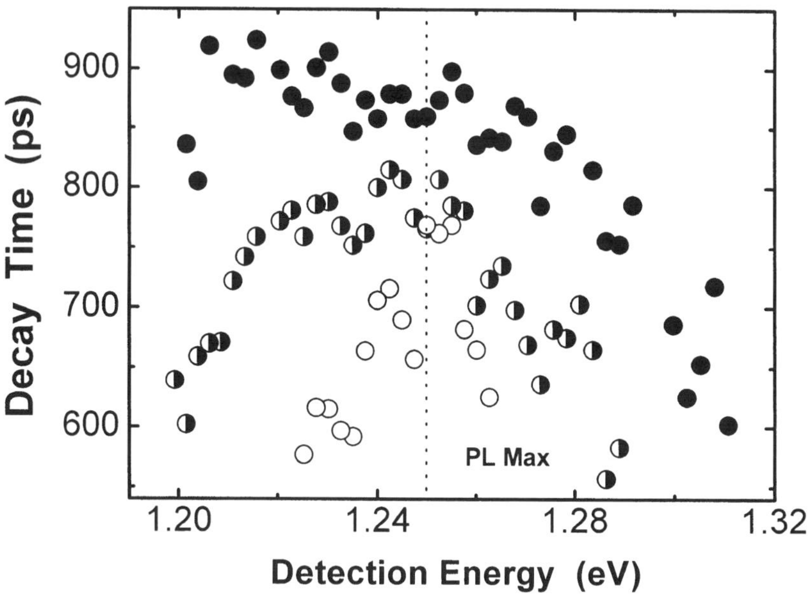

Figure 4. Decay times measured at 10 K under near resonant conditions in the sample containing QDh ensemble: 878 nm (solid circles), 928 nm (semi-solid circles) and 951 nm (hollow circles). The vertical dotted line indicates the PL peak energy.

Conclusion

In summary, it is demonstrated that the different shape and lateral sizes of the nanostructures produce changes in their electronic structure, as revealed from their PL and PLE spectra. The decay time measured within the main part of the PL band is correlated with the inverse of the oscillator strength for ground excitonic optical transitions expected for the size distribution of the nanostructure's ensemble. The exciton lifetime for QC (750 ps) and QR (800 ps) nanostructures is smaller than that for QDh (950 ps), because of the smaller three dimensional size confinement in the later case (lower optical transition energy). For near resonant excitation a strong filtering effect is observed on the number of nanostructures to be populated, which leads to a maximum decay time at the PL energy whose first exciton excited state is being resonantly excited. The decay times measured under near resonant excitation is slightly reduced because of the internal dynamics within the QDh ensemble. Possible non radiative channels at the InAs/GaAs interfaces can contribute to that decay time reduction. On the other hand, such non radiative channels are strongly limiting the PL efficiency in the samples containing QC and QR, because now the InAs/GaAs top interface includes an InGaAs layer acting as a sink for carriers under near resonant conditions.

Acknowledgments
This work was partially supported by Nanoself Project No. TIC2002-04096-C03 and by Nanomat of the EC Growth Program, Contract no. G5RD-CT-2001-00545.

References

[1] J. M. Garcia, G. Medeiros-Ribeiro, K. Schmidt, T. Ngo, and P. M. Petroff, Appl. Phys. Lett. **71**, 2014 (1997); R. Songmuang, S. Kiravittaya, and O. G. Schmidt, J. Cryst. Growth **249**, 416 (2003).
[2] A. Lorke, R. J. Luyken, A. O. Govorov, J. Kotthaus, J. M. Garcia, and P. M. Petroff, Phys. Rev. Lett. **84**, 2223 (2000).
[3] R. J. Warburton, C. Schaflein, D. Haft, F. Bickel, A. Lorke, K. Karrai, J. M. Garcia, W. Schoenfeld, and P. M. Petroff, Nature (London) **405**, 926 (2000).
[4] T. Raz, D. Ritter, and G. Bahir, Appl. Phys. Lett. **82**, 1706 (2003).
[5] D. Granados and J. M. García, Appl. Phys. Lett. **82**, 2401 (2003).
[6] R. Heitz, A. Kalburge, Q. Xie, M. Grundmann, P. Chen, A. Hoffmann, A. Madhukar, and D. Bimberg, Phys. Rev. **B 57**, 9050 (1998).
[7] J.W. Tomm, T. Elsaesser, Yu. I. Mazur, H. Kissel, G. G. Tarasov, Z. Ya. Zhuchenko, and W. T. Masselink, Phys. Rev. **B 67**, 045326 (2003).
[8] W. Yang, R. R. Lowe-Webb, H. Lee, and P. C. Sercel, Phys. Rev. **B 56**, 13314 (1997).
[9] C. Santori, G. S. Solomon, M. Pelton, and Y. Yamamoto, Phys. Rev. **B 65**, 073310 (2002).

Applied Physics (APHYS 2003)
A. Méndez-Vilas (Ed.)

Tissue-equivalent TL Sheet Dosimetry System for Gamma-ray Spatial Dose Distribution Measurement

Nobuteru Nariyama*, Akiko Konnai†, Seiki Ohnishi†, Naoteru Odano†, Akio Yamaji†, Naoto Ozasa¶ and Yuhzoh Ishikawa¶

*Synchrotron Radiation Research Institute, Mikazuki, Sayo, Hyogo 679-5198, Japan
† National Maritime Research Institute, Mitaka, Tokyo 181-0004, Japan
¶ Nemoto & Co., Ltd, Takaido-Higashi, Suginami, Tokyo 168-0072, Japan

Abstract. To measure continuous dose distribution for gamma and x rays, a tissue-equivalent thermoluminescent (TL) thin sheet-type dosimeter and the reader system were developed. The TL sheet is made of LiF:Mg,Cu,P mixed with polymer ETFE and the thickness is 0.2 mm. The energy response calculated showed excellent tissue-equivalent property. For the TL reading, a heating plate of 20-cm square was developed, and the temperature profile was measured with an infrared thermal imaging camera. As a result, it was confirmed that the temperature rose linearly within 2% and the homogeneity was within 3%. The TL signal emitted is detected using a CCD camera with an image intensifier and displayed as spatial dose distribution. From measurement results using synchrotron radiation, the TL sheet dosimetry system can be expected for dose mapping of various purposes.

INTRODUCTION

Recently, accelerator facilities providing beam shape radiation are increasing in number, of which the radiation types are various and the energies are distributed widely: synchrotron radiation is a low-energy x-ray beam, and the proton and heavy ion beams used for medical applications have high energies. When the beam radiation is incident to a human body intentionally or accidentally, the dose distribution is extremely localized and cannot be detected with a conventional point-type dosimeter satisfactorily. Furthermore, the spatial dose distribution changes drastically, and so even if many small dosimeters are set for the measurement, such a localized distribution cannot be measured in detail, and the highest dose also cannot be detected if accidentally irradiated. Thus the dose monitoring necessary much differs from that for the traditional radioisotopes and nuclear facilities.

In the situation, position sensitive two-dimensional dosimeters can be expected as a useful tool; several types dosimeters have been developed until the present. In early stages, thermoluminescence (TL) distribution in a millimeter-size area has been measured using a photograph camera [1]. Taking a step forward, optically stimulated luminescence (OSL) from small natural and artificial materials has been measured using a photomultiplier tube [2,3] and a CCD camera [4-6] for several tens μm resolution. For macroscopic dose distribution, a laser-heating thermoluminescent dosimeter system has been developed [7]. However, it was considered that the heat load was large and the space resolution was worse than OSL: the system has not become popular for the practical dose measurements.

The OSL is certainly a suitable method for the two-dimensional dosimeter because of the availability of narrow laser beam for the high space resolution and no heat loading. Nevertheless, rather than the space resolution, tissue-equivalent property is much more significant for the radiation dosimeters to deal with the diversities of the radiation and energies. For OSL, such a phosphor with satisfactory sensitivity has not been discovered. An imaging plate is a high-sensitive OSL detector developed for imaging; the detector shows the energy response over one hundred times higher for 50-keV photons than for Co-60 gamma rays, according to the calculations using the mass energy-absorption coefficients of BaFBr:Eu^{2+}. Moreover, the phosphor is known to show a strong fading [8]. The other practical used OSL phosphor of Al_2O_3:C also exhibits increase of the energy response in the low energy region.

For the TL, two kinds of tissue-equivalent phosphors are practically used: lithium fluoride (LiF) and lithium borate. Using the phosphors, a tissue-equivalent sheet-type dosimeter is possible. When a CCD camera is used for the TL detection, a conventional ohmic-heating method is applicable, which is easier for the temperature control. Actually, such an equipment is applied to $BaSO_4$:Eu TL sheet [9,10]. Here should it be noted that to make a sheet, TL material powder has to be mixed with a binder material, which inevitably lower the sensitivity than that of the original one, and so high-sensitive TL material is necessary. Therefore, LiF:Mg,Cu,P phosphor was chosen in the present work because of the highest sensitivity among the tissue-equivalent TL phosphors [11].

In this study, a tissue-equivalent thermoluminescent dosimeter (TLD) sheet using LiF:Mg,Cu,P was developed for dose mapping to various radiation. For the heating, a 20-cm-square plate with temperature linearity and homogeneity was also developed, and using the reader system, characteristics of the sheet TLDs were investigated.

MATERIALS AND METHODS

Advantages of the LiF:Mg,Cu,P phosphor are the energy response very similar to soft tissue [12-14], the high sensitivity and the low annealing temperature of 240 °C, which is favorable to the selection of the binder material. On the other hand, the disadvantage is the annealing-temperature sensitivity: heating over 245 °C destroys the TL properties. That is, the baking has to be made below 245 °C and at the same time, the binder material should have the heat resistance at 240 °C at the annealing. This slight 5 °C difference makes the choice of the binder material difficult. Moreover, the binder material itself should also have a tissue-equivalent property.

In general, Teflon is preferable for the binder material because the physical properties such as the mass energy-absorption coefficients are very similar to LiF, and the light transparency is high at the TL wavelength. The problem is the high melting point: 327 °C for PTFE. Hence, low-melting fluorocarbon polymer ETFE (Asahi

Glass Co., Ltd.) was selected instead. The ETFE is a copolymer comprising of tetrafluoroethylene (C_2F_4) and ethylene (C_2H_4) and the melting point is 260 °C.

For the sheet production, LiF:Mg,Cu,P powder NT-250 was newly developed, which has considerably high sensitivity and can be produced at a low cost. The powder was mixed with ETFE and baked into the sheet of 0.2-mm thickness during the press.

For the heating plate, an area of at least 20-cm square was necessary because of the human-body size. Moreover, the plate has to be heated linearly and homogeneously with a precise absolute temperature control: linearity is necessary for reproducibility of the temperature and obtaining the glow curves. Several kinds of heating methods were tested and as a result, ohmic heating was found to be the only method that can satisfy the above conditions.

To develop the linear-heating plate, the resistor pattern was modified to meet the requirements by measuring the temperature profile with an infrared thermal imaging camera NEC-Sanei TH4104MR and five thermocouples set at the different positions in the plate. To attain the homogeneity, a gap was made at the edge of the plate to suppress the radiation of heat and in the center a heat sink was set to promote the heat radiation.

The plate is made of copper in black. To set the sheet, the plate is slid out as shown in Fig. 1, and quartz glass is put on the sheet to smooth out the wrinkles of the thin sheet and improve the contact with the heating plate. For the same purpose, vacuum absorption is also used during the heating, which absorbs the sheet through the minute holes in the plate. The heating rates available are 0.5 and 1 °C/s, and it takes about ten minutes to recover to 50 °C after a measurement by forced air cooling.

FIGURE 1. Sliding heating plate

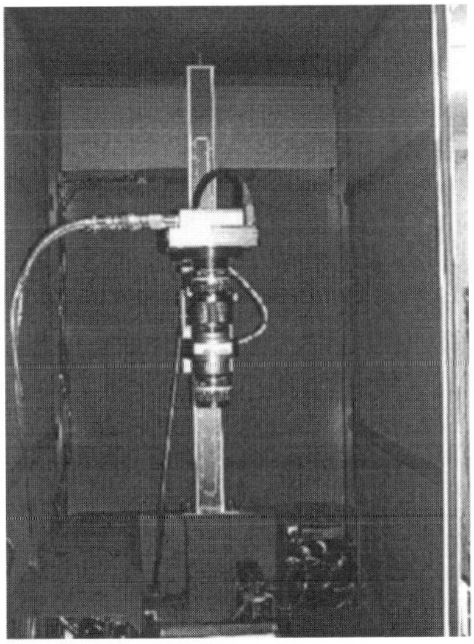

FIGURE 2. TL detection camera

The TL signal is detected with a CCD camera. Performance required for the camera is the high quantum efficiency at the TL wavelength, the wide dynamic range, the large sensitive area and the low dark current. Moreover, the large pixel size increases the sensitivity, while the high readout frequency is not necessary for the integrated measurement. To increase the sensitivity of the camera, an image intensifier was adapted to the camera: in the present system, an Apogy AP8-2 camera was installed with a Proxitronic BV2581EZ image intensifier. Figure 2 shows the TL detection system. The quantum efficiency is 62% at 400 nm and the sensitive area is 24.6-mm square with 1024×1024 pixels. The dynamic range is over 86 dB and 16-bit AD conversion. The heating and data acquisition are controlled with a Microsoft Windows personal computer.

RESULT AND DISCUSSION

Figure 3 shows the temperature profile of the heating plate at 1 °C/s up to 240 °C, measured with the infrared thermal imaging camera. Points of a, b, c and d are situated at different positions on the heating plate. It was confirmed that the position dependence was within 3% and the linearity was within 2%. The temperature homogeneity was also confirmed with the five thermocouples in the heating plate.

The energy response in soft tissue was calculated using the photon mass energy-absorption coefficients [15]. The result is shown in Fig. 4. The used ratio of LiF to ETFE was 2/3. The response became almost constant over 10 keV. The phosphor LiF:Mg,Cu,P, however, is known to show the different response than the calculation [12,14]. The response considering the effect is also indicated in Fig. 4. While the response decreased in the low-energy region, it is maintained over 0.75 above 30 keV. The actual response has to be determined experimentally.

Figure 5 shows the dose distribution for 60-keV synchrotron radiation at 0.165 Gy. The irradiation was made at SPring-8 and the beam size was 6.5 mm by 3.0 mm. The localized distribution was clearly demonstrated.

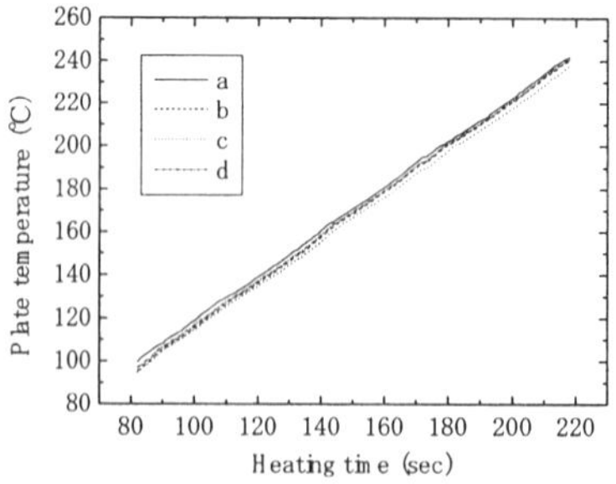

FIGURE 3. Linearity and homogeneity of the temperature rise on the heating plate

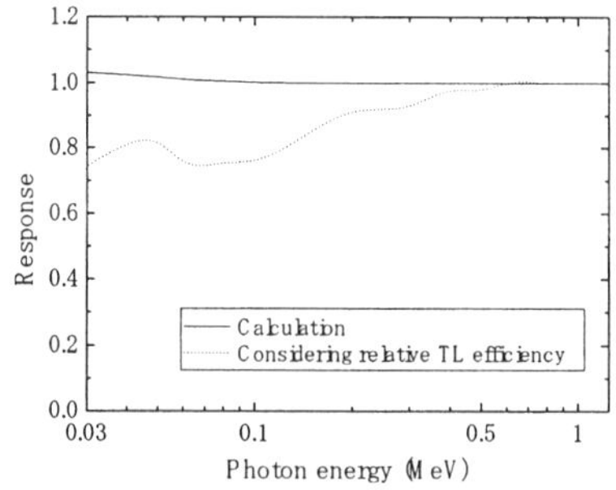

FIGURE 4. Calculated energy response of the sheet TLD in soft tissue

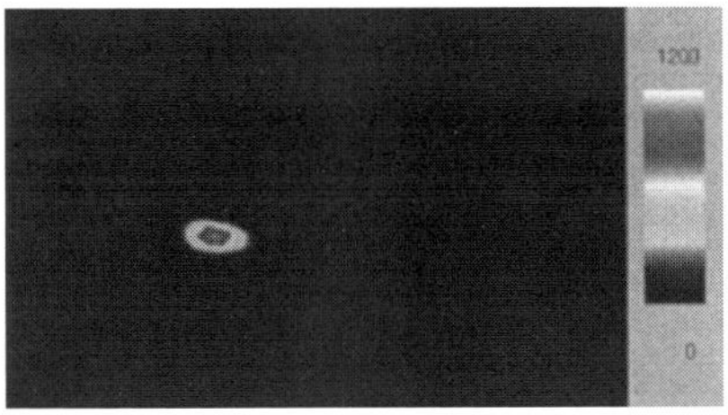

FIGURE 5. Dose distribution for synchrotron radiation at 0.165 Gy

CONCLUSION

A tissue-equivalent TL sheet dosimeter composed of LiF,Mg,Cu,P and ETFE was developed. For the heating, a 20-cm square plate was developed, and the linearity and homogeneity of heating was confirmed with an infrared thermal imaging camera. The present system was found to be promising for the dose mapping of gamma- and x-ray radiation.

ACKNOWLEDGMENTS

This study was financially supported by the Budget for Nuclear Research of the Ministry of Education, Culture, Sports, Science and Technology, based on screening and counseling by the Atomic Energy Commission.

REFERENCES

1. A.J. Walton and N.C. Debenham, *Nature* **284**, 42-44 (1980).
2. L. Dusseau, G. Ranchoux, G. Polge, D. Plattard, F. Saigné, J.C. Bessière, J. Fesquet and J. Gasiot, *IEEE Trans. Nucl. Sci.* **46**, 1757-1761 (1999)
3. I.K. Bailiff and V.B. Mikhailik, *Radiat. Mea.* **37**, 151-159 (2003)
4. N.A. Atari and G.K. Svensson, *Med. Phys.* **13**, 354-360 (1986)
5. G.A.T. Duller, L. Bøtter-Jensen and B.G. Markey, *Radiat. Mea.* **27**, 91-99 (1997)
6. N.A. Spooner, *Radiat. Mea.* **32**, 513-521 (2000)
7. J. Gasiot, P. Braunlich and J.P. Fillard, *J. Appl. Phys.* **53**, 5200-5209 (1981)
8. G.F. Knoll, Radiation Detection and Measurement, John Wiley & Sons, Inc. (2000)
9. K. Imaeda, T. Kitajima, K. Kuga, S. Miono, A. Misaki, M. Nakamura, K. Ninagawa, Y. Okamoto, O. Saavedra, T. Saito, N. Takahashi, Y. Takano, T. Tomiyama, T. Wada, I. Yamamoto and Y. Yamashita, *Nucl. Instru. Meth.* **A241**, 567-571 (1985)
10. K. Iwata, H. Yoshimura, Y. Tsuji, A. Shirai, F. Uto, T. Tamada, I. Yamamoto, T. Tomiyama, T. Wada, H. Ohishi and H. Uchida, *Int. J. Radiat. Oncology Biol. Phys.* **22**, 1109-1115 (1992)
11. T. Nakajima, Y. Murayama, T. Matsuzawa and A. Koyano, *Nucl. Instru. Meth.* **157**, 155-162 (1978)
12. P. Olko, P. Bilski, E. Ryba and T. Niewiadomski, *Radiat. Prot. Dosim.* **47**, 31-35 (1993)
13. T. Kron, L. Duggan, T. Smith, A. Rosenfeld, M. Butson, G. Kaplan, S. Howlett and K. Hyodo, *Phys. Med. Biol.* **43**, 3235-3259 (1998)
14. N. Nari yama, Y. Namito, S. Ban and H. Hirayama, *Phys. Med. Biol.* **46**, 717-728 (2001)
15. Internatio nal Commission on Radiation Units and Measurements, ICRU Report 44 (1989)

Applied Physics (APHYS 2003)
A. Méndez-Vilas (Ed.)

Active Thermography Applied for Quantitative Determination of Stomatal Resistance

G. Klinger*, P. Bajons**, V. Schlosser**

*Department of Meteorology and Geophysics, University of Vienna, Althanstrasse14, A-1090 Vienna, Austria
**Department of Material Physics, University of Vienna, Strudlhofgasse4, A-1090 Vienna, Austria

Abstract. Based on an extension of the theory of energy balance the stomatal resistance of leaves in vivo is calculated by using temperature rise and fall times (thermal time constants) which are caused by a sudden change of irradiation intensity. The change in the irradiation was performed by turning on/off a red laser. To measure the temperature changes a commercial IR-camera was employed. Experiments were performed under laboratory conditions (monitoring of air temperature and humidity, no wind). As model-material leaves of basil were taken. The applicability of the new method is demonstrated.

INTRODUCTION

Surface temperatures of canopies or leaves are often considered as possible indicators of plant health, plant stress or infection. The associated heat transfer between the plant and the environment is -besides meteorological circumstances- affected by plant physiological processes. The aperture of the stomata regulates transpiration and thus the stomatal resistance to evaporation influences heat transfer. Reported passive thermographic methods calculate stomatal resistance from measurements of the temperature of leaves in vivo, in the dry and in the wet state and need the use of artificial reference leaves.

THEORY

In the present study active thermography, i.e. temperature changes due to the action of defined light pulses, is used for the determination of stomatal resistance. The underlying theory is based on an extension of the energy balance equation: At equilibrium the rate of radiation energy absorbed by a leaf equals the heat energy stored in tissues (S) as well as the energy losses by reradiation, by sensible heat exchange between the leaf and the surrounding air (C) and by transpiration (λE, where E is the transpiration rate and λ is the latent heat of vaporisation). Therefore the energy stored in a leaf can be expressed as follows:

$$S = \Phi - C + \lambda E \qquad (1)$$

where Φ is the net heat gain from radiation (absorbed radiation minus reradiation).

When either one of the components of the energy balance changes then the temperature of the leaf is also changed. The rate of this temperature change depends on the heat capacity per unit area of the tissue. In case of an instantaneous change in the incident radiation, it was shown [1,2] that the temperature change follows an exponential law. The thermal time constant τ_{leaf} of this process is given according to

$$\tau_{leaf} = (\rho_l c_l d / \rho_a c_a) / (1 / r_{HR} + \varepsilon_a / (r_{aw} - r_s)) \qquad (2)$$

where ρ_l is the density of leaf tissue, c_l is the specific heat capacity of leaf tissue, d is the leaf thickness (volume to area ratio), ρ_a is the density of the surrounding air, c_a is the specific heat of air and ε_a is the ratio of the increase of latent heat content to increase of sensible heat content of saturated air. The symbol r stands for the resistivity with $1 / r_{HR} = 1 / r_R + 1 / r_{aH}$ where r_R is the 'resistance' to radiative heat transfer, r_{aH} is the mean leaf boundary layer resistance to heat, r_{aw} is the mean leaf boundary layer resistance to water vapor and r_s is the mean stomatal resistance to water vapor.

For the same leaf in the dry state the mean stomatal conductance $1/r_s$, will be zero and on the other hand if the same surface is wet the mean stomatal resistance will equal zero. Thus in case of closed stomata a thermal time constant τ_{dry} and in the wet state a thermal time constant τ_{wet} will be obtained.

Replacement of the non biotic terms of the leaf boundary resistances by the thermal time constants allows to calculate the stomatal resistance of a leaf in vivo according to

$$r_s = (\varepsilon_a \rho_a c_a / \rho_l c_l d) (\tau_{dry})^2 (\tau_{leaf} - \tau_{wet}) / ((\tau_{dry} - \tau_{leaf}) (\tau_{dry} - \tau_{wet})) \qquad (3)$$

EXPERIMENT

The experiments were performed under laboratory conditions, i.e. the temperature and the humidity of the air was controlled and monitored and air movements were restricted (no wind). Temperature measurements were performed by a commercial IR-camera AGEMA-570 (focal plane array with 320 pixel x 240 pixel, spectral range: 7,5 µm - 13 µm). Taking into consideration that the thermal resolution of the infrared camera (with a nominal temperature sensitivity of 0.1 K) is limited by random noise and by permanent parasitic signals, images were – whenever necessary – further processed. In case where random noise reduction was wanted, e.g. for illustration purposes, the values of several consecutive thermal images were recorded, afterwards averaged pixel by pixel and then again displayed. Local temperature changes could be detected by subtracting pixel by pixel one average image from the other. By this also parasitic signals could be suppressed.

Measurements of leaf temperature changes were made in the dark and under illumination by a lamp. To measure the thermal time constants the leaves were additionally illuminated by turning on and off a laser (675 nm, rectangular radiation characteristic). In all the experiments performed within the scope of this study the leaf to air temperature difference resulting from the combined action of the laser and the lamp was kept in the range of 3 to 10 K, while the leaf temperature increase resulting only from the illumination of the laser was kept in the range of 2 to 5 K.

Leaves of Mediterranean type basil (ocimum basilicum L.) with a heavily vaulted surface were studied. To obtain wetted surfaces water was sprayed by means of a diffuser on both sides of the leaf. Dry surfaces were obtained by covering the leaf with a thin coating of petroleum jelly (vaseline).

RESULTS

The thermal behavior of basil leaves was tested over a period of several months. Within this period several plants with different leaf size, at different age and subjected to various degree of water stress were studied. In the following some typical examples of leaf temperature behavior are given. Fig.1a shows a part of an illuminated basil leaf (and part of the cold and thus black background). The image, which is the mean of ten consecutive images, contains the information of an area of 3 cm x 3 cm and is divided into 35 pixels x 35 pixels. The gray scale is chosen such that the whole temperature range between the minimum temperature (black, T_{air} = 293.5 K) and the maximum (white, 304 K) is covered. The temperature in the center of the leaf is higher and decreases to values close to the background temperature at the edges. Besides physiological reasons the temperature distribution on the leaf surface is influenced by the following physical factors: Due to the fact that the surface of the leaf has a strong curvature, the angle of light incidence varies upon the leaf surface. Thus in the given example the absorbed radiation energy decreases from the center to the edge of the leaf. Additionally, the thickness of the boundary layer, which strongly influences the heat loss from the leaf surface to the surrounding

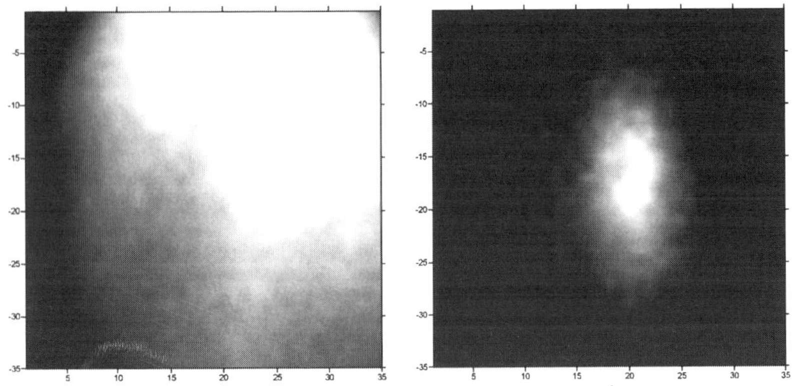

a b

FIGURE 1. a) Thermal image of illuminated leaf b) Differential image of laser action

medium, decreases - at least in principle - from the middle to the edge of the leaf.

In a next step the leaf was further illuminated by the laser, which creates in the plane of the leaf an illumination rectangle of 8 mm x 20 mm. The thermal reaction of the leaf due to the action of the laser light becomes more evident, when the average image of the leaf illuminated only by the lamp (Fig.1a) is subtracted from the new average image. The result is shown in Fig.1b, where the area of laser action is clearly detectable against the black background. The observed shape of the temperature distribution may again be attributed to the leaf curvature and surface waviness, but also long-term physiological reactions of the plant or even slight movements of the leaf towards the source of illumination must be taken into account. These uncertainties, namely the problems in the interpretation of temperature changes arising from geometrical conditions of illumination and from plant long-term thermal rearrangements, strengthen the necessity of time resolved temperature measurements.

Fig.2 visualizes the temporal development of the temperature on a leaf surface, which is obtained by a turn on, turn off respectively, of the laser. It was measured in a similar experiment (T_{air} = 296.2 K) and represents the mean of twelve pixels located in the center of the hot area. The temperature is 302K. Turning on the laser raised the temperature further by 3 K and turning off the laser lowered the temperature again to the former level. The same procedure was repeated with the leaf in the wet and dry (coated) state.

The thermal rise time for the leaf was calculated as 21.1 s and the decay time as 20.9 s. Comparing the rise and the decay time it follows that the action of the laser did not influence the physiological status of the leaf. However, at this point it may be mentioned that repeated measurements of rise and/or decay times often

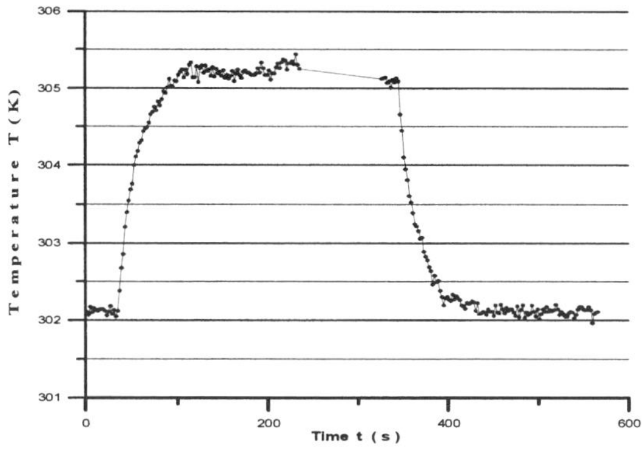

FIGURE 2. Thermal behavior of the leaf due to laser action (the rise and the decay time is 21 s)

yields a scatter of the values up to a range of ± 1.5 seconds. Similarly a mean value of τ_{dry} = 30 s and τ_{wet} =10 s was obtained. The thickness of the leaf was measured as d = 0.2 mm. For the determination of a mean value of ρ_l c_l we followed the arguments given in previous publications [1], i.e. the product of the average leaf density and the average specific heat capacity was assumed to be 2.7 MJ / K m^3. Also the tabulated values given in this publication were used which gave ε_a = 2.56, c_a = 1010 J / kg K and ρ_a = 1.191 kg / m^3. Thus a mean stomatal resistance to water vapor was calculated as r_s = 0.31 s / mm, a value which is in agreement with observations on the microscope.

CONCLUSIONS

The method used for the determination of leaf resistance is active thermography. The non-biotic parameters of heat exchange, i.e. the resistances of the boundary layer, can be eliminated by measuring the thermal time constants of the leaf in the wet state and in the dry state. Especially when these data are determined on similar leaves under similar meteorological conditions, the action of the stomata can be followed by simply measuring the rise and decay times of a leaf in vivo. By this early detection of plant stress ore plant discase will become possible.

REFERENCES

1. H.G. Jones, Plants and Microclimate, Cambridge University Press, Cambridge 1992
2. J.L. Monteith and M.H. Unsworth, Principles of Environmental Physics, ed. E Arnold, London 1990

Applied Physics (APHYS 2003)
A. Méndez-Vilas (Ed.)

Local Characterization Of Multicrystalline Silicon Wafers And Solar Cells

Viktor Schlosser*, Milena Dineva*, Peter Bajons*, Rita Ebner†, Johann Summhammer†, Gerhard Klinger¶

*Institut für Materialphysik, University of Vienna, Strudlhofgasse 4, A-1090 Wien, AUSTRIA
†Atominstiut der Österreichischen Universitäten, Stadionallee 2, A-1020 Wien, AUSTRIA
¶Institut für Meteorologie und Geophysik, University of Vienna, Althanstrasse 14, A-1090 Wien, AUSTRIA

Abstract. The current voltage characteristics of multicrystalline solar cells with individually designed front contact grids under different bias light conditions in the temperature range of 300 K to 340 K were investigated. Cells with a grid placed predominately on grain boundaries exhibit higher conversion efficiencies than cells where a great portion of the grid covers crystallites. The diode loss current observed for the first type of cells is governed by a thermal activation energy which is about 30 meV larger than found for other cells. This is attributed to a local increase of the recombination center density within grain boundaries.

INTRODUCTION

During the preparation of multicrystalline silicon (mc-Si) cells different crystal orientations as well as the region between the grains which has an extremely high density of lattice defects and accumulates impurities causes inhomogeneous chemical surface treatment and different doping conditions during diffusion steps. In the final solar cells especially the grain boundaries exhibit unwanted electrical characteristics. They tend to have a high concentration of electrically active defects which cause high recombination for minority carriers thus reducing the collection efficiency of light generated carriers [1, 2]. Furthermore grain boundaries often act as potential barriers for majority carriers which introduces an additional contribution to internal power losses of the solar cell [3]. Therefore cells of mc-Si suffer from lower conversion efficiencies and batches of cells have larger dispersion of the electrical parameter compared with monocrystalline Si cells. In order to minimize grain boundary effects on the solar cell performance electrically active defects are passivated by the introduction of atomic hydrogen frequently in combination with a thermal treatment to getter metallic impurities [4]. Recently an other approach to minimize efficiency losses due to grain boundary effects was suggested [5]. The metal grid of the front contact of a mc-Si solar cell was applied mainly above grain boundaries. Two effects were expected to take place. First the shadowing of grains with high light generated photocurrent densities is suppressed. Second the series resistance is lowered due to a reduction of current paths across grain boundaries. Previously a statistically elaborated study reported an increase of the output power between 2-5 % [6]. Content of the present work is to further improve this method of individually processing mc-Si wafers to solar cells.

EXPERIMENTAL

The starting material for the preparation of solar cells were p-type mc-Si wafers as delivered from Bayer or Eurosolare. The wafer surfaces were chemically polished and cleaned either by an acetic or by a hydroxide solution. The planar pn-junction was formed by a phosphorous diffusion. The back side of all cells was fully metalized by screen printing a paste containing aluminum and silver particles. In order to compare cells with different front grids pairs or triplets of subsequent wafers from one batch which have almost identical distributions of crystallites were selected. For one of these cells the surface structure was determined by an optical contrast image which was transferred to a computer. A program starts from an initial grid with rectangular lines defined by the user's input of the desired grid parameter such as line spacing, total line length and percentage of the cell area which shall be covered by the metal. Based on the information of the optical contrast image the program distorts the initial grid layout towards a grid which still maintains the inputted parameter but is located predominately (>77 %) along boundaries of high optical contrast caused by the different crystallographic orientations of the grains. Depending on the way the front contact is formed the final lay out of the grid is used to either drive a plotter or to output a mask for photolithography. Currently two methods to apply the front grid are in use: (1) A silver ink is plotted onto the grain boundaries according to the computer driven plot instructions and then burnt in or (2) a masked photoresist covers the wafer's surface leaving regions uncovered where the metal is intended to be deposited by an electrochemical deposition from a liquid solution. The grid properties of the two deposition methods are summarized in Tab. 1. On the first cell of a selected pair the grid was applied above the grain boundaries. These cells will be referred to as "ON cells" further on. The same grid was used for the second cell of the pair but is was rotated by 90 degrees causing an intentional misalignment of grid lines and grain boundaries ("OFF cells"). In the case of triplets a standard H-pattern grid was used for the third cell. Except a simple surface passivation by covering the cell's surface with a polymeric layer no advanced techniques to reduce the electrically active defect density on surfaces or within the grain boundary region were used. The optical and electrical properties of the solar cells were investigated. In Fig. 1 the map of a light beam induced current scan, LBIC, carried out on an "ON cell" with electrochemically deposited contacts illustrates how the metal lines shown as black polygons surround the crystallites shaded in grey (a).

TABLE 1. Properties of the front metal grid.

	Electrochemical deposition	Silver ink plotting
Line width	210 μm	350 μm
Line height	3 – 5 μm (Ni)+10 - 15 μm (Ag)	100 μm
Line resistance	10 mΩ/cm	30 mΩ/cm
Contact resistance	1.5 mΩcm^2	11 - 12 mΩcm^2
Shadowed to total surface	6.0 %- 7.9 %	7.5 % - 11 %

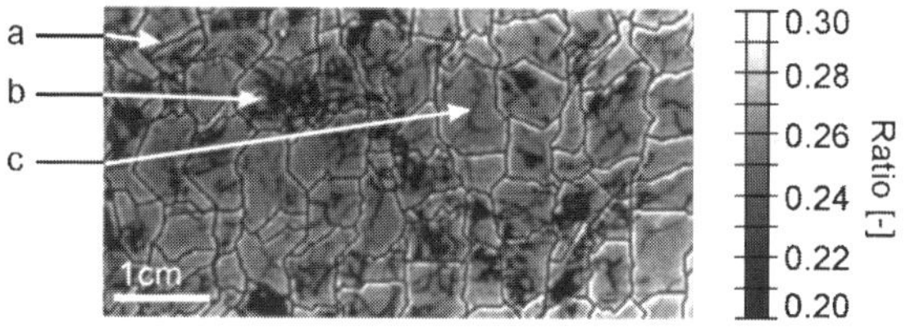

FIGURE 1. Grey scale map of the local variation of a light beam induced current in an "ON cell". The shadowed areas covered by the front contact grid appear as black polygons (a). A grain with an high density of crystal defects is indicated by (b). An example of a grain boundary (c) that was invisible to optical surface contrast.

Two lasers spots with wavelengths at 635 nm and 905 nm respectively were scanned simultaneously over the cell surface and the ratio of the two signals were shown as Grey scale contrast in the image. Displaying the ratio rather than recording the signal arising from a single laser excitation reduces misinterpretation caused by the large spatial variation of surface reflection. In some cases at each step of the solar cell preparation cycle the local light excited charge carrier density of the partly processed cells were characterized with the help of a microwave bridge. The method allowed us to estimate the local minority carrier diffusion length without any additional preparation and without the application of electrical terminals [7] A comparison of optical contrast images which were used to define front contact grids and the maps of the microwave signal indicates that only a portion of boundaries between areas of high optical contrast are electrically active. On the other hand electrically active boundaries were sometimes invisible to an optical surface scan due to low contrast. As a consequence the grid calculation should preferably be based on minority carrier diffusion length mapping.

RESULTS

The electrical characteristics of the photovoltaic devices were investigated in the dark and under different bias light conditions between 300 K and 340 K. The solar cell parameter, short circuit current, I_{SC}, open circuit voltage, V_{OC}, curve fill factor, *CFF*, and maximal outputted power, P_m, were determined and statistically evaluated for batches of cells. As previously reported an increase in P_m of 2-5 % for "ON cells" compared with standard pattern cells was found when the front grid was plotted with silver ink [6]. However P_m for "ON cells" varied stronger with temperature than for the other types. This was ascribed to a lower series resistance, R_S observed for "ON cells". The results reported in this paper refer to cells with electrochemically deposited front contact grid. As can be seen in Tab. 1 the line width is reduced by 40 % resulting in a reduction of shadowed cell area. Moreover the contact resistance to the underlying highly doped emitter could be significantly lowered which results in almost negligible low values for R_S for all types of cells. Nevertheless the temperature dependence of P_m for "ON cells" still was significantly larger. In order to investigate this behavior in more detail linear temperature coefficients were defined. Analogue to Eq. 1 which expresses the coefficient for P_m coefficients for the other solar cell parameter were defined.

$$\alpha_P = \frac{1}{P_m(300K)} \frac{\Delta P_m}{\Delta T} \tag{1}$$

A statistical analysis of these coefficients showed that for P_m, V_{OC} and CFF the coefficients are systematically larger for "ON cells" whereas no differences were found for I_{SC} which suggests that the differences in the temperature dependence is caused by the diode loss currents which are located in the semiconductor below the cell area which is covered by the metal lines. Therefore the current voltage characteristics $I_{cell}(V_{cell})$ of the cells in the dark and under illumination were fitted to a single diode modell as described below. Where the light generated current, I_L, practically equals the measured I_{SC}, I_D is the diode loss current and I_{SH} is a loss current caused by short circuits across the pn-junction.

$$I_{cell}(T) = I_L - I_D - I_{SH} \tag{2}$$

$$I_D(T) = I_0 \left\{ \exp\left(\frac{V_{cell} - R_S I_{cell}}{2k_B T} \right) - 1 \right\} \tag{3}$$

$$I_{SH}(T) = G_{SH}(V_{cell} - R_S I_{cell}) \tag{4}$$

$$I_0 \propto \exp\left(-E_{act} / k_B T \right) \tag{5}$$

I_0 is the diode saturation current thermally activated by an energy, E_{act}. The shunt conductance G_{SH} as well as R_S were found to be almost temperature independent.

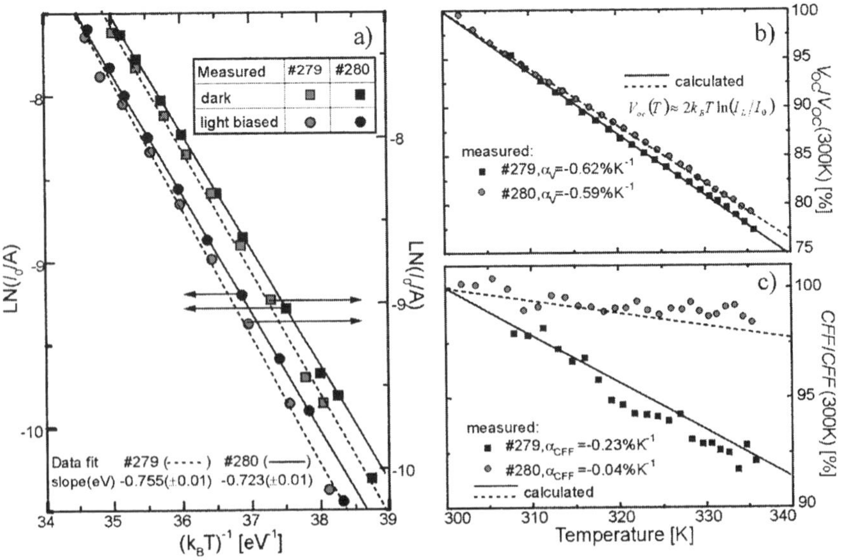

FIGURE 2. Arrhenius plot of the diode saturation current I_0 in order to retrieve it's activation energy E_{act} (a). Calculated and measured temperature dependence of V_{OC} (b) and CFF (c) for a pair of solar cells.

For 100 cm^2 cell area R_S was typically 1 mΩ – 3 mΩ, G_{SH} about 0.1 S. Since the effect of slight variations of these two parameter on the quality of data fitting was negligible no significant differences between cells types could be derived. It appears that "ON cells" have slightly lower values of R_S and slightly higher values of G_{SH}. A typical result of fitting measured $I_{cell}(V_{cell})$ data according to a recombination dominated diode loss current is shown in Fig. 2 for a pair of mc-Si cells. From the arrhenius plot of fitted I_0 values (Fig. 2a) E_{act} was determined which is independent from the illumination level. The shift between results from measurements in the dark and under illumination which can be seen in Fig. 2a is likely to be caused by a change of the active area for the diode current. We found that in "ON cells" E_{act} was about 30 meV larger compared with corresponding "OFF cells" which explains the observed differences in the temperature dependence of V_{OC} very well as can be seen in Fig. 2b. Beside I_D also R_S and, depending on the illumination intensity, G_{SH} determine *CFF*. Therefore the agreement of calculated and measured temperature variations of *CFF* (Fig. 2c) is less accurate than it is for V_{OC}.

CONCLUSIONS

Using an electrochemical deposition process for the application of individually designed front contact grid yields improved electrical output of mc-Si solar cells compared with screen printed front contacts. Cells having a galvanically processed grid covering predominately grain boundaries show an increase of P_m between 5-25 % compared with cells with a geometrical standard grid. This is significantly more than previously found for cells prepared by silver ink plotting [6]. Due to the high density of electrically active defects in the grain boundary region of mc-Si the thermal activation energy of the diode loss current differs from the ones within the crystallites. This was found to be the origin of larger temperature variations of P_m, V_{OC} and *CFF* observed for "ON cells" and could potentially be suppressed by a passivation procedure during solar cell preparation.

REFERENCES

1. Seto, J. Y. W, *J. Appl. Phys.* **46**, 5247-5254 (1975).
2. Edmiston, S. A., Heiser, G., Sproul, A., B., and Green, M., A., *J. Appl. Phys.* **80** 6783-6795 (1996).
3. Landsberg, P. T., and Abraham, M., S., *J. Appl. Phys.* **55**, 4284-4293 (1984).
4. Macdonal, D., and Cuevas, A., *Sol. Energ. Mat. Sol. C.* **65**, 509-516 (2001).
5. Summhammer, J., and Schlosser, V., "Investigations of a novel front contact grid on poly silicon solar cells" in *Proceedings of the twelfth European Photovoltaic Solar Energy Conference*, edited by H. A. Ossenbrink et al., Bedford, UK: H.S. Stephens and Associates, 1994, pp. 734-737.
6. Ebner, R., Radike, M., Schlosser, V., and Summhammer, J., *Prog. Photovolt: Res. Appl.* **11**, 1-13 (2003).
7. Schlosser, V., Markowitsch, W., Klinger, G., Bajons, P., Chancy, S., Ebner, R., and Summhammer, J., "Investigations of the Electro-Optical Properties of Multicrystalline Silicon during Solar Cell Processing " in *Conference Proceedings PV in Europe*, edited by J. L. Bal et al., Munich: WIP, 2002, pp. 32-35.

Applied Physics (APHYS 2003)
A. Méndez-Vilas (Ed.)

Contributions of Steady Heat Conduction to the Rate of Chemical Reaction

Kim Hyeon-Deuk[1] and Hisao Hayakawa

Graduate School of Human and Environmental Studies, Kyoto University, Kyoto 606-8501, JAPAN
Department of Physics, Yoshida-South Campus, Kyoto University, Kyoto 606-8501, JAPAN

Abstract. We have derived the effect of steady heat flux on the rate of chemical reaction based on the line-of-centers model using the explicit velocity distribution function of the steady-state Boltzmann equation for hard-sphere molecules to second order. We have found that the second-order velocity distribution function plays an essential role for the calculation of it. This indicates the significance of the second-order coefficients in the solution of the steady-state Boltzmann equation as terms which reflect the local nonequilibrium effect.

CHEMICALLY REACTING GAS

In the early stage of a chemical reaction between monatomic molecules:

$$A + A \rightarrow products, \tag{1}$$

the rate of chemical reaction is not affected by the existence of products. From the viewpoint of kinetic collision theory[1], the rate of chemical reaction (1) can be described as

$$R = \int d\mathbf{v} \int d\mathbf{v}_1 \int d\mathbf{k}\, f f_1 g \sigma(g), \tag{2}$$

where $\mathbf{v}$ and $\mathbf{v}_1$ are the velocities of the molecules, $g = |\mathbf{v} - \mathbf{v}_1|$ their relative speed, $\mathbf{k}$ the solid angle, $f = f(\mathbf{r}, \mathbf{v})$ and $f_1 = f(\mathbf{r}, \mathbf{v}_1)$ are the distributions of $\mathbf{v}$ and $\mathbf{v}_1$ at $\mathbf{r}$, respectively.

The line-of-centers model proposed by Present has been accepted as a standard model to describe the chemical reaction in gases.[1] It assumes the chemical cross-section as

$$\sigma(g) = \begin{cases} 0 & g < \sqrt{\dfrac{4E^*}{m}} \\ \dfrac{d^2}{4}\left(1 - \dfrac{4E^*}{mg^2}\right) & g \geq \sqrt{\dfrac{4E^*}{m}} \end{cases} \tag{3}$$

with m mass of the molecules and E^* the threshold energy of the chemical reaction. d is regarded as a distance between centers of monatomic molecules at contact.

[1] Present Address: Department of Chemistry, Kyoto University, Kyoto 606-8502, JAPAN

NONEQUILIBRIUM EFFECT ON THE RATE OF CHEMICAL REACTION

In order to calculate the rate of chemical reaction (2), we expand the velocity distribution function to second order as

$$f = f^{(0)} + f^{(1)} + f^{(2)} = f^{(0)}(1+\phi^{(1)}+\phi^{(2)}), \tag{4}$$

around the local Maxwellian, $f^{(0)} = n(m/2\pi\kappa T)^{3/2}\exp[-m\mathbf{v}^2/2\kappa T]$, with n the density of molecules, κ the Boltzmann constant and T the temperature defined from the kinetic energy. Substitution of eq.(4) into eq.(2) leads to

$$R = R^{(0)} + R^{(1)} + R^{(2)}, \tag{5}$$

up to second order. The zeroth-order term of R, the rate of chemical reaction of the equilibrium theory, becomes $R^{(0)} = \int d\mathbf{v}\int d\mathbf{v}_1\int d\mathbf{k} f^{(0)} f_1^{(0)} g\sigma(g) = 4n^2\sigma^2(\pi\kappa T/m)^{1/2}\exp[-E^*/\kappa T]$. The first-order term of R, i.e. $R^{(1)}$, does not appear because $\phi^{(1)}$ is an odd function of $\mathbf{c}$. The second-order term of R, i.e. $R^{(2)}$, is divided into

$$R^{(2,A)} = \int d\mathbf{v}\int d\mathbf{v}_1\int d\mathbf{k} f^{(0)} f_1^{(0)}\phi^{(1)}\phi_1^{(1)} g\sigma(g), \tag{6}$$

and

$$R^{(2,B)} = \int d\mathbf{v}\int d\mathbf{v}_1\int d\mathbf{k} f^{(0)} f_1^{(0)}[\phi^{(2)}+\phi_1^{(2)}] g\sigma(g). \tag{7}$$

Since the integrations (6) and (7) have the cutoff from eq.(3), the explicit forms of $\phi^{(1)}$ and $\phi^{(2)}$ of the steady-state Boltzmann equation for hard-sphere molecules are required to calculate $R^{(2,A)}$ and $R^{(2,B)}$, respectively.

Although Burnett had determined the second-order pressure tensor for the Boltzmann equation, he had not derived the explicit second-order velocity distribution function of the Boltzmann equation.[2,3] Therefore, none has succeeded to obtain the correct reaction rate of Present's model except for Fort and Cukrowski who adopted information theory[4] as the nonequilibrium velocity distribution function to second order.[5,6]

On the other hand, we have recently derived the explicit velocity distribution function of the steady-state Boltzmann equation for hard-core molecules to second order in density and the temperature gradient.[3] This enables us to calculate the effect of steady heat flux on the rate of chemical reaction based on the line-of-centers model.[7]

RESULTS: THE RATE OF CHEMICAL REACTION

Although we have derived the explicit expressions of $R^{(2,A)}$ and $R^{(2,B)}$, we show only the graphical results of $R^{(2)}$ compared with those of $R^{(2,A)}$ in Fig.1. In our calculation, we have confirmed that $R^{(2,B)}$ for the steady-state Boltzmann equation is determined only by the terms in $\phi^{(2)}$ which Burnett[2] had not derived, but we have derived in ref.[3]. We have found that $R^{(2,B)}$ plays an essential role for the evaluation of $R^{(2)}$, and that there are no qualitative differences in $R^{(2)}$ of the steady-state Boltzmann equation, the steady-state Bhatnagar-Gross-Krook(BGK) equation and information theory. It should be mentioned that, however, we have found qualitative differences among these theories in pressure tensor and the kinetic temperature as Table 1 shows.[3] We have also found that the steady-state BGK equation belongs to the same universality class as Maxwell molecules, and that information theory is inconsistent with the steady-state Boltzmann equation.[8]

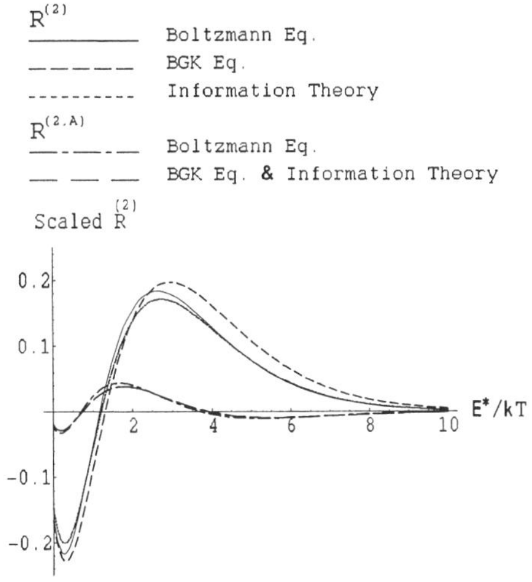

FIGURE 1. Both of $R^{(2)}$ and $R^{(2,A)}$ are scaled by $\pi^{1/2}d^2m^{1/2}J_x^2/\kappa^{5/2}T^{5/2}$. Here J_x means a steady heat flux.

TABLE 1. Numerical Constants for the pressure tensor $P_{ij} = n\kappa T[\delta_{ij} + \lambda_{ij}mJ_x^2/n^2\kappa^3T^3]$ and the each component of the kinetic temperature $T_i = T[1+\lambda_{ii}mJ_x^2/n^2\kappa^3T^3]$. Note that the off-diagonal components of a λ_{ij} re zero, and that $\lambda_{yy} = \lambda_{zz} = -\lambda_{xx}/2$.

	λ_{xx}
Hard-Core Molecules	-4.600×10^{-2}
Maxwell Molecules	0
BGK equation	0
Information Theory	12/25

DISCUSSION

The nonequilibrium effect on the rate of chemical reaction will substantiate significance of the second-order coefficients in the solution of the steady-state Boltzmann equation, although their importance has been demonstrated only for descriptions of shock wave profiles and sound propagation phenomena. This indicates the significance of the second-order coefficients as terms which reflect the local nonequilibrium effect.

We also propose a *thermometer* of a monatomic dilute gas system under a steady heat flux.[7] We mean that we can measure the temperature T around a heat bath at T_0 in the nonequilibrium steady-state system indirectly with the aid of the nonequilibrium effect on the rate of chemical reaction. The nonequilibrium effect in the early stage of chemical reaction around the heat bath can be measured experimentally. Thus, one can compare the experimental result with the theoretical result shown in Fig.1 with $T = T_0$. The difference between the former and the latter will indicate that the temperature T around the heat bath is not identical with T_0, but $T = T_0 + \Delta$ where Δ depends upon the steady heat flux in general. Substituting this temperature expression into the explicit expressions of $R^{(0)}$, we obtain a new correction term

$$R^{(new)} = \frac{2n^2\sigma^2\Delta}{T_0}\left(\frac{\pi\kappa T_0}{m}\right)^{\frac{1}{2}}\left(1+\frac{2E^*}{\kappa T_0}\right)e^{-\frac{E^*}{\kappa T_0}}, \tag{8}$$

as the nonequilibrium effect on the rate of chemical reaction besides $R^{(2)}$ in Fig.1 with $T = T_0$. We can estimate the gap Δ so as to make the new correction term match the experimental result. For example, if Δ in eq.(8) is proportional to the heat flux, we will confirm the relevancy of the slip effect[9]. If Δ in eq.(8) is identical with $2mJ_x^2/5n^2\kappa^3T_0^2$, the experimental result will agree with the theoretical result from the steady-state Boltzmann equation with the *nonequilibrium temperature* $\theta = T_0$ predicted by Jou *et al.*.[4] We show the comparison of the theoretical results in Fig.2. We note that one can obtain the explicit form of $R^{(2)}$ with $\theta = T_0$ expressed as the dashed line in Fig.2 as the sum of $R^{(2)}$ from eqs.(6) and (7) for the steady-state Boltzmann equation with $T = T_0$ and $R^{(new)}$ from eq.(8) with $\Delta = 2mJ_x^2/5n^2\kappa^3T_0^2$. We have found that there is a significant difference between $R^{(2)}$ with $T = T_0$ and that with $\theta = T_0$ for small $E^*/\kappa T_0$.

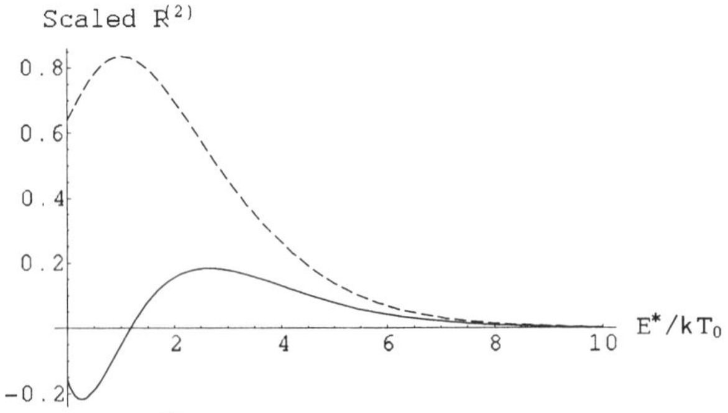

FIGURE 2. Comparison of $R^{(2)}$ from the steady-state Boltzmann equation with $T = T_0$ (the solid line) and that with $\theta = T_0$ (the dashed line) as a function of $E^*/\kappa T_0$. Note that the former is identical with the result shown by the solid line in Fig.1 with $T = T_0$. $R^{(2)}$ is scaled by $\pi^{1/2}d^2m^{1/2}J_x^2/\kappa^{5/2}T^{5/2}$.

ACKNOWLEDGMENTS

This presentation is partially supported by the Sasagawa Scientific Research Grant from The Japan Science Society, Kim Man-Yu Scientific Organization and the Grant-in-Aid of ministry of Education, Science and Culture, Japan (Grant No.15540393).

REFERENCES

1. R. D. Present, *Kinetic Theory of Gases*, New York, Mcgraw-Hill, 1958.
2. D. Burnett, *Proc. Lond. Math. Soc.* **40**, 382- 435(1935).
3. Kim. H.-D. and H. Hayakawa, *J. Phys. Soc. Jpn.* **72**, 1904-1916 (2003).
4. D. Jou, J. Casas-Vazquez and G. Lebon, *Extended Irreversible Thermodynamics*, Berlin, Springer, 2001.
5. J. Fort and A. S. Cukrowski, *Chem. Phys.* **222,** 59-69 (1997).
6. J. Fort and A. S. Cukrowski, *Acta Phys. Pol.* B **29,** 1633-1646 (1998).
7. Kim. H.-D. and H. Hayakawa, *Chem. Phys. Lett.* **372,** 314-319 (2003).
8. Kim. H.-D. and H. Hayakawa, *J. Phys. Soc. Jpn.* **72**, 2473-2476 (2003).
9. C. Cercignani, *Mathematical Methods in Kinetic Theory*, New York, Plenum Press, 1990.

Applied Physics (APHYS 2003)
A. Méndez-Vilas (Ed.)

Dynamic Force Spectroscopy: Looking at the Total Harmonic Distortion

Robert W. Stark[1]

Center for Nanoscience and Ludwig-Maximilians-Universität München, Section Crystallography, Theresienstr. 41, 80333 München, Germany

Abstract. Tapping mode atomic force microscopy is a standard technique for inspection and analysis at the nanometer scale. The understanding of the non-linear dynamics of the system due to the tip sample interaction is an important prerequisite for a correct interpretation data acquired by dynamic AFM. Here, the system response in tapping-mode atomic force microscope (AFM) simulated numerically. In the computer model the AFM microcantilever is treated as a distributed parameter system. With this multiple-degree-of-freedom (MDOF) approach the the total harmonic distortion in dynamic AFM spectroscopy is simulated.

INTRODUCTION

The atomic-force microscopy (AFM) has become an standard inspection and analysis tool in research as well as in industry. The tapping or intermittent contact mode is presently the most widely used imaging modes in practical AFM applications. In this mode of operation, the forced oscillation amplitude of the force sensor is adjusted to a value between 10 nm and 100 nm. During imaging the amplitude is limited by the specimen surface which can be understood as a non-linear mechanical controller limiting the amplitude. Thus, a theoretical description of the system dynamics in this mode requires an understanding of the non-linear system dynamics [1]. The non-linear tip-sample interaction leads to a complicated system behavior. It was shown that the system is well behaved for a large set of parameters but that it also can exhibit a complex dynamics [2-4]. The non-linearity also induces higher harmonics in the system response which are amplified by the higher eigenmodes of the force sensor [5-8].

These higher harmonics can be measured by dynamic force spectroscopy recording the full spectral response of the system. This also allows one to directly measure transient tip-sample interaction forces by signal inversion [9]. In a simplified analysis the individual higher harmonics characterize the system dynamics [10].

The total harmonic distortion is a measure for the degree of the generation of higher harmonics. In the following, the response of the total harmonic distortion to the variation of average tip-sample gap is investigated by numerical simulations. To model the higher eigenmodes of the cantilever a 6-th-order state space model is used.

---

[1] E-mail: stark@nanomanipulation.de. This work was supported by the German Federal Ministry of Education and Research (BMBF) under Grant 03N8706

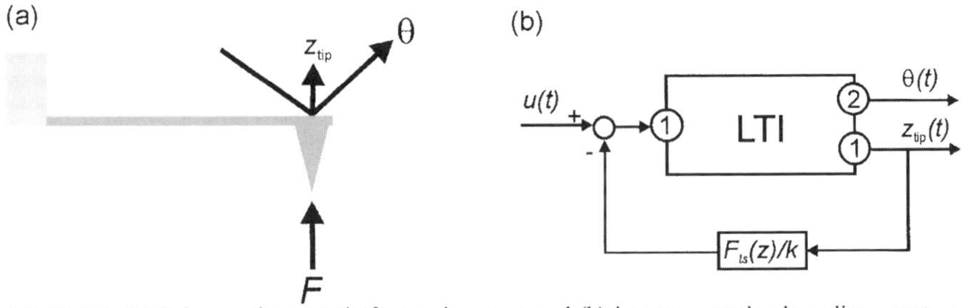

FIGURE 1. (a) Scheme of an atomic force microscope and (b) its representation by a linear system with a non-linear output feedback. The force F acting onto the tip is fed into the linear time invariant system. The tip position and the light lever readout are the system outputs.

MODELLING

The microcantilever in atomic-force microscopy is approximated by a linear and time invariant (LTI) system. In tapping mode, its deflection is typically of the order of a few nanometers, whereas its thickness is in the range of microns. Therefore, the cantilever [Fig. 1 (a)] is modelled as a LTI system with a non-linear output feedback [Fig. 1 (b)]. To investigate the basic phenomena in the following, only forces that act onto the tip at input (1) will be considered..

Approximating the microcantilever by a n degrees-of-freedom (n eigenmodes) LTI system the equations of motion are given in state-space form by

$$\dot{\mathbf{x}} = \mathbf{A}\mathbf{x} + \mathbf{B}u, \tag{1}$$

$$\mathbf{y} = \mathbf{C}\mathbf{x}. \tag{2}$$

The time dependent state-vector $\mathbf{x} = (x_1, \dot{x}_1, \cdots, x_n, \dot{x}_n)$ contains the modal displacements and velocities. The system matrix

$$\mathbf{A} = \begin{bmatrix} \Phi_1 & 0 & 0 \\ 0 & \ddots & 0 \\ 0 & 0 & \Phi_n \end{bmatrix}, \; with\; \Phi_i = \begin{bmatrix} 0 & 1 \\ -\hat{\omega}_i^2 & -2\gamma_i\hat{\omega}_i \end{bmatrix} \tag{3}$$

is a $2n \times 2n$ matrix. It consists of 2×2 submatrices Φ_i along the diagonal. The matrices Φ_i characterize the individual eigenmodes of the weakly damped system. The eigenfrequencies $\hat{\omega}_i = \omega_i / \omega_1$ are normalized to the fundamental resonance frequency, the modal damping is γ_i. In the case of heavy damping as it is the case for example in a liquid environment matrix A also contains non-diagonal elements. The input vector is

$$\mathbf{B} = \left[0, \varphi_1(\xi_{tip})/M_1, \cdots, 0, \varphi_n(\xi_{tip})/M_n\right]^T. \tag{4}$$

It contains the modal deflection at the tip $\varphi_i(\xi_{tip})$ which is normalized by the generalized modal mass $M_i = \int_0^1 m\varphi_i(\xi)^2 d\xi$. Scalar u is the input to the model, i.e. the driving force minus the tip-sample interaction force.

The components of the output vector $\mathbf{y} = [y_1, y_2]^T$, i.e. the tip displacement output y_1 that is used for feedback and the photodiode signal output y_2, are linear combinations of the states as defined in the output matrix

$$\mathbf{C} = \begin{bmatrix} \varphi_1(\xi_{tip})/n_{pos} & 0 & \cdots & \varphi_n(\xi_{tip})/n_{pos} & 0 \\ \varphi_1'(\xi_{sens})/n_{sig} & 0 & \cdots & \varphi_n'(\xi_{sens})/n_{sig} & 0 \end{bmatrix}. \tag{5}$$

The tip deflection output (1) is normalized with n_{pos} to obtain a unit DC gain, i.e. it is normalized to a quasi-static spring constant $\hat{k}_{cant} = 1$ of the system at $\omega = 0$. The optical lever sensor output (2) is normalized by n_{sig} to a unit response at $\omega = 0$.

The tip displacement $y_1 = \sum_{i=1}^{n} x_{2i-1}$ at output (1) is used to calculate the non-linear tip-sample interaction force $F_{ts}(y_1 - z_s)/k$. The resulting force is fed back to input (1) of the model.

The attractive part of the interaction force $(y_1 - z_s \geq a_0)$ is modelled as a van der Waals interaction force. A Derjaguin-Müller-Toporov (DMT) model [?] was used in the repulsive regime $(y_1 - z_s < a_0)$. Thus, the interaction force is

$$F_{ts}(y_1) = \begin{cases} -HR/\left[6(y_1 - z_s)^2\right] & y_1 - z_s \geq a_0 \\ -HR/6a_0^2 + \frac{4}{3}E^*\sqrt{R}(a_0 - y_1 + z_s)^{3/2} & y_1 - z_s < a_0 \end{cases}, \tag{6}$$

where H is the Hamaker constant, R the tip radius, and a_0 an interatomic distance. The effective contact stiffness is given by $E^* = \left[(1-\nu_t^2)/E_t + (1-\nu_s^2)/E_s\right]^{-1}$, where E_t and E_s are the respective elastic moduli and ν_t and ν_s the Poisson ratios of tip and sample.

As numerical parameters typical values for a beam shaped cantilever were used. Three eigenmodes were considered $(n = 3)$ for the computation of the system response. The modal deflection φ_n and deflection angle φ_n were calculated from the well known eigenmodes of a uniform beam [?]. The tip and laser spot were assumed to be collocated at the end of the cantilever beam $\xi_{tip} = \xi_{sens} = 1$. The damping was set to $\gamma_i = 0.0025$ for all modes. Further parameters were: $k = 10\ Nm^{-1}$, $R = 15\ nm$, $E_t = 129GPa$, $\nu_t = 0.28$, $E_s = 70GPa$, $\nu_s = 0.3$, $a_0 = 0.166nm$, and $H = 6.4\times10^{-20} J$. The driving frequency was $\omega = 1.0$, the amplitude of the driving force was $F_{dr} = 0.97\ nN$, resulting in a free amplitude of $A_0 = 20\ nm$. The simulation was implemented in MATLAB RELEASE 13 using SIMULINK (The Mathworks Inc., Natick, MA, USA).

TOTAL HARMONIC DISTORTION

In order to compute the system response in a dynamic AFM spectroscopy experiment the sample position z_s was reduced by ramping. At each approach step the ramp was halted and the system was allowed to equilibrate for more than 1000 cycles before data was extracted for Fourier transform (FFT) analysis. For $z_s \leq -3nm$ data was extracted every 0.5 nm, for larger z_s the distance was decreased to 0.2 nm to capture the complex dynamics at small distances. Figure 2 shows the evolution of the amplitude and phase of the first harmonic (fundamental) together with the total harmonic distortion of the position output (1) and the average force. The harmonic distortion is defined by $THD = \left(\sum_{n=2}^{\infty} |c_n|^2\right)^{-1/2} / \left(\sum_{n=1}^{\infty} |c_n|^2\right)^{-1/2}$, where $|c_n|$ is the FFT amplitude of the n-th harmonic. It gives the fraction of power that is transferred into the higher harmonics as compared to the total power.

Far away from the sample, the oscillation amplitude of the fundamental is $2|c_1| = -20nm$, the phase is at $-90°$. There is only a very small average attractive force and a very small total harmonic distortion. Approaching to $z_s = -20nm$ the system is in the net attractive (low amplitude) regime as can be seen by the net-negative interaction force. With increasing strength of the attractive interaction the THD of the output signal also increases. Between $z_s = -18.5nm$ and $z_s = -18nm$ the system transits to the high amplitude state (arrows). This transition prevails in the phase as well as in the average interaction force. It is also visible in the THD which increases by 50%. Approaching further, the dynamics of the system changes at $z_s = -2.8nm$. The THD decreases significantly and recovers at $z_s = -1nm$ before it drops to zero. This behavior can be explained by the generation of subharmonics where spectral power is transferred into subharmonics.

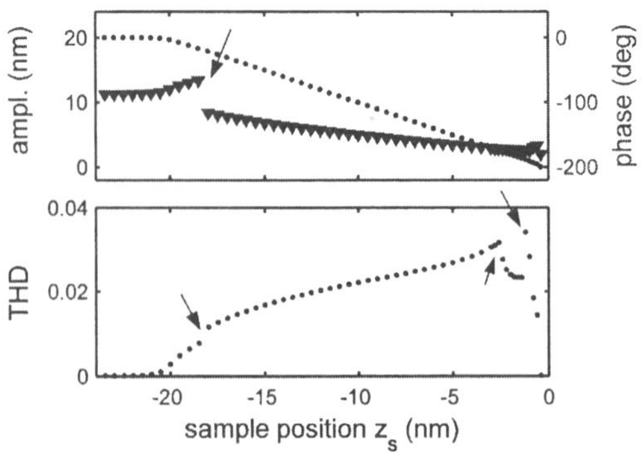

Figure 2. (Above) Amplitude and phase of the first harmonic. The transition from the low amplitude state into the high amplitude state can be identified by the phase jump (arrow). (Below) Total harmonic distortion of the position output. The transition between both states is accompanied by a step in the THD (arrow, left). At a small z_s subharmonics are generated, which first leads to a reduction of the THD followed by a final maximum (arrows, right).

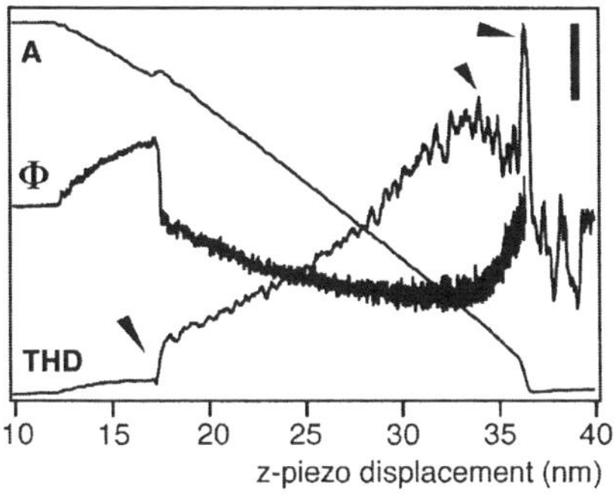

Figure 3. Experimental approach curves on a silicon sample. Amplitude A, phase Φ of the fundamental, and the total harmonic distortion THD. The arrows indicate the transition from the attractive to the repulsive branch (left) and both maxima in the THD (right). Scale bar: A: a.u., Φ: 36°, THD: 4 %. (From Ref. [10]. (c) 2000 AIP, reprinted with permission).

CONCLUSIONS

In comparison with experimental data obtained earlier [10] the characteristics of the response of the simulated *THD* are similar to that of the experimental data in Fig. 3 although the scaling is different. The increase in the *THD* at the transition from the attractive to the repulsive state (arrow) was also observed in the experimental data. Additionally, both maxima in Fig. 3 of the *THD* (arrows) are well reproduced by the numerical simulations. This indicates that the numerical simulations capture basic features of the dynamics in tapping mode AFM. In order to achieve a better match of the numerical simulations to the experimental data a more precise modelling is necessary. This includes a better mathematical model for the cantilever as it can be obtained e.g. by system identification and an more precise model of the contact mechanics.

REFERENCES

1. García, R., and Pérez, R., Surface Science Reports, 47, 197–301 (2002).
2. Hunt, J., and Sarid, D., Appl. Phys. Lett., 72, 2969–2971 (1998).
3. Basso, M., Giarré, L., Dahleh, M., and Mezi´c, I., J. Dyn. Syst. Meas. Control, 122, 240–245 (2000).
4. Rützel, S., Lee, S. I., and Raman, A., Proc. R. Soc. Lond. A, 459, 1925–1948 (2003).
5. Stark, R. W., and Heckl, W. M., Surf. Sci., 457, 219–228 (2000).
6. Sahin, O., and Atalar, A., Appl. Phys. Lett., 79, 4455–4457 (2001).
7. Rodriguez, T. R., and García, R., Appl. Phys. Lett., 80, 1646–8 (2002).
8. Balantekin, A., and Atalar, A., Phys. Rev. B, 67, 193404 (2003).
9. Stark, M., Stark, R. W., Heckl, W. M., and Guckenberger, R., Proc. Natl. Acad. Sci. USA, 99, 8473–8478 (2002).
10. Stark, M., Stark, R. W., Heckl, W. M., and Guckenberger, R., Appl. Phys. Lett., 77, 3293–5 (2000).
11. Derjaguin, B. V., Muller, V. M., and Toporov Yu, P., J. Coll. Interf. Sci., 53, 314–26 (1975).
12. Clough, R., and Penzien, J., Dynamics of structures, McGraw-Hill, Singapore, 1993, 2. edn.

Applied Physics (APHYS 2003)
A. Méndez-Vilas (Ed.)

Optical Fiber Sensors For Nuclear Environments

Gaspar M. Rego[*ξδ], Alberto Fernandez Fernandez[ϒ], José L. Santos[ξα], Henrique M. Salgado[ξδ], Francis Berghmans[ϒ] and Andrei Gusarov[ϒ]

[*]*Escola Superior de Tecnologia e Gestão, Instituto Politécnico de Viana de Castelo, Av. do Atlântico, 4900 Viana do Castelo, Portugal*
[ξ]*INESC Porto, Unidade de Optoelectrónica e Sistemas Electrónicos, Rua do Campo Alegre 687, 4169-007 Porto, Portugal*
[δ]*Dep. de Eng. Electrotécnica e de Computadores, Fac. de Engenharia, Universidade do Porto, Rua Dr. Roberto Frias, 4200-465 Porto, Portugal*
[α]*Dep. de Física, Faculdade de Ciências, Universidade do Porto, Rua do Campo Alegre 687, 4169-007 Porto, Portugal*
[ϒ]*SCK•CEN, Belgian Nuclear Research Center, Instrumentation Department, Boeretang 200, B-2400, Mol, Belgium*

Abstract. We have experimentally studied the effect of ionizing radiation on the properties of LPFGs fabricated in a pure-silica fiber using the arc discharge technique. The goal of the study is to assess the possibility of using such gratings in radiation environments for sensing applications.

INTRODUCTION

The well-known properties of the FBG-based optical fiber sensors (OFS) make them very attractive for the application in the nuclear industry where they can perform, among others, structural integrity and temperature monitoring [1]. The implementation of such sensors is compromised by their sensitivity to radiation. It is, therefore, of interest to study the effect of radiation on FBGs written in pure-silica-core fibers, which are usually expected to be more radiation-resistant as compared with germanium-doped fibers [2]. Gratings fabrication in pure-silica-core fibers requires that the inscription be based on CO_2 lasers [3] or arc-discharges [4]. Those techniques allow a very high temperature stability of the gratings. Recently, the possibility of the use of a 193-nm excimer laser has also been demonstrated [5].

GRATINGS FABRICATION AND CHARACTERIZATION

Long-period fiber gratings (LPFGs) used in the present study have been written in a pure-silica-core fiber (D_{core} = 9 μm and N.A. = 0.11) using the electric arc-discharge technique [6]. The fabrication consists in placing a fiber under a tension between the electrodes of a splicing machine, which produces a short duration arc discharge. Then, the fiber is moved by a grating period Λ. The sequence "arc discharge - fiber displacement" is repeated several times and is computer controlled. It should be noted that the optimal writing parameters are to be defined for each fiber type individually. In the preset case a grating period Λ = 730 μm was chosen so that a resonance peak appeared in the third telecommunication window. Figure 1 shows the transmission spectra of four gratings written in the SiO_2-core fiber under different conditions. As it can be seen, an increase of the tension and a decrease of the arc duration lead to a stronger coupling. The influence of the discharge current was briefly studied and the results indicate that its increase may speed-up the growth of gratings although an increase of the current does not necessarily mean high amplitude of the transmission dip, unless a certain "threshold tension" is also used. For example, when a mass of 1.4 g was applied in combination with a discharge current of 11 mA with duration 1 s or 0.3 s no grating or a weak grating (< 0.7 dB) were formed after 25 periods, respectively. It may be relevant to note that both an increase of the tension and the current also contributes to a higher insertion loss and therefore a compromise is required to obtain a grating with useful parameters [6].

LPFGs were also written in a fiber pre-annealed at 1000 °C during 90 min. Using the same writing conditions as for LPFG #3 only 34 arc discharges were required to obtain a resonance centered at 1633.6 nm with a loss-peak isolation of 23 dB.

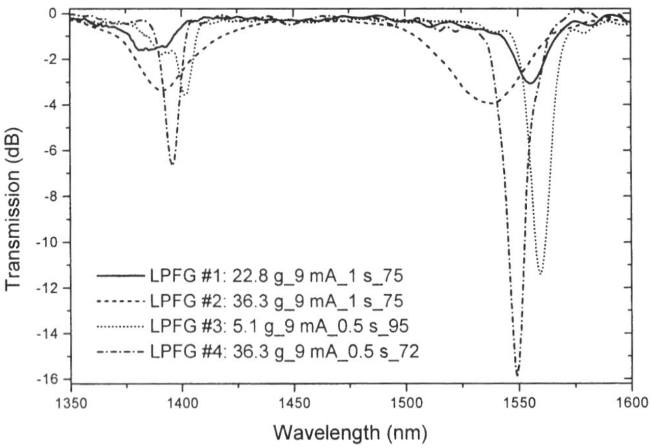

FIGURE 1. Transmission spectra of the gratings written in the SiO_2-core fiber using different parameters (external applied tension: 5.1 g to 36.3 g; current: 9 mA; arc duration: 0.5 to 1 s and number of discharges: 72 to 95).

In order to determine the temperature dependence of the transmission dip position for the LPFGs written in the SiO_2-core fiber two gratings were placed inside a tubular oven under a slight tension (a suspended mass of 1.4 g was used). The temperature of the oven was increased at a low rate and the transmission spectra of the gratings were periodically recorded. Figure 2 shows that the position of the transmission dip moves towards longer wavelengths as the temperature rises up to 350-400 °C. At higher temperatures they move towards shorter wavelengths. In standard fibers such a behavior can only be seen at very high temperatures. The "strain point" for this fiber, i.e., the temperature at which the annealing of intrinsic stresses begins, occurs at temperatures well below a 700-900 °C range observed for other fibers [6]. The temperature sensitivity of the gratings (10 - 30 pm/°C) is also significantly lower as compared with that for gratings written in standard fibers; for latter case it can be as high as 80 pm/°C, for temperatures in a range 0 - 125 °C. It is necessary to note that a difference for the temperature sensitivity of the two LPFGs studied was observed. The effect can be probably related with the difference in the arc duration used for their inscription. A further work is required to address this behavior correctly. Irreversible changes in the shape of the spectrum start above 650 °C.

To measure the strain sensitivity of the grating the fiber was fixed at two points separated by 50 cm, one of them being on a micrometer positioner. The LPFG was then pulled in steps of 0.05 mm with the total displacement up to 0.6 mm. Figure 3 shows the strain measurement results for LPFG #2 and LPFG #4. The strain sensitivity of the LPFG #3 was determined to be 0.34 pm/με after the grating being subject to a temperature of about 450 °C. Therefore, it is possible that the annealing and the writing conditions also affect the strain sensitivity. For comparison, the strain sensitivity of gratings written in standard Ge-doped fibers was found to be in a range 0.8-1.1 pm/με.

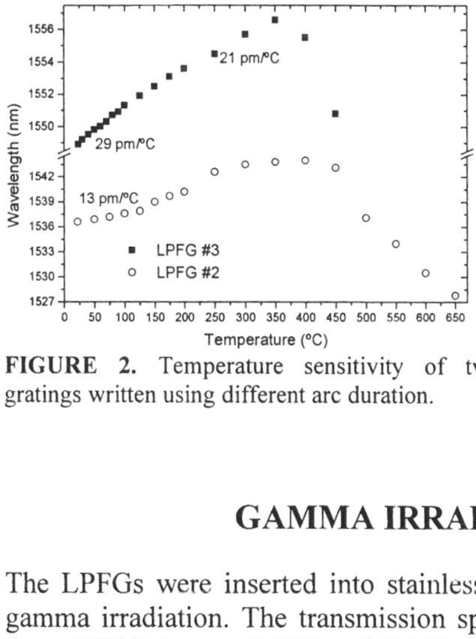

FIGURE 2. Temperature sensitivity of two gratings written using different arc duration.

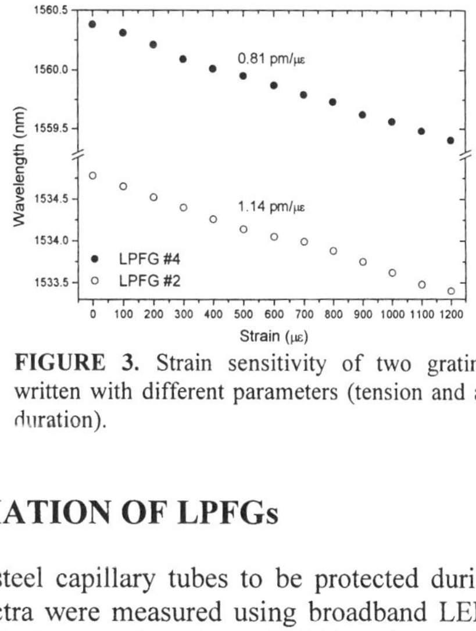

FIGURE 3. Strain sensitivity of two gratings written with different parameters (tension and arc duration).

GAMMA IRRADIATION OF LPFGs

The LPFGs were inserted into stainless steel capillary tubes to be protected during gamma irradiation. The transmission spectra were measured using broadband LEDs from BW-Tek, an ANDO AQ6315B optical spectral analyzer (OSA) and two SM optical switches (OS)(see Figure 4). The emission of the optical sources was continuously monitored via an internal reference fiber. The use of a temperature-stabilized FBG as an external wavelength reference allowed a wavelength measurement stability better than 20 pm throughout the experiment.

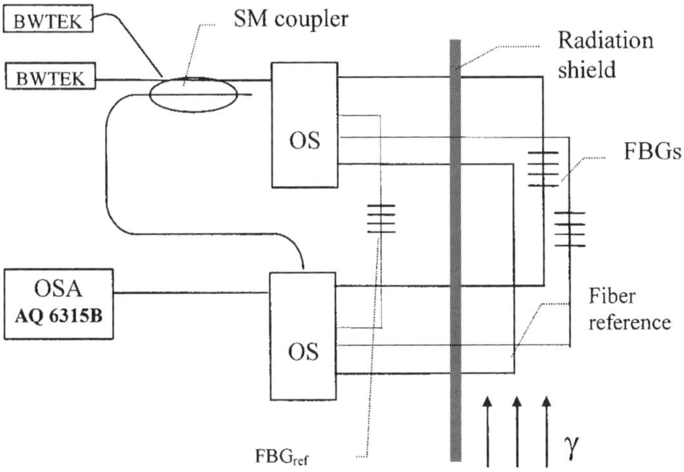

FIGURE 4: Optical measurement set-up

The LPFGs were evaluated under ^{60}Co irradiation in the RITA (Radio Isotope Test Arrangement) facility (SCK•CEN, Belgium), described in details in [7]. For the present experiment the gamma dose rate was about 1 kGy/h [H_2O] with the total dose about 560 kGy. The temperature in the irradiation rig was kept 37.4 ± 0.1 °C.

The effect of gamma-radiation on the transmission spectra strongly depends on the grating fabrication conditions. E.g., Figures 5 and 6 depict results for LPFG #1 and #3, respectively. It can be seen that for LPFG #1 the amplitude of the 1500 nm peak is decreasing, while for LPFG #3 it is increasing. Such a behavior should be attributed to a modal index change under radiation. It is known that the radiation sensitivity of pure silica is very low. However, the refraction index of the F-doped cladding can be changed by radiation. A more detailed analysis of the modal structure, including numerical simulations, is planned for a future.

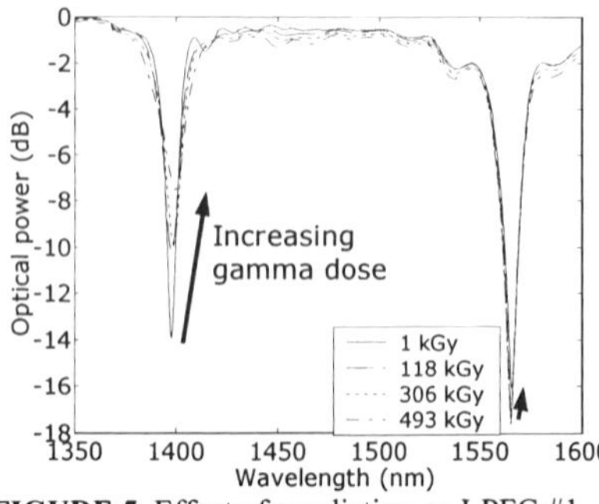

FIGURE 5. Effect of γ-radiation on LPFG #1

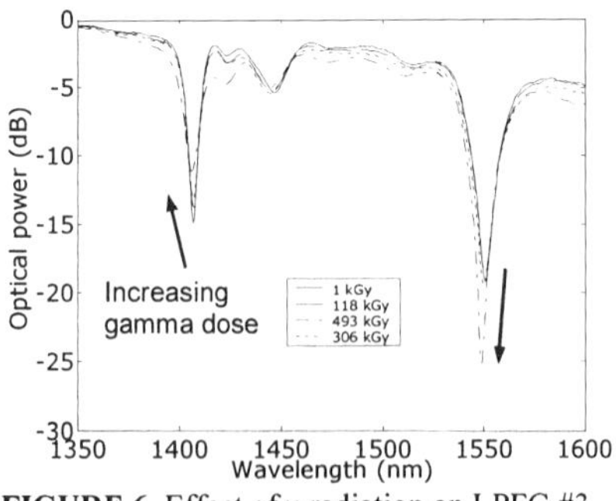

FIGURE 6. Effect of γ-radiation on LPFG #3

DISCUSSION AND CONCLUSION

The basic condition for implementation of a LPFG for strain sensing in a radiation environment is that the grating survives under radiation. Our results show that even under a dose in excess of 0.5 MGy the transmission deeps of the LPFGs are easily detectable. The radiation does change the amplitude and the positions of the peaks. However, the gratings fabricated under different writing conditions have opposite trends in the parameters change. That means that by adjusting the writing conditions it should be possible to fabricate a grating for which the dip position or/and amplitude are not changed by radiation.

In conclusion, LPFGs written in a pure-silica-core fiber were characterized in terms of the temperature and strain sensitivity. The effect of ionizing radiation on their properties has also been investigated. The results of the irradiation experiments enable to conclude that arc-induced gratings in pure-silica-core fibers can be used to perform multi-parameter simultaneous measurement in radiation environments.

ACKNOWLEDGMENTS

G. M. Rego is thankful for the grant through the Program PRODEP III - Medida 5 - Acção 5.3. - Formação Avançada de Docentes do Ensino Superior, integrada no Eixo 3. A. Fernandez Fernandez wish also to acknowledge the financial support of the European Commission – EURATOM Fusion Technology Program. The content of the publication is the sole responsibility of its publishers and it does not necessarily represent the views of the Commission or its services.

REFERENCES

1. Gusarov, A., Defosse, Y., Deparis, O., Blondel, M., Fernandez Fernandez, A., Berghmans, F., and Decréton, M., "Effect of ionizing radiation on the properties of fiber Bragg gratings written in Ge-doped fiber" in Proc. Bragg Gratings, Photosensitivity and Poling in Glass Waveguides, Stresa, Italy, paper BThC30-1, 2001.
2. Kakuta, T., Sakasai, K., Shikama, T., Narui, M., and Sagawa, T., "Absorption and fluorescence phenomena of optical fibers under heavy neutron irradiation", *J. Nucl. Mater.* **258-263**, 1893-1896 (1998).
3. Akiyama, M., Nishide, K., Shima, K., Wada, A., and Yamauchi, R., "A novel long–period fiber grating using periodically released residual stress of pure-silica core fiber" in Proc. Optical Fiber Communications, paper ThG1, 1998, pp. 276-277.
4. Enomoto, T., Shigehara, M., Ishikawa, S., Danzuka, T., and Kanamori, H., "Long–period fiber grating in a pure-silica core fiber written by residual stress relaxation" in Proc. Optical Fiber Communications, paper ThG2, 1998, pp. 277-278.
5. Albert, J., Fokine, M., Margulis, W., "Grating formation in pure silica-core fibers" in Proc. Bragg Gratings, Photosensitivity and Poling in Glass Waveguides, Stresa, Italy, paper BThC9-1, 2001.
6. Rego, G., Okhotnikov, O., Dianov, E., and Soulimov, V., "High Temperature Stability of Long–Period Fiber Gratings Produced Using an Electric Arc", *J. Lightwave Technol.* **19**, 1574-1579 (2001).
7. Fernandez Fernandez, A., Ooms, H., Brichard, B., Coeck, M., Coenen, S., Berghmans, F., and Décreton., M., "SCK·CEN irradiation facilities for radiation tolerance assessment", in Proc. IEEE NSREC 2002, Radiation Effect Data Workshop, Phoenix, USA, 2002.

Measurement Of The Salinity In Water Through Long-Period Gratings Arc-Induced In Pure-Silica-Core Fibers

Gaspar M. Rego[*ξδ], José L. Santos[ξα] and Henrique M. Salgado[ξδ]

[*]*Escola Superior de Tecnologia e Gestão, Instituto Politécnico de Viana de Castelo, Av. do Atlântico, 4900 Viana do Castelo, Portugal*
[ξ]*INESC Porto, Unidade de Optoelectrónica e Sistemas Electrónicos, Rua do Campo Alegre 687, 4169-007 Porto, Portugal*
[δ]*Dep. de Eng. Electrotécnica e de Computadores, Fac. de Engenharia, Universidade do Porto, Rua Dr. Roberto Frias, 4200-465 Porto, Portugal*
[α]*Dep. de Física, Faculdade de Ciências, Universidade do Porto, Rua do Campo Alegre 687, 4169-007 Porto, Portugal*

Abstract. We have investigated the sensitivity of arc-induced gratings to changes of ambient refractive index. Two pure-silica-core fibers and a standard fiber were used in this study. For a 6×10^{-3} change of the refractive index a 240 pm shift of the resonant wavelength was achieved.

INTRODUCTION

In nowadays life, there is an increasing demand for monitoring the water quality. In particular, the salinity is a very important parameter for a broadband of water related topics. The well-known properties of optical fiber sensors (OFS) make them very attractive for the measurement of physical parameters such as temperature and refractive index (which depends on salinity). A basic OFS configuration consists on the use of Bragg gratings (BGs) and/or long-period gratings (LPGs) for which changes on resonant wavelengths or power are related to the parameter to be measured. In opposition to BGs, LPGs are very prone to changes of the external medium since the effective refractive index of the cladding modes depends on the difference between the cladding and ambient refractive indexes. The sensitivity to refractive index changes increases with the reduction of that difference [1]. Therefore, pure-silica-core fibers, where the cladding is doped with fluorine in order to decrease its refractive index, possess an intrinsic advantage over standard fibers. Further increases in sensitivity can be obtained through etching of the fiber cladding or concatenating two LPGs in a Mach-Zenhder topology.

In this paper we present results of the sensitivity of arc-induced LPGs in pure-silica-core fibers to changes of the ambient refractive index simulating typical values for the salinity in the water. These results show a 2-3 times increase of the sensitivity when compared to gratings written in standard fibers.

GRATINGS FABRICATION

LPGs used in this study have been written in two pure-silica-core fibers (Fiber #1: D_{core} = 9 μm, D_{clad} = 125 μm and N.A. = 0.11; Fiber #2: MFD = 12.9 μm, D_{clad} = 150 μm and $\lambda_{cut\text{-}off}$ = 1.36 μm) and in the SMF-28 fiber from Corning using the electric arc-discharge technique [2]. The fabrication consists in placing a fiber under tension between the electrodes of a splicing machine, which produces a short duration arc discharge. Then, the fiber is moved by the grating period Λ. The sequence "arc discharge - fiber displacement" is repeated several times and is computer controlled. It should be noted that the optimal writing parameters are to be defined for each fiber type individually. Concerning the existence a resonance peak at 1.5 μm band, grating periods of Λ = 730 μm and 540 μm, were chosen for fibers #1-2 and for SMF-28 fiber, respectively. Figure 1 shows the transmission spectra of three gratings written in those fibers.

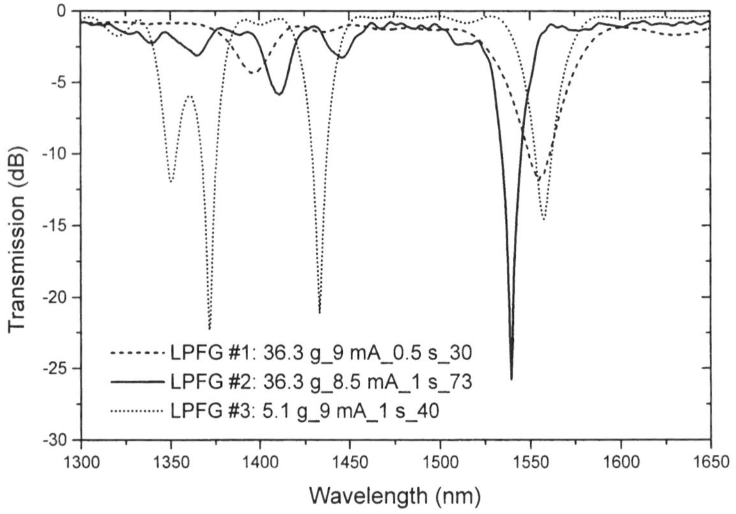

FIGURE 1. Transmission spectra of the gratings under investigation.

REFRACTIVE INDEX MEASUREMENTS

To assess the performance of the fabricated gratings for water salinity measurement, three samples containing mixtures of distilled water and ethylene glycol were prepared such that their respective refractive indexes fall in the 1.3325-1.3385 range. The lowest value corresponds to that of distilled water measured using an Abbe refractometer with a sodium lamp. The refractive index of the samples at 1.5 μm could be calculated using the Cauchy equation being the respective parameters determined illuminating the refractometer with, for instance, the Helium-Neon and Argon lasers. However, the relative difference of the refractive index between different samples is kept constant, when measured at distinct wavelengths [3].

It is well known that there is a linear relation between the salinity and the refractive index of a given sample (see, for instance, Ref. 4) and therefore, from now on, we will only talk about refractive index measurements. In order to perform such measurements fibers were kept under a tension of 22.8 gf to avoid bending of the gratings. A broadband optical source and an optical spectrum analyzer (OSA), set to maximum resolution (0.05 nm), were used to register the gratings spectra in the vicinity of a particular resonance. Spectra were recorded for gratings on air, embedded in water and in the three water-glycol mixture samples (n_1 = 1.3345, n_2 = 1.336 and n_3 = 1.3385). Careful has to be taken while exchanging the liquids to avoid contamination and to guaranty no temperature drift. Fig. 2 shows the spectra of two gratings written in two different fibers on air and immersed in water.

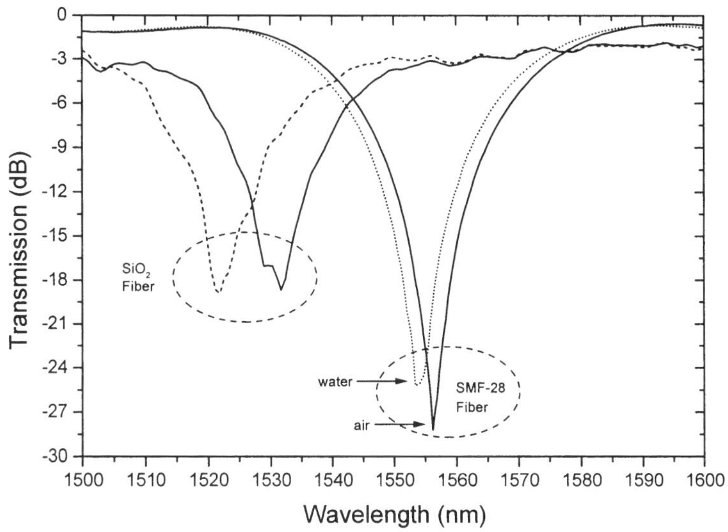

FIGURE 2. LPGs spectra in air and in water for two different fibers.

As it can be seen as the refractive index of the external medium increases the spectrum moves towards shorter wavelengths. This can be understood through an increase of the effective refractive index of the cladding, which according to the resonance condition leads to lower resonant wavelengths [2]. The shift is longer for the pure-silica-core fiber than for the standard fiber. That is a result of doping the cladding with fluorine which lowers its refractive index and therefore reduces the refractive index difference between the cladding and its surroundings increasing the grating sensitivity [1]. The high insertion loss for the pure-silica-core fiber is due to mode-field mismatch.

Figures 3 and 4 shows the change that occurs in the transmission band of the grating induced in Fiber #1 for different external fluids and its respective resonant wavelength shift with their correspondent refractive indexes. As it can be seen the shift is almost linear for the small range of interest. The results for the three fibers investigated are summarized in Table 1.

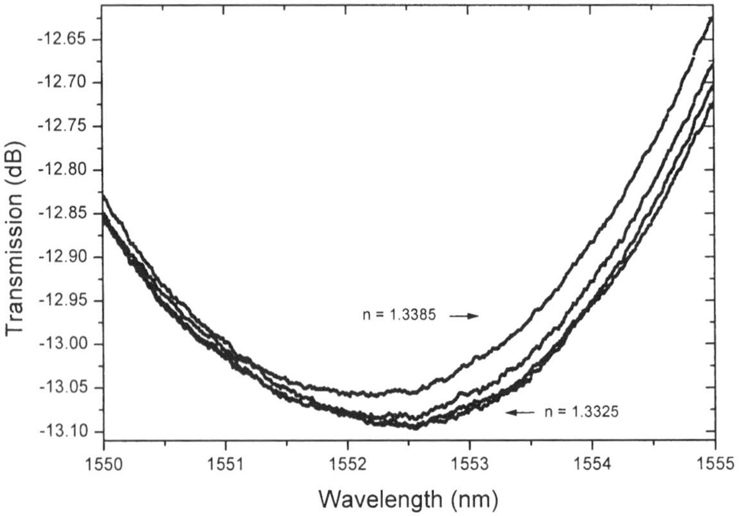

FIGURE 3. Transmission of the resonant band of the grating written in Fiber #1 for different fluids.

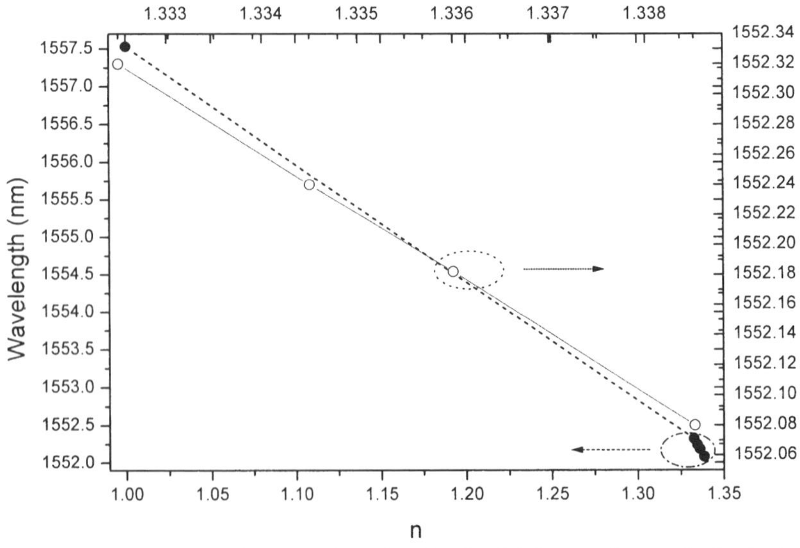

FIGURE 4. Resonant wavelength (grating written in Fiber #1) for several external refractive indexes.

TABLE 1. Resonant wavelength shift upon changes of the surrounding refractive index.

Fiber \ Medium	Δλ/nm Air-water 1-1.3325	Δλ/nm Water-mixture 1.3325-1.3385
SiO_2-125 μm	5.21	0.24
SiO_2-150 μm	8.96	0.11
SMF28-125 μm	2.34	0.10

It is interesting to note that when going from air to water the highest sensitivity was obtained for the SiO_2 fiber with the largest cladding diameter whilst for the small range of interest, corresponding to refractive index values of sea-water, the SiO_2 fiber with standard dimensions revealed to be twice effective. More modest values were achieved with the standard fiber despite a higher order cladding mode has been monitored. For comparison, the Bragg wavelength shift of a grating written in a strongly etched fiber was of only ~50 pm for the same $6x10^{-3}$ change in the refractive index [3].

CONCLUSION

The sensitivity of arc-induced gratings to changes of ambient refractive index was investigated. It was found that LPGs written in pure-silica-core fibers possess higher sensitivity compared with those induced in standard fibers. Sensitivities up to 40 nm/r.i.u. were obtained. Further increase in gratings response can be achieved using higher order cladding modes. Etching of the fiber cladding and/or concatenating two LPGs also lead to improvements of the above result. Therefore, a sensor head based on arc-induced gratings in pure-silica-core fibers can perform the measurement of salinity in water.

ACKNOWLEDGMENTS

G. M. Rego is thankful for the grant through the Program PRODEP III - Medida 5 - Acção 5.3. - Formação Avançada de Docentes do Ensino Superior, integrada no Eixo 3.

REFERENCES

1. James, S.W., and Tatam, R.P., "Optical fibre long-period grating sensors: characteristics and application", *Meas. Sci. Technol.*, **19**, 49-61 (2003).
2. Rego, G., Okhotnikov, O., Dianov, E., and Soulimov, V., "High Temperature Stability of Long–Period Fiber Gratings Produced Using an Electric Arc", *J. Lightwave Technol.* **19**, 1574-1579 (2001).
3. Pereira, D., Frazão, O., and Santos, J.L., "Fibre Bragg grating sensing system for simultaneous measurement of salinity and temperature", accepted for publication in *Opt Engineering.*
4. Zhao, Y., and Liao, Y., "Novel optical fibre sensor for simultaneous measurement of temperature and salinity", *Sens. And Actua. B,* **63**, 608-612 (2002).

Applied Physics (APHYS 2003)
A. Méndez-Vilas (Ed.)

Characterization of Cement Mortars with Ultrasonic Testing

L. Mariano del Río*, Antonio Jiménez*, Margarita Jiménez*, M. Montaña Rufo*, Jesús M. Paniagua* and Felicísima López&

*Department of Physics and &Department of Techniques, Means and Building Components, University of Extremadura, Avda. de la Universidad s/n, 10071 Cáceres, SPAIN

Abstract. This paper reports on an investigation of the application of ultrasonic inspection techniques to characterize construction materials. In this study, we have analyzed the properties of several cement mortars by using ultrasounds at frequencies of 40 kHz, as well as destructive techniques. Our goal is to correlate the properties of cement with respect to the propagation speed of ultrasounds. In particular, we have plotted several histograms of mechanic resistances and speeds of propagation of sonic waves in prismatic recipients (40x40x160 mm) with cement. We have also investigated the existing correlations between speeds of propagation and flexion/compression strengths. The above experiments have allowed us to find a relationship between direct, destructive techniques, and indirect, non-destructive (NDT) ultrasound-based approaches. A detailed analysis of the temporal evolution of the above-mentioned propagation speed with regards to the hardening process of cement samples is also reported. With all these experiments, we elaborate a model for accurate verification of other samples using NDT techniques only, which can be used to quantify the performance of this type of experiments.

INTRODUCTION

In this paper, we analyse the speeds of propagation of ultrasound in various samples of cement mortars, comparing the results with the samples' compressive/flexion strength characteristics as evaluated during destructive testing. Two types of study were performed. The first attempted to relate different physical variables of the cement with the speed of propagation of ultrasound waves. The second analyzed the relationship between the ultrasound speed and the hardening of the cement, i.e., the variation of those same physical variables as a function of the time lapsed after fabrication of the concrete. In all cases, we followed Spanish regulations concerning the inspection of concrete with ultrasound [1].

We prepared 18 specimens for inspection by ultrasound, all with dosages of 1:3:0.5 cement/sand/water. The specimens differed in the cement used: for those labeled A and A′, the cement used was of 42.5 N/mm^2 nominal resistance; for B and B′, 32.5 N/mm^2; and for C and C′, 22.5 N/mm^2. Each specimen was prepared in triplicate. The regulatory norms were followed at all times during the process of fabrication, curing, and conservation [2]. The specimens were manufactured as prisms of 40x40x160 mm^3. After preparation, the specimens were placed in a humidity chamber at 20°C and 90% humidity. The flexural strength F was determined using a testing device capable of applying loads of up to 10 kN with a precision of ±1%, and a load rate of (50±10) N/s. Their compressive strength R was determined using a hydraulic press with a load rate of (2400∀200) N/s. The velocity of the propagation of ultrasound pulses was measured by direct transmission using a Steinkamp BP-V ultrasound device (Germany). This measures the time of propagation of ultrasound pulses in a sample in the range (0.1-9999.9) μs with a precision of 0.1 μs. The transducers used were 28 mm in diameter, and had maximum resonant frequencies, as measured in our laboratory, of 42.5 kHz.

The measurement protocol used to study the relationship between the speed of ultrasound and the physical properties of the mortars was as follows. At 28 days after fabrication [2], the specimens were removed from the humidity chamber, carefully weighed and analysed by ultrasound. For each specimen, 12 speed measurements were made, six through the longitudinal section of the specimen (16 cm) (v), and six through the transversal section (4 cm) (v'). The specimen was then subjected to flexural rupture. As a result of this test, the specimen was spit into two parts allowing subsequent compressive rupture.

The samples used to study the hardening of the concrete were those labeled B and C′. The specimens were removed from the humidity chamber to be weighed and analysed by ultrasound, and then returned to the chamber. Ultrasonic testing was carried out on days 1 (measurements made at different hours), 2, 3, 6, 7, 13, 17 and 28 after fabrication.

RESULTS

Characterization

Figure 1 shows histograms of the nominal (R_N) and experimental (R) compressive strengths, the flexural strength (F), and the longitudinal (v) and transversal (v') speeds of ultrasound. As was expected, the values of R_N and R were similar. The value of F increased with increasing R_N. The values of the speeds v were systematically greater than those of v'. This difference, which was less than 4% in all cases, may be because the transversal path traversed by the wave was only 4 cm, less than the wavelength of the ultrasound which was about 10 cm [3].

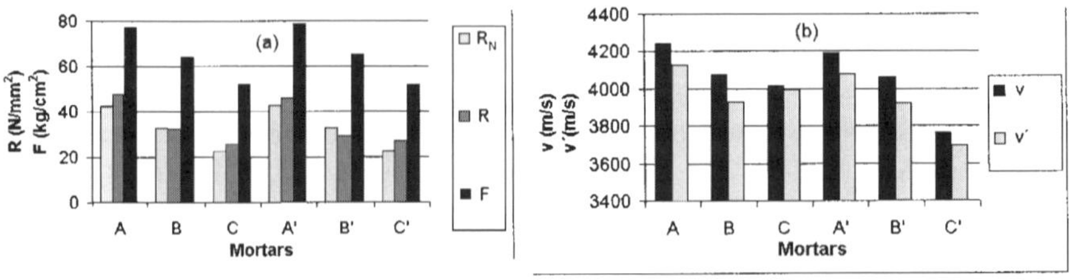

FIGURE 1. Histograms of the values of (a) nominal strength (R_N), compressive strength (R), and flexural strength (F), and (b) longitudinal (v) and transversal (v') speed of ultrasound of the specimens.

One sees in Figure 1 that there is gradual decline in v and v' as the strength of the mortar decreases. Indeed linear regressions applied to the data gave high values of the correlation coefficient r. For the ranges of compressive/flexural strength of the present work, the relationships are given by the expressions of Eqs. (1), (2), (3) and (4):

$$R\left(\frac{\mathrm{N}}{\mathrm{mm}^2}\right) = 0.0395 \cdot v\left(\frac{\mathrm{m}}{\mathrm{s}}\right) - 123.6; \quad r = 0.860. \qquad (1)$$

$$R\left(\frac{\mathrm{N}}{\mathrm{mm}^2}\right) = 0.0445 \cdot v'\left(\frac{\mathrm{m}}{\mathrm{s}}\right) - 141.3; \quad r = 0.727. \qquad (2)$$

$$F\left(\frac{\mathrm{kg}}{\mathrm{cm}^2}\right) = 0.0534 \cdot v\left(\frac{\mathrm{m}}{\mathrm{s}}\right) - 149.9; \quad r = 0.884. \qquad (3)$$

$$F\left(\frac{\mathrm{kg}}{\mathrm{cm}^2}\right) = 0.0609 \cdot v'\left(\frac{\mathrm{m}}{\mathrm{s}}\right) - 174.7; \quad r = 0.868. \qquad (4)$$

Hardening

By way of example, Figure 2 shows the temporal evolution of v and v' in one of the specimens (B1). In general terms, the other samples showed similar trends. One observes that both speeds increase up to 28 days. This is, in principle, indicative of a process of homogeneous hardening in all the layers and directions of the mortar.

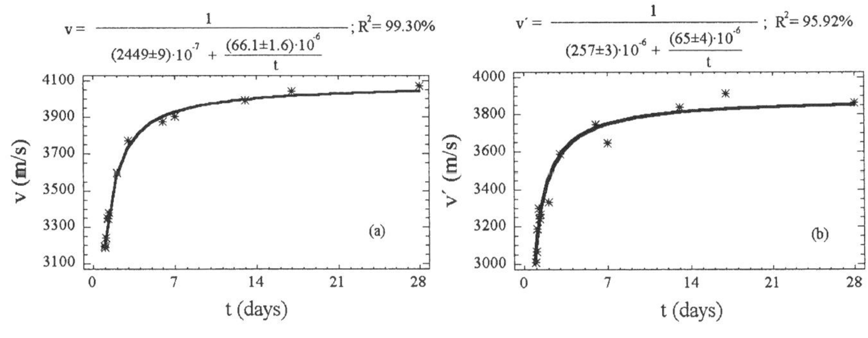

FIGURE 2. Plot of the temporal evolution of the speeds of ultrasound, (a) v and (b) v', with the time of aging, t, in specimen B1, together with the fits from Eqs. (5) and (6).

Figure 2 also shows this evolution together with the least squares fits to the double reciprocal equations characterized by the function of Eqs. (5) and (6) where v (or v') is expressed in m/s and t in days.

$$v = \frac{1}{a + \frac{b}{t}}. \tag{5}$$

$$v' = \frac{1}{c + \frac{d}{t}}. \tag{6}$$

Equations (5) and (6) suggests that, many days after fabrication, v and v' tend asymptotically to the value $1/a$ (m/s) and $1/c$ (m/s) respectively. This type of asymptotic behaviour has been reported for other types of specimens, such as concrete [4]. For the present case of cement mortars, from the fit for n = 28 days (as is established in the Spanish legislation [1]) for the sample B1, Eqs. (5) and (6) give a value of $1/a$ = (4083±15) m/s and $1/c$ = (3890±40) m/s (corresponding to t 6 4). Doing the same for n = 6 days, the estimates for $1/a$ and $1/c$ are (4068±23) m/s and (3870±80) m/s, with relative errors of 0.6% and 2.1%, respectively. Table 1 summarizes the values of $1/a$ and $1/c$ in the fits to Eqs. (5) and (6) with the data up to day 28 and up to day 6 in the entire set of samples analysed in the hardening study.

TABLE 1. Values of $1/a$ and $1/c$ obtained from the fits of Eqs. (5) and (6) respectively to the data up to day 28 and up to day 6 in the entire set of specimens analysed in the hardening study. The mean values and the standard errors of these coefficients are included for each of the two types of mortar.

Sample	$1/a$ (m/s) when t=28 days	$1/a$ (m/s) when t=6 days	$1/c$ (m/s) when t=28 days	$1/c$ (m/s) when t=6 days
B1	(4083±15)	(4068±23)	(3890±40)	(3870±80)
B2	(4098±15)	(4077±22)	(3940±50)	(3920±90)
B3	(4096±15)	(4109±23)	(3900±50)	(3880±110)
Mean Value (B – Type)	(4092±15)	(4085±23)	(3910±50)	(3890±90)
C′1	(3760±14)	(3759±24)	(3680±50)	(3630±100)
C′2	(3780±15)	(3779±24)	(3700±50)	(3710±90)
C′3	(3776±15)	(3776±23)	(3740±50)	(3730±90)
Mean Value (C′ – Type)	(3772±15)	(3771±24)	(3710±50)	(3690±90)

In other words, according to the fitted model given by Eqs. (5) and (6), it would not be necessary to wait for 28 days or more in order to determine with high precision and accuracy the speed of propagation of ultrasound in manufactured specimens, and hence the values of compressive/flexural strength associated with the different mortars from Eqs. (1), (2), (3) and (4). Also, within a confidence interval of ±2σ, the proposed model distinguishes between the longitudinal speeds associated with the mortars labeled B, fabricated with a cement of 32.5 N/mm^2 nominal strength, and those labeled C′, fabricated with another cement of 22.5 N/mm^2. This distinction is feasible even when only 6 days have passed after the preparation of the mortar.

CONCLUSIONS

In the present study, first we confirmed that there exists a linear relationship between the compressive and flexural strengths (*R* and *F*) of the prismatic mortar specimens and the speed of propagation of ultrasound in them (v and v'). Second, from the cement hardening data, we developed a time-dependent model with which it was found to be straightforward to predict the values of v and v' at 28 days after fabrication of a specimen on the basis of the measurement at only 6 days after fabrication.

ACKNOWLEDGMENTS

The support of Isabel Peco and Juan Carlos Cadenas is gratefully acknowledged.

REFERENCES

1. UNE 83-308-86. "Norma Española: Ensayos de Hormigón. Determinación de la Velocidad de Propagación de los Impulsos Ultrasónicos", Spain, 1986.
2. UNE 196-1:94. "Norma Española: Determinación de Resistencias Mecánicas", Spain, 1996.
3. ASTM D 2845-00. "Standard Test Method for Laboratory Determination of Pulse Velocities and Ultrasonic Elastic Constants of Rock", United States, 2000.
4. del Río, L. M., Jiménez, A., López, F., Rosa, F. J., Rufo, M. M., and Paniagua, J. M., "Characterization and Hardening of Concrete with ultrasonic Testing" in *Ultrasonics International 2003*, Abstract Book 1.79 E, Granada (Spain): Elsevier, 2003.

Applied Physics (APHYS 2003)
A. Méndez-Vilas (Ed.)

Topographical investigation of drug-sensitive and drug-resistant lung tumor cells H69 by scanning near-field optical microscopy

Weihong Qiao, Franck H. Lei, Aurélie Trussardi-Regnier, Jean-F. Angiboust, Jean-M. Millot, Ganesh-D. Sockalingum and Michel Manfait

Unité MéDIAN, CNRS-UMR 6142, UFR de Pharmacie, IFR 53, Université de Reims Champagne-Ardenne, 51 rue Cognacq Jay, 51096 REIMS Cédex, France

Abstract. Drug resistance is one of the major obstacle in the success of anticancer therapy. The topography of human small-cell lung-cancer (SCLC) cell lines, both sensitive and resistant to chemotherapy, has been characterized by scanning near-field optical microscopy. The obtained shear force images show that the average spatial height of drug-resistant cells is about 200nm higher than that of drug-sensitive cells. The results suggest that the vertical profile differences may be due to clustering phenomenon at the cell surface. The existence of such clusters in sensitive and resistant H69 cells was observed by the SNOM images. Moreover, the surface structure of the drug-resistant tumor cells is more compact than that of drug-sensitive ones by comparison the images of two-dimension and three-dimension.

INTRODUCTION

Successful chemotherapeutic treatment of tumors is often seriously hampered by their poor response to anticancer drugs[1]. Drug resistance can be observed at the outset of therapy (intrinsic resistance) or after exposure to the anticancer agent (acquired resistance). The mechanism responsible for intrinsic drug resistance are still poorly understood[2].

Scanning near-field optical microscopy (SNOM) techniques is a promising and useful method for observing and analyzing the biological surface structure with high lateral resolution[3-6]. By regulating the tip-sample distance during the scanning, the topographic image is obtained synchronously with the optical data. However, high quality biological imaging is often difficult because of soft, ductile and humid sample surfaces. Sometimes, the sample is immersed in liquid. The SNOM tip may also touch and destroy the cell surface. In these cases, a SNOM with high sensitivity is necessary.

In a previous studies involving the cell surfaces by SNOM, Péter *et al*[4] have studied the large-scale association pattern of erbB2 in quiescent and activated cells labelled with fluorescent anti-erbB2 monoclonal antibodies using SNOM. ErbB2 was found to be concentrated in irregular membrane patches with a mean diameter of approx. 0.5 μm in non-activated SKBR3 and MDA453 human breast tumor cells. The average number of erbB2 proteins in a single cluster on non-activated SKBR3 cells was about 10^3. They concluded that an increase in cluster size may constitute a general phenomenon in the activation of erbB2. Perner *et al*[7] studied the variations in cell surfaces of treated breast cancer cells detected by a combined instrument for far-field and near-field microscopy. They found the surface topography of untreated control cells was regular and smooth with small overall height modulations. In physiological condition the surfaces became increasingly jagged as detected by an increased variation in membrane height. After application of un-physiological medium, the cell surface structures appeared to be smoother again with an irregular fine structure.

The purpose of this work is to study the difference in surface morphology between drug-sensitive and drug-resistant lung tumor cells in vertical profile. The clusters' existence was observed by the near-field results. The cell biological aspects of these observations maybe offer new perspectives to be empirically used in multi-drug resistance characterization of lung cancers.

MATERIALS AND METHODS

The sample preparation

The human small-cell lung-cancer (SCLC) cell lines used were NCI-H69[8]. The multidrug resistant SCLC cell line (NCI-H69/VP) was selected for resistance to VP-16[9]. All cell lines were maintained at 37 °C in RPMI 1640 medium with glutamax and 10% fetal calf serum in humidified atmosphere containing 5% CO_2. Cytospin tumor cell lines preparation were rapidly air dried and fixed at 4 °C during 20 minutes, in para-formaldehyde (3.7%).

Scanning near-field optical microscopy/ Optical microscope

The SNOM set-up used in this study is a homemade shear force near-field microscope coupled with an inverted optical microscope Olympus IX-70. The force sensor of the SNOM is a bimorph-based cantilever incorporating force feedback technique, which allows high force detection sensitivity for biological samples. A detailed description of this SNOM can be found elsewhere [10]. The bimorph cantilever was fabricated by cutting a commercially available bimorph wafer (T215, t = 0.38 mm, $E = 5.2 \times 10^{10}$ N/m^2, and $\rho = 7800$ kg/m^3, Piezo Systems Inc., USA).

The SNOM probes with tip diameters of about 50 nm were made by etching an optical fiber of 125 μm (cladding diameter) in 48% hydrofluoric acid solution.

RESULTS AND DISCUSSION

The shear force imaging of lung tumor cells were performed by SNOM in ambient conditions. The SNOM force sensor was driven with a sine voltage v_i = 0.2 mV, corresponding to a tip vibration amplitude of about 6 nm. The scan rate was fixed to10 μm/s for an imaging area of 23×23 μm^2 and to 7 μm/s for 15×15 μm^2 with a resolution of 256x256 pixels. To avoid the perturbation caused by any instability, the acquisition was repeated 30 times for each point and then averaged.

The obtained topographical images are shown in Fig.1 and Fig. 2, respectively. Figure 1 is a drug- resistant cell image of a 23×23 μm^2 scan area in two-dimension (2-D) [Fig. 1(a)] and in three-dimension (3-D) (Fig. 1b). Figure 2 shows drug- sensitive cell images of 15×15μm^2 in 2-D [Fig. 2(a)] and in 3-D [Fig. 2(b)], and an image of 21×21μm^2 in 2-D [Fig. 2(c)] and in 3-D [Fig. 2(d)]. All the 2-D images show some common features: oblique rhombic structures approximately 5~8×2~4μm^2. Although the samples were deposited at random, and even if the samples were changed so as to

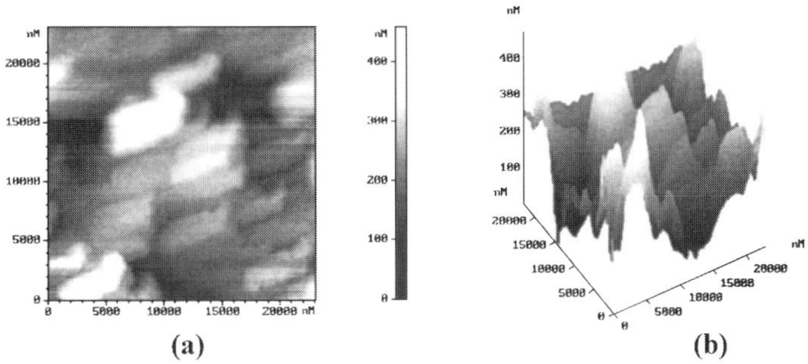

FIGURE 1. Topographic SNOM images of drug-resistant cell lines. (a) 23×23μm^2 scan area in 2-D (b) 3-D image with the same area and same size as in (a)

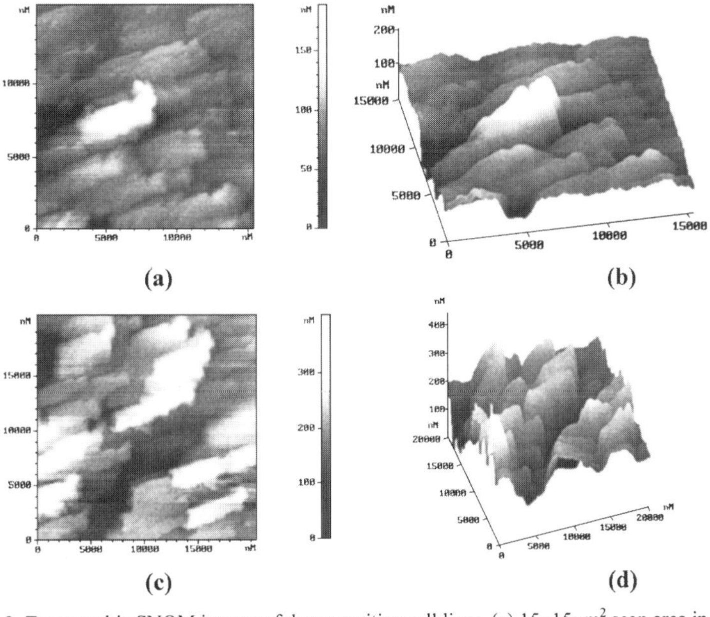

FIGURE 2. Topographic SNOM images of drug-sensitive cell lines. (a) 15×15μm^2 scan area in 2-D (b) 3-D image with the same area and same size as in (a); (c) 21×21μm^2 scan area in 2-D (d) 3-D image with the same area and same size as in (c)

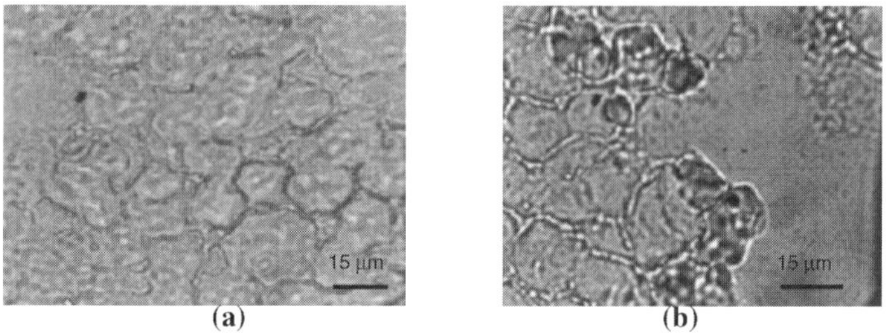

FIGURE 3. Optical microscopy images of (a) drug-resistant cells (b) drug-sensitive cells.

lie in other directions or opposite direction, these common features were still visible. We hope to do further work to test and verify whether there exist some pseudomorphs caused by instrument or software etc. to disturb our surveys.

By comparing the images in Fig. 1 and 2, it can be seen that the variation in vertical height for the drug-resistant cells is about 300-600 nm, while that for drug-sensitive cells is only about 100-400 nm. The average height difference between them is about 200 nm. This indicates the spatial height variation of drug-resistant cells is almost twice that of drug-sensitive cells. These height changes may be caused by clustering phenomenon on the membrane of the drug-resistant cell. The clusters' existence in drug-resistant and drug-sensitive cells was observed by the SNOM images.

In our studies, the cell sizes taken by shear force image and optical microscope were also compared. The average diameter size of drug-resistant and drug-sensitive cells is about 15 microns, as shown in Fig. 3(a) and (b) by optical microscopy. In fact, the shape of these cells is an irregular fine structure, so the average size was almost the same, thus the optical microscope images and 2-D images (except for the mentioned common features) did not reveal any distinct significant morphological differences on their surfaces. However, the 3-D images [Fig. 1(b), Fig. 2(b) and (d)], show in the scope of about $15\times15\mu m^2$ area, a cell surface structure with concave-convex like peaks. In Fig.1(b) and Fig.2(b), the area of about $15\times15\mu m^2$ covers almost a cell, while in Fig.2(d), the same area contains images from two neighboring cells. By comparing these 3-D images, the surface structure of drug-resistant tumor cells seems to appear much more compact with protuberant peaks, while that of drug-sensitive tumor cells looks more dispersed, this may be one of the reasons to form the clustering phenomenon of vertical variation.

CONCLUSION

The topographical structure of drug-sensitive and drug-resistant lung tumor cells H69 was studied by SNOM. The cell surface spatial height variation of drug-resistant cells is about 200 nm higher than that of drug-sensitive cells. The former surface structure exhibits more compact features, while the latter looks more dispersed by comparison of the 2-D and 3-D topographic SNOM images. The vertical profile difference may be caused by clustering phenomenon on the cell membrane. The cluster existence was observed by the SNOM results. As for understanding of the clustering phenomenon, SNOM analysis of living cell membrane will be associated with the analysis of near-field fluorescent spectra in future studies.

REFERENCES

1. Dong Z.,Ward N.E., Fan D., Gupta K., O'Brain C.A., *Molecular Pharmacology,* **39**, 563-569 (1992)
2. Meschini S, Marra M, Calcabrini A, Monti E, Gariboldi M, Dolfini E, Arancia G., *Toxicology In Vitro,* **16 (4)**, 389-98(2002).
3. M.De Serio, R. Zenobi, V.Deckert, *Trends in Analytical Chemistry*, 22(2),70-77(2003)
4. Péter Nagy, Attila Jenei, Achim K. Kirsch, Jànos Szöllösi, Sàndor Damjanovich and Thomas M. Jovin, *Journal of Cell Science*, **112**, 1733-1741(1999)
5. Tomoyuki Yoshino, Shigeru Sugiyama, Shoji Hagiwara, Daisuke Fukushi, Motoharu Shichiri, Hidenobu Nakao, Jong-Min Kim, Tamaki Hirose, Hiroshi Muramatsu, Toshio Ohtani, *Ultramicroscopy*, **97**, 81-87(2003)
6. C.HÖPPENER, D.MOLENDA, H.FUCHS and A.NABER, *Journal of Microscopy*, **210**,288-293(2003)
7. Perner P, Rapp A, Dressler C, Wollweber L, Beuthan J, Greulich KO, Hausmann M., *Anal Cell Pathol.* **24(2-3)**,89-100 (2002)
8. Desmond N.Carney, Adi F. Gazdar, Gerold Bepler, John G.Guccion, Paul J. Marangos, Terry W. Moody, Mark H.Zweig and John D.Minna, *Cancer Research*, **45**,2913-2923(1985)
9. Peter Buhl Jensen, Henrik Roed, Maxwell Sehested, Erland J.F.Demant, Lars Vindelov, Ib Jarle Christensen and Heine Hoi Hansen, *Cancer Chemother Pharmacol*, **31**,46-52(1992)
10. Lei F.H, Nicolas J-L, Troyon M. Sockalingum G-D, Rubin S, Manfait M, *J Appl Phys* **93 (4)**, 2236-2243(2003)

Applied Physics (APHYS 2003)
A. Méndez-Vilas (Ed.)

IRON OXIDE THIN FILMS GROWN BY PULSED LASER DEPOSITION

M.L. PARAMÉS[1], N. POPOVICI[1], P.M. SOUSA[1], A.J. SILVESTRE[2] AND O. CONDE[1]

[1]*Dept. Física, Faculdade de Ciências da Universidade de Lisboa, Campo Grande, Ed. C8, 1749-016 Lisboa, Portugal*

[2]*Instituto Superior de Engenharia de Lisboa, R. Cons. Emídio Navarro, 1749-014 Lisboa, Portugal*

Abstract: Iron oxide thin films have been grown at room temperature onto (100) Si substrates by pulsed laser deposition (PLD). Ablation of commercial Fe_3O_4 targets (contaminated with α-Fe_2O_3) was performed using a KrF excimer laser, in vacuum (1.4×10^{-5} and 3.2×10^{-5} mbar) or with a background gas pressure of argon or oxygen (8×10^{-4} -3×10^{-3} mbar). For each set of experiments different laser fluences were tested, in the range 2–7 Jcm^{-2}. The as-deposited films were characterized by X-ray diffractometry using glancing incidence geometry (GIXRD), micro-Raman spectrometry, scanning electron microscopy (SEM) and energy dispersive spectroscopy (EDS). This study showed that different iron oxide phases are grown depending on the processing parameters.

INTRODUCTION

Iron oxides have been extensively studied because of their technological interest. They present an ensemble of chemical and physical properties that depend on the chemical composition and crystal structure. Concerning magnetic properties, while pure iron is ferromagnetic, Fe_3O_4 and γ-Fe_2O_3 are ferrimagnetic, and α-Fe_2O_3 and FeO are antiferromagnetic [1].

Magnetite (Fe_3O_4) has a predicted half-metallic band structure fully spin-polarised at the Fermi level and a very high Curie temperature (~860 K) [2]. This half-metallic behaviour turned it a promising material for spintronic applications, in which case a low temperature deposition process is required.

A number of deposition techniques have been used to grow Fe_3O_4 thin films on a diversity of substrates, at different temperatures and from different pure Fe_3O_4, α-Fe_2O_3 and Fe targets [3-6]. However, it is still difficult to grow them with a controlled and well defined composition and microstructure. In particular, iron oxide thin films grown at high temperature usually consist of multi-phase domains [1]. Pulsed laser deposition (PLD) is considered one of the most effective methods that allow the control of thin film growth even at room/low temperatures. In PLD processes it is well known that the formation of oxides is favoured by the presence of a small amount of O_2 inside the deposition chamber that interacting with the ablation plume will promote oxygen incorporation in the growing film.

The main goal of this work was to identify possible roots to achieve PLD of pure magnetite thin films at room temperature (RT). Iron oxide films grown at RT by ablation of Fe_3O_4/α-Fe_2O_3 targets under vacuum and passive or active environments are presented and discussed.

EXPERIMENTAL

The experimental set-up consists of a stainless steel high vacuum deposition chamber and a pulsed UV laser (KrF, 248 nm wavelength, 30 ns pulse width) with associated beam delivery optics. The laser beam, with a repetition rate of 5 Hz, was focussed on the target surface under an angle of 45°. A commercial Fe_3O_4 target, contaminated with α-Fe_2O_3, was used. The target was continuously rotated and periodically translated. The deposition time was 100 minutes for all the experiments. The ejected matter was collected on a (100)Si substrate placed in front and at 4 cm from the target, and kept at room temperature. Prior to each deposition the unfocussed laser beam was rastered across the target surface in HV. Films were grown in high vacuum, 1.4×10^{-5} and 3.2×10^{-5} mbar, and in oxygen or argon atmosphere at different working pressures, from 3×10^{-3} to 8×10^{-4}mbar, with laser fluences in the range 2 – 7 J cm^{-2}.

The morphology of the films was analysed by scanning electron microscopy (SEM) and qualitative chemical analyses were performed by energy dispersive spectroscopy (EDS). Phase analysis was carried out by X-ray diffraction with CuKα radiation, using glancing geometry (GIXRD) at 1° angle of incidence to the specimen surface. Micro-Raman spectrometry with an Ar^+ laser, λ=514.5 nm, and a triple monochromator was used for microstructural studies. The thickness of the as-deposited films was measured by optical profilometry. Typically, a value of ~2 μm was obtained.

RESULTS AND DISCUSSION

Figure 1 presents the GIXRD patterns of two films deposited in high vacuum at a residual pressure of 3.2×10^{-5} mbar (Fig. 1a) and in argon environment, $p_{Ar} = 3.0\times10^{-3}$ mbar (Fig. 1b), with laser fluences of 3.0 and 2.3 J cm^{-2}, respectively. Despite the different deposition conditions, similar iron oxide phases can be found in both samples: magnetite (Fe_3O_4), hematite (α-Fe_2O_3), their polymorphs (not labelled in the figure) and stoichiometric and non- stoichiometric wustite ($Fe_{1-x}O$).

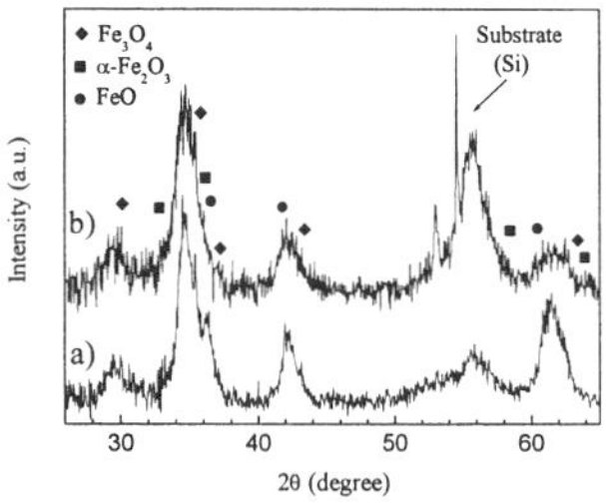

FIGURE 1 - GIXRD patterns of iron oxide films grown in: a) high vacuum, $p = 3.2\times10^{-5}$ mbar, $F = 3$ J cm^{-2}; b) argon atmosphere, $p_{Ar} = 3.0\times10^{-3}$ mbar, F=2.3 J cm^{-2}

By changing the deposition atmosphere to an oxygen environment, the number of iron oxide phases developed in the films tend to decrease, as can be seen in the diffractograms of films grown at a laser fluence of 2.3 J cm^{-2} and with oxygen background pressures of 8×10^{-4} and 3×10^{-3} mbar (Fig. 2). In the film grown with the lowest oxygen pressure, only stoichiometric magnetite and hematite are present (Fig. 2a). The XRD data also show that this film is polycrystalline with no preferred grain orientation. Values of ~50 nm for the mean grain size of magnetite and hematite crystallites were estimated using a Gaussian peak fit and Scherrer's equation [7]. Micro-Raman analysis confirmed the XRD results. The Raman spectra of the films prepared at the lowest oxygen pressure, $p_{O2} = 8\times10^{-4}$ mbar, show weak bands at 665 cm^{-1} and 536 cm^{-1} which are not far from the reference magnetite values of 670 cm^{-1} (A_{1g} mode) and 540 cm^{-1} (T_{2g} mode), and one band around 250 cm^{-1} corresponding to hematite [8]. An increase of the oxygen pressure leads also to the formation of Fe_2O_3 polymorphic phases, as can be deduced by the broadened region centred at ~34.5° (Fig. 2b). The disappearance of FeO in the films deposited with oxygen assisted ablation shows that this reactive process favours the formation of phases with iron at its highest oxidation state (Fe^{3+}), promoting the formation of Fe_2O_3, by contrast to the oxygen deficient phases deposited under vacuum or in argon.

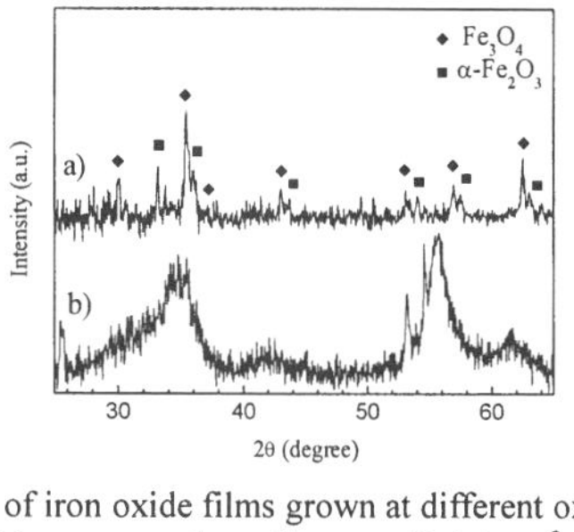

FIGURE 2 - GIXRD patterns of iron oxide films grown at different oxygen pressures: a) $p_{O2} = 8\times10^{-4}$ mbar, b) $p_{O2} = 3\times10^{-3}$ mbar, with a constant laser fluence of 2.3 J cm^{-2}.

Besides ambient environment, other crucial experimental parameter is laser fluence. Fig. 3 shows diffractograms of films synthesised at an oxygen pressure of 8×10^{-4} mbar with the highest and lowest laser fluences considered in this study, i.e. 2 – 7 J cm^{-2}. By comparing both patterns, one can see that an increase of laser fluence induces the formation of FeO besides the Fe_3O_4 and Fe_2O_3 phases. This result may be understood by considering that the amount of ejected material increases with laser fluence. Therefore, if the oxygen content in the gas phase remains constant, the [Fe]/[O] ratio will increase, meaning that laser fluence favours iron oxide formation with iron in the lowest oxidation state (Fe^{2+}).

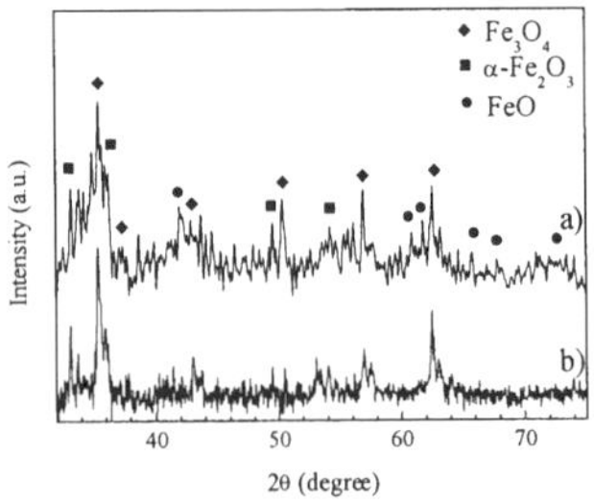

FIGURE 3 - GIXRD patterns of iron oxide films grown at constant oxygen pressure, $p_{O2} = 8\times10^{-4}$ mbar, and laser fluence of 7 J cm^{-2} (a) and 2 J cm^{-2} (b).

SEM analysis of the coating surface shows a very smooth morphology independent of the oxygen pressure and laser fluence used, as can be seen in Fig. 4. Embedded within the films a number of round drop-like features, Fe-enriched, can be observed (Fig.5) as a consequence of the formation of liquid droplets during the ablation process. Also tiny particulates can be seen on these films, due to the flake directly generated from the target by the ablation. The density of both features increases with increasing laser fluence.

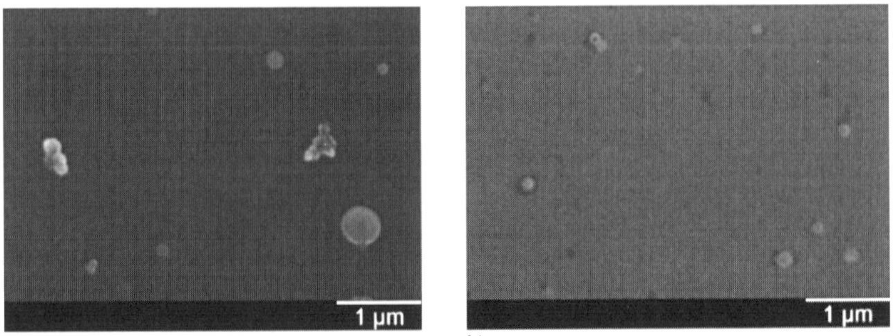

a) b)

FIGURE 4 - SEM micrographs of films deposited at laser fluence of 2.3 J cm^{-2}, with oxygen assistance: a) $p_{O2} = 8\times10^{-3}$ mbar, b) $p_{O2}=8\times10^{-4}$ mbar.

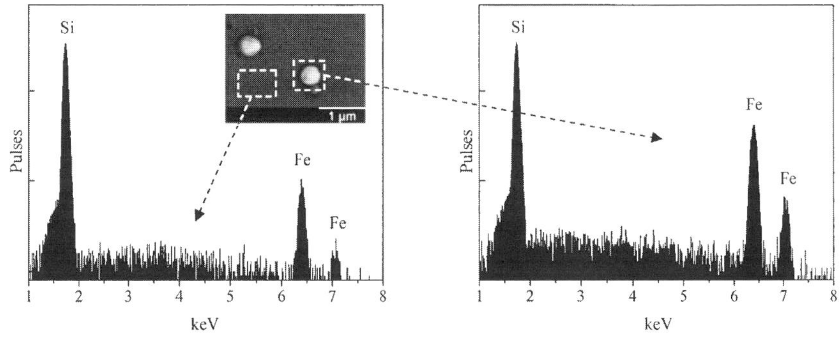

FIGURE 5 - EDS spectra of a film deposited at oxygen pressure of 8×10^{-4} mbar and laser fluence of 2.3 J cm^{-2}. Spectrum a) was recorded over the film surface and spectrum b) over a droplet. Both spectra are plotted with the same vertical scale.

CONCLUSIONS

Ablation of $Fe_3O_4/\alpha\text{-}Fe_2O_3$ targets was performed in high vacuum, argon and oxygen background, at various pressures, using a pulsed UV KrF laser. The ablated material was collected onto (100)Si substrates. The films consist of several iron oxides phases, either stoichiometric or non-stoichiometric, depending on the atmosphere inside the deposition chamber. In O_2 reactive atmosphere, and for some limited conditions, films consist of the same phases as the target. PLD at room/low temperature is currently under investigation aiming at the growth of pure magnetite thin films.

ACKNOWLEDGEMENTS

This work was supported by the EU contract FENIKS: G5RD-CT-2001-00535.

REFERENCES

1 K.J. Kim, D.W. Moon, S.K. Lee, K.H. Jung, Thin Sol. Films **360** (2000) 118.
2 Yu. S. Dedkov, U. Rüdiger, G. Güntherodt, Phys. Rev. B **65** (2002) 64417.
3 S. Kale, S.M. Bhagat, S.E. Lofland, T. Scabarozi, S.B. Ogale, A. Orozco, S.R. Shinde, T. Venkatesan, B. Hannoyer, B. Mercey, W. Prellier, Phys. Rev. B **64** (2001) 205413, and references therein.
4 W.L. Zhou, K.–Y. Wang, C.J. O'Connor, J. Tang, J. Appl. Phys. **89** (2001) 7398.
5 J.P. Hong, S.B. Lee, Y.W. Jung, J.H. Lee, K.S. Yoon, M.H. Jung, Appl. Phys. Lett. **83** (2003) 1590.
6 S.B. Ogale, K. Ghosh, R.P. Sharma, R.L. Greene, R. Ramesh, T. Venkatesan, Phys. Rev. B **57** (1998) 7823.
7 B.D. Cullity, Elements of X-Ray Diffraction, Addiso-Wesley Publ. Co., MA, 1978.
8 L.V. Gasparov, D.B. Tanner, D.B. Romero, H. Berger, G. Margaritondo, L. Forro, Phys. Rev. B **62** (2000) 7939.

Applied Physics (APHYS 2003)
A. Méndez-Vilas (Ed.)

ELECTROCHEMICAL IMPEDANCE SPECTROSCOPY AS A TOOL FOR ELECTRICAL AND STRUCTURAL CHARACTERIZATIONS OF MEMBRANES IN CONTACT WITH ELECTROLYTE SOLUTIONS

J. BENAVENTE
Grupo de Caracterización Electrocinética y de Transporte en Membranas e Interfases. Departamento de Física Aplicada I. Facultad de Ciencias. Universidad de Málaga. E-29071 Málaga (Spain).

Abstract: Different membrane/electrolyte systems have been studied from electrochemical impedance spectroscopy (EIS) measurements. Membranes with different structures (ideally porous, dense and composite) and from different materials (polyetilentereftalate, polyamide and polyamide/polysulfone) have been considered. Measurements were made with the membranes in contact with NaCl solutions at different concentrations ($10^{-3} \leq$ c(M) $\leq 5x10^{-2}$). Membrane electrical parameters (resistance and capacitance) were determined from impedance plots by using equivalent circuits as models. Differences in the equivalent circuits obtained for the studied membranes are related to the different membrane structure. The uniform capillary model was assumed for the ideal porous membrane and good agreement was obtained when experimental and calculated electrical resistance values are compared. This point indicates that electrical measurements could also be used for geometrical characterisation of this kind of membranes. For dense membranes, capacitance values allow the estimation of structural or material characteristic parameters, while for composite membranes EIS measurements allow the characterisation of each sublayer.

INTRODUCTION

Membrane separation techniques are now commonly used in many industrial processes and different membrane structures and materials are used to obtain high fluxes and selectivity, which are the two key factors for membrane performance (1). Structural parameters are important to characterise the membrane retention, and they are usually determined from hydrodynamic measurements, but electrical parameters are also important if the movement of ions or charged macromolecules across membranes is considered (2-4). Electrochemical impedance spectroscopy (EIS) is a non-destructive technique for characterising materials and interfaces, which allows the determination of the electrical properties of heterogeneous systems (i.e. membrane/electrolyte systems) by using equivalent circuits as models. In the case of composite membranes, EIS measurements permit us to characterise sublayers with different electrical/structural properties (5-6). This technique has been used for characterising both artificial and biological (cuticle) membranes in contact with electrolyte solutions (membran working conditions) (7,8).

In this paper, both commercial and experimental membranes with diverse structures and made from different materials have been electrically and structurally characterised by EIS measurements, which were carried out with the membranes in contact with electrolyte solutions at different concentrations ($10^{-3} \leq$ c(M) $\leq 5x10^{-2}$).

The fitting of the experimental data allows the determination of the electrical parameters, while structural parameters for the membrane (or the sublayers forming a composite membrane) such as porosity and/or thickness can also be estimated from these results. Differences found in the equivalent circuits associated to the studied membranes are mainly attributed to differences in the membranes structure.

THEORY

Electrochemical impedance spectroscopy (EIS) is a relatively new a.c. technique for characterizing materials and interfaces that has emerged with the development of instruments capable of measuring impedance in a wide range of frequency (between 10^{-6} Hz and 10^{9} Hz). It is a successful tool to determine the electrical properties of heterogeneous systems formed by a series array of layers with different electrical (and even structural) characteristics such as membrane/electrolyte or composite membrane systems, since it allows a separate evaluation of the electrical contribution of each sublayer (5,8). IS measurements enable us to obtain information about the different sublayers of these heterogeneous systems by using the impedance plots and the equivalent circuits as models, where the different circuit elements are related to the structural/transport properties of the systems (9).

When a linear system is perturbed by a small v(t) voltage, its response, the electric current i(t), is determined by a differential equation of nth order in i(t), or a set of n differential equations of the first order. If v(t) is a sine wave input:

$$v(t)=V_o\sin\omega t \qquad (1)$$

the current intensity i(t) is also a sine wave,

$$i(t)=I_o\sin(\omega t+\phi) \qquad (2)$$

a transfer function, the admittance function, can be defined: $Y^*(\omega) = |Y(\omega)|e^{j\phi}$. The impedance function, $Z(\omega)$, is the inverse of the admittance function: $Z(\omega)=[Y^*(\omega)]^{-1}$. Since both the amplitude and phase angle ($|Y(\omega)|$ and ϕ, respectively) of the output may change with respect to the input values, the impedance is expressed as a complex number: $Z=Z_{real}+jZ_{img}$, where Z_{real} is the real part and Z_{img} the imaginary one.

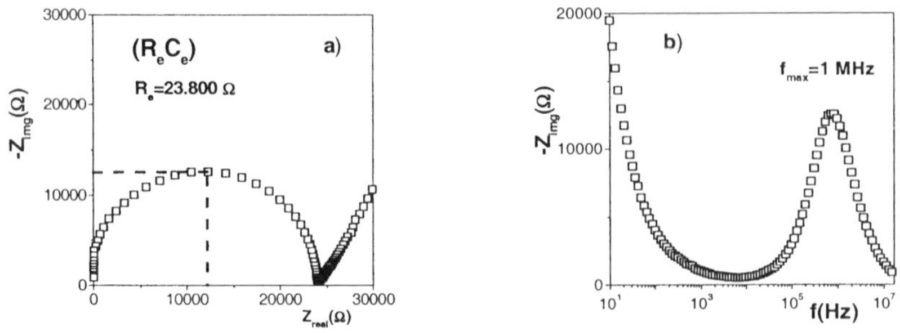

Figure 1: Impedance plots for an electrolyte solution (NaCl). a) Nyquist plot; b) Bode plot.

Analysis of the impedance data is often carried out by complex plane method by using Nyquist plot ($-Z_{img}$ vs Z_{real}). The equation for a parallel R-C circuit gives rise to a semi-circle in the $Z^*(\omega)$ plane as that shown in Figure 1.a for a NaCl solution, which has intercepts on the Z_{real} axis at R_∞ ($\omega \to \infty$) and R_o ($\omega \to 0$), being (R_o-R_∞) the resistance of the system. The maximum of the semi-circle equals $0.5(R_o-R_\infty)$ and occurs at such a frequency that $\omega RC=1$, being RC the relaxation time (10). The Bode plot ($-Z_{img}$ vs frequency) is shown in Figure 1.b, and it allows the determination of the interval of frequency associated to a given relaxation process. For electrolyte solutions a unique relaxation process exists, and the Bode plot shows a symmetrical and well-defined peak, with the maximum frequency around 1 MHz.

Complex systems may present different relaxation times and the resulting plot is a depressed semi-circle, because of which a non-ideal capacitor or constant phase element (CPE) can be considered. Impedance for the CPE is expressed by (10): $Q(\omega) = Y_o(j\omega)^{-n}$, where the admittance Y_o and n are experimental parameters ($0 \leq n \leq 1$).In these cases, an equivalent capacitance (C^{eq}) an be determined (11):

$$C^{eq}=(RY_o)^{(1/n)}/R \qquad (3)$$

A particular case is obtained when n=0.5, then the circuit element corresponds to a "Warburg Impedance" (W), which is associated to a diffusion process according to Fick´s first law.

In addition to impedance there are several other derived can be calculated from IS measurements (10). In order to get a clear picture of the processes taking place in a particular system, one or another of the different dielectric diagrams can be used.

EXPERIMENTAL

Material

The following membranes with different structures and from different materials were studied:

a) an ideally porous membrane obtained by etching the track in a poly(ethylene) terephtalate film caused by irradiation with heavy ions (membrane PET). This membrane was obtained in the Laboratoty of Nuclear Filters, Shubnikov Institute of Cristallography, (Russian Academy of Sciences), Moscow, and submitted by Dr. A. Nechaev. The membrane geometrical parameters (given by suppliers) are: thickness, $\Delta x_m=(10\pm1)$ μm, pore radii, $r_p=(4.5\pm0.5)$ nm and number of pores by membrane area, $N=2\times10^{13}$ m^{-2}.

b) a dense aromatic polyamide membrane with an ionic pendent group (membrane DPA3) obtained in the Polymer Institute (CSIC, Madrid) and submitted by Prof. J. de Abajo. The membrane thickness is $\Delta x_m=(35\pm3)$ μm.

c) a composite polyamide/polysulfone membrane for reverse osmosis (membrane PAC) by PRIDESA (Erandio, Spain) obtained for interfacial polymerisation (1). The membrane basically consists of two subalyers: a dense and very thin polyamide layer and a thick and porous polysulfone support; moreover, the membrane is maintained on an unwoven support for reinforcement reasons. Membrane hydraulic permeability and total thickness are: $L_p=10^{-11}$ m/s.Pa and $\Delta x_m=120$ μm.

Measurements were carried out with aqueous NaCl solution at six different concentrations ($10^{-3} \leq c(M) \leq 5\times10^{-2}$), at a constant temperature t=(25.0±0.3)°C and standard pH=(5.9±0.3). Before use, the membranes were immersed for several hours (around 10 h) in a solution of the appropriate salt concentration.

Electrochemical Impedance Spectroscopy measurements.

The test cell used for impedance measurements is similar to that described elsewhere (12). The membrane was tightly clamped between two glass half-cells by using silicone rubber rings. Electrochemical impedance measurements were carried out by using an Impedance Analyzer (Solartron 1260) controlled by a computer. The experimental data were corrected by both software and the influence of connecting cables as well as by other parasite capacitances. The measurements were carried out using 100 different frequencies in the range 10 Hz-10^7Hz at a maximum voltage of 0.01 V, the solutions at both sides of the membrane having the same concentration.

RESULTS AND DISCUSSION

Typical impedance plots obtained for the different membranes are shown in Figure 2 (a-c); in Figure 2.a the impedance data obtained for the NaCl solution at the same concentration are also drawn. Very good agreement between the part assigned to the electrolyte solution in the heterogeneous membrane/solution plot and that measured for the electrolyte alone can be observed. The equivalent circuits associated to each system are also indicated in Figure 2:

i) a non-ideal capacitor or constant phase element has been considered in the circuit associated to the ideally porous PTF membrane/electrolyte solution, this means: (R_eC_e)- (R_mQ_m).

ii) for the dense PAD3 membrane a sub-circuit formed by the parallel association of a resistance and a capacitor was obtained, and the total circuit is: (R_eC_e)-(R_mC_m).

iii) the equivalent circuit for the composite membrane consists of a series association of two subcircuits, one for each sublayer: a parallel association of a resistance and a capacitor (R_dC_d) for the dense sublayer, and a parallel association of a resistance and a Warburg impedance (R_pW_p) for the porous one; then, the total circuit for the membrane/NaCl solution is: (R_eC_e)-(R_pW_p)-(R_dC_d).

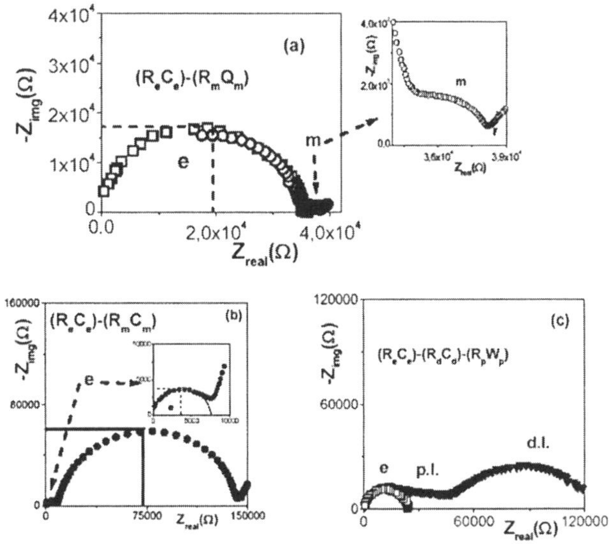

FIGURE 2: Nyquist plots and equivalent circuits for different membrane/electrolyte systems. (a) porous PTF membrane (o); electrolyte (□). (b) dense PA3 membrane (•). (c) composite polyamide/polysulfone PAC membrane (τ); electrolyte (□).d.l.: dense sublayer, p.l.: porous sublayer

Figure 3 shows the Bode plots (-Z_{img} vs f) for the different membranes, which allow the estimation of the interval of frequency for thc relaxation processes associated to each membrane (or sublayer) depending on their structures. For both dense membrane and dense sublayer, a well-defined and symmetric peak with a maximum frequency ranging between 5x102 Hz and 5x103 Hz can be observed, while for porous membrane/sublayer the maximum frequency is not so well established. In all cases, the maximum frequency for the electrolyte solution appears at 1-2 MHz. Plots similar to those indicated in Figures 2 and 3 were obtained for the other concentration studied.

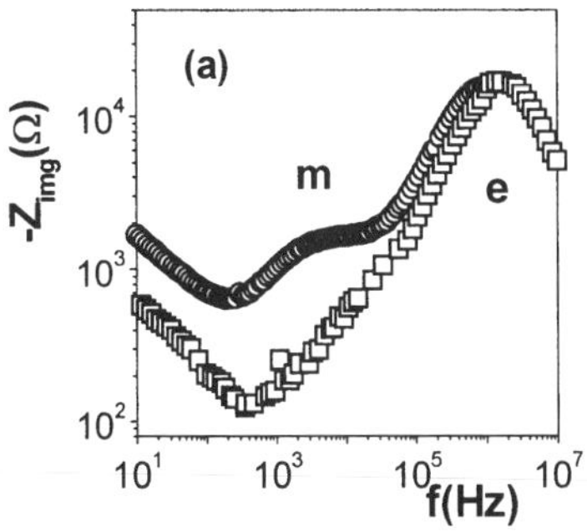

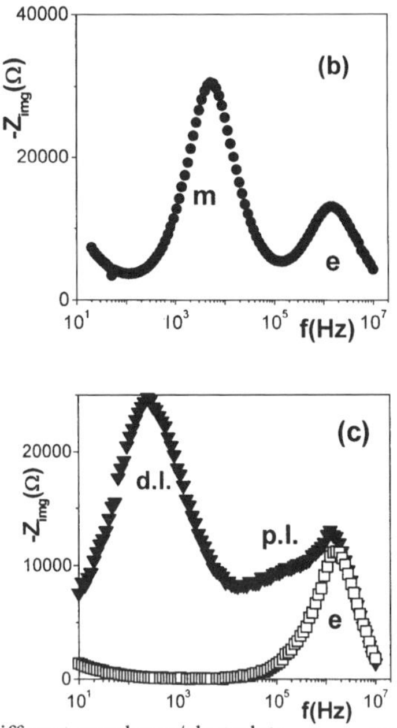

FIGURE 3: Bode plots for different membrane/electrolyte systems. (a) porous PTF membrane (o); electrolyte (□). (b) dense PA3 membrane (•). (c) composite polyamide/polysulfone PAC membrane (τ); electrolyte (□).

The fitting of the experimental points by means of a non-linear program (13) allows the determination of the different circuit parameters (R_m, C_m and W_p). Variation of R_m values with salt concentration is shown in Figure 4. The decrease of membrane resistance when external salt concentration increases is attributed to the concentration dependence on the electrolyte filling the membrane structure (pores or void volume) (7,14).

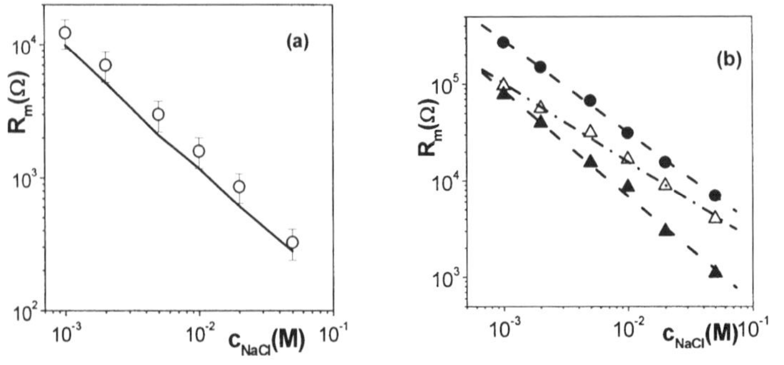

FIGURE 4: Membrane electrical resistance versus external salt concentration for the different membranes. (a) porous PTF membrane: (o) experimental values; (----) theoretical values. (b) dense PA3 membrane: (•); composite PAC membrane: (Δ) porous sublayer, (σ) dense sublayer.

Due to the well-defined geometrical structure of track-etched membranes the uniform capillary physical model is usually adopted. This model assumes that an ideally porous membrane is an assembly of parallel cylindrically shaped pores of uniform dimension (only straight pores without tortuosity or constriction). Then, assuming the electrical resistance of the solid matrix, R_{mx}, is much higher that for the pore, R_p, ($R_{mx}>>R_p$), the electrical resistance of ideal porous membranes can be calculated as that corresponding to a parallel association of pores filled by the electrolyte solution, this means:

$$R_m{}^T=R_p/N=(1/\lambda_o)(\Delta x_m/NS_p) \quad (4)$$

where λ_o is the solution conductivity, while N and S_p represent the number of pores and the pore area and porosity, respectively. A comparison between experimental values for PTF membrane and those calculated using Eq. (4) is also shown in Figure 4.a, and very good agreement between both R_m and $R_m{}^T$ can be seen. This results confirms the possibility of using electrical parameters for geometrical determinations as is indicated in the literature (5,15); it is worth indicating the good agreement obtained is directly related to the well-defined capillary structure of track-etched membranes, their narrow pore distribution and known pore length.

Capacitance values for both dense membrane and dense sublayer of the composite one (dl) were also obtained by the fitting of the experimental points. Capacitance values are practically independent of salt concentration, and the following average values per membrane surface for the whole range of concentrations were determined: $<C_{DPA3}>^*=(5.0\pm0.7)x10^{-6}$ F/m^2 and $<C_{dl}>^*=(4.5\pm0.6)x10^{-4}$ F/m^2. Assuming the dense membrane behaves as a plane-plate capacitor (9-15):

$$<C>=\varepsilon_o\varepsilon S/\Delta x \quad (5)$$

then, the membrane dielectric constant, ε, can be determined by Eq. (5), if its thickness is known (ε_o represents the permittivity of vacuum). The following value was obtained for polyamide (in hydrated state): $\varepsilon=(20\pm2)$. This result was used to determine the thickness of the polyamide dense layer in the composite PAC membrane, and the value $\Delta x_{dl} \approx (0.39\pm0.04)$ μm was obtained. This value agrees with that assumed from polymer diffusion for dense sublayers of composite membranes obtained from interfacial polymerisation ($\Delta x_{dl}< 1$ μm, (1)).

For the porous PET membrane an equivalent capacitance, C^{eq}, was determined by Eq. (3) and its variation with salt concentration is shown in Figure 5. The concentration dependence for the Warburg impedance, W_p associated to the porous sublayer of composite PAC membrane is also shown in Figure 5. As can be seen, an increase of both C^{eq} and W_p values when the concentration increases was obtained, which is attributed to the accumulation of charge on the membrane surfaces (7). On the other hand, differences in the circuit parameters obtained for both porous membrane and porous sublayer of the composite one are attributed to the high salt rejection presented by the dense active layer of reverse osmosis membranes, which could originate a concentration gradient in this sublayer.

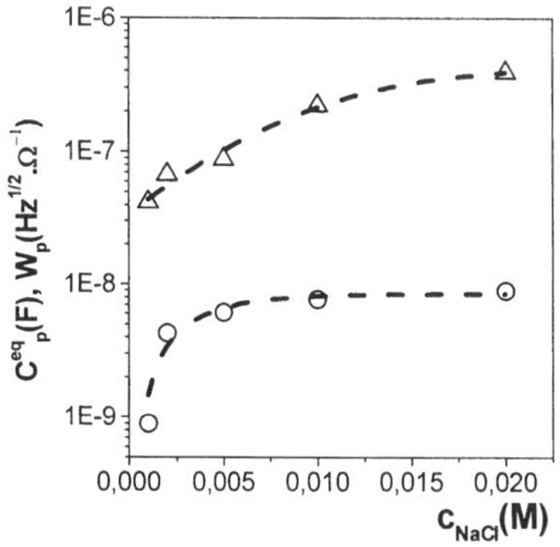

FIGURE 5: Equivalent capacitance, C^{eq}, for porous PTF membrane (o) and Warburg impedance, W_p, for porous sublayer of composite PAC membrane (Δ) versus salt concentration.

CONCLUSIONS

Electrochemical impedance spectroscopy (EIS) measurements allows the characterisation of heterogeneous systems formed by a series array of layers with different electrical/structural properties by using equivalent circuits as models. This technique can be used in the case of membrane/electrolyte systems, which allow the characterisation of membranes in "working conditions" (in contact with electrolyte solutions).

For ideally porous membranes, the uniform capillary model allows the estimation of geometrical parameters from electrical resistance. In the case of composite membranes, EIS also permits us to estimate electrical/geometrical parameters for the different sublayers from measurements carried out for the whole membrane.

The use of equivalent circuits also allows a correlation between the different circuit elements and the processes taking place in the membrane system.

REFERENCES

(1) Mulder, M. Basic Principles of Membrane Technology; Kluwer Acad. Pub.: Dordrecht, The Netherland, 1992.
(2) Tsuru, T., Nakao, S.I. and Kimura, S., J. Chem. Eng. Jap., 23 (1990) 604.
(3) H.U. Demisch, W. Pusch, J.Colloid Interface Sci., 69 (1979) 247.
(4) Y. Kimura, H.J. Lim, T. Iijima, J.Membr. Sci., 18 (1984) 285.
(5) Asaka, K., J. Membr. Sci., 50 (1990) 189.
(6) Ariza, M.J., Cañas, A. and Benavente, J., Surf. Inter. Anal., 30 (2000) 425.
(7) Benavente, J., Ramos-Barrado, J.R. and Bruque, S., in Interfacial Dynamics, Ed. Kallay, N., Marcel Dekker, Inc., New York, USA, 1999.
(8) Benavente, J. Ramos-Barrado, J.R. and Heredia, A., Colloids & Surfaces A, 140 (1998) 333.

(9) Benavente, J., García, J.M., Riley, R. and de Abajo, J., J. Membr. Sci., 32 (2000) 43.
(10) Macdonald, J.R., Impedance Spectroscopy, Wiley, New York, 1987.
(11) Honscher, A.K., Dielectric Relaxation in Solid, Chesea Dielectric Press, London, 1983.
(12) Benavente, J., Muñoz, A. and Heredia, A., Solid State Ionics, 97 (1997) 89.
(13) Boukamp, B.A., Solid State Ionics, 18&19 (1986) 136.
(14) Oleinikova, M., Muñoz, M., Benavente, J. and Valiente, Langmuir, 16 (2001) 716.
(15) Coster, H.G.L., Chilcott, T., Coster, A.C.F., Bioenergetic, 40 (1996) 79.

Applied Physics (APHYS 2003)
A. Méndez-Vilas (Ed.)

Effect of Titanium/Oxygen Compositional Gradient on Adhesion of Titanium-Oxygen System Film deposited onto Titanium-based Alloy by Reactive DC Sputtering

T. SONODA, A. WATAZU, J. ZHU, T. YAMADA, K. KATO AND T. ASAHINA

National Institute of Advanced Industrial Science and Technology(AIST)
2266-98 Anagahora,
Shimoshidami,Moriyama-ku, 463-8560 Nagoya, JAPAN
E-mail: tsutomu.sonoda@aist.go.jp

Abstract. Coating of titanium-based alloy such as Ti-6Al-4V with Ti-O compositionally gradient films was carried out by reactive DC sputtering, in order to improve not only the biocompatibility of the alloy[1)] but also the adhesion between the deposited film and the alloy substrate with preserving the high hardness of Ti-O films[2)]. The effects of Ti/O compositional gradient on adhesion of the film to the alloy substrate were investigated by comparing the adhesion of Ti-O compositionally gradient films to the alloy substrates with that of Ti-O compositionally constant films to the alloy substrates. The compositional gradient was realized by varying continuously the oxygen content in Ar-O_2 sputter gas mixture during the sputter-deposition. According to AES in-depth profiles, the oxygen(O) concentration in the deposited film decreased gradually in-depth direction from the surface toward the substrate, confirming that a film with Ti/O compositional gradient, i.e., a Ti-O compositionally gradient film had formed on the alloy substrate, and thereby expecting that the stress concentrated at the interface between the deposited film and the alloy substrate could be relaxed. On the basis of indentation-fracture tests, it was found the compositionally gradient films were more adhesive to the alloy substrates than the compositionally constant films, concluding that the Ti/O compositional gradient improved the adhesion of the deposited Ti-O films to the alloy substrates

INTRODUCTION

Ti 6Al 4V alloy attracting attention as a biomaterial features excellent mechanical properties and corrosion resistance, and super plasticity, so enables the forming of denture bases of complicated shapes. However, this alloy contains aluminum and vanadium liable to do serious harm to human bodies[1,2], so actual use will require prevention of direct contact with biological tissues. In our previous study, a method of coating the alloy surface with pure titanium film of excellent biocompatibility as a barrier layer was developed by DC sputtering in argon atmosphere, to improve the dental applicability of the alloy[3]. The pure titanium barrier layer prevents such harmful substances from leaching out to biological tissues. However, the pure titanium layer is soft and easily suffers mechanical damages, and thus the contamination of biological tissues with such harmful substances may be caused in case of the alloy being exposed to the tissues by the damages. Therefore, it is necessary to improve the hardness of the barrier layer so that the layer should resist the damages. Hence it can be noticed that the hardness of titanium dissolving oxygen atoms increases with increasing the oxygen concentration, titanium oxide has high hardness and its coatings enhance the bonding of titanium alloy implants to living bone[4,5]. In the present work, we performed coating of the alloy substrate with Ti/O compositional gradient film by reactive DC sputtering in Ar-O_2 gas mixture varying continuously the oxygen content in it, expecting the improvement not only of the hardness and the biocompatibility of the barrier layer but also of the relaxation of the stress concentrated at the interface between the layer and the substrate. Furthermore we investigated the structure and the mechanical properties of the deposited film.

EXPERIMENTAL

A planar magnetron sputtering system(ANELVA Corp. type SPF-210H) with a 200mm-diameter, 130mm-height stainless steel chamber was used. The planar target used for this study was a 100mm-diameter 99.99mass% pure titanium disk. Ti-6Al-4V alloy substrates($13\times9mm^2$, thickness 0.55mm) cleaned with organic solvent were mounted on the water-cooled substrate holder. Ti/O compositional gradient films were deposited onto the substrate by reactive DC sputtering under the Ar-O_2 atmosphere in which the oxygen content was continuously varied with depositing time. Discharge voltage and current for the sputtering were 350V and 1A, respectively. The procedure of the deposition was as follows. First, pure titanium was deposited for 1minute by sputtering in the atmosphere of pure argon. Then, oxygen was gradually introduced into the chamber on condition that the oxygen flow rate was constantly increased for 18.5minutes up to 3.0mL/min, aiming at the formation of Ti/O compositional gradient film by reactive sputtering in the Ar-O_2 gas mixture. Finally, the reactive sputtering under the constant argon flow rate (1.4mL/min) mixed with the constant oxygen flow rate (3.0mL/min) was carried out for 0.5 minute. Schematic change in flow rate of the oxygen gas introduced into the chamber with depositing time is shown in Fig.1. The thickness of deposited film was measured by tracing the substrate-film step using a surface roughness tester. The surface morphology of the film was studied on SEM images. In-depth profiles of the chemical composition (Ti/O ratio) of the film were analyzed by AES with an ion sputter etching method using argon ion beam. Indentation-fracture tests were carried out using a Vickers hardness tester under the load of 50g, to estimate the adhesion of the obtained film to the alloy substrate.

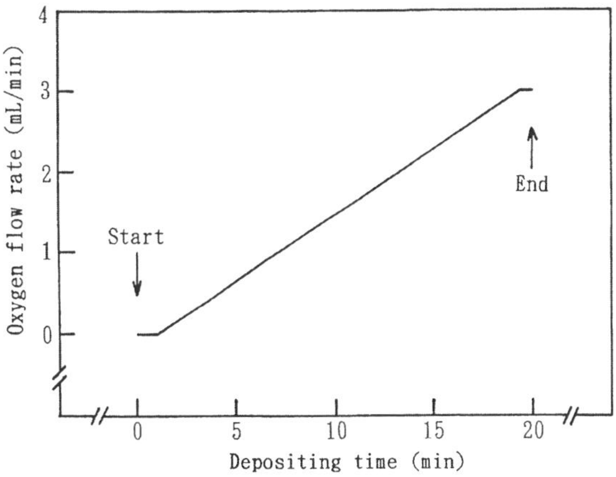

Figure 1. Schematic change in flow rate of oxygen gas introduced into the chamber with depositing time.

RESULTS AND DISCUSSION

The thickness of the obtained film was approximately 3μm. The film appeared to be uniform and adhesive. In regard to the mechanical durability, the film was hard and durable without being damaged by a tip of pincette. The surface of the film was observed to consist of smooth accumulated deposit and fine particles dispersed on the deposit. Under the more detail observation on the topography of the accumulated deposit, some pits and bumps in the order of several microns, which might reflect those existing originally on the substrate surface, were found on the deposit. Thus the obtained film exhibited to be adhesive to the substrate surface.

AES in-depth profiles of titanium, oxygen and nitrogen for the deposited film and the substrate are shown in Fig.2. The AES signals for the analysis were Ti-LMM(480eV), O-KLL(510eV) and Al-KLL(1396eV). A specific increase of aluminum concentration was detected along the depth direction of the sample. This implies the position of the interface between the film and the substrate. It is shown that the oxygen concentration in the film decreases gradually in depth direction from the surface toward the substrate while the titanium concentration increases gradually in contrast with the oxygen. Therefore it was confirmed that a Ti/O compositional gradient film had formed on the alloy substrate. Then, a specific increase of oxygen concentration was also detected in the vicinity of the interface between the film and the substrate. The amount of the oxygen concentration integrated with depth ranged in the vicinity corresponded to that of oxygen concentration integrated with depth ranged in the surface layer of the alloy substrate where oxygen atoms concentrate, which was detected by AES analysis for the surface of the alloy substrate. Therefore it was concluded that the oxygen atoms which had concentrated in the surface layer of the alloy substrate diffused into the titanium layer in the compositional gradient film.

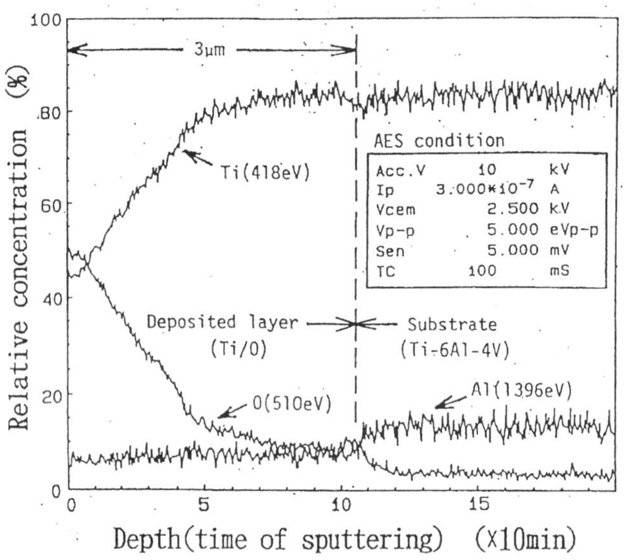

Figure 2. AES in-depth profiles of titanium, oxygen and nitrogen for the deposited film and the Ti-6Al-4V alloy substrate

The typical SEM image of 50g indentation-fracture tests on the surface of the Ti/O compositional gradient film is shown in Fig.3(b), compared with that of the Ti-O compositional constant film [Fig.3(a)], which was deposited by reactive DC sputtering under the same sputtering conditions as in this study except the gas mixing condition, i.e., in 1.4mL/min constant Ar+3.0mL/min constant O_2 mixture. Not only lateral cracks but also radial cracks were observed on the both films. The lateral cracks on the Ti/O compositional gradient film exhibited to be segmental such as circular arcs, while those on the Ti-O compositional constant film exhibited to be continuous such as a circle. Therefore it was found that the compositional gradient film was more adhesive to the alloy substrate than the compositional constant film. Then, the radial cracks on the Ti/O compositional gradient film exhibited to be short propagated and few in number, while those on the Ti-O compositional constant film exhibited to be long propagated and many in number. Therefore it was found that the compositional gradient film was tougher than the compositional constant film.

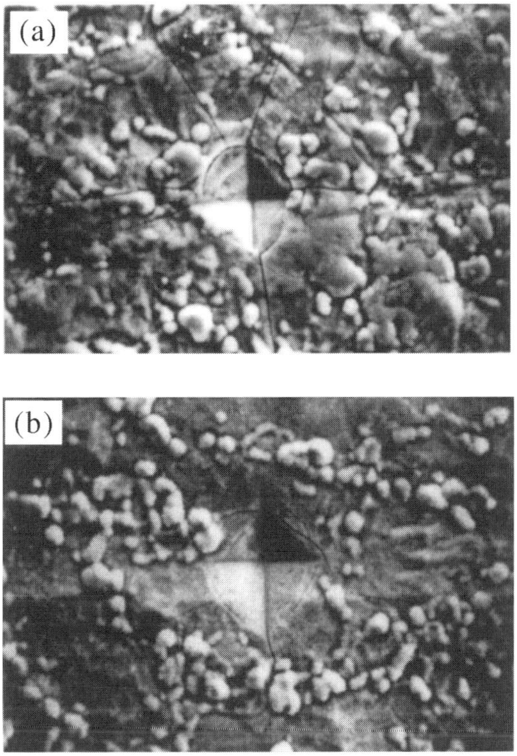

------------ 10μm

Fig.3. SEM image of 50g indentations on the surface of Ti-O compositional constant film deposited under the same sputtering conditions as in this study except the gas mixing condition, i.e., in 1.4mL/min constant Ar+3.0mL/min constant O_2 mixture(a) and the Ti/O compositional gradient film obtained in this study(b).

CONCLUSIONS

Coating of Ti-6Al-4V alloy substrates with Ti/O compositional gradient films was carried out by rcactive DC sputtering, not only to improve the biocompatibility and the surface hardness of the alloy but also to relax the stress concentrated at the interface between the film and the alloy substrate. The compositional gradient was realized by varying continuously the oxygen content in Ar-O_2 sputter gas during depositing. Under SEM, the surface of the deposited film was found to have fine particles dispersed on a smooth accumulated deposit. Under AES, the oxygen concentration in the film decreased gradually in depth direction from the surface toward the substrate, confirming that a Ti/O compositional gradient film had formed on the alloy, and a specific increase of the oxygen concentration was detected in the vicinity of the interface, concluding that the oxygen atoms which had concentrated in the surface layer of the alloy substrate diffused into the titanium layer in the compositional gradient film. Based on indentation-fracture tests, it was concluded that this depositing method improved the adhesion of the coated film to the alloy substrate and the toughness of the film

REFERENCES

[1] S.G.Steinemann and S.M.Perren, Titanium Sci. Technol. 2(1985)1327.
[2] P.Galle, Compt. rend. 299(1984)536.
[3] M.Kato and T.Sonoda, J. Iron and Steel Institute of Japan 77(1991)1206.
[4] A.Dubertret and P.Lehr, Compt. rend. 263(1966)591.
[5] T.Kitsugi et al., J. Biomed. Mater. Res. 32(1996)149.

Applied Physics (APHYS 2003)
A. Méndez-Vilas (Ed.)

Ground Level Air Radioactivity Monitoring in Belgrade Urban Area

Dragana J. Todorovic*, Dragana Lj. Popovic[ζ], Mirjana B. Radenkovic[¶]* and Mirjana D. Tasic[¶]

*Environmental and Radiation Protection Laboratory, Institute for Nuclear Sciences "Vinca", P.O.Box 522, 11000 Belgrade, Serbia and Montenegro
[ζ]Department of Physics and Biophysics, Faculty of Veterinary Medicine, University of Belgrade, Bul.JA 18, 11000 Belgrade, Serbia and Montenegro
[¶]Institute of Physics, Pregrevica 118, 11080 Zemun, Serbia and Montenegro

Abstract. The paper presents the preliminary results of determination of natural and anthropogenic radionuclides (^{7}Be, ^{210}Pb, ^{235}U, ^{238}U, ^{232}Th, ^{40}K, ^{137}Cs) in ground level air, soils and tree leaves in Belgrade urban area. Activities of the radionuclides were determined on an HPGe detector (ORTEC, relative efficiency 23%) by standard gamma spectrometry. The content of the radionuclides in soils were within the average values for the region and the ^{235}U/^{238}U ratio confirmed the natural origin of uranium. The average radionuclides concentrations in ground level air in the urban areas were within the range of the values obtained in the long-term study on the ground level air radioactivity in the city area, exhibiting in general the same seasonal variations pattern. The highest concentrations of ^{210}Pb in aerosols were measured in the very center of the city. There were no significant differences in the content of ^{7}Be and ^{40}K in leaves, whereas further investigations are needed to confirm the differences in the content of ^{137}Cs and ^{210}Pb due to the plant morphology (horse chestnut, linden) or seasonal variations. The study is in progress.

INTRODUCTION

Long-term investigations of the contents of natural and anthropogenic radionuclides in ground level air are of utmost importance in the studies of atmospheric processes on the global level, indicating the migrations of air masses from stratosphere to troposphere and global climate changes [1,2,3]. Some of the radionuclides as ^{210}Pb are today used as tracers in sedimentological and biogeochemical and geochronological studies [4,5,6]. On the other hand, urban areas are of special interest in environmental pollution studies, due to heavy traffic and high population density [7].

Although the radionuclides in ground level air in the Belgrade city area have been the issue of investigations for more than thirty years [8,9,10], there are no data about the content of natural and anthropogenic radionuclides within the Belgrade residential area. Therefore, the aim of the study was to determine the content of radionuclides in soils, plant and ground level air within the very center of the city The results are to be correlated with the results of determing the suspended particles in air [11,12] and enable a prognostic model for transport of pollutants in urban areas to be developed.

MATERIALS AND METHODS

Samples of air, soils and tree leaves were taken on three representative "black spots" in the residential and business center of the city (the city parks) during 2002 and 2003 and analyzed. Aerosols were sampled daily, on filter paper (F/W, relative efficiency 80%, constant rate 600 m^3/day), ashed at 400°C and formed as a composite monthly sample ($15x10^3$ m^3). Soils and leaves were measured in native state; leaves were dried up to 105°C before analyzing.

Activity of the radionuclides was determined on an HPGe detector (ORTEC, relative efficiency 23%) by standard gamma spectrometry. All results are presented as "means ± mean standard error" calculated for the sampling dates. For the radionuclides content in air, the values were calculated for the middle of the sampling period and presented as "monthly means Bq/m3 ± mean standard error". Minimum detectable concentrations (MDC) were: for ^{137}Cs in aerosols $0.1x10^{-5}$ Bq/m3, for ^{235}U in soils 0.2 Bq/kg, and for ^{137}Cs and ^{210}Pb in leaves 0.6 Bq/kg dry weight.

RESULTS AND DISCUSSION

The results of determination of radionuclides content in soils, aerosols and leaves in Belgrade urban area are presented in Tables 1, 2 and 3.

TABLE 1. Radionuclides in soils (Bq/kg) in Belgrade city center

Site	^{226}Ra	^{232}Th	^{40}K	^{137}Cs	^{238}U	^{235}U
S1	39 ± 5	33± 5	402± 40	21± 2	15± 5	2.0± 0.3
S2	34 ± 3	35± 4	402± 32	35± 3	28± 11	1.9± 0.4
	31 ± 3	34± 4	388± 38	26± 3	27± 8	1.3± 0.3
S3	26 ± 3	27± 4	378± 30	35± 2	16± 10	<MDC
Mean	32 ± 4	32± 4	392± 35	29± 2	16± 8	1.7± 0.3

There were no significant differences in the content of radionuclides in soils between the sites and the obtained values were within the range of the average values of the radioactivity in soils in the region [3,9]. The ^{235}U/^{238}U ratio confirmed the natural origin of uranium.

TABLE 2. Radionuclides in ground level air (Bq/m^3) in Belgrade city center

Site	^{7}Be (x10^{-3})	^{137}Cs (x10^{-5})	^{210}Pb (x10^{-3})
S1 - July 2002	3.2 ± 0.7	1.4 ± 0.3	0.6 ± 0.2
- December 2002	1.4 ± 0.3	0.3 ± 0.03	0.5 ± 0.1
- April 2003	2.6 ± 0.5	4.0 ± 1.0	0.5 ± 0.1
- May 2003	3.2 ± 0.4	4.2 ± 0.7	0.4 ± 0.1
- June 2003	2.9 ± 0.5	< MDC	0.5 ± 0.1
S2 - July 2002	3.7 ± 0.6	3.6 ± 0.5	1.4 ± 0.3
S3 - June 2002	5.4 ± 1.0	8.1 ± 1.1	1.1 ± 0.3
- October 2002	2.0 ± 0.3	0.3 ± 0.06	0.5 ± 0.1

There were no significant variations in radionuclides content in air on different sampling sites, except for ^{137}Cs, but to determine expected local climate specificities further investigations are required. The highest concentrations of ^{210}Pb were measured in the very center of the city. Generally, average monthly concentrations in ground level air in the central city were within the range of the values obtained in long-term radioactivity measurements of the ground level air in the Belgrade city area, following the pattern of seasonal variations with pronounced one/two maxims in summer/early fall and a winter minimum for ^{7}Be, one/two maxims in spring/summer and winter for ^{137}Cs, and a maximum in early fall for ^{210}Pb [3,8,9,10].

TABLE 3. Radionuclides in plant leaves (Bq/kg dry weight) in Belgrade city center

SITE	^{7}Be	^{137}Cs	^{40}K	^{210}Pb
S1 May 2002L	86 ± 17	2.0 ± 0.2	393 ± 67	12 ± 2
C	54 ± 12	0.6 ± 0.2	352 ± 63	8 ± 1
S1 Sept.2002 L	19 ± 4	<MDC	DE	< MDC
C	59 ±13	0.5 ± 0.1	297 ± 53	< MDC
S1 May2003 L	< MDC	< MDC	370 ± 74	< MDC
C	37 ± 9	< MDC	274 ± 55	< MDC
S2 May2002 L	22 ± 3	2.6 ± 0.5	358 ± 64	17 ± 2
C	48 ± 11	0.9 ± 0.2	525 ± 89	16 ± 3
S2 Sept2002 L	44 ± 10	2.9 ± 1.3	366 ± 62	< MDC
C	93 ± 18	1.3 ± 0.3	744 ±112	< MDC
S2 May2003 L	7 ± 3	7 ±1	442 ± 88	< MDC
C	20 ± 5	1.2 ± 0.4	293 ± 59	8 ± 2
S3 May2002 L	41 ± 10	< MDC	380 ± 68	< MDC
C	7 ± 6	0.6 ± 0.2	438 ± 79	6 ± 2
S3 Sept.2002 L	112 ± 27	< MDC	315 ± 38	< MDC
C	19 ± 4	< MDC	DE	< MDC
S3 May2003 L	35 ± 8	0.9 ± 0.3	456 ± 82	< MDC
C	10 ± 3	1.1 ± 0.3	386 ± 77	7 ± 3

(horse chestnut C, linden L, DE - detection error, MDC - Minimal Detectable Concentration)

There were no significant differences found in the content of radionuclides among the plant species (horse chestnut, linden) for ^{7}Be and ^{40}K, while no conclusion could be made for ^{210}Pb and ^{137}Cs. The study is in progress.

REFERENCES

1. H.W. Gagelar, *Radiochim. Acta* **70/71,** 345- 353 (1995).
2. A.Baeza et al., *J. of Radioanal.and Nucl.Chem.* **207,** 331-344 (1996).
3. D. Popovic, D. Todorovic, M. Radenkovic, and G. Djuric, *J. of Environm. Protection & Ecology* **S124,** 130- 134 (2000).
4. R. Arimoto, *J.of Geophys.Research* **104**, 301-321(1999).
5. A.A.Peters, et al., *J. of Geochem.Research* **102,** 5971-5978 (1997).
6. El Daoushy F., *Environment International* **14,** 305-319 (1998).
7. F.Tondeur, I.Gerardy, and N.Manderlier, Proceedings Natural Radiation Environment NREVII Symposium, Rhodos (2002) 64-66.
8. D.Todorovic, D. Popovic, and G. Djuric, *Vinca Bull.* **2,** 635-638 (1997).
9. D.Todorovic, D. Popovic, and G. Djuric, *Environment International* **25,** 59-66 (1999).
10. D. Popovic, D. Todorovic, G. Djuric, and M.Radenkovic, *Atmosph. Environm.* **34,** 3245-3248 (2000).
11. G.M.Marcazzan, S.Vaccaro, G.Valli and R.Vecchi, R., *Atmosph.Environm.***35,** 4639- (2001)
12. M. Tasic, S.Rajsic, V. Novakovic, Z. Mijic and M.Tomasevic, Proceedings 5th General Conference of the Balkan Physics Union, V.Banja (2003) CD-SP12-007.

Applied Physics (APHYS 2003)
A. Méndez-Vilas (Ed.)

MODELLING AND SIMULATION OF AN ABSORPTION SOLAR COOLING SYSTEM WITH LOW GRADE HEAT SOURCE IN ALICANTE

Cámara-Zapata, J.M.[1]*; Perea, M.C.[2] and Juan-Igualada, J.M.[1]
[1]: Physics and Computers Architecture Department. [2]: Statistics and Applied Mathematics
Universidad Miguel Hernández, Ctra Beniel sn, km 3.2 03312. Alicante. Spain
**Author to whom correspondence should be addressed*
Tel. 34 96 674 97 30; Fax: 34 96 674 96 71; e-mail: jm.camara@umh.es

Abstract: This paper presents the results of modeling and simulation of an absorption solar cooling system whit low-grade heat source. The proposed model is of a nodal type and was obtained applying mass and energy balances as well as the state equations of the LiBr-H_2O solution. The program's code is written in free format MATLAB 6.5. For the heat supply of the absorption cooling system we selected two single-glazed flat-plate solar collectors with a selective surface that more commercially available. The selection of generating temperature T_G is especially important for this system since it affects not only the COP of the absorption solar cooling system (COP_{ASC}), but also the efficiency of the solar collector η_{SC}. We conclude that the COP values are higher than 0.4 under the investigated conditions. The optimum collector inlet temperature lays between 73 and 75°C depending on the working conditions and the type of collector considered.

INTRODUCTION

In the South – East of Spain, in Alicante, during the summer months the daytime mean temperature reaches 28°C with maximum temperatures sometimes attaining 42°C (CENSOLAR, 2002). Therefore, in order to provide a comfortable environment, there is a need to lower considerably the indoor air temperature (Florides et al. 2002). The possibility of providing cooling and air conditioning by means of solar energy has attracted Man's attention since the early development of solar technology (Tabor 1962). The necessity of air conditioning for thermal comfort in hot areas of the world and the abundance of sunshine in these areas has always intrigued researchers about how to combine the two factors for people's benefit. Furthermore, in contrast with other solar applications such as heating, the greatest demand for cooling occurs when the solar radiation is most intense, thus making its use for this application much more attractive. Growing demand for air conditioning in recent years, particularly in hot climates such as in Mediterranean countries, has imposed a significant increase in demand for primary energy resources. Electric utilities have their peak loads in hot summer days, and Southern European countries are often faced with blackout situations, barely capable of meeting the demand. With suitable technology, solar cooling can help alleviate, if not eliminate, the problem (Grossman 2002).

Experience has shown that closed-cycle systems most suitable for solar cooling are based on absorption cycles. Absorption systems have many advantages: possible operation at wide ranges of heat source and heat sink temperatures; possible multi-staging for improving the coefficient of performance (COP) with high-temperature heat sources; quiet operation with no moving parts; environmentally-safe working fluids. Moreover, absorption systems are amenable to combining available solar heat with back-up heat to meet the cooling demand (Grossman 2002).

Most employed absorption cooling systems use LiBr-H_2O which have a higher efficiency than NH_3-H_2O cycles (Florides et al. 2002). The operation of single-effect absorption chillers is well documented in the literature (ASHRAE, 1988). A single-stage absorption system is not suited to utilize a heat source at temperatures higher than about 100°C. To take advantage of a high temperature heat source, absorption systems must be configured in stages (Alefeld 1982). The principle is to utilize the heat rejected from the condenser to power additional desorbers, thereby approximately doubling or tripling the amount of refrigerant extracted out of the solution spending no extra solar heat. The major disadvantage of those systems is the high purchasing cost. When using solar energy as the energy source, investment cost is higher due to the cost of the auxiliary facility. The first solar-powered absorption systems were carried out in the 1970s, most of which have employed single-effect LiBr-H2O chillers with low-temperature solar collectors. These first projects employed conventional machines originally designed for gas or oil-fired operation.

Ghaddar et al. (1997) presented the modelling and simulation of a solar absorption system for Beirut. The economic analysis performed showed that the solar cooling system is marginally competitive only when it is combined with domestic water heating. Sumathy et al. (2002) showed results obtained with an integrated solar cooling system based on the two-stage chiller designed by Li et al. (1999) and installed on a 24-story building in Jiangmen City, south China (longitude: 113.0°E, latitude: 22.4°N). The solar system supplied hot water to the building for daily use throughout the year and provided air conditioning for an education center, located on the 22nd floor. The resulting COP value oscillated around 0.4 when water temperature was about 65°. Most remarkably, the system performance was better than that of a single effect machine under the same conditions.

In the Mediterranean region, using solar energy for producing hot water is more widespread than solar absorption cooling systems. So, in Alicante one can find relatively frequently solar thermal energy facilities in business buildings, hotels, etc. that produce hot sanitary water almost all year round though during summer time those facilities are underutilized since energetic availability overrides demand. The most commonly employed collector to produce water at a temperature between 50 and 60 ºC is the flat plate collector. During the summer, temperature can increase to 80ºC because energetic availability is higher, thus solar absorption cooling systems of a low – temperature energy source could be employed to use the energy surplus.

Under the economic point of view, one must consider that the cost of the solar energy facility is much bigger than that of the absorption system when both are expressed in terms of climatization energy units (Grossman 2002). One of the common goals of present-day research activities on solar assisted cooling is to find an optimum combination of collector and cooling system that matches the special cooling demands (Lazzarin et al. 1993).

The selection of a generating temperature T_G is especially important for this system since it affects not only the COP of the absorption solar cooling system (COP_{ASC}), but also the efficiency of the solar collector η_{SC}. An increase in T_G increases the COP_{ASC} but decreases the η_{SC}. The overall efficiency is the product of the particular coefficients:

$$COP_O = COP_{ASC} \text{ x } \eta_{SC}$$

The aim of this work is the modelling and simulation of an absorption solar cooling system whit two-stage chillers and low grade heat source, evaluate the η_{SC} for commercially available solar collectors and determine the optimal T_G that corresponds to a maximum COP_O.

DESCRIPTION OF THE SYSTEM

The description of the two-stage absorption refrigeration cycle with water as the refrigerant and $LiBr-H_2O$ as the absorbent is given in the flow chart shown in Figure 1.

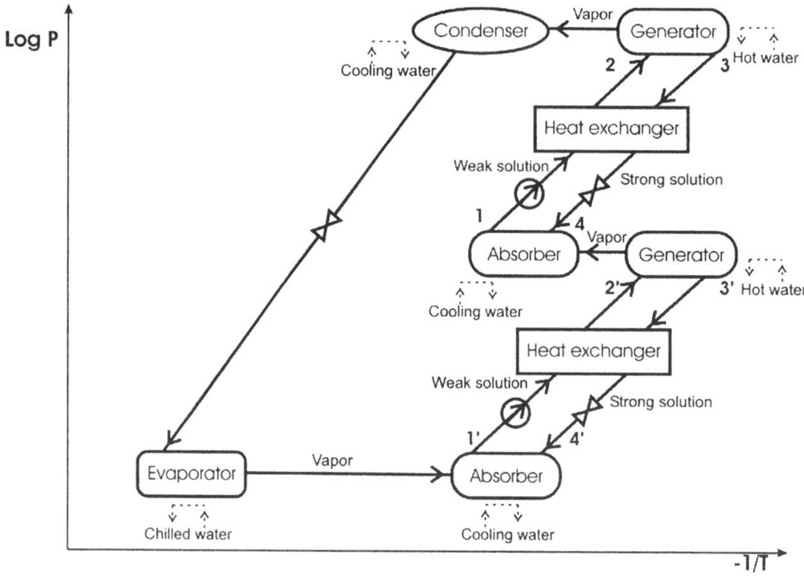

Figure 1. Schematic description of a two-stage absorption refrigeration cycle

In the high-pressure (HP) generator, diluted LiBr solution (2) is heated by hot water to generate water vapor, which is condensed in the condenser at the condensation pressure p_C. The condensed water-refrigerant is then circulated through an expansion valve to the evaporator (in the low pressure LP stage,) where evaporation occurs at the evaporation pressure p_E, producing the desired cooling effect. Later, the evaporated vapor is absorbed in the low-pressure absorber at p_E by the concentrated solution. At the end of this process, the weak solution is circulated from the LP absorber (1') to the LP generator (2') through the heat exchanger for heat recovery. In the LP generator, the weak solution is heated by hot water. During this process, water vapor is generated at intermediate pressure p_M and is circulated to the HP absorber. At the same time, concentrated solution coming form the HP generator (4) is circulated through the HP heat exchanger to the HP absorber to absorb the vapor coming form the LP generator, at the same pressure p_M. After this absorption process, the weak solution is, once again, circulated through the HP heat exchanger to the HP generator, to begin the HP generation process, and thus completing a full cycle of operation (Sumathy et al. 2002).

MODELLING AND SIMULATION OF THE SYSTEM

Absorption cooling system

The proposed model is of a nodal type and was obtained applying mass and energy balances as well as the state equations of the LiBr-H_2O solution. In the heat transfer processes, the components efficiencies were used. The following assumptions were made for the model elaboration: a) head losses are negligible; b) there are no variations in kinetic or potential energy nor heat exchanges with the surrounding environment; c) the liquid phase solution is incompressible; d) at the condenser outlet we obtain saturated liquid at the condensation temperature; e) the values of the liquid state properties are approximated to those of the saturated liquid ; and f) the water vapor is supposed to be at saturation conditions.

The program's code is written in free format MATLAB 6.5. The input or initial data are:

T_{OLG} (ºC) : Outlet temperature at the low-pressure generator,
T_{OHG} (ºC) : Outlet temperature at the high-pressure generator,
T_{OLA} (ºC): Outlet temperature at the low-pressure absorber,
T_{OHA} (ºC): Outlet temperature at the high-pressure absorber,
p_E (kPa) : pressure at the evaporator
p_{LG} (kPa) : pressure at the low-pressure generator
p_{HG} (kPa) : pressure at the high-pressure generator
Q_E (kW) : cooling power
$p_{HG} = p_C$; where p_C (kPa) : pressure in the condenser

Firstly, the approximate solution concentration (ξ) is determined in the most significative states, using the Newton-Raphson method, starting from the relationship between the boiling point of the solution (T, ºC), the dew point of the vapor (D, ºC) and the mole ratio (X) that is stated in Feuerecker et al. (1993):

$$T = \sum_{i=0}^{4} A_i X^{i/2} + D\sum_{i=0}^{4} B_i X^{i/2}$$

where $X = \frac{M_W}{M_S}\frac{\xi}{1-\xi}$; and M_W and M_S are the molar mass of water and salt, respectively, and ξ is defined as: $\xi = \frac{mass\ salt}{mass\ solution}$

ξ_{OLA}, ξ_{OLG}, ξ_{OHA} y ξ_{OHG} have to be determined. After checking the validity of the preceding results, the mass and energy balances are established in order to determine the absorption cycle variables that allow the estimation of the COP.

In the evaporator, the required water vapor mass flow is:

$$m_{VE} = \frac{Q_E}{h_{VE} - h_{LC}} \text{ (kg/s)} \qquad h_{VE} = f(p_E) \qquad h_{LC} = f(p_C)$$

Where h_{VE} and h_{LC} are obtained using interpolation functions of MATLAB and the database (Wexler and Hyland, 1980).

In the low-pressure absorber, the mass flow at the low-pressure absorber, (m_{OLA}) (kg/s), and at the low-pressure generator, (m_{OLG}) (kg/s), are:

$$m_{OLG} = m_{VE} \frac{\xi_{OLA}}{\xi_{OLG} - \xi_{OLA}} \qquad m_{OLA} = m_{VE} + m_{OLG}$$

It is verified that: $m_{VE} = m_C$: refrigerant mass flow that exits the condenser (kg/s)
$m_{OLG} = m_{OHG}$: mass flow that exits the high-pressure generator (kg/s)
$m_{OLA} = m_{OHA}$: mass flow that exits the high-pressure absorber (kg/s)

In the low-pressure heat exchanger, the temperature increment of the mixing that exits the absorber is lower than that exiting the generator because its specific heat is higher (Feuerecker et al. 1993). The enthalpy at the low pressure generator inlet (h_{ILG}, kJ/kg) is estimated from:

$$m_{OLA}(h_{ILG} - h_{OLA}) = \eta\, m_{OLG}(h'_{OLG} - h_{OLA})$$

where η : heat exchanger efficiency.

The solution enthalpies are obtained as a function of their temperature (K) and of ξ (%) according to Feuerecker et al. (1993).

$$h'_{OLG} = \sum_{n=0}^{4} a_n \xi_{OLA}^n + T_{OLG} \sum_{n=0}^{3} b_n \xi_{OLA}^n + T_{OLG}^2 \sum_{n=0}^{2} c_n \xi_{OLA}^n + T_{OLG}^3 d_0$$

$$h_{OLA} = \sum_{n=0}^{4} a_n \xi_{OLA}^n + T_{OLA} \sum_{n=0}^{3} b_n \xi_{OLA}^n + T_{OLA}^2 \sum_{n=0}^{2} c_n \xi_{OLA}^n + T_{OLA}^3 d_0$$

Once h_{ILG} is obtained, the enthalpy at the low-pressure absorber inlet can be obtained (h_{ILA}, kJ/kg)

$$h_{ILA} = h_{OLG} - \frac{m_{OLA}}{m_{OLG}}(h_{ILG} - h_{OLA})$$

where $h_{OLG} = \sum_{n=0}^{4} a_n \xi_{OLG}^n + T_{OLG} \sum_{n=0}^{3} b_n \xi_{OLG}^n + T_{OLG}^2 \sum_{n=0}^{2} c_n \xi_{OLG}^n + T_{OLG}^3 d_0$

In the high-pressure heat exchanger, the same deduction allows the calculation of the enthalpy at the high-pressure generator inlet (h_{IHG}, kJ/kg) and the enthalpy at the high-pressure absorber (h_{IHA}, kJ/kg).

The required heat for the system is

$$Q_G = G_{LG} + G_{HG} = (m_{VE}\,h_{VE} + m_{OLG}\,h_{OLG} - m_{ILG}\,h_{ILG}) + (m_C\,h_{VC} + m_{OHG}\,h_{OHG} - m_{IHG}\,h_{IHG})$$

where $h_{VC} = f(p_C)$ is also obtained using interpolation function of MATLAB and the database (Wexler and Hyland, 1980). Finally COP is calculated as, $COP = \frac{Q_E}{Q_G}$.

The tolerance used throughout the code is 2.220446049250313e-016 and the tolerance used in the Newton-Raphson method is 1.000000000000000e-006.

Solar collector system

For the heat supply of the absorption cooling system we selected two single-glazed flat-plate solar collectors with a selective surface that are commercially available. Type A is a low-cost specially designed and Type B is a conventional one. The difference between them is that type A collector has a 10 cm layer of air space (insulation) beneath the glass cover. The steady-state thermal performance curves of these solar collectors are

$$\eta_{SC} = 0.8 - C_A \frac{(T_i - T_a)}{I_T} \qquad C_A = 3.5\ \text{m}^2/\text{W K (Type A)}$$

$$\eta_{SC} = 0.8 - C_B \frac{(T_i - T_a)}{I_T} \qquad C_B = 5.7\ \text{m}^2/\text{W K (Type B)}$$

where I_T is the incident solar radiation upon the collector aperture (W/m^2); T_i and T_a are the collector inlet and the ambient temperatures, respectively. Values of C_A and C_B were obtained from the manufacturer.

RESULTS AND DISCUSSION

In this study, we first determine the COP of an absorption cooling system for different climatic conditions (condensate temperature, T_C = 24-32°C) in an air conditioning regime (evaporation temperature, T_E = 6°C) for a low range of generation temperatures (T_G = 60-75°C). The cooling capacity was fixed at 20 kW. The results are shown in Figure 2.

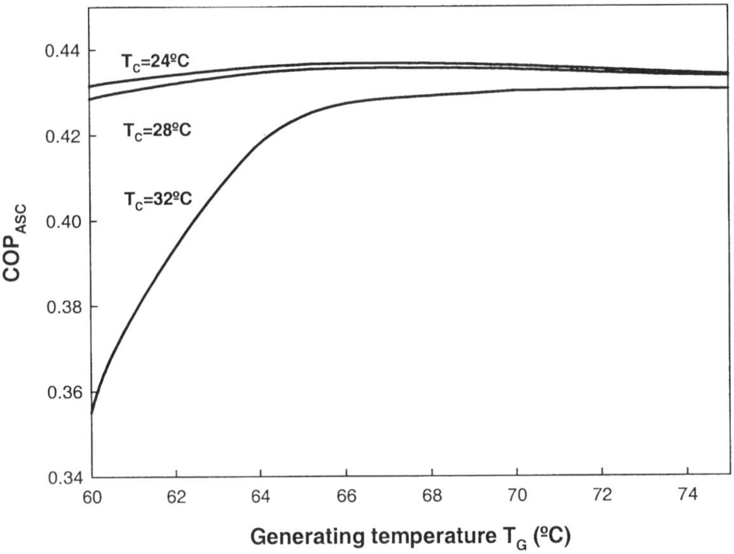

Figure 2. Variation of COP_{ASC} as a function of the generating temperature (T_G) for an evaporator temperature (T_E = 6°C) at different condensate temperatures (T_C = 24, 28 and 32°C)

The values of the COP_{ASC} obtained show that the two-stage system employed int this work is specially adapted for climatization from a low grade heat source, since the system´s efficiency does not increase with increasing TG as it happens in other two-stage systems (Grossman 2002).

The value of the COP_{ASC} hardly changes for the different T_G considered when la T_C is low (24 and 28°C). Under those conditions, the maximum value of COP_{ASC} is reached when T_G = 67°C. Nevertheless, for T_C = 32°C ther is a sharp reduction of COP_{ASC} with decreasing values of T_G. However, the COP_{ASC} values obtained can be considered acceptable if we take into account the low T_G values employed.

According to de performance of the ASC shown in Figure 2, we then evaluated the overall performance. The solar thermal energy is transported to the generator by a heating medium (water). The inlet temperature of the solar collector was assumed to be 8°C higher than the generating temperature of de ASC, i.e. T_I = T_G+8°C. The incident solar radiation was assumed to be I_T = 700 W/m^2 and the ambient temperature T_A= 30°C. The thermal efficiency of a solar collector η_{SC} can be calculated from these data. The overall COP of the absorption solar cooling system COP_O is evaluated for a given T_G, T_C, and T_E.

Figures 3 and 4 show the variation of COP_O with T_G for T_E = 6°C at T_C = 24, 28, and 32°C using type A and type B collectors, respectively. The values of COP_O obtained are higher with type A collector than those obtained with type B regardless of the conditions, which was to be expected due to the better insulation of the former. In both cases, the higher values of COP_O are obtained at T_C = 24 - 28°C, resembling the behavior of COP_{ASC}, though in this case the values of COP_O decrease when T_G increases. For T_C = 32°C, the value of COP_O is the lowest for any value of TG, though its variation shows a maximum at T_G = 67°C for Type A collector and at T_G = 65°C for Type B collector.

The results allow establishing the optimum working temperature of the ASC. When η_{SC} is not taken into account, the higher values of COP_{ASC} at T_C = 24, 28°C are obtained at T_G = 67°C, (T_I = 75°C). For T_C = 32°C, the value of COP_{ASC} increases slowly with increasing T_G. Under those conditions, the optimum value of COP_O is obtained at T_I = 75°C for type A collector, and T_I = 73°C for type B collector.

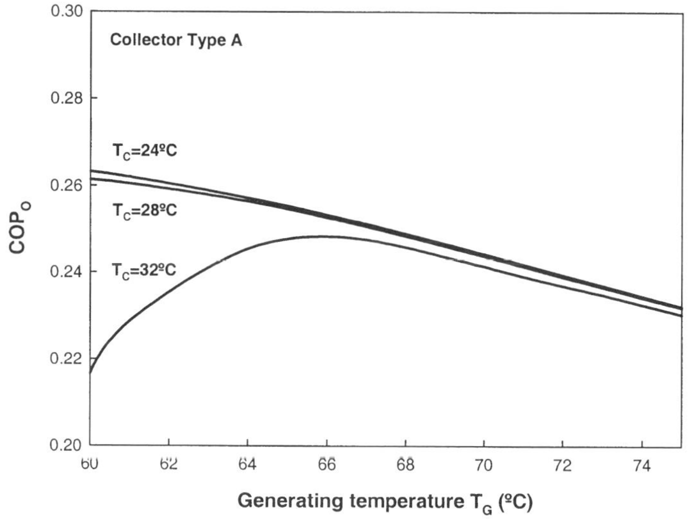

Figure 3. Variation of COP_O as a function of T_G for T_E = 6°C, I_T = 700 W/m^2, T_A = 30°C, at three T_C (collector type A)

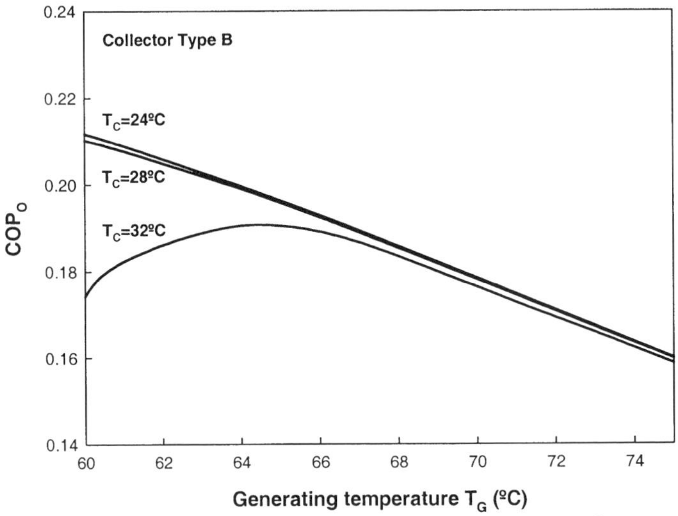

Figure 4. Variation of COP_O as a function of T_G for T_E = 6ºC, I_T = 700 W/m^2, T_A = 30ºC, at different T_C (Type B collector)

The lowering of COP_O values with increasing generation temperatures at low condensation temperatures indicate a strong effect of the collector efficiency on the values of the system's global efficiency. Using other types of high performance collectors (CPC and evacuated tube) can surely improve the overall performance of the system though, under the economic point of view, the system becomes too expensive. As a result, it may be preferable to evaluate the system´s efficiency combining the solar absorption system with domestic water heating (Ghaddar et al. 1997). Furthermore, it has to be considered that the major part of the acquisition costs of an absorption solar cooling system are due to the solar system costs (86% for single-effect and 87% for double-effect) (Grossman 2002).

CONCLUSIONS

In this paper, we have shown the results of the modeling and simulation of an absorption solar cooling system that has an adequate performance operating with a low-grade heat source. The COP values are higher than 0.4 under the investigated conditions. The optimum collector inlet temperature lays between 73 y 75ºC depending on working conditions and the type of collector utilized. Many domestic water heating facilities in Alicante can improve their efficiency if combined with absorption solar cooling systems.

REFERENCES

Alefeld G. 1982. Regeln für den Entwuf von Mehrstufigen Absorbermaschinen (Rules for the design of multistage absorption machines). Brennst-Warme-Kraft 34, 64-73.

ASHRAE Handbook. 1988. Absorption Cooling, Heating, and Refrigeration Equipment. Equipment Volume, Chapter 13.

CENSOLAR 2002. Centro de estudios de la energía solar (Center of Solar energy learning). Curso de instalador de energía solar. Volumen 1.

Feuerecker G., Scharfe J., Greiter I., Frank C. and Alefeld G. Measurement of thermophysical properties of aqueous LiBr-solutions at high temperatures and concentrations. 1993. International Absorption Heat Pump Conference. AES Volume 31.

Florides G.A., Kalogirou S.A., Tassou S.A. and Wrobel L.C. 2002. Solar energy 72(1), 43-51.

Ghaddar N.K., CIAV M. and Bdeir F. 1997. Modelling and simulation of solar absorption system performance in Beirut. Renew. Energy 10(4), 539-558.

Grossman G. 2002. Solar-powered systems for cooling, dehumidification and air-conditioning. Solar energy 72(1), 53-62.

Lazzarin R.M., Romagnoni P. and Casasola L. 1993. Two years of operation of a solar cooling plant. International Journal of Refrigeration 21(2), 89-99.

Li J.H., Ma W.B., Jiang Q., Huang Z.C. and Xia W.H. 1999. A 100 kW solar air conditioning system. Acta Energiae Solaris Sinica (Chinese) 20(3), 239-243.

Sumathy K., Huang Z.C. and Li F. 2002. Solar absorption cooling with low grade heat source – a strategy of development in South China. Solar energy 72(2),155-165.

Tabor H.Z. 1962. Use of solar energy for cooling purposes. Solar energy 6, 395-399.

Wexler A. and Hyland R.W. 1980. A formulation for the thermodynamic properties of the saturated pure ordinary water-substance from 173.15 to 476.15 K (Final Report, ASHRAE Research Project RP 216, Pt 1)

Applied Physics (APHYS 2003)
A. Méndez-Vilas (Ed.)

PRECOOLING TIME ESTIMATION AND MEASUREMENT METHODS IN FORCED AIR PRECOOLING SYSTEMS

J. M. Cámara-Zapata, M. Ferrández-Villena, D. Martínez y S. Castillo

Escuela Politécnica Superior de Orihuela. Universidad Miguel Hernández

Carretera de Beniel, km 3,8 S/N, 03312 Orihuela, Alicante, jm.camara@umh.es

Abstract: In food precooling it is important to know the coefficients of heat transmission by convection to obtain cooling time estimation. In this study, the simulation of a fruit precooling system has been carried out using artificial rubber fruits with a spherical form. The coefficient of heat transfer by convection has been calculated by starting from adimensional relationships, applying Buckingham's pi theorem and adapting the necessary parameters to the characteristics of the precooling process. The cooling time has been determined by measuring the temperature in the different positions of the simulator. Moreover, this parameter has been calculated starting from the coefficient of heat transmission by convection, solving the general equation of heat conduction. The results show that the cooling time obtained starting from the complete resolution of the above-mentioned equation is quite a good estimation. However, when simplified methods are used, the results are less satisfactory, due to the fact that we obtain cooling times that are lower than the ones determined in the simulator.

INTRODUCTION

The production of fresh fruits and vegetables is affected, as much in their quantity as in their quality, by the degradation they go through in conditions and transport between the production areas and the consumption areas. Inadequate handling after harvesting, in transport or storage, increases losses. Precooling constitutes the first stage of post-crop treatment, and as such, it contributes decisively to maintaining the quality of the products (Linke and Geyer 2000). Precooling systems should decrease the temperature of freshly harvested horticultural crops as fast as feasible (Gollete et al. 1996). The most frequent precooling methods used are air, water, vacuum and ice.

The precise determination of the precooling time is essential for minimizing losses in the final quality of the fruit. On one hand, if the fruit does not reach the necessary temperature at the appropriate time, proliferations of fungics and/or bacterial illnesses can occur. On the other hand, if the precooling system acts for a longer time than necessary, the final temperature is excessively low and freezing of the fruit can occur, which means the loss of its commercial quality.

There are many studies in which heat transfer through a channel of particles has been studied (Hermida 2000), although little is known about this process in a tunnel of fruit precooling. A primordial difference between both types of systems resides in the importance of homogeneity of the final condition in a food precooling system, with the purpose of minimizing losses in harvesting. Precooling is complicated but in practice it is possible to establish a relationship between the final quality of the product and its position. Defective cooling is associated to one or more of the following factors: small refrigeration capacity, scarce or inadequate flow of air through the packed product, insufficient and heterogeneous humidification, inadequate temperature control, humidity and/or speed of the air, and finally, incorrect stacking of boxes (Verboven et al. 2000).

The appropriate design of a precooling system requires knowledge of the complex phenomena of heat transfer that takes place. Through simulation of the process we can obtain experimental data with which to determine the coefficients of heat transfer starting from appropriate correlations. Moreover, in this way it is possible to find out the analytical solution of the heat conduction analysis, and therefore, to determine cooling time.

The simplest method to estimate the coefficient of heat transfer by convection is dimensional analysis using Buckingham's π theorem and the method of indexes. The related adimensional groups are Reynolds' number (*Re*), Nusselt's number (*Nu*) and Prandtl's number (*Pr*). In a precooling system, a generic relationship exists among these

$$\mathrm{Nu} = f(\mathrm{Re}, \mathrm{Pr}) \tag{1}$$

being

$$\mathrm{Re} = \frac{v\,\mathrm{L}}{\upsilon} \qquad \mathrm{Nu} = \frac{\mathrm{h\,L}}{\mathrm{k}} \qquad \mathrm{Pr} = \frac{c_P\,\mu}{\mathrm{k}} \tag{2}$$

where v, υ, μ, c_P and k are the characteristic speed, the cinematic and dynamic viscosities, the specific heat and the thermal conductivity of the air, respectively, L is the characteristic length, and finally, h is the coefficient of heat transfer by convection. To evaluate the value of the latter precisely enough, it is necessary to determine the other parameters correctly and to use appropriate relationships among the adimensional groups according to their values.

Starting from the coefficient of heat transmission by convection we can approach to the analytical solution of the general equation for heat conduction. Analytical solutions are used very often in engineering and they constitute useful patterns to verify the accuracy of numerical solution methods (Mills 1995).

In non stationary or transitory conduction, the temperature is a function of both time and spatial coordinates and, if internal heat generation is not considered and a radial distribution of fruit temperature is supposed (Arpaci 1996; Mulet and Bon 1993), this:

$$\frac{\partial T}{\partial t} = \alpha \nabla^2 T = \alpha \frac{1}{r^2}\frac{\partial}{\partial r}\left(r^2 \frac{\partial T}{\partial r}\right) \tag{3}$$

is verified, where T is the temperature, t is the time and α and r are thermal diffusivity and the distance to the centre of the fruit which we want to precool, respectively. The initial and contour conditions are

$$t=0 \qquad T=T_0 \tag{4a}$$

$$r=0 \qquad \frac{\partial T}{\partial r} = 0 \tag{4b}$$

$$-k\left(\frac{\partial T}{\partial r}\right)_{r=R} = h\left((T)_{r=R} - T_S\right) \tag{4c}$$

$$r=R \qquad T=T_S \tag{4d}$$

where R is the radius of the fruit to be precooled, 0 makes reference to the initial temperature and S to air temperature. The solution of equation (3) is simplified if variable changes are carried out. In this way, these:

$$\theta = \frac{T - T_S}{T_0 - T_S} \qquad \eta = \frac{r}{R} \qquad \zeta = \frac{\alpha t}{R^2} = \frac{t}{t_C} \tag{5}$$

are defined, where t_C is a constant named critical time. The variable of adimensional time is commonly known as Fourier's number, Fo. When the changes are carried out, we obtain

$$\frac{\partial \theta}{\partial \zeta} = \frac{1}{\eta^2}\frac{\partial}{\partial \eta}\left(\eta^2 \frac{\partial \theta}{\partial \eta}\right) \tag{6}$$

The initial and contour conditions are

$$\zeta = 0 \qquad \theta = 1 \tag{7a}$$

$$\eta = 0 \qquad \frac{\partial \theta}{\partial \eta} = 0 \tag{7b}$$

$$-\left(\frac{\partial \theta}{\partial \eta}\right)_{\eta=1} = \frac{h\,R}{k}(\theta)_{\eta=1} = Bi\,(\theta)_{\eta=1} \tag{7c}$$

$$\eta = 1 \qquad \theta = 0 \tag{7d}$$

where $Bi=h\,R/k$ is a remarkable adimensional group in heat transfer processes by conduction. To solve the differential equation (6) we apply the method of separation of variables

$$\theta\,(\zeta,\eta) = Z\,(\zeta)\,H\,(\eta) \tag{8}$$

so when substituting in (6) we obtain an equality in which each member is a function of a single independent variable, which fits if both members are equal to a constant named lambda. The general solution of the differential equation is

$$\theta = \sum_{n=1}^{\infty} A_n\, e^{(-\lambda_v^2\, Fo)}\, f_n(\lambda_v\, \eta) \tag{9}$$

where A_n and f_n are constants whose value depends on Bi. The series solution converges quickly for long periods of time and when $Fo>0.2$, it is only necessary to consider the first term of the series to obtain an error inferior to 2% (Mills 1995).

In general, we are interested in knowing the temperature in the centre of the fruit ($r=0$) where the answer is slower. If we denominate $\theta_C= (T_C-T_S)/\,(T_0-T_S)$ to the adimensional centre temperature, where T_C is the centre temperature, and only the first term of the series (9) is considered, we obtain

$$\theta_C = A_1\, e^{(-\lambda_1^2\, Fo)} \qquad Fo>0.2 \qquad (10)$$

as the function is infinite for x=0 (Mills, 1995). If only the first term of the series is considered, the temperature distribution is invariable in time. In this way, θ is related at any point with θ_C simply as

$$\theta = \theta_C\, f_1(\lambda_1\, \eta) \qquad Fo>0.2 \qquad (11)$$

Another very interesting concept is the loss of fractional energy, Φ, which is the relationship between the real energy loss in an interval of time and the total loss that is necessary to reach the ambient temperature. By means of an energy balance we obtain

$$\Phi = 1 - \frac{\bar{T} - T_S}{T_0 - T_S} \qquad (12)$$

where T is the average temperature calculated on the volume to precool, whose value can be obtained starting from (9), resulting

$$\Phi = 1 - \bar{\theta} = 1 - \sum_{n=1}^{\infty} A_n\, e^{(-\lambda_v^2\, Fo)}\, B_n \qquad (13)$$

where B_n is a constant whose value depends on *Bi*. If in (13) only the first term of the series is considered, the result is

$$\Phi = 1 - B_1\, \theta_C \qquad Fo>0.2 \qquad (14)$$

The approximation of a single term is valid during most of the cooling period. However, there are graphs that show the answer to *A* and *B* according to time, carried out with the complete solution to equation (9) and, therefore, they are applicable for any value of *Fo*.

MATERIALS AND METHODS

Precooling chamber

Precooling is used by a forced air simulation system located inside a cold-storage room so that the air temperature has been maintained under *0°C* during the experiment.

The simulator consists of a rectangular chamber, which is *40 cm* long [in which the compartment that contains the product (load area) only occupies *28 cm*], *21.9 cm* wide and *24.2 cm* high, so it has *0.015* m^3 capacity. Inside the box there is a second compartment of *10.2 cm* long, which is divided by a panel perforated with 18 circular holes each of *0.01 m* diameter. Both the air flow and the number of holes in service can be regulated.

The transverse limits of the load area according to the direction of the air flow are plastic grilles, while the rest of the walls are made of wood.

Two aspiration systems have been used, one of low air flow by means of a helical tubular extractor, model TDM-100 of S&P, and another of high flow, model MC-E651C of Panasonic with a consumed power variable between *850* and *1,000 W*. In both cases the connection to the simulator has been carried out through a cylindrical tube of *6 cm* diameter.

Artificial rubber fruits, with a spherical and solid shape, have been used, a total of 173, and with the thermophysic properties as shown in Chart 1.

Material	Shape	Diameter (*m*)	c_P ($J\ kg^{-1}\ {}^oC^{-1}$)	ρ ($kg\ m^{-3}$)	k ($W\ m^{-1}\ {}^oC^{-1}$)	Mass (*kg*)	Total Mass (*kg*)
Rubber	Solid sphere	0.04	1,900	1,150	0.16	0.040	6.930

Chart 1. Properties of the artificial fruits used

The temperature has been determined with a datalogger CR10X of Campbell Scientific, using T type thermocouples. Four of them have been located inside the artificial fruits, distributed in a strategic way, and another at the entrance of the simulator (Figure 1). Data has been obtained in a Windows environment through PC208W software.

On both sides of the regulator grille, there are two tubes joined to a difference of pressure meter, model Testo *25* of Text, with a precision of ±*0.2%*.

The cooling process in the simulator has been maintained until the fruit has reached the temperature of *7/8* ($T_{7/8}$), defined as that equal to the fruit initial temperature minus *7/8* from the difference between the above-mentioned initial temperature and that of the air which produces precooling. The necessary time is denominated 'cooling time' and is represented as $t_{7/8}$.

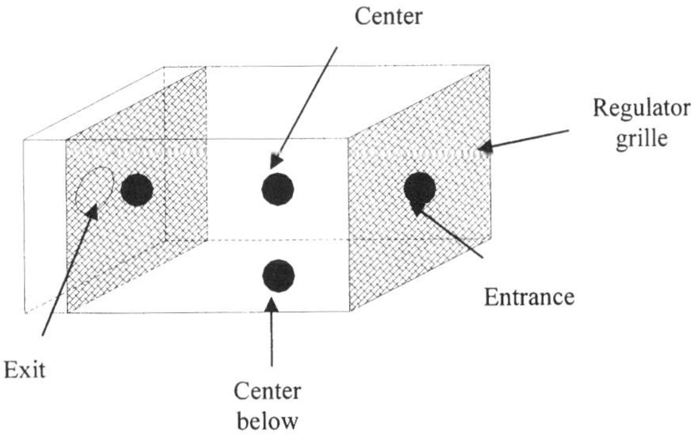

Figure 1. Scheme of the simulator and disposition of the thermocouples inside the load area

Transfer of heat

Convection

The calculus of the adimensional groups related to heat transfer by convection *Re*, *Nu* and *Pr* have been carried out starting from their definitions (1), being

$$v = \frac{\dot{m}}{\rho A_C \varepsilon_V} (m\,s^{-1}) \begin{cases} \dot{m} : air\ mass\ flow (kg\,s^{-1}) \\ \rho : air\ density (1.265\,kg\,m^{-3}) \\ A_C : simulator\ transvers\ area (m^2) \\ \varepsilon_V : porosity = \left(\frac{V_{CHAMBER} - V_{FRUIT}}{V_{CHAMBER}} \right) \end{cases}$$

υ: *air cinematic viscosity* $(13.91 \cdot 10^{-6}\ m^2\ s^{-1})$

μ: *air dynamic viscosity* $(17.60 \cdot 10^{-6}\ kg\,m^{-1}\ s^{-1})$

$$L = D_{FRUIT} \left(\frac{\varepsilon_V}{1 - \varepsilon_V} \right) (m)$$

h: heat transfer mean coefficient by convection

The mass flow is determined as

$$\dot{m} = \rho \dot{V} = A_{WAY} \sqrt{2\,\rho\,\Delta p} \tag{16}$$

where A_{WAY} is the air way area through the regulator panel and Δp is the pressure variation on both sides of this panel.

An appropriate correlation for heat transfer in the case of a gas that flows through a compact channel is (Whitaker, 1972)

$$Nu = (0.5\,Re^{1/2} + 0.2\,Re^{2/3})\,Pr^{1/3}; \qquad 20 < Re < 10^4 \tag{17}$$

Air properties vary with temperature and they have an influence on the value of *Nu*, so it is convenient to correct their value (Mills 1995) through the relationship

$$\frac{Nu}{Nu_S} = \left(\frac{\mu_E}{\mu_S} \right)^{-0.14} \tag{18}$$

where the sub indexes *E* and *S* make reference to the temperature of the mass to precool and to the average temperature of the air in the simulator, respectively. Air properties have been evaluated with an average temperature of *10ºC* and the correction of the variation of properties has been carried out considering $T_E = 20ºC$ and $T_S = 10ºC$.

Conduction

The solution of the general equation of heat conduction has been determined in three different ways. In the first place, we have used the evolution of θ_C (Figure 1) obtained with the complete solution of equation (9). This way, the solution is valid for any value of *Fo* and the result is the most approximate to the real value obtained in the simulator.

On the other hand, it has been proved that under experimental conditions $Fo > 0.2$ becomes real, so we can apply the simplified method which consists of considering only the first term of the series that appears in (9).

Lastly, we have used the evolution of Φ that appears in Figure 2, obtained with the complete solution of equation (9). However, the value of Φ has been determined starting from the first term of the series that appears in (13), that is to say, it has been simplified as shown in the expression (14).

The values of λ_1^2, A_1 and B_1 in function of Bi appear in Chart 2 (Mills, 1995).

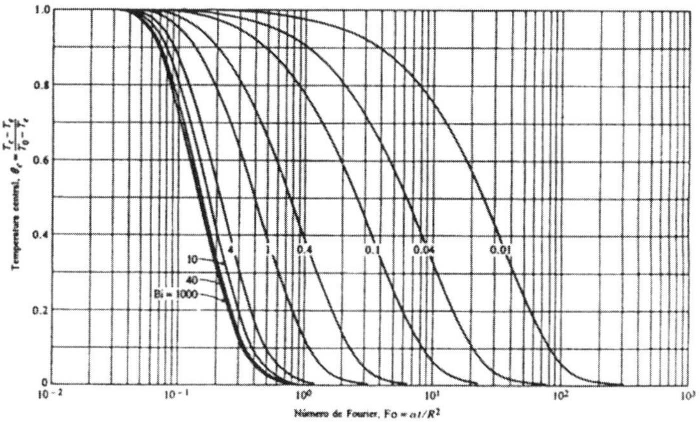

Figure 2. Answer of the temperature of the centre of a sphere cooled by convection

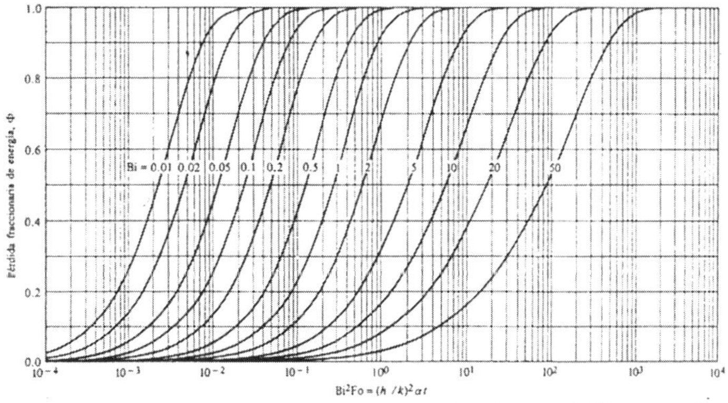

Figure 3. Answer to the loss of fractional energy of a sphere cooled by convection

Bi	λ_1^2	A_1	B_1	Bi	λ_1^2	A_1	B_1
0.02	0.060	1.006	0.994	2	4.116	1.479	0.645
0.04	0.119	1.012	0.988	4	6.030	1.720	0.513
0.06	0.178	1.018	0.982	6	7.042	1.834	0.452
0.08	0.236	1.024	0.977	8	7.647	1.892	0.417
0.1	0.294	1.030	0.971	10	8.045	1.925	0.395
0.2	0.577	1.059	0.944	20	8.914	1.978	0.350
0.4	1.108	1.116	0.894	30	9.225	1.990	0.335
0.6	1.599	1.171	0.849	40	9.383	1.994	0.327
0.8	2.051	1.224	0.809	50	9.479	1.996	0.322
1	2.467	1.273	0.774	100	9.673	1.999	0.313

Chart 2. Coefficients of the approximation of a term for the spheres cooling by convection

RESULTS AND DISCUSSION

Chart 3 shows the experimental results obtained when determining temperature and cooling time ($t_{7/8}$). The $t_{7/8}$ value varies with the position, so it is higher in the air entrance area, very similar in the central situations and lightly inferior to the air exit.

The results obtained in the simulator (Chart 3) show a fall in the cooling speed as cold air advances in its course. The fruit located in the entrance cools down more quickly, probably because the air temperature is lower in this area and it increases as it advances into the simulator. It is evident that these differences can affect to the homogeneity of the final quality of the fruit. These results suggest that the type of air flow (flow and direction) should be studied with the purpose of reducing these differences. In this sense, it is likely that the differences in the fruit cooling speed decreased if the forced air flow was not directly guided toward the fruit, but that it penetrated in the load area in a lateral way and not directly. To succeed in this, corridors could be permitted between lines of fruit containers so that the air can circulate freely through these corridors. However, this configuration could mean an excessive increase of the cooling time with the consequent risk of proliferation of illnesses and loss of commercial quality.

Flow ($m^3 h^{-1}$)	Position	Average T (°C)	Initial fruit T (°C)	$T_{7/8}$ (°C)	$t_{7/8}$ (min)
9.4	Entrance	-0.50	20.14	2.08	68.50
	Center		20.63	2.14	166.00
	Below		21.33	2.23	176.69
	Exit		20.54	2.13	197.53
	Average value		20.66	2.15	152.18
11.1	Entrance	-1.61	17.71	0.81	68.00
	Center		18.46	0.90	148.50
	Below		18.30	0.88	150.00
	Exit		18.55	0.91	192.50
	Average value		18.26	0.87	139.75
35.4	Entrance	-0.70	21.45	2.07	42.65
	Center		22.69	2.22	94.50
	Below		22.98	2.26	102.07
	Exit		21.99	2.14	126.72
	Average value		22.28	2.17	91.49
40.1	Entrance	-2.93	23.90	0.42	40.10
	Center		25.41	0.61	90.59
	Below		23.81	0.41	99.63
	Exit		24.70	0.52	115.18
	Average value		24.50	0.50	86.37
64.2	Entrance	-2.51	20.45	0.36	36.09
	Center		21.84	0.53	66.04
	Below		22.22	0.58	72.63
	Exit		21.59	0.50	83.97
	Average value		21.53	0.50	64.68
	Entrance		16.70	0.55	34.66

77.6	Entrance	-1.76	16.70	0.55	34.66
	Center / Below (overprinted)		17.[illegible]	0.[illegible]	5[illegible]
	Exit		16.35	0.50	66.69
	Average value		16.90	0.57	54.48
82.6	Entrance	-1.21	18.82	1.29	32.45
	Center		19.70	1.40	55.76
	Below		20.81	1.54	60.13
	Exit		18.85	1.30	64.15
	Average value		19.55	1.39	53.12
	Entrance		17.07	0.79	30.25
	Center		18.32	0.95	51.91
	Below		18.06	0.92	58.56
	Exit		18.58	0.98	63.89
	Average value		18.01	0.91	51.15

Chart 3. Temperatures (average of the cold-storage room, initial and of fruit cooling) and cooling time found in different positions of a precooling system simulator

Figure 3 shows the variation of cooling time with the air flow from inside the precooling tunnel. As time increases, a significant decrease of the cooling time occurs, although the relationship is not lineal. That is to say, when the flow is low, small increases in this volumetric flow mean important reduction in the cooling time value, while when the flow is high, an increase in this flow means a small decrease in cooling time. Consequently, an increase in the forced air flow always implies a tendency to reduce cooling time. Although, it is necessary to establish an efficient flow for which the cooling time does not differ significantly from the one obtained with greater flows, that is to say, flow values superior to this efficient flow do not imply a significantly lower cooling time. Under the conditions of our experiments, you could adopt $77.6\ m^3\ h^{-1}$ as this value.

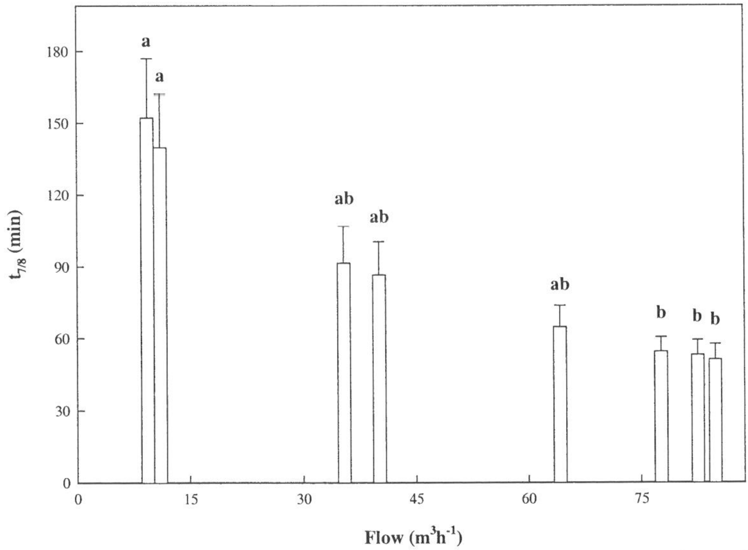

Figure 3. Average values of the cooling time for different flows. The error bars represent the standard error of the mean. The values are averages of four repetitions. Different letters indicate significant differences at a significance level of 95%.

The value of the coefficient of heat transfer by convection is shown in Chart 4, where the results obtained when calculating the related parameters appear. Eight different flows have been used for which the air way area through the regulation grille has varied between 6, 10, 12, 16 and 18 holes in service. The values of ε_V, L, A_C, $\dot{m}$ and v have been obtained starting from their definitions. The flow has been obtained as the relation between $\dot{m}$ and the density of the air, ρ. Under the conditions of our experiments, the value of Re varied between *372* and *3,352*. The coefficient of heat transmission by convection, h, varied between *7.0* and *25.9* $W\ m^{-2}\ K^{-1}$. A high coefficient of determination between the values of Re and those of h ($r = 0.99$) was found.

When the cooling times $t_{7/8}$ determined in the pretender (it Figures 3) and the values of dear h (Chart 4) were correlated, a high correlation was found ($r = -0.99$), which indicates that the adopted relation (17) between Re and Nu has been appropriate.

Flow ($m^3\ h^{-1}$)	ε_V	*L* (*m*)	A_C (m^2)	$\dot{m}$ (*kg* h^{-1})	Δp (*Pa*)	A_{PASO} (cm^2)	*Re*	*v* ($m\ s^{-1}$)	*Nu*	*h* ($W\ m^{-2}\ K^{-1}$)
9.4				12.0	19.6	4.7	372	0.08	17.6	7.0
11.1	0.61	0.06	0.05	14.1	9.8	7.9	439	0.10	19.4	7.7
35.4				44.7	68.7	9.4	1.393	0.30	38.4	15.3
40.1				50.7	88.3	9.4	1.580	0.34	41.4	16.5
64.2				81.3	127.5	12.6	2.532	0.55	54.9	21.9
77.6				98.2	147.2	14.1	3.060	0.67	61.5	24.5
82.6				104.5	166.8	14.1	3.257	0.71	63.9	25.4
85.0				107.6	176.6	14.1	3.352	0.73	65.0	25.9

Chart 4. Coefficient of heat transmission by convection in a precooling system

To estimate the cooling time and to compare it with the one obtained experimentally (Chart 3) three methods have been used. In the first place, Chart 5 shows the results obtained when solving the general equation (9). The critical time is determined starting from its definition in (5). The value of Bi is obtained starting from (7c), taking as coefficient of heat transfer by convection (h) value as determined previously (Chart 4). The value of θ_C results from imposing the condition that the precooling process takes place until the fruit reaches the temperature of *7/8* ($T_{7/8}$). The value of Fo has been obtained by means of Figure 2 starting from the values of Bi and θ_C. Lastly, the cooling time, $t_{7/8}$, has been calculated using the equation (5). The $t_{7/8}$ value varied between *116.53* and *55.53 min*. The correlation between the estimated values of h and those of $t_{7/8}$ starting from the general equation of heat conduction was also very high ($r = -0.98$). However, when comparing the cooling times measured in the simulator and the estimated ones, the estimation is not adequate for low flows, but only for the highest flows of those which have been used under the conditions of our experiments.

In practice, the precooling process is usually maintained until the fruit with lower cooling speed reaches the temperature $T_{7/8}$, so that when the estimated value of $t_{7/8}$ goes slightly above the value of $t_{7/8}$ determined in the simulator, we can consider that the estimation is valid. This reasoning could be applied to the values of $t_{7/8}$ estimated for the flows greater than $64.2\ m^3\ h^{-1}$ under the conditions of our experiments.

On the other hand, since it is determined that $Fo>0.2$, only the first term of the series (9) has been considered and the cooling time has been obtained in a simplified way (Chart 6). The values of A_1 and λ_1^2 are obtained by interpolation of Chart 2 and the value of θ_C as in the previous case (Chart 5). The value of Fo is determined starting from the expression (10).

Caudal ***($m^3\ h^{-1}$)***	t_C ***(min)***	θ_C	***Bi***	***Fo***	$t_{7/8}$ ***(min)***
9.4			0.88	1.28	116.53
11.1			0.97	1.11	101.05
35.4			1.91	0.87	79.20
40.1	91.04	0.13	2.06	0.84	76.47
64.2			2.73	0.71	64.64
77.6			3.06	0.64	58.27
82.6			3.18	0.62	56.44
85.0			3.24	0.61	55.53

Chart 5. Cooling time ($t_{7/8}$) according to the flow and related parameters obtained starting from the complete solution of the equation of heat conduction considering the evolution of the temperature of the centre θ_C

Moreover, the cooling time has been determined starting from the graph that shows Φ throughout time (Figure 3) and from the simplification shown in expression (14). The results obtained appear in Chart 7. The value of B_1 has been obtained interpolating Chart 2. In this case, the value of Fo has been determined in Figure 3.

Flow ***($m^3\ h^{-1}$)***	***Bi***	A_1	λ_1^2	θ_C	t_C ***(min)***	***Fo***	$t_{7/8}$ ***(min)***
9.4	0.88	1.244	2.217			1.04	94.36
11.1	0.97	1.266	2.405			0.96	87.64
35.4	1.91	1.461	3.968			0.62	56.41
40.1	2.06	1.486	4.173	0.13	91.04	0.59	54.01
64.2	2.73	1.567	4.815			0.53	47.81
77.6	3.06	1.607	5.130			0.50	45.32
82.6	3.18	1.621	5.245			0.49	44.48
85.0	3.24	1.628	5.303			0.48	44.07

Chart 6. Cooling time ($t_{7/8}$) according to the flow and related parameters obtained when considering the first term of the solution of the equation of heat conduction

The simplified methods (Charts 6 and 7) offer values of $t_{7/8}$ inferior to those determined in the simulator for all the flows used under the conditions of our experiments. The biggest differences take place when the flows are lower.

Figure 4 shows the relation between the flow and the estimated cooling time value. For small flows, the estimation is insufficient whichever method is used. However, when the value flow goes above of $65.2\ m^3\ h^{-1}$, the estimation of the cooling time starting from the general conduction equation is adequate since it is above the average value of the cooling time obtained by means of the simulator.

Flow ($m^3\,h^{-1}$)	*Bi*	B_1	Φ	θ_C	t_C (min)	*Fo*	$t_{7/8}$ (min)
9.4	0.88	0.795	0.90			1.01	91.70
11.1	0.97	0.779	0.90			0.93	84.24
35.4	1.91	0.657	0.92			0.62	56.73
40.1	2.06	0.641	0.92	0.13	91.04	0.60	54.28
64.2	2.73	0.597	0.93			0.57	52.27
77.6	3.06	0.575	0.93			0.54	48.98
82.6	3.18	0.567	0.93			0.53	47.84
85.0	3.24	0.563	0.93			0.52	47.28

Chart 7. Cooling time ($t_{7/8}$) according to the flow and related parameters obtained starting from the complete solution of the equation of heat conduction considering the evolution of fractional loss of energy, Φ

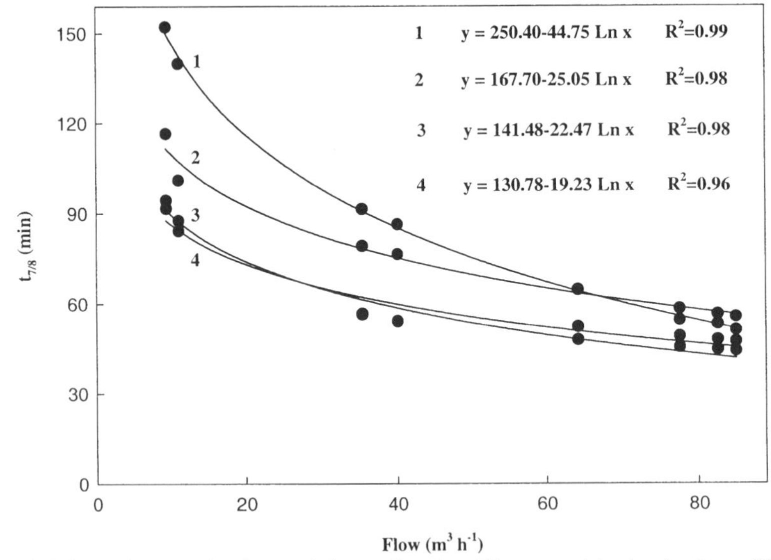

Figure 4. Relation between the flow and the cooling time (1) measured in the simulator, (2) estimated starting from the complete solution of the general equation of heat conduction, (3) estimated starting from the approximation used for Fo>0.2 and (4) estimated starting from the fractional energy with the approximation for Fo>0.2

If all the cooling times determined in our experiments are considered, the relative error made with the estimation starting from the general conduction equation is of *7%*, while with the simplified methods it is greater (*28%* using equation (10) and *25%* starting from equation (14) and Figure 3, respectively). Consequently, it is not advisable to use simplifications in the determination of cooling time.

CONCLUSIONS

The empirical relations between the adimensional groups that take part in the processes of heat transfer by convection enable us to obtain the corresponding coefficient, *h*, in a fruit precooling process. Starting from this point, the general equation of heat transfer by conduction can be solved to estimate cooling time accurately. When the resulting solution is simplified, the results are not satisfactory, so the cooling times estimated are lower than the ones obtained when carrying out the trial, so its application can cause fruit not to reach an appropriate cooling temperature.

ACKNOWLEDGEMENTS

This work has been carried out thanks to the sponsorship of the Valencian Generalitat by means of the Project Ref. GV00-071-13.

BIBLIOGRAPHY

Arpaci, V. S. 1996. Conduction heat transfer. Addison Wesley.

Linke, M., Geyer, M., 2000. Determination of flow conditions close to the produce. Improving post harvest technologies of fruits. Vegetables and ornamentals. Eds. Artés, F., Gil, M.I., Conesa, M.A., IIR Conference. Murcia, p. 872-878.

Mulet, A., Bon. J., 1993. Transmisión de calor por conducción. Universidad Politécnica de Valencia.

Mills, A.F., 1995. Transferencia de calor. McGraw-Hill/Irwin.

Verboven, P., Nguyen, T.A., Hoang, M.L., Nicolaï, B.M., 2000. A variable-porosity porous medium model of transient combined convection air cooling of apples in boxes. Improving post harvest technologies of fruits. Vegetables and ornamentals. Eds. Artés, F., Gil, M.I., Conesa, M. A., IIR Conference. Murcia, p. 843-850.

Whitaker, S., 1972. Forced convection heat transfer correlations for flow in pipes, past flat plates, single cyclinders, single spheres, and for flow in packed beds and tube bundles. AIChE Journal. vol. 18, p. 361-371.

Applied Physics (APHYS 2003)
A. Méndez-Vilas (Ed.)

Cleaning Noising from An ECG

Diaz Calavia Emilio, J.*; Elizalde Soba, Pedro*; Berraondo Lopez, Pedro*;
Teijeira Alvarez, Jose M.;*
Perez Cajaraville, Juan[2]; Ortuño Fernandez-Pedreño, Felipe.[2]

**Biofísica. Dpto. Física y Matemática Aplicada.*
[2]Clinica Universitaria de Navarra.
Universidad de Navarra. 31080 -Pamplona.
ediaz@unav.es

Abstract. Generally, an ECG is analysed by a classical method. Our hypothesis is that an ECG hides information from detection by conventional methods, which nevertheless could be made manifest by mathematical and computerised methods. Investigating an ECG, we try to find out it's true morphology. Cleaning the noising from an ECG without distorting it's signals, we try to find the "ECG-sound" which is the most close to reality. Filters are tested applying them to a sinusoidal signal of linear frequency between 0 and 10 kHz.

INTRODUCTION

In order to work properly with an ECG, we need:
-at first, to eliminate parasitical noising;
-second, not to distort its signal.

Considering each ECG cycle a sinusoidal synthesis (A. Fourier), "cleaning" means to eliminate the elements of Fourier which do not belong to an ECG.

<u>Problem:</u>

1. The true / real ECG of an individual is not known.
2. The noising added to an ECG is not known.
3. The sum is known (noising + ECG).
4. It is known that at short intervals of time an ECG is relatively recurring.
5. The level of frequency which constitutes the ECG is known.
6. With these giving we try to obtain an ECG free of noising.

We apply conventional filters (FIR., Butterworth, ICA., etc.) and another one prepared as moving average. The proof is executed applying the filters to a variable linear sinusoidal frequency starting from 0 Hz and increasing up to much higher frequencies than those which can be detected in an ECG.

We try to find the original ECG sound without distortion by filters, or, at least, the most similar sound.

The problem consists in removing only the sinusoidal components of the signal and not the ECG sound.

MATERIAL

ECG signal, D II lead. Sampling at 20 kHz. Software comercial: **MATLAB** y **HP VEE.**

APPLIED METHODS

Noise cleaning.

Minimum sequence register of: 5 minutes sequentially.

We obtain a long series of sequential ECGs. The cycles, relatively recurring, are very similar and vary during the time of testing. The obtained signal shows noising of low and high frequencies.

Cleaning of Low Frequencies Noising.

Noise <0,1 Hz. being no component of the ECG.
Noise cleaning by conventional methods.

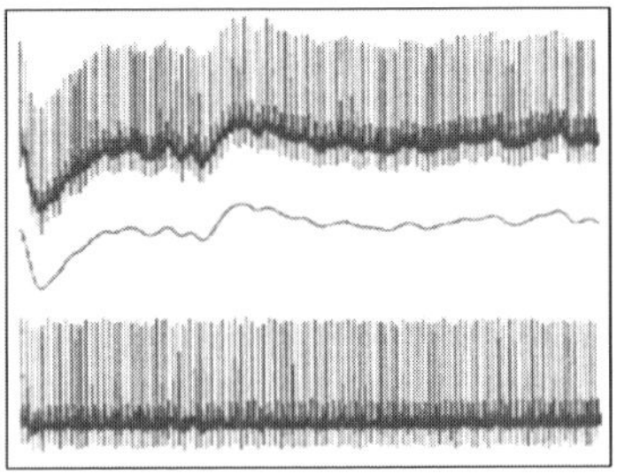

Figure1.- Series of sequential ECGs. Removing the noising of low frequency, we obtain a series of ECGs with a virtual isoelectric base line.

There still persists high frequency noise.

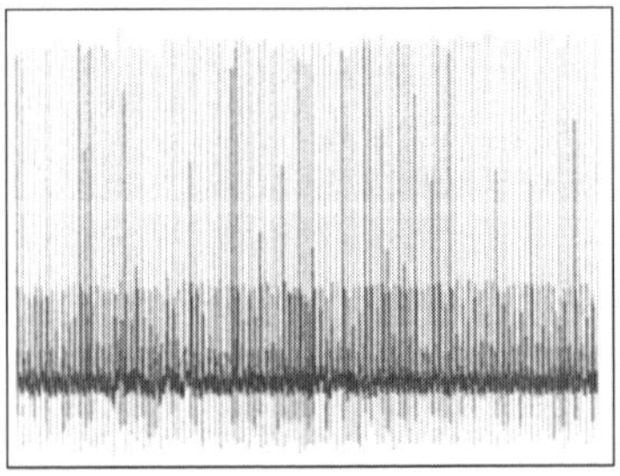

Figure 2 ECG series without low frequency noising, of a (relatively) "high" frequency.

Cleaning of the Noising Remnant of High Frequency.

In the previous graph we obtained virtually a basal reference line. There still persists nevertheless the noise of a higher frequency than the cleaned.

Noise cleaning up to now has been done by conventional filters (FIR., ICA., Butterworth,...) or by the prepared moving average filter.

The ECG as synthesis of sinusoidal signals

We consider the obtained signal (ECG + remnant noising) as a synthesis of Fourier elements.

One of the ECG cycles cleaned of noising of low and high frequencies can be seen in Fig. 3

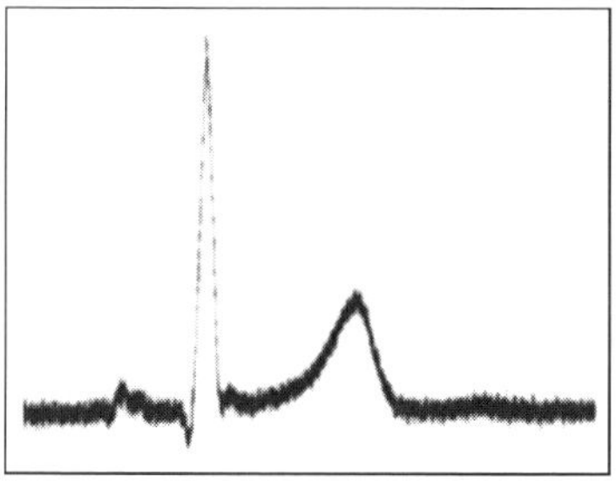

Figure 3. – One cycle of the ECG series of Figure 2. Noising can still be observed.

To clean the noising signal, we use:

1. Conventional methods.
2. The algorithm prepared as moving average filter.

Algorithm for average filter

We prepared an average filter, which is analogous to those referred to in scientific literature.

We talk about an algorithm which function is smoothing the signal depending on the variations of a segment of the ECG. The length of the segment has a number of dots selected empirically in each sampling.

Preparation of the Algorithm

The continuous mathematical function **(V_{ECG}t)** it´s sampled at 20 kHz with a resolution of 16 bits.

We obtain roughly 20 000 samples of cardiac cycles, stored in different memories from 0,...to 19999. The following cycle (if each would last for one second) will/would have stored .from 20 000 to 39999.

We assume the amplitude of point **p** as the sum of his own amplitude as well as the amplitude of all the former ones.

$$\int_{i=0}^{p} V_{ECG}\, i$$

1.- Then we trace a line (see Fig. 4) which roughly represents the "integral" of a cardiac cycle.

2.- Moved that graph **n** points.

3.- Subtract from the "original integral" graph, the other one.

4.- We obtain a series of dots (another graph). In this new graph, the amplitude of dot number **m** is the sum as well of it's own amplitude as of the previous **(n-1)** dots.

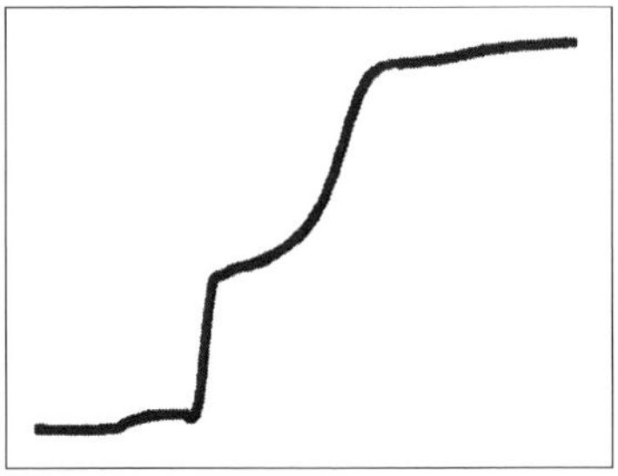

Figure 4.- "Integral" of ECG and noise, of Fig.3.

The amplitude of dot **m** is:

[m + (m - 1) + (m - 2) + ...+ (m - n + 1)]

The amplitude of dot **(m+1)** is:

[(m + 1) + m + (m -1) + (m - n + 2)]

Dividing the amplitude of this range by **n**, a series of dots appear with the following amplitudes:

$$\int_{i=0}^{p} V_{ECG}\, i$$

Figuratively, and taking into account the great number of dots of the series, we can define the amplitudes as

$$Vn_i = \frac{\int_{i=0}^{m} V_{ECG_i}.dt - \left[\int_{i=0}^{m} V_{ECG_i}.dt - \int_{i=m-n}^{m} V_{ECG_i}.dt\right]}{n}$$

Dot **m** of this series has a value which is the average of its own value and of the preceding (**n-1**). In this way, variations are averaged (and also the fluctuation of noising when aleatory).

The first **n** dots do not meet this condition and are eliminated. This operation can be applied to the whole series of ECGs or only to one segment. To average adjacent dots, **n** has to be **n<<m**, otherwise they would interfere in the real

fluctuation of the signal. An averaging of the values is thus obtained for the amplitudes (except for the first **n** dots).

If the process is applied to different dots, less distortion and more cleaning-up will be obtained.

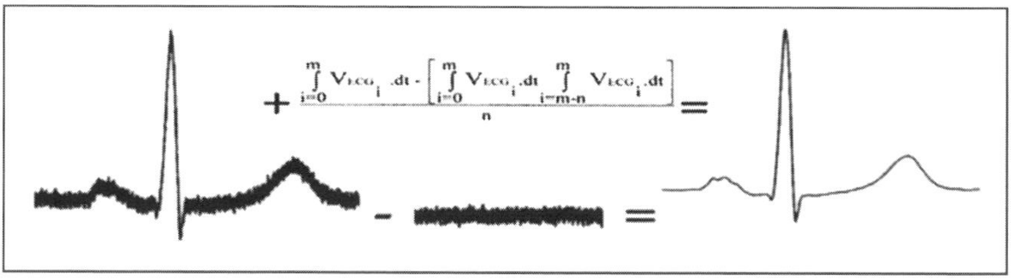

Figure 5.-Clearing noise of ECG.

SENOIDAL WAVES AND THE CHOOSING OF FILTERS

We try to apply filters to a linear sinusoidal wave of a growing frequency between 0 and 10 kHz. We try to apply filters to a linear sinusoidal wave of a growing frequency between 0 and 10 kHz.

There may be a sinusoidal wave of 0 to 2 kHz. We assume that there are no higher frequencies in an ECG. (Elements of a frequency >200 virtually do not appear). Nevertheless, in some cases frequencies of until 500 Hz can be analyzed. Higher frequencies are noising. Which is the distortion produced by the filters (having ruled out the disappearence of the first **n** dots of the series of ECGs)

1.- We generate a sinusoidal of

- a constant amplitude
- a linear variable frequency between 0 Hz and more than 5 kHz.

This sinusoid (of no linear function) includes all the possible elements of the ECG.

A sampling of a constant 20 kHz frequency is made (the same frequency of the ECG sampling).

2.- We move this function **n** points that we think is appropriate.

3.- From the ordinate of the original sinusoid we subtract the corresponding ordinates of the displaced (for the corresponding abscissas).The result will be a graph.

4.- In this graph we discern –both visually and by the numeric values of the ordinates– if there is any variation of amplitudes and for which frequency. In this way we are able to know scientifically/empirically which value of **n** should be chosen.

5.- The most appropriate way of proceeding is to select a small value of **n** and to repeat several times the procedure.

This is the proceeding to test the filters.

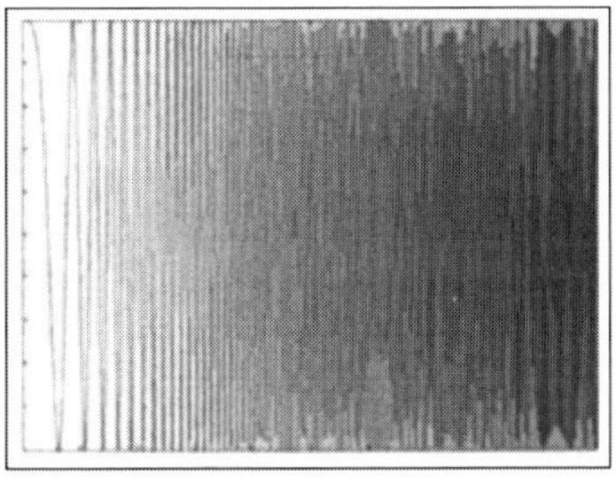

Figure 6 .- Sinusoid of increasing frequency from 0 Hz to 500 Hz.

Possible distortions

This method does not change the frequencies. That get involved in the amplitude, in relation with: sample frequency, **n** number of points for the basal segment and the interactions of the process. It looks very useful for frequencies below 200 Hz.

Also useful to show signals in time (R point for temporal analysis). It is a fast procedure. We consider that it works in real time for high number of interactions.

Figure 7.- Sinusoid of increasing frequency from 0 to 500 Hz. Filter n.4.

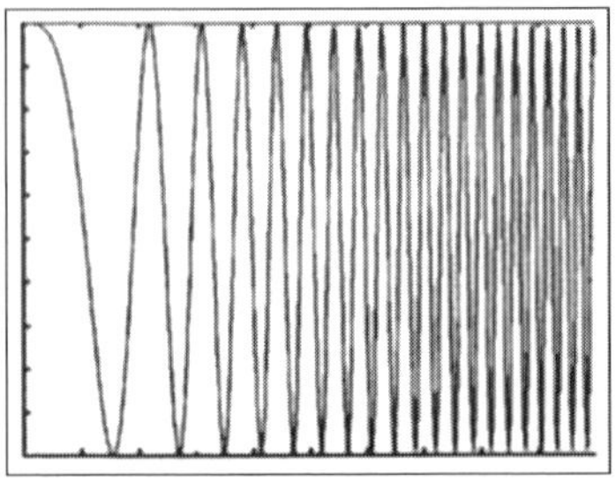

Figure 8.- Amplitude variation on the first 200 cycles; n=4 with 15 interactions.

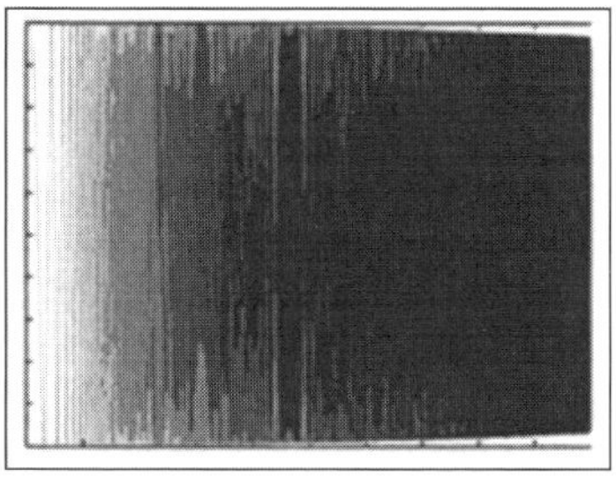

Figure 9.- Sinusoid 0,11kHz.;Amplitude variation n=4 with a interaction.

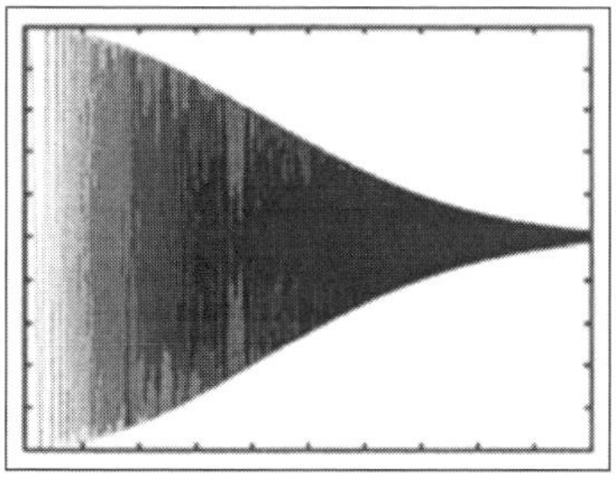

Figure10.- Sinusoid 0 – 1 kHz; 5 interactions; n=3.

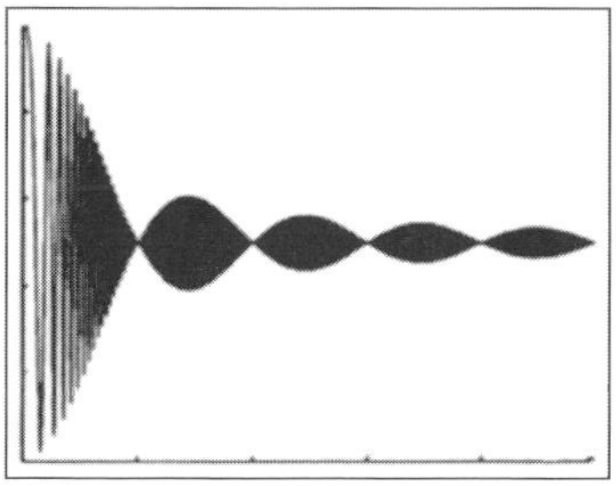

Figure 11 Sinusoid of 0 – 1 kHz. Filtrated for n = 10.

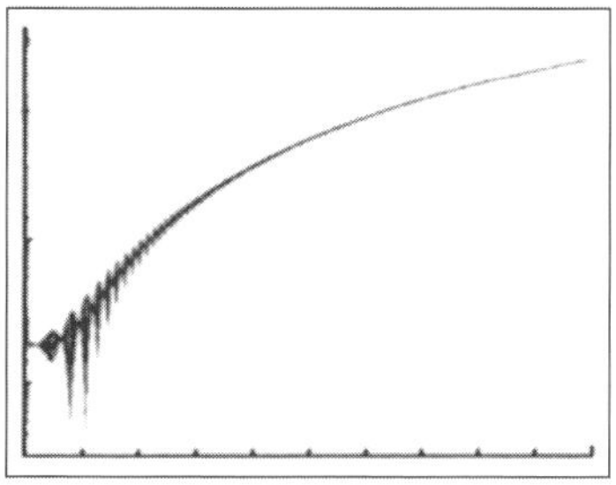

Figure 12.- FFT of sinusoid 0 – 2 kHz

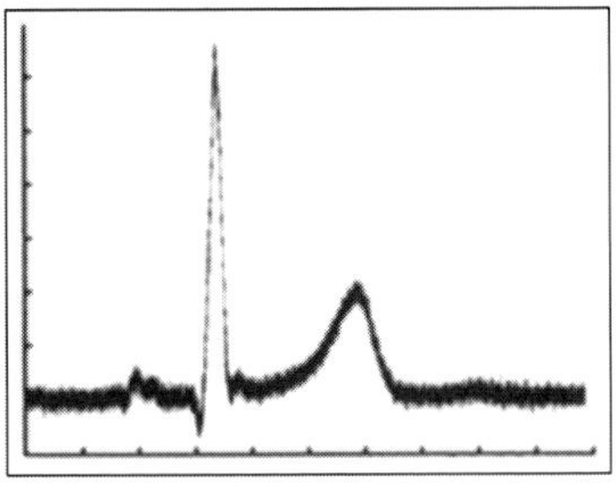

Figure13.- ECG un-filtrated.

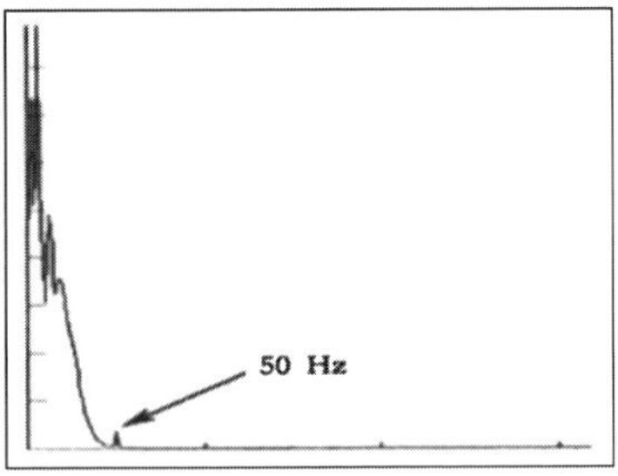

Figure 14.- Fourier elements of a ECG un-filtrated.

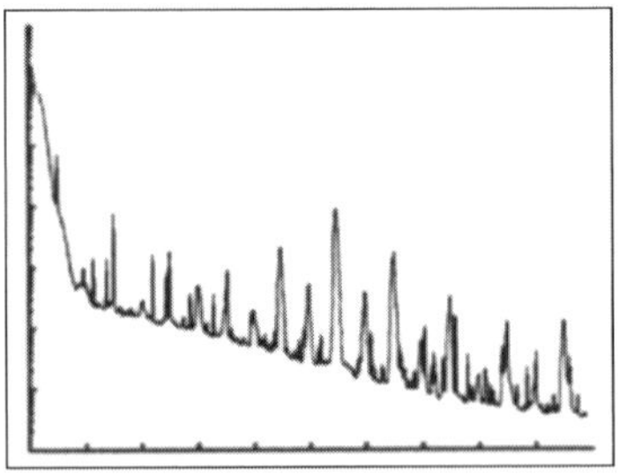

Figure 15.- Semi-logarithm graph FFT of ECG unfiltrated.

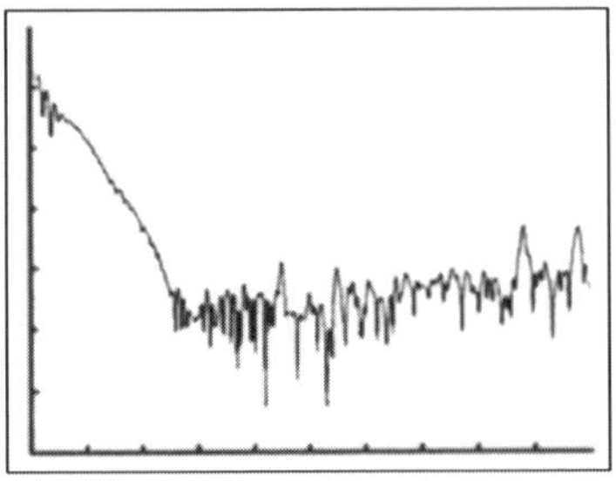

Figure 16.- FFT. Semilogarith graph ECG unfiltrated.

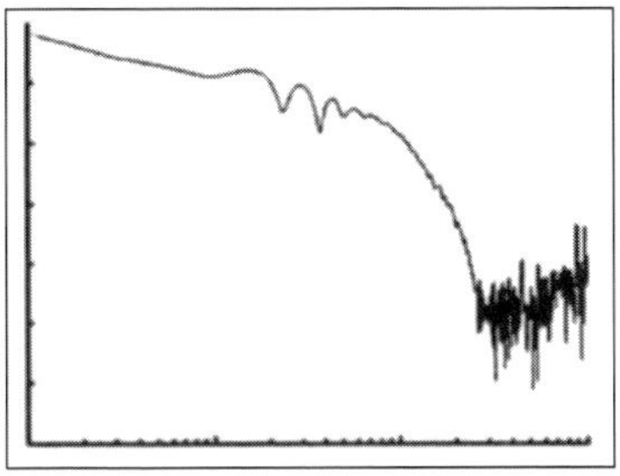

Figure 17- FFT. Logarithm figure. ECG unfiltrated.

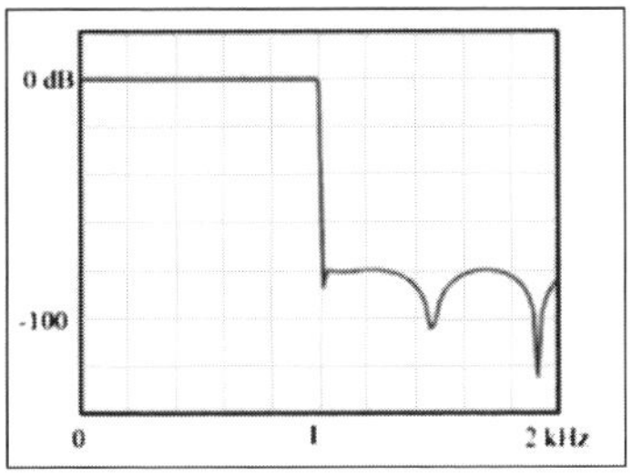

Figure 18.- FIR filter. Result of a constant amplitude of 0 to 1 kHz.

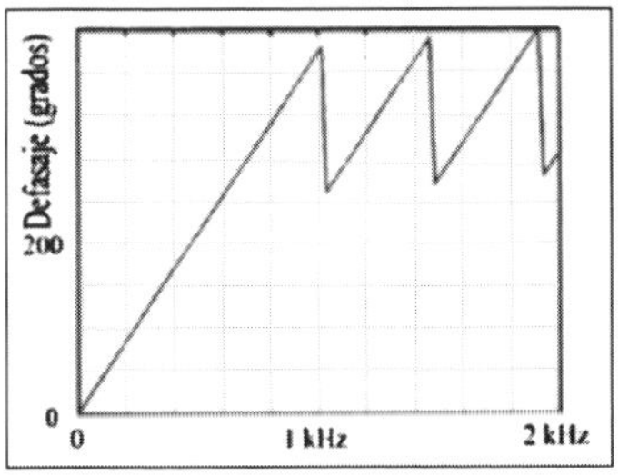

Figure 19.- Phase variation apply to a FIR filter.

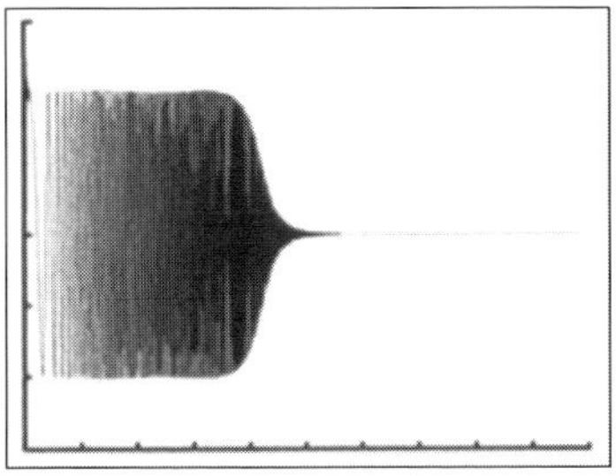

Figure 20.- Butterworth filter. Amplitude variation of 0 – 1 kHz. 18 interactions.

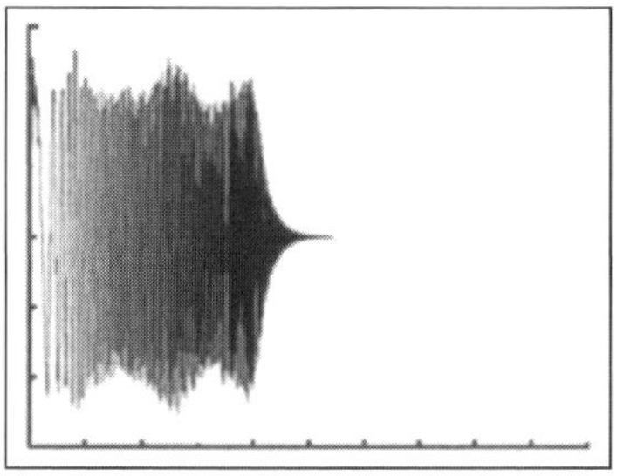

Figure 21.- Butterworth filter. Amplitude variation of 0 – 1 kHz. 20 interactions. Instability.

FIR and Butterworth filters

If we apply the Butterworth filter, we can achieve a very good result in a short period of time. It keeps amplitudes, but there is a risk of non lineal des-phase. It places in a non- stable phase for high level.

FIR is a ripple system, which you can solve if we increase the number of interactions. Spends time in the process and the phase difference is lineal.

CONCLUSIONS

The averaged filter that we present is good if you want to work with fast measurements, with fix time interval, although affects to the amplitude deformation.

If we need to have amplitudes and time, we recommend FIR, even if takes more time.

REFERENCES

[1] www. moving average
[2] www. ecg moving average
[3] www. moving average algorithm
[4] National Institute of Standards and Technology: "Engineering Statistics Handbook" www.itl.nist.gov/div898/handbook/pmc/section4 /pmc4.htm
[5] University of Newcastle upon Tyi.
Chemical and process Engineering. "Dealing with Measurement Noise" Moving average filter. Exponentially Weighted Moving Average Filter
Lorien.ncl.ac.uk/ming/filter/filave.htm
[6] Data Instruments.
"A Closer Look At Advanced CODAS Moving Average Algorithm"
data.com

Applied Physics (APHYS 2003)
A. Méndez-Vilas (Ed.)

Experimental and modelling study on the uptake kinetics of radionuclides by SPM. Discussion on box-models applications.

H. Barros* & J.M. Abril
Dpto. Física Aplicada I, Universidad de Sevilla. EUITA, Ctra. Utrera km 1, C.P. 41014, Sevilla, Spain.

Abstract: The dispersion and fate of radionuclides and other hazardous materials in natural aquatic environments strongly depends on their uptake by suspended particulate matter (SPM). This work presents a mathematical discussion on the box models commonly used to describe the kinetic of that process. We deal with the differences and similarities between the model that describes the uptake in terms of parallel reversible reactions and the model based on consecutive reversible reactions. It will be demonstrated that, under the usual assumptions, both models are mathematically equivalents. Here we combine the experimental and the modelling approaches to illustrate the applications and limitations of such models. The experiments were traced with ^{133}Ba because it is a good indicator of the ^{226}Ra behaviour and can be easily measured by γ-spectrometry. We simulate in the laboratory different situations that actually occurs in the interface water-sediments in environments under the influence of tides and currents: multiple releases in presence of suspended sediments, desorption from the solids after an uptake and decantation processes. Also it was studied the effect of the specific surface area of the particles. Sediments and waters were sampled in the Odiel and Tinto estuaries (Huelva, Southwest of Spain), where two phosphate-fertilizer industries have been enhancing the ^{226}Ra concentration. The uptake is characterized by three different stages, each one with a characteristic time (minutes, hours and days). Another remarkable result is that with both models (parallel and consecutive) a successful prediction is obtained for the different observed processes by using the transfer coefficients obtained by a single release uptake experiment. This shows the applicability of this kind of models for its incorporation into more general models devoted to study the dispersion of radionuclides in aquatic environments, but imposes the need for testing the models against variable conditions prior to these applications.

INTRODUCTION.

Increasing experimental effort has been paid to study the sorption of radionuclides and heavy metals by SPM in aquatic environments. It is essential to better understand the environmental behaviour of such inorganic pollutants in aquatic systems. It becomes also necessary to develop suitable predictive models capable to evaluate the dispersion, sinks and residence times of these hazardous material in the dissolved phase (related with the biodisponibility) since with those information would be possible to asses their impact on the human health and the environment.

A detailed description of the sorption kinetics becomes important to deal with different situations: dispersion in systems with high SPM concentrations or during its episodic increment (storms, heavy rain, remedial actions, etc.), when the input of pollutants affects large areas (e.g., atmospheric deposition), or when the contaminated sediments or particles become a secondary source of pollutants (desorption scenarios).

The linear box model approach has been widely used to successfully describe the uptake kinetics. Most of them are versions of two basic approaches: one containing several parallel reactions (up to three, since rarely the experimental data can justify more complex models) while the other involves consecutive reactions. All the reactions are reversible (irreversibility is included as a particular case) with concentration independent coefficients. The often used approach found in the literature consists in tracing experiments to record the time course of the dissolved phase concentrations. A fitting procedure is generally used to determine both, type of model and the numerical values of the kinetic coefficients. This method also can support consistent discussions on the effect of different parameters (salinity, pH, specific surface area, etc). The scientific literature presents a sparse set of models describing a wide variety of behaviours, as it is summarised below.

Nyffeler *et al.* [1] described the sorption kinetics in sea waters by means of a single reversible reaction model, and for some radioisotopes included a consecutive irreversible channel. Yuan-Hui Li *et al.* [2] used a single reversible reaction to describe the oxidation of metals in seawater as a path way for metal's uptake by solids. Benes & Polliak [3] and Benes *et al.* [4] presented evidences of a two or three-steps sorption kinetic for ^{58}Co, ^{85}Sr and ^{137}Cs and described these behaviour in terms of models with two reactions. Comans *et al.* [5] presented an interesting study about caesium sorption on illite, and used models incorporating a Freundlich isotherm and a 2 or 3-box model plus a consecutive irreversible process. A discussion on the effect of salinity and pH in the sorption of ^{133}Ba was given by Laissaoui *et al.* [6], but due to the low SPM concentration of the studied environment a single reversible reaction was enough to describe the data.

An extensive experimental work presented by Bunker *et al.* [7], includes Co, Sr, Ru and Cs. In this study (made with suspensions of 10 g.L^{-1} of lake's sediments) a linearization approach is compared with a consecutive 2, 3 or 4-box model, and a statistical discussion about the level of improvement between models through the *F*-test is given. Ciffroy *et al.* [8] presented a consecutive 3-box model which satisfactorily described the adsorption/desorption of six metallic radioisotopes by natural river water (with 20 mg.L^{-1} of SPM). Box models can also serve to identify different pathways or channels for the sorption onto the particles. Thus, Comans *et al.* [9] discussed the effect of competitive ions and pointed that the reversibility was affected by the slow kinetic. In an interesting experimental work, Børretzen and Salbu [10] estimated the apparent rate coefficients for parallel and consecutive 3-box models in terms of a post-sorption sequential extraction scheme. And latter the same authors [11] presented an interesting discussion about the application of semi-Markov and Markov stochastic process model coupled with 3-step parallel and consecutive box models in order to explain the residence times of Cs in a seawater-sediment system. El-Mrabet *et al.* [12] presented a 4-box model to explain the uptake kinetics of Pu in natural aqueous suspensions with some 10 mg.L^{-1}.

The present work demonstrates the mathematical equivalence between parallel and consecutive box models under the usual assumptions. This is, both models are producing the same analytical solution for the uptake curve (time course of the concentrations in the dissolved phase). Furthermore, a similar behaviour between the corresponding solid compartments was found.

Under the effect of water movements, the water-sediment interface can behave as a fluid mud layer featured by a high SPM concentration, it is made basically with material from the surface sediments when they are removed by the currents (Eisma [13]). In particular the estuary under study is strongly affected by tides and cycles of resuspension and decantation alternates continuously. This kind of process can radically alter the fate and the description of the dispersion/sorption of the dissolved pollutants.

As it was referred above, different radionuclides shows a complex 2 or 3-stages kinetics in natural suspensions with low SPM concentrations, but in the same conditions other group of tracers (as ^{133}Ba) exhibit a more conservative behaviour. Our results for that kind of suspensions match exactly with those described by Laissaoui *et al.* [6] for the same environment, and therefore it is not discussed here. However, in the presence of high SPM concentrations (> 5 $g.L^{-1}$) the 3-stage kinetics once again is the observed phenomenology.

Thus, the mentioned mathematical equivalence between models will be illustrated using the results of a series of tracing experiments performed with ^{133}Ba and suspended sediments, and covering a wide range of situations. This result unifies the description of the observed behaviour, but it points out the question of the physical meaning of the involved coefficients.

MATERIALS AND METHODS

The samples

The samples were collected in the Huelva estuary (south of Spain) where the intense industrial activity have been increasing the ^{226}Ra concentration during the last decades (Martínez-Aguirre *et al.* [14]). Water (pH≈7 and electrical conductivity 60 $mS.cm^{-1}$) was stored in darkness and was not chemically treated. A portion of the sediment were oven dried and manually disaggregated to separated it in different fractions by dry sieving, the densities range from 2.4 to 2.6 $g.cm^{-1}$. The ^{133}Ba was selected as a tracer because it is relatively easy to measure this artificial radionuclide by gamma spectrometry and it is a good chemical analogous of the ^{226}Ra.

Experimental

It were simulated, in the laboratory, several situations that actually occurs in the water-sediments interface in environments under the influence of tides/currents or planned/accidental wastes releases. Such situations are: single or multiple releases in presence of suspended sediments, decantation processes and desorption of pollutants from the solids after sorption process and the renewal of waters (simulating the current).

The first kind of experiment consists of a single release, in each case 6 or 20 g of sediments were mixed with 1 L of clean (unfiltered) estuarine water obtained by decantation. The suspension was maintained by a mechanical stirrer and it was wait 24 hours for chemical equilibrium. Afterwards the solution was traced with a known amount of ^{133}Ba, which was added like $BaCl_2$, then it was extracted 15 mL aliquots of solution at the different observation times and the solid fraction was separated form the water by centrifugation at 4000 r.p.m. The liquid fraction was directly measured by gamma spectrometry (HPGe and REGe detectors were used). The multiple tracing experiments were performed in the same manner, an additional amount of tracer was added to the single tracing experiment after few days of contact time, once a pseudo-equilibrium was reached and when the fast and intermediate processes were not dominant.

Another type of experiment was a decantation and it's consequent desorption process, it was a continuation of the experiments previously traced. After sorption becomes slow in a release experiment the stirring mechanism was stopped and several aliquots were sampled after the deposition process. Then the major part of the overlaying water replaced with clean water (from the same origin) while at the same time the stirring system was restarted, simulating the water masses displacement in a natural environment.

Additionally there were performed measurements of the grain size distribution of the different sediment fractions by laser diffraction using a MasterSizer device.

RESULTS AND DISCUSSION

Analytical solution of the Parallel Reversible Reaction (*PRR*) Model

Since the models considered here are linear, to deal with an uptake process with three characteristic times it is necessary to use four compartments (Fig. 1a). The *PRR* model assumes that all the reactions occurs parallels, being independent one from each other. The governing equations for the tracer concentration in each compartment can be written as:

$$\begin{aligned}
\dot{a}_w &= -k_{11} \cdot a_w + k_{21} \cdot a_{s1} - k_{12} \cdot a_w + k_{22} \cdot a_{s2} - k_{13} \cdot a_w + k_{23} \cdot a_{s3} \\
\dot{a}_{s1} &= +k_{11} \cdot a_w - k_{21} \cdot a_{s1} \\
\dot{a}_{s2} &= +k_{12} \cdot a_w - k_{22} \cdot a_{s2} \\
\dot{a}_{s3} &= +k_{13} \cdot a_{s2} - k_{23} \cdot a_{s3}
\end{aligned} \tag{1}$$

where a represent the different concentrations: in the liquid phase (a_w), in the more accessible part of the particle (a_{s1}, which corresponds to the faster reaction), in a fraction not so accessible (a_{s2}, pores, free edges, etc.) and where the process is very slow and/or irreversible (a_{s3}). The so called constant rates or kinetic coefficients k_{in} are constant as the physic-chemical conditions of the experiment.

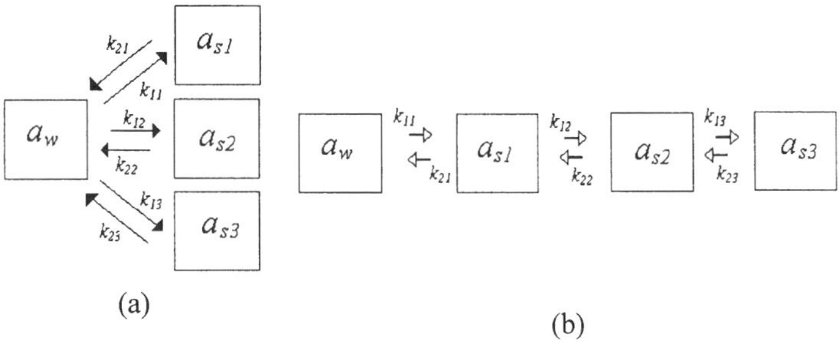

Figure 1. Schematic (a) Parallel and (b) Consecutive Reversible Reaction Models. a_w represents the dissolved phase concentration, while a_{si} the concentration associated to the different sites on the particles. This general approach with four compartments (3-steps kinetics) contents the cases with three or two compartments

The concentration is measured in mol.L^{-1} and mol.g^{-1}; or in Bq.L^{-1} and Bq.g^{-1} in the case of radionuclides. Thus, some authors use the amount of suspended loads m_s (g.L^{-1}) as a multiplicative factor to obtain a bulk concentration for each material compartment. In this work we used bulk concentrations, and the coefficients contains all the corresponding information instead to be the product of two or more constants, then each k_{in} has dimensions of T^{-1}.

Eq. 1. can be written in a matrix form, then the characteristic times are related with the eigenvalues s_i of the matrix in Esq. 2 by $t_i = -s_i^{-1}$. The trivial eigenvalue $s_0 = 0$ just indicate the mass conservation.

$$K = \begin{pmatrix} -k_{11} - k_{12} - k_{13} & k_{21} & k_{22} & k_{23} \\ k_{11} & -k_{21} & 0 & 0 \\ k_{12} & 0 & -k_{22} & 0 \\ k_{13} & 0 & 0 & -k_{23} \end{pmatrix} \qquad (2)$$

Then the solution of Esq. 1. corresponds to a multiple exponential function with those characteristic times:

$$a_w(t) = a_w(0)\left[A_1\left(e^{s_1 \cdot t} - 1\right) + A_2\left(e^{s_2 \cdot t} - 1\right) + A_3\left(e^{s_3 \cdot t} - 1\right) + 1\right] \qquad (3)$$

In the last equations it has been assumed that all the initial concentrations associated to particle's compartments are zero. This simplification is justified since we are focusing on releases of anthropogenic pollutants, but the result can be easily extended for a more general situation.

Then, introducing the initial conditions and solving the system, k_{in} coefficients can be obtained as the solutions of the following equations:

$$\sum_{ij} k_{ij} = -\sum_{l} s_l$$
$$k_{21}k_{22} + k_{21}k_{23} + k_{22}k_{23} + k_{11}(k_{22} + k_{23}) + k_{12}(k_{21} + k_{23}) + k_{13}(k_{21} + k_{22}) = s_1 s_2 + s_1 s_3 + s_2 s_3$$
$$k_{21}k_{22}k_{23} + k_{11}k_{22}k_{23} + k_{12}k_{21}k_{23} + k_{13}k_{21}k_{22} = -s_1 s_2 s_3$$
$$k_{21} + k_{22} + k_{23} = s_1(A_1 - 1) + s_2(A_2 - 1) + s_3(A_3 - 1)$$
$$k_{21}k_{22} + k_{21}k_{23} + k_{22}k_{23} = s_1 s_2 + s_2 s_3 + s_1 s_3 - A_1 s_1(s_2 + s_3) - A_2 s_2(s_1 + s_3) - A_3 s_3(s_1 + s_2)$$
$$k_{21}k_{22}k_{23} = s_1 s_2 s_3 (A_1 + A_2 + A_3 - 1)$$
(4)

Then, once obtained the uptake curve form an adsorption experiment, it is possible to find out the corresponding parameters for the kinetic model in terms of those directly obtained from the fitting procedure (s_i and A_i). On the other hand, it have to be ensured that there are enough experimental points in each stage (minutes, hours and days) to obtain a good description of the curve in the whole temporal range.

Analytical solution of the Consecutive Reversible Reaction (*CRR*) Model

In this model the reactions are consecutives, i.e. there is a fast reaction which retains the dissolved ions near to the particle; then another different mechanism acts causing the fixation of these "weekly adsorbed" ions into the particle and afterwards another one reaction traps the ion even more strongly. In such situation the available amount of ions in each step depends on the previous process, and the scheme should be chain like (see Fig 1b). The kinetics produce a constrain between the different apparent rates: the more exterior channels are faster than the interior ones.

As before, the same procedure and interpretation may be done. The only important difference is the structure of the process and its implications. The differential equations are described in equation 5 and the associated matrix in Eq. 6:

$$\begin{aligned}
\dot{a}_w &= -k_{11} \cdot a_w + k_{21} \cdot a_{s1} \\
\dot{a}_{s1} &= +k_{11} \cdot a_w - k_{21} \cdot a_{s1} - k_{12} \cdot a_{s1} + k_{22} \cdot a_{s2} \\
\dot{a}_{s2} &= +k_{12} \cdot a_{s1} - k_{13} \cdot a_{s2} - k_{22} \cdot a_{s2} + k_{23} \cdot a_{s3} \\
\dot{a}_{s3} &= +k_{13} \cdot a_{s2} - k_{23} \cdot a_{s3}
\end{aligned}$$
(5)

$$K = \begin{pmatrix} -k_{11} & k_{21} & 0 & 0 \\ k_{11} & -k_{21} - k_{12} & k_{22} & 0 \\ 0 & k_{12} & -k_{22} - k_{13} & k_{23} \\ 0 & 0 & k_{13} & -k_{23} \end{pmatrix} \qquad (6)$$

Once again, the eigenvalues indicate the characteristic times; and it is possible to obtain the solution in terms of a multiple exponential curve as in Eq. 3. As before, the kinetic coefficients can be found in terms of amplitudes and frequencies solving the next equations.

$$\sum_{ij} k_{ij} = -\sum_{l} s_l$$
$$k_{11}(k_{12}+k_{13}+k_{22}+k_{23})+k_{22}(k_{23}-k_{12})+(k_{12}+k_{21})(k_{13}+k_{22}+k_{23}) = s_1s_2+s_1s_3+s_2s_3$$
$$k_{11}k_{22}k_{23}+k_{11}k_{12}(k_{13}+k_{23})+k_{21}k_{22}k_{23} = -s_1s_2s_3$$
$$k_{12}+k_{13}+k_{21}+k_{22}+k_{23} = s_1(A_1-1)+s_2(A_2-1)+s_3(A_3-1)$$
$$k_{22}k_{23}-k_{12}k_{22}+(k_{12}+k_{21})(k_{13}+k_{22}+k_{23}) = \sum_{i\neq j} s_is_j - \sum_{l} A_ls_l(s_{l+1}+s_{l+2})$$
$$k_{21}k_{22}k_{23} = s_1s_2s_3(A_1+A_2+A_3-1) \tag{7}$$

Where *i, j* and l = *1, 2* and *3*, and the sum over index "l" is a sum module *3* (i.e, *3+1= 1*, etc.).

Mathematical equivalence between parallel and consecutive kinetic box models

Both, in laboratory experiments and during in situ measurements, it is really complex to measure separately the concentrations related to the different sites on the particle. Thus, often in the scientific literature the fitting parameters are obtained through the water concentration curves, i.e. following the behaviour of a_w. (see references above). But since equation 3 is the solution in both cases, and observing that the righter side of the equations 4 and 7 are identical, it is possible to generate the same analytical function $a_w(t)$ using two different sets of kinetics coefficients regarded to *PRR* and *CRR* models respectively. Then the two models are mathematically equivalents in that approach. In other words, once obtained an uptake curve for water concentration, it is not possible to determine which mechanism is the cause of the phenomenology, and it makes an important question about the physical meaning of the kinetic coefficients.

Moreover, the functions describing the particle-associated compartments are not exactly the same, but their behaviours are similar.

Experimental results and applications of the models

Single tracing experiments with different SSA.

Experiments 1 and 2 were performed with the same amount of suspended particles, 20 $g.L^{-1}$ of sediments from the Odiel river. In the Experiment 1 we have discarded particles larger than 2 mm and in the case of Experiment 2 we have used the fraction with grain size between 63 and 400 μm. The higher specific surface area (SSA = 0.66 $m^2.L^{-1}$) of Experiment 1 corresponds to the stronger uptake, while the slower uptake is due to a lower SSA (0.014 $m^2.L^{-1}$). In both cases ^{133}Ba sorption has been successfully fitted to a third order exponential decay, corresponding either to *CRR* or *PRR* models, Fig. 2. The obtained coefficients and the parameters for all the experiments are given in Table I.

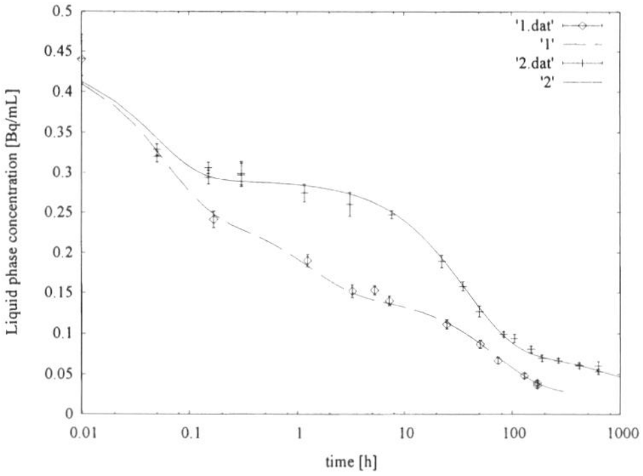

Figure 2. Uptake curves of ^{133}Ba by suspended estuarine sediments. Experiment 1 corresponds to SSA = 0.66 $m^2.L^{-1}$ and Experiment 2 to SSA = 0.014 $m^2.L^{-1}$. In each case a third-order exponential decay was fitted and the corresponding *PRR* and *CRR* coefficients are summarized in Table I.

Table I. Kinetic and distribution coefficients calculated with *PRR* and *CRR* models. Experiments with estuarine sediments in suspension.

Experiment	N° 1		N° 2		N° 3	
Model	*PRR*	*CRR*	*PRR*	*CRR*	*PRR*	*CRR*
k_{11} [h^{-1}]	6.3700	6.9700	6.4500	6.4757	0.4000	0.4100
k_{21} [h^{-1}]	9.0700	8.3271	12.4000	12.3508	2.0000	1.9514
k_{12} [h^{-1}]	0.5674	0.7038	0.0257	0.0492	0.0097	0.0484
k_{22} [h^{-1}]	0.4651	0.4764	0.0081	0.0081	0.0072	0.0072
k_{13} [h^{-1}]	0.0326	0.0277	0.0043	0.0043	0.0003	0.0002
k_{23} [h^{-1}]	0.0024	0.0025	0.0005	0.0005	0.0000	0.0000
k_d [$mL.g^{-1}$]	540		360		320	
χ	1.02		0.99		0.80	
m_s [$g.L^{-1}$]	20		20		6	
Grain size [μm]	$\phi < 2000$		$63 < \phi < 400$		$63 < \phi < 400$	
SSA [$m^2.L^{-1}$]	0.66		0.014		0.013	

Note: - χ corresponds to the exponential fitting to the water concentration
- k_d refers the distribution at the last experimental point available
- SSA was calculated from grain size distribution, assuming spherical shape

For each case separately, it can be seen that the two models are equivalents for the compartment a_w, while the corresponding material compartments (a_{si}) are functions with practically the same behaviour. Figure 3 shows the three particle associated compartments of Experiment 1, as functions of the contact time for *CRR* and *PRR* models. In Experiment 1 we found the highest differences between the two models, in the other experiments that differences are smaller and in some cases practically can not be appreciated.

Experiments with a second tracing and a desorption process.

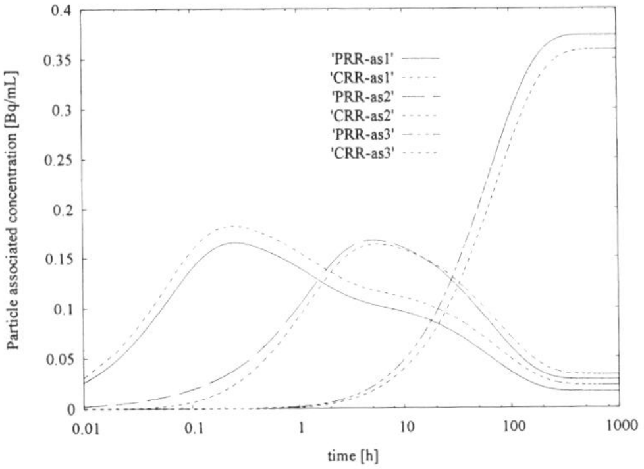

Figure 3. Particle associated concentration (Bq per mL of solution) of Experiment 1. The compartment a_{s1} (electrostatic sorption) grow rapidly during the first 10 minutes, afterwards the second reaction channel (a_{s2}) becomes more important and the ions are transferred there. Finally it can be seen the character weakly reversible of a_{s3} which accumulated the major part of the tracer in the last days of the experiment.

In Experiment 2 the second spike was of the same order of magnitude than the first one, after that a desorption process was performed. In the figure 4.a it is shown the curve fitted to the first uptake data and the prediction that *PRR* and *CRR* models makes for the second tracing experiment and for the desorption. Note that there is only one curve for each set of data, since both models gives exactly the same solution. In this kind of graph the time was reset for each stage of the experiment for an easier comparison, then all the curves starts in t = 0.

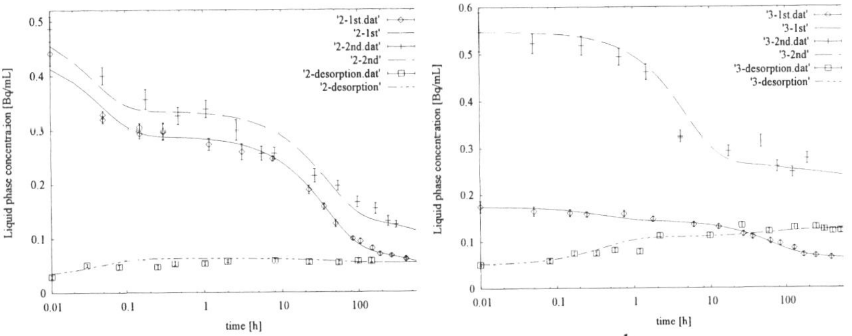

a b

Figure 4. Dissolved concentrations in a) the Experiment 2. and b) the Experiment 3. The first discharge was fitted by using a third-order exponential decay. The coefficients thus calculated form *PRR* and *CRR* models gives the same prediction for the second discharge and for the desorption process, and that solutions agrees reasonably the data of the corresponding parts of the experiment (which were not used in the fitting procedure).

The Experiment 3 was performed with 6 grams of sediments from the Tinto river in one litre of water, it was used the fraction from 63 to 400 μm (detail in Table I). The second spike was near to three times higher than the first one. That's why the desorption (after the second spike) released to the water an amount of ^{133}Ba close to the initial condition of the first spike. In figure 4.b it can be seen the fitting for the first uptake data, and the predictions (prescribed by the two models) for the second tracing and for the desorption. As before, both models gives the same result for a_w in the whole experiment.

Discussion on some ideas to distinguish between *PRR* and *CRR* models.

The use of different SPM (or SSA) concentrations.

In the case of *PRR* it is clear, from the definition, that the direct coefficients $\boldsymbol{k_{11}}$ and $\boldsymbol{k_{12}}$ should be proportional to the available particle's surface and consequently to the load concentration. A similar behaviour should be expected for the first direct coefficient $\boldsymbol{k_{11}}$ in the *CRR* model. Then if we design a sets of experiments (varying SPM or SSA) to search for the suitability of one model over the other one (*CRR* and *PRR*), no conclusion can be obtained.

This is easily explained by analysing the mathematical relationship between the two set of coefficients. From Eqs. 4 and 7 it can be obtained an explicit relationship between both set of coefficients. It is presented the case of a 2-step model to simplify the explanation and avoid unnecessary large mathematical expressions:

$$
\begin{aligned}
k_{11}^{*} &= k_{11} + k_{12} \\
k_{21}^{*} &= \frac{k_{11}k_{21} + k_{12}k_{22}}{k_{11} + k_{12}} \\
k_{22}^{*} &= k_{21}k_{22}\frac{k_{11} + k_{12}}{k_{11}k_{21} + k_{12}k_{22}} \\
k_{12}^{*} &= k_{21} + k_{22} - \frac{k_{11}k_{21} + k_{12}k_{22}}{k_{11} + k_{12}} - k_{21}k_{22}\frac{k_{11} + k_{12}}{k_{11}k_{21} + k_{12}k_{22}}
\end{aligned}
\tag{8}
$$

Where the k_{ij}^{*} are related to the consecutive reactions are expressed in terms of the parallel reactions coefficients. Then it is clear that any increase of k_{11} and k_{12} in a given factor corresponds to the increasing of k_{11}^{*} by the same factor, preserving the others coefficients values and making impossible trough this way the observation of any difference in the description of the water concentration.

The use of chemical treatments to separate the particle associated concentrations

The attempt to obtain information, about the radionuclide's distribution in the inner part of the particles, can be assessed by chemical treatments. However this approach would have inherent problems. Firstly the association between a compartment and the amount of radionuclide extracted by an specific reagent have to be done operationally, and no control on the number of processes involved can be taken since the complex geochemistry of the particles. In second term, the molarity and the contact time with the reagent will also change the amount of tracer extracted from the particles, adding more uncertain to the final value for the associated concentration. And finally, we show that the *PRR* and the *CRR* models predetermine the compartmental concentrations if the curve for dissolved radionuclides is well defined with enough experimental data points, this imposes constrains on the behaviour of particle associated concentrations.

Moreover, in many cases (see for instance Fig. 3) the description of the particle associated compartments by the two models are very similar. Then also under the hypothetical case of finding accurate experimental values for them, it would be a challenge to distinguish between one model and another with a resolution higher than the error bars of the measurements.

CONCLUSIONS.

The uptake of radionuclides by suspended sediments is featured by a complex sorption that actually occurs in different steps, in general three. These process can be described in terms of its characteristics times by multi-exponential functions representing the liquid phase and the particle associated concentrations. That description can also be expressed by means of multi-compartmental *PRR* or *CRR* model, where the boxes represents the amount of radionuclides in the water and in the different sites on the particles.

In the study of the sorption kinetic, the existence of very fast channels implies the necessity of describe the process from the very beginning, recording the process with a good time resolution.

The *PRR* and *CRR* models are equivalents for representing the time evolution of water concentration. Thus, changes in the load concentration or in the SSA, multiple releases and desorptions are equivalently explained with both models, being this description equally good independently of the magnitude of the releases and the intensity of the desorption. It reopen the discussion about the physical meaning of the kinetic coefficients.

Generally, experimental conditions are controlled and fixed in order to find out which one is the main behaviour, and perhaps the main mechanism, of a given process. However natural systems are normally featured by changing conditions, and this aspect have to be incorporated in the models in some way. In other words a such model capable to describe an experimental situation have to be fully tested for its suitability confronting changing conditions before being employed in natural systems.

Finally, the main conclusions of the present work indicates that more care have to be undertaken in the choice of a given model type, it is necessary to include more information in the studies in order to insure the mechanistic hypothesis performed.

ACKNOWLEDGEMENTS:

This work has been partially supported by a Research Contract with the Spanish ENRESA (0770105).

REFERENCES

[1] Nyffler U. P.; Li Y. H. & Santschi P. H. A kinetic approach to describe trace-element distribution between particles and solution in natural aquatic systems. *Geochim. et Cosmochim. Acta.* **48,** 1513 (1984).

[2] Yuan-Hui L., Burkhardt L., Buchholtz M., O'Hara P. and Santschi P. H. Partition of radiotracers between suspended particles and seawater. *Geochim et Cosmochim Acta.* **48,** 2011 (1984).

[3] Benes P. & R. Poliak. Factors affecting interaction of radiostrontium with river sediments. J. of Radioanal. And Nucl. Chem. **141,** (1) 75 (1990).

[4] Benes, P. Picat, P. Cernik, M. Quinault, J. M. Kinetics of radionuclide interaction with suspended solids in modelling the migration of radionuclides in rivers I. Parameters for two-step kinetics. *J. of Radioanal. Nucl. Chem.* **159,** (2) 175 (1992).

[5] Comans R. N. J., Haller M. and De Preter P. Sorption of caesium on illite: Non equilibrium behaviour and reversibility. *Geochim. et Cosmochim. Acta.* **55,** 433 (1991).

[6] Laissaoui, A. Abril, J.M. Periañez, R. García-León & M. García Montaño, E. Kinetic transfer coefficients for radionuclides in estuarine waters: Reference values from ^{133}Ba and effects of salinity and suspended load concentration. *J. of Radioanal. Nucl. Chem.* **237,** (1-2) 55 (1998).

[7] Bunker D. J., Smith J. T., Lievens F. R. and Hilton J. Kinetics of metal ion sorption on lake sediments – approaches to the analysis of experimental data. *Applied Geochem.* **16,** 651 (2001).

[8] Ciffroy, P. Garnier, J.M. Phan, M.K. Kinetics of the adsorption and desorption of radionuclides of Co, Mn, Cs, Fe, Ag and Cd in freshwater systems: experimental and modelling approaches. *J. of Environ. Rad.* **55,** 71 (2001).

[9] Comans R. N. J. & Hockley D. E. Kinetics of cesium sorption on illite. *Geochim. et Cosmochim. Acta.* **56,** 1157 (1992).

[10] Børretzen P. & Salbu B. Estimation of apparent rate coefficients for radionuclides interacting with marine sediments from Novaya Zemlya. *Sci. of the Total Environ.* **262,** 91 (2000).

[11] Børretzen P. & Salbu B. Fixation of Cs to marine sediments estimated by a stochastic modelling approach. *J. of Environ. Rad..* **61,** 1 (2002).

[12] El Mrabet, R. J.M Abril, G. Manj6n and R. García Tenorio. Experimental and modelling study of the Plutonium uptake by suspended matter in aquatic environments from the south of Spain. *Water Research.* **35,** (17) 4184 (2001).

[13] Eisma, D. *Suspended Matter in the Aquatic Environment.* Springer-Verlag, Berlin-Heidelberg. 1993.

[14] Martínez-Aguirre A., Garcia-Le6n M. & Ivanovich M. The distribution of U, Th and ^{226}Ra derived from the phosphate fertilizer industries in an estuarine system in southwest Spain. *J. Environ. Radioactivity.* **22,** 55 (1994).

Applied Physics (APHYS 2003)
A. Méndez-Vilas (Ed.)

Kinetically controlled radionuclide sorption by sediment cores from two different environments. Experimental studies using ^{133}Ba as a tracer.

H. Barros[1*], J.M. Abril[1], El-Mrabet[2] & A. Laissaoui[2].
[1]*Dpto. Física Aplicada I. EUITA. Universidad de Sevilla, Crta de Utrera km 1. DP 41013. Sevilla, Spain.*
[2]*Centre Nat. de L'Energie des Sciencies et des Techniques Nucléaires. 65 Rue Tansift Agdal. Rabat, Morocco.*

Abstract
Considerable efforts have been devoted to experimental studies on the sorption kinetics and the transport of radionuclides in porous media. Motivations: their environmental impact, selection of materials for nuclear waste deposits and pollution in underground aquifers. The present study investigates aspects of non-conservative radionuclide's dispersion and sorption kinetics in natural sediments. Waters and sediments were sampled in two different environments. It was investigated the uptake kinetics of ^{133}Ba by sediments in aqueous suspensions and cores. Samples from the upper most layer (estuary) reveals an important and fast uptake by sediment cores. The depth distribution could be reasonably described by means of an effective diffusion coefficient. The uptake kinetics by suspended sediments could be described by a compartmental kinetic model, but the effective diffusion model (based in a distribution coefficient) shows serious limitations to describe the profiles along the time.

Keywords: Radionuclides dispersion, porous sediment, diffusion/sorption, environment.

1. Introduction

Diverse hazardous materials enters into the environment, either accidentally or following planned releases, and find their pathway towards the aquatic systems. Radionuclides are particle-reactive, then the sorption by sediments influence it's dispersion in the aquatic systems determining the bioavailability of these hazardous elements in the environment. Therefore, water-sediment partitioning of pollutants and their sorption kinetics have been a main focus of research in this field in the last years. Often the experimental studies have been undertaken with aqueous suspensions [1-9], and when the uptake of radionuclides by static sediments is studied on large times (decades) the sediment is often considered as a continuous media under accretion, and the post-depositional redistribution of radionuclides as governed by an advection-diffusion equation [10]. Other authors consider interstitial water and solids separately, but only an equilibrium coefficient is used to describe the uptake [11,12].

Those results have been applied to hydrodynamic models which attempts to describe the radionuclide's mobility in real aquatic environments. Thus, some works included, in marine dispersion models, an active upper layer of sediments with two categories of sizes [13], others included the kinetics through a single reversible reaction [14], and also some transport models [15,16] tried to reproduce the dynamics of suspended loads, including an approach for the interaction with sediments.

Some works [17] provides theoretical studies for the description of the uptake kinetic by suspended particles, however deeper studies still being required to clarify many aspects, as those related to the diffusion through the sediment pores coupled with the sorption of ions at the *interstitial water / pore's surface* interface level. The importance of such studies lie in the necessity of know and predict the mobility and final fate of the referred pollutants, to evaluate it's bioavailability and allow a reliable estimation of it's radiological impact in the environment and the human health.

In this work, a radioactive tracer is used in a series of experiments designed to study the main features of radionuclide sorption kinetic by sediments cores in two different environments. The use of ^{133}Ba does not limit the broad interest, since the basic principles applied to understand the experimental results are of general use. Ba is an artificial gamma emitter which is a good analogue of the environmental behaviour of ^{226}Ra (natural radionuclide of radiological relevance).

2. Materials and methods.

2.1. Sampling and sample treatment

The samples were collected from the estuary of Huelva and the Gergal reservoir (SW of Spain). In the estuary sediments from the first 10 cm were dredged up, and in the reservoir the sediments were carefully extracted with a cylindrical sampler in order to preserve its natural aggregation state. No chemical treatments were performed to the samples, which were stored in dark to avoid algae growth. In the case of the estuary, sediment sub-samples were air-dried in an oven (24 hours at 110°C) gently disaggregated and sieved to remove the particles larger than 2 mm. The sediment samples from the reservoir were stored at 4 °C until it were directly used in the experiments.

2.2. Experimental set up.

The experiment design consisted of two parts: to study the uptake by suspended sediments ("S" experiments hereafter) following an established method [5,8]; and a second part carried out with sediment cores under a water column ("R" experiments), see Table I.

Table I. Experimental set up: Suspended Sediments and Static Cores†

Sample	Experiment ¶	Activity [Bq]	$a_{w(0)}$ [Bq/mL]	Duration days	Sediment porosity§	Sediment density [g.L^{-1}]	E.C.* [mS cm^{-1}]
Gergal	S-A	254 ± 20	0.249 ± 0.012	52	-	2.60 ± 0.03	0,46 ±
	R-A#	267 ± 21	0.109 ± 0.008	41 / 221	0.63		
Tinto	S-B	445 ± 35	0.445 ± 0.035	100	-	2.58 ± 0.02	59 ± 3
	R-B	445 ± 35	0.659 ± 0.046	30	0.67		
Odiel	S-C	445 ± 35	0.445 ± 0.035	7	-	2.40 ± 0.05	63 ± 3
	R-C	445 ± 35	0.630 ± 0.044	30	0.69		

† Water from the same location as the sediment was used in each experiment

¶ Experiments labelled with "S" refers to suspensions and "R" refers to a sediment column at rest. (dry weight)

§ The value corresponds to the first layer of the experiments "R"

* EC, electrical conductivity of the water. In all the cases the water had a neutral (pH ≈ 7)

For the Gergal, two replicates of "R" experiments were made with different contact times

For the "S" experiments, an amount equivalent to 6 or 20 grams of dry sediment were added into one litre of water, the different values of mass were used in order to evaluate the effect of this parameter on the uptake. The sediment was maintained in suspension by a magnetic stirrer (at 25 °C). Aliquots were extracted at determined contact time after tracing, then the solids were separated by centrifugation (20 min. 4000 rpm).

For the estuary R-experiment were performed with 500 grams of dry sediment which were introduced in a PVC cylinder of 10 cm diameter with 1 L of water. We note that the handling of samples force some particle-size stratification in the core, as it will be discussed further. The experimental set up was left at rest during 1 day to achieve sediment saturation, then the water column of 10 cm high (0.75 L) was traced with a known amount of ^{133}Ba. After the corresponding contact time, the supernatant was removed and measured. The sediment column was frozen (–80°C) to stop diffusion processes. Subsequently, the sediment was sectioned into 1 cm thick sections.

For the reservoir R-experiment were performed similarly, but with a natural undisturbed column of sediments (8 cm high) and an overlying water column of 15 cm high (2,44 L) in a plastic cylindrical container with 14,3 cm of diameter. This container was specially designated in order to allow a precise sectioning of the sediment core without modifying it's natural structure. The disc base of the cylinder is a piston like mobile piece, then after the experiment the container is placed in a mechanical device coupled with an hydraulic jack which push the piston upwards making possible to cut a thin layer of the sediment lifted above the upper cylinder's border.

In all the cases, interstitial water was extracted by vacuum filtration and then measured. Water and sediment samples were measured using HPGe, REGe or XTRA detectors attending to the level of radioactivity of each sample. Each sample was measured once and the activities were reported with the corresponding analytical errors.

3. Results and discussion.

3.1. ^{133}Ba uptake by suspended sediments.

The experiment from the Gergal reservoir was performed with a replicate in order to evaluate it's variability, since the sediments were not perturbed and then could be not so homogeneous. In all the experiments it was observed a complex kinetics of three characteristic times. A rapid decrease of the liquid phase concentration was observed during the first minutes, then a moderate reduction occurs in the scale of hours and finally a slow decrease along the days. In the experiment S-A (Gergal) it was observed the fastest sorption reaction, 60% of the tracer was transferred in 2 minutes, after another 10% in the next 2 hours and finally 1 to 4% was transferred slowly. In the experiment S-B (Tinto) in 2 minutes 20% of the tracer was adsorbed, and in the experiment S-C, after 10 minutes 46%. Afterwards the uptake continue at a slower rate, probably due to process related with the ion accessibility into the particle's pore structure [8].

A kinetic compartment model can be used as a practical approach to this complex process [18], where the model's coefficients are found by a fitting procedure to the experimental uptake curves. Despite the local application of the coefficient's values, the approach is of general applicability. In this work, it was successfully used a box model based on a model already applied for Pu uptake [8]. It use two parallel reaction and an additional slow channel which follows the slower of the previous ones. The model distinguishes three sites in the particles through it's bulk activity concentrations: a_{s1}, a_{s2} and a_{s3}, and a_w represents the dissolved concentration (all in Bq mL^{-1}). The mathematical representation of such model is presented in Equation 1 (dots indicates time derivatives and k_{ij} have dimensions of $time^{-1}$).

$$\begin{aligned}
\dot{a}_w &= -(k_{11} + k_{12}) \cdot a_w + k_{21} \cdot a_{s1} + k_{22} \cdot a_{s2} \\
\dot{a}_{s1} &= +k_{11} \cdot a_w - k_{21} \cdot a_{s1} \\
\dot{a}_{s2} &= +k_{12} \cdot a_w - k_{22} \cdot a_{s2} - k_{13} \cdot a_{s2} + k_{23} \cdot a_{s3} \\
\dot{a}_{s3} &= +k_{13} \cdot a_{s2} - k_{23} \cdot a_{s3}
\end{aligned} \qquad (1)$$

A fitting procedure was applied to the experimental curves. In the A experiment the two replicates are shown in Figure 1.a, in Table II the "averaged" coefficients are reported. In the case of the estuary (B and C) the coefficient's errors have been calculated [18]. For the two rivers the experimental curves have a similar shape with similar reaction times, although the transferred fraction in each reaction are different (Figure 1.b).

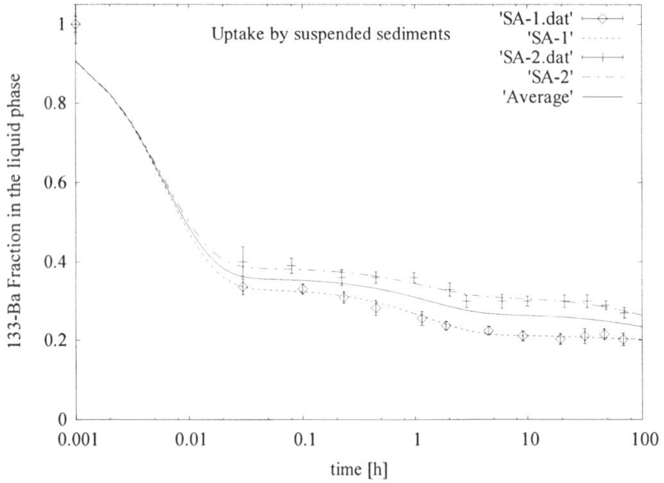

Figure 1.a. Time evolution of dissolved phase activity fraction for experiments "S-A" (Table 1). Activities are normalised to the initial value. The initial condition has been artificially shifted in all the plots (logarithmic time scale) to provide a global view.

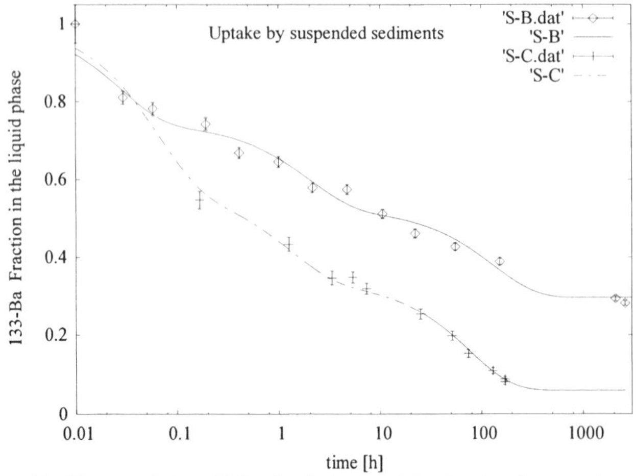

Figure 1.b. Time evolution of dissolved phase activity fraction for experiments "S-B" and "S-C" (Tinto and Odiel rivers). Fitted curves according to Eq. 1. Kinetic and distribution coefficients in Table 2.

Table II. Kinetic coefficients for the uptake experiments with suspended sediments†.

	Sub-experiment		
	S-A*	S-B	S-C
ms [g.L^{-1}]	20	6	20
ϕ [mm]	$\phi < 2$	$\phi < 2$	$\phi < 2$
SSA [m^{2}g^{-1}]	N.R.	0.66	0.66
k_{11} [h^{-1}]	94.4 ± 0.2	9.1 ± 1.5	6.4 ± 1.0
k_{21} [h^{-1}]	53 ± 6	26 ± 6	9.1 ± 1.6
k_{12} [h^{-1}]	0.5 ± 0.3	0.21 ± 0.01	0.60 ± 0.05
k_{22} [h^{-1}]	0.55 ± 0.06	0.35 ± 0.03	0.44 ± 0.04
k_{13} [h^{-1}]	0.007 ± 0.004	0.010 ± 0.006	0.025 ± 0.003
k_{23} [h^{-1}]	0.006 ± 0.003	0.004 ± 0.003	0.0025 ± 0.0008
χ_ν-test	3	1.3	1.2

†For coefficients k_{ij}, model description in Eq. 1. Values within 1σ (fitting) error
ms: concentration of suspended sediments; ϕ: particle's diameter; N.R: non reported
SSA: specific surface area (spherical shape approach)
*Corresponds to the average between the two replicates (details in the text)

The coefficients k_{ij} are of the same order in experiments S-B and S-C (estuarine system) but in experiment S-A (reservoir) the first direct coefficient is about one order of magnitude higher. This effect could be related to the fresh water environment respect to the salty waters from the estuary (see electrical conductivity in Table I), although the differences in the main mineralogical composition between the two environments.

3.2. Kinetics of the uptake by sediment cores.

The dissolved activity in the overlaying water was recorded along the time for the experiments R-A and R-B. Two replicates were carried out in the case of the reservoir (R-A 1 and 2), they were prepared in identical conditions, but one was studied at 41 days of contact time and the other at 221 days.

In the experiments R-A the dissolved concentration remain almost constant during 10 hours, then decrease gradually up to 85%. This process can be understood as the diffusion of the dissolved ion through the porous media and the resulting decreasing in the overlaying water concentration. The two replicate provide the same result (Figure 2a).

The experiment R-B was completely different, it was observed a fast adsorption which, in 3 minutes, transfers the 80% of the pollutant to the static (bulk) sediment. It is shown in the Figure 2b, where after the first interaction the system reach a kinetic equilibrium and the dissolved phase concentration remained constant. The last result can not be explained in terms of diffusion of the tracer through the interstitial water since the reaction was too fast. It is important to note that (during the preparation of these substratum) an unavoidable stratification have set up the fine particles (clay fraction) in the upper most layer of the sediment; and it is well known that these kind of particles are highly reactive [2]. This fact is not a method associated problem because these kind of behaviour actually occurs in the natural studied environment due to currents.

We propose the existence of a direct channel related to adsorption by the particle's surface more accessible from the water column, in these process the complex and irregular and water-sediment interface acts in a similar way as the suspended particles. From a physical point of view the advection-diffusion-uptake model is not complete due to the rupture of symmetry at the interface level.

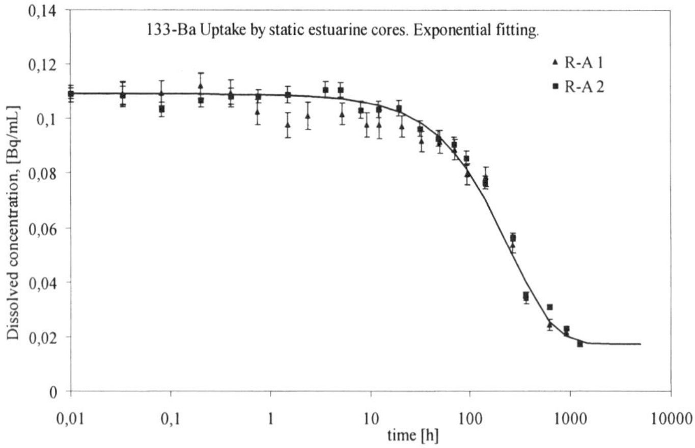

Figures 2. a) Uptake kinetics of ^{133}Ba from overlaying water by a static column of sediment. For the R-A experiments an empirical exponential fit was performed to be used in the description of the corresponding penetration profile. **b)** R-B experiment, the line represents the best fit to a one reversible reaction model (corrected for the extraction of the aliquots).

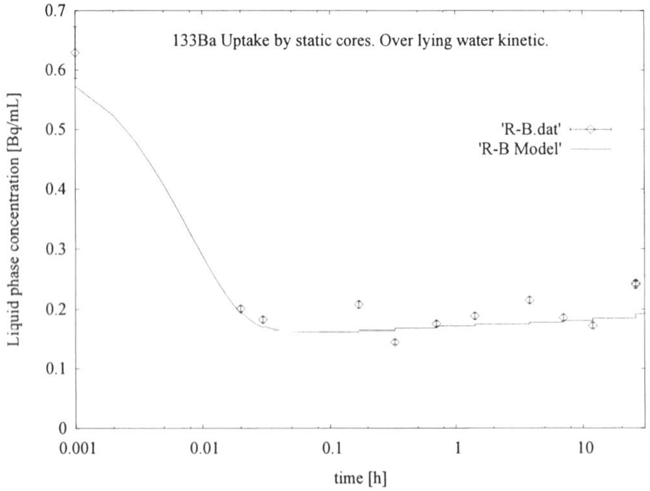

Thus uptake curve could be reasonably fitted by a single reversible reaction model. The obtained coefficients are: $k_{11} = 90 \pm 15$ and $k_{21} = 30 \pm 8$ h^{-1}, this direct uptake dominates the short-term fate of the tracer. Within the substratum it was not possible to detect activity beyond a depth of 2 cm (Table III) which is reasonable since a higher penetration depth cannot be achieved in a short time (30 days). In natural systems the net uptake by sediments is a complex balance between simultaneous vertical diffusion, horizontal transport and dispersion. However, in shallow, well mixed and still waters, the direct uptake of ^{133}Ba by sediments is an important transfer channel as shown in our experiments.

Table III. Activity distribution in sediments and interstitial waters for the "R-" experiments.

	R-A 1	**R-A 2**	**R-B**	**R-C**
Contact time (days)	41	221	30	30
Supernatant (Bq) initial	267 ± 21	267 ± 21	445 ± 30	445 ± 30
Supernatant (Bq mL^{-1}) initial	0.109 ± 0.008	0.109 ± 0.008	0.66 ± 0.05	0.63 ± 0.05
Supernatant (Bq) final	43 ± 2	37 ± 2	70 ± 5	111 ± 7
Supernatant (Bq mL^{-1}) final	0.022 ± 0.001	0.017 ± 0.001	0.22 ± 0.01	0.19 ± 0.01
Layer 1 solid sediment (Bq)	210 ± 15	212 ± 20	330 ± 17	320 ± 20
Layer 1 solid sediment (Bq $g^{-1\#}$)	1.208 ± 0.017	1.237 ± 0.043	5.6 ± 0.3	5.5 ± 0.4
Layer 1 pore water (Bq)	1.83 ± 0.09	0.45 ± 0.02	10 ± 1	-
Layer 1 pore water (Bq mL^{-1})	0.0169 ± 0.0007	0.0053 ± 0.0001	0.088 ± 0.008	-
Layer 2 solid sediment (Bq)	4.30 ± 0.09	20.6 ± 0.8	1.5 ± 0.1	-
Layer 2 solid sediment (Bq $g^{-1\#}$)	0.0289 ± 0.0007	0.172 ± 0.004	0.025 ± 0.002	-
Layer 2 pore water (Bq)	- §	0.018 ± 0.005	-	-
Layer 2 pore water (Bq mL^{-1})	-	0.00058 ± 0.00015	-	-
Layer 3^+ solid sediment (Bq)	0.041± 0.012	2.4 ± 0.3	-	-
Layer 3 solid sediment (Bq $g^{-1\#}$)	0.00033 ± 0.00008	0.017 ± 0.001	-	-
Layer 4^+ solid sediment (Bq)	-	0.24 ± 0.04	-	-
Layer 4 solid sediment (Bq $g^{-1\#}$)	-	0.0013 ± 0.0001	-	-

Dry weight. § Not detected. + Pore water activity was not detected for layers 3 and 4.

3.3. Discussion on the k_d values.

In the case of sediment columns it is possible to define an apparent distribution coefficient as the ratio between the bulk activity concentration in the sediment (1st layer) and the overlaying water concentration. Also it can be defined an intrinsic distribution coefficient as the ratio between the activity concentration in the solid fraction to the corresponding in pore water. The obtained values are shown in Table IV, where the latest available data point was used for the calculation and do not necessarily corresponds to equilibrium.

Table IV. Intrinsic and apparent distribution coefficients for "S-" and "R-" experiments.

Experiment	**A 1**	**A 2***	**B**	**C**
Suspension $k_{d\,sus}$ [L kg^{-1}]	180 ± 30	180 ± 30	416 ± 20	540 ± 110
Core apparent $k_{d\,ap}$ [L kg^{-1}]	56 ± 3	73 ± 6	25 ± 3	30 ± 3
Core intrinsic $k_{d\,sed}$ [L kg^{-1}]	74 ± 4	233 ± 12	63 ± 6	N.C

‡ N.C Not calculated (interstitial water was not measured)
* Second layer in experiment R-A 2 gave $k_{d\,sed} = 296 \pm 83$
For A experiments $k_{d\,sus}$ corresponds to the average value

In the experiments R-B and R-C, the sediment layer was 1 cm depth due to the difficulty found in the sectioning of the frozen sediments. In the experiments R-A the layers can be as thin as 5 mm. The measured concentrations corresponds to the averaged value in the corresponding sediment layer.

Since activity in pore water decreases with depth following diffusion, it is expected that the intrinsic k_d values should be higher than the apparent ones (as confirmed in Table IV). However apparent k_d values depend on the selected sediment thickness, and for the first layer the tracer comes from two different channels, then no quantitative conclusions can be derived about an effective scaling factor of general use.

3.4. Diffusion controlled ^{133}Ba depth distributions in sediment cores.

Diffusion is a slow transport phenomenon, then the transference of the tracer due to this mechanism only affected the upper layers of the static sediment, in our experiments up to a maximum depth of 3 cm. The activities in the rest of the layers are below the detection limit of 5.10^{-5} Bq g^{-1} for sediments and 10^{-4} Bq mL^{-1} for pore water (XTRA detector).

The sorption of ions by pore's surface is relatively fast in comparison with the transit time through the interstitial water, then the effect of these two process can be described by means of an effective diffusion coefficient D_e [12], which depends on the free diffusion coefficient (D^*), the porosity (φ), the intrinsic distribution coefficient (k_d) and the ionic strength of the solution (α). The Eq. 2 shows a mathematical relationship for this effective coefficient, where φ^n ($n \in [1, 2]$) is an empirical expression for the tortuosity.

$$D_e = \frac{\alpha D^* \varphi^n}{1 + k_d} = \frac{D_s}{1 + k_d} \tag{2}$$

Regarding the different border conditions, the diffusion equation can be solve. If the initial activity in the sediment is zero and the activity in the overlying water is maintained constant (A_0), a solution can be found for D_e being constant. Eq. 3 shows the pore water activity, from which the activity's concentration in the sediment can be obtained (Eq. 4)

$$a_{wi}(x,t) = a_0\, erfc(\frac{x}{2\sqrt{D_e t}}) \tag{3}$$

$$a_s(z,t) = k_d\, a_{wi}(z,t) \tag{4}$$

The above border conditions match with our experiment R-B. Where, after the initial fast uptake by the upper sediment surface, the activity remained constant (Fig. 2.b). Thus, radionuclide diffusion and their subsequent uptake by the sediment core can be explained by Eqs. 3 and 4 with $A_0 = 0.22$ Bq mL^{-1} and t = 30 days. Thus, using the found intrinsic k_d value (Table IV) we used the parameter D_s for the interstitial water and the solids, finding a reasonable agreement with the measured values (Fig. 3a, 3b). However, the low penetration due to the accumulation of clays in the upper surface have limited the data. It is important to note that this approach cannot account for the transient regimen when the overlying water activity is varying.

Then to approach the experiment R-A, we performed an empirical fitting of the data showed in Figure 2.a, obtaining an exponential decay for $a_w(t)$. Then we introduced such a function in the integral coming from the Convolution Theorem applied to the diffusion equation and the border conditions (Eq. 5) and it was confirmed (numerically) that in the present case the contribution from the transient was less than 1%, so we avoid unnecessary complications and proceed as in the case of a constant function.

$$a_{wi}(x,t)=\frac{2}{\sqrt{\pi}}\int_{x/2\sqrt{D_e.t}}^{+\infty}e^{-\nu^2}a_w(t-\frac{x^2}{4.D_e.\nu^2})d\nu\approx a_0.\frac{2}{\sqrt{\pi}}\int_{x/2\sqrt{D_e.t}}^{+\infty}e^{-\nu^2}d\nu \qquad (5)$$

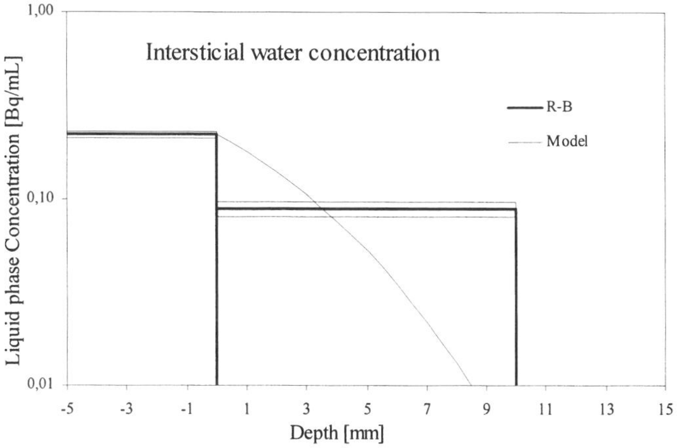

Figures 3. a) Measured and calculated activity concentrations in interstitial water, and **b)** in the solid phase in experiment R-B (Tables 1 and 3). Measured data are represented by step straight lines that corresponds to the depth intervals. The model is given by equations 3 and 4, and is represented by the corresponding curves.

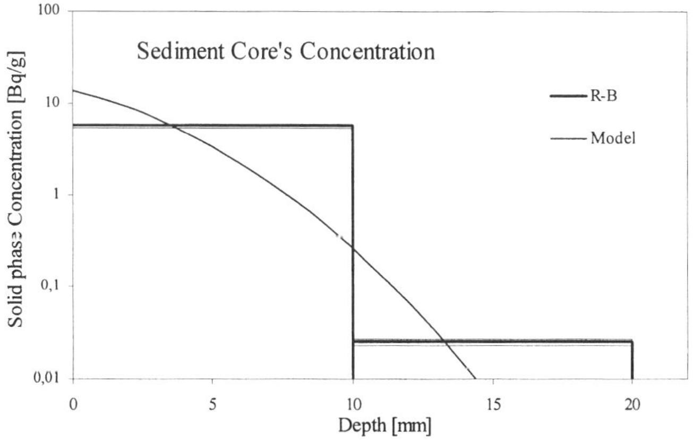

The results for experiments R-A 1 and R-A 2 are shown in the Table 3. Then we used the measured intrinsic k_d, porosity, and the referenced value for the free diffusion coefficient. The Fig. 4.a shows the interstitial water concentration corresponding to the two different contact times, straight lines corresponds to the measured values, and the curves corresponds to the model; Figure 4.b shows the corresponding solid concentrations.

We have considered the kinetic transfer as a very fast process [12] and then we used the Equation 2 to describe the sorption/diffusion process as a diffusion droved by partition coefficient. However in Table 4 it is shown that the intrinsic partition coefficient is not constant, in fact it change from 74 to 230 in six months an introduce a factor three in the effective diffusion coefficient. This is a clear limitation for the proposed approximation.

Consequently the best fit found with these model, despite its agreement with the majority of the data, show serious limitations, as can be seen in the figure 4.a.

Further more the present model can not explain the increment in the intrinsic k_d and then is not suitable to describe any other kinetic aspects related to the process.

4. Conclusions

The uptake of radionuclides (as ^{133}Ba) by suspended sediments is a process with 3 characteristic times that transfer an important fraction of the dissolved ions to the material particles. Therefore, episodic resuspension of sediments can play an important role in controlling the final fate of particle-reactive pollutants. These complex process can be described by means of a compartmental model with two parallel uptake channels plus a slow and weakly reversible reaction. It was also confirmed that reproducibility in the uptake experiments is achieved although we used sediments from different places (within a perimeter of 4-5 meters), it indicates that the coefficients obtained via these kind of experiments can be applied to approximate the general situation in the studied environment.

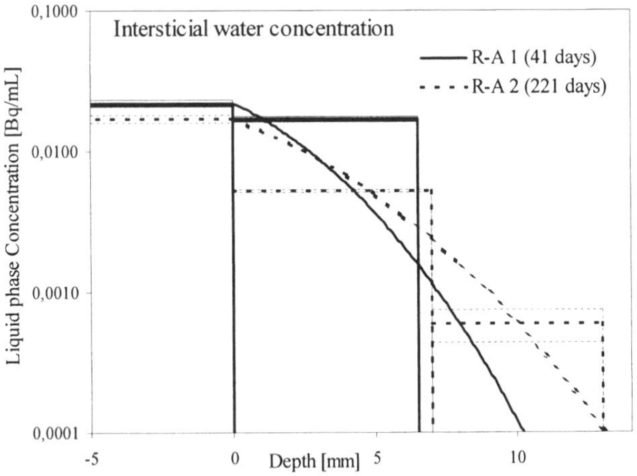

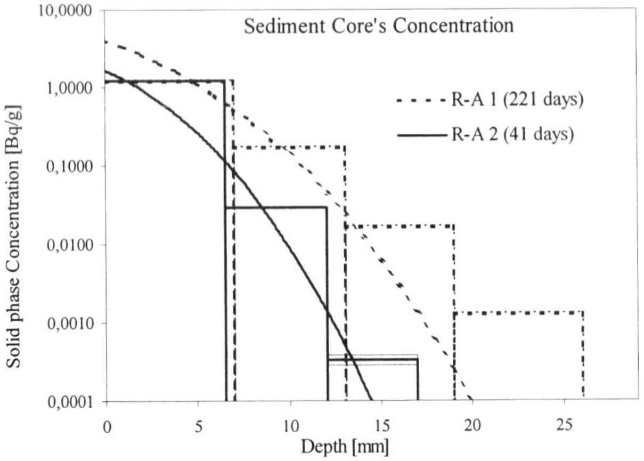

Figures 4. a) Measured and calculated activity concentrations in interstitial water, **b)** the solid phase. Experiment R-A (Tables 1 and 3). The measurements are represented by the step straight lines, and the model by the curve lines.

In some environments the porous water column-sediment interface shows a relatively large surface (regarded to it's irregularity) which is directly accessible for the dissolved radionuclides. It can represent a fast and important channel of uptake, and has to be explicitly considered since the symmetry of the diffusion-uptake process in sediments is broken at this interface. This last shows the importance of the kinetics in the study and modelling of the dispersion of these kind of radionuclides in aquatic environments.

In experiment R-B after the initial uptake, the activity in the overlaying water remained virtually constant, and it was possible to use an analytical solutions for the diffusion-uptake problem, by means of an effective diffusion coefficient. The analytical solution for the model, which uses some of the experimentally obtained parameters, could then reasonably describe the depth distributions of ^{133}Ba in solids and in the interstitial water. On the other hand in the Gergal's experiment, the uptake was controlled by diffusion but in this case serious limitations in the use of distribution coefficient were observed. Then, although more complex, the suitable approximation for detailed studies on sorption/diffusion process between water and sediments have to consider the sorption kinetics at both, interface and porous water level.

^{133}Ba was used as a tracer, but the results of the present work still having a general character since many pollutants are present in solution as positively charged ions and are known to behave in relatively similar way in the presence of solid particles.

Acknowledgements

This work has been partially supported by ENRESA (I+D: 774511 – 0770105) Spanish Public Corporation of Radioactive Residues. We also acknowledge the AMA (Agencia del Medioambiente de Andalucía) for its assistance in carrying out the sampling.

References.

1. Benes P and Poliak, R. Factors affecting interaction of radiostrontium with river sediments. J. Radioanal Nucl Chem 1990; 141:75-90.

2. Comans R N J, Haller M and De Preter P. Sorption of cesium on illite: Non-equilibrium behaviour and reversibility. Geochim Cosmochim Ac 1991; 55: 433-440.

3. Bird G A and Evenden W G. Effect of sediment type, temperature and colloids on the transfer of radionuclides from water to sediment. J Environ Radioactiv 1994; 22: 219-242.

4. He Q and Walling D E. Interpreting particle size effects in the adsorption of ^{137}Cs and unsupported ^{210}Pb by mineral soils and sediments. J Environ Radioactiv1996; 30: 117-137.

5. Laissaoui A, Abril J M, Periañez R, García-León M and García-Montaño E. Determining kinetics transfer coefficients for radionuclides in estuarine waters: Reference values from ^{133}Ba, and effects of salinity and suspended loads concentration. J Radioanal Nucl Chem 1998; 237: 55-66.

6. Smith J T, Commans R N J, Ireland D G, Nolan L and Hilton J. Experimental and in situ study of radiocaesium transfer across the sediment-water interface and mobility in lake sediments. Appl Geochem 2000; 15: 833-848.

7. Ciffroy P, Garnier J M and Phan M K. Kinetics of the adsorption and desorption of radionuclides of Co, Mn, Cs, Fe, Ag and Cd in freshwater systems: experimental and modelling approaches. J Environ Radioactiv. 2001; 55: 71-91

8. El-Mrabet R, Abril J M, Manjón G and García-Tenorio R. Experimental and modelling study of plutonium uptake by suspended mater in aquatic environments from southern Spain. Wat Res 2001; 35(17): 4184-4190.

9. Børretzen P and Salbu B. Estimation of apparent rate coefficients for radionuclides interacting with marine sediments from Novaya Zemlya. Sci Total Environ 2000; 262:91-102.
10. Christensen E R and Bhunia P K. Modeling radiotracers in sediments: Comparison with observations in lakes Huron and Michigan. J Geophys Res 1986; 91 C7: 8559-8571.
11. Robbins J A. A model for particle-selective transport of tracers in sediments with conveyor belt deposit feeders. J Geophys Res 1986; 91: 8452-8558.
12. Krezoski J R, Robbins J A and White D S. Dual radiotracer measurement of zoobenthos-mediated solute and particle transport in freshwater sediments. J Geophys Res 1984; 89: 7937-7947.
13. Abril, J M and García-León M. A 2D-4Phases marine dispersion model for non-conservative radionuclides. Part I: Conceptual and computational model. J Environ Radioactiv 1993; 20:71-88.
14. Periáñez R, Abril J M and García-León M. Modelling the dispersion of non-conservative radionuclides in tidal waters. Part 1: Conceptual and mathematical model J Environ Radioactiv 1996; 31: 127-141.
15. Periáñez R. Three-dimensional modelling of the tidal dispersion of non-conservative radionuclides in the marine environment. Application to $^{239,240}Pu$ dispersion in the eastern Irish Sea. J Marine Syst 1999; 22: 37-51.
16. Abril J M and Abdel-Aal M M. A modelling study on hydrodynamics and pollutant dispersion in the Suez Canal. Ecol Model 2000; 128: 1-17.
17. Abril J M. Basic microscopic theory of the uptake kinetics, transfer and distribution of dissolved radionuclides by suspended particulate matter. Part I: Theory development. J. Environ Radioactiv 1998; 41: 307-324.
18. Benes P, Picat P, Cernick, M and Quinault J M. Kinetics of radionuclides interaction with suspended solids in modelling the migration of radionuclides in rivers. I. Parameters for two-step kinetics. J Radioanal Nucl Chem 1992; 159:175-186.

Applied Physics (APHYS 2003)
A. Méndez-Vilas (Ed.)

Development and Production of Iodine-125 Seeds for Brachytherapy

M. E. C. M. Rostelato; C. P. G. Silva; P. R. Rela; H. T. Casiglia; C. A. Zeituni; A. Feher and V. Lepki
Comissão Nacional de Energia Nuclear – CNEN
Instituto de Pesquisas Energéticas e Nucleares – IPEN
elisaros@ipen.br

Abstract. The number of prostate cancer cases in Brazil is increasing and part of the patients are submitted to brachytherapy treatment using Iodine-125 radioactive seeds, which nowadays are imported at a high cost, restricting their application. The local production of these radioactive sources became a priority in order to reduce the problems of prostate cancer management for end users. Such action will permit to spread the use to a larger number of patients. Due to such reasons, the Nuclear Energy Research Institute established a program in order to produce Iodine-125 radioactive seeds. In brachytherapy, these small seeds with Iodine-125 are implanted into the prostate to irradiate the tumor. The Iodine-125 seeds consist of a welded titanium capsule containing Iodine-125 adsorbed onto a silver rod. Concerning the setup of the local production, the optimization of the following activities have been carried out: superficial treatment of the silver rod, development of a process to absorb the Iodine in the silver rod, welding methodology to seal the seeds, leakage and contamination test and source activity measurement.

INTRODUCTION

Considered a public health problem in Brazil, cancer is the second death cause of disease. An estimate by the National Institute of Cancer – INCA showed that 122.600 people died from cancer and another 337.535 had the disease in the country, in 2002 [1].

The prostate cancer is the second larger death cause among men. Studies by the National Institute of Cancer – INCA estimated that, for the year 2002, the occurrence of new cases of prostate cancer would affect more than 25.600 patients [1].

One of the options for prostate cancer treatment is the brachytherapy. By this technique, small seeds with Iodine-125, a radioactive material, are implanted in the prostate. The advantages of radioactive seed implants are the preservation of healthy tissues and organs near the prostate, besides the low rate of impotency and urinary incontinence, compared to conventional treatments, such as the radical prostatectomy and the external radiation beam [2,3].

The 125-Iodine is produced in a nuclear reactor, from 124-Xenone. It decays by electronic capture and internal conversion to 125-Telurium. In this process, it emits photons of 27keV, 31keV and 35keV, with an average energy of 29keV. Due to its low average energy of emission, its photons have a short penetration. The isotope has a half-life of 60 days.

The seeds, which have microscopic dimensions, consist of a titanium capsule (material inert to human tissue) of 0.8 mm external diameter, 0.05mm wall thickness and 4.5mm long. The inner capsule houses a silver wire, 3mm long and 0.5mm diameter, containing the adsorbed 125-Iodine. The typical seed apparent activity is of 0.4mCi (14.8MBq), with a recommended variation of about 5% at most, in a same lot of seeds [4].

The 125-Iodine seed implants have been carried out in Brazil with imported seeds. Nowadays, the 125-Iodine demand in Brazil is of 2,500 to 3,000 seeds/month.

Taking into account the seeds price and the difficulties to import, the Energy and Nuclear Research Institute – IPEN, which belongs to the Nuclear Energy National Commission – CNEN, established a program for the development of the technique and production of 125-Iodine seeds. The estimate for the 125-Iodine seeds demand if of 8,000 seeds/month and the laboratory to be implanted will need this production capacity.

The project goal is to enable the country for the 125-Iodine seeds production, at a cost meeting the Brazilian reality and, thus, allowing a larger number of patients to access this type of therapy.

The project will be divided in two phases: technological development of a prototype seed and a pilot plant implementation for the production of the 125-Iodine seeds, aiming to meet the medical faculty requests. This paper covers the technological prototype seed development.

METHOD

During the project execution, the following methods were developed: the seed core (Silver) cutting, the titanium tube cutting, the iodine immobilization through its deposition in silver substrate, and the sealing of the seeds through the microplasma welding process, so that the classification of the seeds, as sealed sources, and the leakage tests can be done according to the international norms ISO-2919 e ISO-9978 [5,6].

RESULTS

A model of the 125-Iodine seed was developed, as shown in Fig. 1.

The seed core (silver) and titanium tube cuttings were carried out with a "cut-off" device, with an aluminum oxide disc, then the debris were sandpapered. The visual inspection was done in an optical microscope. The result was a perpendicular cutting, without debris.

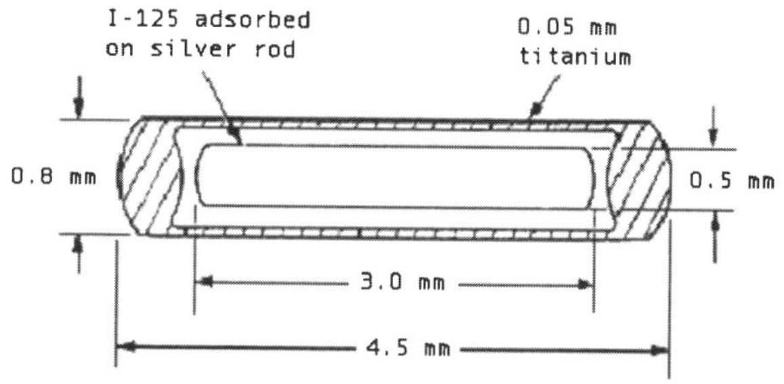

FIGURE 1: Schematic diagram of the 125 – Iodine seed.

In the 125-Iodine deposition on the silver substrate, a reaction yield up to 90% was obtained with an average value of 80% over 500 experiments. The deposition was done for lots of 30 seeds.

The seeds sealing was accomplished by microplasma welding process and the result was an homogeneous weld without inclusions, cracks or fissures, as seen in Fig. 2.

The 125-Iodine seed prototype is shown in Fig. 3.

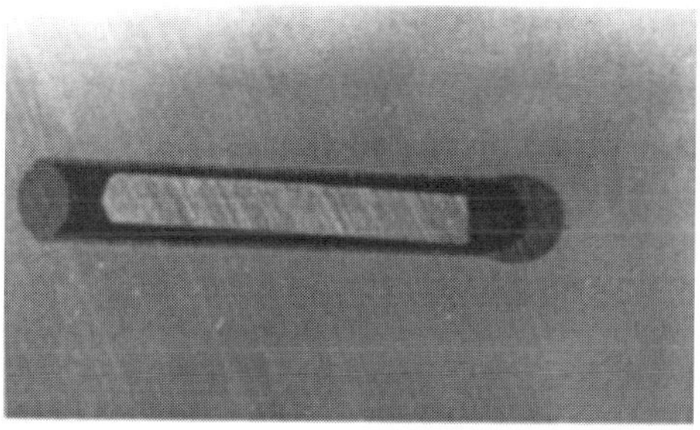

FIGURE 2: Longitudinal cut to view the microplasma welding process.

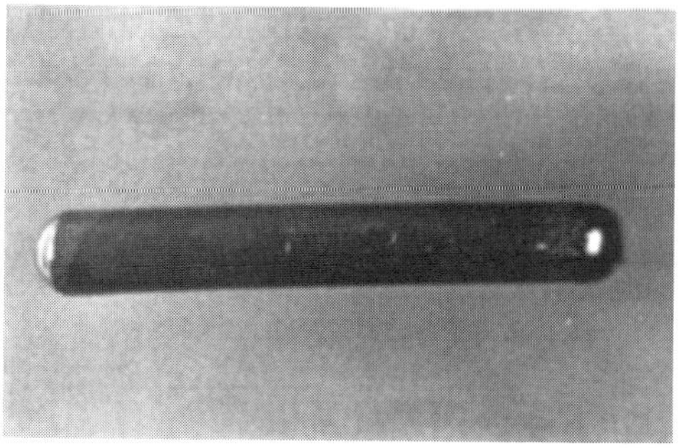

FIGURE 3: IPEN's 125-Iodine seed prototype.

DISCUSSION

Cutting of the seed core: The silver wire has to undergo a perfect cutting, perpendicular to its own axis. Imperfections in the cutting can cause deformation in the isodose lines of the seed.

The titanium tube cutting must be perfect, free of debris and without deformations so that the seed core can be inserted without difficulty.

The yield of the 125-Iodine deposition reaction on the silver is expected to be high, approximately 80%, and with a homogeneous distribution in a lot of seeds. In a 30-seed lot, the permitted variations in the activity of the 125-Iodine was up to 10%. With these requirements met, an almost thorough yield of the radioactive material and a lot of seeds with homogeneous activities were obtained.

The sealing should not present inclusions, cracks or fissures allowing radioactive material leakage. The weld should be as uniform as possible to diminish the anisotropy. IPEN'S seed presented a good weld quality, compatible with that of other seeds in the market. After sealing, the weld integrity was evaluated with the use of an optical microscopy and leakage test, according to the norm ISO-9978 [6].

CONCLUSIONS

According to this paper target, a 125-Iodine seed prototype was developed in Brazil. The seeds showed to be satisfactory as to the 125-Iodine deposition, welding method and leakage tests carried out as requested by the norm ISO-9978. Nowadays, this prototype is being submitted to classification tests, according to the norm ISO-2919.

REFERENCES

1. Ministério da Saúde INCA / Conprev. Estimativa da incidência e mortalidade por câncer no Brasil 2002. Rio de Janeiro, 2002.
2. American Urological Association Prostate Cancer. Clinical Guidelines Panel, "The Management of Localized Prostate Cancer – A Patient's Guide". USA 1998.
3. P. Grimm. Ultrasound-Guided Prostate Permanent Seed Implant Therapy. Swedish Medical Center's Seattle Prostate Institute, USA, 1997.
4. J. Blasko, M. J. Datolli and K. Wallner. Prostate Brachytherapy. Smart Medicine Press, Washington, USA, 1997.
5. International Standard Organization. Radiation protection – Sealed radioactive sources – General requirements and classification. Mar. 08, 1995. (ISO-2919).
6. International Standard Organization. Radiation protection – Sealed radioactive sources – Leakage test methods. Feb.15, 1992. (ISO-9978).

Quartz Crystal Microbalance And Electrical Impedance Characterization Of Nickel Dissolution Process.

J. J. García-Jareño, D. Giménez-Romero, J. Gregori, F. Vicente.

Department of Physical Chemistry, University of Valencia, C/ Dr. Moliner 50, 46100 Burjassot (Spain)
E-mail: proclq@uv.es (F. Vicente)

Abstract. The anodic nickel dissolution in acid media is analysed by means of EQCM and EIS techniques. The experimental impedance spectra have been fitted to the equivalent circuit which corresponds to two consecutive electron transfers followed by a Ni(II) desorption. That way rate constants and surface concentrations of the Ni(0) and Ni(I) species are obtained. EQCM also provides information about the mechanism of deposition and passivation of nickel as well as the hydrogen evolution.

INTRODUCTION

Nickel has been the subject of many researches related with dissolution and passivation mechanism in acid medium by means of different electrochemical techniques[1-4]. Electrochemical impedance spectroscopy (EIS) is a technique which allows us to obtain mechanistic information about the processes which take place on an electrode surface at steady state potential[5,6]. Electrochemical quartz crystal microbalance (EQCM) is a device which allows to frequently measure the mass changes on an electrode surface when an electrochemical reaction takes place[7-11]. EQCM in combination with voltammetry technique provides important information about mechanism and stoichiometry in metal dissolution and deposition process[10,11] by means of the mass/charge ratio defined by:

$$F\frac{\Delta m}{\Delta Q}=\sum_i \frac{MW_i}{n_i}\nu_i \pm \textit{mass changes due to uncharged species} \qquad (1)$$

where MW_i is the molecular mass of a specie i, which interchanges n_i electrons and ν_i is the charge ratio due to process i.

The aim of this work is to correlate the mechanistic information on nickel electrodissolution obtained from EQCM measurements with the shape of the Nyquist plots of impedance spectra. In this sense the theoretical impedance function for the proposed electrodissolution mechanism is obtained and it is related with the equivalent circuit proposed in order to fit the impedance spectra.

EXPERIMENTAL

All the experiments have been carried out in an electrochemical cell of three electrodes. A *SSE* reference electrode and a platinum sheet of large area as auxiliary electrode were used. For impedance measurements the potential was controlled with a Potentiostat-Galvanostat 273A EG&G PAR and the frequency analyser was used with the Lock-in Amplifier 5210 EG&G PAR. The working electrodes, A = 0.25 cm^2, were made from a nickel sheet (99.9%, Johnson&Matthey). The frequency range was $[10^5, 5\cdot10^{-2}]$ Hz with the signal amplitude of 10 mV r.m.s. The fitting of experimental impedance data to the proposed equivalent circuit was carried out by means of a non-linear least squares procedure based on the Marquard algorithm[12]. For EQCM experiments the working electrodes were made from a quartz sheet embedded between two pieces of gold connected to a resonance circuit. The electrical area was 0.228 cm^2 and the effective mass area was 0.196 cm^2. The potential sweep was carried out in the [200,-1500] mV potential range at 20 mV/s. The microbalance was a UPR15/RT0100 (UPR of the CNRS). The resonance frequency of the quartz was measured with a frequenciometer Fluke PM6685. The current was measured with a multimeter Keithley PM2000. The potential was applied with a Potentiostat 263A EG&G PAR. The EQCM was calibrated by means of a galvanostatic Cu deposition[10]. The experimental Sauerbrey constant was $9.50\cdot10^7$ $Hz\cdot g^{-1}$.

EXPERIMENTAL RESULTS AND DISCUSSION

EQCM Measurements

In order to discuss the EQCM experimental results it is convenient to divide in several regions the mass variation domain as can be seen in Fig. 1.

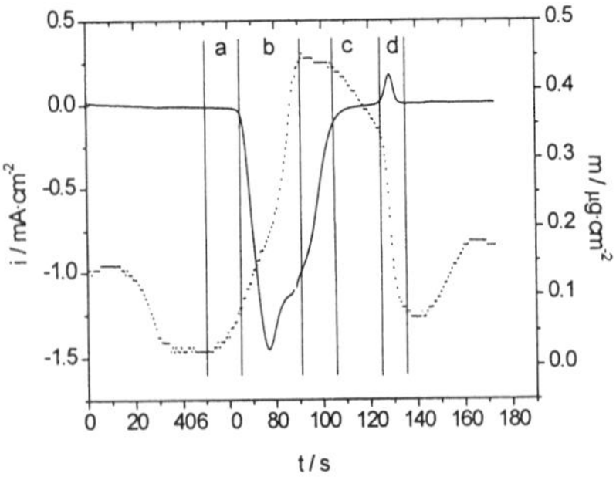

FIGURE 1. Mass changes (dashed line) and current changes (continuous line) vs time. 10^{-3} M $NiSO_4$, 0.245 M K_2SO_4, $5\cdot10^{-3}$ M H_2SO_4. pH = 2.7. T = 298 K. Potential ranges: **a** [-800,-1100] mV, **b** [-1100,-1500] mV in cathodic scan and [-1500,-1400] mV in the anodic one, **c** [-1100,-650] mV, **d** [-650,-500] mV.

In region **a** and **b** a considerable mass increase takes place most probably caused by nickel electrodeposition. In region **b** this process is accompanied by a great cathodic current due to the hydrogen evolution[13]. In regions **c** and **d** a considerable mass decrease takes place. In region **d,** this mass decrease is due to nickel dissolution[8,9].

In region **a** F·Δm/ΔQ value is -25 g·mol^{-1}. It is close to the theoretical value for an electrodeposition process:

$$Ni^{2+} + 2e^{-} \rightarrow Ni \tag{2}$$

and F·Δm/ΔQ$_{theoretical}$ = -29.3 g·mol^{-1}. The difference between experimental and theoretical value is due to hydrogen evolution contribution to ΔQ.

In region **d** F·Δm/ΔQ = -38 g·mol^{-1}. This value is slightly higher than the value which corresponds to a loss of a nickel atom according to a global process:

$$Ni \rightarrow Ni^{2+} + 2e^{-} \tag{3}$$

where F·Δm/ΔQ$_{theoretical}$ = 29.3 g·mol^{-1}. This behavior seems indicate a complex dissolution mechanism where a two consecutive electron transferences can be considered for the early stages of nickel electrodissolution[8].

EIS Measurements

Figure 2 shows the impedance spectra of nickel electrodissolution at different potentials. The experimental impedance spectra are fitted to the equivalent circuit of fig. 2.

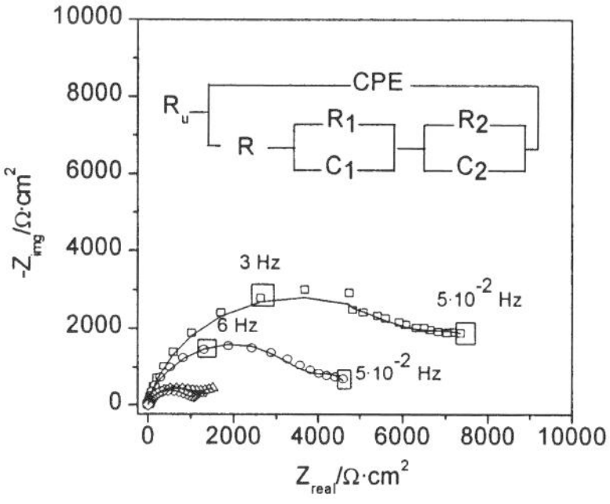

FIGURE 2. Experimental impedance spectra at E_0 = -650 mV (squares), E_0 = -600 mV (circles), E_0 = -585 mV (up triangles), E_0 = -555 mV (down triangles) and E_0 = -535 mV (diamonds). 0.245 M K_2SO_4, $5\cdot10^{-3}$ M H_2SO_4. pH = 2.7. T = 298 K. Solid line indicates the fitting to the equivalent circuit.

Two consecutive electron transferences can be considered as the initial stage for metal electrodissolution in many metals and also for nickel electrodissolution[8,11,14,15]:

$$Ni(0)(\theta_0) \xrightarrow{k1} Ni(I)(\theta_1) + e^{-} \xrightarrow{k2} Ni(II)(\theta_2) + e^{-} \xrightarrow{k3} Ni^{2+} \tag{4}$$

compatible with EQCM results. If we consider that the kinetic constants for the electrochemical steps follow a Butler-Volmer relationship and elementary steps obey a first order kinetic so the rates for each elemental step can be expressed as:

$$r_1 = k_{01}\theta_0 e^{b_1 E}\ ;\ r_2 = k_{02}\theta_1 e^{b_2 E}\ ;\ r_3 = k_{03}\theta_2 \tag{5}$$

and the mass balance and the charge balance at the electrode surface are:

$$\frac{i_F}{F} = r_1 + r_2 \,;\quad \frac{d\theta_0}{dt} = r_3 - r_1 \,;\quad \frac{d\theta_1}{dt} = r_1 - r_2 \qquad (6)$$

where θ_0, θ_1 and θ_2 are the surface concentrations of Ni(0), Ni(I) and Ni(II), respectively, on the electrode surface. The theoretical faradaic impedance for the above reaction mechanism can be calculated:

$$FZ_F = \frac{\frac{\partial r_1}{\partial\theta_0}\frac{\partial r_2}{\partial\theta_1}+(\frac{\partial r_1}{\partial\theta_0}+\frac{\partial r_2}{\partial\theta_1})\frac{\partial r_3}{\partial\theta_2}+(\frac{\partial r_1}{\partial\theta_0}+\frac{\partial r_2}{\partial\theta_1}+\frac{\partial r_3}{\partial\theta_2})j\omega-\omega^2}{2\frac{\partial r_2}{\partial E}\frac{\partial r_1}{\partial\theta_0}\frac{\partial r_3}{\partial\theta_2}+2\frac{\partial r_1}{\partial E}\frac{\partial r_2}{\partial\theta_1}\frac{\partial r_3}{\partial\theta_2}+(\frac{\partial r_2}{\partial E}\frac{\partial r_1}{\partial\theta_0}+2\frac{\partial r_1}{\partial E}\frac{\partial r_2}{\partial\theta_1}+\frac{\partial r_1}{\partial E}\frac{\partial r_3}{\partial\theta_2}+\frac{\partial r_2}{\partial E}\frac{\partial r_3}{\partial\theta_2})j\omega-(\frac{\partial r_1}{\partial E}+\frac{\partial r_2}{\partial E})\omega^2} \qquad (7)$$

In this theoretical faradaic impedance function two time constants are present, one for each reaction intermediate[16], as in equivalent circuit of figure 2. Impedance function for this circuit is:

$$Z_F = \frac{R + R_1 + R_2 + (R(R_1C_1 + R_2C_2) + R_1R_2(C_1 + C_2))j\omega - RR_1R_2C_1C_2\omega^2}{1 + (R_1C_1 + R_2C_2) - R_1R_2C_1C_2\omega^2} \qquad (8)$$

From equations (5), each partial derivative can be obtained. If we compare equation (7) and equation (8) we can obtain an equation system for the partial derivatives which can be numerically solved from the obtained values of the elements of equivalent circuit, table1. In Fig. 3 $\ln(\partial r_i/\partial\theta_i)$ *vs* E is plotted. From the slope and the intercept point of this plot we can obtain b_i and k_{0i} for each electron transference. The obtained values for b_i are very close to values previously published[17]. k_{03} is nearly potential independent as corresponds to a non-electrochemical process. We can also obtain the surface concentration of Ni(0) and Ni(I) on electrode surface, table 1. θ_0 is relatively small if compared with θ_1. In this sense nickel surface begins poorly active in electrodissolution process. Moreover, the k_3 value is smaller than k_2 in all the studied potential range. That way, the low frequencies behaviour is capacitive in these experimental conditions[15].

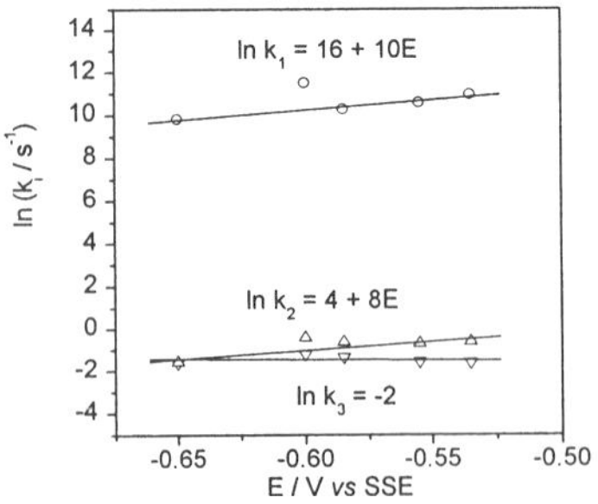

FIGURE 3. Kinetic parameters for the nickel electrodissolution process.

TABLE 1. Elements of the equivalent circuit and surface concentration of Ni(0) species, θ_0, and Ni(I) species, θ_1, on the electrode surface.

E (vs SSE) mV	R $\Omega\cdot cm^2$	R_1 $\Omega\cdot cm^2$	C_1 $\mu F\cdot cm^{-2}$	R_2 $\Omega\cdot cm^2$	C_2 $\mu F\cdot cm^{-2}$	θ_0 $mol\cdot cm^{-2}$	θ_1 $mol\cdot cm^{-2}$
-650	4	2750	750	7000	1	$4.9\cdot10^{-13}$	$8.0\cdot10^{-11}$
-600	2	1100	1250	4000	4	$4.9\cdot10^{-12}$	$1.8\cdot10^{-10}$
-585	6	700	2700	1270	8	$2.0\cdot10^{-12}$	$4.9\cdot10^{-10}$
-555	3	430	4400	1250	12	$4.1\cdot10^{-12}$	$4.8\cdot10^{-10}$
-535	3	300	7600	1000	8	$4.6\cdot10^{-12}$	$5.6\cdot10^{-10}$

CONCLUSION

EQCM results of nickel electrodissolution in acid media are consistent with a two consecutive electron transferences at the early stages of nickel dissolution. The theoretical impedance function can be obtained for such mechanism. The comparison of such function with the faradaic impedance of the equivalent circuit used in the fitting procedure allows to obtain kinetic information for each elementary process as well as the superficial concentration of each reaction intermediate. So, nickel dissolution is limited, in all the studied potential range at these experimental conditions, for the Ni(II) $\rightarrow$ Ni^{2+}_{aq} process, and in this way the low frequencies behavior is capacitive. The concentration of free sites on nickel surface, θ_0, is small compared with the concentration of Ni(I) species, θ_1, and nickel surface begins poorly active in electrodissolution process

ACKNOWLEDGMENTS

This work has been partially supported by CICyT-Mat/2000-011-P4. D. Giménez-Romero acknowledges a Fellowship from the Generalitat Valenciana (FPI program). J. Gregori acknowledges a Fellowship from the Spanish Education Ministery (FPU program). J.J. García-Jareño acknowledges the financial support of the program "Ramón y Cajal" from the Spanish Science and Technology Ministery.

REFERENCES

1. Real, S.G., Vilche, J.R., and Arvía, A.J., *Corros. Sci.* **20,** 563-& (1980).
2. Barbosa, M.R., Real, S.G., Vilche, J.R., and Arvía, A.J., *J. Electrochem .Soc.* **135**, 1077-1085 (1988).
3. Jouanneau, A., Keddam, M., and Petit, M.C., *Electrochim. Acta* **21,** 287-292 (1976).
4. Barbosa, M.R., Bastos, J.A., García-Jareño, J.J., and Vicente, F., *Electrochim. Acta* **44**, 957-965 (1998).
5. Bard, A.J., and Faulkner, L.R., *Electrochemical Methods. Fundamentals and Applications*, 2nd edn., New York: John Wiley & Sons, 2001, pp. 316-370.
6. Macdonald, D.D., *Transient Techniques in Electrochemistry*, New York: Plenum Press, 1977, pp. 229-311.
7. Sauerbrey, G., *Z. Physik* **155**, 206-& (1959).

8. Itagaki, M., Nakazawa, H., Watanabe, K., and Noda, K., *Corros. Sci.* **39**, 901-911 (1997).
9. Lachenwitzer, A., and Magnussen, O.M., *J. Phys. Chem. B* **104**, 7424-7430 (2000).
10. Giménez, D., García-Jareño, J.J., and Vicente, F., *Materiales y Procesos Electródicos I*, València: INSDE, 2002, pp. 65-84.
11. Giménez, D., García-Jareño, J.J., and Vicente, F., *J. Electroanal. Chem.* **558**, 25-33 (2003).
12. Vicente, F., García-Jareño, J.J., and Sanmatías, A., *Procesos Electródicos del NAFIÓN y del Azul de Prusia sobre electrodo transparente de óxido de indio-estaño: un modelo de electrodo multicapa*, Burjassot: Moliner-40, 2000.
13. Song, K.-D., Kim, K.B., Han ,S.H., and Lee H.K., *Electrochem. Commun.* **5**, 460-466 (2003).
14. Shao, H.B., Wang, J.M., Zhang, Z., Zhang, J.Q., and Cao, C.N., *J. Electroanal. Chem.* **549**, 145-150 (2003).
15. Giménez, D., García-Jareño, J.J., and Vicente, F., *Electrochem. Commun.* **4**, 613-619 (2003).
16. Gabrielli, C., and Keddam, M., *Electrochim. Acta* **41**, 957-965 (1996).
17. Jouanneau, A., and Petit, M.C., *J. Chim. Phys.* **73**, 82-88 (1976).

Applied Physics (APHYS 2003)
A. Méndez-Vilas (Ed.)

Electrical Properties Of Poly(Neutral Red) Deposited On Polycrystalline Nickel

J. Agrisuelas, J. J. García-Jareño, J. Gregori, D. Giménez-Romero, F. Vicente

Departament de Química Física, Universitat de València. C/ Dr. Moliner 50, 4610, Burjassot, València, Spain. e-mail address: proclq@uv.es (F. Vicente)

Abstract. A poly(neutral red) surface-layer has been electrogenerated by means of cyclic voltammetry on nickel surface. Electrical impedance measurements have proved that this polymeric material shows insulating or alternatively an intrinsic conducting character, which depends on the applied potential. This material is a colored lacquer that protects the metallic surface of its anodic dissolution.

INTRODUCTION

The poly(neutral red) (PNR) is included in an interesting family of electroactive polymers of phenazine dye [1-5]. Figure 1 shows the neutral red (NR) monomer structure (N^8,N^8,3-trimethylphenazine-2,8-diamine) as well as the scheme for the redox reaction.

During polymerization, the phenazine group of monomer preserves its electroactivity which remains intact [1-8] making this film useful as an electronic mediator catalyzing the reduction of different molecules [3], as redox indicator of bioelectrochemical electronic exchange processes [4,5], as a pH potentiometric sensor [3,5] or as a biosensor [9].

One interesting property of electrogenerated polymers on the surface of metals is their protective role against corrosion. That way, polymers based on aniline and pyrrole have been tested for this purpose [10-15]. This is a very essential tool in the particular case of technological materials based on nickel [16-20], since environmental nickel generates mutagenic and allergic effects on human body [21,22]. Moreover, it is necessary to consider that the mechanisms for the anodic dissolution of this metal are known [23-26].

The aim of this paper is to generate PNR on a polycrystalline nickel electrode and the study of this system by electrochemical impedance spectroscopy.

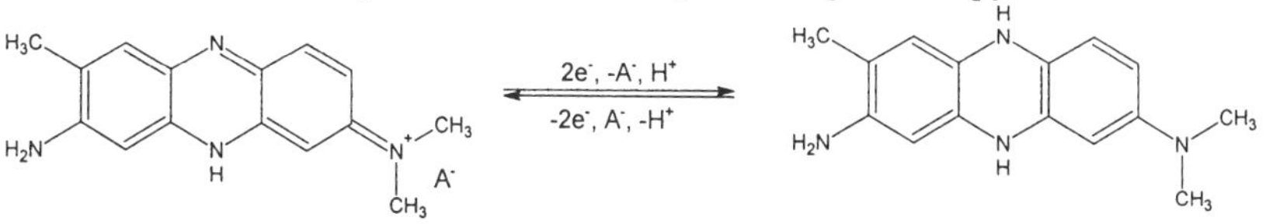

FIGURE 1. Structure and proposed redox reaction for monomer of neutral red [6].

EXPERIMENTAL

A typical three-electrode cell was used for electrogeneration. $Hg|Hg_2SO_4|K_2SO_4$ (sat.) was the reference electrode (all potentials are given *vs* this one) and a platinum mesh was used as counter electrode. The working electrode consisted of a polycrystalline nickel plate (Alfa®, Ni>99%). The cell was filled with 0.25 M K_2SO_4 (analytical grade, Panreac), 10^{-2} M H_2SO_4 (Merk) and $5x10^{-3}$ M NR (for microscopy, Merk) aqueous solution at 298 K (all reagents used as received) in an atmosphere of Argon U-N45 (Air liquide S. A.). The impedance measurements were carried out by means of RCL bridge (mod. PM 6304, Fluke and Phillips) connected up to PGS 81 potentiostat (Bank Elektronik) in the frequency range between 50 Hz and 100 KHz with 50 mV *rms* of signal amplitude. The above mentioned reference electrode was short-circuited with a platinum plate counter electrode (1.57 cm^2) for impedance measurements. The outer solution was the same but free of monomer at 298 K.

RESULTS AND DISCUSSION

Figure 2 shows a voltammogram of a Ni electrode in an aqueous solution containing NR monomer. Peak I can be attributed to the formation of NR radical cations and their polymerization [5]. The peak current associated to this peak increases with the number of cycles. This increase has been attributed to the fact that part of the electrogenerated polymer is oxidized at these potentials [8]. The peak potential displaces to more anodic potentials due to the modification of the electrode surface.

The peak III in Fig. 2 corresponds to the reduction of the polymer since both its current and involved electrical charge increase with cycles indicating the build-up of a surface-bound electroactive material. However, for the latest cycles, the hydrogen evolution has moved to less cathodic potentials and that peak is not detected. This displacement has been attributed to the fact that the polymer catalyzes the hydrogen evolution [4]. Besides, it should be noted that the polymer reduction takes place at the same potentials as the monomer reduction does [2]. It is not clear where the oxidation of the polymer takes place. Looking at Fig. 2, a wave is observed at potentials before peak IV that can be attributed to this oxidation, but that oxidation is likely to complete at potentials where the polymerization takes place.

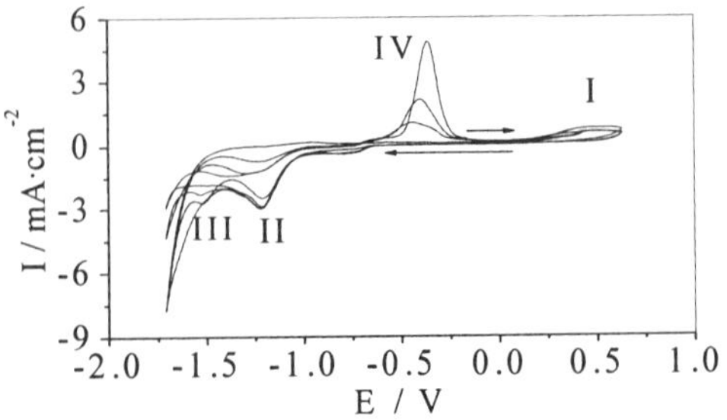

FIGURE 2. Successive voltammetric curves during PNR electrogeneration on polycrystalline nickel electrode in 0.25 M K_2SO_4, 10^{-2} M H_2SO_4 and $5x10^{-3}$ M NR aqueous solution. The arrows indicate the direction of the scans. E_i=-0.18V; E_{max}=-0.66V; E_{min}=-1.7V; v=20mV·s^{-1}; T=298K.

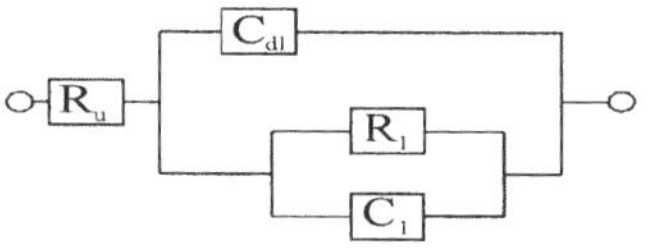

FIGURE 3. Equivalent circuit proposed for the Ni/PNR system.

Finally, peak IV is associated with the nickel electrodissolution where the formation of nickel hydroxide is favored [24] and the peak II to the nickel hydroxide reduction. Figure 2 shows that the peak current for the peak IV decreases quickly with the number of cycles. This decrease can be explained by both the formation of a passive layer of Ni hydroxide and the formation of PNR on the Ni electrode surface which prevents the dissolution. In agreement with other similar systems [14,15], the corrosion potential of nickel has been slightly moved to more anodic potentials due to the metal surface ennobling by the PNR film.

Impedance data were fitted to the equivalent circuit of Fig. 3 by simplification of a circuit that models a coated metal [27], where R_u is the uncompensated resistance due to cell assembly, C_{dl} is the double-layer capacitance at the metal-solution interface, R_1 is the transport and transfer charge resistance and C_1 is the charge transfer capacitance.

Figure 4 shows complex plane plots of capacitance. This capacitance (C) was calculated from raw impedance (Z) data by C=1/Zωj, where ω is the angular frequency and $j = \sqrt{-1}$. The higher frequencies loop (5-100 KHz) in these figures can be attributed to parameters R_u and C_{dl} in the equivalent circuit of Fig. 3. The real part of impedance at high frequencies corresponds to the R_u parameters and from the dependence of the imaginary part of impedance on ω it is possible to calculate the C_{dl}.

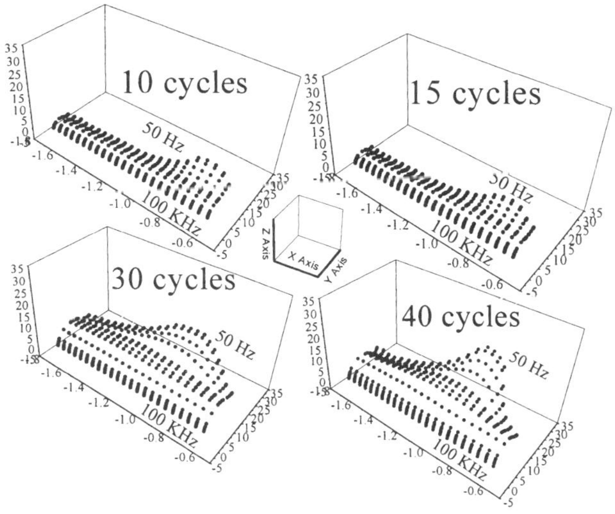

FIGURE 4. Complex plane plots of capacitance for the Ni/PNR electrodes in solution free of monomer at different potentials between 50 and 10^5 Hz. T=298K. C_{real}=-Z_{img}/X and C_{img},=Z_{real}-R_u/X (where X=[$(Z_{real}-R_u)^2+(Z_{img})^2$]ω). X-axis=E / V; Y-axis=C_{real} / μF·cm^{-2}; Z-axis=C_{img} / μF·cm^{-2}.

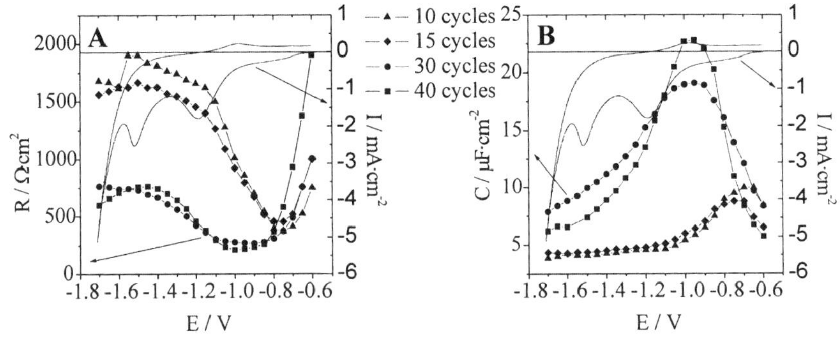

FIGURE 5. Plot of calculated resistance (a) and capacitance (b) values at 50 Hz. $R=[(Z_{real}-R_u)^2+(Z_{img})^2]/(Z_{real}-R_u)$ and $C=C_T-C_{dl}$ where $C_T=Z_{img}/[(Z_{real}-R_u)^2+(Z_{img})^2]\omega$. The continuous line corresponds to voltammetric curve at same interval of potentials from Fig. 2 (cycle 20).

The low frequencies loop (50-2000 Hz), associated to R_l and C_l parameters, is only well observed at potentials where faradaic processes take place. In fact, these parameters are related to these processes. It is also observed that this loop proves larger for the thickest films, and then, it could be attributed to the faradaic processes associated to the electrogenerated polymer.

Figure 5 shows the dependence of R_l and C_l parameters evaluated at 50 Hz with the film thickness. It is shown that at potentials where the polymer oxidation takes place (E=-1.0 to -0.9 V), R_l proves minimum and C_l proves maximum for the thickest films. For more cathodic potentials values of R_l are also smaller and C_l larger for the thickest films, a probable explanation would be the fact that the presence of polymer on the electrode surface catalyzes the hydrogen evolution.

Finally at more anodic potentials than -0.8 V the values for R_l are larger and the values of C_l are smaller for the thickest films. At these potentials where there is no faradaic process associated to the PNR film, these parameters are determined by the Ni oxidation process. Then, it seems that the presence of PNR on the electrode makes difficult the Ni oxidation reaction and causes the increase in the R_l value.

CONCLUSION

The electrogenerated film in aqueous media yields a colored lacquer that cover the nickel surface and that is able to inhibit nickel electrodissolution; this demonstrates a possible use as an anticorrosive coating. Moreover, from the experiments by means of RCL bridge, it has been demonstrated that this material can provide other interesting applications on metallic surface since it shows a dual property as conducting/insulating film.

ACKNOWLEDGMENTS

J. Agrisuelas and D. Giménez-Romero acknowledge a fellowship from Generalitat Valenciana, Consellería d'Educació i Ciència. J. Gregori acknowledges a fellowship from Ministerio de Educación, Cultura y Deporte. J. J. Garcia-Jareño acknowledges the financial support from Ministerio de Ciencia y Tecnología and FEDER (Programa Ramón y Cajal). PN and FEDER (MAT2000-0100-P4-03) have supported part of this work.

REFERENCES

1. Karyakin, A. A., Strakhova, A. K., Karyakina, E. E., Varfolomeyev, S. D., and Yatsimirsky, A. K., *Biolectrochem. Bioenerg.* **32**, 35-43 (1993).
2. Karyakin, A. A., Karyakina, E. E., and Schmidt, H. L., *Electroanalysis* **11**, 149-155 (1999).
3. Chen, S., and Lin, K., *J. Electroanal. Chem.* **511**, 101-114 (2001).
4. Inzelt, G., and Csahók, E., *Electroanalysis* **11**, 744-748 (1999).
5. Benito, D., García-Jareño, J. J., Navarro-Laboulais, J., and Vicente, F., *J. Electroanal. Chem.* **446**, 47-55 (1998).
6. Karyakin, A. A., Bobrova, O. A., and Karyakina, E. E., *J. Electroanal. Chem.* **399**, 179-184 (1995).
7. Benito, D., Gabrielli, C., García-Jareño, J. J., Keddam, M., Perrot, H., and Vicente, F., *Electrochem. Commun.* **4**, 613-619 (2002).
8. Benito, D., Gabrielli, C., García-Jareño, J. J., Keddam, M., Perrot, H., and Vicente, F., *Electrochim. Acta.* In press.
9. Sun, Y., Ye, B., Zhang, W., and Zhou, X., *Anal. Chim. Acta* **363**, 75-90 (1998).
10. Beck, F., Michaelis, R., Schloten, F., and Zinger, B., *Electrochim. Acta* **39**, 229-234 (1994).
11. Sazou, D., and Georgolios, C., *J. Electroanal. Chem.* **429**, 81-93 (1997).
12. Herrasti, P., and Ocón, P., *Appl. Surf. Sci.* **172**, 276-284 (2001).
13. Tallman, D. E., Spinks, G., Dominis, A., and Wallace, G. G., *J. Solid State Electrochem.* **6**, 73-84 (2002).
14. Spinks, G. M., Dominis, A. J., Wallace, G. G., and Tallman, D. E., *J. Solid State Electrochem.* **6**, 85-100 (2002).
15. Santos, J. R., Mattoso, L. H. C., and Motheo, A., *Electrochim. Acta* **43**, 309-313 (1998).
16. Dávila, M. M., Roig, A., Vicente, F., Martínez, E., and Scholl, H., *J. Mater. Sci. Lett.* **13**, 602-606 (1994).
17. Manini, P., Napolitano, A., Camera, E., Caserta, T., Picardo, M., Palumbo, A., and D'Ischia, M., *Biochim. Biophys. Acta* **1621**, 9-16 (2003).
18. Feng, F., Geng, M., and Northwood, D. O., *Int. J. Hydrog. Energy* **26**, 725-734 (2002).
19. Fritz, T., Mokwa, W., and Schnakenberg, U., *Electrochim. Acta* **47**, 55-60 (2001).
20. Bonilla, F. A., Ong, T. S., Skelton, P., Thompson, G. E., Piekoszewski, J., Chmielewski, A. G., Sartowska, B., and Stanislawski, J., *Corros. Sci.* **45**, 403-412 (2003).
21. Denkhaus, E., and Salnikow, K., *Crit. Rev. Oncol./Hematol.* **42**, 35-56 (2002).
22. Thomas, P., Barnstorf, S., Summer, B., Willmann, G., and Przybilla, B., *Biomaterials* **24**, 959-966 (2003).
23. Real, S. G., Vilche, J. R., and Arvía, A. J., *Corros. Sci.* **20**, 563-586 (1980).
24. Barbosa, M. R., Real, S. G., Vilche, J. R., and Arvía, A. J., *J. Electrochem. Soc.* **135**, 1077-1085 (1988).
25. Abd el Rehim, S. S., Abd el Wahaab, S. M., and Abd el Meguid, E. A., *Surf. Coat. Technol.* **29**, 325-333 (1986).
26. Hummel, R. E., and Smith, R. J., *Corros. Sci.* **30**, 849-854 (1990).
27. Silverman, D. C., "Primer on the AC impedance technique" in *Electrochemical techniques for corrosion engineering*, edited by NACE Publication, USA: 1986, pp. 73-79.

Applied Physics (APHYS 2003)
A. Méndez-Vilas (Ed.)

On the topology of two dimensional generalized cell systems

I. Zsoldos*, J. Janik*, T. Réti**
*Szent István University, Engineering Faculty, Godollo, Pater K. u. 1., Hungary
** Faculty of Technology Sciences, Szechenyi Istvan University, Hungary

Abstract: The applicability limits of the Aboav's law for trivalent polygonal systems have been described and this knowledge has been generalized.

INTRODUCTION

The growing interest in cellular pattern studies does not come as a surprise, for the knowledge gained from these studies bears far-reaching implications in a great variety of disciplines. The following scientific fields compile only a short list of the possible applications of studying cellular systems: biology (arrangement of human, animal and vegetable tissues, e.g. human and animal fat-tissue, etc.), geology (e.g. rock structure formations, etc.), agricultural science (e.g. structure of chopped straw, grain arrangement in stores, etc.), metallurgy (e.g. atomic structure of metallic glasses and grain structure of polycrystalline alloys, domain patterns of magnetic materials, etc.), technological applications (e.g. fracture of condensed materials, structure of products of powder metallurgy, internal structure of granulates, clustering of foams, etc.).
All these materials have one characteristic in common: their structures have similar geometric patterns. In our present work we discuss the topological properties of 2D patterns.
We find it suitable to distinguish two sets of the 2D patterns:

- Trivalent polygonal cell systems, where: The cells are polygons, or the system is topologically equivalent with a polygonal system. E.g. the edges of the cells in a 2D soap foam pattern are curved, but the system is topologically equivalent to a polygonal system. The consequence of this condition is, that the number of the boundary edges of a cell (EN) can be $3 \leq n < \infty$. The cells cover the plane without gaps and overlaps. Three edges bisect in every vertex.

Until now trivalent polygonal cell systems (like the examples above) have been mostly dealt with in the relevant scientific literature [1]- [13]. The topological properties and the applicability of several topological and metric laws (e.g. Euler's, Weaire's, Aboav's, Peshkin's, and von Neumann's laws) have been investigated experimentally on different real trivalent polygonal structures and theoretically in different models, computer simulations and other studies, in mechanical equilibrium or in dynamic development. Researchers were published in numerous topics with entirely different systems such as: grain section of polycrystals [1], soap foams [1], Bénard-Marangoni convective structures [2], crack patterns in ceramics [3], magnetic liquid froths [4], cork [5], biological tissues [6], topological models and scaling properties for coarsening in soap froths [7], the shell analysis in random cellular structures [8], the entropic prediction [9], random cellular models generated from Ising ferromagnet [10], structures generated by random fragmentation [11], random Hamiltonian cellular model [12], and models for fractal cellular structures [13].

- Generalized cell systems, where:

The shape of the cells may also be a multiple contiguous area (Fig. 1.), and the boundary edges can be either straight or curved. The points where the curve is non-derivable define the vertices. The number of the boundary edges of a cell (EN) can be n=1 (e. g. circle) or $1<n<\infty$. Two or more edges bisect each other at each vertex. The cells have a complete tessellation, cover a plane without gaps or overlaps. (The actual gap is considered to be a new cell.) There is only one condition: the direct neighbours of a cell can only be other cells, a cell cannot be its own neighbour.

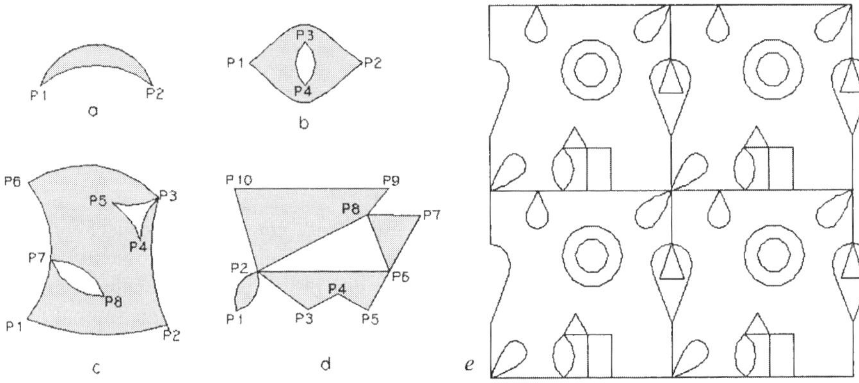

Figure 1. *a-d:* Examples for generalized cells (cells with *a*: n=2, *b*: n=4, *c*: n=9, *d*: n=14 ENs, the grey area is the cell). *e:* A part of a generalized cell system. The pattern has four elementary units. By imagining their parallel shifts we are able to imagine the entire system [14].

We mention only several examples for materials having tricky (not polygonal in their 2D sections) internal structure: a lot of biological tissues (e. g. epidermic tissue of the stomach or the intestines, the kidney, the ovary, or a simple human cell), the chopped straw, the most plastics, a lot of ceramic materials, the ESD magnetic materials. For generalized systems only the applicability of the Weaire's law was investigated [14].
The goal of our present paper is to describe the applicability limits of the Aboav's law for trivalent polygonal systems and to generalize this knowledge.

THE APPLICABILITY OF THE ABOAV'S LAW

Aboav's law declares a linear connection between the EN of the cells (n) and the mean total EN of first-neighbour cells of cells bearing n EN (nm(n)):

$$nm(n) = (\langle n\rangle - a)n + \langle n\rangle a + \mu_2 \qquad (1)$$

- $\langle n\rangle$ is the average EN in the system,
- $\mu_2 = \sum_{i\geq 1} (\langle n\rangle - i)^2 p_i$ is the so-called second moment where p_i is the relative frequency of the cells having i EN,
- 'a' is a system-constant.

In the early 1980 Aboav showed that the law bearing his name is only an approaching connection [1]. He found a very good approximation on real 2D trivalent polygonal systems, for example on the patterns of polycrystalline alloys sections and soap foam. But he pointed out that the approximation can deviate considerably from the principal value.
The accuracy of the linear formula (Eq. 1) is determined by the deviations from the nm(n) values. Cells with the same EN might have many kinds of neighbourhoods. Let

us denote by t(n) the total EN of first-neighbour cells of a particular cell with n EN. t(n) can fluctuate from cell to cell and thus it is a distributed quantity.
t(n) is a random variable and its mean is $\langle t(n)\rangle = nm(n)$.
Note that if a cell with k sides shares i sides with the central n-sided cell, k is counted i times in t(n) for generalized systems.
In the next part we demonstrate that the deviation of t(n) around the linear fit has a limit, which is smaller in the case of trivalent polygonal systems than in that of generalized systems.

Trivalent polygonal systems

The accuracy of the linear regression is characterised by the correlation coefficient (Pearson product) $R\leq 1$, and, as it is well known, the closer the R value is to 1, the more accurate the linear fit is.

Let us determine the minimal R value. Denote the minimal and maximal EN by n_{min}, and by n_{max}, repeatedly:

- The upper limit of the t(n) values is nn_{max}.
- The lower limit of the t(n) values is: $4n+3$ if $n_{min}<5$, nn_{min} if $n_{min}=5$ and $t(6)=nm(6)=36$ if $n_{min}=n_{max}=6$.
- The possible t(n) values plotted in a diagram (Fig.2) are in a domain (in a quadrangle). The actual domain depends on the values of n_{min} and n_{max}, Fig. 2.

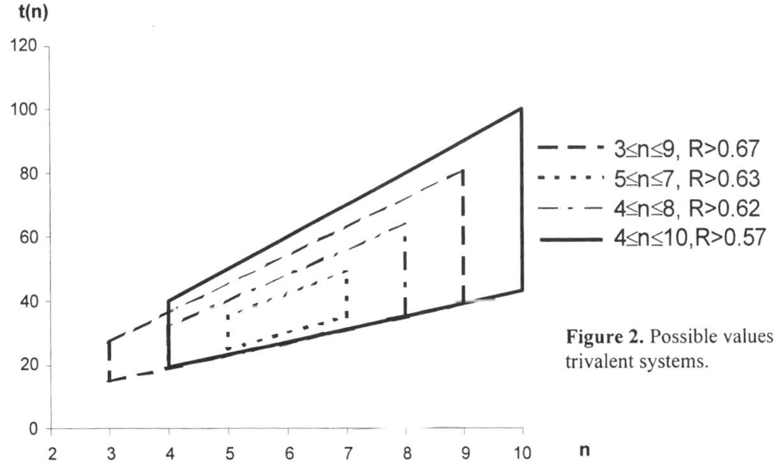

Figure 2. Possible values of t(n) for trivalent systems.

When will the accuracy of a linear regression be the worst (R has its minimum), if the possible values of t(n) are confinable with a quadrangle? In the worst case:

- there are t(n) values only at the boundary of the quadrangle,
- there are t(n) values in every possible point of the boundary.

In this extreme case the theoretical minimum of R (R_{min}) can be computed. We show the values of R_{min} for several case of $n_{min}\leq n\leq n_{max}$ in Figure 2.
Actually the extreme case cannot occur in trivalent cell systems. If we try to approach the extreme case, we obtain the following results: cells having t(n) value at the boundary of the domain can only rarely occur in a cell system. Surrounding these cells there have to be a lot of cells having t(n) value inside the domain in the system, Figure 3. This is a strong topological constraint, its consequence for trivalent polygonal systems is that the R value is significantly higher than R_{min}. In Figure 3, we show a

trivalent pattern in which the extreme values of t(n) occur. But only five t(n) values from the 113 cells are at the boundary. Any larger deviation than this rate cannot be enforced, therefore the value of R is much higher than the theoretical minimum for trivalent polygonal systems.

Figure 3. Left: O ne unit of a trivalent periodical pattern generated from pentagons, hexagons and

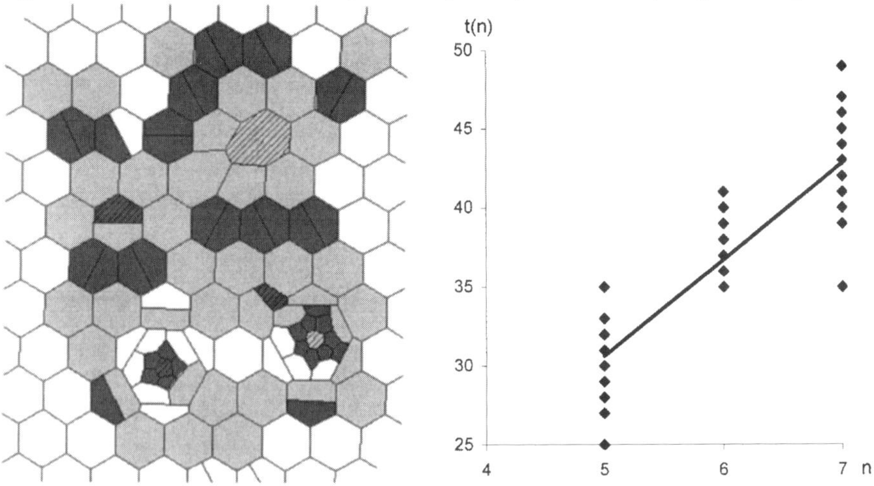

heptagons. The environment of the hatched cells consists of solely pentagons or solely heptagons. There are only five such ones out of 113 cells. Right: The dispersion of the t(n) values is high. Nevertheless, the value of R is significantly higher than the theoretical minimum: $R=0.92>R_{min}=0.63$.

Generalized systems

Values of t(n) have limits in generalized systems as well. The upper and the lower limits are: nn_{max} and nn_{min}. If $n_{min}=1$, then the lower limit is n+1. The possible t(n) values are in a trapeziform domain in this case as well. This domain can be drawn for every case of $n_{min}\leq n\leq n_{max}$, Figure 4.

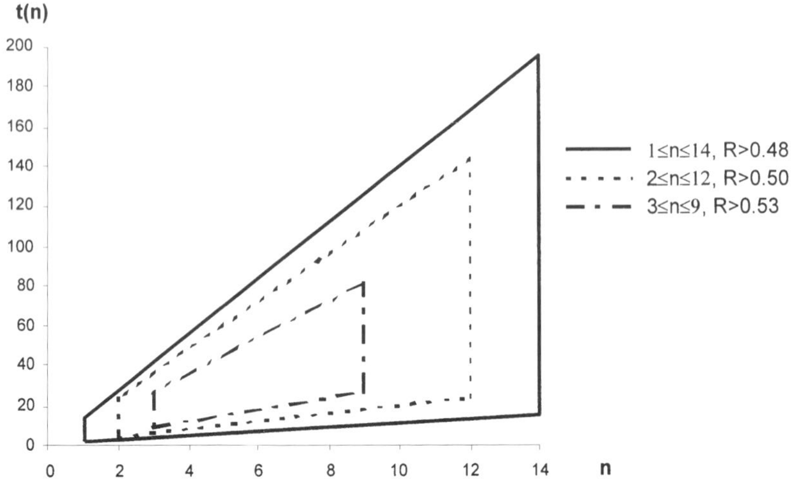

Figure 4. Possible values of t(n) for generalized systems

The lower limit of t(n) values either equals or is smaller for generalized systems than for trivalent polygonal systems. Therefore the theoretical minimum of R equals or it is smaller for generalized systems than for trivalent polygonal systems. E. g., for the $3 \leq n \leq 9$ case R_{min} is 0.67 and 0.53 for trivalent and generalized systems, respectively, Figure 4. On the other hand, for the $5 \leq n \leq 7$ case the theoretical minima of these systems are equal to each other.

That extreme case in which every t(n) value is at the boundary of the domain, cannot occur in generalized systems either. But the number of cells found at the boundary can be much greater than the one observed in the trivalent case. Therefore the actual R value comes closer to the R_{min} value, Figure 5.

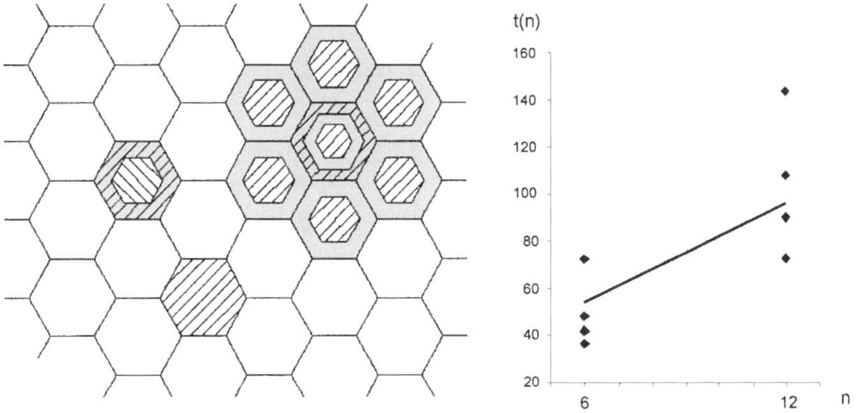

Figure 5. Left: Generalized cell system consists of cells having n=6 and n=12 ENs. The t(n) value of the hatched cells (one third of the cells) is extreme. Right: The dispersion of the t(n) values is high. The R value is only a bit higher than the theoretical minimum: $R=0.78>R_{min}=0.60$.

CONCLUSION

Aboav's law is a good approximation for the trivalent polygonal systems and this approximation is weaker in generalized systems. The different appearance of the topological constraints causes the difference between trivalent and generalized systems.

ACKNOWLEDGEMENTS

This work was supported by OTKA grants T 42717 and the Széchenyi István Fellowship in Hungary.

REFERENCES

[1] D.A.Aboav Metallography 13: 43-58, 1980, 16: 265-273, 1983.
[2] P.Cerisier,S.Rahal,N.Rivier Phys.Review E 54:5086-5094, 1996.
[3] W.Korneta,S.K.Mendiratta,J.Menteiro Phys.Review E 57:3142-3152,1998.
[4] F.Elias,C.Flament,J.C.Bacri,O.Cardoso,F.Graner Phys.Rev.E 56:3310-3318,1997.
[5] P.Pina,M.A.Fortes J.Phys.D 29:2507-2514,1996.
[6] J.C.M.Mombach, R.M.C.deAlmeida, J.R.Iglesias Phys.Review E 47:3712, 48:598, 1993
[7] F.Bolton, D.Weaire Phil.Mag.B **63**:795- 809,1991.
[8] K.Y.Szeto, X.Fu, W.Y.Tam Phys. Rev. Letters 88: 138302, 2002.
[9] M.A.Peshkin,K.J.Strandburg,N.Rivier Phys.Rev.Lett. 67:1803-06,1991.
[10] G. Le Caer, R.Delannay J.Phys.A:Math. 26:3931,1993.
[11] R.Delannay, G. Le Caer Phys. Rev. Letters 73:1554, 1994.
[12] G.Schliecker, S. Klapp Europhys. Letters 48 (2):122, 1999.
[13] G. Le Caer, R.Delannay J.Phys. I France 5:1417, 1995.

[14] T.Réti, I.Zsoldos Proceeding of the 8th Seminar of the International Federation for Heat Treatment and Surface Engineering Dubrovnik, p.381, 2001.

Applied Physics (APHYS 2003)
A. Méndez-Vilas (Ed.)

Effect of Reaction Parameters on Morphology of Synthesized MFI

Phairat Phiriyawirut[a], Rathanawan Magaraphan[a], Alexander M. Jamieson[b] and Sujitra Wongkasemjit[a]

[a]*The Petroleum and Petrochemical College, Chulalongkorn Unitversity, Bangkok 10330, Thailand.*
E-mail: phairat.p@student.chula.ac.th
[b]*The Macromolecular Science Department, Case Western Reserve University, Cleveland, Ohio, USA.*
E-mail: amj@po.cwru.edu Continue Here

Abstract. New synthetic route of MFI using silatrane as the precursor and tetrabutylammonium hydroxide (TBAOH) as a template was studied. Synthesis formulation is focused on the effect of OH^-, Na^+ and template concentration as well as operating conditions as aging, heating time and temperature on the morphology of MFI crystal. The rate of MFI formation decreases with increasing Na^+, decreasing OH^- or lowering template concentration. Higher temperatures correlate to shorter aging and heating times. At fixed temperature, longer aging times lead to shorter heating times. It is found that a longeragi ng time is more important to achieve high crystallinity than the heating time.

INTRODUCTION

MFI is classified as a high silica zeolite, and synthesized by interaction between a silicate precursor and an organic template, acting as guest molecule in the structure of MFI. In some cases, the template can be omitted, for example in a high aluminum MFI synthesis [1]. Several templates have been studied experimentally and via theoretical calculation [2-3]. The most effective organic templates are alkylammonium derivatives [4-6], and tetrapropylammonium salt (TPA) is the most common template for MFI synthesis, utilised in many published studies over the past few years [7-13].

De Moor et. al. (in 1999) have proposed a nucleation and growth mechanism for MFI formation from a mixture of silicic acid and tetrapropylammonium hydroxide (TPAOH) and found that nanometer-scale primary units ("nanoblocks") containing a specific MFI topology are formed [10,13]. Nucleation involves the aggregation of these primary particles. Crystal growth occurs via stepwise addition of primary particles throughout the reaction period. In the same year, Martens et. al studied the reaction of TEOS and concentrated TPAOH [11-12] and observed the same result as reported in the study of De Moor et al [10,13].

We investigate a new synthetic route to MFI, using silatrane as the precursor, which was successfully used as precursor for mesoporous [14] and microporous [15] syntheses. The effect of OH^-, Na^+ and template concentration on the morphology of MFI crystals is explored. Other operating conditions, namely, aging time, heating time and temperature are also investigated

EXPERIMENTAL

Instruments

FTIR spectroscopic analysis was conducted using a Bruker-EQUINOX55 with a resolution of 4 cm^{-1}. Mass Spectrometry was carried out with a Fison VG Autospec model 7070E, using the positive fast atomic bombardment (FAB^+-MS) mode and glycerol as a matrix. Thermogravimetric analysis utilized a Perkin Elmer TGA7 at a scanning rate of 10 °C/min under nitrogen atmosphere and simultaneous thermal analysis (STA) was conducted using a Netzsch-STA 409 at a scanning rate of 20 °C/min under nitrogen and oxygen atmosphere. Crystal morphology was characterised using a JEOL 5200-2AE scanning electron microscope. Crystal structure was determined using a Rigaku X-Ray Diffractometer at a scanning speed of 5 degree/sec and CuK as radiation. Hydrothermal treatment by microwave heating technique was conducted using a MSP1000, CME Corporation.

Methodology

Silatrane (Tris(silatranyloxy-ethyl)amine or SiTEA) was synthesized following Wongkasemjit's method [16] by reacting 0.125 mol triethanolamine with 0.1 mol silicon dioxide using ethylene glycol as solvent at 200 °C under nitrogen atmosphere. The reaction was complete within 10 hr, and the reaction mixture was cooled to room temperature, before removing excess solvent by distilling under vacuum (10^{-2} torr) at 110 °C overnight. The resulting brownish white solid was washed three times with dry acetonitrile to obtain a fine white powder. The product was characterized using XRD, TGA, FTIR and FAB^+-MS.

The silatrane product: FAB^+MS; 100% intensity of 174 m/e ($N[CH_2CH_2O]_3Si^+$), 11.3% of 236 m/e ($H_2^+OCH_2CH_2OSi[OCH_2CH_2]_3N$) and 2.6% of 323 m/e ($H^+[HOCH_2CH_2]_2NCH_2CH_2OSi[OCH_2CH_2]_3N$) . FTIR; 3422 cm^{-1} (ν OH), 2986-2861 cm^{-1} (ν CH), 2697 cm^{-1} (ν N→Si), 1459-1445 cm^{-1} (δ CH), 1351 cm^{-1} (ν CN), 1082 cm^{-1} (ν Si-O-C), 1049 cm^{-1} (ν CO) and 579 cm^{-1} (ν N→Si). XRD; MFI crystalline product. TGA; one mass loss transition at 400°C with 18.36% ceramic yield (close to the theoretical yield of the $Si((OCH_2CH_2)_3N)_2H_2$ structure).

SiTEA equivalent to 0.25 g SiO_2 was dispersed in 5 ml of water and continuously stirred before adding TBAOH. Hydroxide and Na^+ concentrations were controlled by adding calculated amounts of NaOH and NaCl, respectively. The formulation of SiO_2:TBA:OH^-:Na^+:H_2O was fixed at 0.1:0.4:0.4:114 to evaluate the effect of aging time, heating time and temperature. All reaction mixtures were aged at room temperature for various times with continuously stirring prior to heating. Different proportion of TBA, hydroxide ion and sodium ion were studied to observe the effect on morphology of MFI crystals.

RESULTS AND DISCUSSION

Effect of Aging and Heating Times

From the experimental study, to investigate the influence on MFI crystals of increasing the heating time, the aging time and microwave temperature were fixed at 84 hr and 150 °C, respectively. After 5 hr of heating, agglomeration of amorphous material was observed and very few of small crystals of 2 x 1 micron in orthorhombic shape were present, as shown in figure 1a. This was confirmed by XRD analysis which showed a broad amorphous peak around 23 /2theta together with some characteristic peaks of MFI at 7.95, 8.90 and 23.18 /2theta, see figure 2a. The broad amorphous XRD peak largely disappears when the sample mixture has been heated for 10 hr. However, there remains a little amorphous phase as most clearly seen in the SEM micrograph (figure 1b), which shows that crystals up to 6 x 2 micron are present, with small particles of amorphous material on their surfaces. The MFI crystal

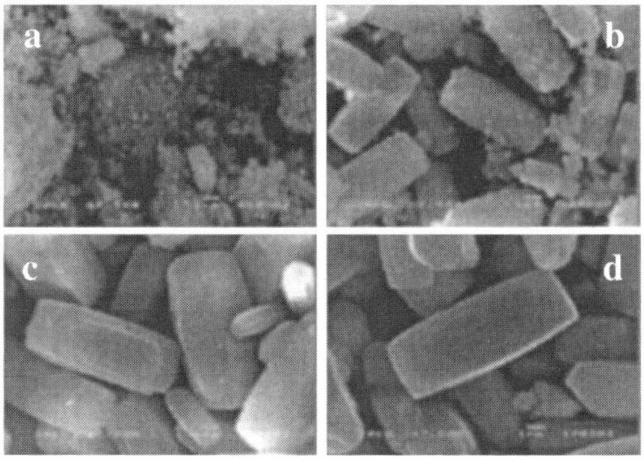

FIGURE 1. Effect of heating time on reaction of SiO_2: 0.1 TBA: 0.4 OH^-: 0.4 Na^+: 114 H_2O at 150°C, after 84 hr aging time a) 5 hr, b) 10 hr, c) 15 hr

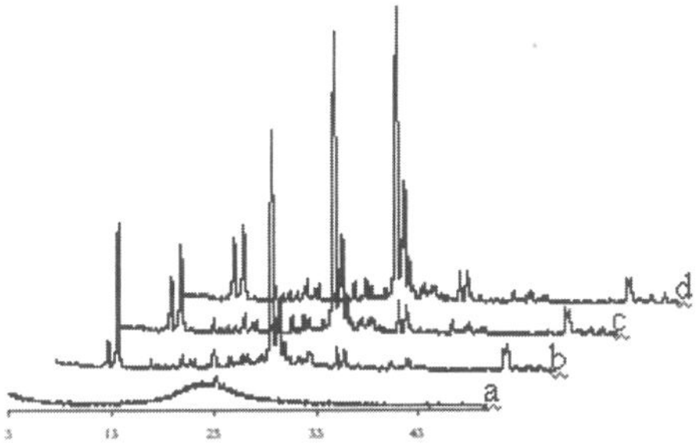

FIGURE 2. XRD patterns of sample in Figure 1. a) 5 hr b) 10 hr c) 15 hr and d) 20 hr

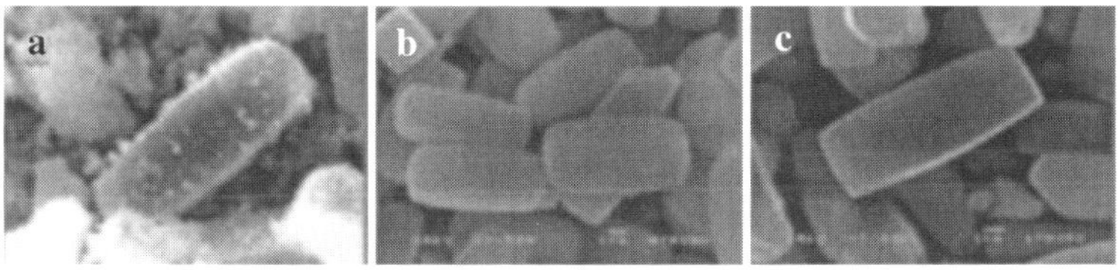

FIGURE 3. Effect of aging time on morphology of products formed from SiO_2: 0.1 TBA: 0.4 OH^-: 0.4 Na^+: 114 H_2O mixtures with heating for 20 hr at 150°C: a) 36 hr, b) 60 and c) 84 hr.

size increases with heating time. However, the heating time does not have any effect on the crystal size after 15 hr of heating, as seen by comparing figures 1c and 1d.

As mentioned earlier, 84 hr aging time is required to produce perfect, fully-grown crystals. This result is illustrated in figure 3. When the mixture is aged for 36 hr, more amorphous phase is obtained (Fig. 3a); as the aging time is increased to 60 hr, no amorphous phase is observed, but the crystals are not fully-grown (Fig. 3b); after aging for 84 hr, fully-grown crystals are formed.

These results demonstrate that the aging process is important for synthesis of MFI from silatrane precursor. It is well known [17] that sol-gel processing via hydrolysis and condensation reaction causes the silicate precursor to polymerize and that the hydrolysis does not involve all the alkyl ligand initially, thus some organic component is present in the obtained polysilicate. This organic component can facilitate homogeneous mixing with the added organic template. The capability to enhance the insertion of organic template into the primary unit of MFI during hydrolysis allows more controllable condensation prior to primary unit formation. Mechanistic studies of MFI formation report that primary particles are always formed despite the fact that different types of precursor, either inorganic silicate [9] or organosilicate compound [11-12], are used. Therefore, we expect that primary particles exhibiting MFI structure are also formed in synthesis from silatrane.

Effect of Temperatrue

It is well known that temperature strongly influences the formation of zeolites, using either organosilicate or inorganic precursors. The optimal temperature range depends on the Si/Al ratio. Pure silicate systems, such as MFI, tend to require higher temperatures, around 140° to 180 °C [18]. We studied the effect of temperature at 120°, 150° and 180 °C, using the sample ratio SiO_2:0.1TBA:0.4OH^-:0.4Na^+:114H_2O, an 84 hr aging time and 10 hr heating time. Selected SEM results are shown in figure 4. Heating at 120°C for 10 hr produces a morphology (Fig. 4a) and XRD result (not shown) similar to that produced by heating for 5 hr at 150 °C (c.f. Fig. 1a and Fig. 2a), i.e. mostly amorphous materials with few MFI crystals. However, when the heating time was extended to 20 hr, the amount of amorphous material decreases, and many small crystals of size around 1 micron are observed (figure 4b). These observations indicate that the temperature is too low for efficient MFI crystallization. At higher temperature (180 °C), larger crystals are obtained in shorter times (10 hr), as seen by comparing figure 4b with figure 4c.

It appears that MFI formation can be attained using a shorter heating and/or aging time with higher temperature. However, use of the microwave technique in zeolite synthesis at high temperature, especially at 180 °C, can degrade the organic template via Hoffman reaction, as reported by Arafat et al [7]. In our experiment, the clear colorless solution after thermal treatment at 180°C became reddish brown. However, perfect MFI crystals were also formed. It is possible that the template is degraded after the primary particles were formed, since, as explained by de Moor et al [9,13] the primary particles are present throughout the heating period and should have

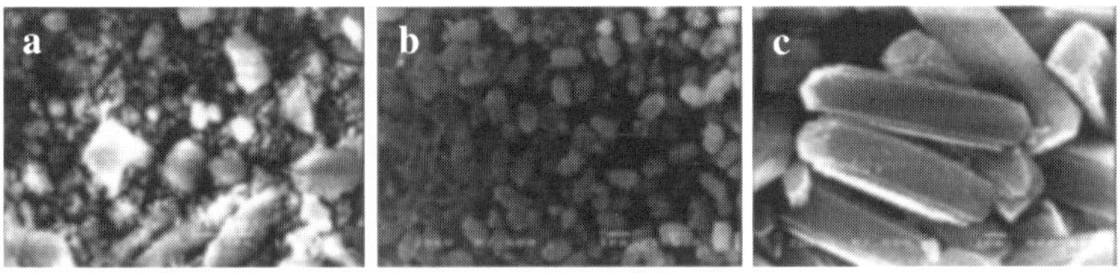

FIGURE 4. Effect of temperature on products formed from SiO_2: 0.1 TBA: 0.4 OH^-: 0.4 Na^+: 114 H_2O after 84 hr aging time and heating at: a) 120°C for 10 hr, b) 120°C for 20 hr and c)180°C for 10 hr.

more stability to withstand the reaction temperature. These authors also found that the crystallization rate of MFI increases with reaction temperature, which correlates with our observation that the same crystal can be synthesized by shorter heating times at higher temperatures.

Role of Organic Template

Template molecules play a key role in zeolite synthesis, especially in the case of pure silicate or high silicate zeolites, such as MFI. Most studies agree that the optimal template for MFI synthesis is tetrapropylammonium (TPA). In the present study, we have succeeded in synthesizing MFI using a TBAOH template. This may be related to the fact that we have also used a different precursor, silatrane. For example, it is possible that, since silatrane generates a large organic ligand during hydrolysis, this makes the polysilicate more homogeneous with organic template.

The effect of template concentration was studied at TBA/Si = 0.017, 0.034, 0.1 and 0.15. Formulation variables were fixed at SiO_2:xTBA:0.4OH^-:0.4Na^+:114H_2O (x = 0.017, 0.034, 0.1

and 0.15), and reaction conditions at 20 hr heating time, 150°C heating temperature and 84 hr aging time. As shown in the figure 5, the crystal size at TBA/Si = 0.017 is esentially the same size for TBA/Si = 0.034. However, at low TBA concentration more agglomerates of amorphous silicate are formed. This result is consistent with our observation that increased aging time enhances the crystallization rate. The concentration of template molecules is directly proportional to that of the primary units formed, which relates in turn to the number of nucleation sites.

Theoretically, complete conversion should be obtained at the stoichiometric equivalent ratio of 0.034 TBA/Si. However, our results indicate that some template may not participate in the formation of MFI, possibly due to interactions between template and organic ligands derived from the silatrane. As a result, MFI has to be synthesized at much higher template concentration. When TBA/Si = 0.1, there is no agglomeration, and the crystal size becomes smaller. Evidently,

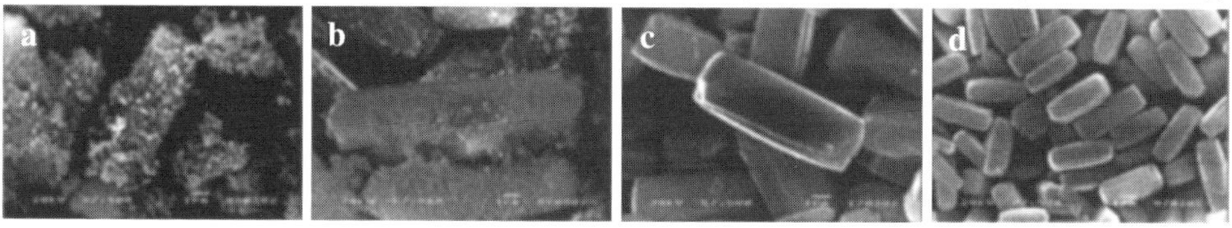

FIGURE 5. Effect of template concentration on products formed from SiO_2: x TBA: 0.4 OH^-: 0.4 Na^+: 114 H_2O after aging for 84 hr and heating at 150°C for 20 hr: TBA/Si ratios, x, are a) 0.017, b) 0.034, c) 0.1 and d) 0.15 with aged.

higher template concentration results in more nucleation events, resulting in smaller crystals. Thus the crystal size strongly depends on the concentration of organic template.

Effect of Sodium Ions

Here, we formulated the reaction mixture as SiO_2:0.1TBA:0.4OH^-:xNa^+:114H_2O where x = 0.3, 0.6, 0.9, with reaction conditions set at 84 hr aging time, 10 hr thermal treatment and 150 °C heating temperature. At 0.3 Na^+/Si, large crystals of MFI are formed together with a small amount of amorphous material. Completely amorphous product was obtained at Na^+/Si = 0.6 and 0.9. This result is consistent with the conclusion of Koegler et al [8] that sodium ion and organic template have opposing effects on the formation of primary units. Sodium ions favor the formation of negatively charged SiO^- groups, whereas the organic template prefers when more hydrophobic Si-O-Si bonds are formed. Excess of sodium ion thus reduces the interaction between template and precursor, causing agglomeration of polysilicate to form dense amorphous phase.

On the other hand, lack of sodium ion in the system also results in less stabilization of nucleation step. We found the bigger crystal and amorphous obtained in the less sodium than 0.3 Na^+/Si. The fact is that in the lower sodium system, less conversion of precursor to crystal was obtained, due to less nucleation.

Effect of Hydroxide Ions

Hydrolysis of precursor occurs in the initial stages of the reaction, and strongly depends on the OH^-/Si ratio. Thus the role of hydroxide ion in zeolite synthesis is a major factor to be considered. We fixed the formulation at SiO_2:0.1TBA:xOH^-:0.4Na^+:114H_2O (x= 0.1, 0.2, 0.3, 0.4 and 0.5), with reaction conditions at 84 and 20 hr aging and heating time, respectively, 150°C. At 0.1 and 0.2 OH^-/Si, no MFI is formed. Crystals are observed from mixtures containing 0.3 to 0.5 OH^-/Si. The crystal size decreases with increasing OH^-/Si ratio.

At 0.1 OH^-/Si, a white cloudy gel was obtained after 84 hr aging time. Only at this condition was a gel formed. At 0.2 OH^-/Si, a white cloudy sol was obtained, and the reaction mixtures became clearer as the OH^-/Si ratio was further increased. The formation of gel becomes unfavorable during the aging period due to fast hydrolysis followed by condensation reaction with precursor, causing fewer interactions between precursor and TBA. Moreover, lower OH^- also causes fewer nucleation events, as can be seen by the larger crystal size and the presence of amorphous material in the SEM micrographs (figure 6a). At 0.5 OH^-/Si, the crystal size is smaller and some crystal fracture

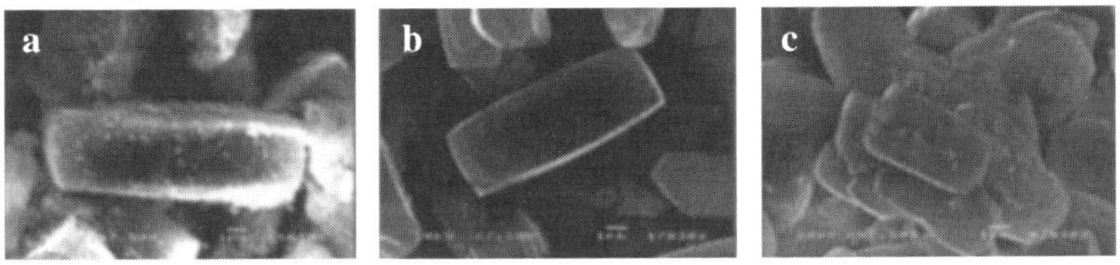

FIGURE 6. Effect of hydroxide concentration on products formed from SiO_2: 0.1 TBA: x OH^-: 0.4 Na^+: 114 H_2O after aging for 84 hr and heating at 150°C for 20 hr: OH^-/Si ratios, x, are a) 0.3, b) 0.4 and c) 0.5.

appears to occur, as shown in figure 6c. Possibly, dissolution of crystalline material under the highly basic conditions may take place under microwave conditions, as indicated in the work of Arafat et al [7]. The OH^-/Si ratio of 0.4 appears to be the optimum since no amorphous is observed (Fig. 6b).

CONCLUSIONS

Silatrane complex SiTEA was synthesized via the OOPS process and used as a silica source for the MFI synthesis. Aging time of the reaction mixture prior to heating plays a very important role in the formation of MFI. Complete conversion of SiTEA to MFI crystals was obtained with longer aging times. However, heating time also has an effect on SiTEA conversion, a longer heating time resulting in better conversion. Higher temperatures can be applied to curtail aging and heating times. As for the effect of OH^-, Na^+ and template concentration, less perfect crystals of MFI are obtained when increasing Na^+, decreasing OH^-, or decreasing TBA concentration.

ACKNOWLEDGMENTS

This research work was fully supported by the Thailand Research Fund (TRF), Postgraduate Education and Research Program in Petroleum and Petrochemical Technology (ADB) Fund and Ratchadapisake Sompote Fund, Chulalongkorn University.

REFERENCES

1. Machado F.J., Lopez C.M., Centeno M.A. and Urbina C., Appl. Cat. A 181, 29-38 (1999).
2. Boyett R.E., Stevens A.P., Ford M.G., and Cox P.A., Zeolites 17, 508-512 (1996).
3. Chatterjee A., J. Mol. Cat. A 120, 155-163 (1997).
4. Moini A., Schmitt K.D., and Polomski R.F., Zeolites 18, 2-6 (1997).

5 Rollmann L.D., Schlenker J.L., Lawton S.L., Kennedy C.L., Kennedy G.J. and Doren D.J., J. Phys. Chem. B 103, 7175-7183 (1999).

6 Rollmann L.D., Schlenker J.L., Kennedy C.L., Kennedy G.J. and Doren D.J., J. Phys. Chem. B 104, 721-726 (2000).

7 Arafat A., Jansen J.C., Ebaid A.R., and Bekkum H., Zeolites. 13, 162 (1993).

8 Koegler J.H., Bekkum H. and Jansen J.C., Zeolites 19, 262-269 (1997).

9 De Moor P-P.E.A., Beelen T.P.M., and Santen R.A., J.Phys. Chem. B.103, 1639-1650 (1999).

10 Ravishankar R., Kirschhock C.E.A., Knops-Gerrits P-P., Feijen E.J.P., Grobet P.J., Vanoppen P., De Schryver F.C., Miehe G., Fuess H., Schoeman B.J., Jacobs P.A., and Martens J.A., J. Phys. Chem B 103, 4960-4964 (1999)

11 Kirschhock C.E.A., Ravishankar R., Verspeurt F., Grobet P.J., Jacobs P.A. and Martens J.A., J. Phys. Chem. B 103, 4965-4971 (1999).

12 Kirschhock C.E.A., Ravishankar R., Looveren L., Jacobs P.A. and Martens J.A., J. Phys. Chem. B 103, 4972-4978 (1999).

13 De Moor P-P.E.A., Beelen T.P.M., Santen R.A., Beck L.W.and Davis M.E., J. Phys. Chem. B 104, 7600-7611 (2000).

14 Cabrera S., Haskouri J., Guillem C., Latorre J., Beltran-Porter A., Beltran-Porter D., Marcos M.D., and Amoros P., Solid State Sci. 2, 405-420 (2000).

15 Sathupunya M., Gulari E., and Wongkasemjit S., J. Eur. Cera, Soc., 22, 2305-2314 (2002).

16 Piboonchaisit P., Wongkasemjit S. and Laine R.M., J. Sci. Soc. Thailand 25, 113 (1999).

17 Klein L.C., Engineered Materials Handbook. Ceramics and Glasses Volume 4, ASM International The Materials Information Society, USA, pp. 209-213.

18 Cundy C.S., Collect. Czech. Chem. Commun. 63, 1699 (1998).

Applied Physics (APHYS 2003)
A. Méndez-Vilas (Ed.)

Ipen Environmental Monitoring Programme: Assessment of the Gamma Radiation Levels with Thermoluminescence Dosimeters

Brigitte R. S. Pecequilo, Marcia P. Campos, Marcos M. Alencar and Marcelo B. Nisti

Environmental RadiometricDivision, Instituto de Pesquisas Energéticas e Nucleares, Av. Prof. Lineu Prestes, 2242,Cidade Universitária,

05508-000,São Paulo, SP, Brazil, E-mail: brigitte@ipen.br

Abstract. The Institute of Nuclear and Energetic Researches, IPEN (Instituto de Pesquisas Energéticas e Nucleares), at São Paulo city, Brazil, is the largest institute in the nuclear research field in Brazil and consists of a number of nuclear and radiative facilities. The external exposure determined by thermoluminescence dosimeters showed that the mean annual background equivalent dose of 1 $mSv.y^{-1}$ for the decade 1993-2002 around IPEN facilities is below the outer TLD dose value and in agreement with the international radiological protection dose limits.

INTRODUCTION

The Institute of Nuclear and Energetic Researches – IPEN (Instituto de Pesquisas Energéticas e Nucleares) is an autarchy of the State of São Paulo Govern, linked to the Science, Technology and Economic Development Bureau. It is managed administrative and financially by the Nuclear Energy National Commission (CNEN) and associated to São Paulo University (USP) for post-graduation teaching purposes.

The Institute, situated in the São Paulo University campus, on the northwest of the city of São Paulo, Brazil, covers an area of 500,000 m^2 and has a major role in several fields of the nuclear activity, such as radiation and radioisotope applications, nuclear reactors, materials and fuel cycle and radiological protection and dosimetry.

The major nuclear and radiactive facilities at IPEN are the IEA-R1 Swimming Pool Nuclear Research Reactor, built by the Babcock & Wilcox Co. and two isochronal Ciclotrons. The oldest one, a 24 MeV cyclotron from TCC (The Cyclotron Corporation) accelerates protons, deuterons, Helium-3 and alpha particles, and, in 1989, started to produce the ^{67}Ga radioisotope for nuclear medicine. Nowadays, this machine is employed in irradiations for basic and applied research in several fields. The new cyclotron, CV-30, a new generation machine, accelerates only high efficiency and reliability protons, being used in the radioisotopes production for nuclear medicine purposes: ^{67}Ga, ^{201}TI, ^{111}In and ^{18}F.

Also, there is a Nuclear Fuel Center producing uranium dioxide pastilles, metallic uranium rods and fuel elements in plates for brazilian research reactors and the Radiopharmacy Center, in charge for the production and distribution of radiopharmaceuticals like ^{153}Sm, ^{131}I, ^{67}Ga, ^{201}TI, ^{111}In, and ^{18}F. Radiation detectors like Geiger-Müller detectors and Ionization chambers are calibrated with ^{60}Co sources and X-ray tubes inside a proper concrete construction called "bunker". At all IPEN facilities, researchers are developing activities in the fields of nuclear physics, radiochemistry, environmental diagnosis and monitoring, nuclear engineering, radiation protection and application of nuclear techniques in industry, radioisotope and radiopharmaceuticals.

The normal operation of nuclear and radiactive facilities involves the production of radioactive waste products. As complete removal of radioactivity from gaseous and liquid effluents discharged to the environment is practically impossible to achieve, all released effluents will always contain small amounts of radioactive waste material. Specially for a nuclear facility we must ensure that, under normal operating conditions, any waste discharged from it will not give rise to radiation exposures of members of the public above the limits stipulated by competent authorities.

In order to estimate the radiological impact to the environment around IPEN facilities, a monitoring programmme was established [1,2,3,4]. Within this programme, samples of ground water, rainwater and filters for air sampling in the influence area of IPEN were measured by using gamma spectrometry, total alpha and beta counting and instrumental neutron activation analysis.

The external exposure was determined by thermoluminescence dosimeters, that can integrate exposures over long period with no attention.

METHODOLOGY

The measurement of the environmental outdoor gamma radiation levels in the surroundings of IPEN was realized by using $CaSO_4$:Dy thermoluminescence dosimeters. Five monitoring stations were placed at points of maximum predicted ground-level concentrations. Ten monitoring stations were placed at more distant points, assumed with no influence from the institute facilities, covering the IPEN area in all directions. An outer dosimeter was placed far-away more that 15 km from the Institute, still inside the city of São Paulo, in the prevailing upwind direction.

Until 1999 each monitoring station was an wood stake with an wood box. One PVC plastic badge, with three $CaSO_4$:Dy detector chips is placed inside the wood box., 1 m above the ground. Each plastic badge is identified with a numbered tag. In 1999, after one year of compared measurements, all wood stakes were substituted by PVC plastic tubes, solidly bounded to ground, as PVC is more resistent to weather changes.

The measuremente were carried out quarterly, observing the four seasons of the year.

RESULTS AND DISCUSSION

The annual external exposure for each monitoring point was obtained by integrating the data for the four seasons of the year. The average annual background radiation for IPEN surroundings was calculated considering all results for the ten environmental monitoring stations. All background radiation values are of the same order as external exposures from different places in the world, measured in the same way [5, 6, 7].

The equivalent dose due to gamma radiation exposure was calculated considering the ICRP 60 [8] procedures through the gamma radiation levels determined with the thermoluminescence dosimeters.

Figures 1, 2, 3, 4 and 5 show the equivalent doses from 1993 to 2002 for the five monitoring points of maximum predicted ground-level concentrations together with the average annual background equivalent dose for IPEN surroundings and the outsider background value.

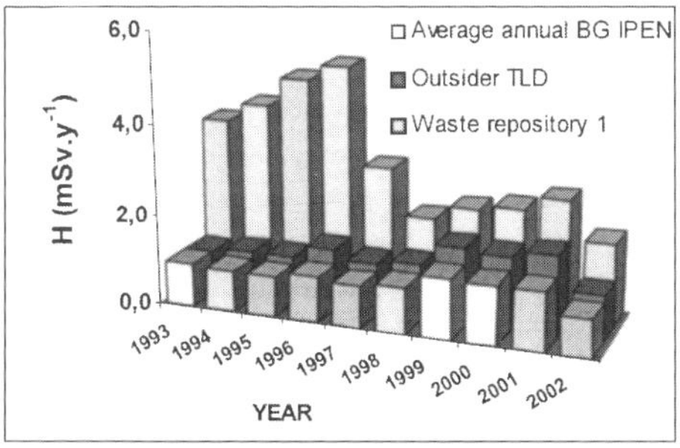

FIGURE 1. Equivalent dose for the decade 1993-2002 in front of the waste repository of IPEN, position 1, keeped away since 1997.

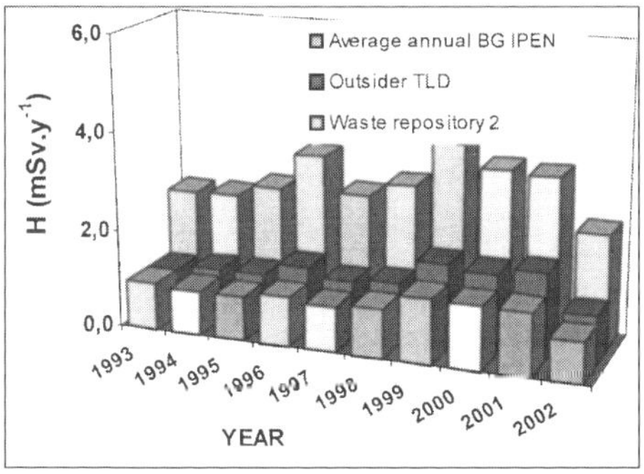

FIGURE 2. Equivalent dose for the decade 1993-2002 near the waste repository of IPEN, position 2.

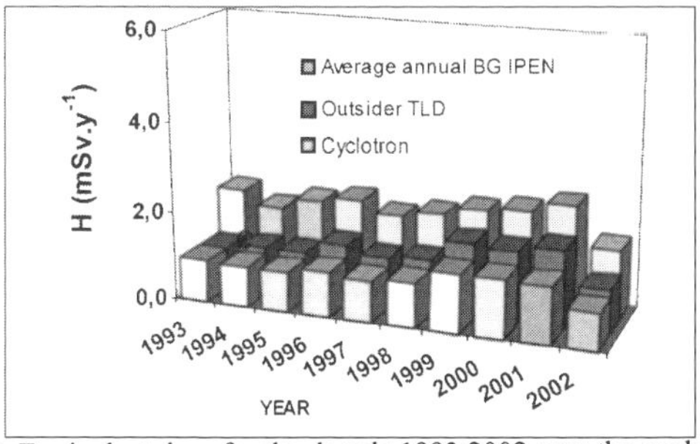

FIGURE 3. Equivalent dose for the decade 1993-2002 near the cyclotrons, IPEN.

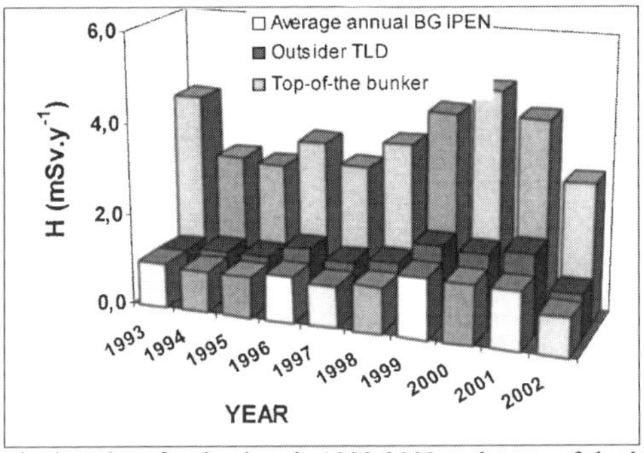

FIGURE 4. Equivalent dose for the decade 1993-2002 at the top-of-the-bunker, IPEN.

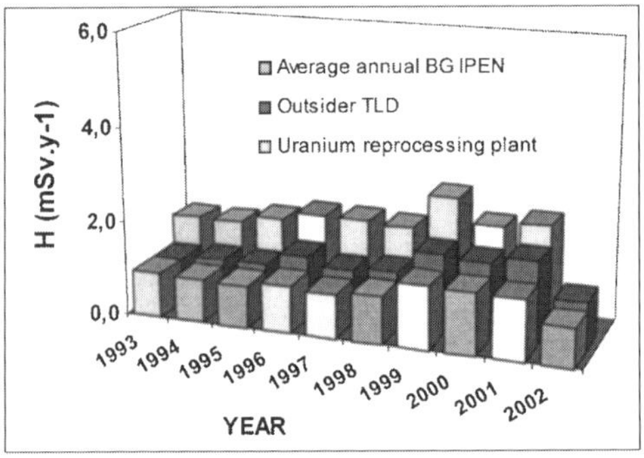

FIGURE 5. Equivalent dose for the decade 1993-2002 near uranium reprocessing plant, IPEN

The highest values were obtained close to the cyclotrons (2.4 $mSv.y^{-1}$), to the solid waste repository (3,9 $mSv.y^{-1}$ and 5 $mSv.y^{-1}$) and at the top of the bunker (4.9 $mSv.y^{-1}$), as a consequence of the normal operation of the cyclotron facility when producing radiopharmaceuticals. As IPEN surroundings are not of free access to the general public, we can consider those values below 20 $mSv.y^{-1}$, suggested as an annual effective dose limit for occupational exposure by ICRP 60 [8], considering a working time of 2000 hours per year.

The average annual IPEN background equivalent dose for the decade 1993-2002 is 1 $mSv.y^{-1}$, in agreement with ICRP 60 dose limit [8] for general public. The equivalent dose for the outsider TLD is below the recommended UNSCEAR value of 2.4 $mSv.y^{-1}$ for natural radiation [9].

ACKNOWLEDGMENT

The authors would like to thanks to the Calibration and Dosimetry Division of IPEN for producing and processing the thermoluminesce dosimeters.

REFERENCES

1. *Manual on environmental monitoring in normal operation.* IAEA SS-16, Vienna, 1966.
2. *Objectives and design of environmental monitoring programmes for radioactive contaminants.* IAEA SS-41, Vienna, 1975
3. Departamento de Radioproteção Ambiental. *Avaliação Radiossanitária do Meio-Ambiente sob Influência do IPEN.* São Paulo, nov., 1995, IPEN-Pub-401.
4. V. M. F. Jacomino and M. F. Maduar, *Monitoração Ambiental nas Imediações de Instalações Nucleares.* São Paulo, fev., 1992, IPEN-Pub-363.
5. T. F. L Daltro, M. A. P. V. de Moraes and B. E. Sartoratto, *Environmental Natural Gamma Radiation in Centro Experimental Aramar (CTMSP-Brazil).* V Regional Congress on Radiation Protection and Safety (Regional IRPA Congress), Recife, Brazil, 29/04 to 4/5/2001. Proceedings.
6. B. M. R. Green, J. S. Hughes, P. R. Lomas and A. Janssens. *Natural Radiation Atlas of Europe.* Radiat. Prot. Dosim. **45**, 491 (1992).
7. K. A. Al-Hussan, and N. F. Wafa, *Environmental Radiation Background Level in Riyadh City.* Radiat. Prot. Dosim. **40**, 59 (1992).
8. ICRP-60. INTERNATIONAL COMMISSION ON RADIOLOGICAL PROTECTION. 1990 *Recommendations of the international commission on radiological protection.* Pergamon Press. Oxford, 1991.
9. UNSCEAR. United Nations Scientific Committee on the Effects of Atomic Radiation. *Sources, effects and risks of ionizing radiation.* New York, U.N. 1988.

Applied Physics (APHYS 2003)
A. Méndez-Vilas (Ed.)

Working With A Neutron Activation Analyzer

G. P. Westphal, F. Grass, H. Lemmel and J. Sterba

TU Wien, Atominstitut der Österreichischen Universitäten, Stadionallee 2, A-1020 Vienna, Austria
westphal@ati.ac.at

Abstract. Dubbed "Analyser" because of its simplicity, a neutron activation analysis facility for short-lived isomeric transitions is based on a low-cost rabbit system and an adaptive digital filter which are controlled by a software performing irradiation control, loss-free gamma spectrometry, spectra evaluation, nuclide identification and calculation of concentrations in a fully automatic flow of operations.

Designed for TRIGA reactors from inexpensive plastic tubing and an aluminium in-core part, the rabbit system features of samples of 5 ml and 10 ml with sample separation at 150 ms and 200 ms transport time or 25 ml samples without separation at a transport time of 300 ms. For the smaller samples a computer-controlled sample changer is available.

By automatically adapting shaping times to pulse intervals the Preloaded Digital Filter gives best throughput at best resolution up to input counting rates of 10^6 c/s. Loss-Free Counting enables quantitative real-time correction of counting losses of up to 99%.

Due to its excellent high-rate performance the facility is specially suited for activation analysis of short-lived nuclides while above average sample sizes compensate for lower fluxes at small research reactors. Its simplicity makes it also an excellent beginner´s system for the first-time introduction of activation analyis at small research reactors in developing countries.

ACTIVATION ANALYSIS AT SMALL RESEARCH REACTORS

Trace and main element analysis are essential in many fields of the sciences: in environmental investigations, in nutrition research, in searching for mineral deposits, to mention only a few. Small research reactors offer the opportunity of investigating these fields effectively. The sensitivity of determinations by NAA depends on the activity induced. This activity being proportional to neutron flux and sample weight, a lower flux can be compensated by irradiation of larger samples with the distinct advantage of much lower radiation damage and better sample representation. In contrast to other methods of trace element analysis a further important advantage is the simple sample manipulation which can be performed by untrained personnel. With larger samples the danger of sample contamination can be neglected. In 2001, 283 research facilities were available in 56 countries around the world. Many of these research reactors, especially in developing countries, could be used for main- and trace-element analysis if a modern system for activation analysis were accessible.

At present, however, most permanent in-core irradiation systems in these research reactors contain metallic parts - mainly aluminium - which are strongly activated. As the transfer of the samples should be fast to allow the measurement of short-lived radio-nuclides in the range of seconds to minutes, the rapid movement of the rabbits involves friction which transfers activity to the sample catcher. The contamination of the transport-rabbit by metallic parts is minimised if re-enforced graphite pipes or quartz- tubes are used for the in-core parts. A much less expensive solution is the automatic separation of the sample from the rabbit.

Without sample separation the rabbit system allows the transfer of samples of up to 20g. Thus, low flux research reactors can attain sensitivities comparable to high-flux reactors where only a few 100 mg samples are routinely activated.

A recently developed technique enables determination of the interesting essential and toxic elements in nutrition research, i.e., F, Cl, Br, Se, Cu, Ca, Mg, I, Mn, Sr, K, Na, Al by short- and medium- lived nuclides, by the registration and evaluation of several hundred loss-corrected and non-corrected γ-spectra, thus allowing the selection of the optimum temporal range for the evaluation of elemental concentrations[1]. The decay curves enable quality assurance of the measurements. A further advantage of a system with sample separation is the possibility to irradiate together with the transfer rabbit a flux-monitor documenting the actual flux the sample has obtained.

A SIMPLE RABBIT SYSTEM

A simple but highly effective pneumatic transport system is made from an aluminium in-core part connected to inexpensive plastic tubing, not at last to avoid costly bending of aluminium tubes. Pressure comes from an industrial compressed air generator or, if available, from a central compressed air supply. Pressures from 5 to 10 bar have been used successfully with our design. Argon exhaust is collected in plastic balloons wherefrom it is removed by the TRIGA reactor´s beam tube ventilation system. Fig. 1 shows the system installed in the F ring of the TRIGA reactor. Fig. 2 shows a large sample container (left) of a volume of 25 ml and a smaller container of 5 ml (right) which separate, inertia driven, from a transport capsule (centre, shown with leaving small container). Containers of 10 ml are not shown here.

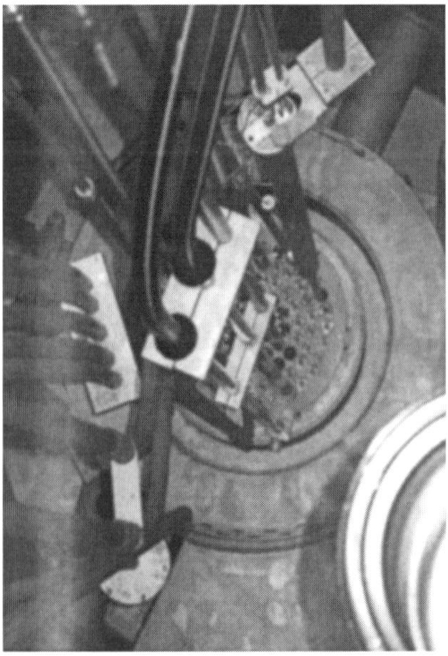

Figure 1 The rabbit system installed in the TRIGA reactor

Fig. 3 shows the sample changer. Rotated electrically, the shooting and measurement position is followed by a sample unload position and in turn by a sample reload position. Here, at any given time, one of a series of stacked samples is dropped into the barrel, to be transferred into the shooting position. From there, the sample is driven into the transport container, waiting upstairs in the sample separation stage. Together, transport container and sample travel to the irradiation position, to be separated again when returning to the sample separation stage. The transport container remains upstairs while the sample, one level lower, is measured by the facing detector. The sample changer as well as the rabbit system are controlled by the computer or manually by switches.

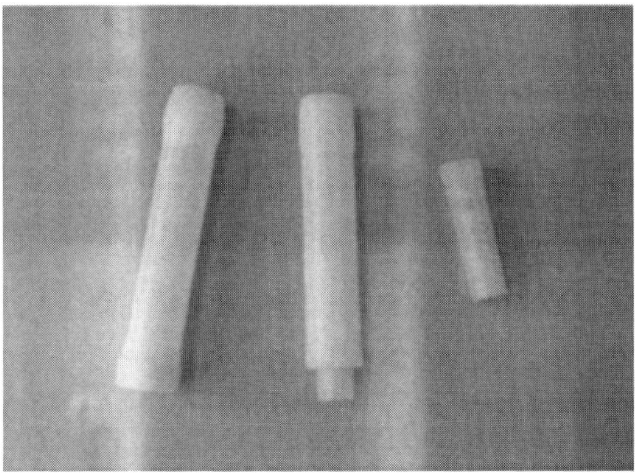

Figure 2 Sample containers and transport capsule (centre)

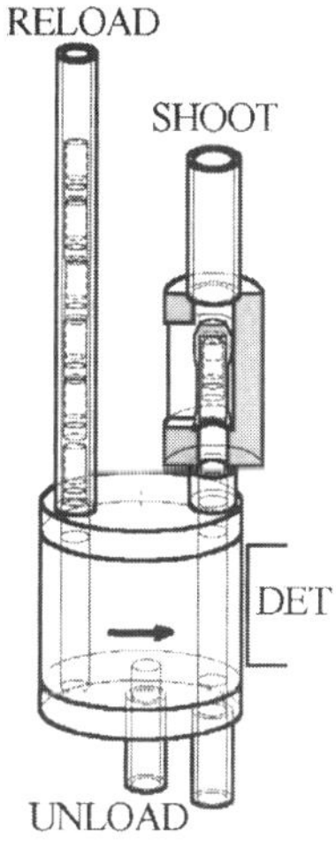

Figure 3 Sample changer

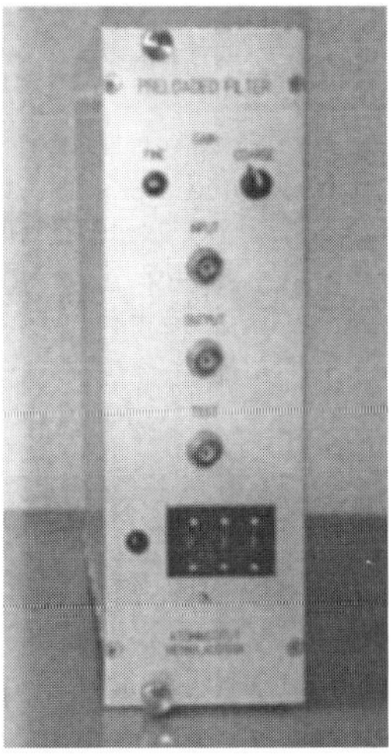

Fig. 4 The Preloaded Digital Filter is a double width NIM module, plug-compatible with MCA´s from Canberra.

THE SPECTROMETRY SYSTEM

As a necessary prerequisite for high-rate gamma spectrometry a high-purity Germanium detector (of 36% relative efficiency) with transistor reset preamplifier is the main building block of the spectrometry system. Pulse processing is performed by the Preloaded Digital

Filter developed at the Atominstitut (Fig. 4) which alternatively may be replaced by (less powerful) plug-compatible digital filters or analog filter-ADC combinations by Canberra.

By automatically adapting the noise filtering time to individual pulse intervals, the Preloaded Digital Filter (PLDF)[2] combines low- to medium-rate resolutions comparable to those of high-quality Gaussian amplifiers with throughput rates of up to 100 kc/s, and high-rate resolutions superior to those of state-of-the-art gated integrator systems. In contrast to commercially available digital filters, the PLDF in its new implementation performs pulse shortening as well as pole zero cancellation in the analog domain to render possible the use of cheap monolithic converters such as the AD9240 by Analog Devices. By presenting to the ADC very short signals as near as possible to the current pulses originating in the detector, pulse pileup at elevated counting rates and the consequent degradation of system resolution is completely avoided. At the same time, much better use may be made of the dynamic range of the converter as allowances for pulse pileup are no longer necessary. Added benefits are a much simpler digital circuit and conventional instead of digital pole zero cancellation.

To further optimise the signal to noise ratio, a pseudo-rectangular pre-filter with a pulse duration of 1.5 μs has been chosen which makes possible a maximum throughput rate after pileup rejection of 92 kc/s.

A specially designed parallel interface connects the digital filter to a fast PC and at the same time serves for the generation of weighting factors for real-time correction of counting losses with digital dead-time extension or according to the recently developed method of computed pileup correction[3].

Running not under WINDOWS as customary commercial products but under the DOS operating system, a software implementation of a Loss-Free Counting multi-channel analyser is programmed in Assembler and C++ and thus does not waste more than 90% of computer power on a fancy graphical user interface but saves it all for uncompromising real-time performance. Storing immediately into the multi-megabyte memory of a Pentium type PC it offers programmable rabbit control and enables the collection of up to 1000 pairs of simultaneously recorded loss-corrected and non-corrected spectra of 16 k channels each, in a true sequence without time gaps in between, at throughput rates of up to 200 kc/s. Counting loss correction is performed by Loss-Free Counting with digital dead-time extension or according to the recently developed method of Computed Pileup Correction[3] which also makes possible the loss-corrected multi-scaling of the system's input counting rate. Intended for FNAA (Fast Neutron Activation Analysis), the system renders possible peak to background optimisations and separations of lines with different half-lives without an a priori knowledge of sample composition by summing up appropriate numbers of spectra over appropriate intervals of time.

The activation analysis system is controlled by the automatic sequence of the programs PREPARE (sample, irradiation and measurement description), MEASURE (irradiation and measurement), and ANALYZE (spectra analysis, nuclide identification and calculation of elemental concentrations).

Interactive control of the multi-channel analyser, a live spectrum display and a number of system test facilities are also provided, together with a suite of spectrum manipulation programs from the (file and spectrum) compatible ACCUSPEC/ASAP system including the spectrum analysis program PEAK, by Nuclear Data/Canberra.

AN AUTOMATIC SYSTEM FOR ACTIVATION ANALYSIS

To facilitate the introduction of NAA at research reactors where appropriate specialists are not (yet) available, an automatic system avoiding some of the procedural obstacles of ordinary activation analysis could be extremely helpful. Such a system is based on irradiations at a stable and reproducible neutron flux as it is available for instance at TRIGA and SLOWPOKE reactors, and on measurements in a fixed and reproducible geometry. Determined under these preconditions, a table of saturation activities per gram of all elements under investigation, together with their half-lives and line energies, is already sufficient for nuclide identification as well as the direct calculation of elemental concentrations, immediately after automatic peak search and net peak area analysis. All that steps are performed on all sequentially measured spectra, producing concentrations for every element found at the same time in a corrected and the corresponding non-corrected spectrum. From these intermediate values a weighted average for every element found is computed as the final result of analysis.

This automatic approach for the analysis of measurements is called ANALYZE, and it is complemented by PREPARE, a program for the preparation of samples, and MEASURE, the performance of measurements. In PREPARE, command files and data file headers are prepared for a number of samples, later on to be fed to the automatic sample changer, recording sample name and weight, irradiation and measurement times for all intended sequential spectra, spectral resolution, then basic peak search sensitivity which by MEASURE is modified automatically for each spectrum according to its average counting loss correction factor, and basic FWHM resolution versus energy, for properly resolving multiplets, which by MEASURE is automatically adapted for each individual spectrum according to its counting rate. MEASURE, finally, controls the sample changer, performs the irradiation of each individual sample, and measures its predetermined sequence of corrected and non-corrected spectra. ANALYZE may follow immediately in a batch sequence, to give essentially what may be called an *Automatic Neutron Activation Analyzer*. Alternatively, the data may be transferred for analysis to an off-line computer which frees the on-line machine for an uninterrupted sequence of measurements.

A TEST OF SYSTEM PERFORMANCE

To investigate system performance in a single, easily reproducible measurement[4], a sample of Dysprosium was activated to give an initial input counting rate in the order of 1 million c/s, which was to be measured together with a reference sample of ^{60}Co fixed to the detector, to give net peak areas for both lines during a counting period of 50 s of 42000 and 37000 counts, respectively. By measuring a succession of loss-corrected and non-corrected spectra together with a multi-scaling spectrum of the loss-corrected input counting rate, the temporal response of the rabbit system, resolution and peak stability of the digital filter and the accuracy of the counting loss correction system was investigated in a single experiment.

From the time histogram of the loss-corrected input counting rate, the transport time of the big 25 ml capsule at a pressure of 5 bar has been found to be 300 ms. For transport rabbits carrying samples of 5ml and 10 ml transport times of 150 ms and 200 ms have been determined.

A peak position change of not more than 0.25 keV and a resolution degradation of less than 0.6 keV have been found and are certainly adequate for a range of input counting rates going up to 1 million c/s. The net peak areas of ^{60}Co remained constant over the whole range within their counting errors which at the uppermost data point corresponding to 1 million c/s and total counting losses of 98,4% are 10 % at 1173 keV and 9% at 1332 keV. The evaluation of these data has been done by Gaussian fitting while counting errors have been determined from the non-corrected spectra.

REFERENCE MATERIALS

As a test of system reproducibility in sample separation geometry, measurements were performed of NBS 1570 Spinach (1.33 g), NBS 1572 Citrus Leaves (2.43 g), NBS Pine Needles (2.42 g), NBS 1547 Peach Leaves (0.62 g), and Bowens Kale (2.30g) [Y. Muramatsu and R. M. Parr, IAEA /RL/128/ Dec. 1985], at an activation time of 10 s each [x]. The data for K, Cl, Mn, Mg, Ca, Sc, V, compare excellently with the consensus values for NIST biological Standard Reference Materials, compiled by I. Roelandts and E. S. Gladney, Fresenius J. Anal. Chem. **1998**, pp. 360. Though some of the consensus data show a rather large scatter, the larger sample size makes for an excellent sample representation, so that the consensus values correspond nicely.

Results Obtained Automatically From NBS 1572 Citrus Leaves

NBS 1572 - Citrus Leaves				
Nuklid	conc	err%	std	err%
Al-28	1.23E-04	0.61	8.70E-05	12.64
Br-80g	9.56E-06	4.82	8.20E-06	4.88
Cl-38a	4.26E-04	1.47	4.90E-04	18.37
Cl-38b	4.38E-04	1.64	4.90E-04	18.37
Cl-38m	3.94E-04	4.78	4.90E-04	18.37
Cu-66	9.41E-06	2.88	1.65E-05	7.88
Dy-165m	4.68E-08	3.08	4.50E-08	8.89
F-20	2.96E-06	12.21	3.20E-06	21.88
I-128	2.48E-06	5.22	1.80E-06	3.89
K-42	1.82E-02	2.25	1.82E-02	3.30
Mn-56a	1.85E-05	6.81	2.24E-05	7.14
Mn-56b	2.27E-05	2.11	2.24E-05	7.14
Na-24a	2.10E-04	4.94	1.60E-04	7.50
Na-24b	2.04E-04	4.53	1.60E-04	7.50
Rb-86m	6.61E-06	6.93	4.60E-06	6.52
Sc-46m	1.37E-08	12.81	1.10E-08	9.09
Sr-87m	1.02E-04	16.77	9.90E-05	5.05
Ti-51	1.11E-05	1.49	5.40E-06	31.48
V-52	2.77E-07	4.01	2.47E-07	5.26

Results Obtained Automatically From NBS 1570 Spinach

NBS 1570 - Spinach				
Nuklid	conc	err%	std	err%
Al-28	8.96E-04	0.32	8.20E-04	10.98
Br-80g	5.35E-05	1.73	4.90E-05	8.16
Br-82b	6.77E-05	44.80	4.90E-05	8.16
Ca-49b	1.36E-02	0.89	1.33E-02	6.02
Cl-38a	7.12E-03	0.69	6.50E-03	13.85
Cl-38b	7.13E-03	0.94	6.50E-03	13.85
Cl-38m	7.87E-03	2.02	6.50E-03	13.85
Co-60m	2.82E-06	14.15	1.49E-06	9.40
Cu-66	7.65E-06	21.90	1.19E-05	5.88
Dy-165m	5.10E-08	24.02	7.90E-08	k/a
Hf179m1	4.04E-08	8.84	4.50E-08	k/a
I-128	1.15E-06	46.05	1.19E-06	8.40
K-42	3.25E-02	2.37	3.54E-02	4.24
Mg-27a	7.80E-03	3.28	8.50E-03	4.71
Mg-27b	7.87E-03	1.38	8.50E-03	4.71
Mn-56a	1.79E-04	2.43	1.64E-04	5.49
Mn-56b	1.53E-04	1.80	1.64E-04	5.49
Na-24a	1.32E-02	1.68	1.39E-02	0.72
Na-24b	1.29E-02	1.23	1.39E-02	0.72
Rb-86m	1.24E-05	21.95	1.21E-05	10.74
Sc-46m	1.28E-07	3.48	1.65E-07	100.00
Sr-87m	9.85E-05	28.85	8.20E-05	8.54
Ti-51	5.38E-05	9.74	2.80E-05	28.57
V-52	1.22E-06	3.67	1.20E-06	14.17

CONCLUSION

A low-cost but powerful neutron activation analysis facility for small research reactors has been described which due to its operational simplicity may be called a *Neutron Activation Analyser*. It is excellently suited to be a beginners AA system as it makes possible immediate analyses without going into details with spectrum evaluation, nuclide identification and the calculation of elemental concentrations. On the other hand, quite sophisticated research is possible: Short-lived nuclides are facilitated by a fast rabbit and a high-rate spectrometry system with real-time correction of counting losses. Sample-rabbit separation and above average sample sizes make for good sensitivity even at lower fluxes. And, finally: A completely programmable system with an ultra-fast processor and the capability of storing virtually unlimited numbers of sequential files is, in the hands of the more experienced user, a powerful tool for performing novel experiments.

REFERENCES

1. G. P. Westphal, H. Lemmel, F. Grass, R. Gwozdz, K. Jöstl, P. Schröder and E. Hausch, „A Gamma Spectrometry System for Activation Analysis," lecture presented at the 5th *Int. Conf. Methods Applications Radionanalytical Chemistry, Kona, HI, 2000, J. Rad. Chem. Vol. 248, No. 1 (2001) pp. 53 - 60*

2. G. P. Westphal, K. Jöstl, P. Schröder and W. Winkelbauer, „Adaptive Digital Filter for High-Rate, High-Resolution Gamma Spectrometry," *IEEE Trans. Nucl. Sci. vol. 48, pp. 461 – 465, 2001*

3. G. P. Westphal and H. Lemmel, „Loss-Free Counting," lecture presented at the *3rd International Symposium on Nuclear Analytical Chemistry, Halifax, Nova Scotia*, 2001, to be published in *J. Rad. Chem.*

4. G. P. Westphal et al., "Large Sample Activation Analysis at Small Research Reactors", lecture presented at the *NAMLS-7 Conference, Antalya, 2002*, to be published in *J. Rad. Chem.*

Applied Physics (APHYS 2003)
A. Méndez-Vilas (Ed.)

Behavior of Irradiated BICMOS Components for Space Applications

D. Codegoni[1], A. Colder[2], N. Croitoru[1,3], P. D'Angelo[1], M. De Marchi[1], G. Fallica[4], A. Favalli[1], S. Leonardi[4], M. Levalois[2], P. Marie[2], R. Modica[4], S. Pensotti[1,5], P.G. Rancoita[1] and A. Seidman[1,3]

[1] *INFN - Istituto Nazionale di Fisica Nucleare, sezione di Milano, Milan, Italy*
[2] *Lermat, Caen, France*
[3] *Department of Physical Electronics, Tel-Aviv University, Ramat Aviv, Israel*
[4] *STMicroelectronics, Catania, Italy*
[5] *and University of Milano-Bicocca, Italy*

Abstract. Bipolar transistors were irradiated by neutrons, ions (or by both of them) and recently by electrons. Fast neutrons, as well as other types of particles, produce defects, mainly by displacing silicon atoms from their lattice positions to interstitial locations, i.e. generating vacancy-interstitial pairs, the so-called *Frenkel pairs*. The experimental results indicate that the gain variation is linearly related to the non-ionizing energy deposition (NIEL) for neutrons, incoming ions and electrons.

INTRODUCTION

The flux of GCR's is low compared to trapped particles but it is a hazard to spacecraft electronics, because GCR's can penetrate shielding materials [1]. After a standard aluminum shield of 100 mils (see, for instance, Sect. 5.4 in [1]) fast energetic charged particles (namely above 22.5 MeV for protons and 90 MeV/a for iron ions, but only above 1.2-1.5 MeV for electrons) will be able to make radiation damages. In this case, each isotopic element of these penetrating charged particles (see for instance, fluxes and energy distributions of protons, helium, carbon and iron shown in Sect. 23 of [2]) releases a dose and loses an amount of energy by a Non Ionizing Energy Loss (NIEL) process ([3] and references therein), which are non negligible fractions of those ones of the proton component. As a consequence, when a 100 mils aluminum shield is taken into account the proton component is expected to contribute with an important, but not dominant, fraction of both the dose and the NIEL deposition. The high energy electron component in GCR's is about 4-10 % of the dominant proton component [4,5] and is larger than any other isotopic component.

Taking into account the space-radiation environment in which the large number of electronic circuits are going to be introduced, investigations of the interdependence effects between the type, intensity, duration of irradiations typical of space environment have to be carried out on basic components of VLSI technologies (see for instance [6-12] and references therein). These basic components are indeed the essential part of any circuit, thus the knowled ge of irradiation effect on them is critical.

IRRADIATION AND MEASUREMENTS

The BJT's were irradiated by neutrons, C, Ar and Kr ions, and recently with electrons. The neutron irradiation was performed at the Triga reactor RC:1 of the National Organization of Alternative Energy (ENEA) at Casaccia, Rome. The obtained fluences were in the range of (1.2×10^{13}-1.2×10^{15}) n/cm^2 and are shown in Table 1. The C ions were made available at GANIL, at two different energies: a) ^{12}C accelerated at 95 MeV/a (High Energy, HE), and b) ^{13}C ions at energy of 11.1 MeV/a (Medium Energy, ME), while ^{36}Ar ions at 13.6 MeV/a and ^{86}Kr ions at 60 MeV/a. Electrons have been obtained with a mean kinetic energy of 9.1 MeV , at the ISOF Institute in Bologna.

Both npn and pnp devices were characterized, using a modular DC source/monitor (Hewlett-Packard HP4142B), controlled by a work station HP9000/C160. The forward voltage applied to the emitter-base junction (VBE), was in the range 0.2-1.2 V, which allows to measure the value of forward gain for collector currents, Ic, within the range $10^{-6}<Ic<10^{-3}$A, with the base and collector grounded.

TABLE 1. Particle fluences in [particles/cm^2]. For neutrons, the fluences are those for fast neutrons with kinetic energy above 10 keV.

n [n/cm^2]	C(HE) [ion/cm^2]	C(ME) [ion/cm^2]	Ar [ion/cm^2]	Kr [ion/cm^2]	e [e/cm^2]
1.35×10^{13}	5.2×10^{10}	1.0×10^{11}	1.0×10^{10}	1.0×10^{9}	2.0×10^{14}
1.35×10^{14}	1.0×10^{11}	5.0×10^{11}	5.0×10^{10}	5.0×10^{9}	7.0×10^{14}
6.75×10^{14}	5.1×10^{11}	1.0×10^{12}	1.0×10^{11}	1.0×10^{10}	2.0×10^{15}
1.35×10^{15}	1.0×10^{12}	5.0×10^{12}		5.0×10^{10}	
	5.0×10^{12}	1.0×10^{13}			
	1.0×10^{13}				

DISPLACEMENT DEFECTS AND GAIN DEGRADATION

Fast neutrons, as well as other types of particles, produce defects mainly by displacing silicon atoms from their lattice positions to interstitial locations, i.e., generating vacancy-interstitial pair, the so-called Frenkel pairs. Displacements produced in silicon during neutron irradiation have been widely studied ([12] and references therein). In case of incoming neutrons, the concentration of Frenkel pairs (*FP*), i.e. the number of Frenkel pairs per cm^3, can be evaluated in the Kinchin-Pease model from $FP = E_{dis}/(2\ E_d)$, where E_d, about 25 eV, is the energy required to displace a silicon atom from its lattice position to an interstitial location and E_{dis} is the energy density, i.e. the energy per cm^3, deposited through atomic displacements by neutrons. It can be computed using the neutron spectral fluence and the damage function [12]. The calculated Frenkel pairs concentration (as used in the present paper) per incoming neutron per cm^2 has to be decreased by about 20% in the modified Kinchin-Pease model [13], i.e., only 0.8 E_{dis} is the energy density available for Frenkel pairs creation. Similarly to an incoming neutron, an incoming ion may interact and displace a silicon atom from its lattice position. This primary displacement is followed by a small cascading process, in which secondary displacements are induced by the recoil silicon atom. The concentration of the displaced atoms has been calculated [11,12] by means of the TRIM Monte Carlo program [14].

For incoming electrons, the NIEL deposition in silicon has been estimated by interpolating the computed NIEL deposition for $E_d = 21$ and 30 eV in [3].

The relationship between the variation of the inverse of the gain and the fast neutron fluence is provided by means of the Messenger-Spratt Equation ([15-17] and references therein):

$$\Delta\left(\frac{1}{\beta}\right) = \frac{1}{\beta_i} - \frac{1}{\beta} = \frac{1}{\omega_T}\frac{\Phi_n}{K} \tag{1}$$

where β and β_i are the common emitter current gain before and after the irradiation, ω_T is 2π times the common emitter gain cut-off frequency, K the relevant damage constant which depends also on the base resistivity and Φ_n the fast neutron fluence in n/cm^2. Eq. (1) can be rewritten [12] as:

$$\Delta\left(\frac{1}{\beta}\right) \propto \frac{\text{NIEL}}{\omega_T} = \frac{\lambda}{\omega_T} FP \tag{2}$$

where λ is independent of the damage function, which is accounted by E_{dis} and, in turn, by FP. To a first approximation, Eq. (2) can be applied to the case of damages induced by ions as already observed for irradiations with neutrons and C ions in [10] and for irradiation with Ar and Kr in [11,12] namely there is a NIEL scaling of the gain variation of bipolar transistors.

BIPOLAR TRANSISTORS BEHAVIOR

In previous papers [10-12], it has already been shown that for the case of small area npn transistors (with the emitter region of 5μm x 5μm and irradiated with C, Ar and Kr ions, and neutrons) there is a similar gain degradation when the concentration of the created Frenkel pairs (CF), interstitial-vacancy, is the same. This occurs in spite of the large difference in the absorbed dose between neutron and ion irradiations for generating an equal amount of FP concentration. Similarly, it was also shown that npn large emitter area transistors, and vertical and lateral pnp transistors have their gain degradation depending on the concentration of the created Frenkel pairs.

The present investigation was carried out with 9.1 MeV electrons and fluences between $2\text{x}10^{14}$ and $2\text{x}10^{15}$ e/cm^2 (see Table 1). The gain degradation was investigated for collector currents Ic between 1 μA and 1 mA. A linear dependence of $\Delta(1/\beta)$ (see for instance Fig. 1 for Ic = 1 mA) as a function of the concentration of Frenkel pairs was found. The experimental data show that the gain degradation depends almost linearly on the Frenkel pairs concentration.

Collected data (including the present data obtained with incoming electrons) show a linear dependence of the bipolar gain degradation on the Frenkel pairs concentration for incoming neutrons (as expected by the Messenger-Spratt equation), electrons and ions. This confirms that the basic mechanism of the bipolar gain variation is mainly related to silicon displacements, namely to the NIEL deposition. In Fig. 2, slopes of the linear fit to data are shown for small and large npn transistors, and vertical and lateral pnp transistors as a function of the collector current. The slope depends on the collector current Ic and on the type of the transistors [12].

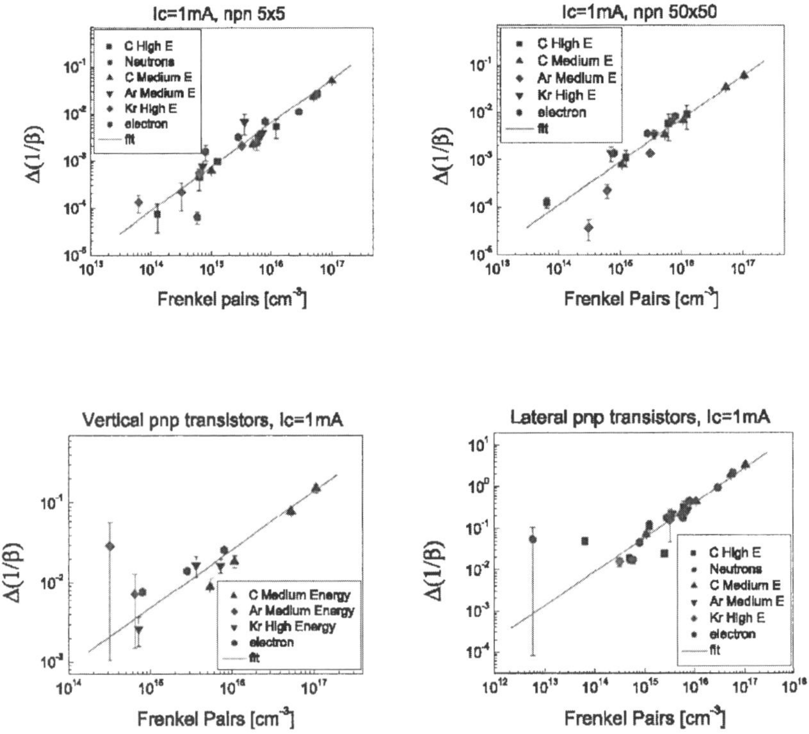

FIGURE 1. Linear dependence of the quantity $\Delta(1/\beta)$ on the concentration of Frenkel pairs for npn small area transistors with the emitter region of 5μm x 5μm, for npn large area transistors with the emitter region of 50μm x 50μm, for pnp vertical transistors, for pnp lateral transistors, and collector currents of 1 mA .

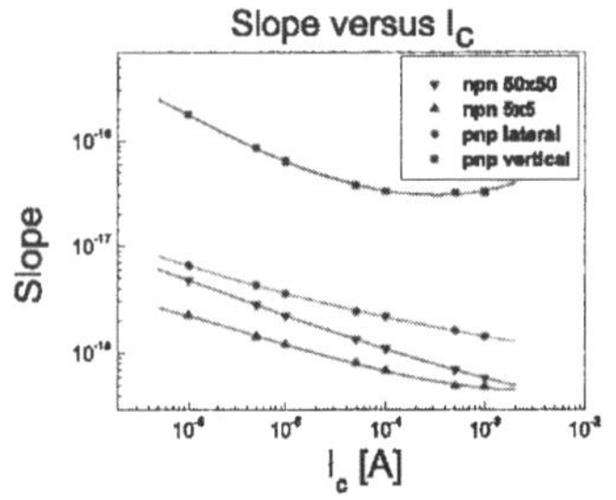

FIGURE 2. Slopes of the linear fit to data are shown for largeemitter area region npn transistors (npn 50x50), for smallemitter area region npn transistors (npn 5x5), for vertical pnp transistors (pnp vertical) and for lateral pnp transistors (pnp lateral) as a function of the collector current Ic.

CONCLUSIONS

The measurements confirm the fact that in bipolar transistors the gain degrades linearly with respect to the Frenkel pair concentration, for incoming electrons, neutrons, and C, Ar, and Kr ions and is mainly due to NIEL. For space application the known and linear (as observed from experimental data) $\Delta(1/\beta)$ dependence on the concentration of Frenkel pairs allows the possibility to evaluate the total amount of the gain degradation of VLSI components due to the flux of charged particles during the full life of operation of any pay-load. In fact, the total amount of expected Frenkel pairs can be computed taking into account the isotopic spectra, and subsequently related to the overall gain degradation. It has to be point out that in cosmic rays there is relevant flux of electrons and of massive particles up to Fe nuclei, which are within the range of particles presently investigated.

ACKNOWLEDGMENTS

The authors would like to thank P. Fuochi and M. Lavalle for their help during the electrons irradiations at the ISOF Institute in Bologna.

REFERENCES

1. Barth, J., Short Course on Applying Computer Simulation Tools to Radiation Effects Problems, IEEE Nuclear and Space Radiations Effects Conference, Snowmass Village (1977).
2. Hagiwara, K. et al.., *Phys. Rev. D*, **6**, 010001.1 - 010001.974 (2002).
3. Messenger, S. R. et al., *IEEE Trans. on Nucl. Sci.*, **46**, 1595 - 1602 (1999).
4. Alcaraz, J. et al., *Nucl. Instr. and Meth. in Phys. Res. B,* **472**, 215 - 226 (2000).
5. Alcaraz, J. et al., *Phys. Lett. B,* **484**, 10 - 22 (2000).
6. Johnston , A. H. et al., *IEEE Trans. on Nucl. Sci.*, **41**, 2427 - 2436 (1994).
7. Fleetwood, D. M. et al., *IEEE Trans. on Nucl. Sci.*, **41**, 1871 - 1883 (1994).
8. Baschirotto, A. et al., *Nucl. Instr. and Meth. in Phys. Res. A,* **362**, 466 (1995).
9. Baschirotto, A.. et al., *Nucl. Instr. and Meth. in Phys. Res. B,* **114**, 327 - 331 (1996).
10. Colder, A. et al., . *Instr. and Meth. in Phys. Res. B,* **179**, 397 - 402 (2001).
11. Colder, A. et al., "Effects of Ionizing Radiation on BiCMOS Components for Space Application", Proc. of the European Space Component Conference (Toulose 24-27 September 2002), 377 – 386 (2002).
12. Codegoni, D. et al., *Nucl. Instr. and Meth. in Phys. Res. B,* **217**, 65 - 76 (2004).
13. Norgett, M. J. et al., *Nucl. Engin. and Des.*, **33**, 50 (1975).
15. Ziegler, J.F., Biersack, J.P.and.Littmark, U., *The Stopping and Range of Ions in Solids,* Pergamon Press, New York (1985).
15. Messenger, G. C., *IEEE Trans. on Nucl. Sci.*, **39**, 468 (1992).
16. Messenger, G. C., *IEEE Trans. on Nucl. Sci.*, **16**, 160 (1969).
17. Messenger, G. C., *IEEE Trans. on Nucl. Sci.*, **13**, 141 - 159 (1966).

Applied Physics (APHYS 2003)
A. Méndez-Vilas (Ed.)

Unexploded Ordnance Discrimination Using Neutrons

Phillip C. Womble, Michael Belbot, Jon Paschal, Kirk Cantrell, and Lindsay Hopper

Applied Physics Institute, Western Kentucky University, 1 Big Red Way, Bowling Green, KY, USA
E-mail: womble@wku.edu

Abstract. In the U.S., the majority of unexploded ordnance (UXO) is often found to pose no risk, since inert and explosive ordinance can easily be distinguished. However, there are items that are not identifiable due to age and condition. These UXO must often be treated as though they contained high explosives or other hazardous material. This leads to unnecessary cost and potential impacts to the environment and communities. For subsurface anomalies, excavation is necessary for visual inspection. This procedure adds to the risk and cost associated with neutralizing UXO. Pulsed fast/thermal neutron analysis (PFTNA) is a technique used for bulk chemical analysis. In PFTNA, neutrons are produced with a pulsed 14 MeV (d-T) neutron generator. Separate gamma-ray spectra from fast neutron, thermal neutron and activation reactions are accumulated and analyzed to determine elemental content. A man-portable explosives detection system called PELAN has been developed using this technique. Recent tests of the PELAN have shown that this system works very well for artillery shells larger than 90 mm with false negative rates less than 1%. For smaller munitions, the false negative increases dramatically to values greater than 50%. This increase is due to the poor signal to noise ratio present when PELAN is used to examine smaller UXO.

INTRODUCTION

As the ability to detect sub-surface metallic objects increases, the organizations which must remediate sites designated for base realignment and closure (BRAC) and at Formerly Used Defense Sites (FUDS) face an increasing problem. The problem is to differentiate between unexploded ordnance (UXO) and other metallic clutter. The problem is exacerbated at former firing ranges where ordnance filled with inert materials was fired to test ballistic characteristics. At the Jefferson Proving Ground in Indiana it is estimated that for every six shells found only two contain high explosives[1]. To treat every piece of ordnance found as "live" increases the cost and time required for the remediation process.

The elemental composition of explosives has been shown[2,3], to be quite different from other innocuous materials and contraband (see Table 1). Neutrons are excellent probes for the nondestructive analysis of elemental content. In particular, 14 MeV neutrons can be used to produce three types of reactions in materials: inelastic scattering, thermal neutron capture, and activation.

TABLE 1. Average elemental weight densities of explosives, illicit drugs, and innocuous materials from Reference 2. Elemental data has been normalized to the average carbon weight density of explosives (0.364 g/cc).

	H	C	N	O	C/H	C/N	C/O	H/O	H/N	N/O
Explosives	0.13	1.00	1.07	2.10	7.93	0.93	0.48	0.06	0.12	0.51
Fabrics and Plastics	0.13	1.00	0.05	0.17	7.49	20.30	5.81	0.78	2.71	0.29
Foods and other Organic Materials	0.22	1.20	0.40	1.00	5.37	2.97	1.19	0.22	0.55	0.40
Illicit Drugs	0.11	1.92	0.16	0.40	17.5	12.35	4.75	0.27	0.70	0.38

The Pulsed Fast/Thermal Neutron Analysis (PFTNA) technique is a method for producing these three reactions in a single system. In PFTNA, a pulsing deuterium-tritium (d-T) neutron generator is utilized to create pulses of neutrons with a duration of a few microseconds and a frequency of several thousand Hertz.

During the neutron pulse, the 14 MeV neutrons produce gamma rays from elements primarily through inelastic scattering (n,n'γ) (e.g C (~200 mb), O (~100 mb)) or N (~20 mb). Between pulses, the fast neutrons are moderated by materials in the environment. These moderated neutrons can measure elements such as H, S, Cl, and N that have substantial (n,γ) cross-sections can be measured. Finally, neutron activation reactions from (n,p), (n,α), etc. are also possible with 14 MeV neutrons. The generator can be turned off to measure gamma rays from the short half-life activation reactions. A single data acquisition system that can switch between memory groups is used to collect and store these spectra at different memory locations. A wide variety of applications including landmine detection and bulk chemical analysis using PFTNA is given in Reference 4.

Recently, the Environmental Security Technology Certification Program in conjunction with the U.S. Navy EOD Technology Division has been testing a PFTNA-based system that our group developed[5,6] called Pulsed ELemental Analysis with Neutrons (PELAN) for non-intrusive filler identification of unexploded ordnance.

The PELAN (shown in Figure 1) consists of a pulsing d-T neutron generator and a bismuth germanate (BGO) gamma-ray detector. PELAN is a small man-portable device composed of two units which interlock to form the shape shown in Figure 1. In one unit the neutron generator along with the computer and the various power supplies are housed. The second unit contains the BGO γ-ray detector and the necessary material to shield the detector from the direct neutrons. The total mass of PELAN is less than 45 kg. PELAN automatically analyzes the three spectra derived from inelastic scattering, thermal capture, and neutron activation to determine whether a threat is present.

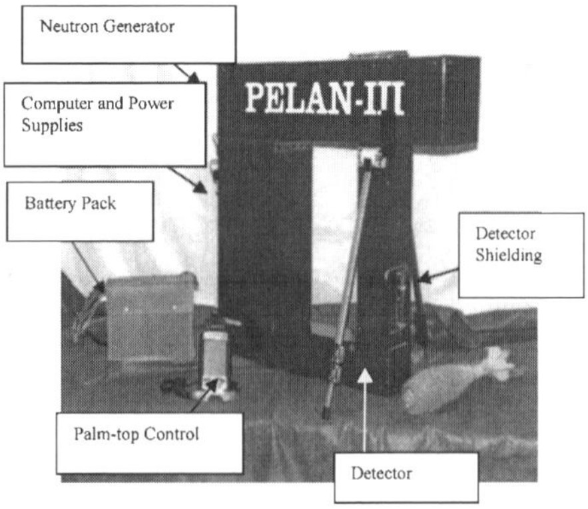

Figure 1. The PELAN system.

FIELD TRIAL, MAY 2002

During a two-week period in May 2002 (May 13-24) at the U.S. Navy Explosive Ordnance Disposal Technology Center at Indian Head, Maryland, an extensive validation test of PELAN was conducted. The purpose of the test was a) to examine PELAN's stability over a long time period, different types of soil, etc. and b) to collect data on different fill materials for shells (both explosive and non-explosive) and other explosives not found in UXO.

The demonstration was performed using:

- Five different environments: on a table and four on different types of soil (gravel, sand, wet soil in 3'x3'x1' boxes and regular soil)
- Shell sizes of 60, 76, 81, 82, 90, 105, 122, and 155 mm.
- Three landmines designed for both anti personnel and anti tank warfare
- The following explosives: TNT, RDX, COMPB, ANFO (ammonium nitrate/fuel oil mixture), PBX-108, PETN, Octal, Semtex, smokeless powder, and mixtures of TNT and RDX
- Shells of inert types: sand, wax, epoxy, plaster of Paris, and empty

A total of 164 cases were examined. All the data were automatically analyzed at the end of each run and recorded in the PELAN computer and on a spreadsheet. Each run was 5 minutes in duration. The data were used to establish such parameters as the validity of decision-trees built on the 164 cases, repeatability of measurements, critical analysis of spectral analysis software, etc.

In the case of PELAN, a decision tree is a chain of logic statements (Boolean or otherwise) which makes a determination of the type of material it is analyzing. In the case of UXO, not only was the presence of a threat returned ("threat" or "no threat") but the type of filler material was also determined (e.g. RDX, wax, etc.).

Three decision trees were written. The first decision tree was created from the data from only the UXO items (shells and mortars). The second decision tree was created from all of the explosives and inert items in the field trials (including landmines and raw explosives). In these decisions, the required inputs were the H, C, O, and N elemental content and the ratios of C/H and C/N. These decision trees are given in Table 2.

The third decision tree, unlike the others, required the user to input the size of the UXO before a decision could be made and was based only on the UXO items. The data set from Indian Head was flawed in that there were limited variety in the fills for certain sizes.

In Table 3, we show the results of these decision trees. FN denotes a false negative i.e. returning a value of "no threat" when a threat is present. FP denotes false positives or false alarms. The shell environment (depending on whether the shell is placed on a table or partially buried) will vary these results.

As shell size decreases, the PELAN's FN rate increases. This is due to the fact that the signal from the small explosive size (300 g) is overwhelmed by that from the environment.

Another trend seen in Table 3 is that additional information on the shell size significantly reduces the FN rate for the smaller shells since the decision tree can be

TABLE 2. Decision trees 1 and 2 in tabular form.

Decision Tree 1						
	TNT	Comp B	Explo-sives	Wax	Sand	
H (cps)						
C (cps)	>2	>2	>2		<2	
O (cps)						
N (cps)			>2	<2		
C/H	>20	<20	<1			
C/N			>1 and <10	<1or >10		
Decision Tree 2						
	ANFO	TNT	Comp B	Explo-sives	Wax	Sand
H (cps)						
C (cps)		>2	>2	>2	>2	<2
O (cps)						
N (cps)	>2			>2	<2	<2
C/H	>.001an d <.0.65	>20 and <40	>1		>1	
C/N				>1 and <10		

adapted, accordingly. Information on the environ-ment (such as soil type) may have a further beneficial result.

In previous papers (e.g. Reference 4), our group has discussed the possibility of utilizing the ratios of chemical elements as a method of discriminating explosives from innocuous materials. In Figure 2, we present the receiver-operator characteristic curve for C, O, and their ratio, C/O. This curve indicates that the ratio of two elements does not enhance the prediction of a threat over using single elements. A possible reason is that averages shown in Table 1 have very large standard deviations (approaching 100% of the average).

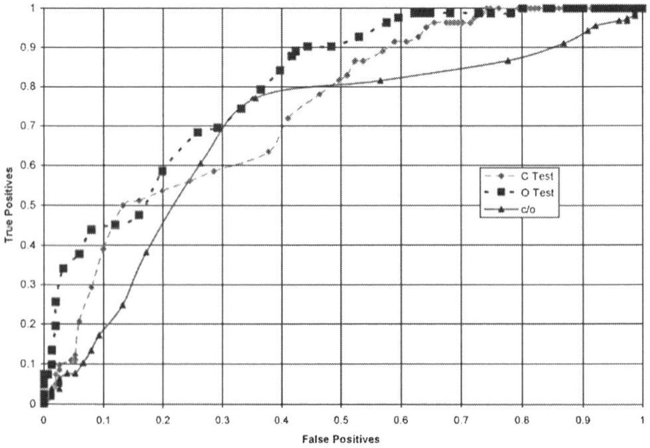

Figure 2. ROC curves for C, O and their ratio.

CONCLUSION

We have shown that a neutron-based system can be used to discriminate UXO from other metallic clutter. While the false negative rate is rather low (3% or less), it has a significant false alarm rate (~20%).

Further work is necessary in developing algorithms for threat detection. While in principle the ratio of elements seems to be an indicator of an explosive, it is difficult in practice to use these ratios because they overlap extensively for each class of objects in Table 1.. An alternate method such as linear discriminant analysis suggested in Reference 2 or neural networks may prove more reliable.

We are also continuing work discussed in References 5 and 6 in changing the physical geometry of the system to increase the signal to noise ratio in the spectra.

TABLE 3. Results of decision trees for the field trial. FN denotes false negative and FP denotes false alarm.

				Decision Tree 1		Decision Tree 2		Decision Tree 3	
Size	Approximate Explosive Mass (g)	Number of Inerts	Number of Explosives	FP %	FN %	FP %	FN %	FP %	FN %
90 mm-155mm	> 900	36	36	22	3	22	0	0	3
76 mm -82 mm	700	28	40	29	20	29	15	18	3
60 mm	300	12	10	8	70	8	70	42	0
TOTAL		76	86	22	19	22	8	20	1.3

ACKNOWLEDGMENTS

This work is supported through Department of Defense Contracts DACA72-01-C-0017 and N0017400C0073.

REFERENCES

[1] U. S. ARMY CORPS OF ENGINEERS – Louisville District, Archive Search Report for the Jefferson Proving Ground at Madison, Indiana, accessed at http://dogbert.mvr.usace.army.mil/military/derp/ir/projects/jefferso/oew/asr/asr.html, July 1999.

2. Shea, P. and Gozani, T., "Decision Processing in Explosive Detection Systems", *Proceedings of Nuclear Techniques for Analytical and Industrial Applications,* G. Vourvopoulos and T. Paradellis, eds. Western Kentucky University, p. 359-379, (1992).

3. Vourvopoulos, G., Chemistry and Industry, 18 April 1994, p. 297-300.

4. Vourvopoulos, G., and Womble, P.C., *TALANTA*, **54**, 2001, pp. 459-468.

5. Womble P.C., Vourvopoulos G., Paschal J., Novikov I., and Chen G., NIM A 505, pp.470-473, 2003.

6. Womble P.C, Paschal J. , Cantrell K., Belbot M., and Hopper L., Evaluation of UXO Discrimination Using PELAN, *Proceedings of Accelerators in a Nuclear Renaissance*, American Nuclear Society, June 2003, in press

Applied Physics (APHYS 2003)
A. Méndez-Vilas (Ed.)

Optical Effects In Gaussian Pulse Reflection And Transmission By Linearly Accelerated Interfaces

Maria Hermínia Marçal

Centro de Electrodinâmica, IST, Technical University of Lisbon,
Avda.Rovisco Pais, 1049-001 Lisboa, PORTUGAL
E-mail: pcmarcal@mail.ist.utl.pt

Abstract. The reflection and transmission of optical pulses by a linearly accelerated interface are investigated theoretically. Although only linear, loss-less and non-dispersive media are considered, several interesting optical effects still occur, due to the accelerated motion of the interfaces. Namely, pulse compression or broadening and a frequency chirp across the pulse are readily put to evidence for an input Gaussian pulse at normal incidence. There is also a Døppler shift in the frequency of the waves after one or more transits through the accelerated slab, an effect that does not occur for constant linear velocity. Optical devices for measuring linear acceleration based on this effect can be considered the linear analogues of optical gyroscopes and are within the range of current technology.

INTRODUCTION

The propagation of light through a medium with a finite proper acceleration has been re-considered by several authors [1-3] in recent times. Here we shall consider this problem in a classical setting, by solving for the reflection and transmission of a classical plane wave, at normal incidence, by a uniformly accelerated semi-infinite dielectric and dielectric slab. Although the interaction of electromagnetic plane waves with accelerating media or interfaces has been studied before, [4,5] no specific solutions for a given incident pulse shape have been presented. The focus of this work is on optical pulse behaviour, and an incident optical Gaussian pulse will be used to better illustrate the various effects in the reflected and transmitted pulses. The results of the research could be used for understanding the phenomena of electromagnetic wave processes in bounded accelerating media, as well as for technological applications as optical moving sensors.

BASIC EQUATIONS

The electromagnetic field in the accelerated medium is formally described by a generally covariant formulation of Maxwell equations as presented in detail in [4].

In this formulation the components of the electromagnetic field tensor F_{ik} and the tensor density G^{ik} are expressed as

$$F_{ik} = \begin{pmatrix} 0 & cB_Z & -cB_Y & E_X \\ -cB_Z & 0 & cB_X & E_Y \\ cB_Y & -cB_X & 0 & E_Z \\ -E_X & -E_Y & -E_Z & 0 \end{pmatrix} \quad G^{ik} = \begin{pmatrix} 0 & H_Z & H_Y & -cD_X \\ -H_Z & 0 & H_X & -cD_Y \\ H_Y & -H_X & 0 & -cD_Z \\ cD_X & cD_Y & cD_Z & 0 \end{pmatrix} \tag{1}$$

As a result, Maxwell equations can be written in the well-known vector form

$$\begin{aligned} \nabla \times \boldsymbol{E} &= -\frac{\partial \boldsymbol{B}}{\partial t}, \qquad \nabla \cdot \boldsymbol{B} = 0 \\ \nabla \times \boldsymbol{H} &= -\frac{\partial \boldsymbol{D}}{\partial t}, \qquad \nabla \cdot \boldsymbol{D} = 0 \end{aligned} \tag{2}$$

Under arbitrary holonomic coordinate transformations $x^i \to \bar{x}^k(x^i)$ the components F_{ik} and G^{ik} transform according to the tensorial law and consequently Eqs. (2) maintain their form in the new coordinates. This simplicity however has its price, since the necessary connections between $\bar{F}_{ik}$ and $\bar{G}^{ik}$, become dependent on the space-time metrics. In the rest frame of the medium they assume the form

$$\bar{G}^{4\alpha} = c\varepsilon(-g)^{1/2}\bar{F}^{4\alpha}, \qquad \bar{G}^{\alpha\beta} = (c\mu)^{-1}(-g)^{1/2}\bar{F}^{\alpha\beta} \tag{3}$$

where ε and μ are the permittivity and the permeability of the medium, g is the determinant of the metric and $\bar{F}^{ik}$ is obtained from $\bar{F}_{ik}$ by the usual process of raising the indices. These constitutive relations are valid in the assumption that the acceleration acting on it does not change the macroscopic properties of the medium, [4], which is considered to be homogeneous, isotropic and non-dispersive.

The coordinate transformation relating the (upper case) Minkowski coordinates of the inertial frame I (X, Y, Z, cT) to the (lower case) coordinates of the linearly accelerated frame L (x, y, z, ct) which is stationary relative to the moving medium can be written as

$$\begin{aligned} X &= (c^2/g)(1+gx/c^2)\cosh\theta \\ Y &= y \qquad Z = z \\ cT &= (c^2/g)(1+gx/c^2)\sinh\theta \end{aligned} \tag{4}$$

where $\theta = gt/c$. In L the line element has the form

$$ds^2 = (1+gx/c^2)c^2dt^2 - dx^2 - dy^2 - dz^2 \tag{5}$$

which, as expected, reduces to the Minkowski form for null acceleration.

REFLECTED AND TRANSMITTED WAVES

It is assumed that the medium is a solid dielectric, which is accelerated as a whole and moves rigidly with constant proper acceleration g along the X-axis, as shown in Fig. 1.

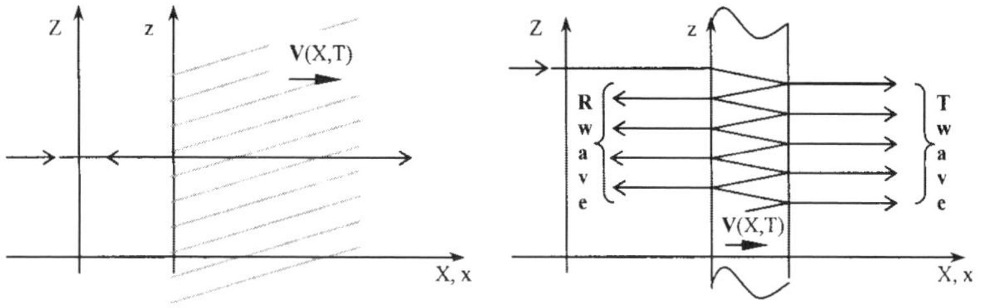

FIGURE 1. Geometry of the problem for a semi-infinite dielectric (left) and a dielectric slab of width h (right).

The waves reflected and transmitted by the accelerated medium were obtained for an incident plane wave of the general form

$$E^{I}{}_{Y}(\tau) = E_o \underbrace{\exp\left(-s(\tau/\tau_o)^2/2\right)\exp\left(-j\omega_o\tau\right)}_{f_i} \qquad (6)$$

where τ=X/c-T. For s=0 this represents a plane monochromatic wave of angular frequency ω_o, and for s=1 an optical Gaussian pulse with τ_o representing the pulse 1/e intensity half-width. The problem was solved in the rest frame of the medium matching the field at the boundaries, [6].

In the case of an accelerated half-space, we can locate the only boundary at x=0 Using (4) and the boundary conditions at x=0, one obtains for the reflected wave, upon transforming back to the inertial frame

$$E^{R}{}_{Y}(\eta) = \tfrac{1-n}{1+n}\tfrac{1-\beta(T_r)}{1+\beta(T_r)}\exp\left(-s\left(\tfrac{1-\beta(T_r)}{1+\beta(T_r)}\right)^2\left(\tfrac{\eta}{\tau_o}\right)^2/2\right)\exp\left(-j\omega_o\tfrac{1-\beta(T_r)}{1+\beta(T_r)}\eta\right) \qquad (7)$$

where η=X/c+T and $\beta(T_r)$ is the instantaneous relative velocity of the interface at the time of the reflection, The continuous Døppler effect in the amplitude, instantaneous frequency and pulse width of the reflected wave is clearly visible.

In the case of the accelerated dielectric slab it is found that the waves reflected and transmitted into the air consist of a sequence of components waves, which have been reflected and transmitted at x=0 and x=h, as shown in Fig. 1.

The solutions were obtained using the multiple reflections method and can be mathematically expressed as

$$E^R{}_Y(\eta) = RE_o(g\eta/c)^{-2} f_i\left((g\eta/c)^{-2}\right) + \\ + TT'E_o(g\eta/c)^{-2}\sum_{p=1}^{\infty} R'^{2p-1}(1+gh/c^2)^{2np} f_i\left((g\eta/c)^{-2}(1+gh/c^2)^{2np}\right) \quad (8)$$

$$E^T{}_Y(\tau) = TT'E_o\sum_{p=1}^{\infty} R'^{2(p-1)}(1+gh/c^2)^{-1+(2p-1)n} f_i\left((1+gh/c^2)^{-1+(2p-1)n}\tau\right)$$

where *R, T* (*R', T'*) are the usual reflection and transmission coefficients at a stationary dielectric/air (air/dielectric) interface. Apart from those coefficients, each term of the series differs from the previous one, in amplitude or phase, by a constant factor,

$$(1+gh/c^2)^{2n} = \exp(2gt_s/c) \quad (9)$$

where t_s is the transit time in a one-way trip in the slab. This term accounts for the change of velocity of the interfaces between two successive reflections or refractions.

Fig. 2 and 3 illustrate the results of applying Eqs. (8) to some particular situations. In these graphics time is normalized such that *c/g* is a unit of time and position is normalized such that c^2/g is a unit of length. *Eo* is a unit of field strength. A *short* incident pulse with 1/e intensity half-length equal to 1/10 the slab's width was used throughout. The incident pulse period was taken equal to 1/5 the pulse length.

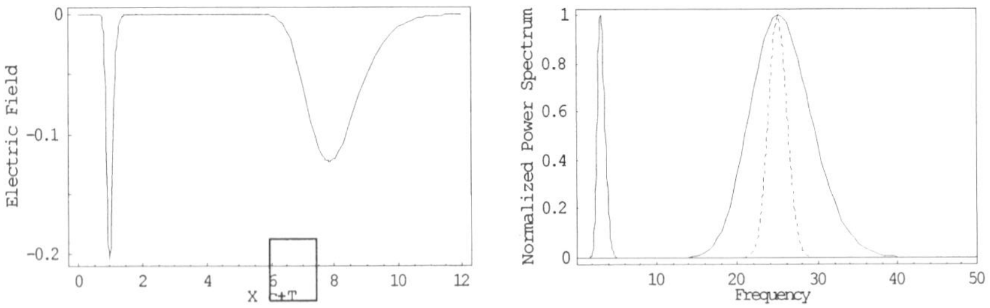

FIGURE 2. Reflection from a free space/dielectric (n=1.5) interface at x=0 and a metallic interface (obtained in the limit n→∞) at x=h: Envelope (left) and power spectrum (right) of the first two reflected pulses. The spectrum of the incident pulse is also shown (dashed line).

The results for the reflected wave depend on the spatial position of the incident wave and the dielectric slab at T=t=0, causing an unusual dependence of the wave parameters on (X, T). Except for *g=0*, the reflected optical pulses do not maintain their Gaussian form. On the other hand, the waves transmitted into the air depend on the change in velocity of the two slab interfaces during the transit of the wave through the slab, but not on the velocity itself. Therefore, the transmitted waves are much easier to analyze than the reflected waves.

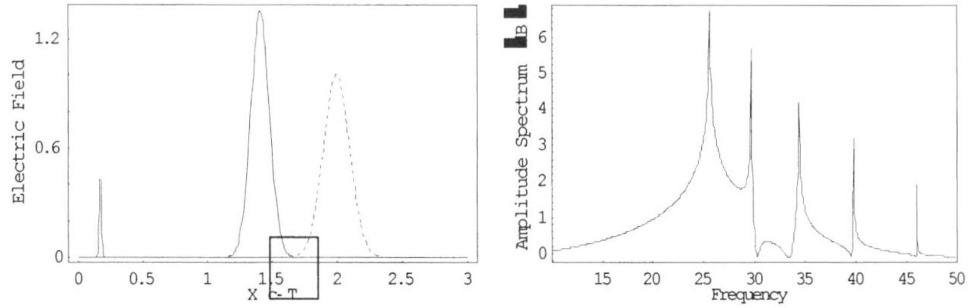

FIGURE 3. Transmission by a dielectric slab (n=1.5) of width h: Envelope of the first two transmitted optical pulses (left) and spectrum of the sum of the first five transmitted monochromatic waves. The incident pulse is also shown (dashed line).

The transmitted pulses maintain their Gaussian form, with the 1/e intensity half-width changing according to

$$\tau_o(p) = \tau_o /(1 + gh/c^2)^{-1+(2p-1)n} \tag{10}$$

A Fourier analysis of the time-dependent output for an incident monochromatic wave was obtained, and the results are shown in Fig. 3. The several lines corresponding to the different signal components are clearly visible. The lines separation is proportional to the acceleration of the sample for $gh/c^2 << 1$.

If the accelerated slab is inserted in one arm of a Mach-Zehnder interferometer, the frequency shift due to the presence of the accelerated medium can be measured, either by recording the output intensity as a function of time, or by Fourier analysis of the output, [3]. Both procedures are within the limits of current technology.

REFERENCES

1. Leonhardt, U. and Piwnicki, P., *Phys. Rev. A* **60**, 4301-4312 (1999).
2. Van Meter, J.R., Carlip, S. and Hartemann, F.V., *Am. J. Phys.* **69**, 783-787 (2001).
3. Neutze, R. and Stedman, G.E., *Phys. Rev. A* **58**, 82-90 (1998).
4. Ryon, J.W. and Anderson, J.L., *Phys. Rev. D***2**, 2745-2755 (1970).
5. Tanaka K., *J. Appl. Phys.* **49**, 4311-4319 (1978); *Phys. Rev. A***25**, 385-390 (1982)
6. Abreu Faro, M. and Marçal, M.H., *Mem. Academia Ciências Lisboa,* ***Tomo XXVI***, 52-70 (1985)

Feasibility Of $(Pb_{1-x}Ca_x)TiO_3$ Thin Films With x ~0.5 For Electronic Applications

J. Mendiola, R. Jimenez, P. Ramos[a] C. Alemany, M. L. Calzada, E. Maurer

Instituto de Ciencia de Materiales de Madrid, CSIC, Cantoblanco, 28049, Madrid, Spain
[a]Dpto. de Electrónica, Universidad de Alcalá, Alcalá de Henares, 28871, Madrid, Spain

Abstract. Calcium modified lead titanate thin films, $Pb_{1-x}Ca_xTiO_3$, with x ~ 0.5 have been deposited onto the crystalline $(Pt/TiO_2/SiO_2/(100)Si$, and $Pt/(100)SrTiO_3$ substrates by spinning-on and crystallisation of several layers of sol-gel precursor solutions. The layers are crystallized by a rapid thermal processing (RTP). Dielectric measurement with temperature and frequency show a wide broadening of the maximum of the permittivity close to room temperature, with high values of dielectric constant (K′ ~ 350), moderate dielectric losses (tgδ ~ 0.05) and positive shift of temperature of the maximum, T_m, for increasing frequencies that follows the empirical Vögel-Fulcher law, which could be interpreted as a relaxor behaviour. The high density of capacitance / μ^2 and low leakage currents lead to high retention of its charge that promises its use in DRAM. Measurements of capacitance with bias voltage (C-V) at different temperatures and frequencies show an interesting non-linear behaviour (varactor effect) with tunabilities about 70% and figure of merit of FOM > 30%. Therefore, these films can be considered as an alternative to traditional materials for voltage-tuneable devices for high frequencies.

INTRODUCTION

New DRAM (dynamic random access memories) generation looks for improving the integration density and the access velocity at lower prices [1,2]. Thus, alternative materials to the traditional SiO_2, such as, ferroelectric thin films $(Ba_xSr_{1-x})TiO_3$, BST, with high permittivity are extensively studied to fabricate memory elements equivalents to few nm of SiO_2 thickness which are unfeasible with this material [3]. On the other hand, at present, semiconductor varactors cannot be used beyond 3-5 GHz and to surpass these values several thin films based on textured $(Ba_xSr_{1-x})TiO_3$ have been intensively studied to fulfil the main requirements which combine a high permittivity with low dielectric losses (see [4] and citation quoted herein). Main drawback of thin films with respect to bulk ceramics is the strong depressed values of permittivity that is reduced below 25 times. Other alternative ferroelectric thin film materials based on the calcium titanate-lead titanate ($CaTiO_3$ - $PbTiO_3$) solid solution are proposed in this work for these applications.

Lead titanate is a well-known ferroelectric compound with tetragonal symmetry, P4mm space group that exhibits a ferroelectric-paraelectric phase transition at 490 ^{0}C [5]. Substitution of Pb^{2+} cations at B sites with isovalent Ca^{2+} cations causes drastic tetragonality changes. The phase transition temperature shifts to lower values and consequently, room temperature permittivity increases and remanent polarization decreases [6,7]. The $PbTiO_3$-$CaTiO_3$ solid solution considered from the $CaTiO_3$ side, where Ca is replaced by Pb, has been studied by Lemanov et al. [8]. They reported that the system $Pb_xCa_{1-x}TiO_3$ behaves as an incipient ferroelectric for a critical value x_0 = 0.28. They also suggested the existence of a morphotropic phase boundary (MPB) around x=0.5. According to powder neutron diffraction studies carried out by Ranjan et al. [9], the crystal structure of $Pb_{.5}Ca_{.5}TiO_3$ bulk ceramic was refined as orthorhombic with Pbnm space group, following the model of Glazer [10] for $CaTiO_3$ with the $a^-a^-c^+$ tilt system of the oxygen octahedral. The present paper is an extension of the work previously reported by the authors [11,12] on $(Pb_{1-x}Ca_x)TiO_3$ (PCT) films with x close to 0.5 that exhibit promising properties to their use in DRAM or varactors, as compare with the former materials.

EXPERIMENTAL PROCEDURES

A precursor solution with nominal composition of $Pb_{0.50}Ca_{0.50}TiO_3$ was prepared by a sol-gel method reported elsewhere [13]. This solution was deposited and crystallized onto $Pt/SrTiO_3(100)$ and

(Pt/TiO_2/SiO_2/(100)Si substrates, named as SrCa and SiCa respectively. One SrCa film was prepared with a thickness of ~ 2160 nm, and two SiCa films were prepared with thickness of ~ 190 nm and ~1400 nm. Thickness was measured by profilemetry. Pt dot electrodes of 0.2 mm^2 were deposited by sputtering on the film surface, by a shadow mask to perform electric measurements. Crystal phase was studied by X-ray diffraction, XRD, on a Siemens D-500 diffractometer with θ-2θ scan experimental configuration, and Cu Kα radiation. An LCR-meter HP 4284A was used to measure dielectric permittivity as a function of temperature at several frequencies. Capacitance variation with bias voltage (C-V) at different temperatures and frequencies has also been measured by an Impedance Analyser HP 4194A. Leakage current density was measured for the electric field range employed, by using a pulse generator HP- 8116A and an electrometer Keithley 6514. Testing of voltage drop of the small capacitor after pulse charging is performed by means a data acquisition hardware.

RESULT AND DISCUSSION

All prepared films show by XRD an unique crystalline phase of pseudocubic symmetry. Dielectric measurements performed in the films have been obtained for temperature range 125 to 425 K in the frequency range 500Hz to 1MHz. K′ values have been corrected by the effect of series resistance, R_s, due to the electric path between capacitor electrodes that were deduced previously from impedance measurements.

In sample SrCa (Fig. 1), a broad maximum is observed for K′ with a shift to higher temperatures as the frequency increases and an important decrease of dielectric dispersion above T_m, that suggest the relaxor nature of the material.

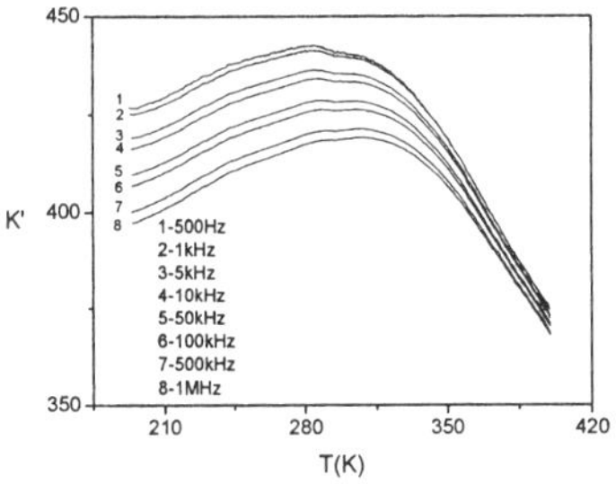

FIGURE 1. Sample SrCa. K′-T corrected values for frequencies 500Hz to 1MHz.

Figure 2 shows the capacitance versus bias voltage, C-V, measured at 10kHz for 258 K, 298 K and 356 K. Although hysteresis still remains well above the temperature of maximum of permittivitty, is small. Calculation of tunabilities for the whole set of C-V measurements were performed using the expression,

$$\text{Tun} = (C_{max} - C_{min})/C_{max} \qquad (1)$$

where C_{max} and C_{min} are the values of maximum and minimum of capacitance. The figure of merit of the capacitor is obtained as follow:

$$\text{FOM} = \text{Tun}/\ \text{tg}\delta \qquad (2)$$

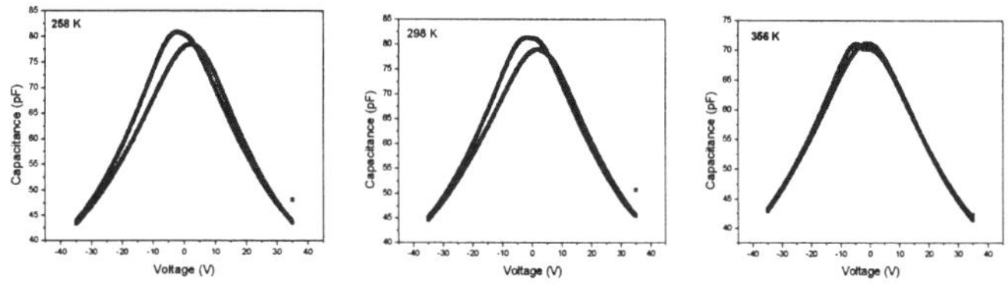

FIGURE 2. Sample SrCa. Capacitance as a function of dc bias voltage for temperatures 258 K, 298 K and 356 K.

tgδ are the dielectric losses corresponding to the C_{max} value. Figure 3 shows the change of tunability and FOM in a broad range of temperature of the SrCa film. Note that tunability remains higher than 60% even above 373 K, being FOM more than 30% at any temperature.

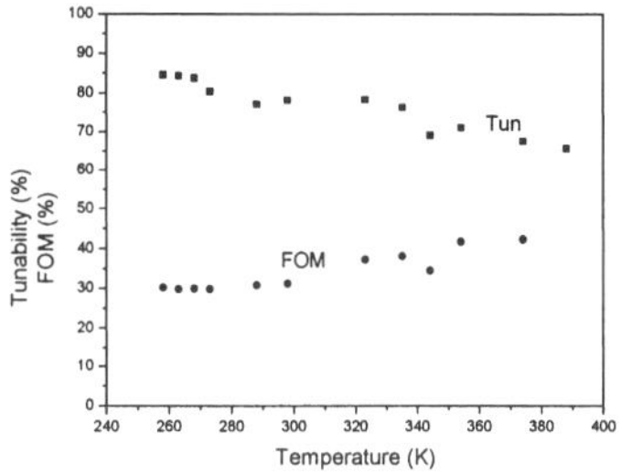

FIGURE 3. Sample SrCa. Changes of Tunability and FOM with temperature.

Dielectric measurements of the two SiCa samples are depicted in Fig. 4. The thicker sample has higher dielectric constant values than the thinner one, appearing an extra maximum in the last, probably due to the mixing of phases in the morphotropic boundary.

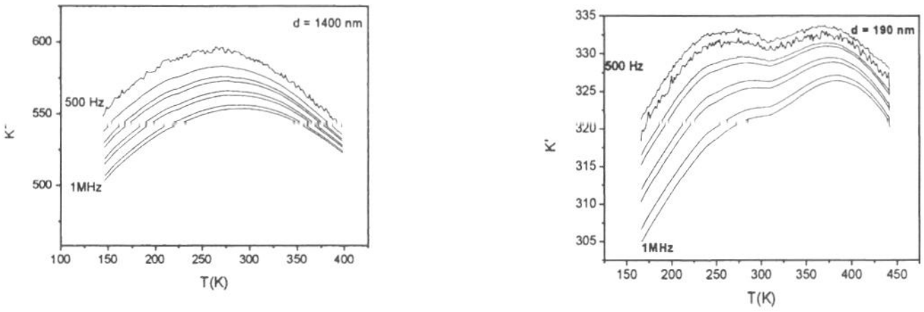

FIGURE 4. SiCa samples. Variation of dielectric constant with temperature for several frequencies.

For the 1400 nm thick SiCa sample, correction for dead layer capacitance has been made [14], assuming, in a series model, a lower permittivity that of the bulk film, being also temperature independent. Figure 5a shows the behaviour of a relaxor ferroelectric compound. Above the temperature of maximum, T_m, an important reduction of dispersion is observed and the maximum shifts to higher temperatures as the frequency increases. Notice the high values of permittivity and the maximum broad around room temperature. The relaxor nature is reinforced by the good fitting of the Vogel-Fulcher law,

$$\omega = \omega_0 \exp(-E_a/R(T_{max} - T_f)) \qquad (4)$$

Figure 5b shows the fitting of experimental data to the equation (4), with the calcutated parameters.

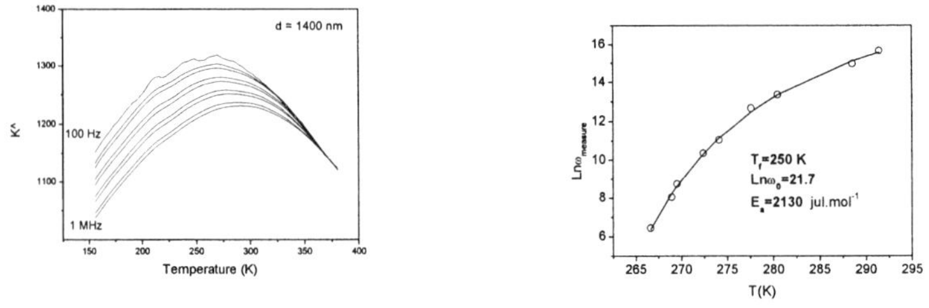

FIGURE 5. Relaxor behaviour of SiCa 1400 nm thick sample. a) K′ values after all corrections. b) Vogel-Fulcher law applied to the frequency dispersive nature of dielectric constant maximum, T_m.

Leakage current density measurements for a wide applied voltage range show values of ~10^{-8} A/cm^2 at ~3V (Fig. 6a) which is of high interest together with large retention of the voltage, after the capacitor is charged. To know the DRAM pulse behaviour of theses capacitors, the voltage retained after its charged up is depicted in Figure 6b. The signal decreases ~ 3.5% after 300ms for a pulse of charge of 3V.

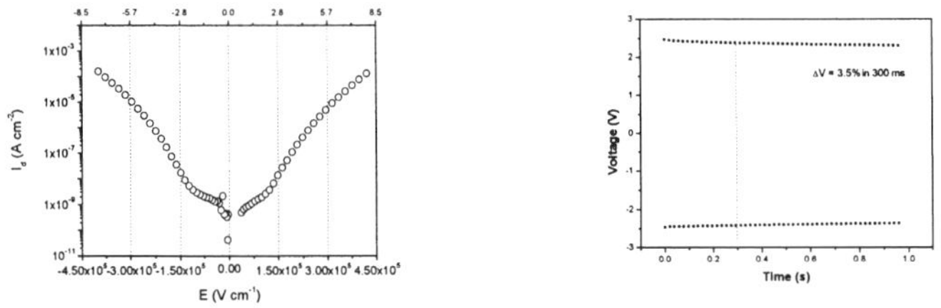

FIGURE 6. DRAM Properties for sample SiCa 190 nm thick. a) Leakage current density; b) Pulse behaviour

Dielectric results here reported have been measured in planar capacitors and at frequencies below 1MHz. However, it is expected that this behaviour can be extrapolated for higher frequencies (> 10 GHz), by measuring with interdigital electrodes. Therefore, these data support the potential application of these films in two types of electronic devices. The important varactor effect (C-V changes) which leads to tunability values around 70% with a 10% deviation in the measured range of temperature and the low losses (tgδ < 0.05) that leads to a figure of merit about 40%, make feasible the use of these materials in tuneable high frequency circuits, as an alternative to the BST materials.

The moderate high permittivity values, in a wide range close to room temperature (variations less than 5% in 100K intervale) and the very low leakage current densities measured (~10^{-8} A/cm^2) at ~ 3V make these films attractive for the fabrication of dynamic random access memories (DRAM). Capacitance density close to 15 fF/μm^2 has been obtained, that means 4 nm of equivalent SiO_2 thickness.

CONCLUSIONS

In $Pb_{0.50}Ca_{0.50}TiO_3$ films, the temperature of the maximum K′ is shifted to around room temperature, obtaining K′ values of 325 for 190 nm thick films. The low leakages for work voltages of

3V and the high retention of the store charge make attractive these compounds for the use in DRAM. Furthermore, in thicker films, the low dielectric losses and the high tunability values in a wide temperature range close to room temperature, make also feasibly their use in high frequency tuneable circuits. Theses previous results do not yet surpass the reported values obtained for BST epitaxial films. However, improvements are possible if the dead layer effect is reduced. This joined to the simplicity of the sol-gel fabrication would make this material a good candidate for the mentioned applications.

ACKNOWLEDGMENTS

This work has been supported by the Spanish project MAT2001-1564. R. Jimenez recognizes the support of Ramon y Cajal contract of the MCyT.

REFERENCES

1. O. Auciello, J. F. Scott and R. Ramesh, *Physics Today* 7, 22-27 (1998).
2. D.E. Kotecki. *Integrated Ferroelectrics*, **16**, 1-19 (1997)
3. A.I. Kingon, J.P. Maria, S.K. Streiffer, *Nature*, **406**, 1020 (2000)
4. M. W. Cole, W. D. Nothwang, C. Hubbard, E. Ngo, M. Ervin, *J. Appl. Phys.*, **93**, 9218-9225 (2003)
5. G. Shirane, R. Pepinsky, B.C. Frazer, *Acta Cryst.*, **9**, 131- 140 (1956)
6. T.Yamamoto, M. Saho, K. Okazaki, *Jpn. J. Appl. Phys.*, 26, Supplem 26-2, 57-60 (1987)
7. J. Mendiola, B. Jiménez, C. Alemany, L. Pardo and L. Del Olmo, *Ferroelectrics*, **94**, 183-188 (1989)
8. A. V. Lemanov, E. P. Sotnikov, M. Smirnova, Weihnacht, *Appl. Phys. Lett.*, **81[5]**, 886 (2002).
9. R. Ranjan, N. Singh, D. Pandey, V. Siruguri, P.S.R. Krishna, S.K. Paranjpe, A. Banaerjee, *Appl. Phys. Lett.*, **70 (24)**, 3221-3223 (1997).
10. A.M. Glazer, *Acta Cryst.*, **A31**, 756-762 (1975)
11. J. Mendiola, R. Jiménez, P. Ramos, C. Alemany, M. L. Calzada, E. Maurer, *Bol. Soc. Es. Cer, Vidrio*, (in press, 2003).
12. R. Jiménez, C. Alemany, M.L. Calzada, J. Mendiola, *Intrgrated Ferroelctrics* (submitted, 2003).
13. I. Bretos, J. Ricote, R. Jiménez, J. Mendiola and M. L. Calzada, *Integrated Ferroelectrics, (submitted, 2003)*
14. M. Tyunina and J. Levoska, *Physical Rev.*, **63**, 224102, 1-8 (2001)

Applied Physics (APHYS 2003)
A. Méndez-Vilas (Ed.)

Acceleration and mixing in the radiometric dating of recent sediments: A further discussion supported by the IMZ model.

J.M. Abril

Department of Applied Physics I. University of Seville. E.U.I.T.A., Carretera de Utrera km 1. D.P. 41013. Seville (Spain)

Abstract. Man-made fallout radionuclides, such as ^{137}Cs, are increasingly used to provide the necessary support to the ^{210}Pb-based dating of recent sediments. In the cases with constant ^{210}Pb activities in the topmost sediments, the presence of a distinct ^{137}Cs peak within the ^{210}Pb plateau has been used as a definitive demonstration of acceleration (increase in the sedimentation rate in recent years) versus fast mixing. This work tries to demonstrate that the simple methods based in the identification of the ^{137}Cs fallout peaks cannot provide a definite support for CRS chronologies. Thus, the incomplete mixing within the top sediment zone (described through the Incomplete Mixing Zone model) can explain quantitatively and simultaneously the ^{137}Cs peak and the flattening in the ^{210}Pb activity profile. This is demonstrated using selected examples from literature data.

INTRODUCTION

Radiometric dating is an important tool to study rates of sedimentation and mixing and to interpret historical records of hazardous chemicals in sediments.

The CRS model [1] is a ^{210}Pb-based radiometric dating which is widely applied. The model assumes a constant rate of supply of unsupported ^{210}Pb and no post-depositional mixing, but the sediment accumulation rate can be variable. Smith [2] argued that ^{210}Pb geochronologies must be validated using at least one independent tracer, which separately provides an unambiguous time-stratigraphic horizon. ^{137}Cs is of fallout origin, with a distinct maximum in 1963. This radionuclide is often used to support ^{210}Pb-based chronologies.

In many cases one can find constant ^{210}Pb activity in the topmost sediments. The presence of a distinct ^{137}Cs peak within the ^{210}Pb plateau has been used as a definitive demonstration of acceleration (increase in the sedimentation rate in recent years) versus fast mixing in these cases [3].

This paper shows that simple methods based on the identification of the ^{137}Cs fallout peaks cannot provide a definitive support for CRS chronologies. Thus, the incomplete mixing within the top sediment zone (described through the Incomplete Mixing Zone model [4]) can explain quantitatively and simultaneously the ^{137}Cs peak and the flattening in the ^{210}Pb activity profile. This is demonstrated using selected examples from literature data.

INCOMPLETE MIXING ZONE MODEL

The advection-diffusion equation for a particle-associated radiotracer under steady-sate compaction can be written in the mass-depth variable as follows [5]:

$$\frac{\partial s}{\partial t} = -\lambda s + \frac{\partial}{\partial m}\left(k\frac{\partial s}{\partial m}\right) - \frac{\partial}{\partial m}(w s) \qquad (1)$$

In Eq. 1 s is the activity concentration [Bq M^{-1}], λ [T^{-1}] the radioactive decay constant, w [M L^{-2} T^{-1}] the sedimentation rate, and k [M^2 L^{-4} T^{-1}] a mixing coefficient.

The IMZ model [4] assumes the existence of a mobile fraction in radionuclide activity (dissolved radionuclides, and those attached to colloids and very small particulate matter), which is rapidly and homogeneously mixed (through the connected pore water) within a top zone in the sediment core. On the other hand, settled particles have an irreversibly bounded fraction of activity, and can undergo aggregation processes in which part of their free surface becomes unavailable for any further radionuclide exchange. The model approaches this situation assuming that a fraction g of the incoming flux, $\phi(t)$, undergoes a fast and complete mixing ($k\to\infty$) within a top sediment zone of mass thickness m_a, while the remaining fraction does not follow any post-depositional redistribution. This model is applicable for ^{210}Pb and for fallout radionuclides with time-dependent fluxes (such as ^{137}Cs). The corresponding analytical solution is: $s = g\, s_1 + (1-g) s_2$, with

$$s_1 = \frac{\phi(t - m/w)}{w} e^{-\lambda m/w}\ ; m/w < t \quad , \quad s_1 = 0\ ; m/w > t$$

$$s_2(m \le m_a) = \frac{\sum_\phi(t)}{m_a};\ \text{with}\ \sum_\phi(t) = \int_0^t \phi(t-u)\, e^{-(\lambda + \frac{w}{m_a})u}\, du$$

$$s_2(m > m_a) = \frac{\sum_\phi\left(t - \frac{m - m_a}{w}\right)}{m_a} e^{-(\lambda \frac{m - m_a}{w})} \qquad (2)$$

CASE STUDIES

The first case study is a sediment core from the Kattegat area (Sweeden), sampled in 1984 at 96 m depth and at 20 km from the coast. The site description and the radiometric data can be found in the work from Abril et al. [4]. The unsupported ^{210}Pb profile showed the typical plateau, while the distinct ^{137}Cs peak appeared within this plateau. The IMZ model, as shown in Figure 1, could explain the whole data set.

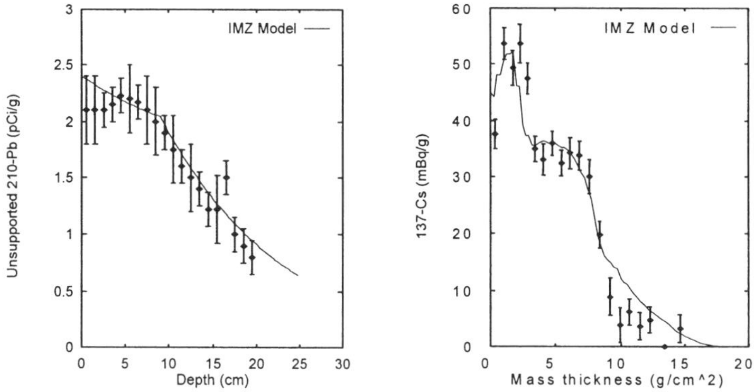

FIGURE 1. ^{137}Cs and unsupported ^{210}Pb profiles in the marine sediment from Goteborg. IMZ model uses the following parameters: w= 0.374 g cm^{-2} y^{-1}, g= 0.7, m_a= 6.0 g cm^{-2}. The flux was constant for ^{210}Pb and time-dependent for ^{137}Cs (Data from Abril et al. [4]).

The second case study corresponds to a sediment core sampled in 1979 by Erten et al.[6] in the Lake Zürich, at 61 m water depth and at approximately 1 km from the shoreline.

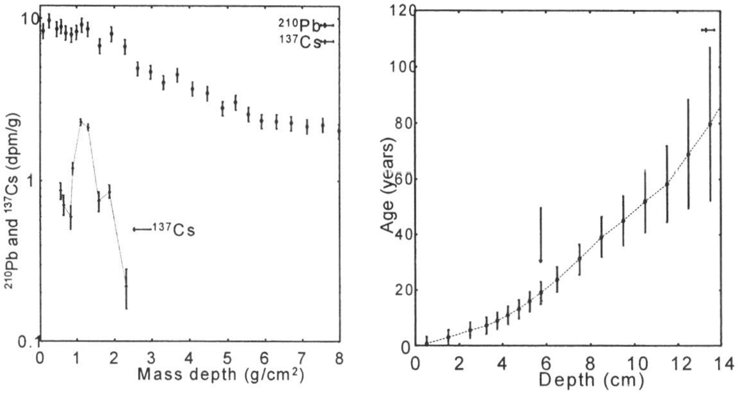

FIGURE 2. ^{137}Cs and unsupported ^{210}Pb profiles in the sediment from Lake Zürich [6]. On the right, the computed CRS ages (age zero corresponds to the time of sampling), in excellent agreement with the ^{137}Cs time-mark (year 1963, see arrow).

The distribution of ^{137}Cs in the cores showed a distinct maximum in activity. ^{7}Be was measured in the uppermost sediment layer from an additional core, while non-detectable activities were found deeper. Finally, the texture analysis showed well-developed and unperturbed laminations. Erten et al. [6] concluded that mixing or bioturbation did not occur at least not for the solid parts of the sediment.

The historical records of ^{137}Cs atmospheric deposition were estimated from the ones measured by the Risø National Laboratory in Denmark and by introducing a normalization factor (the ratio between the measured inventory in sediments and the cumulative atmospheric deposition). Thus, from the estimated ^{137}Cs atmospheric deposition and using the (variable) sedimentation rates resulting from the CRS model, it was possible to compute the expected ^{137}Cs versus depth distribution (provided that no post-depositional mixing occurred, as it is stated in the CRS model). But CRS model could not explain the size of the peak, nor the presence of ^{137}Cs in the deeper layers. Thus, one has to accept that mixing occurs at least for ^{137}Cs.

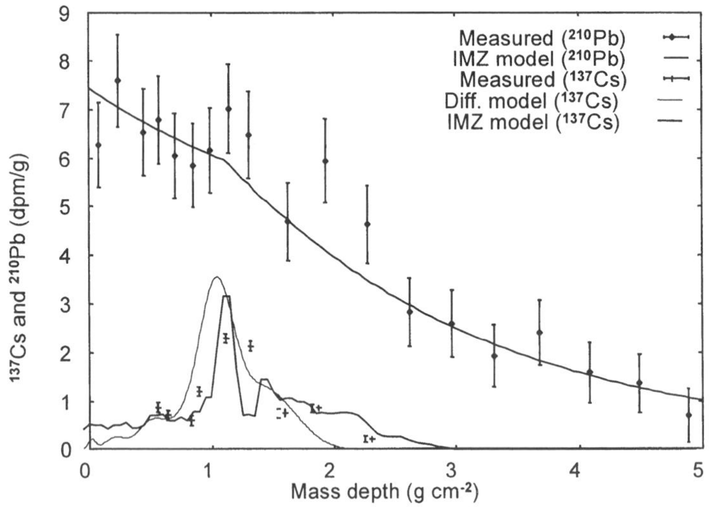

FIGURE 3. ^{137}Cs and unsupported ^{210}Pb activities versus mass depth profiles in the sediment core from Lake Zürich [6]. A model of constant diffusion fits the ^{137}Cs profile. The IMZ model simultaneously explains both ^{137}Cs and ^{210}Pb data, leading to a different chronology (see text).

To test the hypothesis of mixing, the ^{137}Cs profile in the sediment core from Lake Zürich was studied using a model with a constant mixing coefficient and the aforementioned records of ^{137}Cs atmospheric deposition. The model used a mixing coefficient of 0.002 g^2 cm^{-4} y^{-1} and a constant mean value of the mass sedimentation rate of 0.075 g cm^{-2} y^{-1} (from CRS results). It provided a reasonable description of the data, as shown in Fig. 3, although activities in the deepest layers were poorly described. Nevertheless, this particular diffusion model is not compatible with the CRS model since it considers the mixing of the total specific activity, and thus affects solid particles.

The IMZ could reasonably explain both ^{137}Cs and ^{210}Pb specific activities versus mass depth profiles, as shown in Fig. 3. The model used a constant value for the mass sedimentation rate of 0.075 g cm^{-2} y^{-1}, a mixing zone at the top of the sediment with a mass depth of 1.1 g cm^{-2} and a mixing fraction $g = 0.6$ (the same parameter values as for ^{137}Cs and ^{210}Pb).

ACKNOWLEDGEMENTS

This work has been partially supported under contract I+D with ENRESA (Project 774511, Code: 0770105).

REFERENCES

1. P.G. Appleby and F. Olfield. *Catena* **5**, 1-8 (1978)
2. J.N. Smith. *J. Environ. Radioactiv.* **55**,121-123 (2001).
3. P.G. Appleby P.G. *Limnol.* **59** (Suppl.1): 1-14.(2000).
4. J.M. Abril, M. García-León, R. García-Tenorio, C.I. Sánchez and F. El-Daoushy *J. Environ. Radioactiv.* **15**:135-151 (1992)
5. J.M. Abril. *J. Paleolimnology* (in press)
6. H.N. Erten, H.R. von Gunten, E. Rössler and M. Sturm, M. *Schwiz. Z. Hydrol.* **47**(1), 5-1 (1985).

Applied Physics (APHYS 2003)
A. Méndez-Vilas (Ed.)

A new theoretical treatment of sediment compaction: a reviewed basis for the radiometric dating of recent sediments with compaction and time-dependent fluxes.

J.M. Abril
Department of Applied Physics I. University of Seville. E.U.I.T.A., Carretera de Utrera km 1. D.P. 41013. Seville (Spain)

Abstract. Radiometric dating of recent sediments is a powerful tool to interpret historical records of anthropogenic impacts and the behaviour of hazardous chemicals in aquatic systems. Man-made fallout radionuclides are increasingly used to provide the necessary support for the ^{210}Pb-based chronologies. This requires solving an advection-diffusion problem with time-dependent fluxes and under compaction. Classical treatment of mass conservation of solids in growing sediments states an advection-diffusion equation for the bulk density of sediment. Nevertheless, in a gravity field, compaction cannot be treated as diffusion but as mass flow. A *compaction potential energy* is then defined so that its spatial gradients force a mass flow involving a *conductivity* function. Thus, the continuity equation for density involves only an advection term. Typical bulk density profiles, with an asymptotic increase with depth, can be obtained as a steady-state solution under constant sedimentation rate and constant conductivity. From this basis, the advection-diffusion equation for a particle-associated tracer is rewritten and their numerical and analytical solutions are found out for several radiometric-dating models of special relevance. Literature data are used to illustrate different aspects of the theory development and its applications.

INTRODUCTION

Studies on the behavior of particle-associated tracers in aquatic sediments are of widespread interest. Radiotracers can provide valuable information on rates of sedimentation and mixing, which are necessary to interpret historical records and the behavior of hazardous chemicals. As anthropogenic impacts increased with the industrial era their effects on sediments appear in the top zone, usually affected by active compaction.

The use of appropriate mathematical models is a valuable tool for the determination of sedimentation parameters and the sediment chronology. The most complex (and interesting) situation corresponds to time-dependent fluxes (as it is the case for fallout radionuclides) and active compaction. The classical differential equations for the conservation of solids, pore water and the particle-associated tracers, are given by Berner [1]. Based on Berner's equations,

Christensen and Bhunia [2] developed a comprehensive model for the activity of radionuclides in sediments. They stated an advection-diffusion equation for the bulk sediment density. The diffusion term accounts for sub-grid scale processes consisting in exchanges of the extensive property (volume), and resulting in a zero net balance of this property, but with changes in the intensive property (dry mass concentration). In the gravity field the exchanges of solid particles by pore water (or reciprocally) are forced, and they cannot be treated as diffusion but as a mass flow.

Abril [3] presented a new formulation based on the idea of a *compaction potential energy,* which allows rewriting the continuity equation for density and the advection-diffusion equation for a particle-associated tracer. This work presents a further development of their numerical and analytical solutions, which are found out for several radiometric-dating models of special relevance. Literature data will be used to illustrate different aspects of the theory development and its applications.

COMPACTION POTENTIAL ENERGY

One can assume that the sediment is locally homogeneous at any horizontal cross-section. Thus, only the variations with depth are of relevance and the problem can be studied in one-dimension. As result of the accumulation of new material, the sediment-water interface displaces up. From a framework anchored to this boundary, the sediment moves down as a whole with the sedimentation velocity v.

The *bulk density* ρ is the mass of dry matter per unit volume of undisturbed sediment. Due to progressive compaction, typical bulk densities asymptotically increase with depth (the depth z is measured downwards from the water-sediment interface). This obeys to a forcing action. In fact, as density of solids is greater than density of water, a solid particle tends to occupy a water pore under it unless the resultant of the contact forces from other solid particles prevents it. Thus we can, at least conceptually, introduce a specific potential energy for solid particles, ψ, which decreases when water pores are occupied by solids. Let us call ψ the *compaction potential.* It is defined as energy per unit weight; consequently it has dimensions of L. Conceptually, the spatial gradients of ψ can only be upwards directed and they represent a forcing term resulting in a downwards-directed flux of matter:

$$q\rho = -K(z)\frac{\partial\psi}{\partial z} \tag{1}$$

Eq. 1 is similar to the linear transport equations of classical physics, being q the velocity (L T^{-1}) associated to the flow of solids. Thus, *K(z)* can be interpreted as a *conductivity* function. It has dimensions of M L^{-2} T^{-1}. Thus, taking into account the mass flow associated to the sedimentation rate *v*, one can introduce the continuity equation as follows,

$$\frac{\partial \rho}{\partial t} = -\frac{\partial}{\partial z}(\rho v + \rho q). \tag{2}$$

It is more convenient to define the *dry mass sedimentation rate* $w = \rho(v + q)$, with dimensions of M L^{-2} T^{-1}. From our formulation, *v* is not depth-dependent but it can vary with time. Under steady state compaction $\rho(z,t) = \rho(z)$, and the continuity equation for the sediment mass leads to the following result: $w(z,t) = w(0,t)$.

Let us consider the simplest compaction potential $\psi = A - B\rho$, where *A* and *B* are positive constant coefficients. This potential is belittled as ρ increases. Thus, under a constant sedimentation velocity *v*, and constant conductivity, *K*, the steady-state solution of Eq. 2 is the typical asymptotic bulk density profile:

$$\rho = \rho_\infty - \rho_1 e^{-\alpha z} \tag{3}$$

with $\rho_\infty = -\frac{a}{v}$ and $\alpha = \frac{v}{KB}$, where *a* is a constant.

Alternatively, it can be shown that for any given steady-state bulk density profile following Eq. 3, the compaction potential is necessarily given by $\psi = A - B\rho$, if the mass sedimentation rate *w* and the conductivity *K* are assumed to be constant with depth.

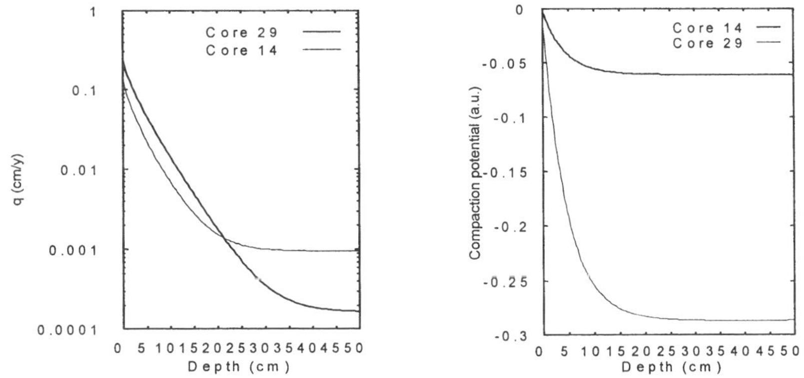

FIGURE 1. Values of *q* and ψ versus depth for core 14 (ρ=0.253-0.156 $e^{-0.228z}$ gcm^{-3}, w = 0.022 $gcm^{-2}y^{-1}$) from Lake Huron and core 29 (ρ=0.561-0.314 $e^{-0.210z}$ gcm^{-3}, w = 0.105 $gcm^{-2}y^{-1}$) Lake Michigan (ρ and *w* data from Christensen and Bhunia [2]).

The complete advection-diffusion equation, including the radioactive decay constant λ for radionuclides, can be written as follows

$$\frac{\partial(\rho s)}{\partial t} = -\lambda \rho s + \frac{\partial}{\partial z}\left(D\rho \frac{\partial s}{\partial z}\right) - \frac{\partial}{\partial z}(ws) . \quad (4)$$

Let us introduce the mass-thickness $m(z) = \int_0^z \rho dz'$. Thus, $\frac{\partial}{\partial z} = \rho \frac{\partial}{\partial m}$ for steady-state compaction and, introducing $k = D\rho^2$, Eq. 4 can be rewritten as:

$$\frac{\partial s}{\partial t} = -\lambda s + \frac{\partial}{\partial m}\left(k \frac{\partial s}{\partial m}\right) - \frac{\partial}{\partial m}(ws) \quad (5)$$

The time-dependent problem can be solved in the Laplace's space as the steady state case of Eq. 5. This allows finding out useful analytical solutions for radiometric dating models accounting for mixing and compaction. Eq. 5 can also be numerically solved applying standard methods.

APPLICATION: SOME RADIOMETRIC DATING MODELS

The *linear model* assumes no post-depostional mixing (k=0) and constant sedimentation rate (w). With constant fluxes ϕ the steady-sate solution of Eq. 5 is: $s = s_0 e^{-\lambda m/w}$. Applying Laplace's transformations we can find out the corresponding solution for the time-dependent problem ($\phi=\phi(t)$):

$$s = \frac{\phi(t - m/w)}{w} e^{-\lambda m/w} \; ; m/w < t \quad , \quad s = 0 \; ; m/w > t \quad (6)$$

The *complete mixing zone* (CMZ) model assumes constant w and homogeneous and instantaneous mixing ($k \rightarrow \infty$) over a top sediment zone of mass thickness m_a, and no post-depostional mixing deeper. The corresponding steady state solution with constant flux is:

$$s(m \le m_a) = \frac{\phi}{w + \lambda m_a} \quad ; \quad s(m > m_a) = \frac{\phi}{w + \lambda m_a} e^{-\lambda(m - m_a)/w} \quad (7)$$

For the time-dependent problem the corresponding solution is:

$$s(m \le m_a) = \frac{\sum_\phi(t)}{m_a}; \text{ with } \sum_\phi(t) = \int_0^t \phi(t-u)\, e^{-(\lambda + \frac{w}{m_a})u} du$$

$$s(m > m_a) = \frac{\sum_\phi(t - \frac{m - m_a}{w})}{m_a} e^{-(\lambda \frac{m - m_a}{w})} \quad (8)$$

The IMZ model [4] assumes the existence of a mobile fraction in radionuclide activity (dissolved radionuclides, and those attached to colloids and very small particulate matter), which is rapidly and homogeneously mixed (through the connected pore water) within a top zone in the sediment core. On the other hand, settled particles have an irreversibly bounded fraction of activity, and can undergo aggregation processes in which part of their free surface becomes unavailable for any further radionuclide exchange. The model approaches this situation assuming that a fraction g of the incoming flux undergoes a fast and complete mixing within a top sediment zone of mass thickness m_a, while the remaining fraction does not follow any post-depositional redistribution. This model is applicable for ^{210}Pb and for fallout radionuclides with time-dependent fluxes (such as ^{137}Cs). The corresponding analytical solution is a linear combination (with coefficients g and $1-g$) of the corresponding solutions for CMZ and *Linear* models.

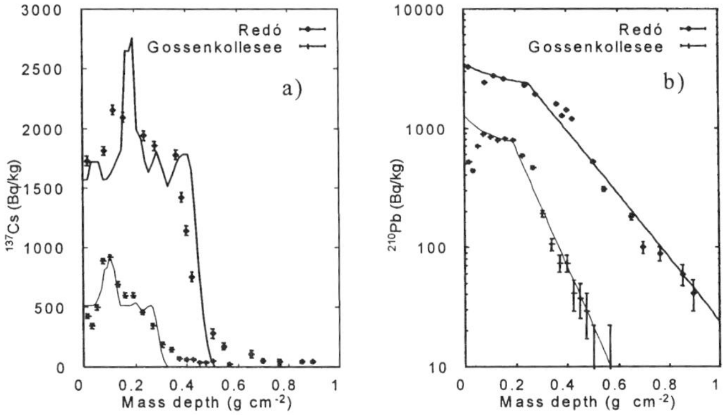

FIGURE 2. ^{137}Cs (fig. a) and unsupported ^{210}Pb (Fig. b) specific activities (in Bq kg^{-1}) versus mass depth profiles. Comparison between measured (data with error bars) and computed values from the IMZ model for Lake Redó (w = 0.0056 g cm^{-2} y^{-1}, m_a = 0.25 g cm^{-2} and g = 0.9) and Lake Gossenköllesse (w = 0.0030 g cm^{-2} y^{-1}, m_a= 0.19 g cm^{-2} and g = 0.8). The parameter values are the same for both two radionuclides. Measured data from Appleby [5].

ACKNOWLEDGEMENTS

This work has been partially supported under contract I+D with ENRESA (Project 774511, Code: 0770105).

REFERENCES

1. R.A.Berner. *Early Diagenesis: A Theoretical Approach.* Princeton University Press. Princeton. N.J. 1982
2. E.R. Christensen and P.K.. *J. Geophys. Res.* **91**: 8559-8571 (1986).
3. J.M. Abril. *J. Paleolimnology* (in press)
4. J.M. Abril, M. García-León, R. García-Tenorio, C.I. Sánchez and F. El-Daoushy *J. Environ. Radioactiv.* **15**:135-151 (1992)
5. P.G. Appleby P.G. *Limnol.* **59** (Suppl.1): 1-14.(2000).

Applied Physics (APHYS 2003)
A. Méndez-Vilas (Ed.)

SOLAR RADIATION MAP OF EXTREMADURA FROM OTHER WEATHER DATA

A. RAMIRO, J.J. REYES, J.F.GONZÁLEZ, E.SABIO, M.L.GONZÁLEZ-MARTÍN, C.M.GONZÁLEZ-GARCÍA AND J.GAÑÁN

Escuela de Ingenierías Industriales, University of Extremadura, Avda. de Elvas s/n, 06071 Badajoz, SPAIN.
Fax: 924-289601 E-mail: aramiro@unex.es

M. NÚÑEZ

Centro Meteorológico Territorial de Extremadura (**I.N.M.**)*, Avda. de Elvas s/n, 06071 Badajoz, SPAIN.*
E-mail: marcelino.nunez@inm.es

Abstract- In a previous work, we have found correlation expressions that permit to estimate the mean monthly values of daily diffuse and direct solar irradiation on a horizontal surface in function of some weather parameters. In this work, the incident radiation on a horizontal surface has been estimated in thirty zones of Extremadura by means of weather data from existing stations located in these zones and its orography. The weather data used have been the monthly average values of the highest temperatures and the sunshine fraction. These monthly average values have been obtained from measurements carried out in the weather stations during the period 1985-2002. The results are presented as interactive maps in Arcview language, associated to a conventional data base.

KEYWORDS: (Solar energy; Estimation solar radiation; Empirical correlations)

1. Introduction

The design of any solar installation requires the knowledge of long-term solar radiation data in the locality in question. Up till now, only data supplied by the weather stations situated in the province capitals were available. Nevertheless, environment temperature, cloud cover, sunshine fraction and pluviometry data are available in many weather stations distributed by all the territory, with different degree of automation. From these data the solar irradiation could be estimated.

Various investigators (Lumb, 1964; Brinsfield et al., 1984) have considered the possibility to find expressions that relate the solar radiation with the cloud cover. The most extensive investigation to this respect was carried out by Haurwitz (1945, 1948), who analyzed the data during eleven years in Blue Hill, Massachussets. Kasten and Czeplak (1979) have completed the work of Haurwitz and recently their results have been applied for the U.K. This class of models is called CRM (cloud cover radiation model) and needs only cloud cover data for the estimation of the solar radiation.

Another type of model is the one called MRM (meteorological radiation model) that requires for the estimation of the solar radiation another weather series of data, such as the temperature environment and the sunshine fraction or quotient among the real hours of sun (time in hours when the irradiance exceeds a certain threshold) and the theoretical hours or day length, expressed in hours. Muneer *et al.* (1996, 1997) have developed this type of models. A comparison between the two models cited has been carried out by Mehreen S. Gul *et al.* (1998).

In a previous work, we have found expressions of correlation that permit to estimate the monthly mean values of daily direct and diffuse irradiation on a horizontal surface ($B_{dm}(0)$ and $D_{dm}(0)$, respectively) in function of some of these weather parameters (Ramiro *et al.* 2003).

2. The Solar Analyst.

The Solar Analyst is a comprehensive geometric solar radiation modelling tool. It calculates insolation maps using digital elevation models (DEMs) as input. Highly optimized algorithms account for the influences of the viewshed, surface orientation, elevation, and atmospheric conditions. The Solar Analyst provides a convenient and effective tool for understanding spatial and temporal variation of insolation at landscape and local scales

The Solar Analyst (Pinde Fu y Paul M.R., 2000) draws from the strengths of both point-specific and area-based models. In particular, it generates an upward-looking hemispherical viewshed, in essence producing the equivalent of a hemispherical (fisheye) photograph for every location, on a DEM (Digital Elevation Models). The hemispherical viewsheds are used to calculate the insolation for each location and produce an accurate insolation map. The Solar Analyst can calculate insolation integrated for any period of time. They account for site latitude and elevation, surface orientation, shadows cast by surrounding topography, daily and seasonal shifts in solar angle, and atmospheric attenuation.

The Solar Analyst calculates direct, diffuse and global (as the sum of direct and diffuse) radiation from atmospheric and topographic parameters. Two of these atmospheric parameters are the transmittivity of the atmosphere (T), defined as the quotient among the direct radiation that arrives to the land and the extraterrestrial radiation, and the diffuse proportion (D), which is the proportion of global normal radiation flux that is diffuse.

The user interface which permits parameters input and calls the calculation engine, was developed using the ArcView Dialog Designer and Avenue.

The sky parameter window (Fig 1) permits specification of parameters relevant to sunmap and skymap calculations.

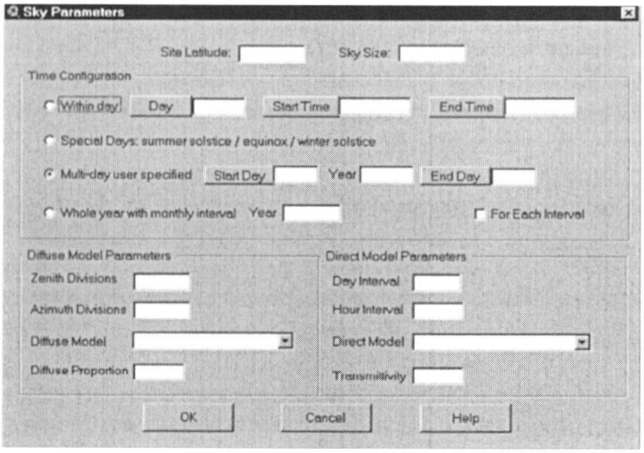

Fig 1. The sky parameter window

The topographic parameter window (Fig 2) permits specification of calculation parameters relevant to the topographic surface being analyzed

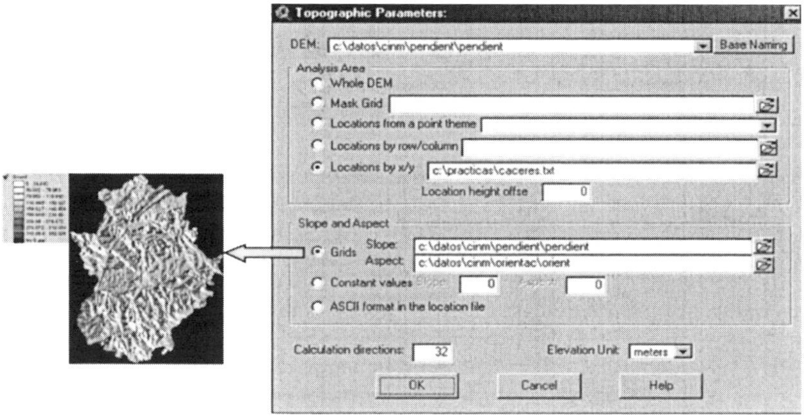

Fig 2. The topographic parameter window

The output dialog window (Fig 3) permits specification of custom output.

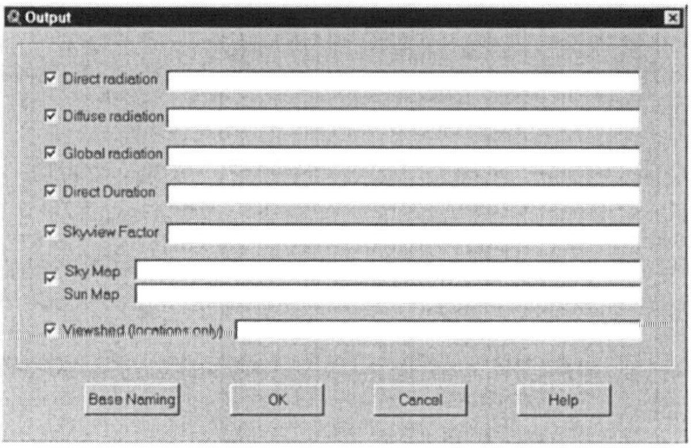

Fig 3. The output dialog window

3 Experimental data and statistical methods

To obtain the functions that relate the sky parameters, T and D, with the monthly average values of maximum temperatures and the sunshine fraction, and the natural day according with the last day of every mouth, the next methodology was followed:

a) Obtaining the T and D values that fit the monthly mean daily irradiation data, by iteration using Solar Analyst. The monthly average values of daily irradiation on a horizontal surface data, measured in the weather station of Cáceres during the period 1985-2002 have been used.

b) Obtaining the correlation functions between T and D values versus the weather variables used and the natural day of every month, by means of the statistical program Statgraphics.

c) Comparison of the experimental and estimated values of irradiation. The comparison was carried out using the monthly average values of experimental and estimated daily irradiation data from the weather stations of Cáceres, Cádiz and Huelva.

3.1. Multiple Regression Analysis.

The output of Statgraphics program shows the results of fitting a multiple linear regression model to describe the relationship between the sky parameters, *diffuse proportion* (D) and *transmittivity* (T), and the independent variables, *day* (natural day according to the last day of every month), *h* (the monthly average value of suhshine fraction of every month) and T_m (the monthly average value of maximum temperatures of every month). The correlation functions obtained using the previous methodology, as well as the values of R^2, standard error of the estimated and P-Values are the following:

$$D = 3,65679 - 3,62679 * h \wedge 0,1 + 0,00145574 * day \wedge 1,1 - 0,00000726726 * day \wedge 2 + \\ +0,0371491 * \cos(T_m) * \sin(T_m * day) \wedge 3 - 0,00145326 * \cos(T_m) \wedge 3 * day - 1,2 * T_m \wedge 1,6 \quad (1)$$

$R^2 = 98,0839$ %
Standard error of the estimated = 0,00554295
P-Values < 0.01

From November to April:

$$T = 0,195747 + 0,00000462478 * day \wedge 2 - 0,0305196 * \sin(T_m) \wedge 2 - 0,0423018 * day \wedge 0,5 + \\ +0,873115 * h \wedge 0,5 \quad (2)$$

$R^2 = 93,647$ %
Standard error of the estimated = 0,16678
P-Values < 0.05

From May to October:

$$T = 0,408136 + 6,67292E-8 * day \wedge 2,5 + 0,000641831 * T_m * h \wedge 3 \quad (3)$$

$R^2 = 97,9318$ %
Standard error of the estimated = 0,00817111
P-Values < 0.01

3.2. Comparison of measured and estimated values of irradiation

Fig 4 shows the comparison between experimental and estimated values of monthly direct (B), diffuse (D) and global (G) irradiation on a horizontal surface for Cáceres. It reveals that the experimental and theoretical values obtained with Solar Analyst are agree with a lower than 5% average error.

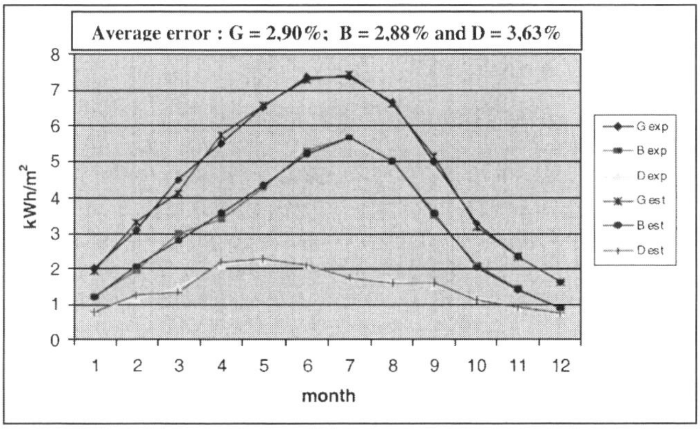

Fig 4. Comparison between experimental and estimated values of irradiation for Cáceres.

In order to validate the model, the irradiation data from two other weather stations in Cadiz and Huelva were used. These data are the corresponding to the 1991-2001 period. Fig 5 shows the comparison between experimental and estimated values of monthly global irradiation on a horizontal surface (G) for Cádiz.

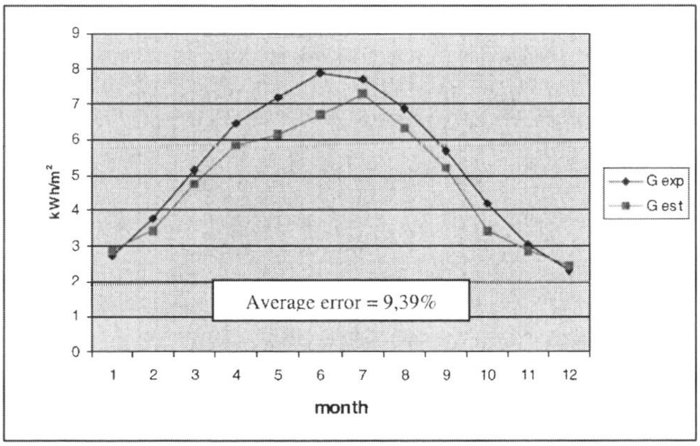

Fig 5. Comparison between experimental and estimated values of global irradiation for Cádiz.

Fig 6 shows the comparison between experimental and estimated values of monthly global irradiation on a horizontal surface (G) for Huelva. In both of the two cases, the average errors are lower than 10%

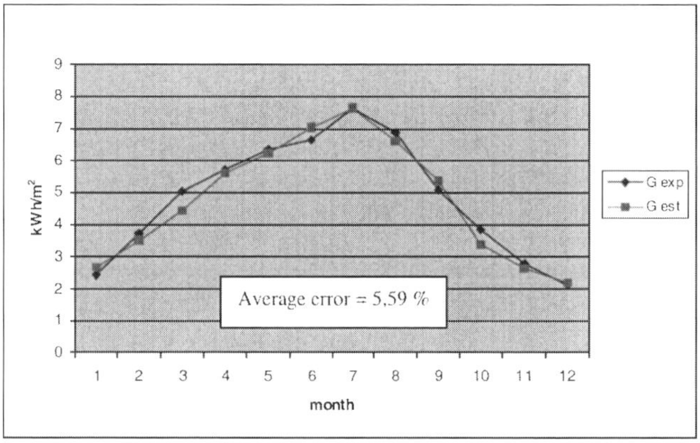

Fig 6. Comparison between experimental and estimated values of global irradiation for Huelva.

3.3. Weather data of Extremadura.

To obtain a solar map of radiation for Extremadura, applying the methodology exposed, is necessary to know topographical data of the region and the sunshine and maximum temperatures data, measured along a sufficiently large period of time so that the results obtained be significant. For this work the measured data in Extremadura weather stations along 1985-2002 have been used The availability of these data will determine the number of same weather zones or surfaces in which Extremadura will be divided. The subsequent application of the area-based model of the Solar Analyst will supply the monthly mean values of direct, diffuse and global irradiation for each cell-DEM or same topography parameters zone.

For instance, Fig 7 shows a same sunshine fraction zones map and Fig 8 shows a same maximum temperature zones map, monthly average values of June and weather stations in Extremadura where these data were measured.

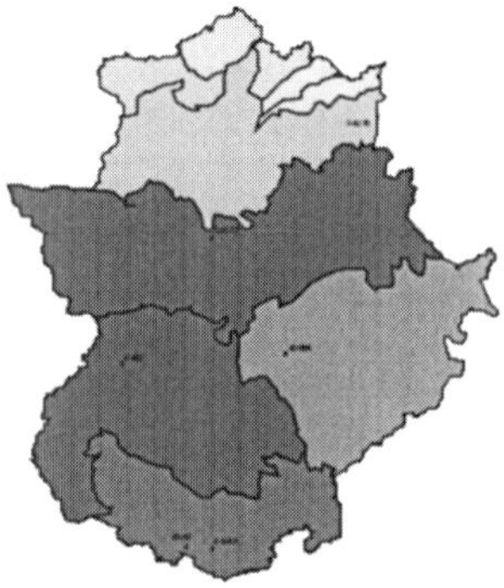

Fig 7. Same sunshine fraction zones map of Extremadura

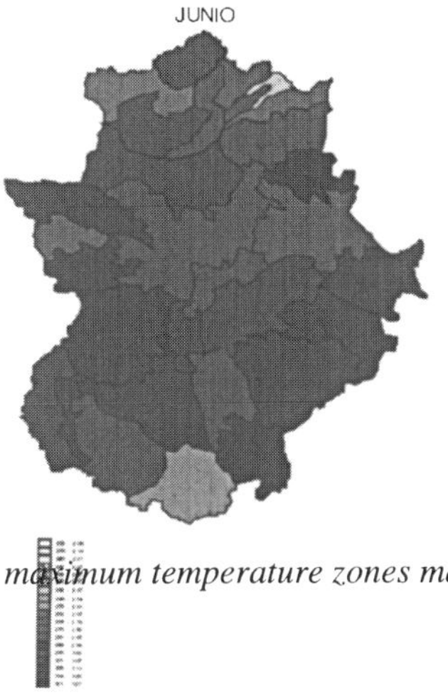

Fig 8. Same maximum temperature zones map of Extremadura

4. Results and discussion.

The results are presented as interactive maps in ArcView language, associated to a conventional data base.

Figs 9, 10 and 11 show three examples of the solar irradiation maps. Monthly diffuse irradiation of January is showed in Fig 9. Monthly direct irradiation of January is showed in Fig 10 and Fig 11 shows the monthly global irradiation of January.

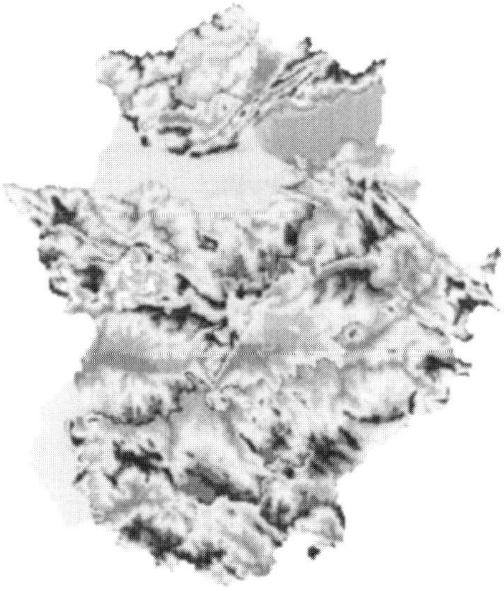

Fig 9. Monthly diffuse irradiation of January.

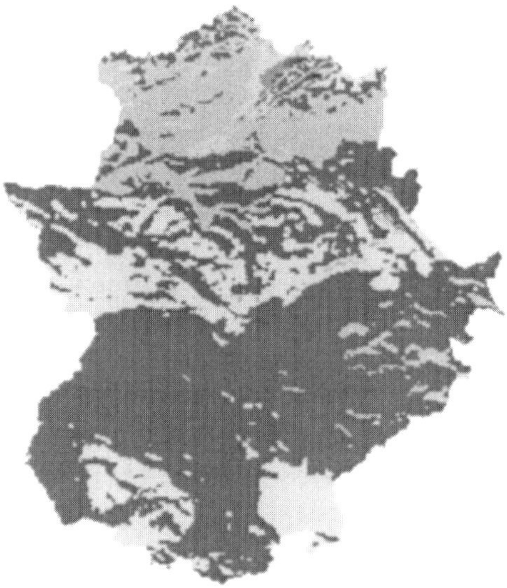

Fig 10. Monthly direct irradiation of January

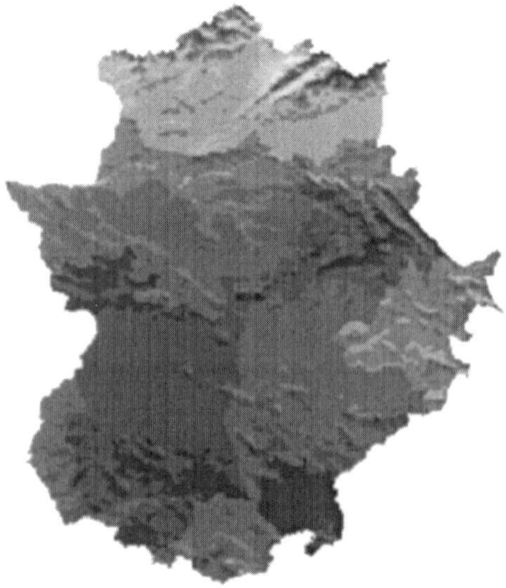

Fig 11. Monthly global irradiation of January

Fig 12 shows an ArcView window with an interactive map of the Guadiana basin. In the figure appears the monthly global irradiation from the cell-DEM indicated (X), its *x* and *y* coordinates and the number of the cells-DEM with the same monthly global irradiation.

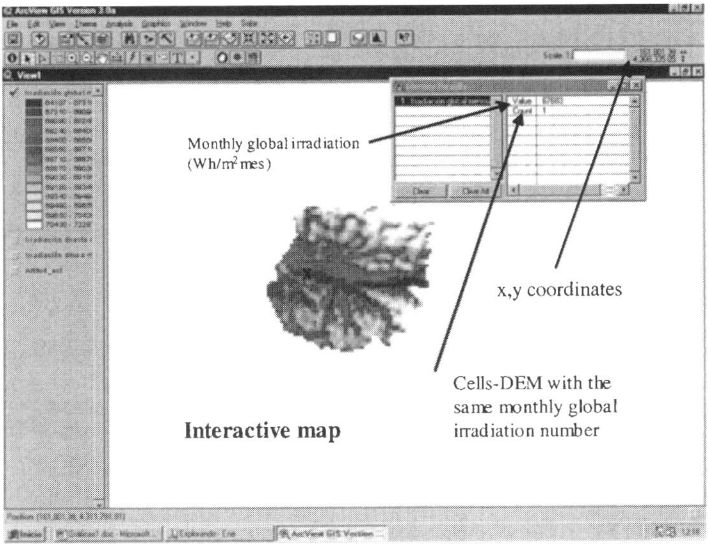

Fig 12. Interactive map example.

5. Conclusions.

A statistical model to estimate global solar irradiation on a horizontal surface, and its direct and diffuse components, from other weather data has been obtained using Solar Analyst.

The Solar Analyst calculates direct, diffuse and global radiation from atmospheric and topographic parameters. Two of these atmospheric parameters are the transmittivity (T), defined as the quotient among the direct radiation that arrives at the land and the extraterrestrial radiation, and the diffuse proportion (D), the proportion of global normal radiation flux that is diffuse.

A multiple linear regression model to describe the relationship between the sky parameters, diffuse proportion (D) and transmittivity (T), and the independent variables, day (natural day according to the last day of every month), h (the monthly average value of sunshine fraction of every month) and T (the monthly average value of maximum temperatures of every month) has been used..

The average error between global irradiation estimated values and experimental values is lower than 10 %.

The results are presented as interactive maps in ArcView language, associated to a conventional data base

References.

Centro Meteorológico Territorial de Extremadura. *Datos de radiación de Cáceres, año 2000.*

Haurwitz B. (1948) Insolation in relation to cloudiness and cloud density. *J. Meteoroly* **2**, 154-156.

Haurwitz B. Insolation in relation to cloud type. *J. Meteorology* 5, 110-113.

Lumb F.E. (1964) The influence of cloud on hourly amounts of total solar radiation at the sea surface. *Quarterly J. Royal Meteorological Society* **90**,383.

Kasten F. and Czeplak G. (1979) Solar and terrestrial radiation dependent on the amount and type of cloud. *Solar Energy* **24**, 177-189.

Manugistics. (1995) *Statgraphics plus. Time-Series Analysis.* Rockville (U.S.A.)

Manugistics (1996) *Statgrpics plus. Advanced regression. Rockville* (U.S.A.)

Mehreen S. Gul, Tariq Muneer and Harry D. Kambezidis (1998) Models for obtaining solar radiation from other meteorological data**.** *Solar Energy* **64,** *99-108*

Muneer T., Gul M. S., Kambezidis H. D. and Allwinkle S. (1996) An all-sky solar meteorological radiation model for the U.K. In *Proceedings of the Joint CIBSE/ASHRAE Conference*, Vol. II, 29 September–1 October, Harrogate, U.K. Chameleon Press, London, pp. 271–279.

Muneer T. Gul M.S. and Kambezidis H.D. (1997) *Longterm evaluation of a meteorological solar radiation model against U.K. data Energy Conversion and Management* **39**,1.

Pinde Fu y Paul M. Rich (2000) *Manual Solar Analyst* 1.0. Helios Enviromental Institute (HEMI)

A. Ramiro, M. Núñez, J.F. González, M.L. González-Martín, E. Sabio, C.M.González-García, J.Gañán y F.J. Fernández (2003). *Estimación de las componentes de la radiación solar a partir de algunos parámetros meteorológicos.* III Jornadas Nacionales de Ingeniería Termodinámica. Valencia.

Applied Physics (APHYS 2003)
A. Méndez-Vilas (Ed.)

Controlling Nano Sized Particles Obtained Via Emulsion Polymerization Using a Polymeric Surfactant and a Water Soluble Initiator

Aida M. Martínez, Carmen González, José M. Gutiérrez, and Montserrat Porras

Department of Chemical Engineering and Metallurgy, University of Barcelona, C/ Martí i Franquès 1, Barcelona 08028, SPAIN

Abstract. A series of emulsion systems based on styrene, mixtures of styrene/butyl acrylate (BA) and styrene/methyl methacrylate (MMA) in the presence of polystyrene-co-maleic anhydride cumene terminated (SMA) as a surfactant and ammonium persulfate as initiator were developed. Experimental results indicate that it is possible to obtain particles below 200 nm with the block copolymer as surfactant. The extent to which varying the monomer concentration and copolymer composition could affect the polymer particle size during the polymerization was examined. The system showed that the particle diameter is influenced by the addition of the hydrophobic monomer. With varying type of monomer particle size decreases in the following order: styrene/BA, styrene and styrene/MMA.

INTRODUCTION

The ability to synthesize macromolecules with complex and controlled architecture is becoming an increasingly important aspect of polymer science [1]. The state of art in nanotechnology is the on-going way to develop a reaction media which can offer compositional and architectural control on a nanometer scale. Organized self-assembled surfactant phases have received a lot of attention as possible reaction and templating media; however, due to their dynamic nature, conventional surfactant structures are of limited use. Polymeric surfactants, on the other hand, offer potential for developing hybrid nanosized reaction and templating media with constrained geometries [2].

Emulsion polymerization process has several advantages. It is easy to control the process, the latex product is often directly valuable and the small particle size allows the attainment of low residual monomer levels [3]. The main mechanism of conventional emulsion polymerization is that particle nucleation can occur either in monomer-swollen micelles (micellar nucleation) or in the aqueous phase (homogeneous nucleation) and also might occur in the monomer droplets (nanoemulsion polymerization) [4]. In emulsion polymerizations, the use, or in situ production, of surfactants is necessary in order to achieve stabilization during polymerization and on the derived products. However, the presence of surfactant is a disadvantage for certain applications of emulsion polymers like coatings, paints, finishes or polishes [5]. On the other hand, today's environmental regulations mandate that volatile organic chemicals (VOCs) are virtually diminished to zero and these surfactants must be removed from the final latex. The removal of surfactant can lead to coagulation or flocculation of the destabilized latex. Moreover the surfactant could remain in the polymer particles after polymerization and may have a deleterious effect on the properties of the polymer.

Polymeric surfactants have attracted considerable attention in recent years as dispersants for solids in liquids and as emulsifiers. The effectiveness of these molecules as emulsifiers is based on the role of stabilization of particles and droplets against flocculation and/or coalescence by a mechanism referred to as steric stabilization [6]. Recent studies are based on a novel class of emulsion stabilizers, namely, the amphiphilic block copolymers [7]. It was recently demonstrated that the block copolymer micellar aggregates offer the unique property of acting as a seed for the creation of particles, provided that they are stable over long periods of time [8]. This is due to the enhanced structural stability of the poly-micelles compared to spherical aggregates or conventional low molecular weight surfactants [9]. In fact, to improve some properties of the polymer latex, one may use surfactants with functional groups that are capable of interacting with the radical polymerization process [10]. Alternating copolymers of SMA offer a great potential to engineer interfaces where control of the morphology on a nanoscale is crucial. This work examines the feasibility of using SMA having a molecular weight of 1900 composed of 74% by weight in styrene and 26% by weight in maleic anhydride as the surfactant in an emulsion polymerization based on a hydrophobic monomers and ammonium persulfate system. This study examines the parameters of monomer concentration and copolymer composition, as well as how they affect the particle sizes.

EXPERIMENTAL

Materials

Prior to use, styrene (99%), BA (98%) and MMA (99%) were passed through a chromatographic column filled with aluminium oxide (Merck) until no inhibitor remained and stored at 4 °C before use. Ammonium persulfate was used as a water soluble initiator. Deionized water was used throughout all the experiments. The other materials were supplied by Sigma-Aldrich and used as received.

Polymer Hydrolysis

An ammonium solution 28%wt was prepared. The block copolymer was then dissolved in water by alkaline hydrolysis using the ammonium solution and heating up to 70 °C for approximately 4 hours until dissolution took place. Thus, the solubilized copolymer gains polyelectrolytic properties and can be ionized to different degrees depending on the pH [7]. The degree of ionization was fixed by the molar ratio of alkali/polymer (82:6).

Emulsion Preparation and Polymerization

Batch polymerizations were run in a 100 mL four-necked flask fitted with a rubber septum, thermometer, nitrogen inlet, reflux condenser, and equipped with a mechanical stirrer. The solubilized copolymer was fed into the reactor and heated up to 70 °C. While stirring, nitrogen gas (Air Liquide) was flushed to remove oxygen. The oil-in-water emulsion was prepared in situ by adding continuously a) styrene b) a mixture of styrene/BA (46:54) c) a mixture of styrene/MMA (20:80, 46:54, 80:20) and afterwards a solution of ammonium persulfate 5%wt. All the polymerizations were carried out at 70 °C using a stirrer speed of 700 rpm during 4 hours and then cooled at room temperature.

Particle Size

Particle sizes were obtained from Dynamic Light Scattering (DLS). The DLS measurements were performed on a photon correlation spectrometer Malvern 4700 Instrument system equipped with an argon laser at a wavelength of 488 nm. The scattering angle was fixed at 90 ° and the temperature of the solutions was maintained at 25 °C.

Differential Scanning Calorimetry

The melting behavior of the particles obtained was studied by means of Differential Scanning Calorimetry (DSC). For the DSC studies, small samples (about 10 mg) were prepared and washed several times with hot ammonium solution (20%wt) until no polymeric surfactant remained. The measurements were carried out by a Mettler Toledo Star System, under a nitrogen atmosphere and a temperature scanning speed of 10 °C/min from 50 to 550 °C. Presence of SMA copolymer in the final particles was determined on the basis of the DSC melting curves comparing the DSC scans of particles made by using SMA block copolymer with the ones produced without the SMA surfactant.

Atomic Force Microscopy

The morphology of the polymer particles was investigated by Atomic Force Microscopy (AFM). The latex samples were applied on thin sheets of mica. AFM was thereafter performed on these latex films after drying at room temperature under an argon chamber. The AFM Instrument used was Nanoscope III Multimode from Digital Instruments. The AFM images of the samples were observed using tapping mode.

RESULTS AND DISCUSSION

The particle size was determined by DLS. The primary information given by DLS data are intensity distributions whereby the relative amount of each particle size is measured by the intensity scattered by all the particles of the considered size. The intensity distributions can be converted into distributions by volume and by number and this leads to obtain different mean average sizes depending on the distribution chosen. The results for polystyrene (PS) by DLS technique given by intensity distributions were compared with experiments made with a mixture of styrene and BA (46:54) at the same conditions as shown in Fig. 1. The PS particles diameter increased with the increasing concentration of the monomer. The PS/BA particles showed the same behavior, though the particles obtained only with styrene were smaller.

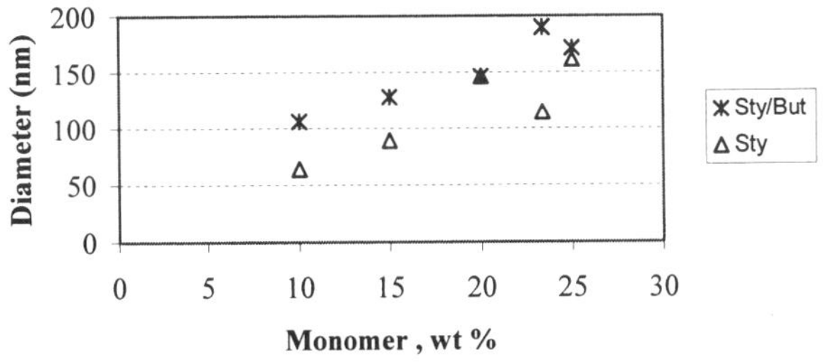

FIGURE 1. Latex Particles Diameter (styrene and styrene/BA (46:54), molar ratio NH_3: surfactant (82:6)) obtained by DLS.

The results of studying the influence of monomer type were made by using different mixtures of styrene and MMA (23%wt of total monomer). The results showed that when the relation of styrene/MMA was 46:54, the average particle size decreased to a minimum and then gradually increased as shown in Fig. 2. We also can notice that the PS/MMA particles are much smaller than the ones obtained only with styrene or with a mixture of styrene/BA.

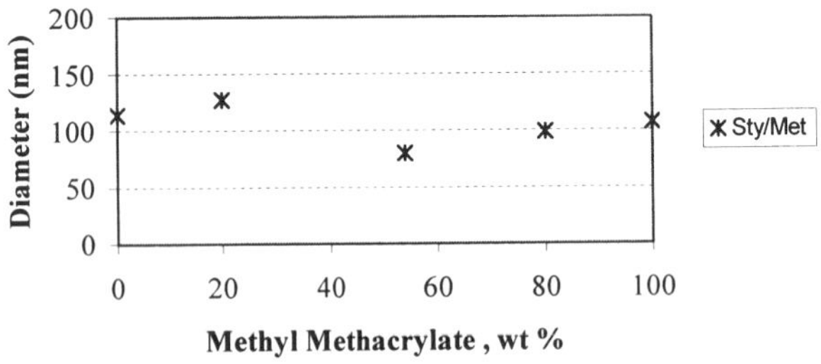

FIGURE 2. Latex Particles Diameter (styrene/MMA at different concentrations (23wt% total monomer), molar ratio NH_3: surfactant (82:6)) obtained with DLS.

The images obtained by AFM show that the particles are uniformly and regularly spherical. However, we detected variation in particle size as shown in Fig. 3. We also notice that the DLS and AFM techniques gave different particle sizes, but when comparing DLS number data with AFM the particle sizes where quite similar. This can be explained due to the polydispersity of the latex samples. We have very tiny particles (DLS number) but we also have bigger particle sizes that scatter a significant amount of light so the diameter given by DLS intensity distributions is higher.

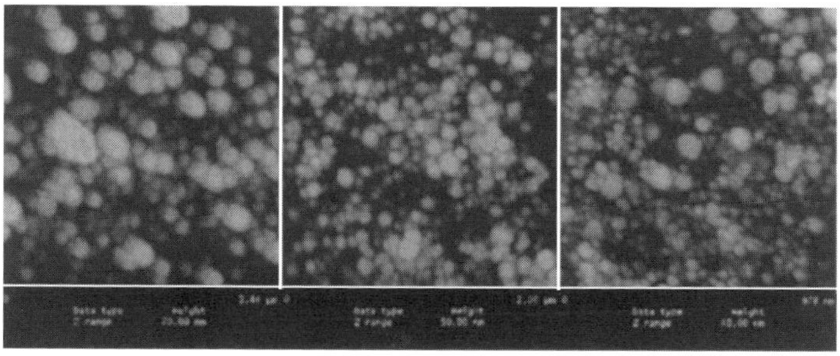

(a) (b) (c)

FIGURE 3. AFM Topographic Images of (a) PS/BA 23%wt (46:54) size 2.44 mm x 2.44 mm (b) PS 15%wt size 2.0 mm x 2.0 mm (c) PS/MMA 23%wt (20:80) size 978 nm x 978 nm dried at room temperature under ambient conditions.

On the other hand we can notice that the average particle diameter is in the order of nanometers. The polymerization was carried out in an emulsion using the typical recipe: monomer, surfactant, water and initiator. However, the presence of a water-insoluble compound (hydrophobe) such as polystyrene in the monomer droplets retards the diffusion of the monomer out of the droplets. The inclusion of a small amount of hydrophobe can effectively retard the diffusional degradation of droplets and maintain a small droplet size [11]. This could explain the fact of obtaining tiny particles.

The thermal properties of PS particles at 23 %wt of monomer obtained with SMA copolymer and without surfactant were determined by DSC. The glass-transition temperature of the sample made with polymeric surfactant was 113 °C meanwhile that of PS was 102 °C. This fact demonstrates that a portion of the amphiphilic block copolymer is quite clearly incorporated into the particles.

CONCLUSIONS

The use of SMA as surfactant allows to obtain nanoparticles in the system studied. As the polymerization was carried out using an amphiphilic polymer as a surfactant, the latexes produced were free of any other surfactant or protective colloid. In fact, a portion of the block copolymer is quite clearly incorporated into the final particles as the results DSC thermogram confirmed this fact. It was found that the particle size slightly increases as the quantity of monomer increases for styrene and styrene/BA monomers. PS/BA particles are bigger than particles made only with styrene. Particle size is smaller when the monomer used is a mixture of styrene/MMA instead of styrene/BA or only styrene. The AFM data confirmed the small particle size given by

DLS. We suggest that smaller and near monodisperse particles could be produced by controlling the molecular weight and composition of the hydrophilic-hydrophobic copolymer.

ACKNOWLEDGMENTS

This work was funded under MCYT Project No. PPQ2002–04514–C03–02. We also grateful acknowledge to CONACYT for the financial support given.

REFERENCES

1. E. S. Park, M. N. Kim, I. M. Lee, H. S. Lee and J. S. Yoon, *J. of Polymer Science* **38**, 2239-2244 (2000).
2. M. Summers and J. Eastoe, *Adv. In Colloid and Interface Sci.*, 1-16 (2002).
3. S. L. Rosen, *Fundamental Principles of Polymeric Materials*, Wiley-Interscience, New York, 1993, p.p. 135-175.
4. M. S. Lim and H. Chen, *J. of Polymer Science* **38**, 1818-1827 (2000).
5. G. G. Odian, *Principles of Polymerization*, Wiley Interscience, New York, 1991, p.p. 335-354.
6. C. Stevens, A. Meriggi, M. Peristeropoulou, P. Christov, K. Booten, B. Levecke, A. Vandamme, N. Pittevils and T. Tadros, *Biomacromolecules* **2**, 1256-1259 (2001).
7. G. Garnier, M. Duskova, R. Vyhnalkova, T. G. Van de Ven and J. F. Revol, *Langmuir* **16**, 3757-3763 (2000).
8. C. Burguière, S. Pascual, C. Bui, J. P. Vairon, B. Charleux, K. A. Davis, K. Matyjaszewski and I. Bétremieux, *Macromolecules* **34**, 4439-4450 (2001).
9. K. Stähler, J. Selb and F. Candau, *Mat. Sci. and Eng. C* **10**, 171-178 (1999).
10. Y. Li, L. Wang and X. Liu, *Langmuir* **14**, 6879-6885 (1998).
11. I. Capek, S. Y. Lin, T. J. Hsu and C. S. Chern, *J. of Polymer Science* **31**, 1477-1486 (1999).

Applied Physics (APHYS 2003)
A. Méndez-Vilas (Ed.)

Extraction Of Informative Features From The Images Of Diagnostic Structures In Dried Drops Of Biological Liquids

Nuidel I., A.Chaikin, A.Tel'nykh, O.Sanina, V.Yakhno, T.Yakhno, L.Karimova,* O.Kruglun,* and N.Makarenko*

Institute of Applied Physics of the Russian Academy of Sciences,46 Ulyanov Str., 603950 Nizhny Novgorod, Russia. E-mail: nuidel@awp.nnov.ru
**Institute of Mathematics, Almaty, Kazakhstan.E-mail: makarenko@math.kz*

Abstract. We processed and analyzed half-tone images of specimens of dried drops of normal blood serum and serum of oncologic patients. We used two procedures: (a) obtaining binary specimens from the initial image using uniform distributed neuron-like systems, followed by calculation of quantitative features for extracted objects and (b) calculation of the values of the Minkowsky functionals from the initial image. The neuron-like algorithms for processing of initial images permit one to extract characteristic regions in the images such as objects of given dimensions. We developed variants of quantitative estimates enabling one to segment characteristic fragments of analyzed images. For this, we use geometric features of objects in the extracted regions, including coordinates of the centers of characteristic regions, perimeter, area, form factor, etc. Reliable characteristics of the difference between specimens of normal and pathologic serum are determined statistically. The developed procedure will be tested for the analysis of specimens of other biological liquids. The Minkowsky functionals (area, Euler characteristic, and total perimeter) were used for processing of different images depending on their binarization threshold level. It is shown that the Euler characteristic of an image taken from a drop of normal blood serum is considerably different from the Euler characteristic of serum from oncologic patients. The information capacity of other Minkowsky functionals as diagnostic features is also discussed.

NEURON-LIKE ALGORITHMS OF DIAGNOSTIC ESTIMATE CALCULATION

To calculate the diagnostic features of dried-drop images of blood serum in healthy donors and oncologic patients, a homogeneous distributed neuron-like system consisting of coupled active elements of the same type was used [1].

Model Of Homogeneous Disributed Neuron-Like Systems For Image Processing

The model is written in the form of one integro-differential equation (1) [1]. It was obtained as a balance equation for spikes in the fibers of excitatory and inhibitory neuron networks in the part of animal brain cortex containing hundreds of thousands of nervous cells in the approximation of uniformity of the part considered.

$$\tau_u \, \partial u / \partial t = -u + F\left[-T + \alpha \int_{-\infty}^{+\infty} \Phi_u(\vec{\xi} - \vec{r}) \cdot u(\vec{\xi}, t) \cdot d\vec{\xi} + u_{ex}(\vec{r}, t)\right] \qquad (1),$$

Here, $u(\vec{r},t) = u(x,y,t)$ describes the excitation in the patterns in a two-dimensional distributed neuron-like system (image byte per point), τ_u is the relaxation time for the initial condition, *T* is the threshold of active elements on a general external signal from the coupled field, $\Phi_u(\vec{r})$ is a coupling function of lateral inhibition type with the positive center and the negative surround (2), and α is the norm constant for the coupling function. The nonlinear function *F[Z]* is written in the piecewise-linear form $\Phi(\vec{r} - \vec{r}_0) = (1 - b(\vec{r} - \vec{r}_0)^2)\exp(-a(\vec{r} - \vec{r}_0)^2)$, $\vec{r} = (x, y)$. (2) $u(x,y)|_{t=0} = u_0(x,y)$ (3)
The distribution of the initial excitation is assigned in the form of a fixed image (3).

In [1], it is shown analytically that autowave solutions exist in system (1) in the form of stable immobile pulses and traveling propagating excitation fronts. On the basis of solutions in the form of immobile pulses, a new approach has been proposed for processing of images in such systems. From an immobile gray-tone image, which is fed in the form of the initial condition of state of two-dimensional system (1), a stationary structure is formed with the time (after a few time steps) in the form of a binary pattern, which is interpreted as a simplified preparation of the initial image. For example, the contour, fragments of different spatial scales, lines of different directions, etc. are extracted.

Description Of Image Processing and Analysis

Processing algorithms are demonstrated using images for five dried drop preparations of blood serum from healthy donors (d4,d5,d6,d7,d8) and seven preparations from oncologic patients (o1,o2,o3,o4,o5,o6) as the examples (Fig.1). The image processing was organized step by step.

1. Gray-scale images in byte-per-point format are presented as a set of one-layer distributed neuron-like systems. A set of simplified binary patterns for different scales is obtained.

This was achieved by processing of the initial and inverted images in one-layer distributed neuron-like systems tuned to extraction of lines of eight different directions (lines at angles 0, 22.5, 45, 67.5, 90, 112.5, 135, and 157.5 deg to the vertical on eight spatial scales. The coupling functions are tuned to extraction of objects with scales 3*3, 10*10, 15*15, 20*20, 30*30, 40*40, 60*60, and 80*80 pixels).

Eight images are added, and characteristic regions for one of the scales are obtained. We obtained 16 gray-scale images characterizing the initial image with different degrees of accuracy (8 regions of different scales for the initial image and the same for the inverted initial image).

2. Among the resulting images we chose the images characterizing the initial image exhaustively. Such images for the donor preparation d4.512 are presented in Fig.2a,b,c,d (the initial image is d4.512 in Fig.2a; characteristic regions for d4.512 on the scale 80*80 pixels are given in Fig.2b; characteristic regions for the inverted initial image on the scale 40*40 pixels are shown in Fig.2c; characteristic regions for the

inverted initial image on the scale 20*20 pixels are shown in Fig.2d). The corresponding images for o6.512 are given in Fig.2m,n,o,p.

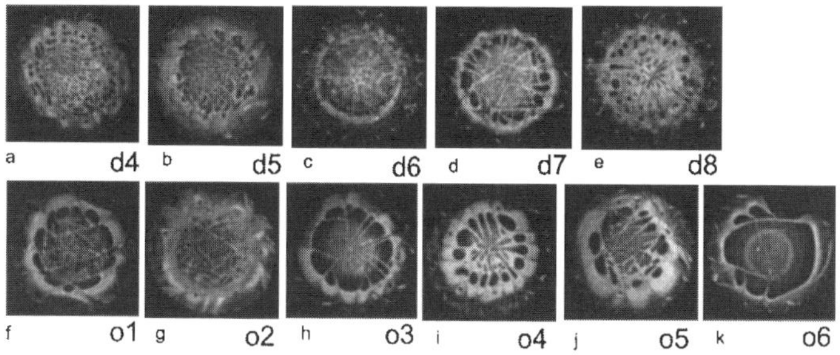

FIGURE 1. Examples of initial images for dried-drop preparations of blood serum from healthy donors and (d4,d5,d6,d7,d8) and oncologic patients (o1,o2,o3,o4,o5,o6).

The gray-scale images obtained by image processing in distributed neuron-like systems with different thresholds are transformed into binary patterns (bitmapped images). The ultimate result of binary image processing is the table of statistical information for objects from the image. Quantitative estimations of the difference between images are found on the basis of number and geometric characteristics of the extracted regions. They are the coordinates of the centers of characteristic regions, perimeter, square, coefficients of form, line sizes, etc.) (4).

$$X_{ч.м.} = \frac{\sum_{i=1}^{k} x_i n_i}{S}; Y_{ч.м.} = \frac{\sum_{i=1}^{k} y_i n_i}{S}; P; S; K_{\phi} = \frac{P}{\sqrt{S}}; L = \sqrt{X_{\max}^2 + Y_{\max}^2}; \Phi = \max(X_{\max}, Y_{\max}) \qquad (4)$$

Here, n_i is the brightness at point I, S is a square, or the number of points in the object. Reliable characteristics of the difference between serum preparations are determined statistically.

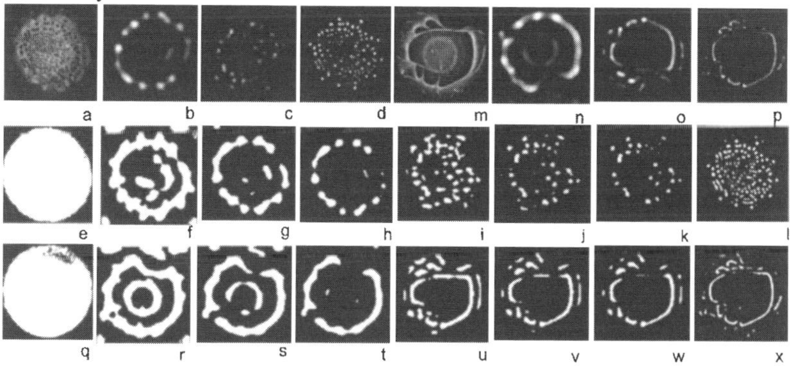

FIGURE 2. Binary preparations obtained from donor (d4.512) and oncologic ones (o6.512).
a,m are the initial images; e,q are binary preparations obtained by initial-image binarization with threshold T=2; b,n are characteristic regions obtained from the initial image on scale 80*80 pixels; f,g,h (r,s,t) are binary preparations obtained from the images b,m by binarization with thresholds T=2;8;100 correspondingly; c,o are characteristic regions obtained from the inverted initial image on scale 40*40 pixels; I,j,k(u,v,w) are binary preparations obtained from the images c,o by binarization with thresholds

T=20;60;80 correspondingly; d,p are characteristic regions obtained from the inverted initial image on scale 20*20 pixels; l,x are binary preparations obtained from the images d,p by binarization with thresholds T=50.

It is proposed to use the following parameters as diagnostic ones:

1. The form coefficient of the drop area (in the norm, the mean form coefficient of areas smaller than in the case of pathology.)
2. The number of large-scale areas for binary images obtained from the initial one near the drop center (the coupling function of a neuron-like system is 80*80 pixels, and the thresholds are T=2,4,8,10). The presence of two ring areas is evidence of pathology. In the norm, the number of large-scale areas near the drop center is smaller than in the case of pathology.
3. The moments for binary images obtained from the initial one in a neuron-like system with coupling function 80*80 pixels with binarization thresholds T=120,140,160,180,200. The moment is equal to the product of the square of the area and interval from the form center of the area to the drop center. The mean dispersion of area moments in the norm is smaller than in the case of pathology.
4. The moments for binary images obtained from the inverted initial one in the neuron-like system with coupling function 40*40 pixels with binarization thresholds T=20,40,60,80,100. The mean dispersion of area moments in the norm is smaller than in the case of pathology.
5. The moments for binary images obtained from the inverted initial one in the neuron-like system with coupling function 20*20 pixels with binarization thresholds T=50. The mean dispersion of area moments in the norm is smaller than in the case of pathology.
6. The number of round areas inside the drop for binary images obtained from the inverted initial ones near the drop center (the coupling function of neuron-like system is 20*20 pixels, and the thresholds are T=50). In the norm, the number of round areas is larger than in the case of pathology. The developed procedure will be used for analyzing preparations of other biological liquids.

MATHEMATICAL MORPHOLOGY METHODS FOR IMAGE FEATURE DISCRIMINATION

The morphological image analysis (MIA) [2, 3] characterizes objects in terms of the form (geometry) and connectivity (topology) with the help of the Minkowski functionals. A digital image is described by functional $I(ij)$ given on a square lattice of pixels (i,j). A cross-section $I=h$ forms an excursion set $A_h = \{(i,j) | I \geq h\}$ as a function of a threshold h. This set consists of black and white pixel clusters. In this case, Minkowski functionals [2, 3] correspond to the area (W_2), the length of a boundary (W_1) and the connectivity (W_0) of the clusters of fixed color. The connectivity or the Euler characteristic is the number of simply connected domains minus the number of holes in those domains. These three functional dependences $W_i(h)$, $i=0,1,2$ can be used for the quantitative description of geometrical and topological complexity of the image. We used them in this work for the purpose of qualitative discrimination of ill and healthy people.

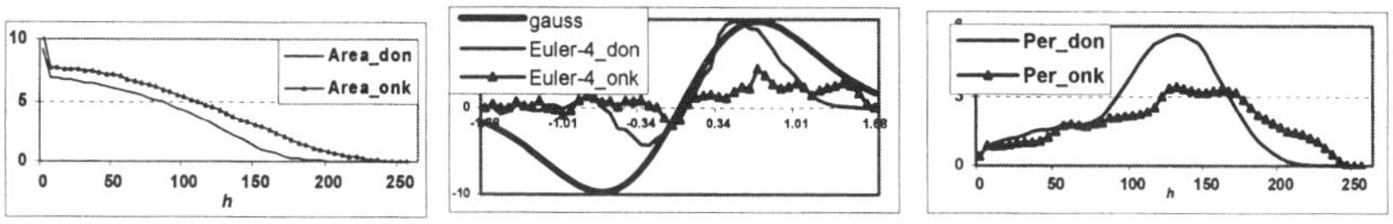

a b c

FIGURE 3. Minkowski functionals: the area W_2 (a), the Euler characteristic W_0 (b), the perimeter W_1 (c) of healthy (d5.jpg) and ill (o5.jpg) people.

The preprocessing of images was performed as follows. First, an RGB image of drops was transformed to a gray-scale picture. Then, the cross-sections on different brightness thresholds were plotted. Thus, the image of each level was represented as a black-and-white picture and the functionals $W_i(h)$, $i = 0,1,2$ were computed.

Figures 3a-c demonstrates the Minkowski functional curves of o5 and d5 images. The results obtained for all available images gives a possibility for making the following conclusion. The W_0 and W_1 functionals appear to be the most significant characteristics for the discrimination of ill and healthy people (Fig. 3b, c). The W_2 functional is less important and does not give any additional information (Fig. 3a).

CONCLUSIONS

It is shown that using neuron-like algorithms and Minkowski functionals for image processing allows one to extract diagnostic characters. Reliable characteristics for the statistic difference of blood-serum preparations in the norm and pathology are determined. The developed procedure is also proposed for analyzing preparations of other biological liquids.

ACKNOWLEDGEMENTS

This work was supported by research grant RFBR, 01-01-00388, RFEBR, 02-04-49342, Russian Academy grant for young scientists of 6th competition – examination (1999), project number 108, INTAS 01-0690.

REFERENCES

1. V.Yakhno, I.Nuidel., A.Ivanov, N.Bellustin, D.Budnikov, A.Tel'nykh, E.Eremin, A.Kogan, M.Kostin, A.Perminov, Yu.Radzhabova, M.Sorokin, S.Shilin, D.Tikhomirov, and A.Chaikin "Investigation of dynamic regims in neuron-like systems," *Information technologies and computing systems,* 2 (2003).
2. K.Michielsen and H.De Raedt Integral-geometry morphological image analysis. // Phys. Rep., 2001. - Vol. 347 №6 – 463-538.
3. L. Karimova and N. Makarenko "Diagnosis of stochastic fields by the mathematical morphology and computational topology methods" in Nuclear Instruments and Methods in Physics Research A, 502, pp. 802-804 (2003).

Applied Physics (APHYS 2003)
A. Méndez-Vilas (Ed.)

Computing the Differential Invariants for Second-Order ODEs

A. Martín del Rey

Department of Applied Mathematics, EPS, University of Salamanca
C/Santo Tomás s/n, 05003-Ávila, Spain
E-mail: delrey@usal.es

J. Muñoz Masqué

Institute of Applied Physics, CSIC
C/ Serrano 144, 28006-Madrid, Spain
E-mail: jaime@iec.csic.es

G. Rodríguez Sánchez

Department of Applied Mathematics, E. P. S., University of Salamanca
Avda. Cardenal Cisneros 34, 49022-Zamora, Spain
E-mail: gerardo@usal.es

Abstract. In this work a novel algorithm to compute differential invariants of second-order ODEs is proposed. Also a characterization of Painlevé transcendents in terms of their invariants is presented.

INTRODUCTION

The goal of this paper is to describe the implementation of a new algorithm for determining the differential invariants of second-order ODEs with respect to the group Aut(p) of all automorphisms of the natural projection p: $\mathbf{R}^2 \to \mathbf{R}$, $p(x,y) = x$, of the plane onto the real line, as well as the invariants of its relevant subgroups. A. R. Tresse at the end of 19th century, and more recently other authors have studied semi-invariants with respect to the group Diff($\mathbf{R}^2$) of all planar diffeomorphisms and an algorithm has been announced for this group (see [1]). Nevertheless, in the modern geometric approach to differential equations, the group Aut(p) seems to be more natural as a second-order ODE is usually understood to be a section of the canonical projection of jet bundles p_{21}: $J^2(p) \to J^1(p)$. The elements Aut(p) are also viewed as planar changes of coordinates $X = \mathsf{f}(x)$, $Y = \mathsf{y}(x, y)$. The group Aut(p) acts on the space of sections of p_{21} (*i.e.*, the space of ODEs) and we consider the notion of invariance corresponding to the prolongation of this action to $J^r(p)$. Indeed, we work with the infinitesimal version of this action: The invariance under the Lie subalgebra aut(p)$\subset X(\mathbf{R}^2)$ of p-projectable vector fields, as this point of view allows us to use the basic tools of Geometric Analysis, although both notions are equivalent. As the subgroup Aut(p) is smaller than the group Diff($\mathbf{R}^2$) of all punctual planar diffeomorphisms, the number of invariants at each order increases considerably. Hence, the computational complexity of the problem of determining Aut(p) -invariants is much larger than that corresponding to the full group of diffeomorphisms. In this sense, we obtain complete results for the group Aut(p), the subgroup of horizontal diffeomorphisms defined by y $(x, y) = y$, and denoted by Aut^h (p), and the subgroup of vertical diffeomorphisms defined by f(x) = x, and denoted by Aut^v (p). In fact, for each order r we have a procedure for determining explicitly a system of functionally independent generators of the algebra of differential invariants of order r, for each of these groups. This is an extremely precise formulation of Lie's asymptotic stability theorem. These results allow one to decide whether two given ODEs are equivalent or not with respect to changes of coordinates: For analytic functions this is the case if and only if they have the same invariants. For example, the elementary types of ODEs equations (such as homogeneous, autonomous, linear, special equations, etc.) can be intrinsically classified in terms of their invariants (see [3]). In fact, once a procedure to obtain a basis of invariants for arbitrary order, is known, almost every intrinsically formulated questions about ODEs (such as infinitesimal symmetries, integrating factors, etc.) can be efficiently computerized. Furthermore, the knowledge of differential invariants allows us to obtain a method to characterize several classes of second-order ODEs. This procedure is computational feasable because once a concrete differential equation is given, we can recognize whether such equation belongs to a specific class by running an algorithm in polynomial time. Basically, the method reduces to check the vanishing of several algebraic expressions written in terms of differential invariants. Both algorithms: the calculus of differential invariants and the characterization of some types of second-order ODEs, are implemented by means of the computer algebra system Mathematica. Nevertheless, with minor modifications these algorithms can be adapted to the most computer symbolic packages, such as Maple, etc. As an example we will focus our attention in the second-order ODEs with Painlevé property.

THE NOTION OF INVARIANCE

In order to be able to use the notion of a differential invariant as a function on a jet bundle that is invariant under the induced action of a certain group of transformations (see [5]), a second-order ODE is defined to be a section s of p_{21}: $J^2(p)\to J^1(p)$, where p: $\mathbf{R}^2\to\mathbf{R}$ is the natural projection $p(x, y) = x$. If (x, y, y', y'') is the natural coordinate system on $J^2(p)$, then s is equivalent to giving the function $s^* y'' = F(x, y, y') \in C^\infty(J^1(p))$. Let Aut($p$) be the group of automorphisms of p, *i.e.*, Aut (p) = $\{F\in\mathrm{Diff}(\mathbf{R}^2): p\circ F = f\circ p, f\in\mathrm{Diff}(\mathbf{R})\}$. This group acts on the space of sections of p_{21} in a natural way; *i.e.*, $s \to F^{(2)} \circ s \circ (F^{(1)})^{-1}$, where $F^{(k)}$ denotes de k-jet prolongation of F to $J^k(p)$:

$$\Phi^{(k)}(j_x^k s) = j_{\varphi(x)}^k\left(\Phi \circ s \circ \varphi^{-1}\right)$$

for every section s: $\mathbf{R}\to\mathbf{R}^2$, $s(x) = (x, f(x))$, of p. The notion of invariance that we consider is the one relative to the prolongation of this group action to $J^r(p_{21})$. In this way, a function $I : J^r(p_{21}) \to \mathbf{R}$ is said to be a differential G-invariant, where $G \subseteq$ Aut (p), if for all $j^r_{z'}\sigma \in J^r(p), z' = j^1_x s \in J^1(p)$, and every F $\in G$ the following equation holds:

$$I\left(\left(\Phi^{(2)}\right)^{(r)}\left(j^r_{z'}\sigma\right)\right) = I\left(j^r_{\Phi^{(1)}(z')}\left(\Phi^{(2)} \circ \sigma \circ \left(\Phi^{(1)}\right)^{-1}\right)\right) = I\left(j^r_{z'}\sigma\right)$$

This notion is not very operative in order to calculate explicitly the differential invariants. That is why the infinitesimal version of this notion is introduced. Every differential invariant with respect to G is also an invariant with respect to the Lie algebra g of p-projectable vector fields whose local flow belongs to G; *i.e.*, $(X^{(2)})^{(r)}I = 0$ for all $X \in g$. Both notions are equivalent (see [4]). Below, we deal with the cases

$$G^{\mathrm{h}} = \mathrm{Aut}^{\mathrm{h}}(p) = \{\Phi \in \mathrm{Aut}(p) : p \circ \Phi = p\},$$

$$G^{\mathrm{v}} = \mathrm{Aut}^{\mathrm{v}}(p) = \{\Phi \in \mathrm{Aut}(p) : \Phi(x, y) = (\varphi(x), y), \varphi \in \mathrm{Diff}(\mathrm{R})\},$$

which we refer to as the "horizontal" and "vertical" subgroups, respectively. The case of total group $G = \mathrm{Aut}(p)$ is also considered.

Also, in [3] the number of functionally independent r-order differential invariants for the horizontal group is proved to be equal to $(r(r+1)(r+5)+1)/6$ on the dense open subset defined by $y'\neq 0$, and for the vertical group, the number of functionally independent r-th order invariants is 1 if $r = 0,1$; 2 if $r = 2$; and $(r(r-2)(r+5)+1)/6$ if $r \geq 3$. Finally, for the case of the total group, the number of functionally independent r-th order invariants is 0 for $r = 0,1,2,3$; and $(r^3+3r^2-16r-12)/6$ for $r \geq 4$.

A basic question is how to obtain invariants of order $r + 1$ starting from invariants of order less or equal than r. The standard procedure goes back to Lie's ideas (see [2]). If the operators D_x, D_y, and $D_{y'}$ stand for the total derivatives on jet bundles with respect to the variables x, y, and y', respectively, then for every positive integer number r, the operators X_1, X_2 and X_3, transform r-th order differential invariants into $(r+1)$-th order differential invariants for the horizontal subgroup, whereas the operators Y_1, Y_2, Y_3 run for the vertical group and Z_1, Z_2, Z_3 are the differential operators related to the total group (see [3]), where:

$$X_1 = D_y \qquad X_2 = y' D_{y'} \qquad X_2 = \frac{1}{y'}\left(D_x + y'' D_{y'}\right)$$

$$Y_1 = \frac{1}{\sqrt{y''_{003}}} D_{y'} \quad Y_2 = D_x + y' D_y + y'' D_{y'} \qquad Y_3 = \frac{2D_y + y''_{001} D_{y'}}{\sqrt{y''_{003}}}$$

$$Z_1 = Y_1(K)^{1/3} Y_1 \qquad Z_2 = \frac{1}{Y_1(K)^{2/3}} Y_2 \qquad Z_3 = \frac{1}{Y_1(K)^{1/3}}\left(Y_3 - \frac{Y_3(K)}{Y_1(K)} Y_1\right)$$

As a consequence the algebras of G^{h}-invariants, G^{v}-invariants and G-invariants are generated by algebraic operations and derivations with respect to the operators D_x, D_y, and $D_{y'}$ above. Precisely for the horizontal case, a system of functionally independent generators of the ring of r-th order is given by:

$$y,$$
$$I_{abc} = \left(Y_1^a \circ Y_2^b \circ Y_3^c\right)(I),\ \ 0 \le a+b+c \le r-1,$$
$$J_{\beta\gamma} = \left(Y_2^\beta \circ Y_3^\gamma\right)(J),\ \ 0 \le \beta+\gamma \le r-1,$$

where $I = \dfrac{y_{010}^{//}}{y'^2}$ and $J = \dfrac{y''-y'y_{001}^{//}}{y'^2}$ are G^{h}-invariants of order 1. For the vertical case a system of functionally independent generators of the ring of r-th order differential invariants is given by:

$$x,$$
$$K_{abc} = \left(Z_1^a \circ Z_2^b \circ Z_3^c\right)(K),\ \ 0 \le a+b+c \le r-2,$$
$$K_{\gamma\beta} = \left(Z_3^\gamma \circ Z_2^\beta \circ Z_1\right)(K),\ \ 1 \le \beta+\gamma \le r-3, \gamma \ne 0,$$
$$Z_1^\alpha(V),\ \ 0 \le \alpha \le r-4,$$

where $K = \dfrac{y_{002}^{//2}}{2} - y^{//}y_{002}^{//} + 2y_{010}^{//} - y^{/}y_{011}^{//} - y_{101}^{//}$ and $V = \dfrac{y_{004}^{//}}{y_{003}^{//3/2}}$ are G^{v}-invariants of order 2 and 4 respectively. Finally, for the total case, a system of functionally independent generators of the ring of r-th order differential invariants is given by $\left\{\left(Z_a^\alpha \circ Z_b^\beta \circ Z_c^\gamma\right)(T_i)\right\}$, where $1 \le i \le 6, 0 \le \alpha+\beta+\gamma \le r-4, (a,b,c) \in S_3$, and:

$$T_1 = \frac{K_{200}}{K_{100}^{1/3}} \qquad T_2 = VK_{100}^{1/3} \qquad T_3 = \frac{K_{101} - K_{10}}{K_{100}^{4/3}}$$
$$T_4 = \frac{2K_{001} - K_{110}}{2K_{100}^{5/3}} \quad T_5 = \frac{K_{10}}{K_{100}^{4/3}} - T_1\frac{K_{001}}{K_{100}^{5/3}} \quad T_6 = \frac{K_{002}K_{200} + K_{10}^2}{K_{200}^4}$$

One can efficiently implement an algorithm by using Mathematica to obtain the invariants in the last formulas recursively.

CHARACTERIZATION OF PAINLEVÉ TRANSCENDENTS

The search for nonlinear ordinary differential equations with solutions without moving critical points (critical points which location depends on the initial conditions to the differential equation) -so named equations with Painlevé property- was an important mathematical problem in 19th century. For the equations of the form $y' = F(x, y, y')$, such that are rational in y', algebraic in y and analytic in x. Painlevé and Gambier found fifty types of differential equations satisfying Painlevé property, six of them have solutions in terms of the Painlevé transcendents. The first three of these equations were due to Painlevé:

$$y' = 6y^2 + x,$$
$$y' = 2y^3 + xy + \alpha,$$
$$y' = \frac{y'^2}{y} - \frac{y'}{x} + \frac{\alpha y^2 + \beta}{x} + \gamma y^3 + \frac{\delta}{y}.$$

The role of such equations in several aspects of Physics (statistical mechanics, theory of solitons and integrable dynamical systems, etc.) is fundamental.

Basic results in order to characterized the first ans second Painlevé transcendents in terms of differential invatiants are the following:

Theorem. An arbitrary ODE of second-order, $y' = F(x, y, y')$ is reducible to first Painlevé transcendent under a change of coordinates of the independent variable if and only if the following relations hold:

$$0 = 2IJ + I_{001}$$
$$0 = 2I + I_{010}$$
$$0 = 2J + J_{10}$$
$$0 = I_{200}$$
$$0 = J_{02} + 6J^3 + 7JJ_{01}.$$

Theorem. An arbitrary second-order ODE, $y' = F(x, y, y')$, is reducible to the second Painlevé transcendent under a change of the independent variable if and only if the following conditions hold:

$$0 = 2I + I_{010}$$
$$0 = 2J + J_{10}$$
$$0 = I_{100} - yI_{200}$$
$$0 = 2JI_{100} + 2I^2 + I_{101}$$
$$0 = y(I_{001} + 2IJ) - 2J^2 - J_{01}$$
$$0 = 1728y^3(I_{001} + 2IJ)^2 - I_{100}^2.$$

ACKNOWLEDGMENTS

This work is supported by Ministerio de Ciencia y Tecnología (Spain), under grant BFM2002-00141.

REFERENCES

1. Berth, M. and Czichowski, G., *Appl. Algebr. Eng. Comm.* **11**, 359-376 (2001).
2. Kumpera, A., *J. Differential Geom.*. **10**, 289-345, 347-416 (1975).
3. Martín del Rey, A., "Differential Invariants of First and Second-Order ODEs," (in spanish), Ph. D. Thesis, Madrid, 2000.
4. Martín del Rey, A., "Invariants of second-order ODEs under changes of the independent variable," in Proceedings of the First Colloquium on Lie Theory and Applications, University of Vigo, Vigo, 2002, pp. 127-134.
5. Olver, P., *Equivalence, Invariants and Symmetry*, Cambridge University Press, 1995.

Applied Physics (APHYS 2003)
A. Méndez-Vilas (Ed.)

The Role Of Electrochemical Etching Conditions In The Growth Of Porous Silicon Layers

P. Fernández-Siles, A. Ramírez-Porras*

Centro de Investigación en Ciencia e Ingeniería de Materiales (CICIMA) and Escuela de Física Universidad de Costa Rica, San Pedro 2060, COSTA RICA

Abstract. Porous silicon (PSi) has been subject to a quite extended research due to its potentiality in the production of optoelectronic and chemical or biological sensor devices. It is widely accepted that morphology and grow rate of PSi layers depends strongly not only on the type and resistivity of the crystalline wafer, but also on the anodic etching conditions such as hydrofluoric acid content, current density and exposure time to the etching solution. Nevertheless, it is not well known which of these parameters dominates. In this study, the grow rate is examined in the case of p-type lowly doped silicon wafers. Comparison with already published results for highly doped wafers suggests a predominance in current density over the doping level in the PSi grow rate control.

INTRODUCTION

Since the discovery of strong luminescence on porous silicon at room temperature when illuminated with UV light [1], a lot of work has been devoted to understand the physics and chemistry of this system. Porous silicon has been firstly produced many years ago [2] as a surface film treatment of crystalline silicon by anodic attack, usually called etching. This material is a candidate to develop diverse light and chemistry-based devices for electronics, optics and biomedical applications [3]. However, the layer growing mechanisms are not yet fully understood. One of the important parameters to control is the porous layer thickness as time passes, in other words, the etching rate. The control of this parameter is crucial for a variety of potential applications. Examples are light-emitting diodes [4,5], Bragg reflectors [6] and optical microcavities [7]. The literature reports at least two behaviors of the etching rate of p-type wafers with respect to the initial etching parameters (HF content, current density, doping level and temperature). The first one estates that the etching rate increases with an increase in current density and conductivity, and decreases with an increase in HF concentration [3]. This has been explained as a process of charge migration onto the pore pits which initiates local electrochemical etching and forms cylindrical pores penetrating into the semiconductor bulk material perpendicularly with respect to the surface. In the second behavior, the etching rate is almost constant for high values of current density and wafer doping (resistivities around 1 Ω·cm), but it raises for an increase in low-level values of current density at same doping level, as stated by the first model [8]. In the high current density case, the etching

* Corresponding author: ramirezp@cariari.ucr.ac.cr

process is faster than the charge migration speed, and therefore a large number of inhomogeneous pores are formed both in the perpendicular and parallel direction. The overall effect is to reduce the etching rate. We will call the case of increased etching rate with increased current density the “linear” regime, and the constant etching rate the “independent” regime. The goal of this article is to discriminate between these two regimes for the samples of our own and from other researchers in order to distinguish the kind of relationships among the porous silicon preparation parameters.

EXPERIMENTAL SETUP

The samples were grown at room temperature on the polished side of boron doped silicon wafer of 15-45 Ω·cm resistivity and 0.25 mm thick. An aqueous solution of HF in ethanol in the proportions $HF:H_2O:EtOH = 1:1:2$ was placed in contact with the silicon polished side and employed as the etching solution where a current density of 100 mA/cm^2 was set through a platinum electrode, immersed in the solution, and a back ohmic contact of Aluminum deposited on the bottom of the silicon wafer. The anodization process was applied from 5 to 40 minutes. After rinsing with deionized water and drying with a flow of nitrogen, the samples were cleaved to expose their cross sections. SEM micrographs were taken to study the porous layers. Figure 1 shows a typical micrograph of such a kind of samples in the case of 30 minutes of anodization. The porous layer is located on the top, with a thickness of 150 μm. It is noticeable the presence of columnar structures in the perpendicular direction with respect to the wafer surface (following the electric field developed between the platinum electrode not shown on the top and the aluminum contact clearly visible on the bottom as a white line). These structures are the remaining semiconductor material after the pores were formed. A very sharp and straight interface between the porous layer and the crystalline region is also noticeable. Comparing the obtained morphology with the pictures reported in reference 8, we conclude that our samples belong to the “linear” regime already pointed out in the previous section.

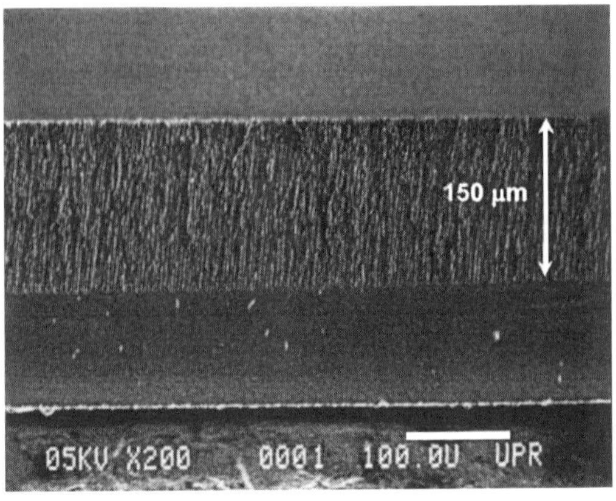

FIGURE 1. p-Type Silicon wafer cross section showing the porous layer growth on the top side after 30 minutes of anodization in an aqueous ethanoic solution of HF.

RESULTS AND DISCUSSION

Figure 2 shows a plot of the porous layer thickness as a function of anodization time for data of our own (curve A) and from other researchers (curves B to G and region H, where no precise correspondence between thickness and time has been reported) for different doping levels, current densities and HF content in the etching solution [9-12]. The effect of ethanol concentration does not appreciably shift the data points position. All points can be fitted to straight lines where the slopes correspond to the etching rates. Table 1 specifies all the values, sorted in decreased etching rates. A rapid inspection of this table points at a fairly direct relation between the etching rate and the current density. This is particularly true for lines F and G, where the resistivity is around 1 Ω·cm and the only change in the parameters is the current density. According to what has been exposed in the introduction, these cases belong to the "linear" regime. Additionally, lines B, C and D correspond to somewhat lowly doped silicon (*p*⁻-type) and also show such linear behavior. Region H can also be grouped here because of its doping level. Taking into account case A (a *p*⁻⁻ lowly doped wafer similar to case E), which already has been recognized as linear, we can infer that, for the present preparation conditions, all *p*, *p*⁻ and *p*⁻⁻ belong to the same regime.

The next step is to determine the relative dominance of preparation parameters. In general, current density shows up as the dominant factor in the etching rate control. This is quite evident in lines A and E, where the doping levels are coincident and the only difference is the current density. The role of HF can be determined from cases D and H (*p*⁻ wafers): a slight increment in current density (23 to 30 mA/cm^2) is not sufficient to prevent a decrease in etching rate because of the decrement in HF concentration (34 to 25%). This can be explained in terms of

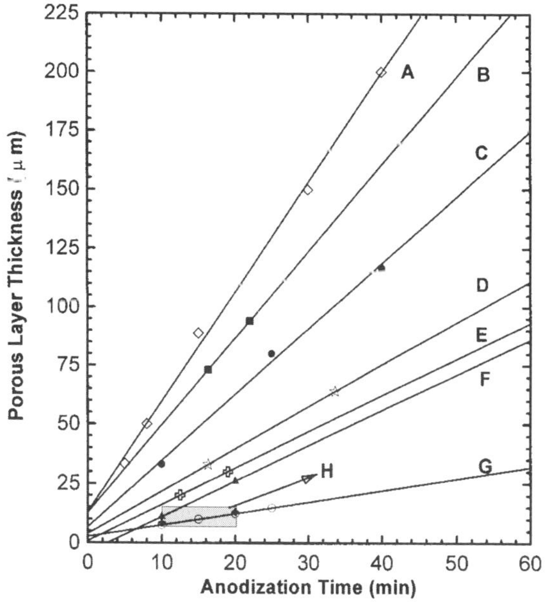

FIGURE 2. Porous layer thickness as function of the anodization time for various preparations (labels A through H specified in Table 1).

TABLE 1. Preparation parameters and Etching rates for different reported cases. The labels correspond to the ones shown in Figure 2.

Label	Doping level	Resistivity (Ω.cm)	Current density (mA/cm^2)	HF concentration (%)	Etching rate (μm/min)	Data source
A	p^{--}	15-45	100	25	4.66	This work
B	p^{-}	8-15	57	34	3.69	Ref. [9]
C	p^{-}	8-15	40	32	2.80	Ref. [9]
D	p^{-}	8-15	23	34	1.79	Ref. [9]
E	p^{--}	10-30	35	25	1.54	Ref. [11]
F	p	1-1.5	32	20	1.50	Ref. [10]
G	p	1-1.5	14	20	0.49	Ref. [10]
H	p^{-}	10	30	25	~ 0.50	Ref. [12]

the requirements for dissolution chemistry: because holes (from the semiconductor) and fluorine ions (from the solution) are necessary to start breaking the silicon covalent bonds [13], a lower concentration of HF should lower the reaction speed, and therefore the etching rate.

The role of doping wafer is not so evident. A shift from case E to case F yields an increase in doping level (resistivity decreases) along with small decrements in current density and HF content. The result is a slight decrement in etching rate, which perhaps could be attributed more to the latter parameters variation instead of the resistivity decrement of one order of magnitude. On the contrary, a shift from case H to case F (doping highly and current density slightly increased, HF content decreased), yields a higher raise in the etching rate. We therefore conclude that a variation in the doping level gives a second order change with respect to current density variation. In other words, current density dominates over doping.

Thus, it is suitable to propose a general relation between the rate at which the thickness layer (ε) grows as a function of current density (J), HF proportion and resistivity (ρ):

$$\frac{d\varepsilon}{dt} = AJ^{\alpha}[\mathrm{HF}]^{\beta}\rho^{-\gamma} \tag{1}$$

where A, α, β and γ are positive constants ($\alpha > \beta, \gamma$). The determination of such constants is a subject of a further study.

CONCLUSIONS

In this paper we have studied the role of some the preparation parameters in the growth of porous silicon layers. We have determined a clearly strong direct relation between the porous layer etching rate with current density. This behavior is characteristic of a linear regime where the pores are formed following the direction of the applied electric field, that is, perpendicular to the semiconductor surface. A slightly weaker direct relation has been found with respect to the concentration of hydrofluoric acid in the anodic solution. Finally, the etching rate seems to be inversely related to the wafer resistivity in a fairly weak proportion.

ACKNOWLEDGMENTS

One of us (A. R-P.) wishes to acknowledge partial support from the Fundación para la Investigación de la Universidad de Costa Rica (FUNDEVI), and the Comisión Nacional de Investigaciones Científicas y Tecnológicas (CONICIT) for providing founds through the Programa de Fondo de Riesgo para la Investigación (FORINVES), under the grant # FV-035-02.

REFERENCES

1. Canham, L.T., *Appl. Phys. Lett.* **57**, 1046 (1990).
2. Uhlir, A., *Bell Syst. Tech. J.* **35**, 333 (1956).
3. Bisi, O., Ossicini, S., and Pavesi, L., *Surface Science Reports* **38**, 97 (2000).
4. Steiner, P., Kozlowski, F., and Lang, W., *Appl. Phys. Lett.* **62**, 2700 (1993).
5. Tsybeskov, L., Duttagupta, S.P., Hirschman, K.D., and Fauchet, P.M., *Appl. Phys. Lett.* **68**, 2058 (1996).
6. Mazzoleni, C. and Pavesi, L., *Appl. Phys. Lett.* **67**, 2983 (1995).
7. Pavesi, L., *La Rivista del Nuovo Cimento* **20**, 1 (1997).
8. Mason, M.D., Sirbuly, D.J., and Buratto, S.K., *Thin Solid Films* **406**, 151 (2002).
9. Benhida, A., Achargui, N., Combette, P., and Foucaran, A., *Journal of Crystal Growth* **186**, 565 (1998)
10. Bessaïs, B., Ben Jounes, O., Ezzaouia, H., Mliki, N., Boujmil, M.F., Oueslati, M., and Bennaceur, R., *J. Luminescence* **90**, 101 (2000).
11. A xelrod, E., Givant, A., Shappir, J., Feldman, Y., and Sa'ar, A., *J. Non-Crystalline Sol.* **305**, 235 (2002).
12. Belogorokhov, A.I., Bologorokhova, L.I., and Gavrilov, S., *Journal of Crystal Growth* **197**, 702 (1999).
13. Le hman, V., and Gösele, U., *Appl. Phys. Lett.* **58**, 856 (1991).

Applied Physics (APHYS 2003)
A. Méndez-Vilas (Ed.)

Future Trends in Nuclear Medical Imaging

Habib Zaidi

Division of Nuclear Medicine, Geneva University Hospital
CH-1211 Geneva, Switzerland

Abstract. Continuous efforts to integrate recent research findings for the design of different geometries and various detector technologies of single-photon emission computed tomography (SPECT) and positron emission tomography (PET) scanners have become the goal of both the academic community and nuclear medicine industry. As PET has become of more interest for clinical practice, several different design trends seem to have developed. Systems are being designed for "low cost" clinical applications, very high-resolution research applications, and just about everywhere in-between. All of these systems are undergoing revisions in both hardware and software components. The development of dual-modality imaging systems is an emerging research field. One of the major advantages is that SPECT/PET data are intrinsically aligned to anatomical information from the X-ray CT without the use of external markers or internal landmarks. On the other hand, combining PET with Magnetic Resonance Imaging (MRI) technology is scientifically more challenging owing to the strong magnetic fields. Nevertheless, significant progress has been made resulting in the design of a prototype small animal PET scanner coupled to three multi-channel photomultipliers via optical fibers so that the PET detector can be operated within a conventional MR system. Thus, there are many different design paths being pursued - which ones are likely to be the main stream of future commercial systems? It will be interesting, indeed, to see what technologies become the most popular in the future. This paper briefly summarizes state-of-the art developments in nuclear and dual-imaging devices. Future prospects will also be discussed

INTRODUCTION

Radionuclide imaging, including planar projection imaging, single-photon emission computed tomography (SPECT) and positron emission tomography (PET), relies on the tracer principle, in which a minute quantity of a radiopharmaceutical in introduced into the body to monitor the patient's physiological function. In a clinical environment, resulting radionuclide images are interpreted visually to assess the physiological function of tissues, organs, and organ systems, or can be evaluated quantitatively to measure biochemical and physiological processes of importance in both research and clinical applications. Nuclear medicine relies on noninvasive measurements performed with external (rather than internal) radiation detectors in a way that does not allow the radionuclide measurement to be isolated from surrounding body tissues or cross-talk from radionuclide uptake in non-target regions.

An important consequence of the cost- and performance-conscious environments of health care today is the constant pressure to minimize the cost of nuclear medicine imaging devices, while at the same time there is also pressure to provide the most accurate diagnostic answers through the highest performance possible. The dilemma is that both approaches can lower the cost of health care. Continuous efforts to integrate recent research findings in detector development for the design of different geometries of nuclear medicine imaging instruments have become the goal of both the medical imaging academic community and nuclear medicine industry.

TRENDS IN SPECT INSTRUMENTATION

Most scintillation cameras used for clinical imaging of single-photon emitters are based on the original design proposed by Anger more than 45 years ago [1]. The first nuclear medical imaging systems (single-photon and positron) were designed with NaI(Tl) scintillation detectors. With exception of the recently marketed NeuroFocus (Neurophysics Corporation, Shirley, Massachusetts) multi-conebeam system with a claimed intrinsic spatial resolution of 3 mm, all commercial imaging devices available nowadays employ NaI(Tl) scintillation detectors coupled to photomultiplier tubes (PMT's), likely in the near future, photodiodes or avalanche photodiodes (APDs) to convert the light output into electrical signals.

The system spatial resolution and sensitivity depends to a larger extent on the type of collimator used and the intrinsic performance of the camera. Independent of the collimator, system resolution cannot get any better than intrinsic resolution. There have been several collimator designs in the past fifteen years, which optimized the resolution/sensitivity inverse relation for their particular design. Converging hole collimators, for example fan-beam and cone-beam have been built to improve the trade-off between resolution and sensitivity. More modern collimator designs, such as half-cone beam, astigmatic and rotating slat, have also been conceived. Sensitivity has seen an overall improvement by the introduction of multi-camera SPECT systems. A typical triple-head camera SPECT system equipped with high resolution fan-beam collimators can achieve a resolution of 7-8 mm in typical brain imaging conditions. Other types of collimators with only one or a few channels, called pinhole collimators, have been designed to image small organs and human extremities, such as the wrist and thyroid gland, in addition to research animals such as rats.

With the exception of coded aperture techniques, all collimators exhibit a limiting detection sensitivity that is inversely proportional to the spatial resolution. This motivated the development of Compton cameras, which provide information about the incoming photon direction electronically without any restriction with respect to the solid detection angle [2]. The development of pixelated detectors allowed improving the spatial resolution at the expense of deteriorating the energy resolution as a result of light losses in the crystals compared to that of a single crystal [3].

An interesting design, which if successful in a clinical environment, will revolutionize the practice of nuclear medicine and result in a quantum jump in the history of nuclear medicine instrumentation, is the SOLid STate Imager with Compact Electronics (SOLSTICE). This system combines the direct gamma ray conversion through solid-state detectors with the better compromise offered by rotating slat collimators (Figure 1) to break the limitations of conventional scintillation cameras-based design [4]. Some promising results have been presented during the last few years but the instrument is still in a development phase and appears to be well suited for high resolution small-animal imaging. The viability and cost-effectiveness of the product still needs to be demonstrated.

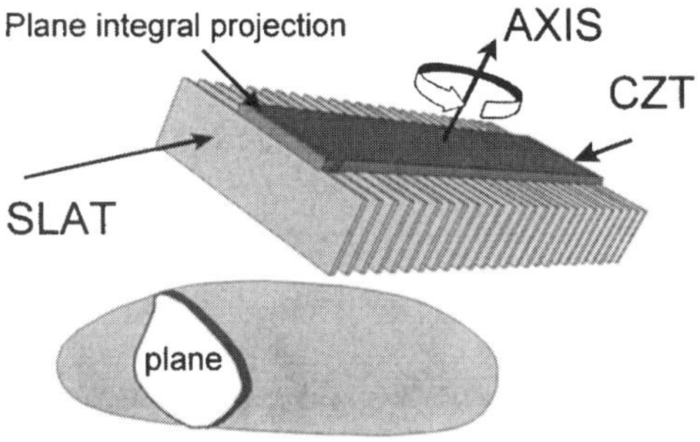

FIGURE 1. Schematic illustration of the SOLSTICE prototype system. Plane integrals are measured by the SOLSTICE imaging system for each projection angle in contrast to line integrals measured in the conventional approach. A specially designed reconstruction algorithm which handles this acquisition geometry has been developed for this purpose.

TRENDS IN PET INSTRUMENTATION

There has been a significant evolution in PET instrumentation from a single ring of bismuth germanate (BGO) detectors with a spatial resolution of 15 mm, to multiple rings of small BGO crystals resulting in a spatial resolution of about 4-6 mm. Improvements in spatial resolution have been achieved by the use of smaller crystals and the efficient use of photomultiplier tubes (PMT's) and Anger logic-based position readout.

Dedicated full-ring PET tomographs have evolved through at least 4 generations since the design of the first PET scanner in the mid 1970s and are still considered the high-end devices. The better performance of full-ring systems compared to camera-based dual or triple-headed systems is due to higher overall system efficiency and count rate capability which provides the statistical realization of the physical detector resolution and not a higher intrinsic physical detector resolution. Obviously, this has some important design consequences since even if both scanner designs provide the same physical spatial resolution as estimated by a point spread function, the full-ring system will produce higher resolution images in patients as a result of the higher statistics per unit imaging time. Figure 2 illustrates possible geometric designs of positron imaging systems used so far. The most important aspect related to the outcome of research performed in the field is the improvement of the cost/performance optimization of clinical PET systems.

The geometric camera design actually affects to a greater extent the solid angle coverage, which has direct consequences on the resulting sensitivity. On the other hand, the reconstruction algorithm used significantly affects the achieved spatial resolution. It has been shown that iterative algorithms outperform conventional analytic methods [5, 6].

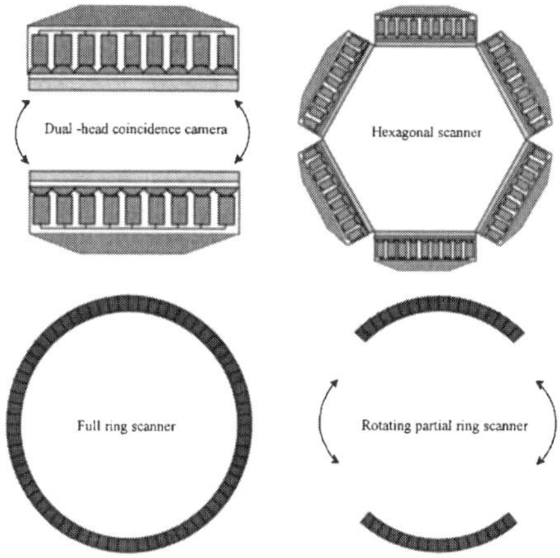

FIGURE 2. Illustration of the range of different geometries of positron volume imaging systems. The dual-head coincidence camera and partial ring tomographs require the rotation of the detectors to collect a full 180° set of projection data.

The critical component of PET tomographs is the scintillation detector. New detection technologies that are emerging include the use of new Cerium doped crystals (LSO:Ce, GSO:Ce and $LaBr_3$:Ce) as alternatives to conventional BGO crystals, the use of layered crystals and other schemes for depth-of-interaction (DOI) determination. In particular phoswich detectors received considerable attention for the design of high resolution scanners dedicated for brain, positron emission mammography (PEM) and small animal imaging. This may be implemented with solid-state photodiode readouts, which also allows electronically collimated coincidence counting. Such a design has been implemented on the ECAT high resolution research tomograph (HRRT) with LSO crystals [7]. Figure 3 illustrates the principle of the conventional detector block and the phoswich approach where two detectors are assembled in a sandwich-like design, the difference in decay time of the light is used to estimate the crystal where the interaction occurred.

The intrinsic physical performance of conventional designs (even with DOI capability) seem to have been reached encouraging the development of innovative approaches capable of providing improved performance at a reduced or comparable cost. This motivated the proposal of a novel design concept which provides full 3D reconstruction with high resolution over the total detector volume, free of parallax errors. The key components are a matrix of long scintillator crystals and Hybrid Photon Detectors (HPDs) with matched segmentation and integrated readout electronics [8]. The HPDs read out the two ends of the scintillator package. Both excellent spatial (x,y,z) and energy resolution is obtained. The concept allows enhancing the detection efficiency by reconstructing a significant fraction of events which underwent Compton scattering in the crystals.

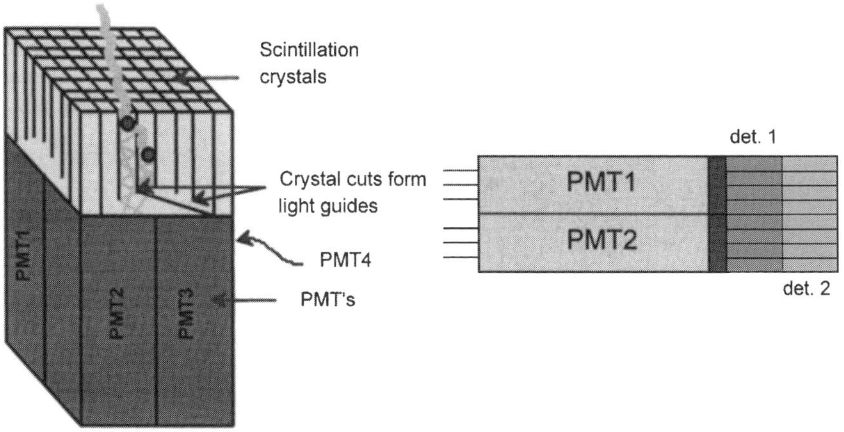

(a) (b)

FIGURE 3. (a) Conventional block detector consisting of a set of crystals having cuts of different depths acting as light guides and segmenting the block into 64 (8x8) detection elements in this example. The block is optically coupled to four photomultiplier tubes at the back, and the crystal in which photoabsorption occurs is identified by comparing the outputs of the four photomultiplier tubes (Anger logic). (b) Detector block consisting of a phoswich (detector 1 and 2) with depth-of-interaction measurement capability.

TRENDS IN IN SMALL ANIMAL IMAGING INSTRUMENTATION

The advent of molecular imaging attracted the interest of biomedical researchers in the small animal imaging technology. The use of animal models has been motivated to a great extent by the availability of modern transgenic and knockout techniques. This interest motivated the development of high resolution research prototype systems dedicated for imaging small animals including projection imaging, SPECT and PET (or both), the latter being the most appealing approach for molecular imaging research owing to ease of incorporation into a wide range of molecules and more advanced tracer development activities using this imaging modality [9].

An important conclusion drawn from simulation studies is that unlike human imaging where both sensitivity and spatial resolution limitations significantly effect the quantitative imaging performance of a tomograph, the imaging performance of dedicated animal tomographs is almost solely based upon its spatial resolution limitations. Thus, different PET designs have been suggested encompassing conventional small ring radius cylindrical block-detector based design with DOI capability and APDs readout, a renewed interest in the 3D high density avalanche chamber (HIDAC) camera that achieves sub-millimeter resolution along with many other designs. Nowadays, high-resolution animal scanner designs are plentiful and are being developed in both academic and corporate settings, with more than three such devices (both SPECT and PET) being offered commercially (Fig. 4). More recently, advanced versions of these same technologies have begun to be used across the breadth of modern biomedical research to study non-invasively small laboratory animals in a myriad of experimental settings. In addition, combined imaging devices such as SPECT/CT, PET/CT and PET/MRI are under development by different research groups and scanner manufacturers, leading the biomedical imaging community to forecast a promising progress during the next few years.

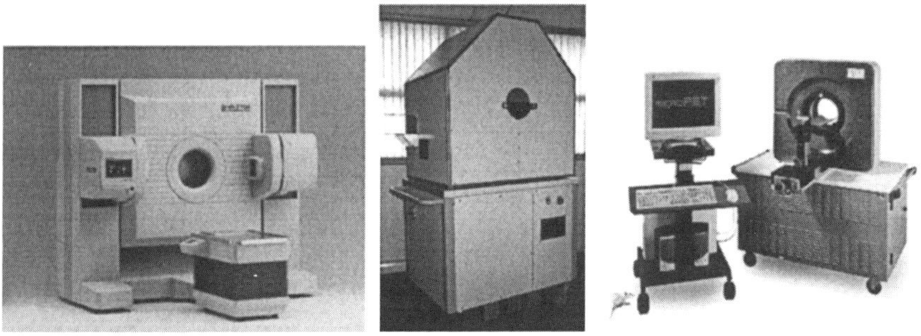

(a) **(b)** **(c)**

FIGURE 4. Photographs of small animal PET scanners showing (a) the Hamamatsu SHR-7700 PET scanner based on BGO detector blocks designed for non-human primate imaging, (b) the multiwire proportional chamber technology-based HIDAC system, and (c) the microPET P4 scanner using LSO scintillation crystals. (Photographs courtesy of Hamamatsu Photonics KK, Japan, Oxford Positron Systems, UK and Concorde Microsystems, USA, respectively).

TRENDS IN DUAL-MODALITY IMAGING INSTRUMENTATION

The principle of dual-modality imaging technology is to combine a functional imaging device (SPECT or PET) with an anatomical imaging instrument (CT or MRI) to acquire coregistered images with a single integrated system. This is a hardware approach to image fusion as opposed to the software approach where manual or automated image registration techniques are used to realign intra-modality images acquired separately on two different imaging systems [10]. This development is an emerging research field driven particularly by whole-body imaging in clinical oncology. One of the major advantages is that SPECT/PET data are intrinsically aligned to anatomical information from the X-ray CT without the use of external markers or internal landmarks. Figure 5 shows a clinical study illustrating the principle of PET/CT image fusion and highlights the value of this technology for better anatomic localization of abnormal tissue metabolism (tumors). Different designs of combined SPECT/CT and PET/CT tomographs were developed for diagnostic purposes in clinical oncology and three such systems are now commercially available from the major vendors. Quantification is also improved by using the low noise CT transmission information during the correction of the PET data for self-attenuation, contamination from scattered photons and for partial volume effects.

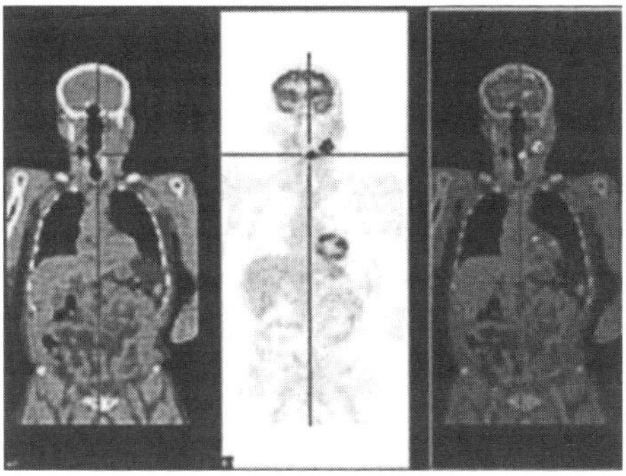

FIGURE 5. Clinical study of a patient with head and neck cancer illustrating the anatomical CT images (left), PET image (center), and the fused PET/CT image (right) allowing better localization of the tumor.

On the other hand, combining PET with MRI technology is scientifically more challenging owing to the strong magnetic fields. Nevertheless, significant progress has been made resulting in the design of a prototype small animal PET scanner with LSO:Ce detector blocks of 3.8 cm ring diameter coupled to three multi-channel photomultipliers via optical fibers [11] so that the PET detector can be operated within a conventional MR system. The authors reported no appreciable artifacts caused by the scintillators in the MR images. A second larger (11.2 cm) prototype is being developed for simultaneous PET/MR imaging of mice and rats at different magnetic field strengths [12]. While the advantages of combined PET/CT could in principle be replicated by combined PET/MRI, the usefulness of MRI for attenuation correction in radionuclide imaging has not been established until recently in brain scanning [13]. Potential applications of the technique in whole-body imaging still need to be demonstrated.

The availability of dual-modality instruments is likely to influence the approaches to image correction and reconstruction, with algorithms that combine anatomical information likely to increase in popularity. These approaches have previously relied on accurate registration and ready access to anatomical data. The advent of dual modality instruments makes this much more practical. In particular, the development of dual-modality imaging systems stimulated the interest in Bayesian-type maximum a posteriori (MAP) reconstruction algorithms. This approach uses the anatomical information from a registered MRI or CT image to guide and tune the noise suppressing prior in a MAP-algorithm, by limiting smoothing to within organ boundaries revealed by the anatomical data.

SUMMARY

In many respects, the field of nuclear medicine has been ahead of other areas of image science in objective assessment and optimization of image quality. Nuclear medicine has poor spatial resolution compared to other imaging modalities in radiology, although its ability to measure physiological processes is unsurpassed.

While the potential for PET in this field is undisputed, the challenges must also be recognized.

A brief overview of current state-of-the art developments in nuclear medicine instrumentation was presented. It has been emphasized that there are many different design paths being pursued in both academic and corporate settings - which ones are likely to be the main stream of future commercial systems? It will be interesting, indeed, to see what technologies become the most popular in the future.

ACKNOWLEDGMENTS

The author would like to acknowledge support from the Swiss National Science Foundation under grant SNSF 3152A0-102143.

REFERENCES

1. Anger H. *Rev. Sci. Instr.* **29**, 27-33 (1958).
2. Meier D, Czermak A, Jalocha P, et al. *IEEE Trans. Nucl. Sci.* **49**, 812-816 (2002).
3. Loudos GK, Nikita KS, Giokaris ND, et al. *Appl. Radiat. Isot.* **58**, 501-508 (2003).
4. Gagnon D, Zeng GL, Links JM, et al. *Proc.* IEEE Nuclear Science Symposium and Medical Imaging Conference, Oct. 4-10, San Diego, CA, 2002, pp. 1156-1160.
5. Qi J and Leahy RM. *IEEE Trans. Med. Imaging* **19**, 493-506 (2000).
6. Moses WW and Qi *J. Nucl. Instr. Meth. Phys. Res.* A **497**, 82-89 (2003).
7. Wienhard K, Schmand M, Casey ME, et al. IEEE Trans. Nucl. Sci. **49**, 104 -110 (2002).
8. Braem A, Chesi E, Joram C, et al. *Conf. Proc.*of the First International meeting on Applied Physics, 13-18th October 2003, Badajoz, Spain. 2003, pp. 86-67.
9. Green MV, Seidel J, Vaquero JJ, et al. *Comput Med Imaging Graph* **25**, 79-86 (2001).
10. Hasegawa BH, Iwata K, Wong KH, et al. *Acad. Radiol.* **9**, 1305-1321 (2002).
11. Shao Y, Cherry SR, Farahani K, et al. *Phys. Med. Biol.* **10**, 1965-1970 (1997).
12. Slates RB, Farahani K, Shao Y, et al. *Phys. Med. Biol.* **44**, 2015-2027 (1999).
13. Zaidi H, Montandon M-L, Slosman DO. Med. Phys. **30**, 937-948 (2003).

Decontamination of ^{137}Cs Radioactive Liquid Wastes by Membrane Technology

J.M. Arnal*, M. Sancho*, J.M. Campayo†, G. Verdú*, J. Lora*

* *Chemical and Nuclear Engineering Department, Polytechnic University of Valencia*
Camino de Vera s/n, 46022 Valencia, Spain
† *Logística y Acondicionamientos Industriales S.A. (LAINSA)*
El Palleter 13, 46008 Valencia, Spain

Abstract. As a consequence of an accidental fusion of a ^{137}Cs source in a stainless steel production factory, part of an oven and the refrigeration circuit were radioactively contaminated. After the decontamination and cleaning of the factory, 40 m^3 of radioactive liquids and 2000 T of ashes of low-medium activity level were generated. The radioactive liquid wastes were treated by the Chemical and Nuclear Engineering Department of the Polytechnic University of Valencia (Spain) and LAINSA company by means of a reverse osmosis plant. This treatment was very successful since more than 36 m^3 of the radioactive liquid waste were declassified after the treatment by reverse osmosis. The remaining 4 m^3 of liquid were treated by evaporation, resulting in a final volume lower than 1 m^3. The radioactive ashes have not been treated yet, but they could be decontaminated by means of a solid-liquid extraction process. This would generate a liquid waste containing most of the original ^{137}Cs which could be treated by the reverse osmosis plant to reduce its volume or to be reused in the extraction process. This paper describes the treatment of the liquid wastes carried out and the proposal for decontaminating the radioactive ashes.

INTRODUCTION

As a result of an accidental fusion of a ^{137}Cs source in a stainless steel production factory, one of the ovens and the refrigeration circuit were radioactively contaminated. LAINSA (Logística y Acondicionamientos Industriales S.A.) company took charge of the plant decontamination process, in which the following wastes were generated:

- 40 m^3 of ^{137}Cs contaminated liquids with an average activity of 300 kBq/L
- 2000 T of radioactive ashes with a low-medium activity level

After the decontamination process, LAINSA contacted the Chemical and Nuclear Engineering Department of the Polytechnic University of Valencia (UPV) to develop a project for ^{137}Cs contaminated liquid treatment. In this project, the proposed solution for treating the radioactive liquids consisted in the application of reverse osmosis, with the main objective of reducing the waste volume to be disposed.

Membrane Technology Applied to Radioactive Liquid Wastes

Membrane processes have become common in the treatment of radioactive liquid wastes or effluents. Reverse osmosis for example is being successfully applied to radioactive liquid waste treatment in many nuclear installations [1,2]. The main advantages of membrane technology in comparison to traditional separation processes applied to radioactive effluents (evaporation, chemical precipitation and ionic exchange) are: moderate operating conditions, simple apparatus, high decontamination factors and low energy consumption [3].

Reverse osmosis (RO) is a process driven by pressure in which the membrane acts as a barrier between two liquids. A reverse osmosis membrane allows the passing of water (solvent) through it, preventing the transference of salts and larger compounds (solutes). By means of reverse osmosis application it is possible to transform a great volume of a feed solution into two streams: a small volume of a very concentrated liquid (retentate) and a stream almost exempt from solutes (permeate).

The application of reverse osmosis to the ^{137}Cs contaminated liquids would result in a treated liquid (permeate) with an activity level lower than the legal discharge limit for the ^{137}Cs radioisotope (300 Bq/L), and a concentrated liquid (retentate) containing most of the original radioisotope, that must be managed as a radioactive liquid waste.

The reverse osmosis process would be also suitable for the treatment of the liquid used for extracting the ^{137}Cs from the radioactive ashes. By means of the membrane treatment, the extraction liquid could be reused in another extraction stage.

This paper describes the treatment of the liquid wastes carried out by reverse osmosis and the possible application of this technology to the treatment of the liquid used for decontaminating the radioactive ashes.

THE TREATMENT OF ^{137}Cs RADIOACTIVE LIQUID WASTES

After some preliminary tests in which the efficiency of reverse osmosis (RO) process in the treatment of ^{137}Cs contaminated liquids was tested [4], the reverse osmosis plant was designed by the Chemical and Nuclear Engineering Department of the Polytechnic University of Valencia (UPV), and built by LAINSA company. Figure 1 shows the flow chart of the RO plant used for treating the ^{137}Cs contaminated liquids. The main components of the treatment plant are the following:

- *Filtration unit* (K). The liquid to be treated by RO is pretreated in a filtration unit which comprises two parallel lines with two filters of different pore size each one: a 5 μm filter and a cartridge filter of 0,45 μm.
- *High pressure pump* (P). The high-pressure pump, which feeds RO membranes, is controlled by means of a speed variator and regulation valves.
- *Reverse osmosis modules* (F, G, H). The plant comprises two kinds of spiral-wound RO aromatic polyamide composite membranes [4,5]: low pressure (LP) and high pressure (HP). The main difference between them is their working pressure limit: 40 bar for the low-pressure membranes and 70 bar for the high-pressure ones.

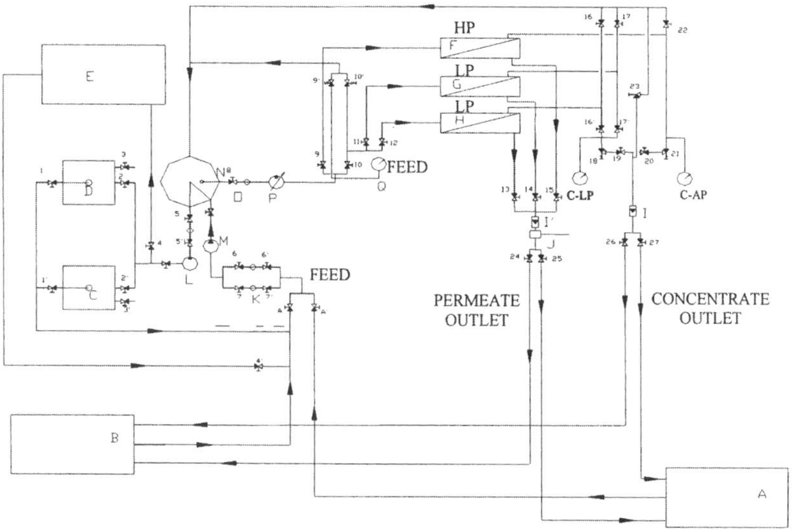

FIGURE 1. Flow chart of the reverse osmosis treatment plant.

The research team UPV-LAINSA carried out the treatment of the ^{137}Cs liquid wastes, which lasted a month, approximately. In a first stage both HP and LP membranes were applied to the liquid wastes. In final stage evaporation had to be applied to the concentrated liquid in order to reduce more its volume. Table 1 summarises the main results in each of the treatment stages.

TABLE 1. Results obtained in each stage of the treatment.

Day of treatment	Stage	Volume to treat	Volume treated
1	Beginning of membrane treatment	40 m^3	0 m^3
20	End of membrane treatment	4 m^3	36 m^3
21	Beginning of evaporation treatment	4 m^3	0 m^3
30	End of evaporation treatment	0,6 m^3	3,4 m^3

At the end of the treatment 39,4 m^3, approximately, were decontaminated and discharged in a purifier, and only 600 litres had to be managed as radioactive wastes.

THE PROPOSAL FOR TREATING THE RADIOACTIVE ASHES

The most suitable way of treating the ^{137}Cs radioactive ashes consists in a solid-liquid extraction process after which the ashes can be considered radioactively declassified [7]. This treatment generates a ^{137}Cs contaminated liquid (50 litres per each gram of ashes, approximately) that must be post-treated in order to reduce its volume or to be reused in the extraction process. This post-treatment can be made by two techniques: selective adsorption using ammonium phosphomolibdate as an additive or by means of membrane technology.

When applying selective adsorption some physic-chemical characteristics of the solution must be modified, for example, pH has to be reduced to 1 by means of sulfuric acid addition. Besides, experimental assays with ^{137}Cs contaminated ashes have shown that between 500 and 1000 ppm of ammonium phosphomolibdate are required per each litre of radioactive liquid to be treated. Therefore, selective adsorption would consume a great amount of reagents, which is a clear disadvantage in comparison to membrane technology.

According to the results obtained in the experimental assays of ^{137}Cs extraction from the contaminated ashes, the values of conductivity and activity of the liquid used in the extraction are within the range of values of the ^{137}Cs contaminated liquids treated by the RO plant. Therefore, it can be said that it is possible to treat the liquid resulted in the extraction by a reverse osmosis process, in a similar way to the treatment described previously in this paper.

CONCLUSIONS

- The application of reverse osmosis in the treatment of the 137Cs contaminated liquids was highly successful, having declassified more than 90% of the original volume by means of membrane treatment, and about 98% using evaporation in the last stage of the treatment.
- Selective adsorption is not a suitable method for treating the liquid used in the extraction of ^{137}Cs from the contaminated ashes because it requires the use of a great amount of reagents.
- According to the conductivity and activity values of the liquid used in the extraction of ^{137}Cs from the contaminated ashes, the most suitable way of treating it is the application of reverse osmosis.

ACKNOWLEDGMENTS

The authors would like to express their gratitude to ACERINOX and LAINSA people for their collaboration during the treatment of the ^{137}Cs liquid wastes, and to all the people that have collaborated in the experiments with contaminated ashes.

REFERENCES

1. Sen Gupta, S.K., and Rimpelainen, S., *Ultrapure Water* **14** (1), 32-39 (1997).
2. Chmielewski, A.G., Harasimowicz, M., Tyminski, B., and Zakrzewskatrznadel, G., *Sep. Sci. Technol.* **36** (5-6), 1117-1127 (2001).
3. Chmielewski, A.G., Harasimowicz, M., Tyminski, B., and Zakrzewskatrznadel, G., *Czech. J. Phys.* **49**, 979-985 (1999).
4. Arnal, J.M., Sancho, M., Verdú, G., Campayo, J.M., and Villaescusa, J.I., *Desalination* **154**, 27-33 (2003)
5. Chmielewski, A.G., and Harasimowicz, M., *Nukleonika* **42** (2), 857-862 (1997).
6. Arnal, J.M., Campayo, J.M. et al., *Desalination* **129**, 101-105 (2000).
7. Arnal, J.M., "Doctoral Thesis", Valencia: Polytechnic University of Valencia, 2002.

Applied Physics (APHYS 2003)
A. Méndez-Vilas (Ed.)

Application of the Unscear 2000 Report in the Valencian Breast Cancer Screening Program

M. Ramos [(1)], S. Ferrer [(1)], J.I. Villaescusa [(2)], G. Verdú [(1)], M.D. Salas [(3)], and M.D. Cuevas [(4)]

[(1)] ***Departamento de Ingeniería Química y Nuclear.*** *Universidad Politécnica de Valencia. Camino de Vera s/n, 46022 Valencia Tel: (+34) 963879631 / Fx: (+34) 963877639 mirapas@iqn.upv.es, silferto@iqn.upv.es, gverdu@iqn.upv.es*
[(2)] ***Servicio de Protección Radiológica.*** *Hospital La Fe de Valencia. Avda Campanar,21, 46009 Valencia, Spain. Tf: +34963868782 / Fax: +34963862770 villaescusa_ign@gva.es*
[(3)] ***Dirección General de Salud Pública.*** *Consellería de Sanidad de Valencia. C/ Micer Mascó, 31, 46021 Valencia, Spain. Tf: +34963869278 salas_dol@gva.es*
[(4)] ***Dirección General de Prestación Asistencial.*** *Consellería de Sanidad de Valencia. C/ Micer Mascó, 31, 46021 Valencia, Spain. Tf: +34963868051 cuevas_dol@gva.es*

Abstract. The breast cancer is one of the most frequent diseases in women, with a high incidence rate. The best fight against breast cancer is the early detection by means of mammography in a screening program. The Valencian Breast Cancer Screening Program (VBCSP) started at 1992, and it is comprised of twenty-four screening centers. The program is targeted towards asymptomatic women from 45 to 69 years old, but this screening has a negative influence in the studied woman, whatever the diagnosis was. The UNSCEAR 2000 report includes new epidemiological models for risks of solid cancer mortality and incidence due to radiation exposures, based on attained-age or age-at-exposure at a given dose level. Different MCNP-4c2 models, simulating the craniocaudal (*CC*) and mediolateral oblique view (*OBL*), for calculating breast glandular doses have been developed. The radiological detriments have been transported from the population under study in the UNSCEAR 2000 report (Life Span Study, Swedish patients irradiated for skin haemangioma in childhood and individual estimates of organ doses from Stockholm and Gothenburg, from others) to the Valencian Community. The fatal breast cancer excess deaths have been obtained considering the Valencian Community mortality and breast physiology, applying the ICRP (1991) morbidity data. Also, the excess absolute risks and the risk of exposure-induced death for breast cancer have been calculated and compared by means of the ASQRAD software, with an older risk projection model, the UNSCEAR 1994.

INTRODUCTION

For the dose and detriment calculation, three sample populatutions of 100 women from each unit of the Valencian Breast Cancer Screening Program were taken during a three year period. By means of the Monte Carlo MCNP-4c2 [(1)] code and the real mammographic exposure conditions of the sample, the induced doses have been calculated at each exposure. The ASQRAD computer code [(2)] and the UNSCEAR 2000 report [(3)] have been used to obtain the excess absolute risk per 100000 women-year and per sample population.

METHOD

The European Protocol on Dosimetry in Mammography methodology [4] and an alternative method have been used for calculating the average mean glandular breast doses per sample population. The photon fluxes calculated with MCNP-4c2 code have been employed to obtain three conversion factors between calculated $ESAK_c$ (entrance surface air kerma) and MGD (mean glandular dose), according to

$$MGD = f\,ESAK_c = f_K f_{TD} f_{GD} ESAK_c \qquad (1)$$

where f_K is the backscatter factor (air-standard breast), f_{TD} is the conversion factor between $ESAK_c$ and total dose (glandular + adipose tissues), and f_{GD} is the total to glandular dose conversion factor. All factors have been calculated with MCNP-4c2.

The average mean glandular breast dose ($AMGD$) per mammogram and per female breast delivered by each unit i has been calculated using the expression

$$AMGD(i) = \sum_{j=1}^{nu(i)} \sum_{k=1}^{nw(i,j)} \frac{\sum_{l=1}^{nv(i,j,k)} MGD(i,j,k,l)}{nv(i,j,k) \sum_{j=1}^{nu(i)} nw(i,j)}$$
$$= \sum_{j=1}^{nu(i)} \sum_{k=1}^{nw(i,j)} \sum_{l=1}^{nv(i,j,k)} \frac{f_K(i,j,k,l) f_{TD}(i,j,k,l) f_{GD}(i,j,k,l) ESAK_c(i,j,k,l)}{nv(i,j,k) \sum_{j=1}^{nu(i)} nw(i,j)} \qquad (2)$$

where $nu(i)$ is the number of units contributing to dose in the sample population i, $nw(i,j)$ is the maximum number of women in the considered unit j in the sample i and $nv(i,j,k)$ is the maximum total number of mammograms for the woman k in the unit j in the sample population i. For simplicity, the additional views, those exposures which are made as a complementary study in the diagnosis, were not included in the dose calculations.

The UNSCEAR 2000 report contains the most recent observed and expected number of cases for breast cancer incidence and mortality. In our study, it has been considered an age at exposure model; therefore the site-specific solid cancer risks estimated as the risk of exposure-induced death ($REID$) is calculated by the expression

$$REID^{VBCSP}(e) = \int_{e+L}^{e+L+P} h_0^{VBCSP}(a')\, ERR^{UNS2000}(e)\, da' \approx ERR^{UNS2000}(e) \sum_{j=e+L}^{e+L+P} \left(\prod_{k=e}^{j} \left(1 - h_{0\,all}^{VC}(k)\right) \right) h_0^{VC}(j) \qquad (3)$$

where e is the age-at-exposure and h_{0all}^{VC} is the mortality rate of the Valencian Community women population, due to all causes, including other breast cancer fatalities. The $ERR^{UNS2000}$ for incidence (total) breast cancer are obtained through the morbidity data presented in the ICRP (1991) [5]. Although the $ERR^{UNS2000}$ has been adjusted to observed data, the $REID^{VBCSP}$ has been also possible to calculate by means of the ASQRAD software [6].

RESULTS AND DISCUSSION

Table 1 shows the average values of age, voltage, tube load, compressed breast thickness and the average mean glandular dose per mammogram in each sample population. If the dose results are closely analyzed, they show a large variance in the average glandular dose between the several units of the Valencian Breast Cancer Screening Program (*VBCSP*), but this deviation is reduced from a sample to the next one.

TABLE 1. Average values in each sample population

Sample population	Age	kV	mAs	s_b (cm)	AMGD (mSv)
1st sample	55.934 ± 1.847	26.732 ± 1.646	107.111 ± 22.686	5.120 ± 0.788	1.686 ± 0.671
2nd sample	55.943 ± 1.995	27.467 ± 1.862	96.401 ± 37.627	5.197 ± 0.638	1.528 ± 0.526
3rd sample	55.438 ± 1.439	27.568 ± 1.754	91.197 ± 31.930	5.569 ± 0.635	1.216 ± 0.439

The total excess absolute risk, $\Omega^{VBCSP}(i)$, quantifies the associated risk in the target population of the Valencian Community, assuming that the average mean dose in the sample population *i* has been delivered to all women. The total detriment has been estimated as

$$\Omega^{VBCSP}(i) = \frac{AMGD(i)}{\sum_{j=45}^{69} nw_j^{VC}} \sum_{j=45}^{69} REID_j^{VBCSP} v_j w_j^{VC} \quad (4)$$

where *AMGD (i)* is the contribution to dose for all units in the sample population *i*, $REID^{VBCSP}(j)$ is the excess absolute risk for fatal breast cancer per *mSv* at the exposed *jth*-age group, v_j is the number of views per woman at age *j* and w_j^{VC} is the number of women at the age *j* in the entire Valencian Community. Furthermore, it has been calculated the radiological detriment with the hypothesis of 50 years old as the initial age of the screening program. In the calculation, it has been considered that each woman receives only one view per breast (two views per woman). The results for each projection risk model (UNSCEAR 94 and UNSCEAR 2000) are shown in Table 2.

TABLE 2. Average REID results per 100000 women-year

Sample population	Risk estimator	UNSCEAR 94		UNSCEAR 2000	
		Initial age (45)	Initial age (50)	Initial age (45)	Initial age (50)
1st sample	Total cancer	1.190	0.814	2.698	2.127
	Fatal cancer	0.595	0.407	1.349	1.063
2nd sample	Total cancer	1.078	0.737	2.445	1.927
	Fatal cancer	0.539	0.369	1.223	0.964
3rd sample	Total cancer	0.858	0.587	1.946	1.534
	Fatal cancer	0.429	0.293	0.973	0.767

Figures 1 and 2 show the evolution of the radiological detriment at each age-at-exposure, under the UNSCEAR 94 and UNSCEAR 2000 models, as the number of fatal breast cancer per 100000 women in each age group. Thus, these results should be considered as qualitative indicators, subjected to all kind of uncertainties.

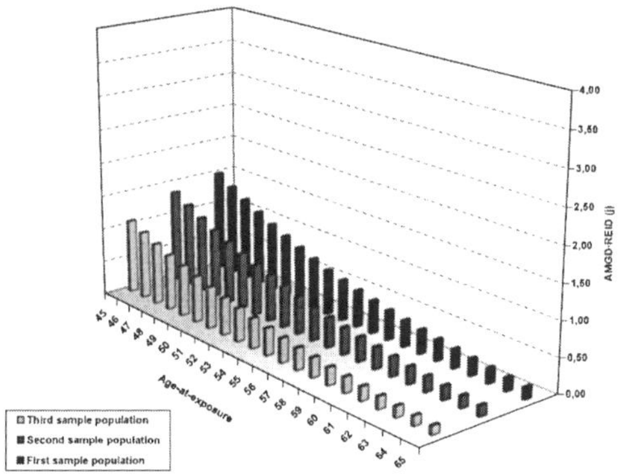

FIGURE 1. Fatal Breast Cancer per 100000 women under the UNSCEAR 94 risk model.

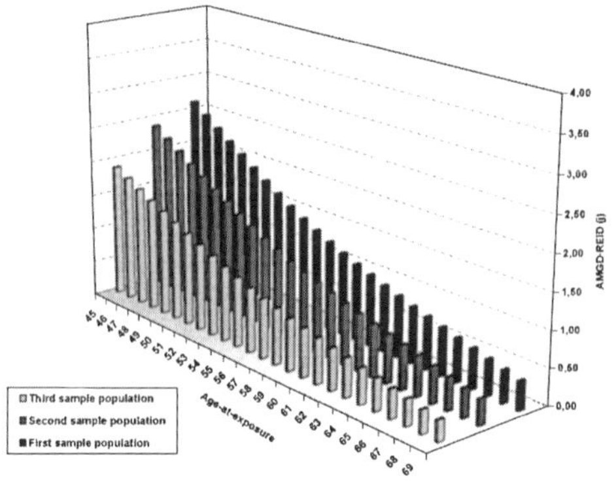

FIGURE 2. Fatal Breast Cancer per 100000 women under the UNSCEAR 2000 risk model.

RESULTS AND DISCUSSION

The large variation in the average glandular breast dose per screening location demonstrates the need of continuous quality control in a breast cancer screening program. As the number of units employed increases, the difficulty of equalizing the procedures becomes unattainable without a centralized control, thus the implementation of a quality control system is very important for any breast cancer screening program.

The excess absolute risk measured as number of fatal cancers number induced by the mammography is lower than 0.6 per 100000 exposed women for the model UNSCEAR 94 and lower than 1.5 per 100000 women for the more conservative model, UNSCEAR 2000. Also the radiological risk could be reduced in 31% for UNSCEAR 94 model and 21% in UNSCEAR 2000 model if the age at which the program starts was 50 years old instead of 45.

These results show that it is possible to reduce the radiological detriment by making reviewing and equalizing procedures. In this line, studies are being carried out to analyze and to optimize the ratio detected cancers per induced cancers, focused to increase the quality image trying a reduction in the average glandular breast dose.

ACKNOWLEDGMENTS

We are grateful with the Valencian Breast Cancer Screening Program and the European Breast Cancer Network (2001 Project 4.3). The work of M. Ramos is funded by the MCYT (Spain) under an FPI grant.

REFERENCES

1. Briesmeister, J.F. 'MCNP^{TM}- A General Monte Carlo N-Particle Transport Code – v.4C' LA-13709-M (Manual) – March 2000
2. Assessment System for the Quantification of Radiation Detriment (ASQRAD) CEPN NRPB ISBN 92-827-5085-X
3. 'Sources and Effects of Ionizing Radiation, vol II. Annex I: Epidemiological evaluation of radiation-induced cancer' United Nations Scientific Committee on the Effects of Atomic Radiation, UNSCEAR 2000 Report
4. European Commission 1996 'European Protocol on Dosimetry in Mammography' European Commission Report EUR 16263 EN (Luxembourg, 1996)
5. ICRP (1991). International Commission on Radiological Protection. 1990 Recommendations of the International Commission on Radiological Protection, ICRP Publication 60, Annals of the ICRP 21 (Pergamon Press, Elmsford, New York)
6. Schneider, T.; Hubert, D.; Degrange, J.; Bertin, M. 'Use of risk projection models for the comparison of mortality from radiation-induced breast cancer in various populations' Health Physics 11/94 452-459

Applied Physics (APHYS 2003)
A. Méndez-Vilas (Ed.)

Single Conductor DC Magnetic Field Reduction

J. R. Riba Ruiz and O. Bertran Cánovas

Department of Physics, EUETI d'Igualada, UPC, Plaça del Rei 15, 08700 Igualada, SPAIN.
E-mail: jordi@euetii.upc.es

Abstract. In this paper, different geometry of shields for the magnetic field created by a single conductor through which a dc current is flowing, are analyzed. The shielding method proposed consists in introducing a ferromagnetic material (a material that is only a conductor is not valid) that works as a shield towards the magnetic field. This shield must possess such a geometry (generally no symmetrical) that causes an asymmetry in the distribution of the magnetic field generated by the single conductor in a way that in the area to be shielded the magnetic field diminishes. It is due to the fact that the magnetic field intensity is increased in another space area. It is shown that the reducing factor of the magnetic field decreases with the distance between the point of measure and the surface of the conductor.

INTRODUCTION

This paper studies how to shield the magnetic field generated by a single, very long conductor through which an already known direct current intensity is flowing. We attempt both to develop a method to determine the most efficient way to shield the magnetic field generated by the conductor under study and also to systematize the process of calculating the most adequate magnetic shield.

In general, in the case of industrial frequency (50 or 60 Hz) the commonly used method to shield the magnetic field consists in causing eddy currents. Therefore, the principle of magnetic field shielding used in the case of alternate current of industrial frequency is not applicable to this problem due to the fact that the magnetic field created by the conductor is absolutely static and its temporal derivative is nil. In this case the shielding method proposed consists in introducing a ferromagnetic material (a means that is single a conductor is not valid) that works as a shield towards the magnetic field. This shield must possess such a geometry (generally no symmetrical) that an asymmetry is caused in the distribution of the magnetic field generated by the single conductor in a way that in the area to be shielded the magnetic field diminishes usually due to the fact that the magnetic field intensity is increased in another space area.

MAGNETIC FIELD SHIELDING CREATED BY A VERY LONG RECTILINEAR CONDUCTOR

We attempt to shield (reduce) the magnetic field created by a very long rectilinear conductor (through which a known direct current intensity is flowing) in a determined area of the space. In this case a void cylinder of conductor or ferromagnetic material does not have the function of magnetic field shielding since the eddy currents are not induced in it. This is due to the fact that in the void cylinder there is no variation of the magnetic flow and therefore the electric currents can't be induced. In addition, according to the Ampere law, $\int_c B.dl = \mu_0 I$, and due to the fact that through a circular radio line external to the shielding (see figure 1) the total current is not modified by having introduced the void cylinder, the magnetic field outside the cylinder is the same as if it was not there.

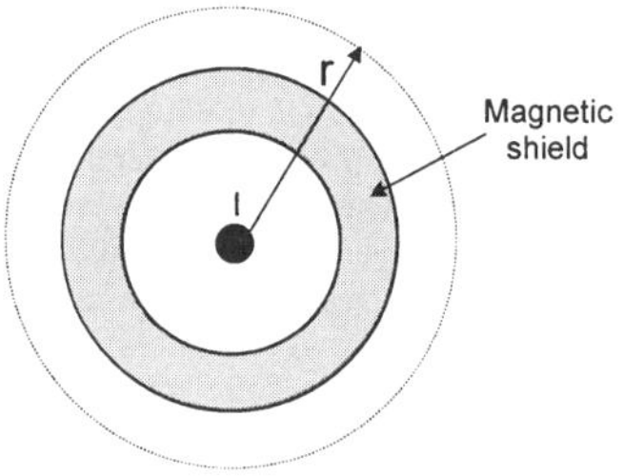

FIGURE 1. Conductor and cylindrical shields

Thus we have to resort to other strategies. We may modify the distribution of the magnetic field created by the conductor by introducing a ferromagnetic shielding with an asymmetric geometry. Hence, the values of the magnetic field in a determined area of the space (where we want to shield it) will decrease to subsequently increase in another region.

Figure 2 shows several types of useful geometry to pursue this end.

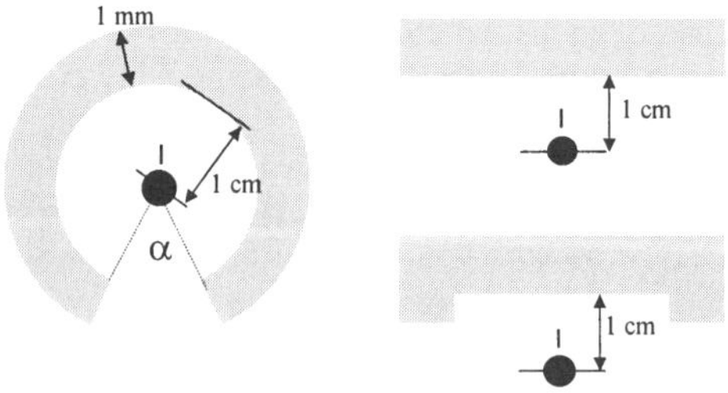

FIGURE 2. Several magnetic field shieding geometries analysed

CALCULATIONS

All numerical results obtained in this paper come from simulations performed with the program FEMM version 3.1. All simulations have assumed a copper conductor of 70 mm^2 section (radio = 4.7 mm) through which a direct current intensity of 100A is flowing. We have always made the hypothesis that the relative permeability of the ferromagnetic shielding is $\mu_r = 5000$ and that its electrical conductivity is 10^7 $(\Omega.m)^{-1}$.

We define the magnetic field reducing factor as:

$$RF_{\%} = 100 \cdot \frac{B_o - B}{B_o} \tag{1}$$

where B_0 is the magnetic field in a determined point in space when there is no ferromagnetic shield, whereas B is the magnetic field in the same point when the shield is present. In the event of having magnetic field reduction due to the effect of the shield, the values of the reducing factor would be positive whereas if there were an increase negative values would result.

Figure 3 shows the axles of reference taken in the study.

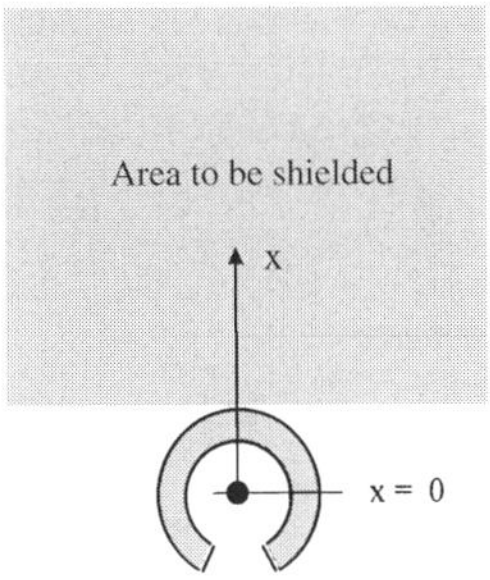

FIGURE 3. Axles of reference for the calculations and area to be shielded

Figure 4 shows the magnetic field averaged in a squared area of 30 cm x 30 cm situated as shows figure 3. The calculation has been for total angle openings ranging from 5° to 300°. The curve representing the magnetic field in accordance with the opening angle has not an outstanding minimum but a quite plain minimum zone delimited by opening angles ranging from 5° to 60°. The calculations made from now on for this type of shielding will assume total opening angles of 35° (it is found in this area of minimum field).

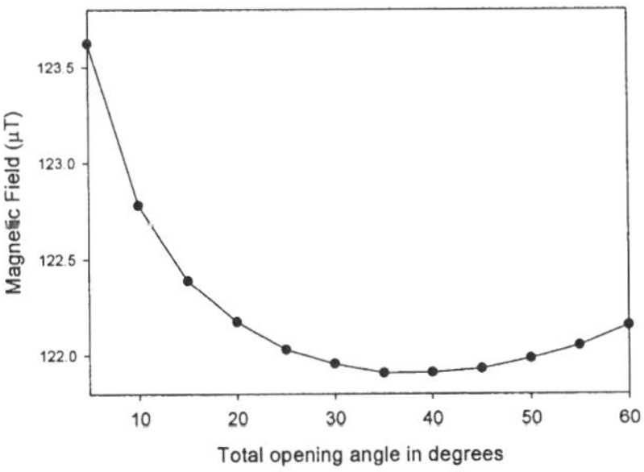

FIGURE 4. Average magnetic field with cylindrical shielding according to opening angle

In the three types of magnetic shields proposed we have supposed the same quantity of ferromagnetic material, namely a 1 mm-thick sheet and 6 cm-long sheet (corresponding to a 1 cm-radioring with an opening angle of 35°) with different geometries.

Figure 5 shows a comparative of the shielding effect of the different geometries supposing in all cases that the shields keep a distance of 1 cm from the center of the conductor.

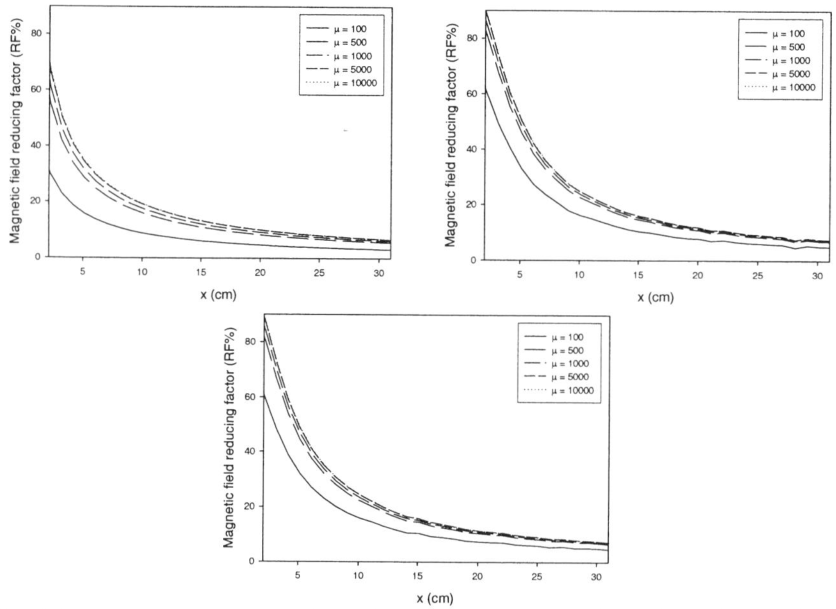

FIGURE 5. Comparative of the shielding effect of the several geometries analysed. Cylindrical shield with $\alpha = 35^{o}$ (up left), flat sheet (up right) and flat sheet with extensions (down).

From figure 5 we deduce that with the same iron, the geometry providing the highest magnetic field reducing factor is the flat sheet. In all cases we observe that magnetic field reducing factor decreases rapidly with the distance to the shielding.

Figure 6 shows a comparative study of the behavior of the flat shield when dealing with different thickness of ferromagnetic shield. From figure 6 we deduce that if we aim to raise significantly the reducing factor of the magnetic field we must raise significantly the thickness of the ferromagnetic shield.

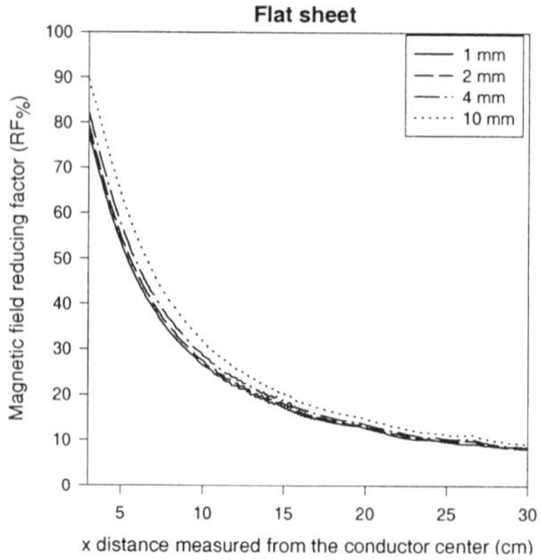

FIGURE 6. $RF_{\%}$ according to x distance for different sheet thicknesses

CONCLUSIONS

From the results obtained in this paper we can summarize the results as follows.

- For a 1mm-thick, the geometry offering a better shielding in the area studied (x distances below 20 cm.) is the flat sheet, followed by the sheet with arms and finally the cylindrical layer with an opening angle of 20°.
- After analyzing several thicknesses for flat sheet, we have proved that when the distances are inferior than the dimensions of the sheet, the magnetic field reducing factor is few dependent on the shield thickness.
- This study has been carried out assuming a ferromagnetic material with a relative magnetic permeability of $\mu_r = 5000$ and an electric conductivity of $\sigma = 10^7\ (\Omega.m)^{-1}$. The effect of the magnetic shielding increases as the relative permeability value μ_r also does so. As we work with direct current the area of the hysteresis cycle of the material used has no influence at all.
- With the same iron thickness, the dimensions of the flat sheet are decisive to determine the magnetic field reducing factor in a point that is relatively distanced from the shield. In order to obtain high reducing factors we must make sure that the distance between the point to be shielded and the screen is lower than the shield width. Therefore, the dimensions of the sheet (width) are the most determining factor in order to obtain a good degree of shielding.

REFERENCES

1. Yaping Du, T.C. Cheng. "Principles of Power-Frequency Magnetic Field Shielding with Flat Sheets in a Source of Long Conductors" in IEEE Transactions on Electromagnetic Compatibility, **38 (3)**, 450-459 (1996).
2. K. Wassef, V. V. Varadan, V. K. Varadan. "Magnetic Field Shielding Concepts for Power Transmission Lines" in IEEE Transactions on Magnetics, **34 (3)**, 649-654 (1998).
3. C. F. Yang, J. S. Yang, J. T. Hwang, T. H. Lee. "Measurements and Simulations of Cable Shielding against Magnetic Field Induction" in IEEE Transactions on Power Delivery, **14 (3)**, 1102-1109 (1999).
4. ICNIRP Guidelines. "Guidelines for limiting exposure to time-varying electric, magnetic and electromagnetic fields (up to 300 GHz)" in Health Physics **74 (4)**, 494-522 (1998).
5. Handbook of Chemistry and Physics, E-128. 70th Edition, edited by CRC Press, 1989-1990.
6. David Meeker. "Finite Element Method Magnetics (FEMM 3.1). User's Manual". http://members.aol.com/dcm3c. October 25, 2001.
7. Amalia Ivanyi. "Magnetic Shielding for Phase Conductors" in IEEE Transactions on Magnetics, **32 (3)**, 1481-1484 (1996).
8. R. B. Schulz, V. C. Plantz, D. R. Brush. "Shielding Theory and Practice" in IEEE Transactions on Electromagnetic Compatibility", **30**, 187-201 (1988).
9. J.R. Moser. "Low-frequency low-impedance electromagnetic shielding" in IEEE Transactions on Electromagnetic Compatibility, **30**, 202-210 (1988).
10. Jordi-Roger Riba Ruiz, Xavier Alabern Morera. "Simulació del camp magnètic generat per les línies aèries d'alta tensió" in Revista de Física, **3(1)**, 38-42 edited by Societat Catalana de Física, Institut d'Estudis Catalans, (2001).
11. Jordi-Roger Riba Ruiz, Xavier Alabern Morera. "Simulació del camp elèctric generat per les línies aèries d'alta tensió. Càlcul i simulació" in Revista de Física, **3(2)**, 47-52 edited by Societat Catalana de Física, Institut d'Estudis Catalans (2002).

Applied Physics (APHYS 2003)
A. Méndez-Vilas (Ed.)

Evaluation of Healthy Bone by a Method Based on Image Analysis

A. Baltasar Sánchez and A. González-Sistal*

*Department of Physiological Sciences II, Faculty of Medicine, University of Barcelona, Feixa Llarga s/n, 08907 Hospitalet de Llobregat, Barcelona, SPAIN.

Abstract. The heterogeneity of healthy bone observed on simple X-rays is a characteristic that often makes difficult the medical diagnosis of bone diseases. The purpose of this study is to characterize the healthy bone by the development of a new automatic method. This method is based on techniques of image processing and analysis, applied on digitalized radiographs. This study is limited to the anatomical and histological bone classifications. This work shows the usefulness of the method to contribute in the differential diagnosis between healthy bone and pathological bone in the future.

INTRODUCTION

On simple X-rays, we can observe different grey levels which are related with different materials: black with air, grey with muscle or fat, and white with bone or metal. The grey levels corresponding to healthy bone (HB) are heterogeneous and they depend on the following parameters: bone mineral density, microarchitecture, size and geometry [1]. Most of these parameters are related to the balance of bone resorption and formation. The osteoblast is the main cell that contributes to bone formation, which produces and secretes the structural components of the bone matrix [2]. The osteoclast is the cell specialized in resorbing the mineralized bone matrix [3]. Bone remodelling is regulated from genetic factors, environmental influences (such as mechanical loading of bone), microfractures, medication and illnesses [4]. Ageing might affect bone expression of skeletal endocrine factors [5].

Bones have a common histological structure: an external cortical bone tissue which serves as a protective covering and surrounds the internal trabecular bone tissue (except on places where it is replaced by bone marrow or an air space, like long bones diaphysis or maxilar senus face respectively), which plays a very important role in bone metabolism duties. The ratio of cortical and trabecular bone combination varies through the skeleton. Some morphologic characteristics constitute the anatomical classification in: long, short and flat bones (see Table 1).

Image techniques are in continuous development. The basic principles of interpretation and knowledge of anatomy and pathology scarcely change. Therefore, we can use different current image techniques or future applications to enhance the differential diagnosis between HB and pathological bone. The simple X-rays of the skeleton are the most requested image technique, mainly due to its low cost. For this reason, in this work we present an image procedure in order to enhance the identification and diagnosis of HB by simple digitalized radiographs (DR).

We performed a methodology based on the basic steps of digital image processing and analysis: acquisition / storage, enhancement / processing, and measurements [6] (see Fig.1). The subsequent application of this methodology allows to obtain results to validate and classify the bone groups. We studied 60 digital monochrome images corresponding to DR which included HB from different patients of both sexes. The images were 512 x 512 array, grey level 8 (it provides 256 levels of grey: 0-255, where 0 is equivalent to black and 255 to white). The diversity of cases employed allows that the grey levels being valuable inside these limits.

We analyzed the bone groups previously knowing the morphologic differential features that constitute the anatomical bone classification in: long (LB), short (SB) and flat bones (FB). Table 1 shows the characteristics and the skeletal distribution of LB, SB and FB. In the LB group we studied the cortical (CO) and trabecular (TR) components because they are clearly delimited on simple X-rays.

The aim of the enhancement techniques was to process the original images (HB_{OR}) in order to minimize the noise and the scattering of data [7]. The noise was improved using a filter. In earlier tests (with a gaussian, lowpass and median filters) it was shown an appropriate filtering [8]. The filtering applied was a smoothing filter lowpass size 3x3 with weights 1, 1, 1. This filter was applied a mean of 20 times to every image.

The HB_{OR} and filtered images (HB_{FT}) were analyzed measuring its brightness. We calculated the mean value of grey level (MGL: 0-255) of an area of interest (AOI) of 7000 pixels. We made the histogram (X: grey level; Y: pixels) in every AOI to obtain the following parameters: average, standard deviation, maximum, minimum and sum. This procedure provided the MGL. This parameter is equivalent to information about anatomical, physiological or malignant processes.

When we had defined the methodology, we created a macro with Visual Basic language in order to automatize the process of image processing and analysis. Therefore, we created an alternative automatic method.

TABLE 1. Anatomical Classification of Healthy Bone.

Long Bones	Short Bones	Flat Bones
Characteristics		
Tubulars: Length > Width Diaphysis (occupied by bone marrow or empty) Epiphysis or bone extremes (occupied by trabecular bone)	Shape of cube, trapeze,... Thin cortical supported by a trabecular internal bone	Flats A internal layer of trabecular bone (diploe) closed by the external and internal cortical sheets.
Skeletal Distribution		
Extremities: humerus, tibia, femur, .. Metacarpals, metatarsals and phalanges	Carpus and tarsus	Skull, sternum, ribs, ...

OTHER SPECIFICATIONS

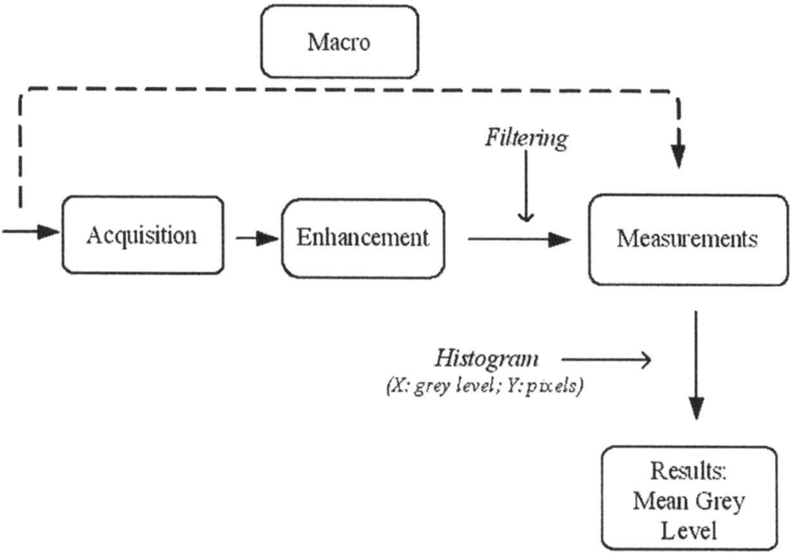

FIGURE 1. Scheme of the process developed to characterize the Healthy Bone.

TABLE 2. Mean Value of Grey Level (MGL) and Standard Deviation (SD) of Healthy Bone.

Groups	n	MGL	SD
HB_{OR}	60	207.43	±28.05
HB_{FT}	60	197.26	±6.84
LB	40	199.97	±7.01
SB	10	178.90	±6.36
FB	10	204.80	±6.61
CO	25	215.32	±6.15
TR	15	174.40	±8.45

Table 2 lists the results obtained for all DR for MGL and standard deviation (SD), as well as information on number of cases studied (n). The MGL of HB_{OR} differs from HB_{FT} (207.43 ± 28.05 vs 197.26 ± 6.84). The MGL of every anatomical group is: LB 199.97 ± 7.01, SB 178.9 ± 6.36 and FB 204.80 ± 6.61. The CO component of LB group differs from TR (215.32 ± 6.15 vs 174.40 ± 8.45). Figures 1-2 show the MGL of every group.

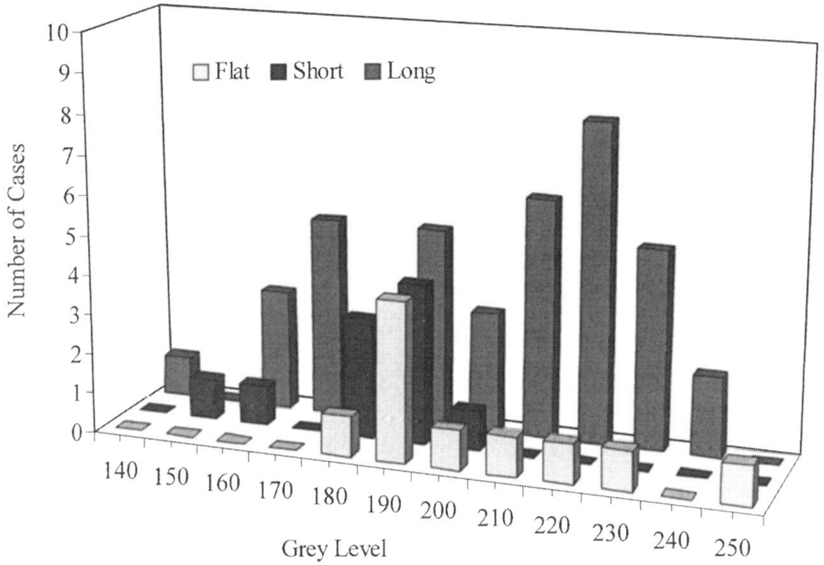

FIGURE 2. Representative histogram which shows the distribution of grey levels between: long, short and flat bones (see Table 1).

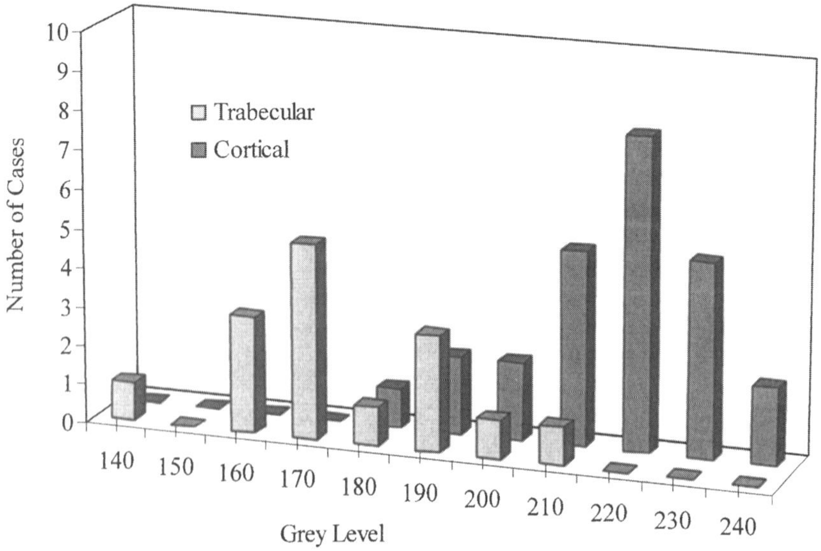

FIGURE 3. Representative histogram which shows the distribution of grey levels between long bones for the cortical and trabecular components (see Table 1).

This study is the presentation of a complementary method to analyze the heterogeneity of HB. There is a clear differentiation between the groups studied into the anatomical and histological classification. The MGL in the group FB is bigger than LB and SB. The same is applicable for the CO component in respect of TR in the LB group.

This study proves that the HB is heterogeneous. The simplicity of the method reduces the time of computation, which allows to obtain an automatic diagnosis. The methodology posed allows to enunciate future applications: to raise a helpful method in case that the simple radiological exploration is insufficient and to compare the HB with pathological bone.

ACKNOWLEDGMENTS

This work was partly financed by the MCyT, DGESIC Project (BFI 2001-3331) of the Spanish Government.

REFERENCES

1. J. A. Kanis, *Lancet.* **359**, 1929-1936 (2002)
2. E. J. Mackie, *IJBCB.* **35**, 1301-1305 (2003)
3. T. Komori, *Cell.* **89**, 755-764 (1997)
4. T. L. Noman and Z. Wang, *Bone.* **20**, 375-379 (1997)
5. D. Goltzman, A. C. Karaplis, R. Kremer and S.A. Rabbani, *Cancer.* ***88***, 2903-2908 (2000)
6. J.C. Russ, *The Image Processing Handbook*, edited by CRC Press, Boca Raton, Florida, fourth edition, 2002
7. R.C. González and R.E. Woods, *Digital Image Processing*, edited by Addison-Wesley, 1993
8. A.González-Sistal, *Study of the Renal Function from Dynamic Sequences of Gammagraphic Images,* Ph.D Thesis, edited by Publicacions Universitat de Barcelona, 1991

Applied Physics (APHYS 2003)
A. Méndez-Vilas (Ed.)

A Method to Improve the Characterization of Bone Tumors from Digitalized Radiographs

A. Baltasar Sánchez and A. González-Sistal*

**Department of Physiological Sciences II, Faculty of Medicine, University of Barcelona, Feixa Llarga s/n, 08907 Hospitalet de Llobregat, Barcelona, SPAIN.*

Abstract. Firstly, the diagnosis of bone tumors depends on simple X-ray where osteoblastic or osteolytic changes are observed in respect of the surrounding healthy bone. This study sought to characterize bone tumors by applying an automatic method based on image processing and analysis. The method distinguish between healthy bone and bone tumors. The results will allow to raise a useful method in the future in case that the radiological exploration is insufficient.

INTRODUCTION

The primary bone tumors are classified depending on the cell or normal bone tissue from which are originated (see Table 3 and 4). In the general bone classification all cellular types that constitute the bone are included. The incidence of the different benign primary bone tumors are unknown, because for most of the lesions a biopsy is not performed. However, these are hundred times higher than the malignant primary bone tumors. In the USA, there is a diagnosis of 2,100 bone sarcomas (malignant primary bone tumors) per year with an annual mortality of 1,300 patients [1].

The secondary metastasic bone tumors are originated in neoplasic cells or tissues out of bone. The incidence in population (100,000 inhab/year) are: a) adenocarcinoma of breast (108.8), lung (37.8), and kidney (5.6) in women and, b) prostate (92.2), lung (82.5), and kidney (11.6) in men [2].

The biological potential of bone tumors varies from those totally benign to those lethal. This diversity forces to make an accurate diagnosis. Firtsly, the study of the morphology of bone tumors depends on simple X-ray, completed with the Computerized Axial Tomography (TAC) and the Nuclear Magnetic Ressonance (RMN). An accurate radiological diagnosis requires a relation between the radiological data (size, location and morphology of the tumor), history, symptoms, age and sex of the patient [3]. On simple X-ray bone tumors are visualized as osteoblastic (bone formation, as an area of radiolucency) or osteolytic (associated with a geographic bone destruction, with well-defined lucent defects) changes in respect of the surrounding healthy bone (HB) [4].

The purpose of this study is to use an image procedure in order to enhance identification and diagnosis of bone tumors by simple digitalized radiographs (DR). The methodology was previously developed, based on image processing and analysis techniques, in order to characterize the heterogeneity of HB and in this way enhance the differential diagnosis between HB and bone illnesses [5]. We studied 200 DR corresponding to HB, primary bone tumors and secondary metastasic bone tumors (see Table 1). The images were 512 x 512 array, grey scale 8 (it provides 256 levels of grey: 0-255, where 0 is equivalent to black and 255 to white).

We filtered the original images with a smoothing filter lowpass size 3x3 with weights 1, 1, 1. This filter was applied a mean of 20 times to every image. The filtered images were analyzed measuring its brightness. We calculated the mean value of grey level (MGL: 0-255) of an area of interest (AOI) of 7000 pixels. We made the histogram (X: grey level; Y: pixels) in every AOI. We created a macro with Visual Basic language in order to automatize the process [5].

TABLE 1. Bone Groups Considered in this Work.

Groups	n	Types
Healthy Bone (HB)	60	
Benign Primary Bone Tumors (BPT)	50	Chondroma, Osteochondroma, Osteoid Osteoma, Non ossifying Fibroma and Giant Cell Tumor
Malignant Primary Bone Tumors (MPT)	50	Chondrosarcoma, Osteosarcoma, Fibrosarcoma, Multiple Myeloma, Ewing's Sarcoma
Osteolytic Metastasic Secondary Bone Tumors (LMT)	20	Lung, Breast and Kidney
Osteoblastic Metastasic Secondary Bone Tumors (BMT)	20	Lung, Breast and Prostate

TABLE 2. Skeletal Locations Considered in this Work.

	Types	n	Location
BPT	Chondroma	10	Tibia, Humerus
	Osteochondroma	10	Tibia, Fibula and Humerus
	Osteoid Osteoma	10	Femur
	Non ossifying Fibroma	10	Femur, Tibia
	Giant Cell Tumor	10	Radius, Humerus and Ulna
MPT	Chondrosarcoma	10	Femur
	Osteosarcoma	10	Femur
	Fibrosarcoma	10	Humerus, Radius and Tibia
	Multiple Myeloma	10	Skull, Femur
	Ewing's Sarcoma	10	Humerus, Pelvis

TABLE 3. Benign Primary Tumors.

Tumors	Features
Chondroma	Lobulated hyaline cartilage separated by nornal medullary bone
Osteochondroma	Smooth or lobulated hyaline cartilage Central portion: medullary bone and calcifications
Osteoid Osteoma	Mass (nidus) of immature bone (osteoid)
Non Ossifying Fibroma	Fibroblastic proliferation Cortical shows scalloped border
Giant Cell Tumor	Stromal mononuclear cells and multinucleated giant cells proliferation Cortical is immature bone (osteoid)

TABLE 4. Malignant Primary Tumors*.

Tumors	Features
Chondrosarcoma	Lobulated hyaline or myxoid cartilage Cortical destruction and thickening
Osteosarcoma	Immature bone (osteoid) Types: osteoblastic, fibroblastic or chondroblastic
Fibrosarcoma	Spindle cells proliferation Non-specific aspect, similar to soft tissue tumors
Multiple Myeloma	Small nodules of densely packed plasma cells
Ewing's Sarcoma	Round cells proliferation Cortical breakthrough

* Extension to the soft tissues and cortical destruction is common in all these tumors.

OTHER SPECIFICATIONS

Table 5 lists the results obtained for all DR for MGL and standard deviation (SD), as well as information on number of cases studied (n). The MGL of HB and bone tumors differ: HB (197.26 ± 6.84), BPT (132.58 ± 6.53), MPT (104.36±8.08), LMT (107.40±5.98) and BMT (208±6.16). Figures 1 shows the value of MGL of every group.

TABLE 5. Mean Value of Grey Level (MGL) and Standard Deviation (SD) of Groups Studied.

Groups	N	MGL	SD
HB	60	197.26	±6.84
BPT	50	132.58	±6.53
MPT	50	104.36	±8.08
LMT	20	107.40	±5.98
BMT	20	208	±6.16

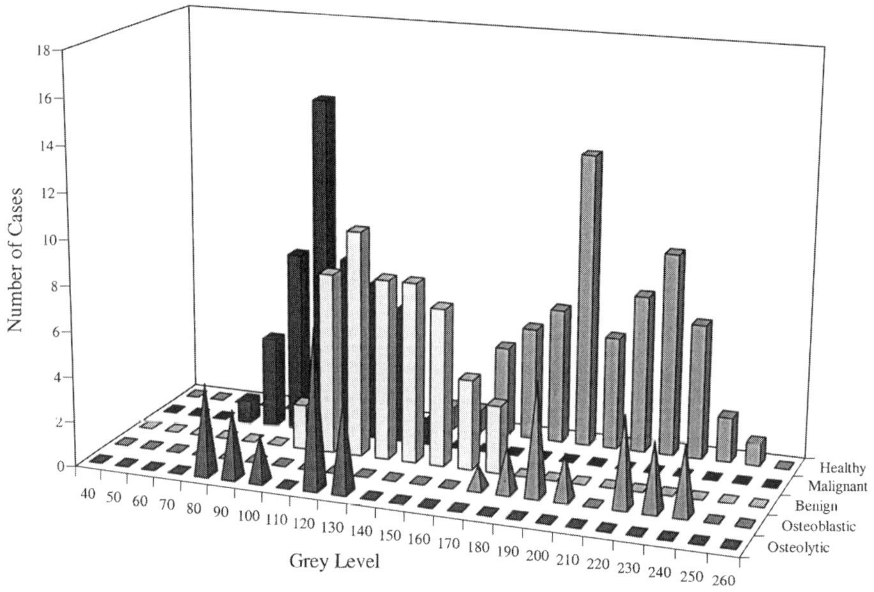

FIGURE 1. Representative histogram which shows the distribution of the grey level to: Healthy Bone, Benign Primary Bone Tumors, Malignant Primary Bone Tumors, Osteolytic and Osteoblastic Metastasic Secondary Bone Tumors (see Table 1).

We can observe a clear differentiation of MGL among the groups studied. We can distinguish the value of MGL for HB from bone tumors. Some values of MGL are the same in the groups BPT and MPT. There is no difference between MPT and LMT because the evolution of these tumors is the destruction of bone. We can distinguish between the osteolytic and osteoblastic metastases. The MGL of HB and BMT are very similar because this kind of metastases corresponds to new bone formation.

The methodology presented allows to enunciate future applications: 1) to raise a helpful method in case that simple radiological exploration is insufficient, 2) to study the evolution of some tumors under medical treatment, specially the metastasic bone tumors and, 3) to analyze the origin of different types of secondary metastasic bone tumors.

ACKNOWLEDGMENTS

This work was supported in part by the MCyT, DGESIC Project (BFI 2001-3331) from the Spanish Government.

REFERENCES

1. P. Bulloughs, *Orthopedic Pathology*, edited by Times Mirror International Publishers Limited, third edition, London, 1997
2. H.D. Dorfman and B. Czerniak, *Cancer.* **75**, 203-210 (1995)
3. A. Giudici, *Radiologic Clinics of North America.* **31**, 237-259 (1993)
4. M. Krikun, *Imaging of Bone Tumors*, edited by W.B. Saunders, Philadelphia, 1993
5. A. Baltasar Sánchez and A. González-Sistal. "Evaluation of healthy bone by a method based on image analysis", *Recent Advances in Interdisciplinary Applied Physics,* New York, American Institute of Physics, 2003

Applied Physics (APHYS 2003)
A. Méndez-Vilas (Ed.)

Neutronic Time-step Size And Direct Heating Influence On Power Peak Obtained By TRAC/BF1-NOKIN Coupled Code

Ana Mª Sánchez-Hernández, Gumersindo Verdú and Rafa Miró

Departamento de Ingeniería Química y Nuclear de la Universidad Politécnica de Valencia, Camino de Vera s/n, 46022 Valencia, SPAIN.

E-mail: ansanher@iqn.upv.es; gverdu@iqn.upv.es; rmiro@iqn.upv.es

Abstract. In this work we have made a sensitivity study analyzing separately the neutronic time-step and direct heating influence on power peak. The extreme scenario simulated is a turbine trip without scram, with bypass system relief failure and without safety and relief valves opening. NOKIN is a code developed at the Polytechnic University of Valencia. NOKIN code uses the best estimate code TRAC/BF1 to give account of the heat transfer and thermal-hydraulic processes. NOKIN code uses a one-step backward discretization of the neutron diffusion equation.
The objective of this work is to find the optimum time-step size to perform the analyzed case with an acceptable accuracy and an available CPU time. Also, we have calculated three cases with different direct moderator heat values.

INTRODUCTION

This paper has the purpose of analyzing the influence on the obtained results of some variables. We have analyzed separately the neutronic time-step and direct heating influence on power peak. The extreme case simulated is a turbine trip without scram, with bypass system relief failure and without safety and relief valves opening.

This work has been developed within the resolution of the benchmark "Boiling Water Reactor Turbine Trip Benchmark"[3].

TRAC-BF1/NOKIN COUPLING

NOKIN is a nodal neutronic code that uses the best estimate code TRAC/BF1 to give account of the heat transfer and thermal-hydraulic processes. NOKIN code uses a one-step backward discretization of the neutron diffusion equation.

TRAC/BF1 is a thermal-hydraulic code that uses an axial-radial heat transfer equation to model the heat transfer in the fuel, while the thermal-hydraulics processes are modelled solving six balance equations of mass, momentum and energy for liquid and vapor phases.

The nuclear cross sections associated to each neutronic node are obtained interpolating the values of multiple entries tables in terms of the thermal-hydraulic variables and the control rods insertion pattern. These nuclear cross sections are used to calculate the power distribution with the neutronic module. This power distribution is used as an input for TRAC/BF1 to obtain the thermal-hydraulic variables, which are utilized to get a second set of cross sections. After this, TRAC/BF1 uses the nodal power distribution provided by the neutronic module in the PREP and OUTER stages. Both sets of cross sections are used in the implicit integration of the nodal equations. The cross sections at intermediate time-steps are calculated by linear interpolation from both sets of cross sections. In this way, we obtain an explicit coupling between the neutronic module and TRAC/BF1.

NUMERICAL RESULTS

Time-Step Size

The transient has been calculated with different neutronic time-steps as shown in table 1:

TABLE 1. Time-Step Size.

Time-step	Time (s)
Δt_1	0.005
Δt_2	0.010
Δt_3	0.015
Δt_4	0.020

Figure 1 shows the total power distribution during the transient, calculated for the different time-steps.

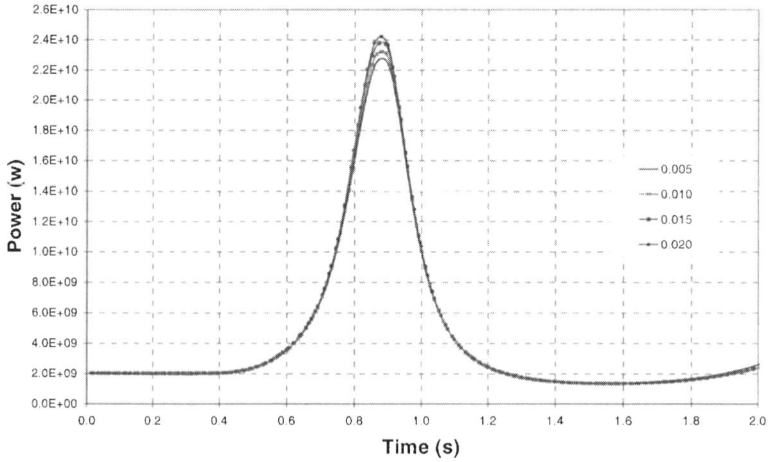

FIGURE 1. Total power distribution.

We can observe in figure 1 that the power evolution is practically the same in all the transients, but on the power peak changes with the different time-steps, the higher time-step the higher power peak.

Table 2 contains the power peak values and the CPU times for all the simulated cases. We can observe that using the lowest time-step we can obtain a power peak 6.4% lower than using the highest time-step, but the CPU time is a 47.8% bigger.

TABLE 2. Power Peak and CPU Time

Time-step	Power Peak (w)	CPU Time (s)
0.005	2.2749E+10	12179.78
0.010	2.3223E+10	9196.36
0.015	2.3771E+10	7745.45
0.020	2.4218E+10	6339.08

Figure 2 shows that the power peak changes practically linearly with the neutronic step size between Δt_1 y Δt_4.

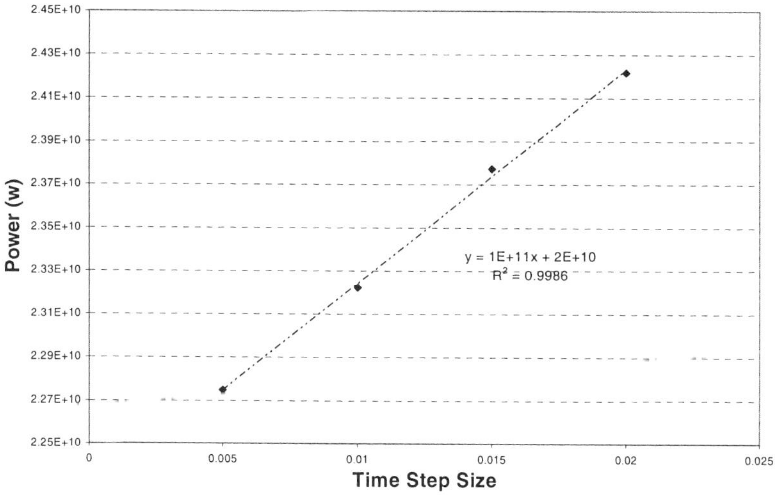

FIGURE 2. Power Peak.

Figure 3 presents the CPU time used for calculating the scenario. We can represent the CPU time dependence of the time-step using an exponential function between the interval of time-steps used:

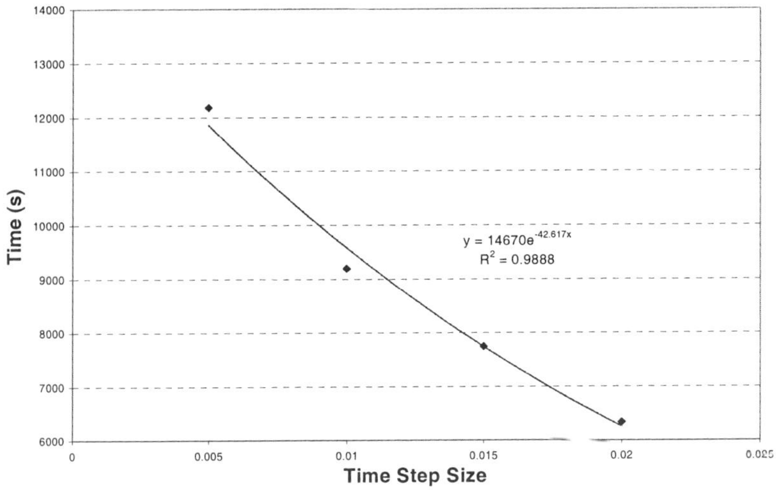

FIGURE 3. CPU Time.

Direct Moderator Heat

Direct moderator heat is the part of the total reactor thermal heat released to the moderator. In this section we have analyzed the influence of this heat on power peak. We have calculated three cases, a case without considering the moderator heat, a case with a 2% of the thermal heat to the moderator and a third case with a 4%.

The results obtained with the different cases are very similar as we can see in figure 4, where is represented the power distribution. Table 3 shows the power peak values for the different options.

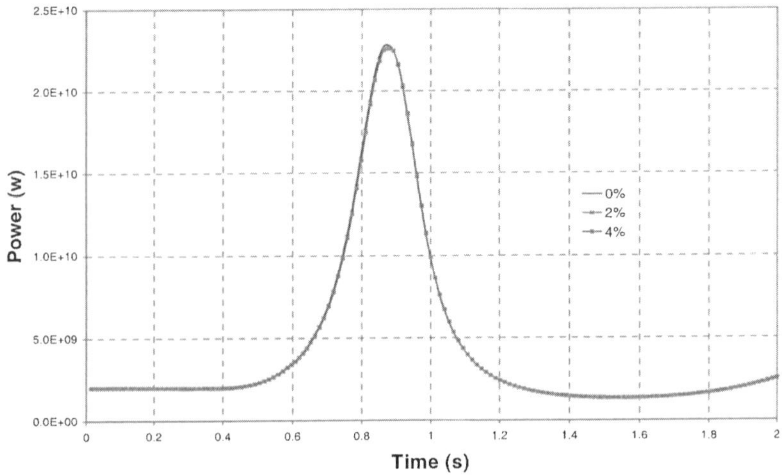

FIGURE 4. Total power distribution.

TABLE 3. Power Peak

Moderator Heat (%)	Power Peak (w)
0%	2.2772E+10
2%	2.2667E+10
4%	2.2545E+10

The greatest divergence is on the peak, but we can observe that results are very similar in the three cases. The difference of the results obtained for the case without direct moderator heat and with a 4% of the heat to the moderator is lower than 1.2% of the power.

CONCLUSIONS

We can obtain an acceptable accuracy choosing an appropriated neutronic time-step size using a reasonable CPU time.

Moderator heat has a small influence on power.

The highest divergences are obtained on power peak, but in the rest of the transient the results are practically the same.

ACKNOWLEDGMENTS

The authors are indebted to the 'Programa de incentivo de investigación de la UPV' (PPI-00-02).

REFERENCES

1. G. Verdú, R. Miró, D. Ginestar, V. Vidal, "The Implicit Restarted Arnoldi Method, an Efficient Alternative to Solve the Neutron Diffusion Equation". Ann. Nucl. Energy, 26, 7, pp.579-593, (1999).
2. J.A. Borkowski, N.L. Wade, "TRAC-BF1/MOD1: An Advanced Best-Estimate Computer program for Boiling Water Reactor Accident Analysis. Model Description". NUREG/CR-4356, EGG-2626, Vol 1, (1992).
3. J. Solís, K.N. Ivanov, B. Sarikaya, A.M. olson, K.W. Hunt, "Boiling Water Reactor Turbine Trip (TT) Benchmark. Volume I: Final Specifications". NEA/NSC/DOC(2001)1.
4. D. Ginestar, G. Verdú, V. Vidal, R. Bru, J. Marín, J.L. Muñoz-Cobo, "High Order Backward Discretization of the Neutron Diffusion Equation", Ann. Nucl. Energy, 25, 1-3, pp. 47-64, (1998).
5. R. Miró, D. Ginestar, G. Verdú, D. Hennig, "A Nodal Modal Method for the Neutron Diffusion Equation. Application to BWR Instabilities Analysis". Ann. Nucl. Energy, 29, pp.1171-1194 (2002).

Applied Physics (APHYS 2003)
A. Méndez-Vilas (Ed.)

Uranium-Isotopes Determinations In Waters From Almonte-Marismas Aquifer (Southern Spain)

J.González-Labajo[1], J.P. Bolívar[1] and R. García-Tenorio[2]

1 Department of Applied Physics, University of Huelva, campus El Carmen, 21005-Huelva, Spain. E-mail: labajo@uhu.es
2 Department of Applied Physics II, University of Sevilla, ETSA, Avenida Reina Mercedes 2, 41012-Sevilla-Spain.

Abstract. In this work, we have analysed the geographical distribution of the U-isotopes concentrations in the waters of the Almonte-Marismas aquifer, located at the south of Spain. During the year 2000, a systematic sampling was carried out, being all the samples analysed for U isotopes by alpha-particle spectrometry, both in the dissolution and suspended phases. From the spatial distribution obtained in this detritic aquifer for the dissolved Uranium and for the U equilibrium distribution coefficients, it has been possible to delimit its main oxidised and reduced zones. It has been then demonstrated, how the determination of Uranium isotopes concentrations over a detritic aquifer is a powerful and elegant tool to understand some basic processes that govern the geochemical behaviour of these natural systems.

INTRODUCTION

A great amount of studies have been historically conducted in different aquifer systems distributed over the world in order , a) to analyse the levels and dispersion of trace and contaminant elements in them, and b) to perform its geochemical characterisation.

In particular, in some of these aquifers the spatial distribution of the concentrations of several natural radionuclides have been determined in different water phases (dissolved and suspended matter phases). These radioactive determinations have allowed as a first approach the characterisation of the analysed aquifers from the radiological point of view, but, additionally, in some cases, it has been possible its geochemical and hydrological characterisation through the analysis of the gradients observed inside the aquifer for several natural radionuclides and through the analysis of the spatial distribution of several daughter/parent pair activity ratios (Ivanovich and Osmond, 1991). In this sense, it is known the utility of several radionuclides from the Uranium series, and their activity ratios, as isotopic signatures of different processes affecting the analysed groundwater ecosystem (Osmond and Cowart, 1982).

According to these features, in this paper we will show specifically how it is possible the identification of the different redox areas in an Spanish aquifer (i.e. the characterisation of its oxidised and reduced regions, as well as the fixation of the location of the redox barrier in its interior) through the construction of two detailed spatial distribution maps: one reflecting the U-levels in dissolution in the analysed aquifer, and the other showing their associated $K_{d(U)}$ values.

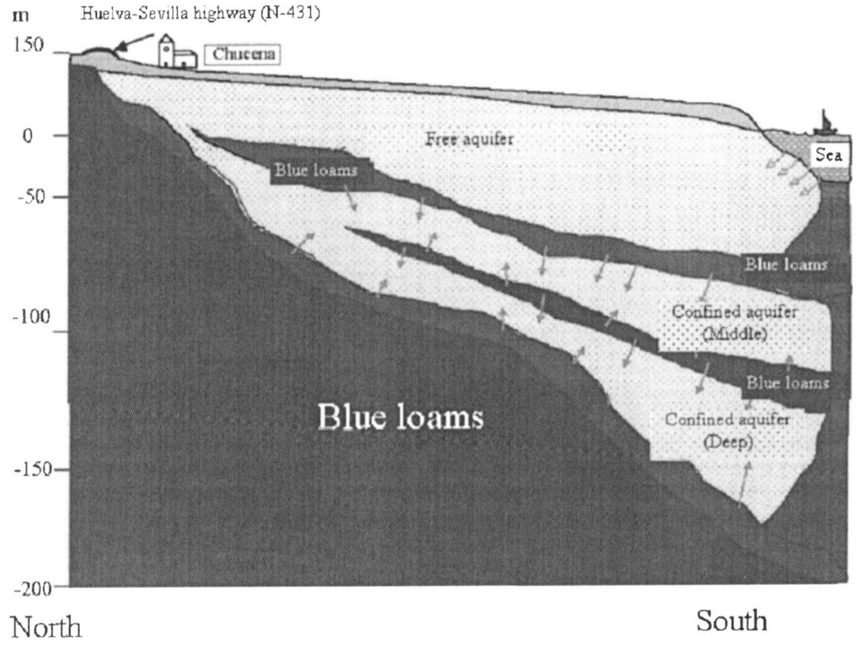

FIGURE 1: Simplified transverse profile of the Almonte-Marismas aquifer. The arrows indicate the salt contribution to the waters. (Romero. 1998)

The Spanish aquifer elected for this study, called Almonte-Marismas is located at the Southwest of the Iberian Peninsula (in the region of Andalusia) and can be considered as an idealised artesian groundwater system that is smoothly sloped from its recharge area placed in the north towards the south (see Figure 1).

The extension of this aquifer is 3400 km^2 and is formed mainly by three different sandy layers separated between them by aquitards formed by blue loams (Figure 1). The uppermost layer, called free aquifer, it has been the analysed one in this work and as the aquifer as a whole, shows a soft inclination from the recharge area, placed in the north, to the south making it suitable for studies of flow motion. The thickness of this layer is variable (1 to 30-40 meters), increasing also in the direction north to south.

SAMPLING AND METHODS

A total of 28 groundwater samples from the uppermost layer of the Almonte-Marismas aquifer were collected for the performance of this study. The sampling points were distributed in a quite homogeneous way over all the extension of the aquifer (see Figure 2) collecting at each place a total amount of 20 litres.

In each sampling location, in-situ measurements of different parameters like temperature, pH and conductivity were carried out in aliquots of the collected waters. These samples were immediately transported to our laboratory, and then filtered through a 0.45 μm Millipore filter in order to separate their dissolved particulate fractions. The filters were then dried at room temperature, and the amount of suspended material determined by weighing, while the dissolved phase was acidified to pH=1 with nitric acid for its preservation before the radiochemical analysis.

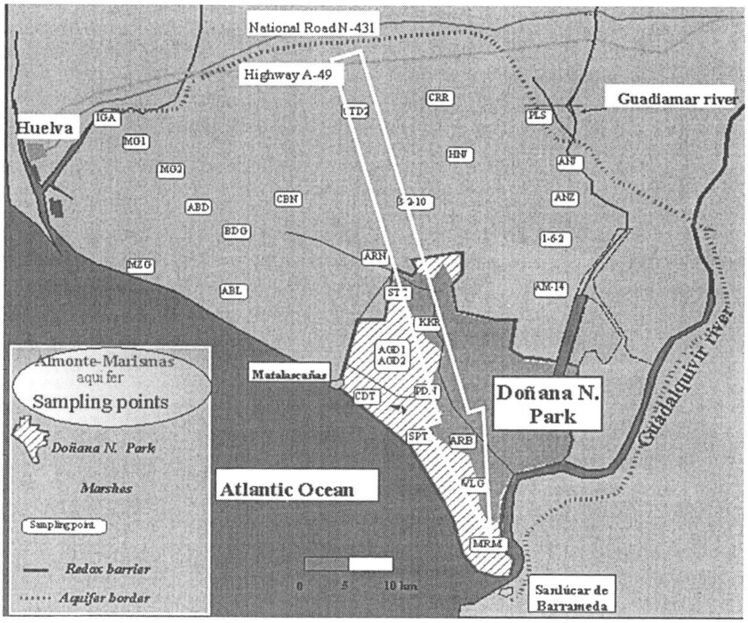

FIGURE 2: Location of sampling wells and the flowline studied in the Almonte-Marismas aquifer.

Uranium-isotopes determinations have been carried out in the particulate and dissolved fractions of the collected samples by applying the alpha-particle spectrometry technique with PIPS semiconductor detectors. Previously, each particulate and dissolved aliquot was radiochemically treated in order to isolate the Uranium isotopes from chemical and radioactive interferents, and to adapt the final Uranium source to the conditions required for the measurement. More details of these radiochemical steps can be found elsewhere (González-Labajo J. 2003).

RESULTS AND DISCUSSION

In Table 1, it is possible to find for the twenty eight analysed groundwater samples the following results: the ^{238}U concentration in dissolution (Bq/m3), the ^{238}U concentration in the particulate fraction (Bq/kg) and, finally, the associated Uranium equilibrium distribution coefficient (l/kg).

TABLE 1. ^{238}U **in dissolution and suspended matter, distribution coefficients of U, K_U in groundwater.**

Sample	Dissolution ^{238}U (Bq/m^3)	Suspended matter ^{238}U (Bq/kg)	$K_{dU} \cdot 10^3$ (l/kg)	Sample	Dissolution ^{238}U (Bq/m^3)	Suspended matter ^{238}U (Bq/kg)	$K_{dU} \cdot 10^3$ (l/kg)
Oxic zone				Anoxic zone			
IGA	25 ± 2	28 ± 12	1.1 ± 0.5	ABD	2.5 ± 0.3	31 ± 9	13 ± 4
1.6.2	16 ± 1	18 ± 5	1.1 ± 0.5	ABL	1.8 ± 0.3	22 ± 6	12 ± 5
3.2.10	15 ± 1	35 ± 18	2.4 ± 1.2	BDG	1.2 ± 0.3	15 ± 4	13 ± 5
ANJ	37 ± 4	14 ± 4	0.4 ± 0.1	CBN	2.3 ± 0.3	18 ± 4	7.8 ± 2.0
ANZ	11 ± 1	15 ± 4	1.4 ± 0.4	MG1	8.4 ± 0.9	41 ± 18	4.9 ± 2.3
CRR	37 ± 4	N.M.	-----	MG2	4.7 ± 0.6	69 ± 23	15 ± 5
CTD1	24 + 2	N.M.	-----	MZG	5.2 ± 0.5	N.M.	-----
CTD2	30 ± 3	48 ± 9	1.6 ± 0.4	AGD1	1.1 ± 0.1	3.8 ± 1.0	3.5 ± 1.0
HNJ	20 ± 2	17 ± 5	0.9 ± 0.5	AGD2	4.5 ± 0.4	4.6 ± 1.6	1.0 ± 0.4
PLS	15 ± 1	32 ± 8	2.2 ± 0.6	ARB	2.8 ± 0.3	26 ± 4	9.5 ± 1.8
Media	23 ± 9	26 ± 11	1.4 ± 0.6	ARN	2.3 ± 0.3	15 ± 3	6.5 ± 1.9
				CDT	1.5 ± 0.1	26 ± 10	18 ± 7
				KKR	1.0 ± 0.1	31 ± 7	31 ± 7
				MRM	2.0 ± 0.2	88 ± 18	44 ± 10
				PDN	1.1± 0.1	100 ± 22	92 ± 21
				SPT	1.3 ± 0.2	4.1 ± 1.4	3.2 ± 1.2
				STC	3.9 ± 0.4	27 ± 6	6.9 ± 1.8
				VLG	1.4 ± 0.2	14 ± 3	10 ± 3
				AM-14	1.2 ± 0.3	12 ± 3	10 ± 4
				Media	2.6 ± 1.9	30 ± 27	17 ± 21

By other hand, the spatial distribution in the aquifer obtained for the ^{238}U concentrations in dissolution is shown in the Figure 3. From the observation of the data reflected in this Figure an immediate conclusion can be outlined: the Uranium levels in dissolution are not homogeneous over all the aquifer, existing two main and clearly differenced geographical zones. The Uranium in dissolution is clearly higher in the water samples collected from the north of the free aquifer, and experiments a sudden and appreciable decrease when following the direction of the water flow we move from the north to the south and the waters cross an imaginary interface line/band (which is also plotted in Figure 3). In fact, applying the t-student test to the ^{238}U concentrations in dissolution, we have found the existence of significant differences between the values found at the north and at the south to the level of 0.01.

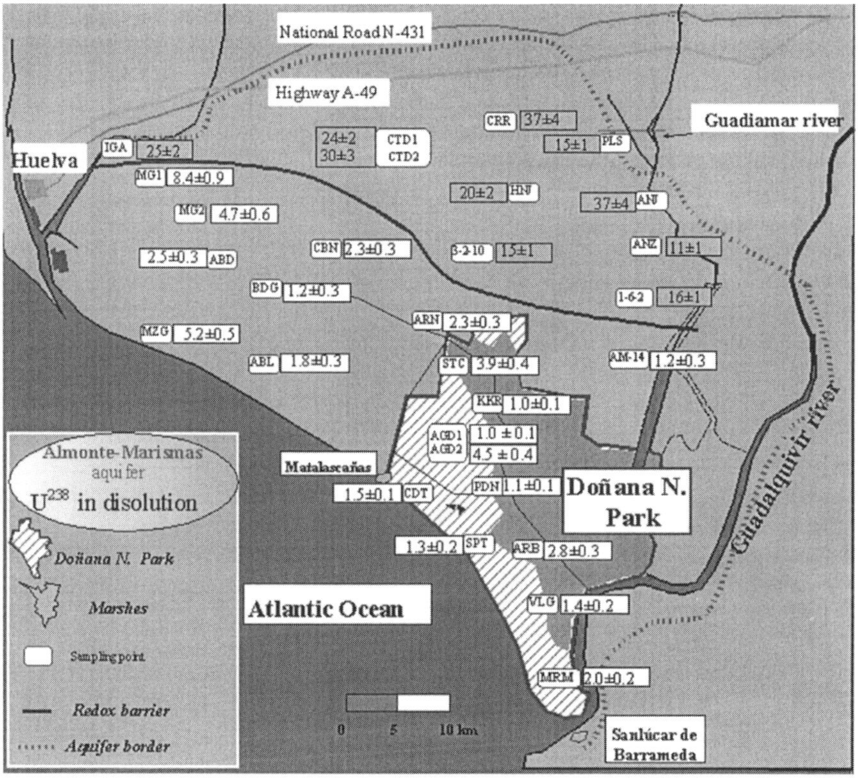

FIGURE 3: Distribution of ^{238}U concentration in dissolution in samples collected in Almonte-Marismas aquifer.

Several arguments can be found, as a first approach, to explain the heterogeneous dissolved Uranium distribution over the free aquifer. As a first hypothesis, the obtained results could be simply a reflect of the heterogeneous composition of the substrata where the waters of the aquifer are embedded. In this case, the higher levels of Uranium in dissolution in the northern region of the aquifer would be associated to the presence in this region of a substratum with higher concentrations of this element that in the south.

But, as a second hypothesis, and for a proper interpretation of the data compiled in Figure 3, it is necessary to have also in consideration the no simple behaviour of the Uranium in aqueous systems. The amount of Uranium present in dissolution in aqueous solutions can be affected by different processes like precipitation, redissolution, coprecipitation and/or sorption, being in any specific case the magnitude and predominance of some of these processes clearly influenced by the geochemical status of the waters. In this sense, we can indicate for example the higher tendency of the Uranium to remain in dissolution under oxidised conditions (when the Uranium is in the form of U(VI)) than under reduced/anoxic conditions (when the Uranium is in the form of U(IV)). The reduced uranium has in fact a high tendency to be associated to the particulate fraction in aqueous solutions, and/or coprecipitate associated to solid material.

Under this second hypothesis, the higher levels of dissolved Uranium found in the northern region of the Almonte-Marismas aquifer could be clearly related with the existence of predominant oxidised conditions in this zone (in agreement with the fact that the main recharge area of the aquifer is placed in this region of the aquifer) while the lower dissolved Uranium levels found in the south could be simply an indication of the predominance of anoxic conditions in this area. Consequently, the sudden decrease in the Uranium dissolved concentrations observed at the interface band/line which was plotted in Figure 3 could be associated to the existence of quite well defined redox-barrier that delimits clearly the oxidised and reduced areas of the analysed aquifer.

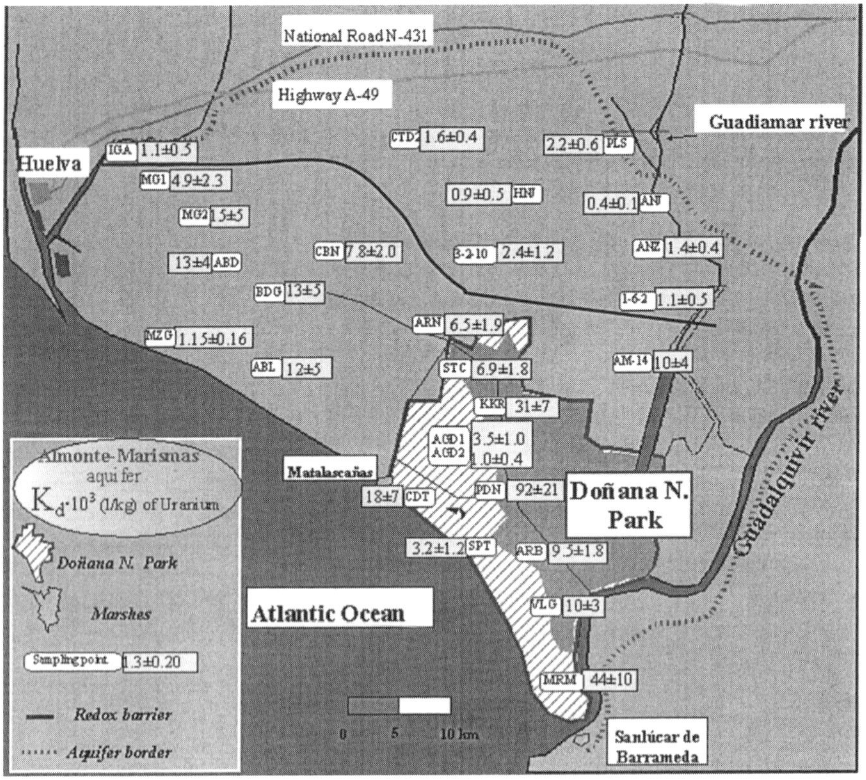

FIGURE 4: Distribution of K_d values of U from samples of Almonte-Marismas aquifer.

In order to identify which of the two hypothesis, briefly described below, cause the obtained spatial dissolved Uranium distribution in the Almonte-Marismas aquifer, additional radiometric determinations were used. Uranium concentrations measured in the particulate fractions of all the collected samples allowed the obtention of the spatial aquifer map of the equilibrium distribution coefficients for Uranium ($K_{d(U)}$). This coefficient is simply defined as the ratio between the Uranium levels per unit mass in the particulate fraction (Bq/kg) and the Uranium levels per unit volume remainig in dissolution (Bq/l), and its value (expressed in l/kg) informs about the distribution of this element between the two mentioned aqueous phases: dissolution.

The Uranium K_d map obtained in the study area is plotted in the Figure 4. From the data shown in this Figure it is possible to differentiate clearly the same two zones (north and south) that in the Figure 3, supporting as a consequence the hypothesis that the differences in the dissolved Uranium concentrations between the two zones is produced by the redox processes affecting the aquifer waters. In fact, we can affirm that in the north zone exists oxidised conditions because we found the higher values of dissolved Uranium and the smaller Uranium K_d values, reflecting the trend of the Uranium to remain in dissolution under oxidised forms. On the contrary, in the southern zone we can affirm that predominates the anoxic conditions because we found very low values of dissolved Uranium and the higher Uranium K_d values, reflecting the greater trend of the reduced Uranium to precipitate and/or to be associated to the particulate fractions.

At the same time, the hypothesis indicating that the observed differences in the dissolved Uranium concentrations between the northern and southern regions of the aquifer are mainly caused by differences in the composition of the substratum in the two zones, should be neglected according to the obtained Uranium K_d map. If we assume quite stable geochemical conditions over all the aquifer, the Uranium K_d distribution map should be quite homogeneous with independence of the enrichment in some places of the Uranium in dissolution concentrations due to the characteristics of the substratum. In these places the enrichment should be also noted in the Uranium concentrations associated to the particulate fraction, and consequently, and according to its definition, would be obtained similar K_d values than in the zones with lower Uranium concentrations in dissolution.

We can conclude, after all the discussion reflected in this section, indicating that the determination of the levels of some specific natural radionuclides in groundwater systems are not only carried out with radiological purposes. These radionuclides can be used as excellent tracers of the different processes (geochemical, hydrological, etc) affecting the analysed ecosystem, as it has been demonstrated at least partially in this work.

CONCLUSIONS

In this work it has been demonstrated the utility of the Uranium determinations performed in the different water phases (dissolution and suspended matter) as an isotopic signature to analyse and understand the extension and importance of some basic geochemical processes affecting the analysed aquifer. In particular, and taken an aquifer located at the Southwest of Spain as a case study, we have shown how through the construction of the dissolved Uranium concentration and the Uranium distribution coefficient maps it is possible to delimit in the analysed aquifer the areas whose waters are under oxidised or anoxic conditions as well as to locate its main redox barrier.

REFERENCES

1. Ivanovich and Osmond.,"Application of U-series disequilibrium concepts to rock/water interaction studies-A review", *Nuclear Geopphysics,* 1991.
2. Osmond J.K. & Cowart J.B., "Groundwater", In *Uranium Series Disequilibrium: Applications to Environmental Problems in Earth Sciences*, Edited by M. Ivanovich and R.S. Harmon, Clarendon Press, 1982, pp. 202-245.
3. Romero E., *Caracterización de las salinidades en el Preparque Norte del Parque Nacional de Doñana*, Doctoral thesis, Huelva: Universidad de Huelva, 1998.
4. González-Labajo J., *Radionuclidos Naturales en el Parque Nacional de Doñana y su Entorno,*. Doctoral thesis, Sevilla: Universidad de Sevilla, 2003.

Applied Physics (APHYS 2003)
A. Méndez-Vilas (Ed.)

TRACK LIKE STRUCTURES IN CR-39 DETECTOR FROM ALPHA PARTICLES WITH INCIDENT ANGLES CLOSE TO AND BELOW THE CRITICAL ANGLES

C.W.Y.Yip, D. Nikezic, J.P.Y. Ho, K.N. Yu

Department of Physics and Materials Science, City University of Hong Kong,
Tat Chee Avenue, Kowloon Tong, Kowloon, Hong Kong. P.R. CHINA

ABSTRACT

The critical angles for the CR-39 detector were studied. In the first step, CR-39 detectors were irradiated with alpha particles with normal incidence to the detector surfaces, and with alpha particles in the energy range from 1 to 5 MeV. The detectors were etched in 6.25 N NaOH solutions kept at 70 °C. The track diameters were fit by a V function, which has a maximum value of 2.87, so the critical angle should be 20.4°. Therefore, at incident angles at or smaller than 20°, no tracks should be observed. The next step was the irradiation of the detectors under small incident angles, namely, 20° and 10°. Track-like structures were indeed observed under the optical microscope. A model taking into consideration the finite lateral dimension of the latent track was proposed to explain such track development.

INTRODUCTION

The response of solid-state nuclear track detectors (SSNTDs) to fast heavy charged particles is characterized by the V_t function, which is the rate at which the etching progresses along the particle track. The ratio $V(R') = V_t/V_b$ is usually used instead, where R' is the residual range of the particles, and V_b is bulk etch rate, i.e., the etching rate of undamaged surfaces. Models of the track growth developed in the past have been founded on these two parameters.

Based on the two rates, V_t and V_b, it is possible to define the so-called critical angle as $\theta_{crit} = \text{asin}(1/V)$. The critical angle is the angle below which the bulk etch rate overcomes the component of the etching rate along the tracks in the direction normal to the detector surface. Tracks will not be developed in such cases. The critical angle has been a subject for many measurements, and has also been used as a base in many calculations. For example, Barrilon et al. [1] measured the critical angle in CR-39 detectors irradiated by alpha particles. For their etching condition, the maximum value of the critical angle is a little bit smaller than 30° (with respect to the detector surface).

The critical angle depends on the incident particle energy as well as on the thickness of the removed layer during etching. If the detector is exposed to radon and its progeny, alpha particles with energies from 0 to 7.69 MeV would strike the detector (if thoron progeny are present in air, the maximum alpha energy can reach 8.78 MeV). Under such circumstances, it is difficult to define a unique critical angle.

Therefore, the concept of a single critical angle is not very useful if one calculates the detector efficiency to radon and its progeny. Instead, the functional dependence of the critical angle on the incident energy should be established for the employed etching condition [2].

In fact, it is possible to define the theoretical critical angle as:

$$\theta_{theor} = \text{asin}(1/V_{max}) \tag{1}$$

where V_{max} is the maximum value of the V function. Particles striking the detector at angles smaller than θ_{theor} will not leave visible tracks. If the incident angle is larger than θ_{theor}, tracks may be developed under proper etching conditions.

METHODS AND RESULTS

Irradiation of CR-39 detectors with alpha particles with normal incidence

In the present paper, the critical angle for the CR-39 detector will be examined. As the first step, the V function for our CR-39 detectors and for our etching conditions were derived. CR-39 detectors were irradiated with alpha particles with normal incidence to the detector surfaces, and with alpha particles in the energy range from 1 to 5 MeV with steps of 0.5 MeV. The detectors were etched in 6.25 N NaOH solutions kept at 70°C. Under these etching conditions, the bulk etch rate as V_b was previously found to be 1.1 μm/h [3]. Etching was performed for 15 h. The source of alpha particles was ^{241}Am (main initial energy E_o = 5.49 MeV). The stopping medium between the source and the detectors was air. The alpha particle energy E incident on the detector at a distance x from the source was determined from the $E(x)$ curve, which was measured beforehand with a surface barrier detector and a multi-channel analyser. The track diameters were measured by an image analyser with a magnification of 1000.

Modified V_t function

In addition to the measurements, we also calculated the track diameters using the following V function given by Brun et al. [4] (hereafter referred to as the Brun's function) in the form

$$V = 1 + e^{-a_{1B}R' + a_{4B}} - e^{-a_{2B}R' + a_{3B}} + e^{a_{3B}} - e^{a_{4B}} \tag{2}$$

with the constants $a_{1B} = 0.1$, $a_{2B} = 1$, $a_{3B} = 1.27$ and $a_{4B} = 1$.

Since the etching condition and the readout procedure were different between our experiments and the experiment of Brun et al. [4], these constants were modified to obtain the best fit for the data for normal incidence. The modified constants are

$$a_{1m} = 0.29,\ a_{2m} = 1.21,\ a_{3m} = 1.23 \text{ and } a_{4m} = 1.05. \tag{3}$$

The more significant modifications are for a_1 and a_2, while a_3 and a_4 are only slightly modified. Comparison between the experimental data for the track openings and the computational results using the modified Brun's function are given in Fig. 1. Very good agreement is observed.

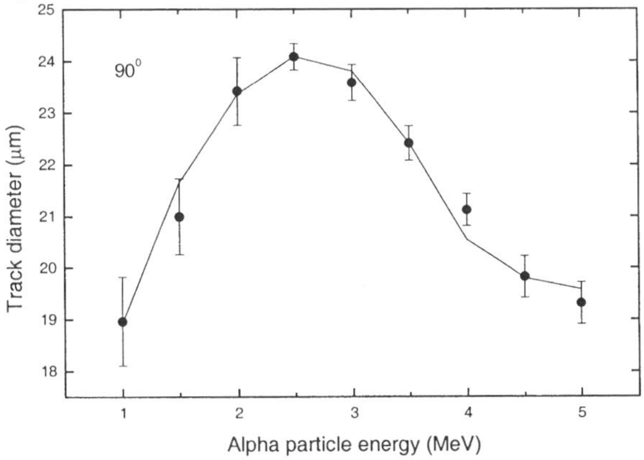

Fig. 1. Comparison between the experimental data for the track diameters and the computational results using the modified Brun's function. Solid circles: experimental data; solid line: values calculated using the modified Brun's function.

Discrepancies for small incident angles

The next step was the irradiation of the detectors under small incident angles, namely, 20° and 10°. Track-like structures were observed, sometimes even after only a few hours of etching. For example, Fig. 2 shows track-like structures on a piece of CR-39 detector resulted from irradiation of 5 MeV alpha particles with an incident angle of 10° with respect to the detector after 7 h of etching. However, if the modified constants for the Brun's function are employed, the maximum value for V is V_{max} = 2.87, so the critical angle is $\theta_{crit} = \text{asin}(1/V_{max}) = 20.4°$. Therefore, at incident angles at or smaller than 20°, no tracks should be observed. Therefore, it can be concluded that the development of tracks formed by particles with small incident angles is not well understood. An attempt to model such track development will be outlined in the next section.

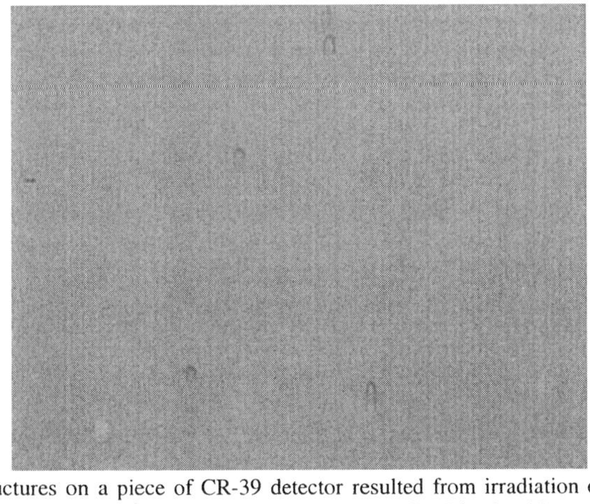

Fig. 2. Track-like structures on a piece of CR-39 detector resulted from irradiation of 5 MeV alpha particles with an incident angle of 10° with respect to the detector after 7 h of etching.

MODEL FOR TRACK DEVELOPMENT FOR INCIDENT ANGLES BELOW THE CRITICAL ANGLE

The discrepancies described in the preceding section has prompted us to look for some overlooked information in existing track development models. The parameter we would like to study in the present paper is the finite diameter of the latent track, which may become important if the incident angle is small. All models of track growth consider the particle trajectory as a line with an infinitesimally small lateral dimension. Etching progresses along that line with the track etch rate V_t. In all other directions, etching progresses with the bulk etch rate V_b. However, it is known that the lateral track has a finite lateral dimension between 10 and 20 nm.

In the following, track etching when the incident angle is below the critical one and the latent track has a diameter *d* will be examined. The model is shown schematically in Fig. 3. Here, V_t is assumed to be the same in all directions in the track core. The part of the detector to the left of point *A* is etched with V_b and the part of detector to the right of *C* is also etched with V_b. Etching from point *B* progresses vertically down with V_t until etching reaches the opposite border of the damaged area. Etching along the line *L* and along the line *R* progress with mixed rates. From L_1 to L_2, the etching has the rate V_b and from L_2 to L_3 the rate V_t. The opposite is along line *R*; from R_1 to R_2 the etching has the rate V_t and from R_2 to R_3 the rate V_b. Such an etching scheme will create the track profile *TT'T"*. This is a consequence of the finite dimension of the track core. If the dimension of the track core is neglected, no tracks are formed when $\sin\theta < 1/V$.

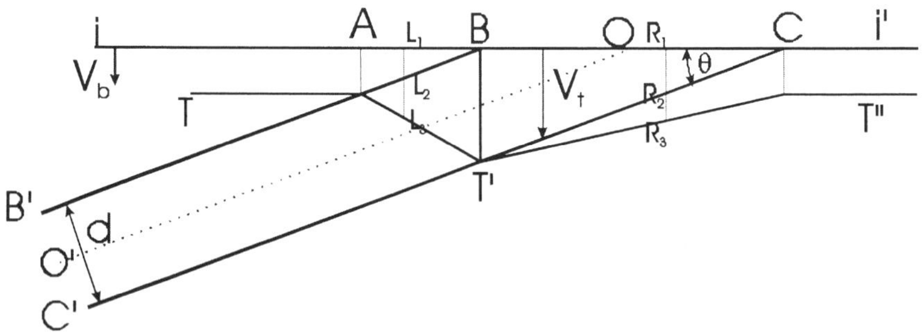

Fig. 3. Schematic diagram showing the model of track growth for incident angles below the critical angle by considering the finite diameter of the latent track. Notations are as follows: *ii'* is the initial surface of the detector; *O* the entrance point of the alpha particle with the trajectory shown as the dotted line *OO'*; *BB'* is the border on the left hand side of the damaged area while *CC'* is the border on the right hand side; θ the incident angle; *d* the diameter of the latent track, i.e. the diameter of the damaged area; V_b the bulk etch rate, V_t the track etch rate; and *TT'T"* gives the shape of the surface obtained after the initial phase of etching.

If *V* is a constant (which is used for illustration purposes and does not represent a realistic situation), further etching will not increase the track depth, but will significantly expand the dimension of the created entity. Whether the created track is seen under a microscope or not is a question of magnification and quality of the microscope.

A simple program has been prepared for simulation of the track growth when $\sin\theta < 1/V$. As inputs, $V_b = 1$ μm/h, $V_t = 2$ μm/h and $\theta = 20^{\circ}$ have been adopted. Under these conditions, the critical angle is θ_{crit} = asin(1/2) = 30°, so the incident angle is well below the critical angle. The diameter of the core is further assumed to be 20 nm. Fig. 4 shows the track development at different time, with the lines corresponding to etching time intervals of 0.001 h. A shallow track is formed, of which the diameter increases with the etching duration. The result for prolonged etching depends on the V_t/V_b ratio, the incident angle and the core diameter, but some track-like structures would remain on the detector surface.

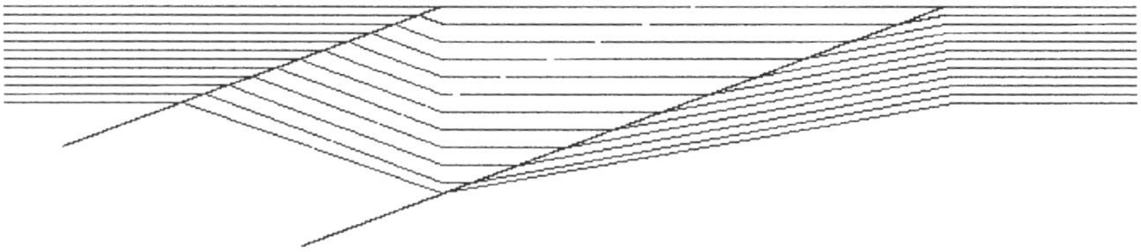

Fig. 4. Computer simulation of track development when the incident angle is below the critical angle. The finite diameter of the latent track has been taken into account, which is a necessary condition for track development. Input parameters: V_b = 1 μm/h, V_t = 2 μm/h, θ = 20° and d = 20 nm. Successive profiles of the formed track are obtained with time steps of 0.001 h for the etching duration.

The current model, which takes into consideration the latent track diameter, is still a simplified one. For example, the variable V function has not been taken into account. Nevertheless, it has already illustrated the mechanism behind the formation of tracks for incident angles below the critical angle. The revised model which can generate realistic numerical results will be published in the future.

ACKNOWLEDGMENT

The present research is supported by the CERG grant CityU1081/01P from the Research Grant Council of Hong Kong.

REFERENCES

1. Barillon, R., Fromm, M. and Chambadet, A. *Varitaion of the critical registration angle of alpha particles in CR-39: Implications for radon dosimetry.* Radiat. Meas. **25** 631-634 (1995).
2. Calamosca, M. , Penzo, S. and Gualdrini, G. *Experimental determination of CR-39 counting efficiency to α particles to design the holder of a new radon gas dosemeter.* Radiat. Meas. **36** 217-219 (2003).
3. Ho, J.P.Y., Yip, C.W.Y., Nikezic, D. and Yu, K.N. *Effects of stirring on the bulk etch rate of CR-39 detector.* Radiat. Meas. **36** 141-143 (2003).
4. Brun, C., Fromm, M., Jorffroy, M., Meyer, P., Groetz J.E., Dörschel, B., Hermsdorf, D., Bretschneideer, R., Kadner K. and Kühne H. *Inter-comparison study of the detection characteristics of the CR-39 SSNTD for Light ions: Present status of the Besancon-Dresden approaches.* Radiat. Meas. **31** 89-98 (1999).

Applied Physics (APHYS 2003)
A. Méndez-Vilas (Ed.)

Quality Control of a Pencil Ionization Chamber

Ana F. Maia and Linda V.E. Caldas

Instituto de Pesquisas Energéticas e Nucleares (IPEN)
Comissão Nacional de Energia Nuclear
Av. Lineu Prestes, 2.242
05508-000, São Paulo, SP
Brazil
afmaia@ipen.br, lcaldas@ipen.br

Abstract. This work presents a set of quality control tests that were applied to a Victoreen pencil ionization chamber. An acrylic support was developed for the repeatability and reproducibility tests since no supports are commercially available. The highest coefficients of variation obtained for the repeatability and reproducibility tests were 0.32% and 2.74% respectively, both within the limits of 1% and 3% of international recommendations. The maximum variation obtained for the calibration coefficients was 1.3% in the energy dependence test (32 – 46 keV). In the case of the linearity test, measurements were taken at several air kerma rates, and a linear fit was obtained. The uncertainty obtained in the angular coefficient was ±0.07 %. For the angular dependence test, the pencil ionization chamber was rotated around its central axis. The maximum variation obtained was 0.65%, far below the limit of 3% of international recommendations.

INTRODUCTION

Pencil ionization chambers have a special design, presenting some particular properties, because they were developed for computed tomography (CT) dosimetric purposes [1,2]. A typical characteristic of those chambers is their partial volume response, i.e., the chamber reading is proportional to the irradiated length. Moreover, these chambers present uniform response around their central axis, which is very important since X-ray tubes rotate around the patients during CT medical procedures [3,4].

The external aspect of those chambers is very similar to a thimble chamber; they are just longer and thinner. The sensitive length of a typical pencil ionization chamber is about 10–15cm, its external diameter is about 9 mm, and its sensitive volume is about 3 cm^3 [5].

As all instruments used for dosimetric purposes, pencil ionization chambers have to be submitted to a quality control program to assure their good performance. This work presents a set of quality control tests that were applied to a pencil ionization chamber.

MATERIALS AND METHODS

A Victoreen pencil ionization chamber, model 660-6, was coupled to a Victoreen electrometer, model 660. The sensitive volume of this chamber is 3.2 cm^3, the sensitive length is 10 cm, and it is filled with atmospheric air. The physical quantity measured by this chamber is the exposure in air length product, and the electrometer readout is in the old units R.cm or R.cm/min, with a range from 0.01 R.cm/min (0.001 R.cm) to 999 R.cm/min (99.9 R.cm).

A $^{90}Sr + ^{90}Y$ check source, Physikalisch-Technische Werkstätten (PTW; 5.77 MBq, 2003), was used to perform the repeatability and reproducibility tests. An X-ray system, diagnostic radiology level, Medicor Mövek Röntgengyara, model Neo-Diagnomax, that operates from 40 to 125 kV at the radiographic mode and from 45 to 100 kV at the fluoroscopic mode, was used for the energy dependence test. Diagnostic qualities defined by the International Electrotechnical Commission, IEC 61267 [6], were used in this system, and their parameters are listed in Table 1. The reference system for these qualities was a parallel plate ionization chamber with 1 cm^3 of sensitive volume, PTW, model 77334, with a PTW electrometer, model UNIDOS 10001. This chamber was calibrated by PTW, with traceability to Physikalisch-Technische Bundesanstalt (PTB), Germany. Another X-ray system, therapy level, Pantak, model HF320, which operates up to 320 kV, was used for the linearity and angular dependence tests. For the linearity test, the air kerma rates were determined using a reference ionization chamber. This reference system was a cylindrical ionization chamber NE, model 2505/3 (0.6 cm^3 sensitive volume) with a PTW electrometer, model UNIDOS 10001. This chamber was calibrated in air kerma by the Brazilian Laboratory for Ionizing Radiation Metrology, Rio de Janeiro, Brazil; the calibration is traceable to the Bureau International des Poids et Mesures (BIPM).

TABLE 1. IEC diagnostic radiology qualities in the *Medicor Mövek Röntgengyara* X-ray equipment.

Radiation Quality	Tube Voltage (kV)	Total Filtration (mmAl)	Half-Value Layer (mmAl)	Effective Energy (keV)
RQR3	52	2.5	1.82	32.0
RQR5	70	2.5	2.45	39.2
RQR7	90	2.5	3.10	46.0

RESULTS AND DISCUSSION

For most types of ionization chambers, supports for repeatability and reproducibility tests are commercially available. However, this is not the case of supports for pencil ionization chambers. Therefore, an acrylic support (Figure 1) was developed for those two tests.

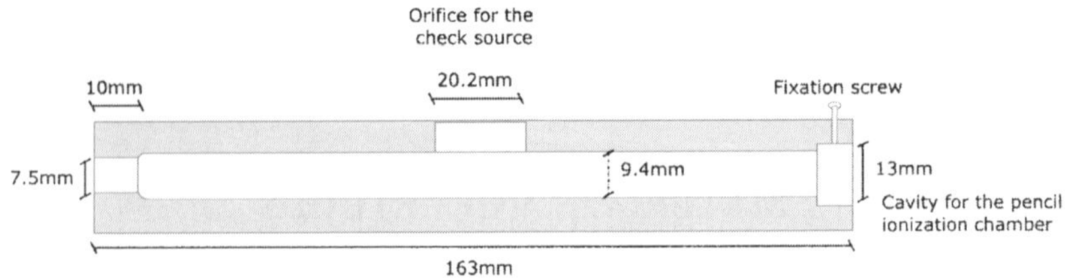

FIGURE 1. Acrylic support for repeatability and reproducibility tests of the Victoreen pencil ionization chamber.

The repeatability test was performed by taking several measurements with the chamber exposed to a check source under reproducible conditions. By IEC 61674 [7], the maximum acceptable coefficient of variation is 1% for CT specific chambers. The highest coefficient of variation obtained was 0.32%, far below the recommended limit.

The reproducibility test was obtained by plotting the results of the repeatability test in function of time, so the long-term stability of the ionization chamber could be observed. By IEC 61674 [7], the medium value obtained in each repeatability test must not differ from the reference value more than 3%. Figure 2 shows the results obtained for the pencil ionization chamber. All values obtained in this work were within the recommended limit.

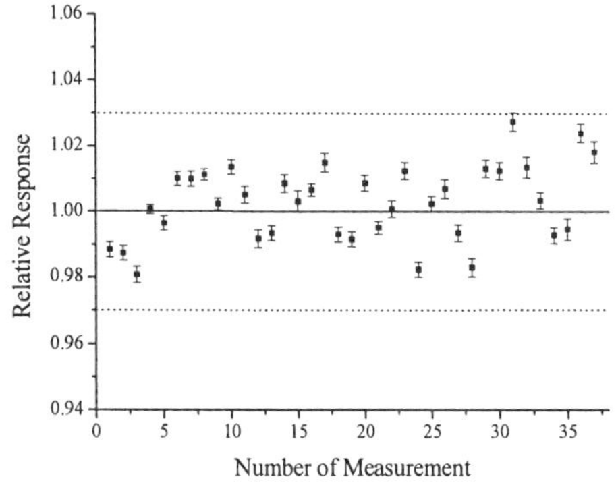

FIGURE 2. Reproducibility test of the Victoreen pencil ionization chamber, using a ^{90}Sr + ^{90}Y check source.

For the energy dependence test, the chamber was calibrated in the beam qualities described in Table 1. For those qualities, the maximum variation obtained for the calibration coefficients was 1.3%, as shown from data in Table 2.

TABLE 2. IEC Calibration coefficients for the Victoreen pencil ionization chamber in diagnostic radiology standard beams.

Radiation Quality	Calibration Coefficient (dimensionless)
RQR3	1.044 ± 0.026
RQR5	1.058 ± 0.026
RQR7	1.052 ± 0.024

In the case of the linearity test, the pencil ionization chamber was exposed to several air kerma rates. In order to provide the air kerma rate variation, the nominal current was varied between 1 and 25 mA at the fixed potencial of 100 kV, half-value layer (HVL) of 4.027 mmAl of the Pantak X-ray system. The air kerma rates were determined using the reference system calibrated for this quality beam. Figure 3 shows the chamber response variation, normalized for the reading using a current of 1 mA, in function of the air kerma rate. A linear fit was provided, and the uncertainty obtained in the angular coefficient was ±0.07 %.

For the angular dependence test, the pencil ionization chamber was exposed to the same standard beam of the Pantak X-ray system. The chamber was rotated around its central axis from −180° to +180°, in steps of 30°. By IEC 61674 (7), the value obtained in each angle must not differ from 0° more than 3%. The maximum variation obtained was only 0.65%, as shown in Figure 4.

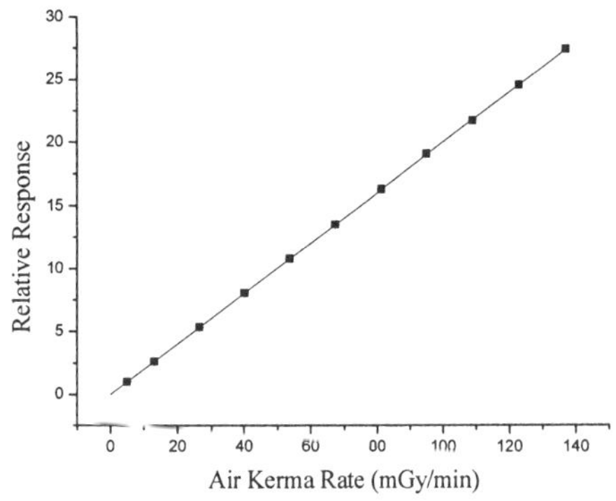

FIGURE 3. Linearity test of the Victoreen pencil ionization chamber, in standard X-ray beams (100 kV, HVL of 4.027 mmAl). Normalization of the chamber response was performed in relation to 1mA current.

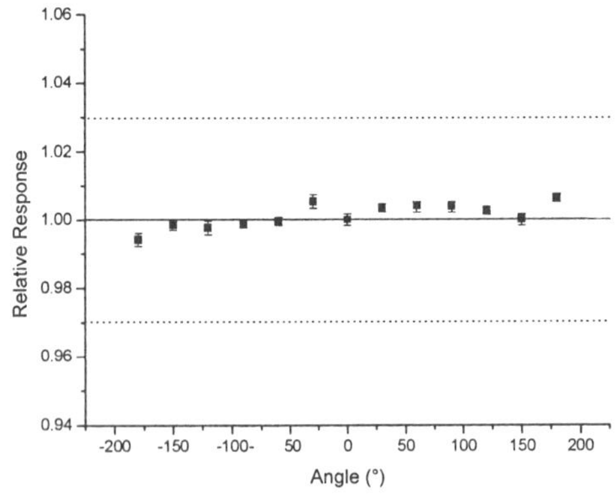

FIGURE 4. Angular dependence test results of the Victoreen pencil ionization chamber, in standard X-ray beams (100 kV, 5 mA, HVL of 4.027 mmAl). Normalization of the chamber response was performed in relation to 0°.

It was not possible to measure the leakage current in the chamber response because of its manual zero adjustment.

CONCLUSIONS

It is very important to keep pencil ionization chambers in a quality control program to assure their proper performance and the high reliance of the measurements. All quality control tests realized in this work showed results within the acceptable limits.

Most of the users of pencil ionization chambers do not perform repeatability and reproducibility tests because of the lack of commercial supports. This work shows that this deficiency can easily be fulfilled, since the developed acrylic support is very simple, presents low cost, and is adequate to perform this kind of tests.

ACKNOWLEDGEMENTS

The authors wish to acknowledge the partial financial support of Fundação de Amparo à Pesquisa do Estado de São Paulo (FAPESP) and Conselho Nacional de Desenvolvimento Científico e Tecnológico (CNPq), Brazil.

REFERENCES

1. Jucius, R.A. and Kambic, G.X., *SPIE Proc.* **127,** 286-295 (1977).
2. Poletti, J.L., *Phys. Med. Biol.* **29,** 725-731 (1984).
3. Pavlicek, W., *Health Phys.* **37,** 773-774 (1979).
4. Bochud, F.O., Grecescu, M. and Valley, J.F., *Phys. Med. Biol.* **46**, 2477-2487 (2001).
5. Suzuki, A. and Suzuki, M.N., *Med. Phys.* **5**, 536-539 (1978).
6. International Electrotechnical Commission. *Medical diagnostic X-ray equipment - Radiation conditions for use in determination of characteristics.* **IEC 61267** (1994).
7. International Electrotechnical Commission. *Medical electrical equipment - Dosimeters with ionization chamber and/or semi-conductor detectors as used in X-ray diagnostic imaging.* **IEC 61674** (1997).

Applied Physics (APHYS 2003)
A. Méndez-Vilas (Ed.)

Gamma-ray measurements of naturally occurring radioactive isotopes in Lesvos Island Igneous rocks (Greece)

A. B. Petalas, S. Vogiannis, S. Bellas & C. P. Halvadakis

Waste Management Laboratory, Department of Environmental Studies, University of the Aegean, GR-81100 Mytilene, Lesvos, Greece

Abstract. Thick igneous rock-successions of mid-crustal plutonic and volcanic origin partly characterize the geology of Lesvos Island. Since radiation of higher levels is mainly associated with igneous than sedimentary rocks, this work attempts to correlate radioactive isotopes levels among various igneous rock types (formations) exposed on the island. In Situ γ-ray spectrometry measurements at 65 geologically well-defined sites were carried out, by using a '3 x 3 in.' NaI (Thallium) portable detector. High values in K-40 were found in several geological formations of Lesvos. (Polychnitos & Skopelos Ignimbrites, Undivided Lower Lavas etc.). High values in Th-232 were counted not only in Polychnitos & Skopelos Ignimbrites, but also in Skalohorion formation and high values in U-238 were found in Skalohorion formation and Kapi rhyolite formation.

INTRODUCTION

The study of naturally occurring radioactive materials (NORM) is of great interest to many fields of Earth sciences, most of them involving the determination of U, Th and K content of rocks. Among the different analytical techniques, field γ-ray scintillation spectrometry offers a number of advantages. In situ analysis is relatively fast and provides immediate and inexpensive results over wide areas. Large rock volumes are measured so that the calculated values are more representative than smaller samples analysed in the laboratory [3].

This study follows the work done by our team who mapped Lesvos island for measured background gamma dose rate using a Geiger-Müller instrument (Micro Sievert by Bicron) and Kriging technique for map configuration [1]. Figure 1 shows the adsorbed dose in the island from the background radiation. This work was the first radiological survey carried out in the island. This survey showed, that the north part of the island has increased values of background radiation. This founding is in accordance to the geological background. This area consists of acid volcanic rocks. Figure 2 shows the geological map of the island, and the formations Sykaminea, Skalohorion, Skoutaros, Kapi rhyolite, Polichnitos & Skopelos Ignimbrites and undivided Lower Lavas. The Skoutaros formation comprises andesite and basalt flows, generally lacking hydrous mineral phases and with a shoshonite geochemistry. In the upper part of the formation, Skoutaros lavas include more dacite and interbed with felsic pyroclastic rocks of the Sigri pyroclastic formation, which in turn is overlain by the Polychnitos ignimbrite. The Polichnitos ignimbrite is thicker near Mantamados, where the upper sheets correlate with those near Polychnitos [7]. The Skopelos ignimbrite south - west of the Gulf of Geras is petrologically quite distinct from the Polychnitos ignimbrite, but is of similar age (18.1 Ma). Extensive ignimbrites in western Anatolia north of Lesvos are associated with the Kestanbol granite in pluton. The term Acid Volcanic unit has been used for these varied felsic volcanic rocks.

In western Lesvos, the Sigri pyroclastic formation includes spectacular silicified tree trunks. Eastward, both pyroclastic and sedimentary rocks become coarser grained and interbedded lavas are more common. Many of the pyroclastic rocks appear derived from a caldera near Vatousa, which shows widespread alternation and mineralisation. Hybrid lavas of the Skalohorion formation, occuring only in the centre of the island, overlie the Sigri pyroclastic formation. They are overlain by thick stratovolcanoes of the Sykaminea formation. Prominent centres occur at Lepetimnos in the north of the island and between Kalloni and Agra in the South. The Sykaminea lavas differ from

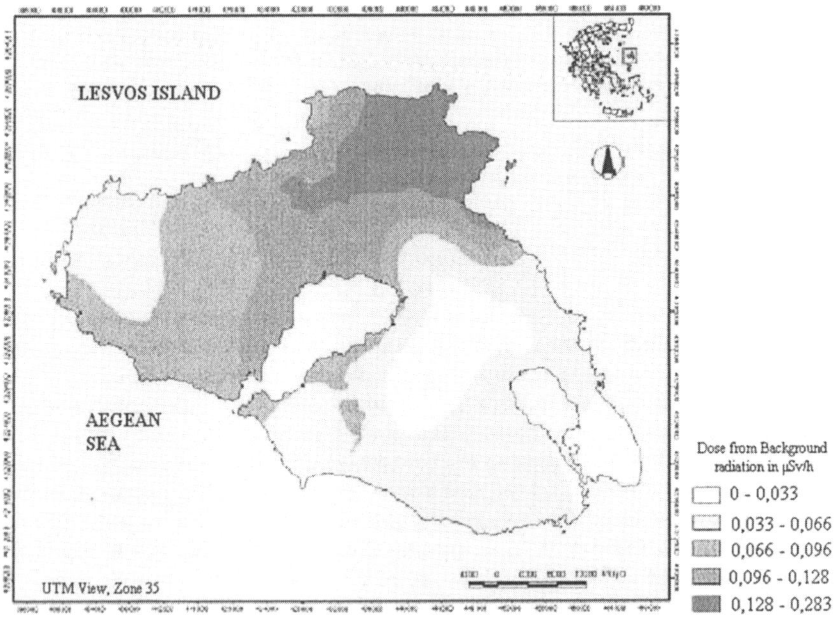

FIGURE 1. Map of the Lesvos Island, with estimated absorbed Dose from background radiation.

Skoutaros lavas in greater abundance of hornblende and biotite. In the type area, the Sykaminea formation comprises dacites with minor interbedded pyroclastics. Near Agra, there is andesite and minor interbedded agglomerate. South of Kalloni, most andesitic lavas have a typical hydrous mineral assemblage (including hornblende and biotite), but many in addition contain pyroxenes and are thus similar to some Skoutaros formation lavas. Near Polichnitos, the Sykaminea formation consists of andesite and overlying basalt. Most rocks of the Sykaminea formation are K-rich (shoshonitic), but some individual flows have lower K values and classify as calc-alkaline [7].

Measurements in this study were taken from these and other formations, such as Post Miocene sediments, to indicate existing differences. The aim of the present work was to relate the results of the above-mentioned study, using in-situ γ-ray spectrometry for the isotopes ^{40}K, ^{238}U and ^{232}Th. The relation between the measured activities and the geological formation in each area was under consideration.

MEASUREMENT TECHNIQUE

During the present radiometric survey, carried out in July and August 2003, γ-ray spectra at 65 sites were acquired. Particular care was taken to avoid outcrop surfaces showing evidence of weathering or not presenting flat geometry. Any measurement suspected of being affected by vegetation or detrital cover was rejected. Many sectors of the island were inaccessible because of the very rough topography of the volcanic edifices. During the field survey, a flat surface about 20 cm x 20 cm was prepared at each location. The detector was then placed on the flat surface and encased within a 2 cm thick cylindrical lead shield. The energy range was set at 3000 keV and a live time of 15 min was used for the measurements. Energy calibration was performed in the laboratory, using standard point sources, type D from Isotopen Laboratory Products, of ^{22}Na, ^{54}Mg, ^{57}Co, ^{60}Co, ^{109}Cd, ^{133}Ba and ^{137}Cs.

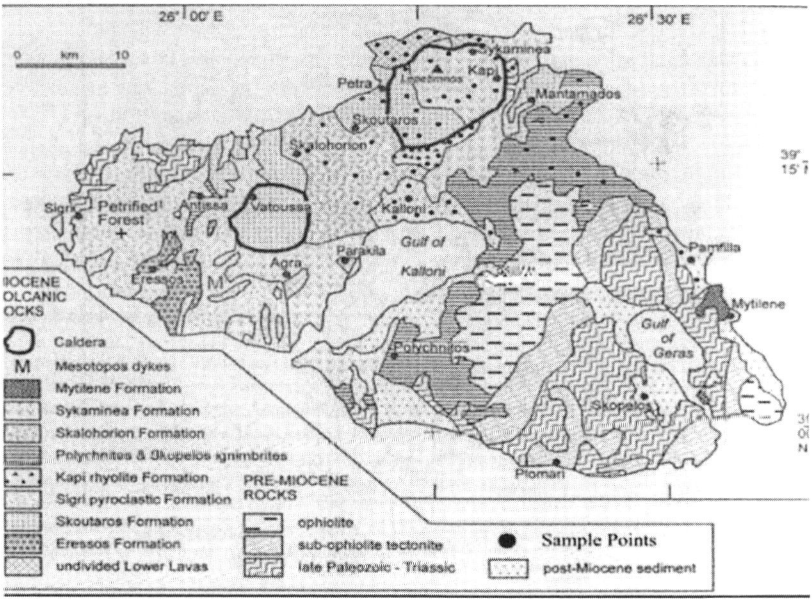

FIGURE 2. Geological map of Lesvos Island (Greece) with the study sites indicated.

The lead shield served to reduce gamma ray contributed from cosmic rays and the surrounding environment as well as to ensure a consistent detection angle and sample geometry during the field survey. Based on the attenuation efficiency of lead against gamma-ray, the lead shield should have eliminated about 70 % of the environmental contribution [2]. To further increase sensitivity, background measurements were taken at the same locations with the detector shielded and placed on 3 cm thick lead bricks. The measured background was subtracted from the field measurements to obtain the "true readings". All measurements carried out with the same protocol of measurements that has been described above.

The in-situ measurements were carried out by using an apparatus consisting of a 76.2 mm x 76.2 mm thallium-activated sodium iodide scintillation detector (Model 802, by Canberra) and a 1024-channel spectrometer unit (NaI Inspector Portable unit by Canberra). The detector is enclosed in a single integral unit with a photomultiplier tube, a high-voltage supply and signal preamplifier. It is thermally insulated and housed in an aluminium cylinder. During the experiment, a reference isotopic source of ^{137}Cs was used, with an initial activity of 37 kBq, to control automatically the system gain and prevent from gain shifting caused by temperature effect or component ageing. The 65 γ-ray spectras were processed with Genie PC 2000 software, by Canberra.

For the determination of the concentration of U, Th and K, three energy windows were investigated, by recording γ-rays associated with characteristic photo-peaks in the decay spectra of ^{40}K, ^{238}U and ^{232}Th. The determinations of U and Th are based on measured γ-radiation from the decay of ^{214}Bi (1.76 MeV) in the ^{238}U decay series and from ^{208}Tl (2.62 MeV) in the ^{232}Th series. The primary decay of ^{40}K (1.46 MeV) is measured directly. The energy window of total count is set from 0.12 to 3.00 MeV [2],[3].

Figure 3 shows a typical spectrum analysed by Genie PC 2000. The three regions of interest (ROI), which are highlighted in the spectra, are corresponding to the three isotopes of interest, respectively (ROI 1 to ^{40}K, ROI 2 to ^{238}U and ROI 3 to ^{232}Th). Measurements are carried out under the assumption of a flat surface and a halfspace with uniform distribution of radionuclides. However, for natural radionuclides, ground roughness effect is negligible, whereas it may be a problem for fallout nuclides like ^{137}Cs. The photon mass attenuation coefficients are very accurate for the range used in this study (0.05-3 MeV), although there might be some uncertainties below 10 KeV.

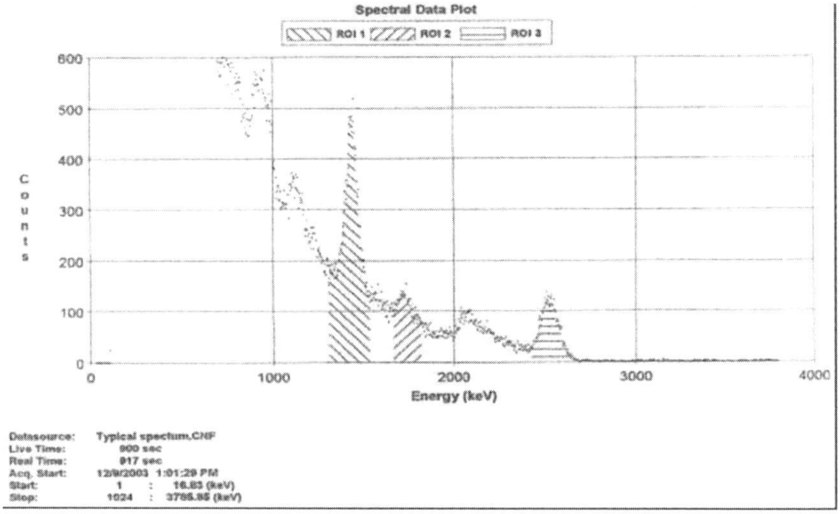

FIGURE 3. Typical spectrum from Genie PC, by Canberra

RESULTS AND DISCUSSION

The values of each isotope, ^{40}K, ^{238}U and ^{232}Th, in the formations that were studied, are presented in table 1.

	Type of Rock	Number of Spectras	K-40 (cpm)			U-238 (cpm)			Th-232 (cpm)		
			Min	Max	Range	Min	Max	Range	Min	Max	Range
p1	Polychnitos & Skopelos Ignimbrites	9	5799	6901	1102	378	512	134	1971	2658	687
p2	Undivided Lower Lavas	9	5345	6321	976	218	411	193	1281	2044	763
p3	Kapi rhyolite formation	10	4862	6014	1152	589	901	312	1895	2578	683
p4	Sykaminea formation	10	3488	5920	2432	321	558	237	1426	2244	818
p5	Skalohorion formation	8	3758	6275	2517	399	518	119	1658	2197	539
p6	Skoutaros formation	8	1987	2826	839	299	389	90	874	1219	345
p7	Post Miocene Sediments	11	1367	1762	395	139	221	82	287	462	175

TABLE 1. Results of γ-ray spectrometry studies

From the table it can be seen, that there are big differences among the values of each isotope from formation to formation. Even in the same formation the values of the isotopes have a significant range. As far as the isotope K-40 concerns, Polichnitos & Skopelos Ignimbrites (p1) show the maximum values for K-40 and the range for these values is 1102 counts. Undivided Lower Lavas (p2), which appear in the northwest part of the island (Fig. 2), follow, having the second higher value in K-40, after p1. Skalohorion formation (p5), Kapi rhyolite formation (p3) and Sykaminea formation (p4), are the formations that follow in the values of K-40. The last two formations, Skoutaros (p6) and Post Miocene Sediments (p7) have the lower values in K-40 and the range of them, especially for p7 is only 395 counts. Attention should be paid in the great range of values, the formations p5 and p4 appear. This is happening, because of the big geographical distribution the formations have. Figure 4 shows the values of each formation in K-40.

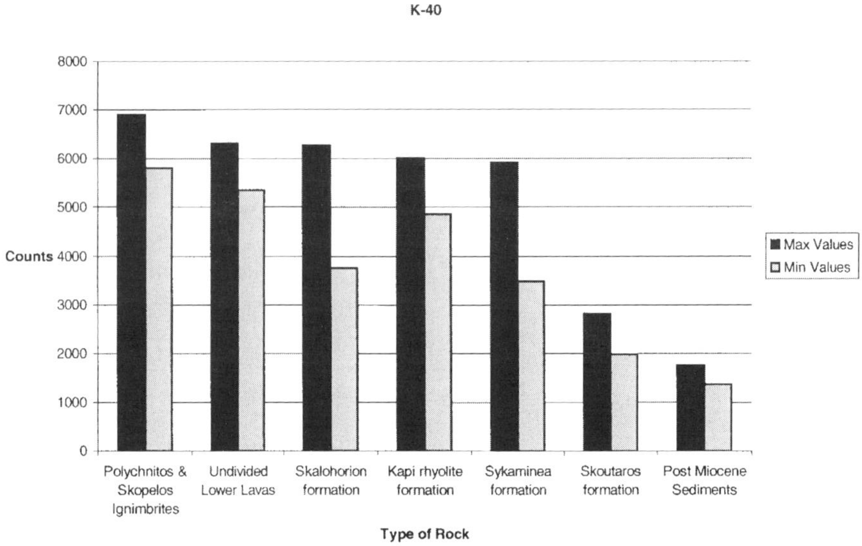

FIGURE 4. Values from the studied formations in K-40

The second isotope was U-238, which was studied through its daughter element Bi-214. For this isotope, p5 is the formation that has the higher values and the biggest range. The formation p3, p4 and p1 are coming after p5. P3 formation is the one that has the bigger range of the three. P2 formation, although it had high values in K-40, in Th-232 its values are rather low, than the other formations. Figure 5 shows the values of the formations in U-238.

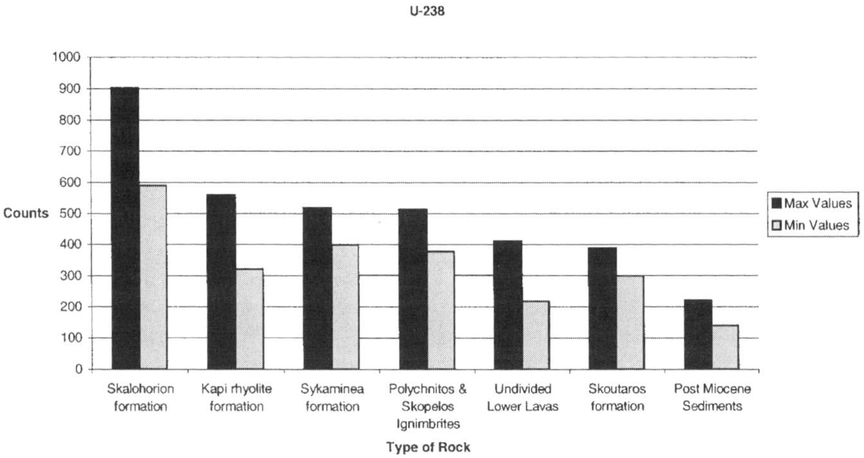

FIGURE 5. Values from the studied formations in U-238

The last isotope that was studied was Th-232. P1 and p8 formations have high values, as they have in K-40. It is not accurate to say that there is connection between these two elements, because this relation is not obvious to all the formations that were studied. Formations p6 and p7 have the lower values in the three isotopes. Post Miocene sediments, is the youngest formation of all. Weathered rocks created it, by transportation of materials, mainly in flat lands. Figure 5 shows the distribution of Th-232 in the studied formations.

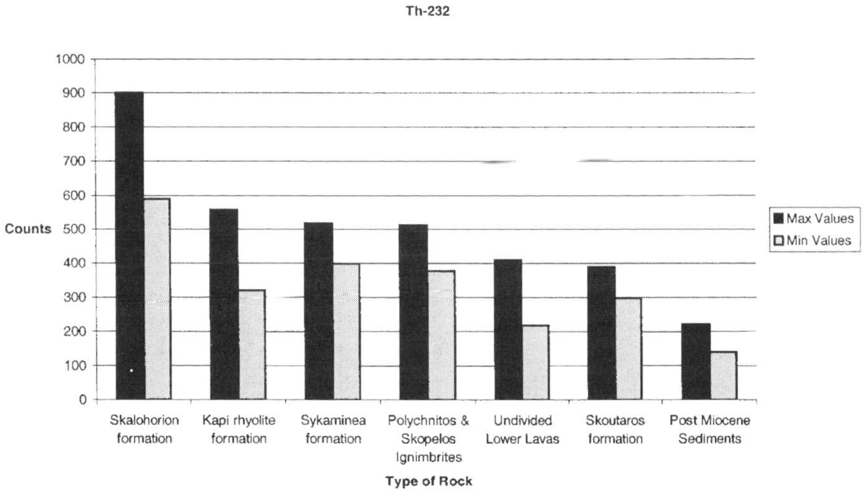

FIGURE 6. Values from the studied formation in Th-232

The area, that was selected to be studied, was the one that had the higher values in the absorbed dose (Fig 1). Formations that are high in K, U and Th are dominant in the north part of the island. Post Miocene sediments were studied, to help our team to see the differences in the values of the three isotopes. These differences are obvious in background dose rates. This dose comes from both cosmic and terrestrial γ-radiation. The altitude, which affects the levels of γ-rays cosmic, has not big variations in the studied area.

This study is a first step to indicate the relation between the geological background of an area and absorbed dose. Our future aim is to find out, the exact concentrations of the three isotopes in Lesvos Island. It is interesting to know the distribution of these isotopes, for our department (Dept of Environmental Studies), because it can be combined with other applications, such as the passage of these elements in the food chain and related environmental problems.

REFERENCES

1. B. Boulgaraki and H. Tsetoura, *Background radiation measurements in Lesvos Island, Greece*. Batchelor Thesis, Dept of Environmental studies, University of the Aegean, Mytilene, 2003 (*In Greek*)
2. F. Q. M. Chen and L. S. Chan, *In-situ gamma-ray spectrometric study of weathered volcanic rocks in Hong Kong*. J Earth Surf. Process. Landforms 27, 2002, pp. 613-625
3. P. Chiozzi, V. Pasquale and M. Verdoya, *Naturally occurring radioactivity at the Alps-Apennines transition,* J. Radiation Measurements 35, 2002, pp. 147-154
4. M. Eisenbud and T. Gesell, *Environmental Radioactivity*, Fourth Edition, Academic Press, 1997, ISBN 0-12-235154-1
5. H. S. Haralampous, *Introduction to Nuclear Physics*, Aristotelian University of Thessaloniki, Thessaloniki Greece, 1973 (In Greek)
6. G. Pe-Piper and D. J. W. Piper, *The igneous rocks of Greece – The anatomy of an orogen*, Gebrüder Borntraeger, Berlin – Stuttgart, 2002, pp. 284-296
7. G. Pe-Piper and D.J.W. Piper, *Geochemical variation with time in the Cenozoic high-K volcanic rocks of the island of Lesvos, Greece: significance for shoshonite petrogenesis*, Journal of Volcanology and Geothermal Research 53, 1992, pp. 371-387.

Applied Physics (APHYS 2003)
A. Méndez-Vilas (Ed.)

MEASUREMENTS OF TRACK PARAMETERS IN CR-39 DETECTOR USING SURFACE PROFILOMETRY

F.M.F. Ng, C.W.Y. Yip, J.P.Y. Ho, D. Nikezic, K.N. Yu*

Department of Physics and Materials Science, City University of Hong Kong,
Tat Chee Avenue, Kowloon Tong, Kowloon, Hong Kong. P.R. CHINA

*Corresponding author: phone: (852) 27887812 fax: phone: (852) 27887830
email: peter.yu@cityu.edu.hk

Abstract. A method based on surface profilometry is proposed to determine track lengths in CR-39 detectors. Tracks from α-particles with an incident energy of 4 MeV have been chosen to demonstrate the method. After irradiation and chemical etching, the tracks were measured by the Form Talysurf PGI Profilometer with a 90° conisphere shaped stylus. Considering the geometry of the stylus, true track lengths will be obtained when the ratio S between the recorded track diameter and the recorded track length is larger than 2. For CR-39 detectors etched in a 6.25 N NaOH at 70 °C, tracks from 4 MeV α-particles will give S = 2 for an etching duration of 15 h, and S = 2.2 for an etching duration of 18 h. In the latter case, the track length is given as 15.6 μm and the track-opening diameter as 34 μm, which should reflect the true values.

INTRODUCTION

The problem of track development has attracted much attention for a long time [1–9]. A method for calculating track parameters based on analytical and three-dimensional consideration was presented [10]. However, consideration was restricted to tracks in the first phase of development, where etching does not reach the end point of the particle range and the track is conical in shape. More recently, Nikezic and Yu [11] extended the consideration to tracks for which etching has passed the end point of the particle track. These over-etched tracks are rounded or spherical in shape. All these track development models should be examined with experimental data.

Measurements of these alpha track parameters are usually performed by optical methods. While measurements of track-opening diameters are relatively straightforward, direct measurements of track lengths are relatively difficult. One approach involves the breaking of SSNTDs to reveal the lateral images of the tracks for direct measurements [12]. In the present work, we propose a new method based on surface profilometry to determine the lengths of tracks in CR-39 detectors through measurements. However, the geometry of tracks and the geometry of the stylus of the surface profilometry equipment should be carefully considered, since the stylus may not reach the bottom of the tracks if the geometries do not match.

METHODOLOGY

For the sake of demonstrating the applicability of surface profilometry in revealing the lengths of alpha tracks in the CR-39 detectors, only alpha particles with an incident energy of 4 MeV and with normal incidence on the detectors are studied. The CR-39 detectors used in the present study were purchased from Page Mouldings (Pershore) Limited (Worcestershire, England).

The CR-39 detectors were irradiated with alpha particles with an energy of 4 MeV under normal incidence through a collimator. For validation of the track development models, irradiation with normal incidence will be sufficient. The alpha source employed in the present study was a planar ^{241}Am source (main alpha energy = 5.49 MeV under vacuum). Normal air was used as the energy absorber to control the final alpha energies incident on the detector. A relationship between the alpha energy and the air distance traveled by an alpha particle (with initial energy of 5.49 MeV from ^{241}Am) was therefore needed. This relationship was obtained by measuring the energies for alpha particles passing different distances through normal air using α spectroscopy systems (ORTEC Model 5030) with Passivated Implanted Planar Silicon (PIPS) detectors of areas of 300 mm^2.

After irradiation, the detectors were etched in a 6.25 N aqueous solution of NaOH maintained at 70 °C by a water bath, for 15 and 18 h. These two etching periods gave tracks which were used to demonstrate whether the stylus of the surface profilometry equipment could reach the bottom of the tracks. The track lengths were then measured by the Form Talysurf PGI Profilometer (Taylor Hobson, Leicester, England), which was a contact stylus instrument based on a Phase Grating Interferometric (PGI) transducer. The PGI transducer is within a gauge and converts the movements of the stylus into an electrical signal. This type of transducer employs a laser diode in conjunction with a fine optical grating to detect very small movements of the stylus and provides a high-resolution output of both amplitude and direction. During measurements, the stylus was set to scan the studied surface many times to ensure that the stylus would pass across any one track at least 10 times.

RESULTS AND DISCUSSION

Fig. 1 shows a two-dimensional image of a piece of CR-39 detector, with tracks from 4 MeV alpha particles, etched in a 6.25 N aqueous solution of NaOH at 70 °C for 15 h, given by the Form Talysurf PGI Profilometer. It is noted that the track characteristics will vary with etching and irradiation conditions. Moreover, conditions of polymerization and hardening can also produce batch to batch differences, although these can be somewhat controlled or minimized.

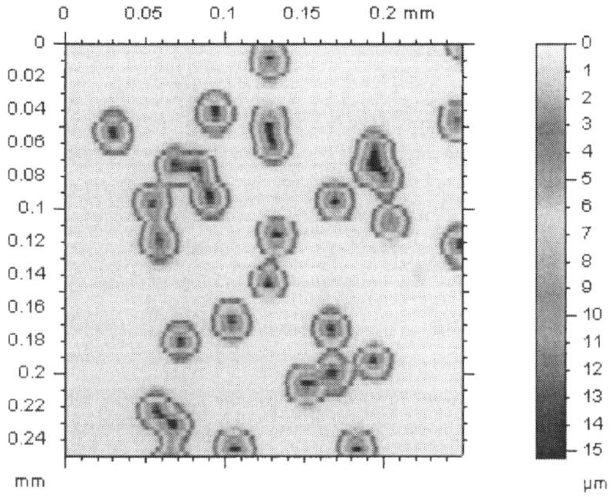

Fig. 1. A two-dimensional image of tracks on a piece of CR-39 detector etched in a 6.25 N aqueous solution of NaOH at 70 °C for 15 h with tracks from 4 MeV alpha particles. The lengths of the tracks are shown by different colors.

The dimensions as well as the lengths of the tracks are conveniently read from the figure. The software can also generate the lateral views of those tracks shown in Fig. 1 as protruding objects for more convenient visualization, which are shown in Fig. 2. Here, the track lengths are more clearly given as 14.4 µm. The highest peaks will be taken to calculate the track lengths since the stylus may not be able to scan across the deepest point of all tracks. The track opening diameters are clearly given as 29 µm.

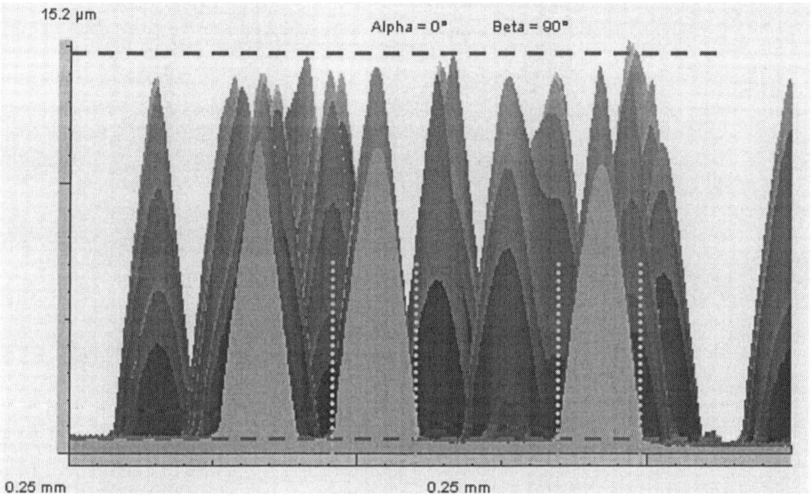

Fig. 2. The lateral views of the tracks shown in Fig. 1 expressed as protruding objects for more convenient visualization. The track length is given by the difference between the two dashed lines, which is (14.8 – 0.4) = 14.4 μm. The track opening diameter of a peak is given by the difference between the two dotted lines containing the peak, which is (121 – 92) or (198 – 169) = 29 μm.

In fact, for the tracks just described, the lengths were not truly reflected. For our surface profilometer, the stylus was a 90° conisphere shaped stylus with a tip radius of 2 μm. Therefore, for all tracks with an opening angle smaller than 90°, the stylus cannot reach the bottom of the track, and the recorded profile is in fact only the shape of the stylus itself. Under such circumstances, the ratio *S* between the recorded track diameter and the recorded track length should be equal than 2. This is the case for our tracks from 4 MeV alpha particles in a CR-39 detector etched in a 6.25 N aqueous solution of NaOH at 70 °C for 15 h. In other words, the true track lengths should be larger than 14.4 μm.

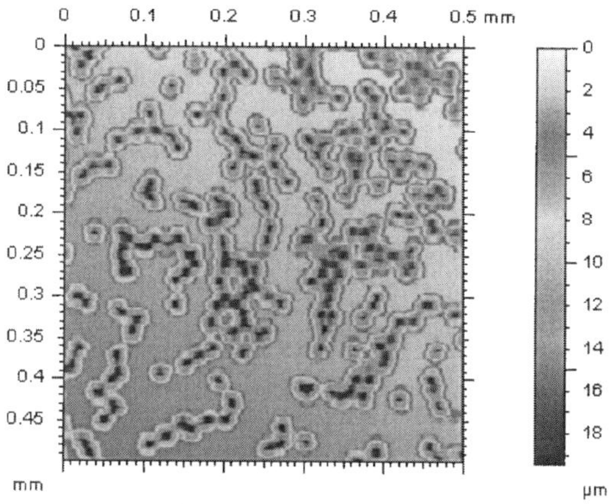

Fig. 3. A two-dimensional image of a piece of CR-39 detector etched in a 6.25 N aqueous solution of NaOH at 70 °C for 18 h with tracks from 4 MeV alpha particles.

Fig. 3 shows a two-dimensional image of a piece of CR-39 detector, with tracks from 4 MeV alpha particles, etched in a 6.25 N aqueous solution of NaOH at 70 °C for 18 h. The dimensions as well as the lengths of the tracks can be read from the figure. Again, the software can also generate the lateral views of tracks shown in Fig. 3 as protruding objects for more convenient visualization, which are shown in Fig. 4. Here, the track lengths is clearly given as 15.6 µm while the track opening diameter is clearly given as 34 µm. For this track, S = 2.18 which is larger than 2. This over-etched track is also expected to be rounded in shape [11], so we conclude that the stylus has reached the bottom of the track, and the length of the track has been truly reflected. However, we should note that the shape of the track might not be truly reflected. Fortunately, the shape is usually not needed in validating the track development models.

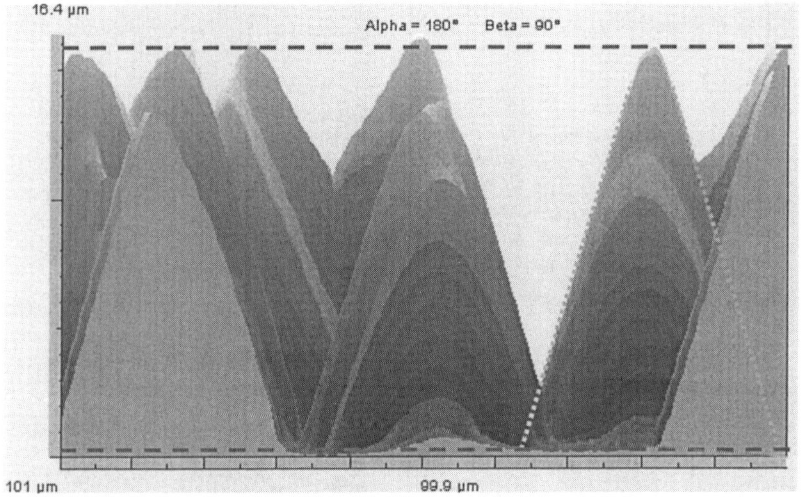

Fig. 4. The lateral views of some tracks shown in Fig. 3 expressed as protruding objects for more convenient visualization. The track length is given by the difference between the two dashed lines, which is (15.8 – 0.2) = 15.6 µm. The track opening diameter of a peak is given by the intersection between the two dotted lines with the lower dashed line, which is (98 – 64) = 34 µm.

ACKNOWLEDGMENT

The present research is supported by the CERG grant CityU 102803 from the Research Grant Council of Hong Kong.

REFERENCES

1. Ditlov, V. *Calculated tracks in plastics and crystals.* Radiat. Meas. **25**(1/4), 89-94 (1995).

2. Hatzialekou, U. Henshaw, D.L. and Fews A.P. *Automated image analysis of alpha–particle autoradiographs of human bone.* Nucl. Instr. Meth. A **263** 504-514 (1988).

3. Henke, P.R. and Benton, E. *On geometry of tracks in dielectric nuclear track detector.* Nucl. Instr. Meth. **97** 483-489 (1971).

4. Fromm, M., Chambaudet, A. and Membrey, F. *Data bank for alpha particle tracks in CR39 with energies ranging from 0.5 to 5 MeV recording for various incident angles.* Nucl. Tracks Radiat. Meas. **15** 115-118 (1988).

5. Meyer, P., Fromm, M., Chambaudet, A., Laugier, J. and Makovicka, L. *A computer simulation of n, p conversion and resulting proton tracks etched in CR39 SSNTD.* Radiat. Meas. **25**(1/4) 449-452 (1995).
6. Nikezic, D. and Kostic, D. *Simulation of the track growth and determination the track parameters.* Radiat. Meas. **28** 185-190 (1997).
7. Paretzke, G.H., Benton, E. and Henke, P.R. *On the particle track evolution in dielectric track detectors and charge identification through track radius measurements.* Nucl. Instr. Meth. **108** 73-80 (1973).
8. Somogyi, G. and Szalay, A.S. *Track diameter kinetics in dielectric track detector.* Nucl. Instr. Meth. **109** 211-232 (1973).
9. Somogyi, G. *Development of etched nuclear tracks.* Nucl. Instr. Meth. **173** 21-42 (1980).
10. Nikezic, D. *Three dimensional analytical determination of the track parameters.* Radiat. Meas. **32** 277-282 (2000).
11. Nikezic, D. and Yu, K.N. *Three-dimensional analytical determination of the track parameters. Over-etched tracks.* Radiat. Meas. **37** 39-45 (2003).
12. Dörschel, B., Fülle, D., Hartmann, H., Hermsdorf, D., Kadner, K. and Radlach, Ch. *Measurement of track parameters and etch rates in proton-irradiated CR-39 detectors and simulation of neutron dosemeter responses.* Radiat. Prot. Dosim. **69**(4) 267-274 (1997).

Applied Physics (APHYS 2003)
A. Méndez-Vilas (Ed.)

Non linear effects of harmful agents: application of α irradiation and taxol on human cancer cells in vitro

Jesús Soto[a], Carlos Sainz[a], Samuel Cos[b], Domingo Gonzalez Lamuño[c]

[a] Departamento de Física Médica.
[b] Departamento de Fisiología y Farmacología.
[c] Departamento de Genética y Pediatria
Facultad de Medicina, Universidad de Cantabria, 39011 Santander (Spain).
Correspondence to C. Sainz (Tel. 942201974; Fax 942201903; E-mail: sainzc@unican.es)

Abstract. Alpha radiation is the main component of man irradiation from natural sources. Its biological effects on the low doses range still remain unknown, especially in the case of combination with other physical or chemical agents. Taxol is a compound obtained from the bark of the Pacific Yew tree (taxus brevifolia) which is used in cancer treatment, mainly in mammary and ovarian diseases.

To analyse the combined action of these environmental agents a study was carried out to test the possibility that breast cancer cells show increased sensitivity to the chemotherapeutic agent taxol when they have been treated with low radiation doses of alpha particles. To this end, the cells were cultivated in a medium containing dissolved radon (^{222}Rn) and then in a second medium containing a concentration of taxol (20, 50, 75 or 100 nM). After the culture phase the surviving cells were counted and their viability was assessed.

The results obtained indicate that the cells treated with low doses of alpha particles exhibit increased sensitivity when treated with certain concentrations of taxol; in particular, a significantly lower survival rate and lower viability were observed in cells treated with alpha radiation and 50 nM of taxol than in cells treated with the same concentration of taxol alone. These effects could be the result of the influence of the alpha particles on the expression of apoptosis-related genes, which is complementary to the action of taxol on bcl-2 related genes.

INTRODUCTION

The biological effects due to the combined action of environmental factors usually show several non-additive characteristics. This fact implicates a wider understanding manner of the relationships between life and environment which includes explanation of several phenomena like hormesis, therapeutic drug enhancement and influence on immune system.

Exposure to alpha particles is the most usual way of man irradiation from natural sources of ionising radiation, mainly radon gas. In the same way of all kinds of ionizing radiation, the biological responses induced from alpha particles exposure are dose dependent. The most typical effects of low doses are those related to protein stress and adaptive response.

On the other hand, the main target of chemotherapeutical treatment is to reach the maximum or, if possible, the absolute removal of malignant cells, leading the minimum harm to adjacent healthy cell groups. That means a limit for the therapy related with the restriction of the maximum admissible dose of citotoxic drug.

Moreover, the appearance of drug resistance, present in some types of tumour, supposes an another challenge. Normally this kind of phenomena is associated to the expression of apoptosis related genes [9,10]. For example, a correlation between the expression of apoptosis.inhibitor gene bcl.2 and multidrug resistance mechanisms in several tumour cell lines has been observed [3, 6, 11].

Low doses of alpha particles can also be used as sensitizer agent to the later action of chemotherapeutic drugs. We have observed significant influences on the growth of MCF-7 breast cancer cell populations by using radioactive gas radon previously dissolved in culture medium as irradiation source [17]. Under the same conditions, we have found that irradiated cells express apoptosis related genes to a different degree with respect to non irradiated cells [16]. In this paper we show the results from a set of experiments carried out to study whether the modifications produced by irradiating MCF-7 cells with low doses of alpha particles modify the sensitivity of the cells to different concentrations of the agent taxol.

METHODS

To study the modification produced by alpha particles in the sensitivity of the cancer cells to taxol we used the stable MCF-7 line of human breast cancer cells. Cells were purchased from the American Type Collection (Rockville, MD) and maintained as monolayer cultures in 75-cm^2 plastic culture flasks in Dulbecco's Modified Eagle's Medium (DMEM) (Sigma Chemical Co., St Louis, MO) supplemented with 5% fetal bovine serum (FBS) (GIBCO, France), penicillin (20 units/ml) and streptomycin (20 μg/ml) (Sigma Chemical Co., St Louis, MO) at 37°C in a humid atmosphere containing 5% CO_2. Cells were subcultured every 3-4 days by suspension in 5 mM Na2-EDTA in phosphate buffered saline (PBS) (pH 7.4) at 37°C for 5 min.

Before the experiment, cells from stock subconfluent monolayers (80%) were incubated with 5 mM Na2-EDTA in PBS (pH 7.4) at 37°C for 5 min, resuspended in DMEM supplemented with 5% FBS and passed repeatedly through a 25-gauge needle to produce a single cell suspension. Cell number and viability were determined by staining a small volume of cell suspension with 0.4% trypan blue saline solution and examining the cells in a hemocytometer. MCF-7 cells (3 x 10^5 cells/dish) were seeded into 60 x 15 mm tissue culture dishes in DMEM supplemented with 5% FBS, penicillin (20 units/ml) and streptomycin (20 μg/ml). After the cells were firmly attached to the dishes (4 h), the culture media were aspirated and the dishes with and without radon were refilled with fresh media.

Cell irradiation with alpha particles was due to the presence of ^{222}Rn and its short life daughters, previously dissolved in culture medium. The radon containing culture medium was obtained by dissolving the gas from a liquid radium (^{226}Ra) sample [18]. The experimental device for doing so is shown in Fig. 1. The radon produced by the radium sample diffuses through the air in a closed system, and is dissolved in the culture medium in a sterile vial. The concentration of radon in the culture medium depends on the time the gas has been diffusing from the source. After the time of exposure to radium, the vials were tightly closed and the radon concentration was measured with a gamma spectrometry semiconductor GeHP (Canberra) coupled to a multichannel analyzer. The short-life radon daughter gamma emitters were determined under the 352 keV ^{214}Pb and 612 keV ^{214}Bi photopeaks. Successive measurements showed a progressive radioactive equilibrium between radon and its daughters, and a plateau was reached after 3 h. This measurement system was used to determine the radon concentration in equilibrium with its daughters in the culture.

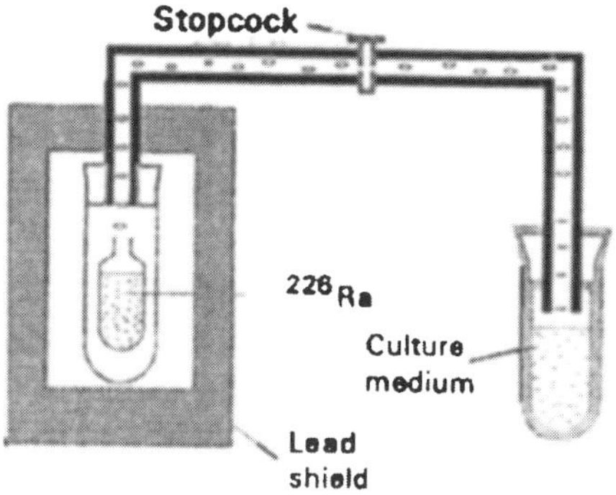

FIGURE 1: Experimental device for dissolution of radon in culture medium

In each case, the dose absorbed by the cell cultures was calculated from the initial radon concentration in radioactive equilibrium with its descendants. The radon gas initially dissolved in culture medium diffuses to the air because cultures were incubated in open Petri dishes. Therefore, only his short life daughters ^{218}Po and ^{214}Po are responsible of alpha dose. Beta and gamma emissions scarcely deposits energy in the small dimensions of culture plates and are not taken into account for dosimetric purposes. In the same way, the probability of alpha emission of another daughters such as ^{218}At, ^{214}Bi or ^{210}Bi is very low and its contribution to total dose results negligible. Under these conditions, total absorbed dose in cell culture can be calculated bearing in mind initial populations of each isotope as well as their alpha energy of emission (6 MeV- ^{218}Po and 7.68 MeV-^{214}Po). Total doses in the experiments performed were between 1.7 and 15 mGy.

Fig. 2 shows the relation between dose rate and time. It can be observed that dose rate is not constant during irradiation period. Because of the absence of radon, dose rate become zero after 200 minutes, being the decrease curve a combination of ^{218}Po and ^{214}Po decay ones.

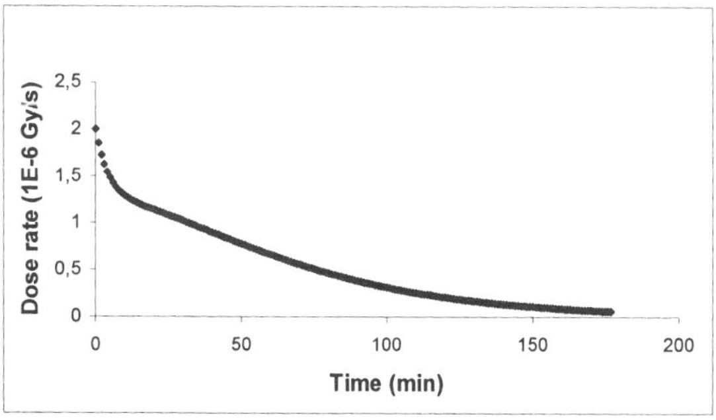

FIGURE 2: Dose rate evolution in presented experiments. Due to radon absence, the shape of the curve results a combination of ^{218}Po and ^{214}Po decay ones.

A total of six experiments were performed with different alpha radiation doses and/or different concentrations of taxol. The explanation diagram of each experiment is shown in Fig. 3. Every experiment was divided into two stages. In the first, 12 x 5 ml portions of cell culture were placed in Petri dishes. Four of them contained medium with dissolved radon, and the other eight ones standard culture medium only. In both cases the media were heated al 37 ªC and filtered in order to ensure their sterility. After three days incubation, the twelve culture media were aspirated and replaced with fresh media in four of them, and with fresh media with a concentration of taxol (20, 50, 75 or 100 nM) in the remaining eight dishes. Thus, we had four irradiated cultures with taxol (R-T), four unirradiated cultures with taxol (C-T) and another four dishes with standard culture medium only (C-C). After three days, survival and viable cells were counted. For doing so, cells were harvested by treatment with 5 mM Na2-EDTA in PBS (pH 7.4) at 37°c for 5 min, passed through a 25-gauge sterile needle to produce a single cell suspension and counted with a hemocytometer. Viability was determined by the trypan blue exclusion test.

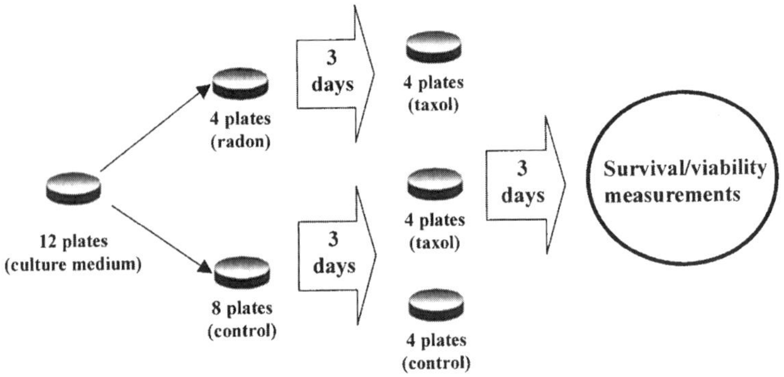

FIGURE 3: Flow diagram of the experimental design. MCF-7 were incubated in either radon-containing or control culture media for three days. After this time, media in the dishes cultured with radon were aspirated and replaced with fresh media containing taxol. The media in dishes not cultured with radon were also replaced with fresh media containing either taxol or the taxol diluent ethanol. The concentrations of taxol used were 20, 50, 75 and 100 nM. After three days, cells were harvested and counted with a hemocytometer. Viability was determined by the trypan blue exclusion test.

RESULTS

Three sets of experiments (one of them with each taxol concentration, 20, 50 and 75 nM) were performed with different alpha radiation doses. The results are given in tables I, II and III. All data were analysed by one way analysis of variance (ANOVA) followed by the Student-Newman-Keuls multiple comparisons test. Differences between groups were considered significant at $p<0.05$ and $p<0.001$. Each table gives the mean values of cell survival, viability and long-term survival (survival plus viable cells), expressed as a percentage of control cell values. The mean survival and viability values are expressed with their standard deviation. Also the conditions of cell culture are shown in tables: C-C cultures in standard medium, C-T cells cultured for three days in standard medium and three days in medium with taxol, and R-T cells incubated for three days in radon containing medium and afterwards in a medium with taxol.

Using a taxol concentration of 20nM and alpha radiation doses from 1.7 to 3.2 mGy we performed two experiments. Values in table I indicate that there were not significant differences between cells treated with taxol only and those previously irradiated.

Culture conditions	Survival (% control)	Viability (%)	Long-term survival (%)
C-C	100	87 ± 2	87
C-T	90 ± 7	83 ± 5	75
R-T	95 ± 6	76 ± 3	95

TABLE I: Mean values of survival rate and viability for doses from 1.7 to 3.2 mGy and 20 nM of taxol. (Significant differences: [a] vs control, $p < 0.05$; [b] vs control, $p<0.001$; [c] vs taxol 20 nM, $p<0.05$)

In another three experiments we used a taxol concentration of 50 nM and alpha doses from 2.7 to 3.1 mGy. Results are given in table II. In this case, taxol produced a significant decrease in cell survival in comparison with control cells and this decrease was greater for cells treated with radon and taxol. Also, the differences on survival between cells treated with taxol only and those treated with alpha particles and taxol were significant in all cases.

Culture conditions	Survival (% control)	Viability (%)	Long-term survival (%)
C-C	100	83 ± 2	83
C-T	32 ± 2 [b]	59 ± 3 [b]	20
R-T	18 ± 1 [b c]	47 ± 4 [b c]	9

TABLE II: Mean values of survival rate and viability for doses from 2.7 to 3.1 mGy and 50 nM of taxol. (Significant differences: [a] vs control, $p < 0.05$; [b] vs control, $p<0.001$; [c] vs taxol 50 nM, $p<0.05$)

Table III shows the results obtained irradiating cells with 3 mGy and afterwards treated with 75 nM of taxol. A large decrease in cell survival was observed for taxol treated cells in comparison with control ones. This decrease was greater than in the previous experiences. However, the small decrease produced by radiation and taxol with regard to taxol alone was not significant.

Culture conditions	Survival (% control)	Viability (%)	Long-term survival (%)
C-C	100	83 ± 2	83
C-T	14 ± 2 [b]	57 ± 5 [a]	8
R-T	12 ± 2 [b]	50 ± 5 [a]	6

TABLE III: Mean values of survival rate and viability for doses of 3.0 mGy and 75 nM of taxol. (Significant differences: [a] vs control, $p < 0.05$; [b] vs control, $p<0.001$; [c] vs taxol 75 nM, $p<0.05$)

One further experiment was done with a taxol concentration of 100 nM (results not tabulated), in which the survival of cells treated with taxol and with alpha particles and taxol was nil.

DISCUSSION

Taxol has shown antitumour activity in the treatment of advanced ovarian cancer and metastatic breast cancer. Its mode of action is related to the stabilisation of the structure of the microtubules preventing these from acting as a mitotic axis and stopping cell division in the M phase of the cell cycle. This mechanism leads to cellular apoptosis through increased expression of p53 or by inactivation of bcl-2. Thus taxol can control the growth of tumour cell populations with the highest mitotic activity [7].

Because of the relatively low number of studies about the action mechanism of alpha particles [14, 8] it is difficult to explain how low alpha particle doses can contribute to taxol action. Irradiation of cell culture is not homogeneous when the concentration of alpha emitters in culture medium is low. This is because of the dimensions of cells are of the same order of magnitude as the range of the particles. Thus it is generally accepted that the analysis of the response of cells to alpha particle radiation requires microdosimetric methods [13]. However, several mechanisms such as free oxygen radicals production [12] and the action over another cellular structures different from DNA can explain in part the activation of processes dependent of the expression of cellular damage related genes such as p53.

Although the main findings are those related to taxol concentrations that considerably reduce growth of the cancer cell populations, the results obtained at lower taxol concentrations are also of interest. In the two experiments with taxol concentration of 20 nM, the survivals of cells treated with taxol alone and with irradiation and taxol combined yielded inconsistent results, ranging from 77 % to small insignificant increases in comparison with control cells. The errors associated with the percentage of surviving cells, which are mainly due to different growth in different dishes, are high, demonstrating a small and partly random effect of taxol and of taxol plus radiation. The viability of the surviving cells showed values between 62 and 92 %, which did not differ, significantly from control values.

In the three experiments at taxol concentration of 50 nM, the percentages of surviving cells compared to control cells were 38, 34 and 25 % with small standard deviations, and the viability of the surviving cells was 53 %, thus giving long-term survival rates of 20, 18 and 13 %. These results would be moderately adequate in chemotherapy and the taxol concentration used can be considered as close to therapeutic concentrations.

When 50 nM taxol administration was preceded by alpha irradiation at doses around 3 mGy, the results showed a marked improvement and gave survival rates of 16, 24 and 15 % which were significantly lower than those for taxol alone. This difference in the percentage of surviving cells represents and improvement of about 190 % in favour of treatment with alpha irradiation. Similarly, the application of alpha particles modified the viability of surviving cells, which was reduced to 56, 50 and 15 %. Thus, long term cell survival showed and improvement of 220 % in comparison with taxol alone.

Increasing the taxol concentration to 75 nM did not improve the results of the interaction between alpha radiation and taxol. There was a decrease in survival rates as compared with lower taxol concentrations for cells treated with taxol alone, and there was also a decrease in viability. However, no significant decrease in comparison with taxol alone was observed when cells were first treated with alpha radiation.

In order to explain the different combined effects with the concentrations of taxol used, there must be taken into account the balance between the sublethal injuries, induced by radiation and increased by the later action of taxol, and the efficacy of cellular damage repair mechanisms. Before the treatment with taxol, the results found on irradiated cultures indicate that the injuries produced by low doses of alpha particles are not enough for inducing significant changes in survival rate and viability. In the same way, the lowest taxol concentration (20 nM) does not significantly change the results with regard to control cultures. And the set of injuries induced by radiation and taxol together neither are enough to change the survival and viability values. On the other hand, taxol concentration of 75 nM produces a sufficient quantity of injuries in cell culture for producing a large decrease in survival rate and viability values. For this reason, the effect of alpha particles is minimised by the action of taxol, and not significant differences arise between irradiated and unirradiated cultures. Finally, the greatest taxol concentration 100 nM is able to destroy the whole cell culture without the possibility of observing any effect induced by alpha particles.

In conclusion, these results indicate that low doses of alpha particles can enhance the action of intermediate taxol concentration (50 nM) remaining its effect negligible in combination with low (20 nM) or very high (75, 100 nM) concentrations of taxol.

A more detailed explanation could be a complementary action of alpha particles and taxol that is related to apoptosis, since this is the main mechanism of action of taxol and may be caused by inactivation of bcl-2 by the drug [4]. Within the bcl-2 gene family, bcl-x is translated into two peptides, bcl-x_L and bcl-x_S, with the latter giving rise to apoptosis [2, 19], and possibly to increased mutagenesis [5]. Alpha radiation can contribute to apoptosis by inducing oxidative stress due to the production of free oxygen radicals, and also altering the equilibrium between apoptosis inductor/inhibitor genes. The latter mechanism could be explained by the creation of chemical species that break bcl-x into its two transcripts and thereby enhance the action of taxol.

These kind of in vitro studies constitute the basis for a possible clinical use of low doses of alpha particles. To this end, it would be necessary to study the effects of low doses on healthy cells. In this sense, there is evidence that the low doses of alpha particles from radon and its daughters used to modify the growth of MCF-7 cells but do not change the growth of populations of fibroblasts [17]. Additionally, due to gaseous character of radon, it would add the possibility of irradiating the whole body by inhalation, blood transport and dissolution in tissues, with potential to act on metastases distant from the primary tumour site.

Finally, the results also could be of interest for the risk estimation of people exposed to natural sources of alpha particles. Relative biological effectiveness (EBR) is not accurately determined for alpha particles in the low doses range, because it depends on the biological response studied [1]. The doses used in our experiments are of the same order of magnitude to those received by people who live in areas of high natural radiation for a year, and those received over a shorter period by patients and staff in some spas [15] where radon is dissolved in the water in high concentrations.

REFERENCES

1. Baverstock K., Thorne M. *Int. J. Radiat. Biol.* **74**(6), 799-804 (1998).
2. Boise L.H., Gonzalez-Garcia M., Postema C.E., Ding L., Lindsten T., Turka L.A., Mao X., Nuñes G., Thompson C.B. *Cell.* **74**, 597-608 (1993).
3. Campos L., Rouault J.P., Sabido O., Oriol P., Roubi N., Vasselon C., Archimbaud E., Magaud J.P., Guyotat D. *Blood.* **81**, 3091-3096 (1993).
4. Chen M., Quintans J., Fucks Z., Thompson C., Kufe D.W., Weichselbaum R.R. *Cancer Res.* **55**, 991-994 (1995).
5. Cherbonnel-Laserre C., Gauny S., Kronenberg A. *Oncogene* **13**, 1489-1497 (1996).
6. Dole M., Nuñez G., Merchant A.K., Maybaum J., Rode C.K., Bloch C.A., Castle V.P. *Cancer Res.*, **54**: 3253-3259 (1994).
7. Holmes F.A., Walkers R.S., Theriault R.L., Forman A.D., Newton L.K., Raber M.N., Buzdar A.V., Frue D., Hortobagyi G.N. *J Natl Cancer Inst* **83**, 1797-1805 (1991).
8. Kadhim M.A., Macdonald D.A., Goodhead D.T., Lorimore S.A., Marsden S.J. Wright E.G. *Nature*, **355**, 738-740 (1992).
9. Lowe S.W., Ruley H.E., Jacks T., Housman D.E. *Cell* ,**74**, 975-967 (1993).
10. Lowe S.W., Bodis S., McClatchey A., Remington L., Ruley H.E., Fisher D.E., Housman D.E., Jacks T. *Science,* **266** , 807-810 (1994).
11. Miyashita T., Reed J. *Blood*, **81**, 151-157 (1993).
12. Narayanan P. K., Goodwin E. H., Lenhert B.E. *Cancer Res.* **57 (18)**, 3963-3971 (1997).
13. Polig E. *Cur. Top. Radiat. Res.* **13**, 189-327 (1978).
14. Robertson J.B., Keehler A., George J., Little J.B. *Radiat Res* **96**, 261-274 (1983).
15. Soto J., Gómez J. *Health Phys* **76**, 398-401 (1999).
16. Soto J., Martín A., Cos S., González-Lamuño D. *IAEA-CN-67/190*, 647-650 (1997).
17. Soto J., Quindós L.S., Cos S., Sánchez-Barceló E.J. *Sci Total Environ* **181**, 181-185 (1996).
18. Soto J., Sainz C., Cos S. , Gonzalez Lamuño D. Nuc. Inst. and Meth. B **197**, 310-316 (2002).
19. Yang E., Zha J., Jockel J., Boise L.H., Thompson C.B. *Cell* **80**, 285-291 (1995).

Applied Physics (APHYS 2003)
A. Méndez-Vilas (Ed.)

ENVIRAD: A COLLABORATION BETWEEN INFN AND SECONDARY SCHOOL FOR THE STUDY OF RADON

M. Pugliese*°", M. Ambrosio°, E. Balzano*°, A.M. Esposito°, L. Gialanella°, V. Roca*°, M. Romano*°, C. Sabbarese^°

*)Dipartimento di Scienze Fisiche, Università Federico II , Napoli, Italy
^)Dipartimento di Scienze Ambientali, Seconda Università di Napoli, Caserta, Italy
°)INFN, Napoli
")corresponding author: Dipartimento di Scienze Fisiche, Università Federico II, Complesso di Monte S.Angelo, via Cintia, 80126 Napoli, Italy, fax:+39081676346, E-mail: pugliese@na.infn.it

Abstract. The problem of radon mitigation is certainly a problem of technical order but also a cultural problem. In fact, the first step towards its solution envisages a complete knowledge of the dependence of radon concentration on a number of parameters and on a correct pattern of use of housing. As a matter of fact, these topics are not present in the typical cultural background and into ordinary scholastic curricula.
The ENVIRAD project aims at facing such deficiency in school programs by proposing to underline these topics and setting up a net of schools in which measurements of radon in the ground and indoor will be carried out in order to contribute to the construction of an archive about the natural radioactivity in Campania region. The students will be responsible for the organization of indoor surveys, a the management of a continuous radon monitoring station in the ground and the transmission of the data to a database server.
The method has been set during one year activity carried out with a pilot school.

INTRODUCTION

Radon exposure increases the risk of lung cancer, and represents the most relevant naturally occurring health hazard [(1)]. Hence, radon concentrations have been closely monitored, with particular attention to cohorts living or working in radon-abundant areas. Both scientific knowledge and technical handling of this issue is now adequate insofar as they can provide useful information for risk assessment. In fact, over the past 30 years many tools have been developed that aim at reducing radon levels in buildings. However, effective prevention and mitigation require a suitable pattern of use of housing, schools and workplaces. Rather than due to technical failures, the issue at stakes is the necessity of educating the general public making them aware of the existence of natural radioactivity and of the means by which this can be measured, without feeding unjustified radiophobia.
A particularly fertile environment to operate in is the secondary school, for which has been proposed the ENVIRAD project, supported by Istituto Nazionale di Fisica Nucleare (INFN). Its goal is the involvement in the data collecting process of "non-conventional" operators through a collaboration with schools which goes beyond the usual divulgation schemes

The experimental work on which students will be implicated after a training period will consist on indoor radon surveys and on in-soil radon monitoring.

Indoor radon concentrations

The measurement of indoor radon concentrations is the typical activity in the field of radon study, being the first step toward health radon risk valuation. In Italy a national survey [(2)] was carried out in the 90s , devoted to measure national and regional radon concentrations. In the South of Italy, at the moment, there lacks ofthe organisation of campaigns dedicated to particular communities as the children of primary schools or the inhabitants of the old towns, where building typology is favourable to rising of radon concentrations. Our proposal is a way of reducing, at least partially, this deficiency, encouraging the distributed presence on the territory of atypical but skilled subjects able to perform this job.

As known, measurements of radon concentrations are meaningful only if they are averaged over a one-year period to take into account periodical fluctuations.

Techniques routinely used are based mainly on SSTD and E-Perm. Annual exposition can be carried out on shorter, consecutive periods if the observation of the fluctuations is also of interest.

Another important factor in the planning of a survey is the dimension and the statistical significance of the sample. If one is interested solely in the mean concentration, a random sample must be used. If one particular aspect of radon issues is to be studied, for example the dependence of the concentration on a particular parameter, this has to be the only free variable

Monitoring of radon gas in the soil

Being the soil the main source of radon, the monitoring of radon in the environment is the first step in radon prevention: in fact the knowledge of these concentrations can drive indoor survey towards most interesting situations. Furthermore, , these values could be used in many fields, such as geologic studies or mineralogic research. Hence, the objective of this part of the work is the collection of reliable data, which could be utilized by experts of other disciplines. Also in this case, clearly, a dense presence of monitored sites will facilitate the finding of useful correlations.To perform this type of measurements, the RaMoNa system (Radon Monitoring Naples), which was set up in our Radioactivity Laboratory and is based on alpha-particles spectroscopy of ionised radon daughters collected on a silicon detector [(3)], will be used.

It consists of a collection chamber containing the detector, a set of probes to measure enviromental parameters (pressure, humidity and temperature) and an electronic module providing functions of treatment of detector and probes signals and their acquisition.

Since the monitor efficiency as well the radon emanation from soil depend on P, H and T, the knowledge of these parameters is important to normalize detector response and radon concentration to standard conditions. For this reason a program for the characterization of the chamber respect to these parameters has been carried out [(4)].

RaMoNa was conceived for laboratory usage, so a more friendly version that students can use has been developed. A prototipe is currently being tested and one system, after calibration, will be installed in each of the eleven schools thus far involved in the ENVIRAD project.

THE PROJECT

ENVIRAD aims at introducing students of secondary schools to the issue of natural radioactivity by building a network based on schools in the Campania region where some indoor surveys will be organized in the schools and nearby territory and radon in the soil will be monitored. The students' participation will consist in accomplishing the measurements and forwarding the data to a central database. To do that in a fully aware way, students have to understand properly the general problematic of the environmental radioactivity and have to be familiar with the experimental techniques used. Therefore, it is crucial both the elaboration of a suitable road map to set up programs and laboratory activities that may help the understanding of topics with which young students are not familiar.

To assess the potentialities of this project, in the first year we focussed on the following points:

- Basic elements of radioactivity and radiation
- Interaction of radiation with matter and health effects
- Natural radioactivity and radon
- Parameters that influence radon concentration
- Radon measurement methods
- Criteria to plan a radon survey
- Data analysis.

A program developed according to the above points was proposed, during the last year, to students of a pilot school, the Liceo Scientifico "E.Torricelli" in Somma Vesuviana, a town of the volcanic area surrounding the Vesuvio. Within this framework a series of meetings and lectures with teachers and students were performed while the experimental activity commenced. To facilitate the management of the project and the interaction among all involved components, a dedicated web page has been constructed and is continuously update (www.na.infn.it/envirad). After this first year, the project will be developed on two years, in the first of which training and the indoor measurements will be planned and ,will start; in the second one the continuous monitoring of radon in soil will start and the indoor measurements will be completed. In case of success of the work, the program will be re-proposed for the next years and to other schools.In fact, such a project is suitable to produce many objectives from various points of view. First of all educational objectives, which can be summarised as the experimentation of new didactic schemes, the employment of not usual instrumentation and, mainly, the integration of different disciplines.

We have already indicated the scientific objectives, devoted to realise an archive of indoor and in-soil radon concentrations, useful for health, scientific and industrial purposes . But the most important outcomes of Envirad, is of cultural order, in fact it will be suitable introduce students and teachers to the world of scientific research, which is usually well separated from the world of the school. Secondarily, it will increase the knowledge of the territory, and relationships between students with other institutions and within themselves.

RESULTS

The training initially envisaged lectures and discussions, but after the first month, jointly with the teachers we decided to start some preliminary laboratory activities to introduce the students to the scientific method. On the whole, within a few months the students in few months reached a good level in understanding the whole issue and were encouraged to plan a survey into their school: they had to find useful criteria to project it in optimal manner . To meet these criteria students organised and carried out a series of short-term measurements (six days) using charcoal canisters that they analyzed in our laboratory by gamma - ray spectrometry to verify the dependence of radon level on various parameters, such as height of the floor and ventilation rate.

The students showed a particular interest in various aspects of this work to the extent that after the early training period, they where able to draw up a well thought out survey in their school and in a primary school of Somma Vesuviana. In this case, it was them who explained radon issues to teachers. The survey was carried out leaving in well-studied places SSTD detectors for two periods of six months each, starting from April 2003. The exposed detectors are currently being replaced for the second period measurement.

At the end of May, a meeting was organized where students illustrated the project to other schools, presented Envirad and showed the results of their activities, demonstrating a very high degree of understanding of the whole problematic and the capacity to achieve the proposed objectives.

Their work is now a PowerPoint presentation available on the ENVIRAD web page

CONCLUSIONS

The first year of work with the pilot school demonstrates that the model proposed can be applied to 12 schools partecipating to the network. Some aspects of the original road map have been changed on the basis of this experience, and now a more appropriate program idoneus to our goal is ready, also if it will be subjected to other changes on the basis of suggestions from the real work with the student and the teachers.

REFERENCES

1. Nazaroff, W.W and Nero, A.V. *Radon and its decay products in indoor air.* (New York: John Wiley and sons) 1988.

2. Bochicchio F. et al, *Indagine nazionale sulla radioattività naturale nelle abitazioni.* Rapporto finale, Roma 1994.

3. Pugliese, M., Baiano, G., Boiano, A., D'Onofrio, A., Roca, V., Sabbarese, C., Vollaro, P., *A Compact Multiparameter Acquisition System for Radon Concentration Studies.* Applied Radiation and Isotopes. **53** (2) (2000).
4. Roca, V., De Felice, P., Esposito, a.M., Pugliese, M., Sabbarese, C., Vaupotich, J. *The influence of environmental parameters in electrostatic cell radon monitor response.* Proceedings of ICRM Conference on Low-level radioactivity Measurement Techniques 2003, Vienna, October 13-17, 2003

Applied Physics (APHYS 2003)
A. Méndez-Vilas (Ed.)

Calculation of Scattered Radiation Around a Patient Subjected to the X-Ray Diagnostic Examination

Srpko Marković, Vladan Ljubenov, Olivera Ciraj, and Rodoljub Simović

VINČA Institute of Nuclear Sciences
P.O.Box 522, 11001 Belgrade, Serbia and Montenegro

Abstract. An efficient analytic approach for the calculation of scattered radiation around a patient subjected to angiography, the contrast X-ray diagnostic technique, is presented here. Adopting the well justified physical assumptions that (i) the spectrum of the X-ray tube is known in advance, (ii) the scattering function do not change significantly through the scattering medium and (iii) the linear attenuation coefficients of the initial and single scattered photons are equal for the incident photon energy range from 30 keV to 60 keV, an expression for the space-energy distribution of scattered radiation is derived and the ANGIO computer routine is written. Calculated values are compared with the experimental data originally obtained as well as the Monte Carlo simulation results.

INTRODUCTION

As a part of artificial ionizing radiation sources, X-ray diagnostic techniques irradiate global population with more than ninty percents. From the radiation protection point of view, X-ray diagnostics are the only area with high possibilities for significant collective dose reduction. The exposure of the medical diagnostic staff is mainly related to unnecessary irradiation, while in this respect the angiography is the most critical X-ray contrast technique. The actual extensive technical improvements of the X-ray diagnostic procedures and new devices have not resulted in the expected collective dose reduction except in some computer tomographies and dental techniques [1].

In X-ray diagnostics, a qualified assessment of medical staff exposure is the indispensable foundation for radiation protection optimisation, shielding calculation and other radiation protection measures. A good knowledge of space-energy distribution of scattered radiation around a patient is a necessary condition for this evaluation. Starting from the Compton single scattering model and using suitable physical modelling and mathematical approximations for the actual range of photon energies, the expressions for the space-energy distribution of scattered radiation are derived and ANGIO computer routine is developed [2,3]. The calculated values are compared with detailed dosimetric measurements of the scattered radiation field around a patient (i.e., the water phantom) and with the results of Monte Carlo simulation.

DETERMINATION OF SCATTERED RADIATION

An analytic procedure for calculating the scattered radiation space-energy distribution around the laying patient and the ANGIO computer routine are presented in this section of paper.

Analytic approach

Backscattered radiation from a patient, i.e., the radiation albedo, is the main irradiation source for the physician and the other members of the medical team. We consider the differential intensity of scattered photons, the most suitable physical quantity which is directly related with the exposure, the quantity we can measure.

By formulating probabilities that the initial photon:

(*i*) penetrate through the scattering medium to depth z without interactions;

(*ii*) experience the Compton scattering at depth z and scatter into unit space angle Ω defined with the elevation angle θ and the azimuth angle ϕ; and

(*iii*) hit the target at the distance $l(\theta)$ without any further interaction,

one can obtain the starting expression for the space-angular distribution of the scattered radiation [4]

$$dI_R(\Omega) = I_0 B n_e \frac{d_e\sigma(\theta_R)}{d\Omega} \exp\left[-z\left(\frac{\mu_1}{\cos\theta_0} + \frac{\mu_2}{\cos\theta}\right)\right] \frac{dV}{l^2(\theta)} \tag{1}$$

where: μ_1 and μ_2 are linear attenuation coefficients for the initial photon and the single scattered photon, respectively; θ_0 is the initial photon incident angle; dV is the elementary volume around the scattering point; n_e is the number of electrons in a unit volume of the scatterer; $d_e\sigma(\theta_R)/d\Omega$ is the scattering function given in form of the Klein-Nishina formula where θ_R is the scattering angle [5]; B is the buildup factor, a correction for the photon multiply scattering [6]; I_0 is the initial radiation intensity.

For the sake of brevity, the energy variable E is omitted in previous definitions.

Equation (1) defines the differential intensity of scattered photons $dI_R(\Omega)$ escaped from volume dV to distance $l(\theta)$ in the unit space angle around θ and ϕ. To calculate the intensity $I_R(\Omega)$ at the target position, expression (1) should be integrated over the active part of the scattering volume.

Additionaly, the following simplifications are adopted:

(i) $\cos\theta_0 = 1$, acceptable approximation for the actual angle $\theta_0 \leq 5^\circ$;

(ii) $\mu_1 = \mu_2$, justified for the actual photon energy range from 30 keV to 60 keV. In reality, the photon energy change in the single Compton scattering amounts to only a couple of keV;

(iii) the scattering function $d_e\sigma(\theta_R)/d\Omega$ and the distance $l(\theta)$ do not change significantly with the depth z and can be considered as constants.

With these simplifications, the intensity $I_R(\Omega)$ becomes [3]

$$I_R(\Omega) = C_1 \frac{d_e\sigma(\theta_R)}{d\Omega} \sin^2\theta\, \Im(\theta) \tag{2}$$

where C_1 is the constant

$$C_1 = \frac{I_0\, B\, n_e\, \pi\, R_0^2}{h_c^2\, b^2} \tag{3}$$

b is the target lateral distance from the stream of X-rays, h_c is the distance from the X-ray source to the surface of the scattering medium, while R_0 is the radius of irradiated surface. Moreover, $\Im(\theta)$ stands for the integral

$$\Im(\theta) = \int_0^{z_0} (z + h_c)^2 \exp[-\mu\, z\, (1 + \sec\theta)]\, dz \tag{4}$$

In expression (4) z_0 represents the depth of the scatterer.

The scattered intensity $I_R(\Omega)$ can be written in an even simpler analytic form by using the procedure which takes into account that the beam of incident photons is very homogeneous and narrow in practice, and that the truncated cone of the relevant scatterer can be substituted by an equivalent cylinder [3]

$$I_R(\theta) = \frac{C_2}{l^2(\theta)} \frac{d_e\sigma(\theta_R)}{d\Omega} \frac{1 - \exp[-\mu\, z_0\, (1 + \sec\theta)]}{\mu\, (1 + \sec\theta)} \tag{5}$$

In equation (5), the constant C_2 is defined as $C_2 = I_0\, B\, n_e\, \pi\, R_0^2$.

The ANGIO code

It is well known that the energy aspect of photon interaction probabilities is specially emphasised for the energy range from 20 keV to 60 keV, which is the region of practical interest in contrast X-ray diagnostics. In this range, a small initial photon energy change causes significant variation in the Compton scattering probabilities $d_e\sigma(\theta_R)/d\Omega$, the values of the linear attenuation coefficients μ and the buildup factor B. For these reasons, a multigroup technique in treating the photon spectrum was used and the four energy group ANGIO code was created based on the demonstrated analytic approach [2].

The input for the photon buildup factor for the low-energy region (Fig. 1) is obtained by interpolation of the results from the literature [4,2]. The ANGIO code delivers the exposure rate which can directly be converted to the equivalent dose rate expressed in the units μSv/h.

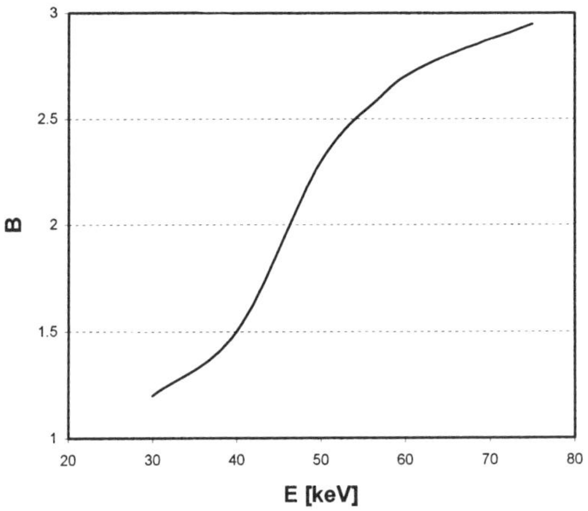

FIGURE 1. Buildup factor for a monoenergetic source of photons embedded in water

RESULTS

In order to check the results obtained by the computer programme, detailed dosimetry measurements of the scattered radiation field around the water phantom of rectangular geometry 25 cm x 25 cm x 15 cm were carried out and Monte Carlo simulations were performed.

A Philips MG-320 X-ray generator was used as the X-ray source [7]. It is equipped with a set of remote filters and operates between 30 kV and 320 kV. For measurements, the following dosimetric instruments were utilized: a standard free air X-ray ionization chamber, a set of cavity chambers for exposure measurements, a tissue equivalent ionization chamber as well as numerous secondary standard dosimeters and radiation detectors. The source spectrum in the experiment was equivalent to the original spectra typically used in the X-ray diagnostic techniques. Two sets of experiments were carried out: the first, characterized by high voltage U=60 kV, output current I=10 mA, filter of 4 mm Al + 0.3 mm Cu and the beam quality defined by the half value layer HVL=0.18 mm Cu; and the second one, defined by high voltage U=80 kV, output current I=10 mA, filter of 4 mm Al + 0.5 mm Cu and the beam quality defined by the half value layer HVL=0.35 mm Cu.

The MCNP-4C code [8] was applied with the standard photon nuclear data library MCPLIB in order to determine the scattered radiation around the water phantom already used in the measurement. Spectrum of incident photons was modeled in 27 energy groups, while the equivalent dose rate values dH/dt were obtained from the photon flux collected in seven group via MCNP F5 point detector tally. Simulations included $2\ 10^7$ photon histories resulting in statistical uncertainty of less than 1% for each point detector. Also, the photon flux-to-dose-rate conversion factors were applied according to ICRP-21 recommendations [9].

The calculated, measured and Monte Carlo results for several target positions (defined by lateral b and vertical h distances from the point of stream incidence) are presented in Figs. 2 and 3. Here, the full line represents the values calculated by the ANGIO code, while the triangles and open circles stand for the measured values and the results of simulation, respectively. It is evident that the computed values are in good agreement with the measured and Monte Carlo data, except for the target position defined by b = 60 cm and h = 60 cm and the high voltage U=60 kV, when the discrepancy of results is higher than 10%.

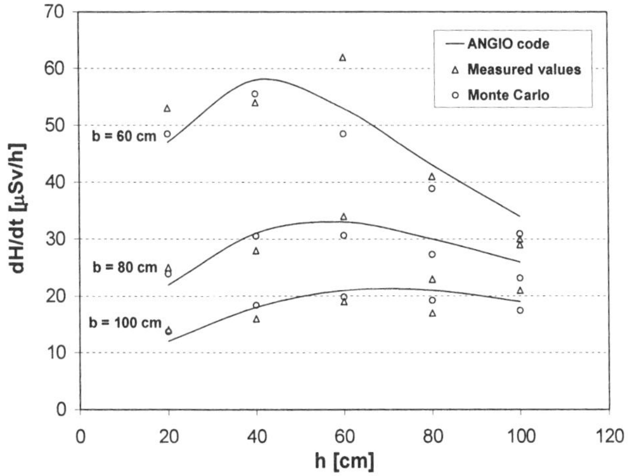

FIGURE 2. Equivalent dose rate of scattered radiation around the water fantom for 60 kV high voltage of the X-ray generator

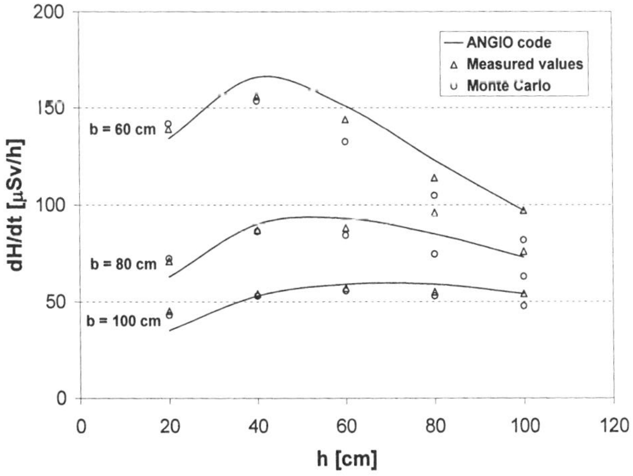

FIGURE 3. Equivalent dose rate of scattered radiation around the water fantom for 80 kV high voltage of the X-ray generator

CONCLUSION

Comparison of the computer programme results with the experimental and Monte Carlo data confirms that the analytic approach is sufficiently accurate for the medical team irradiation assessment. The expressions derived for scatter radiation intensity around a patient are quite general and can be used in calculating of the dose equivalent for the whole set of the existing X-ray techniques. This analytic procedure simplyfies potentialy a collective dose calculation which is an inevitable base for the radiation protection optimization. However, a limitation of this calculating procedure lies in the fact that the spectrum of X-ray tube has to be known in advance.

ACKNOWLEDGMENTS

The Ministry of Sciences, Technologies and Development of the Republic of Serbia supported work on this topic through the Projects No. 1958 (Transport Processes of Particles in Fission and Fusion Systems) and No. 2016 (Radiation Protection Physics).

REFERENCES

1. Sources and effects of ionizing radiation. UNSCEAR 1988, Report of the United Nations Scientific Committee on effects of atomic radiation, Supl. 45, New York, 1988.
2. Marković, S., *Population irradiation reduction in X-ray diagnostics* (in Serbian). MSc thesis, Unuversity of Belgrade, 1993.
3. Marković, S., and Ciraj, O., *Nuklearna tehnologija* **16** (2), 55-59 (2001).
4. Bulatov, B. P., Efimenko, B. A., Zolotuhin, V. G., Klimanov, V. A., and Maskovic, V. P. *Gamma radiation albedo*, Moskow: Atomizdat, 1968.
5. Shultis, J. K., and Faw, R. E., *Radiation shielding*, Upper Saddle River: Prentis Hall PTR, 1996.
6. Chilton, A. B., Shultis, J. K., and Faw, R. E., *Principles of radiation shielding*. Englewood Cliffs: Prentice-Hall, Inc., 1984.
7. Secondary standard dosimetry laboratory "Vinča", Report of the VINČA Institute, 1996.
8. Briesmeister, J. F. (Ed.), *MCNP™ - A general Monte Carlo N-Particle transport code*. Version 4C. LA-13709-M, Manual, LANL, 2000.
9. Data for protection against ionizing radiation from external sources: supplement to ICRP publication 15. ICRP-21. International Commission on Radiological Protection, Graz: Pergamon press, 1971, pp. 319-320

Applied Physics (APHYS 2003)
A. Méndez-Vilas (Ed.)

A PULSED FAST NEUTRON ANALYSIS FOR THE DETECTION OF PLASTIC EXPLOSIVES

SEUNG-KOOK KO, SU-YEOL PARK, BO-YOUNG LEE, and HEE-SEOCK LEE*

Department of Physics, University of Ulsan, Ulsan, 680-749, Korea
**Pohang Accelerator Laboratory, POSTECH, Pohang, 790-784, Korea*

Abstract. The feasibility study to develop the detection techniques of plastic explosives has been carried out using electron linac of the Pohang Accelerator Laboratory. Neutrons produced from thick Bi target irradiated by 2 GeV electrons were used for the pulsed fast neutron analysis (PFNA). The neutron source spectrum and the neutron spectrum transmitted in interesting materials were measured by the time-of-flight method. It is the developing method to inspect explosives to compare both spectra and find out a constituent. Pure graphite, liquid nitrogen and liquid oxygen have been selected for the experiments. The total reaction cross-sections of C, N, and O obtained in these measurements agreed well with ENDF-VI library. Especially the resonance cross-sections between 1 MeV and 8 MeV regions agreed with those within 14 %. The resonance peaks are the key-point to detect plastic explosives in unknown substances.

INTRODUCTION

Among the various non-destructive techniques that are presently developed for the detection of hidden explosives, neutron detection techniques have been shown to possess high qualities that are necessary for an effective detection system. They can examine large volume object with speed and have no memory effects. A number of these developing techniques are presented in many papers[1-4]. It is well known that plastic explosives contain high contents of nitrogen (N), hydrogen (H) and oxygen (O) elements but low carbon(C) concentration. Because hydrogen has no resonance reaction region, resonance reactions of C, N and O elements have been observed through energy-sensitive measurements of the attenuation of a neutron-energy continuum. The three elements C, N, O could be identified through the characteristic resonance reaction emitted from the interaction of neutrons with the corresponding nuclei. C and O are identified through the (n, n′ γ) reaction while N is identified through the (n, γ) reaction[5]. A neutron continuum is produced by a nanosecond-pulsed electron beam slowing down in a thick bismuth target. The neutron spectra transmitted in samples have been measured by the time-of-flight method.

EXPERIMENTAL SETUP

The high-energy neutron facility of Pohang Accelerator Laboratory (PAL) has been used for this experiment. It was conducted at the 10° beam line of the 2.5 GeV electron linac, which was operated in the accelerating energy of 2 GeV, the pulse width of 1 nsec, and the pulse repetition rate of 10 Hz. As shown in Fig. 1, the 2 GeV electrons strike the Pb radiator with 5 mm thickness. The generated photons strike the thick Bi target with 10 cm diameter and 8.1 cm thickness and then the fast neutrons are produced. The fast neutron source spectrum and the attenuation spectrum are measured by a time-of-flight (TOF) technique with the time resolution of 1 nsec and the flight length of 10.4 m and 9.3 m. The fast plastic scintillator, BC418, was used for the neutron detector. A TDC (LeCroy 3377) and a MCS (Fast ComTech 7886A) were used as the time spectral counter. The details of the TOF measurement system were presented in H.S. Lee's papers[6]. The elements of C, N, and O were placed in the neutron path at the front of the Pb collimator as shown in Fig. 1. The graphite (ρ=1.8 g/cm^3), liquid nitrogen and liquid oxygen were selected as the elements, C, N, and O, respectively.

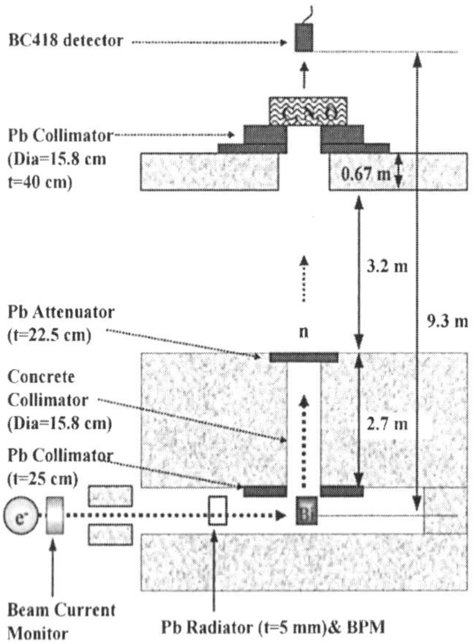

FIGURE 1. Experimental arrangement for measuring neutron spectrum.

The thick Pb attenuator of 22.5 cm was placed at the exit side of a large concrete collimator to reduce huge X-ray flash, which cause spectral noise in the TOF spectrum. This may be most important shortcoming in this detection method to be treated properly in practical state. Lower limit of the neutron energy was set to 1 MeV. In this experiment, the three configurations of elements with neutron detectors were considered as shown in Fig. 2 in order to study the spectral dependence on the material thickness and the geometrical distribution as well as the ability to identify different elements in the layered environment.

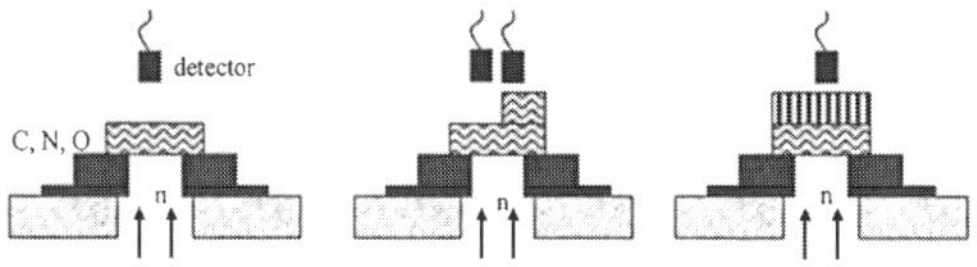

(a) (b) (c)

FIGURE 2. Configuration of elements and detectors for the case study. (a) Attenuation. (b) Geometry dependence. (c) Detection for two elements.

The neutron spectrum has been found by measuring the flight time of neutron. In this time-of-flight method, flight time is the sum of the reach time of gamma flash, the time difference of the peak time of gamma flash and neutron peak as following equation(1).

$$t = l / c + (C_n \tau - C_r \tau), \tag{1}$$

where l is the flight time of neutron, c is light velocity, C_n , C_r are the channel numbers of neutron peak, gamma flash, respectively. τ is the channel width.

EXPERIMENT

5-cm and 10-cm thick carbons have been detected coincidentally with two neutron detectors. These count rates are shown in Fig. 3. Detector BC418 was used for 10 cm thick carbon, Pilot-U detector for 5 cm thick carbon. The differences of total cross-section and the energy with ENDF-VI[7] at interested peak result from the statistical uncertainty due to low count rate. Total reaction cross-section can be calculated by comparing the count rates of neutron detector as shown in equation (2).

$$\sigma_{tôt}(E_n) - \frac{1}{Nt} \ln\left(\frac{C_0(E_n)}{C(E_n)} \right), \tag{2}$$

where E_n, C_0, C are neutron energy in the neutron spectrum, the count rate of the channel in source spectrum and the count rate of the channel in attenuation spectrum, respectively. N is the atomic number density and t is the sample thickness.

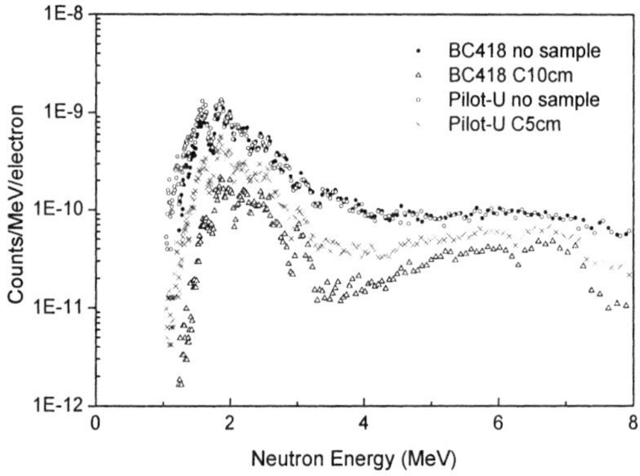

FIGURE 3. Neutron spectrum.

The neutron source spectrum and the attenuation spectrum for carbon have been measured and total cross-section has been calculated by the equation (2). These results were compared with ENDF- VI data, which showed good agreement as shown in Fig. 4. Liquid nitrogen was used for total reaction cross-section of N and liquid oxygen for O. Total reaction cross-sections of N, O are shown in Fig. 5, Fig. 6.

Resonance region of carbon is from 2 MeV to 7 MeV, peaks of resonance cross section appear at 2.074 MeV, 6.296 MeV which values agree with the ENDF- VI data as shown in Fig. 4. Figure 5 shows the resonance region of nitrogen from 1 MeV to 4 MeV, the energy of resonance peak fitting well with the ENDF-VI data is 1.594 MeV. Figure 6 shows the resonance region of oxygen from 1 MeV to 6 MeV, energies of resonance peak fitting well with the ENDF-VI data are 1.318 MeV, 1.652 MeV.

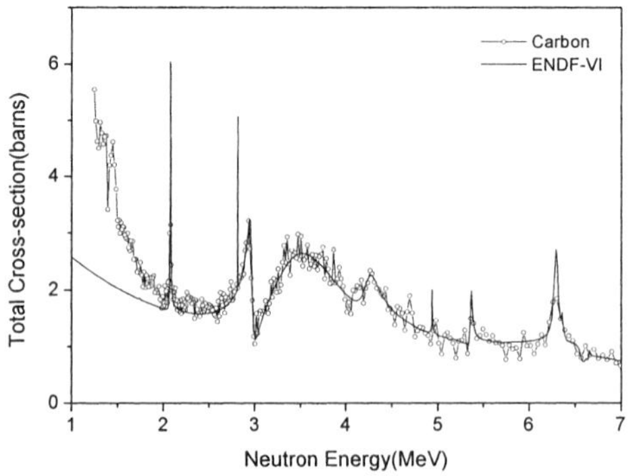

FIGURE 4. . Total reaction cross-section of carbon compared to ENDF-VI library.

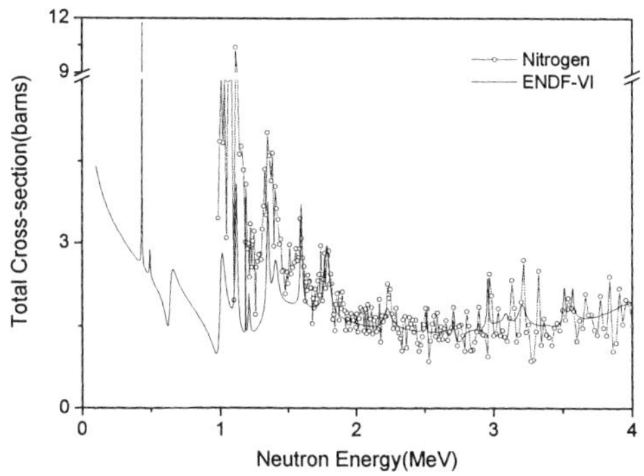

FIGURE 5. . Total reaction cross-section of nitrogen compared to ENDF-VI library.

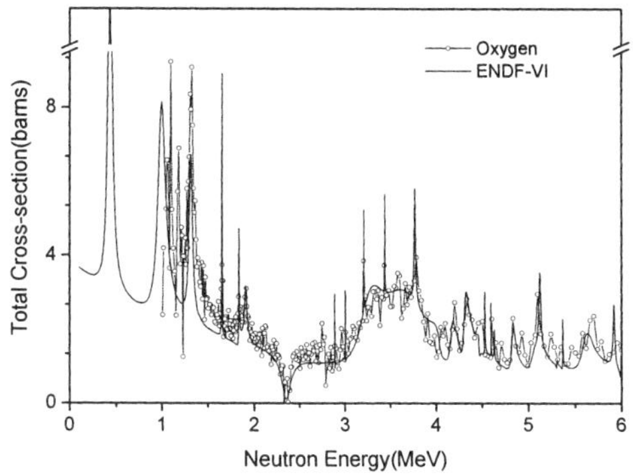

FIGURE 6. . Total reaction cross-section of oxygen compared to ENDF-VI library.

5-cm and 10-cm thick carbon have been detected coincidentally with two neutron detectors, and then total reaction cross-sections were in Fig. 7. Also, detector BC418 was used for 10-cm thick carbon, Pilot-U detector for 5-cm thick carbon. Resonance energies of these samples with different thickness are same with the original at resonant peaks for two different detectors. Resonance energies for distinguishing the sample are summarized in Table 1.

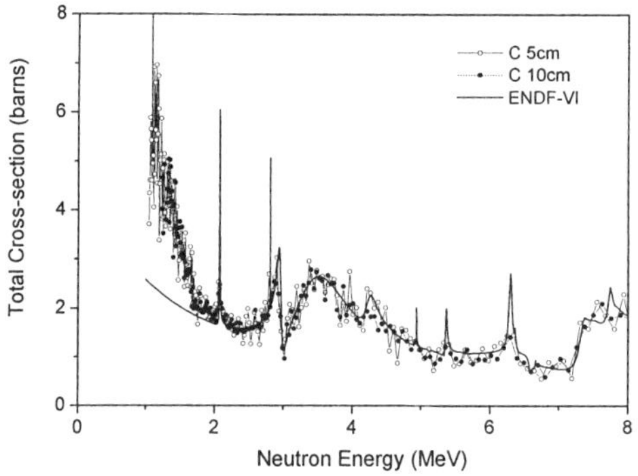

FIGURE 7. Total cross-section of carbon compared to ENDF-VI library by two detectors.

TABLE 1. Resonance energies of C, N and O elements.

	Measurement		ENDF-VI	
	Neutron energy (MeV)	**FWHM (keV)**	**Neutron energy (MeV)**	**FWHM (keV)**
Carbon	2.074	34	2.078	11
	6.296	166	6.296	90
Oxygen	1.318	84	1.315	60
	1.652	31	1.652	6
Nitrogen	1.594	26	1.598	15

RESULTS

Double layered material with 9.75-cm liquid nitrogen and 5-cm carbon has been measured and calculated as shown in Fig. 8. Comparing to the single sample, resonance peak of nitrogen agrees well at 1.594 MeV and energies of carbon fit well at 2.074 MeV, 6.296 MeV.

Double layered material with 10.83-cm liquid oxygen and 10-cm carbon also has been experimented as shown in Fig. 9. 11-cm thick liquid oxygen and 10-cm thick carbon have been used for measurement. Resonance energies of carbon agree well at 2.074 MeV, 6.296 MeV. And energies of resonance peak for liquid oxygen are 1.318 MeV, 1.652 MeV.

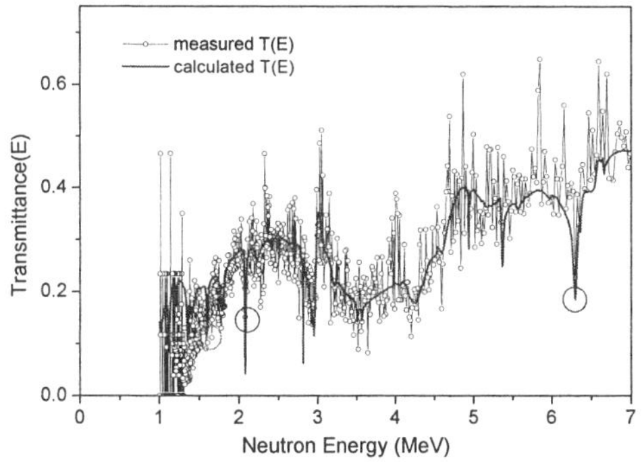

FIGURE 8. Comparison of transmittance for double layered carbon and nitrogen.

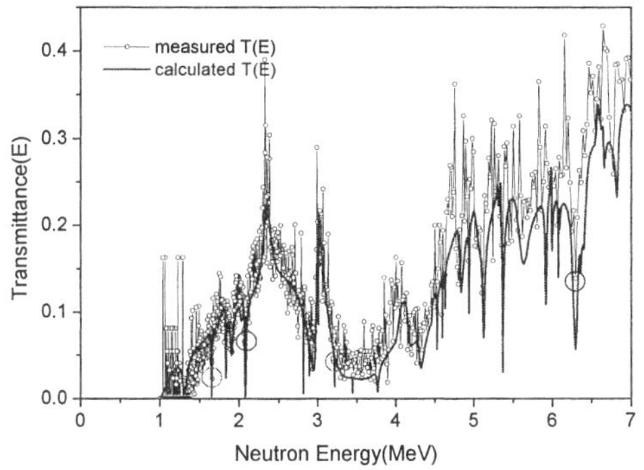

FIGURE 9. Comparison of transmittance for double layered carbon and oxygen.

REFERENCES

1. Esam M. A. Hussein and Edward J. Waller, *Review of one-side approaches to radiographic imaging for detection of explosive and narcotic*, Radiation Measurements. **Vol.29**, No. 6 , 581-591 (1998)
2. Hsiao-Hua Hsu, Kimberlee J. Kearfott, *Effects of neutron source selection on land-mine detection efficiency*, Nucl. Instr. Meth. **A 422**, 914-917 (1999)
3. J.C Overley, M.S. Chmelik, R.J. Rasmussen. R.M.S. Schofield, H.W. Lefevre, *Explosives detection through fast-neutron time-of-flight attenuation measurements*. Nucl. Instr. Meth. **B99**, 728-732 (1995)
4. C.L. Fink, B.J. Micklich, T.J. Yule, P. Humm, L. Sagalovsky and M.M. Martin, *Evaluation of neutron techniques for illicit substance detection*, Nucl. Instr. Meth. **B99**, 748-752 (1995)
5. G. Vourvopoulos and F.J. Schultz, *A pulsed fast-thermal neutron system for the detection of hidden explosives*. Nucl. Instr. Meth. **B79**, 585-588 (1993)
6. H.S. Lee, et al., *Photoneutron Spectra from Thin Targets bombarded with 2.0 GeV Electrons*. Nucl. Sci. Tech. Suppl. l, 207-211 (2000)
7. http://www.nndc.bnl.gov/nndc/endf

Applied Physics (APHYS 2003)
A. Méndez-Vilas (Ed.)

Study Of Two Sequential Extraction Methods And Its Application To Environmental Radioactivity Measures

V. Peña[1], J.C. Nalda[1], C. Cazurro[2] and R. Pardo[2]

1 LIBRA laboratory. I+D Building. University of Valladolid. Avda. Belén 47011. Valladolid. SPAIN
E-mail: victorp@libra.uva.es
2 Departament of Analytical Chemistry. Sciences Faculty. C Real de Burgos s/n. Valladolid. SPAIN
E-mail: ccp@qa.uva.es

Abstract. Determination of background and reference levels of radionuclides in environmental is essential for studies of dynamics of pollutants. For best determination of kinetics characteristics is important know mobility of pollutants from the soil to the other environmental cells. Determination of metals mobility by sequential extraction methods is great known method in analytical chemistry. Its application in radiochemistry had been used with independence of chemical research. In this study possibility of application of two extraction methods for heavy metals (BCR method and NIST method) to Uranium and Thorium measures in environment has been evaluated. Both methods has been applied to reference samples to discern the best method. In a second part it will be used for determination of U and Th contents in samples collected round a coal-mining zone (Guardo, Spain) when we has studying mobility of actinides in environment. U and Th isotopes have been measured by alpha spectrometry in all fractions of sequential method for each sample. This measure provides us with important information for mobility studies.

INTRODUCTION

Pollution by heavy metals and radionuclides is today one of the more pressing environmental concerns. The heavy metals are a very hazardous group of pollutants because of their toxic and accumulative characteristics, whereas the radionuclides add their radioactivity hazards. There are several alternatives to determine their contents in environmental soils: total, extractable, bio-available etc. Background and reference levels are determined on the basis of total contents, but in order to determine the real environmental impact of soils, the extractable or bio-available levels are preferred, because the more mobile and bio-available the metals and actinides are, the more hazardous become. That mobility is related to the form of association of the metals in the soil and can be studied by extraction procedures that, despite uncertainties such as the selectivity of the reagents, provide qualitative evidence regarding the forms of association of metals. It is important know mobility of pollutants from soils to other environmental cells when concentration of metals and radioactive isotopes in the environment is studied.

Most important advances in pollutants extraction has been performed in Analytical Chemistry for the study of pollution by heavy metals as Cd, Cr, Cu, Ni, Pb and Zn. The methodology is simple but there are two main alternatives: single extractants or sequential extraction schemes [1] [2] [3]. These last ones become more useful and imply the sequential treatment of the soil with a series of chemical reagents with increasing reactivity. The metals present in the sample are thus released in order of decreasing mobility, being the more dangerous those present in the first fractions. Metals in the extracts are then determined by standard analytical methods.

Application of these techniques in radioactivity studies is not usual [4]. Only a few references could de founded in bibliography. For this reason we established a collaboration between a low radioactivity measurement laboratory and investigation group in analytical chemistry for the application of sequential extraction techniques in radioactivity analysis that has been carried out by us in environment round coal-mining zone in Guardo, (Palencia – Spain). More exact knowledge can be taken from information provided by these techniques, in special for the study of environmental transport methods that happen in nature. The application of these techniques is also important for the study origin of the different fractions on sediments.

This work look for two clear objectives: first, knowledgement the applicability of sequential chemical extraction techniques for measures of radioactivity isotopes in each extract; and second, determination of optimum method in the laboratory.

METHODOLOGY

There are different methods of sequential extraction depending on number of extracts, chemical reagents are used and their quantities. The common objective of all there is progressive extraction in different phases of pollutants. In present work two specific methods known as B.C.R [5] [6](Standards Measurements and Testing programme of the European Commission) and N.I.S.T [7] [8](National Institute of Standards and Technology, USA) schemes had been compared. These methods are illustrated in figure 1 and 2. The BCR scheme (also named common scheme) is a derivation of Tessier's scheme and is mainly used by European researchers of the analytical chemistry area, working on heavy metal pollution. However, radiochemical and American researchers use a Tessier-type [3] scheme endorsed by N.I.S.T. The number of extractants, experimental conditions and fractions are different in both of the schemes, thus hindering the comparison of results obtained by the two schemes.

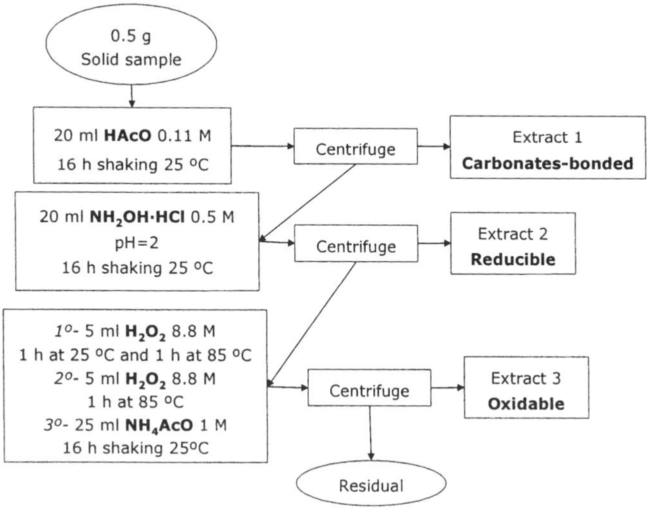

FIGURE 1: BCR sequential extraction scheme

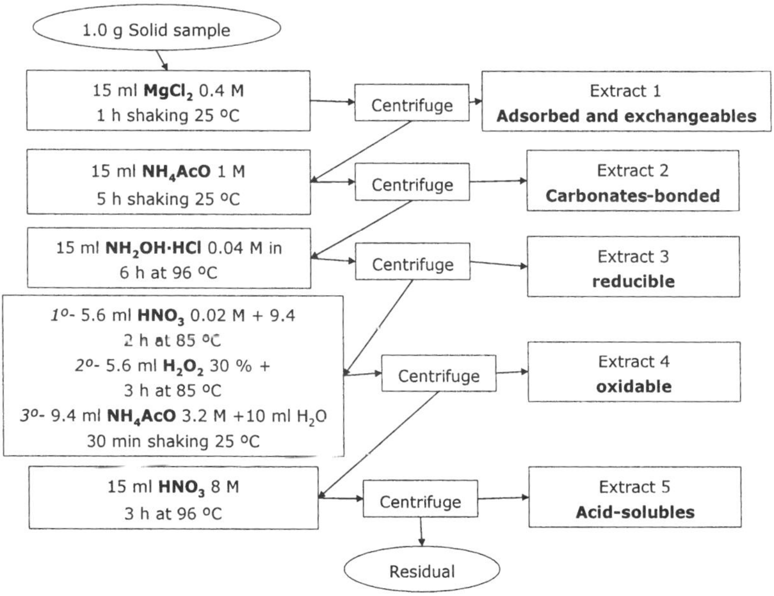

FIGURE 2: NIST sequential extraction scheme

For the comparison of the obtained results, mean comparison tests (significance level $\alpha=0.05$) were carried out on similar fractions of the two schemes as is explained on table 1. For the comparison these two methods were applied for two references sources: CRM 601 (used in the validation of B.C.R. scheme) and N.I.S.T. SRM 4357 (environmental radioactivity standard).

TABLE1: Hypothesis for comparison between NIST and BCR methods.

BCR method	NIST method	Reason
Fraction 1	Fractions 1 + 2	First fraction of NIST scheme doesn't have any similar fraction at BCR scheme
Fraction 2	Fraction 3	Similar reagents
Fraction 3	Fraction 4 + 5	Similar reagents in 3 BCR and 4-NIST

Before the application of these techniques for these samples we determine concentration of Cd, Cr, Cu, Ni, Pb and Zn in one hand, and ^{238}U, ^{234}U, ^{232}Th, ^{230}Th y ^{228}Th in each obtained extract. Total contents are also determined because they are necessary to obtain the quantities of metals and radionuclides in the residual fraction. The procedure for metals consists of a microwave-assisted digestion of the soil sample with concentrated HNO3 according to the E.P.A. 3051 norm. In the case of radio nuclides a HNO3 - HF (1:1) mixture is used for digestion. These elements are determined in the liquid fractions originating from digestions and extraction steps by standard analytical methods: ICP - AES in the case of non-radioactive elements, and α spectrometry for Uranium and Thorium isotopes with a Alpha Analyst Silicon detectors spectrometer. For that analysis TBP [9] (Tri-n-butyl phosphate) liquid-liquid extraction for the isotopic separation of U and Th had been performed. Samples containing Th were purified after with a UTeva (Eichrom INC.) ionic interchange column with the method exposed by Pilvio [10]. Samples were electroplated according to Talvitie schema [11].

RESULTS AND DISCUSSION

In figures 3 the results for determination of heavy metals contents on each extraction solution and the residue for reference sources after application of cited sequential schemes are presented. The residue contents had been calculated as the difference between total contents and the sum of contents in each fraction for each sample. Great differences in the application of both methods are observed. A priori proposed comparison is not corroborated by results in general. The election of optimum method must considerate this fact.

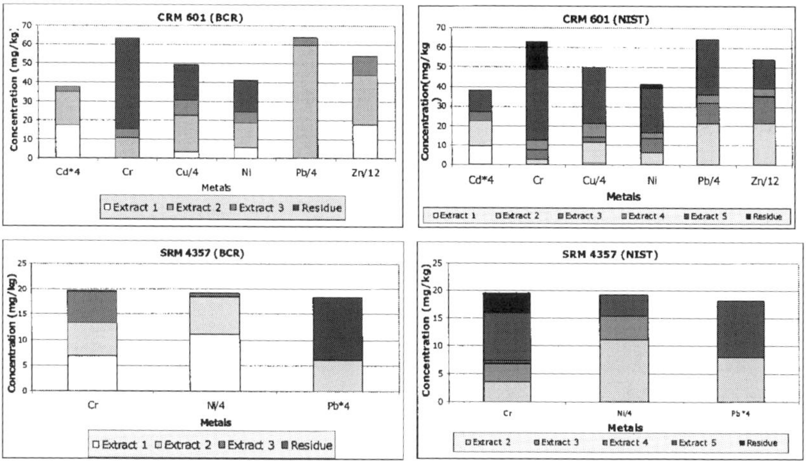

FIGURE 3: Heavy metals contents on extracts from CRM 601 and SRM 4357 samples after application of BCR and NIST methods.

In tables 2 and 3 we present the results obtained for the measures of radionuclides in each extract and the residue of reference samples. Another time, results are great different depend on used method. The most important part of activity remains in residual phase and/or last extract, overcoat in BCR scheme. The most important difference is the measured activity in the second extract of NIST method. More possible information is needed in each analysis when we applied these techniques for real samples. For this reason, NIST method provides us best results despite longer number and time of analysis.

TABLE 2. Activity in Bq/Kg (σ) in fractions for SRM 4357 sample after application of sequential extraction methods.

SRM 4357		^{238}U	^{234}U	^{232}Th	^{230}Th	^{228}Th
Total contents		11,8 (1,1)	11,9 (1,1)	15,3 (1,5)	13,6 (1,4)	10 (2)
BCR	Extract 3	0,84 (0,09)	0,8 (0,08)	1,12 (0,12)	1,13 (0,12)	0,97 (0,09)
	Extract 2	0,27 (0,05)	0,31 (0,04)	0,42 (0,05)	0,39 (0,03)	0,38 (0,03)
	Extract 1	ND	ND	ND	ND	ND
NIST	Extract 5	2,4 (0,3)	3,1 (0,3)	6,3 (0,7)	5,4 (0,6)	5,3 (0,7)
	Extract 4	ND	ND	ND	ND	ND
	Extract 3	ND	ND	ND	ND	ND
	Extract 2	0,39 (0,04)	0,42 (0,05)	0,57 (0,06)	0,51 (0,06)	0,54 (0,06)
	Extract 1	ND	ND	ND	ND	ND

TABLE 3. Activity in Bq/Kg (σ) in fractions for CRM 601 sample after application of sequential extraction methods.

	SRM 4357	^{238}U	^{234}U	^{232}Th	^{230}Th	^{228}Th
	Total contents	50 (5)	50 (5)	62 (6)	56 (5)	61 (7)
BCR	Extract 3	4,1 (0,4)	4,9 (0,5)	4,2 (0,3)	4,1 (0,4)	5 (0,8)
	Extract 2	2 (0,2)	2 (0,2)	0,37 (0,07)	0,48 (0,09)	0,41 (0,08)
	Extract 1	0,27 (0,05)	0,43 (0,06)	ND	ND	ND
NIST	Extract 5	8,5 (0,9)	8,3 (0,9)	43 (4)	37 (3)	42 (4)
	Extract 4	1,2 (0,2)	1,4 (0,2)	ND	ND	ND
	Extract 3	2,6 (0,3)	2,9 (0,3)	0,82 (0,09)	0,76 (0,08)	0,79 (0,08)
	Extract 2	9,4 (1,1)	11,5 (1,3)	1,3 (0,2)	1,2 (0,2)	1,3 (0,2)
	Extract 1	0,34 (0,08)	0,36 (0,08)	ND	ND	ND

CONCLUSIONS

A comparison between two chemical sequential extraction methods and applied their for measurements of heavy metals and radionuclides had been performed These conclusions has been obtained:

- Application of extraction techniques for measurements of radioactivity is possible and advisable because provide us necessary information of the bio-availability of the considered isotopes. These results are very important in environmental studies.

- There are many different extraction schemes and the results change depending on used method because we don't know comparables schemes. Use of different reagents modified the results clearly.

- In most cases the most important part of activity remains in residual extracts. The hazard is less for this reason.

- NIST method provide us more information that BCR scheme for measures of radioactive materials in environment.

Note that SRM 4357 is a radioactive reference sample meanwhile CRM 601 isn't it.
This work must be completed with the application of these methods for different type of real samples (like soils, sediments, ground waters, etc…).

ACKNOWLEDGMENTS

This work is part of a important study that is being realized in LIBRA laboratory in collaboration with Analytical Chemistry department supporting by "Junta de Castilla y Leon", Spain Science and Technology Ministry (Project PPQ2002/01054) and Iberdrola S.A.

REFERENCES

(1) Ure A.M., Mikrochim. Acta, 11, 49 (1991).
(2) A. Kot, J. Namiesnik, Trends in Anal. Chem., 19, 69 (2000).
(3) A.Tessier. P.G.C. Campbell, A. Bisson, Anal. Chem., 51, 844 (1979).
(4) Ivanovich and Harmon, *Uranium Series disequilibrium. Applications to Environmental Problems.* Oxford Science Publications (1982).
(5) A.M. Ure, P.H. Quevauviller, H. Montau and B. Griepink, Intern. J. Environ. Anal. Chem., 51, 135 (1993).
(6) R. Pardo, M. Vega, E. Barrado, Y. Castrillejo and F. Prieto, Química Analítica, 20, 63 (2001).
(7) M.K. Schultz, K.G.W. Inn, Z.C. Lin, W.C. Burnett, G. Smith, S.R. Biegalski, J. Filliben, Appl. Radiat. Int, 49, 1289 (1998).
(8) D.J. Greeman, A.W. Rose, J.W. Wasinghton, R.D. Dobos, E.J. Ciolkosz, App. Geochem., 14, 363 (1999).
(9) Holm, Rioseco and Garcia-Leon, Nucl. Instr. Meth. 223, 204 (1984).
(10) R. Pilviö, M. Bickel, Appl. Rad. Isot. 53, 273 (2000).
(11) Talvitie, Anal. Chem. 44, 280 (1972).

Applied Physics (APHYS 2003)
A. Méndez-Vilas (Ed.)

Investigation On The Soil Profile Around DU Projectile Three Years After Contamination

Mirjana Radenković, Jasminka Joksić, Dragana Todorović, Milojko Kovačević and Jagoš Raičević

Radiation and Environmental Protection Department,
Institute of Nuclear Sciences VINČA, P.O.Box 522, Belgrade, Serbia and Montenegro

Abstract. During the war conflict at former Yugoslavia territory in 1999, depleted uranium (DU) ammunition was used by NATO air forces and the Bratoselce site have been contaminated. Three years later decontamination of the location had been undertaken. During cleaning up the site, samples of projectile and soil samples around it, along the depth profile and parallel to the projectile centerline were collected and analyzed by gamma and alpha spectrometry. Measurements were performed by gamma spectrometry system (HP Ge detector, efficiency 23%) and high-resolution alpha spectrometry systems (PIPS detectors, efficiency 7% and 15%). Concentrations of present uranium, radium and plutonium isotopes were determined and compared with naturally occurring levels of the location investigated. Results indicate that high DU contamination of the soil layers next to the penetrator decrease rapidly with the distance to the 1% of initial values that is about double background levels for the investigated location.

INTRODUCTION

During the war conflict at former Yugoslavia territory, in 1999, depleted uranium (DU) ammunition was used by NATO air forces and some areas of South Serbia have been contaminated. About 1300 DU projectiles[(1)] have been fired into the field of $5400 m^2$ of the Bratoselce site (Fig.1.) in May and June 1999.

This terrain was one half wooded and a half is meadow, lightly upland and very exposed to the local meteorological conditions. Three years later, in September 2002, the clean up operations of the contaminated area have been started. During the first phase of this action about 200 projectile remains with surrounding contaminated soil have been detected and removed.

The measurable contamination levels found at the location have shown that there is no widespread contamination over the surface soil but localized points of concentrated contamination around the spots of projectile impacts into the ground. It could be the indication of a small number of projectiles that hit the hard target and produce airborne radioactivity. The most of them have penetrated almost intact into the ground and remain there. They are found without the aluminum jacket, mostly as not fragmented penetrators coated with the thick layer of uranium oxides.

The contamination levels of the soil profiles around the projectile buried in the ground will be investigated here by the means of gamma and alpha spectrometry methods. This study should provide insight into the downward transport of DU through the soil at the investigated location, based on the state three years after contamination.

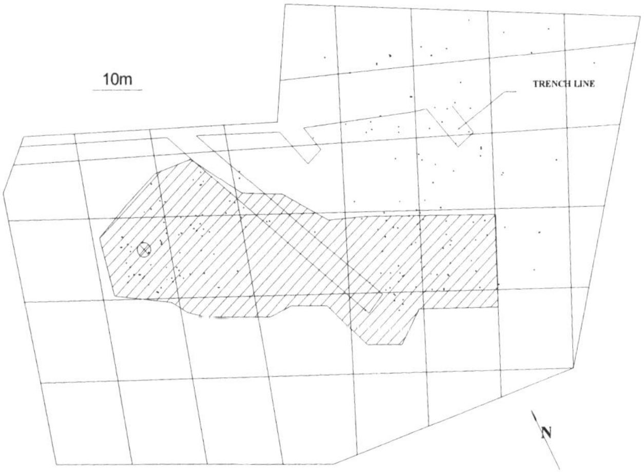

FIGURE 1. Contaminated area of the Bratoselce site. Legend: • -projectile found, ⊗ -sampling location, //// -decontaminated section

It is also important from the aspect of the contaminated soil displacement from the environment in the clean up operations.

MATERIALS AND METHODS

DU penetrator and the soil samples along the downward and sideward profile were collected at the marked position of the contaminated area. (Fig.1.) After the surface soil was removed to the depth of 0.2-0.3 m, the aluminum jacket belonging to the projectile was found. It was crumbled in the projectile entrance corridor path. About 0.5 m deeper the penetrator was found. It was buried into the ground, turned up again and after lost of kinetic energy stopped in almost vertical position.

The penetrator was corroded and covered with the uranium oxides allover. Samples of surrounding soil were collected at the following positions: projectile entrance at the ground surface, along the projectile path through the subsurface ground, the layer just next to DU penetrator, a few layers down along the depth profile and a few at different distances sidewise. About 90 g of each sample was homogenized, sealed for one month and analyzed by gamma spectrometry using HP Ge detector with relative efficiency of 23%.

The topsoil samples from the depth of 0-0.15 m were taken randomly inside the contaminated area, and a few to determine the natural uranium levels outside the contamination zone. Samples were analyzed in the Marinelli geometry, by gamma spectrometry method.

The quantity of 10 g of a few noted samples was taken for alpha-spectrometry analysis. Prior radiochemical procedure was performed before alpha spectrometry measurements. Uranium and plutonium isotopes separation was done by ion-exchange procedure [(2)], followed by Talvitie's procedure [(3)] of radioactive source preparation by electrodeposition on stainless still plates. Tracer solutions of ^{232}U and ^{236}Pu were used to determine radiochemical yield. High-resolution alpha spectrometry was performed using alpha vacuum chambers with PIPS detectors of 100 and 300 mm^2 surfaces and efficiency 7 and 15% respectively.

RESULTS AND DISSCUSSION

The uranium isotopes activities in the analyzed samples of soil around DU penetrator are presented in the Table 1. After the radioactive equilibrium achievement, the ^{235}U activity of the samples was determined on the basis of gamma energy of 163 keV and the ^{238}U activity on the basis of 1000 keV gamma energy. Concentration of ^{226}Ra was determined from daughter's gamma energies and presented in the Table 1, too. The ^{232}Th content in the samples was 72±12 Bq/kg.

TABLE 1. Radionuclides Content In The Soil Samples Around DU Penetrator.

N°	Location	^{235}U (Bq/kg)	^{238}U (KBq/kg)	$^{235}U/^{238}U$ (Bq/Bq)	^{226}Ra (Bq/kg)
1	Entrance into the ground	776±124	90.12±14.42	0.0086	142±20
2	Projectile path through the soil	580±87	78.49±7.85	0.0074	122±17
3	Soil layer just next to penetrator	2174±370	263.40±26.34	0.00825	255±41
4	1 cm downwards	72±13	8.52±0.85	0.0085	107±17
5	2 cm downwards	24±6	2.87±0.29	0.008	111±14
6	4 cm downwards	7.9±1.2	0.82±0.20	0.0096	141±20
7	10 cm downwards	1.2±0.2	0.16±0.06	0.0077	151±26
8	2 cm to the side	25±5	3.44±0.62	0.0073	116±20
9	8 cm to the side	1.1±0.2	0.15±0.05	0.0069	154±15
10	15 cm to the side	0.40±0.07	0.10±0.04	0.0034	141±20

Results have shown high DU contamination in the very vicinity of the penetrator and rapid devolution with the distance through the soil layers both downward and sideward of its centerline. According to the obtained results, the distance of about 150 mm is enough to decrease DU contamination level to 1% of the initial value. The contamination gradient is uniform in two studied directions around the projectile. It will be more apparent at the Fig.2, representing the ^{238}U content in the layers of the profiles around DU penetrator.

Although strongly localized, there is still relatively high contamination of the surface ground. The main reasons should be resuspension and insolubility of present chemical forms of depleted uranium, processes depending mainly on the meteorological conditions and geomorphologic characteristics of the terrain.

Alpha spectrometry analysis of topsoil samples had shown the natural occurring ^{238}U values to be 50-70 Bq/kg at Bratoselce location, known as granite petrology regions. It means that after about 150 mm distance to DU penetrator, uranium activity levels in the soil become double of the background levels for the investigated location. Taking into account this fact, the quantity of the soil that should be removed during the clean-up operations can be considered.

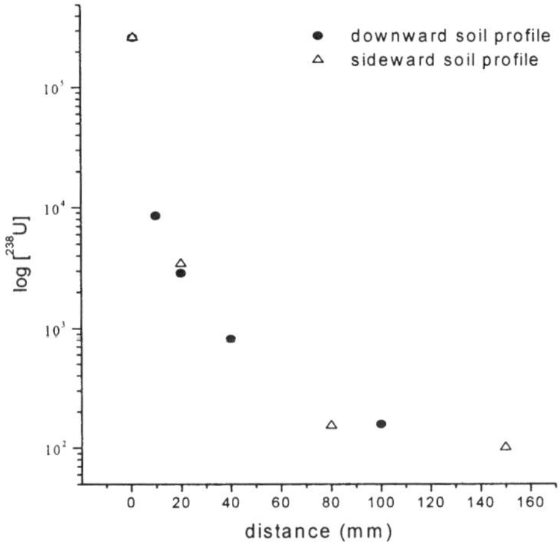

FIGURE 2. Content of ^{238}U in the layers of the profiles around DU penetrator

The penetrator analysis from this location was already done by alpha spectrometry method and reported earlier (4). Traces of ^{236}U, ^{239,240}Pu and ^{237}Np have been found, indicating that "dirty" uranium was a part of depleted uranium ammunition used at the site. In the same way, the sample 5 was analyzed to determine a content of transuranic element ^{239,240}Pu at the 0.4 m distances but its concentration in the sample was not detectable.

CONCLUSION

Study on the contamination levels of the soil around DU penetrator buried in the ground of Bratoselce site had shown strongly localized but still significant surface contamination. Resuspension process and the certain weathering conditions and geomorphologic characteristics of the site will play a major role in DU behavior at the ground surface.

Possible air and groundwater contamination subsequently implies the health risk of DU inhalation and/or ingestion by the local population. (5)

Decrease of DU activity through the soil profiles around penetrator is uniform in two investigated directions and rapidly decrease with a distance. Transport of DU through the soil will depend on the penetrator state, too. Corrosion process already started and it will rapidly go on to the uranium oxidation products formation. That could imply possible future contamination of drinking water supplies. (5)

Due to the named reasons, first of all possible internal human population, DU penetrators with surrounding soil layers should be removed from the contaminated areas. The uranium activity levels in the polluted environment should be systematically controlled by the monitoring system.

REFERENCES

1. UNEP, Depleted Uranium in Serbia and Montenegro, Post-Conflict Environmental Assessment in the Federal Republic of Yugoslavia, Geneva: UNEP (2002).
2. Radenković, M., Vuković, D., Šipka, V., and Todorović, D., *J.Radioanal.Nucl.Chem.* **208/2**, 467-475 (1996).
3. Talvitie, N. A., Anal. Chem. **44**, 280-283 (1972).
4. Šipka, V., Radenković, M., Paligorić, D., and Djurić J., Proceedings of XXI Symposium of Yugoslavian Radiation Protection Association, Belgrade: Institute of Nuclear Sciences Vinča and YRPA, 2001, pp.69-72.
5. Fetter, S., and Hippel, F.N., *Science and Global Security* **8/2**,125-161 (1999).

Applied Physics (APHYS 2003)
A. Méndez-Vilas (Ed.)

Radioactive Disequilibrium of Naturally Occurring Radionuclides in Mineral Waters of Metamorfic Rock Balkan Area

Jasminka Joksić[1], Mirjana Radenković[1], Branislav Potkonjak[2], Snežana Pavlović[1], Dragana Todorović[1]

[1] *Vinča Institute of Nuclear Sciences, Radiation and Environmental Protection Laboratory P.O.BOX 522, 11001 Belgrade,Serbia and Montenegro*

[2] *IHTM-ITR, Njegoševa 12, 11000 Belgrade, Serbia and Montenegro*

Abstract. Uranium series disequilibria in the hydrosphere occur due to geochemical differentiation processes resulting with different mobility of the radionuclides from the same series. Radioactive equilibrium in the mineral waters Čibutkovica and Studenica, originated from metamorphic rock area of the South Serbia was the object of investigation.
Enhanced total alpha/beta activity found in examined natural mineral waters with similar chemical composite and origin indicated high content of naturally occurring radionuclides, members of ^{238}U and ^{232}Th series. Gamma and alpha spectrometry analysis have been undertaken and present natural radionuclides quantified. It was found that content of ^{226}Ra exceeds the value that expected from content of ^{238}U, in Studenica water. Rise in the activity level coresponds to the ^{226}Ra.
In Čibutkovica mineral water disequilibria exist, both in the uranium-238 and thorium-232 series. Content of ^{226}Ra, that should be in equilibrium with ^{238}U, much exceeds the value which is expected. In thorium-232 series there exists a disequilibrium and concentrations of ^{228}Th, ^{228}Ac and ^{228}Ra are much higher than expected from present ^{232}Th. High content of radium isotopes in analysed natural mineral waters indicates contribution from other sources than water, probably environmental sediment.

INTRODUCTION

Mineral waters Čibutkovica and Studenica, originated from metamorphic rock area of the South Serbia. These are carbon acid waters, with similar chemical composite and the same pH value (6.5). Sample preparation and chemical procedure for alpha counting.

EXPERIMENTAL

Alpha spectrometry procedure for uranium and thorium isotopes included: sampling, preliminary samples treatment, ion-exchange chemical separation, ion-exchange purification of separated elements, thin-layer source preparation and alpha spectrometry measurements.[(1)]

Sample preparation and chemical procedure for alpha counting

After collection water samples have been acidified to pH=2, to keep trace elements in the solution, then evaporated, dried and ashed at 550°C.
^{232}U and ^{229}Th have been aded in a quantity of about 0.1 Bq as a tracer to each sample, for radiochemical yield recoveries.
After the preconcentracion treatment, samples have been leached by HNO_3, HCl (3:1), in order to attain complete dissolution of present elements. The initial steps of chemical procedure were filtration, oxidations (H_2O_2, $NaNO_2$), precipitation (NaOH, pH >10), dissolutions and evaporations to dryness.
Selective adsorption and desorption of the uranium and thorium on the ion-exchange resin (DOWEX 1x8, 100-200 mesh) was done. In order to remove the ferum ions, expected to follow the uranium fraction, we have done di-isopropilether extraction (a few if necessary).
Electroplating of purified fractions, was used to made thin-layer radioactive sources.After electrodeposition, a thorium source was covered by vyns-foil.

Sample preparation and chemical procedure for gamma spectrometry

Samples have been concretrated by evaporating 10 l water to 450 ml for gamma analysis. After preparation, samples have been saled in the acril containers for 4 weeks to establish radioequilibrium between ^{228}Ra and ^{228}Ac and ^{226}Ra and ^{222}Rn.

MEASUREMENTS

Low-level activity measurements have been done by the use of Canberra 2004 alpha-spectrometry counting system, including vacuum chamber (20 mbar), PIPS-detector (300 mm^2 surface), with: counting efficiency 15%, at 25mm distance; multichannel energy scale 9.1 keV/ch, resolution 24 keV for ^{241}Am.

The counting time required had to be a few days, that is long enough to ensure an accurate result.

The gamma activity of the samples has been counted using high purity Ge detector,with counting relative efficiecy 23 %.

RESULTS AND DISCUSSION

Enhanced total alpha/beta activity found in examined natural mineral waters with similar chemical composite and origin indicated high content of naturally occurring radionuclides, members of ^{238}U and ^{232}Th series. Gamma and alpha spectrometry analysis have been undertaken and present natural radionuclides quantified.[(2)]

The uranium and thorium series nuclides in these waters should come from the leaching of the metamorphic rock . It is sorption / desorption process of the elements between water and rock. According to the age of the rock, both uranium and thorium series nuclides shoud reach radioactive equilibrium, which means that the radioactivities of a radionuclide and its daughter nuclide are equal. If radionuclide and its daughter nuclide are of the same isotopic elements, it means that they have the same chemical properties and their activities should remain equal after leaching.[(3)]

TABLE 1. Uranium and thorium isotopes activites in water samples

Water Samples	U-238 Bq/l	U-234 Bq/l	Th-232 Bq/l	Th-230 Bq/l	Th-228 Bq/l	Ra-228 Bq/l	Ra-226 Bq/l
Čibutkovica	0,0022	0,0021	0.0009	0,0024	0,0489	0.6085	0,0924
Studenica	0.0038	0.0047	0.00044	0.00035	0.006	/	2.2

Degree of the radioactivity disequilibria for U-234 /U-238 is not great in both mineral waters Čibutkovica and Studenica because the ratios of U-234/U-238 are 0.95 and 1.24, reapectively. Obtained results shows that rise in the activity level coresponds to the ^{226}Ra, in Studenica water. Ratio of Ra-226/Th-230 has been 50000. This ratio for Čibutkovica water has been 38.5. In Studenica water was not found disequilibria in Th-232 series.

In Čibutkovica mineral water exist disequilibria, both in the uranium-238 and thorium-232 series. Content of ^{226}Ra, that should be in equilibrium with ^{238}U, much exceeds the value which is expected. In thorium-232 series there exists a disequilibrium and ^{228}Th, ^{228}Ac and ^{228}Ra concentrations are much higher than expected from present ^{232}Th. Ratios Th-228/Th-232, Ra-226/Th-230 and Ra-228/Th-228 are 54.3, 38.5 and 12.4. In the decay process from Th-232 to Th-228 there is a Ra-228 with $T_{1/2}$ = 5.75 y. Because the chemical properties of radium and thorium isotopes are different , high continent Th-228 may derive from Ra-228.

Radioactive disequilibrium of uranium and thorium series observed in nature, in most of the cases originated from hydrosphere, especially from ground water and aquifer environments. There is three differentiation processes at liquid/solid phase boundary:

1. Solution and precipitation, some of the daughters of a series are more soluble than others;
2. Diffusion, radon diffusion can produce disequilibria;

3. Alpha recoi of daughter products during the process of alpha decay, which can produce disequilibria at phase boundaries.[4]

TABLE 2. Radioactivity ratios of U-234 / U-238 and Th-228 / Th-232 in mineral waters

Water Samples	U-234 / U-238	Th-228 / Th-232	Ra-226/ Th-230	Ra-228 / Th-228
Čibutkovica	0.95	54.3	38.5	12.4
Studenica	1.24	13.6	50000	/

The Th-228 / Th-232 activity ratio varies betwen 5 and 30 in surface and deep waters of the ocean. Thus, only small part of the measured Th –228 derives from Th-232 in the water, and confirms that the rast must derive from the Ra-228 input to the water from the bottom sediments.[4]

ACKNOWLEDGMENTS

Uranium series disequilibria in the hydrosphere occur due to geochemical differentiation processes resulting with different mobility of the radionuclides from the same series. We have investigated radioactive equilibrium in the mineral waters Čibutkovica and Studenica, originated from metamorphic rock area of the South Serbia.

High content of radium isotopes in analysed natural mineral waters indicates contribution from other sources than water, probably environmental sediment.

REFERENCES

1. Radenković,M., Vuković, D., ŠipkaV., Todorović, D., *J.Radioanal.Nucl.Chem.*, **208/2**, 467-475 (1996)
2. Choukri, A., Moutia, Z., Hakam., O.K., Guessous A., Lferde, M., Cherkaoui, R.,Semghouli, S., Mouhiddine M., and Reyss, *M.J.Environ.Radioactivity*, **57**, 175-189 (2001 a)
3. Tieh-Chi Chu, Jeng-Jong Wang, *J.Nuc.Rad.Sciences*, **1/1**, 5-10, (2000)
4. Ivanovich, M., Russel, S., *Uranium Series Disequilibrium Applications to Environmental Problems*, Oxford: Clarendon, 1982

Applied Physics (APHYS 2003)
A. Méndez-Vilas (Ed.)

Absolute Activity Measurement Of Thallium-204 By Efficiency – Tracing Method

L. Mo, M. Smith, M.I. Reinhard, J. Davies and D. Alexiev

Australian Nuclear Science and Technology Organisation
PMB 1 Menai 2234 NSW Australia
E-mail: lmx@ansto.gov.au

Abstract: The Australian Nuclear Science and Technology Organisation (ANSTO) participated in the year 2002 international intercomparison of activity measurements of a thallium-204 solution organised by the Bureau International des Poids et Mesures (BIPM). $4\pi\beta-\gamma$ coincidence counting efficiency - tracing extrapolation method was used for the standardisation of this radionuclide. Cobalt-60 was used as the tracer. VYNS, aluminium and nickel foils were used to attenuate the β – emission in order to vary the β- detection efficiency. The activity concentration of the BIPM original solution was found to be 101.73 ± 0.55 kBq/g, which is in agreement to better than 0.17% with the mean value of the International Intercomparison results.

INTRODUCTION

The year 2002 international intercomparison of activity measurements of a thallium-204 solution is the third comparison on Tl-204 organised by the Bureau International des Poids et Mesures (BIPM). The main purpose of the thallium-204 comparison was to allow BIPM to check its capability to use its liquid scintillation spectrometer to extend the Systeme International de Reference pour la Mesure d'Activite d'Emetteurs γ (SIR) to β – emitting radionuclides.

Thallium-204 decays by 97.1% beta emission to the ground state of lead-204 and 2.9% electron capture to the ground state of mercury-204. No γ radiation is emitted in the decay. The maximum beta energy is 763.72 keV. The half-life of thallium-204 is 1381 ± 1 days (1).

The standardisation of this pure beta radionuclide was carried out by ANSTO using $4\pi\beta-\gamma$ coincidence counting efficiency - tracing technique (2) (3) (4) (5). Cobalt-60 was used as the tracer to establish the beta - counting efficiency.

Cobalt-60 is a $\beta-\gamma$ emitter and has a simple decay scheme. It decays by beta emission, predominantly to the 2505.75 keV excited state of nickel-60. This beta β_1 branch has an emission probability of 99.89% (6). The maximum beta energy of β_1 is 317.9 keV. The nickel-60 excited level promptly de-excites to the ground state via emission of gamma rays of energy $E_{\gamma1}$=1173.22 keV and $E_{\gamma2}$=1332.51 keV. Another beta branch named β_2 has much lower emission probability, of 0.08%. β_2 leaves nickel-60 in the 1332.51 keV excited state, which again promptly de-excites by emitting a gamma ray of energy 1332.51 keV. The half-life of cobalt-60 is 1925.48±0.51 days (7). Because the emitted beta and gamma radiations are prompt, the 4πβ–γ coincidence counting technique can be used for direct determination of the activity and β detection efficiency.

With the efficiency - tracing technique, an accurately known amount of a suitable β–γ radionuclide, known as a tracer, is mixed with the pure β emitter. The β - count rate, γ - count rate and coincidence count rate of the mixed source are then measured in a 4πβ–γ coincidence counting system for a range of different β - detection efficiencies. The total β - count rate of the mixed source, as a function of the β - detection efficiency is then extrapolated to an efficiency of 100% to yield an absolute total activity of the mixed source. The efficiency extrapolation method in coincidence counting was described by Baerg (8). The activity of the pure β emitter is determined by a subtraction of the activity of the tracer radionuclide from the absolute total activity. In this work, VYNS, aluminium and nickel foils were used sequentially to attenuate the β – emission in order to vary the β- detection efficiency.

SAMPLE PREPARATION

An ampoule containing thallium-204 produced by the Bureau National De Metrologie – Laboratoire National Henri Becquerel (BNM - LNHB), France and dispatched by BIPM for the international intercomparison arrived in April 2002. This ampoule contained 3.6273 grams of the thallium-204 solution that had an approximate activity concentration of 100kBq/g (at 0 h UT, 15th of March 2002). The chemical composition of the aqueous solution was 31 μg/g of TlCl contained in 0.1 M HCl.

The cobalt-60 tracer solution selected had an activity concentration of 288.757 ± 0.681 kBq/g (at 0 h UT, 1st of July 2002). An activity contribution of 60% cobalt 60 and 40% thallium-204 was adopted for the production of the thallium-cobalt master solution. To achieve this, 1.0058 g of thallium-204 was taken from BIPM ampoule and mixed, in a vial, with 0.6586 g of cobalt-60, to produce a thallium-cobalt mater solution. The solution was well mixed and then centrifuged before dispensing, so that no activity was lost onto the lid or glass wall. Eleven 4π counting sources were dispensed from the thallium-cobalt master solution with mass ranged between 10.637g to 36.264 g and total activity per source aimed to be in the range of 2-10 kBq. The mass deposited onto a source mount was determined by weighing the syringe pycnometer before and after dispensing. The wet sources were then dried in an oven at 45°C.

The source mount used was VYNS thin copolymer film held by a circular brass mounts. In order to make VYNS conductive, a thin coating of gold/palladium was evaporated onto the foil in a vacuum. Before radioactive solution was deposited catanac was added to the centre of the film and spread to the very edges of the mount and left to dry in an oven at 45°C. Catanac acts as a wetting agent.

$4\pi\beta$–γ COINCIDENCE COUNTING EFFICIENCY – TRACING EXTRAPOLATION METHOD

A conventional $4\pi\beta$–γ coincidence counting system was used. The system consists of a 4π gas-flow counter for the β detection, a 3 by 3 inch thallium activated sodium iodide NaI(Tl) scintillator for the γ-detection and an array of electronics used to process the signal. A 10% Methane in Argon (P-10) counting gas was used in this work.

A non-extendable dead time of 8.00 ± 0.04 μs was imposed on both channels using paralysis units. The coincidence - mixer resolving time was set at about 1 μs and subsequently measured to be 1.081 ± 0.010 μs.

The gamma channel single channel analyser (SCA) was set to gate the 1173 and 1332 keV photopeaks from cobalt-60 decay. The beta channel SCA was left wide open so all of the beta particles were counted.

For a thallium-cobalt mixed source, the total number of events $N_{\beta_{Total}}$, in terms of count rate, detected by the proportional counter is given by

$$N_{\beta_{total}} = N_{0_{Co}}[\varepsilon_{\beta_{Co}} + (1-\varepsilon_{\beta_{Co}})(\frac{\alpha\varepsilon_{ce} + \varepsilon_{\beta\gamma}}{1+\alpha})] + N_{0_{Tl}}[b\varepsilon_{\beta_{Tl}} + (1-b)\varepsilon_{XA}] \quad (1)$$

where $N_{0_{Co}}$ is the absolute activity of the cobalt 60 source, $\varepsilon_{\beta_{Co}}$ is the efficiency of detecting β particles in the β detector from disintegrations of cobalt-60, α is the total internal conversion coefficient (ratio of the number of conversion electrons to the number of γ rays) of cobalt-60, ε_{ce} is the efficiency of detecting conversion electrons in the β detector following disintegrations by cobalt-60, $\varepsilon_{\beta\gamma}$ is the efficiency of detecting γ rays in β detector following disintegrations of cobalt-60, $N_{0_{Tl}}$ is the absolute activity of the thallium-204 source, b is the β branching ratio of thallium-204, 1-b is the electron capture branching ratio of thallium-204, $\varepsilon_{\beta_{Tl}}$ is the efficiency of detecting β particles from disintegrations of thallium -204 in the β detector, ε_{XA} is the probability of detecting at least one X-ray or Auger electron (9) following electron capture in the β detector.

The gamma count rate N_γ detected in the NaI detector is given by

$$N_\gamma = N_{O_{Co}}(\varepsilon_{\gamma 1_{Co}} + \varepsilon_{\gamma 2_{Co}}) \quad (2)$$

Where, $\varepsilon_{\gamma 1_{Co}}$ and $\varepsilon_{\gamma 2_{Co}}$ are the efficiencies of detecting 1173 keV and 1332 keV gamma-rays of cobalt-60 in the γ detector respectively.

The coincidence count rate N_c registered in the coincidence mixer is given by

$$N_C = N_{O_{Co}} \varepsilon_{\beta_{Co}} (\varepsilon_{\gamma 1_{Co}} + \varepsilon_{\gamma 2_{Co}}) \tag{3}$$

From equation (2) and (3), we have

$$1 - \frac{N_c}{N_\gamma} = 1 - \varepsilon_{\beta_{Co}} \tag{4}$$

With the efficiency – tracing method, it is assumed that the efficiency of beta detection for thallium -204 and the one for cobalt-60 tracer are linearly interrelated and that they approach unity simultaneously (4). Therefore we have $(1-\varepsilon_{\beta_{Tl}}) = k(1-\varepsilon_{\beta_{Co}}) = K(1-\frac{N_c}{N_r})$, where k and K are constants. The equation (1) then becomes

$$N_{\beta_{total}} = N_{0_{Co}}[1 + G(1 - \frac{N_c}{N_r})] + N_{0_{Tl}}[b - H(1 - \frac{N_c}{N_r}) + (1-b)\varepsilon_{XA}] \tag{6}$$

Where, G = $(\frac{\alpha\varepsilon_{ce} + \varepsilon_{\beta\gamma}}{1+\alpha}) - 1$ and H = bK (constant). It is reasonable to assume that ε_{XA} approaches unity simultaneously as $N_c/N_\gamma \to 1$. Extrapolation of the 4π β count rate to $N_c/N_\gamma \to 1$ thus yields the total 4π beta count rate of thallium - cobalt mixed source to be $N_{\beta_{total}} = N_{0_{Co}} + N_{0_{Tl}}$. The extrapolation method eliminates the influence of conversion electrons and gamma sensitivity of the β detector. If $\varepsilon_{\beta\gamma}$ and ε_{ce} are constants (in other words, do not vary with β detection efficiency) the curve fitting is linear, otherwise it shows a polynomial fit. The activity of thallium-204 may then be determined by a simple subtraction of the tracer activity from the absolute total β count rate.

RESULTS AND DISCUSSION

A gamma spectrometry measurement showed no impurities present in either the original thallium-204 source or the thallium-cobalt mixed source.

Eleven 4π counting sources were counted in ANSTO's 4πβ–γ coincidence counting instrument generating eleven sets of observed β count rate $N'_{\beta_{total}}$, γ count rate N'_γ and coincidence count rate N'_c. As mentioned before the β- detection efficiency was varied by using VYNS, aluminium and nickel foils to attenuate the β – emission. The measured β efficiencies ranged from 0.55 to 0.95. $N'_{\beta_{total}}$ and N'_γ were corrected for background, dead times using Wyllie's equation (10) and decay as well. N'_c was corrected for dead times and accidental coincidence using Campion's equation (11) and decay as well. For each of the eleven sets of data, a curve $N_{\beta_{Total}}$ as a function of $1\text{-}N_c/N_\gamma$ was plotted. All plots showed linear fitting of the measurement data. A typical data set is shown in Figure 1.

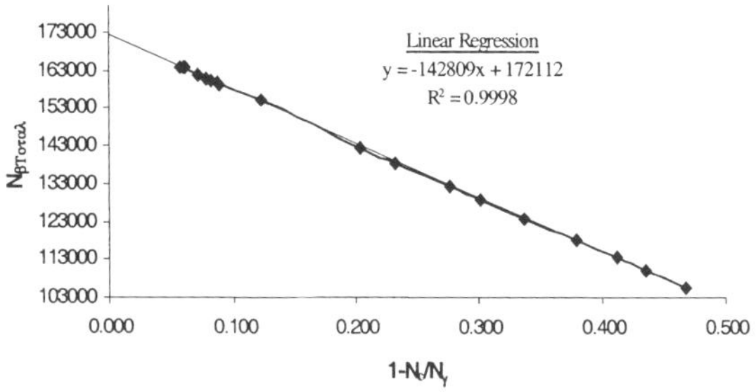

Figure 1. Variations of $N_{\beta_{Total}}$ as a function of $1\text{-}N_c/N_\gamma$. Extrapolation of $N_{\beta_{Total}}$ to unity beta detection efficiency yields total activity of a Tl-204 and Co-60 mixed source. Data obtained from the measurement of the source #7354.

By subtraction of the tracer activity from the extrapolated total activity, the activity of thallium-204 is obtained. By averaging the eleven results, the activity concentration of BIPM thallium-204 source was found to be 101.73 ± 0.55 kBq/g at 0 h UT, 1st of July 2002. The final uncertainty was determined on the basis of the exact evaluation formulas. The basic uncertainty elements are counting statistics on $N_{\beta_{Total}}$ (0.05%), N_γ (0.24%) and N_c (0.22%), standard uncertainties associated with mass weighing (0.14%), dead time (0.50%), resolving time (0.89%), half-life of thallium-204 (0.51%) and cobalt-60 (0.03%). These basic elements are propagated with the formulas of background correction, dead time correction, accidental coincidence correction, decay correction, linear curve fitting and tracer activity subtraction in accordance with the Guide to the Expression of Uncertainty in Measurement (12).

The result deviates only 0.17% from the mean activity concentration of 32 results submitted by 20 participants. Even so, it is generally understood that the use of cobalt-60 as a tracer may not be an ideal method due to the possibility of re-crystallisation of thallium and cobalt when a source was drying in the oven. It results in poor homogeneity of activity and mass distribution.

CONCLUSION

Although cobalt-60 may not be an ideal tracer to be used for the absolute measurement of activity of thallium-204, the activity concentration of 101.73 ± 0.55 kBq/g determined by using 4πβ–γ coincidence counting efficiency - tracing extrapolation technique with cobalt-60 as a tracer agree to better than 0.17% of the mean value for the 2002 International Intercomparison of Activity Measurements of a Solution of thallium-204. It demonstrates the accuracy of Australian Primary Standard.

REFERENCE

1. Bc, M., Coursol, M. N., *Table de Radionucleides*, BNM – CEA/LNHB 01 December 1983 – 18 April 2002.
2. Campion, P. J., Taylor, J. G. V. and Merritt, J. S., *International Journal of Applied Radiation and Isotopes,* **8,** 8-19 (1960).
3. Baerg, A. P., Meghir, S. and Bowes, G. C., *International Journal of Applied Radiation and Isotopes,* **15,** 279 -287 (1964).
4. Steyn, J., *Nuclear Instruments and Methods,* **112,** 157-163 (1973).
5. Lowenthal, G. C., *Nuclear Instruments and Methods*, **112,** 165-168 (1973).
6. Legrand, J., Perolat, J., Lagoutine, F., Gallic, Y., Le *Table de Radionucleides*, Laboratoire de Metrologie des Rayonnements Ionisants (1975).
7. Nichols, A L and Hunt, E., *Nuclear Data Table*, AEA Technology plc, Analytical Services, 551 Harwell, Didcot, Oxfordshire, OX11 0RA.
8. Baerg, A. P., *Nuclear Instruments and Methods,* **112,** 143-150 (1973).
9. Perolat, J. P., *Nuclear Instruments and Methods,* **112,** 179-185 (1973).
10. Wyllie, H. A., *Applied Radiation Isotopes,* **38**, No. 5, 385-389 (1987).
11. Campion, P. J., *International Journal of Applied Radiation and Isotopes,* **4,** 232-248 (1959).
12. Guide to the Expression of Uncertainty in Measurement, International Organisation for Standardisation, 1993.

Applied Physics (APHYS 2003)
A. Méndez-Vilas (Ed.)

A Method For C-14 Specific Activity Detection In Gas – Graphite Reactor Moderators Based On CO_2 "in situ" Generation And Trapping.

A. A. Porta*, F. Campi*, L. Garlati*, M. Caresana*

*CeSNEF – Department Of Nuclear Engineering, Polytechnic Of Milan
Via Ponzio 34/3 20133 Milan Italy

Abstract. A method for C-14 specific activity detection, characterized by simplicity and speed of sampling, was developed by CeSNEF – Department of Nuclear Engineering, Polytechnic of Milan. The method is designed to sample graphite reactor moderator elements at a rate up to 1 sample every 5 minutes without drilling or generating dusts of carbon which could be disperse in the air. It is able to sample about 1g of graphite in 90 seconds using a voltaic arc, simply leaning a particular miniaturized combustion chamber against the graphite surface. The produced CO_2 is sucked and trapped in a saturable chemical solution and then analyzed by liquid scintillation counting (LSC). The detection limit is a specific activity of 327 Bq g^{-1} which typically represents 1:150 – 1:300 of irradiated graphite C-14 activation level.

INTRODUCTION

Graphite, which consists mainly of the element carbon, is used as a moderator in a number of nuclear reactor designs, such as the Magnox and AGR in the United Kingdom and the RBMK design in Russia. At the end of reactor life, the graphite moderator, weighing typically about 2000 t, is a form of radioactive waste which requires eventual management. After neutron irradiation the graphite will contain stored Wigner energy and it will also contain significant quantities of radionuclides from neutron induced reactions in the graphite itself and in the minor impurities contained. The radioisotope content can conveniently be divided into two groups: short-lived isotopes (such as Co-60) and long-lived isotopes (principally C-14). Short-lived isotopes make the graphite difficult to handle immediately after reactor shutdown, but they decay in a few tens of years, while long-lived isotopes requires fifty – sixty thousands of years representing a potential problem for an environmental safe-release. Therefore it is important to know accurately the specific activity of C-14 in preparation to its disposal or dispersion through gasification.

MATERIALS AND METHODS

The system allows the measurement of C-14 specific activity in graphite reactor moderators both as a single measurement and as a sampling campaign, that it can be used to create an activity map up to the single graphite element. The system configuration is schematised as follows (Fig. 1): miniaturized burning chamber (A), air filter (B), HNO_3 – Drechsel (C), saturable NaOH – Trap (D), NaOH– Drechsel (E), Phenolphthalein indicator – Trap (F), diaphragm suction pump (G), flowmeter (if required to monitor the air flow).

The miniaturized burning chamber (Fig. 2-A) is made up of AISI – 310 stainless-steel with a suitable drilled finning designed to remove the heat generated by a voltaic arc, which is set between a graphite electrode and the graphite surface under sampling.

The burning chamber is inserted into a suitable slot milled in an insulating-fireproof surface (Fig. 2-D) which guarantees electric insulation between the graphite element and the chamber itself in order to set the voltaic arc only between the graphite element (Fig. 2-E) and a graphite electrode (Fig. 2-G). The flameless combustion of activated graphite generates CO_2 (contaminated with $^{14}CO_2$) which is trapped in a suitable NaOH-Trap. The exhaust gas passes through an air filter to remove particulates, then it is bubbled in a Drechsel filled with 250ml of HNO_3 0.1M to condense HTO and damp down Cl-36 (which may be present in neutron irradiated graphite). Forced to reach the saturable trap, the CO_2 contained in the exhaust gas reacts with NaOH reaching a saturation point at which is known the carbon content.

An aliquot of solution drawn from the saturable trap can be analyzed by liquid scintillation counting (LSC) to determine the C-14 activity. Division by the relative carbon content of the aliquot gives the C-14 specific activity of the graphite sample.

After the NaOH-Trap the radioactive outflow is damped down through a scrubber filled with 250ml of NaOH 5M and, in order to prevent its saturation with a subsequent possible $^{14}CO_2$ escape, it is also added with a small NaOH-Trap 0.05M with phenolphthalein (Fig. 1). Phenolphthalein is a basic indicator whose colour changes from violet to colourless (across pH 9.1) warning all the operators about the possible danger of undesired C-14 release in the air. The NaOH trap is designed to operate for 10 – 11 samplings, i.e. for 10 – 11 combustions. A diaphragm suction pump is connected at the end of the measuring chain and sets a stable suction pressure anywhere along the chain itself, assuring a flow of about $3\ l\ min^{-1}$.

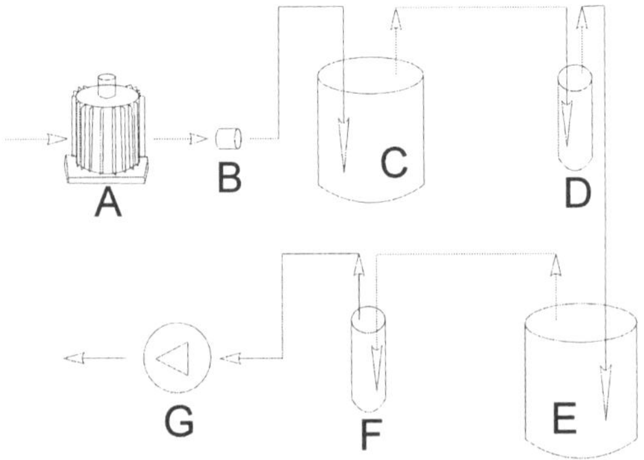

FIGURE 1. Chain measuring elements: (A) miniaturized burning chamber, (B) air filter, (C) 250ml HNO3 0.1M Drechsel-trap, (D) 30ml NaOH 0.05M plastic scrubber trap, (E) 250ml NaOH 5M Drechsel-trap, (F) 30ml NaOH 0.05M + Phenolphthalein plastic scrubber trap, (G) diaphragm suction pump

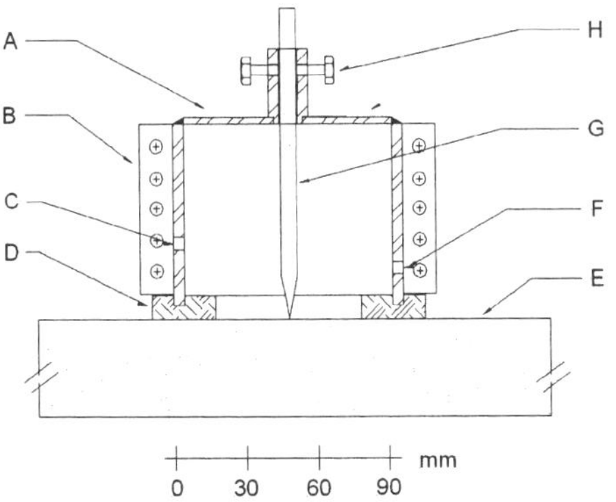

FIGURE 2. Miniaturized burning chamber, list of elements: (A) Burning chamber – AISI 310, (B) Drilled finning, (C) Air intake, (D) Insulating - fireproof surface, (E) PGA/PGB graphite element, (F) Gas outflow, (G) Commercial graphite electrode (high purity), (H) Locking bolts

RESULTS AND DISCUSSION

The sampling cycle is made up of 90 s of flameless combustion and 60 s of pure air suction to collect all the exhaust gas and cleaning the combustion chamber. The total cycle length is 150 s and, during this time, the pump continuously maintains an air flow of 3 l min^{-1}. In the 90 s of combustion the voltaic arc is started up by an high frequency generator and is supplied by a suitable current following a linear ramp starting from 150A dc and reaching, in 10 s, 200A dc to last again 80 s at this intensity. The current profile was optimised to avoid the disruption of the electrode tip at the start up. During these 90 s it is possible to remove a total of 1116.8 +/- 68.2 mg of carbon from graphite elements with a good repeatability (Table I – column 1&2).

The saturable trap receives CO_2 both from the element and from the electrode giving, at the saturation point, a mixture in which C-14 activity is "diluted" by the CO_2 coming from the electrode burning (interference). This fact leads to an underestimation of C-14 specific activity up to –47% (Table I – column 4). The percentual interference (I%) depends on the temperature profile of the sampled surface due to the plasma arc. The graphite electrode keeps a quite constant temperature because of its small dimensions while the graphite element, varying from small to big bars, presents both several thermal capacities and I%. The lower is the temperature profile the higher is I% allowing a major increase in CO_2 production from the graphite electrode. However, as described in the following, it is possible to set a two-stages calibration procedure which allows to determine the interference amount introducing a corrective factor.

One way to determine the corrective factor is to measure the weight loss of the electrode, $\Delta P1_{EL}$, and of a little specimen of the same material to be sampled, $\Delta P1_{SP}$, (stage 1) to create a correlation with the new weight loss of the electrode, $\Delta P2_{EL}$, and that of graphite bar, $\Delta P2_{BAR}$, which is unknown (stage 2).

TABLE I

Combustion number	Electrode weight loss [mg]	Graphite element weight loss [mg]	Electrode interference [%]	Trap saturation [%]
1	499	735	40.44	93
2	501	563	47.09	94
3	433	625	40.93	93
4	538	667	44.65	92
5	390	651	37.46	94
6	441	625	41.37	92
7	501	676	42.52	94
8	510	698	42.22	93
9	384	702	35.36	92
10	521	668	43.82	92
11	508	649	43.91	91
12	397	650	37.92	94
13	414	619	40.08	91
14	428	605	41.43	92
15	511	592	46.33	93
16	451	626	41.88	91
17	521	691	42.99	94
18	452	704	39.10	92
19	387	628	38.13	94
20	458	716	39.01	92

Stage 1, sampling of a specimen: at the sampling cycle finish it is possible to know the $\Delta P1_{EL}$, $\Delta P1_{SP}$ and $V1_{PHE}$, this latter being the volume of the acid solution used to titrate a suitable NaOH trap absorbing all the CO_2 produced. The NaOH trap has to be sufficiently concentrated to form only Na_2CO_3 and to be titrated with phenolphthalein indicator to give V_{PHE} volumes.

Stage 2, sampling of a radioactive graphite bar: as previously exposed, it is possible to determine $\Delta P2_{EL}$, $V2_{PHE}$ but not $\Delta P2_{BAR}$ due to its relevant weight. Anyway, the total CO_2 produced and captured in the NaOH trap will be proportional both to the total weight losses $\Delta P1_{EL}+\Delta P1_{SP}$, $\Delta P2_{EL}+\Delta P2_{BAR}$ and to the $V1_{PHE}$, $V2_{PHE}$ allowing to write:

$$\frac{\Delta P2_{EL}+\Delta P2_{BAR}}{\Delta P1_{EL}+\Delta P1_{SP}} \cong \gamma = \frac{V2_{PHE}}{V1_{PHE}} \tag{1}$$

The knowledge of γ factor, remaining a constant value for all the sampling cycles, allows to neglect $\Delta P2_{BAR}$ value obtaining, anyway, the interference value I% through the following approximation:

$$I\% = 100 * \frac{\Delta P2_{EL}}{\Delta P2_{EL}+\Delta P2_{BAR}} \cong 100 * \frac{\Delta P2_{EL}}{\gamma * (\Delta P1_{EL}+\Delta P1_{SP})} \tag{2}$$

In our experiences it was used a NON radioactive set of pile grade A graphite bricks (PGA graphite with a density of 1700 kg m^{-3}) of 20x45x150 mm and a NaOH 2.075M trap with two drops of phenolphthalein in the calibration procedure.

During sampling cycle the saturable trap is not saturated up to 100% of its total capacity but only to its 93%. Its stability is in the range 93 +/- 1 % (confidence interval (CI): 95%) allowing to keep this percentile value as a corrective factor (Table I – column 5).

To perform LSC analysis 12ml of Packard[1] OPTI-FLUOR® scintillating cocktail were mixed with 3ml of saturable-trap solution (essentially made up of $NaHCO_3$) in a 15ml plastic vial and than were counted for 5 minutes.

In the 12ml+3ml mixture it was put 1.674 +/- 0.144 mg of total carbon (already corrected with the 93 +/- 1 % corrective factor and expressed with CI: 95%) which has still to be corrected by the electrode interference.

Sampling some irradiated graphite test pieces coming from the Magnox nuclear power plant of Latina (Italy) it was found a C-14 specific activity of 67600 Bq g^{-1} +/- 15% (CI: 95%) matching the 76000 Bq g^{-1} +/- 20% (average value) calculated in the "Inventario nazionale dei rifiuti radioattivi – ENEA – 2000" for the same plant. It was found, for a 5 min LSC measurement, a lower limit detection (LLD) of 10.39 CPM (CI: 99%) and a minimum detectable activity (MDA) of 0.3 Bq/vial (CI: 99%) which lead to a minimum detectable specific activity (MSDA) of 327 Bq g^{-1} . The MSDA value has to be compared with the usual C-14 specific activity that can be found sampling neutron irradiated graphite bricks, coming from nuclear power plants (at end-life), which are in the range of 50-100 kBq g^{-1} , that is a sensitivity in the range of 1:150 – 1:300.

CONCLUSIONS

The method seems to be fast and precise.

Its repeatability is good within our experimental experiences.

There are no release of exhaust gases into the atmosphere and/or work-environment.

The associated chemistry is very simple and easy to approach.

As it can be inferred, the system provides a good relative sensitivity, typically 1 part over 150-300, with an MSDA of about 327 Bq g^{-1} . Anyway, the use of LS-analysers with low background and longer counting times can improve the MSDA value.

ACKNOWLEDGMENTS

The authors remember the great importance of all the ideas and scientific activity of Professor Sergio Terrani, deceased on September-25-2002, and with gratitude thank him for years of precious advices.

The authors thank the Sintec Europe s.r.l. (www.sinteceurope.it) and in particular Mr. Radini and Mr. Polat for their helpfulness and promptness in designing and realizing the miniaturized combustion chamber.

---

[1] Packard Instruments Company, Inc. – 800 Research Parkway, Meriden, CT 06450 USA Tel. 1-203-238-2351

Applied Physics (APHYS 2003)
A. Méndez-Vilas (Ed.)

Radioactive Iodine Waste Treatment using Electrodialysis with an Anion Exchange Paper Membrane

Hiroyoshi Inoue* and Mayumi Kagoshima

Radioisotope Institute for Basic and Clinical Medicine, Kurume University School of Medicine, 67 Asahi-machi, Kurume, 830-0011, Japan
**Corresponding Author: Fax: +81-942-317584, Tel: +81-942-317584*
e-mail: hiroin@med.kurume-u.ac.jp

Abstract. In order to simply and safely treat radioactive iodine waste, a study of the removal of the iodide ion from radioactive waste using electrodialysis with an anion exchange paper membrane, in which trimethylhydroxylpropylammonium groups were homogeneously dispersed at a high density. In a $Na^{125}I$ and $Na^{36}Cl$ concentration-cell system, phenomenological coefficients including electric ion and water conductance, have been experimentally determined on basis of nonequilibrium thermodynamics. Prepared paper membranes had a higher permselectivity for ^{125}I ions than ^{36}Cl ions by approximately 21%. On the other hand, water flux, accompanied by an ionic transference in prepared paper membranes was greatly larger than that in typical synthesized membranes. This observation suggested that a depression of water mobility is important to practice an ideal radioactive iodide waste electrodialysis system with a novel anion exchange paper membrane.

INTRODUCTION

Radioactive iodine is used extensively in medical treatment, diagnosis, industry, and laboratorial investigations. Radioactive iodine is used for treating thyroid cancer, after which a significant amount of radioactive waste is generated from patients. More than 70% of the radioactive iodine administered is excreted into the urine, after three days of radioactive iodine treatment [1]. One of the most important radiation management areas is the safe and effective treatment of radioactive iodine waste. Several researchers investigated the prospect of reducing of the volume of stored radioactive waste by filtering radioactive iodine from the general radioactive waste using columns packed with either anion exchange resin or activated charcoal. Some of these techniques provided removal of more than 90% of the radioactive iodine from the general waste [2, 3]. However, column-based methods require frequent regeneration or exchange of the packing material. New radioactive waste is generated in the column regeneration process, and the end result is the generation of large amounts of solid waste. Thus, in order to minimize the waste generated, radioactive iodine concentration treatment using anion exchange paper membranes have been proposed to conveniently and safely deal with radioactive iodine waste. The use of anion exchange paper membranes for radioactive waste disposal has several advantages over more conventional methods.

Radioactive iodine has been reported to bind more strongly to anion sites than ^{36}Cl ions by a factor of more than two [4]. In addition, the membrane/solution distribution of the ^{125}I ion in the presence of electroneutral solutes such as glucose and urea was reported to increase [5]. Furthermore, the membrane separation system could be controlled in this method by changing the kinds of exchange sites within the paper membrane [6]. Anion exchange paper membranes were reported to exhibit high ^{125}I ion permselectivity due to the electrostatic effects in paper membranes bearing trimethylhydroxypropylammonium anion exchange groups [7]. Electrodialysis with a compact apparatus design is an effective method to practically utilize these paper membranes that have properties including high radioactive iodine permselectivity and high ionic fluxes.

In the present investigation, the transport of ^{125}I through the novel anion exchange paper membrane bearing a trimethylhydroxypropylammonium group was systematically investigated in the presence or absence of the electric current in contrast to the transport of ^{36}Cl in a single electrolyte solution system. The transport of water coupled with ionic movements within the paper membrane were simultaneously measured since transport was important to the design the ionic selective separation system. A phenomenological equation based on non-equilibrium thermodynamics with a compact apparatus design was adopted to quantitatively evaluate the permaselectivities of ^{125}I and ^{36}Cl and the simultaneous transport of water.

EXPERIMENTAL

Preparation of anion exchange paper membrane

The method of introduction of trimethylhydroxypropylammonium anion exchange groups onto cellulose, and the preparation of an anion exchange paper membrane was previously reported [7, 8]. However, further quarternizing of the prepared paper membrane using methyl iodine/hexane (40/60 wt/wt) at 30°C for 24 hr was performed. In a prepared anion exchange paper membrane; the ion exchange capacity is 0.30 meq/g dry membrane, the water content is 1.28 g H_2O/g dry membrane, and the thickness is 0.28 mm in the dry form.

Iodide Ion Transport

Membrane potential and ion and water fluxes were measured in the absence and presence of electric current in order to evaluate the transport of ^{125}I ions and ^{36}Cl ions across the prepared paper membranes using a four compartment assembly as shown in Fig 1. The experimental procedure was previously described[4, 7]. Electrodialysis was performed at a current density of 10 mA cm^{-2}. Total fluxes, which summarized the ionic flow and the water flow were measured by moving the mercury droplet in the glass pipette connected to the experimental assembly.

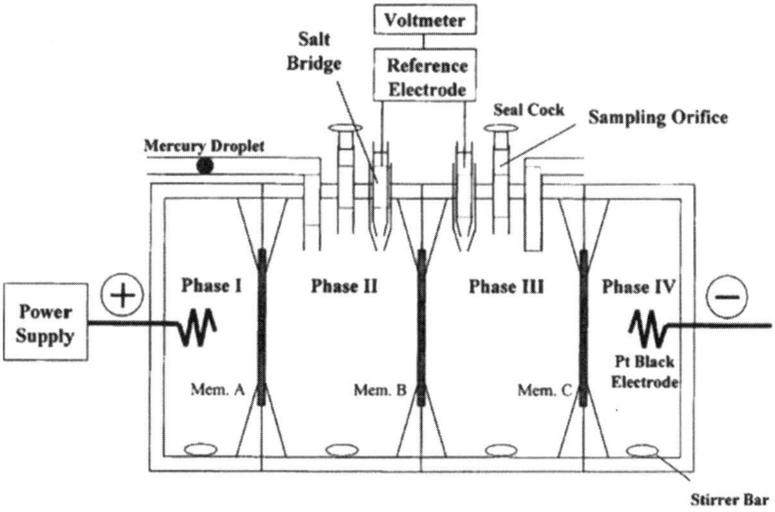

Fig. 1 Schematic representation of the experimental assembly for electrodialysis. $Na^{125}I$ or $Na^{36}Cl$ media solutions were placed into phases II and III. The hypertonic solution was kept in phase III. 2×10^{-1} mol dm^{-3} of NaOH solution was used as a catholyte and an anolyte in phases I and IV, respectively. Mem. A: cation exchange membrane (Neocepta CM-2; Tokuyama Corp.), Mem. B: prepared anion exchange paper membrane, Mem. C: anion exchange membrane (Neocepta AM-2; Tokuyama Corp.). Effective membrane area 4.9 cm^2; Compartment capacities: 15 cm^3 for phases I and IV and 25 cm^3 for phases II and III. All measurements were carried out at 25 $\pm$0.5℃, with all media solution phases stirred at 90 rpm in all cases.

In phases II and III of the Fig. 1, the concentration of $Na^{125}I$ or $Na^{36}Cl$ was varied from 1×10^{-3} to 1×10^{-1} mol dm^{-3}in the one media solution, whereas concentration of the ions in the other media solution of was kept constant at 1×10^{-2} mol dm^{-3}. The radioactive aqueous $Na^{125}I$ solution was prepared by mixing ultra-pure water (>17 MΩ cm), ^{125}I in NaOH (Amersham Pharmacia Biotech Ltd., Japan), and extra-pure NaI (Nacalai Tesque Ltd., Japan). Similarly, the $Na^{36}Cl$ solution was prepared from ultra-pure water, $Na^{36}Cl$ (Amersham Pharmacia Biotech Ltd., Japan), and extra-pure NaCl (Nacalai Tesque Ltd., Japan). The activity coefficients of the $Na^{125}I$ or $Na^{36}Cl$ solutions were estimated using the Debye-Hückel equation and the parameters reported by Kielland were used [9, 10].

RESULTS AND DISCUSSION

The membrane permeability matrix theory reported by Nagata et al. [11] has been proven useful for simultaneous quantitative analysis of the permselective transport of ions and water across a charged membrane. According to this theory, the membrane conductance matrix for the sodium iodide and sodium chloride concentration-cell systems containing a sodium ion;

Na+, anion; A-, and water; W can be expressed using membrane conductance elements, $g_{\alpha\cdot\beta}$ (α or β = Na, A, or W), as:

$$\begin{bmatrix} i_{\mathrm{Na}} \\ i_{\mathrm{A}} \\ Fj_{\mathrm{W}} \end{bmatrix} = -\begin{bmatrix} g_{\mathrm{Na\cdot Na}} & g_{\mathrm{Na\cdot A}} & g_{\mathrm{Na\cdot W}} \\ g_{\mathrm{A\cdot Na}} & g_{\mathrm{A\cdot A}} & g_{\mathrm{A\cdot W}} \\ g_{\mathrm{W\cdot Na}} & g_{\mathrm{W\cdot A}} & g_{\mathrm{W\cdot W}} \end{bmatrix} \begin{bmatrix} V - V_{\mathrm{Na}} \\ V - V_{\mathrm{A}} \\ -V_{\mathrm{W}} \end{bmatrix} \quad (1)$$

where i is the ionic current, F is Faraday's constant, j_{W} is the water flux, V is the transmembrane potential, V_{Na} and V_{A} are the pseudo equilibrium potentials of the sodium ion and the anion, respectively, V_{W} is the effective driving potential of water. Subsubscripts Na, A, and W refer to the sodium ion; Na+, anion ; A-, and water, respectively.

The equation for membrane current, I, is derived from Eq. (1) as follows:

$$I = -g_{\mathrm{Na}}(V - V_{\mathrm{Na}}) - g_{\mathrm{A}}(V - V_{\mathrm{A}}) + g_{\mathrm{W}}V_{\mathrm{W}} = -G_{\mathrm{m}}(V - V_0)$$
$$t_\alpha = \frac{g_\alpha}{G_{\mathrm{m}}} \quad (2)$$

where G_{m} is membrane conductance, V_0 is membrane potential at zero membrane current, and t_α is the apparent transport number of species α.

The fluxes of anions and water as a function of the potential in the presence and absence of the membrane current were analyzed, which enabled the membrane conductance elements to be estimated Then, the following equation was derived from Eqns. (1) and (2).

$$i_\alpha - i_\alpha^0 = t_\alpha I = -g_\alpha(V - V_0)$$
$$F(j_{\mathrm{W}} - j_{\mathrm{W}}^0) = t_{\mathrm{W}} I = -g_{\mathrm{W}}(V - V_0) \quad (3)$$

where g_α (α=I or Cl) and g_{W} denote the electric ion conductance and the electric water conductance, respectively. The superscript, 0, an absence of electric current.

g_α and g_{W} are expressed as a function of the membrane conductance elements in Eqn. (1). Therefore, g_α and g_{W} combinatorially represent to the relationship between the three driving forces, and potentials in the system and the resulting electric flows and electric currents.

The transmembrane potential in the absence of an electric current, V_0, and the apparent transport numbers of ^{125}I and ^{36}Cl ions relative to Na ion are shown in Fig. 2. In both the Na^{125}I and Na^{36}Cl concentration-cell systems, the transmembrane potentials deviated significantly from the Nernstian response (Fig. 2-A). Transmembrane potentials in Na^{125}I system were always larger than those in the Na^{36}Cl system. When examining the apparent transport numbers estimated from the transmembrane potential data, the ^{125}I ions, average; 0.772, were slightly larger than the ^{36}Cl ions, average; 0.722 (Fig. 2-B).

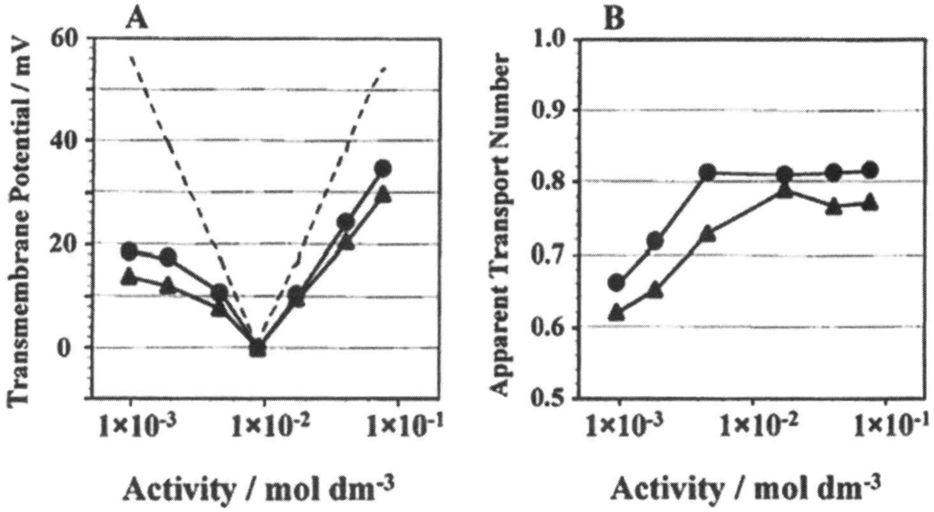

Fig. 2. Transmembrane potential (A) and apparent transport number (B) as a function of the varied solution concentration. ● and ▲ indicate $Na^{125}I$ and $Na^{36}Cl$ systems, respectively. The broken lines in A refer to data derived from the Nernst equation.

The fluxes of ^{125}I and ^{36}Cl ions through the prepared paper membrane were evaluated by measuring the time course of the radioactivity changes in phase II compartment. There is one pathway for the transport of ^{125}I and ^{36}Cl ions through the prepared paper membrane, since these anions move from phase III into phase II and the exit of these anions to catholyte is interfered with by a cation exchange membrane. The results are presented in Fig. 3 (A-I and B-I), as the molar permeations of the ^{125}I and ^{36}Cl ions against electrodialysis time. The fluxes of ^{125}I and ^{36}Cl ions through the prepared paper membrane were calculated as 0.282 and 0.227 mmol h^{-1} cm^{-2}, respectively, in the 1×10^{-2} - 1×10^{-1} mol dm^{-3} system. The fluxes were 0.056 and 0.045 mmol h^{-1} cm^{-2}, respectively, for the 1×10^{-3} - 1×10^{-2} mol dm^{-3} system. Furthermore, the electric current results in not only transference of ions but also convection of pore water, thus transference of water, called "electroosmosis". The pore water carries a net electric charge with the same charge as the counter-ions. In the present study, a large quantity of water flux was observed as shown in Fig. 3 (A-II and B-II). The water fluxes through the prepared paper membrane in the 1×10^{-2} - 1×10^{-1} mol dm^{-3} systems, were calculated as 57.8 and 49.4 mmol h^{-1} cm^{-2} for the ^{125}I system and ^{36}Cl system,_respectively. In the 1×10^{-3} - 1×10^{-2} mol dm^{-3} system, the values were 25.0 and 20.4 mmol h^{-1} cm^{-2}, respectively. The molar permeations of water were approximately 200 to 500 times those of ^{125}I and ^{36}Cl ions. These magnifications in the paper membrane differed greatly from those in typically synthesized membranes [12]. This differences may attributed to the solvation shell conditions within the membrane phases. No water is transferred across the membrane phases unless it is inside an aqueous solvation shell. It has been previously reported that the ions within typical synthesized membranes behave as un-hydrated ions, whereas the ions within the paper membrane used in this study behave as hydrated ions [8]. Unfortunately, the increase in electroosmosis reduces the efficiency of electrodialytic demineralization because the applied electric current is utilized to remove not only ions but also water from the liquid waste.

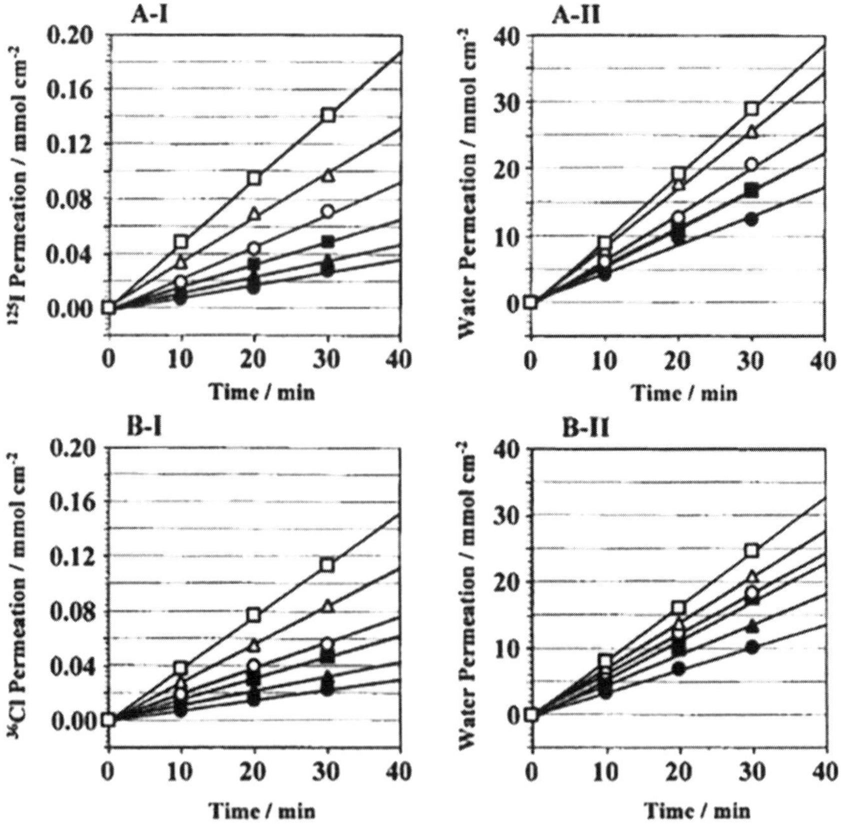

Fig. 3 Ion permeation (A-I and B-I) and water permeation (A-II and B-II) for the $Na^{125}I$ (A; upper panel) and $Na^{36}Cl$ (B; lower panel) concentration-cell systems. ●, ▲, ■, ○, △, and □ indicate 1×10^{-3}, 2×10^{-3}, 5×10^{-3}, 2×10^{-2}, 5×10^{-2}, and 1×10^{-1} mol dm^{-3} for phase II or III, respectively. Each measurement point represents the mean of six samples, without standard error.

The electric ion and water conductance were calculated using Eqn. (3) as shown in Fig. 4. The electric ion conductance of ^{125}I and ^{36}Cl, g_I and g_{Cl}, increased moderately with an increasing concentration of the media, 1×10^{-3} to 1×10^{-1} mol dm^{-3} (Fig 4-A). g_I was larger than g_{Cl}, and the ratios of g_I/g_{Cl} remained almost constant at 1.21 at all concentration combinations. Similarly, the electric water conductance of $Na^{125}I$ and $Na^{36}Cl$ system, $g_{W(NaI)}$ and $g_{W(NaCl)}$, increased moderately with an increasing concentration of the media (Fig 4-B). $g_{W(NaI)}$ was larger than $g_{W(NaCl)}$, and the ratios $g_{W(NaI)}/g_{W(NaCl)}$ remained almost constant at 1.41 at all concentration combinations.

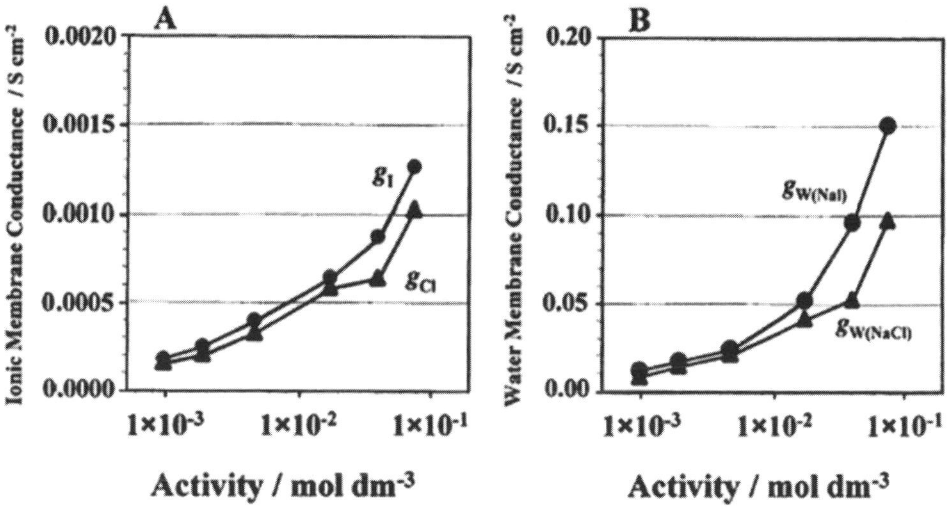

Fig. 4 Electric ion (A) and water (B) conductance as a function of the activity of varied solution concentrations. ● in A: g_I, ● in B: $g_{W(NaI)}$ for $Na^{125}I$ system, ▲ in A: g_{Cl}, ▲ in B: $g_{W(NaCl)}$ for $Na^{36}Cl$ system.

As described above, the ionic and water fluxes (Fig. 3) as well as the electric ion and water conductance (Fig. 4) increased with increasing external media concentration. In a typical synthesized membrane, however, these indices were predicted to remain constant or decreased with increasing external media solution concentration [12-14]. Furthermore, in typical synthesized membranes with high ion exchange capacities, the Donnan exclusion of co-ions becomes less efficient and also the water content of the membrane decreases when the concentration of the treated solution is increased. The incremented concentration co-ions within the membrane results in a decrease in counter-ion transport. On the other hand, when comparing the prepared paper membrane used in this study with a typical synthesized anion exchange membrane, the ion exchange capacity of the former was almost one order of magnitude smaller than that of the latter. In addition, the water content of the prepared paper membrane was more than five times that of a typical synthesized anion exchange membrane, and was not dependent on the external media solution concentration. In the present study, the increase in the Donnan effect and the decrease in membrane the swelling effect due to high external solution concentration had very little influence on the transport of ^{125}I and ^{36}Cl through the prepared paper membrane. Therefore, in the prepared paper membrane, the higher the concentration of the external media solution, the more effectively the applied electric current within membrane phase transferred counter-ions. This is thought to be because the transport of the electric current is accomplished almost exclusively by the counter-ions.

Results suggest that the paper membrane prepared with a trimethylhydroxypropylammonium group has the drawback of large water flux which reduces the electrodialysis efficiency, but has ^{125}I selective removal of 21% in comparison to ^{36}Cl. In order to practically utilize the radioactive waste electrodialysis system with the novel anion exchange paper membrane, the water mobility within the paper membrane must be depressed.

ACKNOWLEDGEMENT

This work supported in part by a Grant-in-Aid for Scientific Research from the Japan Society for the Promotion of Science (No. 14580543) and a Grant-in-Aid for Scientific Research in Priority Areas from the Ministry of Education, Science, Sports and Culture (No. 15020260). The author would like to thank Tokuyama Corp, Japan for supplying the Neocepta CM1 and AM1 membranes.

REFERENCES

1. Leung, P.M.K. and Nikolic, M. *Disposal of therapeutic ^{131}I waste using a multiple holding tank system*, Health Phys. **75(3),** 315-321 (1998).
2. Bohner, K.E., Miller, K.L., Finnegan, J.J., and Erdman, M.C. *Radioiodine in liquid waste – Volume reduction techniques*. Proceedings of the Ninth Biennial Conference of Campus Radiation Safety Officers, Columbia, 23-25 (1983).
3. Edwards, B.E., Couch, N.W., Myers, K.D., Blanchard, S.G., Chandra, G., and Parr, A.F. *^{125}I aqueous waste volume reduction at a pharmaceutical research laboratory*, Health Phys. **71(3),** 379-383 (1996).
4. Inoue, H. and Kagoshima, M. *Removal of ^{125}I from radioactive experimental waste with anion exchange paper membrane*, Appl. Radiat. Isot. **52,** 1407-1412 (2000).
5. Inoue, H. *Influence of glucose and urea on ^{125}I transport across an anion exchange paper membrane*, Appl. Radiat. Isot. **54,** 595-602 (2001).
6. Inoue, H. *Transport of ^{125}I and ^{36}Cl across an anion-exchange paper membrane*, Appl. Radiat. Isot. **56**, 659-665 (2002).
7. Inoue, H. *Radioactive Iodine and Chloride Transport across a Paper Membrane bearing Trimethylhydroxypropylammonium Anion Exchange Groups*. J. Membrane Sci. **222,** 53-57 (2003).
8. Inoue, H. *Effects of Co-ions on Transport of Iodide Ions through a Non Conventional Anion Exchange Paper Membrane*. J. Membrane Sci., in press.
9. Robinson, R.A. and Stokes, R.H. *Electrolyte Solutions*. (London: Butterworths Ltd.).

(1965).
10. Kielland, J. *Individual activity coefficients of ions in aqueous solutions*. J. Am. Chem. Soc. **59,** 1675-1678 (1937).
11. Katchalsky, A. and Curran, P.F. *Nonequilibrium Thermodynamics in Biophysics*. (Harvard Univ. Press). (1964).
12. Nagata, Y., Kohara, K., Yang, W.K., Yamauchi, A., and Kimizuka, H. *Transport Properties of Cation-Exchange Membrane-Aqueous Electrolyte System*, Bull. Chem. Soc. Jpn. **59,** 2689-2693 (1986).
13. Rosenberg, N.W., George, J.H.B., and Potter, W.D. *Electrochemical properties of a cation-transfer membrane*. J. R. Electrochem. Soc. **104,** 111-115 (1957).
14. Helfferich, F. *Ion Exchange*. (New York: Dover Publications, Inc.). (1995).

Applied Physics (APHYS 2003)
A. Méndez-Vilas (Ed.)

Determination Of Naturally Occurring Ra Isotopes In Ubatuba-SP, Brazil To Study Coastal Dynamics And Groundwater Input

Joselene de Oliveira[1*], Barbara P. Mazzilli[1], Washington E. Teixeira[1], Cátia H. Saueia[1], Willard Moore[2], Elisabete de Santis Braga[3], Valdenir V. Furtado[3]

[1]*Divisão de Radiometria Ambiental, Centro de Metrologia das Radiações*
Instituto de Pesquisas Energéticas e Nucleares
Av. Prof. Lineu Prestes, 2242 Cidade Universitária São Paulo SP
*CEP 05508-900 Brazil *e-mail: jolivei@ipen.br*

[2]*Department of Geological Sciences*
University of South Carolina Columbia SC 29208 USA

[3]*Departamento de Oceanografia Química e Geológica*
Instituto Oceanográfico da Universidade de São Paulo
Praça do Oceanográfico s/n Cidade Universitária São Paulo SP
CEP 05508-900 Brazil

Abstract. The four naturally occurring Ra isotopes were measured in seawater, river and groundwater samples collected in a series of small embayments of Ubatuba coastal region, covering latitudes between 23°26'S and 23°46'S and longitudes between 45°02'W and 45°11'W, in order to estimate coastal mixing rates and groundwater discharge fluxes. During the period of this investigation, the activity concentrations of ^{223}Ra in surface seawater varied from 2.7 to 41 mBq 100 L^{-1}, ^{224}Ra in excess from 8.5 to 624 mBq 100 L^{-1}, ^{226}Ra from 131 to 187 mBq 100 L^{-1} and ^{228}Ra from 109 to 409 mBq 100 L^{-1}. The activities of ^{223}Ra, ^{224}Ra, ^{226}Ra and ^{228}Ra in a surface river water sample (Escuro River), which reaches the coast on the local area, were 46 mBq 100 L^{-1}, 954 mBq 100 L^{-1}, 229 mBq 100 L^{-1} and 745 mBq 100 L^{-1}, respectively. Groundwater samples from monitoring wells presented activity concentrations up to 2,033 mBq 100 L for ^{223}Ra, 72,540 mBq 100 L^{-1} for ^{224}Ra in excess, 2,722 mBq 100 L^{-1} for ^{226}Ra and 35,688 mBq 100 L^{-1} for ^{228}Ra. The ^{223}Ra/^{224}Ra activity ratios observed in seawater samples ranged from 0.7×10^{-1} to 0.46, whereas ^{228}Ra/^{226}Ra AR varied in the interval from 0.6 to 1.9. These results seems to indicate that Ra isotopes from ^{232}Th series prevail in a major number of samples, when compared with Ra isotopes from ^{238}U and ^{235}U series. Based on this data, shore-perpendicular profiles of ^{223}Ra and ^{224}Ra in surface waters along the coast were modeled to yield eddy diffusion coefficients. These coefficients allow an evaluation of cross-shelf transport and provide further insight on the importance of groundwater to coastal regions.

INTRODUCTION

The fate of contaminants and natural compounds in estuarine and coastal ocean is determined by a set of biological, geochemical and physical interactions. Although scientists have a basic understanding of the major sources, sinks, and transformations for many substances, to assess offshore fluxes of dissolved materials we need to know coastal water residence times. Until today to quantify residence times in coastal areas remains as a difficult task, since few methodologies are available to assess water exchange in this dynamic region, where currents, waves, tides, river flow and groundwater discharge usually play together a complex role.

The cycling of Ra in oceans can be considered as the most interesting phase of radium geochemistry. The fact that in oceans ^{226}Ra and ^{228}Ra exist in excess of their respective parents, ^{230}Th and ^{232}Th (more than tenfold in excess for ^{226}Ra), led some authors to hypothesize that the excess radium was being supplied by diffusion from deep sea sediments. This hypothesis was later confirmed by deficiencies of the radium isotopes observed in deep sea sediments. It was recognized the potential of these isotopes for studying oceanic circulation because they are the only ones amongst other natural tracers which are injected directly into bottom water from underlying sediments.

The first measurements of radium in the ocean water were made due not so much to its radiological significance, but to the promising role of oceanic radium as a tool for understanding marine geochemical processes. It was Koczy (1958) who first proposed using Ra as a natural tracer in the ocean for the calculation of vertical diffusion coefficients in the water column. This was extended later by Koczy & Szabo (1962) for estimating the renewal time of water masses in the Pacific and Indian Oceans. Numerous investigations have been conducted since then to obtain a fuller understanding of ^{226}Ra distributions in the major oceans of the world, with considerable attention devoted to the variability of ^{226}Ra concentrations with regard to depth and latitude, correlation with barium, silica concentrations, salinity, etc. (Broecker *et al.,* 1967).

^{226}Ra concentrations in surface seawater (0-500 m) appear to be in a narrow range, and nearly uniform in the Pacific (0.74-3.7 mBq L^{-1}) and Atlantic Oceans (0.74-2.96 mBq L^{-1}). However, the Indian Ocean has levels of ^{226}Ra which cover a narrower range (1.11-2.22 mBq L^{-1}). At the increasing depths, a trend of increasing concentrations is observed uniformly in all of the oceans, which, according to most investigators, results from the injection of ^{226}Ra from thorium (^{230}Th) bearing sediments in the ocean floor.

Reports of an interesting correlation of silica and barium concentrations with those of Ra in collateral seawater samples have been made by some researchers, calling attention to the parallel geochemical behavior among these elements. It appears that radium behavior in the marine environment closely parallels that of barium, both of which are carried by marine diatoms, which are highly siliceous organisms.

^{226}Ra concentrations in coastal waters are slightly higher relative to open ocean water. Radium is released by coastal sediments, which form an important source of ^{226}Ra migration to the ocean. This was corroborated by measurements in South Carolina (USA) by Elsinger & Moore (1980) who concluded that a desorption mechanism can quantitatively explain the increase of ^{226}Ra in brackish water. Moore (1981) in his interesting investigations at Chesapeake Bay, demonstrated the clear influence of salinity on the desorption of ^{226}Ra from particulates: at salinities lower than 0.5 ‰, about 12 % of ^{226}Ra in the water is in the soluble phase, while above 5 ‰ salinity, over 80 % is in the dissolved phase.

The spread of ^{226}Ra in coastal waters is more or less uniform (0.74-6.6 mBq L^{-1}), except for estuarine and shelf water in west central Florida. There the ^{226}Ra levels can vary from 1.83 to 53.8 mBq L^{-1} for surface water and from 1.83 to 22.0 mBq L^{-1} for deep water. These high concentrations are attributed to the geology of that area, known for its rich phosphate deposits, seepage of groundwater and to the existence of active geothermal springs.

It is generally observed that nearshore water, particularly from restricted bays and sounds, has higher ^{228}Ra levels than does open ocean surface water. However, these concentrations tend to become lower as one moves away from the coast, thus showing a close correlation between concentration and proximity to land (coastal activities ranging from 0.7 to 11.5 mBq L^{-1}). Some nearshore Indian coastal waters displays significant ^{228}Ra levels (13.7-38.1 mBq L^{-1}) due to significant occurrence of monazite sands.

Moore (1969) has discussed the use of ^{228}Ra in the oceans as a natural tracer for studying marine processes occurring within a 3-30 year time-scale. He has also described the applicability of $^{228}Ra/^{226}Ra$ activity ratios as tracers for studying lateral and vertical movements within the ocean.

Nearshore water, such as surface water close to continents and coastal water in contact with terrigenous sediments were it is observed limited circulation with open ocean, appear to have very high $^{228}Ra/^{226}Ra$ activity ratios. For example, some of the highest $^{228}Ra/^{226}Ra$ ratios (7.1) have been measured in the brackish water of Mississippi Sound. Similar enhanced ratios have also been observed in Chesapeake Bay (1.8); Long Island Sound (1.4); Davis Bay, Mississippi (2.5); Narragansett Bay, Rhode Island (2.0); Wellington Harbour, New Zealand (1.7); Penang Harbour, Malaysia (2.0) and few other places (Moore, 1969).

When nearshore water mixes with oceanic water, the $^{228}Ra/^{226}Ra$ ratio goes down. However, within surface water, large variations in this activity ratio have been observed by Moore (1969). Atlantic surface water generally has higher ratios than Pacific surface water. For example, the Atlantic has activity ratios of 0.09-2.41 and the Pacific 0.011-0.2. According to Moore, the $^{228}Ra/^{226}Ra$ ratio is determined by a balance between the supply of ^{228}Ra to the water body and the lateral and vertical mixing rates. The activity ratios in the Atlantic and Pacific Oceans indicate that the mixing rate of the surface water is considerably longer than ten year mean life of ^{228}Ra.

The $^{228}Ra/^{226}Ra$ activity ratios in the surface water of the Mediterranean, Caribbean and Black Seas are all less than unity and are also in a close range. Inversely, the surface water of the Indonesian seas show typically enhanced ratios, from 0.26 to 3.8, which is due to considerable diffusion of ^{228}Ra from the continental shelves. Other regions such as New York Bight (0.39-1.99), Chesapeake Bay (1.17-4.08), Kalpakkam, India (2.4-4.9) and Bombay, India (1.6-1.9) further support the fact that estuarine and coastal water is rich in ^{228}Ra owing to diffusion from nearshore and estuarine sediments, thereby leading to enhanced $^{228}Ra/^{226}Ra$ ratios, as observed. Moore (1981), following his studies in Chesapeake Bay on ^{226}Ra and ^{228}Ra flux rates and comparing his data with those from other bay and estuarine regions, has proposed that the flux rate of ^{228}Ra from the sediments should be greater than that of ^{226}Ra owing to the faster growth of ^{228}Ra from its parent ^{232}Th and the effects of bioturbation.

This paper reports the determination of ^{223}Ra, ^{224}Ra, ^{226}Ra and ^{228}Ra in a series of small embayments of Ubatuba, São Paulo State-Brazil. The main aim of this preliminary work was to apply these isotopes as tracers to assess coastal mixing rates and groundwater input. Concurrent analysis of ^{226}Ra in sediment, seawater and sediment physical properties, besides nutrients improved the evaluation carried out in this coastal region.

MATERIAL AND METHODS

The field work was carried out during January 2002 (summer), in a series of small embayments of Ubatuba, covering latitudes between 23°26'S and 23°46'S and longitudes between 45°02'W and 45°11'W. The main embayments selected to be studied in this project are Flamengo Bay (Ubatuba Marine Laboratory site), Fortaleza Bay, Mar Virado Bay and Ubatuba Bay **(Fig.1).**

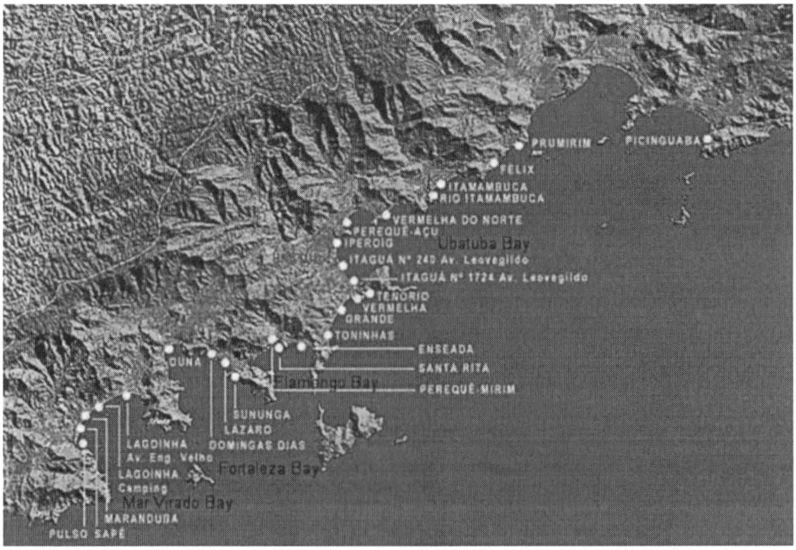

FIGURE 1. Location of the four embayments studied at Ubatuba coastal area: Flamengo, Fortaleza, Mar Virado and Ubatuba. Ubatuba county is located 250 km north from São Paulo city, southeastearn Brazil.

The study area comprises the northernmost part of São Paulo Bight, southeastern Brazil, and is considered a tropical coastal area. The geological/geomorphologic characteristics of the area are strongly controlled by the presence of granites and migmatites of a mountain chain locally called Serra do Mar (altitudes up to 1,000 meters), which reaches the shore in almost all of the study area, and limits the extension of the drainage systems and of Quaternary coastal plains (Mahiques, 1995). In most of the area, the sediments contain mainly silt and very fine sand, and few samples show coarse sand or a clay modal distribution. Wave action is the most effective hydrodynamic phenomenon responsible for the bottom sedimentary processes in the coastal area as well as in the adjacent inner continental shelf. Two main wave directions affect the area. Waves coming from S-SE are associated to the passage of cold fronts over the area and are the most important in terms of reworking of sediments previously deposited. Waves coming from E-NE are mainly generated by trade winds and also during post-frontal periods and are believed to be important to the bottom dynamics. The interaction of wave directions with the extension and orientation of bay mouths and the presence of islands in the inner shelf lead to the occurrence of sensible variations in the dynamics characteristics of the bays, despite that they can all be considered as enclosed bays. The terrestrial input of sediments is strongly dependent on the rainfall regime, leading to a higher contribution of sediments during summer season. During this period, the advance of the South Atlantic Central Water (SACW) over the coast leads to the displacement of the Coastal Water (CW) (Castro Filho *et al.*, 1987), rich in continental suspended materials, and to the transportation of these sediments to the outer portions of the continental shelf. During winter, the retreat of the SACW and the decreasing of the rainy levels restrict the input of sediments from the continental areas. The mean annual rainfall is roughly 1,803 mm, the maximum rainfall rates being observed in February. Sea level varies from 0.5 to 1.5 m, the highest values occurring in months August/September due to greater volume of warm waters of Brazil Current (Mesquita, 1997).

The Ubatuba coastal area is known to be oligo-mesotrophic, because the primary production is limited by the lack of inorganic compounds of nitrogen and phosphorous (Braga & Muller, 1998). The region has been reported to receive nutrient inputs by atmospheric contribution mainly in nitrogenous compounds, and in minor degree by terrestrial contribution, which limit the local primary production. However, from time to time, intrusions of nutrient and oxygen-rich South Atlantic Central Water (SACW) from the open ocean thermocline may reach the shelf edge, and may further be transferred by coastal upwelling, that is driven by northeasterly winds, providing a third source of nutrients for primary production.

For the purposes of pre-concentration of Ra isotopes from large volume of seawater samples described in this paper, Acrylic fibre (Cia. Sudamericana do Brasil, 3.0 denier) was treated with a hot solution of saturated $KMnO_4$ for approximately 10 minutes. The $KMnO_4$ oxidizes specific sites on the acrylic molecule and deposits MnO_2 at these sites. The prepared fibre was washed with purified water free of radium and was kept in plastic bags for use. This produces Mn-fibre having sub-micrometric sized particles of MnO_2 chemically bonded to the fibre. The MnO_2 constitutes 8-10% by mass of the Mn-fibre. This procedure was conducted in a 5 L beacker scale.

To assess the spatial distribution of the Ra tracers, five shore-perpendicular transects were sampled in January 2002, the horizontal profiles were collected up to about 30 km from shore **(Fig.2)**. Large volume seawater samples (196 L) were pumped from 5 m below the surface into plastic drums on the R/V Velliger II. The sample volume was recorded and the seawater was percolated through a column of manganese coated acrylic fibre to quantitatively remove radium from seawater (Moore, 1996). Temperature and salinity profiles were obtained at each station using a 2.00" Micro CTD, from Falmouth Scientific Inc. Samples for salinity and nutrients were also collected in each station.

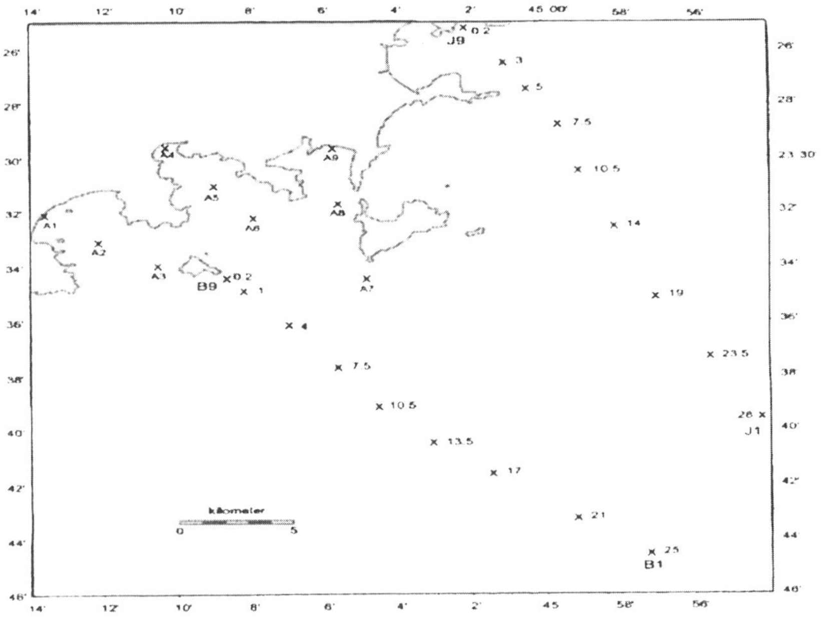

FIGURE 2. Location of the horizontal profiles established for Ra isotopes in Ubatuba region (Jan. 2002).

In each profile, samples for the determination of phosphate, silicate, ammonium, nitrite and nitrate were also collected. Water samples were frozen until time of analysis. The analytical procedures adopted for these determinations were vanadium reduction followed by chemiluminescence detection of NOx for nitrate-nitrite, phenate method for ammonia and ascorbic acid method for phosphate.

In an onshore laboratory each Mn fibre sample was partially dried with a stream of air and placed in an air circulation system described by Moore & Arnold (1996). Helium was circulated over the Mn fibre to sweep ^{219}Rn and ^{220}Rn generated by ^{224}Ra and ^{223}Ra decay in a 1.1 L scintillation cell. The alpha particles from the decay of radon and its daughters were recorded by a photomultiplier tube (PMT) attached to the scintillation cell described previously. Signals from the PMT were routed to a delayed coincident counter system adapted for Ra measurements (Moore & Arnold, 1996). The delayed coincidence system utilizes the difference in decay constants of the short-lived Po daughters of ^{219}Rn and ^{220}Rn to identify alpha particles derived from ^{223}Ra and ^{224}Ra captured on the Mn fiber. The expected error of the short-lived Ra isotope measurement is 10%.

After completing the ^{224}Ra and ^{223}Ra measurements, the Mn fibre samples were aged for 5 weeks to allow excess ^{224}Ra to equilibrate with ^{228}Th adsorbed to the Mn fibre. The samples were measured again to determine ^{228}Th activity and this value was used to correct the ^{224}Ra activity to its unsupported activity.

Following these analyses, the Mn fibre was leached in a beaker with 200 mL HCl under controlled heating, to quantitatively remove the longer-lived Ra isotopes. For the radiochemical separation of ^{226}Ra and ^{228}Ra, carriers of stable barium (20mg) and lead (20 mg) were added to the water sample in the presence of 5 mL of 1 M citric acid and 5 mL of 40% hydroxylamine hydrochloride solutions. The radium was co-precipitated as $Ba,Pb(Ra)SO_4$ by adding 50 mL of 3 M H_2SO_4. The precipitate was dissolved with alkaline EDTA. When the pH is adjusted to 4.5 with glacial acetic acid, $Ba(Ra)SO_4$ is re-precipitated, while interfering elements remain in the solution. The $Ba(Ra)SO_4$ precipitate was transferred to a 2 mL polypropylene tube and sealed to avoid the escape of ^{222}Rn. ^{226}Ra and ^{228}Ra were measured by gamma spectrometry of a $Ba(Ra)SO_4$ precipitate in a WeGe well germanium detector, after 21 days from the precipitation. The detector is a 78 cm^3 coaxial intrinsic germanium crystal with a 1 cm diameter and 4 cm deep well produced by Princeton Gamma Tech.. The ^{226}Ra activities were determined by taking the mean activity of three separate photopeaks of its daughter nuclides: ^{214}Pb at 295.2 keV and 351.9 keV, and ^{214}Bi at 609.3 keV. The ^{228}Ra content of the samples was determined from the 911 keV and 968 keV gamma-ray peaks of ^{228}Ac. Both measurements were performed at the Radioisotope Geochemical Laboratory of the University of South Carolina.

Radium Isotope Disequilibrium to Delineate Coastal Dynamics and Groundwater Input

In the natural radioactive series, there are four radium isotopes: ^{226}Ra ($t_{1/2}$ = 1620 years); ^{228}Ra ($t_{1/2}$ = 5.75 years); ^{223}Ra ($t_{1/2}$ = 11.3 days); ^{224}Ra ($t_{1/2}$ = 3.66 days). Each isotope is produced from the decay of a thorium parent: ^{230}Th ($t_{1/2}$ = 7.54 $\times 10^4$ years); ^{232}Th ($t_{1/2}$ = 1.40$\times 10^{10}$ years); ^{227}Th ($t_{1/2}$ = 18.7 days); ^{228}Th ($t_{1/2}$ = 1.91 years), respectively.

Because thorium remains tightly bound to particles while radium daughters are mobilized into the marine environment, sediments provide a continuous source of Ra isotopes to seawater, at rates set by their respective decay constants. Measurements of the Th isotope activities in the sediments and the distribution coefficient of Ra between the sediments and water provide a means of quantifying the potential input of each isotope to the ocean.

Two short-lived radium isotopes ^{223}Ra and ^{224}Ra can be used as tracers to measure the rate of exchange of coastal waters (Moore, 1998). Shore-perpendicular profiles of ^{223}Ra and ^{224}Ra in surface waters along the coast may be modeled to yield eddy diffusion coefficients. Coupling the exchange rate with offshore concentration gradients, the offshore fluxes of dissolved materials are estimated. For systems in steady-state, the offshore fluxes must be balanced by new inputs from rivers, groundwater, sewers or other sources. Also, it was observed that barium and ^{226}Ra contents can be powerful indicators of groundwater input in marine systems, since they have high relative concentrations in the fluids and low reactivity in the coastal ocean. An estimate of the ^{226}Ra offshore flux is made applying the eddy diffusion coefficients to the ^{226}Ra offshore gradient. Complementary data of ^{226}Ra in subsurface fluids provides a mean of calculate the fluid flux necessary to support the ^{226}Ra concentrations found in the marine environment.

The Ra distribution may be expressed by a simple one-dimensional horizontal diffusion model, in which the distribution is in balance between eddy diffusion and radioactive decay (Moore, 1998):

$$\frac{dA}{dt} = K_h \frac{\partial A}{\partial x^2} - \lambda A \qquad (1)$$

At the steady-state, this expression can be written as the following:

$$A_x = A_0 \exp\left[-x\sqrt{\frac{\lambda}{K_h}}\right] \qquad (2)$$

Where, A_x is the activity at distance x from coast, A_0 is the activity at distance zero from coast, λ is the decay constant and Kh is the horizontal eddy diffusion coefficient.

RESULTS AND DISCUSSION

The location of all stations studied as well as the results of the cruise carried out in January 2002 are shown in **Tab. 1 to 8.**

During the period of this investigation, the activity concentrations of ^{223}Ra in surface seawater varied from 2.7 to 41 mBq 100 L^{-1}, ^{224}Ra in excess from 8.5 to 624 mBq 100 L^{-1}, ^{226}Ra from 131 to 187 mBq 100 L^{-1} and ^{228}Ra from 109 to 409 mBq 100 L^{-1}. The activities of ^{223}Ra, ^{224}Ra, ^{226}Ra and ^{228}Ra in a surface water sample from Escuro River, which reaches the coast on Fortaleza bay were 46 mBq 100 L^{-1}, 954 mBq 100 L^{-1}, 229 mBq 100 L^{-1} and 745 mBq 100 L^{-1}, respectively. Groundwater samples from monitoring wells installed in Flamengo bay (in front of the Marine Lab.) presented activity concentrations up to 2,033 mBq 100 L^{-1} for ^{223}Ra, 72,540 mBq 100 L^{-1} for ^{224}Ra in excess, 2,722 mBq 100 L^{-1} for ^{226}Ra and 35,688 mBq 100 L^{-1} for ^{228}Ra. The $^{223}Ra/^{224}Ra$ activity ratios observed in seawater samples ranged from 0.7×10^{-1} to 0.46, whereas $^{228}Ra/^{226}Ra$ AR varied in the interval from 0.6 to 1.9. These results seems to indicate that Ra isotopes from ^{232}Th series prevail in a major number of samples, when compared with Ra isotopes from ^{238}U and ^{235}U series.

The ^{226}Ra activity concentrations found in surface seawater samples studied at Ubatuba region are of the same order of magnitude than those reported by other author in the southeastern coast of United States (typical values from 133 to 283 mBq 100 L^{-1}) (Moore, 1999).

TABLE 1. Location of the samples from Transect B, between Mar Virado and Fortaleza embayments, Jan. 2002.

Sample	Date	Volume (L)	Latitude	Longitude	Distance from shore (km)	Salinity (psu)	Temperature (°C)
B-1	22/Jan/02	196	S23°44.592	W44°57.259	30.6	35.337	24.1
B-2	22/Jan/02	196	S23°43.265	W44°59.154	27.9	35.393	24.5
B-3	22/Jan/02	196	S23°41.684	W45°01.435	23.5	35.348	23.5
B-4	22/Jan/02	196	S23°40.445	W45°03.006	20.4	34.917	24.0
B-5	22/Jan/02	196	S23°39.163	W45°04.395	17.4	35.054	24.3
B-6	22/Jan/02	196	S23°37.713	W45°05.697	14.4	35.003	25.0
B-7	22/Jan/02	196	S23°26.233	W45°06.888	11.5	35.003	25.0
B-8	22/Jan/02	196	S23°34.976	W45°08.645	9.3	35.025	25.0
B-9	22/Jan/02	196	S23°34.316	W45°08.645	8.4	35.018	25.0

TABLE 2. Concentration of the natural Ra isotopes, Ra^{223}/Ra^{224} and Ra^{228}/Ra^{226} activity ratios and nutrients determined in the samples from Transect B.

Sample	^{223}Ra mBq 100 L^{-1}	^{224}Ra mBq 100 L^{-1}	$^{223}Ra/^{224}Ra$	^{226}Ra mBq 100 L^{-1}	^{228}Ra mBq 100 L^{-1}	$^{228}Ra/^{226}Ra$	^{228}Th mBq 100 L^{-1}	Nitrate µmol L^{-1}	Nitrite µmol L^{-1}	Silicate µmol L^{-1}	Phosphate µmol L^{-1}
B-1	6.5	29	0.22	153	159	1.0	4.2	0.016	0.096	1.72	0.43
B-2	2.7	22	0.12	147	133	0.90	6.2	0.607	0.098	2.39	0.86
B-3	1.0	8.5	0.12	119	109	0.91	2.3	-	-	5.96	0.33
B-4	5.8	45	0.13	131	123	0.94	6.3	0.241	0.127	2.30	0.51
B-5	13	126	0.10	163	244	1.5	10	0.014	0.049	3.13	0.60
B-6	9.3	107	0.09	168	268	1.6	8.5	0.299	0.069	3.77	0.60
B-7	12	132	0.09	158	305	1.9	12	0.130	0.059	2.78	0.43
B-8	14	106	0.13	141	255	1.8	7.5	0.288	0.059	4.73	0.62
B-9	15	152	0.10	173	285	1.7	8.5	0.028	0.098	2.70	0.51

TABLE 3. Location of the samples from Transects A, inside Mar Virado, Fortaleza and Flamengo embayments, Jan. 2002.

Sample	Date	Volume (L)	Latitude	Longitude	Distance from shore (km)	Salinity (psu)	Temperature (°C)
A-1	23/Jan/02	196	S23°32.089	W45°13.380	1	34.798	25.0
A-2	23/Jan/02	196	S23°33.094	W45°12.015	4	35.136	25.0
A-3	23/Jan/02	196	S23°33.994	W45°10.196	7	34.947	25.5
A-4	23/Jan/02	196	S23°29.823	W45°10.196	1	34.221	26.0
A-5	23/Jan/02	196	S23°31.027	W45°09.057	4	35.114	25.0
A-6	23/Jan/02	196	S23°32.303	W45°07.912	7	34.885	25.0
A-7	23/Jan/02	196	S23°34.397	W45°04.705	13	35.163	26.0
A-8	23/Jan/02	196	S23°34.347	W45°29.629	6	34.820	25.0
A-9	23/Jan/02	196	S23°29.629	W45°05.758	1	-	27.0

TABLE 4. Concentration of the natural Ra isotopes, Ra^{223}/Ra^{224} and Ra^{228}/Ra^{226} activity ratios and nutrients determined in the samples from Transects A.

Sample	^{223}Ra mBq 100 L^{-1}	^{224}Ra mBq 100 L^{-1}	$^{223}Ra/^{224}Ra$	^{226}Ra mBq 100 L^{-1}	^{228}Ra mBq 100 L^{-1}	$^{228}Ra/^{226}Ra$	^{228}Th mBq 100 L^{-1}	Nitrate µmol L^{-1}	Nitrite µmol L^{-1}	Silicate µmol L^{-1}	Phosphate µmol L^{-1}
A-1	43	624	0.07	165	409	2.5	19	0.425	0.036	6.08	0.62
A-2	5.8	34	0.17	162	211	1.3	7.3	0.323	0.048	2.63	0.32
A-3	13	121	0.10	150	258	1.7	8.8	0.004	0.108	3.25	0.33
A-4	27	441	0.06	166	370	2.2	14	0.252	0.096	2.63	0.56
A-5	17	125	0.14	149	247	1.7	9.5	0.008	0.048	7.31	0.50
A-6	8.7	109	0.08	167	307	1.8	7.5	0.193	0.144	3.83	0.58
A-7	6.8	11	0.60	142	167	1.2	6.3	1.064	0.060	1.85	0.64
A-8	16	133	0.12	167	313	1.9	6.0	0.286	0.096	6.02	0.81
A-9	41	513	0.08	163	392	2.4	17	0.126	0.132	3.91	0.46

TABLE 5. Location of the samples from Transect J. Ubatuba bay, Jan. 2002.

Sample	Date	Volume (L)	Latitude	Longitude	Distance from shore (km)	Salinity (psu)	Temperature (°C)
J-1	25/Jan/02	196	S23°31.874	W45°09.898	30	35.728	25.0
J-2	25/Jan/02	196	S23°37.426	W44°55.658	25	35.550	26.0
J-3	25/Jan/02	196	S23°35.073	W44°57.099	20	35.368	26.0
J-4	25/Jan/02	196	S23°32.521	W44°58.186	15	35.526	26.0
J-5	25/Jan/02	196	S23°30.453	W44°59.618	11	35.653	26.0
J-6	25/Jan/02	196	S23°28.842	W44°59.618	8	35.581	26.0
J-7	25/Jan/02	196	S23°27.399	W45°00.448	5	35.339	26.0
J-8	25/Jan/02	196	S23°26.545	W45°01.128	3	35.667	26.0
J-9	25/Jan/02	196	S23°25.201	W45°02.246	1	35.752	26.0

TABLE 6. Concentration of the natural Ra isotopes, Ra^{223}/Ra^{224} and Ra^{228}/Ra^{226} activity ratios and nutrients determined in the samples from Transect J.

Sample	^{223}Ra mBq 100 L^{-1}	^{224}Ra mBq 100 L^{-1}	$^{223}Ra/^{224}Ra$	^{226}Ra mBq 100 L^{-1}	^{228}Ra mBq 100 L^{-1}	$^{228}Ra/^{226}Ra$	^{228}Th mBq 100 L^{-1}	Nitrate µmol L^{-1}	Nitrite µmol L^{-1}	Silicate µmol L^{-1}	Phosphate µmol L^{-1}
J-1	5.5	23	0.24	151	160	1.1	2.0	0.166	0.040	2.03	0.28
J-2	9.7	55	0.18	158	235	1.5	7.7	0.134	0.098	3.43	0.59
J-3	11	93	0.11	149	271	1.8	15	0.176	0.098	5.55	0.66
J-4	19	131	0.15	158	298	1.9	12	0.587	0.108	10.87	0.67
J-5	13	107	0.12	159	221	1.4	6.0	0.539	0.067	4.01	0.36
J-6	11	82	0.14	148	234	1.6	8.2	0.070	0.058	2.42	0.46
J-7	16	104	0.16	187	109	0.6	6.0	0.383	0.029	2.60	0.35
J-8	10	106	0.10	148	256	1.7	8.0	0.038	0.058	4.00	0.57
J-9	16	35	0.46	-	-	-	7.0	0.353	0.115	0.40	0.90

TABLE 7. Location of the groundwater and river samples from Ubatuba region, Jan. 2002.

Sample	Date	Volume (L)	Latitude	Longitude	Salinity (psu)
P-1	24/Jan/02	20	S23°29.952	W45°07.093	0.75
P-2	24/Jan/02	20	S23°31.893	W45°09.891	0.079
P-3	24/Jan/02	20	S23°31.886	W45°09.873	0.060
P-4	24/Jan/02	20	S23°31.872	W45°09.901	0.059
PM-01	24/Jan/02	8	S23°30.009	W45°07.113	-
PM-03	24/Jan/02	15	S23°30.008	W45°07.105	-
PM-04	25/Jan/02	8	S23°30.013	W45°07.095	-
PM-05	25/Jan/02	10	S23°30.018	W45°07.085	-
PM-06	24/Jan/02	8	S23°29.999	W45°07.107	-
PM-08	25/Jan/02	10	S23°30.007	W45°07.093	-
PM-09	25/Jan/02	8	S23°30.012	W45°07.084	-
Escuro River	26/Jan/02	40	S23°29.466	W45°09.830	30.6

TABLE 8. Concentration of the natural Ra isotopes, Ra^{223}/Ra^{224} and Ra^{228}/Ra^{226} activity ratios and nutrients determined in groundwater and river samples from Ubatuba region.

Sample	^{223}Ra mBq 100 L^{-1}	^{224}Ra mBq 100 L^{-1}	^{223}Ra/^{224}Ra	^{226}Ra mBq 100 L^{-1}	^{228}Ra mBq 100 L^{-1}	^{228}Ra/^{226}Ra	^{228}Th mBq 100 L^{-1}	Nitrate µmol L^{-1}	Nitrite µmol L^{-1}	Silicate µmol L^{-1}	Phosphate µmol L^{-1}
P-1	16	609	0.03	224	-	-	21	146	13	57.5	0.57
P-2	58	1,561	0.04	314	-	-	63	12.9	11	12.1	0.42
P-3	36	334	0.11	168	-	-	27	90	11	7.1	0.61
P-4	13	801	0.02	-	-	-	32	-	-	-	0.20
PM-01	19	1,211	0.02	479	-	-	70	-	-	-	-
PM-03	931	16,244	0.06	242	14,533	60	433	96	8	9.6	0.25
PM-04	1,408	72,540	0.02	1,192	35,688	30	1451	16.4	14	54.7	1.63
PM-05	525	34,659	0.02	2,722	23,995	8.8	826	-	-	-	-
PM-06	59	2,821	0.02	-	-	-	238	-	-	-	-
PM-08	2,033	35,405	0.06	933	15,398	17	624	-	-	-	-
PM-09	473	38.009	0.01	1.035	21,167	20	666	-	-	-	-
Rio Escuro	46	954	0.05	229	745	3.3	27	50.2	0.57	-	-

The ^{226}Ra distribution in surface seawater samples from Transect B showed a narrow range along the coast. Since the half-life of ^{226}Ra (1.620 y) is comparable to the mean ocean circulation time established from 750 a 1000 years (Broecker & Peng, 1982), the ^{226}Ra should be very well mixed in seawater (uniform concentrations along the coast). Deviation of this behavior is expected to occur only close to the margins or the bottom (in this case one positive ^{226}Ra signal could be used to identify a groundwater input). In the case of ^{228}Ra, as it has the half-life of 5.7 years, the activities are higher close to the margins, decreasing with distance from shore.

In coastal areas the short-lived Ra isotopes, ^{223}Ra and ^{224}Ra, are flushed from the sediments and are regenerated from their thorium parents in sediments on a time scale of days. This provides a continuous source of ^{223}Ra and ^{224}Ra activity to the overlying seawater that is not accompanied by large additional ^{226}Ra and ^{228}Ra, which are regenerated more slowly. Groundwater discharge directly into sea water may also provide significant additions of Ra activity to the estuaries and ocean shelf bottom water (Moore, 1996; Rama & Moore, 1996).

Fig. 3 shows the distribution of ^{224}Ra in excess vs. ^{223}Ra along the Transect B. This horizontal profile was sampled from the surroundings of Mar Virado Island until 30 km offshore **(see Fig.2)**. This results demonstrate clearly the influence of a second water mass entering the coastal area during the studied period **(Fig. 5-9)**.

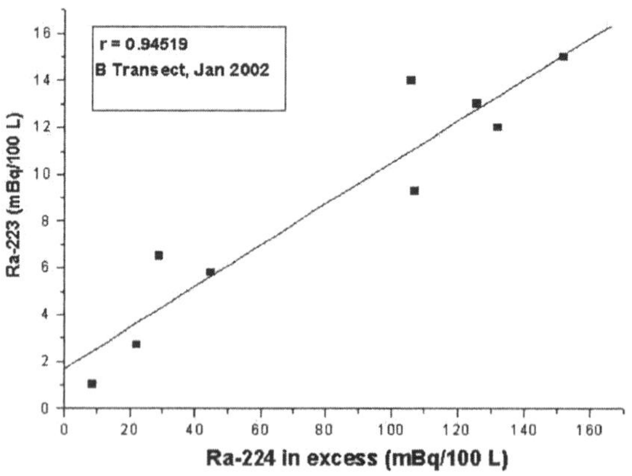

FIGURE 3. ^{224}Ra in excess vs. ^{223}Ra observed in Transect B.

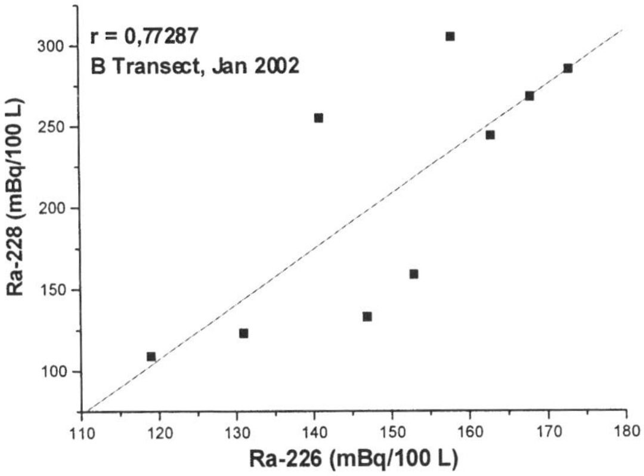

FIGURE 4. Concentrations of ^{226}Ra vs. ^{228}Ra observed in Transect B.

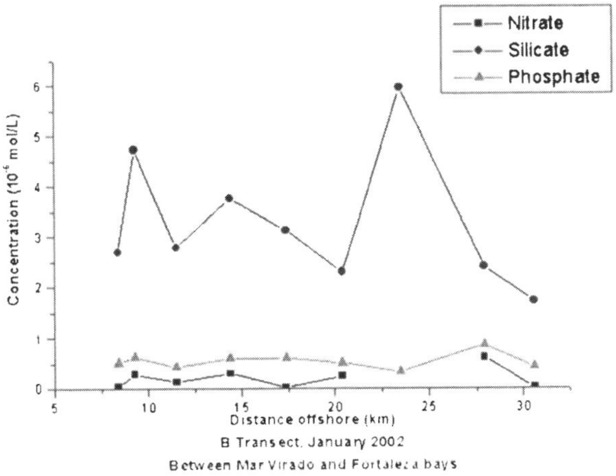

FIGURE 5. Distribution of nutrients along the coast, Transect B.

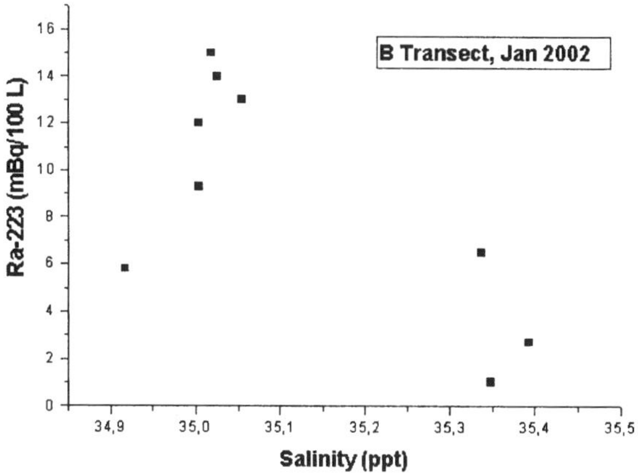

FIGURE 6. A plot of salinity vs. concentration of ^{223}Ra, Transect B.

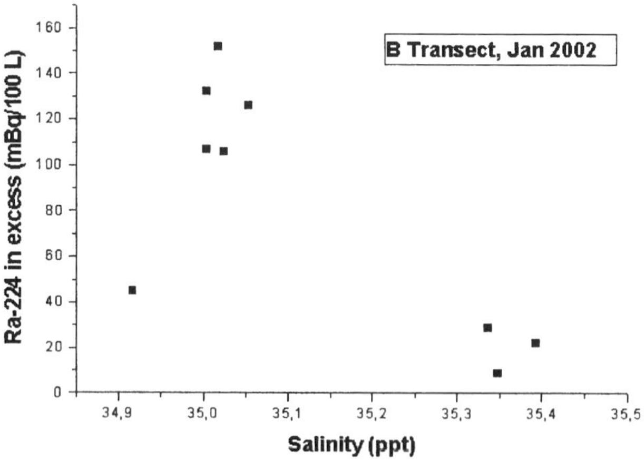

FIGURE 7. A plot of salinity vs. concentration of ^{224}Ra in excess, Transect B.

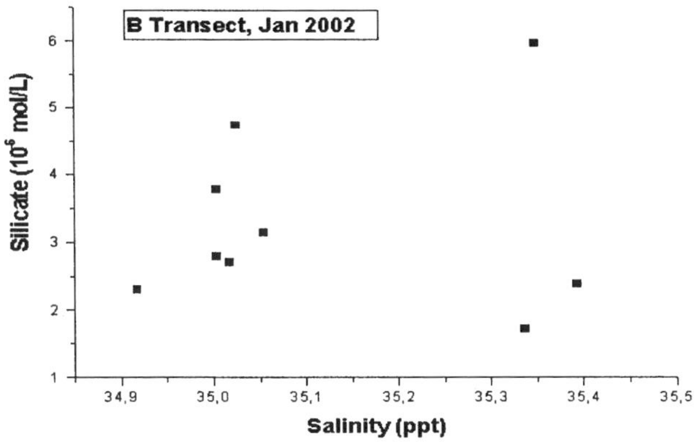

FIGURE 8. A plot of salinity vs. silicate, Transect B.

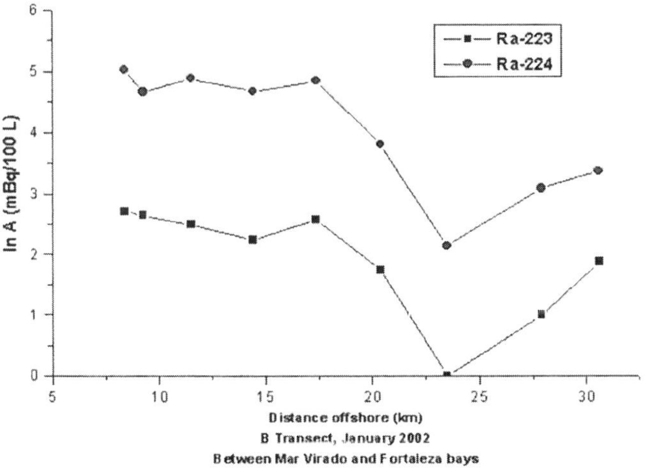

FIGURE 9. The distribution of $\ln^{223}$Ra and $\ln^{224}$Ra in excess along the coast, Transect B.

As it was presented in **Fig. 3**, there is a strong correlation between the short-lived Ra isotopes in Transect B (r = 0.94519). However, the same pattern was not observed for all the samples in a correlation plot of ^{228}Ra vs. ^{226}Ra (r = 0.77287) presented in **Fig. 4.**

Regarding the nutrients, it was found increased silicate concentrations in surface seawater around 25 km offshore **(Fig. 5)**, which can indicates upwelling of a bottom water mass or submarine groundwater discharge. Although, nitrate and phosphate concentrations measured are comparable to the values usually found in Ubatuba region, classified as oligo-mesotrophic following to the content of (N-P) available to the marine ecosystem productivity (Braga & Muller, 1998). Is was not observed any anomalous nitrate concentration in the Transect B samples, which would indicate infiltration of groundwater contaminated by domestic sewage.

It is very well known that ocasionally, the upwelling of SACW can bring nutrients for the local embayments of Ubatuba. Thus, the silicate signal noticed at 25 km offshore can be with great probability due to the passage of SACW close to the coast during the summer.

Fig. 9 shows the distribution of ln^{223}Ra and ln^{224}Ra in excess vs. distance from shore. Theoretically, these concentrations should decrease exponentially with distance offshore. However, it was observed a slight increase of ^{224}Ra and ^{223}Ra activities after 25 km. This fact demonstrates again the presence of a second water mass with distinct characteristics, which could be attributable to SACW upwelling or the circulation pattern of the studied embayments or groundwater discharge. In this case, the existence of an anti-clockwise eddy in opposition to the passage of ocean currents parallel to the coast could carry more fast the short-lived Ra isotopes to this location. The distribution of ln^{226}Ra and ln^{228}Ra along the coast for Transect B is presented in **Fig. 10.**

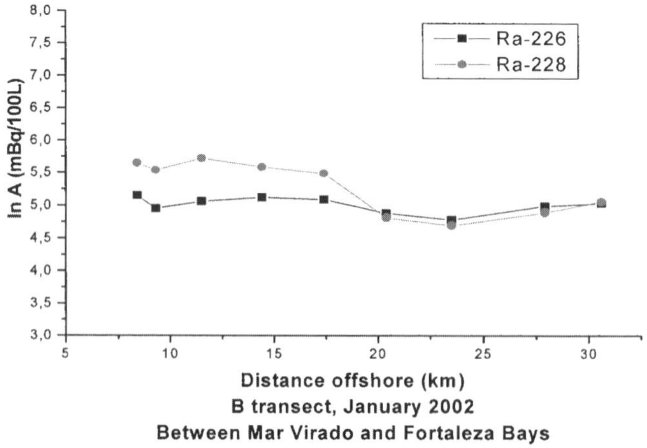

FIGURE 10. The distribution of ln^{226}Ra and ln ^{228}Ra along the coast, Transect B.

The ^{228}Th measurements reveal a marked deficiency of ^{228}Th relative to its parent, ^{228}Ra. This indicates the rapid removal of Th from the water column.

Considering the results obtained in the summer 2002, the exchange time of the surface water masses inside Flamengo, Fortaleza and Mar Virado embayments were estimated using the activity concentrations of ^{223}Ra and ^{224}Ra (small A transects). Apparent ages calculated by this method reflect the time elapsed since the water sample became enriched in Ra and was isolated from the source. This calculations assumed that 100% of the initial concentration of the Ra isotope present in the sediments in the near shore region was transferred to the seawater (end-member fraction=1).

The exchange time obtained using the ^{223}Ra activities were 29.4 days for Flamengo Bay, 18.8 days for Fortaleza Bay and 19.7 days for Mar Virado Bay. Taking into account the ^{224}Ra activities, these apparent ages were respectively: 19.9 days for Flamengo, 7.3 days for Fortaleza and 8.5 days for Mar Virado. However, it is recommended to use the ages determined from the ^{223}Ra profile across the shelf, since the lowest half-life of 3.6 days for ^{224}Ra may be less convenient for this purposes.

The water exchange time in Flamengo Bay was verified using the activities of ^{223}Ra normalized to ^{228}Ra activities to correct for mixing, because its half life is long with respect to the mixing time of near shore waters. This final calculation resulted in an exchange time of 19.4 days for Flamengo Bay. This ages imply that exchange times of the coastal waters across the bays are rapid, of the order of 20 days during the period of investigation.

CONCLUSIONS

The application of the four naturally occurring Ra isotopes and an one-dimensional advection-diffusion model was shown as a tool to assist in the interpretation of coastal ocean circulation and biogeochemistry at Ubatuba region. Since they do not require steady-state conditions with respect to mixing, this isotopic technique can supply useful data which coupled with salinity or any other tracer distributions, provide powerful constraints to follow the circulation patterns and calculate fluxes of several dissolved materials to the ocean. Obviously, additional work during different conditions shall be carried out to estimate average exchange times and seasonal variations.

Once it was indicated and quantified in a previous research work, carried out in Ubatuba using ^{222}Rn as a tracer, that there is a significant inflow of subsurface fluids at rates in excess of several cm per day in the same embayments studied here (Oliveira *et al.*, 2003), we intend to use the Ra data set and the residence times obtained to perform a mass balance (integrating river and groundwater end-member concentrations) to quantify the groundwater input for the same area. This is a research in progress.

ACKNOWLEDGMENTS

This research work was funded by Fundação de Amparo à Pesquisa no Estado de São Paulo – FAPESP, Project nº 2000/10993-7. We would like to tank the crew of

R/V Velliger II, as well as Ana Carolina Garofalo Masini and Patrícia da Costa, who assisted in the sample collection and processing.

REFERENCES

1. Braga, E.S., Muller, T.J. *Continental Shelf Research* **18,** 915-922 (1998).
2. Broecker, W.S.; Li, Y.H.; Cromwell, J. *Science,* **158,**1307-1310 (1967).
3. Broecker, W.S.; Peng, T.H. *Tracers in the Sea.* Lamonh-Doherty Geological Observatory, Palisades, NY, 690 pp (1982).
4. Castro Filho, B.M., Miranda, L.B., Miyao, S.Y. *Boletim do Instituto Oceanográfico,* São Paulo, **35(2),** 135-151 (1987).
5. Castro, B.M., Miranda, L.B. In: Robinson AR and Brink KH (eds), *The Sea,* **Vol.11,** 209-252 (1998), New York, John Wiley & Sons.
6. Elsinger, R.J.; Moore, W.S. *Earth Planet.Sci. Lett.,* **48,** 239-249 (1980).
7. Koczy, F.F. Natural radium as a tracer in the ocean. *Peaceful uses of Atomic Energy* (Proc.2nd Int. Conf. Geneva, 1958), **Vol.1,** United Nations, Geneva (1958).
8. Koczy, F.F.; Szabo, B.J. *J.Oceanogr.Soc.Jpn,* 20th Anniversary volume, 590-599 (1962).
9. Mahiques, M.M. *Boletim do Instituto Oceanográfico, São Paulo,* **43(2),** 111-122 (1995).
10. Mesquita, A.R. Marés, circulação e nível do mar na Costa Sudeste do Brasil. *Relatório Fundespa,* São Paulo (1997).
11. Moore, W.S. *Earth Planet.Sci. Lett.,* **6(2),** 437-446 (1969).
12. Moore, W.S. *Estuar.Coast.Shelf Sci.,* **12,** 713-723 (1981).
13. Moore, W.S. *Nature* **380,** 612-614 (1996).
14. Moore, W.S., Arnold, R. *Journal of Geophysical Research* **101 (C1),** 1321-1329 (1996).
15. Moore, W.S. *Earth Planet.Sci.* **107(4),** 343-349 (1998).
16. Moore, W.S. and Shaw, T.J. *J.Geophys.Res.Oceans* **103,** 21543-21552 (1998).
17. Moore, W.S. *Marine Chemistry* **65,** 111-125 (1999).
18. Oliveira, J.; Burnett, W.C.; Mazzilli, B.P.; Braga, E.S.; Farias, L.A.; Chistoff, J.; Furtado, V.V. *Journal of Environmental Radioactivity* **69,** 37-52 (2003).
19. Rama and Moore, W.S. *Geochim.Cosmochim. Acta* **60,** 4245-4252 (1996).

Applied Physics (APHYS 2003)
A. Méndez-Vilas (Ed.)

Calibration Procedures for Hand-Foot Contamination Monitors

Oscar B. Alvarez, Ana F. Maia and Linda V. E. Caldas

Instituto de Pesquisas Energéticas e Nucleares (IPEN)
Comissão Nacional de Energia Nuclear
Av. Lineu Prestes 2.242
05508-000, São Paulo, SP
Brazil
otbalvar@ipen.br, afmaia@ipen.br, lcaldas@ipen.br

Abstract. Hand-foot contamination monitors are widely used for quick contamination measurements on hands and feet of workers in radioactive environments. The ISO 7503-1 standard recommends using reference sources large enough to cover the entire sensitive areas of the probes for the efficiency determination of this kind of monitor. However, most of hand-foot contamination monitors are built in compliance with the IEC 504 standard having too big foot probes (525 cm^2) compared with the conventional reference sources (150 cm^2). This problem can be solved following the suggestion of the publication HS(G)49, but it presents the disadvantage of being a time consuming procedure. The aim of this study was to establish alternative methods for the calibration of these monitors. The final results showed that it is possible to determine the monitor efficiency using a faster, and even more realistic, method. Therefore, a method is suggested to be used for workplace calibrations, keeping the HS(G)49 method only for periodic calibrations.

INTRODUCTION

Contamination monitoring is one of the most important procedures for radiation protection as well as radiation dose monitoring. For implementation of the proper contamination monitoring, radiation measuring instruments should not only be suitable for the purpose of monitoring, but they should also be properly calibrated [1].

Hand-foot contamination monitors are widely used for quick contamination measurements on hands and feet of workers in radioactive environments. They are available with several kinds of counter tubes such as gas flow proportional counters (for alpha and beta radiation), Xenon filled proportional counters (for beta radiation and low energy X-rays, etc), scintillation counters with ZnS (Ag) (for alpha radiation) or dual phosphors (for alpha, beta and gamma radiation), Geiger-Müller detectors (for beta and gamma radiation), etc.

To comply with IEC 504 standards [2], often the detection areas of hand-foot contamination monitors are 250 cm^2 for hands and 525 cm^2 for feet. As the reference sources [3] available are smaller, usually 15 cm x 10 cm (150 cm^2) in dimensions, the calibration procedure is difficult because the ISO 7503-1 standard [4] recommends the use of reference sources large enough to cover the entire sensitive probe. Even so, the same standard recommends that, in extreme cases, when sources of such dimensions are not available, sequential measurements with smaller distributed sources of at least 100 cm^2 of active area shall be utilized. In compliance with this recommendation, the British standard HS(G)49 [5] suggests a procedure to determine the efficiency taking a measurement and then moving the source to an adjoining area, and repeating the process until covering all the sensitive detector window.

In this paper, the objective was to find an adequate method to determine the efficiency of a hand-foot contamination monitor. For the hand probes, that are much smaller than the feet probes, just one calibration procedure was evaluated because the utilization of a rectangular source (150 cm^2) in the central area of the probe simulates adequately the real positioning of the user hands. However, for the feet probes, four different methods were evaluated: using an available source with an active area smaller than the sensitive detector window (HS(G)49 method); a method recommended by Los Alamos National Laboratory (LANL)(6) that uses a circular source; and also other two alternative methods established in this work. The procedure chosen as most adequate was applied for the routine tests of a hand-foot contamination equipment in its quality control program.

In this work, periodic calibration signifies the calibration made at a metrology laboratory, and routine calibration, the workplace calibration.

MATERIALS AND METHODS

A hand-foot contamination monitor for alpha and/or beta radiation Eurisys/Sirius was used in this study. This equipment has six large proportional detectors, 4 detectors for the hands and 2 detectors for the feet. The detection areas are 250 cm^2 for the hands and 525 cm^2 for the feet.

The equipment efficiency was determined for five beta sources (C-14, Tc-99, Cs-137, Cl-36 and Sr/Y-90), and one alpha source (Am-241). The characteristics of the beta sources are presented in Table 1. The alpha source has the following characteristics: half-life of 432.2 years and surface emission rate of 442 s-1. All these sources present areas of 15 cm x 10 cm, and in the cases of Am-241 and Cl-36 circular sources of 36 mm in diameter were also utilized. The sources are from Amersham, "anodized" variety, and have calibration certificates from Physikalisch-Technische Bundesanstalt (PTB), with the activity nearly uniformly distributed (better than 6%).

TABLE 1. Main characteristics of the beta sources utilized.

Source	Approximate half-life (years)	Surface emission rate (s^{-1})	Average energy (keV)
C-14	5700	406	50
Tc-99	211100	572	85
Cs-137	30.07	586	185
Cl-36	300000	643	246
Sr/Y-90	28.79	1300	934

The instrument efficiency, ε_i, on the reference source is given by the following equation [4]:

$$\varepsilon_i = \frac{n - n_B}{q_{2\pi}} \qquad (1)$$

where

n is the total measured count rate from the reference source plus background, in reciprocal seconds;

n_B is the background count rate, in reciprocal seconds;

and $q_{2\pi}$ is the surface emission rate, in reciprocal seconds, of the reference source.

The calibration coefficient, N, for surface contamination monitors is given by the following equation [7]:

$$N = \frac{q_{2\pi}}{n - n_b} \qquad (2)$$

The calibration coefficient is the inverse of the efficiency.

In all cases, the measurements were taken by placing each source on the bar grid above the left and right foot probes in a reproducible geometry, and for each case the measurements were repeated ten times.

RESULTS AND DISCUSSIONS

The study was divided in two parts. In the first one, the best calibration procedure for the feet detectors was established among four methods. In the second part the chosen procedure was applied for the calibration of the monitor.

The aim of the first part of the study was to establish an alternative method to determine the efficiency when the recommended source is not available, as happen at the

Calibration Laboratory of IPEN. Therefore, studies were performed using circular sources of 36 cm diameter and rectangular sources of 15 cm x 10 cm.

Four methods were tested for the determination of the monitor efficiency. The method A followed the HS(G)49 standard, by covering all the whole sensitive detector window. Figure 1A shows how the source was positioned. Special care was taken in order to avoid overlapping of the covered areas.

In the second test (method B), the efficiency was determined following the method used at LANL (6). In this case, the circular source of 36 mm in diameter was positioned in the central area of the probe (Figure 1B). In method C, the circular source was substituted by the rectangular source to determine the efficiency (Figure 1C).

Lastly, the method D was applied using the rectangular source twice to simulate the area usually covered by the user foot (Figure 1D).

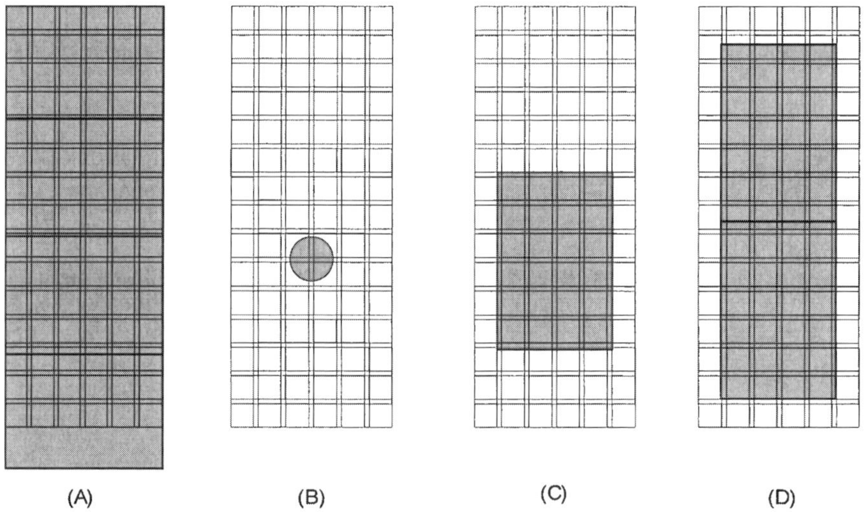

Figure 1. Source positioning in each one of the four calibration methods tested in this work: (a) method A, following the HS(G)49 suggestion; (B) method B, using a circular source as in Los Alamos National Laboratory; (C) method C, using a rectangular source at the central area of the detector; and (D) method D, using the rectangular source twice to simulate the area usually covered by the user foot.

Table 2 shows the efficiencies obtained by the different methods. Although method B shows a higher efficiency, it does not mean that it is the best method, because the covered area represents only 2% of the sensitive window area. Observing the results of the other methods, there is not a great difference between the obtained efficiencies. In the case of the left foot probe, the maximum variation for Am-241 is 10%, and 5.4% for Cl-36; for the right foot probe, the maximum variation for Am-241 is 15%, and 7% for Cl-36.

The main difference among the four methods is the spent time to perform each test; considering that there are two feet detectors and 10 measurements are taken for each one of the six sources, 480 measurements are necessary for method A, 240 measurements for method B and 120 measurements for methods C and D.

TABLE 2. Efficiencies (%) obtained for the left and right foot probes of a Eurisys/Sirius hand-foot monitor using four different methods.

Source	Method A	Method B	Method C	Method D
		Left foot		
Am-241	16.9	20.5	18.9	17.5
Cl-36	24.4	27.6	25.8	24.4
		Right foot		
Am-241	16.3	19.2	17.2	16.7
Cl-36	23.8	25.6	24.5	23.3

Method A is recommended by ISO [4] and HS(G)49 [5], because is the only one that consider the use of a source covering the whole area of the sensitive window, but it is a time consuming procedure, and it increases the uncertainty in the results, due to the multiple source positioning. Method B shows the best efficiency. However, this result is probably overestimated since the covered area is very small compared to the whole sensitive window. Comparing methods C and D, method C is faster, but method D simulates better the real situation; furthermore, the method D is twice faster than the HS(G)49 recommended method. This is the reason why method D was chosen for workplace calibrations at the Laboratory of IPEN.

Finally, the established procedure was applied to the whole monitor, hands and feet. For the hands, the rectangular sources of 15 cm x 10 cm were placed in the central part of the window. For the feet, the rectangular sources were positioned twice, as described in method D.

The results obtained for the efficiency versus the average beta energy of each source are represented in Figure 2. Poor efficiency for beta radiation can be observed below 100keV. Between Tc-99 and Sr/Y-90, the energy response variation is about 30%, that is in agreement with the data published by LANL(6).

Table 3 shows the calibration coefficients obtained, and their related uncertainties. For this determination, A and B uncertainty types were considered. The relative combined uncertainties determined were inferior to 10%.

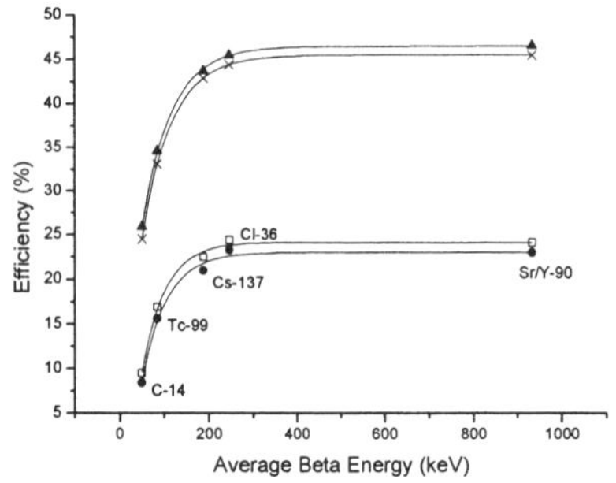

FIGURE 2. Energy dependence curve of a Eurisys/Sirius hand-foot monitor for several beta sources for the left foot probe (∀), right foot probe (,), left hand probe (□) and right hand probe (7).

TABLE 3. Calibration coefficients (dimensionless), with their respective relative combined uncertainty in percentage, obtained for a Eurisys/Sirius hand-foot monitor, using method D.

Source	Left foot	Right foot	Left hand	Right hand
Am-241	5.7 (5.5)	6.0 (5.7)	5.9 (5.5)	6.1 (5.5)
C-14	10.5 (7.2)	11.8 (8.8)	3.9 (4.2)	4.1 (4.1)
Tc-99	5.9 (5.5)	6.4 (5.9)	2.9 (3.8)	3.0 (3.7)
Cs-137	4.4 (5.6)	4.8 (5.8)	2.3 (3.9)	2.3 (3.9)
Cl-36	4.1 (5.5)	4.3 (5.5)	2.2 (3.7)	2.2 (3.7)
Sr/Y-90	4.2 (5.2)	4.3 (5.3)	2.1 (3.6)	2.2 (3.6)

CONCLUSION

The results show that it is possible to determine the monitor efficiency using a faster and even more realistic method by using a smaller source than the monitor sensitive area. However, as the recommended method is the method A, HS(G)49, we suggest using it just for the periodic calibrations, and the method D for the workplace calibrations.

ACKNOWLEDGMENTS

The authors acknowledge the partial financial support of Conselho Nacional de Desenvolvimento Científico e Tecnológico (CNPq), Brazil.

REFERENCES

1. Yoshida, M., *JAERI – Conf 97-008, Conf- 9610322 – IAEA*, Vienna (1997).
2. International Electrotechnical Commission. *Moniteurs et signaleurs de contamination des mains ou des pieds ou des deux*, **IEC-504** (1975).
3. International Organization for Standardization, *Reference sources for the calibration of surface contamination monitors – Beta-emitters (maximum beta energy greater than 0.15 MeV) and alpha-emitters*, **ISO 8769** (1988).
4. International Organization for Standardization. *Evaluation of surface contamination – Part 1: Beta-emitters (maximum beta energy greater than 0,15 MeV) and alpha-emitters*, **ISO 7503-1** (1988).
5. Health and Safety Executive. *The examination and testing of portable radiation instruments for external radiation health and safety series booklet*, **HS (G) 49, HMSO** (1990).
6. Bjork, W. C., *Beta contamination monitors energy response*, **LA-UR-98-2942, CONF-980756** (1998).
7. International Atomic Energy Agency. *Calibration of radiation protection monitoring instruments*. **Safety Reports Series No. 16** (2000).

Applied Physics (APHYS 2003)
A. Méndez-Vilas (Ed.)

ANALYSIS OF WATER, SOIL AND FRUIT QUALITY FROM ECO-LOCATIONS IN SERBIA USING NUCLEAR AND CHEMICAL METHODS

S. Cupic, M. Stojanovic, V. Andric, A. Onjia, N. Stojanovic and A. Kandic

Institute of Nuclear Sciences "VINCA"
P.O. Box 522, 11000 Belgrade, Serbia and Montenegro
E-mail: marcosun@EUnet.yu

Abstract. This paper presents results obtained by using analytical methods and instrumental techniques, such as: energy-depressive x-ray fluorescence spectrometry (EDXRF), gamma spectrometry, inductively coupled plasma-atomic emission spectroscopy (ICP-AES), flame atomic absorption spectrometry (F-AAS), graphite furnace atomic absorption spectrometry (GF-AAS) and hydride-generation/cold vapor atomic absorption spectrometry (HG/CV-AAS), based on standard procedures (AOAC, ISO, EPA etc). Physical/chemical and radiological investigations were performed on soil, water and fruit samples form an ecological micro locations Vlasina and Topola, Serbia. The purpose of those investigations was to determine possible presence of heavy metals and radionuclides, which could have negative influence on the quality of fruit and other food products from that location. Our research included various instrumental techniques in order to obtain more reliable results, and due to the fact that certain techniques can be used to measure only some of the parameters needed for the analysis. Some of those techniques are non-destructive (EDXRF and gamma spectrometry), and others are not. Results presented in this paper indicate very high quality of EMPOZ organic food produced in those ecological areas.

INTRODUCTION

The quality of plants produced for human nutrition, fruits and vegetables in the first place, is not defined by their size and appearance, but by its content of essential matters in natural quantities characteristic for given sort, which strengthen our immune system, inhibit cancer-related processes, prevent cardio-vascular diseases and high blood pressure and help body fight other vicious diseases to which we are exposed nowadays.
Science has determined a connection between human nutrition and many forms of cancer, heart conditions and other dangerous diseases. Many causes of such diseases are found, but there are no known averters. Therefore, averters are intensively looked for. It is known that many of them, if not a large majority of them, are present in food and nutrition. However, it does not apply to food produced using extensive agriculture methods, such as using artificial fertilizers and pesticides. Such production first led to extreme consumption of biogenic elements and soil deterioration, especially in essential microelements, and then to debalancing of concentration and even functions of

microorganisms in rhysosphere of the plants' root systems. It affects the fundamental processes of photosynthesis, as well as the content and quantity of organic matter in the fruits, leaves and root of plants people consume. Such food is often "empty" or even filled with the excess degraded or degenerated cells, heavy metals and other toxic substances that in human organism start reacting in a way different than naturally expected. The surrogate-hormones and false enzymes appear, as well as a lot of free radicals, acting as regular cell-killers. Science has found answers to free radicals, substances that de-stimulate them (anti-oxidants), such as: vitamins C and E, beta-carotene, selenium, zinc, manganesium and some amino acids.
These paper presents physical/chemical methods which can be used to achieve production of the exclusive EMPOZ (Ecological Natural Mineral Enriched) organic food, and to determine it's quality. These methods can successfully be used to perform measurements of the quality of air, water and soils on which EMPOZ food is grown. They can also be used to control the conditions and characteristics of the snvironment in which the food processing is carried out, in order to determine whether the product is in fact EMPOZ food or not [1, 2, 3, 4].

EXPERIMENTAL

Experimental work included measurements and analysis of the samples taken from two ecological microlocations in Serbia (Vlasina – south of Serbia, and Topola – central Serbia). On these locations there is monitoring of ecological parameters related to metheorological, hydrological and geo-morphological aspects. Our complex measurements included measurements of water, air, soils, fruits and vegetable parts (both below and above ground). This paper presents some of the results, which shed light on the quality of these locations, as well as the food produced on them (apples and pears). Measurements included several methods: X-fluorescence measurements, γ-ray spectrometry, atomic absorption spectrometry (AAS), inductively coupled plasma atomic emission spectroscopy (ICP-AES) and gas chdomatography (GC).
The X-fluorescence measurements were performed using the EDXRF spectrometer with Canberra Si(Li) detector (model SL30170 with Be window) and two radioisotopes, ^{109}Cd and ^{241}Am as excitation sources. Fruit samples (apple and pear) are dried at 105°C and meshed. Then pastilles of calibrated geometry and mass (0.5000 g) have been made and measured on the Mylar holders. Gained spectra did not show presence of any heavy metals in significant amount, which can be determined by this method. Peaks in the spectra are due to presence of elements in the surrounding of the instrument.
Radioactivity was measured by low-background γ-ray spectrometer with Canberra HP Ge coaxial detector (model GC 1318-7500SL) having relative efficiency of 14.7%, energy resolution of 1.7 keV and background count rate in the 20-2900 keV energy interval of 0.560 cps. Water sample was concentrated ten times by lamp evaporation and measured. Determined γ radioactivity in the fruit samples originated from natural radioactive

elements in the very low concentrations and results are given as the triple value of the error of the peak determination.

The determination of trace metals was performed in either flame or graphite-furnace mode with a Perkin-Elmer Model 5000/HGA-400 atomic absorption spectrometer (AAS), under optimized measurement conditions using suitable hollow cathode lamps. The signals were measured with background correction (deuterium lamp) at the optimized operating parameters. The concentrations of the different elements in these samples were determined using corresponding standard calibration curves obtained by using standard solutions of the elements.

Inductively coupled plasma atomic emisssion spectroscopy (ICP-AES) measurements were made directly at optimal wavelenghts with a standard configuration of the Perkin-Elmer ICP/6500 system. Milli-Q treated water and chemicals of reagent grade were used throughout the study. The following ICP-AES parameters were optimized for each element: forward power, viewing height, sample flow rate, nebulizer gas flow rate, auxiliary gas flow rate, and plasma gas flow rate. The trace element concentrations were determined in the same way as with AAS, from standard calibration lines.

Analyses of organic micropolutants have been performed by gas chromatography (GC) with various detectors. Solid samples were extracted in a Soxhlet apparatus and measured, while water samples were prepared by standard liquid-liquid or solid-phase extraction prior to GC. Chromatographic data were obtained using Spectra-Physics model SP7100 GC operating with appropriate temperature programs. Nitrogen or helium at a flow rate of 1.0 ml/min were used as the carrier gases. The screening of PAHs, phenols and various pesticides in solid and liquid samples were performed.

RESULTS AND DISCUSSION

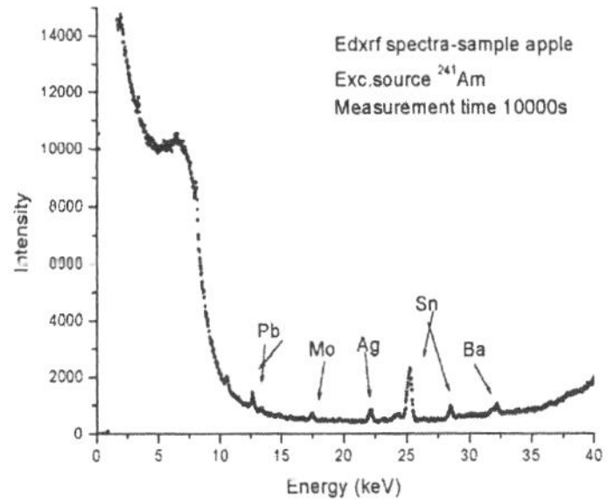

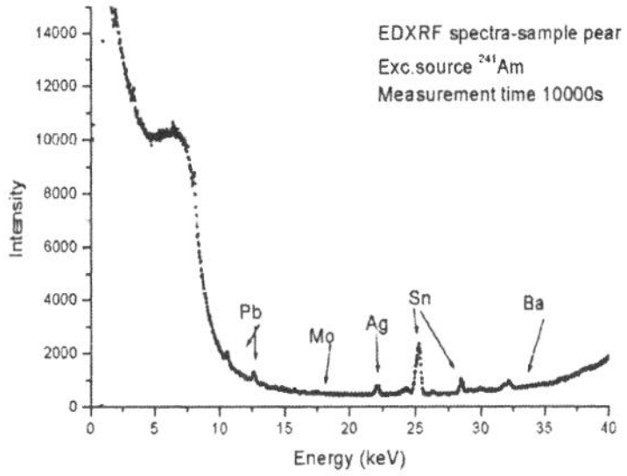

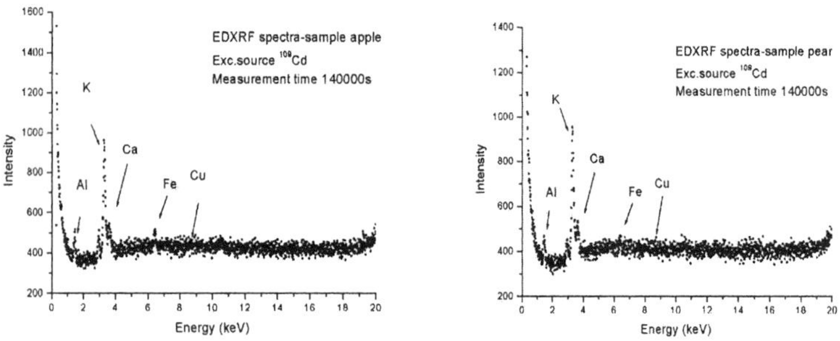

Fig. 1. EDXRF spectra of apple and pear samples.

In EMPOZ food production, it is very important to ensure the presence of all essential matters characteristic for certain fruit kinds, vegetables and parts of vegetables used in human and animal nutrition. It is even more important to avoid the presence of heavy metals, such as Ra, Cd, Rb, Hg, Cr, As etc, which endanger plant growth and health of humans and animals. These elements and their compounds enter the living environment through soils and waters (where they are naturally present), via mineral composts and pesticides. Therefore, methods are being developed, some of which were used in our research, which detect their presence/absence in samples of fruit, vegetables, soils, waters etc.

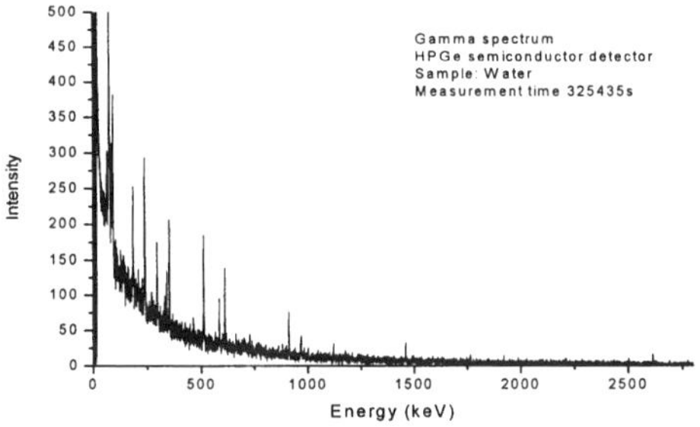

Fig. 2. Gamma spectometry of a water sample.

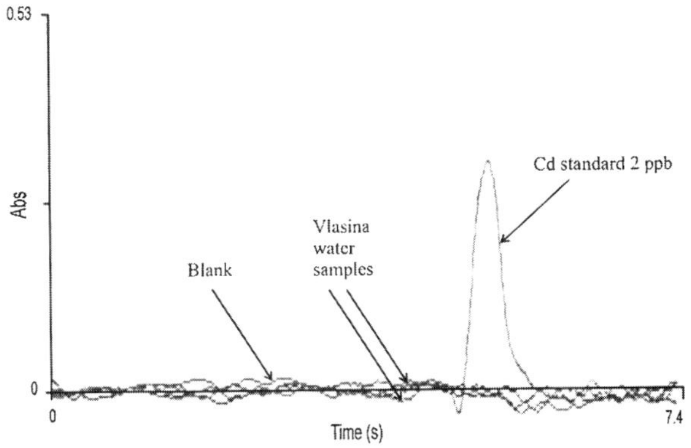

Fig. 3. AAS signal of Cd determination.

Figures 1 – 5 show the results of the analyses of the characteristics of apples, pears and spring waters. These results show that the samples used in our measurements do not contain these matters in quantities dangerous for human health. Also, Tables 1 and 2 show results of the analysis of underground waters from two locations, proving them to be of extraordinary quality, which is a guarantee for possible production of the EMPOZ organic food in those places.

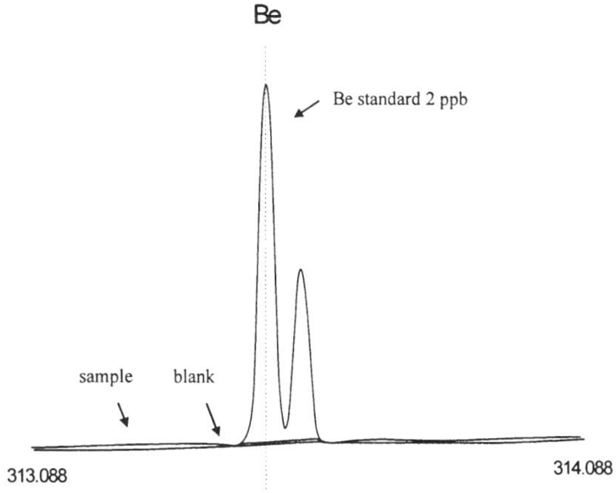

Fig. 4. ICP-AES of Be in a water sample.

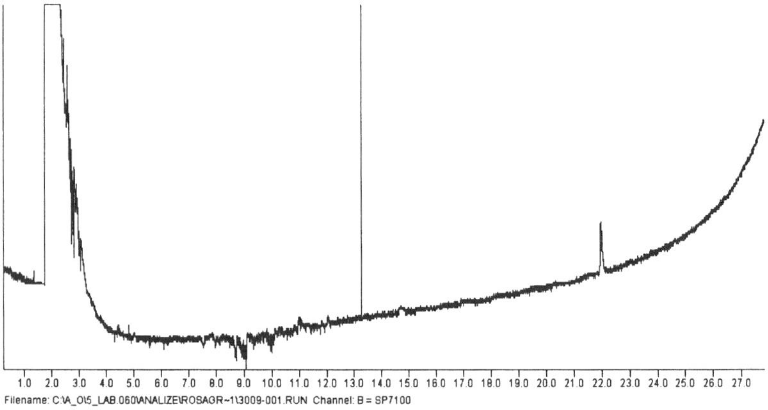

Fig. 5. GC screening of PAHs in water. No peak of any PAH was detected.

Table 1. Water analysis – sample from Topola

No:	Parameter	Sample 1.	MDK
1.	Water temperature, °C	13.0	-
2.	Air temperature, °C	22.1	-
3.	Colour, °Co-pt	0	5
4.	Taste and smell	without	without
5.	Clearness, NTU	43	< 1
6.	pH value	6.73	6.8 – 8.5
7.	Consumption of $KmnO_4$, mg/l	8.53	do 8
8.	Evap. residue on 105 °C, mg/l	645	-
9.	Electrical conditivity, μScm^{-1}	500	< 1000
10.	Ammonia, mg/l	0.00	0.1
11.	Chlorides, Cl^-, mg/l	40.5	200
12,	Nitrites, mg/l	< 0.03	0.03
13.	Nitrates, mg/l	4.0	50
14.	Iron, Fe, mg/l	3.35	0.3
15.	Mangan, Mn, mg/l	< 0.01	0.05
16.	Anj. detergents, mg/l	0.00	0.1
17.	Phenoles, mg/l	< 0.001	0.001
18.	Cadmium, Cd, μg/l	< 0.001	0.005
19.	Sulphates, SO_1^{-2}, mg/l	96.5	-
20.	Total oil and fat, mg/l	without	without

21.	Arsenic, As, µg/l	< 0.001	0.05
22.	Mercury, Hg, µg/l	< 0.001	0.001
23.	Lead, PB, mg/l	0.003	0.05
24.	Copper, Cu, mg/l	0.035	0.1
25.	Zinc, Zn, mg/l	0.067	0.1
26.	Selen, Se, µg/l	< 0.001	0.01
27.	Magnesium, Mg, mg/l	36.79	50.0
28.	Calcium, Ca, mg/l	87.9	200.0
29.	Sodium, Na, mg/l	14.6	150.0
30.	Potassium, K, mg/l	8.14	12.0
31.	Total chromium, Cr, mg/l	< 0.001	0.05
32.	Nickel, Ni, mg/l	< 0.001	0.02
33.	Thorium, ^{232}Th, mBq/l	< 30	19 Bq/l
34.	Uranium, ^{238}U, mBq/l	117	7.4 Bq/l
35.	Cezijum, ^{157}Cs, mBq/l	< 4.43	74 Bq/l
36.	Potassium, ^{40}K, mBq/l	< 137	-

Tabele 2. Water analysis – sample from Vlasina

No:	Parameter	Sample 1.	MRL
1.	Evap. residue on 105 °C, mg/l	52.0	
2.	Colour, °Co-pt	< 5	(5.0)
3.	pH value	7.2	(6.8 – 8.5)
4.	Consumption of $KmnO_4$, mg/l	2.8	(8.0)
5.	Ammonia, mg/l	< 0.01	(0.1)
6.	Electrical conditivity, μScm^{-1}	65.0	(1000)
7.	Nitrites, mg/l	< 0.003	(0.03)
8.	Nitrates, mg/l	1.77	(50.0)
9.	Clearness, NTU	0.3	(1.0)
10.	Chlorides, Cl^-, mg/l	2.5	(200.0)
11.	Total oil and fat, mg/l	< 0.01	(0.1)
12.	Sulphates, SO_1^{-2}, mg/l	3.5	(250.0)
13.	Fluorides, mg/l	0.11	(1.2)
14.	Alcality, as in $CaCO_3$ mg/l	31.5	
15.	Bicarbonates, HCO_3^-, mg/l	36.0	
16..	Calcium, Ca, mg/l	7.1	(200)
17.	Magnesium, Mg, mg/l	0.8	(50.0)
18.	Arsenic, As, µg/l	n.d.	(0.01)
19.	Sodium, Na, mg/l	1.54	(150.0)
20.	Potassium, K, mg/l	0.44	(12)
21.	Cadmium, Cd, µg/l	n.d.	(0.003)
22.	SI SI SI	1.47	

* n.d. – not detected
* n.p. – not performed

CONCLUSION

Results presented in this paper come from the analysis of the data obtained using aforementioned techniques (X-fluorescence measurements, γ-ray spectrometry, atomic absorption spectrometry (AAS), inductively coupled plasma atomic emission spectroscopy (ICP-AES) and gas chdomatography - GC). Those results indicate very high quality of the fruit products produced in these locations. The results also prove these locations to be very suitable for EMPOZ food production.
Serbia is still in great part undamaged by horrible side-effects of artificial fertilizers, pesticides and other chemical compuonds, thanks to the isolation and it was subjected to and general poverty of both the country and individual citizens. Serbian land also posseses many advantages regarding the climate, richness in underground and surface waters etc, so it can be said that Serbia represents an European biological reservoir for natural production of essential matters for human nutrition through EMPOZ organic food. Using aforementioned nuclear and chemical methods provides the necessary analysis and control of such food production.

REFERENCES

[1] G.A.F. Hendry, Ed., *Natural Food Colorants*, Blackie Academic and Professional, 1996.
[2] G.A. Burdock, Ed., *Encyclopedia of Food and Color Aditives*, Vol. II, CRC Press, Boca Raton, 1997.
[3] AOAC Official method 967.09, *ZDBT Colorimetric Method*, Official method of analysis of AOAC Int., 16th ed., Arlington, USA, 1997.
[4] M. Stojanovic, *Ecology and EMPOZ Organic Food*, Institute of Nuclear Sciences "Vinca", Belgrade, 2002.

MODELING IRRADIATION-INDUCED CHARGING-ANNEALING DYNAMICS IN METAL-OXIDE-SEMICONDUCTOR DEVICES

Faigón, A. Cedola, E. G. Redin, G. Kruszenski, J. Lopez, M. Maestri, J. Lipovetzky and A. Docters[2]

Laboratorio de Física de Dispositivos - Microelectrónica, Dpto de Física, Facultad de Ingeniería, Universidad de Buenos Aires.
E-mail: afaigon@fi.uba.ar
2 Planta de Irradiación Semi-Industrial - Centro Atómico Ezeiza - CONEA.

Abstract. The experimental evolution of the threshold voltage of MOS devices during irradiation at changing bias voltages, exhibiting charging and neutralization of the oxide traps, was well fitted by a two-traps dynamic balance model. The values and field dependence of the extracted physical parameters are in close agreement with previously reported ones.

INTRODUCTION

Numerous experimental studies were devoted to the effects of ionizing radiation on metal-oxide-semiconductor (MOS) structures of different size and oxide nature [1]. Ionizing irradiation results mainly in the creation of a net positive charge in the bulk of the oxide and of surface states at the Si/SiO_2 interface.

The dynamics of defects creation has received a great amount of experimental and theoretical work, for understanding the physics underlying the phenomenon and for modeling [2]-[5].

The cancellation of these effects --or defects annealing-- have been studied for both practical and theoretical purposes. It has been found that charges and states anneal thermally and through charge injection from the electrodes [6,7]. Radiation-induced positive charge annealing was reported by applying zero or negative bias to the gate, or a switched bias alternating the polarity [8-11]. Improvements in the modeling of both phenomena have direct application in MOS dosimetry [12, 13].

In this work we describe experimental observations of radiation induced charge neutralization under positive bias, and show the need of at least two kind of traps if an exponential model is accepted [14]. Two or more types of trap or trapped charge were often proposed in the literature in order to explain different phenomena [15, 16]. A two-exponential model, compatible with a dynamic balance between trapping and de-trapping in the two trap centers, fit quite accurately the experimental data.

MEASUREMENTS AND RESULTS

The threshold voltage, Vt, of commercial DMOS transistors was measured each 10 second during gamma irradiation from a Co^{60} source at a dose rate of 26.7 KGy

(SiO_2)/hr. Outside the 70 msec interval of Vt measurements the device was biased with positive voltage to the gate. The repeated switching from bias to measurement did not affect the Vt evolution for intervals between measurements longer than 3 sec. Three devices were tracked simultaneously in each irradiation.
The main results are described in the following:

i) The threshold voltage has a quasi-steady state --as reported in [17]--, which depends on the bias applied during irradiation –Fig. 1-.

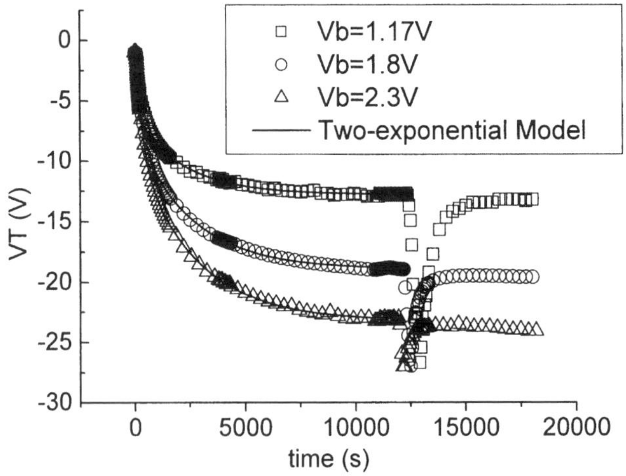

Fig.1. Threshold voltage (Vt) vs. time for different gate applied biases (Vb) during irradiation. After the devices reached a steady state, at about 12000 sec, Vb's were raised, driving the Vt's to more negative values, and then returned to the original biases causing the devices recovery towards the same previous steady states.

ii) Recovery towards higher steady-state values, or positive charge annealing, occurs after a bias change to a less positive voltage --Fig.1 at 12000 sec, and Fig.2--.

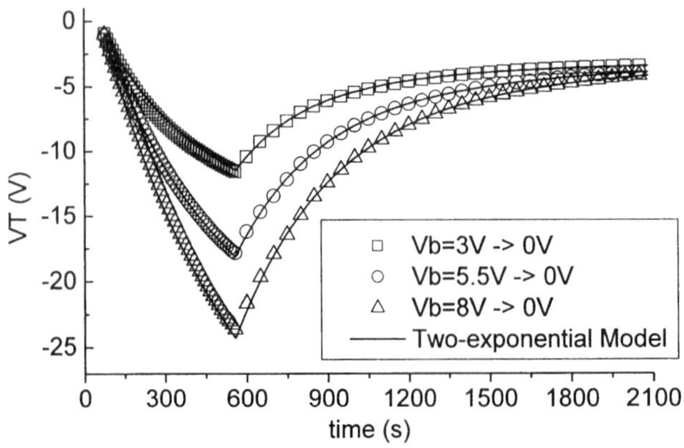

Fig.2. Vt evolution after a bias switch (at about 500 sec). With zero volt applied to the gate electrode the positive oxide charge accumulated anneals during irradiation.

iii) The evolution of Vt during the annealing may exhibit a turnaround as in Fig.3.

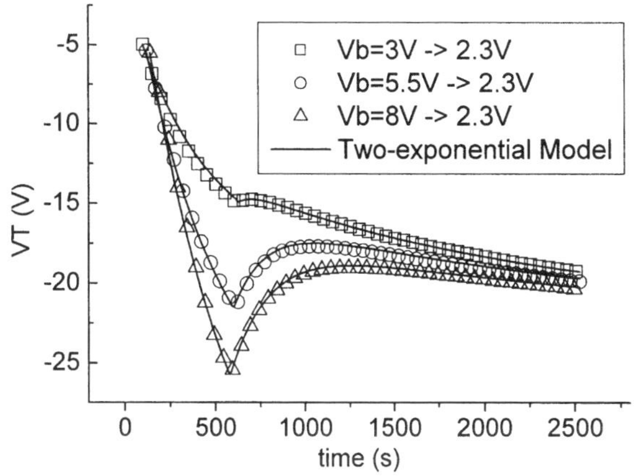

Fig.3. Vt evolution after a gate bias switch (at about 500 sec). The turnaround exhibited by the curves following bias change suggests the contribution of more than one center to the oxide charge trapping-annealing dynamics.

MODELING

Fig.3 shows that the knowledge of Vt is not enough information to predict the device behavior at a given bias. The internal state of the device must be described by at least two internal variables. Capacitance-voltage measurements show no significant contribution from surface states to the changes being described. The quasi-steady Vt value corresponding to a given bias, suffers from a very long-term drift towards lower voltages but is quite stable during the times involved in our experiences. Thus, trap creation is disregarded, and the two internal chosen variables are the population fractions of two kinds of traps. The described behavior is consistent with a dynamic balance model (DBM) in which the occupation of each one of the two classes of trap centers reaches a dynamical equilibrium between charge trapping and de-trapping. A simple such description for a single trap is found in [17]

$$\frac{\partial p_t}{\partial t} = \sigma_p \cdot J_p \cdot (N_t - p_t) - \sigma_n \cdot J_n \cdot p_t \qquad (1)$$

where p_t is the density of trapped holes, $J_{p,n}$ current density, $\sigma_{p,n}$ capture cross section for a hole in a trap and for an electron in a trapped hole, N_t hole traps density. Eq. (1) solves to

$$p_t = p_{t0} + N_t \cdot \frac{\sigma_p \cdot J_p}{\sigma_p \cdot J_p + \sigma_n \cdot J_n} \cdot \{1 - \exp[-(\sigma_p \cdot J_p + \sigma_n \cdot J_n) \cdot t]\} \qquad (2)$$

Assuming for simplicity a sheet of traps at a distance x from the Si-SiO_2 interface, a model with two different traps yield the following simple expression for the threshold voltage evolution

$$Vt = Vt_{f1} + Vt_{f2} + A_1 \cdot \exp(-t/\tau_1) + A_2 \cdot \exp(-t/\tau_2) \quad (3)$$

where Vt_{fi} is the asymptotic value of Vti.

$$A_i = \frac{q}{C_{ox}} \cdot \left(\frac{t_{ox} - x}{t_{ox}}\right) \cdot N_{ti} \cdot \frac{\sigma_{pi} \cdot J_p}{\sigma_{pi} \cdot J_p + \sigma_{ni} \cdot J_n} \cdot \quad (3a)$$

i is 1 or 2; and

$$\tau_i = \frac{1}{\sigma_{pi} \cdot J_p + \sigma_{ni} \cdot J_n} \cdot \quad (3b)$$

Fig. 4 shows the decomposition of different experimental curves into the two exponentials of eq. (3). All the four experimental curves in that figure were measured under the same applied bias, on devices differing only in their initial states. The curves are well fitted by forcing to share the final values Vt_{fi} determined in the long measurement shown in Fig. 4. The four remaining parameters, A_1, A_2, τ_1, τ_2, were independently adjusted for each curve.

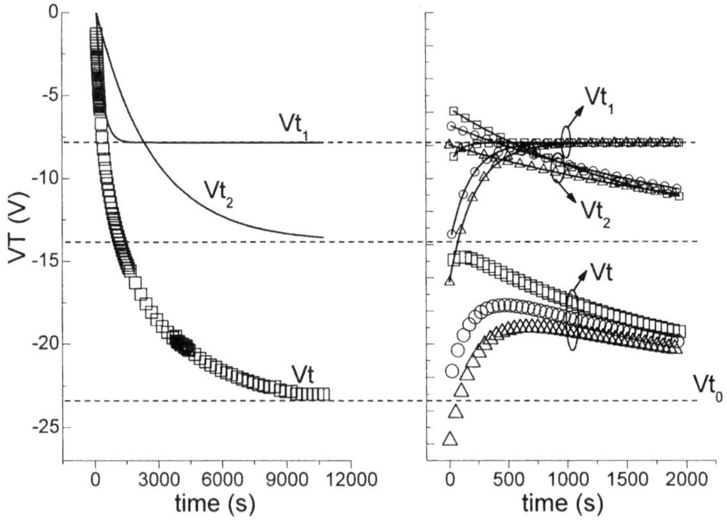

Fig.4. Two-exponential decomposition of the Vt evolution for the above presented curves corresponding to Vb=2.3V.

FIELD DEPENDENCE OF PHYSICAL PARAMETERS

Analog parameter extraction was performed for the different curves presented above, and the field dependence of the obtained parameters was investigated. The amplitudes A_1 and A_2 and typical times for decay, τ_1 and τ_2, were plotted against the mean field at the beginning of the irradiation or following a bias change, as shown in Fig. 5.

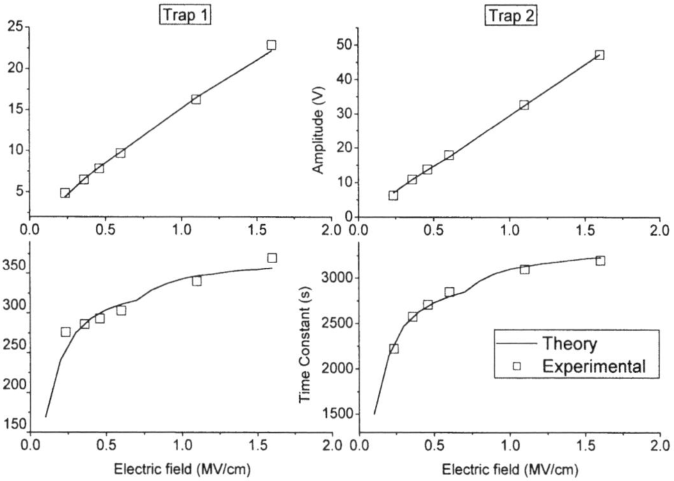

Fig.5. The field dependence of the two-trap model parameters A1, A2, τ_1 and τ_2.

The field dependence of these empirical parameters was reconstructed using accepted field dependences of the physical parameters:
-for the capture cross sections, [17,18,19]

$$\sigma_p(E) = \sigma_{p0} \cdot \left(1 + 1.9 \cdot 10^{-4} \cdot E^{0.55}\right)^{-1} \tag{4}$$

$$\sigma_n(E) = \sigma_{n0} \cdot 10^{15.265} / E^{2.865} \quad , for \quad E > 7 \cdot 10^5 \quad V/cm \tag{5a}$$

$$\sigma_n(E) = \sigma_{n0} \cdot \left(1 + 1.1 \cdot 10^{-11} \cdot E^{2.16}\right)^{-1} \quad , for \quad E < 7 \cdot 10^5 \quad V/cm \tag{5b}$$

-for the fraction Y of carriers escaping geminate recombination in the evaluation of the current density J [20]:

$$Y(E) = \left(1 + 0.55 \cdot 10^6 / E\right)^{-0.7} \tag{6}$$

$$J_p = g_0 \cdot D' \cdot Y(E) \cdot (t_{ox} - x) \tag{7a}$$

$$J_n = g_0 \cdot D' \cdot Y(E) \cdot x \tag{7b}$$

Using the nominal generation rate in SiO_2 g_0=7.6*10^{12} pairs.cm^{-3}.rad^{-1}, a typical value of 5 nm for the distance from the sheet of traps --or traps centroid-- to the interface, x, and dose rate D'=742 rad(SiO_2).sec^{-1} --as determined by PMMA dosimetry (+/- 5%)--, the only parameters to be adjusted are the capture cross sections at E=0 for the exponential time constants, and the density of traps N_t for the amplitudes. The fitting values are given in Table I. For the correct fitting of the amplitudes an empirical field correction was implemented,

$$N_t = N_{t0} \cdot (1+E)^n \qquad (8)$$

with n=1.19 for the fast traps, and n=1.56 for the slower traps.

Table I

	$\sigma_{po}(cm^2)$	σ_{no} (cm^2)	N_{to} (cm^{-2})
Trap 1 (fast)	$1.95*10^{-13}$	$1.00*10^{-11}$	$3.52*10^{12}$
Trap 2 (slow)	$2.18*10^{-14}$	$1.13*10^{-12}$	$5.26*10^{12}$

DISCUSSION

The two traps model was proposed as an analytical tool accounting for the fact that threshold voltage alone, i.e. the amount of trapped charge for a given traps distribution, does not suffice for predicting the evolution of the device at a given bias. In its simplicity, several features of the actual evolution were disregarded. The amount of trapped charge at such high total doses is sufficient for deviate significantly the local fields inside the oxide from the applied field. It is, therefore, surprising the reasonable values and field dependencies found for the capture cross sections: The fitting value for σ_{po2} corresponding to the trap governing the long transient is very close to $1.4*10^{-14}$ obtained in [18], and the short transient cross section for electrons is identical to that obtained in [17] fitting Ning data [19]. The smallest cross section for electrons is almost one order of magnitude smaller as reported for example in ref [21]. This is not the case for the N_t values, governing the amplitudes in our model. Instead of a simple N_t value, we need a field dependent function to fit the experimental results. This is so because the saturation of the Vt evolution is aided by the weakening of the field driving the hole current whereas that driving electrons becomes stronger with positive charge trapping. Ignoring in our model the varying internal fields, requires the semi-empirical correction above mentioned.

CONCLUSIONS

-Voltage dependent quasi-steady-states for oxide charge were observed in gamma irradiated MOS devices.

-Radiation induced charge neutralization was observed at positive gate bias as a consequence.

-The transient towards a bias dependent quasi-steady-state depends on the device initial state. The threshold voltage may describe a non-monotonic path if the initial state is not itself a steady state.

-A double exponential function, consistent with a two-trap dynamic balance model, fit well the experimental results.

-The physical parameters corresponding to the dynamic balance model have quite reasonable values and coincident field dependencies as compared with previously obtained ones through different ways.

ACKNOWLEDGMENTS

The authors are indebted with the technical staff of the PISI - CAE - CNEA, which gave us all the technical assistance needed in performing the irradiations. The work received financial support from the ANCyT (Agencia Nacional de Ciencia y Tecnología) under grant PICT6822, and from UBACyT (Secretaria Ciencia y Tecnica Universidad de Buenos Aires) under grants I603 and I903.

REFERENCES

1. For a review see for example T. P. Ma and P. V. Dressendorfer, *Ionizing Radiation in MOS Devices and Circuits,* Wiley Interscience, New York, 1989.
2. P.M. Lenahan and J.F. Conley, Jr., "A Comprehensive Physically Based Predictive Model for Radiation Damage in MOS Systems", IEEE Transactions on Nuclear Science, vol. 45, pp 2413-2423 (1998)
3. Nicolas T. Fourches "Charging in gate oxide under irradiation: A numerical approach", Journal of Applied Physics, vol 88-9, p. 5410, (2000)
4. D. M. Fleetwood, P. S. Winokur, R. A. Reber, Jr., T. L. Meisenheimer, J. R. Schwank, M. R. Shaneyfelt, and L. C. Riewe "Effects of oxide traps, interface traps, and border traps on metal-oxide-semiconductor devices", Journal of Applied Physics, Vol. 73, pp. 5058-5074, (1993)
5. P. M. Lenahan, J. F. Conley Jr and B.D. Wallace "A model of hole trapping in SiO2 films on Si", Journal of Applied Physics, Vol 81-10, p. 6822, (1997)
6- J. M. Benedetto, H. E. Boesch, F. B. McLean and J. P. Mize, "Hole removal in thin gate MOS-FETs by tunneling", IEEE Trans. Nucl. Sci., NS-32, p. 3916, (1985)
7. P. J. McWorther, S. L. Miller and W. M. Miller, "Modeling the anneal of radiation induced trapped holes in a varying thermal environment", IEEE Trans. Nucl. Sci., NS-37, p. 1682, (1990)
8. R. K. Freitag, C. M. Dozier and D. B. Brown, "Growth and annealing of trapped holes and interface states using time-dependent bias", IEEE Trans. Nucl. Sci., NS-37, p. 1682, (1990)
9. V. S. Pershenkov, V.V. Belyakov, A.V. Shalnov, " Fast Switched-Bias Annealing of Radiation-Induced Oxide-Trapped Charge and its Application for Testing of Radiation Effects in MOS Structures", IEEE Trans. Nucl. Sci., NS-41, p. 2593, (1994)
10. D. M. Fleetwood "Radiation-induced charge neutralization and interface-trap buildup in metal-oxide-semiconductor devices", J. of Appl. Phys., Vol. 67-1, pp. 580-583, (1990)
11. O. Quittard et al., " Use of the Radiation-Induced Charge Neutralization Mechanism to Achieve Annealing of 0.35UM SRAMs", IEEE Trans. Nucl. Sci., NS-46, p. 1633, (1999)
12. A. Holmes-Siedle, "The space charge dosimeter", Nucl. Instrum. Methods 121, p. 169, (1974).
13. C. Conneely, B. O'Connell, P. Hurley, W. Lane and L. Adams, "Strategies for millirad sensitivity in PMOS dosimeters", IEEE Trans. Nucl. Sci., NS-45, p.1475, (1998)
14. P. M. Lenahan et al., " Predicting Radiation Response from Process Parameters: Verification of a Physically Based Predictive Model", IEEE Trans. Nucl. Sci., NS-46, p.1534, (1999)
15. R. K. Freitag, D. B. Brown and C. M. Dozier, "Evidence for two types of radiation-induced trapped positive charge", IEEE Trans. Nucl. Sci., NS-41, p. 1828, (1994)
16. V. S. Pershenkov et al. "Proposed two-level acceptor-donor (QD) center and the nature of switching traps in irradiated MOS structures", IEEE Trans. Nucl. Sci., NS-43, p.2579, (1996)
17. H. E. Boesch, F. B. McLean, J. M. Benedetto and J. M. McGarrity, "Saturation of threshold voltage shift in MOSFETS at high total dose, IEEE Trans. Nucl. Sci., 33-6, p.1191, (1986)
18. R. J. Krantz, L. W. Aukerman, and T. C. Zietlow, "Applied Field and Total Dose Dependence of Trapped Charge Buildup in MOS Devices", IEEE Trans. Nucl. Sci., NS-34, p.1196, (1987)
19. T. H. Ning, "High field capture of electrons by Coulomb-attractive centers in silicon dioxide", J. App. Phys. 47, p. 3207, (1976)
20. C. M. Dozier, D. M. Fleetwood, D. B. Brown, P. S. Winokur, "An evaluation of low energy X-ray and Co-60 irradiations of MOS transistors", IEEE Trans. Nucl. Sci., NS-34, p.1535, (1987)
21. J. M. Aitken and D. R. Young "Electron trapping by radiation-induced charge in MOS devices", Journal of Applied Physics, Vol. 47-3, pp. 1196-1198, (1976)

Applied Physics (APHYS 2003)
A. Méndez-Vilas (Ed.)

New Simplified Technique for Determination of lead-210 in Environmental Samples Using ^{212}Bi as Tracer of Chemical Yield

Oksana Blinova[*], Ramiz Aliev[*,†] and Yury Sapozhnikov[*]

**Radiochemistry Division, Department of Chemistry, Lomonosov Moscow State University, Leninskie Gory, 119992, Moscow, Russia*

[†] Skobeltsyn Institute of Nuclear Physics, Lomonosov Moscow State University, Leninskie Gory, 119992, Moscow, Russia

Abstract. New technique of determination of ^{210}Pb in environmental samples was developed. Method is based on solvent extraction of diethyldithiocarbamate complex of daughter ^{210}Bi, and liquid scintillation determination of its radioactivity. It is suggested to apply short-lived ^{212}Bi from thorium series as a tracer for chemical yield determination. The advantages of this method are: the convenience of the tracer preparation and its short half-life, which leads to make background value lower. Another benefit is the possibility of the determination of both radionuclides (^{210}Bi and ^{212}Bi) using the same method, e.g. the liquid scintillation spectrometry. Thus, the total time of the analysis is decreased.

INTRODUCTION

Radionuclides responsible for internal exposure are lead-210 and its daughter radionuclides. Besides these radionuclides are useful tools for the environmental research. There are several approaches to ^{210}Pb determination in environmental samples[1,2,3] such as high-resolution gamma spectrometry of ^{210}Pb gamma quanta (46.5 keV), measuring of alpha particles of ^{210}Po or beta particles of ^{210}Bi. In the last case ^{210}Bi is counted using liquid scintillation (LS) or proportional counter after separation from parent ^{210}Pb. This separation requires chemical yield of bismuth to be determined. Usually gamma emitter ^{207}Bi or stable bismuth is applied as a tracer. In these cases, determination of chemical yield is an additional step of analysis. Besides, conversion electrons of ^{207}Bi make contribution to blank measurement results. It is suggested to apply short-lived ($T_{1/2}$=1.01 hr) ^{212}Bi from thorium series as a tracer for chemical yield determination in ^{210}Pb analysis. Isotope generator, based on separation of ^{212}Bi by solvent extraction from parent ^{232}Th solution was used for preparing of ^{212}Bi.

EXPERIMENTAL

Alpha- and beta- particles were registered using LS spectrometer Tri-Carb 2700 TR (Packard, USA) and scintillator OptiPhase 'HiSafe-3' (Wallac, Finland). Gamma-spectrometric measurements were carried out using Canberra Packard spectrometer with high pure germanium detector (HPGe) GC-3020 (coaxial detector, 60 mm in diameter, 44 mm in length, 30% relative efficiency, 1.8 keV energetic resolution with 1.3 MeV scale). The software Genie PC-400F was used for data analysis.

The solution of ^{210}Pb in equilibrium with daughter radionuclides, prepared from ^{226}Ra was used to develop the solvent extraction procedure. The developed procedure was tasted using standard sample of marine sediment IAEA-315. The preliminary experiments have proved, that quantitative separation of ^{210}Bi from parent ^{210}Pb may be achieved by solvent extraction of ^{210}Bi as diethyldithiocarbamate from ammonia solution in the presence of the excess of EDTA (Fig. 1 a, b).

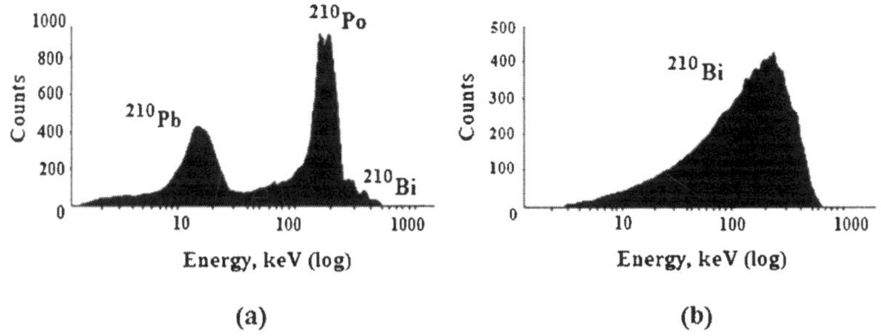

FIGURE 1. (a) LS-spectrum of ^{210}Pb in equilibrium with ^{210}Bi and ^{210}Po (before extraction); (b) LS-spectrum of ^{210}Bi (after extraction).

Tracer preparation

^{212}Bi was prepared by solvent extraction from $^{232}Th(NO_3)_4$ in potassium biphthalate solution (pH 1.7). Radiochemical purity of tracer samples was proved by gamma-ray measurements and LSC method (via ^{212}Bi decay). Gamma-ray spectrum of ^{212}Bi tracer is shown in Fig 2. It is shown that the difference between calculated and tabular half-life values of both radionuclides is negligible and the remained activity of samples corresponded to background activity value in the counting error limits.

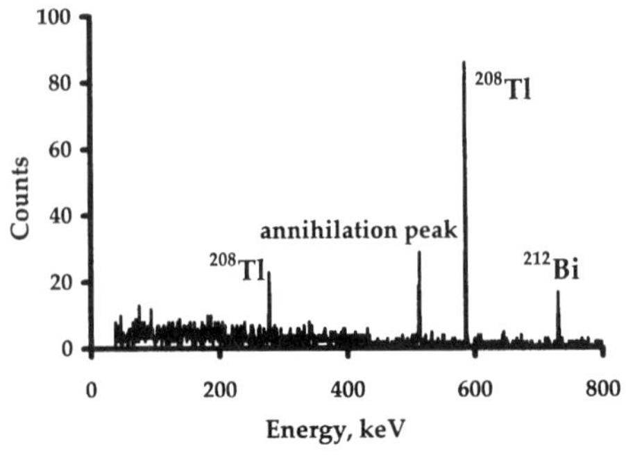

FIGURE 2. Gamma-ray spectrum of ^{212}Bi.

Environmental sample analysis

About 5 g of marine bottom sediment was leached with concentrated HNO_3 for 6 hours, than the solution was separated from insoluble residue. The aliquot of ^{212}Bi solution and excess of EDTA solution (40 g/L) in aqueous NH_3 (1:1) was added to obtain pH of 9-10. Then 5 ml of 1.5% aqueous solution of sodium diethildithiocarbamate (DDTC-Na) was added and bismuth was twice extracted with CCl_4. Then CCl_4 was evaporated and organic residue was digested by boiling with concentrated HNO_3. After HNO_3 was evaporated down to several drops, about 0.5 ml of distilled water was added and sample was transferred to 22 ml LSC plastic vial and was mixed with 10 ml of LS cocktail. The chemical yield was determined via ^{212}Bi by two measurements of 3 min each in the freshly prepared sample and after 1 – 2 hrs. The activity of ^{210}Pb was determined by ^{210}Bi decay curve that was obtained by number of counting during 15 days. The time of each measurement (counting window 10-2000 keV) was kept 3 hrs.

Results and discussion

The method was used to determine ^{210}Pb in the number of environmental samples (marine bottom sediments). Analysis of the certified reference material IAEA-315 resulted to specific activity of ^{210}Pb equal to 28±5 Bk/kg (the reference value is 30 Bk/kg). So, the blank sample result decreases to 10 – 30 cpm in comparison with application of long-lived ^{207}Bi. Besides, the efficiency of beta particles counting using LSC is much higher than in case of gamma quanta registration with semiconductor. Thus, the uncertainty of chemical yield determination considerably decreases. Another advantage of suggested method is a possibility to use single beta-counting device. As a result, the duration and cost of analysis is decreased.

CONCLUSIONS

The new simplified and cost effective technique of determination of ^{210}Pb in environmental samples is presented with minimal detectable activity of about 0.1 Bq.

REFERENCES

1. R.N. Moser, *J. Radioanal. Nucl. Chem. Articles* **173** (2), 283 – 292 (1993).
2. C. Dovlete, Gy. Rusa, F. Basiu and O. Sima, *Environ. Internat.* **22,** S319 – S321 (1996).
3. N. Vajda, J. LaRosa, R. Zeisler, P. Danesi and Gy. Kis-Benedek, *J. Environ. Radioactivity* **37** (3), 355 – 372 (1997).

Applied Physics (APHYS 2003)
A. Méndez-Vilas (Ed.)

Neutron Flux Measurements in Radiation Damage Experiments of ND-FE-B magnets by High Energy Electrons

Hee-Seock Lee*, Dong-En Kim*, Chinwha Chung*, Teruhiko Bizen†, Hideo Kitamura†, Yoshihiro Asano¶

** Pohang Accelerator Laboratory, POSTECH, Nam-gu, Pohang, 790-784, South Korea*
†Japan Synchrotron Radiation Research Institute,SPring-8, Sayo-kun, Hyokoi, 679-5198, Japan
¶Synchrotron Radiation Research Center, JAERI, SPring-8, Sayo-kun, Hyoko, 679-5198, Japan

Abstract. The demagnetization of permanent magnets used for a insertion device, due to 2 GeV electrons and secondary particles, has been studied and the reduction of those magnetic field strengths was observed in several magnet conditions. Coincidently, the neutron flux measurements was conducted using a threshold activation method for the study of interrelation between neutron flux and the demagnetization. The neutron flux from 1.2 MeV to 140 MeV was measured by using Bi, Ti, and In foils. It was also validated that Bi(n,xn) reactions is applicable as a measurement tool of an energy-informed neutron flux in high energy electron accelerator. The interrelation was discussed based on the obtained neutron flux.

INTRODUCTION

The demagnetization of the permanent magnets due to radiation damage is one of hot issues at the 3rd and the 4th generation synchrotron radiation facilities because the insertion device consisting of permanent magnets, especially an in-vacuum undulator, can be irradiated by high energy electron or secondary particles [1,2,3,4]. The demagnetization characteristics of Nd-Fe-B magnets located at the downstream of thick beam stoppers (Cu or Ta), which were irradiated by 2 GeV electrons, has been studied using the electron linac of the Pohang Light Source since 2000 [5,6,7,8]. Many results of demagnetization property depending geometrical condition and manufacturing process were obtained. The demagnetization might be mainly due to neutrons produced through electromagnetic showers and photo-nuclear reactions in the upstream thick beam stopper or the magnets [3]. Intensive gamma field might also be the source. In other experiments to observe the microscopic change inside the Nb-Fe-B magnet, it was found no appreciable change in local structure by the measurements of X-ray absorption property [9]. It is important to know the radiation fields around magnets in order to find out the demagnetization characteristics.

The active detector system such as the time-of-flight technique was known as a good tool to measure high energy neutron spectrum. However, it is limited hardly in high energy electron accelerator because the gamma background is too high. So a passive-type detector is used still at many places. The threshold foil activation is one of them. Recently the method using Bi(n,xn) reactions was developed and used as a neutron detector at high energy proton accelerator [10,11] and rarely at electron accelerator [12]

In this study, the measurement of neutron flux around the beam stopper and the magnet has been carried out. The Bi(n,xn) reactions were chosen as its main detection method and the feasibility was discussed.

EXPERIMENTS

The neutron flux measurements and the demagnetization study have been carried out at 2.5 GeV electron linac of Pohang Light Source. The same target geometry to one in the demagnetization study was used for these measurements shown in Fig. 1. The 2.0 GeV electrons were struck at the 1 mm-thick beam profile monitor, the 10 mm-thick backing bakelite, thick Cu beam stopper and the Nb-Fe-B magnet in series, and then produced electromagnetic gamma shower and neutrons in those. The thicknesses of the Cu beam stopper target and the Nd-Fe-B magnet were 40 mm and 6 mm, respectively. The beam parameters were these; 10 Hz pulse repetition rate, 1 nsec pulse length, and 5 x 10 mm cross-sectional dimention. Its intensity was about 1E+11 electrons per second. The Nd-Fe-B magnet with a holder shown in Fig. 2 is the same to one used for the undulator and tested in the demagnetization study.

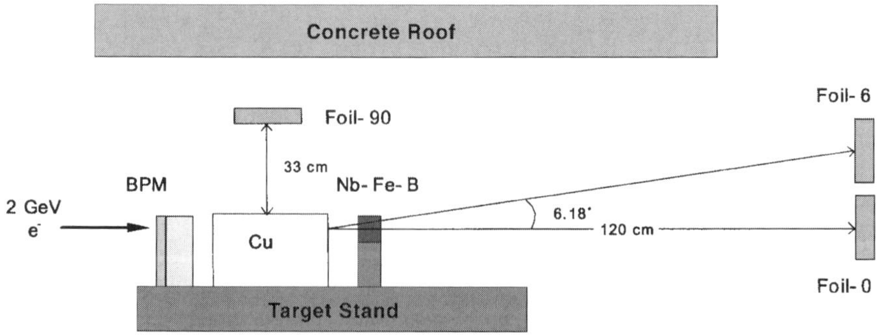

FIGURE 1. Experimental setup for neutron flux measurements during demagnetization study of Nb-Fe-B magnet.

The threshold activation method was adapted to measure the neutron flux. Especially, by using activation foils with several threshold energies, the energy-informed flux could be obtained. Normal In, and Ti(1/2" diameter), and Bi (25 x 25 x 1 mm) were used. The 10 types of reactions, (n,xn), of Bi foils are available with each threshold energy. Those reaction cross-sections were presented in Fig. 3, which were from ENDF/HE-IV library [13]. The other details are listed in Table 1. Those foils were installed at 0, 6, and 90 degrees relative to the incident electrons. The irradiation time of the electron beam was about 1 ~ 2 hours. During the irradiation the electron beam intensity was controlled within about 5 % deviation of average value. The gamma spectra from all activated foils were measured by the ϕ3" x 3" HPGe gamma spectroscopy system after some cooling time. Because of variety of decay times, the gamma spectra were measured repeatedly with some time intervals in order to confirm the reactions and obtain the precise intensities from spectra. The neutron flux was obtained through general activation analysis process.

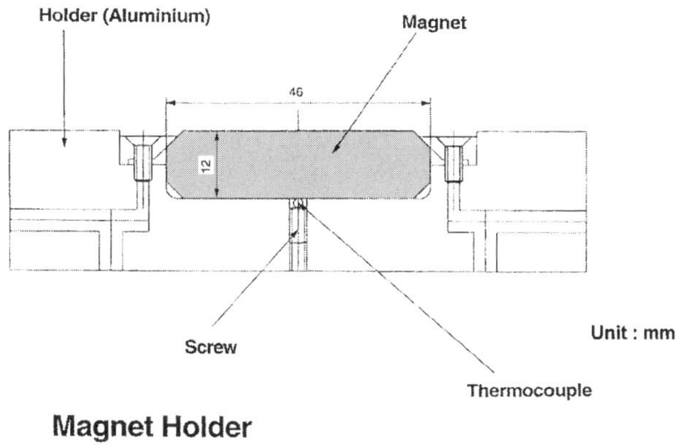

FIGURE 2. Detail structure of magnet sample for irradiation (t= 8 mm).

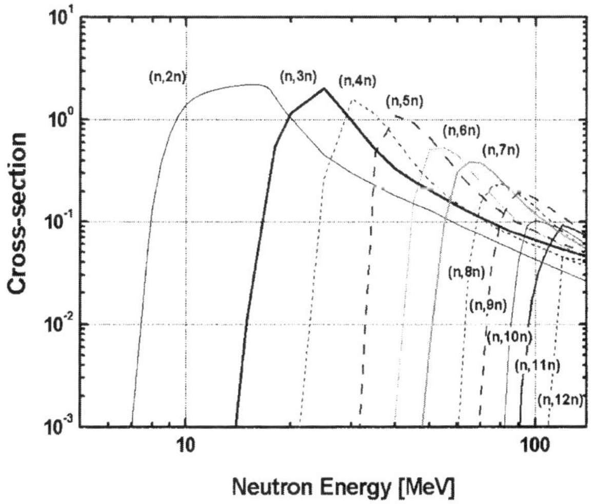

FIGURE 3. Cross-sections of Bi-209 (n, xn) reactions.

TABLE 1. Characteristics of neutron-induced threshold reactions for bismuth, titanium and indium.

Reaction	Half life	Threshold Energy [MeV]	Gamma rays [keV]	Cross-section [barn]
Bi^{209}(n, 12n)Bi^{198}	693s	88.17	562.4, 1063.4	1.72
Bi^{209}(n, 11n)Bi^{199}	27m	78.74	425.3, 841.7	3.96
Bi^{209}(n, 10n)Bi^{200}	36.4m	70.79	462.4, 1026.5	4.1
Bi^{209}(n, 9n)Bi^{201}	108m	62.58	629.1, 936.2	8.02
Bi^{209}(n, 8n)Bi^{202}	1.72h	53.99	422.1, 657.5, 960.6	9.45
Bi^{209}(n, 7n)Bi^{203}	11.76h	45.32	820.2, 825.2, 1847.6	14.5
Bi^{209}(n, 6n)Bi^{204}	11.22h	37.99	899.2, 983.9	16.8
Bi^{209}(n, 5n)Bi^{205}	15.31d	29.62	703.5, 1764.3	27.2
Bi^{209}(n, 4n)Bi^{206}	6.243d	22.56	803.1, 881	28
Bi^{209}(n, 3n)Bi^{207}	31.55y	14.42	569.7	32.5
Bi^{209}(n, 2n)Bi^{208}	3.68E+5y	7.5	1063.6	31.7
Ti^{48}(n, p)Sc^{48}	43.67h	7.6	983.5, 1037.5, 1312.1	0.303
Ti^{47}(n, p)Sc^{47}	3.345d	2.2	159.4	21.4
Ti^{46}(n, p)Sc^{46}	83.79d	3.9	889.3, 1120.5	10
In^{115}(n, n')In^{115m}	4.486h	1.2	336.2, 497.3	170

* Foil size : 25x25x1(t) mm for Bi, ϕ1/2" x 0.1(t) mm for Ti, In

* *Energy range of cross-section : ≤ 140 MeV for Bi
≤ 20 MeV for Ti, In

RESULTS

In these measurements, the neutron flux was analyzed as following Table 2. The results obtained from Bi(n,2n), and Bi(n,3n) reactions were not clear because many photons make (γ, xn) reactions. Especially, in the case of 0 and 6 degree conditions, huge number of photons from electromagnetic shower is expected. That was also found in the same measurement using Zr foil; so high activities were observed from the foils evenly low reaction cross-section of (n,2n). The threshold reaction over Bi(n, 8n) reaction could not be measured due to very short half-life. In this analysis, the cross-sections up to 140 MeV were used in the case of Bi foils and up to 20 MeV in the case of In and Ti foils.

In the case of 0 degree, the neutron flux was obtained down to about 1.2 MeV from In foils. It was about 1.74E+12 neutrons per cm^2 per second at 1 m from the Cu beam stopper. But in the cases of 6 and 90 degrees, neutrons below about 22 MeV were not measured. This might results from the low neutron flux at those directions in the comparison with the results at 0 degree. The flux at 90 degree was about 7 times higher than results by other measurement using well collimated experimental structure and the time-of-flight method [14]. The target thickness was 14 cm in those measurements.

For three directions the energy information of all fluxes followed the normal differential neutron yield curve even though there was some fluctuation such as higher flux at higher threshold energy condition.

TABLE 2. Neutron flux at the 1 m distance from Cu beam stopper produced by 2 GeV electrons.

Reaction	Lower Energy [MeV]	Neutron Flux* [neutrons/cm^2/sec]		
		0 degree	6 degree	90 degree
Bi^{209}(n, 8n)Bi^{202}	53.99	1.48E+8 (1.95)	6.18E+7 (1.92)	6.62E+5 (5.44)
Bi^{209}(n, 7n)Bi^{203}	45.32	5.11E+8 (2.67)	1.47E+8 (4.06)	5.93E+5 (7.53)
Bi^{209}(n, 6n)Bi^{204}	37.99	5.37E+8 (0.55)	1.48E+8 (1.41)	6.01E+5 (4.15)
Bi^{209}(n, 5n)Bi^{205}	29.62	1.44E+10 (0.44)	4.08E+8 (1.92)	1.41E+6 (14.5)
Bi^{209}(n, 4n)Bi^{206}	22.56	1.50E+10 (0.38)	4.69E+8 (1.20)	9.27E+5 (4.09)
Ti^{48}(n, p)Sc^{48}	7.6	7.92E+10 (4.66)	-	-
Ti^{46}(n, p)Sc^{46}	3.9	2.40E+11 (3.76)	-	-
Ti^{47}(n, p)Sc^{47}	2.2	3.50E+11 (0.71)	-	-
In^{115}(n, n')In^{115m}	1.2	2.85E+11 (4.71)	-	-

* The fluence of incident 2 GeV electrons is 3.26E+10 sec^{-1}.

** Numbers in parenthesis is counting error (%) in gamma spectroscopy system x

After long time (above 6 month) from these irradiation and gamma-ray measurements, the gamma spectroscopy for the irradiated Nd-B-Fe magnets and the SmCo magnets irradiated by the same method were carried out using ϕ1"x1" NaI(Tl) detector. Mn-54 and Co-58 were found as main radioactive elements of the Nd-B-Fe magnet and Co-58 as ones of the SmCo magnet after long cooling period.

DISCUSSION AND SUMMARY

The environmental condition, an energy-informed neutron flux, around the magnet with Cu beam stopper in its demagnetization study was observed by the threshold activation measurements. These results gave the key idea to understand the demagnetization phenomena. The measure flux at 90 degree was higher than other measured results. However the target geometry in this experiment is unique for simulating insertion device and many scattered neutrons are estimated. The energy information of neutron flux has some fluctuation, which might results from lack of cross-section data. These results are still preliminary ones. Only the neutron flux around the magnet was measured. Still there is no enough information of particles inside the magnet, which is very important in order to understand the demagnetization phenomena. More precise measurements and the normal unfolding analysis to get energy spectrum will be done in near future.

In this measurements the feasibility of Bi(n,xn) reactions and the uses of Ti and In foil in high energy neutron measurements were also validated again. The gamma contamination for all of threshold reactions will be checked more.

REFERENCES

1. W.V. Hassenzahl, T.M. Jenkins, Y. Namito, W.R. Nelson, and W.P. Swanson, *Nucl. Instr. & Meth.* **A291**, pp378-382 (1990).

2. P. Colomp, T. Oddolaye, and P. Elleaume, *Demagnetization of permanent magnets to 180 MeV electron beam*, ESRF/MACH-ID/93-09, ESRF, (1993).
3. J. Alderman, P.K. Job, R.C. Martin, C.M. Simmons, and G.D. Owen, *Nucl. Instr. & Meth.* **A281**, pp9-28 (2002).
4. Y. Ito, K. Yasuda, R. Ishigami, S. Hatori, O. Okada, K. Ohashi, and S. Tanaka, *Nucl. Instr. & Meth.* **B183**, pp323-328 (2001).
5. T. Bizen, T. Tanaka, Y. Asano, D.E. Kim, J.S. Bak, H.S. Lee, and H. Kitamura, *Nucl. Instr. & Meth.* **A467-468**, pp185-189 (2001).
6. T. Bizen, Y. Asano, T. Hara, X. Marechal, T. Seike, T. Tanaka, H. Kitamura, H.S Lee, and D.E. Kim, "Improvement of radiation resistance of NdFeB magnets by thermal treatment", *Eighth International Conference on Synchrotron Radiation Instrumentation* (*SRI 2003*) , San Francisco, August 2003 (To be published at proceedings).
7. T. Bizen, Y. Asano, T. Hara, X. Marechal, T. Seike, T. Tanaka, H. Kitamura, H.S Lee, and D.E. Kim, "Introduction of the high radiation resistance of undulator magnet", *Eighth International Conference on Synchrotron Radiation Instrumentation* (*SRI 2003*) , San Francisco, August 2003 (To be published at proceedings).
8. T. Bizen, Y. Asano, T. Hara, X. Marechal, T. Seike, T. Tanaka, H.S Lee, and D.E. Kim, C.W. Chung, and T. Kitamura, *Nucl. Instr. & Meth.* **A515**, pp850-852 (2003).
9. T.Y. Goo, Pohang Accelerator Laboratory, POSTECH, (private communication).
10. E. Kim, T. Nakamura, A. Konno, Y. Uwamino, N. Nakanishi, M. Imamura, N. Nakao, S. Shibata, and S. Tanaka, *Nucl. Sci. & Eng.* **129**, pp209-223 (1998).
11. N. Nakao, T. Shibata, T. Ohkubo, S. Sato, Y. Uwamino, Y. Sakamoto, and D.R. Perry, *Shielding experiment at 800 MeV proton accelerator facility*, KEK Preprint 97-251, KEK (1998).
12. Y. Asano and N. Sasamoto, *J. Nucl. Sci. Tech.* **Suppl. 1**, pp535-539 (2000).
13. ENDF/IV, http://www.nndc.bnl.gov/, National Nuclear Data Center, Brookhaven National Lab.
14. H.S. Lee, S. Ban, K. Shin, T. Sato, S. Maetaki, C.W. Chung, and H.D. Choi, *J. Nucl. Sci. Tech.* **Suppl. 2**, pp1228-1231 (2002).

Applied Physics (APHYS 2003)
A. Méndez-Vilas (Ed.)

MEASUREMENT OF RADON CONCENTRATION IN DIFFERENT CONSTRUCED HOUSES AND TERRESTRIAL GAMMA RADIATION IN ELAZIG, TURKEY

Mahmut Doğru[1], Cumhur Canbazoğlu[1], Nilgün Çelebi[2] and Güner Kopuz[2]

1 Department of Physics, University of Firat, TR-23169 Elazig,TURKEY
E-mail: mdogru@firat.edu.tr
2 Çekmece Nuclear Research and Training Centre, Department of Health Physics, P.K 1 Atatürk Havalimani, Istanbul,TURKEY
E-mail: celebin@nukleer.gov.tr

Abstract. In this study, we have measured the terrestrial γ-radiation in the Elazig, which is the eastern city of Turkey by using portable NaI γ-ray detector. The measurements were performed in 21 critical places to determine the variation of the gamma radiation level in air in the city. The γ-radiation levels were found to be between 38.1 pGy/s and 55.7 pGy/s. We have also measured indoor radon (^{222}Rn) concentration in different constructed houses of the city by using CR-39 passive radon detector. The detectors were obtained from the Çekmece Nuclear Research and Training Centre laboratory and they remained for five months in selected houses rooms. All detectors were evaluated and etched in the same laboratory. The obtained results were interpreted according to the house constructions and the geologic formation of the soil.

Keywords: Nuclear Track Detectors, Radiation, Radon concentration, Elazig.

INTRODUCTION

Humans are continuously exposed to natural radiation based on the earth surface, which includes the radioactive nuclide in its crusts and the cosmic ray sources. Radon and its daughters is the largest contributor to the environmental radiation and to dose accumulation on the population respiratory organs. The important natural radioactive sources for indoor radon are mainly its soil beneath, house building material and the household water. The radiation dose originated by indoor radon and its progenies level is a major percentage of the background received by the population in over the world.

Generally, the natural radioactive sources can be divided in two groups namely the earth and cosmic origin sources. The earth origin radiation sources are coming through the solar system formation (e.g. uranium, thorium and potassium) (1). The radionuclides activity concentration changes according to the soil and rock type. The radioactivity concentration represents the absorbed dose strength at 1m levels in air (2) and the dose intensity is strongly related to the radionuclides at the earth surface. Although the sedimentary and phosphate rocks have higher radioactivity concentration, the radioactivity concentration in the volcanic rocks is relatively higher than the sedimentary ones. Metamorphism rocks have the same radioactivity concentration as in its contents have (3). The earth origin includes the terrestrial radioelement of ^{238}U, ^{232}Th and their series, and ^{40}K have gamma radiation in the crystal rocks regions are relatively higher than the sedimentary rocks regions (4).

The building materials are one of the natural radioactive sources, which have an additive radon concentration to the indoor livings. The most of the outdoor radiation comes from the gamma radiation caused by uranium and thorium decreasing series and from potassium radioelement. Most of the domestic radiation exposure, especially affecting the respiratory tract, is due to radon and its daughters that are coming from the building materials and beneath soil (5). ^{222}Rn is produced from the ^{238}U decay series (from ^{226}Ra), it is noble and radioactive gas with half-life of 3.82 days decays into a series of solid and short-lived radioisotopes. The most additive factor to indoor radon concentration is coming from the soil beneath of the building (6).

The radon and its daughter's exposure may increase the cancer risks on human being (e.g. lung cancer). The International Radiological Protection Commission (IRPC) has recommended upper limit for indoor radon concentration is 200-600 Bqm^{-3} per a year. According to the Commission, the lung cancer risk increases about 6% for the people who live in the 400 Bqm^{-3} exposure media (7, 8).

MATERIAL AND METHODS

We have measured environmental radiation doses of the city by using NaI probe (SPA8) supported by portable radiation monitoring system (ASP2e) supplied by Eberline Instrument (9) with homogeneous distribution in all over the city centre at standard 1m high from the ground. The NaI crystal was in 2.5cmX2.5cm dimensions and it sensitive to the gamma energy range between 40 keV and 1.3 MeV. The detector set to collects the lowest detectable energy up to highest value and all energies between this ranges was integrated. The measurements have been done in suitable time by 5 repeating times at least in 10 positions which was selected in about 20m radius of circumference with the same altitude and similar soil composition for each determined place. The results were obtained by averaged the measured data. The standard deviation of measurements was obtained by using 50 measurements for each position and it has been found to be 5%-7% of the averaged value.

Different home's constructions had not the same contribution to the houses radon concentration. It should be very interesting to evaluate this contribution to the radon concentration in homes. The obtained data may help the researchers to interpret the radon concentration related to different constructed homes. Therefore, to obtain the contribution of different home's construction to indoor radon concentration, the homes were mainly divided into four categories. Since some homes have the same construction but built on different geological formation, we have analysed five sets of homes in four categories. All houses drinking water were supplied by main city water network. The four categories of homes were selected as follows:

First type: Houses were newly (a few years old) built from concrete.

Second type: The houses built from the concrete and at least 30 years old apartment.

Third type: The wall thickness at least 50cm and they built from mud and stones and at least 40 years old (generally called sun dried brick houses). The ground of those houses is completely soil and their ceilings constructed by mud and wooden.

Forth type: Prefabricate houses completely built from wooden and their ground is soil and have big garden.

We have used CR-39 nuclear track detector in order to measure indoor radon concentration in city centre. The thickness of the used detectors was 0.50mm and they made of allyl-diglycol carbonate plastics. The CR-39 foil pieces are cut in dimension of 20 x 20 x 0.50 mm and they have placed in the radon diffusion cup in vertical position. The cups were closed by cloth pieces to avoid dust particles and aerosols other than radon gas to get inside. The calibration was made by using 225 l radon calibration room. The room radon equilibrium concentration was 3.2 kBq. The track detectors were left for 1 to 5 days in this room and than the chemical etching were applied. The tracks was counted by using optical microscope and the calibration factor was found to be 5 (kBq/m^3) / (net tracks / h). The efficiency (tracks $cm^{-2}Bq^{-1}m^3h^1$) of the used track detectors was about 2%. Each detector's background was determined before deliveries to homes.

Two detectors are distributed to the each previously selected home according to their types for leaving in living room and bedroom. They were collected after 5 month and their air holes were covered by nylon band and hence, they could not be influenced by radon and its progeny during the transport to the laboratory (Çekmece Nuclear Research and Training Centre (ÇNAEM), Health Physics Division) located in Istanbul. Detectors immediately submitted to the etching process when arrived to the laboratory.

Detector foils were etched in 30% NaOH solution at 62 ^{0}C for 15 hours and then they washed in double distilled water, and dried. After the etching process ten regions were selected from the each foil and the tracks were counted by using optical microscope. Each counted region had 2.5mm^2 surface area.

THE RESULTS AND DISCUSSIONS

The mean of the environmental gamma radiation dose were determined individually in 21 different places of the city centre which geological formation rich by copper, zinc, chrome and manganese (see Table 1). The lowest mean gamma radiation (38.1 pGy/s) was determined in the 3rd, 8th and 20th regions, and the highest mean value (55.7 pGy/s) were determined in 13th region. The mean gamma radiation dose for the city centre was found to be 46.8 ± 2.9 pGy/s. This value is about three times bigger than the gamma radiation dose (15.8 pGy/s) measured in Istanbul (10). This difference is attributed to the different soil formation and altitude of the cities.

Table 1: The mean γ radiation doses measured in 21 different places of the Elazig city centre

Region Number	The name of the places	γ Radiation dose (pGy/s)
1	Hankendi quarter	52.8
2	Hilalkent quarter	52.8
3	Hilalkent quarter around the lime mine	38.1
4	Abdullahpaşa quarter	44.0
5	Around Mollakendi quarter	48.2
6	Mollakendi quarter	46.9
7	Around Fırat University's Hospital	46.9
8	Rızaiye quarter	38.1
9	Around Harput Dabakhane	44.0
10	Harput Centre	49.9
11	Doğukent quarter	41.1
12	Sanayi quarter	49.9
13	Sanayi quarter (around GSM station)	55.7
14	Aksaray quarter	49.9
15	Sürsürü quarter	49.9
16	Cumhuriyet quarter	44.0
17	Cumhuriyet quarter (around GSM station)	49.9
18	Bahçelievler quarter	49.9
19	Zafran quarter	52.8
20	Gazi Street (city centre)	38.1
21	Around Dental hospital	41.1

The obtained data from the locations around the city centre were as follows.

- 1-4 numbered regions were in the southwest of the city and the mean gamma radiation dose was 46.9 pGy/s.
- 5th and 6th regions located in southeast of the city and the mean gamma radiation dose was 47.7 pGy/s.
- 7-20 numbered regions located in the city centre and the mean gamma radiation dose was 47.2 pGy/s.
- 21st region located in the south of the city and the mean gamma radiation dose was 41.1 pGy/s.

Although the first, second and third regions have almost the same altitude, the third region's radiation dose was found 0.72 times higher than the others, i.e. the gamma radiation dose reduces about 28% between those regions. The difference was attributed to the different soil formation. On the other hand, the 9th and 10th regions have the same soil formation but their altitudes were different. Since the cosmic ray contribution was relatively higher in the higher altitude, difference on their gamma radiation (5.9 pGy/s) was attributed to the altitude differences.

The radon concentration in homes chosen to be represented a different construction building were shown in Table 2. Our aim in this preference was to show the affects of the building material on indoor radon concentration. The radon concentrations in all the city's homes are undergoing projects to be completed. As we have shown in Table 2, considering the living rooms, the highest radon concentration (145 Bqm^{-3}) was measured in second type homes, on the contrary to the lowest radon concentration (71 Bqm^{-3}) in the first type home's living room. The lowest radon concentration, 37 Bqm^{-3}, was measured in the first type and the highest radon concentration, 154 Bqm^{-3}, was measured in the third type home's bedrooms. The averaged values for the living rooms and bedrooms were found to be 137 Bqm^{-3} and 54 Bqm^{-3} in third and first home's type, respectively. The highest mean radon concentration in third type homes, 137 Bqm^{-3}, was not surprising us since they were constructed from mud and stones and also their ground was soil. The people who live in those homes were smoker. Although, the third type homes had the same construction with the second and fifth numbered homes, the radon concentration was 60 Bqm^{-3} different to each other. The reason was attributed to the smoke additive in fifth home. The second highest radon concentration (96 Bqm^{-3}) was measured in the forth type home. Those homes were built directly on the soil and the measured data was agreed with the literature (6). If we compare the first and third homes (see Table 2), we could easily see that the radon concentration in third home is 1.74 times bigger than the first home type. This difference was attributed to the building age. The third home type was very older than the other.

Table 2: The radon concentration in some selected type of homes in Elazig city centre

Home Number	Home type	Living room ($Bq.m^{-3}$)	Bedroom ($Bq.m^{-3}$)	The mean radon concentration ($Bq.m^{-3}$)
1	First type	71	37	54
2	Third type	77	76	77
3	Second type	145	43	94
4	Fourth type	100	92	96
5	Third type	119	154	137

As a result of the measurement of the indoor radon concentration in the selected home types, it has been seen that all of the values was under the permitted upper limit (200-600 Bqm^{-3}) which determined by the International Radiological Protection Commission (IRPC) (11).

ACKNOWLEDGEMENT

This work is supported by Firat University Research Units projects FUBAP-591 and FUBAP-631. We would like to thanks to TÜBİTAK BAYG for their financial support.

REFERENCES

1. Al-Jundi, J. *Population Doses from Terrestrial Gamma Exposure in Areas Near to Old Phosphate Mine, Russaifa, Jordan*, Radiat. Meas. **35** 23-28 (2002).
2. Beck, H.L. The Natural Radiation Environment II. *The Physics of Environmental Gamma Radiation Fields*. USERDA Conf.-720805-P2, 101-104 (1982).

3. National Council on Radiation Protection and Measurements. *Environmental Radiation Measurement*. (NCRP Report 50). ISBN. 0-913392-32-4. (1977)
4. Buchli, R. and Burkart, W. *Correlation Among The Terrestrial γ Radiation, The Indoor Air ^{222}Rn, and The Tap Water ^{222}Rn in Switzerland*, Health Phys. **57**(5) 753-759 (1989).
5. Sharaf, M., Mansy, M., El Sayed, A., Abbas, E. *Natural Radioactivity and Radon Exhalation Rates in Building Materials Used in Egypt*, Radiat. Meas. **31** 491-495 (1999).
6. Farid, M.S. *Indoor and Soil Radon Measurements in Swaziland by Track Detectors*, J. Environ. Radioactivity, **34**(1) 29-36 (1997).
7. Selçuk, A.B., Yavuz, H., Köksal, E.M., Özçinar, B. *Radon Concentration in Elazig Houses and Factories*, Radiat. Prot. Dosim. **77**(3) 211-212 (1998).
8. Srivastava, A., Zaman, M.R., Dwivedi, K.K., Ramachandran, T.V. *Indoor Radon Level in The Dwellings of The Rajshahi and Chuadanga Regions of Bangladesh*, Radiat. Meas. **34** 497-499 (2001).
9. Eberline Instruments. *ASP-2/2e Portable Radiation Monitor*, Technical Manual, Santa Fe, (1995).
10. Karahan, G. *İstanbul'un Çevresel Doğal Radyoaktivitesinin Tayini ve Doğal Radyasyonların Yıllık Etkin Doz Eşdeğeri*. Doktora Tezi, İstanbul Teknik Üniversitesi Nükleer Enerji Enstitüsü, İstanbul, (1997).
11. International Commission on Radiological Protection (ICRP). *Protection Against Radon at Home and at Work*. (Oxford: Pergamon Press). ICRP Publication 65, (1993)

Applied Physics (APHYS 2003)
A. Méndez-Vilas (Ed.)

THE STATISTICAL ANALYSIS OF THE RADIOACTIVITY CONCENTRATION OF THE WATER DATA IN MALATYA CITY, TURKEY

Mahmut Doğru, Mesut Yalçın, Fatih Külahcı, Cumhur Canbazoğlu and Oktay Baykara

Department of Physics, University of Firat, TR-23169 Elazig, TURKEY
e-mail: mdogru@firat.edu.tr

Abstract. The natural radioactivity data of 306 water samples, which were obtained from various 51 sampling points of Malatya that is the eastern city of Turkey, was studied by applying the factor and cluster statistical analysis to interpretation and determination of the general radioactivity in the city. The obtained data from the statistical analysis were evaluated according to the geological formation of the state. The varimax factor rotation was applied to the data. Two factors were extracted from the radioactivity of the water supplies, tap water, and water depots data, which account for about 95.52% of the total variance. Factor 1 and factor 2 explained 48.5% and 47% of the variance, respectively.

INTRODUCTION

The importance of the healthy drinking water for human could not be denied by anyone who lives on the Earth. However, it is very important to determine the chemical and physical properties of usage and drinking water especially consumed by public and any other living creature. The security of the healthy drinking water can be continuously obtained by applying standards on important chemical and physical parameters (e.g. gross alpha and beta radioactivity guidelines). Since, the amounts of the internal exposure dose are strongly depending on the amount of the radionuclides taken through drinking water, the radio-nuclides determination in drinking water is very important to estimate the human internal exposure especially cause by radionuclides [1].

Water, generally, contains various radioisotopes from very well known radioactive series of ^{235}U and ^{232}Th [2]. The soil formation of the water sources surrounds is very effective parameter on the radioactive elements, which occurs in the radioactive decay series (e.g. ^{226}Ra, ^{223}Ra and ^{222}Rn from uranium, and ^{224}Ra and ^{228}Ra radioisotopes from thorium are more probable to be found in ground water [3]). It is also determined that the amount of ^{226}Ra and ^{225}Ra in ground water depends on the distribution of thorium and uranium in the environment and the geo-chemical formation of the places [4]. Drinking water is mostly supplied from ground water by drilling wells. Hence, distribution of the radioactive minerals in the environment and soil formation is very important parameters for obtaining healthy drinking water in acceptable circumstances.

Gross alpha and beta radioactivity measurements, generally, are particular routine monitoring purpose of interest. For the practical purposes, without specified the natural and man-made radionuclides the recommended guideline activity concentration are 0.1 Bql^{-1} and 1 Bql^{-1} for gross alpha and beta in drinking water, respectively [2, 5]. Institution of Turkish

Standards (ITS), Turkey, recommends guidelines for gross-α and gross-β activity in drinking water to be 0.037 Bql^{-1} and 0.37 Bql^{-1}, respectively [6].

Malatya city's drinking water is obtained from Derme water source in Yeşilyurt district, which is 23 km far away of the city center. This water falls into main tunnel and then separates into two tunnels individually. Most of the city center taps and usage water obtains their requirements from the water stores, which their water is supplied from the tunnels [2]. This source also supplies Yeşilyurt and Battalgazi counties' drinking water (the other districts drinking water is supplied by various drilled well around the places). The tunnels, which collect the water from the source, are completely surrounded by glass. The water flow rate is about 2000 l/s and the attitude of the source is about 140 m. Water completely transported by natural flows.

Working on the Malatya city center's drinking water is interesting since most of the water stores are supplied from one spring water source. We have previously studied on the water radioactivity of the stores without making any statistics on the obtained data [1, 7]. In this work, we have added many water samples were obtained from taps (all over the city) and applied the statistics on them. From the point of the view of the evaluation of the water stores effects on the tap water, the results are very interesting. We have tried to conclude this situation through this manuscript.

THE STUDIED FIELD

THE SAMPLING AREA

Malatya is the one of the East Anatolian city, which locates in the Upper Firat catchments basin and it has 12.313 km^2 areas (Fig. 1). The samples were obtained randomly from all over the city center.

THE GEOLOGICAL STRUCTURE OF THE STUDIED AREA

The geological formation of the area might be summarized as below:

- The geological formation of the city center is based on alluvia, limestone. Beydağı town, which is many water stores are accumulated in this place and the water spring area, Yeşilyurt district, have similar geological formation; sand gravel, limestone and conglomerate,

FIGURE 1. The location of Malatya city in Turkey.

- Battalgazi and Yeşilyurt districts, which are both of them close to the city center, have sedimentary rocks.
- Pütürge and Doğanyol districts mainly have volcanic rocks (marble and granite).
- The other districts geologic formation is mainly similar to Yeşilyurt district.

EXPERIMENTAL STUDIES

The water samples were collected into sterilized 1 l capacity glass bottles. 0.5 ml 3*N* nitric acid added to prevent precipitation and absorption of minerals by container walls. The water samples radioactivity measurements used ZnS(Ag) scintillator supported by photo multiplier tube with 7286 low-level alpha counters from NE Technology Inc. which for instrumentation to count gross alpha. At the end of these processes the samples were ready to be counted in counting systems [8].

The Factor Analysis

The purpose of the factor analysis is the description of the observe variables in complex environmental compartments by finding summarizing factors, which are often causally explainable. The extracted factors reflect the main part of information of the data set [9].
Axis rotation might be done to get the clarity of the factors obtained from the analysis. Therefore, the factors load would be effective on the factors identities. There are two rotation techniques called as orthogonal and oblique techniques. We have used varimax technique, which is one of the orthogonal axis rotational techniques in this study.

THE RESULTS OF THE FACTOR ANALYSIS

When we have examined the factor analysis data, we have obtained six data structure, which collected under two factors (Table 1). These two factors explain the variance depending on the scale and the variance is found to be %95.525. It is obtained that the shared variance (communalities) of the two factors, which defined relevant to the structure, is varying between 0.924 and 0.977. Therefore, in our analysis, we have obtained that the important appearance of the two factors together determined the total variance in their structure and the variance related to the scale.

When we conclude the Component Matrix Table, the first three (beta radioactivity) of the six data structure is explained in factor 1 (Table 2). The alpha radioactivity is also explained in the factor 2. Due to the proximity between the "Beta (3)" and "Alpha (1)" we have allowed to use the rotation on the data.
The explanation of the variances, which is explained by the factors after the rotation, was to be 48.54%, and 46.9% for the factor 1 and factor 2, respectively.

TABLE 1. Principal Factor Analysis.

	Factor 1	Factor 2
Beta (1)	0.797	0.585
Beta (2)	0.794	0.574
Beta (3)	0.788	0.595
Alfa (1)	-0.746	0.607
Alfa (2)	-0.735	0.645
Alfa (3)	-0.730	0.636

TABLE 2. Rotated Component Matrix

	Factor 1	Factor 2
Beta (3)	**0.982**	-0.102
Beta (1)	**0.982**	-0.114
Beta (2)	**0.972**	-0.121
Alfa (2)	-0.09	**0.973**
Alfa (3)	-0.101	**0.963**
Alfa (1)	-0.132	**0.952**

When we have examined the rotated Component Matrix Table, we have obviously determined that the alpha and beta radioactivity concentration of the water of Malatya city is well separated from each other. Hence, factor 1 and factor 2 are named as alpha radioactivity and beta radioactivity factors, respectively.

Cluster Analysis

The cluster analysis is the method, which is used for the data structure, which the relationship between the individuals is not clearly known. The method aims to accumulate the unrelated individuals in different sets. The analyses aimed to obtain the homogenous sub sets structure by dividing the individuals (the obtained data) of the data set.

In this work, the method used for the clustering (grouping) the data is

"Agglomerative Hierarchical Cluster Analysis". In this method considering the similarity of the units, it is aimed to combine them in the similar state. It is also different approximation is used for the combination of the units together. We have used Ward (Squared Euclidean Distance) Method in this work. The theory of the method can be found elsewhere [10].

THE RESULTS OF THE CLUSTER ANALYSIS

We have obtained two main clusters on the data. Most of the data included in the first cluster. The first cluster individually is divided to the sub clusters (Table 3). The 60% of the tap waters data accumulated in the cluster A and the rest of the 40% of the data represent the storage water samples in cluster B. These results show that the data in different clusters represent the samples, which is obtained from neighbor places, which have similar geological formation (basalt, clay).

In cluster B the data shows the storage water of the city. Those waters were collected from different places of the main state and the samples alpha and beta radioactivity shows the similarity. The other clusters data accumulation shows the samples results, which have obtained from the similar geological formation places.
The common characteristics of the samples, which accumulated in cluster A, are obtained from the area, which has uranium minerals. The cluster C contains the data obtained from the mostly thorium mineral founds area. The cluster E contains the data were obtained from the samples which collected from the area which has sedimentary rocks and cluster F data shows very similarity on the places which has limestone, conglomerate formation.

CONCLUSION

The gross alpha and beta radioactivity of the main tunnels, which supplies the most of the city center's drinking water, is respectively, 0,117 $\pm$ 0.017 and 0.220 $\pm$ 0.019 Bq/l. The gross alpha and beta radioactivity of the water samples obtained from the storages constructed in different places of the city were found to be 0.170 $\pm$ 0.044 and 0.257 $\pm$ 0.007 Bq/l, respectively. The both radioactivity is relatively higher in storage water and the main source. This difference may be attributed to additive effect of ^{222}Rn in the storage water. The radioactivity level of the city is shown in Fig. 2. The places, which contain more sedimentary rocks, show more radioactivity than the places contains volcanic rocks in their geological formation. The reason for these differences may be attributed to the relatively harder structure of the volcanic rocks than the sedimentary ones.

TABLE 3. Dendrogram using Ward Method (Squared Euclidean Distance) Rescaled Distance Cluster Combine

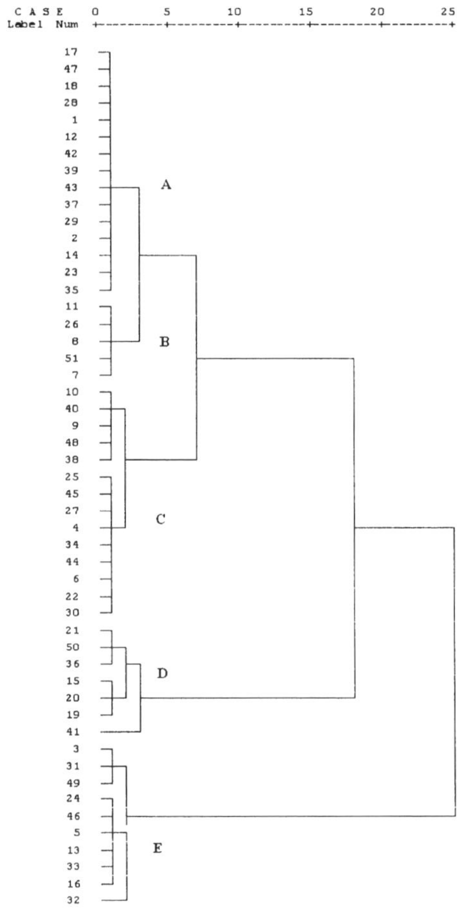

More detailed statistical analysis made on the obtained whole data. As a conclusion of the application two statistics (factor and cluster analyses) on the obtained data very clearly shows that increasing the number of the samples make easier to apply statistics on the data and easily conclude the results individually. It is clearly seen that, application the statistical methods on the radioactivity data gives an opportunity to see the detail on the relationship between radioactivity and studied places. By using the statistical methods on the radioactivity data also helps to conclude the results individually in many related factors.

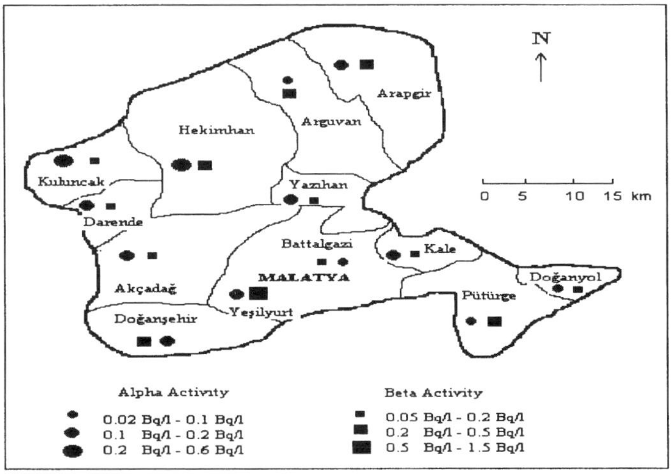

FIGURE 2. Distribution of the radioactivity over the Malatya city.

ACKNOWLEDGEMENT

This work is supported by Firat University Research Units project FUBAP-591. We would like to thanks to TÜBİTAK BAYG for their financial support.

REFERENCES

1. G. Karahan, N. Öztürk, and A. Bayülken, *Water Research* **34**, 4367-4370 (2000).
2. M. Yalçın, M. Doğru, C. Canbazoğlu, *Balkan Physics Letters* **11**, 1-4 (2003).
3. C. Duenas, M.C. Fernandez, J.A. Gonzalez, J. Carretero and M. Perez, *Toxicological and Environ. Chem* **39,** 71-79 (1993).
4. M. Dogan and M. Soylak, *Su Kimyasi*, Erciyes Universitesi Yayinlari, Kayseri, 2000.
5. World Health Organization. *Guidelines for drinking water quality*. Recommendations **1**, Geneva, 1993.
6. Institution of Turkish Standards, Annual Progress Report, 1997.
7. M. Doğru, C. Canbazoğlu, M. Yalçın, O. Baykara and F. Külahcı, *Doğu Anadolu Bölgesı Araştırmaları Dergisi* **1,** 96-99 (2003).
8. F. Külahcı, M. Doğru, O. Baykara, S. Şahin, C. Canbazoğlu, *Balkan Physics Letters* **10**, 220-225 (2002).
9. M. Bakaç, *Environ. Geochem. and Health* **22,** 93-111 (2000).
10. K. Özdamar, *Paket Programlar ile Istatistiksel Veri Analizi*, Kaan Kitapevi, Eskişehir, 1999.

Applied Physics (APHYS 2003)
A. Méndez-Vilas (Ed.)

Dosimetric Characterization Of A Novel Ternary Crystal With Europium

R. Rodríguez-Mijangos and R. Pérez-Salas.

Centro de Investigación en Física, Universidad de Sonora, A. P. 5-88, 83190, Hermosillo, Sonora , México

Abstract. Recently, the optical properties of an europium-doped ternary mixed crystal containing $KCl_{0.5}$, $KBr_{0.25}$, and $RbBr_{0.25}$ (in molar composition) have been reported. In this work, the thermoluminescence properties of this material under β-radiation exposure is investigated. The glow curve shows light emissions for two temperature regions around 140 and 332 °C which increase with the time of exposure, in each region The light emission of the second region increase almost linearly with time. In each region the fading is so small that this material shows high energy storage capacity and a possible application in dosimetry being the main dosimetric glow region found at 332 °C.

Introduction

Since the finding of ternary alkali halide mixed crystals with crystalline characteristics that could behave as well as the single alkali halide crystals [1], renewed research in multiple mixed dielectric crystals with defects has started. [2] As is well known, Eu^{2+}-doped single alkali halide crystals have good TL dosimetric characteristics for β, X and UV radiations [3]. This is based on its greater trapping efficiency of radiation induced electron-hole pairs, where electrons are trapped at the anion vacancies, while holes are trapped around the Eu^{2+}-cation vacancy (I-V) dipoles [4]. However such efficiency is also modified by other parameters such as the content of dislocations or interstitials spaces created by the impurity ion. Dosimetric characteristics of Eu^{2+}-doped mixed alkali halides, particularly that of $KCl_{1-x}Br_x:Eu^{2+}$,[5] have been studied under UV irradiation. These studies show a higher TL efficiency for the middle composition, which is explained by a higher concentration of dislocations than in the end compositions x = 0 or 1. On the other hand as it has been indicated above, the ternary mixed alkali halides with single crystal characteristics have its own physical properties and similar behavior as alkali halide crystals of one component. The difference on the ion size of the different constituents must led to a much rich content of dislocations than in binary mixed alkali halide crystals thus increasing the trapping sites for electrons and holes induced by irradiation, as is clearly shown by the widening of the F band in a ternary crystal in comparison to the F band width observed in its components.[6] In the ternary crystal $KCl_{0.5}KBr_{0.25}RbBr_{0.25}$, member of a novel crystallographic family of mixed alkali halide crystals with great miscibility [2] was found in detailed optical studies in this ternary crystal doped with europium a physical behavior that resembles the single phase condition of a solid solution clearly shown by X ray difractometry. [7] Therefore we will refer to these type of crystals as solid solutions.

A detailed study of the thermoluminescent behavior of a complete set of $KCl_{1-x}Br_x:Eu^{2+}$ crystals has been done [8], and a single thermoluminescent peak has been found which shifts when x changes being a typical characteristic of a solid solution or single phase with lattice constant dependence. Though the precise recombination mechanisms are not well understood it is believed that an energy transfer must be occur because the emission spectrum consists mainly of europium emission bands.

Linking the TL phenomena and his behavior in this work the TL dosimetric characteristics of an Eu^{2+}-doped mixed crystal $KCl_{0.5}KBr_{0.25}RbBr_{0.25}$ is studied.

Experimental

The $KCl_{0.5}KBr_{0.25}RbBr_{0.25}:Eu^{2+}$ crystal was grown by the Czochralsky technique in the Crystal Growth Laboratory of the Instituto de Física de la UNAM, Mexico. $EuCl_2$ was added in the melt of the ternary mixture, previously reduced from $EuCl_3$. Samples were exposed to a ^{90}Sr β radiation source at room temperature. TL measurements were carried out in a RISO TL/OSL-DA-15 system at 5 °Cs^{-1}.

Results and conclusions

A typical TL glow curve of a thermally treated crystal is shown in Fig 1. The most intense glow peaks appears at 140 and 332 °C for

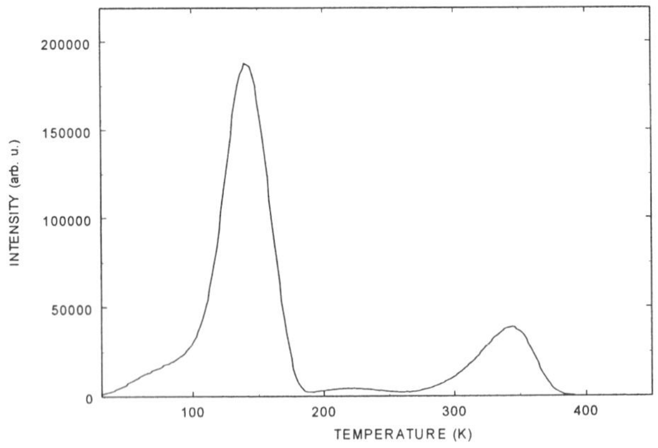

FIGURE 1. Glow curve of $KCl_{0.5}KBr_{0.25}RbBr_{0.25}:Eu^{2+}$ crystal exposed to beta radiation at RT.

all the irradiation doses given at times lower than 2400s. At this time the form of the glow curve does not change appreciably. Figure 2 shows the time dependence of the 140 and 332°C glow peaks intensity. At times lower than 2400s increase linearly. The emission

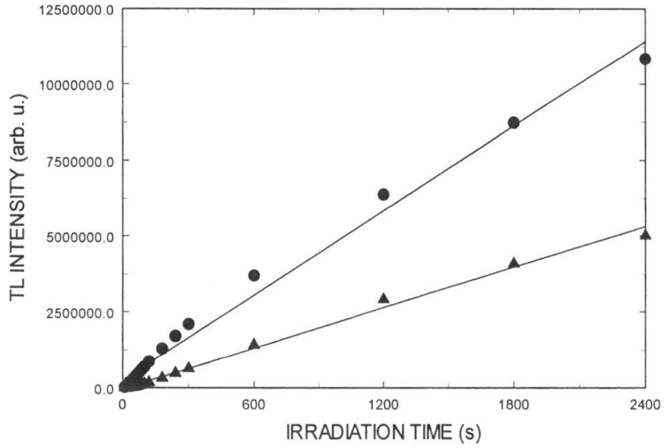

FIGURE 2. Integrated thermoluminiscence of $KCl_{0.5}KBr_{0.25}RbBr_{0.25}:Eu^{2+}$ after irradiation at RT (140C : circles, 332C : triangles).

spectra consist of only a component centered at 420nm which corresponds to the $4f^{6}5d$-$4f^{7}$ Eu^{2+}-electronic transition. This suggests that Eu^{2+}-ion plays an important role in the luminescent process as occur in other Eu^{2+}-doped alkali halide crystals but up today the mechanism is not so clear.

The glow peak found at 332 C is considered as the main dosimetric glow peak The storing effect, shown in Figure 3, indicates that decayment of the two glow peaks is so small that the material can retain almost all the defects induced by irradiation. From these results we can see that also it is a good material for optical memory.

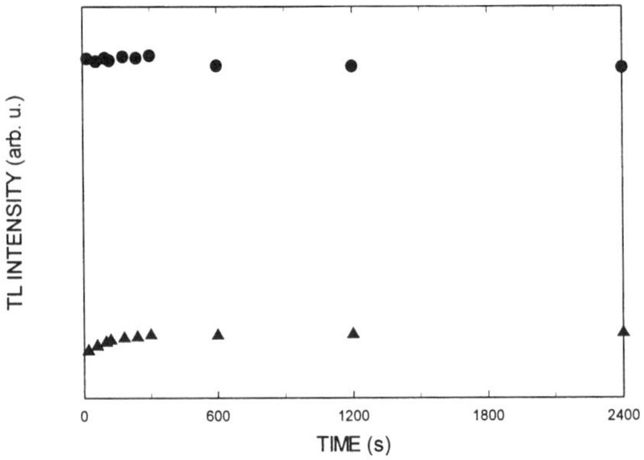

FIGURE 3. Fading of $KCl_{0.5}KBr_{0.25}RbBr_{0.25}:Eu^{2+}$ after irradiation at RT (140C : circles, 332C : triangles).

In Fig 4, the glow curve of europium doped ternary is compared with the glow curves of the ends KCl:Eu and KBr:Eu single crystals, which were obtained under similar conditions than ternary crystal. The glow curves of the ends show the main dosimetric glow peaks located around 200°C and 100°C. The temperatures at the emission maxima of the glow peaks result to be different than those of the emission observed on KCl:Eu and KBr:Eu and RbBr:Eu however the glow peak at 140°C of the ternary crystal is found between the glow peaks of KCl:Eu and KBr:Eu which could be related with the composition of the solid solution in a similar form to that observed in $KCl_{1-x}Br_x:Eu^{2+}$ crystals as function of the molar composition x. This result suggest that a single phase character of the novel ternary crystal. This is congruent with the results obtained for KCl_{1-x} -KBr_x solid solutions. Taking into account that ternary crystal has a lattice constant given by the Generalized Vegard's law of 6.515A, the position of the glow peak is expected to be found around 140°C.

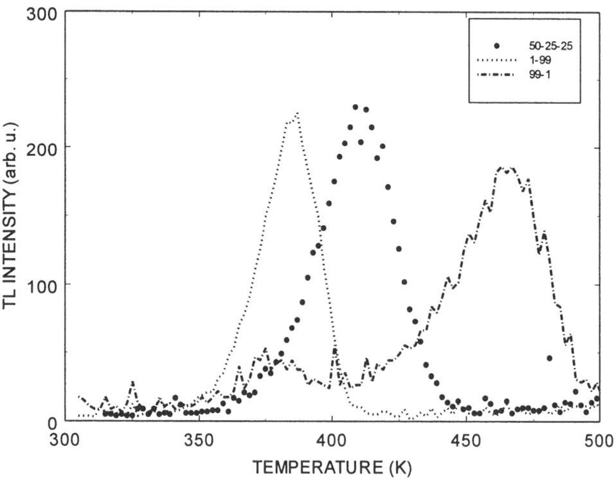

FIGURE 4. Glow curves of (1) $KCl_{0.5}KBr_{0.25}RbBr_{0.25}:Eu^{2+}$, (2) $KCl:Eu^{2+}$, (3) $KBr:Eu^{2+}$ crystals exposed to beta radiation at room temperature

An important conclusion of this work is the potential application of a novel ternary crystal doped with europium in optical and dosimetric applications and the need to make more investigation around the TL dosimetric properties of ternary alkali halide crystals.

ACKNOWLEDGEMENTS

Whe are thankful to Professor H. Riveros and R. Guerrero (IFUNAM,.México) for growing the crystal. This work is supported by CONACyT, MEXICO under Project 2002-C01-40497.

REFERENCES

1. Mijangos R. R., Cordero-Borboa A., Alvarez E and Cervantes M. *Phys. Letters* A, **282**, 195, (2001).
2. Mijangos R. R., Riveros H., Alvarez E., Atondo M., Vazquez-Polo G. and Gonzalez G. M. *Rad. Eff. and Def. Solids*, **158**, 513, (2003).
3. Melendrez R., Perez-Salas R., Pashchenko L. P., Aceves R., Piters T. M. and Barboza-Flores M., *Appl. Phys Lett.* **68**, 3398 (1996).

4. Rubio O. J., Flores M.C., Murrieta S. H., Hernández A. J., Jaque F., and López F. J. *Phys. Rev. B* **26**, 2199 (1982).
5. Castañeda B., Aceves R., Piters T. M., Barboza-Flores M., Melendrez R., and Perez-Salas R. *Appl. Phys. Lett.* **69**, 1388, (1996).
6. Mijangos R. R., Cordero-Borboa A., Camarillo E., and Castaño V. *Phys. Lett.* A, **245**, 123, (1998).
7. Mijangos R. R., Alvarez E., Perez-Salas R., and Duarte C. *J. Opt. Mat.* **25**, 279, (2004).
8. Perez-Salas, R. Aceves, R. Rodriguez-Mijangos, H. G. Riveros and C. Duarte- *J. Phys. Condens. Matter* **16**, 491 (2004)

Applied Physics (APHYS 2003)
A. Méndez-Vilas (Ed.)

Radon Data From Different Laboratories: An Italian Intercomparison

F. Campi[*], M. Caresana[*], M. Ferrarini[*], L. Garlati[*], M. Palermo[¶], R. Rusconi[¶], L. Salvatori[§], L. Verdi[#]

[*] *Dipartimento di Ingegneria Nucleare, Politecnico di Milano, via Ponzio 34/3, 20133 Milano, ITALY*

[¶] *ARPA Lombardia, Dipartimento Città di Milano, via Juvara 22, 20129 Milano, ITALY*

[§] *FGM Ambiente, via delle Margherite 28, 20070 Dresano, ITALY*

[#] *APPA Bolzano, via Amba Alagi 5,39100 Bolzano, ITALY*

Abstract. The Italian act D. Lgs. 241/2000 requires the evaluation of natural radioactivity, among which there is the contribution of radon gas, in certain workplaces. To put these directives into practice it is necessary to have a complete characterization of the radon detectors, also regarding the related uncertainties estimate. Furthermore the directive establishes to identify the radon prone areas. In order to do this, in Italy the Regional or Provincial Agencies for Environmental Protection (ARPA or APPA) are charged to perform indoor radon measurements. This means examining and comparing the data elaborated in various laboratories.
To evaluate the reliability of different measurement devices and procedures, an intercomparison among the ARPA Lombardia, APPA Bolzano and FGM Ambiente has been carried out with nuclear track detectors. We have exposed two sets of LR115 and two sets of CR-39 at different concentrations of radon, into the radon chamber of Politecnico di Milano, so to simulate concentrations from 25 to 800 $Bq{\cdot}m^{-3}$ for a six months measure.
The results are presented, as the comparison among the results of different laboratories for the same type of dosimeter and the comparison and the comparison between the responses of the two different nuclear track dosemeters.

INTRODUCTION

The directive 96/29/EURATOM, acknowledged from Italy trough the act D. Lgs 241/2000, requires the evaluation of natural radioactivity, among which there is the contribution of radon gas in underground workplaces. Moreover, radon concentration must be measured in sites with certain geological characteristics (radon prone areas). The mapping of national territory to locate these areas is entrusted to Italian Regional or Provincial Agencies for Environmental Protection (ARPA or APPA). Each laboratory has own instruments and characterizes them. This implies that each laboratory chooses the detector and drafts a protocol to carry out the measure. To verify the goodness of its protocol, the laboratories are invited to participate in national and international intercomparison (e.g. NRPB radon intercomparison).

Until some time ago, the dosemeter more used by ARPA and APPA laboratories was the cellulose nitrate dosemeter, know with its trade name LR115 detector, in the configuration designed by Tommasino [1]. To standardize and compare the results obtained in different laboratories, in 1990 the ENEA (the Italian national agency for new technologies, energy and the environmental) issued a protocol for the radon measurement with LR115 detectors [2]. Actually many laboratories also use the detector based on the poly-allyl diglycol carbonate plastic (known with its trade name CR-39) and the electret ionisation chamber, even if LR115 is still used.

Furthermore, at present, in Italy there are not directives about how to do the radon measurements and with which detector, drafted by technical commission. There are only the rules drawn up by regional coordination [3]. A technical commission would must give the indications to recognize the laboratories and institutes authorized for doing the radon measurements. This means that a quality system of radon measurement is absent and the results are often given without the uncertainty.

A laboratory that adopt a new measure method must characterize the detector and draft a internal protocol to do the measurements. If another measure system is used, we will make a set of measure with both the systems, to know the differences and the mistakes of the procedures. The calibration just before, the intercomparison are a useful tools to have a good method of measure.

INTERCOMPARISON

In order to evaluate the reliability of different measurement devices and procedures and to estimate the uncertainty of measure, an intercomparison among the ARPA Lombardia, the APPA Bolzano (Bz) and the FGM Ambiente (FGM) has been carried out with nuclear track detectors. The ARPA and APPA laboratories used the LR115 dosemeter, while the APPA Bolzano and the FGM Ambiente laboratories also adopted the CR-39 detector with Radosys system.

We have drafted a intercomparison protocol to expose two sets of LR115 and two sets of CR-39 at different radon concentrations into the radon chamber of Politecnico di Milano.

Intercomparison Protocol

We want to simulate four radon concentrations from 25 to 800 $Bq{\cdot}m^{-3}$ for a six months measure with nuclear track detectors. Each concentration is chosen to be useful both for the calibration and for the characterization of a new measurement protocol. Care is taken for the concentration around 500 $Bq{\cdot}m^{-3}$, that is the action level required by Italian act D. Lgs 241/2000.

The concentration values chosen are listed on end with the number of detector for each set:

1 – 10 dosemeters, exposed at about $3.60{\cdot}10^8$ $Bq{\cdot}s{\cdot}m^{-3}$, simulate 25 $Bq{\cdot}m^{-3}$ for six months. This is the value of environmental background;

2 – 10 dosemeters, exposed at about $1.98{\cdot}10^9$ $Bq{\cdot}s{\cdot}m^{-3}$, simulate 130 $Bq{\cdot}m^{-3}$ for six months;

3 – 10 dosemeters, exposed at about $7.20 \cdot 10^9$ $Bq \cdot s \cdot m^{-3}$, simulate 450 $Bq \cdot m^{-3}$ for six months. This exposure is useful to evaluate the uncertainty of method at the action level;

4 – 10 dosemeters, exposed at about $1.24 \cdot 10^{10}$ $Bq \cdot s \cdot m^{-3}$, simulate 800 $Bq \cdot m^{-3}$ for six months. For the LR115 detector at this concentration, saturation phenomena are important.

Furthermore, each laboratory also send a set of 10 dosemeter as transit control.

Radon Chamber

Radon exposure was performed at Politecnico of Milano in a radon chamber consisting in a glove box of about 0.8 m^3 of volume (Fig. 1). Its walls are made by polycarbonate and the floor is in stainless steel. A system of fans is mounted into the chamber for the atmosphere homogenisation. The radon source is a vial containing radium salt. The radon is introduced in the radon chamber through a volumetric dosage system (five vials of $5 \cdot 10^{-5}$, $5 \cdot 10^{-4}$ and $15 \cdot 10^{-4}$ m^3), that permits to obtain different concentrations. In the radon chamber, the obtainable concentrations are from 10^3 $Bq \cdot m^{-3}$ to 10^5 $Bq \cdot m^{-3}$. Inside the radon chamber, an Alphaguard PQ2000 works as reference instrument and surveys environmental temperature, pressure and humidity.

The samples are introduced through a double air lock door (SAS). The appropriate conditions of exposure are obtained in the chamber before introducing the detectors. These are hanging by a clip on a grille.

FIGURE 1. Radon chamber of Politecnico di Milano.

Once the detector has been received from participants, we labelled each dosemeter and divided the whole group in five sets (one for each exposure and one for transit control). Each set was sealed in a double radon proof bag and stored in a low-radon area, until the exposure. At the end of exposure, the detector are placed in a low-radon laboratory to allow radon to diffuse out. After two days, the detectors were double-bagged and sent to the laboratories.

RESULTS

The dosemeters sets (two for LR115 and two for CR-39) were exposed at the exposure levels summarised in Table 1.

TABLE 1. Exposure levels used for the intercomparison.

N. of exposure	Duration (s)	Radon exposure 10^9 (Bq s m^{-3})	Uncertainty[(1)] 10^9 (Bq s m^{-3})
1	157968	0.406	0.013
2	114300	1.809	0.056
3	403488	6.835	0.206

(1) Estimated uncertainty at 68% confidence level

Among the laboratories which took part, the ARPA Lombardia used the exposures as calibration to develop the uncertainty protocol [4] according to the ISO Guide [5]. Therefore a comparison can be made among three dosemeter sets: two CR-39 sets and one LR115 detectors. In Table 2 we report the mean data of each exposure with the standard deviation over the data and the percentage value of difference between the reference and the mean value. In Fig. 2 we represent the ratio between the mean and the reference value with the uncertainty calculated considering the uncertainty of both values. The CR-39 detectors overestimate the exposure value, while the LR115 detectors underestimate it. The CR-39 dosemeters wonder from reference value more for low track density; instead the LR115 detectors wonder more for higher value.

TABLE 2. Exposure results obtained from three laboratories in the intercomparison.

Dosemeter	Exp. 1 10^9 (Bq s m^{-3})	% SD	% Diff	Exp. 2 10^9 (Bq s m^{-3})	% SD	% Diff	Exp. 3 10^9 (Bq s m^{-3})	% SD	% Diff	Exp. 4 10^9 (Bq s m^{-3})	% SD	% Diff
LR115 Bz	0.414	34	-2	1.649	8	9	6.060	4	11	10.494	9	12
CR-39 Bz	0.512	20	-26	2.253	8	-25	7.720	8	-13	13.008	2	-9
CR-39 FGM	0.506	16	-24	2.003	10	-11	7.587	10	-11	12.873	3	-8

% SD is the percentage standard deviation.
% Diff is the percentage difference between the reference value and the mean reported.

The reasons of deviations from reference values should be carefully analysed; for example in the LR115 case, on basis of our experience [4], we guess that deviations from reference values for high track density is due to saturation effects of spark counter used to count tracks. However, the analysis of differences among measure protocols adopted by different laboratories was not the aim of this work.

Instead, the responses of the two CR-39 detector sets are similar.

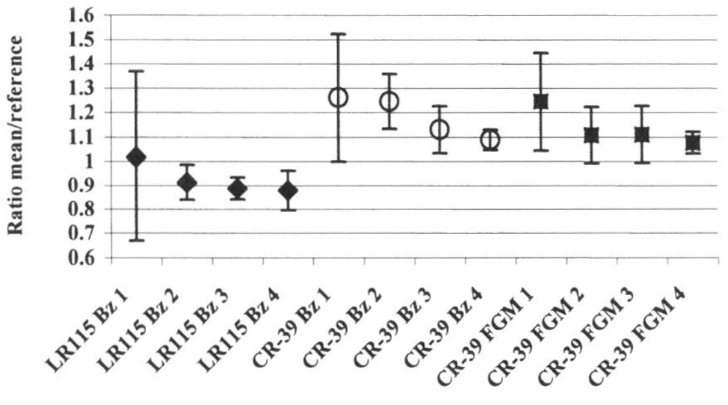

FIGURE 2. The ratio between the mean value and the reference value of Italian intercomparison.

The data obtained in this intercomparison can be used to improve the reading method.

Regarding the ARPA Lombardia, a direct check of the intercomparison is not possible. A check of calibration and method quality is given by the analysis of 2003 NRPB intercomparison results. The APPA Bolzano laboratory also took part at the NRPB intercomparison in 2003. Therefore, the comparison between the results of the laboratories will be made.

Both laboratories used LR115 track detectors, but different methods of correction for spark counter saturation effects were used.

The analysis was carried out in two steps: first we considered only the uncertainties given by laboratory, then we analysed the data with their own standard deviation.

The ARPA Lombardia gave the results with uncertainty obtained considering all uncertainty causes (spark counter, etching time, film thickness, track saturation); the APPA Bolzano supplied only the exposure values of single dosemeters.

In Table 3 and in the Fig. 3 we report the NRPB reference value and the mean values with percentage uncertainty.

Instead in Table 4 and in Fig. 4 we report again the mean values, but the uncertainty is given as standard deviation, which takes care of the data dispersion.

Furthermore the value of the standard deviation and the difference between the reference and the mean value are the parameters that the NRPB laboratory use to define the category the laboratories, which take part at the intercomparison, belong to.

TABLE 3. Intercomparison NRPB: reference value and results obtained with LR115 detector.

N. exposure	Reference value (NRPB) 10^9 (Bq s m^{-3})	Un. %	ARPA Lombardia mean value 10^9 (Bq s m^{-3})	Un.[(1)] %	APPA Bolzano mean value 10^9 (Bq s m^{-3})	Un.[(2)] %
1	0.421	3	0.443	6.27	0.408	20
2	1.058	3	1.126	4.46	1.030	20
3	8.244	3	8.135	3.65	7.380	20

[(1)] Uncertainty calculated from ARPA Lombardia protocol in reference 4.

[(2)] Uncertainty generally assumed from ARPA laboratories.

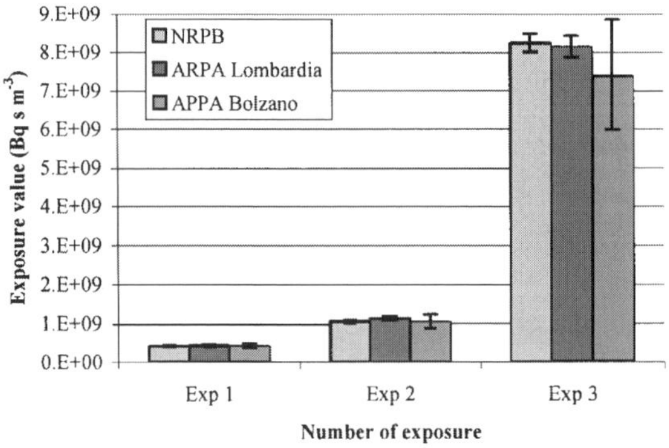

FIGURE 3. 2003 NRPB reference value and the mean values obtained by ARPA Lombardia and APPA Bolzano.

The Figure 5 represent the ratio between the mean and the reference value. We note that the ratios, obtained from the mean values of ARPA Lombardia, are around the unit with a low percentage difference. This is a proof of good quality of calibration into the radon chamber of Politecnico. Concerning the other set of dosemeters, there is a slight underestimate at highest exposure value of 2003 NRPB intercomparison.

TABLE 4. Intercomparison NRPB: standard deviation and difference of results.

	Exp. 1 10^9 (Bq s m^{-3})	**% SD**	**% Diff**	**Exp. 2 10^9 (Bq s m^{-3})**	**% SD**	**% Diff**	**Exp. 3 10^9 (Bq s m^{-3})**	**% SD**	**% Diff**
ARPA Lombardia	0.443	14.8	-5.2	1.126	15.2	-6.3	8.135	12.8	1.3
APPA Bolzano	0.408	20.3	3.1	1.030	12.7	2.7	7.380	12.0	10.5

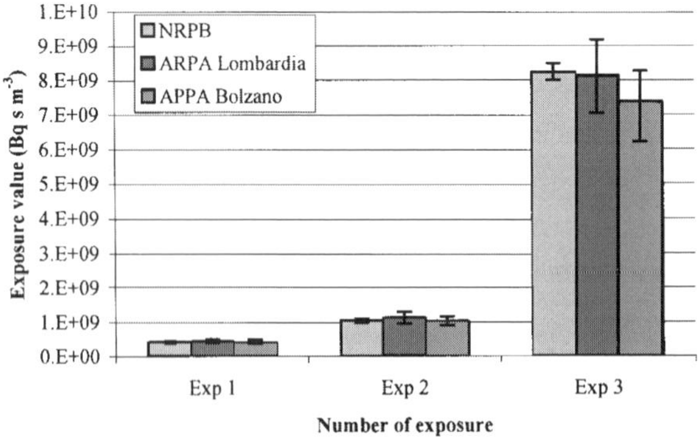

FIGURE 4. 2003 NRPB reference value and the mean values obtained by ARPA Lombardia and APPA Bolzano with standard deviation uncertainty.

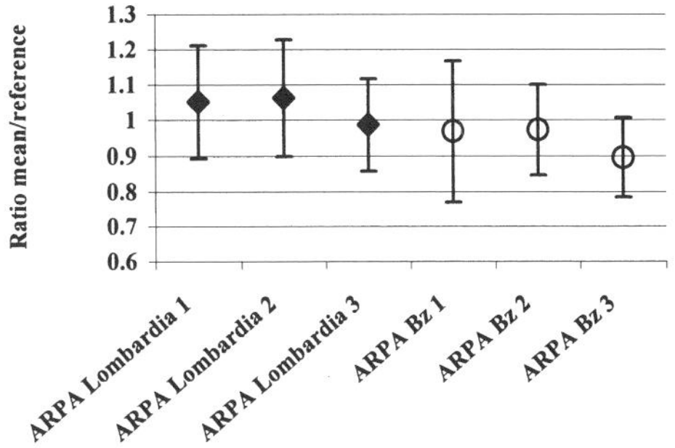

FIGURE 5. The ratio between the mean value and the reference value of 2003 NRPB intercomparison.

The data of both dosemeter sets are rather dispersed around the mean value (SD>10%). However, the standard deviation can not be used as uncertainty because it does not consider the uncertainty of single measurement.

Moreover, when we make the radon measurement we use only one or two dosemeters, not ten as in calibration. Then the study of single measurements uncertainty is desirable so to have measurements of good quality owing to the radioprotection law.

CONCLUSIONS

The radioprotection national law imposes the evaluation of radon concentration in workplaces. To have reliable measurements, one needs to do periodic calibrations or intercomparisons from the laboratories that make these measurements. This is useful to verify the quality of one's own work method. Furthermore it is important to study the measurement uncertainties associated with the method. A priori, from the results of Italian intercomparison, it is impossible to consider one detector better the other. We can say that if a system is well characterised, we obtain good results (e.g. LR115 detectors in 2003 NRPB intercomparison).

REFERENCES

1. D. Azimi-Garakani, B. Flores, S. Piermattei, A.F. Susanna, J. L. Seidel, L. Tommasino, G. Torri, *Rad. Prot. Dos.* **24**, 267-272 (1988).
2. G. Torri, *Metodo delle tracce nucleari su film sottili (LR115): misura della concentrazione di radon 22,* ANPA technical report (1990).
3. Conferenza dei presidenti delle regioni e delle province autonome di Trento e Bolzano, *Linee guida per le misure di concentrazione di Radon in aria nei luoghi di lavoro sotterranei* (2003).
4. F. Campi, M. Caresana, L. Garlati, M. Ferrarini, M. Palermo, R. Rusconi, *Uncertainties evaluation for a LR115 nuclear track dosimeter for radon detection*, Aphys2003, Badajoz (2003).
5. ISO, *Guide to the expression of uncertainty in measurement,* ISBN 92 67 10188 9 (1995).

Applied Physics (APHYS 2003)
A. Méndez-Vilas (Ed.)

Peculiarities of cesium sorption-desorption on bottom sediments

G. Lujanienė*, B. V. Šilobritienė* and K. Jokšas†

*Institute of Physics, Savanoriu ave 231, Vilnius, LT-02300, Lithuania
†Institute of Geology and Geography, T. Sevcenkos 13, Vilnius, LT-2600, Lithuania

Abstract. The Curonian Lagoon has no analogues in the world and influences the radionuclide migration processes in the Lithuanian coastal zone of the Baltic Sea due to special hydro meteorological, geochemical and geographical conditions. The activity concentration of cesium in the Curonian Lagoon is lower by two orders of magnitude as compared with that in the Baltic Sea water. The sorption-desorption processes of cesium in disturbed bottom sediments - fresh and sea water systems - were investigated in long-term experiments (up to 375 days) in order to better understand cesium behavior during seawater flooding events.

INTRODUCTION

Cesium is one of the most important anthropogenic radionuclides, which were introduced in the environment through nuclear weapon testing and nuclear accidents, its inventories in radioactive waste are significant, in addition, ^{135}Cs is extremely long-lived. Its harmful effects on animate nature are obvious from its properties: unlimited solubility, potential mobility and high bioavailability. The sorption-desorption processes of radiocaesium on various sorbents have been the subjects of many recent studies [1]. The numerous publications concerning the binding of Cs by clays especially illite, reflect its importance as radionuclide relevant to safety assessment and on the other hand the searching for low–cost barriers for isolation of radioactive wastes. These studies are mostly dealing with a description of the cesium sorption on homogenous particles and there are difficulties in extrapolation of laboratory sorption data to field conditions. However, the various soil components never contribute independently to the soil sorption ability because the knowledge of its component is insufficient to predict sorption behavior of radionuclides in a complex heterogeneous system as soil or sediments [2]. It is now generally recognized that the sorption of ^{137}Cs by soil and sediments is mainly determined by specific sorption on illite clay minerals. Actually, a trace amount of illite can effectively immobilize cesium even in the presence of a high content of organic substances [3]. However, organic matter and iron oxides have indirect effect on the ^{137}Cs affinity to the clay minerals. The adsorption of macromolecules on clay minerals influences their affinity for cesium, and clay-humic substance complexes adsorb less cesium than uncoated clay minerals. The coatings of iron oxides block the cesium uptake sites as well. The organic matter and iron oxide coatings on soil and sediment particles can serve as intermediate phases when the exchangeable ^{137}Cs permeates towards the mineral core of clay particles.

Moreover, desorption of exchangeable ^{137}Cs from clay minerals can be inhibited by such coatings. As a result, the value of exchangeable ^{137}Cs measured during the extraction procedures can be reduced. In addition, organic matter enhances a "fixation" of Cs+ ions in the clay lattice. The mechanisms of these interactions are poorly understood.

The objective of this study is to focus on the sorption and desorption behavior of Cs in the complex heterogeneous system of bottom sediments.

EXPERIMENTAL

A sample of bottom sediments collected in the Curonian Lagoon was used for sorption experiments. The selective properties of this sample are summarized in Table 1. Total carbon (TC) and TOC were determined using the LECO CS-125 analyzer. After drying two subsamples of approximately 30 mg for each bulk sample, TC was directly measured in the first subsample, whereas TOC was measured in the second subsample following treatment with 0.25-N HCl to remove inorganic carbon. Inorganic carbon (IC) was calculated and converted to weight percent $CaCO_3$ by using equations 1 and 2:

$$IC(wt\%) = TC(wt\%) - TOC(wt\%), \tag{1}$$

$$CaCO_3(wt\%) = IC(wt\%) \cdot 8.333. \tag{2}$$

is where 8.333 is the stoichiometric calculation factor for $CaCO_3$. The precision of the TC and TOC measurements is within ±0.25 %.

The grain size distribution was determined by the gravimetric pipette method. Stable Cs concentration was determined using ICP–MS, and clay minerals were identified by X–ray diffraction. The sample was passed through a 1–mm mesh sieve and homogenised. Then it was divided into equal portions, placed into the plastic vessels and 40 ml of filtered sea water of 7.0 salinity labelled with ^{134}Cs was mixed with 4 g of sediments and sealed by Parafilm TM. The total concentration of Cs in solution was 0.04 ppb. The contents of vessels were mixed 5 min/day using a magnetic stirrer. Samples were stored in the dark at 4°C at different contact times between solution and sediments. Another sorption experiment was performed in the dark, and samples were stored in the room where temperature in wintertime was about 5–7°C and in summertime about 20°C. The experiment was started in November, and the amount of ^{134}Cs activity added to the samples was reduced by factor of two as compared to the first experiment. The solids were separated by centrifugation at 4000 rpm for 15 min. The sample was carefully mixed and split into two equal parts. One of those was used for sequential extraction analyses, and the other one for the desorption experiment.

TABLE 1. Physical-chemical parameters of the sediments.

Sampling location	IC, %	$CaCO_3$, %	TC, %	Mineral, %	TOC, %	Sand, %	Silt, %	Clay, %
Curonian Lagoon 7	3.5	29.5	9.0	58. 9	5.5	15	71	14

The modified Tessier [1] sequential extraction method was used to study association of Cs in sediments. The following extracting agents were used and subsequent characterisations are:

1M $MgCl_2$, pH 7.0, (exchangeable);
1M NH_4Cl, pH 7.0, (exchangeable);
1M $NH_4C_2H_3O_2$ pH 5, (carbonate bound);
0.1M $Na_2P_2O_7$ (organically bound);
0.04 M $NH_2OH{\cdot}HCl$, (oxide bound);
Residue was measured directly by gamma spectrometry.

Fractions were separated using centrifugation and filtration through the membrane filter 0.2 μm.

Desorption experiments were performed using natural river water. Samples were poured with 20 millilitres solution mixed 5 min/day by a magnetic stirrer left to stand overnight, mixed again 5 min and centrifuged at 4000 rpm. The procedure was repeated during 10 days.

^{137}Cs and ^{134}Cs activities were measured with an intrinsic germanium detector (resolution 1.9 keV/1.33 Mev and efficiency 42 %). An efficiency calibration of the system was performed using calibration sources of different densities and geometry that were close to real measured samples. For the source preparation, the Reference solution from AMERSHAM, UK. Solution No. FD 998 was used. The measurement accuracy was tested in intercomparison runs, organized by the Riso National Laboratory, Denmark. The precision of ^{134}Cs measurements by gamma spectrometry was <3 % at $\pm 1\ \sigma$

RESULTS AND DISCUSSION

Data from two sorption experiments of radiocaesium from labelled ^{134}Cs sea water are presented in Figs. 1 and 2. It can be seen that the results obtained from two experiments with different concentration of cesium differ insignificantly. Results of fraction distributions of ^{134}Cs in sediments after different contact time indicated that about 70 % of ^{134}Cs tracer was sorbed during the first three days; however, the decrease in exchangeable fraction was determined only after 11 days of contact time (Fig. 1). A similar result was obtained from another experiment in which the activity of sea water spiked with ^{134}Cs was lower by factor of two (Fig. 2). The decrease in activity of about 15 % in contacting solution was observed during the whole experiment. The exchangeable fraction extracted with $MgCl_2$ solution reveals insignificant variations. The most interesting picture of the sorption process was found for caesium distribution between exchangeable (NH_4Cl), carbonate and residual fractions. The observed distribution in exchangeable fraction to a great extent deals with sorption process on illite mineral when after ion exchange process Cs ions penetrate to interlayer sites exposed along the edges of the clay platelets – “frayed edge sites” [4.5]. The Cs ions then lose their hydration waters and the ions move into the hexagonal indentations on the face of the layer. This repairs a local charge defect; waters due to osmosis migrate outside causing the junction of adjacent layers and isolate Cs from the surrounding solution by fixating it in the clay lattice. This process is responsible for the rather high sorption rate determined at the first stage of the sorption experiment. The decrease in the cesium amount in the exchangeable fraction possibly corresponds to the decrease in the number of frayed edge sites available for Cs sorption.

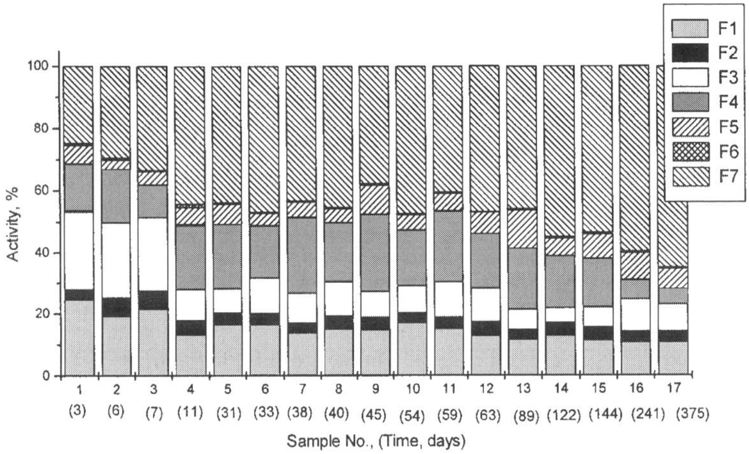

FIGURE 1. Fraction distribution (1 experiment), F1- contact solution, F2- exchangeable ($MgCl_2$), F3- exchangeable (NH_4Cl), F4 - carbonates, F5 - organic, F6 - oxides, F7-residue.

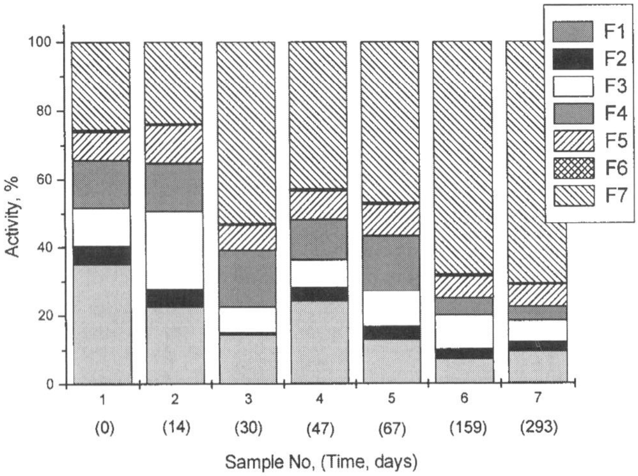

FIGURE 2. Fraction distribution (2 experiment), F1- contact solution, F2- exchangeable ($MgCl_2$), F3- exchangeable (NH_4Cl), F4 - carbonates, F5 - organic, F6 - oxides, F7-residue.

The decrease in the sorption rate accompanied by an increase in association of ^{134}Cs with carbonate fraction can be attributed to the effect of coatings that is normally observed in natural heterogeneous sediments. The particles of different sizes are usually present in the sediments, the mineralogy of particles is different, moreover their surfaces could be covered fully and/or partly with the coating of different origin, e.g., iron compounds and organic substances. There are some publications concerning the role of surface coatings in which its effect was studied by removing iron or organic substances from sediment and later used for sorption experiments [6]. These studies confirmed that surface coatings had a marked effect on the ability of clay to sorb Cs.

In the sediment sample used for the sorption experiment a high content of carbonate was determined (Table 1). This sample was collected in the Curonian Lagoon where the precipitation of $CaCO_3$ was observed and attributed to the high concentrations of Ca in the Nemunas river water. The deposition of $CaCO_3$ on sediment particles and the formation of carbonate counting are possible in this case. Thus the increase in percentage of caesium associated with the carbonate fraction can be interpreted as the effect of carbonate coatings on Cs^+ sorption in the carbonate-rich sediments (Fig. 1). It seems that carbonate coatings inhibit caesium uptake by clay minerals. The caesium ions need time to diffuse through carbonate coating to the clay particles where they can be sorbed on the available sorption sites. The comparative analyses of ^{134}Cs fraction distribution after 241 and 375 days of the sorption experiment with that of the ^{137}Cs distribution, determined in the same sediment sample before the sorption experiment, indicated that after 375 days of sorption the equilibrium was not fully reached, but distribution was found to be close to the equilibrium (Fig.3). In addition, the decrease in association of ^{134}Cs with carbonate fraction was found on the 241st and the 375th days of sorption that indicated the reduction of the coatings effect when the equilibrium was reached.

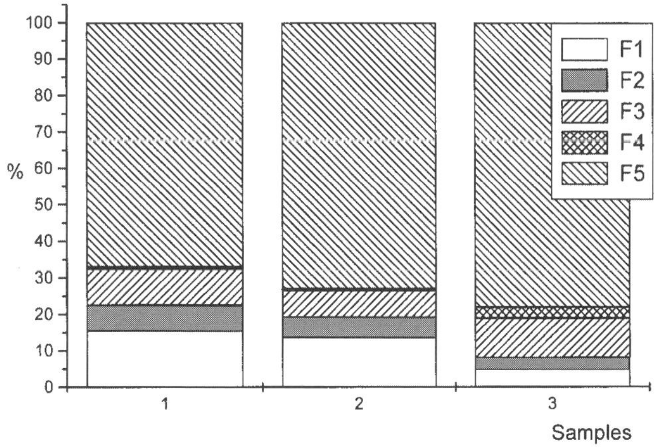

FIGURE 3. Comparison of fraction distribution in samples from 1 and 2 experiments and original sample (3), F1- exchangeable, F2 - carbonates, F3 - organic, F4 - oxides, F5-residue.

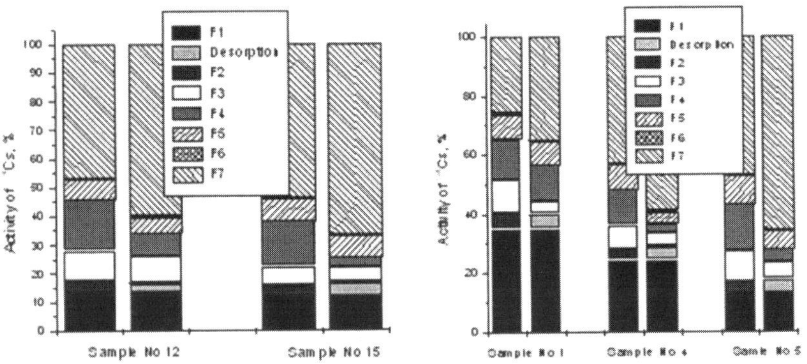

FIGURE 4. Comparison of fraction distribution before and after desorption (1 experiment in the top, 2 experiment in the bottom), F1- contact solution, F2- exchangeable ($MgCl_2$), F3- exchangeable (NH_4Cl), F4 - carbonates, F5 - organic, F6 - oxides, F7-residue.

Results of desorption experiments presented in Fig.4 reveal the insignificant amount of Cs desorbed from the regular exchange sites [7], from which Mg^{2+} can easily displace Cs^+. However, the higher content of caesium tracer held in the residual fraction after desorption experiments clearly demonstrates the redistribution of ^{134}Cs between geochemical fractions during the desorption procedure. A possible explanation is the quick exchange reactions and another possibly deals with the coatings which can inhibit desorption. The redistribution found in samples evidenced the caesium transfer to the frayed edge sites on the layered clay minerals with its further fixation in crystal lattice. Thus the short-term desorption experiments indicated an insignificant release of Cs^+ from regular exchange complex. The remobilization of "fixed" caesium was not observed, on the contrary the redistribution of caesium towards its fixation in clay minerals was observed and on a time scale its fixation kinetics was comparable with the sorption experiment. This study indicated that flooding events of seawater contaminated with caesium can cause the significant accumulation of caesium in the bottom sediments of the Curonian Lagoon.

ACKNOWLEDGMENTS

This work was partly performed under the auspices of IAEA under project LIT/7/002.

REFERENCES

1. Comands, N.T., Geochim. Cosmochim. Acta 56, 1157–1164 (1992).
2. Ohnuki, T., Radiochim.Acta 65, 70–76 (1994).
3. Dumat, C and Staunton, S., Journal of Environmental Radioactivity 46, 187–195 (1999).

4. Tessier, A., Camphell, P.G.C. and Bisson M., Anal. Chem. 51, 844-851 (1979).
5. Evans, D.W. Alberts, J.J. and Clark III, R.A., Geochim Cosmochim. Acta 47, 1041–1049 (1987).
6. Dumat, C, Cheshire, M.V., Fraser, A., Shand, C., Staunton, S., *European Journal of Soil Science* **48,** 675–683 (1997).
7. Walters, J., Vidal, M., Elsen, A. and Cremers, A., *Applied Geochemistry* **11**, 595–599 (1996).

Applied Physics (APHYS 2003)
A. Méndez-Vilas (Ed.)

Influence of particle size distribution on the behavior of ^{137}Cs in the Baltic Sea

G. Lujanienė*, B. V. Šilobritienė* and K. Jokšas[†]

*Institute of Physics, Savanoriu ave 231, Vilnius, LT-02300, Lithuania
[†]Institute of Geology and Geography, T. Sevcenkos 13, Vilnius, LT-2600, Lithuania

Abstract. In order to gain a better understanding of the fate and transport of radioactive contaminants in the marine environment, it is necessary to obtain information on the physical and chemical parameters of radionuclide carriers. The associations of radionuclides may change dramatically in the zone of interaction between fresh and saline waters due to sorption-desorption processes. In addition, the particle separation occurring during settling can affect the properties of suspended matter (size distribution, density and chemical composition), and as a result affects the activity of ^{137}Cs associated with particles. Activity concentrations and speciation of ^{137}Cs in the Baltic Sea suspended matter and bottom sediments of different size particles were determined. The modified Tessier method was used to study cesium associations with geochemical phases of bottom sediments.

INTRODUCTION

Radioactive contamination of the Baltic Sea was caused by three main factors such as global fallout, discharges from the reprocessing plants and the fallout after the Chernobyl accident in April 1986. After the Chernobyl accident, deposition as well as partly river and overland runoff were responsible for an increase in ^{137}Cs concentrations in the Baltic Sea. Although concentration of the mentioned nuclide has about half decreased since 1986, but remains the highest in comparison with the worldwide concentrations. At present, the average activity concentration of ^{137}Cs in surface water of the Baltic Sea was estimated to be about 60 Bq/m^3, while the worldwide average concentration due to global fallout is about 2 Bq/m^3. The Baltic Sea is semi-enclosed, largest brackish water and unique shallow sea in which self-cleaning processes are slow and dissolved substances remain there for a long time. Measurements of total activity concentrations of ^{137}Cs as well as other technogenic and natural radionuclides in the Baltic Sea water have been carried out in many riparian countries and in international cooperation with the group MORS and under umbrella of the HELCOM. However, there is a lack of information on the speciation of radionuclides in the Baltic Sea, but just this parameter in most cases conditions bioavailability, migration and self-cleaning processes.

At present there is no doubt that migration regularities of radionuclides, processes of self-cleaning and redistribution in the environment depend on their specific chemical forms or type of binding rather than on the total element or nuclide content. Due to the very low concentration of radionuclides typically found, there are only some examples of direct determination of their speciation in the chemist's sense in the environmental samples (oxidation state, coordination numbers, and charge). The determination of speciation or binding forms of radionuclides in the environment is difficult and often hardly possible. Therefore, the determination of physicochemical forms or fractions in practice, using sequential extraction methods, is a reasonable compromise to evaluate the associations of radionuclides in the environmental samples (e.g. with carbonate minerals, Fe/Mn oxides and organic substances). Although the results obtained by a sequential extraction procedure are operationally defined and cannot be used as input data for thermodynamic equilibrium models, they provide useful information about the behavior of radionuclides. Knowledge about an association of radionuclides with geochemical phases is important from the point of view of radiation protection as the binding of radionuclides and the stability of geochemical phases provide data on bioavailability and migration ability of the radionuclides. At present, the only available tool to study radionuclide associations at low-level concentrations is sequential extraction methods.

In some papers the criticism is given to these methods, which deals with incomplete selectivity of reagents used to dissolve one particular phase without the additional attack of other geochemical phases and re-adsorption of analytes after release. The results obtained by Kheboian and Bauer from synthetic models and their interpretation were intensively discussed since the publication appeared in 1987. Re-adsorption processes of some elements were observed for iron and organics rich sediments, non-selective extraction was found during extraction of elements from anoxic sediments [1]. The obvious advantages were achieved in analyses of speciation of heavy metals [2]. The most widely used sequential procedure proposed by Tessier et al. [3] was modified in order to avoid these problems by changing the extraction time, the extractant-sample ratio, the reagent concentration, the extraction temperature.

The aim of this study was to determine speciation of ^{137}Cs suspended particles and bottom sediments in order to better understand the behavior, redistribution and sink of the mentioned nuclide in the Lithuanian coastal area of the Baltic Sea.

EXPERIMENTAL

Water and bottom sediment samples were collected during the expedition in the Lithuanian economical zone of the Baltic Sea, the Curonian Lagoon and near the seashore of the Baltic Sea and the Curonian Lagoon in 1999–2001. The sampling locations are presented in Fig. 1, and ranges of activities are presented in Table 1. Surface water samples were collected with the 20 l vessel with a 2 m long handle. The amount of sampled water depending on the lower detection limit of the radionuclide varied from 40 to 100 liters. After sampling, the water samples in the 20 and 50 l polyethylene bottles were transported to the Preila background station laboratory and were filtered as quickly as possible to separate suspended matter. The separation of the suspended matter was achieved using membrane filtration. Water was filtered through Nuclepore (Dubna) membrane 0.2 μm and Filtrak 388 prefilters using the filtration equipment (Millipore) consisting of the Dispensing Pressure Vessel of 10 l and a Stainless Steel Filter Holder of 293 mm.

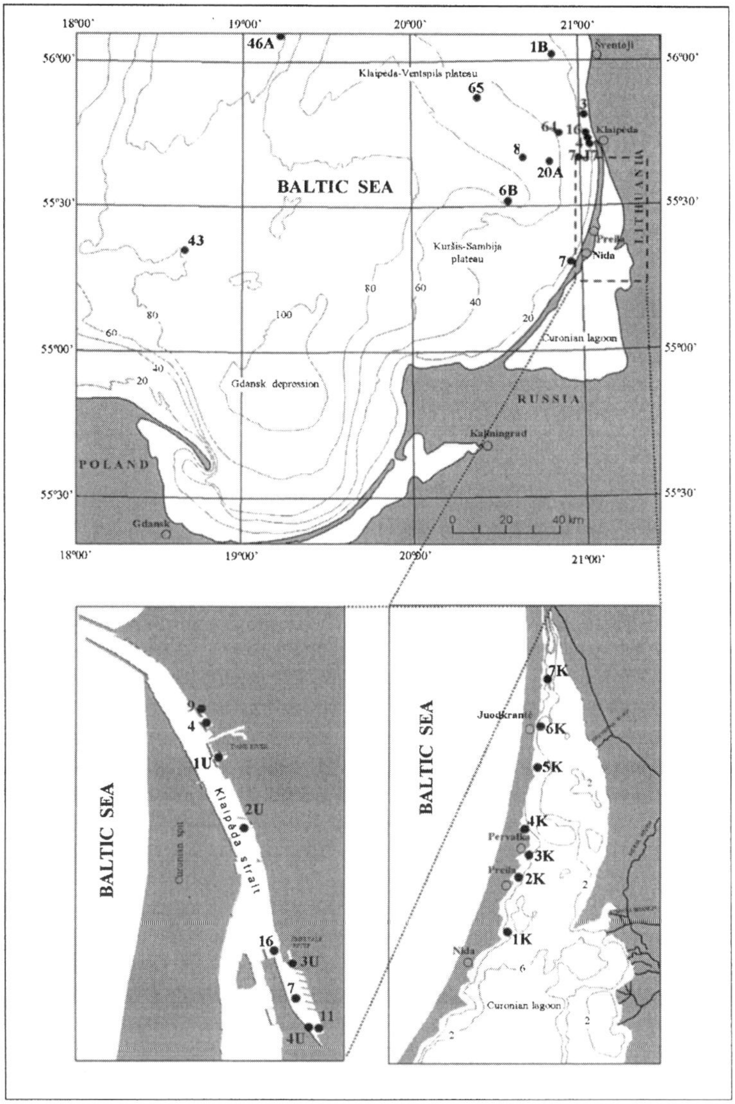

FIGURE 1. Sampling locations in the Baltic Sea and Curonian Lagoon

TABLE 1. Ranges of total ^{137}Cs activities in seawater (Bq/m^3) in sediments (Bq/kg) in the Curonian Lagoon and the Baltic Sea along Lithuanian coast.

The Baltic Sea area	^{137}Cs, Bq/m^3	Sites, n	^{137}Cs, Bq/kg	Sites, n
Open sea water	14 - 130	12	4 - 450	19
Near shore water	50 - 120	12		
Klaipeda port	51 - 52	2	8 - 140	13
Curonian Lagoon	1 - 87	5	0.4 - 210	15

In addition, the suspended particle samples were collected in situ by filtering a large volume of water (~400-600 L) through two consecutive 1 μm polypropylene Sediment Filter Cartridges (US Filter Plymouth Products), particles were filtered from solution using a pumping system. Water was filtered from a depth of about 0.2-0.5 m below the surface and pumping times were 1-1.5 h. The filters and sediment filter cartridges were dried in a desiccator for 2 days before they were weighed. After sampling cartridges containing suspended particles were dried and weighed again using the same procedure to determine the amount of suspended solids. Then the filters and cartridges containing suspended solids were ashed at 380°C.

The specific activity of ^{137}Cs in suspended matter in water samples of 50–100 l volume was close to detection limits of gamma spectrometric methods. For this reason, cesium was separated radiochemically. The samples were digested using HF/HNO_3 and HCl, then cesium was precipitated as $Cs_3Sb_2I_9$ and measured using the proportional Emberline FHT 770 T MULTI–LOW–LEVEL–COUNTER. Chemical yield was determined gravimetrically.

After filtration water samples were acidified. ^{137}Cs was precipitated using the Fe ferrocyanide method and the residue was filtered, dried and placed into the standard plastic containers of 50 or 150 ml volume for the determination of ^{137}Cs with the HPGe detector. For chemical yield determination, ^{134}Cs tracer was used.

Bottom sediment samples were collected during different sampling companies in 1999-2001. The bottom sediments in the Baltic Sea were collected using a Van Veen grab sampler available on the ship "Vejas". The Bottom Sampler acc. Ekman-Birge with an effective grasping area of 225 cm^2 and weight of 3.5 kg was used for the bottom sediment sampling in the Curonian Lagoon. Bottom sediments were sectioned on board the ship and the upper 0-2 cm layer of sediments was selected for radionuclide speciation analyses. Sediments were mixed, placed into the 0.5 l polyethylene containers and transported to the laboratory.

Bottom sediment samples were usually stored frozen. Samples were dried in the air and/or using ovens (100°C). Samples selected for the speciation determination were stored in a fridge as short as possible in the nitrogen atmosphere. Fractionation of sediments was also performed in the nitrogen atmosphere as quickly as possible. Samples were homogenized before measurements. Two standard geometries of 50 and 150 ml volumes were used.

^{137}Cs activities were measured with an intrinsic germanium detector (resolution 1.9 keV/1.33 Mev and efficiency 42 %). Measuring times varied according to sample activities.

An efficiency calibration of the system was performed using calibration sources of different densities and geometry that were close to measured samples. For the source preparation, the Reference solution from AMERSHAM, UK. Solution No. FD 998 was used. Measurement accuracy was tested in intercomparison runs, organized by the Riso National Laboratory, Denmark (the mineral, sea and lake water matrices).

Precision of ^{137}Cs measurements by gamma spectrometry was <10 %. Activity concentrations derived from beta counting had uncertainties from less than 10 to 20 %, depending on the activity of the measured source.

For the determination of physical and chemical association of radionuclides, bottom sediment particle size separations were performed using wet sieving and column

TABLE 2. Physical-chemical parameters of sediments in the Curonian Lagoon and the Baltic Sea.

Station	Location	Clay, %	Loss on ignition, %	CO_2, %	$CaCO_3$, %	CaO, %	Mineral, %	Organic, %	^{137}Cs activity concentration, Bq/kg
16	Baltic Sea 55°45' 21°02'	2	1.93	0.79	1.79	1	97.07	1.14	76
3	Baltic Sea	28	19.46	9.27	21.06	11.79	68.75	10.19	259
4	Baltic Sea 55°44' 21°03'	2.5	2.3	1.66	3.77	2.11	95.59	0.64	10
6b	Baltic Sea 55°31' 20°34'	3	1.96	0.28	0.65	0.36	97.68	1.68	39
65	Baltic Sea 55°53' 20°20'	6	3.25	0.75	1.71	0.96	95.79	2.5	71
4	Curonian Lagoon	12	15.32	6.93	15.75	8.82	75.86	8.39	160
7	Curonian Lagoon	14	24.6	12.98	29.49	16.51	58.89	11.62	136
9	Curonian Lagoon	22	18.29	8.68	19.73	11.05	70.66	9.61	207
11	Curonian Lagoon	16	26.16	13.16	29.9	16.74	57.1	13	165

settling techniques. The fractions of > 50 μm, 50 – 4 μm and <4 μm were separated and characterized as sand, silt and clay particles. In addition, a limited number of samples were separated into four fractions of > 50 μm, 50 – 4 μm, 4 – 1 μm and 1 – 0.2 μm. The separation of 1 – 0.2 μm fraction was achieved using membrane filtration.

The selective properties of sediments selected for speciation analyses are presented in Table 2. Total carbon (TC) and TOC were determined using a LECO CS-125 analyzer. After drying two subsamples of approximately 30 mg for each bulk sample, TC was directly measured in the first subsample, whereas TOC was measured in the second subsample following treatment with 0.25-N HCl to remove inorganic carbon. Inorganic carbon (IC) was calculated and converted to weight percent $CaCO_3$ by using equations 1 and 2:

$$IC(wt\%) = TC(wt\%) - TOC(wt\%), \tag{1}$$

$$CaCO_3(wt\%) = IC(wt\%) \cdot 8.333. \tag{2}$$

where 8.333 is the stoichiometric calculation factor for $CaCO_3$. The precision of the TC and TOC measurements is within ±0.25%.

Stable Cs concentration in seawater was determined using ICP–MS, grain size distribution - by the gravimetric pipette method and clay minerals were identified by X–ray diffraction.

In the investigation of radionuclide chemical forms, the modified Tessier sequential extraction method was used. The following extracting agents were used and subsequent characterizations are:

F1. H_2O, (water-soluble);

F2. 1M $MgCl_2$, pH 7.0, (exchangeable);
F3. 1M NH_4Cl, pH 7.0, (exchangeable);
F4. 1M $NH_4C_2H_3O_2$ pH 5 CH_3COOH, (carbonate bound);
F5. 0.04 M $NH_2OH \cdot HCl$ in 25% CH_3COOH, (oxide bound);
F6. 30% H_2O_2 at pH 2 (HNO_3), then 3.2 M $NH_4C_2H_3O_2$ in 20% HNO_3 (organically bound);
F7. 7M HNO_3, (acid soluble);
F8. 40% HF, HNO_3 (residual).

RESULT AND DISCUSSION

Ranges of activity concentrations of ^{137}Cs in seawater, fresh water and specific activities of ^{137}Cs in bottom sediments are presented in Table 1. It can be seen that activity concentrations of cesium in seawater are by about two orders of magnitude higher as compared to those of the Curonian Lagoon. It should be noted that sampling was mainly performed in the zone of input of fresh waters from the Curonian Lagoon and the correlation between the ^{137}Cs and salinity was observed ($r = 0.76$). Data of monitoring performed at stations 4, 7, 46a and 65 in 1997-2002 by the Ministry of Environment (Fig. 2) also indicated the relation of activity concentration of ^{137}Cs with salinity ($r=0.73$). Specific activities of ^{137}Cs in bottom sediments of the Baltic Sea and Curonian Lagoon ranged from 4 to 450 and from 0.4 to 596 Bq/kg, respectively. Studies performed on the particle size distribution and activity concentrations in bottom sediments collected in the Baltic Sea and the Curonian Lagoon are presented in Fig. 3.

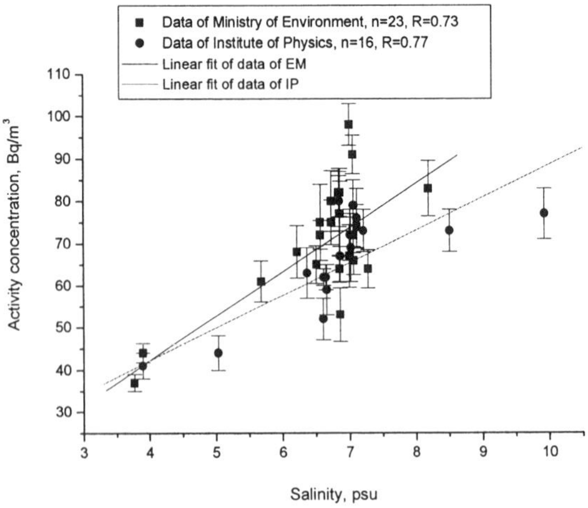

FIGURE 2. Activity concentration of sea water versus water salinity in the Baltic Sea, data of expedition October 1999 and data of 1997-2002, from Stations No. 4, 7, 46a and 65, presented by Ministry of Environment of the Republic of Lithuania and data from Institute of Physics.

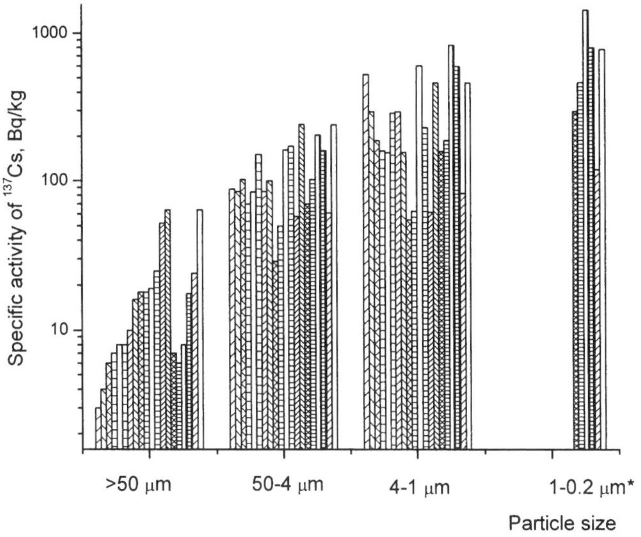

FIGURE 3. Specific activities of ^{137}Cs in bottom sediments samples collected in the Curonian Lagoon and the Baltic Sea.
*In limited number samples particles of 1-0.2 μm size were separated by filtering using membrane filtration.

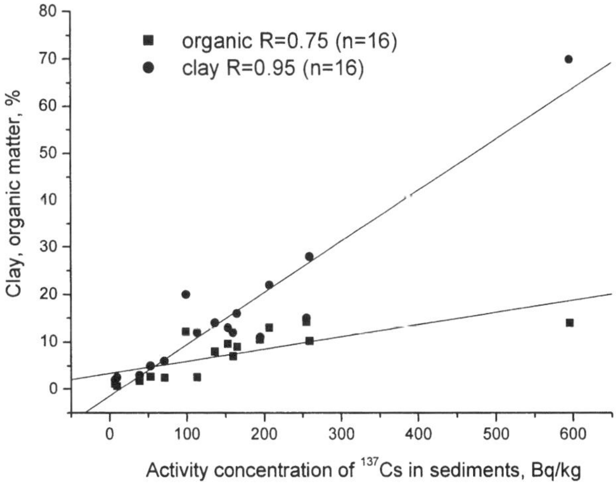

FIGURE 4. Organic and clay matter (%) versus ^{137}Cs activity concentration in bottom sediments.

An increase in the specific activities of ^{137}Cs (over 1000 Bq/kg) observed in fine particles can be due to the increase in the specific surface area of sediments participating in sorption processes. Another possible explanation of this deals with the variation in the mineralogical content of different size particles. Moreover, a comparatively smaller amount of fine particles of < 4 μm was found in samples from the Baltic Sea, it ranged from 1 to 28 % (on average – 6 %) as compared to the Curonian Lagoon bottom sediment samples where it varied from 11 to 70 % (with average value of 20 %). The correlation of total specific activities of ^{137}Cs in sediments with the amount of clay and organic substances was observed (Fig 4).

Studies of ^{137}Cs geochemical fractionation performed on bottom sediments collected in the Curonian Lagoon indicated insignificant variations in partitioning of cesium between different geochemical fractions in the samples collected at different stations (Fig. 5). In samples collected in the Baltic Sea, a wider range of fraction distribution of cesium was determined. In these samples higher percentage of ^{137}Cs associated with the exchangeable and the residual fraction was found. The highest activity of 259 Bq/kg, a small amount of ^{137}Cs in the exchangeable fraction and peculiarities of mineralogical composition (Table 2) of sample collected at station 3 make it possible to characterize this sample as possibly originated from the Curonian Lagoon. The mentioned station is located in the near-shore zone and suspended particles due to the specific hydrometeorological conditions could be deposited at this site. The higher percentage of residual fraction in the mentioned sample could be due to the peculiarities of exchange process, contact time with seawater and the amount of illitic clay.

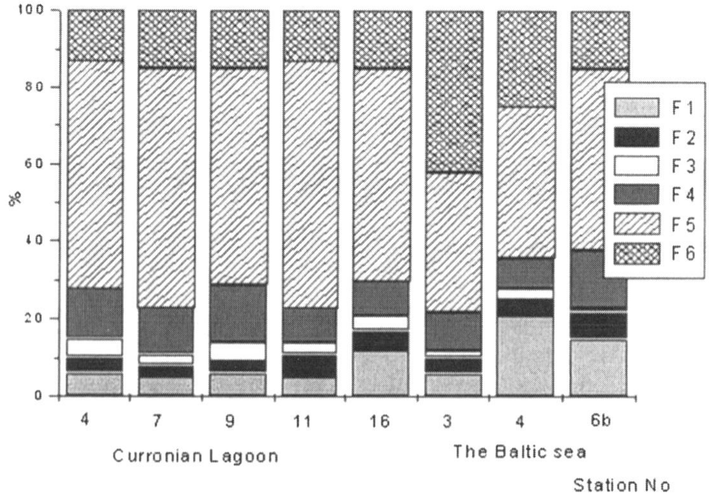

FIGURE 5. Speciation of ^{137}Cs in bottom sediment of the Baltic Sea and the Curonian Lagoon samples, F1 – exchangeable, F2 – carbonates, F3 – oxides, F5 – organic, F6 – acid soluble, F – 7 residue.

In general, the fraction distribution studies indicated the strong association of ^{137}Cs with sediment particles – about 70 % of ^{137}Cs was found in the acid-soluble and residual fractions (Fig. 5), whereas the organic fraction comprised a small part of total activity of ^{137}Cs in the sediments. However, a strong correlation was found both between specific activity of ^{137}Cs and the content the clay particles ($r = 0.95$, $n = 16$) or the amount of organic substances ($r = 0.75$, $n = 16$). This result is in good agreement with the data of other researchers indicating that clay minerals, especially illite, can effectively immobilize cesium even in the presence of a high content of organic substances [4]. Organic substances can inhibit sorption of cesium on clay minerals increasing its mobility in the environment, but perhaps they can enhance the binding of cesium by minerals. Correlation between total activities of cesium and organic substances usually found in the environmental samples confirms this. However, the mechanism of this phenomenon is not clear.

Fig. 6 presents the association of ^{137}Cs with various geochemical fractions of particles of different sizes collected in the Baltic Sea. The wide ranges of geochemical partitioning of ^{137}Cs were determined. Nevertheless, the highest amount of ^{137}Cs bound with organic, oxides and carbonate fractions was found in particles of >50 μm size, they make up to 34, 16 and 12 % of total cesium activity, respectively. It is characteristic of studied samples that the lowest activities of exchangeable cesium (e.g., up to 1 % in the sample collected at station 3) and its highest activities bound with residual and acid soluble fractions (e.g., up to 82 % in the sample from station No 4) were found associated with the smallest particles of < 4μm.

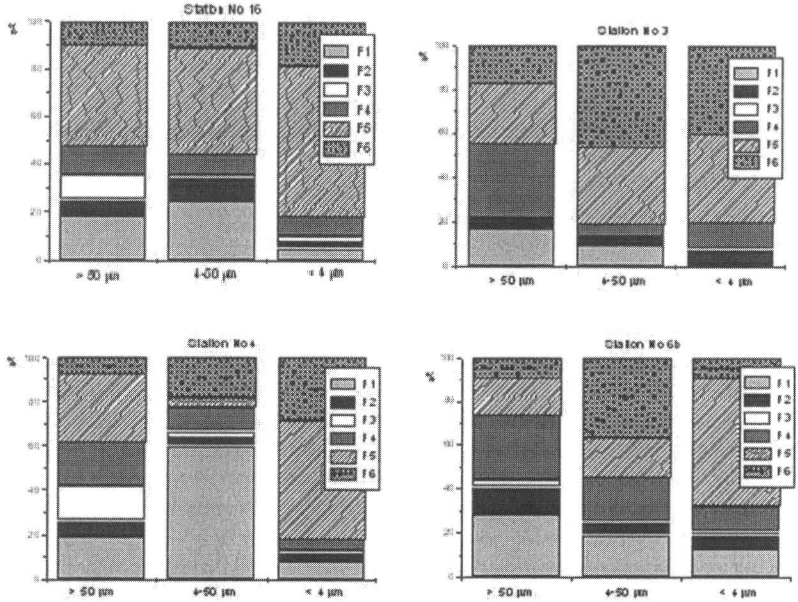

FIGURE 6. Physicochemical characteristics of ^{137}Cs in bottom sediment of the Baltic Sea samples F1 – exchangeable, F2 – carbonates, F3 – oxides, F5 – organic, F6 – acid soluble, F7 – residue.

It is obvious that partitioning of ^{137}Cs determined in the fine < 4 μm size fraction can be attributed to the binding pattern of ^{137}Cs by clay minerals. Therefore the highest activities of ^{137}Cs determined in the fine fraction can be interpreted as cesium binding by clay minerals in brackish waters of the Baltic Sea. The explanation of a comparatively high content of ^{137}Cs in the acid–soluble and especially in the residual fraction that is usually related to the fixation of cesium in the illite crystal lattice can be the coagulation or sticking of fine clay to the coarse sand and silt particles. The sticking particles present on the coarse ones can affect size-dependent distribution of radionuclide activities in soil and sediments observed in most cases [5].

Suspended particles in the near-shore and open seawaters of the Baltic Sea and the Curonian Lagoon were collected during three expeditions. Specific activities of ^{137}Cs in suspended particle samples collected in the Curonian Lagoon and separated using the 0.22μm membrane ranged from 16 to 254 Bq/kg. Insignificantly higher activities of ^{137}Cs from 15 to 372 Bq/kg were found in the near-shore zone of the Baltic Sea waters. The activity concentrations of ^{137}Cs in the surface and bottom waters of the Baltic Sea differ from 84 to 632Bq/kg and from 89 to 965Bq/kg, respectively (Table 3). The increase in ^{137}Cs specific activities accompanied by an increase in water salinity and a decrease in concentration of suspended particles were observed. At all sampling stations the specific activities of particulate ^{137}Cs in surface samples were lower as compared to the samples collected near bottom. Moreover, a strong relation of specific activities of ^{137}Cs with depth was found with the exception of the sample collected at station 6b, which possibly could be explained by water circulation in this area. In samples collected using the sediment filter cartridges with the nominal pore size of 1 μm the comparatively lower specific activities of ^{137}Cs (from 20 to 61 Bq/kg) were determined (Table 4). It seems that suspended particles of the 0.22 – 1 μm size, which were lost during this sampling, contribute considerably to the total particulate activity of ^{137}Cs. Moreover, the opposite relation between specific activities, salinity and concentration of total suspended particles in later samples allow us to suppose not only the fractionation of suspended particles according to the sizes but to the chemical/mineralogical composition. Different behavior of cesium in connection with different nature of particles can be expected. Possibly, the fine fraction of suspended particles collected in May 1999 consists mainly of clay minerals which can more effectively adsorb Cs even from brackish seawater containing high concentrations of potassium. The increase in the specific activity with salinity and sampling depth is a result of sorption process during particles transport and settling in the seawater. In sampling performed in October 2001, due to different cut-off filters the collected suspended matter exhibited rather different sorption ability. The decrease in the specific activity with an increase in water salinity observed in the mentioned samples can be interpreted as desorption of cesium from suspended particles due to exchange reactions with the highly abundant potassium in sea waters (Table 4). It seems that stable cesium had no effect on the behavior of ^{137}Cs because its concentrations were lower by 7 orders of magnitude as compared with that of potassium.

Some studies were performed on the association of ^{137}Cs in the bottom sediments collected in the Curonian Lagoon. The data presented in Fig. 7 indicated that only 1 % of ^{137}Cs could be extracted using 1M $MgCl_2$ and about 10 % - using 1M NH_4Cl from bottom sediments collected in the Curonian Lagoon.

TABLE 3. Specific activity, activity concentration and concentration of suspended particles in the Baltic Sea water (May 1999).

Sampling station	Specific activity, Bq/kg	Total ^{137}Cs, Bq/m^3	Suspended particles, mg/l	Salinity, psu	Depth, m
4 (surface)	84	41	11	3.9	
4 (bottom)	89	52	5.7	6.61	12
16 (surface)	99	44	7.8	5.03	
16 (bottom)	120	59	4	6.66	14
1b (surface)	166	62	3.6	6.65	
1b (bottom)	224	69	2.3	7.02	25
64 (surface)	202	63	2	6.37	
64 (bottom)	299	72	1.7	7.01	30
65 (surface)	91	73	2.6	7.21	
65 (bottom)	368	67	2	6.86	44
6b (surface)	207	62	2.2	6.62	
6b (bottom)	224	73	2.3	8.51	65
46A (surface)	632	76	1,2	7.11	
46A (bottom)	965	77	1	9.93	70
7 (surface)	102	80	2	6.85	
43 (surface)	90.5	79	1.6	7.06	

TABLE 4. Parameters of the Baltic Sea water samples.

Station	Cs, ppb	K, ppb	Salinity, psu	^{137}Cs SM, Bq/kg
4	0.03	56776	6.1	46
J7	0.04	50949	4.5	61
7	0.04	64590	7.0	53
8	0.05	62426	6.9	33
20A	0.04	75674	7.0	20

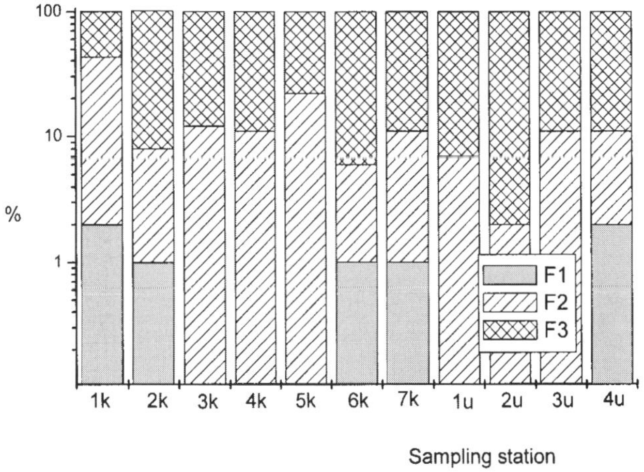

FIGURE 7. Exchangeable ^{137}Cs in bottom sediment samples collected in the Curonian Lagoon, F1 – exchangeable ($MgCl_2$), F2 – exchangeable (NH_4Cl), F3 – residual.

Thus, 1 % of ^{137}Cs can be displaced when such particles containing ^{137}Cs are transported from the Curonian Lagoon to the Baltic Sea under normal conditions and about 10 % under anoxic ones. Studies performed on sorption and release of cesium from particulate matter of the Baltic coastal zone indicated effective sorption of ^{137}Cs (up to 80 %) by bottom sediments, in which the illite was identified as dominant clay minerals. About 2 and 30 % of cesium was desorbed from two different sediments. It was pointed out that fresh water sediments were more susceptible to exchange with ions present in seawater [5]. The results obtained from experiments with living plankton cells and dead plankton show a similar sorption of cesium as compared to soft sediments. However, more rapid and intensive (up to 55 %) desorption from dead plankton was determined [6]. Thus, large variation in sorption–desorption ability of suspended particles of different origin and composition can be expected, the more so that the suspended matter transported to the Baltic from the Curonian Lagoon can contain dead plankton and clay minerals with coatings of various origin.

Our study indicated complicated sorption–desorption behavior of ^{137}Cs in the Baltic Sea water as a result of which it can be mobilized by suspended particles or released to the seawater. Thereby the fate of ^{137}Cs can be considerably affected by mineralogical composition of suspended and bottom sediments, by characteristic geochemistry of sea water and the presence of other pollutants.

From the results obtained it can be seen that fine particles are responsible for ^{137}Cs binding in the bottom sediments and suspended matter in the Baltic Sea. Fine particles of 1–0.2 μm can remain in the water column for a long time and can be transported for long distances and result in the redistribution of ^{137}Cs activity in the bottom sediments from one area to another. Results of sequential extraction analyses performed on the bottom sediments of the Baltic Sea show the strongest binding of ^{137}Cs by fine particles of < 4 μm. Thus, the fine clay particles are responsible for ^{137}Cs binding in the bottom sediments. However, the analysis of granulometric types of sediments performed in the Institute of Geology indicated that the content of aleuritic and pielic silt in bottom sediments of Lithuanian Economic Zone was rather low [8]. Thus, one of the possible reasons for slow sink of ^{137}Cs activities can be peculiarity of mineralogical composition of bottom sediments in the Baltic Sea as well (e.g., low content of clay minerals).

ACKNOWLEDGMENTS

This work was performed under the auspices of IAEA under project LIT/7/002.

We would like to express our gratitude to Ms. Rasa Morkuniene, Ms. Narciza Spirkauskaite and Mr. Nikolaj Tarasiuk for technical assistance.

REFERENCES

1. Khebonian, C. and Bauer, C. F., *Anal. Chem.* **59,** 1417–1423 (1987).
2. Sahuquillo, A., Rigol A. and Rauret G., *Trends in Anal. Chem.* **22,** 152–159 (2003).
3. Tessier, A., Camphell, P.G.C. and Bisson M., *Anal. Chem.* **51**, 844-851 (1979).

4. Evans, D.W., Alberts, J.J. and Clark III R.A,. *Geocim Comochim. Acta.* **47**, 1041–1049 (1987).
5. He, Q. and Walling, D.E., *J. Environ. Radioactivity*. **30**, 117–137 (1996).
6. Knapinska–Skiba, D. Bojanowski, R. and Radecki Z., *Netherlands Journal of Aquatic Ecology* **28,** 413–419 (1994).
7. Knapinska–Skiba, D., Bojanowski, R., Radecki Z. and Lotocka M., *Netherlands Journal of Aquatic Ecolog*, **29**, 283–290 (1994).
8. Radzevičius, R.. *Abstract of doctoral dissertation*, Vilnius, 25 p. (2001).

Modelling Angular Distribution Of Light Emission In Granular Scintillators Used In X-Ray Imaging Detectors

I.Kandarakis[1], D. Cavouras[1], D.Nikolopoulos[1], P.Liaparinos[2], A. Episkopakis[1], K. Kourkoutas[1], N. Kalivas[2], N. Dimitropoulos[3], I.Sianoudis[1], C.Nomicos[1], G.Panayiotakis[2]

[1]Departments of Medical Instruments Technology, Physics and Electronics, Technological Educational Institution (TEI) of Athens, Ag. Spyridonos, Aigaleo, 122 10, Athens, Greece
[2]Department of Medical Physics, Medical School, University of Patras, 265 00, Patras, Greece
[3]Medical Center "Euromedica" Alexandas, Athens, Greece Author's Affiliation

Abstract. The aim of this study was to examine the angular distribution of the light emitted from radiation excited scintillators in medical imaging detectors. This distribution diverges from Lambert's cosine law and affects the light emission efficiency of scintillators. Hence it also affects the dose burden to the patient. In the present study the angular distribution was theoretically modeled and was used to fit experimental data on various scintillator materials. Results of calculations revealed that the angular distribution is more directional than that predicted by Lambert's law. Divergence from this law is more pronounced for high values of light attenuation coefficient and thick scintillator layers (screens). This type of divergence reduces light emission efficiency and hence it increases the incident x-ray flux required for a given level of image brightness.

INTRODUCTION

Scintillator based radiation detectors are employed in a large variety of medical imaging systems (x-ray radiography and fluoroscopy, computed tomography, gamma camera, positron emission tomography). In all cases scintillators are directly or indirectly coupled to optical sensors (films, photocathodes, photodiodes)[1-4]. In designing and evaluating radiation detectors it is often important to accurately determine the fraction of scintillator emitted light collected by the optical sensor. The larger this fraction the lower is the radiation dose burden to the patient combined with better image quality. In many cases this fraction is largely determined by the angular distribution of light emission. In addition the shape of the angular distribution affects the performance of an image detector, e.g. emission efficiency, spatial resolution etc. In most practical applications it is assumed that light emission follows closely the distribution of Lambertian light sources [5]. However experimental data do not agree with this assumption [6].

In the present study the angular distribution of light emission from scintillating screens was modeled as a function of screen thickness and intrinsic physical properties of the scintillator material. The model was used to fit experimental data obtained for various scintillator materials. Fitting allowed the determination of the values of optical attenuation coefficients. Finally the effect of the shape of angular distribution on the x-ray luminescence efficiency (XLE) was examined.

METHOD

Model For Angular Distribution Of A Scintillating Screen

The scintillator was considered to be in the form of a layer (scintillating screen or phosphor screen) of thickness T, irradiated by a parallel x-ray beam. The layer was subdivided into elementary thin layers of thickness dt as depicted in figure 1. These elementary layers absorb x-rays and produce light photons. The surface of each elementary thin layer was assumed to emit light following Lambert's cosine law. The luminous intensity produced by a thin layer at depth t and directed at an angle ϑ, with respect to the normal, is given as [6]:

$$dI(\vartheta) = dI(0)\cos\vartheta \qquad (1)$$

$dI(0)$ is the luminous intensity directed perpendicular with respect to the surface of the thin layer (Figure 1). Luminous intensity is defined as $dI = d\Psi_\lambda / d\Omega$, where $d\Psi_\lambda$ is the luminous flux (light energy flux in watts per m^2) emitted within a solid angle $d\Omega$. The light energy flux may be expressed in terms of the incident x-ray energy flux and the x-ray absorption and conversion properties of the scintillator[7,8,9]:

$$d\Psi_\lambda = \int_0^{E_{max}} \overline{\psi}_x(E)\overline{\eta}(E,T)\eta_C x_R(E,t)dtdE \qquad (2)$$

where $\overline{\psi}_x(E)$ is the incident x-ray energy flux expressed in terms of spectral density (elementary flux per energy interval of the polychromatic x-ray spectrum)[8]. E is the energy of an x-ray photon. $\overline{\eta}(E,T)$ is the x-ray energy absorption efficiency of the scintillating screen. η_C is the x-ray to light conversion efficiency of the scintillator expressing the fraction of absorbed x-ray energy that is converted into light within the scintillator [7,9]. $x_R(E,t)$ is a function giving the relative probability of x-ray absorption within an elementary thin layer of thickness dt, situated at depth t (see appedix). Integration is performed over the entire x-ray energy spectrum. Mean values are considered over scintillator's area.

$dI(\vartheta)$ is the luminous intensity generated within an elementary thin layer. Due to light attenuation effects only a fraction of $dI(\vartheta)$ is transmitted through the rest of the screen to escape towards a direction ϑ. This fraction is expressed by the light transmission efficiency, $G(\vartheta,\sigma,t)$, describing the light attenuation within the screen and giving the fraction of luminous intensity transmitted through the rest of the screen to the emitting surface [7,9,10] (see appendix). σ denotes the light attenuation coefficient of the scintillator material. Thus the luminous intensity emitted from the screens surface is denoted as $dI^e(\vartheta)$ and is given by the relation:

$$dI^e(\vartheta) = dI(\vartheta)G(\vartheta,\sigma,t) = [dI(0)\cos\vartheta]G(\vartheta,\sigma,t) \quad (3)$$

By taking into account the definition of luminous intensity and relation (2), it can be written:

$$dI^e(\vartheta)d\Omega = \int_0^{E_{max}} \overline{\psi}_x(E)\overline{\eta}(E,T)\eta_C x_R(E,t)\cos\vartheta G(\vartheta,\sigma,t)dtdE \quad (4)$$

A relation corresponding to the total luminous intensity emitted by the screen towards ϑ may be obtained after integration of (4) over screen thickness:

$$I^e(\vartheta)d\Omega = \int_0^{E_{max}} \overline{\psi}_x(E)\overline{\eta}(E,T)\eta_C \int_0^T [x_R(E,t)\cos\vartheta G(\vartheta,\sigma,t)]dtdE \quad (5)$$

The angular distribution of the emitted luminous intensity $I^e(\vartheta)$ depends upon the product of $\cos\vartheta G(\vartheta,\sigma,T)$. Thus the angular distribution of $I^e(\vartheta)$ depends on the form and the magnitude of $G(\vartheta,\sigma,t)$. The latter is lower than unity thus reducing lateral light emission and distorting the shape of the angular distribution with respect to the Lambert's cosine law.

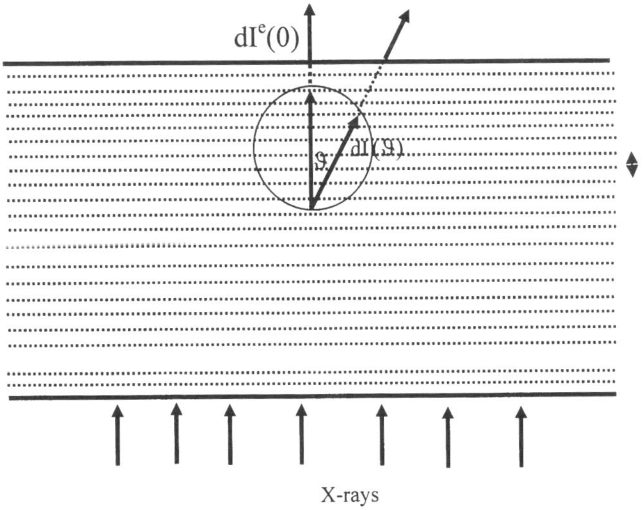

FIGURE 1. Cross section of a scintillating screen subdivided in elementary thin layers. Depicted is the lambertian emission of an elementary thin layer.

In $I^e(\vartheta)$ calculations the effect of K-fluorescence emission was also taken into account. Characteristic K-fluorescent x-rays may be produced if the energy of the K-absorption edge of the heavier element in the scintillator material is lying within the energy spectrum of the incident x-ray beam. These rays may be absorbed within the scintillator mass and create light photons far from the site of primary x-ray absorption. Thus the correction for the K-fluorescence emission was performed by adding to (5) a second term corresponding to the angular distribution of the light photons produced by the K-x rays absorbed in the scintillator ($I(\vartheta)+I_K(\vartheta)$).

Light Signal Loss And Effect Of Angular Distribution On XLE

The non-Lambertian shape of the angular distribution is expected to affect the number of light photons emitted by a scintillating screen. In the following, the effect of angular distribution on scintillator's light emission efficiency will be examined. The light emission efficiency may be expressed either by the x-ray luminescence efficiency (η_λ), defined as follows:

$$\eta_\lambda = \Psi_\lambda^e / \Psi_X \qquad (6)$$

where, Ψ_X is the incident x-ray energy flux (energy per unit of area and time). XLE is suitable for diagnostic radiology systems evaluation since the response of x-ray imaging detectors depends on the x-ray energy absorbed within the scintillator.

Equation (5) gives the luminous flux $\Psi_\lambda^e(\vartheta)$ emitted by the scintillating screen within a solid angle element $d\Omega$. Since $d\Omega = 2\pi \sin\vartheta d\vartheta$, the total light energy flux emitted from the surface of the screen is written as:

$$\Psi_\lambda^e = 2\pi \int_0^{\pi/2} I^e(\vartheta) \sin\vartheta d\vartheta \qquad (7)$$

or

$$\Psi_\lambda^e = 2\pi \int_0^{\pi/2} I^e(0) \sin\vartheta G(\vartheta,\sigma,t) d\vartheta \qquad (7b)$$

$$\Psi_\lambda^e = 2\pi \int_0^{E_{\max}} \overline{\psi}_x(E)\overline{\eta}(E,T) \int_0^{\pi/2}\int_0^{T} [x_R(E,t)\eta_C \cos\vartheta G(\vartheta,\sigma,t)] dt dE d\vartheta \qquad (7c)$$

If the angular distribution follows Lambert's cosine law, equation (1) is taken into account and the light flux may be written as follows:

$$\Psi_{\lambda,L}^e = 2\pi \int_0^{\pi/2} I_L^e(0) \cos\vartheta \sin\vartheta d\vartheta \qquad (8)$$

where the index L signifies that the corresponding luminous intensity and light energy flux follows lambertian distribution. $I_L^e(0)$ is the luminous intensity along the normal to the scintillator surface. Since $I_L^e(0)$ does not depend on ϑ, equation (8) leads to:

$$\Psi_{\lambda,L}^e = 2\pi I_L^e(0) \int_0^{\pi/2} \cos\vartheta \sin\vartheta d\vartheta = \pi I_L^e(0) \qquad (9)$$

To correct x-ray luminescence efficiency for the non-Lambertian distribution, equation (7) was first divided by equation (9). Thus the ratio ψ_λ of the actually emitted total energy flux over the total energy flux emitted by Lambertian surfaces is obtained. ψ_λ is often lower than unity and expresses the degree of optical signal loss due to deviation from Lambertian distribution. Considering the angular distribution normalized to $\vartheta = 0$ (i.e. $I_N^e(\vartheta) = I^e(\vartheta)/I^e(0)$) and assuming that $I^e(0) = I_L^e(0)$, we may write:

$$\psi_\lambda = \frac{2}{I_L^e(0)} \int_0^{\pi/2} I^e(\vartheta) \sin\vartheta d\vartheta = 2 \int_0^{\pi/2} I_N^e(\vartheta) \sin\vartheta d\vartheta \qquad (10)$$

$I_N^e(\vartheta)$ is a function of the angle ϑ and of the optical attenuation properties of the scintillator, expressed by the light transmission efficiency $G(\vartheta,\sigma,t)$. Thus ψ_λ is also a function of $G(\vartheta,\sigma,t)$. X-ray luminescence efficiency may then be corrected as follows:

$$\eta_\lambda = \eta_{\lambda L} \psi_\lambda \qquad (11)$$

where $\eta_{\lambda L}$ corresponds to lambertian distribution.

Measurements And Calculations

The model equations given in previous section were applied to fit experimental data obtained on various scintillator materials. Fitting was performed by the Levenberg-Marquard method [6,7,10]. Data on x-ray absorption properties of these materials and parameters related to the K-fluorescence emission effects were obtained and from the literature [11,12,13]. The function $\overline{\eta}(E,T)$, in (2) and (5), was calculated by considering exponential x-ray absorption. The function $G(\vartheta,\sigma,t)$ describing the light transmission efficiency was modeled according to previous studies [6,7,10]. Data concerning the intrinsic conversion efficiency, η_C in relation (4), (5) and the light attenuation coefficients (σ) of the scintillators were also obtained from previous studies [6,7,13,14,15,16]. Fitting was refined by allowing η_C and σ to vary slightly from their initial values. Measurements were performed on scintillating screens prepared in our laboratory from various materials. The latter were supplied in powder form by Derby Luminescent Ltd., Lumilux and Phosphor Technology Ltd. Screens of various thicknesses were prepared using sedimentation techniques. X-ray excitation was performed on a Philips Optimus radiographic unit incorporating a tungsten target x-ray tube. Experimental data were obtained according to previous studies using an EMI 9592 B photomultiplier equipped with an S-10 photocathode coupled to a Cary 401 electrometer with angular translation on a Rigaku-Denki SG-9D horizontal goniometer equipped with a 0.05° accuracy step scan controller

RESULTS AND DISCUSSION

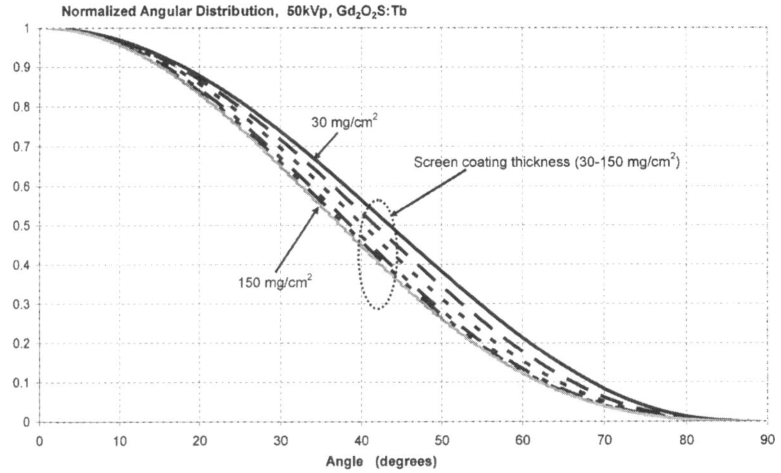

FIGURE 2. Normalized angular distribution of light emitted by granular scintillating screens (Gd_2O_2S:Tb) of various coating weights at 50 kV.

Shown in figure 2 are normalized light emission angular distribution curves (i.e. $I_N^e(\vartheta) = I^e(\vartheta)/I^e(0)$) derived by fitting relation (A4), in appendix, to experimental data. Results correspond to Gd_2O_2S:Tb scintillating screens of various coating thickness, from 30 mg/cm^2 to 150 mg/cm^2 . Data were obtained at 50 kV x-ray tube voltage and tungsten target x-ray spectra. It should be noted that variation in x-ray tube voltage was not found to affect significantly the shape of normalized angular distribution. As it may be observed screen-coating thickness affects the shape of the light emission angular distribution. Thick screens exhibit lower normalized luminous intensity values in the range from 20° to 70°. For the 150 mg/cm^2 screen a decrease, in relative values, of up to 30% was observed at 50°. This effect results in a more directional angular distribution shape, with respect to what was expected from the corresponding lambertian distribution, as screen thickness increases. Increased directionality of light emission may ameliorate image spatial resolution. In absolute luminous intensity values, however, this directionality accounts (corresponds) for a fractional loss in absolute optical signal level (see ψ_λ in equation 9). Hence resulting in a relative decrease of x-ray luminescence efficiency (with respect to lambertian screens). This decrease is mainly due to the increasing amount of optical scattering effects caused by additional phosphor layer accumulation in thick screens. These results are in agreement with image quality considerations stating that spatial resolution is improved if laterally emitted light photons are highly attenuated.

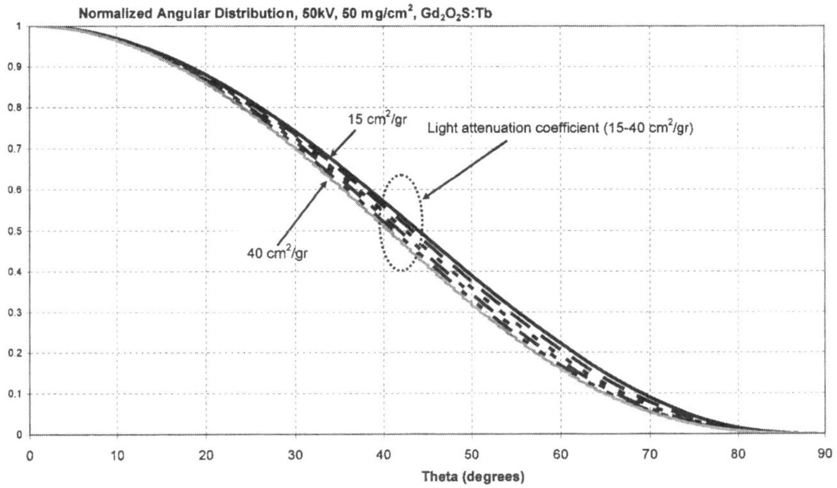

FIGURE 3. Normalized angular distribution of light emitted by granular scintillating screens for various values of the light attenuation coefficient.

Figure 3 shows calculated normalized angular distribution curves obtained for the same host material (e.g. Gd_2O_2S) but to different values of the light attenuation coefficient. The shape of the angular distribution becomes more directional with increasing light attenuation coefficient. At $\vartheta = 45^0$ the difference between the upper curve, corresponding to $\sigma = 15 cm^2 / g$, and the lower curve, corresponding to $\sigma = 40 cm^2 / g$, was approximately 10%. The light attenuation coefficient is a function of both light wavelength and phosphor grain size. For a given scintillator material host, e.g. Gd_2O_2S, and equal grain size, this coefficient is affected by the emitted light wavelength spectrum of the scintillator. This spectrum is determined by the type of ion activator (e.g. Tb^{3+}, Eu^{3+}, Ag, Cu etc) incorporated within the host material. Such a property should be taken into account when a high absorption efficiency scintillator is to be employed in various radiation detectors with different spectral sensitivity optical sensors. In these cases scintillators are activated with various activators suitably modifying their scintillator emission wavelength spectrum to match the optical sensor sensitivity. However this could alter angular distribution thus reducing light collection and detection efficiency. For identical scintillator host and activator, changes in the attenuation coefficient are due to different size of the grains. Small grains increase scattering effects resulting in higher attenuation coefficient. This gives an angular distribution of better directionality, which ameliorates spatial resolution. However small grains are expected to reduce the sensitivity of a detector, which thus requires higher amounts of incident radiation for a given level of image brightness.

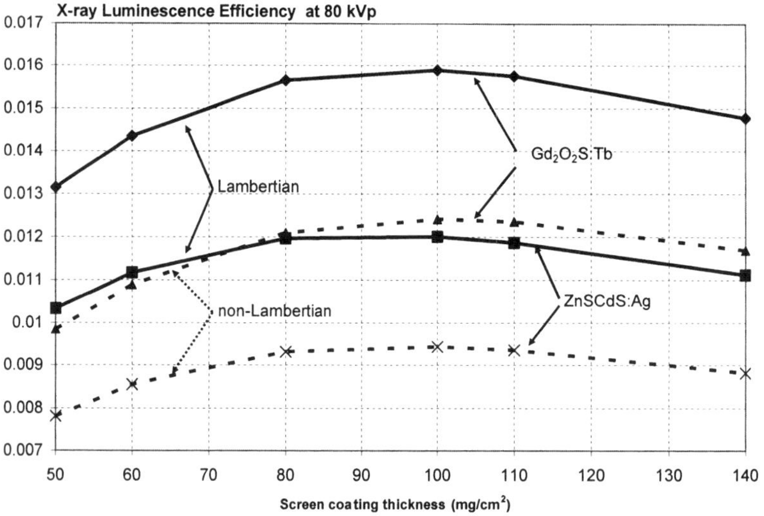

FIGURE 4. Variation of x-ray luminescence efficiency with scintillator coating weight for lambertian and non-lambertian distribution .

Figure 4 shows calculated results on the variation of x-ray luminescence efficiency with coating thickness considering lambertian and non-lambertian angular distribution. XLE is given in unitless values expressing the fraction of incident x-ray energy that is converted into light energy emitted from the screen. Calculations were performed for 80-kVp x-ray tube voltage. Transmission mode or front screen configuration data are shown in both figures, e.g. light flux emitted from the non-irradiated side was calculated. This configuration corresponds to most scintillator-optical sensor combinations: digital radiology detectors, front screen in radiographic cassettes etc. Values corresponding to non-lambertian distribution were found significantly lower with respect to values from Lambertian one. The difference was found to vary from approximately 15 to 30% depending on scintillator material and screen coating thickness.

Model and data presented in this study indicate that the angular distribution of the light emitted by granular scintillators shows a divergence from the well known lambertian distribution. This divergence is mainly determined by the light attenuation properties of the materials, expressed by the light transmission efficiency. Scintillator light emission shows higher directionality than lambertian sources. This effect may improve light collection to the optical sensor and ameliorate image resolution. However x-ray luminescence efficiency was found reduced with respect to lambertian light sources, thus requiring higher levels of incident x-ray flux to obtain a given level of detector sensitivity.

APPENDIX

The light signal emitted by a scintillator, may be expressed by the emitted light energy flux Ψ_λ (in W.m^{-2}) [7,9,10,14,15,16]:

$$\Psi_\lambda^e = \int_0^{E_{max}} \overline{\psi}_x(E)\overline{\eta}(E,t)\eta_C \int_0^T x_R(E,t)G_\lambda(\sigma,t)dtdE \qquad (A1)$$

$\overline{\psi}_x(E)$ is the x-ray energy spectral density distribution $[d\Psi_x(E)/dE]$ of the incident x-ray beam. E is the energy of x-ray photons. The second integral in relation (A1) expresses the light transmission efficiency. The function x_R gives the relative probability of x-ray absorption at depth t and was expressed by the relation:

$$x_R(E,t)dt = \frac{\mu(E)\exp[-\mu(E)t]dt}{\int_0^{w_O} \mu(E)\exp[-\mu(E)t]dt}$$

(A2)

where $\mu(E)$ represents the total mass energy absorption coefficient of the phosphor material. The numerator gives the probability for an x-ray photon to be absorbed at depth t within the phosphor and the denominator gives the total probability of x-ray absorption within the whole phosphor screen.

The function G_λ, gives the fraction of light photons, created within an elementary thin layer at depth t, that escape from screen surface. This function was modeled by considering exponential light attenuation determined by the light attenuation coefficient σ (relation (A1)) described in previous studies [7,9,10,14,15,16].

The energy luminous intensity, emitted within a solid angle element $d\Omega$, is given by the relation

$$I^e(\vartheta) = \frac{1}{4\pi} \int_0^{E_{max}} \overline{\psi}_x(E)\overline{\eta}(E,t)\eta_C \cos\vartheta \int_0^T x_R(E,t)G_\lambda(\vartheta,\sigma,t)dtdE \qquad (A3)$$

This quantity normalized to zero-degree angle is written as follows:

$$I_N^e(\vartheta) = \frac{\int_0^{E_{max}} \overline{\psi}_x(E)\overline{\eta}(E,t)\eta_C \cos\vartheta \int_0^T x_R(E,t)G_\lambda(\vartheta,\sigma,t)dtdE}{\int_0^{E_{max}} \overline{\psi}_x(E)\overline{\eta}(E,t)\eta_C \int_0^T x_R(E,t)G_\lambda(\vartheta=0,\sigma,t)dtdE} \qquad (A4)$$

REFERENCES

1. Wieczorek., B H. , *Rad. Meas.* **33,** 541-545 (2001)
2 Hell, E., Knüpfer, W. and Mattern, D., *Nucl. Instr. Meth. Phys. Res.A* 454 40-48 (2000).
3. van Eijk C.W.E., *Phys. Med. Biol.* **47** R85-R106 (2002).
2. H.J. Besch., *Nucl. Instr. Meth. Phys. Res. A* **419** 201-216 (1998).
5. Giakoumakis, G.E. and Miliotis, D. M., *Phys. Med. Biol.* **30** 21-29 (1985).
6. Begunov, B.N. Zakaznov, N.P., Kiryushin, S.I. and Kuzichev. V.I., *Optical instrumentation. Theory and design,* Mir publishers, Moscow. 1988.
7. Kandarakis, I.. Cavouras, D.. Ventouras, E. and Nomicos, C., *Radiat. Phys. Chem.* **66** 257-267 (2003).
8. Storm., E., *Phys. Rev. A* **5**, 2328-2338 (1972).
9. Ludwig G. W., *J. Electrochem. Soc.* **118 (7)** 1152-1159 (1971).
10. Swank, R.K., *Appl. Opt.* **12 (8)** 1865-1870 (1973)...
11. Storm, E., H. and Israel, H., *Photon cross-sections from 0.001 to 100 MeV for elements 1 through 100 Report LA-3753* Los Alamos Scientific Laboratory of the University of California (1967).
12. Hubbel, J., H. and Seltzer, S., M., *Tables of X-ray mass attenuation coefficients and mass energy absorption coefficients 1 keV to 20 MeV for elements Z=1 to 92 and 48 additional substances of dosimetric interest*, U.S. Department of commerce, NISTIR 5632 (1995).
13. Hubbel, J., H., Trehan, P., N., Singh, N., Chand, B., Mehta, D., Garg, M., L., Garg, R.,R., Singh,S. and Puri, S., *J. Phys. Chem. Ref. Data.* **23 (2)**, 339-364 (1994).
14. Kandarakis, I., Cavouras, D., Panayiotakis, G.S.and Nomicos, C.D., ***Phys. Med. Biol.*** **42**,1351-1373 (1997).
15. Kandarakis, I. and Cavouras. D., *Nucl. Instr. Meth. Phys. Res. A* **460**, 412-423 (2001).
16. Kandarakis, I. and Cavouras, D., *Appl. Rad. Isot.* **54**, 821-831 (2001).

Applied Physics (APHYS 2003)
A. Méndez-Vilas (Ed.)

THE FORMATION OF PEROXIDE COMPOUNDS AS ONE OF THE WAYS OF TRANSFORMING OXYGEN-CONTAINING ANIONS UNDER RADIOLYSIS

Vladimir A. Anan'ev

Department of Analytical Chemistry, Kemerovo State University, Krasnaya str. 6, Kemerovo, 650043, Russia.

Abstract. Optical absorption spectra of γ - irradiated caesium nitrate crystals have been studied. It is shown that γ - irradiation of caesium nitrate crystals at 310 K results in the peroxynitrite ion - $ONOO^-$. The presence of peroxynitrite is supposed to give a systematical error in the nitrite determination after dissolving irradiated samples. Alkali nitrates can be used for dosimetry under certain conditions.

INTRODUCTION

The advances in radiation technologies would have been impossible without determining the value of absorbed dose in a particular material. That is why there is a continuing search of substances which may be used for dosimetry. Johnson [1] stated that radiation yields of nitrite in solid nitrates do not depend on the absorbed dose rate ($10^{15} \div 10^{20}$ eV$\bullet$g$^{-1}\bullet$c^{-1}). On the basis of this fact and that the nitrite accumulation is linear in a wide range of absorbed doses some authors suggest using alkali nitrates as dosimeters. However, different values of the radiation yield of nitrite for the same nitrate presented by different authors (see, Review [2]) do not allow one to do it. At the same time these differences make one suggest there should be a factor resulting in a systematical error in the nitrite determination techniques.

That peroxynitrite is a precursor of nitrite under photolysis was stated in [3] where it was stated that nitrite is formed from peroxynitrite during the dissolving of UV - irradiated samples. R.C. Plumb and J.O. Edwards found out that the common method of determining the content of nitrite involved in the dissolving of the UV – irradiated nitrate samples when pH = 7 is not reliable [4]. Recent studies [5] state that in UV – irradiated solid alkali nitrates nitrite exists as [NO_2^-...O] complexes, most of them being formed from peroxynitrite.

It is stated that the peroxynitrite ion can be formed under the γ - irradiation of solid alkali nitrates [6]. It is accounted for by the fact that after dissolving the irradiated samples in alkaline solution the resulting products react with the permanganate ion like the peroxynitrite ion. The peroxynitrite ion in solution is transformed either into the nitrite ion or into the nitrate ion and which reaction will occur depends on pH and the presence of admixtures in the solution [7]. Thus, it is peroxynitrite that could be the product of radiolysis of alkali nitrate crystals that can be result in a systematical error in nitrite determination techniques. However, whether peroxynitrite is formed in solid state is not evident because the irradiation of solid nitrates at room temperature, as is known [6], results in O_2^-, and probably NO^-, that react with the permanganate ion like the peroxynitrite ion. The goal of the present paper is to analyze the optical spectra of γ - irradiated caesium nitrate crystals to determine the formation of peroxynitrite in solid state. The choice of the above crystals was prompted by the fact that the radiation yield of nitrite in $CsNO_3$ crystals is ~2.5 times as much as the one in $RbNO_3$ crystals and the optical absorption in $CsNO_3$ crystals is of an isotropic character contrary to the optical absorption in potassium nitrate crystals [8].

EXPERIMENTAL

$CsNO_3$ crystals both pure and doped with the nitrite ions were grown by slow evaporation of saturated aqueous solutions. The nitrate used was a.r. grade (three times crystallized from redistilled water before use). The nitrite ions were added to nitrate solutions in the form of the nitrite salt. The shape of doped $CsNO_3$ crystals was the same as the shape of undoped crystals. The procedure for the chemical analysis of nitrite was described earlier [9]. The concentrations of nitrite in doped caesium nitrate crystals was 1.42 mol.%.

The crystals were irradiated with ^{60}Co γ - rays at ~ 310 K. The dose rate was measured using a Fricke dosimeter, taking the radiation yield of Fe^{3+} as 15.6 (100 eV)$^{-1}$. The dose absorbed by the sample was calculated using the mass energy absorption coefficients. The dose rate was in the 1.1 – 2.5 Gy/s range. After irradiation the samples were stored at room temperature during at least 1 hour and analyzed as soon as possible. If necessary thermal annealing of the samples after irradiation was conducted in the thermostat at ~100 ^{0}C (the accuracy was ±1 ^{0}C). The nitrite ions formation in non - irradiated crystals was not observed at this temperature.

Optical measurements were carried out at room temperature. The experiments were made on polished crystals cut parallel to growth sides of the $CsNO_3$ crystals (trigonal type). The samples were 1×0.5 cm and their thickness varied from 0.0050 to 0.5000 cm. The method for the overlapping optical spectra resolution was described earlier [8]. To increase the reliability of the data 3 - 5 crystals were studied.

EXPERIMENTAL RESULTS AND DISCUSSION

The absorption spectra of $CsNO_3$ crystals both pure and doped with the nitrite ions are presented in Fig.1. Only a single band is observed in the doped crystal (half – width is 2400 cm^{-1}, molar absorptivity is ~20 M^{-1}•cm^{-1}). The peak position (355 nm) coincides with the low-energy maximum of the nitrite ion absorption band observed in aqueous solution [10].

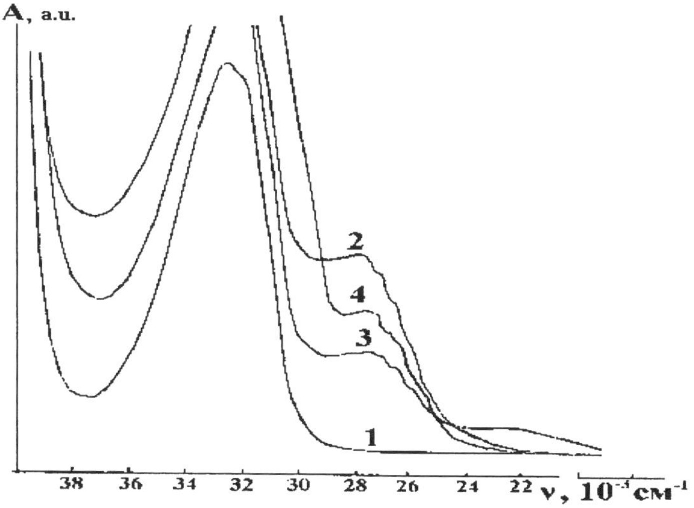

Figure 1. Optical spectra of $CsNO_3$ crystal (pure, γ - irradiated and thermal annealed at 100 ^{0}C at 1 hour – 1, 2, 3 respectively). Optical spectra of $CsNO_3$ crystal doped with the nitrite ions (4).

The absence of the high-energy band at ~287 nm in optical spectra is due to the value of molar absorptivity (three times as less as for the band at 355 nm) and to its strong overlapping with the absorption band of the host crystal. The band at 355 nm has a vibrational structure with a progression which has frequency intervals of ~ 600 cm^{-1}.

The optical spectrum of the γ - irradiated pure CsNO3 crystal is shown in Fig.1. As seen, the spectrum has a vibrational structure like the spectrum of the doped crystal. The optical density of γ - induced band (at 355 nm) after thermal annealing of the crystal at ~100 ^{0}C during 1 hour is twice as less than that of the untreated sample (Fig.1) and a new absorption band at 450 nm appears which in terms of the data available can be referred to the $O_3^{\bullet-}$ species. Their formation pathways do not depend on the nitrite ion and the peroxynitrtite ion formation pathways because the $O_3^{\bullet-}$ species are not formed during thermal annealing of UV – irradiated samples [5]. The continuation of annealing of the crystal at this temperature does result in the change of the optical spectrum in the low – energy region (< 300 nm). As known [5] the decay of peroxynitrite is observed at this temperature which is confirmed both by the chemical analysis data and optical measurements. The annealing of nitrite does not exceed 10 % under these conditions. Since no other products of radiolysis of alkali nitrates absorb light in the low – energy region (< 300 nm), the results obtained prove that peroxynitrite can be formed in γ - irradiated CsNO3 crystals in solid state.

The estimation of the concentration of peroxynitrite in γ - irradiated samples is carried out by comparing the spectra (due to peroxynitrite and the parameters of absorption band reported in в [5]) of photolysed crystals and the spectra (due to the nitrite ions) of doped crystals with the spectra of the γ - irradiated samples.

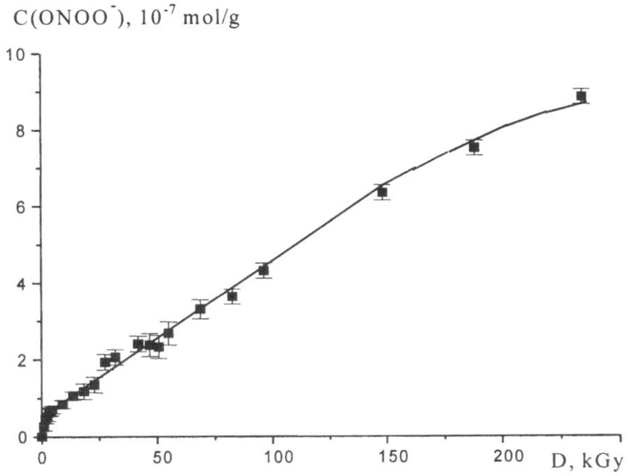

Figure 2. The curves of peroxynitrite accumulation vs. absorbed doses for $CsNO_3$ crystals γ - irradiated at 310 K.

The peroxynitrite accumulation estimated on the basis of optical measurements during the radiolysis of $CsNO_3$ crystals are shown in Fig.2. It is assumed that the nitrite accumulation is linear for the absorbed doses studied. As seen the peroxynitrite accumulation curve can be subdivided into three regions. The radiation yield of peroxynitrite in the first low – absorbed doses region is 0.30±0.05 $(100\ eV)^{-1}$; in the second linear - 0.038±0.004 $(100\ eV)^{-1}$. The radiation yield of the peroxynitrite ions is much less than the radiation yield of the nitrite ions – 1.61 $(100\ eV)^{-1}$ [2].

Thus, the formation of nitrite and peroxynitrite under radiolysis have independent pathways, while under photolysis most of the nitrite ions are formed under the decomposition of the peroxynitrite ions.

Hence, depending on the conditions (pH, presence of heavy metal ions as catalysts of peroxynitrite decomposition, etc.) under which the chemical analysis of irradiated nitrates is carried out it is possible to suppose a systematical error in determining the radiation yield of nitrite.

CONCLUSION

The above data allow us to state that alkali nitrates can be used for dosimetry only when one and the same technique for determine nitrite with the reference reagents and the same pH will be used by all investigators concerned.

ACKNOWLEDGMENTS

This work was supported by the Grant of President of Russian Federation NSH-20.2003.4 and "Universities of Russia" UR.06.01.007.

REFERENCES

1. E.R.Johnson, J.Phys.Chem., **66**, 755-761 (1962).
2. E.R.Jonson, *The radiation – induced decomposition of inorganic molecular ions*. Gordon&Breach Science Publishers, N.-Y. – London – Paris, 1970, pp. 33 – 85.
3. T. Yrmazova, L.Koval and Serikov, Chim. Vys. Energ (Rus) **17,** 151-155 (1984).
4. J.O.Edwards and R.C.Plumb, Prog. Inorg. Chem., **41**, 599-635 (1994)
5. V.Anan'ev , M.Miklin, N.Nelyubina and M.Poroshina, Photochem.& Photobiol.: Part A. Photochem. **162**, 67-72 (2003).
6. V.Anan'ev, L.Kriger and M.Poroshina, Chem.Phys.Lett. **365**, 554-558 (2002).
7. R.Kissner and W.H.Koppenol, J.Am.Chem.Soc. **124**, 234-239 (2002).
8. V.Anan’ev and M.Miklin, Opt.Mat. **14**, 303-311 (2000).
9. M.Miklin, L.Kriger, V.Anan’ev and V.Nevostruev, Chim. Vys. Energ (Rus) **23**, 506-509 (1989) .
10. Drago, R.S. *Physical Methods in Chemistry*, W.B.Saunders Company, Philadelphia, 1977.

Applied Physics (APHYS 2003)
A. Méndez-Vilas (Ed.)

Determination of ^{90}Sr impurities in ^{89}Sr solutions, Intended for Medical Use

Yury A. Sapozhnikov

Chemical Department, Moscow State University, Moscow, 119992, RUSSIA
E-mail: yas@radio.chem.msu.ru

Abstract. The solutions of strontium-89, intended for medical use, may contain the extremely undesirable impurities of long-lived strontium-90. The majority of techniques for determination of strontium-90 is based on separation and determination of its daughter - yttrium-90, which identification is carried out frequently via time consuming half-life period measurement. The aim of this work was to develop the procedure for the rapid determination of strontium-90 microimpurities (to $n \times 10^{-4}$ %) in strontium-89 solutions. Procedure is based on the measurement of three above mentioned radionuclides (^{89}Sr, ^{90}Sr and ^{90}Y) with cherenkov radiation, excited by β-particles, emitted by their nuclei. The application of wavelength shifter makes it possible to increase substantially the registration efficiency of these radionuclides and changes the shapes of their spectra. This enables determination and identification of strontium-90 (via yttrium-90) in the presence of high strontium-89 concentrations.

INTRODUCTION

The liquid scintillation (LS) counting is the most widely applied method for measurement of "pure" β-emitters and their mixtures. Radioactive substances are dissolved in the organic cocktail, capable to scintillate under an impact of β-radiation. Registration efficiency of β-radiation with maximal energy > 200 keV by a LS method is close to 100 % [1].

The cherenkov radiation, arising during the movement of high-energy β-particles in the transparent dielectric media with velocity (v) exceeding the velocity of light (c') in the given media, LS equipment can register as well [1].

The requirement $v>c'$ is equivalent to $c'/c = 1/n$, where n is the index of refraction. For water $n = 1,33$, accordingly $1/n = 1/1,33 = 0,75$, and for electrons, moving in water, the condition $v/c' > 0,75$ is carried out at energies above 263 keV. The intensity of cherenkov radiation grows quickly with increasing of β-particles energy.

Values of the cherenkov radiation registration efficiency for radionuclides of interest in water solutions are presented in Table 1.

TABLE 1. The cherenkov radiation registration efficiency for radionuclides of interest in water solutions [2].

$E_{\beta max}$, keV	Radionuclide	Registration efficiency, %
1492	^{89}Sr	40
546	^{90}Sr	1,5
2280	^{90}Y	70

Shapes of the cherenkov radiation spectra for the majority of high-energy β-emitters in water solutions occupy rather narrow part on a scale of the pulse amplitudes registered by LS spectrometers. It essentially complicates the analysis of mixtures of various β-emitters by cherenkov radiation. Quanta of cherenkov radiation are emitted within a cone, which axis is directed along the trajectory of β-particle movement. It reduces the probability of simultaneous registration of cherenkov radiation by two photomultiplier tubes, usually used in LS systems.

Widely used component LS of cocktails - 2,5 diphenyl oxazole (PPO) is used frequently at measurement cherenkov radiations as wavelength shifter, absorbing quanta of cherenkov radiation with the high efficiency and isotropically re-emitting the light quanta of more long-wave region [2-4].

Application of a wavelength shifter raises the registration efficiency of cherenkov radiation and expands a range of the amplitudes, registered by the LS spectrometer. As a result it becomes easier to distinguish high-energy region of the cherenkov radiation spectra of yttrium-90 and tails of rather low energy spectra of strontium-89.

The aim of this work was to develop the procedure for the rapid determination of strontium-90 microimpurities (to $n \times 10^{-4}$ %) in strontium-89 solutions, intended for medical use.

EXPERIMENTAL

In this work the radionuclides Sr-89 and Sr-90, delivered by the Central Scientific Research Institute of Nuclear Reactors (Dimitrovgrad, Russia) were used.

Measurements were carried out using liquid scintillation spectrometer Tri-Carb-2700TR (Packard, USA).

Absolute activity measurements of the initial solutions were carried out by LS methods with the cocktail OptiPhase "Hi-Safe" III (Wallac, Finland).

Typical cherenkov radiation spectra of Sr-89 and Sr-90 with Y-90 in water solutions are submitted on fig. 1.

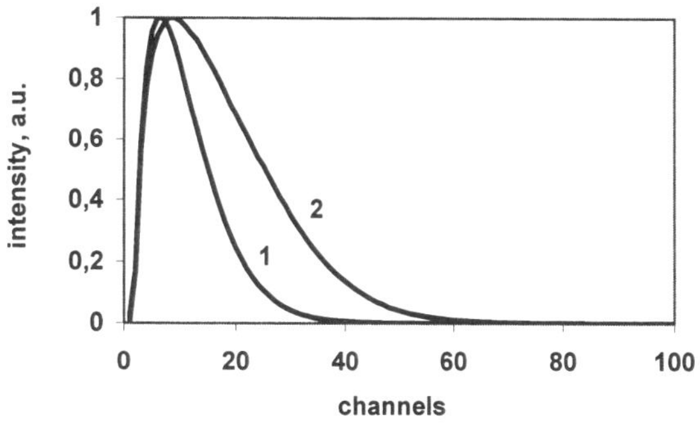

FIGURE 1. Cherenkov spectra of ^{89}Sr (1) and $^{90}Sr + ^{90}Y$ in water solution.

As a wavelength shifter the PPO solution in ethanol (3 g/l) was used.

At the presence of a wavelength shifter the range of the amplitudes, registered by the LS equipment, essentially extended (fig. 2), that detection of small impurity of strontium-90 in a solution of strontium-89 (via yttrium-90) become possible. The best results were obtained, when the basic component of a liquid phase was the PPO

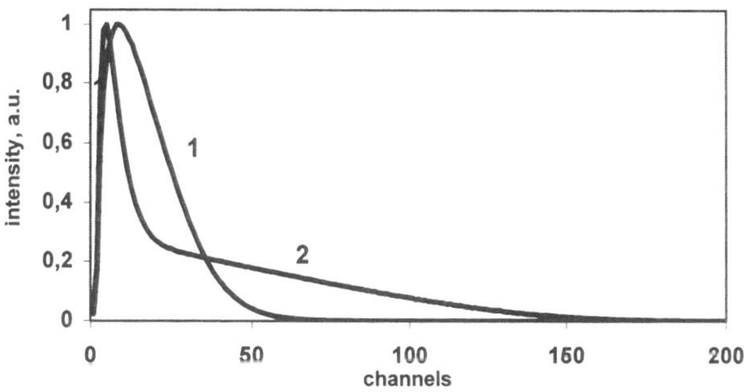

FIGURE 2. Cherenkov spectra of $^{90}Sr + ^{90}Y$ without (1) and with (2) wavelength shifter.

solution in ethanol, and water solutions of radionuclides occupied only insignificant part of the system volume. The high-energy parts of spectra of the mixes $^{89}Sr + ^{90}Sr$ (with ^{90}Y) are presented in fig. 3.

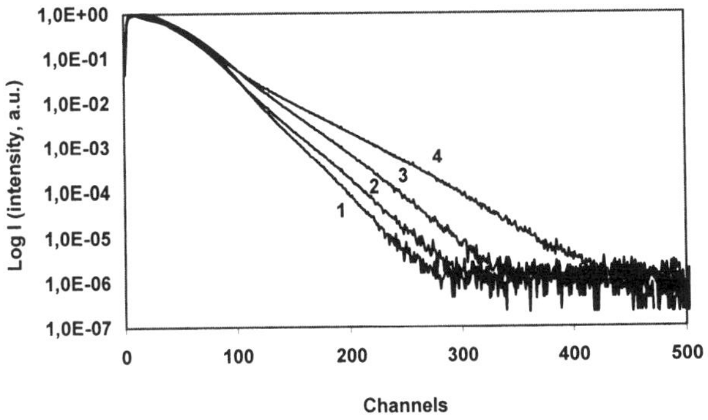

FIGURE 3. Cherenkov radiation spectra of ^{89}Sr solutions with microamounts of ^{90}Sr+^{90}Y (dependence of the slope on ^{90}Sr/ ^{89}Sr ratio: 1- 0; 2 - 8,6*10-4%; 3 - 3,7*10-3%; 4 - 1,1*10-2%).

DISCUSSION

The important feature of the cherenkov radiations is that it is raised only under an impact of the charged particles moving with velocities, exceeding velocity of light in the given media.

In real conditions of these experiments such particles are the electrons (mainly β-particles, emitted by nuclei of the radionuclides of interest) and high-energy protons of cosmic radiation, which intensity is rather low. Therefore the background in the high-energy part of the cherenkov radiation spectra is much below, than in case of LS measurements, where, for example, the role of the muons is more important.

Practically no other components of background are significant in such measurements.

A number of authors [2-4] has established, that addition of wavelength shifters, containing aromatic compounds, results in essential change of efficiency of cherenkov radiation registration. Apparently, with growth of the wavelength shifters concentration the process of an energy transfer in the system, sharply raising efficiency of registration of rather low-energy particles essentially increases.

CONCLUSION

Measurement of cherenkov radiation with wavelength shifter allows to obtain ^{90}Sr microimpurities in ^{89}Sr solutions, intended for medical use.

ACKNOWLEDGEMENT

Author thanks the Central Scientific Research Institute of Nuclear Reactors (Dimitrovgrad, Russia) for presented radionuclides for the performance of this work.

REFERENCES

1. M.J. Kessler (Ed.) *Liquid scintillation counting*. Science and technology. Packard Instrument Co., Publ. No 169-3052, Inc. (1989).
2. S.C. Scarpitta, I.M. Fisenne, *Appl. Radiat. Isot.* **47**, 795-800 (1996).
3. H. Fujii, M Takiue. *Nucl. Instrum Methods* **A265**, 558-560 (1988).
4. G.A. Peck, J.D. Smith, M.B. Cooper, *J. Radioanal. Nucl. Chem.* **238** 163-165 (1998).

Applied Physics (APHYS 2003)
A. Méndez-Vilas (Ed.)

MICROWAVE EFFECTS UPON VEGETAL CELL CULTURES

Fl.`M. Tufescu, D.E. Creanga

Al. I. Cuza University, Faculty of Physics,11 Carol Blvd.,700506 Iaşi , Romania,
tel.:+40-32-201177, fax +40-32-210775, e-mail: ftufescu@uaic.ro

Abstract. Daily exposures to microwaves from X-band were carried out aiming to reveal putative biological effects in vegetal tissue cultures (callus) intended for plant cloning. Low power density irradiation in free space was chosen in order to evidence non-thermal effects of natural exposures from atmospheric sources. Unexpected enhance of chlorophyll level was noticed together with the diminution of nucleic acid level. Possible utilization of microwaves as biotechnological tool in plant micro-propagation is revealed.
Key words: electromagnetic irradiation, poppy callus, photosynthesis

I. INTRODUCTION

Non-ionizing electromagnetic waves represent a ubiquitous environmental component of atmosphere and related media (water and soil). Microwaves, either from cosmic sources (see Crab constellation) or artificial ones (electric power stations and transmission lines, radio and TV stations, Radar stations etc.) are more and more studied by biologists and biophysicists due to their biological effects, already reported in the specialty literature.

Microwave influence upon living bodies was noticed in different species of animals and plants (Tambiev A.H., Kirikova N.N, 1994, [1], Riipulk J, Hiinrikus, H.,1999, [2]). Since low intensity microwave flows are susceptible of specific, non-thermal effects in the last decades, the interest of several research groups was directed toward the study of such biological effects (Ruzic R., Jerman I., Jelig A., Fefer D., 1992, [3], Polk Ch and Postow E., 1996, [4], Vlahovici et al, 2000, [5]). Experiments were designed in different configurations, using different power density beams and irradiation schedules. Effects were evaluated by means of various types of measurements, at the level of different biophysical and biochemical parameters, concerning animal metabolism, microorganism survival, plant photosynthesis. Stimulation or inhibition of plant germination and growth was found as dependent of plant species or ontogenetic stage, on a hand, and on the physical parameters of microwave flow (modulation, power density, frequency) on the other hand.

However there is a significant lack of information regarding microwave action in vegetal tissue cultures, i.e. in callus developed from plant somatic tissue samples (not from embryo cell). Plant cloning in sterile environment is primarily based on such non-differentiated cell tissues (callus), extremely important in plant biotechnology, intended for micro-propagation of plants free of viruses or for the conservation of very rare exemplars from disappearing species or from fossil ones. During various protocol in plant biotechnology physical factors are used as control tools in modeling plant features at the level of the non - differentiated callus that substitutes the embryo. The microwaves from the X band may be such biotechnological tool if used properly, biological effects being putative useful consequences upon plant photosynthesis. Thermal and non-thermal microwave effects may be involved as well.

Thermal microwave effect is well known and widely applied in many fields. Thermal effect of microwaves involves 2 distinct mechanisms- dielectric and ionic.

The dielectric component of the thermal microwave effect is based mainly on water molecule polar character. Water molecular dipoles are able to oscillate up to high frequencies of the electrical external field, these oscillations generating water heating.

The ionic component of thermal microwave effect is originating in the ions oscillatory migration under the action of the external oscillating microwave field, these migrations underlying the conversion of electrical energy into thermal one. Data and Anantheswaran[6-9] proposed the next relation to describe the rate of heat generation per unit volume, W, at a certain location into an aqueous medium:

$$W = 2\pi\nu\varepsilon_0\varepsilon_r E^2 \qquad (1),$$

where $\boldsymbol{E}$ and ν are the electrical microwave field strength and frequency while ε_0 and ε_r are the electrical permittivity values in the vacuum and in the substance. Consequences of the physical effect are the denaturation of enzymes, proteins, nucleic acids, or other vital components.

Less studied than thermal effects, non-thermal microwave effects remain an actual challenge for the biophysicist. Many experiments have been carried out on aqueous cell cultures based on microorganisms – bacteria and fungi.

Four major hypotheses have been formulated to explain non-thermal inactivation of microorganisms by microwaves: selective heating, electroporation, cell membrane rupture, and magnetic field coupling (Kozempel and others 1998)[10].

The selective heating hypothesis states that solid microorganisms are heated more effectively by microwaves than the surrounding medium and are thus killed more readily.

Electroporation is caused when pores form in the membrane of the microorganisms due to the variation of the electrical potential across the membrane, resulting in leakage.

Cell membrane rupture is related in that the voltage drop across the membrane causes it to rupture.

In the fourth hypothesis, cell lysis occurs due to coupling of electromagnetic energy with critical molecules within the cells, disrupting internal components of the cell. Heddleson and Doores (1994)[12], Khalil and Villota (1988; 1989a;b[13-14]) The experimental findings also showed that microwave-injured cells recovered better when microwave heating was carried out anaerobically. This effect was not seen with conventional heating. The mentioned authors speculated that the microwaves catalyzed oxidative reactions, possibly in membrane lipids, decreasing recovery of exposed cells. In another study, Khalil and Villota (1989b) demonstrated that while both conventional and microwave heating destroyed the 16S subunit of RNA of sublethally-heated S. aureus FRI-100, only microwave heating affected the integral structure of the 23S subunit [14].

Brunkhorst [11] subsequently designed a new system that was capable of isolating thermal and non-thermal effects of microwave energy. The system was a double tube that allowed input of microwave energy but removed thermal energy with cooling water. With this system, the researchers found no inactivation of Enterobacter aerogenes, E. coli, Listeria innocua, Pediococcus, or a yeast in various fluids including water, egg white, whole egg, tomato juice or beer at sublethal temperatures. They concluded that, in the absence of other stresses such as pH or heat, microwave energy did not inactivate microorganisms; however, they did suggest that microwave energy may complement or magnify thermal effects.

Researchers focused on plant responses to microwaves reported that germination was inhibited during microwave exposure, fact which can not be assigned to thermal effect because heating stimulates germination [15].

As one can see most of the experimental findings and suppositions concerning non-thermal microwave effects consider disruptive actions as in the case of the thermal effect. Electromagnetic waves effects in biological membrane were mathematically modeled by Francescetti & Pinto [16] as well as by Bussang [17]. Starting from the hypothesis that the cell may be represented by a homogeneous sphere into an infinite homogeneous field, the membrane potential (difference between the electrical potential of the inner and outer membrane faces) was written as:

$$\Delta V = V_0 + \delta V \tag{2}$$

where ΔV_0 is the membrane potential in the lack of the external constraint, δV. The corresponding membrane current density is:

$$j_m = \sum_{k=1}^{\infty} j^{(k)}{}_m \text{ , with } j_m{}^{(k)} = \int_{-\infty}^{t} d\tau_1 ... \int_{-\infty}^{t} d\tau_k \gamma_k(\tau_1,...\tau_k)\delta V(t-\tau_1)...\delta V(t-\tau_k) \tag{3}$$

This represents a Volterra functional expansion (memory Taylor series), able to describe the most general relation non-linear, non-instantaneous, between the current j_m and the external electrical field, the terms γ_k being Volterra nuclei and the first term being the usual linear response. The Volterra nuclei can be obtained resolving the Hodgkin-Huxley equations by means of Volterra series.

The membrane ion current $I(t)$ is given also by the equations proposed by Hodgkin and Huxley [1952] are already very well known in biophysics [18]:

$$C_m \frac{dV_m}{dt} + g_k n^4 (V_m - V_k) + g_{Na} m^3 h(V_m - V_{Na}) + g_l (V_m - V_l) = I(t) \tag{4}$$

$$\frac{d\xi}{dt} = \alpha\xi(1-\xi)\beta\xi, \xi = m,n,h \tag{5}$$

where V_m, V_{Na}, V_K, V_l are: membrane potential and respectively, electrical potentials given by Na+, K+ and other ions circulating through the membrane; g are the values of the electrical conductance due to Na+, K+ and other ions (mainly chloride but generally considered as electrical leakages); α, β are exponential functions depending only on the membrane potential while **m, n** and **h** are non-dimensional parameters ranging between 0 and 1 depending on the membrane potential variation; C_m is the membrane electrical capacity.

The electrical model of the biological cell membrane is generally expressed by the scheme from figure 1a corresponding to the linearization of the Hodgkin-Huxley equations around the equilibrium state (in the absence of external stimuli).

The theoretical analysis of membrane non-linear response to external electrical fields

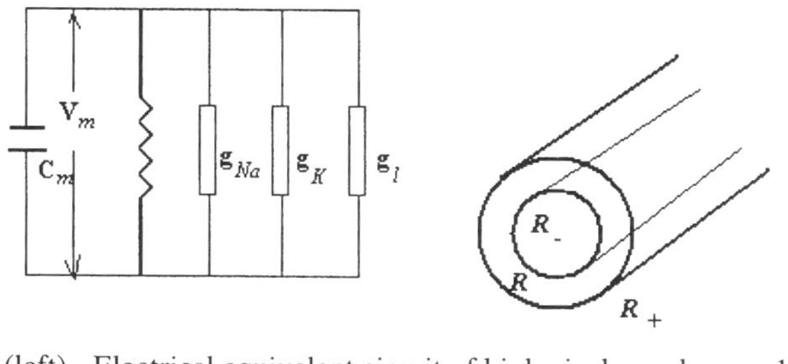

Figure 1 a (left) - Electrical equivalent circuit of biological membrane ; 1b (right) – Spherical cell membrane

with frequency up to 10 GHz – the upper limit of this range representing also the frequency of the experimental investigation reported in this paper - may be accomplished adopting polar coordinates $\{r,\phi,\theta\}$ system originating in the center of the spherical cell having the radius R and the membrane width R_+-R_- (figure 1b) as well as considering a plane electromagnetic wave incident normally to the cell. Such mathematical model is based on the assumption that all electrical fields and currents are derived from the same scalar potential $\Phi(\vec{r},t)$. If the electrical potential induced in the cell is $\Phi_S(\vec{r},t)$ then:

$$\Phi(\vec{r},t)=\Phi_S(\vec{r},t)-e^i(t).r.\cos\theta \tag{6}$$

$$\nabla^2\Phi(\vec{r},t)=0 \tag{7}$$

with boundary conditions:

$$S_e.\frac{\partial\Phi}{\partial r}\Big]_{r=R_+}=S_i.\frac{\partial\Phi}{\partial r}\Big]_{r=R_-}=J_m \tag{8}$$

where S_e and S_i are the inverse Fourier transforms of the conductivities of the extracellular and intracellular media.

To solve the above equation the authors of this modeling expressed the electrical potential induced into the cell as a Volterra series:

$$\Phi(\vec{r},t)=\Phi_S(\vec{r},t)-Ae^i(t).r.\cos\theta \tag{9}$$

$$\Phi_S(\vec{r},t)=\Phi_0+\sum_{m=1}^{\infty}A^m\Phi_S^{(m)}(\vec{r},t) \tag{10}$$

where $\Phi_0=-V_0, r<R$ and $\Phi_0=0, r>R$, $\Phi_S^{(m)}$ is a homogeneous linear function (grade m) relative to the incident field and A is a calculation variable tending toward unit. In the next step it is found that the above equation admits solutions of the form:

$$\Phi_S^{(m)}=\sum_{n=0}^{\infty}C_n^{(m)}r^nP_n(\cos\theta), r<R \tag{11}$$

$$\Phi_S^{(m)}=\sum_{n=0}^{\infty}D_n^{(m)}r^{-n-1}P_n(\cos\theta), r>R \tag{12}$$

where $P_n(x)$ is Legendre polynomial and the unknown expansion coefficients $C_n^{(m)}$ and $D_n^{(m)}$ are functions depending only on the time, t.

The model was validated on experimental NaCl solutions at the physiological temperature of 37 ^{0}C which revealed the next:

-for cell radius equal to 1 mm, a decrease of the membrane response to the frequency enhance occurs for frequencies higher than 10^4Hz;

-for cell radius equal to 1 micrometer, the membrane response is lower, with a decrease for frequencies higher than 10^8 Hz.

All this are showing that cell membrane can be affected by electromagnetic exposure and mathematical tools are able to describe such phenomena.

In the next we present our attempt of revealing the influence of microwaves of low power density on several biological and biochemical parameters of in vitro cell cultures of Papaver pseudo-orientale- a pharmaceutical plant. Not only inhibitory but also stimulatory influences may be assumed when the experimental findings are interpreted.

II. MATERIALS AND METHODS

II.1. Biological material

Papaver pseudo-orientale (poppy) in vitro cultures, meaning vegetal tissues, non-differentiated, no older than three months were obtained from explants of leaves and flowers provided by adult individuals, grown in the Botanic garden of AL. I. Cuza University from Iasi (fig. 2 a). This plant species is important for its pharmaceutical uses.

Murashige Skoog agarized medium with a suitable hormone balance was used to conduct cell culture development for three months. Then, series of five vials, three with leave explants and two with flower explants have been exposed to six different microwave doses, corresponding to irradiation times equal to 2; 4; 6; 8; 10 and 12 hours. Exposures have been performed daily for a week.

II.2. Exposure system

Microwave generator was assembled in our laboratories being based on an IMPATT diode. It is able to deliver at the level at its horn antenna (Fig. 2 b) a radiation flow characterized by a constant level power density of 0.9 mW/cm^2, in a frequency range of 9. 75 - 10.75 GHz. A probe detector was used to measure the spatial distribution of microwave energy in order to assure a geometrical arrangement compatible with a quasi-uniform irradiation of the sample. Microwave flow was let to irradiate biological samples in open atmosphere (free space) in order to simulate better natural irradiation. A thermostat tank (25 ^{0}C) was utilized to keep callus vials during electromagnetic exposure. Absorbed doses were adjusted by means of exposure time to what they are proportional so that any quantitative mentions regarding microwave exposure are in terms of irradiation duration.

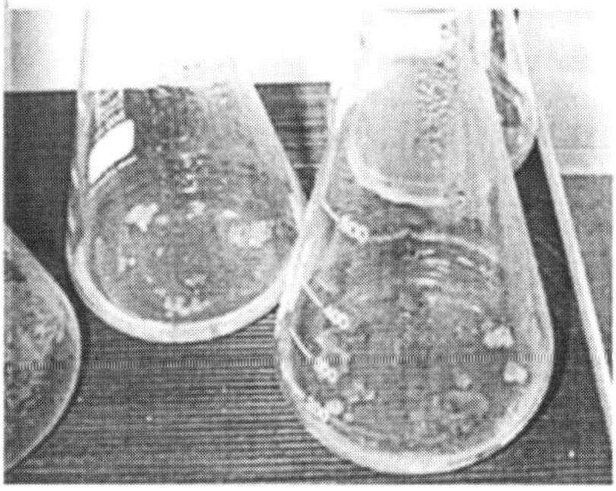

Figure 2a: Vials with plant cell cultures (callus) – the spots visible on the bottom of the vial

Figure 2b: Microwave generator: 1-electric power supply, 2-horn antenna, 3-thermostate with callus vials

II. 3. Photosyntetic pigment extraction and assay were accomplished following standard procedure - extraction solvent was acetone 90% in distilled water, light wavelengths for reading extinction coefficients were 440.5 nm, 644 nm and 662 nm. Nucleic acid extraction was accomplished in perchloric acid 6% and light extinction at the wavelengths of 270 nm and 290 nm were measured. For pigment assay the Meyer-Bertenrath [19] method while for average nucleic acid the Spirin's method were used [20].

II. 4. Spectrocolorimetric assay was carried out using a JASCO laboratory device, computer assisted, with quartz cells.

III. RESULTS AND DISCUSSION

In the vials with vegetal tissue cultures derived either from leaves or flowers callus was obtained during three months of growth in sterile environment. From initial explants, picked up from poppy flowers or leaves, new cells developed following growth stimulation with adequate hormone balance in the sterile culture medium (callusogenesis). In the first developmental stages the callus (non-differentiated vegetal cells) is not green, but, later, chlorophylls are synthesized and callus color shifts to greenish. In time, from callus small roots derived and the all types of tissues are generated until the whole plant is restored (organogenesis). This is plant cloning through in vitro micro-propagation. As well as embryos tissue, extremely sensitivity to external factors characterizes callus cells, allowing biotechnologists to conduct plant growth by means of various physical and chemical agents.

III.1.Biological parameters

In experimental samples corresponding to small exposure times (2; 4; 6; 8 hours) no significant modification of callus aspect, color and volume were noticed.

Neither callusogenesis nor organogenesis was influenced by microwave treatment, i.e. the same developmental stages were observed in control and exposed samples. In samples corresponding to 10 and 12 hours exposure daily exposure the next changes were noticed (in comparison to control samples):

- color shifted, step-by-step, from green - white - cream to grey;
- more pronounced increased of volume in comparison to the surface increase (with consequences on the friability) was remarked;
- the diminution of proliferation rate of the vegetal tissue (probably in the favor of morphogenesis, hyperhydricity of re-differentiated material) was observed;
- in the sub-cultures derived further from the tissues directly exposed to microwave flow a graduate restoration of initial look was observed;
- it was not remarked the appearance of roots after microwave treatment.

III.2. Biochemical parameters

Assimilatory pigment levels are presented in Figs. 3-4. For short exposure times (2; 4; 6 h) all three pigments present averaged values ranging around control values, i.e. small enhancements or small diminutions, without statistical significance (see Table I).). For longer exposure times (8; 10 and 12 h) significant stimulatory effect was revealed, especially in chlorophyll b and chlorophyll a. We noticed that chlorophyll b level is higher than that of chlorophyll a in contrast with the situation in adult plants (chlorophyll b, as secondary photosynthesis pigment, has an important role in photosynthesis sustaining).

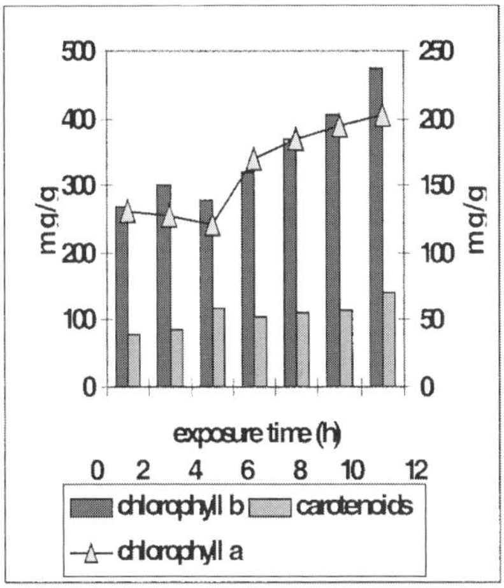

Figure 3: Assimilatory pigment content

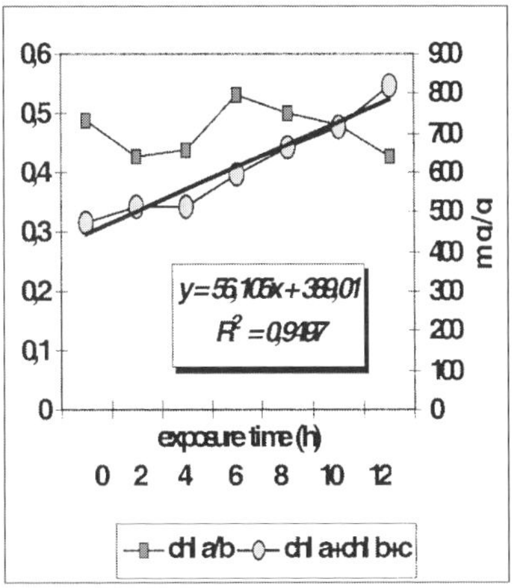

Figure 4: The ratio of chlorophylls contents and the total pigment content

In Figure 5 the average content of DNA and RNA is given. Significant decrease is observed for relatively longer exposure times.

The interpretation of these experimental results need to take into account both thermal and non-thermal effect of microwaves. In the present case power density of microwave flow was rather low so that a thermal effect could be only a weak one.

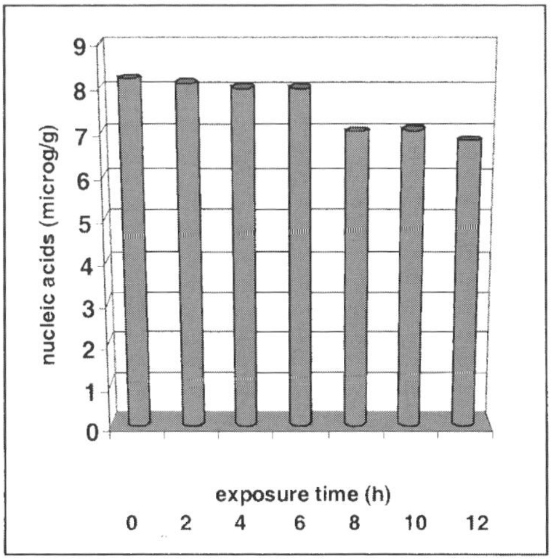

Figure 5: Average content of DNA and RNA

The results obtained for assimilatory pigments are different to those obtained for the average content of nucleic acids.

The levels of chlorophyll a, chlorophyll b and carotenoid pigments appeared as increased after microwave exposure and so is their sum that seems to be enhanced proportionally to the microwave exposure time (correlation coefficient of the regression line is over 0.9). However the physiological parameter which is the chlorophyll ratio presents some values lower than in control sample: for 2; 4; 10 and 12 hours.

The correspondent values of average nucleic acid content is significantly diminished for relatively longer exposure time: 8; 10; 12 hours.

We outline that no difference was observed between the behavior of samples obtained from leaves and those obtained from flowers so that the type of explant is not significantly involved in the cell culture sensitivity to microwave treatment. In Table I the result of Student t-test application is presented, the significance criterion being $p<0.05$. The comparison of the five control samples series to every of the five exposed samples revealed several non significant differences (n.s.) especially for shorter exposure times.

Table I. Statistic significance of assimilatory pigment and nucleic acid assays

Level of significance	Chll a	Chll b	Carotenoids	(DNA+RNA)/2
2 h	n.s.	n.s.	n.s.	n.s
4 h	n.s.	n.s.	n.s.	n.s
6 h	n.s.	n.s.	P<0.05	n.s
8 h	n.s.	P<0.05	P<0.05	P<0.05
10 h	P<0.05	P<0.05	P<0.05	P<0.05
12 h	P<0.05	P<0.05	P<0.05	P<0.05

The expected behavior of callus need to be understood as intermediate between cell cultures aqueous suspensions (such as in experiments with microorganisms) and young vegetal organisms derived from normal embryos. The cells from callus interact with each other much more than microorganisms cells while their capacity of differentiate in various types of cells is missing to microorganisms. In contrast with young plant organisms cells are still of the same kind and physiological interactions among them are much more simple.

Callus exposure to low power density microwaves reduced very much thermal effect (the power density we used is less than 1 mW cm^{-2} which is associated with a temperature increase of 0.1 ^{0}C).

Putative *molecular effects* in pigments and nucleic acid need to be judged first in energetic terms. The energy level of a microwave photon is approximately a million times lower than the energy required to break a covalent bond. Based on this fact, it has been stated in the literature that "microwaves are incapable of breaking the covalent bonds of DNA" [21-22]. However bacterial DNA molecules appeared mach damaged (in electron microscopy) in the Kakita's [23] experiment, but this may be only an indirect, complex effect of the microwaves acting not on isolated DNA molecules but on aqueous media containing nucleic acids. Still, no theory currently exists to explain the phenomenon of DNA fragmentation by microwaves although research is ongoing which may elucidate the mechanism.

In our experiment nucleic acid molecular damages may explain the behavior to longer exposure times of the average content of DNA and RNA. Regarding chemical bonds assuring pigment molecule primary structure, it is not impossible that electromagnetic fields are able to destroy them also by similar indirect mechanism as in the case of nucleic acids. Only this hypothesis is insufficient to explain the obvious increase in assimilatory pigment of microwave exposed callus though the diminished chl a/b ratio for several exposure times can be evoked in this frame.

Molecular effects of microwaves in exposed plant tissues able of photosynthetic activity may be related also to the role of the light harvesting protein complex of photosynthetic system II (LHC II). Encoded in the nucleus it is known to exhibit remarkable structural flexibility upon changes in the environmental conditions (such as, in the present case, the electromagnetic environmental component); so, environmental factors are considered able to regulate LHC II like phytochromes, redox, diurnal cycle etc. are doing too.

LHC II is slightly involved in plant capacity of light harvesting as well as in regulatory process such as via phosphorilation and non photochemical quenching. Consequently the chlorophyll a/b ratio, much dependent upon the LHC II content in the thylakoid membranes, can be affected by microwave exposure, at least during very early ontogenetic stages of vegetal organisms.

In this frame, we need to mention also the enzyme capacity to capture and transmit free energy from oscillating electric fields, as assumed by Westerhoff [24] which may be related to putative cellular effects of low power density microwaves.

Cellular effects resulted from the non-thermal microwave effect, may consist in any of the four damaging processes invoked above, in the introductory part regarding the cell membrane. However other influences of the electromagnetic exposure may occur in the complex structure of a biological cell. Mainly the rate of biochemical reactions, all controlled by specific enzymes, could be influenced by microwaves and consequently the biosynthesis mechanisms may be affected: not necessarily inhibit but also stimulate in some cases. This may happen by means of some enzyme activity enhancement of by the perturbation of ion channels functions. Mainly ion transport through thylakoid membranes can be invoked in this case, though ion channels of plasma membrane and other cytoplasmatic structures are also supposed to experience microwave action [25]. Ample variations in ion conductance may eliminate one of the terms from equation (4) if, for instance, voltage dependent ion channels are blocked following electromagnetic exposure.

Perturbation of ion transport may lead to perturbations of various biochemical reactions which are controlled by ion messengers (especially calcium ions are known as intra-cellular messengers); so, it is possible that biochemical reactions involved in the assimilatory pigment biosynthesis are stimulated for the duration of microwave action.

This is, of course, only an attempt to explain some apparently stimulatory effects of low power density microwaves, by means of the spectral methods presented above. More deep investigations are necessary to formulate a valuable hypothesis about cellular effects of electromagnetic exposure. In further biophysical projects we have planned to do so using also mathematical models for experimental results interpretation.

IV. CONCLUSION

Microwaves of low power density were able to influence tissue cultures developed from somatic cells of Papaver pseudo-orientale after relatively long exposure times. The low level thermal effect, ubiquitous for these electromagnetic radiations, possible sustained by non-thermal effects able to damage cell membrane, induced notable changes in the aspect, color, volume and friability of exposed callus. In the same time quantitative increasing in the photosynthesis pigment levels could be detected with statistical significance. The enhancement of chlorophyll b level, but also of chlorophyll a and carotene, remarkable especially for long exposure times, suggested the stimulation of bioshynthesis.

REFERENCES

[1] Tambiev A.H., Kirikova N.N, 1994, in Biological aspects of Low Intensity Millimeter Waves, Ed. Deviatkov N.D. and Betskii O.V., Moscow

[2] Russello V., Tamburello C., Scialabba A., 1996, Microwave effects on germination and growth of Brassica drepanensis seeds. Proceedings of 3rd Internat. Congress of the European Bio Electromagnetics association, p. 89
[2] Riipulk, J., Hiinrikus, H., 1999, Microwave radiometry for medical applications, Medical and Biological Engineering and Computing, 37 (1):99-103
[3] Ruzic R., Jerman I., Jelig A., Fefer D., 1992, Electromagnetic stimulation of buds of Castanea sativa Mill. in tissue culture, Electro and Magneto Biology, 11(2): 145-153
[4] Polk Ch and Postow E., 1996, Handbook of Biological Effects of Electromagnetic Fields, Ed. CRC Press Ltd., Boca Raton, New York, 1996
[5] Vlahovici, Al., Gasner, P., Pavel, A., Trifan, M., Creanga, D., 2000, Microwave exposure influence on fluorescence spectra of assimilatory pigments in Chelidonium majus, Ann. Sci. Univ. Besancon., B40-48
[6] Datta, A. K. and Hu, W. 1992. Quality optimization of dielectric heating processes. Food Technol. 46(12):53-56
[7] Datta, A. K. and Liu, J. 1992. Thermal time distributions for microwave and conventional heating of food. Trans I Chem E. 70(C):83-90
[8] Datta, A. K., Sun, E. and Solis, A. 1994. Food dielectric property data and its composition-based prediction. M. A. Rao and S. S. H. Rizvi(eds.). Engineering Properties of Food. New York. Marcel Dekker. 457-494.
[9] Datta, A. K. .2000. Fundamentals of heat and moisture transport for microwaveable food product and process development. A. K. Datta and R. C. Anatheswaran. (eds.). Handbook of Microwave Technology for Food Applications. Marcel Dekker, Inc. New York.
[10] Brunkhorst, C., Ciotti, D., Fredd, E., Wilson, J.R., Geveke, D.J., Kozempel, M. Development of process equipment to separate nonthermal and thermal effects of RF energy on microorganisms. Journal Microwave Power Electromagnetic Energy. 2000. v. 35(1). p. 44-50.
[11] Kozempel, M. F., Annous, B. A., Cook, R. D., Scullen, O. J. and Whiting, R. C. 1998. Inactivation of microorganisms with microwaves at reduced temperatures. J Food Protect. 61(5):582-585
[12] Heddleson, R. A. and Doores, S. 1994. Factors affecting microwave heating of foods and microwave induced destruction of foodborne pathogens - a review. J Food Protect. 57(11):1025-1037
[13] Khalil, H. and Villota, R. 1989a. A comparative study of the thermal inactivation of *B.* stearothermophilus spores in microwave and conventional heating. Food Engineering and Process Applications. New York, NY. Elsevier Applied Science Publishers. 1. 583-594
[14]Khalil, H. and Villota, R. 1989b. The effect of microwave sublethal heating on the ribonucleic acids of Staphylococcus aureus. J Food Protect. 52(8):544-548
[15] Russello V., Tamburello C., Scialabba A., 1996
[16] Franceschetti, G., Pinto, I, 1984, Cell membrane nolinear response to an applied electromagnetic field, IEEE Trans on MTT, 653
[17] Busang, J., Non linear system withmultiple inputs, ,Proc. IEEE, 1974, 66, 1088
[18] Hudgkin, A.L., Huxley, A.F., 1952, ,A quantitative ededscription of membrane current and its application to conductance and excitation in nerves, J. Physiol., 17, 500
[19] Hager, A., Meyer-Bertenrath, T., 1966, Die Isolierung und quantitative Bestimmung der Carotenoide und Chlorophylle von Blattern, Algen und isolierten Chloroplasten mit Hilfe dunnschichtchromatographischer Methodes, *Planta* 69, 128-217,

[20] Spirin, A., 1958, Spektrofotometriceskoe opredelenie summarnovo kolicestva nuclienovih kislot, *Biochimia,* Ed. Mir , Moscow, p. 656

[21] Jeng, D. K. H., K. A. Kaczmarek, A. G. Woodworth, G. Balasky, 1987, Mechanism of microwave sterilization in the dry state, *Applied and Environ. Microbiol.* 53, 2133-2137.

[22] Fujikawa, H., H. Ushioda and Y. Kudo, 1992, Kinetics of *Escherichia coli* destruction by microwave irradiation, *Applied and Environ. Microbiol.* 58, 920-924

[23] Kakita, Y., N. Kashige, K. Murata, A. Kuroiwa, M. Funatsu and K. Watanabe,1995. "Inactivation of Lactobacillus bacteriophage PL-1 by microwave irradiation." Microbiol. Immunol. 39: 571-576.

[24] Westerhoff, H.V., Tsong, T. Y., Chock, P.B., Chen, Y.D., Astumian, R.D., 1986, How enzymes can capture and transmit free-energy from an oscillating electric field, *Proceed. of the National Academy of Sciences of the United States of America*, 83 (13), 4734-4738

[25]Hinrikus, H., 1999, Low-level Microwave Field Effects on Living Systems, *Med. Biol. Eng. Comput.*, 37 (1), 45-49

Applied Physics (APHYS 2003)
A. Méndez-Vilas (Ed.)

The Critical Nature of Electromigration

E. Dalton, D. Corcoran, G.Gooberman, A.Arshak

Department of Physics, University of Limerick, Plassey Technological Park, Limerick, Ireland.

Abstract. High-resolution resistance measurements of electromigration in an Al thin-film reveal complex temporal fluctuations. Early fluctuations are consistent with internal fabrication stress relaxation, which in some cases is stimulated by electromigration. Later fluctuations at current densities ≥ 0.5 MA/cm^2 appear to be caused by electromigration alone. Spatial and temporal scale-invariance of the later fluctuations are indicative of a critical origin.

INTRODUCTION

Electromigration (EM) is the driven diffusion of atoms/vacancies due to the momentum imparted by electrons, and it has become of increasing importance as scales in the microelectronic industry have reduced [1]. Divergence of the atom/vacancy flux occurring at grain boundaries, for example, in an integrated circuit metal conductor can lead to heterogeneous atomic accumulations and depletions, eventually producing spurs and voids that cause line shorts and breaks respectively. To date the study of the phenomenon has principally been motivated by practical considerations, such as predicting when the detrimental breakdown or shorting will occur. Yet EM is also of fundamental research interest because of the dynamic and spectral complexity it exhibits. Studies have shown, for instance, an array of EM resistance noise features thought to be related to underlying changes in microstructure, such as void formation or dislocation dynamics [2]. Indeed, many researchers have attempted to study EM noise using low frequency power spectra, with a view to characterizing EM damage [3]. In this work, we study the temporal complexity and spectral signature of EM, and suggest that these can arise because the associated microstructural rearrangements are occurring close to a critical point.

EXPERIMENTAL PROCEDURE

We have conducted resistance measurements on an unannealed Al thin film of dimension 300x6x0.5 μm^3. The test structure used was a resistance bridge to minimize the effects of ambient thermal fluctuations. The structure was prepared on a Si, SiO_2 substrate (SiO_2 thickness ~10 nm). An Al film was RF sputtered onto the substrate, and using a mask of the test structure, a negative pattern was exposed onto the film using standard lithographic techniques. The sample was then hard-baked and the Al

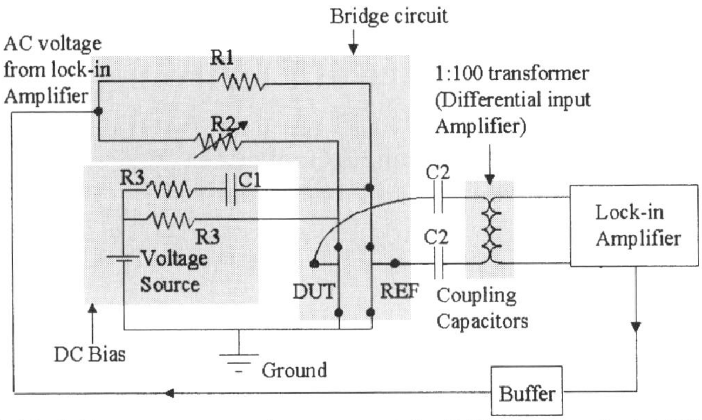

FIGURE 1. The bridge test structure and measurement circuit. The device under test (DUT) and the reference resistor (REF) are matched to remove the effect of ambient thermal fluctuations. A DC bias bridge was used to apply current stress to the DUT.

film wet-etched using hot sulphuric acid. The resulting test structure was connected to the measurement circuit shown in Fig. 1, and contained within a thermal/Faraday enclosure to minimize extraneous noise. Resistance changes were detected by a lock-in amplifier (EEG Model 124), which was interfaced to a computer for data recording.

A 10 V_{rms} AC probe voltage was applied to the sample and the sample substrate raised from room temperature to 50 ^{o}C in approximately 3 hours. The substrate temperature was set to 50 ^{o}C for all subsequent experiments. Resistance changes were initially recorded for a period of 12 hours (experiment 1 or E1). After a further 12 hours, a constant damaging current density of 0.5 MA/cm^2 was applied to the sample for 36 hours while continually recording (E2). The DC current was then reduced to zero and the resistance monitored for a 48-hour period (E3) for further possible events. Potential events were found only in the initial 12-hour period.

Subsequently, resistance fluctuations were recorded over 24 hour periods for applied current densities of 0.16 (E4), 0.33 (E6), 0.5 (E8), and 0.8 MA/cm^2 (E10), with monitored periods of 12 hours at zero applied DC current in between.

RESULTS

Resistance fluctuations were detected in experiment 1, even though no damaging current was applied (see Fig. 2a). The activity was observed to be intermittent and consisted of predominantly positive gaussian-shaped pulses. At the largest scales the pulses were asymmetric in shape beginning with a slow build-up in resistance. The small-scale behaviour also consisted of pulses, though the slow build-up was not observed. The lack of build-up is probably due to the increased activity at these scales, the increasing background resistance imposed by the larger event, and the limitations of the measurement system resolution (~10^{-7} $\Delta\Omega/\Omega$ and 5 s respectively). Remarkably, the pulses spanned more than 3 decades in size ($\Delta R/R$).

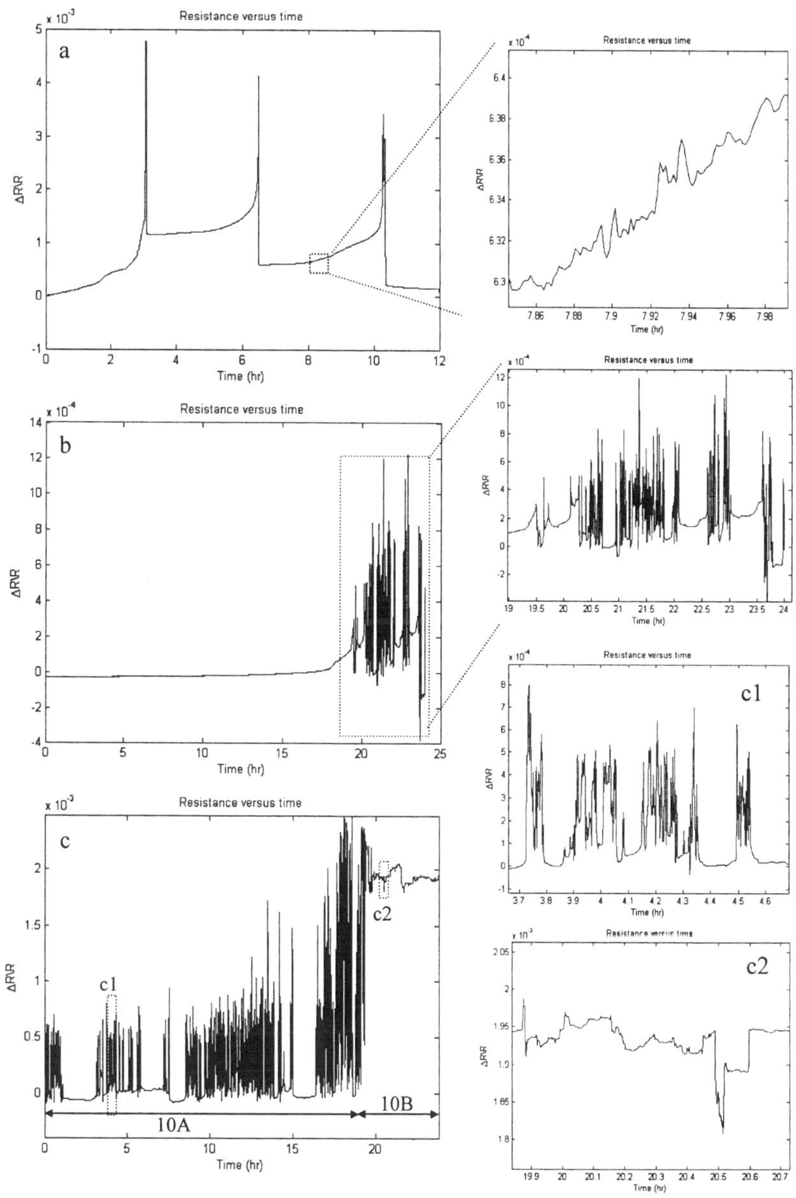

FIGURE 2. The time series of resistance for experiments (a) E1 with 0 applied DC current density, (b) E8 with applied current density of 0.5 MA/cm^2 and (c) E10 with applied current density of 0.8 MA/cm^2.

Similar resistance changes to those of experiment 1 were noted for all experiments with current densities ≤ 0.5 MA/cm^2, though it was observed that with increased current density the build-up time for a given pulse size decreased. In Fig. 2b, the time series of resistance for experiment 8 is shown. The presence of a long period of quiescence at the beginning of the experiment is notable. Such quiescence periods were also found in the preceding experiments 4 and 6, and interestingly the period increased in size with increasing applied current density.

In experiment 10 (see Fig. 2c) at an applied current density of 0.8 MA/cm^2, resistance fluctuations were observed from the beginning of the experiment and the early onset marks a transition in the dynamic behaviour of the sample. Experiment 10 can be divided into 2 intervals labeled A and B. Interval A at the beginning of experiment 10 [1 s, $7x10^4$ s], has very similar morphology to the end of experiment 8, with pulses appearing to cluster on a range of temporal scales. Interval B [$7x10^4$ s, $8.8x10^4$ s] is unlike any other temporal observation. The resistance changes are less frequent in number, appear more asymmetric, and have no preferred initial orientation.

To describe the pulsed resistance changes or fluctuations quantitatively, we have defined an event to be a continuous abrupt resistance change decreasing at a rate above 3.0 x 10^{-8} $\Delta\Omega/\Omega$ s^{-1}. In Fig. 3, we present the probability density distributions (PDDs) of event sizes for experiments 8 and 10A. The PDDs are power law decays, $p(s) \sim s^{-\alpha}$, α=1.1, with an excess of large events. The PDDs for the intervals between events are shown in the inset to Fig. 3 and reveal power laws without an excess, $P(i) \sim i^{-\gamma}$, γ=1.6. The spectra for these experiments also reveal power law decay $1/f^{\beta}$, with β in the range 1.0 to 1.2.

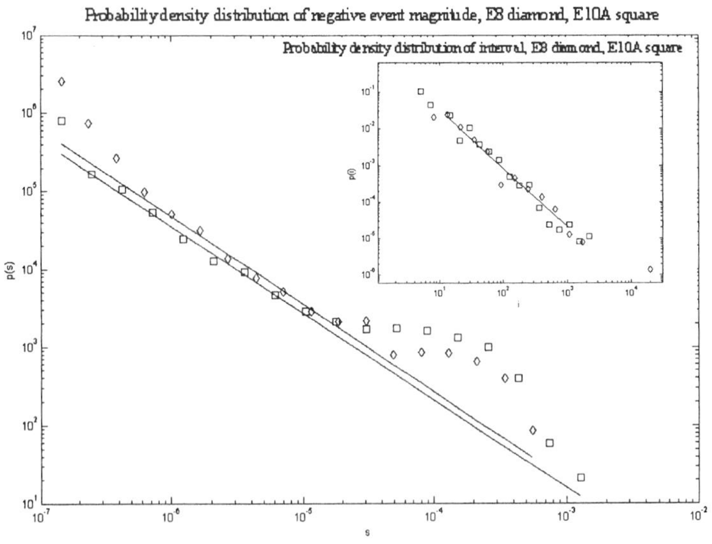

FIGURE 3. The probability density distribution of negative event magnitude and the interval (insert) for E8 and E10A.

ANALYSIS AND CONCLUSION

As mentioned in the introduction resistance fluctuations are thought to be related to micro-structural stress changes. Following Shingubara et al. [2] the resistance pulses observed here for experiment 1 are identified with stress relaxation from void formation or dislocation dynamics. If the gaussian-like pulses at the small scales have the same origin, then these relaxations span several decades in size.

Since no DC current was applied to the sample in experiment 1, the observed resistance fluctuations must have arisen from thermally activated relaxation from a pre-existing state, most probably a result of the fabrication process. The time-series of resistance for experiments 1 to 8, at current densities $\leq$ 0.5 MA/cm^2, are also consistent with relaxation of stresses formed during fabrication. The increasing quiescence period, in particular, even though the supposed damaging current was increasing, points toward a progressively more relaxed state from which the fluctuations emerged. Nevertheless, as the zero DC current intervening experiments did not exhibit significant resistance fluctuations, it is clear that the relaxations in the applied DC current experiments with current density $\leq$ 0.5 MA/cm^2 were being initiated by the electromigration process.

In Experiment 10, given the sudden removal of the quiescence period, it most likely marks a transition from fabrication dominated stress relaxations to an electromigration damaged stress state from which relaxation occurs. The pulsed nature of the events in interval A suggests that they also arise from void formation or dislocation dynamics. The PDDs of events for experiments 8 and 10A, are similar, and it is probable that the microstructural state has been retained in going from experiment 8 to 10A, and that the transition spans both experiments.

A system at a critical point exhibits a divergence of its underlying correlation time and length scales, manifested by temporal and spatial scale-invariance. A 1/f noise spectral signature, as has been observed here would be consistent with a diverging timescale, as would the power law PDDs observed for the intervals between events. The power law decays in event size, obtained for the EM fluctuations in experiments 8 and 10A, are a reflection of underlying spatial scale invariance. The excess of large events however, suggests that the system is not exactly at a critical point but that it has been overdriven to supercriticality.

ACKNOWLEDGMENTS

The authors would like to acknowledge project funding from Enterprise Ireland.

REFERENCES

1. J.R. Lloyd, *J. Phys. D.* **32,** R109-R118 (1999).
2. S. Shingubara, K. Fujiki, A. Sano, K. Inoue, H. Sakaue, M. Saitoh and Y. Horiike, *Jpn. J. Appl. Phys.* **34**, 1030-1036 (1995).
3. C. Ciofi and B. Neri, *J. Phys. D.* **33,** R199-R216 (2000).

Applied Physics (APHYS 2003)
A. Méndez-Vilas (Ed.)

CORRELATIONSHIP BETWEEN MICROSCOPIC OBSERVATIONS AND ELECTROCHEMICAL BEHAVIOUR OF DIFFERENT KIND OF GALVANIZED STEEL.

J.J. García-Jareño, D. Giménez-Romero and F. Vicente*

Departament de Química Física. Universitat de València. C/ Dr Moliner, 50, 46100, Burjassot, València (Spain), E-mail proclq@uv.es.

**To whom correspondence should be addressed.*

Abstract. Zinc anodic dissolution has been studied according to the steel galvanized method by means of the electrochemical impedance spectroscopy (EIS) and microscopic observations. Relevant information on the galvanized method is provided by the analysis of experimental data. The galvanized method has no influence on the kinetics parameters of the zinc anodic dissolution process. The galvanized method only changes the surface texture of the working electrode. Thus, the EIS fitting allows to calculate the fractal dimension of the surface of the working electrode.

INTRODUCTION

One of the most important applications of zinc is its use as a protective coating against corrosion on steel for outdoor use [1]. Thus, the zinc corrosion process has been the subject of numerous publications [2-4].

Electrochemical impedance spectroscopy (EIS) has become highly relevant when applied to the study of electrode reactions [5]. So, the impedance spectra of the dissolution reactions of metals usually show one or several relaxations (capacitive or inductive), which have been assigned to absorbed intermediates participating in the reaction [6]. The deviation between the idealized model and the real system data is a very common problem. The irregularity of a real interface frequently means an adequate explanation for that deviation [7, 8].

The texture of images is an important tool used in pattern recognition to characterize the arrangement of basic constituents of a material on a surface [9]. One way to determine the texture is by analysing the surface intensity by means of the fractal dimension. Pentland [10] provided the first theory in this respect by stating that fractal dimension correlates quite well with human perception of smoothness versus roughness. Some methods for the estimation of the fractal dimension are: the "blanket" method [11]; box counting method [12]; differential fractal Brownian motion method and the frequency domain method [13-15].

The aim of this work is to find a correlationship between EIS, optical observations and observations of electronic microscopy. This purpose will be established by means of the study of the relationship between the fractal dimension and the EIS fitting.

EXPERIMENTAL

The electrochemical experiments were carried out by means of a typical three electrodes cell where a platinum plate was the counter electrode and a Ag|AgCl|KCl_{sat} the reference electrode. The working electrodes were steel sheets coated by galvanized zinc plate supplied by GALESA with a geometrical surface equal to 0.5 cm^2. The galvanized plates used as working electrodes were: shooping plate, hot dip galvanized steel, continuous hot-dipped galvanized steel and electro-galvanized steel.

The working solution was composed of H_3BO_3 (R.P. Normapur®) 0.32 M, NH_4Cl (p. a. Panreac) 0.26 M and Na_2SO_4 (GEHE&Co.A.G. Dresden) 1.33 M, pH=4.4 [16]. The cell was thermostatized at 24.4ºC by means of a HETO DENMARK bath and bubbled with Ar (AIR LIQUIDE) for about five minutes before the experiment.

These experiments were carried out by means of a potentiostat-galvanostat PAR 273A and the frequency analyser was used with the lock-in-amplifier PAR 5210. The impedance measurements were made at stabilization potential of –0.975 V and in the frequency range from 0.05 to 10^4 Hz, with signal amplitude of 5 mV *rms*. Fitting of experimental data to the theoretical models proposed was done by means of a least squares 'home-made' software based on the Marquardt method for optimisation of functions [17, 18].

The electrodes images were realized by means of the scanner Genius Color Page-HR6X at a resolution of 2400 ppi and over a geometrical surface of 50 mm^2. On the other hand, the electronic microscopy (SEM) images were created by means of XL-30 ESEM (Philips) over 0.36 mm^2 and the Scanning Tunnelling Microscopy (STM) images by means of the METRIS-1000 (Burleigh Instrumentations) over 0.0049 mm^2.

RESULTS AND DISCUSSION

Zinc anodic dissolution can be considered as two consecutive monoelectron transfers [19, 20]:

$$Zn(\theta_0) \xrightarrow{k_1} Zn(I)(\theta_1) + 1e^- \qquad \text{(i)}$$

$$Zn(I)(\theta_1) \xrightarrow{k_2} Zn(II)(\theta_2) + 1e^- \qquad \text{(ii)}$$

$$Zn(II)(\theta_2) \xrightarrow{k_4} Zn^{2+}(\theta_2) \qquad \text{(iii)}$$

The EIS spectra of this electrodic process (Figure 1) present two consecutive loops in these experimental conditions, a capacitive loop at high frequencies and an inductive loop at low frequencies [21].

Figure 1 shows the experimental EIS spectra of the zinc anodic dissolution process under kinetics control for the different types of galvanized steel. In this Figure, it is observed that these spectra show both loops for all kind of galvanized. Thus, it is possible to consider that the transfer mechanism is the same for all these sheets.

The EIS spectra are fitted to the theoretical expression of the faradaic impedance function of two consecutive monoelectron transfers, Eq. (1) [20].

$$FA\frac{dE}{di}=\frac{\left(-\omega^2+R\right)\left(S+T\omega^2\right)+Y\phi\omega^2}{\omega^2\phi^2+\left(T\omega^2+S\right)^2}+j\frac{\left(TY+\phi\right)\omega^3+\omega\left(SY-\phi R\right)}{\omega^2\phi^2+\left(T\omega^2+S\right)^2} \quad (1)$$

$$R=\left(k_2^0+k_1^0\right)k_4+k_1^0k_2^0 \quad (2)$$

$$S=\left(\left(k_2^0+k_1^0\right)k_4+k_1^0k_2^0\right)\frac{k_4}{k_1^0k_2^0+k_1^0k_4+k_2^0k_4}\theta^0\left(k_1^0b_1k_2^0+k_2^0b_2k_1^0\right)- \quad (3)$$

$$-2k_2^0k_1^0k_2^0b_2\frac{k_1^0k_4}{k_1^0k_2^0+k_1^0k_4+k_2^0k_4}\theta^0$$

$$T=-\left(k_1^0b_1\frac{k_2^0k_4}{k_1^0k_2^0+k_1^0k_4+k_2^0k_4}\theta^0+k_2^0b_2\frac{k_1^0k_4}{k_1^0k_2^0+k_1^0k_4+k_2^0k_4}\theta^0\right) \quad (4)$$

$$\phi=\left(k_1^0+k_2^0+k_4\right)\frac{k_4}{k_1^0k_2^0+k_1^0k_4+k_2^0k_4}\theta^0\left(k_1^0b_1k_2^0+k_2^0b_2k_1^0\right)- \quad (5)$$

$$-k_1^0k_2^0b_2\frac{k_1^0k_4}{k_1^0k_2^0+k_1^0k_4+k_2^0k_4}\theta^0$$

$$Y=\left(k_1^0+k_2^0+k_4\right) \quad (6)$$

where A is the electroactive surface of the working electrode, F the Faraday constant, ω the frequency signal, E the applied potential, i the system intensity, j the square root of -1, k_i is the kinetics constant of the ith transfer, b_i the exponential factor of the ith kinetics constant and θ^0 is the number of the initial active centres..

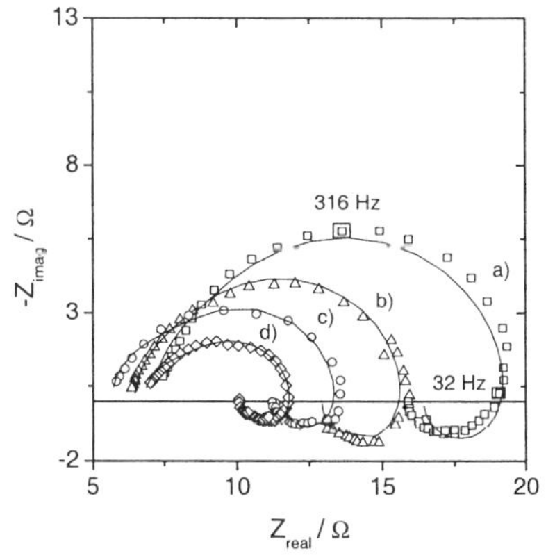

FIGURE 1. Nyquist plots for the zinc anodic dissolution for different textures of the working electrode at E = -0.975 V with regard to the Ag|AgCl|KClsat electrode. Experimental conditions were H_3BO_3 0.32M, Na_2SO_4 1.32 M, NH_4Cl 0.26M, pH=4.4 and T 297.5±0.1 K. The working solution was deaerated by bubbling Ar for 5 min. (O) is the spectrum for the hot dip galvanized steel, (◊) is for the shooping plate, (□) for the continuous hot-dipped galvanized steel and (Δ) for the electro-galvanized steel. The continuous lines are the EIS simulations, where the kinetics parameters are shown in Table 1.

To fit the experimental data, it is taken into account that the electrical impedance is composed of: the faradaic impedance, Eq. (1), and the impedance of the cell, uncompensated resistance, R_u, and the double layer capacitance, C_{dl}. In a previous paper [21], the exponential factors have been evaluated by using the following values: $b_1 = 16$ and $b_2 = 14$. Subsequently, these values are also considered in the EIS fitting.

Table 1 shows the kinetics parameters obtained from these fittings. In this Table, it is possible to see that the surface of the working electrode changes with the type of working electrode, whereas the kinetics parameters can be considered constant with respect to the galvanized method. The variations amongst these parameters are only due to fluctuations in the experimental data since the more accurate kinetic parameters correspond to the product $k_1\theta_0$ and these ones are identical in all sheets. Thus, it is possible to say that the variations of the EIS spectra are only due to the modification of the surface texture of the working electrode. The values of surface area of the galvanized steel calculated by means of the EIS fittings increases as the roughness of the working electrode also increases.

The quantitative relationship between microscopic observations and the EIS fitting is established by means of the fractal dimension calculation. The fractal dimension is related with the surface area of the working electrode, A, through the scaling law [22]:

$$A(\lambda) = A_0 \lambda^{-(fd-2)} \tag{10}$$

where λ is the scaling ratio, fd is called the fractal dimension of the surface and A_0 is the geometrical surface of the working electrode.

The fractal dimension in this work, Table 1, is calculated by means of the Differential Box Counting (DBC) method [23, 24], using, as mathematical algorithm of calculation, the algorithm established by Biswas *et al.* [12]. In this method, the λ value is 1/4. The fractal dimension, calculated in this manner, increases as the working electrode increases its roughness, Table 1.

The Table 1 shows that similar values for the fd are obtained regardless of the image analyzed. This way, the validity of this sheets as a fractal object model in the self-similarity [25] range established by these images. Although, this self-similarity gets lost in the STM measurements since the fractal dimensions calculated from these images are different from the previous ones.

Figure 2 shows the relationship of the surface values of the different type of galvanized with the fractal dimension measured in the self-similarity range. This relationship is logarithmic as shows the Eq. (1), since the scaling ratio is only dependent on the mathematical algorithm used for its calculation. The experimental relationship between the fractal dimension and the logarithm of the surface is linear (Figure 4) and with a slope similar to the napierian logarithm of the scaling factor (theoretical value 1.4 and experimental value 1.7). On the other hand, the ordinate intercept of this representation corresponds to -13.7 whereas its expected value corresponds to -12.7.

Accordingly, it is possible to say that this Figure makes evident the dependence of the faradaic impedance on the surface texture of the working electrode. The galvanized method only modifies the texture of the working electrode whereas the kinetics parameters can be considered constant between the different types of sheets since the different packing degree of zinc does not affect the value of these parameters.

Table 1. Kinetics parameters of the zinc anodic dissolution in deareated sulphate medium.

Working Electrode	fd_{50mm^2} *Scanner*	$fd_{0.36mm^2}$ *SEM*	$fd_{0.0049mm^2}$ *STM*	R_u / Ω	$C_{dl} / \mu F$	A / cm^2	k_1^0 / s^{-1}	k_2^0 / s^{-1}	k_4 / s^{-1}	$\theta^0 / \mu mol\ m^{-2}$
electro-galvanized steel	2.48	2.44	2.24	6.5	62.3	0.98	3329	4	12	223
continuous hot-dipped galvanized steel	2.33	2.34	2.19	7.1	26.4	0.74	6071	14	30	73
hot dip galvanized steel	2.51	2.53	2.05	5.7	32.8	0.99	4224	7	14	182
shooping plate	2.66	2.55	2.05	6.8	34.3	1.34	5656	6	19	212

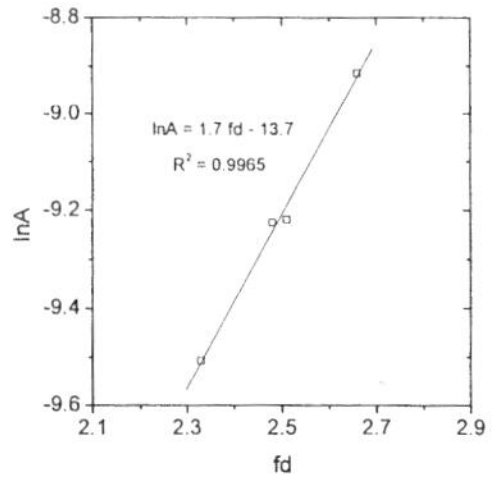

FIGURE 2. Curve of the experimental surface of the working electrode vs the fractal dimension of this surface. The continuous line is the linear fit.

CONCLUSION

The EIS measurements are a characteristic of the working electrode highly dependent on the surface area. These spectra depend directly on the texture of the working electrode and, thus, are related exponentially with the fractal dimension of the surface of these electrodes. It is also proved that the packing degree of zinc does not affect the value of the kinetics parameters of the mechanism of zinc anodic dissolution. These parameters are independent of the roughness of the working electrode.

ACKNOWLEDGEMENTS

This work has been partially supported by CICyT-MAT/2000-0100-P4. D. Giménez-Romero acknowledges a Fellowship from the Generalitat Valenciana, Program FPI. J.J. García-Jareño acknowledges the financial support of the program "*Ramón y Cajal*" from Ministerio de Ciencia y Tecnología (Spain).

REFERENCES

1. E. Ahlberg and H. Anderson. Acta Chem. Scand. 46 (1992) 15.
2. E.B. Yousfi, J. Fouache and D. Lincot. Appl. Surf. Sci. 153 (2000) 223.
3. O. A. Ashiru and J. Shirokoff. Appl. Surf. Sci. 103 (1996) 159.
4. C. Cachet, F. Ganne, G. Maurin, J. Petitjean, V. Vivier and R. Wiart. Electrochim. Acta 47 (2001) 509.
5. T. Tsai, Y. Wu and S. Yen. Appl. Surf. Sci. 214 (2003) 120.
6. A. Ahlberg and H. Anderson. Acta Chem. Scand. 47 (1993) 1162.
7. G. A. McRae, M. A. Maguire, C. A. Jeffrey, D. A. Guzonas and C. A. Brown. Appl. Surf. Sci. 191 (2002) 94.
8. S. A. M. Refaey and G. Schwitzgebel. Appl. Surf. Sci. 135 (998) 243.
9. S. Chesters, H. Y. Wen, M. Lundin and G. Kasper. Appl. Surf. Sci. 40 (1989) 185.
10. A.P. Pentland. IEEE T. Pattern Anal. 6 (1984) 661.
11. S. Peleg, J. Naor, R. Hartley and D. Avnir. IEEE T. Pattern Anal. 6 (1984) 518.
12. M.K. Biswas, T. Ghose, S. Guha and P.K. Biswas. Pattern Recogn. Lett. 19 (1998) 309.
13. K.L. Chan. IEEE T. Bio.-Med. Eng. 42 (1995) 1033.
14. C.C. Chen, J.S. Daponte and M.D. Fox. IEEE T. Med. Imaging 8 (1989) 133.
15. K. Liao, P. Cavalieri and J. Pitts. T. ASAE 33 (1990) 298.
16. T. Ohtsuka and A. Komori. Electrochim. Acta 43 (1998) 3269.
17. J.R. Macdonald. Solid State Ionics 58 (1992) 97.
18. F. Vicente, A. Roig, J.J. García Jareño and A. Sanmatías. Procesos electródicos del Nafión y del Azul de Prusia/Nafion sobre electrodo transparente óxido de Indio-Estaño: Un modelo de electrodos multicapa. Ed. Moliner 40, Burjassot, 2001.
19. R. Wiart. Electrochim. Acta 35 (1990) 1587.
20. D. Giménez-Romero, J.J. García-Jareño and F. Vicente. J. Electoanal. Chem. In Pres.
21. D. Giménez-Romero, J.J. García-Jareño and F. Vicente. Electrochem. Commum. 5 (2003) 722.
22. Y.B. Wang, R.K. Yuan and M. Willander. Appl. Phys. A 63 (1996) 481.
23. N. Sarkar and B.B. Chaudhuri. Pattern Recogn. 25 (1992) 1035.
24. N. Sarkar and B.B. Chaudhuri. IEEE T. Syst. Man Cyb. 24 (1994) 115.
25. J. Navarro-Laboulais, J. Trijueque and F. Vicente. Materiales y Procesos Electródicos (II), Ed. INSDE, Burjassot, 2003, Ch. 15.

Applied Physics (APHYS 2003)
A. Méndez-Vilas (Ed.)

Actin- & Tubulin-Based Structures Under Low Frequency Noise Stress

Mariana Alves-Pereira*, João Joanaz de Melo*, and Nuno A. A. Castelo Branco

*Dept. Environmental Sciences & Engineering, DCEA-FCT, New University of Lisbon, 2829-516 Caparica, Portugal mariana.pereira@oninet.pt
Center for Human Performance, Scientific Board, Apartado 173
2615 Alverca Codex, Portugal

Abstract. Low frequency noise (LFN) (≤500 Hz, including infrasound) is a genotoxic agent of disease which impinges on the respiratory tract. In LFN-exposed rat models, the respiratory tract cellular populations become dramatically altered: a) tubulin-based structures, such as tracheal cilia, lose tension becoming wilted, or appear sheared, as if clipped, despite an intact internal axoneme structure, and b) actin-based structures, such as brush cell microvilli, appear fused. In the cochlea of LFN-exposed rats, actin-based stereocilia are also fused, and in human pericardia of LFN-exposed individuals, no cilia have been identified, as would be expected. Other actin-based structures of LFN-exposed rats are currently under study and first results will be presented. The biomechanical properties of these biological structures will be reviewed in order to provide preliminary hypotheses that might explain the behavior of these cellular components in the presence of acoustical energy in the form of LFN.

BACKGROUND

Actin and tubulin are proteins that can be found in numerous mammalian tissues and cellular structures. Microtubules (MT) are polar structures, consisting of 13 linear protofilaments, aligned in parallel and forming long hollow cylinders, each composed of alternating α- and β-tubulin molecules. With an outer diameter of 25 nm, MT are the primary organizers of the cytoskeleton. Actin protein forms 2-stranded helical polymers of 5-9 nm in diameter. Actin filaments (AF) can, in turn, form linear bundles, 2D networks and 3D gels. Within the cytoskeleton, AF are highly concentrated underneath the plasma membrane, at the cell cortex, they are more rigid than MT, and are highly polarized structures. AF restructuring of the cell cortex can be triggered by extracellular signalling, and re-organization of the actin cortex can influence the plasma membrane above. The actin cortex can also integrate movements of the animal cell over the entire surface.

Ciliary axonemes, with an outer diameter of 25 μm, are composed of 9 MT doublets, organized in a ring centered on an additional pair of MT. Axoneme MT are associated with several proteins that form cross-links maintaining MT bundles together, and generating the driving force for cilliary bending motion. MT also form mechanically-driven relay systems that control motion. Cilia allow for the movement of fluid over a surface, for the propelling of cells through a fluid, and are also secretory structures (1). Cilia distribution in the human respiratory epithelia: $10^9/cm^2$.

Microvilli, are finger-like, cell-surface projections that increase the active surface area of cells. At the core of each microvillus is a rigid bundle of parallel AF, cross-linked by actin-binding proteins. Laterally, the AF core is connected to the plasma membrane through motor proteins (myosin). Auditory hair cells in the cochlea, or stereocilia, are also formed by actin bundles.

Since 1992, and within the scope of vibroacoustic disease (2-4), animal models have been used to study the effects of low frequency noise ($\leq$500 Hz, including infrasound) (LFN) on biological structures. Through electron microscopy, the respiratory tract lining (respiratory epithelium) was studied in LFN-exposed rodents (1). The results raised many more questions than they answered. Some of these questions will be put forth herein, in light of physical and biomedical principles.

The goal of this report is to a) qualitatively analyze the changes of actin and tubulin-based structures induced by acoustic exposure; b) tentatively advance possible bio-mechanisms responsible for these changes; and c) bring this topic to the forum of biophysicists, bioengineers, and bio-mathematicians in an attempt to gain a deeper understanding of these biological processes.

FIGURE 1. Non-exposed rat bronchial epithelium. A brush cell (arrow) under a compact cilliary field featuring multiple cilliary vesicles.

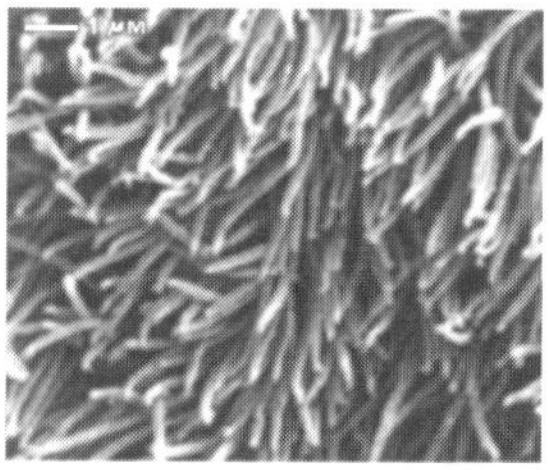

FIGURE 2. Non-exposed rat bronchial epithelium. A compact cilliary field featuring multiple cilliary vesicles.

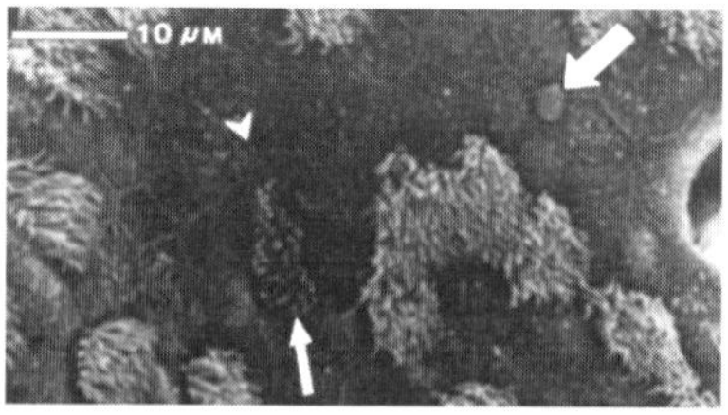

FIGURE 3. Rat tracheal epithelia, gestated and born in LFN & exposed to 235 more hours of occupationally-simulated LFN. Cilia are depleted, shaggy or entirely sheared (small arrow). Rosetta structures, centered on a brush cell (large arrow) are observable, although in some, the brush cell seems to have disappeared (arrowhead).

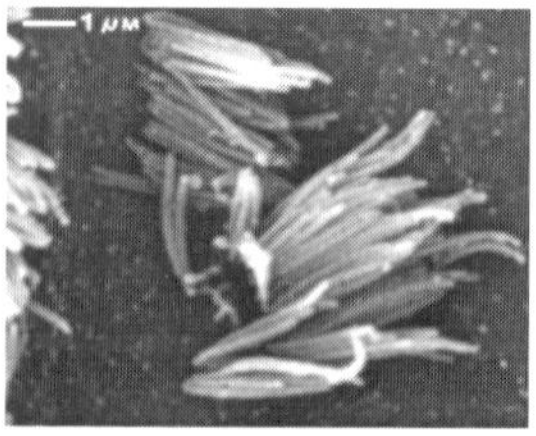

FIGURE 4. Rat Bronchial epithelium exposed to 2160 hours of continuous LFN, followed by 8 days of silence. A tuft of cilia, with some seemingly severed strands that appear to lie fallen on the epithelium surface. Multiple cilliary vesicles are visible.

METHODOLOGY

Data analyzed herein has been already been presented in medical and biological forums, hence animal treatment and electron microscopy techniques are described in detail elsewhere (1). In summary, Wistar rats were exposed to LFN during varying lengths of time: from 24 continuous hrs to 5304 hrs of occupationally-simulated exposure (8hrs/day, 5days/week, weekends in silence). After sacrifice, fragments of several organ tissues were removed for ultrastructural observation under scanning (SEM) electron microscopy. All ultramicrographs herein are coated with gold-palladium. The qualitative analysis of the response of biological tissue to LFN is approached herein from a biomechanical standpoint, drawing upon previously published results of both LFN-exposed human populations and animal models.

RESULTS

In LFN-exposed rodents, ciliary fields become depleted (Fig. 3). A distinct pattern of destruction, though, was not identifiable. Within the same micrograph, some areas had sheared cilia while other adjacent, ciliated cells exhibited longer, non-sheared strands, albeit shaggy (Fig 3). Shaggy cilia were seen in many micrographs although axoneme structures seemed intact. Sheared cilia seemed as though they had been clipped, and strands of seemingly clipped cilia were observed lying on the epithelial surface (Fig. 4). This is in stark contrast to control rats, where the respiratory epithelium possesses many cilliary fields (Figs. 1-2, 5).

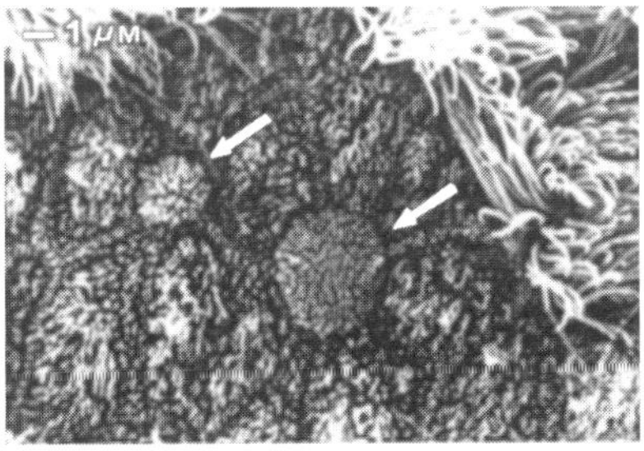

FIGURE 5. Non-exposed rat tracheal epithelium. Two brush cell (BC) (arrows), at different stages of development, are surrounded by secretory cells (SC) in rosetta-shaped structures. Cilia are long, uniform, and exuberant. BC microvilli are uniformly distributed. Rosetta-forming SC have exuberant microvilli at different lengths indicating that these SC are at different stages of life cycles.

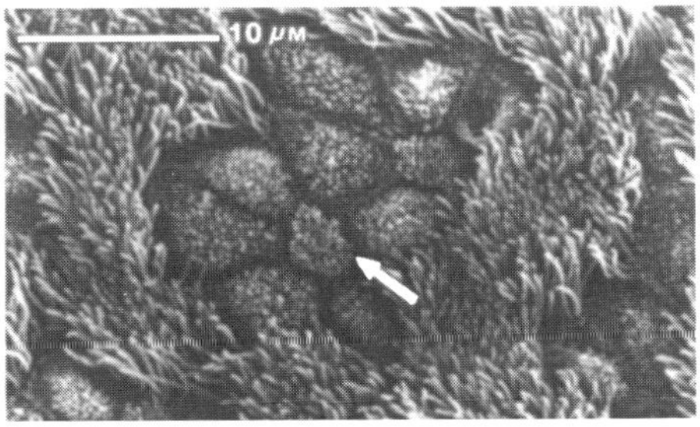

FIGURE 6. Tracheal epithelium of rat exposed to 1864 hours of occupationally-simulated LFN. Swelling is uniformly distributed. Cell borders are well defined as valleys. The rosetta centered on the BC (arrow) is conserved. BC microvilli are clustering into groups losing their uniform appearance. SC microvilli are still in clearly different stages of growth, but are visibly shorter than in controls.

FIGURE 7. Tracheal epithelium of rat exposed to 4399 hours of occupationally-simulated LFN. Amplification of a tracheal BC. Microvilli are clearly grouped together and, in some locations, appear almost fused. Surrounding SC microvilli are uniformly short and stubby.

FIGURE 8. Tracheal epithelium of rat exposed to 4399 hours of occupationally-simulated LFN. BC (arrow) with fused microvilli, seemingly spreading outward from the center, forming a type of indentation. Around the edges, individual microvilli are still identifiable. Rosetta structures are difficult to identify. Cilia of different lengths are slightly shaggy, and SC microvilli are short and disorganized. A few vesicles budding from cilia are visible.

Brush cells (BC) possess microvilli uniformly distributed over the apical surface that is open to the airway (Fig. 5). In LFN-exposed rodents, microvilli clustered together, and with increasing exposure time, became fused (Figs. 6-8). Secretory cell microvilli in exposed specimens appeared stunted (Figs. 6-8).

Cochlear stereocilia also appeared fused in LFN-exposed rats, both among themselves as well as with the upper tectorial membrane (Figs. 9,10).

DISCUSSION

Ciliary depletion of the epithelium does not appear to be solely due to some biochemical process. In fact, sheared cilia adjacent to long-stranded cilia suggest that a mechanical aggression may be at play. Considering the structure of cilia, severing of the axoneme would imply a fairly significant amount of external force. Severed ciliary strands, as seen in Fig. 4, would be physically easier if the bending properties were inexistent.

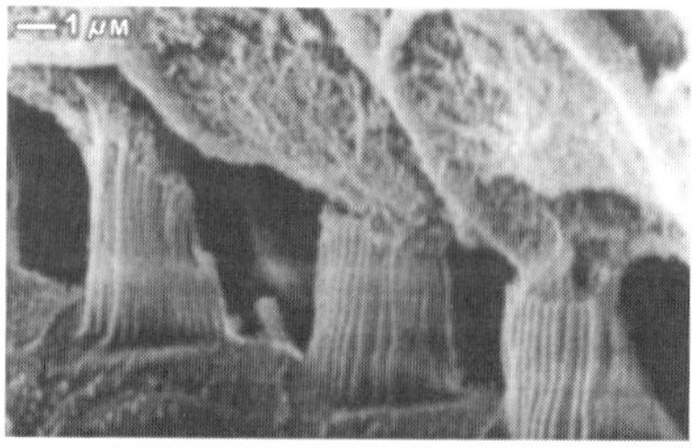

FIGURE 9. SEM of rat cochlear stereocilia exposed to 4399 hours of occupationally-simulated LFN. Cochlear stereocilia are fused with the upper tectonic membrane.

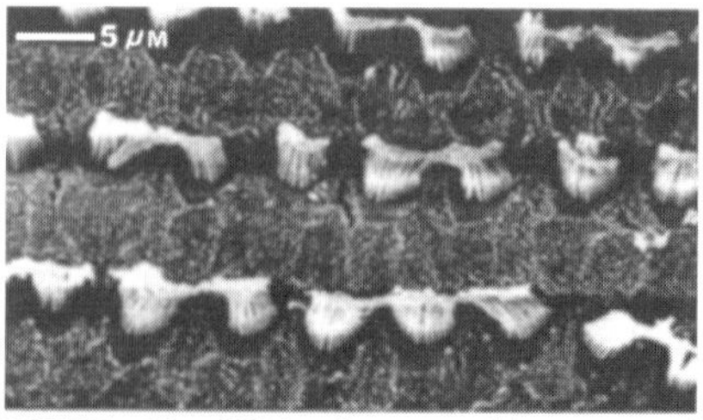

FIGURE 10. SEM of rat cochlear stereocilia exposed to 4399 hours of occupationally-simulated LFN. Cochlear stereocilia after removal of the tectonic membrane, portions of which remain fused to the stereocilia forming bridges between adjacent cells .

Since acoustic phenomena are involved, the concept of resonance is immediately present. It is possible that the LFN-induced vibration transmitted to the tracheal epithelia may resonate cilia at precisely the frequency that severs them. However, long-stranded cilia exist adjacent to severed cilia indicating that if resonance phenomena are at play, they are local events. The possibility of locally different vibratory nodes allied with the possibility of resonance phenomena, could provide a preliminary hypothesis for adjacent sheared and long-stranded ciliated cells.

Why BC microvilli respond to prolonged LFN stress by fusing is unknown. However, the fact that AF can form both rigid (but flexible) bundles as well as gel-like networks, taken together with the fact that motor proteins connect the AF core to the plasma membrane, microvilli fusion does not seem to be such a remote possibility, given the right triggering events. Fusion of cochlear stereocilia AF has also been observed, and corroborates this notion.

ACKNOWLEDGMENTS

The authors thank the Portuguese Ministry of Defense (CIMO) for animal facilities, FLAD (Luso-American Foundation for Development) for continuous support, Prof. Nuno Grande (ICBAS) and Prof. Carlos Sá (CEMUP) for electron microscopy, Daniela Sousa Silva (CEMUP), António Costa e Silva and Emanuel Monteiro (ICBAS), and Carlos Lopes (CIMO) for technical support, and Pedro Castelo Branco for image treatment. Alves-Pereira & Joanaz de Melo also thank IMAR (Instituto do Mar) for hosting project POCTI/MGS/41089/2001 and FCT (Fundação para a Ciência e Tecnologia) for its funding.

REFERENCES

1. Castelo Branco, N. A. A., Alves-Pereira, M., Martins dos Santos, J., and Monteiro, E., "SEM and TEM Study of Rat Respiratory Epithelia Exposed to Low Frequency Noise," in *Science and Technology Education in Microscopy: An Overview*, edited by A. Mendez-Vilas, Badajoz, Spain: Formatex, 2003, pp. 505-533.1.
2. Castelo Branco, N. A. A., and Rodriguez Lopez, E., , *Aviat. Space Environ. Med.* **70**, A1-A6 (1999).
3. Castelo Branco, N. A. A., *Aviat. Space Environ. Med.* **70**, A32-A39 (1999).
4. Castelo Branco, N. A. A., Rodriguez Lopez, E., Alves-Pereira, M., and Jones, D. R., *Aviat. Space Environ. Med.* **70**, A145-A151 (1999).

Applied Physics (APHYS 2003)
A. Méndez-Vilas (Ed.)

Low Frequency Noise Exposure and Biological Tissue: Reinforcement of Structural Integrity?

Mariana Alves-Pereira*, João Joanaz de Melo*, and Nuno A. A. Castelo Branco

*Dept. Environmental Sciences & Engineering, DCEA-FCT, New University of Lisbon, 2829-516 Caparica, Portugal mariana.pereira@oninet.pt
Center for Human Performance, Scientific Board, Apartado 173
2615 Alverca Codex, Portugal

Abstract. Low frequency noise (LFN) (≤500 Hz, including infrasound) exposure causes different types of damage to different organs. In the pericardium – a translucid sac that surrounds the heart - it causes the formation of a new layer of tissue, increasing pericardial thickness 100 to 300%. In the blood vessels, LFN exposure induces thickening of the vessel wall, resulting in a continuous, blanket-like layer throughout the vessel. This is in opposition to artherosclerotic plaques, for example, which attach to vessel walls in clumps. In the respiratory tract, LFN exposure causes the growth of diffuse fibrosis in the trachea and focal fibrosis in the lung. Simultaneously, it is responsible for the shearing and depletion of cillary fields and fusion of brush cell microvilli. How the transmission of acoustical energy induces these biological effects is still greatly unknown. Although biochemical factors must contribute greatly, biomechanical effects due to the direct presence of acoustical phenomena cannot be discarded. The need for biological tissue to maintain structural integrity given the probable acoustical disturbance of biological tension and compression factors, will be discussed within the scope of vibroacoustic disease and the principles of biotensegrity.

BACKGROUND

Since 1980, researchers at the Center for Human Performance (CPH), Alverca, Portugal, have been studying the effects of low frequency noise (LFN) (≤ 500 Hz, including infrasound) exposure on humans. In 1992, small rodents began to be exposed to similar acoustic environments in order to investigate the effects of LFN on the respiratory tract. Today, 11 years past, although many of the initial questions have been answered, many more have been raised, challenging classical biomedical concepts.

Vibroacoustic disease (VAD) is a whole-body (systemic) pathology, characterized by the abnormal growth of the extra-cellular matrices, and caused by long-term exposure (years) to LFN (1-3). (See Table 1) VAD has been diagnosed in aircraft technicians, commercial and military aircraft pilots, flight attendants and cabin crewmembers, restaurant workers, disk-jockeys, ship-machinists, and in the general population exposed to environmental LFN. Until the 1987 autopsy of a deceased VAD patient (4), LFN-induced pathology was thought to be restricted to the realm of neuropsychology. However, the involvement of the cardiovascular, digestive and respiratory systems was confirmed in autopsy (4), and later in animal models (5).

TABLE 1. Clinical stages of VAD. Data from a group of 140 aircraft technicians (selected from an initial group of 306), occupationally exposed to LFN. Exposure time (in years) refers to the amount of time it took for 70 individuals (50%) to develop the corresponding sign or symptom (2). Multiple signs and symptoms appear in each patient- they are not mutually exclusive.

Clinical Stage	Sign/Symptom
Stage I – Mild (1-4 years)	Slight mood swings, Indigestion & heart-burn, Mouth/throat infections, Bronchitis
Stage II-Moderate (4-10 years)	Chest pain, Definite mood swings, Back pain, Fatigue, Fungal, viral and parasitic skin infections, Inflammation of stomach lining, Pain and blood in urine, Conjunctivitis, Allergies
Stage III – Severe (> 10 years)	Psychiatric disturbances, Hemorrhages of nasal, digestive and conjunctive mucosa, Varicose veins and hemorrhoids, Duodenal ulcers, Spastic colitis, Decrease in visual acuity, Headaches, Severe joint pain, Intense muscular pain, Neurological disturbances

The purpose of this report is twofold: 1) to qualitatively analyze what is known to date on the response of living tissue to LFN exposure, and 2) bring this topic to the forum of biophysicists, bioengineers, and biomathematicians in an attempt to gain a deeper understanding of these biological processes.

METHODOLOGY

Data analyzed herein has been already been presented in medical and biological forums, hence animal treatment and electron microscopy techniques are described in detail elsewhere (5). In summary, Wistar rats were exposed to LFN during varying lengths of time: from 24 continuous hrs to 5304 hrs of occupationally-simulated exposure (8hrs/day, 5days/week, weekends in silence). After sacrifice, fragments of several organ tissues were removed for ultrastructural observation under scanning (SEM) electron microscopy. All ultramicrographs are coated with gold-palladium. The qualitative analysis of the response of biological tissue to LFN is approached herein from a biomechanical standpoint, drawing upon previously published results of both LFN-exposed human populations and animal models.

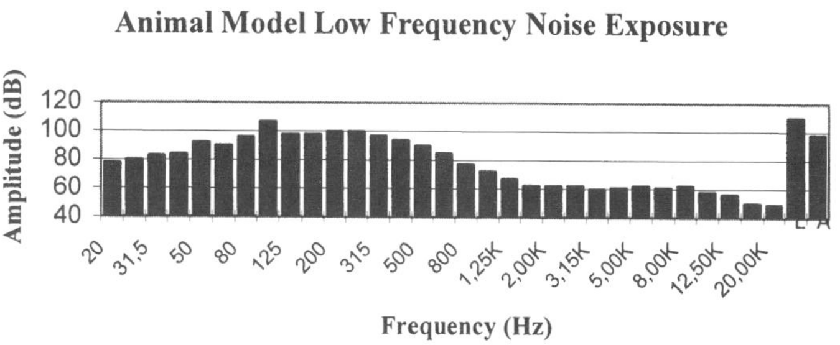

FIGURE 1. Overall linear (L) and A-weighted (A) noise levels, as well as the spectral analysis of the excitation signal collected at the position near the rat test group inside the chamber. Acoustic energy was highly concentrated in the lower frequency bands: in the frequency bands ranging from 50 Hz to 500 Hz, noise levels exceeded 90dB. Overall levels registered above 109dB, with A-weighted levels around 98dB (A).

RESPONSE OF BIOLOGICAL TISSUES TO LFN EXPOSURE

Cardiovascular Structures

LFN-exposed organisms exhibit thickened arteries, cardiac valves and pericardium. Pericardial thickening is the hallmark of VAD (6). No inflammatory process is present in VAD patients.

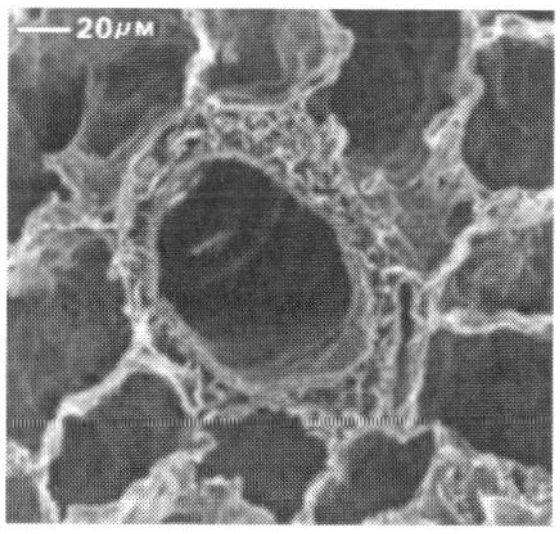

FIGURE 2. SEM of rat alveolar structure exposed to 2160 hours of continuous LFN. Amplification of small artery with an adjacent vein, both with thickened walls. Intima is thickened. Internal and external elastic laminae are also thickened and exhibit disruptions. This response is also observed in humans, as seen through autopsy. Note that this thickening is not similar to artherosclerotic plaques because it continuously lines the lumen and thus does not occur in clumps.

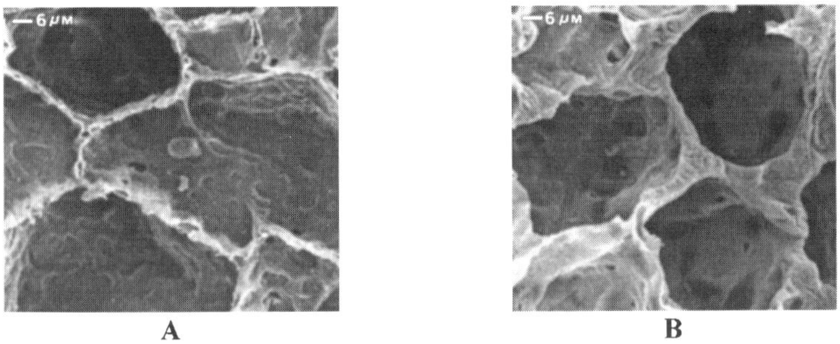

FIGURE 3. SEM of rat alveolar structure. (**A**) Non-exposed alveolar walls are thin and wall structure is visible. (**B**) Exposed to 2160 hours of continuous LFN. Alveolar walls are greatly thickened, and wall structure appears effaced.

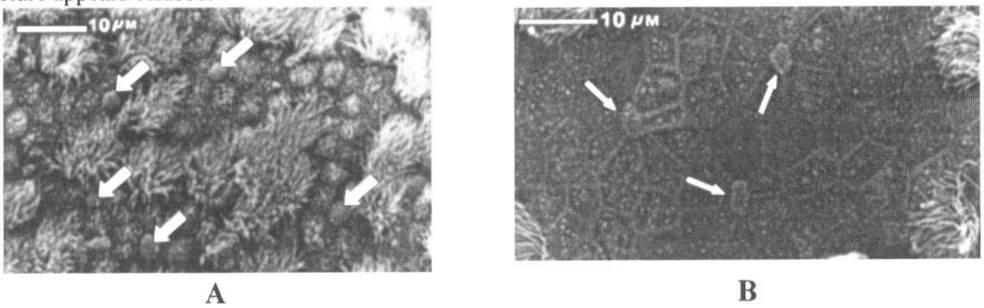

FIGURE 4. SEM of rat tracheal epithelium. All arrows point to brush cells (BC). (**A**) Non-exposed exhibiting several BC with tufted microvilli, surrounded by a ring of secretory cells (SC) in rosetta-shaped formations. SC microvilli are in different stages of development indicating that these cells are at different life cycles. Microvilli of both BC & SC sprout uniformly with the same density and shape in all the cells. Some intercellular junctions are fairly visible. Exuberant cilia, most of which in metachronal coordination, are also present. (**B**) Gestated and born in LFN, and subsequently exposed to 2213 hours of occupationally-simulated LFN. The amount of cilia is greatly reduced. Rosetta structures are visible, but SC are irregularly shaped. BC are distinctly visible in the center of the rosettas. BC microvilli have lost their uniform distribution, and SC microvilli are mostly short and stubby. Intercellular junctions are thick and prominent, and SC surfaces are flat or slightly sunken.

Kidney & Stomach

In the rat kidney structure, LFN exposure causes swelling of structures (See Fig. 5), that disappears if the animal is kept in silence for 1 week after exposure. This also occurs in the respiratory system structures. In the stomach, the areolar (loose tissue) layer of the wall thins considerably in LFN-exposed rodents (8).

DISCUSSION

All data presented herein strongly suggest that LFN poses a threat to the structural integrity of biological structures. LFN exposure consists of airborne acoustic phenomena that impinge on visco-elastic tissue. The transduction of low frequency acoustic energy to tissues may be processed through acoustically-induced vibration. The end result, however, seems to be a form of adaptation that is guided by the need to maintain structural integrity. The fact that artery thickening occurs uniformly throughout the vessels (effectively decreasing lumen size), insinuates that the process is not localized.

LFN-exposed respiratory epithelium evidences an overall depletion of cilia, and cell de-differentiation (Fig 4B) is suggestive of precancerous lesions. It has already been demonstrated, in both human (9) and animal (10) models, that LFN is a genotoxic agent. Thus it is not surprising that many VAD patients develop respiratory tract tumors. What is striking is that, to date, all respiratory system tumors have been of one, unique type: squamous cell carcinomas (11). Similarly, all CNS tumors have been gliomas. The glia is the connective tissue of the brain. Blood vessel walls are also connective tissue, as is the areolar layer in the stomach wall.

As a final comment, the notion of tensegrity systems (12) cannot be excluded from this analysis given the overall response of the organism: connective tissue targeting, thickening in some locations and thinning in others. The authors invite bio-physicists, -engineers and –mathematicians to consider the challenges of these findings.

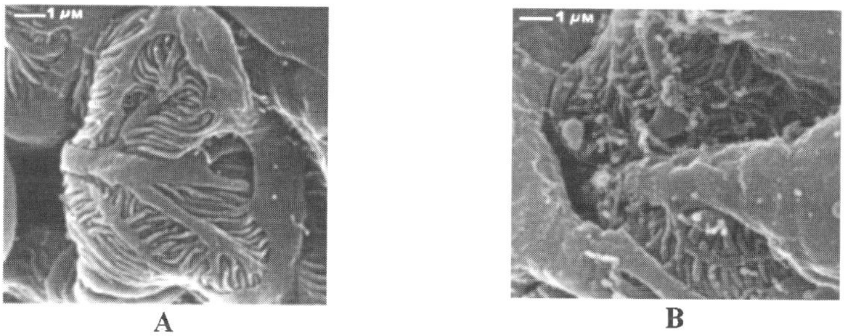

FIGURE 5. SEM of rat kidney. **A.** Non-exposed exhibiting neatly organized epithelial cell extensions and podocytes. Scattered strands of microvilli are visible. **B.** Exposed to 2160 hours of continuous LFN and sacrificed right after exposure: visible overall swelling of structures with loss of organization, more microvilli.

ACKNOWLEDGMENTS

The authors thank the Portuguese Ministry of Defense (CIMO) for animal facilities, FLAD (Luso-American Foundation for Development) for continuous support, Prof. Nuno Grande (ICBAS) and Prof. Carlos Sá (CEMUP) for electron microscopy, Daniela Sousa Silva (CEMUP) António Costa e Silva and Emanuel Monteiro (ICBAS), and Carlos Lopes (CIMO) for technical support, and Pedro Castelo Branco for image treatment. Alves-Pereira & Joanaz de Melo also thank IMAR (Instituto do Mar) for hosting project POCTI/MGS/41089/2001 and FCT (Fundação para a Ciência e Tecnologia) for its funding.

REFERENCES

1. Castelo Branco, N. A. A., and Rodriguez Lopez, E., , *Aviat. Space Environ. Med.* **70**, A1-A6 (1999).

2. Castelo Branco, N. A. A., *Aviat. Space Environ. Med.* **70**, A32-A39 (1999).
3. Castelo Branco, N. A. A., Rodriguez Lopez, E., Alves-Pereira, M., and Jones, D. R., *Aviat. Space Environ. Med.* **70**, A145-A151 (1999).
4. Castelo Branco, N. A. A., *Aviat. Space Environ. Med.* **70**, A27-31 (1999).
5. Castelo Branco, N. A. A., Alves-Pereira, M., Martins dos Santos, J., and Monteiro, E., "SEM and TEM Study of Rat Respiratory Epithelia Exposed to Low Frequency Noise," in *Science and Technology Education in Microscopy: An Overview*, edited by A. Mendez-Vilas, Badajoz, Spain: Formatex, 2003, pp. 505-533.
6. Holt, B. D., "The Pericardium," in *Hurst's The Heart*, edited by Furster, V., Wayne Alexander, R., and Alexander, F., New York: McGraw-Hill, 2000, pp. 2061-82.
7. Reis Ferreira, J. M., Couto, A. R., Jalles-Tavares, N., Castelo Branco, M. S. N., and Castelo Branco, N. A. A., *Aviat. Space Environ. Med.* **70**, A63-69.
8. Fonseca, J., Martins dos Santos, J., Castelo Branco N. A. A., and Grande, N., *40th Ann. Meet. Port. Soc. Anatomy*, Lisbon, Portugal, 2003, pp. 13.
9. Silva Reis Silva, M. J., Carothers, A., Castelo Branco, N. A. A., Dias, A., Boavida, M. G., *Mutuation Research* **44**, 129-134 (1999).
10. Silva, M. J., Dias, A., Barreta, A., Nogueira, P. J., Castelo Branco, N. A. A., Dias, A., and Boavida, M. G., *Teratogen Carcinogen Mutagen* **22**, 195-203 (2002).
11. Castelo Branco N. A. A., 8th *Inter Cong Sound Vib.* Hong Kong, P.R. China, July 2001,pp. 1501-8.
12. Ingberg, D. E., *J Cell Sci* **104**, 613-627 (1993).

AUTHOR INDEX

D

E

F

G

H

I

J

K

L

M

N

O